THE KEW RECORD OF TAXONOMIC LITERATURE

1973

AF768863

ROYAL BOTANIC GARDENS, KEW

The Kew Record of Taxonomic Literature

relating to vascular plants for

1973

LONDON: HER MAJESTY'S STATIONERY OFFICE

© *Crown copyright 1975*
First published 1975

ISBN 0 11 241109 6*

Contents

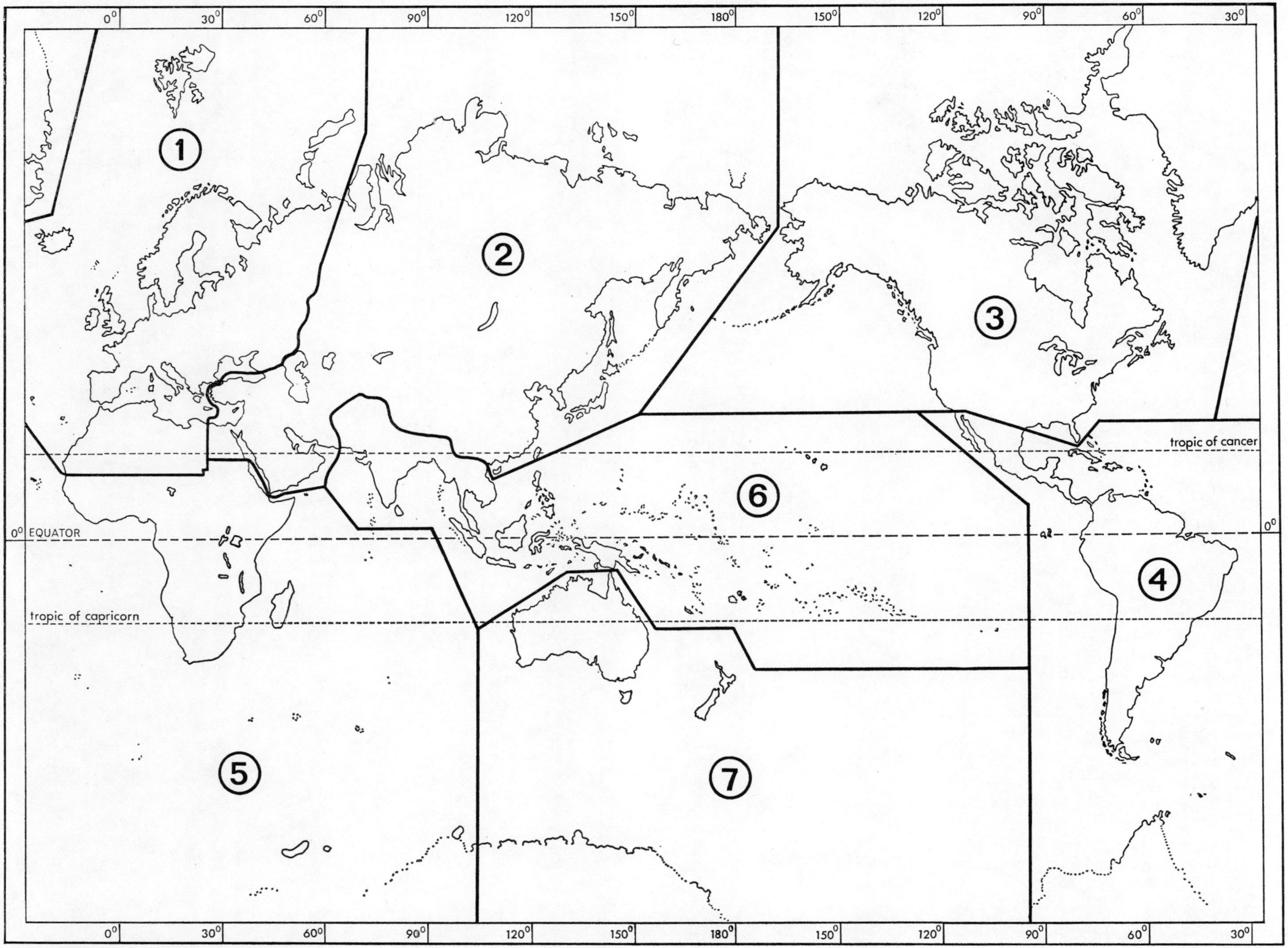

Geographical subdivisions of the World (see page xiii)

Introduction

The third volume of the *Kew Record of Taxonomic Literature* contains references for 1973, together with any references for 1971 and 1972 which were received too late for inclusion in the previous issues. Papers published in 1971–1973 but received after 31st March 1974 will be included in the *Kew Record of Taxonomic Literature 1974.*

All relevant literature received at Kew, the British Museum (Natural History) and the Commonwealth Forestry Institute has been scanned, but we would be grateful for notification of omitted papers, reprints of which will be most acceptable.

Scope and format

As in previous issues all papers relevant to the taxonomy of flowering plants, gymnosperms and ferns have been included, but those in related fields such as pure genetics or physiology have usually been omitted. Entries are arranged, wherever possible, in systematic groups, with more general papers contained in the sections on nomenclature, chromosome surveys, chemotaxonomy, anatomy and morphology, palynology, embryology and reproductive biology.

The format of each entry follows the pattern established in *Kew Record 1971* and *1972.* The author's name, whether personal or corporate, precedes the title of the article, which may be followed by a translated title in parentheses where one is given with the article. When there are several translated titles, preference is given to the English version. A translated title enclosed in square brackets indicates that the title is one given by a secondary source and that the original has not been seen by the indexers.

The abbreviations used for periodical titles follow a system devised at Kew in the absence of any agreed international scheme. The list of abbreviations on p. 347 supplements the lists published in *Kew Record 1971* and *1972.*

Where the date of publication is given in the following way: (1972 publ. 1973), the former date is the printed date of publication, and the latter is the effective date.

The language of an article, where it is not clear from the title, is shown in an abbreviated form. A summary or abstract in another language is shown by a language abbreviation enclosed in parentheses. A list of the language abbreviations is given on p. xv. Transliteration of Cyrillic languages has been in accordance with the British Standard for Transliteration of Cyrillic and Greek Characters (BS 2979: 1958). Consultations about the use of this Standard are now in progress with Russian and other authorities and, after the consensus of opinion is known, its continued use will be reconsidered.

A dash separates the author, title and periodical reference from notes recording the presence of illustrations, portraits, maps, dot maps, chromosome numbers, keys, bibliographies (occurring mainly in the Personalia section) and any explanatory information.

Author Index

A change has been made in the presentation of the Author Index. In cases of joint authorship, cross-references are no longer used, so that the name of each author appears only once, followed by the serial numbers of all relevant references. Where there has been doubt regarding the citation of names in the bibliographical references, national usage has, in general, been observed, as outlined in Chaplin, A. H. and Anderson, D. *Names of Persons: National Usages for Entry in Catalogues.* IFLA/FIAB, 1967.

Geographical analysis

Where relevant, a geographical number has been added to all entries in order to assist botanists with regional interests. The boundaries of the seven regions into which the world has been divided are shown in the map. Each is represented by a code number which appears in the left-hand margin for every entry. The codes and regions are as follows:

1. Europe and North Africa
2. Temperate Asia
3. North America
4. Central and South America
5. Tropical and South Africa
6. Tropical Asia and the Pacific
7. Australia and New Zealand

The code 0 indicates that the reference relates to plants known only from cultivation.

New Names

All new names at every rank have been included, with the exception of cultivars. The requirements of the International Code of Botanical Nomenclature have been adhered to in questions of validity, with reference to the suggestions in *Regnum Vegetabile* 60: 42–57 (1969) where interpretation has been in doubt. The significance of the typography of the new name entries is explained on p. xv.

In distributions given for new specific and infraspecific taxa, all countries cited by the author are given, the first being the country of origin of the type specimen.

Once again this bibliography aims to provide a comprehensive unified coverage of the world's taxonomic literature in botany for international use, and has been compiled and edited by members of the Library and Herbarium staff.

Professor J. Heslop-Harrison
Director

Introduction

The third volume of the *Kew Record of Taxonomic Literature* contains references for 1973, together with any references for [illegible] and [illegible] which were received too late for inclusion in the previous issues. Papers published in 1973 [illegible] after [illegible] March [illegible] will be included in the *Kew Record of [illegible]* [illegible] 1974.

All [illegible] literature received at Kew [illegible] [illegible] [illegible] [illegible] [illegible] Commonwealth [illegible] [illegible] [illegible] [illegible] [illegible] would be grateful for [illegible] [illegible] [illegible] papers [illegible] of which will be [illegible].

Scope and format

As in previous issues all papers relevant to the taxonomy of flowering plants, [illegible] [illegible] [illegible] are [illegible] [illegible] [illegible] [illegible] [illegible] [illegible] [illegible] [illegible] [illegible] [illegible] [illegible] [illegible] [illegible] [illegible] [illegible] [illegible] [illegible] [illegible] [illegible] taxonomic groups [illegible] [illegible] general papers [illegible] [illegible] [illegible] [illegible] [illegible] [illegible] [illegible] [illegible] [illegible] and [illegible] [illegible] palynology, embryology and [illegible] [illegible].

The format of each entry follows the pattern established in *Kew Record 1971* and *1972*. The author's name, whether personal or corporate, precedes the title of the [illegible] which may be followed by a translated title in parentheses [illegible] [illegible] [illegible] [illegible] [illegible] [illegible] [illegible] [illegible] [illegible] [illegible] [illegible] given in the English [illegible]. A translated title enclosed in square brackets indicates that the title is one given by a secondary source and that the original has not been seen by the indexers.

The abbreviations used for periodicals [illegible] [illegible] [illegible] [illegible] in the absence of [illegible] [illegible] [illegible] [illegible] The list of [illegible] [illegible] [illegible] supplements the [illegible] [illegible] *Kew Record 1971* and *1972*.

Where the date of publication is given in the [illegible] [illegible] [illegible] [illegible] [illegible] [illegible] [illegible] [illegible] [illegible] and the latter is the effective date.

The [illegible] of an article, where [illegible] not clear from the [illegible] is shown in an abbreviated form. A summary or abstract in another language is shown by a language abbreviation enclosed in parentheses. A list of the languages [illegible] [illegible] [illegible] [illegible] [illegible] of Cyrillic languages has been [illegible] [illegible] [illegible] *British Standard for the Transliteration of Cyrillic* [illegible] [illegible] [illegible] [illegible] [illegible] [illegible] [illegible] [illegible] [illegible] [illegible] [illegible] [illegible] [illegible] [illegible] [illegible] [illegible] [illegible] [illegible] common [illegible] [illegible] [illegible] will be reconsidered.

[illegible] the author, title and periodical reference a [illegible] note [illegible] the presence of illustrations, [illegible] maps, [illegible] tion maps, chromosome counts, [illegible] [illegible] [illegible] [illegible] [illegible] [illegible] [illegible] [illegible].

[illegible] [illegible]

[illegible] [illegible] [illegible] [illegible] [illegible] [illegible] [illegible] Author Index [illegible]

Geographical analysis

[illegible] [illegible] [illegible] geographical [illegible] [illegible] [illegible] [illegible] added to all [illegible] [illegible] [illegible] [illegible] [illegible] [illegible] [illegible] [illegible] [illegible] [illegible] [illegible] The [illegible] [illegible] [illegible] [illegible] [illegible] [illegible] [illegible] [illegible] [illegible] [illegible] divided are shown in the map, each is represented by a code number which appears in the left-hand margin [illegible] [illegible] [illegible] [illegible] [illegible] [illegible] [illegible].

1. Europe and North Africa
2. [illegible] Europe and Asia
3. North America
4. [illegible] and [illegible] America
5. [illegible] and South Africa
6. Tropical Asia and the Pacific
7. Australia and New Zealand

[illegible] code indicates that the [illegible] refers to plants known only from cultivation.

[illegible] of names

All new names [illegible] [illegible] [illegible] [illegible] been included, with the exception of [illegible] [illegible] [illegible] [illegible] of the International Code of [illegible] [illegible] [illegible] [illegible] [illegible] [illegible] have been adhered to [illegible] has been [illegible] [illegible] [illegible] of the [illegible] [illegible] of the new name [illegible] [illegible] [illegible] [illegible] [illegible] [illegible].

In distributions given for new specific and infraspecific taxa, [illegible] [illegible] [illegible] [illegible] [illegible] [illegible] are given, the first being the country or [illegible] of the type specimen.

[illegible] [illegible] [illegible] [illegible] [illegible] [illegible] [illegible] provided comprehensive [illegible]

[illegible] [illegible] [illegible] [illegible] [illegible]

[illegible]

Abbreviations

Key to symbols, abbreviations and typography

* reference has been taken from a secondary source, usually an abstracting journal, and consequently the bibliographical details have not been verified.

publ. published

Ordinary roman type: used for names proposed as new but not validly published.

Italic type: used for names which are not proposed as new.

Bold face type: used for names proposed as new and apparently validly published.

The language of an article or an abstract is indicated by the following language abbreviations:

Af	Afrikaans	En	English	In	Indonesian	Pol	Polish
Al	Albanian	Es	Estonian	It	Italian	Por	Portuguese
Ara	Arabic	Fi	Finnish	Ja	Japanese	Rum	Rumanian
Arm	Armenian	Fl	Flemish	Ka	Kazakh	Rus	Russian
Az	Azerbaijani	Fr	French	Ko	Korean	Se	Serbo-Croatian
Be	Belorussian	Ge	German	La	Latin	Si	Siamese
Bu	Bulgarian	Geo	Georgian	Latv	Latvian	Sl	Slovenian
Bur	Burmese	Gre	Greek	Li	Lithuanian	Sp	Spanish
Ch	Chinese	He	Hebrew	Mac	Macedonian	Sw	Swedish
Cz	Czech	Hi	Hindi	Mong	Mongolian	Ta	Tadzhik
Da	Danish	Hu	Hungarian	No	Norwegian	Tu	Turkish
Du	Dutch	Ic	Icelandic	Pe	Persian	Uk	Ukrainian

Taxonomic literature 1973

General

1 Altschul, S. von R.
Drugs and foods from little-known plants: notes in Harvard University Herbaria. Cambridge, Mass., Harvard University Press, 1973. xii,366p.—Illustration.

2 Bobrov, E. G.
Introgressive hybridisation, Sippenbildung und Vegetationsänderung. *Feddes Repert.*, 84(4):273–293 (1973)—(En,Rus); maps.

3 Coats, A. M.
The book of flowers: four centuries of flower illustration. London, Phaidon, 1973. 208p.—Illustrations, coloured illustrations.

4 European Weed Research Council:
Parasitic Weeds Research Group *and* Malta:
Royal University
Symposium on parasitic weeds, 11–13 April 1973, Royal University of Malta, Malta. [Wageningen], European Weed Research Council, 1973. 259p.

5 Fosberg, F. R.
Temperate zone influence on tropical forest land use: a plea for sanity. *In* Meggers, B. J., Ayensu, E. S. *and* Duckworth, W. D., *eds*. Tropical forest ecosystems in Africa and South America: 345–350(1973).

6 Frankel, O. H., *ed.*
Survey of crop genetic resources in their centres of diversity: 1st report. FAO and IBP, 1973. xiv,164p.—Illustration, maps, dot maps.

7 Gilmour, J. S. L.
And never the twain shall meet: horticulture and botany—allies not enemies. *In* Green, P. S. *ed.* Plants: wild and cultivated: 13–17(1973).

8 International Union for Conservation of Nature and Natural Resources
Convention on international trade in endangered species of wild fauna and flora . . . held at Washington, D.C. from 12 February to 2 March, 1973. *I.U.C.N. Bull.*, Suppl. 4(3). 12p. (1973).

9 Johnson, H.
The international book of trees: a guide and tribute to the trees of our forests and gardens. London, Mitchell Beazley, 1973. 288p.—Illustrations, coloured illustrations, portraits, maps.

10 Kîvilaan, A. *and* Bandurski, R. S.
The ninety-year period for Dr Beal's seed viability experiment. *Amer. J. Bot.*, 60(2):140–145(1973).

11 Kondo, K.
Carnivorous plants. *Lasca Leaves*, 23(2):77–84 (1973)—Illustrations.

12 Landolt, E.
Current research in plant biosystematics and taxonomy in Switzerland. *Int. Organ. Pl. Biosyst. Newsl.*, no.7:14–20(1973).

13 McLean, R. C. *and* Ivimey-Cook, W. R.
Textbook of theoretical botany: vol. 4. London, Longman, 1973. vii,596p.—Illustration, maps.

14 Malagarriga Heras, R. de P.
Las subespecies y la variacion geográfica. [Barcelona], 1973. 28p.

15 Marks, G. C. *and* Kozlowski, T. T., *eds*
Ectomycorrhizae: their ecology and physiology. New York, London, Academic Press, 1973. xvi,444p. (Physiological Ecology)—Illustrations.

16 Martins, E. M. Occhioni
Glossário de termos botânicos: 2. *Leandra*, 2(3): 61–71(1972).

17 Mathew, B.
Dwarf bulbs. London, Batsford, 1973. 240p.—Illustrations, coloured illustrations.

18 Meikle, R. D.
The development of garden plants from wild species. *In* Green, P. S. *ed.* Plants: wild and cultivated: 114–123(1973).

19 Odum, E. P.
Fundamentals of ecology. 3rd ed. Philadelphia, etc., Saunders, 1971. xiv,574p.—Illustrations, coloured illustrations, maps.

20 Parr, C.
A glossary of botanical terms. *Bull. Afr. Succ. Pl. Soc.*, 8(5):188–191(1973).

21 Raghavan, R. S., Wadhwa, B. M. *and* Kulkarni, B. G.
Notes on plant teratology. *Indian Forester*, 99(11): 659–662(1973)—Illustrations.

22 Rob, C. M.
Garden escapes and naturalized plants. *In* Green, P. S. *ed.* Plants: wild and cultivated: 146–151(1973).

23 Rohweder, O.
Angiospermen-Morphologie-Ergebnis oder Ausgangspunkt phylogenetischer Hypothesen? *Bot. Jahrb.*, 93(3):372–403(1973)—(En). Also publ. in *Mitt. Bot. Mus. Univ. Zürich*, no. 263.

24 Sander, R. *and* Rohumets, L.
Milliseid taimeliike kaitsta ja kuidas? (Which plants need protection and how to provide it?). *Eesti Loodus*, 16(5):276–279(1973)—(Rus,En); illustrations.

25 Satter, R. L. *and* Galston, A. W.
Leaf movements: Rosetta Stone of plant behavior? *BioScience*, 23(7):407–416(1973).

26 Schultes, R. E. *and* Hofman, A.
The botany and chemistry of hallucinogens. Springfield, Thomas, 1973. xxiv,267p. (American Lecture Series, no. 843)—Illustrations, portraits.

27 Sharma, A. K.
A quarter century with chromosomes. *Proc. Indian Sci. Congr.*, 60(2):123–132(1973).

28 Shetler, S. G. *and* Read, R. W., *eds*
International index of current research projects in plant systematics: no. 7. Washington, Smithsonian Institution, 1973. xxii,118p. (Flora North America Report 71).

29 Short, L. L.
The first International Congress of Systematics and Evolutionary Biology. *Taxon*, 22(5–6):641–646 (1973).

30 Stern, W. L.
The wood collection—what should be its future? *Arnoldia*, 33(1):67–80(1973).

31 Ukhacheva, V. N.
O tipe rastitel'nosti 'kriofil'nȳe podushechniki'. (On the type of vegetation 'cryophytic cushion plant formation'). *Vestn. Leningrad. Univ., Biol.*, no. 9(2): 58–64(1973)—(En).

32 Waisel, Y.
Biology of halophytes. New York, London, Academic Press, 1972. xiv,395p. (Physiological Ecology)—Illustrations.

33 Willis, J. C.
A dictionary of the flowering plants and ferns. 8th ed. rev. by H. K. Airy Shaw. Cambridge, University Press, 1973. xxii,1246,lxvip.

Theory

34 Andrade da Mata Rezende, C. M. *and* Gottlieb, O. R.
Xanthones as systematic markers. *Biochem. Syst.*, 1(2):111–118(1973).

35 Atsatt, P. R.
Parasitic flowering plants: how did they evolve? *Amer. Nat.*, 107(956):502–510(1973).

36 Barkley, F. A.
Outline classification of organisms. (ed. 5). [Boston, Mass., Northeastern University?], 1973. iii,48p.

37 Baum, B. R.
The concept of relevance in taxonomy with special emphasis on automatic classification. *Taxon*, 22(4): 329–332(1973).

38 Becker, K. M.
A comparison of angiosperm classification systems. *Taxon*, 22(1):19–50(1973).

39 Berlin, B.
Folk systematics in relation to biological classification and nomenclature. *Ann. Rev. Ecol. Syst.*, 4:259–271(1973).

40 Bidault, M.
Où en est la systématique végétale? *Bull. Féd. Soc. Hist. Nat. France-Comté*, 74(1):16–21(1972 publ. 1973)—Illustrations; chromosome numbers.

41 Canny, M. J.
Where is botany going? *Search*, 4(10):438–444 (1973)—Maps.

42 Clifford, H. T. *and* Williams, W. T.
Classificatory dendrograms and their interpretation. *Austral. J. Bot.*, 21(1):151–162(1973).

43 Cowan, R. S.
The herbarium as a data bank. *Arnoldia*, 33(1):3–12 (1973).

44 Daniels, G. S.
The botanist and the computer. *Arnoldia*, 33(1): 26–36(1973).

45 Fosberg, F. R.
Reflections after the Symposium on Systematic Biological Collections, July 6–8, 1972. *Pl. Sci. Bull.*, 19(1):4–6(1973).

46 Golubev, V. N.
K probleme évolyutsii zhiznennȳkh form rasteniĭ. (On the problem of evolution of plant life forms). *Bot. Zhurn.*, 58(1):3–10(1973)—(En).

47 Gray, J. R.
The origin of weediness. *Trans. Illin. Acad. Sci.*, 66(1–2):94–98(1973).

48 Guinochet, M.
Phytosociologie et systématique. *In* Heywood, V. H., *ed.* Taxonomy and ecology: 121–140(1973)—(En).

49 Heywood, V. H.
Ecological data in practical taxonomy. *In* Heywood, V. H., *ed.* Taxonomy and ecology: 329–347(1973).

50 Heywood, V. H., *ed.*
Taxonomy and ecology: proceedings of an International Symposium held at the Department of Botany, University of Reading. London, etc., Academic Press, 1973. x,370p. (Systematics Association Special Volume no. 5)—Illustrations, map, dot maps.

51 Heywood, V. H.
Taxonomy in crisis? or Taxonomy is the digestive system of biology. *Acta Bot. Acad. Sci. Hung.*, 19 (1–4):139–146(1973).

52 Hutchinson, J.
The families of flowering plants: arranged according to a new system based on their probable phylogeny. 3rd ed. Oxford, Clarendon Press, 1973. xx,968p.—Illustrations, maps; keys.

53 Inger, R. F.
Numerical taxonomy. *Caldasia*, 11(52):7–28(1973).

54 Jones, K.
The relevance of genetics. *In* Green, P. S., *ed.* Plants: wild and cultivated: 72–78(1973).

55 Kaul, M. L. H.
Mutagens and mutations: a molecular approach. *Sci. Cult.*, 38(5):236–239(1972).

56 Kubitzki, K.
Systematik und Evolution der Samenpflanzen. *Fortschr. Bot.*, 35:245–288(1973).

57 Latorre, F. *and* Cardenas, G.
Programa Fortran para taxonomia numerica. *Cienc. Nat.*, 13(1–2):20–24(1972).

58 Lewis, D. H.
The relevance of symbiosis to taxonomy and ecology, with particular reference to mutualistic symbioses and the exploitation of marginal habitats. *In* Heywood, V. H., *ed.* Taxonomy and ecology: 151–172(1973).

59 Khokhryakov, A. P.
Osnovnȳe zakonomernosti évolyutsii ontogeneza u rasteniĭ. (Basic patterns of the evolution of ontogenesis in plants). *Byull. Mosk. Obshch. Ispȳt. Prir., Biol.*, 78(1):59–69(1973)—(En).

60 Kubitzki, K.
Probleme der Grosssystematik der Blütenpflanzen. (Problems of the macro-systematics of the flowering plants). *Ber. Deutsch. Bot. Ges.*, 85(5–6):259–277 (1972)—(En).

61 Lambert, J. M., Meacock, S. E., Barrs, J. *and* Smartt, P. F. M.
Axor and Monit: two new polythetic-divisive strategies for hierarchical classification. *Taxon*, 22(2–3): 173–176(1973).

62 Long, D. G.
Serological studies in higher plant taxonomy. *Trans. Bot. Soc. Edinburgh*, 42(1):119(1973). Short note.

63 Mayr, E.
Znachenie sistematiki v sovremennoj biologii. (The importance of systematics in modern biology). *Zhurn. Obshch. Biol.*, 34(3):323–330(1973).

64 Meeuse, A. D. J.
Anthecology, floral morphology and angiosperm evolution. *In* Heywood, V. H., *ed.* Taxonomy and ecology: 189–200(1973).

65 Meeuse, A. D. J.
Co-evolution of plant hosts and their parasites as a taxonomic tool. *In* Heywood, V. H., *ed.* Taxonomy and ecology: 280–316(1973).

66 Merxmüller, H.
Systematic botany—an unachieved synthesis. Lecture given at the Annual General Meeting of the Systematics Association, November 1971. *Biol. J. Linn. Soc.*, 4(4):311–321(1972 publ. 1973)—Chromosome numbers.

67 Meyen, S. V.
Plant morphology in its nomothetical aspects. *Bot. Rev.*, 39(3):205–260(1973)—Illustration.

68 Raynal, J.
La méthode taxinomique numérique. *Encyclopaedia Universalis*, 15:683–685(1973).

69 Renfrew, J. M.
Palaeoethnobotany: the prehistoric food plants of the Near East and Europe. London, Methuen, 1973. xviii,248p.—Illustrations, maps.

70 Rowley, G. D.
Save the succulents! a practical step to aid conservation. *Cact. Succ. J.* (U.S.A.), 45(1):8–11(1973)—Illustrations.

71 Scott, P. J.
A consideration of the category in classification. *Taxon*, 22(4):405–406(1973).

72 Serebryakova, T. I.
K voprosu ob évolyutsionnȳkh vzaimootnosheniyakh drevesnȳkh i travyanistȳkh zhiznennȳkh form tsvetkovȳkh rasteniĭ. (On the evolutional relations of arborescent and herbaceous life forms in flowering plants). *Byull. Mosk. Obshch. Ispȳt. Prir., Biol.*, 78(3): 76–88(1973)—(En).

73 Sitnik, K. M.
Tendentsiyi rozvȳtku botanikȳ. (Tendencies of botany development). *Ukr. Bot. Zhurn.*, 30(4):409–417 (1973)—Uk(En).

74 Snaydon, R. W.
Ecological factors, genetic variation and speciation in plants. *In* Heywood, V. H., *ed.* Taxonomy and ecology: 1–29(1973).

75 Sneath, P. H. A. *and* Sokal, R. R.
Numerical taxonomy: the principles and practice of numerical classification. San Francisco, Freeman, 1972. xvi,573p.

76 Sporne, K. R.
The survival of archaic dicotyledons in tropical rain-forests. *New Phytol.*, 72(5):1175–1184(1973).

77 Stebbins, G. L.
Evolutionary trends in the inflorescence of angiosperms. *Flora* (Germany), 162(6):501–528(1973)—(Ge).

78 Takhtadzhyan, A. L.
Evolution und Ausbreitung der Blütenpflanzen. Jena, G. Fischer, 1973. 189p.—Illustrations, maps. Originally: Proiskhozhdenie i rasselenie tsvetkovȳkh rasteniĭ. Leningrad, 1970.

79 Tsitsin, N. V.
Nekotorȳe voprosȳ zvolyutsii i formoobrazovaniya u rasteniĭ. (Some questions on the evolution and form-development in plants). *Izv. Akad. Nauk SSSR, Biol.*, no. 2:165–176(1973)—(En); illustrations; chromosome numbers.

80 Tsvelev, N. N.
O znachenii stepeni spetsializatsii taksonov dlya ikh dal'neĭsheĭ évolyutsii. (On the importance of the degree of specialization for the further evolution of taxa). *Byull. Mosk. Obshch. Ispȳt. Prir., Biol.*, 78(2): 71–81(1973)—(En).

81 Valentine, D. H.
The taxonomic treatment of infraspecific variation in relation to floristic studies. *Int. Organ. Pl. Biosyst. Newsl.*, no. 7:6–13(1973).

82 Van Valen, L.
Are categories in different phyla comparable? *Taxon*, 22(4):333–373(1973). Zoological classification.

83 Van Valen, L.
Body size and numbers of plants and animals. *Evolution*, 27(1):27–35(1973).

Practice

84 Anon.
Herbaria: data processing. *Nature*, 246(5428):62 (1973). Report of an International meeting at the Royal Botanic Gardens, Kew, October 3–6, 1973.

85 Brenan, J. P. M.
The value of herbaria for cultivated plants. *In* Green, P. S., *ed.* Plants: wild and cultivated: 38–47 (1973).

86 Charpin, A. *and* Monthoux, O.
L'emploi de l'ordinateur pour la cartographie floristique de la Haute-Savoie. *Bull. Soc. Bot. France*, 118(9):793–800(1971 publ. 1973)—(En,Ge).

87 Eisner, T., Eisner, M., Hyypio, P. A., Aneshansley, D. *and* Silberglied, R. E.
Plant taxonomy: ultraviolet patterns of flowers visible as fluorescent patterns in pressed herbarium specimens. *Science*, 179(4072):486–487(1973).

88 Gómez-Pompa, A.
The thrust of present and future research in the lowland tropics of Mexico. *Ann. Missouri Bot. Gard.*, 60(2):169–173(1973).

89 Gómez-Pompa, A. *and* Nevling, L. I.
The use of electronic data processing methods in the Flora of Veracruz program. *Contrib. Gray Herb.*, no. 203:49–64(1973).

90 Hall, A. V.
The use of a computer-based system of aids for classification. *Contrib. Bolus Herb.*, no. 6.[ii]110p. (1973).

91 Jeffers, J. N. R., *ed.*
Mathematical models in ecology: the 12th Symposium of the British Ecological Society, Grange-over-Sands, Lancashire, 23–26 March 1971. Oxford, etc., Blackwell, 1972. viii,398p.—Illustrations.

92 Jorgensen, V.
The preparing, pressing and mounting of bromeliads. *J. Bromeliad Soc.*, 23(6):211–214 (1973).

93 Kowal, T.
The role of the number of characters in distinguishing taxa in taxonomic investigations performed by the dendrite method. (Rola liczby cech przy wyróznianiu taksonów w badaniach systematycznych prowadzonych metodą dendrytowa). *Acta Soc. Bot. Pol.*, 42(1):103–119(1973)—(Pol).

94 Krauss, H. M.
The use of generalized information processing systems in the biological sciences. *Taxon*, 22(1):3–18 (1973).

95 Nevling, L. I.
Report of the Committee for Recommendations in Desirable Procedures in Herbarium Practice and Ethics: 2. *Brittonia*, 25(3):307–310(1973).

96 Perring, F. H.
Rare plant recording and conservation in Great Britain. *B.S.B.I. Welsh Region Bull.*, 16:5–14(1972).

97 Radford, G. L. *and* Pankhurst, R. J.
A conservation data base. *New Phytol.*, 72(5): 1191–1206(1973).

98 Saenz de Rivas, C.
Aplicación del microscopio electrónico 'scanning' a la taxonomía vegetal. *An. Inst. Bot. A.J. Cavanilles*, 30:99–107(1973)—(En).

99 Shetler, S. G. *and others*
An introduction to the Botanical Type Specimen Register. *Smithsonian Contrib. Bot.*, no. 12. 186p. (1973)—Illustrations.

100 Stant, M. Y.
The role of the scanning electron microscope in plant anatomy. *Kew Bull.*, 28(1):105–115(1973).

101 Stolze, R. G.
Inadequacies in herbarium specimens of large ferns. *Amer. Fern J.*, 63(2):25–27(1973)—Illustrations.

102 Vasil'chenko, I. T.
O ratsional'noĭ organizatsii gerbariev. (On the rational organization of herbariums). *Bot. Zhurn.*, 58(12):1723–1726(1973).

103 Wright, R.
Remote sensing techniques in vegetation mapping—a review of some uses and limitations. *Trans. Bot. Soc. Edinburgh*, 42(1):69–81(1973).

Systematics

PTERIDOPHYTA

104 Adamson, J. A. *and* Telford, I. R. H.
An introduction to ferns. *Austral. Pl.*, 7(56):158,
1–7 180–187(1973).

105 Atkinson, L. R. *and* Stokey, A. G.
The gametophyte of some Jamaican thelypteroid ferns. *Bot. J. Linn. Soc.*, 66(1):23–36(1973)—
4 Chromosome numbers.

106 Brownsey, P. J. *and* Jermy, A. C.
A fern collecting expedition to Crete. *Brit. Fern Gaz.*, 10(6):331–348(1973)—Illustrations, coloured
1 illustration, map; chromosome number.

107 Brunkener, L.
Beiträge zur Kenntnis der frühen Sporangienentwicklung der Pteridophyten und der Gymnospermen. (Contributions to the knowledge of the early development of the sporangium in pteridophytes and gymnosperms). *Svensk Bot. Tidskr.*, 67(4):333–400
1–7 (1973)—(En).

108 Caixinhas, M. L.
Selaginellaceae. *Anuár. Soc. Brot.*, 38:87–88(1972)—
1 Por; key.

109 Carvalho e Vasconcellos, J. de
Azollaceae. *Anuár. Soc. Brot.*, 38:99–100(1972)—
1 Por; key.

110 Carvalho e Vasconcellos, J. de
Equisetaceae. *Anuár. Soc. Brot.*, 38:89–91(1972)—
1 Por; key.

111 Carvalho e Vasconcellos, J. de
Hymenophyllaceae. *Anuár. Soc. Brot.*, 38:97–98
1 (1972)—Por; key.

112 Carvalho e Vasconcellos, J. de
Lycopodiaceae. *Anuár. Soc. Brot.*, 38:81–85(1972)—
1 Por; key.

113 Carvalho e Vasconcellos, J. de
Ophioglossaceae. *Anuár. Soc. Brot.*, 38:93–94
1 (1972)—Por; key.

114 Carvalho e Vasconcellos, J. de
Osmundaceae. *Anuár. Soc. Brot.*, 38:95(1972)—
1 Por.

115 De la Sota, E. R.
Adiciones a flora pteridofitica del noroeste de
4 Argentina. *Darwiniana*, 18(1–2):264–265(1973).

116 De la Sota, E. R.
La distribución geográfica de las Pteridofitas en el cono sur de América meridional. *Bol. Soc. Argent. Bot.*,
4 15(1):23–34(1973).

117 De la Sota, E. R.
Sinopsis de las pteridófitas del noroeste de Argentina: 2. (A synopsis of fern flora of north western Argentina: 2). *Darwiniana*, 18(1–2):173–263
4 (1973)—(En); illustration; keys.

118 Debray, M.
Eléments pour la connaissance de la flore du Havre et de ses environs. *Rev. Soc. Sav. Haute Normandie*, no. 64:29–42(1971 publ. 1973)—Chromosome
1 numbers.

119 Dorn, R. D. *and* Dorn, J. L.
The ferns and other pteridophytes of Montana, Wyoming, and the Black Hills of South Dakota. Laramie, University of Wyoming, 1972. 94p.—
3 Illustrations.

120 Exell, A. W. *and* Wild, H.
A statistical analysis of a sample of the Flora Zambesiaca: 2. Gramineae and Pteridophyta. *Kirkia*,
5 9(1):87–93(1973).

121 Hasegawa, M. *and* Taneyama, M.
Chicoric acid from *Onychium japonicum* and its distribution in the Ferns. *Bot. Mag. Tokyo*, 86(1004):
1–7 315–317(1973).

122 Hill, R. H.
The comparative ecology of several pteridophytes of southeastern Michigan. *Diss. Abstr. Int., B*, 33(11):
3 5173–5174(1973).

123 Holttum, R. E.
The identity of three type specimens in the Willdenow Herbarium. *Amer. Fern J.*, 63(3):81–84
6 (1973). Thelypteridaceae.

124 Holttum, R. E.
Some comments on the ferns on Mt. Kinabalu. *In* Singapore: University. Mount Kinabalu Field Trip, April/May 1973 souvenir programme. [2]p. (1973)—
6 Key.

125 Huynh, K.-L.
L'arrangement des spores dans la tétrade chez les Ptéridophytes. *Bot. Jahrb.*, 93(1):9–24(1973)—
1–3 (Ge,En).

126 Inobe, T.
(The news of ferns in Tokushima Prefecture: 4). *J.*
2 *Geobot.*, 20(1):25(1972)—Ja.

127 Ito, H.
Notes on ferns of the Bonin Islands. *J. Jap. Bot.*,
2 48(5):159–160(1973)—(Ja).

128 Iwatsuki, K.
Phytogeography of the pteridophytes in peninsular Thailand. *Amer. Fern J.*, 63(3):129–134(1973)—
6 Map.

129 János, D.
Páfrányok a lakásban és a kertben. Budapest, Mezögazdasági Kiadó, 1973. 157p.—Illustrations,
1 coloured illustrations.

130 Kato, M.
(Taxonomical evaluation of the articulated hairs found in the Athyriaceae). *Acta Phytotax. Geobot.*,
6 25(4–6):119–126(1973)—Ja(En); illustrations.

131 Khullar, S. P. *and* Mehra, P. N.
Cytotaxonomy of W. Himalayan ferns: 3. Polypodiaceae (sensu stricto). *Nucleus*, 15(3):156–162
6 (1972)—Illustrations; chromosome numbers.

132 Klekowski, E. J.
Genetical features of ferns as contrasted to seed plants. *Ann. Missouri Bot. Gard.*, 59(2):138–151(1972
1–7 publ. 1973)—Illustrations.

133 Knobloch, I. W.
The present status of hybridity among the pterido-
1–7 phytes. *Taiwania*, 18(1):29–34(1973).

134 Laínz, M.
In Florae Europaeae Atlantem chorologicae animadversiones. *Anuár. Soc. Brot.*, 39: 117–123(1973)
1 —Sp.

135 LeCoq, C., Guervin, C. *and* Laroche, J.
Morphologie de quelques spores de Polypodiacées sensu lato en microscopie électronique à balayage. *Pollen Spores*, 15(1):109–115(1973)—(En). *Gymno-*
4 *gramme* and *Scolopendrium*.

136 Lewalle, J.
Flore d'Afrique centrale (Zaïre-Rwanda-Burundi): Pteridophytes. Osmundaceae. 1973. [ii]4p.—Illustra-
5 tions, dot map; chromosome number.

137 Lugardon, B.
La structure fine de l'exospore et de la perispore des Filicinées isosporées: 1. Généralités: Eusporangiées et Osmundales. *Pollen Spores*, 14(3):227–261 (1972 publ. 1973)—(En). Ophioglossales, Marattiales,
1–7 Osmundales.

138 Mitui, K.
A cytological survey on the Pteridophytes of the Bonin Islands. *J. Jap. Bot.*, 48(8):247–253(1973)—
2 (Ja); chromosome numbers.

139 Ohba, H.
Additional notes on Pteridophytes of the Bonin and Volcano Islands (1). *J. Jap. Bot.*, 48(10):318–319
2 (1973)—(Ja).

140 Page, C. N.
Ferns, polyploids, and their bearing on the evolution of the Canarian flora. *Monogr. Biol. Canar.*,
1 no. 4:83–88(1973)—Chromosome number.

141 Pande, P. C.
Pteridophytic flora of Ranikhet. *Indian Forester*,
6 99(1):49–52(1973).

142 Payne, W. W. *and* Peterson, K. M.
Observations of the hypodermises of ferns. *Amer.*
1–7 *Fern J.*, 63(2):34–42(1973).

143 Pichi-Sermolli, R. E. G.
Fragmenta Pteridologiae: 3. *Webbia*, 27(2):389–460
5 (1972 publ. 1973)—En(It); illustrations.

144 Pichi-Sermolli, R. E. G.
Fragmenta Pteridologiae: 4. *Webbia*, 28(2):445–477
1–7 (1973)—En(It).

145 Sehnem, A.
Flora ilustrada catarinense: Pteridáceas. 1972.
4 244p.—Illustrations, dot maps; keys.

146 Serizawa, S.
(Notes on the Pteridophyte flora of the Ryukyu: 1.).
2 *J. Geobot.*, 20(1):5–9(1972)—Ja,En.

147 Shaw, G. A.
1 Ripon ferns. *Naturalist*, no. 926: 99–100(1973).

148 Sossountzov, L.
Structure et fonctionnement du méristème apical des Ptéridophytes: présent et avenir. (Structure and functioning of fern apices: present and future). *Bull. Soc. Bot. France*, 119(5–6):341–352(1972 publ. 1973)—
1–7 (En).

149 Tagawa, M.
New combinations of fern names. *Acta Phytotax.*
6 *Geobot.*, 25(4–6):180(1973).

150 Tardieu-Blot, M. L.
Flore de Madagascar et des Comores: 13e famille Lycopodiacées (Lycopodiaceae): famille 13 bis Huperziacées (Huperziaceae): 3–48(1971)—
5 Illustrations, map; keys.

151 Telford, I.
Australian native ferns. *Austral. Pl.*, 7(56):159–167, 188–190(1973)—Illustrations, coloured illustrations.
7 Tree ferns only.

152 Tosco, U.
Catalogo floristico del Parco Nazionale del Gran Paradiso: prima parte. Tallofite, Briofite, Pteridofite. *Webbia*, 28(2):227–322(1973)—Map. Pteridophyta
1 pp.306–322.

153 Vega, I. S.
Catalogo de las especies de Polypodiaceae de la Province de Cajamarca. *Bol. Soc. Bot. Libertad*, 3(2):
4 137–144(1971)—(En).

154 Walker, T. G.
Additional cytotaxonomic notes on the pteridophytes of Jamaica. *Trans. Roy. Soc. Edinburgh*, 69(5): 109–135(1973)—Illustrations, dot map; chromosome
4 numbers.

155 Wilmanns, O. *and* Rasbach, H.
Observations on the pteridophytes of São Miguel, Açores. *Brit. Fern Gaz.*, 10(6):315–329(1973)—
1 Illustrations, maps, dot maps.

156 Windisch, P. G.
Filices novae Austroamericanae. *Bradea*, 1(37):371–
4 376(1973)—(En); illustrations.

FILICOPSIDA

ACROSTICHUM

157 Rao, T. A., Mukherjee, A. K. *and* Banerjee, L. K.
Is *Acrostichum aureum* L. truly a mangrove fern?
2–7 *Curr. Sci.*, 42(15):546–547(1973).

ADIANTOPSIS

minutula Sehnem in Fl. Ilustr. Catar., 1 (Pteridác.):
4 73(1972)—Brazil.

ADIANTUM

158 Fosback, J.
Two Oregon plant records. *Madroño*, 22(4):213
3 (1973). *Adiantum jordanii.*

159 Lyster, E.
Plant portrait: *Adiantum pedatum. Quart. Bull. Alp.*
3 *Gard. Soc.*, 41(1):14–15(1973)—Illustration.

160 Sledge, W. A.
Native and naturalised species of *Adiantum* in Ceylon. *Ceylon J. Sci., Biol. Sci.*, 10(2):144–154(1973)—
6 Illustration; key.

ALEURITOPSIS

afra Pichi-Sermolli in *Webbia*, 27(2):415(1972 publ.
5 1973)—Tanzania.

ALSOPHILA

humilis (Hieron.) Pichi-Sermolli in *Webbia*, 27(2):
5 405(1972 publ. 1973): *Cyathea humilis.*

AMPHINEURON

terminans (Hooker) Holttum in *Amer. Fern J.*,
6 63(3):82(1973): *Nephrodium terminans.*

ANGIOPTERIS

161 Chang, Chih-yü
The morphology of *Archangiopteris* Christ et Gies. and its relationship with *Angiopteris* Hoffm. *Acta Bot.*
6 *Sin.*, 15(2):261–276(1973)—Ch(En); illustrations.

ANOGRAMMA

162 Eberle, G.
Beobachtungen am Frühlings-Nacktfarn (*Anogramma leptophylla*) am Wuchsort und in der Kultur. *Nat. Mus.* (Frankfurt), 103(3):93–97(1973)—
1 Illustrations.

caespitosa Pichi-Sermolli in *Webbia*, 27(2):429
5 (1972 publ. 1973)—Tanzania.

ANTROPHYUM

163 Gómez P., L. D.
An unusual population of *Antrophyum lanceolatum.*
4 *Amer. Fern J.*, 63(1):22(1973)—Illustrations.

ARACHNIODES

164 Serizawa, S.
Arachniodes chinensis. J. Jap. Bot., 48(7):218–219
2, 6 (1973)—Ja,En.

holttumii (Tagawa) Serizawa in *J. Jap. Bot.*, 48(7):
6 219(1973): *Rumohra holttumii.*
tripinnata (Goldm.) Sledge in *Bull. Brit. Mus.* (*Nat.*
6 *Hist.*), *Bot.*, 5(1):41(1973): *Polystichum tripinnatum.*
yaoshanensis (Wu) Serizawa in *J. Jap. Bot.*, 48(7):
2 219(1973): *Polystichum amabile* var. *yaoshanense.*

ARCHANGIOPTERIS

165 Chang, Chih-yü
The morphology of *Archangiopteris* Christ et Gies. and its relationship with *Angiopteris* Hoffm. *Acta Bot.*
6 *Sin.*, 15(2):261–276(1973)—Ch(En); illustrations.

ASPLENIUM

166 Faden, R. B.
Some notes on the gemmiferous species of *Asplenium* in tropical East Africa. *Amer. Fern J.*,
5 63(3):85–90(1973)—Illustration.

167 Harborne, J. B., Williams, C. A. *and* Smith, D. M.
Species-specific kaempferol derivatives in ferns of the Appalachian *Asplenium* complex. *Biochem. Syst.*,
3 1(1):51–54(1973).

168 Key, J. S. *and* Redfearn, P. L.
Additions and a deletion to the fern flora of Missouri. *Amer. Fern J.*, 63(1):1–2(1973). *Asplenium*
3 *montanum*, *A. rhizophyllum* f. *auriculatum*.

169 Mendes, E. J.
Additiones et adnotationes florae Mozambicanae: 4. (Pteridophyta). *Garcia de Orta, Sér. Bot.*, 1(1–2):19–20
5 (1973)—(En).

170 Nardi, E.
Ad floram italicam notulae taxonomicae et geobotanicae: 9. *Asplenium adulterium* Milde nei pressi di Stresa (Lago Maggiore). *Webbia*, 28(1):37–47
1 (1973)—(En); chromosome numbers.

171 Reichstein, T., Lovis, J. D., Greuter, W. *and* Zaffran, J.
Die Asplenien der Insel Kreta. (The *Asplenium* species of Crete). *Ann. Mus. Goulandris*, 1:133–163
1 (1973)—(En,Gre); illustrations, dot maps.

172 Rivas-Martinez, S., Izco, J. *and* Costa, M.
Asplenium cunefolium Viv. (*A. serpentini* Tausch) en Sierra Bermeja (Málaga): comentarios sobre la flora y vegetación serpentínicola y dolomitícola. *Trab. Dep. Bot. Fisiol. Veg. Univ. Madrid*, 6:23–30(1973)—
1 (En).

173 Sutter, R.
Über Vorkommen und Verbreitung von *Asplenium petrarchae* (Guérin) DC. und die Assoziationen Phagnalo-Asplenietum petrarchae B.-Bl. und Parietarietum lusitanicae ass. nov. prov. *Bot. Jahrb.*, 93(4):568–602(1973)—(En,Fr); illustrations, maps, dot
1 maps; chromosome numbers.

174 Wróbel-Stermińska, W.
Zdeformowane liście paproci *Asplenium nidus* L.
1 *Wiad. Bot.*, 17(1):53–55(1973)—Illustrations.

aegaeum Lovis, Reichstein & Greuter in *Ann. Mus.*
1 *Goulandris*, 1:141(1973)—Crete.
centrafricanum Pichi-Sermolli in *Webbia*, 27(2):
5 436(1972 publ. 1973)—Burundi, Zaïre.
creticum Lovis, Reichstein & Zaffran in *Ann. Mus.*
1 *Goulandris*, 1:145(1973)—Crete.
erectum Bory ex Willd.
var. **mutilatum** (Kaulf.) Pichi-Sermolli in *Webbia*,
5 28(2):468(1973): *Asplenium mutilatum*.
var. **sphenolobium** (Kunze) Pichi-Sermolli *op. cit.*
5, 6 468(1973): *Asplenium lunulatum* var. *sphenolobium*.
fontanum (L.) Bernh.
subsp. **forisiense** (Le Grand) Malagarriga, Las
1 Subesp. y Variac. Geogr.: 4(1973): *A. forisiense*.
lepidum C. Presl
var. **fissoides** Ritter & Schumacher ex Reichstein, Lovis, Greuter & Zaffran in *Ann. Mus. Goulandris*,
1 1:138(1973)—Crete.
lilloanum de la Sota in *Darwiniana*, 18(1–2):198
4 (1973)—Argentina.
pocsii Pichi-Sermolli in *Webbia*, 27(2):432(1972 publ.
5 1973)—Tanzania.
scolopendrium L.
subsp. **antri-jovis** (Kümm.) Brownsey & Jermy in *Brit. Fern Gaz.*, 10(6):346(1973): *Biropteris*
1 *antri-jovis*.

ATHYRIUM

lewalleanum Pichi-Sermolli in *Webbia*, 27(2):440
5 (1972 publ. 1973)—Burundi.

AZOLLA

175 Ott, F. D. *and* Petrik-Ott, A. J.
Azolla and its occurrence in Virginia. *Brit. Fern Gaz.*,
3 10(6):305–309(1973).

BELVISIA

176 Rao, A. R. *and* Srivastava, P.
On the morphology and anatomy of *Belvisia spicata* (L. Fil.) Mirbel. *Proc. Indian Acad. Sci.*, *B*, 77(1):25–30
5, 6 (1973).

BLECHNUM

177 Rios, R. R.
Morfologia de los protalos y esporofitos jovenes de algunas especies chilenas de *Blechnum* (Polypodiaceae s.l.). *Gayana, Bot.*, no. 22. 30p.(1973)—(En);
4 illustrations.

rosenstockii de la Sota in *Darwiniana*, 18(1–2):254
4 (1973)—Brazil.
chambersii Tindale in N. C. W. Beadle, O. D. Evans & Carolin, Fl. Sydney Reg.: 86(1972), nom. nov.:
7 *B. lanceolatum* (R.Br.) Sturm.

BOTRYCHIUM

178 Abeywickrama, B. A. *and* De Fonseka, R. N.
The Ceylonese Ophioglossaceae. *Ceylon J. Sci., Biol. Sci.*, 10(2):132–142(1973)—Illustration;
6 keys.

179 Buzanov, V. A.
Nakhodka *Botrychium virginianum* (L.) Sw. v Udmurtii. (The find of *Botrychium virginianum* (L.)
2 Sw. in Udmurtia). *Bot. Zhurn.*, 58(5):694–695(1973).

180 Holub, J.
A note on the classification of *Botrychium* Sw. s.l. (Poznámka ke klasifikaci rodu *Botrychium* Sw. s.l.).
1–4, 6 *Preslia*, 45(3):276–277(1973)—(Cz).

BOTRYPUS

chamaeconium (Bitter & Hieronym.) Holub in *Preslia*, 45(3):277(1973), as 'chamaeciconium':
5 *Botrychium chamaeciconium*.
cicutaria (Savigny) Holub *op. cit.* 277(1973):
4 *Botrychium cicutaria*.
lanuginosus (Hooker & Greville) Holub *op. cit.* 277
2, 6 (1973): *Botrychium lanuginosum*.
subsp. **leptostachyus** (Hayata) Holub *op. cit.* 277
2 (1973): *Botrychium leptostachyum*.
strictus (Underwood) Holub *op. cit.* 277(1973), as
2 'striatus': *Botrychium striatum*.
parvus (Ching) Holub *op. cit.* 277(1973): *Botrychium*
2 *parvum*.
virginianus (L.) Holub *op. cit.* 277(1973): *Osmunda*
1–4, 6 *virginiana*.
subsp. **europaeus** (Angstr.) Holub *op. cit.* 277
1–3 (1973): *Botrychium virginianum* subsp. *europaeum*.

CERATOPTERIS

181 Hickok, L. G. *and* Klekowski, E. J.
Abnormal reductional and non-reductional meiosis in *Ceratopteris*: alternatives to homozygosity and hybrid sterility in homosporous ferns. *Amer. J. Bot.*, 60(10):1010–1022(1973)—Chromosome
2, 6 numbers.

182 Lloyd, R. M.
Systematics of the genus *Ceratopteris* (Parkeriaceae): 1. Sexual and vegetative reproduction in Hawaiian *Ceratopteris thalictroides*. *Amer. Fern J.*, 63(1):12–18
6 (1973).

CETERACH

183 Piperon, G.
Asarum europaeum et *Ceterach officinarum* à Boigneville. *Bull. Ass. Nat. Vallée Loing Massif*
1 *Fontainebleau*, 48(7–8):89(1972).

CHEILANTHES

184 Reichstein, T. *and* Vida, G.
Cheilanthes corsica Reichst. & Vida spec. nov. *Candollea*, 28(1):83–91(1973)—Ge(En,Fr); illustra-
1 tions; chromosome numbers.

corsica Reichstein & Vida in *Candollea*, 28(1):85
1 (1973)—Corsica, Spain, Portugal, Turkey, Italy, France, Tunisia.
erythraea Pichi-Sermolli in *Webbia*, 27(2):420(1972
2, 5 publ. 1973)—Ethiopia.

CHRISTELLA

dentata (Forssk.) Brownsey & Jermy in *Brit. Fern*
1, 2, 5, 6 *Gaz.*, 10(6):338(1973): *Polypodium dentatum*.
prolixa (Willd.) Holttum in *Amer. Fern J.*, 63(3):83
6 (1973): *Aspidium prolixum*.

COPELANDIOPTERIS B. C. Stone in *Webbia*,
6 28(2):492(1973).

185 Stone, B. C.
Copelandiopteris, a new genus of Philippine ferns.
6 *Webbia*, 28(2):491–494(1973)—(It); key.

calocarpa (Copel.) B. C. Stone in *Webbia*, 28(2):
6 493(1973): *Schizostege calocarpa*.
pachysora (Copel.) B. C. Stone *op. cit.* 493(1973):
6 *Schizostege pachysora*.

CRYPSINUS

bakeri (Luerss.) Tagawa in *Acta Phytotax. Geobot.*,
6 25(4–6):180(1973): *Polypodium bakeri*.
hagerupii (C. Chr.) Tagawa *op. cit.* 180(1973):
6 *Polypodium hagerupii*.

CRYPTOGRAMMA

186 Söderbäck, O.
Cryptogramma crispa funnen i Östergötland. (*Cryptogramma crispa* found in Östergötland). *Svensk*
1 *Bot. Tidskr.*, 67(1):67(1973).

CTENITIS

aquilina (Thouars) Pichi-Sermolli in *Webbia*, 28(2):
5 468(1973): *Polypodium aquilinum*.

CTENOPTERIS

insidiosa (Slosson) Pichi-Sermolli in *Webbia*, 28(2):
4 468(1973): *Polypodium insidiosum*.
trifurcata (L.) Pichi-Sermolli *op. cit.* 469(1973):
4 *Polypodium trifurcatum*.

CYATHEA

187 Anon.
(Materials for the distribution of vascular plants in Japan: 82). *J. Jap. Bot.*, 48(8):246(1973)—Ja. *Cyathea*
2 *spinulosa*.

DICRANOPTERIS

elegantula Pichi-Sermolli in *Webbia*, 27(2):402 (1972 publ. 1973)—Zaïre, Sierra Leone, Liberia,
5 Ivory Coast, Ghana, Gabon.

DIDYMOGLOSSUM

curtii (Rosenst.) Pichi-Sermolli in *Webbia*, 28(2):469
4 (1973): *Trichomanes curtii*.
gourlianum (Grev. ex J. Smith) Pichi-Sermolli *op.*
4 *cit.* 469(1973): *Trichomanes gourlianum*.
pinnatinervium (Jenm.) Pichi-Sermolli *op. cit.*
4 469(1973): *Trichomanes pinnatinervium*.
rhipidophyllum (Slosson) Pichi-Sermolli *op. cit.*
4 469(1973): *Trichomanes rhipidophyllum*.

DIPHASIUM

hastulatum Siplivinskiĭ in *Novosti Sist. Vȳssh. Rast.*,
2 10:348(1973)—U.S.S.R. (Siberia).

DIPLAZIUM

humbertii (C. Chr.) Pichi-Sermolli in *Webbia*,
5 27(2):444(1972 publ. 1973): *Athyrium humbertii*.
megistophyllum (Copel.) Tagawa in *Acta Phytotax.*
6 *Geobot.*, 25(4–6):180(1973): *Athyrium megistophyllum*.
velaminosum (Diels) Pichi-Sermolli in *Webbia*,
5 27(2):443(1972 publ. 1973): *Hemitelia velaminosa*.

DORYOPTERIS

188 Backes, A.
Contribuição ao estudo do gênero *Doryopteris* J. Smith (Polypodiaceae). *An. Acad. Brasil. Cienc.*,
4 43(3–4):767–790(1971)—(En); illustrations, maps.

angustata Sehnem in Fl. Illustr. Catar., 1 (Pteridác.):
4 140(1972)—Brazil.
multipartita (Fée) Sehnem *op. cit.* 176(1972):
4 *D. raddiana* var. *multipartita*.

DRYOPTERIS

189 Benl, G. *and* Eschelmüller, A.
Über *Dryopteris remota* und ihr Vorkommen in Bayern. *Ber. Bayer. Bot. Ges.*, 44:101–141(1973)—
1 Illustrations, map; chromosome numbers.

190 Dale Thomas, R., Wagner, W. H. *and* Mesler, M. R.
Log fern (*Dryopteris celsa*) and related species in Louisiana. *Castanea*, 38(3):269–274(1973)—
3 Illustrations; chromosome numbers.

191 Eschelmüller, A. *and* Bär, A.
Dryopteris assimilis S. Walker—ein übersehener Farn im Allgäu. *Mitt. Naturwiss. Arbeitskr. Kempten*
1 *(Allgäu)*, 17(2):33–44(1973)—Chromosome number.

192 Key, J. S. *and* Redfearn, P. L.
Additions and a deletion to the fern flora of Missouri. *Amer. Fern J.*, 63(1):1–2(1973). *Dryopteris*
3 *celsa*.

193 Piękoś, H. *and* Passakas, T.
The chromosome number in *Dryopteris assimilis* S. Walker from the west Bieszczady Range (East Carpathian Mts.). (Liczba chromosomów u *Dryopteris assimilis* S. Walker z Bieszczadów Zachodnich). *Fragm. Flor. Geobot.*, 19(3):305–308
1 (1973)—(Pol); illustrations.

194 Sledge, W. A.
The Dryopteroid ferns of Ceylon. *Bull. Brit. Mus. (Nat. Hist.), Bot.*, 5(1). 43p.(1973)—Illustrations;
6 key.

195 Steinwendtner, R.
Dryopteris filix mas Agg. *Mitt. Bot. Arbeit. Oberösterr.*
1–6 *Landesmus. Linz*, 4(1–2):45(1972)—Ge.

196 Widén, C.-J., Faden, R. B., Lounasmaa, M., Vida, G., Euw, J. von *and* Reichstein, T.
Die Phloroglucide von neun *Dryopteris* Arten aus Kenya sowie der *D. oligodonta* (Desv.) Pic.-Serm.und *D.* '*dilatata*' von den Canarischen Inseln. *Helv. Chim. Acta*, 56(7):2125–2151(1973)—(En); chromosome
5 numbers.

197 Widén, C.-J., Fraser-Jenkins, C. R., Lounasmaa, M., Euw, J. von *and* Reichstein, T.
Die Phloroglucide von *Dryopteris caucasica* (A.Br.) Fraser-Jenkins et Corley. *Helv. Chim. Acta*, 56(3):
2 831–838(1973)—(En).

ambigua Sledge in *Bull. Brit. Mus. (Nat. Hist.), Bot.*,
6 5(1):23(1973)—Sri Lanka.
approximata Sledge *op. cit.* 11(1973)—Sri Lanka,
6 India.
barbigera (Moore) O. Kze.
var. falconeri (Hk.) R. R. Stewart, Annot. Cat. Vasc. Pl. W. Pakistan & Kashmir: 11(1972),
6 without basionym ref.: *Nephrodium falconeri.*
dilatata A. Gray
var. **chanterae** (Moore) Soó in *Acta Bot. Acad. Sci. Hung.*, 18(1–2):177(1973), as 'chanteriae':
1 *Lastraea dilatata* var. *chanteriae.*
var. **collina** (Newm.) Soó *op. cit.* 177(1973):
1 *Lastraea dilatata* var. *collina.*
var. **dumetorum** (Sm.) Soó *op. cit.* 177(1973):
1 *Aspidium dumetorum.*
gracillima Ching
var. **prolongata** Sledge in *Bull. Brit. Mus. (Nat.*
6 *Hist.), Bot.*, 5(1):19(1973)—Sri Lanka.
var. **triangularis** Sledge *op. cit.* 19(1973)—Sri
6 Lanka.

ELAPHOGLOSSUM

198 Gómez P., L. D.
Contribuciones à la pteridología costarricense. 3. Dos nuevas especies y una nota sobre *Neurocallis* Fée. *Rev. Biol. Trop.*, 20(2):181–187(1972)—(En); illustrations of type specimens of new species of
4 *Elaphoglossum.*

yuhgense de la Sota in *Darwiniana*, 18(1–2):246
4 (1973)—Argentina.

GYMNOGRAMME

199 Guervin, C. *and* Le Coq, C.
La tige et l'organisation post-embryonnaire du *Gymnogramme sulphurea* Desv.: 1. Données anatomiques. (The stem and the post embryonic organisation of *Gymnogramme sulphurea* Desv. 1. Anatomical data). *Bull. Soc. Bot. France*, 119(5–6):
4 383–398(1972 publ. 1973)—(En).

GYMNOPTERIS

200 Kaur, S.
Morphology of the prothallus of *Gymnopteris vestita. Phytomorphology*, 22(1):46–49(1972 publ.
4, 6 1973).

HELMINTHOSTACHYS

201 Abeywickrama, B. A. *and* De Fonseka, R. N.
The Ceylonese Ophioglossaceae. *Ceylon J. Sci.*,
6 *Biol. Sci.*, 10(2):132–142(1973)—Illustration; keys.

HICRIOPTERIS

202 Holttum, R. E.
On the typification of *Mertensia* Willd. non Roth (Gleicheniaceae) with notes on *Sticherus* Presl and
6 *Hicriopteris* Presl. *Taxon*, 22(4):447–450(1973).

HYMENOPHYLLUM

203 Mendes, E. J.
Additiones et adnotationes florae Mozambicanae: 4. (Pteridophyta). *Garcia de Orta, Sér. Bot.*, 1(1–2):19–20
5 (1973)—(En).

HYPOLEPIS

204 Imaichi, R. *and* Nishida, M.
(Studies on the extra-axillary buds of *Hypolepis punctata*: 1. External morphology and anatomy).
2 *J. Jap. Bot.*, 48(9):268–280(1973)—Ja(En); illustra tions.

KUNIWATSUKIA Pichi-Sermolli in *Webbia*,
2, 6 28(2):455(1973).
cuspidata (Bedd.) Pichi-Sermolli *op. cit.* 455(1973):
6 *Lastrea cuspidata.*

LASTREA

205 Chandra, P.
Tetrahedral spores in another species of *Lastrea.*
6 *Amer. Fern J.*, 63(1):9–11(1973). *L. tenericaulis.*

LECANOLEPIS Pichi-Sermolli in *Webbia*, 28(2):
4 449(1973).
membranacea (L.) Pichi-Sermolli *op. cit.* 449(1973):
4 *Trichomanes membranaceum.*

LEPISORUS

206 Bir, S. S. *and* Trikha, C. K.
Taxonomic revision of the polypodiaceous genera of India: 6. *Lepisorus excavatus* group. *Proc. Indian*
6 *Sci. Congr.*, 60(3):402–403(1973). Abstract only.

amaurolepida (Sledge) Bir & Trikha
var. longifolius Bir & Trikha in *Proc. Indian Sci. Congr.*, 60(3):403(1973), without latin descr.—
6 India.
excavatus (Bory) Ching
var. himalayensis Bir & Trikha *op. cit.* 403(1973),
6 without latin descr.—India.
var. mortonii Bir & Trikha *op. cit.* 403(1973),
6 without latin descr.—India.

LINDSAEA

207 Brownlie, G.
The genus *Lindsaea* in Fiji. *Amer. Fern J.*, 63(3):
6 91–98(1973)—Key.

pulchra (Brackenr.) Carr. ex Seem.
var. **protracta** (Copel.) Brownlie in *Amer. Fern J.*,
6 63(3):97(1973): *L. protracta.*
quadrangularis Raddi
var. **pallecens** Sehnem in Fl. Ilustr. Catar., 1
4 (Pteridác.): 38(1972)—Brazil.

LOXOSCAPHE

theciferum (H.B.K.) Moore
var. **cornutum** (Alston) Pichi-Sermolli in *Webbia*,
5 28(2):469(1973): *Asplenium cornutum.*

MARGINARIA

bombycina (Maxon) Pichi-Sermolli in *Webbia*,
4 28(2):469(1973): *Polypodium bombycinum.*
tridens (Kunze) Pichi-Sermolli *op. cit.* 469(1973):
4 *Polypodium tridens.*

MARSILEA

208 Launert, E.
Tropical African plants: 33. An overlooked synonym of *Marsilea minuta* (Marsileaceae). *Kew Bull.*,
5 28(2):319(1973).

209 Mendes, E. J.
Additiones et adnotationes florae Mozambicanae: 4. (Pteridophyta). *Garcia de Orta, Sér. Bot.*, 1(1–2):
5 19–20(1973)—(En).

MATTEUCCIA

210 Lloyd, R. M.
Facultative apomixis and polyploidy in *Matteuccia orientalis*. *Amer. Fern J.*, 63(2):43–48(1973)—
2, 6 Chromosome numbers.

MECODIUM

capense (Schrad.) Pichi-Sermolli in *Webbia*, 27(2):
5 404(1972 publ. 1973): *Hymenophyllum capense*.
endiviifolium (Desv.) Pichi-Sermolli in *Webbia*,
4 28(2):469(1973): *Hymenophyllum endiviaefolium*.
trichomanoides (v. d. Bosch) Pichi-Sermolli *op. cit.*
4 469(1973): *Hymenophyllum trichomanoides*.

MERTENSIA

211 Holttum, R. E.
On the typification of *Mertensia* Willd. non Roth (Gleicheniaceae) with notes on *Sticherus* Presl and
6 *Hicriopteris* Presl. *Taxon*, 22(4):447–450(1973).

MICROGRAMMA

212 De la Sota, E. R.
A new species of *Microgramma* from Argentina.
4 *Amer. Fern J.*, 63(3):61–64(1973).

× **mortoniana** E.R. de la Sota in *Amer. Fern J.*, 63(3):64(1973)—Argentina. (*M. squamulosa* ×
4 *M. vacciniifolia*).

MICROLEPIA

fadenii Pichi-Sermolli in *Webbia*, 27(2):406(1972
5 publ. 1973)—Tanzania.

MICROSORIUM

213 Anon.
Microsorium pteropus (Blume) Ching. *Inform. ZAG Wasserpflanzen*, no. 3, *Beilage*. [2]p.(1973)—Ge;
2, 6 illustration.

NEGRIPTERIS

tricholepifera Pichi-Sermolli in *Webbia*, 27(2):411
5 (1972 publ. 1973)—Ethiopia.

NEPHELEA

214 Gastony, G. J.
A revision of the fern genus *Nephelea*. *Contrib. Gray Herb.*, no. 203:81–148(1973)—Illustrations, dot
4 maps; keys.

erinacea (Karst.) Tryon
var. **purpurascens** (Sod.) Gastony in *Contrib. Gray*
4 *Herb.*, 203:122(1973): *Cyathea purpurascens*.
fulgens (C.Chr.) Gastony *op. cit.* 104(1973): *Cyathea*
4 *fulgens*.
imrayana (Hook.) Tryon
var. **basilaris** (Christ) Gastony *op. cit.* 127(1973):
4 *Cyathea basilaris*.
incana (Karst.) Gastony *op. cit.* 137(1973): *Cyathea*
4 *incana*.
sternbergii (Sternb.) Tryon
var. **acanthomelas** (Fée) Gastony *op. cit.* 135
4 (1973): *Cyathea acanthomelas*.
tryoniana Gastony *op. cit.* 118(1973)—Guatemala,
4 Honduras, Nicaragua.
woodwardioides (Kaulf.) Gastony *op. cit.* 107(19 73):
4 *Cyathea woodwardioides*.

var. **cubensis** (Maxon) Gastony *op. cit.* 110(1973):
4 *Cyathea cubensis*.
var. **hieronymi** (Brause) Gastony *op. cit.* 112
4 (1973): *Cyathea hieronymi*.

NEUROCALLIS

215 Gómez P., L. D.
Contribuciones à la pteridología costarricense. 3. Dos neuvas especies y una nota sobre *Neurocallis* Fée. *Rev. Biol. Trop.*, 20(2):181–187(1972)—(En); illustrations of type specimens of new species of
4 *Elaphoglossum*.

NOTHOLAENA

216 Knobloch, I. W., Tai, W. *and* Ninan, T. A.
The cytology of some species of the genus *Notholaena*. *Amer. J. Bot.*, 60(1):92–95(1973)—
3, 4 Chromosome numbers.

perlanata Pichi-Sermolli in *Webbia*, 27(2):417
5 (1972 publ. 1973)—Zaïre.

OPHIOGLOSSUM

217 Abeywickrama, B. A. *and* De Fonseka, R. N.
The Ceylonese Ophioglossaceae. *Ceylon J. Sci.*,
6 *Biol. Sci.*, 10(2):132–142(1973)—Illustration; keys.

218 Goswami, H. K. *and* Khandelwal, S.
Abnormalities in *Ophioglossum*. *Brit. Fern Gaz.*,
1–7 10(6):311–314(1973).

219 Mesler, M. R.
Sexual reproduction in *Ophioglossum crotalophor-*
3, 4 *oides*. *Amer. Fern J.*, 63(2):28–33(1973)—Illustrations.

ORTHIOPTERIS

brasiliensis (Pr.) Sehnem in Fl. Ilustr. Catar.,
4 1(Pteridác.):5(1972): *Microlepia brasiliensis*.

OSMUNDA

220 Avidsson, L., Borén, L., Hallingbäck, T., Sahlin, E. *and* Wendelbo, P.
Osmunda regalis funnet i Hardanger. (*Osmunda regalis* found in Hardanger, west Norway). *Blyttia*, 31(4):195–198(1973)—(En); illustration, dot
1 maps.

221 Gibson, H. L.
An unusual royal fern. *New York Bot. Gard. Newsl.*, no. 12:3(1973)—Illustrations. *Osmunda regalis* in
3 America.

222 Klekowski, E. J.
Genetic load in *Osmunda regalis* populations. *Amer.*
1–6 *J. Bot.*, 60(2):146–154(1973).

223 Shimura, Y.
(Study of reproduction of *Osmunda* × *intermedia* Sugimoto). *J. Geobot.*, 20(2):38–42(1972)—Ja(En);
2 illustrations.

PELLAEA

224 Rigby, S. J.
Chromosome pairing in obligately apogamous ferns; *Pellaea atropurpurea* and *Pellaea glabella* var.
3 *glabella*. *Rhodora*, 75(801):122–131(1973).

PELTAPTERIS

225 Gómez P., L. D.
Una nueva especie del género *Peltapteris* Link de los Andes peruanos. *Rev. Biol. Trop.*, 18(1–2):217–
4 220(1970 publ. 1971)—(En); illustration.

PERANEMA

226 Sledge, W. A.
The Dryopteroid ferns of Ceylon. *Bull. Brit. Mus. (Nat. Hist.), Bot.*, 5(1). 43p.(1973)—Illustrations;
6 key.

PHLEGMARIURUS

sooanus Lawalrée in *Acta Bot. Acad. Sci. Hung.*,
5 19(1–4):195(1973), as 'sooianus'—Zaïre.
5 **tournayanus** Lawalrée *op. cit.* 197(1973)—Zaïre.

PHYLLITIS

227 Löve, Å. *and* Löve, D.
Cytotaxonomy of the boreal taxa of *Phyllitis*. *Acta Bot. Acad. Sci. Hung.*, 19(1–4):201–206(1973)—
1–4 Chromosome numbers.

PHYMATOPTERIS Pichi-Sermolli in *Webbia*,
2, 6 28(2):460(1973), nom. nov.: *Phymatopsis* J. Smith.
albido-squamata (Blume) Pichi-Sermolli *op. cit.*
6 461(1973): *Polypodium albido-squamatum*.
albopes (C. Chr. & Ching) Pichi-Sermolli *op. cit.*
2 461(1973): *Polypodium albopes*.
chenkouensis (Ching) Pichi-Sermolli *op. cit.* 461
2 (1973): *Phymatopsis chenkouensis*.
chrysotricha (C. Chr.) Pichi-Sermolli *op. cit.* 461
2 (1973): *Polypodium chrysotrichum*.
conjuncta (Ching) Pichi-Sermolli *op. cit.* 461(1973):
2 *Phymatopsis conjuncta*.
conmixta (Ching) Pichi-Sermolli *op. cit.* 461(1973):
2 *Phymatodes conmixta*.
connexa (Ching) Pichi-Sermolli *op. cit.* 461(1973):
2 *Phymatodes connexa*.
crenato-pinnata (Clarke) Pichi-Sermolli *op. cit.* 461
2, 6 (1973): *Polypodium crenato-pinnatum*.
cruciformis (Ching) Pichi-Sermolli *op. cit.* 462
2 (1973): *Polypodium cruciforme*.
cunea (Ching) Pichi-Sermolli *op. cit.* 462(1973):
2 *Phymatopsis cunea*.
dactylina (Christ) Pichi-Sermolli *op. cit.* 462(1973):
2 *Polypodium dactylinum*.
digitata (Ching) Pichi-Sermolli *op. cit.* 462(1973):
2 *Phymatodes digitata*.
ebenipes (Hook.) Pichi-Sermolli *op. cit.* 462(1973):
2, 6 *Polypodium ebenipes*.
echinospora (Tagawa) Pichi-Sermolli *op. cit.* 462
2 (1973): *Phymatodes echinospora*.
engleri (Luerss.) Pichi-Sermolli *op. cit.* 462(1973):
2 *Polypodium engleri*.
erythrocarpa (Mett. ex Kuhn) Pichi-Sermolli *op. cit.*
6 462(1973): *Polypodium erythrocarpum*.
fukienensis (Ching) Pichi-Sermolli *op. cit.* 462(1973):
2 *Phymatopsis fukienensis*.
glaucopsis (Franch.) Pichi-Sermolli *op. cit.* 462(1973):
2 *Polypodium glaucopsis*.
griffithiana (Hook.) Pichi-Sermolli *op. cit.* 462(1973):
2, 6 *Polypodium griffithianum*.
hainanensis (Ching) Pichi-Sermolli *op. cit.* 462(1973):
2 *Phymatodes hainanensis*.
hastata (Thunb.) Pichi-Sermolli *op. cit.* 462(1973):
2 *Polypodium hastatum*.
hirtella (Ching) Pichi-Sermolli *op. cit.* 462(1973):
2 *Phymatopsis hirtella*.
hunyaensis (Ching) Pichi-Sermolli *op. cit.* 462(1973):
2 *Phymatopsis hunyaensis*.
intermedia (Ching) Pichi-Sermolli *op. cit.* 462(1973):
2 *Phymatodes intermedia*.
kingpingensis (Ching) Pichi-Sermolli *op. cit.* 463
2 (1973): *Phymatopsis kingpingensis*.
kwangtungensis (Ching) Pichi-Sermolli *op. cit.*
2 463(1973): *Phymatodes kwangtungensis*.
laciniata (Presl) Pichi-Sermolli *op. cit.* 463(1973):
6 *Phymatodes laciniata*.
lagunensis (Christ) Pichi-Sermolli *op. cit.* 463(1973):
6 *Polypodium lagunense*.
laipoensis (Ching) Pichi-Sermolli *op. cit.* 463(1973):
2 *Phymatopsis laipoensis*.
likiangensis (Ching) Pichi-Sermolli *op. cit.* 463(1973):
2 *Phymatopsis likiangensis*.
longisquamata (Tagawa) Pichi-Sermolli *op. cit.* 463
2 (1973): *Phymatodes longisquamata*.
macrochasma (Bak.) Pichi-Sermolli *op. cit.* 463(1973):
6 *Polypodium macrochasmum*.
majoensis (C. Chr.) Pichi-Sermolli *op. cit.* 463
2 (1973): *Polypodium majoense*.
malacodon (Hook.) Pichi-Sermolli *op. cit.* 463
2, 6 (1973): *Polypodium malacodon*.
montana (Sledge) Pichi-Sermolli *op. cit.* 463(1973):
6 *Crypsinus montanus*.
nigrovenia (Christ) Pichi-Sermolli *op. cit.* 463(1973):
2 *Polypodium shen-siense* var. *nigrovenium*.
obtusa (Ching) Pichi-Sermolli *op. cit.* 463(1973):
2 *Phymatopsis obtusa*.
omeiensis (Ching) Pichi-Sermolli *op. cit.* 463(1973):
2 *Phymatopsis omeiensis*.
oxyloba (Wall. ex Kunze) Pichi-Sermolli *op. cit.*
2, 6 464(1973): *Polypodium oxylobum*.
palmata (Blume) Pichi-Sermolli *op. cit.* 464(1973):
6 *Polypodium palmatum*.
pellucidifolia (Hayata) Pichi-Sermolli *op. cit.* 464
2 (1973): *Polypodium pellucidifolium*.
philippinensis (Ching) Pichi-Sermolli *op. cit.* 464
6 (1973): *Phymatopsis philippinensis*.
pingpienensis (Ching) Pichi-Sermolli *op. cit.* 464
2 (1973): *Phymatopsis pingpienensis*.
quasidivaricata (Hayata) Pichi-Sermolli *op. cit.* 464
2 (1973): *Polypodium quasidivaricatum*.
rhynchophylla (Hook.) Pichi-Sermolli *op. cit.* 464
2, 6 (1973): *Polypodium rhynchophyllum*.
roseo-marginata (Ching) Pichi-Sermolli *op. cit.* 464
2 (1973): *Phymatodes roseo-marginata*.
rotunda (Ching) Pichi-Sermolli *op. cit.* 464(1973):
2 *Phymatopsis rotunda*.
shen-siensis (Christ) Pichi-Sermolli *op. cit.* 464
2 (1973): *Polypodium shen-siense*.
simplicifolia (Ching) Pichi-Sermolli *op. cit.* 464
2 (1973): *Phymatopsis simplicifolia*.
stewartii (Bedd.) Pichi-Sermolli *op. cit.* 464(1973):
2, 6 *Pleopeltis stewartii*.
stracheyi (Ching) Pichi-Sermolli *op. cit.* 464(1973):
2 *Phymatodes stracheyi*.
subebenipes (Ching) Pichi-Sermolli *op. cit.* 464
2 (1973): *Phymatopsis subebenipes*.
suboxyloba (Ching) Pichi-Sermolli *op. cit.* 464(1973):
2 *Phymatopsis suboxyloba*.
taeniata (Swartz) Pichi-Sermolli *op. cit.* 465(1973):
2, 6 *Polypodium taeniatum*.
taiwanensis (Tagawa) Pichi-Sermolli *op. cit.* 465
2 (1973): *Phymatodes taiwanensis*.
tenuipes (Ching) Pichi-Sermolli *op. cit.* 465(1973):
2 *Phymatopsis tenuipes*.
triloba (Houtt.) Pichi-Sermolli *op. cit.* 465(1973):
2, 6 *Polypodium trilobum*.
trisecta (Bak.) Pichi-Sermolli *op. cit.* 465(1973):
2 *Polypodium trisectum*.
veitchii (Bak.) Pichi-Sermolli *op. cit.* 465(1973):
2, 6 *Polypodium veitchii*.
yakuinsularis (Masam.) Pichi-Sermolli *op. cit.* 465
2 (1973): *Polypodium yakuinsulare*.
yakushimensis (Mak.) Pichi-Sermolli *op. cit.* 465
2 (1973): *Polypodium engleri* var. *yakushimense*.

PHYMATOSORUS Pichi-Sermolli in *Webbia*,
2, 5–7 28(2):457(1973).

commutatus (Blume) Pichi-Sermolli *op. cit.* 459
6 (1973): *Polypodium commutatum.*
diversifolius (Willd.) Pichi-Sermolli *op. cit.* 459
6 (1973): *Polypodium diversifolium.*
ensiformis (Thunb.) Pichi-Sermolli *op. cit.* 459
5 (1973): *Polypodium ensiforme.*
longissimus (Blume) Pichi-Sermolli *op. cit.* 459
2, 6 (1973): *Polypodium longissimum.*
lucidus (Roxb.) Pichi-Sermolli *op. cit.* 459(1973):
2, 6 *Polypodium lucidum.*
nigrescens (Blume) Pichi-Sermolli *op. cit.* 459(1973):
6 *Polypodium nigrescens.*
novae-zealandiae (Bak.) Pichi-Sermolli *op. cit.* 459
7 (1973): *Polypodium novae-zealandiae.*
powellii (Bak.) Pichi-Sermolli *op. cit.* 459(1973):
6 *Polypodium powellii.*
scandens (G. Forst.) Pichi-Sermolli *op. cit.* 459(1973):
7 *Polypodium scandens.*
scolopendria (N.L. Burm.) Pichi-Sermolli *op. cit.*
2, 5–7 460(1973): *Polypodium scolopendria.*
suisha-stagnalis (Hayata) Pichi-Sermolli *op. cit.*
2 460(1973): *Polypodium suisha-stagnale.*
variabilis (Ching) Pichi-Sermolli *op. cit.* 460(1973):
2 *Phymatodes variabilis.*

PITYROGRAMMA

rupicola Pichi-Sermolli in *Webbia*, 27(2):425(1972
5 publ. 1973)—Zaïre, Burundi.

PLEOPELTIS

loriformis (Wall ex Mett.) Moore
var. steniste (Clarke) R. R. Stewart, Annot. Cat.
Vasc. Pl. W. Pakistan & Kashmir: 20(1972),
without basionym ref.: *Polypodium lineare* var.
6 *steniste.*

subconfluens (Ching) Pichi-Sermolli in *Webbia*,
2 28(2):469(1973): *Lepisorus subconfluens.*

PNEUMATOPTERIS

228 Holttum, R. E.
Studies in the family Thelypteridaceae: 5. The
genus *Pneumatopteris* Nakai. *Blumea*, 21(2):293–325
5–7 (1973)—Key.

afra (Chr.) Holttum in *Blumea*, 21(2):306(1973):
5 *Dryopteris afra.*
angusticaudata Holttum *op. cit.* 308(1973)—New
6 Guinea (Papua).
auctipinna Holttum *op. cit.* 310(1973)—Celebes,
6 Java, Lesser Sunda Is., Mollucas.
6 **basicurtata** Holttum *op. cit.* 309(1973)—Sumatra.
brooksii (Copel.) Holttum *op. cit.* 308(1973):
6 *Dryopteris brooksii.*
bryanii (C. Chr.) Holttum *op. cit.* 317(1973):
6 *Dryopteris bryanii.*
caudata (Holttum) Holttum *op. cit.* 321(1973):
6 *Pseudocyclosorus caudatus.*
cheesmaniae Holttum *op. cit.* 307(1973)—New
6 Guinea (W. Irian).
christelloides Holttum *op. cit.* 311(1973)—Borneo
6 (Sabah).
comorensis Holttum *op. cit.* 318(1973)—Comoro
5 Archipelago.
costata (Brack.) Holttum *op. cit.* 305(1973):
6, 7 *Goniopteris costata.*
deficiens Holttum *op. cit.* 321(1973)—Lesser Sunda
6 Is.
dicranogramma (v. A. v. R.) Holttum *op. cit.* 301
6 (1973): *Dryopteris dicranogramma.*
ecallosa (Holttum) Holttum *op. cit.* 310(1973):
6 *Cyclosorus ecallosus.*
egenolfioides Holttum *op. cit.* 306(1973)—New
6 Guinea (W. Irian).
excisa (Holttum) Holttum *op. cit.* 321(1973):
6 *Pseudocyclosorus excisus.*
glaberrima (Richard) Holttum *op. cit.* 318(1973):
6 *Aspidium glaberrimum.*
glabra (Copel.) Holttum *op. cit.* 301(1973):
6 *Cyclosorus glaber.*
glandulifera (Brack.) Holttum *op. cit.* 302(1973):
6 *Goniopteris glandulifera.*
hudsoniana (Brack.) Holttum *op. cit.* 317(1973):
6 *Nephrodium hudsonianum.*
5 **humbertii** Holttum *op. cit.* 313(1973)—Malagasy.
6 **imbricata** Holttum *op. cit.* 322(1973)—Mollucas.
6 **incisa** Holttum *op. cit.* 317(1973)—Mollucas.
inclusa (Copel.) Holttum *op. cit.* 306(1973):
6 *Dryopteris inclusa.*
jermyi Holttum *op. cit.* 310(1973)—New Guinea
6 (Terr. New Guinea).
6 **kerintjiensis** Holttum *op. cit.* 312(1973)—Sumatra.
keysserana (Rosenst.) Holttum *op. cit.* 320(1973), as
6 'keysseriana': *Dryopteris keysserana.*
laevis (Mett.) Holttum *op. cit.* 308(1973): *Aspidium*
6 *laeve.*
laticuneata Holttum *op. cit.* 312(1973)—New Guinea
6 (Terr. New Guinea).
latisquamata Holttum *op. cit.* 305(1973)—New
6 Guinea (Terr. New Guinea).
ligulata (Presl) Holttum *op. cit.* 320(1973): *Lastrea*
6 *ligulata.*
longipes (Bl.) Holttum *op. cit.* 306(1973): *Aspidium*
6 *longipes.*
lucida (Bak.) Holttum *op. cit.* 307(1973): *Nephrodium*
6 *lucidum.*
macroptera (Copel.) Holttum *op. cit.* 316(1973):
6 *Dryopteris macroptera.*
magnifica (Copel.) Holttum *op. cit.* 319(1973):
6 *Dryopteris magnifica.*
mesocarpa (Copel.) Holttum *op. cit.* 303(1973):
6 *Dryopteris mesocarpa.*
michaelis Holttum *op. cit.* 313(1973)—Philippine Is.,
6 Borneo (Sabah).
microauriculata Holttum *op. cit.* 311(1973)—
6 Borneo (Sabah).
microloncha (Chr.) Holttum *op. cit.* 308(1973):
6 *Dryopteris microloncha.*
micropaleata Holttum *op. cit.* 319(1973)—Borneo
6 (Sabah).
nephrolepioides (C. Chr.) Holttum *op. cit.* 320
6 (1973): *Dryopteris nephrolepioides.*
nitidula (Presl) Holttum *op. cit.* 318(1973):
6 *Nephrodium nitidulum.*
novae-caledoniae Holttum *op. cit.* 313(1973)—
6 New Caledonia.
6 **obliqua** Holttum *op. cit.* 309(1973)—Celebes.
obstructa (Copel.) Holttum *op. cit.* 302(1973):
6 *Dryopteris obstructa.*
oppositifolia (Hook.) Holttum *op. cit.* 304(1973):
5 *Polypodium oppositifolium.*
oxyoura (Copel.) Holttum *op. cit.* 315(1973):
6 *Dryopteris oxyoura.*
papuana Holttum *op. cit.* 311(1973)—New Guinea
6 (Papua).
parksii (Ballard) Holttum *op. cit.* 302(1973):
6 *Dryopteris parksii.*
patentipinna Holttum *op. cit.* 307(1973)—New
6 Guinea (Terr. New Guinea).
pennigera (Forst.) Holttum *op. cit.* 305(1973):
7 *Polypodium pennigerum.*
pergamacea Holttum *op. cit.* 315(1973)—New
6 Ireland.
petrophila (Copel.) Holttum *op. cit.* 322(1973):
6 *Dryopteris petrophila.*

prismatica (Desv.) Holttum *op. cit.* 316(1973):
5 *Nephrodium prismaticum.*
regis (Copel.) Holttum *op. cit.* 322(1973)—
6 *Dryopteris regis.*
remotipinna (Bonap.) Holttum *op. cit.* 311(1973):
5 *Dryopteris remotipinna.*
rodigasiana (T. Moore) Holttum *op. cit.* 319(1973):
6 *Nephrodium rodigasianum.*
rotumaensis (St. John) Holttum *op. cit.* 318(1973):
6 *Cyclosorus rotumaensis.*
sandwicensis (Brack.) Holttum *op. cit.* 303(1973):
6 *Stegnogramma sandwicensis.*
6 **sibelana** Holttum *op. cit.* 301(1973)—Moluccas.
sogerensis (Gepp) Holttum *op. cit.* 315(1973):
6, 7 *Dryopteris sogerensis.*
stokesii (E. Brown) Holttum *op. cit.* 302(1973):
6 *Dryopteris stokesii.*
subappendiculata (Copel.) Holttum *op. cit.* 303
6 (1973): *Dryopteris subappendiculata.*
subpennigera (C. Chr.) Holttum *op. cit.* 304(1973):
5 *Dryopteris subpennigera.*
sumbawensis (C. Chr.) Holttum *op. cit.* 323(1973):
6 *Dryopteris sumbawensis.*
superba (Brause) Holttum *op. cit.* 303(1973):
6 *Dryopteris superba.*
6 **tobaica** Holttum *op. cit.* 316(1973)—Sumatra.
transversaria (Brack.) Holttum *op. cit.* 319(1973):
6 *Nephrodium transversarium.*
truncata (Poir.) Holttum *op. cit.* 314(1973):
2, 6 *Polypodium truncatum.*
var. **loyalii** Holttum *op. cit.* 314(1973)—India,
6 Assam.
unita (Kunze) Holttum *op. cit.* 304(1973):
5 *Gymnogramma unita.*
usambarensis Holttum *op. cit.* 312(1973)—
5 Tanzania.
vaupelii (C. Chr.) Holttum *op. cit.* 317(1973):
6 *Dryopteris vaupelii.*
venulosa (O. Kuntze) Holttum *op. cit.* 315(1973):
5 *Nephrodium venulosum.*
versteeghii Holttum *op. cit.* 321(1973)—New Guinea
6 (W. Irian).
walkeri Holttum *op. cit.* 323(1973)—New Guinea
6 (Terr. New Guinea).

POLYPODIUM

229 Grzybek, J.
Biological and phytochemical investigations on *Polypodium vulgare* L.: 1. Anatomical analysis. *Acta*
1–3, 5, 6 *Biol. Cracov., Bot.*, 16(1):29–35(1973).

230 Novruzova, Z. A. *and* Askerov, A. M.
Osobennosti anatomicheskoj struktury̆ novykh taksonov paporotnikov iz Azerbaĭdzhana. *Izv. Akad. Nauk Az. SSR, Biol. Nauk*, no. 1:3–6(1973)—(Az).
2 *Polypodium vulgare* subsp. *issaevii* and var. *zuvandicum.*

231 Planderová, E.
Morphological differences among spores of some species of the genus *Polypodium* L. as showed by the study of their sexine microstructure. (Morfologické rozdiely spór niektorých druhov rodu *Polypodium* L. ną základe študia mikroštruktúry ich sexiny). *Biológia, A* (Czechoslovakia), 28(10):813–817(1973)—(Cz,
1, 3–5 Rus).

232 Rutherford, A. *and* Stirling, A. McG.
Observations on *Polypodium australe* Fée in Scotland.
1 *Watsonia*, 9(4):357–361(1973).

233 Wróblewski, A.
Paprotka zwyczajna *Polypodium vulgare* jako epifyt. (The common polypody as an epiphyte). *Chrońmy Przyr. Ojczystą*, 29(4):47–48(1973)—
1 Illustration.

POLYSTICHUM

234 Anon.
Sur deux stations de *Polystichum cristatum* du Massif de Fontainebleau. *Bull. Ass. Nat. Vallée Loing Massif*
1 *Fontainebleau*, 48(7–8):88–89(1972).

235 Becherer, A.
Sulla distribuzione di *Polystichum setiferum* (Forskal) Th. Moore nella Svizzera transalpina e nelle zone italiane di confine. *Boll. Soc. Ticin. Sci. Nat.*,
1 63:22–31(1972–73 publ. 1973).

236 Chojnacki, W.
Stanowiska *Polystichum lobatum* (Huds.) Sw. na zboczach nadmorskich kolo Ustki. (Localities of *Polystichum lobatum* (Huds.) Sw. on the sea cliffs near Ustka (Pomerania)). *Fragm. Flor. Geobot.*, 19(3):285–
1 289(1973)—(En); illustrations, dot map.

237 Daigobo, S.
Chromosome numbers of the fern genus *Polystichum*:
1–3, 6 1. *J. Jap. Bot.*, 48(11):337–344(1973).

238 Nakaike, T.
Studies in fern genus *Polystichum*: 1. Observations on the section *Metapolystichum* at Gobô-sawa, Pref. Chiba. *Bull. Nation. Sci. Mus.* (Tokyo), 16(3):437–457 (1973)—Illustrations, dot maps; chromosome
2 number.

239 Nakaike, T. *and* Sakakibara, M.
(Re-examination of the specimen cited in *Catalogue of Japanese Plants in the Herbarium of the Natural History Department, Tokyo Imperial Museum*: 1. *Polystichum*). *J. Geobot.*, 20(1):10–13(1972)—Ja,
2 En.

240 Novruzova, Z. A. *and* Askerov, A. M.
Osobennosti anatomicheskoj struktury̆ novykh taksonov paporotnikov iz Azerbaĭdzhana. *Izv. Akad. Nauk Az. SSR, Biol. Nauk*, no. 1:3–6(1973)—(Az).
2 *Polystichum kadyrovii, P. × safarovii.*

241 Sleep, A. *and* Synnott, D.
Polystichum × illyricum: a hybrid new to the British Isles. *B.S.B.I. News*, 2(1):10(1973). Full report
1 in *Brit. Fern Gaz.*, 10(5):281–282(1972).

242 Wagner, W. H.
Reticulation of holly ferns (*Polystichum*) in the western United States and adjacent Canada. *Amer. Fern J.*, 63(3):99–115(1973)—Illustrations, dot map;
3 chromosome numbers.

anomalum (Hook. & Arn.) J. Smith
var. **travancoricum** (Bedd.) Sledge in *Bull. Brit. Mus. (Nat. Hist.), Bot.*, 5(1):33(1973): *P. aculeatum*
2, 6 var. *travancoricum.*
kilimanjaricum Pichi-Sermolli in *Webbia*, 27(2):
5 445(1972 publ. 1973)—Tanzania.
setiferum (Forssk.) T. Moore ex Woynar
var. **nigropaleaceum** (Christ) Sledge in *Bull. Brit. Mus. (Nat. Hist.), Bot.*, 5(1):37(1973): *P.*
5, 6 *aculeatum* var. *nigropaleaceum.*
walkerae (Hook.) Sledge *op. cit.* 29(1973): *Polypodium*
6 *walkerae.*
var. **bipinnatum** (Bedd.) Sledge *op. cit.* 30(1973):
6 *Polypodium walkerae* var. *bipinnatum.*

POLYTAENIUM

elongatum (Cav.) Pichi-Sermolli in *Webbia*, 28(2):
4 469(1973): *Hemionitis elongata.*

PTERIDIUM

243 Klekowski, E. J.
Genetic endemism of Galapagos *Pteridium*. *Bot. J.*
4 *Linn. Soc.*, 66(2):181–188(1973)—Map.

PTERIS

bavazzanoi Pichi-Sermolli in *Webbia*, 27(2):422
5 (1972 publ. 1973)—Tanzania.

PYRROSIA

abbreviata (Zoll.) Tagawa in *Acta Phytotax. Geobot.*,
6 25(4–6):180(1973): *Niphobolus abbreviatus.*
winkleri (Rosenst.) Tagawa *op. cit.* 180(1973):
6 *Cyclophorus winkleri.*

SADLERIA

244 Holbrook-Walker, S. G. *and* Lloyd, R. M.
Reproductive biology and gametophyte morphology of the Hawaiian fern genus *Sadleria* (Blechnaceae) relative to habitat diversity and propensity for
6 colonization. *Bot. J. Linn. Soc.*, 67(2):157–174(1973).

SALVINIA

245 Kępczyński, K. *and* Fabiszak, S.
Salvinia natans (L.) All. i zespól Spirodelo-Salvinietum Slavnič 1956 na terenie województwa bydgoskiego. (*Salvinia natans* (L.) All. und Spirodelo-Salvinietum Slavnič 1956 in der Woiwodschaft Bydgoszcz). *Zesz. Nauk. Uniw. Mikol. Kopernika*
1 *Toruniu, Biol.*, 30:33–40(1972)—(Ge); maps.

246 Kleinschmidt, H. E.
Salvinia is a plant to watch. *Queensl. Agric. J.*,
4–7 99(2):73–74(1973)—Illustrations.

247 Shilov, M. P.
K ekologii nekotorȳkh redkikh vodnȳkh rasteniĭ na severnom predele ikh rasprostraneniya. (To the ecology of some rare aquatic plants on the northern limits of their area). *Bot. Zhurn.*, 58(10):1554–1559
1, 2 (1973). *Salvinia natans.*

SCEPTRIDIUM

alabamense (Maxon) Holub in *Preslia*, 45(3):277
3 (1973): *Botrychium alabamense.*
formosanum (Tagawa) Holub *op. cit.* 277(1973):
2 *Botrychium formosanum.*
negeri (Christ) Holub *op. cit.* 277(1973): *Botrychium*
4 *negeri.*
nipponicum (Makino) Holub *op. cit.* 277(1973):
2 *Botrychium nipponicum.*

SCHELLOLEPIS

benguetensis (Copel.) Pichi-Sermolli in *Webbia*,
6 28(2):470(1973): *Polypodium benguetense.*
korthalsii (Mett.) Pichi-Sermolli *op. cit.* 470(1973):
6 *Polypodium korthalsii.*
percussa (Cav.) Pichi-Sermolli *op. cit.* 470(1973):
6 *Cyathea percussa.*
persicifolia (Desv.) Pichi-Sermolli *op. cit.* 470(1973):
6 *Polypodium persicifolium.*
tomentella (C. Chr.) Pichi-Sermolli *op. cit.* 470
6 (1973): *Polypodium tomentellum.*

SPHAEROCIONIUM

adiantoides (v. d. Bosch) Pichi-Sermolli in *Webbia*,
4 28(2):470(1973): *Hymenophyllum adiantoides.*
angustum (v. d. Bosch) Pichi-Sermolli *op. cit.* 470
4 (1973): *Hymenophyllum angustum.*
caparaoense (Brade) Pichi-Sermolli *op. cit.* 470
4 (1973): *Hymenophyllum caparaoense.*
consanguineum (Morton) Pichi-Sermolli *op. cit.*
4 470(1973): *Hymenophyllum consanguineum.*
delicatulum (Sehnem) Pichi-Sermolli *op. cit.* 470
4 (1973): *Hymenophyllum delicatulum.*
dependens (Morton) Pichi-Sermolli *op. cit.* 471
4 (1973): *Hymenophyllum dependens.*
dimorphum (Christ) Pichi-Sermolli *op. cit.* 471
4 (1973): *Hymenophyllum dimorphum.*
fragile (Hedw.) Pichi-Sermolli *op. cit.* 471(1973):
4 *Trichomanes fragile.*
glaziovii (Bak.) Pichi-Sermolli *op. cit.* 471(1973):
4 *Hymenophyllum glaziovii.*
horizontale (Morton) Pichi-Sermolli *op. cit.* 471
4 (1973): *Hymenophyllum horizontale.*
latifrons (v. d. Bosch) Pichi-Sermolli *op. cit.* 471
4 (1973): *Hymenophyllum latifrons.*
lindenii (Hook.) Pichi-Sermolli *op. cit.* 471(1973):
4 *Hymenophyllum lindenii.*
lobato-alatum (Klotz.) Pichi-Sermolli *op. cit.* 471
4 (1973): *Hymenophyllum lobato-alatum.*
maxonii (Christ ex Morton) Pichi-Sermolli *op. cit.*
4 471(1973): *Hymenophyllum maxonii.*
molle (Morton) Pichi-Sermolli *op. cit.* 471(1973):
4 *Hymenophyllum molle.*
multialatum (Morton) Pichi-Sermolli *op. cit.* 471
4 (1973): *Hymenophyllum multialatum.*
prionema (Kunze ex Sturm) Pichi-Sermolli *op. cit.*
4 471(1973): *Hymenophyllum prionema.*
roraimense (Morton) Pichi-Sermolli *op. cit.* 471
4 (1973): *Hymenophyllum roraimense.*
silvaticum (Morton) Pichi-Sermolli *op. cit.* 471
4 (1973): *Hymenophyllum silvaticum.*
simplex (Morton) Pichi-Sermolli *op. cit.* 472(1973):
4 *Hymenophyllum simplex.*
superbum (Morton) Pichi-Sermolli *op. cit.* 472
4 (1973): *Hymenophyllum superbum.*
tenerrimum (v. d. Bosch) Pichi-Sermolli *op. cit.*
4 472(1973): *Hymenophyllum tenerrimum.*
urbanii (Brause) Pichi-Sermolli *op. cit.* 472(1973):
4 *Hymenophyllum urbanii.*

SPHAEROPTERIS

248 Hoshizaki, B. J.
A tree fern cultivar. *Amer. Fern J.*, 63(1):21(1973)—
7 Illustration. *Sphaeropteris cooperi.*

bradei Windisch in *Bradea*, 1(37):372(1973)—
4 Colombia.
cyatheoides (Desv.) Windisch *op. cit.* 372(1973):
4 *Hemitelia cyatheoides.*
dissimilis (Morton) Windisch *op. cit.* 372(1973):
4 *Hemitelia dissimilis.*
intramarginalis Windisch *op. cit.* 373(1973)—
4 Venezuela.
macrosora (Baker) Windisch *op. cit.* 372(1973):
4 *Alsophila macrosora.*
var. **reginae** Windisch *op. cit.* 374(1973)—Colombia,
4 Venezuela, Guyana.
var. **vaupensis** Windisch *op. cit.* 374(1973)—
4 Colombia, Venezuela.
rufescens (Mett.) Windisch *op. cit.* 372(1973):
4 *Hemitelia rufescens.*

STEIROPTERIS (Christensen) Pichi-Sermolli in *Webbia*, 28(2):449(1973): *Dryopteris* subgen.
4 *Steiropteris.*
clypeolutata (Desv.) Pichi-Sermolli *op. cit.* 451
4 (1973): *Nephrodium clypeolutatum.*
crassiuscula (Maxon) Pichi-Sermolli *op. cit.* 451
4 (1973): *Dryopteris crassiuscula.*
deltoidea (Swartz) Pichi-Sermolli *op. cit.* 451(1973):
4 *Polypodium deltoideum.*
densisora (C. Chr.) Pichi-Sermolli *op. cit.* 451(1973):
4 *Dryopteris densisora.*

fendleri (D.C. Eat.) Pichi-Sermolli *op. cit.* 451(1973):
4 *Aspidium fendleri.*
gardnerana (Bak.) Pichi-Sermolli *op. cit.* 451(1973),
4 as 'gardneriana': *Nephrodium gardnerianum.*
glandulosa (Desv.) Pichi-Sermolli *op. cit.* 451(1973):
4 *Polypodium glandulosum.*
incana (Christ) Pichi-Sermolli *op. cit.* 451(1973):
4 *Aspidium incanum.*
insignis (Mett.) Pichi-Sermolli *op. cit.* 451(1973):
4 *Aspidium insigne.*
leprieurii (Hook.) Pichi-Sermolli *op. cit.* 451(1973):
4 *Nephrodium leprieurii.*
lonchodes (D.C. Eat.) Pichi-Sermolli *op. cit.* 451
4 (1973): *Aspidium lonchodes.*
praetervisa (Kuhn) Pichi-Sermolli *op. cit.* 451(1973):
4 *Aspidium praetervisum.*
valdepilosa (Bak.) Pichi-Sermolli *op. cit.* 451(1973):
4 *Nephrodium valdepilosum.*
wrightii (Mett. ex D.C. Eat.) Pichi-Sermolli *op. cit.*
4 451(1973): *Aspidium wrightii.*

STENOLOMA

catharinae (Hook.) Sehnem in Fl. Ilustr. Catar.,
4 1(Pteridác.):48(1972): *Lindsaea catharinae.*

STICHERUS

249 Holttum, R. E.
On the typification of *Mertensia* Willd. non Roth
(Gleicheniaceae) with notes on *Sticherus* Presl and
6 *Hicriopteris* Presl. *Taxon*, 22(4):447–450(1973).

inflexus Pichi-Sermolli in *Webbia*, 27(2):397
5 (1972 publ. 1973)—Rwanda, Zaïre.

TECTARIA

trichotoma (Fée) Tagawa in *Acta Phytotax. Geobot.*,
6 25(4–6):180(1973): *Aspidium trichotomum.*

THELYPTERIS

250 Smith, A. Reid
The Mexican species of *Thelypteris* subgenera
Amauropelta and *Goniopteris*. *Amer. Fern J.*, 63(3):116–
127(1973)—Illustrations; chromosome numbers;
4 keys.

251 Tryon, A. *and* Tryon, R.
Thelypteris in northeastern North America. *Amer.
Fern J.*, 63(3):65–76(1973)—Dot map; chromosome
3 numbers.

subgen. **Amauropelta** (Kunze) A. Reid Smith in
Amer. Fern J., 63(3):121(1973): genus
4 *Amauropelta.*
catamarcensis de la Sota in *Darwiniana*, 18(1–2):216
4 (1973)—Argentina.
cheilanthoides (Kze.) Proctor
var. **mucosa** A. Reid Smith in *Amer. Fern J.*,
4 63(3):123(1973)—Mexico.
hatchii A. Reid Smith *op. cit.* 118(1973)—Guatemala,
4 Mexico.
jujuyensis de la Sota in *Darwiniana*, 18(1–2):222
4 (1973)—Argentina.
laterepens (Trotter & Hope) R. R. Stewart, Annot.
Cat. Vasc. Pl. W. Pakistan & Kashmir: 16(1972):
6 *Polypodium laterepens.*
mortonii A. Reid Smith in *Amer. Fern J.*, 63(3):125
4 (1973)—Mexico.
4 **munchii** A. Reid Smith *op. cit.* 120(1973)—Mexico.
nubicola de la Sota in *Darwiniana*, 18(1–2):225
4 (1973)—Argentina.
oaxacana A. Reid Smith in *Amer. Fern J.*, 63(3):125
4 (1973)—Mexico, El Salvador.

resiliens (Maxon) A. Reid Smith *op. cit.* 121(1973):
4 *Dryopteris resiliens.*
subpubescens (Blume) R. R. Stewart, Annot. Cat.
Vasc. Pl. W. Pakistan & Kashmir: 17(1972),
6 without basionym ref.: *Aspidium subpubescens.*
tablana (Christ) A. Reid Smith in *Amer. Fern J.*,
4 63(3):127(1973): *Aspidium tablanum.*
toganetra A. Reid Smith *op. cit.* 118(1973)—Mexico,
4 British Honduras, Guatemala.

TRICHOMANES

252 Hébant-Mauri, R.
Fonctionnement apical et ramification chez
quelques fougères du genre *Trichomanes* (L.)
(Hyménophyllacées). *Adansonia*, 13(4):495–526
1–7 (1973)—(En).

VANDENBOSCHIA

martinezii (Rovir.) Pichi-Sermolli in *Webbia*, 28(2):
4 472(1973): *Trichomanes martinezii.*

WOODSIA

253 Becherer, A.
Über die Verbreitung von *Woodsia glabella* R. Br.
1 in der Schweiz. *Bauhinia*, 5(1):17–20(1973).

254 Lafontaine, J. D.
Range extension of the blunt-lobed *Woodsia*,
Woodsia obtusa (Spreng.) Torr. (Polypodiaceae), in
3 Canada. *Canad. Field-Nat.*, 87(1):56(1973).

XIPHOPTERIS

aethiopica Pichi-Sermolli in *Webbia*, 27(2):450
5 (1972 publ. 1973)—Ethiopia.
aphelolepis (Morton) Pichi-Sermolli in *Webbia*,
4 28(2):472(1973): *Grammitis aphelolepis.*
limula (Christ) Pichi-Sermolli *op. cit.* 472(1973):
4 *Polypodium limula.*
strangeana Pichi-Sermolli in *Webbia*, 27(2):453
5 (1972 publ. 1973)—Kenya.

LYCOPSIDA

HUPERZIA

afromontana Pichi-Sermolli in *Webbia*, 27(2):394
(1972 publ. 1973)—Burundi, Uganda, Kenya,
5 Tanzania.
arctica (Tolm.) Siplivinskiĭ in *Novosti Sist. Vȳssh. Rast.*,
2 10:347(1973): *Lycopodium selago* subsp. *arcticum.*
dura Pichi-Sermolli in *Webbia*, 27(2):392(1972
5 publ. 1973)—Kenya.
petrovii Siplivinskiĭ in *Novosti Sist. Vȳssh. Rast.*,
2 10:346(1973)—U.S.S.R. (Siberia).

ISOETES

255 Fijałkowski, D.
Stanowiska *Isoëtes lacustris* L. w województwie
lubelskim. (Stations of *Isoëtes lacustris* L. in the Lublin
Voivodship). *Fragm. Flor. Geobot.*, 19(1):5–7(1973)—
1 (En); map.

256 Landry, G. *and* Thieret, J. W.
Isoetes lousianensis (Isoetaceae), a new species from
3 Louisiana. *Sida*, 5(2):129–130(1973).

257 Michaux, N.
La ligule chez l'*Isoetes setacea* Lam. *Compt. R.
Hebd. Séances Acad. Sci., D* (Paris), 277(1):41–44
1 (1973)—Illustrations.

258 Robert, D., Roland-Heydacker, F., Denizot, J.,
Laroche, J., Fougeroux, P. *and* Davignon, L.
La paroi mégasporale de l'*Isoetes setacea* Bosc et
Delile. Étude en microscopies photonique et
électroniques. Localisation et nature de la silice
entrant dans sa constitution. *Adansonia*, 13(3):333–340
1 (1973)—(En).

asturicense (M. Laínz) M. Laínz in *Anuár. Soc. Brot.*,
1 39:118(1973): *I. boryanum* subsp. *asturicense*.
louisianensis Thieret in *Sida*, 5(2):129(1973)—
3 U.S.A. (Louisiana).

LYCOPODIUM

259 Anon.
Materials for the distribution of vascular plants in
Japan (80). *Lycopodium casuarinoides* Spring. *J. Jap.*
2 *Bot.*, 48(5):146(1973)—Ja.

260 Dostál, L.
Plavúň pučivý (*Lycopodium annotinum* L.) aj v
Slánskom pohorí. (*Lycopodium annotinum* L. im
Gebirge Slánske pohorie). *Zprávy Česk. Bot. Spol.*,
1 8(2):119–121(1973)—(Ge).

261 Levin, D. A. *and* Crepet, W. L.
Genetic variation in *Lycopodium lucidulum*: a
phylogenetic relic. *Evolution*, 27(4):622–632(1973)—
3 Dot maps.

262 Kuvaev, V. B. *and* Rudskiĭ, V. V.
O rasprostranenii plauna *Lycopodium inundatum* L.
na territorii Azii. (On the distribution of the *Lyco-
podium inundatum* L. in Asia). *Bot. Zhurn.*, 58(6):
2 880–882(1973)—Dot map.

263 Primack, R. B.
Growth patterns of five species of *Lycopodium*.
1–3 *Amer. Fern J.*, 63(1):3–5(1973).

PALHINHAEA

veigae Vasconcellos in *Anuár. Soc. Brot.*, 38:83
1 (1972)—Madeira.

PHLEGMARIURUS

264 Lawalrée, A.
Deux *Phlegmariurus* (Lycopodiaceae) nouveaux du
Zaïre. *Acta Bot. Acad. Sci. Hung.*, 19(1–4):195–199
5 (1973)—(En); illustrations, dot map.

SELAGINELLA

265 Cour, P.
Cycles végétatifs et phases sporangifères distribution
raisonée de *Selaginella denticulata* (L.) Spring.
(Vegetative cycles and sporangifer phases, geographical
distribution of *Selaginella denticulata* (L.) Spring). *Bull.
Soc. Bot. France*, 119(5–6):399–404(1972 publ. 1973)—
1 (En).

266 Crabbe, J. A. *and* Jermy, A. C.
Seven new species of *Selaginella* from tropical
South America. *Amer. Fern J.*, 63(3):135–144(1973)—
4 Illustrations.

267 Grubov, V. I. *and* Ulziĭkhutag, N.
Selaginella sanguinolenta (L.) Spring.—édifikator.
(*Selaginella sanguinolenta* (L.) Spring as edificator).
2 *Bot. Zhurn.*, 58(12):1807–1808(1973).

268 Robert, D.
Edification de structures réticulées à partir du
feuillet interne de la membrane nucléaire dans les
microspores d'une Sélaginelle, *Selaginella selaginoides*
(L.) Link. *Trav. Lab. La Jaysinia*, no. 4:123–130
1 (1972 publ. 1973)—(En).

269 Robert, D.
Le gamétophyte femelle de *Selaginella denticulata*
(L.) Spring. *Rev. Cytol. Biol. Vég.*, 35(3–4):243–280
1, 2 (1972 publ. 1973)—Illustrations.

270 Robert, D.
Le gamétophyte femelle de *Selaginella kraussiana*
(Kunze) A. Br.: 3. Ultrastructure et développement
des archégones. *Rev. Cytol. Biol. Vég.*, 35(3–4):
1, 5, 6 165–241(1972 publ. 1973)—Illustrations.

271 Robert, D.
Quelques particularités de la reproduction chez les
sélaginelles. (Particularities of the reproduction in
Selaginella). *Bull. Soc. Bot. France*, 119(5–6):373–382
1–3, 5, 6 (1972 publ. 1973)—(En).

272 Tagawa, M.
The Selaginellas of Taiwan. *Acta Phytotax. Geobot.*,
2 25(4–6):170–180,192(1973)—Key.

anaclasta Alston ex Crabbe & Jermy in *Amer. Fern J.*,
4 63(3):135(1973)—Venezuela.
chionoloma Alston ex Crabbe & Jermy *op. cit.*
4 137(1973)—Peru, Bolivia.
hirsuta Alston ex Crabbe & Jermy *op. cit.* 138(1973)—
4 Venezuela.
mortoniana Crabbe & Jermy *op. cit.* 139(1973)—
4 Colombia.
palmiformis Alston ex Crabbe & Jermy *op. cit.*
4 141(1973)—Colombia.
pseudonipponica Tagawa in *Acta Phytotax. Geobot.*,
2 25(4–6):177(1973)—Taiwan.
schultesii Alston ex Crabbe & Jermy in *Amer. Fern
J.*, 63(3):141(1973)—Colombia.
4
tanyclada Alston ex Crabbe & Jermy *op. cit.* 143
4 (1973)—Colombia.

PSILOTOPSIDA

PSILOTUM

273 Lugardon, B.
Sur les parois sporales de *Psilotum triquetrum* Sw. et
leur structure fine. *Compt. R. Hebd. Séances Acad.
Sci., D* (Paris), 276(8):1277–1280(1973)—
1–7 Illustrations.

274 Montz, G. N.
Psilotum in New Orleans. *Castanea*, 38(2):207–208
3 (1973).

275 Sinha, B. M. B. *and* Verma, A. K.
A hexaploid *Psilotum nudum* (L.) Beauv. *Proc.
6 Indian Sci. Congr.*, 58(3):431(1971). Abstract only.

276 Swails, L. F.
The recent discovery of *Psilotum nudum* in South
3 Carolina. *Amer. Fern J.*, 63(1):8(1973).

TMESIPTERIS

277 Braithwaite, A. F.
Tmesipteris in the Solomon Islands. *Brit. Fern Gaz.*,
6 10(6):293–303(1973)—Illustrations.

solomonensis A. F. Braithwaite in *Brit. Fern Gaz.*,
6 10(6):302(1973)—British Solomon Is.

SPHENOPSIDA

EQUISETUM

278 Bilderback, Diane E., Bilderback, D. E., Jahn, T. L.
and Fonseca, J. R.
The release mechanism and locomotor behavior of
Equisetum sperm. *Amer. J. Bot.*, 60(8):796–801
1–6 (1973).

279 Cotthem, W. van
Les aires partielles d'*Equisetum telmateia* Ehrh. en Belgique et quelques stations nouvelles intéressantes.
1 *Nat. Belg.*, 5(4):162–169(1973)—Dot maps.

280 Dayanandan, P. *and* Kaufman, P. B.
Stomata in *Equisetum*. *Canad. J. Bot.*, 51(9):1555–
1–6 1564(1973)—(Fr).

281 Duckert-Henriod, M.-M.
Un critère de détermination peu connu pour *Equisetum pratense* Ehrh. *Bull. Soc. Neuchâtel. Sci. Nat.*,
1–3 96:37–40(1973).

282 Duckett, J. G.
Comparative morphology of the gametophytes of the genus *Equisetum*, subgenus *Equisetum*. *Bot. J. Linn.*
1–3 *Soc.*, 66(1):1–22(1973).

283 Laroche, J. *and* Vo Thi Dao
Peut-on admettre un déterminisme génétique du sexe chez certaines espèces d'*Equisetum*? (Does genetic determination of sex exist in certain species of *Equisetum*?) *Bull. Soc. Bot. France*, 119(5–6):363–371
1–3 (1972 publ. 1973)—(En).

284 Øllgaard, B.
Om klitformen af *Equisetum hyemale* i Nordjylland.
1 *Bot. Tidsskr.*, 67(4):333–336(1973)—(En).

285 Page, C. N.
Two hybrids in *Equisetum* new to the British flora.
1 *Watsonia*, 9(3):229–237(1973)—Illustrations.

286 Sen, T. *and* Sen, U.
Morphology and anatomy of *Equisetum diffusum* D. Don and *E. ramosissimum* Desf. subsp. *debile* (Roxb.) Hauke with a discussion on their taxonomy. *Israel J. Bot.*, 22(3):166–174(1973)—Illustrations;
1–6 chromosome numbers; key.

287 Trivedi, B. S. *and* Kher, U.
Abnormal stomata of *Equisetum*. *Curr. Sci.*, 42(12):
6 436–437(1973)—Illustrations.

288 Webster, T. R.
Abnormal strobili in a stand of *Equisetum arvense*.
1–3 *Amer. Fern J.*, 63(1):19–20(1973)—Illustration.

× **rothmaleri** C. N. Page in *Watsonia*, 9(3):229
1 (1973)—Great Britain. (*E. arvense* × *palustre*).

GYMNOSPERMAE

289 Aubréville, A.
Déclin des genres de conifères tropicaux dans le temps et l'espace. *Adansonia*, 13(1):5–35(1973)—
1–7 Illustrations, maps.

290 Aubréville, A.
Distribution des conifères dans la Pangée: Essais.
1–7 *Adansonia*, 13(2):125–133(1973)—Maps.

291 Brunkener, L.
Beiträge zur Kenntnis der frühen Sporangienentwicklung der Pteridophyten und der Gymnospermen. (Contributions to the knowledge of the early development of the sporangium in pteridophytes and gymnosperms). *Svensk Bot. Tidskr.*, 67(4):333–400
1–7 (1973)—(En).

292 De Laubenfels, D. J.
Flore de Madagascar et des Comores: Gymnospermes: 17e famille Cycadacées (Cycadaceae), 18e famille Podocarpacées (Podocarpaceae). 1972. 22p.—
5 Illustrations, map; key.

293 Gaussen, H.
Les Gymnospermes actuelles et fossiles: fasc. 12. *Trav. Lab. Forest. Toulouse*, tome 2. 1(2(3)).
4–7 [144] p. (1973)—Illustrations.

294 Greenberg, B.
Cycads for the succulent collector. *Cact. Succ. J.*
4–7 (U.S.A.), 45(2):63–70(1973)—Illustrations.

295 Hieke, K.
Poznámky k druhové skladbě jehličnanů v zámeckých parcích Čech. (Notes on the specific structure of conifers in the parks of Bohemian castles).
1–3, 6 *Acta Pruhoniciana*, 27:1–16(1972)—(En,Rus,Ge).

296 Jalas, J. *and* Suominen, J., *eds*
Atlas florae europaeae: distribution of vascular plants in Europe: 2. Gymnospermae (Pinaceae to Ephedraceae). Helsinki, Committee Mapping Flora of Europe and Societas Biologica Fennica Vanamo, 1973.
1 40p.

297 Keng, H.
On the family Phyllocladaceae. *Taiwania*, 18(2):
4, 6, 7 142–145(1973)—Map.

298 Lanner, R. M.
Conifers of the Bear Lake area and mountains south of the Great Salt Lake. *Great Basin Nat.*, 31(2):
3 85–89(1971).

299 Lanner, R. M. *and* Warnick, R.
Conifers of the San Francisco mountains, San Rafael Swell, and Roan Plateau. *Great Basin Nat.*,
3 31(3):177–180(1971).

300 Rouane, P.
Etude comparée de la répartition des ramifications au cours de l'ontogenèse de quelques Cupressacées. *Trav. Lab. Forest. Toulouse*, 9 Art.3.[iv]277p. (1973)—
1–7 Illustrations.

301 Sykes, W. R.
Araucariaceae in New Zealand. *Roy. New Zealand*
7 *Inst. Hort. Ann. J.*, no. 1:49–52(1973)—Illustrations.

PHYLLOCLADACEAE (Pilger) E. L. Core ex H. Keng in *Taiwania*, 18(2):142(1973): Podocar-
2 paceae subfam. Phyllocladoideae.

ABIES

302 Budaragin, V. A. *
Kariotip pikhty sibirskoi Kazakhstanskogo Altaya. (Karyotype of *Abies sibirica* in Kazakhstan Altai). *Tsitologiya*, 14(1):130–133(1972)—(En); illustration.
2 Abstr. in *Biol. Abstr.*, 55(9):47354(1973).

303 Juo, P.-S. *and* Stotzky, G.
Electrophoretic analysis of isozymes from seeds of *Pinus*, *Abies* and *Pseudotsuga*. *Canad. J. Bot.*, 51(11):
1–4, 6 2201–2205(1973)—(Fr).

304 Kantor, J. *and* Chira, E.
On the possibility of crossing certain species of the genus *Abies*. (K otázkám křižitelnosti mezi nêkterými druhy rodu *Abies*). *Acta Univ. Agric., Ser. C* (Brno),
1–4, 6 40(1):15–27(1971)—(Cz).

305 Larsen, J. A.
Growth of spruce at Dubawnt Lake, Northwest
3 Territories. *Arctic*, 25(1):59(1972).

306 Liu, T.-S.
The genus *Abies*: 4. *Quart. J. Taiwan Mus.*, 25
2 (3–4):231–295(1972)—Illustrations, maps; key.

307 Liu, T.-S.
The genus *Abies*: 5. *Quart. J. Taiwan Mus.*, 26(1–2):
2 167–215(1973)—Illustrations, map, dot maps; keys.

308 Liu, T.-S.
A monograph of the genus *Abies*. Taipei, National
Taiwan University, 1971. xxx,608p.—(Fr,Ge,Ja,Ch);
1–4, 6 illustrations, maps; keys.

309 Mitchell, A.
Conifers in the western United States: some
observations on bark. *J. Roy. Hort. Soc.*, 98(10):449–
452(1973)—Coloured illustrations. *Abies concolor*,
3 *A. magnifica*.

310 Orsomando, E.
Nuova stazione di abete bianco (*Abies alba* Mill.)
sui Monti della Laga nelle Marche. *Arch. Bot. Biogeogr.*,
1 48(3–4):124–130(1972)—Map.

311 Zavarin, E., Snajberk, K. *and* Critchfield, W. B.
Monoterpene variability of *Abies amabilis* cortical
oleoresin. *Biochem. Syst.*, 1(2):87–93(1973)—Dot
3 maps.

sect. **Vejarianae** Liu, Monogr. Abies: 257 (1971
4 publ. 1973).
cephalonica Loud.
var. **graeca** (Fraas) Liu *op. cit.* 222,354(1971 publ.
1, 2 1973): *A. pectinata* var. *graeca*.
chensiensis van Tiegh.
var. **ernestii** (Rehd.) Liu in *Quart. J. Taiwan Mus.*,
2 25(3–4):236(1972): *A. ernestii*.
delavayi (van Tiegh.) Franch.
var. **smithii** (Vig. & Gauss.) Liu *op. cit.* 247(1972):
2 *A. forrestii* var. *smithii*.
fargesii Franch.
var. **faxoniana** (Rehd. & Wils.) Liu *op. cit.* 260
2, 6 (1972): *A. faxoniana*.
sibirica Ledeb.
var. **semenovii** (Fedtsch.) Liu, Monogr. Abies:
2 188(1971 publ. 1973): *A. semenovii*.
vejarii Mart.
var. **mexicana** (Mart.) Liu *op. cit.* 261,377(1971
4 publ. 1973): *A. mexicana*.

AMENTOTAXUS

312 Miller, H. J.
The wood of *Amentotaxus*. *J. Arnold Arbor.*, 54(1):
2 111–119(1973).

ARAUCARIA

313 Mattos, J. R.
O pinheiro brasileiro. São Paulo, 1972. 618p.—
4 Illustrations, coloured illustrations, maps.

BIOTA

314 Lemoine-Sébastian, C.
A propos de morphoses induites chez *Biota
orientalis* (L.) Endl. et chez *Cryptomeria japonica* (L.)
Don. *Bull. Soc. Sci. Bretagne*, 47(1–2):17–24(1972)—
2 Illustrations.

CALOCEDRUS

315 Phengklai, C.
Cupressaceae. *Thai Forest Bull., Bot.*, no. 7:5–6
2, 6 (1973)—En; illustration, map.

CATHAYA

316 Greguss, P.
Two very rare conifers . . . *Cathaya argyrophylla*. *Int.
2 Dendrol. Soc. Year Book 1972*:52–54(1973).

CEPHALOTAXUS

317 Phengklai, C.
Cephalotaxaceae. *Thai Forest Bull., Bot.*, no. 7:7–8
6 (1973)—En; illustration, map.

CHAMAECYPARIS

318 Henley, J. W.
Port-Orford-cedar: an American wood. *U.S. Dep.
Agric. F.S.*, 228. 6p.(1973)—Illustrations, map.
3 *Chamaecyparis lawsoniana*.

319 Ho, R. H. *and* Sziklai, O.
Fine structure of the pollen surface of some Taxo-
diaceae and Cupressaceae species. *Rev. Palaeobot.
3 Palynol.*, 15(1):17–26(1973). *Chamaecyparis lawsoniana*.

CHRYSOLARIX

320 Moore, H. E.
Chrysolarix renounced—a comedy of restoration.
2 *Taxon*, 22(5–6):587–589(1973).

CRYPTOMERIA

321 Ho, R. H. *and* Sziklai, O.
Fine structure of the pollen surface of some
Taxodiaceae and Cupressaceae species. *Rev. Palaeobot.
2, 3 Palynol.*, 15(1):17–26(1973). *Cryptomeria japonica*.

322 Lemoine-Sébastian, C.
A propos de morphoses induites chez *Biota orientalis*
(L.) Endl. et chez *Cryptomeria japonica* (L.) Don. *Bull.
2 Soc. Sci. Bretagne*, 47(1–2):17–24(1972)—Illustrations.

323 Lemoine-Sébastian, C.
Vascularisation des ramifications axillaires chez
Cryptomeria japonica (L.) Don. (The vascular system
of axillary shoots in *Cryptomeria japonica* (L.) Don).
Flora (Germany), 162(1–2):45–52(1973)—(En);
2 illustrations.

CUNNINGHAMIA

324 Ho, R. H. *and* Sziklai, O.
Fine structure of the pollen surface of some
Taxodiaceae and Cupressaceae species. *Rev.
Palaeobot. Palynol.*, 15(1):17–26(1973). *Cunninghamia
2, 3 lanceolata*.

CUPRESSUS

325 Thomas, G. E. *and* Goggans, J. F. *
A karyotypic study of cypresses indigenous to the
southwestern United States. *Alab. Agric. Exp. Stat.
Circ.*, 201:1–24(1972). Ref. in *Bull. Torrey Bot. Club*,
3 100(1):72(1973).

CYCAS

326 Burkart, A.
Un ejemplar masculino de *Cycas revoluta* Thunb.,
floreció en el jardin de Darwinion. *Darwiniana*,
2 18(1–2):277–279(1973)—Illustrations.

DACRYDIUM

327 Phengklai, C.
Podocarpaceae. *Thai Forest Bull., Bot.*, no. 7:9–20
2, 6 (1973)—En; illustrations; keys.

DIOON

328 Dossaji, S. F. *and* Bell, E. A.
Biflavones of *Dioon*. *Phytochemistry*, 12(2):371–373
5 (1973).

ENCEPHALARTOS

329 Dyer, R. A.
Cycads. *Cact. Succ. J.* (U.S.A.), 44(5):209–216
5 (1972)—Illustrations.

EPHEDRA

330 Bernard, V. V. *and* Meĭer, N. R.
Pȳl'tsa *Ephedra, Welwitschia* i *Gnetum*. (Pollen grains of *Ephedra, Welwitschia* and *Gnetum*). *Vestn. Mosk. Univ. Nauch. Zhurn.*, *6*, 27(6):86–88(1972)—
1, 2, 6 (En).

331 Danin, A. *and* Hedge, I. C.
Contributions to the flora of Sinai: 1. New and confused taxa. *Notes Roy. Bot. Gard. Edinburgh*, 32(2):
2 259–271(1973)—Key. *Ephedra* pp.264–265.

332 Laudi, G. *and* Bonatti, P. M.
Ultrastructure of chloroplasts of some Chlamidospermae (*Ephedra twediana, Gnetum montana, Welwitschia mirabilis*). *Caryologia*, 26(1):107–114 (1973)—(It); illustrations. Discussion of taxonomic
1–4, 6 relationships.

333 Viţalariu, G.
Ephedra distachya L. în flora şi vegetaţia Moldovei. *An. Şti. Univ. Al. I. Cuza Iaşi, Biol.*, 19(1):203–206
1, 2 (1973)—Illustration.

334 Voronin, N. S., Voronkina, N. V. *and* Lodkina, M. M.
Stroenie apeksa koriya Gnetales. (Root apex structure in Gnetales). *Bot. Zhurn.*, 58(11):1589–1596
1–4, 6 (1973)—(En).

GINKGO

335 Anon.
Les ginkgos de France. *Jard. France*, no. 7:10–11
1 (1973)—Illustrations.

336 Winter, L.
Doit-on écrire *Ginkgo* ou *Ginkyo* ou *Ginkyô* ou *Ginkjô*? *Jard. France*, no. 7:14–16(1973)—
2 Illustration.

337 Winter, L.
Les utilisations du *Ginkgo biloba* L. *Jard. France*,
2 no. 7:17–19(1973)—Illustration.

GLYPTOSTROBUS

338 Butala, J. R. *and* Cridland, A. A.
Juniperus washingtonensis, a synonym of *Glyptostrobus nordenskioldi*. *Taxon*, 22(5–6):577–578(1973).
3

GNETUM

339 Bernard, V. V. *and* Meĭer, N. R.
Pȳl'tsa *Ephedra, Welwitschia* i *Gnetum*. (Pollen grains of *Ephedra, Welwitschia* and *Gnetum*). *Vestn. Mosk. Univ. Nauch. Zhurn.*, *6*, 27(6):86–88
4–6 (1972)—(En).

340 Laudi, G. *and* Bonatti, P. M.
Ultrastructure of chloroplasts of some Chlamidospermae (*Ephedra twediana, Gnetum montana, Welwitschia mirabilis*). *Caryologia*, 26(1):107–114 (1973)—(It); illustrations. Discussion of taxonomic
4–6 relationships.

341 Phengklai, C.
Gnetaceae. *Thai Forest Bull., Bot.*, no. 7:21–38 (1973)—En; illustrations, map; keys for male or
6 female plants.

342 Voronin, N. S., Voronkina, N. V. *and* Lodkina, M. M.
Stroenie apeksa koriya Gnetales. (Root apex structure in Gnetales). *Bot. Zhurn.*, 58(11):1589–1596
4–6 (1973)—(En).

JUNIPERUS

343 Adams, R. P.
Re-evaluation of the biological status of *Juniperus deppeana* var. *sperryi* Correll. *Brittonia*, 25(3):284–289
3 (1973)—Illustrations.

344 Banthorpe, D. V., Davies, H.ff.S., Gatford, C. *and* Williams, S. R.
Monoterpene patterns in *Juniperus* and *Thuja*
1–3 species. *Pl. Med.*, 23(1):64–69(1973).

345 Butala, J. R. *and* Cridland, A. A.
Juniperus washingtonensis, a synonym of *Glypto-*
3 *strobus nordenskioldi*. *Taxon*, 22(5–6):577–578(1973).

346 Hall, M. T., Mukherjee, A. *and* Crowley, W. R.
Chromosome counts in cultivated junipers. *J.*
0 *Arnold Arbor.*, 54(3):369–376(1973).

347 Merihein, A.
Kadajas ja tema sugulased. (The common juniper and its relatives). *Eesti Loodus*, 16(7):413–419(1973)—
1–3 Illustration.

348 Powell, R. A. *and* Adams, R. P.
Seasonal variation in the volatile terpenoids of *Juniperus scopulorum* (Cupressaceae). *Amer. J. Bot.*,
3 60(10):1041–1050(1973).

349 Roof, J. B.
Native dwarf junipers. *Four Seasons*, 4(3):2–9(1973)
3 —Illustrations. California.

350 Ward, L. K.
The conservation of juniper: 1. Present status of juniper in southern England. *J. Appl. Ecol.*, 10(1):
1 165–188(1973)—Dot maps. *Juniperus communis*.

deppeana Steud.
f. **sperryi** (Correll) R. P. Adams in *Brittonia*, 25(3):
3 289(1973): *J. deppeana* var. *sperryi*.

LARIX

351 Christiansen, H.
On the development of pollen and the fertilization mechanisms of *Larix* and *Pseudotsuga menziesii*. *Silvae*
1–3, 6 *Genet.*, 21(5):165–174(1972)—(Ge).

352 Kotok, E. S.
Western larch: an American wood. *U.S. Dep. Agric. F.S.*, 243. 6p.(1973)—Illustrations, map. *Larix*
3 *occidentalis*.

353 Milkina, L. I.
Shche raz pro pokhodzhennya modrȳnȳ pol's'koi (*Larix polonica* Racib.) v urochȳshchi Manyava-Skit ta yiyi okhoronu. (Once more on origin of *Larix polonica* Racib. in the locality Manyava-Skit and its protection). *Ukr. Bot. Zhurn.*, 30(5):597–601(1973)—
1 Uk(En).

354 Mills, J. S.
Diterpenes of *Larix* oleoresins. *Phytochemistry*,
1–3 12(10):2407–2412(1973).

LIBOCEDRUS

355 McDonald, P. M.
Incense-cedar: an American wood. *U.S. Dep.*
Agric. F.S., 226. 7p.(1973)—Illustrations, map.
3 *Libocedrus decurrens.*

METASEQUOIA

356 Mennega, E. A.
Metasequoia glyptostroboides. Wachendorffia, 1(2):
2 19–20(1973)—Du; illustration.

MICROBIOTA

357 Van Hoey-Smith, J. R. P.
Two very rare conifers . . . *Microbiota decussata. Int.*
Dendrol. Soc. Year Book 1972: 50–51(1973)—
2 Illustration.

PICEA

358 Anon.
Picea sitchensis (Bongard) Carriere. Sitka spruce,
member of family Pinaceae. *Davidsonia*, 4(4):41–45
3 (1973)—Illustrations.

359 Fanta, J.
Die vegetative Vermehrung der Fichte in der
Lüneburger Heide. (Vegetative reproduction of
Norway spruce in the Lüneburger Heide). *Mitt.*
Deutsch. Dendrol. Ges., no. 66:39–48(1973)—(En);
1 illustrations. *Picea abies.*

360 Mišić, V., Popović, M. *and* Čolić, D.
Individualni varijabilitet smrče (*Picea excelsa* L.) na
Kopaoniku. (Individual variability of spruce (*Picea*
excelsa L.) on Kopaonik mountain). *Glasn. Prir. Muz.*
1 *Beogradu, B*, 27:49–59(1972)—(En).

361 Schroeder, F.-G.
Westerhof, ein natürliches Fichtenvorkommen
westlich des Harzes. (A native occurrence of spruce
west of the Harz mountains). *Mitt. Deutsch. Dendrol.*
1 *Ges.*, no. 66:9–38(1973)—(En); map. *Picea abies.*

362 Schroeder, F.-G. *and* Bayer, R.
Zur Nomenklatur von *Picea polita* und *Pinus*
thunbergii. Mitt. Deutsch. Dendrol. Ges., no. 66:61–68
2 (1973)—(En); illustration.

363 Woodfin, R. O.
Sitka spruce (*Picea sitchensis* (Bong.) Carr.). *U.S.*
3 *Dep. Agric. F.S.*, 265. 7p.(1973)—Illustrations, map.

PINUS

364 Argus, G.
Pinus banksiana Lamb. or *Pinus divaricata* (Aiton)
3 Dumont? *Taxon*, 22(2–3):318–319(1973).

365 Beasley, R. S.
Bristlecone pine (*Pinus longaeva*) in relation to
environmental factors and soil properties in east-
central Nevada. *Diss. Abstr. Int., B*, 33(2):679–680
3 (1972).

366 Carpenter, E. M. *and* Perala, D. A.
Jack pine: an American wood. *U.S. Dep. Agric.*
F.S., 252. 6p.(1973)—Illustrations, map. *Pinus*
3 *banksiana.*

367 Chałupkowa, A. *and* Chałupka, W. R.
Interesująca forma sosny pospolitej—*Pinus*
silvestris L. (An interesting form of Scots pine—*Pinus*
silvestris L.) *Roczn. Sekc. Dendrol. Pol. Tow. Bot.*,
1 27:229–231(1973)—(En); illustrations.

368 Eliçin, G.
Türkiye Sariçam (*Pinus silvestris* L.) 'larinda
morfogenetik araştirmalar. (Recherches morpho-
génétiques sur les pins sylvestres de Turquie). *Istanbul*
Üniv. Orman Fak. Yayinl., no. 1662. 149p.(1971)—
1, 2 (Fr); illustrations, map.

369 Gabilo, E. M. *and* Mogensen, H. L.
Foliar initiation and the fate of the dwarf-shoot
apex in *Pinus monophylla. Amer. J. Bot.*, 60(7):671–677
3 (1973).

370 Hicks, R. R.
Evaluation of morphological characters for use in
identifying loblolly pine, shortleaf pine and loblolly ×
shortleaf hybrids. *Castanea*, 38(2):182–189(1973).
3 *Pinus taeda* and *P. echinata.*

371 Iacob, T.
Cercetări privind cunoaşterea şi ocrotirea speciei
Pinus cembra L. în parcul naţional Retezat. (Research
of *Pinus cembra* L. in the national park Retezat).
1 *Ocrot. Nat.*, 16(2):207–211(1972)—(En); illustrations.

372 Jacquemoud, F.
Le pin à crochets du Salève: apprécié des varappeurs,
méconnu des botanistes. *Mus. Genève*, 136:19–22
1 (1973)—Illustrations. *Pinus montana.*

373 Juo, P-S. *and* Stotzky, G.
Electrophoretic analysis of isozymes from seeds of
Pinus, *Abies* and *Pseudotsuga. Canad. J. Bot.*, 51(11):
1–3, 6 2201–2205(1973)—(Fr).

374 Karpinski, Z. *
(Un étrange phénomène de ramification dans les
paisses de pin). *Sylwan*, 115(10):75–77(1971)—Pol.;
Ref. in *Bull. Signal., Biol. Physiol. Vég.*, 33(6):5368
1 (1972).

375 Koch, P.
Utilization of the southern pines. *U.S. Dep. Agric.*,
Agric. Handb., no. 420. 2v.(1972)—Illustrations, dot
3 maps. *Pinus* spp. pp. 5–51.

376 Kocięcki, S. *
(Sur la ramification des pousses de pin). *Sylwan*,
115(10):78(1971)—Pol; abstr. in *Bull. Signal., Biol.*
1 *Physiol. Vég.*, 33(6):5364(1972).

377 Kotok, E. S.
Ponderosa pine: an American wood. *U.S. Dep.*
3, 4 *Agric. F.S.*, 254. 6p.(1973)—Illustration, map.

378 Laguardia, A. M. *and* Bottazzi, J. A.
Histotaxonomia foliar de pinos cultivados en el
Uruguay. *Bol. Fac. Agron.* (Montevideo), no. 119. 18p.
1–4, 6 (1971)—(En); key.

379 Lamb, A. F. A., comp.
Fast growing timber trees of the lowland tropics:
no. 6. *Pinus caribaea*: vol. 1. Oxford, University of
Oxford Department of Forestry, 1973. [xxii],254p.—
4 Illustrations.

380 Lanner, R. M. *and* Hutchinson, E. R.
Relict stands of pinyon hybrids in northern Utah.
Great Basin Nat., 32(3):171–175(1972)—*Pinus edulis*
3 and *P. monophylla.*

381 Mihăilescu, A. *and* Dalu, M.
Comparative karyotype study in different proven-
ances of *Pinus silvestris* and *Pinus nigra. Rev. Roum.*
Biol., Bot., 17(5):343–353(1972)—Chromosome
1, 2 numbers.

382 Mitchell, A.
Conifers in the western United States: some observations on bark. *J. Roy. Hort. Soc.*, 98(10):449–452(1973)—Coloured illustrations. *Pinus ponderosa*,
3 *P. jeffreyi*.

383 Munson, R. H.
The dwarf and slow-growing pines. *Arbor. Leaves*,
1–3 15(4):22–24(1973)—Illustrations.

384 Phengklai, C.
Pinaceae. *Thai Forest Bull., Bot.*, no. 7:1–4(1973)—
6 En; illustration, map; key. *Pinus* species in Thailand.

385 Saylor, L. C.
Karyotype analysis of the genus *Pinus*—subgenus *Pinus*. *Silvae Genet.*, 21(5):155–163(1972)—(Ge); illustrations. Karyotypes for all species of subgen.
1–3 *Pinus* except subsec. *Sylvestres* (*Lariciones*).

386 Saylor, L. C. *and* Ke Won Kang
A study of sympatric populations of *Pinus taeda* L. and *Pinus serotina* Michx. in North Carolina. *J. Elisha Mitchell Sci. Soc.*, 89(1–2):101–110(1973)—Illustra-
3 tions, map.

387 Schramm, R. W.
Obecny stan występowania kosodrzewiny (*Pinus montana* s.l.) w Karpatach Polskich na wschod od Tatr. (The actual state of the dwarf mountain pine (*Pinus montana* s.l.) in the Polish Carpathians eastwards of the Tatras). *Fragm. Flor. Geobot.*, 19(1):23–27
1 (1973)—(En).

388 Schroeder, F.-G. *and* Bayer, R.
Zur Nomenklatur von *Picea polita* und *Pinus thunbergii*. *Mitt. Deutsch. Dendrol. Ges.*, no. 66:61–68
2 (1973)—(En); illustration.

389 Smouse, P. E. *and* Saylor, L. C.
Studies of the *Pinus rigida-serotina* complex: 1. A study of geographic variation. *Ann. Missouri Bot. Gard.*,
3 60(2):174–191(1973)—Dot maps.

390 Smouse, P. E. *and* Saylor, L. C.
Studies on the *Pinus rigida-serotina* complex: 2. Natural hybridization among the *Pinus rigida-serotina* complex, *P. Taeda* and *P. echinata*. *Ann. Missouri Bot. Gard.*, 60(2):192–203(1973)—Dot
3 maps.

391 Zavarin, E. *and* Snajberk, K.
Variability of the wood monoterpenoids from *Pinus aristata*. *Biochem. Syst.*, 1(1):39–44(1973)—
3 Map.

392 Zielski, A.
Klucz do określania gatunków rodzaju *Pinus* na podstawie cech anatomicznych. (Arten-Bestimmungschlüssel der Gattung *Pinus* nach anatomischen Merkmalen). *Zesz. Nauk. Uniw. Mikol. Kopernika Toruniu, Biol.*, 30:63–72(1972)—(Ge); illustrations;
1 key. Polish species only.

silvestris L.
subsp. **kochiana** (Klotzsch.) Eliçin in *Istanbul Ünv. Orman. Fak. Yayinl.*, 1662 (Türk. Sariç. Larinda
2 Morfogen. Araştir.): 93(1971): P. kochiana.
subsp. **pannonica** (Schott) Soó in *Acta Bot. Acad. Sci. Hung.*, 18(1–2):177(1973): (Basionym not
1 stated).

PODOCARPUS

393 Duarte, A. Pereira
Tentativa para explicar o ocorrência de duas espécies de *Podocarpus* no Brasil. *Arq. Jard. Bot. Rio de*
4 *Janeiro*, 19:199–215(1973)—Key.

394 Morvan, J.
Sur la répartition et la structure des stomates à la surface des organes foliaires appartenant au cône femelle chez *Podocarpus macrophyllus* (Thunb.) Don var. *angustifolius* Blume. *Compt. R. Hebd. Séances Acad. Sci., D* (Paris), 276(21):2891–2894(1973)—
2 Illustrations.

395 Phengklai, C.
Podocarpaceae. *Thai Forest Bull., Bot.*, no. 7:9–20
2, 6 (1973)—En; illustrations; keys.

396 Woltz, P.
Systématique et anatomie comparée de feuilles de *Podocarpus* malgaches. *Trav. Lab. Forest. Toulouse*,
5 1(8) art.1. 14p.(1973)—(En,Ge).

PSEUDOTSUGA

397 Christiansen, H.
On the development of pollen and the fertilization mechanisms of *Larix* and *Pseudotsuga menziesii*. *Silvae*
2 *Genet.*, 21(5):165–174(1972)—(Ge).

398 Durrieu, A.
Etude chromosomique des espèces du genre *Pseudotsuga*. *Bull. Soc. Hist. Nat. Toulouse*, 107(1–2):
2, 3 58–78(1971)—(En); chromosome numbers.

399 Juo, P.-S. *and* Stotzky, G.
Electrophoretic analysis of isozymes from seeds of *Pinus*, *Abies* and *Pseudotsuga*. *Canad. J. Bot.*, 51(11):
2, 3 2201–2205(1973)—(Fr).

400 Strullu, D.-G.
Répartition des stomates à la face supérieure des feuilles chez *Pseudotsuga* (Abiétacées). *Bull. Soc. Sci. Bretagne*, 47(3–4):209–213(1972 publ. 1973)—
2, 3 Illustrations.

SCIADOPITOPHYLLUM

401 Christophel, D. C.
Sciadopitophyllum canadense gen. et sp. nov: a new conifer from western Alberta. *Amer. J. Bot.*, 60(1):
3 61–66(1973)—Illustrations, dot map.

SCIADOPITYS

402 Ho, R. H. *and* Sziklai, O.
Fine structure of the pollen surface of some Taxodiaceae and Cupressaceae species. *Rev. Palaeobot.*
2, 3 *Palynol.*, 15(1):17–26(1973). *Sciadopitys verticillata*.

SEQUOIA

403 Ho, R. H. *and* Sziklai, O.
Fine structure of the pollen surface of some Taxodiaceae and Cupressaceae species. *Rev. Palaeobot.*
2, 3 *Palynol.*, 15(1):17–26(1973). *Sequoia sempervirens*.

TAIWANIA

404 Ho, R. H. *and* Sziklai, O.
Fine structure of the pollen surface of some Taxodiaceae and Cupressaceae species. *Rev. Palaeobot.*
2, 3 *Palynol.*, 15(1):17–26(1973). *Taiwania cryptomerioides*.

TAXODIUM

405 Groth, R.
Taxodium mucronatum: Ursymbol eines Baumriesen.
4 *Kurzmitt. Deutsch. Dendrol. Ges.*, no. 9:21–22(1972).

406 Kennedy, H. E.
Baldcypress: *Taxodium distichum*. *U.S. Dep. Agric.*
3 *F.S.*, 218. 5p.(1973)—Illustrations, map.

GYMNOSPERMAE

TAXUS

407 Kučera, L.
Wundgewebe in der Eibe (*Taxus baccata* L.). *Vierteljahrschr. Naturforsh. Ges. Zürich*, 116(4):445–466
1 (1971)—(En).

THUJA

408 Banthorpe, D. V., Davies, H.ff.S., Gatford, C. *and* Williams, S. R.
Monoterpene patterns in *Juniperus* and *Thuja*
2, 3 species. *Pl. Med.*, 23(1):64–69(1973).

TSUGA

409 Davidson, D.W., Ahlberg, R. A., Koch, R. G. *and* Lahti, D. J.
A westward extension of hemlock in Wisconsin. *Mich. Bot.*, 12(4):209–211(1973)—Map. *Tsuga*
3 *canadensis.*

410 Nemeth, J. C.
A mountain disjunct hemlock stand in the Piedmont of Virginia. *Castanea*, 38(2):171–175(1973). *Tsuga*
3 *canadensis.*

WELWITSCHIA

411 Bernard, V. V. *and* Meĭer, N. R.
Pȳl'tsa *Ephedra, Welwitschia* i *Gnetum*. (Pollen grains of *Ephedra, Welwitschia* and *Gnetum*). *Vestn. Mosk. Univ. Nauch. Zhurn.*, *6*, 27(6):86–88(1972)—
1, 5 (En).

412 Butler, V., Bornman, C. H. *and* Evert, R. F.
Welwitschia mirabilis: morphology of the seedling.
5 *Bot. Gaz.*, 134(1):52–59(1973)—Illustrations.

413 Butler, V., Bornman, C. H. *and* Evert, R. F.
Welwitschia mirabilis: vascularization of a four-week-old seedling. *Bot. Gaz.*, 134(1):59–63(1973)—
5 Illustrations.

414 Butler, V., Bornman, C. H. *and* Evert, R. F.
Welwitschia mirabilis: vascularization of a one-year-old seedling. *Bot. Gaz.*, 134(1):63–73(1973)—
5 Illustration.

415 Laudi, G. *and* Bonatti, P. M.
Ultrastructure of chloroplasts of some Chlamidospermae (*Ephedra twediana, Gnetum montana, Welwitschia mirabilis*). *Caryologia*, 26(1): 107–114(1973)—(It); illustrations. Discussion of
5 taxonomic relationships.

416 Voronin, N. S., Voronkina, N. V. *and* Lodkina, M. M.
Stroenie apeksa koriya Gnetales. (Root apex structure in Gnetales). *Bot. Zhurn.*, 58(11):1589–1596
5 (1973)—(En).

ANGIOSPERMAE

417 Chang, C. P. *and* Mabry, T. J.
The constitution of the Order Centrospermae: *r*RNA-DNA hybridization studies among betalain- and anthocyan-producing families. *Biochem. Syst.*,
3, 4, 6, 7 1(4):185–190(1973). Bataceae excluded.

418 Ehrendorfer, F.
Adaptive significance of major taxonomic characters and morphological trends in angiosperms. *In* Heywood, V. H., *ed.* Taxonomy and ecology:
1–7 317–327(1973).

419 Holub, J.
New names in Phanerogamae: 2. *Folia Geobot.*
1 *Phytotax.*, 8(2):155–179(1973)—(Ge).

420 Hultén, E.
The circumpolar plants: 2. Dicotyledons. *Kungl. Svenska Vetenskapsakad. Handl.*, 13(1). 463p.(1970
1–3 publ. 1971)—Dot maps.

421 Mehra, P. N. *and* Sareen, T. S.
Cytology of some Himalayan trees: Thalamiflorae. *Silvae Genet.*, 22(3):66–70(1973)—Illustrations; chromosome numbers. *Berberis, Bombax, Cocculus, Elaeocarpus, Eurya, Flacourtia, Gordonia, Grewia, Kydia, Michelia, Pterospermum, Pterogota, Shorea,*
6 *Sterculia, Xylosma.*

422 Raynal, J.
Joncales. *Encyclopaedia Universalis*, 9:501–502
1–7 (1971)—Illustrations.

423 Rougé, P.
Sur l'existence de protéines solubles à propriétés antigéniques communes dans les feuilles de diverses espèces d'angiospermes. *Compt. R. Hebd. Séances*
1–7 *Acad. Sci., D* (Paris), 276(19):2645–2648(1973).

424 Rowley, G. D. *and* Newton, L. E., *eds*
Repertorium plantarum succulentarum: 22—1971. Utrecht, International Association for Plant
1–7 Taxonomy, 1973. 24p. (Regnum Vegetabile 87).

ACANTHACEAE

425 Agostini, G.
Taxonomic bibliography for the neotropical flora: Acanthaceae. *Acta Bot. Venezuel.*, 9(1–4):255–263
4 (1974 publ. 1973).

426 Hossain, A.B.M.E.
Notes on Asiatic Acanthaceae. *Notes Roy. Bot.*
2 *Gard. Edinburgh*, 32(3):405–410(1973).

427 Meyer, P. G.
Neue und wenig bekannte Acanthaceen aus dem Kaokoveld (Südwestafrika). *Mitt. Bot. Staatssamml.*
5 *München*, 11:101–113(1973)—Illustration.

APHELANDRA

428 Wasshausen, D. C.
Monograph of the genus *Aphelandra* (Acanthaceae).
4 *Diss. Abstr. Int., B*, 33(8):3516(1973).

429 Wasshausen, D. C.
New species of *Aphelandra* (Acanthaceae). *Phytologia*,
4 25(7):465–502(1973)—Illustrations.

430 Wasshausen, D. C.
Two additional new species of *Aphelandra* (Acanthaceae). *Phytologia*, 26(6):393–396(1973)—
4 Illustrations.

attenuata Wasshausen in *Phytologia*, 25(7):487
4 (1973)—Ecuador.
4 **benoistii** Wasshausen *op. cit.* 489(1973)—Ecuador.
campii Wasshausen *op. cit.* 467(1973)—Ecuador,
4 Peru.
4 **chrysantha** Wasshausen *op. cit.* 483(1973)—Ecuador.
4 **cinnabarina** Wasshausen *op. cit.* 485(1973)—Ecuador.
4 **cuscoensis** Wasshausen *op. cit.* 469(1973)—Peru.
4 **darienensis** Wasshausen *op. cit.* 480(1973)—Panama.
dasyantha Wasshausen *op. cit.* 474(1973)—Ecuador,
4 Peru.

diffusa Wasshausen *op. cit.* 477(1973)—Ecuador,
4 Peru.
4 **dodsonii** Wasshausen *op. cit.* 486(1973)—Ecuador.
4 **dukei** Wasshausen *op. cit.* 475(1973)—Panama.
4 **fasciculata** Wasshausen *op. cit.* 490(1973)—Venezuela.
4 **ferreyrae** Wasshausen *op. cit.* 465(1973)—Peru.
4 **flammea** Wasshausen in *Phytologia*, 26(6):394
4 (1973)—Ecuador.
galba Wasshausen in *Phytologia*, 25(7):479(1973)—
4 Ecuador.
4 **guayasii** Wasshausen *op. cit.* 495(1973)—Ecuador.
guerrerensis Wasshausen in *Phytologia*, 26(6):393
4 (1973)—Mexico.
hapala Wasshausen in *Phytologia*, 25(7):466(1973)—
4 Peru.
4 **hintonii** Wasshausen *op. cit.* 476(1973)—Mexico.
4 **juninensis** Wasshausen *op. cit.* 471(1973)—Peru.
4 **kingii** Wasshausen *op. cit.* 481(1973)—Colombia.
4 **latibracteata** Wasshausen *op. cit.* 468(1973)—Peru.
4 **nephoica** Wasshausen *op. cit.* 483(1973)—Colombia.
4 **paulensis** Wasshausen *op. cit.* 482(1973)—Brazil.
4 **peruviana** Wasshausen *op. cit.* 470(1973)—Peru.
4 **phaina** Wasshausen *op. cit.* 492(1973)—Ecuador.
4 **reticulata** Wasshausen *op. cit.* 493(1973)—Venezuela.
4 **rubra** Wasshausen *op. cit.* 494(1973)—Bolivia, Peru.
steyermarkii Wasshausen *op. cit.* 488(1973)—
4 Venezuela.
4 **tillettii** Wasshausen *op. cit.* 473(1973)—Peru.
4 **wurdackii** Wasshausen *op. cit.* 472(1973)—Peru.

ASYSTASIELLA

chinensis (S. Moore) E. Hossain in *Notes Roy. Bot.*
2, 6 *Gard. Edinburgh*, 32(3):405(1973): *Asystasia chinensis.*

BLEPHARIS

431 Gutterman, Y.
Delayed seed dispersal and rapid germination as survival mechanisms of the desert plant *Blepharis persica* (Burm.) Kuntze. *Oecologia*, 10(2):145–149
2 (1972).

432 Gutterman, Y., Witztum, A. *and* Heydecker, W.
Studies of the surfaces of desert plant seeds: 2. Ecological adaptations of the seeds of *Blepharis*
2 *persica. Ann. Bot.* (U.K.), 37(153):1051–1055(1973).

CHAMPIONELLA

coreana (Lévl.) E. Hossain in *Notes Roy. Bot. Gard.*
2 *Edinburgh*, 32(3):405(1973): *Strobilanthes coreana.*
labordei (Lévl.) E. Hossain *op. cit.* 405(1973):
2 *Strobilanthes labordei.*

CHROESTHES

silvicola (W. W. Sm.) E. Hossain in *Notes Roy. Bot.*
2 *Gard. Edinburgh*, 32(3):405(1973): *Asystasia silvicola.*

CROSSANDRA

433 Hepper, F. N.
Crossandra friesiorum. Curtis's Bot. Mag., 179(3):
5 tab. 647(1973)—En; illustration, coloured illustration.

DIFLUGOSSA

shweliensis (W. W. Sm.) E. Hossain in *Notes Roy. Bot. Gard. Edinburgh*, 32(3):406(1973): *Strobilanthes*
2 *shweliensis.*

ERANTHEMUM

434 Karlström, P.-O.
Embryological studies in the Acanthaceae: 2. The genera *Pseuderanthemum* Radlk., *Ruspolia* Lindau, *Eranthemum* L. and *Lankesteria* Lindl. *Svensk Bot.*
5, 6 *Tidskr.*, 67(3):257–280(1973).

GOLDFUSSIA

chaffanjonii (Lévl.) E. Hossain in *Notes Roy. Bot. Gard. Edinburgh*, 32(3):406(1973): *Strobilanthes*
2 *chaffanjonii.*
equitans (Lévl.) E. Hossain *op. cit.* 406(1973):
2 *Strobilanthes ? equitans.*
feddei (Lévl.) E. Hossain *op. cit.* 407(1973):
2 *Strobilanthes feddei.*
mahongensis (Lévl.) E. Hossain *op. cit.* 407(1973):
2 *Strobilanthes mahongensis.*

GUTZLAFFIA

lyi (Lévl.) E. Hossain in *Notes Roy. Bot. Gard.*
2 *Edinburgh*, 32(3):407(1973): *Ruellia lyi.*

HYGROPHILA

435 Bruggen, H. W. E. van
De vaantjesplant—*Hygrophila difformis. Aquarium*,
6 43(8):188(1973)—Illustration.

436 Bruggen, H. W. E. van
Synnema triflorum heisst jetzt *Hygrophila difformis. Inform. ZAG Wasserpflanzen*, no. 4:3–4(1973)—
6 Illustration.

JUSTICIA

437 Wasshausen, D. C.
New combinations in cultivated *Justicia*
4 (Acanthaceae). *Baileya*, 19(1):1–3(1973)—Illustrations.

kouytcheensis (Lévl.) E. Hossain in *Notes Roy. Bot. Gard. Edinburgh*, 32(3):407(1973): *Ruellia repens*
2 var. *kouytcheensis.*
leonardii Wasshausen in *Baileya*, 19(1):3(1973),
4 nom. nov.: *Jacobinia incana* (Nees) Hemsley.
lepida (Nees) Wasshausen *op. cit.* 3(1973): *Jacobinia*
4 *lepida.*
rizzinii Wasshausen *op. cit.* 3(1973), nom. nov.: *J.*
4 *pauciflora* (Nees) Grisebach.

LANKESTERIA

438 Karlström, P.-O.
Embryological studies in the Acanthaceae: 2. The genera *Pseuderanthemum* Radlk., *Ruspolia* Lindau, *Eranthemum* L. and *Lankesteria* Lindl. *Svensk Bot.*
5, 6 *Tidskr.*, 67(3):257–280(1973).

LEPIDAGATHIS

incurva Ham. ex D. Don
var. ustulata (Nees) R. R. Stewart, Annot. Cat. Vasc. Pl. W. Pakistan & Kashmir: 677(1972),
6 without basionym ref.: *L. ustulata.*

LEPTOSIPHONIUM

venustum (Hance) E. Hossain in *Notes Roy. Bot. Gard.*
2 *Edinburgh*, 32(3):408(1973): *Ruellia venusta.*

MONECHMA

serotinum P. G. Meyer in *Mitt. Bot. Staatssamml.*
5 *München*, 11:112(1973)—South West Africa.

OPLONIA

tetrasticha (Wr. ex Griseb.) Stearn
subsp. **polyece** (Stearn) Borhidi in *Acta Bot. Acad. Sci. Hung.*, 19(1–4):45(1973): *Oplonia tetrasticha*
4 var. *polyece.*

PARARUELLIA

cavaleriei (Lévl.) E. Hossain in *Notes Roy. Bot. Gard.*
2 *Edinburgh*, 32(3):409(1973): *Ruellia cavaleriei.*
delavayana (Baill.) E. Hossain *op. cit.* 409(1973):
2 *Ruellia delavayana.*

ACANTHACEAE

PETALIDIUM

ohopohense P. G. Meyer in *Mitt. Bot. Staatssamml.*
5 *München*, 11:108(1973)—South West Africa.

PHAULOPSIS

439 Hepper, F. N.
Tropical African plants: 33. A new combination in
5 *Phaulopsis*. (Acanthaceae). *Kew Bull.*, 28(2):320(1973).

ciliata (Willd.) Hepper in *Kew Bull.*, 28(2):320
5 (1973): *Origanum ciliatum.*
semiconica P. G. Meyer in *Mitt. Bot. Staatssamml.*
5 *München*, 11:104(1973)—South West Africa.

PHLEBOPHYLLUM

440 Kallarackal, J.
Morphological and embryological studies on *Phlebophyllum kunthianum* Nees (Acanthaceae).
6 *Botanica* (Delhi), 23(4):152–153(1973).

PSEUDERANTHEMUM

441 Karlström, P.-O.
Embryological studies in the Acanthaceae: 2. The genera *Pseuderanthemum* Radlk., *Ruspolia* Lindau, *Eranthemum* L. and *Lankesteria* Lindl. *Svensk Bot.*
5, 6 *Tidskr.*, 67(3):257–280(1973).

RUELLIA

442 Gibson, D. N.
Studies in American plants: 5. *Phytologia*, 25(3):
4 113–115(1973).

443 Long, R. W.
A biosystematic approach to generic delimitation in *Ruellia* (Acanthaceae). *Taxon*, 22(5–6):543–555
3, 4 (1973)—Illustrations.

444 Wasshausen, D. C. *and* Austin, D. F.
A *Ruellia* (Acanthaceae) species in *Exogonium* (Convolvulaceae). *Phytologia*, 25(7):433–437(1973)—
4 Illustrations.

molinae D. N. Gibson in *Phytologia*, 25(3):113
4 (1973)—Honduras.
velutifolia (House) Wasshausen & D. F. Austin in
4 *Phytologia*, 25(7):433(1973): *Exogonium velutifolium.*

RUSPOLIA

445 Karlström, P.-O.
Embryological studies in the Acanthaceae: 2. The genera *Pseuderanthemum* Radlk., *Ruspolia* Lindau, *Eranthemum* L. and *Lankesteria* Lindl. *Svensk Bot.*
5, 6 *Tidskr.*, 67(3):257–280(1973).

SOOIA Pócs in *Acta Bot. Acad. Sci. Hung.*, 19(1–4):
5 461(1973).

446 Pócs, T.
Soóia ulugurica genus et species nova Acanthacearum. *Acta Bot. Acad. Sci. Hung.*, 19(1–4):461–466(1973)—
5 (En); illustrations.

ulugurica Pócs in *Acta Bot. Acad. Sci. Hung.*, 19(1–4):
5 461(1973)—Tanzania.

SYNNEMA

447 Bruggen, H. W. E. van
Synnema triflorum heisst jetzt *Hygrophila difformis*. *Inform. ZAG Wasserpflanzen*, no. 4:3–4(1973)—
6 Illustration.

TARPHOCHLAMYS

darrisii (Lévl.) E. Hossain in *Notes Roy. Bot. Gard.*
2 *Edinburgh*, 32(3):410(1973): *Strobilanthes darrisii.*

TETRAGOGA

esquirolii (Lévl.) E. Hossain in *Notes Roy. Bot. Gard.*
2 *Edinburgh*, 32(3):410(1973): *Strobilanthes esquirolii.*

TETRAMERIUM

448 Agostini, G. *and* Wasshaussen, D.
Tetramerium (Acanthaceae), un género neuvo para la flora de Venezuela. *Acta Bot. Venezuel.*, 8(1–4):
4 163–166(1973)—Illustrations.

ACERACEAE

ACER

449 Drenkovski, R.
Novi prilozi kon poznavanjeto na vidovite *Acer campestre* L. i *Acer marsicum* Guss. vo Makedonija. (Neue Beitrage zur Kenntnis die Arten *Acer campestre* L. und *Acer marsicum* Guss.). *God. Zborn. Biol.*, 25:
1 185–203(1973)—(Ge).

450 Fowden, L. *and* Pratt, H. M.
Cyclopropylamino acids of the genus *Acer*: distribution and biosynthesis. *Phytochemistry*, 12(7):
1, 2 1677–1681(1973).

451 Harris, J. G. S.
Maples in Taiwan and Hong Kong. *Int. Dendrol.*
2 *Soc. Year Book 1972*: 59–62(1973)—Illustration.

452 Walther, H.
Studien über tertiäre *Acer* Mitteleuropas. *Abh. Staatl. Mus. Mineral. Geol., Dresden*, 19. 309p.(1972)—
1 Illustrations.

ACTINIDIACEAE

453 Agostini, G.
Taxonomic bibliography for the neotropical flora: Actinidiaceae. *Acta Bot. Venezuel.*, 9(1–4):265–266
4 (1974 publ. 1973).

ACTINIDIA

454 Karpowiczowa, L.
Aktinidia chińska *Actinidia chinensis* Planch. *Wiad.*
2 *Bot.*, 17(1):48–50(1973)—Illustrations.

455 Wilson, S. J. *and* Richards, R. R.
Chinese gooseberries. *Tasman. J. Agric.*, 44(4):254–
2 256(1973)—Illustration. *Actinidia chinensis.*

AGAVACEAE

456 Daniel, J. *and* Rowland, D.
An introduction to some cacti and succulent plants of New Mexico. *Cact. Succ. J. Gr. Brit.*, 35(3):53–54
3 (1973).

AGAVE

457 Blunden, G., Yi Yi *and* Jewers, K.
The comparative leaf anatomy of *Agave*, *Beschorneria*, *Doryanthes* and *Furcraea* species (Agavaceae: Agaveae). *Bot. J. Linn. Soc.*, 66(2):157–
3, 4 179(1973).

458 Bos, J. J. *and* Lensing, F. H. G.
A new cultivar in sisal from East Africa: *Agave sisalana* Perr. ex Engelm. cv. Hildana. *E. Afr. Agric. Forest. J.*, 39(1):17–25(1973)—Illustrations;
5 chromosome numbers.

459 Freeman, C. E.
A new record of lechuguilla (*Agave lecheguilla*: Agavaceae) in New Mexico. *Southwest. Nat.*, 17(4):
4 423(1973).

460 Graaf, A. de
Agave ferox C. Koch. *Succulenta*, 52(11):207–210
4 (1973)—Du; illustrations.

461 Horich, C. K.
Post aus Costa Rica: *Agave wercklei* Weber ex Trelease. *Kakt. And. Sukk.*, 24(7):158–159(1973)—
4 Illustration.

univittata Haw.
var. **angustifolia** (Berg.) Jacobsen in *Nation. Cact. Succ. J.* (U.K.), 28(1):4(1973): *A. lophantha* var.
4 *angustifolia*.
var. **brevifolia** (Jac.) Jacobsen *op. cit.* 4(1973):
4 *A. lophantha* var. *brevifolia*.
var. **caerulescens** (S.D.) Jacobsen *op. cit.* 4(1973):
4 *A. caerulescens*.
var. **gracilior** (Jacobi) Jacobsen *op. cit.* 4(1973):
4 *A. lophantha* var. *gracilior*.
var. **subcanescens** (Jacobi) Jacobsen *op. cit.* 4
4 (1973): *A. lophantha* var. *subcanescens*.
var. **tamaulipasana** (Berg.) Jacobsen *op. cit.* 4
4 (1973): *A. lophantha* var. *tamaulipasana*.

BESCHORNERIA

462 Blunden, G., Yi Yi *and* Jewers, K.
The comparative leaf anatomy of *Agave*, *Beschorneria*, *Doryanthes* and *Furcraea* species (Agavaceae: Agaveae). *Bot. J. Linn. Soc.*, 66(2):157–
4 179(1973).

DORYANTHES

463 Blunden, G., Yi Yi *and* Jewers, K.
The comparative leaf anatomy of *Agave*, *Beschorneria*, *Doryanthes* and *Furcraea* species (Agavaceae:
7 Agaveae). *Bot. J. Linn. Soc.*, 66(2):157–179(1973).

FURCRAEA

464 Blunden, G., Yi Yi *and* Jewers, K.
The comparative leaf anatomy of *Agave*, *Beschorneria*, *Doryanthes* and *Furcraea* species (Agavaceae:
4 Agaveae). *Bot. J. Linn. Soc.*, 66(2):157–179(1973).

POLIANTHES

465 Trueblood, E. W. E.
'Omixochitl' the tuberose (*Polianthes tuberosa*).
4 *Econ. Bot.*, 27(2):157–173(1973)—Illustrations.

YUCCA

466 McCleary, J. A. *and* Wagner, K. A.
Comparative germination and early growth studies of six species of the genus *Yucca*. *Amer. Midl. Nat.*,
3, 4 90(2):503–508(1973)—Illustrations, map.

467 Munz, P. A.
Record of an unusually tall *Yucca schidigera*. *Aliso*,
3 8(1):13–14(1973)—Illustration.

468 Rusanov, F. N.
Mezhvidovaya gibridizatsiya u vidov gibiskusa i yukki kak primer akklimatizatsii. (Interspecific hibridization of *Hibiscus* and *Yucca* species as an example of the acclimatization). *In* U.S.S.R.: Akademiya Nauk SSSR: Sovet Botanicheskikh Sadov: Uspekhi introduktsii rasteniĭ: 224–231(1973)—
3, 4 Coloured illustrations.

469 Subramaniam, D.
Cytological investigations in *Haemanthus coccineus* L., *Asparagus racemosus* Willd. and *Yucca gloriosa* Linn. *Proc. Indian Sci. Congr.*, 60(3):323–324(1973)—
3 Chromosome number. Abstract only.

AIZOACEAE

470 Agostini, G.
Taxonomic bibliography for the neotropical flora: Aizoaceae. *Acta Bot. Venezuel.*, 9(1–4):267–268(1974
4 publ. 1973).

471 Bent, P.
More notes on the Conophytums. *Bull. Afr. Succ.*
5 *Pl. Soc.*, 8(1):42–43(1973).

472 Dupont, S.
Étude du pollen de Mésembryanthemacées très succulentes. *Compt. R. Hebd. Séances Acad. Sci., D*
5 (Paris), 277(21):2373–2376(1973).

473 Fernandes, R. B.
Aizoaceae. *Anuàr. Soc. Brot.*, 38:123–126(1972)—
1 Por; key.

474 Fernandes, R. B.
Mesembryanthemaceae. *Anuàr. Soc. Brot.*, 38:127–
1 136(1972)—Por; keys.

475 Fernandes, R. B.
Molluginaceae. *Anuàr. Soc. Brot.*, 38:119–122
1 (1972)—Por; keys.

476 Fernandes, R. B.
Novitates et notulae systematicae et nomenclaturales ad florae Lusitanicae, Azoricae et Madeirensis
1 pertinentes. *Anuàr. Soc. Brot.*, 38:61–64(1972).

477 Fernandes, R. B.
Tetragoniaceae. *Anuàr. Soc. Brot.*, 38:137–138
1 (1972)—Por.

478 Hofmann, U.
Morphologische Untersuchungen zur Ümgrenzung und Gliederung der Aizoaceen. *Bot. Jahrb.*, 93(2):247–324(1973); also publ. in *Mitt. Bot. Mus. Univ. Zürich*, no. 262(1973) (Centrospermen-Studien 6)—(En);
1–7 illustrations.

479 Lousley, J. E.
'Mesembryanthemums' established in the Isles of Scilly. *In* Green, P. S. *ed.* Plants: wild and cultivated:
1 83–85(1973)—Illustration.

480 Nasir, Y.
Flora of West Pakistan: no. 40. Molluginaceae.
6 1973. 6p.—Illustrations, map; key.

481 Nasir, Y.
Flora of West Pakistan: no. 41. Aizoaceae. 1973.
6 12p.—Illustrations, map; key.

ALOINOPSIS

482 Hewitt, T.
Aloinopsis schooneesii L. Bol. (Aizoaceae). *Ashingtonia,* 1(1):10(1973)—En; coloured illustration.
5 Erratum in *Ashingtonia,* 1(3):35(1973).

ARGYRODERMA

483 Hartmann, H.
New combinations and a key for the genus *Argyroderma* N.E. Br. (Mesembryanthemaceae Fenzl). *Nation. Cact. Succ. J.* (U.K.), 28(2):48–50(1973)—
5 Illustrations.

subgen. **Roodia** (N.E. Br.) H. Hartmann in *Nation.*
5 *Cact. Succ. J.* (U.K.), 28(2):49(1973): gen. *Roodia.*
framesii L. Bol.
subsp. **hallii** (L. Bol.) H. Hartmann *op. cit.* 50
5 (1973): *A. hallii.*

CARPOBROTUS

484 Hewitt, T.
Carpobrotus edulis (L.) N.E. Br. (Aizoaceae). *Ashingtonia,* 1(1):7, 8(1973)—En; coloured
1, 3–5, 7 illustration.

CHEIRIDOPSIS

pillansii L. Bol.
var. **crassa** (L. Bol.) Rowley in *Nation. Cact. Succ. J.*
5 (U.K.), 28(1):6(1973): *C. crassa.*

CONOPHYTUM

485 Boom, B. K.
Het geslacht *Conophytum* N.E.Br.: 1–10. *Succulenta,* 52(2):22–26, (3):42–45, (4):62–65, (5):82–86, (6):102–105, (7):122–126, (8):142–146, (9):162–166, (10):182–187, (11):202–203(1973)—Illustrations;
5 key.

486 Rawe, R.
Conophytum altum L. Bolus. *Kakt. And. Sukk.,*
5 24(11):248–250(1973)—Ge; illustrations, map.

487 Rawe, R.
Conophytum elishae (N. E. Brown) N. E. Brown. *Kakt. And. Sukk.,* 24(12):285–286(1973)—Ge;
5 illustrations.

488 Rawe, R.
Conophytum laetum L. Bolus. *Kakt. And. Sukk.,* 24(9):197–199(1973)—Ge; illustrations, coloured
5 illustration, map.

flavum N. E. Br.
var. **luteum** (N. E. Br.) Boom in *Succulenta,*
5 52(6):103(1973): *C. luteum.*
laetum L. Bolus
var. **extractum** (Tischer) Rawe in *Kakt. And. Sukk.,*
5 24(9):199(1973): *C. extractum.*
minusculum N. E. Br.
var. **reticulatum** (L. Bol.) Rawe ex Rowley in Rep. Pl. Succ., 22 (Regnum Veg., 87): 16(1973):
5 *C. reticulatum.*
f. **roseum** (Rowley) Rowley *op. cit.* 17(1973):
5 *C. reticulatum* f. *roseum.*
minutum (Haw.) N. E. Br.
var. **nudum** (Tischer) Boom in *Succulenta,* 52(8):
5 143(1973): *C. nudum.*
var. **pearsonii** (N. E. Br.) Boom *op. cit.* 143(1973):
5 *C. pearsonii.*
var. **sellatum** (Tischer) Boom *op. cit.* 143(1973):
5 *C. sellatum.*
saxetanum N. E. Br.
f. **hallianum** Rowley in *Nation. Cact. Succ. J.* (U.K.), 28(1):6(1973)—South Africa (Cape
5 Prov.).

DELOSPERMA

489 Anon.
Delosperma aberdeenense (L. Bol.) L. Bol. (Ficoidaceae). *Ashingtonia,* 1(3):34(1973)—En; coloured
5 illustration.

FENESTRARIA

aurantiaca N. E. Br.
f. **rhopalophylla** (Schlecht. & Diels) Rowley in *Nation. Cact. Succ. J.* (U.K.), 28(1):6(1973):
5 *Mesembryanthemum rhopalophyllum.*

GLOTTIPHYLLUM

longum (Haw.) N. E. Br.
var. **heterophyllum** (Haw.) Rowley in *Nation. Cact. Succ. J.* (U.K.), 28(1):6(1973):
5 *Mesembryanthemum angustum γ heterophyllum.*

LAPIDARIA

490 Bent, P.
The genus *Lapidaria* Dint. et Schw. *Bull. Afr. Succ.*
5 *Pl. Soc.,* 8(2):73–74(1973).

LITHOPS

491 Cole, D. T.
Lithops gulielmi L. Bol.—dismythed. *Nation. Cact.*
5 *Succ. J.* (U.K.), 28(3):89–92(1973)—Illustrations.

492 Helm, D.
Probleme der Lithopspflege. *Stachelpost,* 9(46):
5 108–110(1973)—Illustrations.

MONILARIA

chrysoleuca (Schlechter) Schwant.
f. **purpurea** (L. Bol.) Rowley in *Nation. Cact. Succ. J.* (U.K.), 28(1):6(1973): *M. chrysoleuca* var.
5 *purpurea.*

RUSCHIA

haworthii Jacobsen & Rowley in *Nation. Cact. Succ. J.* (U.K.), 28(1):6(1973), nom. nov.: *Mesembryan-*
5 *themum dilatatum* Haw.

SCHWANTESIA

herrei L. Bol.
f. **major** Rowley in *Nation. Cact. Succ. J.* (U.K.),
5 28(1):6(1973), nom. nov.: *S. loeschiana* Tischer.

ALANGIACEAE

493 Ghafoor, A.
Flora of West Pakistan: no. 50. Alangiaceae. 1973.
6 3p.—Illustration, map.

ALANGIUM

494 Bishnoi, S., Rajeswari, V. M. *and* Rani, S.
Stomata of *Alangium salviifolium. Proc. Indian Sci.*
6 *Congr.,* 60(3):289–290(1973). Abstract only.

495 Espirito Santo, J. do
Flora de S. Tomé e Príncipe: Alangiaceae. 1973. 5p.
2, 5, 6 *Alangium* only.

ALISMATACEAE

496 Agostini, G.
Taxonomic bibliography for the neotropical
flora: Alismataceae. *Acta Bot. Venezuel.*, 9(1–4):269–
4 272(1974 publ. 1973).

497 Charlton, W. A.
Studies in the Alismataceae: 2. Inflorescences of
1–7 Alismataceae. *Canad. J. Bot.*, 51(4):775–789(1973).

498 Leins, P. *and* Stadler, P.
Entwicklungsgeschichtliche Untersuchungen am
Androeceum Alismatales. *Österr. Bot. Zeitschr.*,
1–7 121(1–2):51–63(1973)—(En).

ALISMA

499 Libbey, R. B. *and* Swann, E. L.
Alisma gramineum Lej.: a new county record. *Nat.*
1 *Cambridge.*, no. 16:39–41(1973).

500 Libbey, R. P. *and* Swann, E. L.
Extension of range of *Alisma gramineum* in Britain.
1 *Watsonia*, 9(4):429(1973).

BALDELLIA

repens (Lam.) van Ooststroom ex Lawalrée in Nouv.
Fl. Belg., Luxemb., N. France etc.: 759(1973):
1 *Alisma repens.*

ECHINODORUS

501 Hansen, J. *and* Hansen, P.
The dwarf sword plant. *Aquarist Pondkeeper*, 38(5):
189–190(1973)—Illustration, map. *Echinodorus*
4 *magdalenensis.*

502 Sadilek, V.
Een nieuwe *Echinodorus*-soort voor grote aquaria.
Aquarium, 43(8):180–182(1973)—Illustrations.
4 *E. osiris.*

LIMNOPHYTON

503 Rangaswami Ayyangar, K.
Cytological investigations in *Limnophyton*
tobusifolium (L.) Miq. *Proc. Indian Sci. Congr.*, 59(3):
6 351I(1972)—Chromosome numbers. Abstract only.

RANALISMA

504 Charlton, W. A.
Studies in the Alismataceae: 4. Developmental
morphology of *Ranalisma humile* and comparisons
with two members of the Butomaceae, *Hydrocleis*
nymphoides and *Butomus umbellatus*. *Canad. J. Bot.*,
5 51(5):899–910(1973)—(Fr).

505 Charlton, W. A. *and* Ahmed, A.
Studies in the Alismataceae: 3. Floral anatomy of
Ranalisma humile. *Canad. J. Bot.*, 51(5):891–897
5 (1973)—(Fr).

SAGITTARIA

506 Pop, I.
Sagittaria subulata (L.) Buch. O nouă hidrofita în
apele termale ale băilor I Mai-Oradea. (Jud. Bihor).
(*Sagittaria subulata* (L.) Buch. A new water-plant in
the thermal waters of Oradea (Bihor District)).
1 *Contrib. Bot.*, 115–118(1973)—(En); illustrations.

507 Rataj, K.
Revision of the genus *Sagittaria*: part 1. (Old
World species). *Annot. Zool. Bot., Slov. Narod. Muz.*
1, 2, 5–7 *Bratislava*, 76. 31p.(1972)—Illustrations.

508 Rataj, K.
Revision of the genus *Sagittaria* part 2. (The species
of the West Indies, Central and South America).
Annot. Zool. Bot., Slov. Narod. Muz. Bratislava, 78.
4 61p.(1972)—Illustrations; key.

509 Singh, V. *and* Sattler, R.
Nonspiral androecium and gynoecium of
Sagittaria latifolia. *Canad. J. Bot.*, 51(6):1093–1095
3, 4 (1973)—(Fr).

510 Wooten, J. W.
Taxonomy of seven species of *Sagittaria* from
eastern North America. *Brittonia*, 25(1):64–74(1973)—
3 Illustrations, dot map; key.

ALSTROEMERIACEAE

511 Hunziker, A. T.
Notas sobre Alstroemeriaceae. *Kurtziana*, 7:133–135
4 (1973)—(En).

ALSTROEMERIA

512 Slob, A.
Tulip allergens in *Alstroemeria* and some other
4 Liliiflorae. *Phytochemistry*, 12(4):811–815(1973).

englerana (Kraenzlin) A. T. Hunziker in *Kurtziana*,
4 7:133(1973), as 'engleriana': *Bomarea engleriana.*
involucrosa (Herb.) A. T. Hunziker *op. cit.* 133
4 (1973): *Collania involucrosa.*
macrocephala (Pax) A. T. Hunziker *op. cit.* 133
4 (1973): *Bomarea macrophala.*
ovallei (R. A. Phil.) A. T. Hunziker *op. cit.* 134
4 (1973): *Leontochir ovallei.*

AMARANTHACEAE

513 Agostini, G.
Taxonomic bibliography for the neotropical flora:
Amaranthaceae. *Acta Bot. Venezuel.*, 9(1–4):273–
4 279(1974 publ. 1973).

514 Khoshoo, T. N. *and* Pal, M.
The probable origin and relationships of the garden
cockscomb. *Bot. J. Linn. Soc.*, 66(2):127–141(1973)—
6 Illustrations; chromosome numbers.

515 Laroche, R. C.
Catálogo de nervação foliar das Amaranthaceae e
Annonaceae da Caatinga: 3. *Arq. Jard. Bot. Rio de*
4 *Janeiro*, 19:269–279(1973)—(En); illustrations; keys.

516 Mittra, R.
Cytology of Amaranthaceae. *Proc. Indian Sci.*
Congr., 58(3):470–471(1971)—Chromosome numbers.
6 Abstract only.

517 Pal, M. *and* Khoshoo, T. N.
Evolution and improvement of cultivated
Amaranths: 6. Cytogenetic relationships in grain
types. *Theor. Appl. Genet.*, 43(5):242–251(1973)—
4–6 Chromosome numbers.

518 Smith, L. B. *and* Downs, R. J.
Flora ilustrada catarinense: Amarantáceas. 1972.
4 110p.—Illustrations, dot maps; keys.

519 Townsend, C. C.
Notes on Amaranthaceae: 1. *Kew Bull.*, 28(1):141–
5 146(1973).

ACHYRANTHES

520 Thakur, V.
Cytology of some members of the genus *Achyranthes* in Bihar. *J. Bihar Bot. Soc.*, 1(1–2):12–21
6 (1972)—Chromosome numbers.

ALLMANIA

521 Behera, B. *and* Patnaik, S. N.
A note on the cytotaxonomy of *Allmania nodiflora* Br. and its variety *roxburghii* Hk.f. *Proc. Indian Sci.*
6 *Congr.*, 59(3):360–361(1972). Abstract only.

ALTERNANTHERA

ficoidea (L.) R. Br.
var. **amoena** (Lem.) L. B. Smith & Downs in Fl. Ilustr. Catar., 1(Amarantác.): 62(1972):
4 *A. sessilis* var. *amoena*.
var. **spathulata** (Lem.) L. B. Smith & Downs
4 *op. cit.* 61(1972): *A. spathulata*.
var. **versicolor** (Lem.) L. B. Smith & Downs
4 *op. cit.* 63(1972): *Telanthera ficoidea* var. *versicolor*.

AMARANTHUS

522 Asai, Y.
(Miscellaneous notes on the naturalized flora of Japan: 3). *J. Jap. Bot.*, 48(3):67–72(1973)—Ja(En);
2 illustrations. *Amaranthus galii*.

523 Grüll, F. *and* Priszter, S.
Amaranthus × *sopronìensis* Priszter et Kárpáti, eine interessante Hybride auch in der Tschechoslowakei. (*Amaranthus* × *sopronìensis* Priszter et Kárpáti, zajímavy hybrid také v CSSR). *Preslia*, 45(2):151–153
1 (1973).

524 Sebastian, K. T. *and* Despande, B. D.
Inflorescence anatomy and floral morphology of *Amaranthus leucocarpus* S. Wat. *Acta Bot. Acad. Sci.*
6 *Hung.*, 18(3–4):355–361(1973).

525 Silverwood, J. H.
The genus *Amaranthus* in Malta. *Maltese Nat.*,
1 no. 4:3–6(1973)—Key.

526 Thakur, V.
Origin of *Amaranthus* × *caturus* Heyne ex Hook. f.
6 *J. Bihar Bot. Soc.*, 1(1–2):71(1972). Abstract.

CHARPENTIERA

527 Sohmer, S. H.
Microsporogenesis in *Charpentiera* (Amaranthaceae).
6 *Chromosome Inform. Serv.*, no. 14:14–16(1973).

528 Sohmer, S. H.
A preliminary report of the biology of the genus *Charpentiera* (Amaranthaceae). *Pac. Sci.*, 27(4):399–
6 405(1973).

DIGERA

muricata (L.) Mart.
var. **macroptera** C. C. Townsend in *Kew Bull.*,
5 28(1):142(1973)—Kenya, Tanzania.
var. **patentipilosa** C. C. Townsend *op. cit.* 142
5 (1973)—Socotra, Kenya, Tanzania.
subsp. **trinervis** C. C. Townsend *op. cit.* 141(1973)
5 —Tanzania.

IRESINE

diffusa H. & B. ex Willd.
f. **lindenii** (Van Houtte) L. B. Smith & Downs in Fl. Ilustr. Catar., 1(Amarantác.): 93(1972):
4 *I. lindenii*.

PHILOXERUS

529 Murata, G.
(Short note on *Philoxerus wrightii* Hook. f.). *Acta*
2 *Phytotax. Geobot.*, 25(4–6):187(1973)—Ja(En).

PTILOTUS

530 Benl, G.
Ptilotus wilsonii spec. nov., eine weitere xerophytische Amaranthacee aus Westaustralien. *Mitt. Bot. Staatssamml. München*, 11:57–63(1973)—
7 (En); illustrations; key.

wilsonii Benl in *Mitt. Bot. Staatssamml. München*,
7 11:57(1973)—Australia (W. Australia).

AMARYLLIDACEAE

531 Calderon, C. V.
Amaryllidaceae I have collected. *Pl. Life*, 29(1–4):
4 9–12(1973)—Illustrations.

532 Cardenas, M.
Bolivian Amaryllids. *Pl. Life*, 29(1–4):36–41
4 (1973)—Illustrations.

533 Geerink, D.
Flore d'Afrique centrale (Zaïre—Rwanda—Burundi): Spermatophytes. Amaryllidaceae. 1973.
5 [ii]23p.—Illustrations; keys.

534 Ravenna, P.
Contribuciones sobre Amaryllidaceae de América del Sur: 5. *Mus. Nac. Hist. Nat. Not. Mens.* (Chile),
4 16(189):8(1972).

AMARYLLIS

535 Blossfeld, H.
Amaryllis calyptrata Ker-Gawler. *Pl. Life*, 29(1–4):
4 30–34(1973)—En; illustration.

536 Moldenke, H. N.
Amaryllis traubii forma *doraniana*. *Pl. Life*, 29(1–4):
4 29–30(1973)—En.

leopoldii
f. **whitakeri** Cardenas in *Pl. Life*, 29:36(1973)—
4 Bolivia.
4 **paquichana** Cardenas *op. cit.* 38(1973)—Bolivia.
traubii Moldenke
4 f. **doraniana** Moldenke *op. cit.* 29(1973)—Peru.

CRINUM

537 Raicu, P., Radu, M., Bogdan, D. *and* Kirillova, M.
Cercetări citogenetice la cîteva specii de plante floricole bulboase. (Cytogenetic researches of some bulbous ornamental plants). *An. Univ. București, Biol. Veg.*, 20:49–60(1971)—(Rus,En); chromosome
2 numbers. *Crinum asiaticum*.

538 Rangaswami Ayyangar, K.
Observations on the cyto-taxonomy of *Crinum powellii* Hort. and *Crinum defixum* Ker. *Proc. Indian Sci. Congr.*, 60(3):302–303(1973)—Chromosome
5, 6 number. Abstract only.

539 Verdoorn, I. C.
Crinum crassicaule. *Flow. Pl. Afr.*, 42(3–4):
5 pl.1676(1973)—En; coloured illustration.

540 Verdoorn, I. C.
The genus *Crinum* in Southern Africa. *Bothalia*, 11(1–2):27–52(1973)—Illustrations, coloured
5 illustrations; key.

EMPODIUM

elongatum (Nel) B. L. Burtt in *Notes Roy. Bot. Gard.*
5 *Edinburgh*, 32(3):311(1973): *Forbesia elongata.*
gloriosum (Nel) B. L. Burtt *op. cit.* 312(1973):
5 *Forbesia gloriosa.*
monophyllum (Nel) B. L. Burtt *op. cit.* 313(1973):
5 *Forbesia monophylla.*
occidentale (Nel) B. L. Burtt *op. cit.* 313(1973):
5 *Forbesia occidentalis.*

GALANTHUS

541 Nutt, R.
'Wild' *Galanthus* in the British Isles. *In* Green, P. S.
ed. Plants: wild and cultivated: 129–138(1973)—
1 Illustrations.

HABRANTHUS

542 Hunziker, A. T. *and* Fulvio, T. E. di
Una nueva especie de *Habranthus* (Amaryllidaceae)
de la Provincia de Buenos Aires. *Kurtziana*, 7:255–259
4 (1973)—(En); illustration; chromosome number; key.

barrosianus A. T. Hunziker & Fulvio in *Kurtziana*,
4 7:255(1973)—Argentina.
schulzianus Ravenna in *Mus. Nac. Hist. Nat. Not.*
4 *Mens.* (Chile), 189:8(1972)—Argentina.

HAEMANTHUS

543 Rangaswami Ayyangar, K. *and* Subramaniam, D.
Meiotic studies in *Haemanthus coccineus* Linn. *Proc.*
Indian Sci. Congr., 59(3):350–351(1972)—
6 Chromosome number. Abstract only.

544 Subramaniam, D.
Cytological investigations in *Haemanthus coccineus*
L., *Asparagus racemosus* Willd. and *Yucca gloriosa* Linn.
Proc. Indian Sci. Congr., 60(3):323–324(1973)—
5 Chromosome number. Abstract only.

HAYLOCKIA

545 Cardenas, M.
The genus *Haylockia* in Bolivia. *Pl. Life*, 29(1–4):
4 41–45(1973)—Illustrations.

americana (Hoffgg.) Traub in *Pl. Life*, 29:47(1973),
3 without basionym date: *Sternbergia americana.*
4 **chihuahuayu** Cardenas *op. cit.* 44(1973)—Bolivia.
4 **cochabambensis** Cardenas *op. cit.* 44(1973)—Bolivia.

HIERONYMIELLA

546 Fulvio, T. E. di
Contribucion al conocimiento cariologico de
Amaryllidaceae: 1. Estudio cromosomico en
Hieronymiella y otros generos afines. *Kurtziana*,
4 7:117–131(1973)—(En); chromosome numbers.

HIPPEASTRUM

547 Kozlova, A. G.
Rost i razvitie gippeastruma gibridnogo. (Growth
and development of the hybrid *Hyppeastrum*). *Izv. Akad.*
4 *Nauk Turkm. SSR, Biol. Nauk*, 3:41–45(1973)—(En).

LEUCOJUM

aestivum L.
subsp. **pulchellum** (Salisb.) Malagarriga, Las
1 Subesp. y Variac. Geogr.: 26(1973): *L. pulchellum.*

NARCISSUS

548 Ellis, E. A.
Narcissus pseudonarcissus L. *Trans. Norfolk Norwich*
Nat. Soc., 22(6):437[1973]—En; illustration. Short
1 note on the species in Norfolk.

549 Fernandes, A.
Um novo hibrido de *Narcissus*. *Anuár. Soc. Brot.*,
1 39:15–18(1973)—Illustrations; chromosome number.

550 Fernandez Casas, J., Laínz, M. *and* Ruiz Rejon, M.
Narcissus cuatrecasasii, stat. et nom. nov. *Cuad.*
Cienc. Biol., (Granada), 2(1):3–5(1973)—Sp(En);
1 chromosome number.

551 Isépy, I. *and* Priszter, Sz.
Chorologische und phänologische Untersuchungen
an mediterranen Geophyten: 1. *Narcissus*. *Ann. Univ.*
Budapest. Rolando Eötvos Nom., *Biol.*, 14:105–117
1, 2 (1972 publ. 1973)—Illustrations, maps.

552 Meikle, R. D.
Narcissus obesus. *Curtis's Bot. Mag.*, 179(4):tab.650
1 (1973)—En; illustration, coloured illustration.

553 Smythies, B. E.
A note on *Narcissus hispanicus* Gouan (*N. major*
Curt.) in the Sierra de las Nieves, Malaga. *Lagascalia*,
1 3(1):49–52(1973)—(Sp).

554 Vakhtina, L. I.
Kariologicheskaya izmenchivost prirodnoĭ
populyatsii *Narcissus angustifolius* Curt. (Karyological
variability in the local population of *Narcissus*
angustifolius Curt.) *Bot. Zhurn.*, 58(9):1367–1370
1 (1973)—Chromosome numbers.

cuatrecasasii Fernandez Casas, M. Laínz & Ruiz
Rejon in *Cuad. Cienc. Biol.* (Granada), 2(1):4(1973),
nom. nov.: *N. rupicola* subsp. *pedunculatus* M. Laínz
1 ex Meikle.
× **hannibalis** A. Fernandes in *Anuár. Soc. Brot.*, 39:16
1 (1973)—Portugal. (*N. concolor* × *N. hispanicus*).

PANCRATIUM

555 Björnstad, I. N.
A revision of the genus *Pancratium* L. (Amaryl-
lidaceae) in Africa south of the Sahara. *Norw. J. Bot.*,
5 20(4):281–291(1973)—Illustrations, dot maps; key.

PUCARA Ravenna in *An. Mus. Hist. Nat. Valparaíso*,
4 5:85(1972).

556 Ravenna, P.
Pucara, genero nuevo de Amaryllidaceae del norte
de Peru. *An. Mus. Hist. Nat.* (Valparaíso), no. 5:85–
4 89(1972)—(En); illustrations.

leucantha Ravenna in *An. Mus. Hist. Nat. Valparaíso*,
4 5:86(1972)—Peru.

ZEPHYRANTHES

557 Singh, V. K. *and* Roy, S. K.
Somatic chromosomes of *Zephyranthes* Herb. *Rev.*
4 *Biol.*, 9(1–4):141–149(1973).

albolilacinus Cardenas in *Pl. Life*, 29:39(1973)—
4 Bolivia.
americana (Hoffmsgg.) Ravenna in *Mus. Nac. Hist.*
Nat. Not. Mens. (Chile), 189:8(1972): *Sternbergia*
4 *americana.*
4 **cutleri** Cardenas in *Pl. Life*, 29:38(1973)—Bolivia.

ANACARDIACEAE

558 Agostini, G.
Taxonomic bibliography for the neotropical flora:
Anacardiaceae. *Acta Bot. Venezuel.*, 9(1–4):281–285
4 (1974 publ. 1973).

559 Arrillaga-Maffei, B. R., Ziliani, G. *and* Ren, J.
Anacardiaceas de Uruguay. *Bol. Fac. Agron.* (Montevideo), no. 126. 33p.(1973)—Illustrations;
4 keys.

560 Dames e Silva, J.
Catálogo de nervação foliar das Anacardiaceae da Caatinga: 1. *Arq. Jard. Bot. Rio de Janeiro*, 19:249–256
4 (1973)—(En); illustrations; key.

ANACARDIUM

561 Ascenso, J. C. *and* Mota, M. I.
Phylogenetic derivation of the cashew flower. *Bol. Soc. Brot.*, 46:253–257(1972)—(Por); illustrations.
5 *Anacardium occidentale.*

562 Mota, M. I.
Anomalias florais do cajueiro (*Anacardium occidentale* L.). (Flower abnormalities in the cashew (*Anacardium occidentale* L.)). *Agron. Moçamb.*, 7(1):
5 21–35(1973)—(En,Fr); illustrations.

PISTACIA

563 En'kova, A. P.
Zhiznesposobnost' pȳl'tsȳ fistashki nastoyashcheĭ *Pistacia vera* L. *Izv. Akad. Nauk Tadzh. SSR, Biol.*
1, 2 *Nauk*, 1(50):20–26(1973)—Ta.

564 Popov, K. P.
O vskrȳvanii kostochek *Pistacia vera* L. (On the opening of nuts of *Pistacia vera* L.). *Bot. Zhurn.*,
1, 2 58(3):446–453(1973).

PSEUDOSPONDIAS

565 Fernandes, R. B.
Estudos nas Anacardiaceae africanas: 6. Revisão dos géneros *Pseudospondias* Engl. e *Trichoscypha* Hook. f. de Angola. *Garcia de Orta, Ser. Bot.*, 1(1–2):13–18
5 (1973)—(En); key.

RHUS

566 Merxmüller, H. *and* Roessler, H.
Eine aussergewöhnliche neue Anacardiacee aus Südwestafrika. *Mitt. Bot. Staatssamml. München*,
5 11:65–82(1973)—Illustrations. *Rhus problematodes.*

problematodes Merxmüller & Roessler in *Mitt. Bot. Staatssamml. München*, 11:66(1973)—South
5 West Africa.

SCHINOPSIS

567 Meyer, T. *and* Barkley, F. A.
Revision del genero *Schinopsis* (Anacardiaceae). *Lilloa*, 33(11):207–258(1973)—(En); illustrations, dot
4 maps; key.

glabra (Engl.) F. A. Barkley & T. Meyer in *Lilloa*,
4 33(11):219(1973): *S. brasiliensis* var. *glabra.*

SCHINUS

568 Barkley, F. A.
Schinus meyeri, una nueva especie de Anacardiaceas.
4 *Lilloa*, 33(14):293–297(1973)—En; illustrations.

TRICHOSCYPHA

569 Fernandes, R. B.
Estudos nas Anacardiaceae africanas: 6. Revisão dos géneros *Pseudospondias* Engl. e *Trichoscypha* Hook. f. de Angola. *Garcia de Orta, Ser. Bot.*, 1(1–2):13–18
5 (1973)—(En); key.

ANCISTROCLADACEAE

ANCISTROCLADUS

570 Hallé, N.
Crochets de lianes du Gabon: *Ancistrocladus* et *Anacolosa* (Ancistrocladacées et Olacacées). *Adansonia*,
5 13(3):299–306(1973)—(En); illustrations.

ANISOPHYLLEACEAE

see

RHIZOPHORACEAE

ANNONACEAE

571 Laroche, R. C.
Catálogo de nervação foliar das Amaranthaceae e Annonaceae da Caatinga: 3. *Arq. Jard. Bot. Rio de Janeiro*, 19:269–279(1973)—(En); illustrations;
4 keys.

572 Tanaka, R. *and* Okada, H.
Karyological studies in four species of Annonaceae, a primitive angiosperm. *J. Sci. Hiroshima Univ.*, *B2*,
2 14(2):85–105(1972)—Chromosome numbers.

ANNONA

573 Loureiro, A. A. *
Contribuição ao estudo anatômico da madeiro de Anonáceas da Amazónia: 3. *Annona sericea* Dun., *Annona paludosa* Aubl. e *Guatteria paranesis* R. E. Fries. *Acta Amazonica*, 1(2):85–90(1971 publ. 1972). Ref. in
4 *Bull. Torrey Bot. Club*, 99(6):331(1972).

CANANGA

574 Yeim Yok Siv *and* Paris, R. R.
Sur quelques plantes à flavonoïdes du Cambodge appartenant aux genres *Cananga, Colona, Grewia, Leea* et *Melastoma*. *Pl. Méd. Phytothér.*, 6(4):299–305 (1972)—also publ. in *Trav. Lab. Matière Méd. Pharm.*
6 *Galénique Fac. Pharm. Paris*, 57(2)(1973).

ENANTIA

575 Leboeuf, M. *and* Cavé, A.
Alcaloïdes des Annonacées sur l'isolement de la L(+)-isocorydine à partir des feuilles de l'*Enantia polycarpa* Engl. et Diels. *Pl. Méd. Phytothér.*, 6(2): 87–90(1972). Also publ. in *Trav. Lab. Matière Méd.*
5 *Pharm. Galénique Fac. Pharm. Paris*, 57(2)(1973).

FISSISTIGMA

576 Nguen T'en Ban
Chto takoe *Fissistigma hainanense* Merr. (Annonaceae)? (What is the *Fissistigma hainanense* Merr.
2 (Annonaceae)?). *Bot. Zhurn.*, 58(8):1146–1148(1973).

GONIOTHALAMUS

577 Blunden, G., Kyi, A. *and* Jewers, K.
The comparative leaf anatomy of *Goniothalamus andersonii*, *G. macrophyllus*, *G. malayanus* and *G. velutinus*. *Bot. J. Linn. Soc.*, 67(4):361–376(1973)—
6 Illustrations.

GUATTERIA

578 Loureiro, A. A. *
Contribuição ao estudo anatômico da madeiro de Anonáceas da Amazónia: 3. *Annona sericea* Dun., *Annona paludosa* Aubl. e *Guatteria paranesis* R. E. Fries. *Acta Amazonica*, 1(2):85–90(1971 publ. 1972).
4 Ref. in *Bull. Torrey Bot. Club*, 99(6):331(1972).

MEIOGYNE

hainanensis (Merr.) T'en Ban in *Bot. Zhurn.*, 58(8):
2 1148(1973): *Fissistigma hainanense*.

MONODORA

579 Leboeuf, M., Parello, J. *and* Cavé, A.
Alcaloïdes des Annonacées: 11. Présence de l'isoboldine dans les feuilles de *Monodora angolensis* Welw. *Pl. Méd. Phytothér.*, 6(2):112–114(1972). Also publ. in *Trav. Lab. Matière Méd. Pharm. Galénique*
5 *Fac. Pharm. Paris*, 57(2)(1973).

RHOPALOCARPUS

suarezensis R. Capuron ex Bosser in *Adansonia*,
5 13(1):55(1973)—Malagasy.

UVARIOPSIS

580 Bouquet, A.
Plantes médicinales du Congo-Brazzaville, *Uvariopsis, Pauridiantha, Diospyros. Trav. Lab. Matière Méd. Pharm. Galénique Fac. Pharm. Paris*,
5 57(1).112p.(1973).

APIACEAE

see

UMBELLIFERAE

APOCYNACEAE

581 Markgraf, F.
Espèces et combinaisons nouvelles d'Apocynacées malgaches: 4. *Adansonia*, 12(4):585–591(1972 publ.
5 1973)—(En).

582 Tsiang, Ying *and* Li, Ping-t'ao
Praecursores florae Apocynacearum sinensium. *Acta Phytotax. Sin.*, 11(4):343–397(1973)—Ch;
2 illustrations.

583 Valente, M. da C.
Catálogo de nervação foliar das Apocynaceae da Caatinga: 2. *Arq. Jard. Bot. Rio de Janeiro*, 19:257–267
4 (1973) (En); key.

ALAFIA

perrieri Jum.
var. **parvifolia** (Pich.) Markgraf in *Adansonia*,
5 12(4):591(1972 publ. 1973): *A. parvifolia*.

ALSTONIA

584 Duret, S. *and* Paris, R. R.
Chimiotaxonomie des polyphénols des Apocynacées. Sur les flavonoides et les acides-phénols de deux *Voacanga* (*Voacanga africana* Stapf et *V. thouarsii* Röm. et Schult.) et de deux *Alstonia*: *A. deplanchei* Heurck. et Muell. Arg. et *A. undulata* Guill. *Pl. Med. Phytothér.*, 6(1):58–65(1972). Also publ. in *Trav. Lab. Matière Méd.*
5 *Pharm. Galénique Fac. Pharm. Paris*, 57(2)(1973).

marquisensis M. L. Grant ex Fosberg & Sachet in
6 *Micronesica*, 8(1–2):46(1972)—Marquesa Is.

ALYXIA

funingensis Tsiang & P. T. Li in *Acta Phytotax.*
2 *Sin.*, 11(4):362(1973)—China.
jasminea Tsiang & P. T. Li *op. cit.* 362(1973)—
2 China.

kweichowensis Tsiang & P. T. Li *op. cit.* 360(1973)—
2 China.
menglungensis Tsiang & P. T. Li *op. cit.* 364(1973)—
2 China.
2 **villilimba** C. Y. Wu *op. cit.* 366(1973)—China.

AMSONIA

sinensis Tsiang & P. T. Li in *Acta Phytotax. Sin.*,
2 11(4):368(1973)—China.

ANODENDRON

affine (Hook. & Arn.) Druce
var. **pingpienense** Tsiang & P. T. Li in *Acta*
2 *Phytotax. Sin.*, 11(4):379(1973)—China.
fangchengense Tsiang & P. T. Li *op. cit.* 378(1973)—
2 China.
salicifolium Tsiang & P. T. Li *op. cit.* 379(1973)—
2 Taiwan, China.

ASKETANTHERA

obtusifolia Alain in *Phytologia*, 25(5):271(1973)—
4 Hispaniola (Dominican Republic).

ASPIDIOSPERMA

585 Duarte, A. Pereira
Nova contribuição para o conhecimento de uma nova espécie do gênero *Aspidosperma* da flora goiana.
4 *Sellowia*, no. 24:67–69(1972)—Illustration.

586 Duarte, A. Pereira
Uma nova espécie de *Aspidosperma* da Amazônia.
4 *Acta Amazonica*, 2(1):29–32(1972)—(En); illustrations.

587 Kulkarni, J. D., Ramstad, E., Rowson, J. M. *and* Trease, G. E.
The pharmacognosy of the *Aspidosperma* barks of
4 Brasil. *Pl. Med.*, 23(1):23–34(1973).

divaricatum Duarte in *Sellowia*, 24:67(1972)—
4 Brazil.
formosanum A. P. Duarte in *Arq. Jard. Bot. Rio*
4 *de Janeiro*, 19:218(1973)—Brazil.
williamii Duarte in *Acta Amazonica*, 2(1):29(1972),
4 as 'williamsii'—Brazil.

BAISSEA

588 Duret, S. *and* Paris, R. R.
Chimiotaxonomie des polyphénols des Apocynacées. Sur les flavonoïdes et les acides-phénols de quelques Echitoïdées (appartenant aux genres *Baissea, Echitella, Nerium* et *Strophanthus*). *Pl. Méd. Phytothér.* 6(3): 210–215(1972). Also publ. in *Trav. Lab. Matière Méd. Pharm. Galénique Fac. Pharm. Paris*,
5 57(2)(1973).

BEAUMONTIA

yunnanensis Tsiang & W. C. Chen in *Acta Phytotax.*
2 *Sin.*, 11(4):380(1973)—China.

CARISSA

yunnanensis Tsiang & P. T. Li in *Acta Phytotax.*
2 *Sin.*, 11(4):347(1973)—China.

CATHARANTHUS

589 Guédès, M. *and* Gourret, J.-P.
Architecture des phyllomes végétatifs et périanthaires: cas de *Catharanthus roseus* (L.) G. Don (*Vinca rosea* L.). (Architecture of foliage and perianth phyllomes in *Catharanthus roseus* (L.) G. Don (*Vinca rosea* L.)). *Flora* (Germany), 162(4):309–334(1973)—
2–6 (En); illustrations.

590 Rabaron, A., Plat, M. *and* Potier, P.
Alcaloïdes de *Catharanthus longifolius* (Pichon)
Pich. Isolement d'un alcaloïde dimère: la vindolicine.
5 *Pl. Méd. Phytothér.*, 7(1):53–58(1973).

CHONEMORPHA

floccosa Tsiang & P. T. Li in *Acta Phytotax. Sin.*,
2 11(4):387(1973)—China.
parviflora Tsiang & P. T. Li *op. cit.* 389(1973)—
2 China.

CONOPHARYNGIA

591 Boiteau, P.
Caducité du genere *Conopharyngia* G. Don
exclusion des Apocynaceae de *C. longifolia* (Lam.)
5 G. Don. *Adansonia*, 13(2):249–250(1973).

ECHITELLA

592 Duret, S. *and* Paris, R. R.
Chimiotaxonomie des polyphénols des Apocynacées. Sur les flavonoïdes et les acides-phénols de quelques Echitoïdées (appartenant aux genres *Baissea, Echitella, Nerium* et *Strophanthus*). *Pl. Méd. Phytothér.*, 6(3):210–215(1972). Also publ. in *Trav. Lab. Matière Méd. Pharm. Galénique Fac. Pharm. Paris*, 57(2)
5 (1973).

EPIGYNUM

auritum (Schneid.) Tsiang & P. T. Li in *Acta Phytotax. Sin.*, 11(4):397(1973): *Trachelospermum*
2 *auritum.*

ERVATAMIA

sect. **Pseudervatamia** Tsiang & P. T. Li in *Acta*
2 *Phytotax. Sin.*, 11(4):371(1973).
puberula Tsiang & P. T. Li *op. cit.* 371(1973)—
2 China.

HAZUNTA

membranacea (DC.) Pich.
f. **pilifera** Markgraf in *Adansonia*, 12(4):591(1972
5 publ. 1973)—Malagasy.

HOLARRHENA

593 Paris, R. R. *and* Duret, S.
Sur les polyphénols de diverses espèces d'*Holarrhena.*
5, 6 *Pl. Méd. Phytothér.*, 7(2):145–150(1973)—(En).

KOPSIA

594 Markgraf, F.
Florae Malesianae praecursores: 53. Apocynaceae:
2. 6. *Urnularia*, 7. *Willughbeia*, 8. *Kopsia. Blumea*,
6, 7 20(2):407–425(1972 publ. 1973)—Illustrations; keys.

595 Markgraf, F.
Identification lists of Malaysian specimens: 42.
Apocynaceae: 2. 6. *Urnularia*, 7. *Willughbeia*, 8.
Kopsia. Leyden, Foundation Flora Malesiana, 1972.
6 pp.727–734.

lancifolia Markgraf in *Blumea*, 20(2):425(1972 publ.
6 1973)—Borneo (Sabah).
officinalis Tsiang & P. T. Li in *Acta Phytotax. Sin.*,
2 11(4):356(1973)—China.
profunda Markgraf in *Blumea*, 20(2):424(1972 publ.
6 1973)—Malaya.
sleeseniana Markgraf *op. cit.* 421(1972 publ. 1973)—
6 Borneo (Sarawak).

LANDOLPHIA

myrtifolia (Poir.) Markgraf in *Adansonia*, 12(4):
5 591(1972 publ. 1973): *Echites myrtifolia.*

MASCARENHASIA

angustifolia DC.
var. **keraudreniana** Markgraf in *Adansonia*, 12
5 (4):586(1972 publ. 1973)—Malagasy.
arborescens (Boj.) DC.
var. **boivinii** (Dub.) Markgraf *op. cit.* 588(1972
5 publ. 1973), as 'boivini': *M. boivini.*
var. **comorensis** Markgraf *op. cit.* 588(1972 publ.
5 1973)—Comoro Archipelago, East Africa.
var. **gracilis** Markgraf *op. cit.* 589(1972 publ.
5 1973)—Malagasy.

MELODINUS

axillaris W. T. Wang in *Acta Phytotax. Sin.*, 11(4):
2 349(1973)—China.
tenuicaudatus Tsiang & P. T. Li *op. cit.* 353(1973)—
2 China.
yunnanensis Tsiang & P. T. Li *op. cit.* 355(1973)—
2 China.

MICRECHITES

formicina Tsiang & P. T. Li in *Acta Phytotax. Sin.*,
2 11(4):385(1973)—China.
malipoensis Tsiang & P. T. Li *op. cit.* 381(1973)—
2 China.
var. **parvifolia** Tsiang & P. T. Li *op. cit.* 383
2 (1973)—China.

NEIOSPERMA

brownii Fosberg & Sachet in *Micronesica*, 8(1–2):
6 49(1972)—Marquesa Is.
oppositifolia (Lam.) Fosberg & Sachet *op. cit.* 48
5, 6 (1972): *Cerbera oppositifolia.*

NERIUM

596 Duret, S. *and* Paris, R. R.
Chimiotaxonomie des polyphénols des Apocynacées. Sur les flavonoïdes et les acides-phénols de quelques Echitoïdées (appartenant aux genres *Baissea, Echitella, Nerium* et *Strophanthus*). *Pl. Méd. Phytothér.*, 6(3):210–215(1972). Also publ. in *Trav. Lab. Matière Méd. Pharm. Galénique Fac. Pharm. Paris*,
5 57(2)(1973).

OCHROSIA

597 Boiteau, P., Allorge, L. *and* Sévenet, T.
Revision des *Ochrosia* de Nouvelle Calédonie.
Adansonia, 12(4):625–629(1972 publ. 1973)—
6 Illustrations.

598 Bruneton, J. *and* Cavé, A.
Plantes de Nouvelle-Calédonie: 8. Alcaloïdes des feuilles d'*Ochrosia balansae* Apocynacées. (Plants of New Caledonia: 8. Alkaloids of the leaves of *Ochrosia balansae* Guillaumin (Apocynaceae). *Ann. Pharm. Franç.*, 30(9):629–636(1972)—En. Also publ. in *Trav. Lab. Matière Méd. Pharm. Galénique Fac. Pharm. Paris*,
6 57(2)(1973).

brevituba Boiteau in *Adansonia*, 12(4):627(1972
6 publ. 1973)—New Caledonia.
fatuhivensis Fosberg & Sachet in *Micronesica*,
6 8(1–2):48(1972)—Marquesa Is.
nukuhivensis Fosberg & Sachet *op. cit.* 48(1972)—
6 Marquesa Is.
sevenetii Boiteau in *Adansonia*, 12(4):627(1972 publ.
6 1973), as 'seveneti'—New Caledonia.

PACHYPODIUM

599 Rauh, W.
The genus *Pachypodium*. *Cact. Succ. J.* (U.S.A.),
5 44(1):7–31(1972)—Illustrations.

lealii Welw.
subsp. **saundersii** (N. E. Br.) Rowley in *Nation.*
5 *Cact. Succ. J.* (U.K.), 28(1):4(1973): *P. saundersii.*
rosulatum Baker
var. **drakei** (Cost. & Bois.) Markgraf in *Adansonia,*
5 12(4):590(1972 publ. 1973): *P. drakei.*
var. **horombense** (Poiss.) Rowley in *Nation. Cact.*
5 *& Succ. J.* (U.K.), 28(1):4(1973): *P. horombense.*

PANDACA

mauritiana (Poiret) Markgraf & Boiteau in
Adansonia, 13(2):244(1973), as 'mauritania':
5 *Tabernaemontana mauritiana.*
persicariifolia (Jacquin) Markgraf & Boiteau *op. cit.*
247(1973), as 'persicariaefolia': *Tabernaemontana*
5 *persicariaefolia.*

PARAVALLARIS

yunnanensis Tsiang & P. T. Li in *Acta Phytotax. Sin.,*
2 11(4):373(1973)—China.

PAREPIGYNUM Tsiang & P. T. Li in *Acta*
2 *Phytotax. Sin.,* 11(4):394(1973).
funingense Tsiang & P. T. Li in *Acta Phytotax. Sin.,*
2 11(4):395(1973)—China.

PLECTANEIA

longisepala Markgraf in *Adansonia,* 12(4):585(1972
5 publ. 1973)—Malagasy.

PLUMIERIA

600 Anon.
Plumieria acutifolia Poir. (Apocynaceae). *Ashingtonia,*
4 1(3):36(1973)—En; coloured illustration.

RAUWOLFIA

601 Banerjee, R.
Karyotype analyses of *Rauvolfia serpentina* Benth.
ex Kurz. *Proc. Indian Sci. Congr.,* 59(3):350(1972).
6 Abstract only.

602 Trivedi, B. S. *and* Upadhyay, N.
Studies of stomata and trichomes of *Rauwolfia*
serpentina L. *Curr. Sci.,* 42(11):401–403(1973)—
6 Illustration.

STROPHANTHUS

603 Duret, S. *and* Paris, R. R.
Chimiotaxonomie des polyphénols des Apocynacées. Sur les flavonoïdes et les acides-phénols de quelques Echitoïdées (appartenant aux genres *Baissea, Echitella, Nerium* et *Strophanthus*). *Pl. Méd. Phytothér.,* 6(3):210–215(1972). Also publ. in *Trav. Lab. Matière Méd. Pharm. Galénique Fac. Pharm. Paris,* 57(2)
5 (1973).

TABERNAEMONTANA

604 Cavé, A., Bruneton, J. *and* Paris, R. R.
Étude des graines de *Tabernaemontana undulata* Vahl (Apocynacées). *Pl. Med. Phytothér.,* 6(4):228–231 (1972). Also publ. in *Trav. Lab. Matière Méd. Pharm.*
4 *Galénique Fac. Pharm. Paris,* 57(2)(1973).

TRACHELOSPERMUM

cathayanum Schneid.
var. **tetanocarpum** (Schneid.) Tsiang & P. T. Li
in *Acta Phytotax. Sin.,* 11(4):390(1973):
2 *T. tetanocarpum.*

gracilipes Hook. f.
var. hupehense Tsiang & P. T. Li *op. cit.* 390(1973),
2 without latin descr. & without type—China.
yunnanense Tsiang & P. T. Li *op. cit.* 393(1973)—
2 China.

URNULARIA

605 Markgraf, F.
Florae Malesianae praecursores: 53. Apocynaceae: 2. 6. *Urnularia,* 7. *Willughbeia,* 8. *Kopsia. Blumea,* 20(2):407–425(1972 publ. 1973)—Illustrations;
6 keys.

606 Markgraf, F.
Identification lists of Malaysian specimens: 42. Apocynaceae: 2. 6. *Urnularia,* 7. *Willughbeia,* 8. *Kopsia.* Leyden, Foundation Flora Malesiana, 1972.
6 pp.727–734.

beccarii (Pierre) Markgraf in *Blumea,* 20(2):409
6 (1972 publ. 1973): *Ancylocladus beccarii.*
lanceolata Markgraf *op. cit.* 409(1972 publ. 1973)—
6 Borneo (Sabah).

VALLARIS

indecora (Baill.) Tsiang & P. T. Li in *Acta Phytotax.*
2 *Sin.,* 11(4):375(1973): *Beaumontia indecora.*

VINCA

607 Aynilian, G. H., Trojánek, J. *and* Farnsworth, N. R.
The use of alkaloids in determining the taxonomic position of *Vinca libanotica* (Apocynaceae). *In* Nobel Symposium 25: Chemistry in botanical classification,
1, 2 August 20–25, 1973: abstracts. 1973.[1]p.

608 Czarnota, K.
Stanowisko zimoziołu północnego *Linnaea borealis* i barwinka pospolitego *Vinca minor* w Puszczy Białej na Mazowszu. (The localities of the twin-flower and the periwinkle in the Baiła Primeval Forest in Masovia). *Chrońmy Przyr. Ojczystą,* 29(2):
1 69–70(1973).

609 Guédès, M. *and* Gourret, J.-P.
Architecture des phyllomes végétatifs et périanthaires: cas de *Catharanthus roseus* (L.) G. Don (*Vinca rosea* L.). (Architecture of foliage and perianth phyllomes in *Catharanthus roseus* (L.) G. Don (*Vinca rosea* L.)). *Flora* (Germany), 162(4):309–334(1973)—
2–6 (En); illustrations.

610 Stearn, W. T.
A synopsis of the genus *Vinca* including its taxonomic and nomenclatural history. Taylor, W. I. *and* Farnsworth, N., *eds.* The *Vinca* alkaloids; botany, chemistry and pharmacology: 19–94(1973)—Maps;
1, 2 keys.

611 Taylor, W. I. *and* Farnsworth, N. R., *eds.*
The *Vinca* alkaloids: botany, chemistry, and pharmacology. New York, Dekker, 1973. xii,357p.—
1, 2 Maps; keys.

VOACANGA

612 Duret, S. *and* Paris, R. R.
Chimiotaxonomie des polyphénols des Apocynacées. Sur les flavonoides et les acides-phénols de deux *Voacanga* (*Voacanga africana* Stapf et *V. thouarsii* Röm. et Schult.) et de deux *Alstonia*: *A. deplanchei* Heurck. et Muell. Arg. et *A. undulata* Guill. *Pl. Med. Phytothér.,* 6(1):58–65(1972). Also publ. in *Trav. Lab. Matière Méd. Pharm. Galénique Fac. Pharm. Paris,* 57(2)
5 (1973).

WILLUGHBEIA

613 Markgraf, F.
Florae Malesianae praecursores: 53. Apocynaceae: 2. 6. *Urnularia*, 7. *Willughbeia*, 8. *Kopsia*. *Blumea*, 20(2):407–425(1972 publ. 1973)—Illustrations;
6 keys.

614 Markgraf, F.
Identification lists of Malaysian specimens: 42. Apocynaceae: 2. 6. *Urnularia*, 7. *Willughbeia*, 8. *Kopsia*. Leyden, Foundation Flora Malesiana, 1972.
6 pp.727–734.

angustifolia (Miq.) Markgraf in *Blumea*, 20(2):414
6 (1972 publ. 1973): *Vahea angustifolia*.
var. **gracilior** Markgraf *op. cit.* 414(1972 publ.
6 1973)—Borneo (Sarawak).
anomala Markgraf *op. cit.* 415(1972 publ. 1973)—
6 Borneo (Sarawak).
gigantea (Boerl.) Markgraf *op. cit.* 413(1972 publ.
6 1973): *Leuconotis gigantea*.

APONOGETONACEAE

APONOGETON

615 Anon.
Aponogeton appendiculatus van Bruggen. *Inform. ZAG*
6 *Wasserpflanzen*, no. 4. [2]p.(1973)—Ge; illustration.

616 Anon.
Aponogeton distachyus L.f. *Inform. ZAG*
1, 3–5, 7 *Wasserpflanzen*, no. 4. [2]p.(1973)—Ge; illustration.

617 Bruggen, H. W. E. van
Revision of the genus *Aponogeton* (Aponogetonaceae): 6. The species of Africa. *Bull. Jard. Bot. Nation. Belg.*, 43(1–2):193–233(1973)—Illustrations, dot
5 maps; key.

bogneri H. W. E. van Bruggen in *Bull. Jard. Bot.*
5 *Nation. Belg.*, 43(1–2):207(1973)—Zaïre.

AQUIFOLIACEAE

618 Mendes, E. J.
5 Flora de Moçambique: 47. Aquifoliaceae. 1973. 3p.

ILEX

619 Baas, P.
The wood anatomical range in *Ilex* (Aquifoliaceae) and its ecological and phylogenetic significance.
1–7 *Blumea*, 21(2):193–258(1973)—Illustrations.

620 Bjørndalen, J. E.
Ilex aquifolium L. funnet i Bamble, Telemark. (*Ilex aquifolium* L. recorded from Bamble, Telemark,
1 S.E. Norway.). *Blyttia*, 31(1):1–2(1973)—(En).

621 Eisenbeiss, G. K. *and* Dudley, T. R.
International checklist of cultivated *Ilex*: part 1. *Ilex opaca*. Washington, United States Department of Agriculture, 1973. 85p. (National Arboretum
1–7 Contribution no. 3).

622 Espirito Santo, J. do
Flora de S. Tomé e Príncipe: Aquifoliaceae. 1973.
5 5p. *Ilex* only.

623 Krinard, R. M.
American holly: an American wood. *U.S. Dep. Agric. F.S.*, 242. 5p.(1973)—Illustrations. *Ilex*
3 *opaca*.

624 Santamour, F. S.
Anthocyanins of holly fruits. *Phytochemistry*,
1–7 12(3):611–615(1973).

ARACEAE

625 Bogner, J.
Die Araceae Madagaskars: 2–3. *Palmengarten*, 37(1):
5 10–13, 37(2):37–39(1973)—Illustrations.

626 Marchant, C. J.
Chromosome variation in Araceae: 5. Acoreae to Lasieae. *Kew Bull.*, 28(2):199–210(1973)—
1, 4–6 Chromosome numbers.

627 Ross-Craig, S.
Drawings of British plants: being illustrations of the species of flowering plants growing naturally in the British Isles: part 30. Juncaceae, Typhaceae, Sparganiaceae, Araceae. London, Bell, 1973. 41pl.
1 Araceae pl. 39–40B.

subtribus **PYCNOSPATHINAE** Bogner in
6 *Österr. Bot. Zeitschr.*, 122(4):202(1973).

AMORPHOPHALLUS

628 Gill, L. S. *and* Chinnappa, C. C.
A note on the karyology of *Amorphophallus rivieri* (Araceae). *Baileya*, 19(1):42–43(1973)—Chromosome
5–7 numbers.

629 Meijer, W.
Endangered plant life. *Biol. Conserv.*, 5(3):163–167
5, 6 (1973)—Illustrations. *Amorphophallus decus sylvae*.

ARISAEMA

vexillatum Hara & Ohashi in *J. Jap. Bot.*, 48(4):
6 99(1973)—Nepal.

AROPHYTON

630 Bhattacharya, G. N.
Study of chromosomes in *Arophyton simplex* Buchet (Araceae). *Proc. Indian Sci. Congr.*, 60(3 :304
5 (1973)—Chromosome number. Abstract only.

ARUM

631 Dostál, L. *and* Kollár, A.
Arum orientale M.B. aj na východnom Slovensku. (*Arum orientale* M.B. even in East Slovakia). *Biológia, A*
1 (Czechoslovakia), 28(7):599–602(1973)—(En,Rus).

632 Prime, C. T.
Arums—lords-and-ladies. *In* Green, P. S. *ed.* Plants:
1 wild and cultivated: 167–171(1973)—Dot map.

633 Terpó, A.
Kritische Revision der *Arum*-Arten des Karpatenbeckens. *Acta Bot. Acad. Sci. Hung.*, 18(1–2):215–255 (1973)—(En); dot map; chromosome numbers;
1 key.

alpinum Schott & Kotschy
subsp. **danicum** (Prime) Terpó in *Acta Bot. Acad. Sci. Hung.*, 18(1–2):238(1973): *A. maculatum*
1 subsp. *danicum*.
subsp. **gracile** (Unverricht) Terpó *op. cit.* 238
1 (1973): *A. gracile*.
var. **intermedium** (Schur) Terpó *op. cit.* 235
1 (1973): *A. intermedium*.
1 f. **javorkae** Terpó *op. cit.* 235(1973)—Hungary.
var. **pannonicum** Terpó *op. cit.* 234(1973)—
1 Hungary.

maculatum L.
f. **immaculatum** (Mutel) Țopa in Fl. Republ.
Social. Român., 12:779(1972): *A. maculatum*
1 subvar. *immaculatum*.
×**sooi** Terpó in *Acta Bot. Acad. Sci. Hung.*, 18(1–2):
1 240(1973)—Hungary. (*A. alpinum* × *A. maculatum*).

CYRTOSPERMA

634 N'tepe, C.
Cyrtosperma senegalense (Schott) Engler (Aracée du
Cameroun). *Inform. Ann. Caryosyst. Cytogénét.*, no. 6:
5 21–24(1972 publ. 1973)—Illustrations.

DRACUNCULUS

635 Filatova, N. S. *
Kariosistematika polȳneĭ podroda Rydb.
Dracunculus (Bess.). *Vestn. AN KazSSR*, no. 3238.
24p.(1971). Abstr. in *Ref. Zhurn. Biol.*, 2(1):v538
2 (1972).

MANGONIA

636 Bogner, J.
Otra especie de *Mangonia* (Araceae) del Uruguay.
Darwiniana, 18(1–2):70–79(1973)—(En); illustrations;
4 key.

uruguaya (Hicken) Bogner in *Darwiniana*, 18(1–2):
4 70(1973): *Felipponia uruguaya*.

PISTIA

637 Chaudhuri, J. B.
Cytotaxonomy of *Pistia stratiotes* L. *Proc. Indian
Sci. Congr.*, 60(3):306(1973)—Chromosome number.
2–7 Abstract only.

PROTARUM

638 Bogner, J.
Protarum sechellarum Engl. *Palmengarten*, 37(2):
5 40(1973)—Ge; illustrations.

TYPHONIUM

639 Rangaswami Ayyangar, K.
Karyological affinities of *Typhonium flabelliforme*
Bl. and *Typhonium trilobatum* Schott. *Proc. Indian Sci.
Congr.*, 60(3):303(1973)—Chromosome number.
6 Abstract only.

ZANTEDESCHIA

640 Letty, C.
The genus *Zantedeschia*. *Bothalia*, 11(1–2):5–26
5 (1973)—Illustrations, coloured illustrations; key.

albomaculata (Hook.) Baill.
subsp. **macrocarpa** (Engl.) Letty in *Bothalia*,
5 11(1–2):23(1973): *Z. macrocarpa*.
subsp. **valida** Letty *op. cit.* 25(1973)—South Africa
5 (Natal).

ARALIACEAE

641 Guha, S.
Cytotaxonomy of Araliaceae. *Proc. Indian Sci.
Congr.*, 58(3):471(1971)—Chromosome numbers.
6 Abstract only.

642 Kerrigan, W. J. *and* Blackwell, W. H.
The distribution of Ohio Araliaceae. *Castanea*,
3 38(2):168–170(1973)—Dot maps.

643 Thorne, R. F.
Inclusion of the Apiaceae (Umbelliferae) in the
Araliaceae. *Notes Roy. Bot. Gard. Edinburgh*, 32(2):
1–7 161–165(1973).

6 **ANAKASIA** Philipson in *Blumea*, 21(1):87(1973).

644 Philipson, W. R.
Anakasia, a new genus of Araliaceae from west New
6 Guinea. *Blumea*, 21(1):87–89(1973)—Illustrations.

simplicifolia Philipson in *Blumea*, 21(1):87(1973)—
6 New Guinea (W. Irian).

ARALIA

645 Noack, H.
Das Porträt: *Aralia elata* (Miq.) Seem. *Kurzmitt.
Deutsch. Dendrol. Ges.*, no. 11:20–22(1973)—Ge;
2 illustration.

BRASSAIA

646 Fosberg, F. R.
The name of the octopus tree. *Baileya*, 19(1):45–46
7 (1973).

BRASSAIOPSIS

sect. **Sinopanax** (Li) Yamazaki in *J. Jap. Bot.*, 48(5):
2 141(1973): gen. *Sinopanax*.
formosana (Hayata) Yamazaki *op. cit.* 141(1973):
2 *Oreopanax formosana*.

CUSSONIA

647 Dadoun, H. *and* Cavé, A.
Etude chimique préliminaire de *Cussonia corbisieri*
de Wild. (Araliacées). *Pl. Méd. Phytothér.*, 6(3):223–
227(1972). Also publ. in *Trav. Lab. Matière Méd.
5 Pharm. Galénique Fac. Pharm. Paris*, 57(2)(1973).

648 Strey, R. G.
Notes on the genus *Cussonia* in South Africa.
Bothalia, 11(1–2):191–201(1973)—Illustrations, dot
5 maps.

5 sect. **Capitata** Strey in *Bothalia*, 11(1–2):193(1973).
5 subgen. **Paniculata** Strey *op. cit.* 193(1973).
5 subgen. **Protocussonia** Strey *op. cit.* 193(1973).
arenicola Strey *op. cit.* 194(1973)—South Africa
5 (Natal), Mozambique.
nicholsonii Strey *op. cit.* 197(1973)—South Africa
5 (Natal).
sphaerocephala Strey *op. cit.* 199(1973)—South
5 Africa (Natal).
zuluensis Strey *op. cit.* 195(1973)—South Africa
5 (Natal), Swaziland, Mozambique.

DENDROPANAX

649 Steyermark, J. A.
Novedades de los Andes Venezolanos. *Pittieria*,
no. 4:9–25(1972)—Sp,En; illustration. *Dendropanax
4 veillonii*.

•
veillonii Steyermark in *Pittieria*, 4:9(1972)—
4 Venezuela.

DIDYMOPANAX

650 Assumpção, W. R. C. d', Pedersoli, J. L. *and*
Gavilanez, M. L.
Estudo anatomo-ecológico de *Didymopanax
macrocarpum* (Cham. & Shl.) Seem. *Oréades*, 3(5):20–26
4 (1972)—(En).

ECHINOPANAX

651 Flȳagina, I. A.
Novȳe dannȳe ob *Echinopanax elatum* Nakai.
Byull. Glavn. Bot. Sada (Moscow), 87:29–30
2 (1973).

ELEUTHEROCOCCUS

652 Sandberg, F.
Two glycoside-containing genera of the Araliaceae family: *Panax* and *Eleutherococcus*. *Pl. Med.*, 24(4):
2 392–396(1973).

HARMSIOPANAX

653 Philipson, W. R.
A revision of *Harmsiopanax* (Araliaceae). *Blumea*,
6 21(1):81–86(1973)—Map; keys.

ingens Philipson in *Blumea*, 21(1):84(1973)—New
6 Guinea (Terr. New Guinea, W. Irian, Papua).
subsp. **moniliformis** Philipson *op. cit.* 86(1973)—
6 New Guinea (Terr. New Guinea, Papua).

HEDERA

654 Juhász, G. D. *and* Bart, E.
Angaben zur Entwicklungsmorphologie des Blütenstandes von *Hedera helix* L. *Ann. Univ. Budapest. Rolando Eötvos Nom., Biol.*, 14:119–126
1, 2 (1972 publ. 1973)—Illustrations.

PANAX

655 Hoo, Gin *and* Tseng, Chiang-jiang
(On the Chinese species of *Panax* Linn.). *Acta*
2 *Phytotax. Sin.*, 11(4):431-438(1973)—Ch; key.

656 Sandberg, F.
Two glycoside-containing genera of the Araliaceae family *Panax* and *Eleutherococcus*. *Pl. Med.*, 24(4):
2 392–396(1973).

pseudo-ginseng Wall.
var. **elegantior** (Burkill) Hoo & C. J. Tseng in *Acta Phytotax. Sin.*, 11(4):436(1973): *Aralia*
2 *quinquefolia* var. *elegantior.*
var. **japonicus** (C. A. Mey.) Hoo & C. J. Tseng
2 *op. cit.* 437(1973): *P. japonicum.*
var. **notoginseng** (Burkill) Hoo & C. J. Tseng *op. cit.* 435(1973): *Aralia quinquefolia* var.
2 *notoginseng.*
var. **wangianus** (Sun) Hoo & C. J. Tseng *op. cit.*
2 436(1973): *P. wangianum.*

PLERANDRA

657 Tseng, C. C.
Systematic palynology of *Tupidanthus* and *Plerandra*
6 (Araliaceae). *Grana*, 13(1):51–56(1973)—Key.

PSEUDOPANAX

658 Edgar, E.
Names in *Pseudopanax* C. Koch (Araliaceae). *New*
7 *Zealand J. Bot.*, 11(1):171–172(1973).

SCHEFFLERA

659 Bamps, P.
Schefflera evrardii P. Bamps, Araliacée nouvelle du Zaïre. *Bull. Jard. Bot. Nation. Belg.*, 43(3–4):425–426
5 (1973).

660 Bernardi, L.
Araliacearum Madagascariae et Comores epilogus: 3. Species nova Scheffler arum. *Candollea*, 28(1):7–11
5 (1973)—(Fr,En,Ge); illustrations.

661 Grushvitskiĭ, I. V. *and* Skvortsova, N. T.
Kriticheskie zametki o nekotorȳkh vidakh roda *Schefflera* Forst. et Forst. f. sev. V'etnama i yugo-zap. Kitaya. (Notae criticae de generis *Schefflera* Forst. et Forst. f. speciebus nonnullis e Vjetname boreali et China austro-occidentali). *Novosti Sist. Vȳssh. Rast.*,
2, 6 10:211–217(1973)—Illustration, dot map.

evrardii Bamps in *Bull. Jard. Bot. Nation. Belg.*,
5 43(3–4):425(1973)—Zaïre.
moratii Bernardi in *Candollea*, 28(1):7(1973)—
5 Malagasy.

SCHEFFLEROPSIS

662 Grushvitskiĭ, I. V. *and* Skvortsova, N. T.
Novȳĭ dlya florȳ Severnogo V'etnama rod *Scheffleropsis* Ridl.—Predstavitel drevneĭsheĭ tribȳ sem. Araliaceae. (*Scheffleropsis* Ridl.—the new genus for the North Vietnam flora and representative of the oldest tribe of Araliaceae). *Bot. Zhurn.*, 58(10):1492–1503
6 (1973)—Illustrations, dot map; key.

sect. **Integrifoliae** Grushvitskiĭ & N. T. Skvortsova
6 in *Bot. Zhurn.*, 58(10):1495(1973).
angkae (Craib) Grushvitskiĭ & N. T. Skvortsova
6 *op. cit.* 1496(1973): *S. subintegra* var. *angkae.*
hemiepiphytica Grushvitskiĭ & N. T. Skvortsova
6 *op. cit.* 1496(1973)—Vietnam.

TIEGHEMOPANAX

663 Mohana Rao, P. R.
Morphology and embryology of *Tieghemopanax sambucifolius* with comments on the affinities of the family Araliaceae. *Phytomorphology*, 22(1):75–87
6, 7 (1972 publ. 1973).

TUPIDANTHUS

664 Tseng, C. C.
Systematic palynology of *Tupidanthus* and *Plerandra* (Araliaceae). *Grana*, 13(1):51–56(1973)—
2, 6 Key.

ARISTOLOCHIACEAE

ARISTOLOCHIA

665 Ahumada, L. Z.
Una neuva especie de *Aristolochia* del NE Argentino.
4 *Sellowia*, no. 24:63–66(1972)—(En); illustration.

albertiana Ahumada in *Sellowia*, 24:64(1972)—
4 Argentina.

ASARUM

666 Piperon, G.
Asarum europaeum et *Ceterach officinarum* à Boigneville. *Bull. Ass. Nat. Vallée Loing Massif*
1 *Fontainebleau*, 48(7–8):89(1972).

667 Schönfelder, P.
Asarum ibericum—eine übersehene Sippe unserer
1 Flora? *Göttinger Flor. Rundbr.*, 7(2):29(1973).

DIPLOCYATHA

668 Walfried, B.
Diplocyatha ciliata Thnbg. *Succulenta*, 52(6):115–116
5 (1973)—Du; illustration.

ASCLEPIADACEAE

669 Brack, S.
Key to the succulent Asclepiadaceae. *Bull. Afr.*
2, 5, 6 *Succ. Pl. Soc.*, 7(6):230–233(1973).

670 Kozjek, F., Jay, M. *and* Nétien, G.
Recherches chimiotaxonomiques sur les plantes vasculaires: 29. Contribution biochimique à l'étude des flavonoïdes des Asclépiadacées. *Candollea*, 28(2):
1–7 237–247(1973)—(En,Ge).

671 Leach, L. C.
Stapelieae (Asclepiadaceae) from south tropical
Africa: 7. *Bothalia*, 11(1–2):133–137(1973)—
5 Illustrations; chromosome numbers.

672 Reese, G.
The structure of the highly specialized carrion-
flowers of stapeliads. *Cact. Succ. J.* (U.S.A.), 45(1):
1, 2, 5, 6 18–29(1973)—Illustrations, coloured illustrations.

ACERATES

673 Reveal, J. L.
Comments on two names in an early Utah flora.
Great Basin Nat., 32(4):221–222(1972). *Acerates*
3 *decumbens.*

BRACHYSTELMA

674 Dyer, R. A.
Brachystelma cathcartense. Flow. Pl. Afr., 42(3–4):
5 pl.1667(1973)—En; coloured illustration.

675 Dyer, R. A.
Brachystelma discoideum. Flow. Pl. Afr., 42(3–4):
5 pl.1668(1973)—En; coloured illustration.

676 Dyer, R. A.
Brachystelma tenellum. Flow. Pl. Afr., 42(3–4):
5 pl.1664(1973)—En; coloured illustration.

677 Sathyanarayanachar, M. B.
A new species of *Brachystelma* from India. *Proc.*
6 *Indian Sci. Congr.*, 58(3):435(1971). Abstract only.

discoideum R. A. Dyer in *Flow. Pl. Afr.*, 42(3–4):
5 t.1668(1973)—South Africa (Transvaal).
elenaduensis Char. ex Sathyanarayanachar in *Proc.*
Indian Sci. Congr., 58(3):435(1971), without latin
6 descr. and without type—India.
tenellum R. A. Dyer in *Flow. Pl. Afr.*, 42(3–4): t.
5 1664(1973)—South Africa (Natal).

CARALLUMA

678 Bally, P. R. O.
Notes on the flora of tropical East Africa. *Bull. Afr.*
5 *Succ. Pl. Soc.*, 8(4):157–160(1973)—Illustrations.

679 Bent, P.
Notes on the stapeliads. *Bull. Afr. Succ. Pl. Soc.*,
2, 5 7(6):225,226(1973).

680 Dyer, R. A.
Caralluma dioscoridis. Flow. Pl. Afr., 42(3–4):
5 pl.1662(1973)—En; coloured illustration.

681 Gilbert, M. G.
Notes on Ethiopian stapeliads: 1. The rediscovery of
Caralluma furta. Nation. Cact. Succ. J. (U.K.), 28(2):
5 43–44(1973).

682 Kalmbacher, G.
The story of *Caralluma kalmbacherana. Cact. Succ. J.*
2, 5 (U.S.A.), 45(1):16–17(1973)—Illustration.

683 Leach, L. C.
Stapelieae from south tropical Africa: 1. *Bull. Afr.*
5 *Succ. Pl. Soc.*, 8(3):116–120(1973)—Illustrations.

caudata N. E. Br.
subsp. **rhodesiaca** Leach in *Bothalia*, 11(1–2):134
5 (1973)—Rhodesia.

CEROPEGIA

684 Biester, K.
Ceropegia haygarthii (Schlechter) H. Huber. *Kakt.*
And. Sukk., 24(12):283–284(1973)—Ge;
2 illustration.

685 Lavranos, J.-J.
Sur une nouvelle Asclépiadaceae de Madagascar.
5 *Adansonia*, 13(1):71–73(1973)—(En); illustration.

ampliata E. Meyer
subsp. **madagascariensis** Lavranos in *Adansonia*,
5 13(1):71(1973)—Malagasy.

CYNANCHUM

686 Ohwi, J. *and* Ohashi, H.
Cynanchum yamanakas, a new species of Asclepiad-
aceae from Shikoku, Japan. *J. Jap. Bot.*, 48(12):370–
2 373(1973)—Illustrations.

687 Spellman, D. L.
New combinations in Asclepiadaceae. *Phytologia*,
4 25(7):438(1973).

apocynellum (Gl. & Mold.) Spellman in *Phytologia*,
4 25(7):438(1973): *Tassadia apocynella.*
recurvum (Rusby) Spellman *op. cit.* 438(1973):
4 *Tassadia recurva.*
yamanakae Ohwi & Ohashi in *J. Jap. Bot.*, 48(12):
2 370(1973)—Japan.

DECABELONE

688 Herre, H.
Notes on South African succulents. *Cact. Succ. J.*
5 (U.S.A.), 45(5):238–239(1973)—Illustration.

689 Rowley, G. D.
Decabelone. Ashingtonia, 1(3):28,30(1973)—En;
5 illustrations, coloured illustrations; key.

DOLICHOPETALUM Tsiang in *Acta Bot. Sin.*,
2 15(1):137(1973).

690 Tsiang, Ying
Dolichopetalum genus novum Asclepiadacearum.
Acta Bot. Sin., 15(1):136–139(1973)—Ch,La;
2 illustrations.

kwangsiense Tsiang in *Acta Bot. Sin.*, 15(1):137
2 (1973)—China.

DUVALIA

691 Bent, P.
The genus *Duvalia* Haworth. *Bull. Afr. Succ. Pl.*
5 *Soc.*, 8(5):186–188,191(1973).

692 Lavranos, J.-J.
Notes on the succulent flora of north-east Africa
and southern Arabia. *Cact. Succ. J.* (U.S.A.), 44(6):
257–261(1972)—Illustrations, coloured illustration.
2, 5 *Duvalia eilensis.*

eilensis Lavranos in *Cact. Succ. J.* (U.S.A.), 44(6):260
5 (1972)—Somali Republic.

HOYA

693 Walfried, B.
Hoya bella Hook. *Succulenta*, 52(7):137–138(1973)—
6 Du; illustration.

694 Wills, A. J.
Notes on *Hoya imperalis* and others of the genus.
J. Roy. Hort. Soc., 98(5):224–225(1973)—Coloured
2, 6, 7 illustrations.

LEPTADENIA

695 Pathan, T. S.
Developmental studies in Asclepiadaceae. *Proc. Indian Sci. Congr.*, 60(3):314(1973)—Chromosome
6 numbers. Abstract only.

MATELEA

696 Spellman, D. L.
New combinations in Asclepiadaceae. *Phytologia*,
4 25(7):438(1973).

697 Spellman, D. L. *and* Dwyer, J. D.
A new *Matelea* from Panama. *Ann. Missouri Bot.*
4 *Gard.*, 60(2):568–570(1973)—Illustration.

brasiliensis (Schltr.) Spellman in *Phytologia*, 25(7):
4 438(1973): *Fimbristemma brasiliensis.*
panamensis Spellman & Dwyer in *Ann. Missouri Bot.*
4 *Gard.*, 60(2):568(1973)—Panama.

NEMATURIS

698 Pereira, J. Fontella *and* Silva, N. M. Ferreira da
Estudos em Asclepiadaceae: 3. Sôbre a identidade da *Nematuris volubilis* Turcz. *Arq. Jard. Bot. Rio de*
4 *Janeiro*, 19:223–226(1973)—(En); illustration.

OXYPETALUM

699 Occhioni, P.
O gênero *Oxypetalum* R. Br. na restinga do sudesta Brasileiro. *Leandra*, 2(2):105–120(1972)—Illustrations;
4 key.

700 Valente, M. da C., Pereira, J. Fontella *and* Alencastro, F. M. M. R. de
Contribuição ao estudo das Asclepiadaceae brasileiras: 9. Estudos taxonômico e anatômico de: *Oxypetalum appendiculatum* Mart., *Oxypetalum pilosum* Gardn. e *Oxypetalum sublanatum* Malme. *An. Acad. Brasil. Cienc.*, 45(1):121–149(1973)—(En);
4 illustrations.

STAPELIA

701 Isherwood, D.
Stapelia erectiflora N. E. Br. *Ashingtonia*, 1(1):7,8
5 (1973)—En; coloured illustration.

702 Nakamura, T.
The chromosome numbers and karyotype analysis in some species of the genus *Stapelia* (Asclepiadaceae). *Chromosome Inform. Serv.*, no. 12:5–7(1971)—
2, 4–6 Illustrations.

pulvinata Masson
f. **margarita** (Sloane) Rowley in *Nation. Cact.*
5 *Succ. J.* (U.K.), 28(1):4(1973): *S. margarita.*

TRICHOCAULON

703 Innes, C. F.
Trichocaulon pillansii N. E. Brown. *Ashingtonia*,
5 1(2):19,21(1973)—En; coloured illustration.

VINCETOXICUM

704 Pringle, J. S.
The spread of *Vincetoxicum* species (Asclepiadaceae) in Ontario. *Canad. Field-Nat.*, 87(1):27–33(1973)—
3 Illustrations, dot map.

ser. **Alboviana** Pobedimova in *Novosti Sist. Vŷssh.*
1, 2 *Rast.*, 10:340(1973).
2 ser. **Congestiflora** Pobedimova *op. cit.* 340(1973).
1, 2 ser. **Media** Pobedimova *op. cit.* 340(1973).
1, 2 ser. **Nigra** Pobedimova *op. cit.* 340(1973).
1, 2 ser. **Officinalia** Pobedimova *op. cit.* 341(1973).

albovianum (Kuzn.) Pobedimova *op. cit.* 341(1973):
2 *Cynanchum albovianum.*
stepposum (Pobedimova) Pobedimova *op. cit.* 341
1, 2 (1973): *Antitoxicum stepposum.*

ASTERACEAE
see
COMPOSITAE

AVICENNIACEAE

705 Jaeri, S. M. H.
Flora of West Pakistan: no. 49. Avicenniaceae.
6 1973. 4p.—Illustration, map.

706 Moldenke, H. N.
A fifth summary of the Verbenaceae, Avicenniaceae, Stilbaceae, Dicrastylidaceae, Symphoremaceae, Nycanthaceae and Eriocaulaceae of the world as to valid taxa, geographic distribution and synonymy:
3, 6 Supplement 2. *Phytologia*, 25(4):225–245(1973).

707 Moldenke, H. N.
Flora of Panama: part 9. Family 168a. Avicenniaceae. *Ann. Missouri Bot. Gard.*, 60(1):149–154(1973)—
4 Illustration; key.

708 Villiers, J.-F.
Avicenniacées. Flore du Gabon, 22:63–66(1973)—
5 Illustrations.

AVICENNIA

709 Bird, E. C. F.
Mangroves on the Australian coast. *Austral. Nat.*
7 *Hist.*, 17(5):167–171(1972)—Map.

BALANOPHORACEAE

710 Hansen, B.
Balanophoraceae. Flore du Cambodge, du Laos et du Viêt-nam, fasc. 14:49–58(1973)—Illustration;
6 keys.

BALSAMINACEAE

HYDROSERA

711 Srinivasa Rao, R. V.
Cytological studies in the South Indian Balsaminaceae. *Proc. Indian Sci. Congr.*, 59(3):356–357(1972)—
6 Chromosome numbers. Abstract only.

IMPATIENS

712 Arisumi, T.
Chromosome numbers and interspecific hybrids among New Guinea *Impatiens* species. *J. Hered.*,
6 64(2):77–79(1973).

713 Barkworth, M. E.
Impatiens parviflora in British Columbia. *Madroño*,
3 22(1):24(1973).

714 Bhaskar, V. *and* Razi, B. A.
A new kind of exine sculpturing in *Impatiens* L. (Balsaminaceae) from South India. *Curr. Sci.*, 42(14):
6 510–512(1973).

715 Huang, T.-C.
A revision of Formosan *Impatiens* (Balsaminaceae).
6 *Taiwania*, 18(1):49–54(1973)—Illustrations; key.

716 Shimizu, T.
Chromosome number of *Impatiens ridleyi* Hook. f.
Acta Phytotax. Geobot., 25(4–6):126(1973)—
6 Illustration.

717 Srinivasa Rao, R. V.
Cytological studies in the South Indian Balsamin-
aceae. *Proc. Indian Sci. Congr.*, 59(3):356–357(1972)—
6 Chromosome numbers. Abstract only.

718 Srinivasa Rao, R. V.
Further studies on the cytology of South Indian
Balsaminaceae. *Proc. Indian Sci. Congr.*, 60(3):317
6 (1973)—Chromosome numbers. Abstract only.

719 Srinivasa Rao, R. V.
Karyotype analyses in the varieties of *Impatiens
balsamina* L. *Proc. Indian Sci. Congr.*, 60(3):316–317
6 (1973)—Chromosome numbers. Abstract only.

720 Winters, H. F.
New *Impatiens* from New Guinea. *Amer.
Horticulturist*, 52(3):16–22(1973)—Illustrations,
6 coloured illustrations.

ser. **Circinatae** Pobedimova in *Novosti Sist. Vÿssh.*
2 *Rast.*, 10:339(1973).
1, 2 ser. **Macropetalae** Pobedimova *op. cit.* 339(1973).
1, 2 ser. **Micropetalae** Pobedimova *op. cit.* 340(1973).
devolii T. C. Huang in *Taiwania*, 18(1):49(1973)—
2 Taiwan.

BASELLACEAE

721 Fernandes, R. B.
Basellaceae. *Anuár. Soc. Brot.*, 38:139–140(1972)—
1 Por.

BASELLA

722 Fathima, T., Boraiah, G. *and* Shivashankar, G.
Cytotaxonomy of *Basella*. *Proc. Indian Sci. Congr.*,
58(3):470(1971)—Chromosome numbers. Abstract
5 only.

BEGONIACEAE

723 Barkley, F. A.
Key to the sections of the Begoniaceae. *Buxtonian*,
2, 4–6 1 Suppl. 3. 7p.(1972).

724 Doorenbos, J.
D'où viennent nos Bégonias? *Rev. Hort.*, no. 2317:
3, 4–6 23–26(1973)—Coloured illustrations.

725 Uribe Uribe, L.
Begoniaceae. Catalogo ilustrado de las plantas de
4 Cundinamarca, vol. 5:43–59(1972)—Illustrations.

BEGONIA

726 Baranov, A. I.
A comparative study of the gross morphology of
the inflorescences and flowers of *Begonia*: announce-
ment for the defense of the M.S. thesis. [Boston],
2, 4–6 Northeastern University, 1973. [8]p. Abstract only.

727 Baranov, A. I. *and* Barkley, F. A.
The combination of *Begonia fuchsiaefolia* (A.DC.)
4 Warburg. *Phytologia*, 26(4):217–220(1973).

728 Barkley, F. A.
Additional species of *Begonia*. *Buxtonian*, 1 Suppl.
4, 5 7. 4p.(1973).

729 Barkley, F. A. *and* Lee, Y. S.
Begonia monophylla Pavon ex A. DC. *Buxtonian*,
4 1 Suppl. 6. 7p. (1972)—En; illustrations.

730 Boghdan, K. S.
Comparative developmental anatomy of the
foliage leaves in three species of *Begonia*: announce-
ment for the defense of the Ph.D. dissertation.
2, 4–6 [Boston], Northeastern Univ., 1973. 4p.—Abstract.

731 Burtt, B. L.
The name *Begonia sinuata*. *Notes Roy. Bot. Gard.
5, 6 Edinburgh*, 32(2):273–274(1973).

732 Creighton, H. B.
Epidermal outgrowths in *Begonia*. *Begonian*, 40(1):
2, 4–6 7–10,15–18, (2):31–34,39–42(1973)—Illustrations.

733 Doorenbos, J.
Additional information about *B. parva*. *Begonian*,
5 40(7):155(1973). Nomenclatural note.

734 El Tigani, K. B.
Development anatomy of the inflorescences of the
species of *Begonia*: announcement for the defense of the
Ph.D. dissertation. [Boston], Northeastern University,
2, 4–6 1973. 4p. Abstract.

735 Golding, J.
Begonia conchifolia or what? *Begonian*, 40(8):173–
4 179,188–190(1973)—Illustrations.

736 Humbert, H.
Quelques *Begonia* nains de Madagascar. *Bull. Soc.
Bot. France*, 118(9):733–744(1971 publ. 1973)—(En);
5 illustrations; key.

737 Kondo, K.
Chromosome numbers of five taxa. *Chromosome
Inform. Serv.*, no. 15:33–34(1973). *Begonia
4 sutherlandii.*

738 Lee, Y. S.
Study of stem anatomy in *Begonia*: announcement
for the defense of the M.S. thesis. [Boston], North-
2, 4–6 eastern University, 1973. 4p. Abstract.

739 Smith, L. B.
Herbarium notes: 4 *Phytologia*, 25(6):418–419
4 (1973)—Illustration.

740 Villiers, J.-F.
Ontogénie et signification morphologique des
éléments du trichome épinervaire et laminaire de
Begonia × *vitichotoma* Hort. *Adansonia*, 12(4):593–
2, 4–6 602(1972 publ. 1973).

741 Ziesenhenne, R.
Begonia bogneri: a new species from the Malagasy
Republic. *Begonian*, 40(4):76–80(1973)—
5 Illustrations.

742 Ziesenhenne, R.
Begonia bowerae family: 1–2. *Begonian*, 40(11):244,
4 267, (12):286–288(1973)—Illustrations.

anjuanensis Humbert in *Bull. Soc. Bot. France*, 118(9):
5 735(1971 publ. 1973)—Malagasy.
ankaranensis Humbert *op. cit.* 735(1971 publ. 1973)—
5 Malagasy.
antsingyensis Humbert *op. cit.* 735(1971 publ. 1973)
5 —Malagasy.
bagotiana Humbert *op. cit.* 737(1971 publ. 1973)—
5 Malagasy.

var. **acutialata** Humbert *op. cit.* 737(1971 publ.
5 1973)—Malagasy.
barkleyana L. B. Smith in *Phytologia*, 25(6):418
4 (1973), nom. nov.: *B. dusenii* Brade.
bogneri Ziesenhenne in *Begonian*, 40(4):80(1973)—
5 Malagasy.
bowerae Ziesenhenne
var. **major** Ziesenhenne in *Begonian*, 40(12):288
4 (1973)—Mexico.
var. **nigramarga** Ziesenhenne *op. cit.* 288(1973)—
4 Mexico.
var. **roseflora** Ziesenhenne *op. cit.* 288(1973)—
4 Mexico.
conchifolia Dietrich
var. **rubrimacula** Golding in *Begonian*, 40(8):189
4 (1973)—Costa Rica.
decaryana Humbert in *Bull. Soc. Bot. France*, 118
5 (9):737(1971 publ. 1973)—Malagasy.
flagellaris Hara in *J. Jap. Bot.*, 48(12):358(1973)—
6 Nepal.
fuchsiiflora (A. DC.) A. I. Baranov & F. A. Barkley
in *Phytologia*, 26(4):220(1973): *Casparya fuchsiae-*
4 *flora.*
isalensis Humbert in *Bull. Soc. Bot. France*, 118(9):
5 739(1971 publ. 1973)—Malagasy.
kalabenonensis Humbert *op. cit.* 739(1971 publ.
5 1973)—Malagasy.
leandrii Humbert *op. cit.* 741(1971 publ. 1973)—
5 Malagasy.
leptoptera Hara in *J. Jap. Bot.*, 48(4):98(1973)—
6 Nepal.
neoperrieri Humbert in *Bull. Soc. Bot. France*, 118(9):
5 741(1971 publ. 1973)—Malagasy.
sambiranensis Humbert *op. cit.* 741(1971 publ.
5 1973)—Malagasy.
trujillensis L. B. Smith in *Phytologia*, 25(6):418
4 (1973)—Venezuela.

BERBERIDACEAE

BERBERIS

743 Hunt, D. R.
Berberis mucrifolia. Curtis's Bot. Mag., 179(3):
tab.643(1973)—En; illustration, coloured
6 illustration.

744 Laar, H. J. van de
Berberis. Dendroflora, no. 9:9–37(1972)—Du(En);
1–7 illustrations.

745 Marroquin de la Fuente, J. S.
A monographic study of the genus *Berberis* L. in
3, 4 Mexico. *Diss. Abstr. Int., B*, 33(8):3511(1973).

LEONTICE

746 Geideman, T. S. *and* Nikolaeva, L. P. *
(Nouvel habitat de la plante rare *Leontice odessana*).
Izvest. Akad. Nauk Moldav. SSR, Biol. Khim. Nauk,
no. 1:79–80(1973)—Rus. Ref. in *Bull. Signal., Biol.*
1 *Physiol. Vég.*, 34(10):9527(1973).

MAHONIA

747 Durand, H. K.
Texas *Mahonia*—a neglected economic plant. *Econ.*
3, 4 *Bot.*, 26(4):319–325(1973).

PODOPHYLLUM

748 Selivanova-Gorodkova, E. A.
Osobennosti morfologii i biologii pobegov
Podophyllum hexandrum Royle i *P. peltatum* L.
(Morphology and biology peculiarities of shoots of
Podophyllum hexndrum Royle and *P. peltatum* L.).
3, 6 *Bot. Zhurn.*, 58*a*(2):273–284(1973)—Illustrations.

BETULACEAE

749 Korchagina, I. A.
Osobennosti morfogeneza tsvetka Betulaceae.
(Flower morphogenesis peculiarities in Betulaceae).
1–4 *Bot. Zhurn.*, 58(7):1037–1043(1973).

750 Mehra, P. N. *and* Sareen, T. S.
Cytology of West Himalayan Betulaceae and
Salicaceae. *J. Arnold Arbor.*, 54(3):412–418(1973)—
6 Chromosome numbers.

ALNUS

751 Grootendorst, H. J.
Alnus: keuringsrapport van de Keuringscommissie
Sierbomen N.A.K.-B. *Dendroflora*, no. 9:3–8(1972)—
1–3 Illustrations.

752 Tallantire, P. A.
Some data on the history of alder in Trøndelag,
Norway. *Grana*, 13(1):18–24(1973). *Alnus incana,*
1 *A. glutinosa.*

BETULA

753 Barȳkina, R. P. *and* Kudryashev, L. V.
Anatomicheskoe issledovanie gipoarkticheskikh
kustarnikov *Betula exilis* Sukacz. i *Betula nana* L.
(Anatomical study of hypoarctic shrubs *Betula exilis*
Sukacz. and *Betula nana* L.). *Bot. Zhurn.*, 58(3):
1–3 421–428(1973).

754 Brisbin, R. L. *and* Sonderman, D. L.
Birch: an American wood. *U.S. Dep. Agric. F.S.*,
3 221. 11p.(1973)—Illustrations, dot maps.

755 Forbes, J. C. *and* Kenworthy, J. B.
Distribution of two species of birch forming stands
on Deeside, Aberdeenshire. *Trans. Bot. Soc. Edinburgh*,
1 42(1):101–110(1973)—Maps.

756 Geideman, T. S. *and* Osadchii, V. M. *
O vidakh borezy v Moldavii (*Betula* species in
Moldavia). *Izv. Akad. Nauk Mold. SSR, Biol. Khim.*
Nauk, no. 2:3–8(1972)—Illustration. Abstract in
1 *Biol. Abstr.* 55(9)47248(1973).

757 Kudryashev, L. V., Barȳkina, R. P. *and*
Pugacheva, L. N.
Formirovanie stlanikovoĭ formȳ kusta u
gipoarkticheskikh kustarnikov *Betula exilis* Sukacz. i
Betula nana L. (Formation of the prostrate form of
bushes in hypoaractic *Betula exilis* Sukacz. and *Betula*
nana L.). *Bot. Zhurn.*, 58(1):53–64(1973)—
1, 2 Illustrations.

758 Makhnev, A. K. *and* Rozhdestvenskiĭ, Yu. F.
O vnutrividovoĭ izmenchivosti i sistematike vidov
Betula lyapinskogo Urala. (On the intraspecific
variability in the taxonomy of *Betula* species from the
subpolar Urals). *Bot. Zhurn.*, 58(8):1107–1117
1, 2 (1973)—(En); chromosome numbers.

759 Payne, R. C. *and* Fairbrothers, D. E.
Disc electrophoretic study of pollen proteins from
natural populations of *Betula populifolia* in New
3 Jersey. *Amer. J. Bot.*, 60(2):182–189(1973)—Map.

760 Pork, K. *and* Sander, R.
Maarjakase levikust lääne-eestis. (The occurrence
of the curly birch in west Estonia). *Eesti Loodus*,
1 16(6):332–335(1973)—(Rus,En).

761 Radics, F.
A nyírhibrid-kutatás története. (Investigational
history of birch-hybrids). *Stud. Bot. Hung.*, 8:99–110
1–3 (1973)—(En).

762 Taper, L. J. *and* Grant, W. F.
The relationship between chromosome size and DNA content in birch (*Betula*) species. *Caryologia*, 26(2):263–273(1973)—Chromosome numbers. North
3 American species.

763 Tuley, G.
A taxonomic history of the British birch tree. *Trans.*
1 *Bot. Soc. Edinburgh*, 42(1):31–42(1973).

DUSCHEKIA

viridis (Chaix) Opiz
subsp. **brembana** (Rota) Holub in *Folia Geobot.*
1–3 *Phytotax.*, 8(2):176(1973): *Alnus brembana.*

BIGNONIACEAE

764 Ferguson, I. K. *and* Santisuk, T.
Notes on the pollen morphology of some Asiatic
2, 6 Bignoniaceae. *Kew Bull.*, 28(2):187–194(1973).

765 Gentry, A. H.
An eco-evolutionary study of the Bignoniaceae of southern Central America. *Diss. Abstr. Int., B*, 33(12):
4 5694–5695(1973).

766 Gentry, A. H.
Generic delimitations of Central American
4 Bignoniaceae. *Brittonia*, 25(3):226–242(1973).

767 Leinfellner, W.
Das Gynözeum der Bignoniaceen: 3. Crescentieae (*Amphitecna, Colea, Rhodocolea, Ophiocolea, Phyllarthron, Phylloctenium, Parmentiera, Enallagma* und
4, 5 *Crescentia*). *Österr. Bot. Zeitschr.*, 122(1–2):59–73(1973).

768 Leroy, J.-F.
La genèse d'un genre chez les Bignoniacées-Crescentiées de Madagascar. *Compt. R. Hebd. Séances Acad. Sci., D* (Paris), 275(23):2675–2678
5 (1972)—Illustrations.

769 Williams, L. O.
Bignoniaceae of tropical North America.
4 *Fieldiana, Bot.*, 36(4):21–29(1973).

AMPHITECNA

770 Gentry, A. H.
Studies in Bignoniaceae: 7. *Dendrosicus, Enallagma*
4 and *Amphitecna. Taxon*, 22(5–6):637–640(1973).

donnell-smithii (Sprague) L. O. Williams in *Fieldiana, Bot.*, 36(4):22(1973): *Crescentia donnell-*
4 *smithii.*
molinae L. O. Williams *op. cit.* 22(1973)—
4 Honduras.
montana L. O. Williams *op. cit.* 22(1973)—
4 Guatemala, Mexico.
oblanceolata L. O. Williams *op. cit.* 23(1973)—
4 Guatemala.
obovata (Bentham) L. O. Williams *op. cit.* 25(1973):
3, 4 *Crescentia obovata.*
sessilifolia (Donn.-Sm.) L. O. Williams *op. cit.* 25
4 (1973): *Tabebuia sessilifolia.*
4 **silvicola** L. O. Williams *op. cit.* 25(1973)—Mexico.

ARRABIDAEA

costaricensis (Kränzl.) A. H. Gentry in *Brittonia*,
4 25(3):231(1973): *Saldanhaea costaricensis.*
pubescens (L.) A. H. Gentry *op. cit.* 239(1973):
4 *Bignonia pubescens.*

BARNETTIA Santisuk in *Kew Bull.*, 28(2):172
6 (1973).

771 Santisuk, T.
Notes on Asiatic Bignoniaceae. *Kew Bull.*, 28(2): 171–185(1973)—Key. *Barnettia*, a new genus from
6 Thailand.

kerrii (Barnett & Sandwith) Santisuk in *Kew Bull.*,
6 28(2):174(1973): *Radermachera kerrii.*
pagetii (Craib) Santisuk *op. cit.* 173(1973): *Rader-*
6 *machera pagetii.*

CUSPIDARIA

floribunda (DC.) A. H. Gentry in *Brittonia*, 25(3):
4 232(1973): *Adenocalymma floribundum.*
weberbaueri (Spr.) A. H. Gentry *op. cit.* 233(1973):
4 *Arrabidaea weberbaueri.*

CYDISTA

aequinoctialis (L.) Miers
var. **hirtella** (Benth.) A. H. Gentry in *Brittonia*,
4 25(3):231(1973): *Bignonia sarmentosa* var. *hirtella.*

DENDROSICUS

772 Gentry, A. H.
Studies in Bignoniaceae: 7. *Dendrosicus, Enallagma*
4 and *Amphitecna. Taxon*, 22(5–6):637–640(1973).

773 Gentry, A. H.
Studies in Bignoniaceae: 9. new species of *Dendrosicus* and *Pachyptera. Phytologia*, 26(6):439–450
4 (1973).

donnell-smithii (Sprague) A. H. Gentry in *Taxon*,
4 22(5–6):640(1973): *Crescentia donnell-smithii.*
isthmicus A. H. Gentry in *Phytologia*, 26(6):442
4 (1973)—Panama, Costa Rica.
kennedyi A. H. Gentry *op. cit.* 441(1973)—Panama,
4 Costa Rica.
latifolius (Mill.) A. H. Gentry in *Taxon*, 22(5–6):
4 638(1973): *Crescentia latifolia.*
macrophyllus (Seem.) A. H. Gentry *op. cit.* 639
4 (1973): *Crescentia macrophylla.*
molinae (L. Wms.) A. H. Gentry *op. cit.* 640(1973):
4 *Amphitecna molinae.*
montanus (L. Wms.) A. H. Gentry *op. cit.* 640(1973):
4 *Amphitecna montana.*
sessilifolius (Donn. Sm.) A. H. Gentry *op. cit.* 640
4 (1973): *Tabebuia sessilifolia.*
silvicola (L. Wms.) A. H. Gentry *op. cit.* 640(1973):
4 *Amphitecna silvicola.*
spathicalyx A. H. Gentry in *Phytologia*, 26(6):439
4 (1973)—Panama.
steyermarkii A. H. Gentry *op. cit.* 445(1973)—
4 Guatemala.

DISTICTIS

buccinatoria (DC.) A. H. Gentry in *Brittonia*, 25(3):
4 237(1973): *Pithecoctenium buccinatorium.*

DOXANTHA

774 Sistrunk, D. R.
A study of leaf development in *Doxantha unguis-cati*
4 Rehder. *Diss. Abstr. Int., B*, 33(8):3514(1973).

ECCREMOCARPUS

775 Mennega, E. A.
Een weinig bekende klimplant.
Wachendorffia, 1(3):32–33(1973)—Illustration.
4 *Eccremocarpus scaber.*

ENALLAGMA

776 Gentry, A. H.
Studies in Bignoniaceae: 7. *Dendrosicus, Enallagma*
4 and *Amphitecna. Taxon*, 22(5–6):637–640(1973).

JACARANDA

777 Prentice, J. R.
Two spectacular trees in Australia. *J. Roy. Hort. Soc.*, 98(6):270–271(1973)—Coloured illustration.
4, 7 *Jacaranda acutifolia.*

MACFADYENA

unguis-cati (L.) A. H. Gentry in *Brittonia*, 25(3):
4 236(1973): *Bignonia unguis-cati.*

MELLOA

quadrivalvis (Jacq.) A. H. Gentry in *Brittonia*, 25(3):
4 237(1973): *Bignonia quadrivalvis.*

NYCTOCALOS

778 Santisuk, T.
Notes on Asiatic Bignoniaceae. *Kew Bull.*, 28(2): 171–185(1973)—Key. Additional records of *Nyctocalos*
6 *shanica* in Thailand.

PACHYPTERA

779 Gentry, A. H.
Studies in Bignoniaceae: 9. new species of *Dendrosicus* and *Pachyptera*. *Phytologia*, 26(6):439–450
4 (1973).

alliacea (Lam.) A. H. Gentry in *Brittonia*, 25(3):
4 236(1973): *Bignonia alliacea.*
hymenaea (DC.) A. H. Gentry *op. cit.* 236(1973):
4 *Bignonia hymenaea.*
kerere (Aublet) Sandwith
var. **incarnata** (Aublet) A. H. Gentry *op. cit.* 235
4 (1973): *Bignonia incarnata.*
parvifolia A. H. Gentry in *Phytologia*, 26(6):447 (1973)—Colombia, Honduras, Panama, Peru,
4 Bolivia, Brazil.
standleyi (Steyerm.) A. H. Gentry in *Brittonia*, 25(3):
4 236(1973): *Pseudocalymma standleyi.*

PARMENTIERA

aculeata (HBK.) L. O. Williams in *Fieldiana, Bot.*,
4 36(4):27(1973): *Crescentia aculeata.*
millspaughiana L. O. Williams *op. cit.* 28(1973)—
4 Mexico.

PSEUDOCALYMMA

sagotii (Bur. & Schum.) Sandwith
var. **macrocalyx** (Sandwith) L. O. Williams in *Phytologia*, 25(7):458(1973): *P. alliaceum* var.
4 *macrocalyx.*

PSEUDOCATALPA A. H. Gentry in *Brittonia*,
4 25(3):241(1973).
caudiculata (Standl.) A. H. Gentry in *Brittonia*,
4 25(3):241(1973): *Petastoma caudiculatum.*

SARITAEA

780 Santisuk, T.
Notes on Asiatic Bignoniaceae. *Kew Bull.*, 28(2): 171–185(1973)—Key. The status of the genus *Saritaea*
4 Dugand.

SCHLEGELIA

781 Gentry, A. H.
Schlegelia costaricensis: a familial transfer to
4 Boraginaceae. *Phytologia*, 26(2):67–68(1973).

782 Leinfellner, W.
Das Gynözeum der Bignoniaceen: 2. Die U-förmige Plazenta von *Schlegelia* (Crescentieae).
4 *Österr. Bot. Zeitschr.*, 121(1–2):13–22(1973).

SPATHODEOPSIS

783 Santisuk, T.
Notes on Asiatic Bignoniaceae. *Kew Bull.*, 28(2): 171–185(1973)—Key. *Spathodeopsis* recorded from
6 Thailand.

STEREOSPERMUM

784 Santisuk, T.
Notes on Asiatic Bignoniaceae. *Kew Bull.*, 28(2): 171–185(1973)—Key. Nomenclatural and taxonomic notes on Indian spp. of *Stereospermum* and new
2, 6 synonyms for *S. neuranthium* Kurz.

785 Theobald, W. L.
Notes on the Bignoniaceae. *Ceylon J. Sci., Biol. Sci.*,
6 10(1):75(1972). (Critical Notes on Ceylon Plants 9).

personatum (Hassk.) Chatterjee
var. **angustifolium** (Haines) Santisuk in *Kew Bull.*,
6 28(2):181(1973): *S. chelonoides* var. *angustifolium.*

TABEBUIA

786 Dugand, A.
Bignoniaceae: el genero *Tabebuia* en Colombia.
4 *Cespedesia*, 1(1–2):103–126(1972)—(En); key.

787 Maciel Pacheco, J. *
Contribuição ao estudo botânico da *Tabebuia longiflora* Greenm. (*Tecoma longiflora*). Fam. Bignoniaceae. (Contribution à l'étude botanique de *Tabebuia longiflora*, Greenm. (*Tecoma longiflora*). Fam. Bignoniaceae). *Atas Soc. Biol. Rio de Janeiro*, 16(1): 1–4(1972). Ref. in *Bull. Signal., Biol. Physiol. Vég.*,
3, 4 34(10):8742(1973).

anafensis Urb.
subsp. **munizii** Borhidi in *Acta Bot. Acad. Sci. Hung.*,
4 18(1–2):41(1973)—Cuba.
perfae Alain in *Phytologia*, 25(5):274(1973)—
4 Hispaniola (Dominican Republic).

TECOMA

788 Pillai, A. *and* Mehra, A. D.
Some aspects of the anatomy of *Tecoma stans* (L.)
4–6 H.B.K. *Proc. Indian Acad. Sci., B*, 77(2):64–77(1973).

BIXACEAE

789 Villiers, J.-F.
Bixacées. Flore du Gabon, 22:59–62(1973)—
5 Illustrations.

BOMBACACEAE

790 Huynh, K.-L.
Étude de l'arrangement du pollen dans la tétrade chez les angiospermes sur la base de données cytologiques: 6. Lythraceae et Bombacaceae. *Bol. Soc. Brot.*,
4–6 46:171–180(1972)—(En).

791 Villiers, J.-F.
Bombacacées. Flore du Gabon, 22:31–54(1973)—
5 Illustrations; keys.

DURIO

792 Heaslett, E. A.
Durio malaccensis Planch. with pink-tinged
6 flowers. *Malayan Nat. J.*, 25(1):39(1972).

BORAGINACEAE

793 Delay, J.
Boraginacées (suite). *Inform. Ann. Caryosyst. Cytogénét.*, no. 6:9–12(1972 publ. 1973)—
1 Illustrations; chromosome number.

794 Fernandes, A. *and* Leitão, M. A.
Contribution à la connaissance cytotaxinomique de spermatophyta du Portugal: 5. Boraginaceae. *Bol.*
1 *Soc. Brot.*, 46:389–405(1972)—Chromosome numbers.

795 Higgins, L. C.
The Boraginaceae of Utah. *Brigham Young Univ. Sci. Bull., Biol.*, 16(3). [iv]83p.(1972)—Dot maps;
3 keys.

796 Roth, I.
Estructura anatómica de la corteza de algunas especies arbóreas venezolanas de Boraginaceae. (Anatomical structure of the cortex of several arboreous Venezuelan species of Boraginaceae.). *Acta*
4 *Biol. Venezuel.*, 8(2):131–153(1973)—(En).

AEGONYCHON

calabricum (Ten.) Holub in *Folia Geobot. Phytotax.*, 8(2):164(1973), without basionym date:
1 *Lithospermum calabricum*.
gastonii (Benth.) Holub *op. cit.* 164(1973):
1, 2 *Lithospermum gastonii*.
goulandriorum (Rech. f.) Holub *op. cit.* 165(1973):
1 *Lithospermum goulandriorum*.
purpurocaeruleum (L.) Holub *op. cit.* 165(1973), as
1, 2 'purpurocoeruleum': *Lithospermum purpurocaeruleum*.
zollingeri (A. DC.) Holub *op. cit.* 165(1973):
2, 6 *Lithospermum zollingeri*.

ALKANNA

bracteolata Greuter in *Exsicc. Genav.*, 3:38(1972),
2 nom. nov.: *Anchusa tuberculata* Forsskål.
tuberculata Greuter *op. cit.* 38(1972), nom. nov.:
2 *Anchusa bracteolata* Viv.

ANCHUSA

barrelieri (All.) Vitm.
lusus **peterfii** Soó in *Acta Bot. Acad. Sci. Hung.*,
1 18(1–2):174(1973)—Rumania.

BOURRERIA

costaricensis (Standl.) A. H. Gentry in *Phytologia*,
4 26(2):67(1973): *Schlegelia costaricensis*.

BUGLOSSOIDES

797 Fernandes, R. B.
Notes sur le genre *Buglossoides* Moench. *Acta Bot. Acad. Sci. Hung.*, 19(1–4):93–101(1973)—(En);
1, 2 chromosome numbers.

arvensis (L.) I. M. Johnston
f. **cyanea** R. Fernandes in *Acta Bot. Acad. Sci. Hung.*, 19(1–4):100(1973), nom. nov.: *Lithospermum*
1, 2 *arvense* f. *coeruleum* Borza.
f. **incrassata** (Guss.) R. Fernandes *op. cit.* 96(1973):
1 *Lithospermum incrassatum*.

CORDIA

798 Agostini, G.
Cordia umbellifera, una nueva especies venezolana de la seccion *Gerascanthus* (*Cordia*, Boraginaceae). *Acta Bot. Venezuel.*, 9(1–4):291–294(1974 publ. 1973)—
4 Illustration; key.

799 Agostini, G.
A new Venezuelan species of *Cordia* sect. *Gerascanthus* (Boraginaceae). *Brittonia*, 25(2):174–176
4 (1973)—Illustration; key.

800 Nowicke, J. W. *and* Ridgway, J. E.
Pollen studies in the genus *Cordia* (Boraginaceae).
1–7 *Amer. J. Bot.*, 60(6):584–591(1973).

thaisiana Agostini in *Brittonia*, 25(2):174(1973)—
4 Venezuela.
umbellifera Killip ex Agostini in *Acta Bot. Venezuel.*,
4 9(1–4):292(1974 publ. 1973)—Venezuela.

CYNOGLOSSUM

801 Biswas, M.
Meiotic studies in *Cynoglossum denticulatum* A. DC. *J. Bihar Bot. Soc.*, 1(1–2):30–34(1972)—Chromosome
6 numbers.

ECHIUM

802 Bramwell, D.
Studies in the genus *Echium* from Macaronesia. *Monogr. Biol. Canar.*, no. 4:71–82(1973)—Illustra-
1 tions, map, dot map; chromosome numbers.

803 Fritsch, B.
Karyologische Untersuchungen in der Gattung *Echium* L. *Bot. Notiser*, 126(4):450–458(1973)—(En);
1, 2 chromosome numbers.

804 Vereshchagina, V. A. *and* Ermishina, O. M.
Izuchenie meĭoza v mikrosporotsitakh u polovȳkh form tsvetkov *Echium russicum* J. F. Gmelin (Boraginaceae). (The study of meiosis in microsporocytes of sex flower forms of *Echium russicum* J. F. Gmelin (Boraginaceae)). *Bot. Zhurn.*, 58(12):1760–1764
1, 2 (1973)—Chromosome number.

creticum L.
subsp. grandiflorum (Lap.) Malagarriga, Las Subesp. y Variac. Geogr.: 20(1973), without
1 basionym date: *E. grandiflorum*.
subsp. maritimum (W.) Malagarriga *op. cit.* 20
1 (1973), without basionym date: *E. maritimum*.
subsp. plantagineum (L.) Malagarriga *op. cit.* 20
1 (1973), without basionym date: *E. plantagineum*.
subsp. prostratum (Delile) Malagarriga *op. cit.* 20(1973), without basionym page or date:
1 *E. prostratum*.

ERITRICHIUM

805 Chukavina, A. P.
Novȳe i redkie vidȳ sem. burachnikovȳkh Pamira. *Izv. Akad. Nauk Tadzh. SSR, Biol. Nauk*, 4(49):
2 30–37(1972 publ. 1973)—(Ta).

806 Mathew, B.
Plant awards 1972–73: *Eritrichium howardii* (Gray) Rydberg P. C. *Quart. Bull. Alp. Gard. Soc.*, 41(4):
3 332(1973).

HACKELIA

807 Chukavina, A. P.
Novȳe i redkie vidȳ sem. burachnikovȳkh Pamira. *Izv. Akad. Nauk Tadzh. SSR, Biol. Nauk*,
2 4(49):30–37(1972 publ. 1973)—(Ta).

murgabica Chukavina in *Izv. Akad. Nauk Tadzh. SSR, Biol. Nauk*, 4(49):32(1972 publ. 1973)—
2 U.S.S.R. (Middle Asia).
popovii Chukavina *op. cit.* 31(1972 publ. 1973)—
2 U.S.S.R. (Middle Asia).

HELIOTROPIUM

808 Asai, Y.
(Miscellaneous notes on the naturalized flora of
Japan: 3). *J. Jap. Bot.*, 48(3):67–72(1973)—Ja(En);
2 illustrations. *Heliotropium curassavicum.*

809 Ponert, J.
Species nova iranica *Heliotropium. Folia Geobot.*
2 *Phytotax.*, 8(3):337–338(1973)—(En).

810 Sahay, S. K.
Pollen morphology of *Heliotropium. Proc. Indian*
1, 2, 5, 6 *Sci. Congr.*, 58(3):455(1971). Abstract only.

gossypii Ponert in *Folia Geobot. Phytotax.*, 8(3):337
2 (1973)—Iran.

LAPPULA

occidentalis (Wats.) Greene
var. **cupulata** (Gray) Higgins in *Brigham Young*
Univ. Sci. Bull., Biol., 16(3):62(1972): *Echino-*
3 *spermum redowskii* var. *cupulatum.*

LITHOSPERMUM

811 Kováts, D.
Anatomical investigations on the vegetative system
of *Lithospermum purpureo-coeruleum* L. *Ann. Hist.-*
1 *Nat. Mus. Nation. Hung.*, 65:109–125(1973).

LYCOPSIS

cretica (Mill.) Holub in *Preslia*, 45(4):359(1973):
1 *Anchusa cretica.*

MORITZIA

812 Gentry, J. L.
Moritzia DC. (Boraginaceae): a genus new to
North America. *Fieldiana, Bot.*, 36(2):13–16(1972)—
3 Illustration; key.

MYOSOTIS

813 Berg, C. C. *and* Kaastra, R. C.
Myosotis palustris en *M. laxa* in Nederland. *Gorteria*,
6(9):141–150(1973)—Du(En); illustration, dot maps;
1 chromosome numbers; key.

814 Moore, L. B. *and* Simpson, M. J. A.
A new *Myosotis* from north-west Nelson. *New*
7 *Zealand J. Bot.*, 11(1):163–170(1973)—Illustrations.

815 Saiki, Y. *and* Okegawa, O.
(On *Myosotis baltica* Sam. found newly in Japan).
2 *J. Geobot.*, 20(3):61–64(1972)—Ja(En); illustration.

brockiei L. B. Moore & M. J. A. Simpson in *New*
7 *Zealand J. Bot.*, 11(1):163(1973)—New Zealand.
micrantha Pallas
subsp. minutiflora (Bss. R.) Malagarriga, Las
Subesp. y Variac. Geogr.: 20(1973), without
1 basionym date: *M. minutiflora.*
palustris (L.) L.
subsp. **nemorosa** (Besser) C. C. Berg & Kaastra in
1 *Gorteria*, 6(9):145(1973): *M. nemorosa.*

NONEA

alba DC.
subsp. ventricosa (Gris.) Malagarriga, Las Subesp. y
Variac. Geogr.: 20(1973), without basionym
1, 2 date: *N. ventricosa.*
micrantha Bss. R.
subsp. bourgaei (Cosson) Malagarriga *op. cit.* 20
1 (1973), without basionym date: *N. bourgaei.*

ONOSMA

816 Ponert, J.
A new species of *Onosma* from Iran. (Nový druh
rodu *Onosma* z Íránu). *Preslia*, 45(2):140–141(1973)—
2 (Cz).

817 Tissot-Daguette, M.
Recherches cytotaxonomiques sur le genre *Onosma.*
Ber. Schweiz. Bot. Ges., 82(4):291–299(1972)—(En,Ge);
1 chromosome numbers.

818 Tissot-Daguette, M.
Sur quelques *Onosma* (Borraginacées) d'Europe
occidentale. *Bull. Soc. Neuchâtel. Sci. Nat.*, 96:41–49
(1973)—(Ge,En); map; chromosome numbers.
1 *O. fastigiatum, O. vandense, O. tricerospermum.*

borragoidinum Ponert in *Preslia*, 45(2):140(1973)—
2 Iran.
tuberculatum (Jáv.) Soó, Magyar Fl. Veg., 5:661
(1973), without basionym ref.: *O. arenarium* subsp.
1 *tuberculatum.*

PARDOGLOSSUM Barbier & Mathez in *Candollea*,
1 28(2):305(1973).

819 Barbier, E. *and* Mathez, J.
Contribution à l'étude des Cynoglossées
(Boraginacées): *Pardoglossum*, genre nouveau du
bassin méditerranéen occidental. *Candollea*, 28(2):
1 281–323(1973)—(En,Ge); key.

atlanticum (Pitard) Barbier & Mathez in *Candollea*,
1 28(2):306(1973): *Solenanthus atlanticus.*
cheirifolium (L.) Barbier & Mathez *op. cit.* 306
1 (1973): *Cynoglossum cheirifolium.*
lanatum (L.) Barbier & Mathez *op. cit.* 307(1973):
1 *Anchusa lanata.*
tubiflorum (Murbeck) Barbier & Mathez *op. cit.*
1 307(1973): *Solenanthus tubiflorus.*
watieri (Batt. & Maire) Barbier & Mathez *op. cit.*
1 308(1973): *Solenanthus watieri.*

PULMONARIA

820 Sauer, W.
Beitrag zur Kenntnis von *Pulmonaria stiriaca* und
Verwandten (Boraginaceae). *Mitt. Bot. Staatssamml.*
München, 11:1–32(1973)—(En); illustrations, dot map;
1 chromosome numbers; key.

821 Steinwendtner, R.
Arealgrenzen von *Pulmonaria kerneri* Wettst. in
Oberösterreich. *Mitt. Bot. Arbeit. Oberösterr. Landesmus.*
1 *Linz*, 4(1–2):45(1972).

carnica W. Sauer in *Mitt. Bot. Staatssamml. München*,
1 11:16(1973)—Austria.

SCHLEGELIA

822 Gentry, A. H.
Schlegelia costaricensis: a familial transfer to
4 Boraginaceae. *Phytologia*, 26(2):67–68(1973).

SYMPHYTUM

823 Gadella, T. W. J. *and* Kliphuis, E.
Cytotaxonomic studies in the genus *Symphytum.*:
5. Some notes on W. European plants with the
chromosome number 2n = 40. *Bot. Jahrb.*, 93(4):
1 530–538(1973)—(Ge); illustrations.

824 Wcisło, H.
Karyological studies in *Symphytum* L. *Acta Biol.*
Cracov., Bot., 15(2):153–163(1972)—Map;
1, 2 chromosome numbers.

TOURNEFORTIA

parvifolia Alain in *Phytologia*, 25(5):272(1973)—
4 Hispaniola (Dominican Republic).

BRASSICACEAE
see
CRUCIFERAE

BROMELIACEAE

825 Benzing, D.
The monocotyledons: their evolution and comparative biology: 1. Mineral nutrition and related phenomena in Bromeliaceae and Orchidaceae. *Quart.*
4 *Rev. Biol.*, 48(2):277–290(1973).

826 Beutelspacher B., C. R. *and* Lopez de Beutelspacher, M. G.
Las bromeliáceas de México. *Cact. Suc. Mex.*, 18(2):
4 46–49(1973)—(En); illustrations.

827 Brown, M. L.
Bromeliad trichomes. *J. Bromeliad Soc.*, 22(5):
4 111–118(1972).

828 Ehler, N. *and* Schill, R.
Die Pollenmorphologie der Bromeliaceae. *Pollen*
4 *Spores*, 15(1):13–45(1973)—(En,Fr).

829 Gilmartin, A. J.
Transandean distributions of Bromelaiceae in
4 Ecuador. *Ecology*, 54(6):1389–1393(1973)—Map.

830 Gómez P., L. D.
Karatophyllum bromelioides L. D. Gómez (Bromeliaceae) nov. gen. et sp., del Terciario Medio de Costa Rica. *Rev. Biol. Trop.*, 20(2):221–229(1972)—(En); illustration. Suggests possible affinity with
4 modern species of *Aechmea* and *Bromelia*.

831 Innes, C. F.
Terrestrial bromeliads: a preamble. *Ashingtonia*,
3, 4 1(3):32–33(1973).

832 Kent, L.
Variegated bromeliads. *J. Bromeliad Soc.*, 22(4):
4 99–103(1972)—Coloured illustration.

833 Pereira, E.
Species novae in Brasilia Bromeliacearum: 5.
4 *Bradea*, 1(29):313–318(1973)—Illustrations.

834 Rauh, W., Ehler, N. *and* Schill, R.
A preliminary note on the submicroscope structure of the pollen-grain membrane of some bromeliads.
4 *J. Bromeliad Soc.*, 21(4):86–95(1971).

835 Rauh, W., Schill, R., Ehler, N. *and* Barthlott, W.
Some remarks on the water supply of bromeliads.
4 *J. Bromeliad Soc.*, 23(3):89–111(1973)—Illustrations.

ABROMEITIELLA

836 Brown, J. R.
Interesting succulents. *Cact. Succ. J.* (U.S.A.), 45(2):54(1973)—Illustrations. *Abromeitiella*
4 *brevifolia*.

AECHMEA

837 Anon.
Aechmea cylindrata Lindm. *J. Bromeliad Soc.*,
4 22(3):70(1972)—En; coloured illustration.

838 Anon.
Aechmea fulgens var. *discolor* Brongn. *J. Bromeliad*
4 *Soc.*, 22(3):72(1972)—En; coloured illustration.

839 Anon.
Aechmea fulgens v. *discolor* (E. Morr.) (Bromeliaceae).
4 *Ashingtonia*, 1(2):24(1973)—En; coloured illustration.

840 Anon.
Aechmea pineliana var. *minuta*. *J. Bromeliad Soc.*,
4 22(4):104(1972)—En; coloured illustration.

841 Foster, M. B.
How to recognize an *Aechmea*. *J. Bromeliad Soc.*,
4 22(3):71(1972).

842 Harden, N.
Aechmea cylindrata. *Bromeliads*, no. 16:72–73(1973)—
4 En.

843 Wall, W. F.
Aechmea chantinii Baker. *Bromeliads*, no. 16:69
4 (1973)—En.

844 Wall, W. F.
Aechmea marmorata Mez. *Bromeliads*, no. 16:70
4 (1973)—En.

trigona E. Pereira in *Bradea*, 1(29):313(1973)—
4 Brazil.

ARAEOCOCCUS

845 [Padilla, V.]
Araeococcus flagellifolius Harms. *J. Bromeliad Soc.*,
4 21(5):120(1971)—En; coloured illustration.

BILLBERGIA

846 Abendroth, A.
Billbergia tweedeana in Paqueta, Gb. (Rio). *J.*
4 *Bromeliad Soc.*, 22(2):37–38(1972).

4 **nana** E. Pereira in *Bradea*, 1(29):316(1973)—Brazil.
richterana E. Pereira *op. cit.* 315(1973), as
4 'richteriana'—Brazil.
4 var. **bicolor** E. Pereira *op. cit.* 315(1973)—Brazil.

BROMELIA

847 Innes, C. F.
Bromelia balansae Mez 1873 (Bromeliaceae).
4 *Ashingtonia*, 1(2):15(1973)—En; coloured illustration.

DYCKIA

848 Kondo, K.
Chromosome numbers of five taxa. *Chromosome Inform. Serv.*, no. 15:33–34(1973). *Dyckia altissima* ×
4 *D. brevifolia*.

FASCICULARIA

849 Anon.
Fascicularia pitcairniifolia Mez. (Bromeliaceae).
4 *Ashingtonia*, 1(3):27(1973)—En; coloured illustration.

GUZMANIA

850 Anon.
Guzmania gloriosa. *J. Bromeliad Soc.*, 21(3):72
4 (1971)—En; coloured illustration.

851 Innes, C. F.
Guzmania lingulata minor Mez 1896 (Bromeliaceae).
4 *Ashingtonia*, 1(1):3(1973)—En; coloured illustration.

852 Oeser, R.
A note on *Guzmania sanguinea*. *J. Bromeliad Soc.*,
4 23(6):223(1973). On seed dispersal.

853 Padilla, V.
Guzmania sanguinea (Andre) Mez. *J. Bromeliad Soc.*,
4 23(4):156–158(1973)—En; coloured illustrations.

854 Temple, P.
An interesting *Guzmania*. *J. Bromeliad Soc.*, 22(1):
4 10–12(1972)—Illustrations.

HECHTIA

855 Padilla, V.
The Hechtias of Texas. *J. Bromeliad Soc.*, 22(2):34
3 (1972).

NEOREGELIA

856 Wall, W. F.
Neoregelia tristis L. B. Smith. *Bromeliads*, no. 16:
4 70(1973)—En.

NIDULARIUM

857 [Padilla, V.]
Plant portrait of the month: *Nidularium seidelii* L. B. Smith. *J. Bromeliad Soc.*, 22(3):51(1972)—
4 Coloured illustration.

innocenti Lem.
var. **bicolor** E. Pereira in *Bradea*, 1(29):316(1973)—
4 Brazil.

ORTHOPHYTUM

858 Foster, M. B.
Orthophytum navioides L. B. Smith. *J. Bromeliad*
4 *Soc.*, 23(6):244(1973)—Coloured illustration.

PUYA

859 Ortiz-Crespo, F. I.
Field studies on pollination of plants of the genus *Puya*. *J. Bromeliad Soc.*, 23(1):3–7,(2):54–57(1973)—
4 Illustrations.

RONNBERGIA

860 Smith, L. B.
Ronnbergia petersii. *J. Bromeliad Soc.*, 23(2):46–48
4 (1973)—En; illustration.

petersii L. B. Smith in *J. Bromeliad Soc.*, 23(2):46
4 (1973)—Panama.

THECOPHYLLUM

861 Salas, D. S.
Una bromeliacea costarricense polinizada por murcielagos. *Brenesia*, no. 2:5–10(1973)—(En).
4 *Thecophyllum irazuense* Mez & Wercklé.

TILLANDSIA

862 Anon.
Portrait of the month: *Tillandsia violacea* (Beer) Baker. *J. Bromeliad Soc.*, 22(1):3(1972)—Coloured
4 illustration.

863 Anon.
Tillandsia monadelpha (E. Morren). *J. Bromeliad Soc.*,
4 22(1):24(1972)—En.

864 Abendroth, A.
Tillandsia usneoides var. *minor* in Paqueta. *J.*
4 *Bromeliad Soc.*, 22(4):153–155(1972).

865 Blass, A.
Three silver-leaved Tillandsias. *J. Bromeliad Soc.*, 23(6):224–225(1973)—Illustrations. *T. montana*,
4 *T. tectorum*, *T. heteromorpha*.

866 Blass, A.
Tillandsia portraits. *J. Bromeliad Soc.*, 21(1):23–24,
4 (2):47–48(1971)—Illustrations, coloured illustrations.

867 Blass, A.
Tillandsias—a portfolio. *J. Bromeliad Soc.*, 22(1):
4 21(1972)—Illustration. *Tillandsia orogenes*.

868 Foster, M. B.
How to recognize a *Tillandsia*. *J. Bromeliad Soc.*,
4 22(1):20(1972).

869 Harris, D.
The Tillandsias of Texas. *J. Bromeliad Soc.*, 22(2):
3 33–34(1972)—Illustration.

870 Innes, C. F.
Tillandsia stricta Soland., 1813 (Bromeliaceae).
4 *Ashingtonia*, 1(1):12(1973)—En; coloured illustration.

871 Matuda, E.
New additions to *Tillandsia* (Bromeliaceae) from Mexico. *Cact. Succ. J.* (U.S.A.), 45(4):186–189
4 (1973)—Illustrations.

872 Matuda, E.
Nuevas Tillandsias de México. *Cact. Suc. Mex.*,
4 18(2):50–52(1973)—(En); illustrations.

873 Oeser, R.
Tillandsias—their great adaptability. *J. Bromeliad*
3, 4 *Soc.*, 21(3):53–56(1971)—Illustration.

874 Pereira, E. *and* Pereira, C.
Bromeliáceas da Guanabara: 1. Gênero *Tillandsia*.
4 *Leandra*, 2(2):41–72(1972)—Illustrations; key.

875 Rauh, W.
Notes on Peruvian Tillandsias. *J. Bromeliad Soc.*,
4 23(4):143–148(1973)—Illustrations.

876 Rauh, W.
Tillandsia califani Rauh spec. nov., a new species from Mexico. *J. Bromeliad Soc.*, 21(3):65–67(1971)—
4 Illustration.

877 Rauh, W.
Tillandsia hildae Rauh, a striking new species from northern Peru. *J. Bromeliad Soc.*,21(6):139–143(1971)
4 —Illustrations.

878 Rauh, W. *and* Rauh, H.
Tillandsia marnieri-lapostollei Rauh. *J. Bromeliad Soc.*,
4 22(2):40–42(1972)—En; illustrations.

879 Rauh, W. *and* Smith, L. B.
Tillandsia esseriana Rauh et L. B. Smith. *J. Bromeliad*
4 *Soc.*, 21(1):3–6(1971)—En; illustrations.

880 Subils, R.
Poliembronía en especies argentinas de *Tillandsia* (Bromeliaceae). *Kurtziana*, 7:266–267(1973)—
4 Illustrations.

881 Swalley, B.
Variation of growth on *Tillandsia circinata*. *J.*
4 *Bromeliad Soc.*, 23(6):215–217(1973)—Illustration.

882 Thiele, P.
Sammelnswerte Klein-Tillandsien. *Stachelpost*,
4 9(47):145–149(1973)—Illustrations.

883 Watson, M. M.
Observations on collecting *Tillandsia fasciculata* in
3 Florida. *J. Bromeliad Soc.*, 21(2):27–29(1971).

carlos-hankii Matuda in *Cact. Succ. J.* (U.S.A.)
4 45(4):186(1973)—Mexico.
carnosa L. B. Smith
var. longispicata Rauh in *J. Bromeliad Soc.*, 23(4):
4 148(1973), without latin descr.—Peru.
circinnatioides Matuda in *Cact. Succ. J.* (U.S.A.),
4 45(4):187(1973)—Mexico.
marnieri-lapostollei Rauh in *J. Bromeliad Soc.*, 22(2):
4 41(1972), without latin descr.—Ecuador.
palmasolana Matuda in *Cact. Suc. Mex.*, 18(2):51
4 (1973)—Mexico.
sierrajuarezensis Matuda *op. cit.* 50(1973)—
4 Mexico.
superinsignis Matuda in *Cact. Succ. J.* (U.S.A.),
4 45(4):189(1973)—Mexico.
4 **vriesioides** Matuda *op. cit.* 187(1973)—Mexico.

VRIESEA

884 Anon.
Vriesea imperialis Carr. *J. Bromeliad Soc.*, 22(4):168
4 (1972)—En; coloured illustration.

885 Anon.
Vriesea malzinei E. Morr. *J. Bromeliad Soc.*, 22(2):
4 48(1972)—En; coloured illustration.

886 Anon.
Vriesea platynema var. *variegata* (Guillon) Reitz.
J. Bromeliad Soc., 22(2):46(1972)—En; coloured
4 illustration.

887 Padilla, V.
Two Vrieseas of unusual interest. *J. Bromeliad Soc.*,
23(6):218–219(1973)—Coloured illustrations.
4 *V. patula*, *V. regina*.

888 Smith, L. B.
Vriesea duarteana. *J. Bromeliad Soc.*, 23(1):16–17
4 (1973)—En; coloured illustration.

WITTROCKIA

889 Anon.
Plant portrait *Wittrockia superba* Lindm. *J. Bromeliad*
4 *Soc.*, 23(1):40(1973)—Coloured illustration.

BRUNELLIACEAE

BRUNELLIA

890 Cuatrecasas, J. *and* Porter, D. M.
A new species of *Brunellia* from Panama. *Phytologia*,
4 26(6):485–486(1973).

darienensis Cuatrecasas & D. M. Porter in *Phytologia*,
4 26(6):485(1973)—Panama.

BUDDLEJACEAE
see
LOGANIACEAE

BURMANNIACEAE

891 Hatschbach, G. *and* Guimarães, O.
Burmanniáceas do estado do Paraná. *Bol. Mus. Bot.*
Mun. (Curitiba), no. 5. 17p.(1972)—Illustrations,
4 dot maps; key.

BURMANNIA

892 Arekal, G. D. *and* Ramaswamy, S. N.
Embryology of *Burmannia pusilla* (Wall. ex Miers)
THW. and its taxonomic status. *Beitr. Biol. Pfl.*,
6 49(1):35–46(1973)—(Ge).

BURSERACEAE

893 Abedin, S. *and* Ali, S. I.
Flora of West Pakistan: no. 26. Burseraceae. 1972.
6 6p.—Illustrations, map.

COMMIPHORA

894 Gillett, J. B.
Commiphora Jacq. (Burseraceae)—Englerian species
5 which 'disappear'. *Kew Bull.*, 28(1):25–28(1973).

895 Lisowski, S., Malaisse, F. *and* Symoens, J. J.
Le genre *Commiphora* Jacq. (Burseraceae) au Zaire
(ex Congo-Kinshasa). *Bol. Soc. Brot.*, 46:81–91(1972)—
5 Illustrations; key.

896 Van der Walt, J. J. A.
Die geslag *Commiphora*. *Veld Flora*, 3(1):4–6(1973)—
5 Illustrations. Popular notes.

897 Van der Walt, J. J. A.
A new species of *Commiphora* from the Kaokoveld
(South West Africa). *Dinteria*, no. 9:23–28(1973)—
5 Illustrations.

898 Van der Walt, J. J. A.
The South African species of *Commiphora*. *Bothalia*,
5 11(1–2):53–102(1973)—Illustrations, dot maps; keys.

899 Van der Walt, J. J. A. *and* Van der Schijff, H. P.
The anatomy of the petiole as an aid to the
identification of South African *Commiphora* species.
5 *Kirkia*, 9(1):95–108(1973).

giessii Van der Walt in *Dinteria*, 9:23(1973)—South
5 West Africa.
thermitaria Lisowski, Malaisse & Symoens in *Bol.*
5 *Soc. Brot.*, 46:83(1972 publ. 1973)—Zaïre.

GARUGA

900 Ghosh, R. B.
A note on the gross anatomy of axillary bud in
Garuga pinnata Roxb. *Broteria, Sci. Nat.*, 42(1–2):
6 15–18(1973).

HAPLOLOBUS

901 Leenhouts, P. W.
A revision of *Haplolobus* (Burseraceae). *Blumea*,
20(2):283–310(1972 publ. 1973)—Illustrations;
6 key.

902 Leenhouts, P. W. *and* Widodo, S. H.
Some notes on the seedling of *Haplolobus*
(Burseraceae). *Blumea*, 20(2):311–314(1972 publ.
6 1973)—Illustration.

boswezensis Leenhouts in *Blumea*, 20(2):293(1972
6 publ. 1973)—New Guinea (W. Irian).
canarioides Leenhouts *op. cit.* 294(1972 publ. 1973)—
6 British Solomon Is.
floribundus (K. Sch.) H. J. Lam
var. **hirsutus** Leenhouts *op. cit.* 298(1972 publ.
1973)—Celebes, Mollucas, New Guinea (W.
Irian), British Solomon Is., Santa Cruz Is.
[South Pacific], New Hebrides, Fiji, Samoa,
6 Caroline Is.

subsp. **microphyllus** (Husson) Leenhouts *op. cit.*
6 297(1972 publ. 1973): *H. microphyllus.*
subsp. **moluccanus** (H. J. Lam) Leenhouts *op. cit.*
6 297(1972 publ. 1973): *H. moluccanus.*
subsp. **salomonensis** (C. T. White) Leenhouts
6 *op. cit.* 298(1972 publ. 1973): *H. salomonensis.*
furfuraceus (Laut.) H. J. Lam
var. **glandulosus** (Husson) Leenhouts *op. cit.* 304
6 (1972 publ. 1973): *H. glandulosus.*
lanceolatus H. J. Lam ex Leenhouts *op. cit.* 305 (1972 publ. 1973), without latin descr.—New Guinea
6 (W. Irian).
pubescens H. J. Lam ex Leenhouts *op. cit.* 307(1972 publ. 1973), without latin descr.—New Guinea
6 (W. Irian).

BUTOMACEAE

903 Charlton, W. A. *and* Ahmed, A.
Studies in the Alismataceae: 4. Developmental morphology of *Ranalisma humile* and comparisons with two members of the Butomaceae, *Hydrocleis nymphoides* and *Butomus umbellatus. Canad. J. Bot.*,
1, 2, 4, 6 51(5):899–910(1973)—(Fr).

904 D'Orey, J.
Flora da Guiné Portuguesa: Butomaceae. 1972.
5 5p.—Chromosome number.

HYDROCLEIS

905 Sattler, R. *and* Singh, V.
Floral development of *Hydrocleis nymphoides. Canad.*
4 *J. Bot.*, 51(12):2455–2458(1973).

BUXACEAE

BUXUS

906 Abramson, D., Goad, L. J. *and* Goodwin, T. W.
Triterpenes and sterols of *Buxus sempervirens* and local variations in their levels. *Phytochemistry*, 12(9): 2211–2216(1973). *Buxus* samples from England,
1 Wales and Scotland only.

907 Fosberg, F. R.
Type specimens of *Buxus sempervirens* Linnaeus.
1 *Boxwood Bull.*, October:18–21(1973)—Illustrations.

908 Kakkar, L.
Leaf vasculature of *Buxus* sp. *Proc. Indian Sci. Congr.*,
1–6 60(3):293(1973). Abstract only.

909 Razafindrambao, R. S.
Étude d'une plante médicinale malgache: *Buxus madagascarica* Baill. et ses variétés. Paris, O.R.S.T.O.M., 1973. 96p. (Travaux et Documents de
5 l'O.R.S.T.O.M. no. 25)—Illustrations, dot map.

baracoënsis Borhidi & Muñiz in *Acta Bot. Acad. Sci.*
4 *Hung.*, 18(1–2):34(1973)—Cuba.
crassifolia (Britt.) Urb.
var. **oblongata** Borhidi & Muñiz *op. cit.* 36
4 (1973)—Cuba.

SIMMONDSIA

910 Kakkar, L.
Leaf vasculature in *Simmondsia californica. Proc.*
3 *Indian Sci. Congr.*, 60(3):293–294(1973). Abstract only.

CACTACEAE

911 Assink, F.
Copiapoa's. *Succulenta*, 52(2):27–30(1973)—Du;
4 illustrations.

912 Bleck, M.
Tribe 2: Opuntieae, exclusive of *Opuntia. Cact.*
3, 4 *Succ. J.* (U.S.A.), 45(6):255–256(1973)—Illustrations.

913 Bourdoux, P.
Considerations sur la nomenclature des cactées.
4 *Cactus* (Brussels), 4(2):34–40(1972)—Illustrations.

914 Buining, A. F. H.
Interessante planten uit de hoge Andes van W.
4 Rausch. *Succulenta*, 52(2):33–34(1973).

915 Castellanos, A.
Cactaceas. [Buenos Aires, Librart], 1972. xxiv, 385p.—Illustrations, portrait, dot maps; keys.
4 Reprints from various journals.

916 Daniel, J. *and* Rowland, D.
An introduction to some cacti and succulent plants of New Mexico. *Cact. Succ. J. Gr. Brit.*, 35(3):53–54
3 (1973)—Illustrations.

917 Daniel, J. *and* Rowland, D.
Some cactus and succulent plants of New Mexico. *Cact. Succ. J. Gr. Brit.*, 35(4):86–89(1973)—
3 Illustrations.

918 Gibson, A. C.
Comparative anatomy of secondary xylem in
3, 4 Cactoideae (Cactaceae). *Biotropica*, 5(1):29–65(1973).

919 Gibson, A. C.
Comparative anatomy of secondary xylem in Cactoideae (Cactaceae). *Diss. Abstr. Int., B*, 33(9):
3, 4 4147–4148(1973).

920 Kench, G.
Studies in the Colorado cacti. *Diss. Abstr. Int., B*,
3 33(4):1421–1422(1972).

921 Lau, A. B.
Baja California: eine abenteuerliche Reise Niederkalifornien. *Kakt. And. Sukk.*, 24(3):64–67(1973)—
4 Illustrations.

922 McCarten, N.
A visit to the valley of Little Hell. *Cact. Succ. J.* (U.S.A.), 45(3):103–107(1973)—Illustrations.
4 Presadel Infernillo, Balsas Valley.

923 Meixner, A.
Die Bedeutung der Bodenzusammensetzung für mexikanische Arten und Gattungen. *Kakt. And. Sukk.*,
4 24(2):39–41(1973)—Map.

924 Mottram, R.
Some observations on 'Schumann and Buxbaum
4 reconciled'. *J. Mammillaria Soc.*, 13(2):18–20(1973).

925 Pinkava, D. J., McLeod, M. G., McGill, L. A. *and* Brown, R. C.
Chromosome numbers in some cacti of western
3 North America—2. *Brittonia*, 25(1):2–9(1973).

926 Robinson, H.
New combinations in the Cactaceae subfamily
3, 4 Opuntioideae. *Phytologia*, 26(3):175–176(1973).

927 Sanchez-Mejorada, H.
Nuevas Cactáceas de la Nueva Galicia. *Cact. Suc.*
4 *Mex.*, 18(4):87–95(1973)—(En); illustrations, map.

928 Schill, R., Barthlott, W. *and* Ehler, N.
Cactus spines under the electron scanning microscope. *Cact. Succ. J.*, (U.S.A.), 45(4):175–185(1973)—
4 Illustrations.

929 Schreier, K.
Vom Rio Nazas zum Rio Balsas: 8. *Kakt. And. Sukk.*,
4 24(2):42–46(1973)—Illustration.

APOROCACTUS

930 Ridder, M. M. de
Aporocactus flagelliformis. Succulenta, 52(10):189–191
4 (1973)—Du; illustrations.

ARIOCARPUS

931 Skarupke, E.
Ariocarpus (*Roseocactus*) *kotschoubeyanus* var.
elephantidens v.n. Skarupke. *Stachelpost*, 9(44):33–34
4 (1973)—Ge; illustration.

932 Voldan, M.
Ariocarpus trigonus (Weber) K. Schumann
Ariocarpus elongatus (Salm-Dyck) M. H. Lee. *Kakt. And.*
4 *Sukk.*, 24(12):275–278(1973)—Ge; illustrations.

kotschoubeyanus Lem.
var. **elephantidens** Skarupke in *Stachelpost*,
4 9(44):33(1973)—Mexico.

ARROJADOA

933 Buining, A. F. H.
Arrojadoa theunisseniana Buining et Brederoo.
4 *Kakteen*, 52:CVb(1973)—Illustrations.

934 Buining, A. F. H. *and* Brederoo, A. J.
Arrojadoa aureispina Buining et Brederoo spec. nov.
Chileans, 6(23):68–69(1972)—En. Translated by
4 H. Middleditch from *Kakt. And. Succ.*, 23(4)1972.

935 Buining, A. F. H. *and* Brederoo, A. J.
Arrojadoa dinae Buining et Brederoo spec. nov.
Kakt. And. Sukk., 24(5):99–101(1973)—Ge; illustra-
4 tion, coloured illustration.

936 Buining, A. F. H. *and* Brederoo, A. J.
Arrojadoa eriocaulis Buining et Brederoo spec. nov.
Kakt. And. Sukk., 24(11):241–243(1973)—Ge;
4 illustrations, coloured illustration.

937 Buining, A. F. H. *and* Brederoo, A. J.
Arrojadoa canudosensis Buining & Brederoo sp. nov.
Cact. Succ. J. (U.S.A.), 44(3):111–113(1972)—En;
4 illustrations, coloured illustration.

938 Buxbaum, F.
The genus *Arrojadoa. Chileans*, 6(23):62–65(1972)—
Illustrations. Translated by H. Middleditch from
4 *Kakteen*, 1. 7. 1968.

939 Mace, T.
The genus *Arrojadoa. Nation. Cact. Succ. J.* (U.K.),
4 28(2):37–38(1973)—Illustrations.

canudosensis Buining & Brederoo in *Cact. Succ. J.*
4 (U.S.A.), 44(3):111(1972)—Brazil.
dinae Buining & Brederoo in *Kakt. And. Sukk.*,
4 24(5):99(1973)—Brazil.
eriocaulis Buining & Brederoo in *Kakt. And. Sukk.*,
4 24(11):241(1973)—Brazil.
theunisseniana Buining & Brederoo in *Kakteen*,
4 52:CVb(1973)—Brazil.

ARTHROCEREUS

940 Mace, T.
Arthrocereus, Pygmaeocereus and *Setiechinopsis*.
Nation. Cact. Succ. J. (U.K.), 28(4):102–103(1973)—
4 Illustrations.

ASTROPHYTUM

941 Schütz, B.
Astrophytum capricorne, niveum und *crassispinum*.
4 *Stachelpost*, 9(43):4–7(1973)—Illustrations.

942 Schütz, B.
Astrophytum senile und *aureum. Stachelpost*, 9(44):
4 41–43(1973)—Illustration.

AUSTROCEPHALOCEREUS

943 Buining, A. F. H. *and* Brederoo, A. J.
Austrocephalocereus albicephalus Buining et Brederoo
spec. nov. *Kakt. And. Sukk.*, 24(4):73–75(1973)—Ge;
4 illustrations, coloured illustration.

albicephalus Buining & Brederoo in *Kakt. And.*
4 *Sukk.*, 24(4):73(1973)—Brazil.

BACKEBERGIA

944 Allen, S. F., Prior, D. A. *and* Stubbings, H. G.
The flowers of *Backebergia militaris* (Audot)
H. Bravo. *Nation. Cact. Succ. J.* (U.K.), 28(3):66–67
4 (1973)—Illustrations.

945 Sánchez-Mejorada, H.
The correct name of the Grenadier's Cap. *Cact.*
Succ. J. (U.S.A.), 45(4):171–174(1973)—Illustrations.
4 *Backebergia militaris.*

militaris (Audot) Bravo ex Sánchez-Mejorada in
Cact. Succ. J. (U.S.A.), 45(4):174(1973), combina-
tion already made by Bravo in *Cact. Suc. Mex.*,
4 6(4):90(1961): *Cereus militaris.*

BOLIVICEREUS

946 Till, H.
Bolivicereus samaipatanus Cardenas: ein dankbarer
Blüher. *Kakt. And. Sukk.*, 24(4):80(1973)—Coloured
4 illustration.

BORZICACTUS

947 Buxbaum, F.
Gattung *Borzicactus. Kakteen*, 55–56:CVb(1973)—
3, 4 Illustrations.

subgen. **Bolivicereus** (Card.) F. Buxbaum in
4 *Kakteen*, 55–56:CVb(1973): genus *Bolivicereus.*

BUININGIA

948 Buining, A. F. H. *and* Brederoo, A. J.
Buiningia purpurea Buining et Brederoo spec. nov.
Kakt. And. Sukk., 24(6):121–123(1973)—Ge;
4 illustrations, coloured illustration.

949 Wageringel, H.
Buiningia brevicylindrica var. *elongata* eine neue
Pflanze voller Rätsel. *Kakt. And. Sukk.*, 24(4):88–89
4 (1973)—Illustration.

purpurea Buining & Brederoo in *Kakt. And. Sukk.*,
4 24(6):121(1973)—Brazil.

CARNEGIEA

950 Brum, G. D.
Ecology of the saguaro (*Carnegiea gigantea*):
phenology and establishment in marginal populations.
3 *Madroño*, 22(4):195–204(1973).

951 Mitich, L. W.
The saguaro—a history. *Cact. Succ. J.* (U.S.A.),
44(3):118–129(1972)—Illustrations, portraits, map.
4 *Carnegiea gigantea.*

CASTELLANOSIA

952 Moser, G.
Castellanosia spec. *Kakt. And. Sukk.*, 24(4):86–88
4 (1973)—Ge; illustrations.

CEREUS

953 Buining, A. F. H.
Cereus crassisepalus Buining et Brederoo spec. nova.
4 *Kakteen*, 53:CIVb(1973)—Ge; illustrations.

954 Krähenbühl, F.
Aus meinem mexikanischen Tagebuch. Am Rio Papagayo. *Kakt. And. Sukk.*, 24(7):152–154(1973)—
4 Illustrations.

955 Moser, G.
Cereus tortuosus. Kakt. And. Sukk., 25(10):224–225
4 (1973)—Ge; illustrations.

· **crassisepalus** Buining & Brederoo in *Kakteen*, 53:
4 CIVb(1973)—Brazil.

CLEISTOCACTUS

956 Klenk, P.
Cerei for the window sill: the genus *Cleistocactus*. *Chileans*, 6(22):35–36(1972). Translated by E. W.
4 Bentley from *Kakt. And. Sukk.* Dec. 1971.

COCHEMIEA

957 Feiler, H. *and* Feiler, W.
Die Gattung *Cochemiea* (K. Brandegee) Walton und ihre Kultur. *Kakt. And. Sukk.*, 24(2):30–32(1973)—
4 Illustrations.

COPIAPOA

958 Bleck, M.
Copiapoa Br. and R. *Cact. Succ. J.* (U.S.A.), 44(6):
4 239–242(1972)—En; illustrations.

959 Broeckhoven, M.
Copiapoa cinerea (Phil.) Br. & R. *Cactus* (Brussels),
4 5(2):37–38(1973)—Fr; illustration.

960 Krainz, H.
Copiapoa hypogaea Ritter. *Kakteen*, 53:CVIb(1973)—
4 Ge; coloured illustration.

961 Krainz, H.
Copiapoa pendulina Ritter. *Kakteen*, 52:CVIb
4 (1973)—Ge; coloured illustration.

962 Milkuhn, G.
Copiapoa cinerea (Phil.) Br. & R. *Copiapoa rupestris*
4 Ritt. *Stachelpost*, 9(45):86–87(1973)—Ge.

963 Milkuhn, G.
Copiapoa megarhiza Br. & R. *Stachelpost*, 9(46):
4 107–108(1973)—Ge,En.

CORYPHANTHA

964 Arp, G.
Studies in the Colorado cacti: 3. *Coryphantha*, *Escobaria* and *Neobesseya*. *Cact. Succ. J.* (U.S.A.),
3 45(1):30–31(1973)—Illustration.

965 Bremer, L.
Descripciones de *Coryphantha*: 3. *Cact. Suc. Mex.*, 18(2):54–56(1973)—(En); illustration, coloured
4 illustration. *Coryphantha elephantidens.*

966 Girard, V. *and* Farwig, S.
Two forms of *Coryphantha* (*Neobesseya*) *missouriensis* found in Arizona. *Cact. Succ. J.* (U.S.A.), 45(3):
3 125–131(1973).

967 Zimmerman, D. A.
Comments on certain south-western Coryphanthas of the subgenus *Escobaria*. *Cact. Succ. J.* (U.S.A.),
3 44(4):155–158(1972)—Illustrations.

968 Zimmerman, D. A.
The fruits of *Coryphantha hesteri* Wright. *Cact.*
3 *Succ. J.* (U.S.A.), 44(2):85(1972).

969 Zimmerman, D. A.
A new species of *Coryphantha* from New Mexico. *Cact. Succ. J.* (U.S.A.), 44(3):114–117(1972)—
3 Illustrations, coloured illustration.

albicolumnaria (Hester) D. A. Zimmerman in *Cact. Succ. J.* (U.S.A.), 44(4):157(1972), without
3 basionym page: *Escobaria albicolumnaria.*
orcuttii (Rose ex Orcutt) D. A. Zimmerman *op. cit.*
3 156(1972): *Escobaria orcuttii.*
organensis D. A. Zimmerman in *Cact. Succ. J.*
3 (U.S.A.), 44(3):114(1972)—U.S.A. (New Mexico).

CRYPTOCEREUS

970 Hewitt, T.
Cryptocereus anthonyanus Alex (Cactaceae).
4 *Ashingtonia*, 1(1):3(1973)—En; coloured illustration.

CYLINDROPUNTIA

wigginsii (L. Benson) H. Robinson in *Phytologia*,
3 26(3):175(1973): *Opuntia wigginsii.*

DISCOCACTUS

971 Buining, A. F. H.
Discocactus horstii Buining et Brederoo spec nov.
4 *Kakteen*, 52:CVIf(1973)—Ge; illustrations.

972 Buining, A. F. H.
Discocactus horstii Buining et Brederoo. *Succulenta*,
4 52(12):229–230(1973)—Du; illustration.

973 Buining, A. F. H.
One of several paradises in Minas Gerais. *Cact. Succ. J.* (U.S.A.), 45(5):240–241(1973)—Illustrations.
4 Only known habitat of *Discocactus horstii.*

974 Wageringel, H.
Discocactus horstii—ein grosses Wunder der Natur.
4 *Kakt. And. Sukk.*, 24(1):8–9(1973)—Illustrations.

horstii Buining & Brederoo in *Kakteen*, 52:CVIf
4 (1973)—Brazil.

DOLICHOTELE

975 Parijs, V.
Dolichotele longimamma Br. & R. *Cactus* (Brussels),
4 4(6):118–119(1972)—Fr.

ECHINOCACTUS

976 Boom, B. K.
The correct name of *Echinocactus cumingii*. *Chileans*, 7(25):18–20(1973). Translated by H. Middleditch
4 from *Succulenta*, no. 9(1962).

977 Daniel, J. *and* Rowland, D.
Some cacti and other succulents of the American southwest: no. 1. *Echinocactus horizonthalonius*. *Nation. Cact. Succ. J.* (U.K.), 28(3):68(1973)—
3, 4 Illustrations.

978 Hopffer, C.
A new *Echinocactus*. *Chileans*, 7(25):23(1973).
4 Translated from *Allgemeine Gartenzeitung*, 29 July 1843.

979 Middleditch, H.
Searching for *Echinocactus cumingii*. *Chileans*, 7(26):
4 102–103(1973).

980 Salm-Reifferscheid-Dyck, J.M.F.A.H.I. zu, *Prinz*
Echinocactus cummingii Salm-Dyck. *Chileans*, 7(25):
22(1973)—En. Translated by G. J. Swales from Salm-
4 Dyck: *Cacteae in horto Dyckensi cultae anno* 1849.

ECHINOCEREUS

981 Anon.
Echinocereus fitchii Br. & R. *Cactus* (Brussels),
4 5(1):8–10(1973)—Fr.

982 Arp, G.
Studies in the Colorado cacti: 1. The ecology of
two varieties of *Echinocereus triglochidiatus*. *Cact. Succ.*
3 *J.* (U.S.A.), 44(2):62–63(1972)—Illustrations.

983 Arp, G.
Studies in the Colorado cacti: 5. The spineless
hedgehog. *Cact. Succ. J.* (U.S.A.), 45(3):132–133
3 (1973). *Echinocereus triglochidiatus* var. *inermis*.

984 Bravo, H.
Echinocereus tulensis, una nueva cactácea de
Tamaulipas. *Cact. Suc. Mex.*, 18(4):110–111(1973)—
4 (En).

985 Bravo, H.
Triglochidiata, una nueva sección del genero
Echinocereus. *Cact. Suc. Mex.*, 18(4):108–109(1973)—
3, 4 (En); illustration.

986 Cheetham, G. E.
Choice Echinocerei: *Echinocereus blanckii* (Poselger)
F. Palmer. *Nation. Cact. Succ. J.* (U.K.), 28(1):12
3, 4 (1973)—Illustration.

987 Cheetham, G. E.
Choice Echinocerei: *Echinocereus gentryi* Clov.
Nation. Cact. Succ. J. (U.K.), 28(4):115(1973)—
4 Illustration.

988 Donald, J. D.
Echinocereus pentalophus (D.C.) Rumpler. *Ashing-
tonia*, 1(2):22(1973)—En; coloured illustration.
4 Correction in *Ashingtonia*, 1(3):35(1973).

989 Moran, R.
Echinocereus ferreirianus H. E. Gates. *Cact. Succ. J.*
(U.S.A.), 44(4):162–167(1972)—En; illustrations,
3 coloured illustration, dot map.

sect. **Triglochidiata** H. Bravo in *Cact. Suc. Mex.*,
3, 4 18(4):109(1973).
triglochidiatus Eng.
var. **inermis** (K. Schum.) Rowley in Rep. Pl.
Succ., 22 (Regnum Veg., 87): (8 May 1973),
without basionym page; also Arp in *Cact. Succ.
J.* (U.S.A.), 45(3):132(20 June 1973): *E.
3, 4 phoeniceus* var. *inermis*.
tulensis H. Bravo in *Cact. Suc. Mex.*, 18(4):110
4 (1973)—Mexico.

ECHINOFOSSULOCACTUS

990 Meyran, J.
Echinofossulocactus anfractuosus. *Cact. Suc. Mex.*,
4 18(4):101–108(1973)—Sp(En); illustrations, dot map.

991 Meyran, J.
Echinofossulocactus lamellosus. *Cact. Suc. Mex.*,
4 18(2):40–46(1973)—Sp(En); illustrations, map.

ECHINOMASTUS

992 Gröner, G.
Eine vermeintliche *Mammillaria spinosissima* entpuppt
sich: *Echinomastus macdowellii* (Reb) Br. u. R.
3, 4 *Stachelpost*, 9(46):112–114(1973)—Illustration.

993 Kladiwa, L.
Echinomastus erectocentrus (Coulter) Britton et Rose.
3, 4 *Kakteen*, 53:CVIIIb(1973)—Ge; illustrations.

994 Kladiwa, L.
Echinomastus mapimiensis Backeberg. *Kakteen*,
4 52:CVIIIb(1973)—Ge; illustrations.

ECHINOPSIS

995 Anon.
Echinopsis cinnabarina Lab. *Chileans*, 7(25):45
4 (1973)—En. From *Bluhende Kakteen* Oct. 1900.

EPIPHYLLUM

996 Anon.
Epiphyllum chrysocardium Alex (Cactaceae).
Ashingtonia, 1(3):26,31(1973)—En; coloured
4 illustration.

997 Horich, C. L.
Post aus Costa Rica: *Epiphyllum pittier* (Weber)
Britton & Rose. *Kakt. And. Sukk.*, 24(9):202–203
4 (1973)—Illustrations.

ERIOCACTUS

998 Bourdoux, P.
N.D.L.R.—Ne voilà-t-il pas l'occasion de vous
étendre quelque peu au sujet du genre *Eriocactus*
Backbg.? *Cactus* (Brussels), 4(6):105–108(1972)—
4 Illustration.

999 Dorsselaer, R. van
Eriocactus leninghausii Hge. *Cactus* (Brussels), 4(6):
4 105(1972)—Fr.

1000 Ritter, F.
Ein neuer '*Eriocactus*'. *Bradea*, 1(34):353–355(1973)—
4 (En,Ge); illustrations.

warasii F. Ritter in *Bradea*, 1(34):353(1973)—
4 Brazil.

ESCOBARIA

1001 Arp, G.
Studies in the Colorado cacti: 3. *Coryphantha*,
Escobaria and *Neobesseya*. *Cact. Succ. J.* (U.S.A.),
3 45(1):30–31(1973)—Illustration.

FEROCACTUS

1002 Innes, C. F.
Ferocactus acanthodes (Lem.) Br. & R. *Ashingtonia*,
3, 4 1(1):12(1973)—En; coloured illustration.

FRAILEA

1003 Bleck, M.
Frailea Br. & R. *Cact. Succ. J.* (U.S.A.), 44(1):
4 36–37(1972)—En; illustrations.

1004 Buining, A. F. H.
Frailea albiareolata Buining et Brederoo spec. nova.
4 *Kakteen*, 53(CVIe(1973)—Ge; illustrations.

1005 Buining, A. F. H.
Frailea deminuta Buining et Brederoo spec. nov.
4 *Kakteen*, 54:CVIe(1973)—Ge; illustrations.

1006 Buining, A. F. H.
Frailea lepida Buining et Brederoo spec. nov.
4 *Kakteen*, 54:CVIe(1973)—Ge; illustrations.

1007 Buining, A. F. H. *and* Brederoo, A. J.
Frailea fulviseta Buining et Brederoo spec. nov.
Kakt. And. Sukk., 24(8):170–172(1973)—Ge; illustrations,
4 coloured illustration.

albiareolata Buining & Brederoo in *Kakteen*, 53:
4 CVIe(1973)—Uruguay.
deminuta Buining & Brederoo *op. cit.* CVIe
4 (1973)—Brazil.
fulviseta Buining & Brederoo in *Kakt. And. Sukk.*,
4 24(8):170(1973)—Brazil.
lepida Buining & Brederoo in *Kakteen*, 54:CVIe
4 (1973)—Brazil.

GRUSONIA

bulbispina (Engelmann) H. Robinson in *Phytologia*,
4 26(3):176(1973): *Opuntia bulbispina.*
clavata (Engelmann) H. Robinson *op. cit.* 176(1973):
3 *Opuntia clavata.*
grahamii (Engelmann) H. Robinson *op. cit.* 176
3, 4 (1973): *Opuntia grahamii.*
pulchella (Engelmann) H. Robinson *op. cit.* 176
3 (1973): *Opuntia pulchella.*
schottii (Engelmann) H. Robinson *op. cit.* 176
3, 4 (1973): *Opuntia schottii.*
stanlyi (Engelmann) H. Robinson *op. cit.* 176(1973):
3, 4 *Opuntia stanlyi.*
vilis (Rose) H. Robinson *op. cit.* 176(1973): *Opuntia*
4 *vilis.*

GYMNOCACTUS

1008 Glass, C. *and* Foster, R.
Gymnocactus aguirreanus: a new species from southern Coahuila, Mexico. *Cact. Succ. J.* (U.S.A.),
4 44(2):80–81(1972)—Illustrations.

aguirreanus Glass & R. A. Foster in *Cact. Succ. J.*
4 (U.S.A.), 44(2):80(1972)—Mexico.

GYMNOCALYCIUM

1009 Aerschot, F. van
Gymnocalycium nidulans Bckbg. *Succulenta*, 52(2):
4 37(1973)—Du; illustration.

1010 Backeberg, C.
Gymnocalycium oenanthemum Backbg. *Chileans*, 7(26):69(1973)—En. Abstracted from *Blätter für*
4 *Kakteenforschung*, 9(1934).

1011 Broekhoven, M.
Gymnocalycium marsoneri. Chileans, 6(23):84–85 (1972)—En. Translated from *Cactus* (Brussels),
4 3(8)1971 by H. Middleditch.

1012 Frank, G.
A beautiful interesting new discovery from Argentina: *Gymnocalycium horridispinum* sp. nov. *Chileans*, 6(22):7–9(1972)—Illustrations. Translated
4 by E. W. Bentley from *Kakt. And. Sukk.* 14(1)1963.

1013 Frank, G.
Gymnocalycium castellanosii Backeberg. *Kakteen*,
4 52:CVIf(1973)—Ge; illustration.

1014 Frank, G.
Gymnocalycium hyptiacanthum (Lemaire) Britton et
4 Rose. *Kakteen*, 55–56:CVIf(1973)—Ge; illustrations.

1015 Jeggle, W.
Gymnocalycium striglianum Jeggle spec. nov. *Kakt.*
4 *And. Sukk.*, 24(12):267–268(1973)—Ge; illustrations.

1016 Schutz, B.
Gymnocalycium—of the group Microsemineae Fric. *Chileans*, 7(26):74–76(1973). Translated by K.
4 Wood Allun from *Friciana*, no. 16(1963).

1017 Simon, W.
Der Formenkreis des *Gymnocalycium lafaldense* Vaupel. *Kakt. And. Sukk.*, 24(8):184–186(1973)—
4 Illustrations.

1018 Till, H.
De verwantschapsgroep van *Gymnocalycium monvillei* (Lem.) Br. et R. beschouwd van vanuit de praktijk. *Succulenta*, 52(9):173–175(1973)—
4 Illustrations.

1019 Till, H. *and* Schatzl, S.
De verwantschapsgroep van *Gymnocalycium monvillei* (Lem.) Br. et R. beschouwd vanuit de praktijk. *Succulenta*, 52(10):193–195(1973)—
4 Illustrations.

1020 Vliet, D. J. van
Gymnocalycium tillianum Rausch. *Chileans*, 7(26): 67–69(1973)—En. Translated from *Succulenta*, 50(3)
4 1971 by H. Middleditch.

striglianum Jeggle in *Kakt. And. Sukk.*, 24(12):267
4 (1973)—Argentina.

HAAGEOCEREUS

1021 Buxbaum, F.
Gattung *Haageocereus* (Schluss). *Kakteen*, 54:CVa
4 (1973).

1022 Cullmann, W.
Haageocereus Backeb. *Stachelpost*, 9(45):61(1973)—
4 Ge; coloured illustration.

1023 Krainz, H.
Gattung *Haageocereus. Kakteen*, 53:CVa(1973)—
4 Ge; illustrations.

1024 [Middleditch, H.]
The genus *Haageocereus. Chileans*, 6(24):124–128
4 (1972).

HAMATOCACTUS

1025 Gelder, B. van
De oudjes doen het nog best. *Succulenta*, 52(8):
3, 4 148–153(1973)—Illustrations.

HELIOCEREUS

1026 Horich, C. K.
Post aus Costa Rica: *Heliocereus heterodoxus* Standley et Steyermark. *Kakt. And. Sukk.*, 24(12):
4 273–274(1973)—Illustrations.

1027 Meyran, J.
Heliocereus amecamensis en Michoacán. *Cact. Suc.*
4 *Mex.*, 18(3):67–69(1973)—Ge(En); illustrations.

HOMALOCEPHALA

1028 Martin, M. J.
Cactus album: no. 10. *Homalocephala texensis. Nation. Cact. Succ. J.* (U.K.), 28(2):38(1973)—
3, 4 Illustration.

ISLAYA

1029 Buxbaum, F.
Gattung *Islaya*. *Kakteen*, 53:CVIb(1973)—
4 Illustrations, map.

1030 Milkuhn, G.
Islaya bicolor Akers & Buin. *Stachelpost*,
4 9(46):97–99(1973)—Ge,En.

LEUCHTENBERGIA

1031 Wery, H.
Leuchtenbergia principis Hooker und ihre Kultur.
4 *Kakt. And. Sukk.*, 24(2):37–38(1973)—Illustrations.

LOBIVIA

1032 Braun, B.
Lobivia fungiflora spec. nov. nom. prov. *Stachelpost*,
4 9(46):101–102(1973)—Ge, En.

1033 Braun, B.
Lobivia zecheri v. *fungiflora* var. nov. Braun.
4 *Stachelpost*, 9(47):138–139(1973)—Ge.

1034 Cardenas, M.
Lobivia oligotricha Card. sp. nov. *Chileans*, 7(25):
48–50(1973)—En. Translated by H. Middleditch
4 from *Cactus* (France), 18(78)1963.

1035 Donald, J. D.
More thoughts on *Lobivia* flowers. *Chileans*, 7(26):
4 110–111(1973).

1036 Hopkins, J.
Bolivian Lobivias: the *cinnabarina* group. *Chileans*,
4 7(25):38–43(1973)—Illustrations.

1037 Hopkins, J.
A further look at *Lobivia* seed. *Chileans*, 6(22):55–56
4 (1972).

1038 Hopkins, J.
The genus *Lobivia*—a seed study. *Chileans*, 6(23):
4 76–84(1972)—Illustrations.

1039 Hopkins, J.
A survey of *Lobivia* flowers. *Chileans*, 6(24):157–
4 159(1972)—Illustrations.

1040 Magnin, J. C. A.
Lobivia pentlandii. *Succulenta*, 52(11):205–206(1973)
4 —Du; illustration.

1041 Rausch, W.
Lobivia akersii Rausch spec. nov. *Kakt. And. Sukk.*,
4 24(2):25(1973)—Ge; illustration.

1042 Rausch, W.
Lobivia cinnarbarina var. *grandiflora* Rausch sp. nov.
Chileans, 7(25):55(1973)—En. Translated by H.
4 Middleditch from *Kakt. And. Sukk.*, 23(11)1972.

1043 Rausch, W.
Lobivia draxleriana sp. nov. *Chileans*, 7(25):52–53
(1973)—En. Translated by H. Middleditch from
4 *Succulenta*, 50(10)1971.

1044 Rausch, W.
Lobivia echinata Rausch spec. nov. *Kakt. And. Sukk.*,
4 24(8):169–170(1973)—Ge; coloured illustration.

1045 Rausch, W.
Lobivia fricii Rausch spec. nov. *Kakt. And. Sukk.*,
24(10):220–222(1973)—Ge; illustration, coloured
4 illustration.

1046 Steif, H.
Neu und interessant! *Lobivia glauca* Rausch var.
paucicostata Rausch. *Kakt. And. Sukk.*, 24(8):187
4 (1973)—Illustration.

1047 Wilhelm, K.
Mein Hobby—Lobivien. *Stachelpost*, 9(44):37
4 (1973)—Illustration. *Lobivia vatteri* var. *robusta*.

aguilarii Vasquez in *Kakt. And. Sukk.*, 25(1):1
4 (1974), as 'aguilari'—Bolivia.
akersii Rausch in *Kakt. And. Sukk.*, 24(2):25(1973)—
4 Peru.
echinata Rausch in *Kakt. And. Sukk.*, 24(8):169
4 (1973)—Peru.
fricii Rausch in *Kakt. And. Sukk.*, 24(10):220(1973)—
4 Argentina.
fungiflora B. Braun in *Stachelpost*, 9(46):101(1973),
4 without latin descr. or type—S. America.
zecheri Rausch
var. **fungiflora** B. Braun in *Stachelpost*, 9(47):
4 138(1973)—Peru.

MACHAEROCEREUS

1048 Martin, V. F.
Los dos *Machaerocereus* de Baja California. *Cact.*
4 *Suc. Mex.*, 18(3):73–77(1973)—(En); illustrations.

MAIHUENIA

1049 Bleck, M.
Pereskia and *Maihuenia*. *Cact. Succ. J.* (U.S.A.),
4 45(5):214–215(1973)—En; illustrations.

MAMMILLARIA

1050 Bostoen, M.
Mammillaria guelzowiana—Wedermann. *Cactus*
4 (Brussels), 5(1):18–20(1973)—Fr; illustration.

1051 Cocker, W. de
Mammillaria saboae—Glass. *Cactus* (Brussels),
3, 4 5(3):41–45(1973)—Fr; coloured illustration.

1052 Cocker, W. de
Mammillaria shurliana Gates. *Cactus* (Brussels),
4(2):21–25(1972)—Fr; illustration, coloured
4 illustration.

1053 Feiler, H. *and* Feiler, W.
Mammillaria mundtii K. Schumann. *Kakt. And Sukk.*,
4 24(9):204(1973)—Ge; illustration.

1054 Fiedler, M.
Mammillaria miegiana Earle spec. nov. *Saguaroland*
Bull., 26(7):77(1972). Also publ. in *Kakt. And. Sukk.*,
4 24(9):212(1973)—Ge; illustration.

1055 Fittkau, H.
Mammillaria spinosissima var. *flavida* SD. *Cact.*
4 *Succ. J.* (U.S.A.), 44(2):71–72(1972)—En.

1056 Glass, C. *and* Foster, R.
Mammillaria melaleuca. *Cact. Succ. J.* (U.S.A.),
4 44(2):86(1972)—En; illustration.

1057 Gröner, G.
Eine vermeintliche *Mammillaria spinosissima*
entpuppt sich: *Echinomastus macdowellii* (Reb) Br. u.
3, 4 R. *Stachelpost*, 9(46):112–114(1973)—Illustration.

1058 Hunt, D. R.
Review of *Mammillaria* names in current usage:
parts 29–34. *J. Mammillaria Soc.*, 13(1):10–14,
(2):23–27, (3):35–39, (4):51–56, (5):60–67, (6):76–81
3, 4 (1973).

1059 Köhler, U.
So blühte *Mammillaria herrerae* Werdermann. *Kakt.*
4 *And. Sukk.*, 24(7):165(1973)—Illustration.

1060 Krähenbühl, F.
Aus meinem mexikanischen Tagebuch. Am Rio
Papagayo. *Kakt. And. Sukk.*, 24(7):152–154(1973)—
4 Illustrations.

1061 Krähenbühl, F.
Mammillaria kraehenbuehlii (Krainz) Krainz. *Kakt.*
And. Sukk., 24(6):124–125(1973)—Ge; coloured
4 illustrations.

1062 Krähenbühl, F. *and* Krainz, H.
Mammillaria mathildae Kraehenbuehl et Krainz
spec. nov. *Kakt. And. Sukk.*, 24(12):265–266(1973)—
4 Ge; coloured illustration.

1063 Krainz, H.
Mammillaria armillata Kath. Brandegee. *Kakteen*,
4 55–56:CVIIIc(1973)—Ge; coloured illustration.

1064 Krainz, H.
Mammillaria mathildae Kraehenbuehl et Krainz sp.
4 n. *Kakteen*, 55–56:CVIIIc(1973)—Ge; illustrations.

1065 Krainz, H.
Mammillaria schiedeana Ehrenberg. *Kakteen*, 52:
CVIIIc(1973)—Ge; illustrations, coloured
4 illustration.

1066 Maddams, W. F.
Interesting newer *Mammillarias*. Worcester Park,
Surrey, Mammillaria Society, 1973. ii,39p.—
4 Illustrations.

1067 Meininger, A.
Mammillaria sonorensis Craig. *Kakt. And. Sukk.*,
4 24(10):223(1973)—Ge; illustration.

1068 Mullard, J. W. P.
Mammillarias of merit: 21. *Mammillaria dodsonii*
Bravo-Hollis. *Nation. Cact. Succ. J.* (U.K.), 28(3):
4 79(1973)—Illustration.

1069 New South Wales Mammillaria Study Group
Mammillaria camptotricha and *Mammillaria decipiens.*
4 *J. Mammillaria Soc.*, 13(3):30–32(1973).

1070 Patzen, G.
Mammillaria pseudoperbella Qhehl. *Stachelpost*,
4 9(43):1–2(1973)—Ge; coloured illustration.

1071 Prajer, Z.
Przypadek grzebieniastej formy kwiatu u
Mammillaria zeilmanniana Böd. f. *cristata*. *Wiad. Bot.*,
4 17(3):173–178(1973)—Illustrations.

mathildae Kraehenbuehl & Krainz in *Kakt. And.*
4 *Sukk.*, 24(12):265(1973)—Mexico.

MAMMILLOYDIA

1072 Krainz, H.
Mammilloydia candida (Scheidweiler) F. Buxbaum
ssp. *ortiz-rubiona* (H. Bravo) Krainz comb. nov.
4 *Kakteen*, 55–56:CIVf(1973)—Ge; illustrations.

candida (Scheidweiler) F. Buxbaum
subsp. **ortiz-rubiona** (H. Bravo) Krainz in
Kakteen, 55–56:CIVf(1973): *Neomammillaria*
4 *ortiz-rubiona.*
f. **rosea** (Salm-Dyck) F. Buxbaum *op. cit.* CIVf
4 (1973): *Mammillaria sphaerotricha* var. *rosea.*

MATUCANA

1073 Buxbaum, F.
Gattung *Matucana*. *Kakteen*, 54:CVb(1973)—
4 Illustrations, coloured illustration.

1074 Donald, J. D.
Variability in the subgenus *Matucana*: some pink
flowered forms. *Ashingtonia*, 1(2):16–18(1973)—
4 Coloured illustrations, map.

1075 Krainz, H.
Matucana paucicostata Ritter. *Kakteen*, 54:CVb
4 (1973)—Ge; coloured illustration.

1076 Rauh, W.
Matucana breviflora Rauh & Backbg. spec. nov.
Chileans, 7(26):89–90(1973)—En. Translated by
H. Middleditch from 'Contribution to the knowledge
4 of the Peruvian cactus flora'.

1077 Rauh, W.
Matucana hystrix Rauh & Backbg. sp. nov. *Chileans*,
7(26):90–91(1973)–; illustration. Translated by H.
Middleditch from 'Contribution to the knowledge
4 of the Peruvian cactus flora'.

1078 Skarupke, E.
Matucana calocephala Skarupke spec. n. *Stachelpost*,
4 9(46):99–101(1973)—Ge,En; illustration.

1079 Skarupke, E.
Zu den Farbbildern (S.44 u 45). *Stachelpost*, 9(44):
44–45,50–51(1973)—Coloured illustrations.
4 *Matucana breviflora, Matucana weberbaueri.*

aurantiaca (Vaupel) F. Buxbaum in *Kakteen*, 54:
4 CVb(1973): *Echinocactus aurantiacus.*
calocephala Skarupke in *Stachelpost*, 9(46):99
4 (1973)—Peru.
calvescens (Kimnach & Hutchison) F. Buxbaum in
4 *Kakteen*, 54:CVb(1973): *Borzicactus calvescens.*
humboldtii (H.B.K.) F. Buxbaum *op. cit.* CVb
4 (1973): *Cactus humboldtii.*
icosagona (H.B.K.) F. Buxbaum *op. cit.* CVb
4 (1973): *Cactus icosagonus.*
madisoniorum (Hutch.) Rowley in Rep. Pl. Succ.,
22 (Regnum Veg., 87): 10(1973), without
4 basionym page: *Borzicactus madisoniorum.*
var. pujupatii (Don. & Lau) Rowley *op. cit.* 10
(1973), without basionym page: *Borzicactus*
4 *madisoniorum* var. *pujupatii.*
myriacantha (Vaupel) F. Buxbaum in *Kakteen*, **54:**
4 CVb(1973): *Echinocactus myriacanthus.*

MEDIOLOBIVIA

1080 Köhler, U.
Mediolobivia aureiflora var. *rubriflora* forma
kesselringiana (Cullmann) Köhler. *Kakt. And. Sukk.*,
4 24(9):206–207(1973)—Ge; illustrations.

MELOCACTUS

1081 Buining, A. F. H.
Melocactus albicephalus Buining et Brederoo spec.
4 nova. *Kakteen*, 52:CVId(1973)—Ge; illustrations.

1082 Buining, A. F. H.
Melocactus conoideus Buining et Brederoo spec. nov.
4 *Kakteen*, 55–56:CVId(1973)—Ge; illustrations.

1083 Buining, A. F. H.
Melocactus ferreophilus Buining et Brederoo spec.
4 nova. *Kakteen*, 52:CVId(1973)—Ge; illustrations.

1084 Buining, A. F. H.
Melocactus oreas Miquel. *Kakteen*, 55–56:CVId
4 (1973)—Ge; illustrations.

1085 Buining, A. F. H.
Melocactus salvadorensis Werdermann. *Kakteen*,
4 54:CVId(1973)—Ge; illustrations.

1086 Buining, A. F. H. *and* Brederoo, A. J.
Melocactus cremnophilus Buining et Brederoo species nova. *Cact. Succ. J.* (U.S.A.), 44(1):3–5(1972)—
4 En; illustrations, coloured illustration.

1087 Buining, A. F. H. *and* Brederoo, A. J.
Melocactus deinacanthus Buining et Brederoo spec. nov. *Kakt. And. Sukk.*, 24(10):217–219(1973)—Ge;
4 coloured illustration.

1088 Buining, A. F. H. *and* Brederoo, A. J.
Melocactus erythracanthus Buining et Brederoo spec. nova. *Cact. Succ. J.* (U.S.A.), 45(5):223–226(1973)—
4 En; illustrations, coloured illustration.

1089 Buining, A. F. H. *and* Brederoo, A. J.
Melocactus giganteus: Buining et Brederoo spec. nova. *Cact. Succ. J.* (U.S.A.), 45(5):227–230(1973)—
4 En; illustrations, coloured illustration.

1090 Buining, A. F. H. *and* Brederoo, A. J.
Melocactus glaucescens Buining et Brederoo spec. nova. *Cact. Succ. J.* (U.S.A.), 44(4):159–161(1972)—
4 En; illustrations, coloured illustrations.

1091 Buining, A. F. H. *and* Brederoo, A. J.
Melocactus levitestatus Buining et Brederoo spec. nova. *Cact. Succ. J.* (U.S.A.), 45(6):271–274(1973)—
4 En; illustrations, coloured illustration.

1092 Buining, A. F. H. *and* Eerkens, G.
Welke *Melocactus* species groeit in Suriname? *Succulenta*, 52(3):53–57, (5):93–95, (6):116–119, (7):133–136, (8):155–157, (9):169–172(1973)—
4 Illustrations, map, portrait of A. F. H. Buining.

1093 Riha, J.
Cuba and its *Melocactus*. *Cact. Succ. J.* (U.S.A.),
4 45(6):281–285(1973)—Illustrations.

albicephalus Buining & Brederoo in *Kakteen*, 52:
4 CVId(1973)—Brazil.
conoideus Buining & Brederoo in *Kakteen*, 55–56:
4 CVId(1973)—Brazil.
cremnophilus Buining & Brederoo in *Cact. Succ. J.*
4 (U.S.A.), 44(1):3(1972)—Brazil.
deinacanthus Buining & Brederoo in *Kakt. And.*
4 *Sukk.*, 24(10):217(1973)—Brazil.
erythracanthus Buining & Brederoo in *Cact. Succ. J.*
4 (U.S.A.), 45(5):223(1973)—Brazil.
ferreophilus Buining & Brederoo in *Kakteen*, 52:
4 CVId(1973)—Brazil.
giganteus Buining & Brederoo in *Cact. Succ. J.*
4 (U.S.A.), 45(5):227(1973)—Brazil.
glaucescens Buining & Brederoo in *Cact. Succ. J.*
4 (U.S.A.), 44(4):159(1972)—Brazil.
levitestatus Buining & Brederoo in *Cact. Succ. J.*
4 (U.S.A.), 45(6):271(1973)—Brazil.

MICRANTHOCEREUS

1094 Backeberg, C.
Micranthocereus polyanthus Werd. Backbg. *Chileans*, 6(22):15(1972)—En. Translated by
4 H. Middleditch from 'Die Cactaceae' vol. 3.

1095 Buining, A. F. H.
Micranthocereus violaciflorus Buining spec. nov. *Chileans*, 6(22):15–16(1972)—En. Translated by
4 H. Middleditch from *Kakt. And. Sukk.*, July 1969.

1096 Buining, A. F. H. *and* Brederoo, A. J.
Micranthocereus auri-azureus Buining & Brederoo sp. nov. *Cact. Succ. J.* (U.S.A.), 45(3):120–123
4 (1973)—En; illustrations, coloured illustrations.

1097 Buxbaum, F.
The genus *Micranthocereus*. *Chileans*, 6(22):17–19 (1972). En. Translated by R. Moreton from Krainz,
4 *Kakteen*, 1. 7.1968.

1098 Mace, T.
The genus *Micranthocereus*. *Nation. Cact. Succ. J.*
4 (U.K.), 28(3):87–88(1973)—Illustrations.

auri-azureus Buining & Brederoo in *Cact. Succ.*
4 *J.* (U.S.A.), 45(3):120(1973)—Brazil.

NEOBESSAYA

1099 Arp, G.
Studies in the Colorado cacti: 3. *Coryphantha*, *Escobaria* and *Neobesseya*. *Cact. Succ. J.* (U.S.A.),
3 45(1):30–31(1973)—Illustration.

1100 Kladiwa, L.
Neobesseya cubensis (Britton et Rose) Hester.
3, 4 *Kakteen*, 53:CVIIIc(1973)—Ge; illustrations.

1101 Riha, J.
Neobesseya cubensis (Britton et Rose) Hester. *Kakt. And. Sukk.*, 24(6):134–136(1973)—Ge; illustration,
3, 4 map.

NEOBUXBAUMIA

1102 Bravo H, H., Scheinvar, L. *and* Sánchez-Mejorada, H.
Estudio comparativo del genero *Neobuxbaumia*: 4. *Neobuxbaumia multiareolata*. *Cact. Suc. Mex.*, 18(3): 59–67(1973)—(En); illustrations, coloured
4 illustration.

NEOCHILENIA

1103 Knize, K.
Neochilenia-Arten der chilenischen Provinz Atacama. *Kakt. And. Sukk.*, 24(3):51–56(1973)—
4 Illustrations.

1104 Strigl, F.
Eine schöne und seltene form der *Neochilenia occulta*. *Kakt. And. Sukk.*, 24(3):57(1973)—
4 Illustration.

1105 Dorsselaer, R. van
Neochilenia aricensis Ritt. *Cactus* (Brussels), 5(3):
4 52–53(1973)—Fr.

NEOLLOYDIA

1106 Riha, J.
Neolloydia cubensis (Britton & Rose) Backbg. *Cact. Succ. J.* (U.S.A.), 44(1):32–33(1972)—En;
4 illustrations.

NEOPORTERIA

1107 Jahn, E.
Neoporteria subgibbosa (Haworth) Britton & Rose: eine Leicht zu pflegende Art. *Kakt. And. Sukk.*,
4 24(3):60(1973)—Illustration.

CACTACEAE

1108 Ritter, F.
Neoporteria wagenknechtii Ritter spec. nov. *Chileans*, 6(22):50–51(1972)—En. Translated by W. W.
4 Atkinson from *Succulenta*, Jan. 1973.

1109 Schumann, K.
Neoporteria clavata. *Chileans*, 6(24):174(1972)—En. Translated by H. Middleditch from 'Gesamtbeschreibung der Kakteen'.
4 chreibung der Kakteen'.

NEOWERDERMANNIA

1110 Donald, J. D.
Neowerdermannia chilensis Backeb. *Ashingtonia*,
4 1(1):10(1973)—En; coloured illustration.

NOPALXOCHIA

1111 Rowley, G. D.
Nopalxochia macdougallii. *Ashingtonia*, 1(1):11
4 (1973)—En; illustration.

NOTOCACTUS

1112 Buining, A. F. H.
Notocactus alacriportanus en aanverwante soorten. *Succulenta*, 52(11):204–205(1973)—Coloured
4 illustration.

1113 Buining, A. F. H.
Notocactus linkii (Lehmann) Herter. *Succulenta*, 52(9):161,167–169(1973)—Du; coloured
4 illustration.

1114 Buining, A. F. H.
Notocactus multicostatus Buining et Brederoo sp. nov.
4 *Kakteen*, 55–56:CVIc(1973)—Ge; illustrations.

1115 Buining, A. F. H. *and* Brederoo, A. J.
Notocactus carambeiensis Buining et Brederoo. *Kakt. And. Sukk.*, 24(1):1–3(1973)—Ge; coloured
4 illustration.

1116 Haas, R.
Notocactus werdermannianus Herter noch immer eine Kostbarkeit. *Kakt. And. Sukk.*, 24(1):12(1973)—
4 Illustration.

1117 Krainz, H.
Notocactus erinaceus (Haw.) var. *tephracanthus* (Link et Otto) Krainz comb. nov. *Kakteen*, 53:CVIc
4 (1973)—Ge; coloured illustration.

1118 Krainz, H.
Notocactus erinaceus var. *tephracanthus* (Schluss).
4 *Kakteen*, 54:CVIc(1973)—Illustration.

1119 Lewis, D. J.
Notocactus seeds in close-up. *Chileans*, 6(22):42–43
4 (1972).

1120 Mace, A. W.
Notocacti: report of a talk by Dr. A. W. Mace at the meeting in the R.H.S. Hall on 21st February 1973. *Cact. Suc. J. Gr. Brit.*, 35(2):35–37(1973)—
4 Illustrations.

1121 Schütz, B.
Notocactus ottonis var. *vencluianus* Schütz der rotblühende *Notocactus ottonis*. *Kakt. And. Sukk.*,
4 24(1):4–5(1973)—Coloured illustration.

1122 Schütz, B.
Zur Gattung *Notocactus*. *Kakt. And. Sukk.*, 24(1):
4 6–8(1973)—Illustrations.

1123 Simon, W.
Notocactus für Anfänger: 4. *Stachelpost*, 9(46):104,
4 106–107(1973)—Illustrations.

1124 Vliet, D. J. van
Notocactus en *Wigginsia* één geslacht?:3–5. *Succulenta*, 52(1):2–7, (3):49–52, (4):66–70(1973)—
4 Illustrations.

1125 Vliet, D. J. van
Notocactus roseoluteus species nova. *Succulenta*, 52(6):101,108–113(1973)—Du; illustrations, coloured
4 illustration.

carambeiensis Buining & Brederoo in *Kakt. And.*
4 *Sukk.*, 24(1):1(1973)—Brazil.
erinaceus (Haw.) Krainz
var. **tephracanthus** (Link & Otto) Krainz in *Kakteen*, 53:CVIc(1973): *Echinocactus tephra-*
4 *canthus*.
multicostatus Buining & Brederoo in *Kakteen*,
4 55–56:CVIc(1973)—Brazil.
roseoluteus van Vliet in *Succulenta*, 52(6):108
4 (1973)—Uruguay.
schlosseri van Vliet in *Succulenta*, 53(1):12(1974)—
4 Uruguay.

OBREGONIA

1126 Broeckhoven, M.
Obregonia denegrii. *Cactus* (Brussels), 4(2):29–30
4 (1972)—Fr; illustration.

1127 Hönig, D.
Obregonia denegrii Fric. *Kakt. And. Sukk.*, 24(6):
4 136–137(1973)—Ge; coloured illustration.

OPUNTIA

1128 Anon.
Cactus album: no. 9. *Opuntia cylindrica*. *Nation.*
4 *Cact. Succ. J.* (U.K.), 28(1):7(1973)—Illustration.

1129 Arp, G.
The Galapagos Opuntias: another interpretation.
4 *Notic. Galapagos*, no. 21:33–37(1973).

1130 Arp, G.
Studies in the Colorado cacti: 4. *Opuntia compressa* and *Oputnia macrorhiza*. *Cact. Succ. J.* (U.S.A.),
3 45(2):56–57(1973).

1131 Arp, G.
Studies in the Colorado cacti: 6. *Cact. Succ. J.*
3 (U.S.A.), 45(5):219–221(1973)—Illustrations.

1132 Conde, L. F.
Anatomical comparisons of some cylindro- and platyopuntias. *Diss. Abstr. Int.*, *B*, 33(6):2503
3, 4 (1972).

1133 Kalmbacher, G.
The mystery of the red 'berry' of *Opuntia caribaea*. *Cact. Succ. J.* (U.S.A.), 45(6):257–258(1973)—
4 Illustrations.

1134 Kleiner, E.
Opuntia rafinesquei Eng. *Stachelpost*, 9(45):72
3, 4 (1973)—Ge; illustration.

1135 Leighton-Boyce, G.
The Airampoae: a background note. *Cact. Succ. J.*
4 *Gr. Brit.*, 35(1):6–8(1973)—Coloured illustrations.

1136 Leighton-Boyce, G. *and* Iliff, J.
The sub-genus *Tephrocactus*: a historical survey with notes on cultivation. Morden, Succulent Plant
1–7 Trust, 1973. 106p.—Illustrations.

1137 McLeod, M. G.
Biosystematics of juicy-fruited prickly-pears of Arizona (Cactaceae: *Optuntia*, series Opuntiae). *Diss. Abstr. Int., B*, 33(10):4693(1973). 3

ser. Nopaleae (S.D.) Rowley in Rep. Pl. Succ., 10(1973), without basionym page: genus *Nopalea*. 4
floccosa S. D.
var. cardenasii (Marn.-Lap.) Rowley *op. cit.* 22 (Regnum Veg., 87): 11(1973): *Tephrocactus floccosus* var. *cardenasii*, without type. 4
nuda (Back.) Rowley *op. cit.* 11(1973), without basionym page: *Nopalea nuda*. 4

OREOCEREUS

1138 Buxbaum, F.
Gattung *Oreocereus*. *Kakteen*, 55–56:CVc(1973)—Illustrations. 4

rettigii (Quehl) F. Buxbaum in *Kakteen*, 55–56: CVc(1973): *Echinocactus rettigii*. 4

OROYA

1139 Rauh, W.
Oroya laxiareolata Rauh & Backeberg sp. nov. *Oroya subocculta* Rauh & Backeberg spec. nov. *Chileans*, 6(24):148–151(1972)—En; illustrations. Translated by H. Middleditch from 'A contribution to the knowledge of Peruvian cactus vegetation'. 4

PARODIA

1140 Bleck, M.
Parodia Spegazzini. *Cact. Succ. J.* (U.S.A.), 44(3): 106–107(1972)—En; illustrations. 4

1141 Brandt, F. H.
The genus *Parodia* Spegazzini: parts 1–2. *Chileans*, 6(23):105–108, (24):167–168(1972)—Illustrations. Translated by R. Moreton. 4

1142 Brandt, F. H.
Interessante und neue Parodien. *Stachelpost*, 9(45): 81–86(1973)—Illustrations. 4

1143 Brandt, F. H.
Parodia friciana Brandt sp. nov. *Stachelpost*, 9(45): 68–71(1973)—Ge(En); illustrations, coloured illustration. 4

1144 Brandt, F. H.
Parodia lauii Brandt spec. nov. *Kakt. And. Sukk.*, 24(11):244–245(1973)—Ge; illustrations, coloured illustration. 4

1145 Brandt, F. H.
Parodia neglecta Brandt spec. nov. *Kakt. And. Sukk.*, 24(3):49–51(1973)—Ge; illustrations, coloured illustration. 4

1146 Brandt, F. H.
Parodia neglectoides Brandt sp. nova. *Stachelpost*, 9(47):129–134(1973)—Ge; illustration. 4

1147 Brandt, F. H.
Parodia rostrum-sperma Brandt sp. nova. *Stachelpost*, 9(43):2–4(1973)—Ge; illustrations. 4

1148 Brandt, F. H.
Parodia salmonea Brandt spec. nov. *Kakt. And. Sukk.*, 24(5):97–98(1973)—Ge; coloured illustration. 4

1149 Brandt, F. H.
Parodia salmonea var. *carminata* Brandt var. nov. *Kakt. And. Sukk.*, 24(5):97,99(1973)—Ge; coloured illustration. 4

1150 Brandt, F. H.
Parodia zaletaewana Brandt sp. nova. *Stachelpost*, 9(44):30–33(1973)—Ge; illustrations. 4

1151 Buxbaum, F.
The genus *Parodia*. *Chileans*, 6(23):109–112(1972)—Key. Translated by P. Waterman from *Kakteen*, 1. 9. 1966. 4

1152 Ritter, F.
Parodia comosa Ritter spec. nov. *Chileans*, 6(24): 163–164(1972)—En. Translated by H. Middleditch from *Cactus* (France), 17(75)1962. 4

1153 Rowley, G. D.
Parodia maxima Ritter. *Ashingtonia*, 1(2):24(1973)—En; coloured illustration. 4

1154 Skarupke, E.
Parodia maassii (Heese) Berg. *Stachelpost*, 9(44): 29–30(1973)—Ge; coloured illustration. 4

1155 Weskamp, W.
Die Gattung *Parodia* Spegazzini: 1–4. *Kakt. And. Sukk.*, 24(5):102–105, (6):125–127, (7):149–151, (8):173–175(1973)—Illustrations. 4

friciana F. H. Brandt in *Stachelpost*, 9(45):68(1973)—Argentina. 4
gibbulosoides Brandt in *Stachelpost*, 7:414–416 (1971); cf. Rowley in Rep. Pl. Succ., 22 (Regnum Veg., 87):11(1973)—Locality not stated. 4
laui F. H. Brandt in *Kakt. And. Sukk.*, 24(11):244 (1973), as 'lauii'—Bolivia. 4
microsperma (Weber) Spegazzini
var. **erythrantha** (Spegazzini) Weskamp in *Kakt. And. Sukk.*, 24(6):127(1973): *P. erythrantha*. 4
var. **rubriflora** (Backeberg) Weskamp *op. cit.* 127(1973): *P. aureispina* var. *rubriflora*. 4
neglecta F. H. Brandt in *Kakt. And. Sukk.*, 24(3): 49(1973)—Bolivia. 4
pluricentralis Back. ex Brandt in *Stachelpost*, 7: 365–367(1971); cf. Rowley in Rep. Pl. Succ., 22 (Regnum Veg., 87):11(1973)—Locality not stated. 4
rostrum-sperma F. H. Brandt in *Stachelpost*, 9(43): 2(1973)—Bolivia. 4
salmonea F. H. Brandt in *Kakt. And. Sukk.*, 24(5): 97(1973)—Bolivia. 4
var. **carminata** F. H. Brandt *op. cit.* 99(1973)—Bolivia. 4
spegazziniana Brandt in *Stachelpost*, 7:367–368 (1971); cf. Rowley in Rep. Pl. Succ., 22 (Regnum Veg., 87):11(1973)—Locality not stated. 4
zaletaewana F. H. Brandt in *Stachelpost*, 9(44): 30(1973)—Bolivia. 4

PEDIOCACTUS

1156 Anon.
Revision du genre *Pediocactus*. *Cactus* (Brussels), 5(6):112–119(1973)—Illustrations. 3

1157 Arp, G.
A revision of *Pediocactus*. *Cact. Succ. J.* (U.S.A.), 44(5):218–222(1972). 3

1158 Arp, G.
Studies in the Colorado cacti: 2. Notes on the ecology of *Pediocactus simpsonii*. *Cact. Succ. J.* (U.S.A.), 44(3):108–109(1972). 3

1159 Kladiwa, L.
Pediocactus knowltonii L. Benson. *Kakteen*, 52: CVIIIb(1973)—Ge; illustrations. 3

1160 Kladiwa, L.
Pediocactus sileri (Engelmann) L. Benson. *Kakteen*,
3 55–56:CVIIIb(1973)—Ge; illustrations.

glaucus (K. Schum.) Arp in *Cact. Succ. J.* (U.S.A.),
3 44(5):221(1972): *Echinocactus glaucus*.
mesae-verdae (Boissevain) Arp *op. cit.* 222(1972):
3 *Coloradoa mesae-verdae.*
polyancistrus (Engelm. & Big.) Arp *op. cit.* 222
3 (1972): *Echinocactus polyancistrus.*
robustior (Coulter) Arp *op. cit.* 222(1972):
3 *Echinocactus simpsonii* var. *robustior.*
whipplei (Engelm. & Big.) Arp *op. cit.* 222(1972):
3 *Echinocactus whipplei.*
wrightiae (L. Benson) Arp *op. cit.* 222(1972):
3 *Sclerocactus wrightiae.*

PENIOCEREUS

cuixmalensis Sánchez-Mejorada in *Cact. Suc. Mex.*,
4 18(4):91(1973)—Mexico.
tomentosus (Bravo) F. Buxbaum in *Kakteen*, 55–56:
CIIa(1973), without basionym page: *Wilcoxia*
4 *tomentosa.*

PERESKIA

1161 Bleck, M.
Pereskia and *Maihuenia*. *Cact. Succ. J.* (U.S.A.),
4 45(5):214–215(1973)—En; illustrations.

1162 Crespo, J. H.
La madera de *Pereskia conzattii* Britton y Rose.
Cact. Suc. Mex., 18(2):31–36(1973)—(En);
4 illustrations.

PORFIRIA

1163 Cartier, A.
4 *Porfiria. Cactus* (Brussels), 5(2):39–40(1973)—Fr.

PSEUDOLOBIVIA

1164 Dorsselaer, R. van
Pseudolobivia kermesina Kr. *Cactus* (Brussels),
4 5(1):7(1973)—Fr.

PSEUDOPILOCEREUS

1165 Buining, A. F. H.
Did we indeed rediscover *Pseudopilocereus Bradei*?
Cact. Succ. J. (U.S.A.), 45(6):288–290(1973)—
4 Illustrations.

1166 Buining, A. F. H. *and* Brederoo, A. J.
Pseudopilocereus fulvilanatus Buining et Brederoo
spec. nov. *Kakt. And. Sukk.*, 24(7):145–147(1973)—
4 Ge; illustrations, coloured illustration.

1167 Buining, A. F. H. *and* Brederoo, A. J.
Pseudopilocereus magnificus Buining & Brederoo sp.
nov. *Cact. Succ. J.* (U.S.A.), 44(2):66–70(1972)—
4 En; illustrations, coloured illustrations.

fulvilanatus Buining & Brederoo in *Kakt. And.*
4 *Sukk.*, 24(7):145(1973)—Brazil.
magnificus Buining & Brederoo in *Cact. Succ. J.*
4 (U.S.A.), 44(2):66(1972)—Brazil.

PTEROCEREUS

1168 Glass, C. *and* Foster, R.
A word about the ptaxonomy of the genus
Pterocereus. Cact. Succ. J. (U.S.A.), 45(1):34(1973)—
4 Coloured illustration.

PYGMAEOCEREUS

1169 Mace, T.
Arthrocereus, *Pygmaeocereus* and *Setiechinopsis*.
Nation. Cact. Succ. J. (U.K.), 28(4):102–103(1973)—
4 Illustrations.

PYRRHOCACTUS

1170 Ritter, F.
Pyrrhocactus crispus Ritter sp. nov. *Chileans*, 7(25):
56–57(1973)—En. Translated by H. Middleditch from
4 *Succulenta*, no. 11.(1959).

REBUTIA

1171 Donald, J. D.
Rebutia albopectinata Rausch. *Ashingtonia*, 1(2):
4 23(1973)—En; illustration.

1172 Donald, J. D.
Rebutia heliosa Rausch. *Ashingtonia*, 1(2):13–14
4 (1973)—En; coloured illustration.

1173 Donald, J. D.
Rebutia narvaecense (Card.) Donald nov. comb.
Ashingtonia, 1(2):19,21(1973)—En; coloured
4 illustration.

1174 Kaiser, G.
4 Neue Rebutien? *Stachelpost*, 9(46):110–111(1973).

1175 Kiesling, R.
Especies nuevas o críticas de la Provincia de Jujuy:
5. Una nueva especie de Cactaceas. *Bol. Soc. Argent.*
Bot., 15(1):132–135(1973)—(En); illustration.
4 *Rebutia* (*Mediolobivia*) *gonjianii.*

1176 Rausch, W.
Rebutia (*Aylostera*) *heliosa* Rausch. *Succulenta*,
4 52(5):90–91(1973)—Du.

1177 Rausch, W.
Rebutia (*Aylostera*) *jujuyana* Rausch spec. nov. *Kakt.*
And. Sukk., 24(7):147–148(1973)—Ge; coloured
4 illustration.

gonjianii Kiesling in *Bol. Soc. Argent. Bot.*, 15(1):132
4 (1973)—Argentina.
jujuyana Rausch in *Kakt. And. Sukk.*, 24(7):147
4 (1973)—Argentina.
narvaecense (Card.) Donald in *Ashingtonia*, 1(2):21
4 (1973): *Aylostera narvaecense.*

SCHLUMBERGERA

1178 Anon.
Schlumbergera truncata v. *delicata* (N. E. Br.) Moran
n. comb. (Cactaceae). *Ashingtonia*, 1(3):32(1973)—
4 En; coloured illustration.

SETIECHINOPSIS

1179 Mace, T.
Arthrocereus, *Pygmaeocereus* and *Setiechinopsis*.
Nation. Cact. Succ. J. (U.K.), 28(4):102–103(1973)—
4 Illustrations.

STENOCEREUS

1180 Beutelspacher B., C. R. *and* Ramirez Martinez, M.
Polinización en *Stenocereus marginatus*. (D.C.)
Britton y Rose. *Cact. Suc. Mex.*, 18(3):80–84(1973)—
4 (En); illustration.

fricii Sánchez-Mejorada in *Cact. Suc. Mex.*, 18(4):
4 89(1973)—Mexico.

STEPHANOCEREUS

1181 Mace, T.
Stephanocereus leucostele (Gürke) Berger. *Nation. Cact. Succ. J.* (U.K.), 28(1):8(1973)—En;
4 illustrations.

STROMBOCACTUS

1182 Gewissler, B.
Strombocactus disciformis (De Candolle) Britton & Rose. *Kakt. And. Sukk.*, 24(12):270–271(1973)—Ge;
4 illustration.

SUBMATUCANA

1183 Skarupke, E.
Submatucana intertexta Ritter. *Stachelpost*, 9(46):
4 103,105(1973)—Ge,En; coloured illustration.

1184 Skarupke, E.
Submatucana (subgenus *Incaica*) *aureiflora* Ritter.
4 *Stachelpost*, 9(46):102–103(1973)—Ge,En.

1185 Skarupke, E.
Zu den Farbbildern (S.44 u 45). *Stachelpost*, 9(44): 44–45,50–51(1973)—Coloured illustrations. *Submatucana* sp. *de Balsas*, *Submatucana* sp. n. *Corral*
4 *Quemado*.

SULCOREBUTIA

1186 Donald, J. D.
Distribution of the genus *Sulcorebutia* Backeb.
4 *Ashingtonia*, 1(3):35(1973).

1187 Donald, J. *and* Brederoo, A. J.
Sulcorebutia's. *Succulenta*, 52(10):192–193(1973)—
4 Du.

1188 Köhler, U.
Sulcorebutia arenacea (Cardenas) Ritter. *Kakt. And.*
4 *Sukk.*, 24(11):246–247(1973)—Ge; illustrations.

1189 Rausch, W.
Sulcorebutia rauschii Frank. *Succulenta*, 52(4):
4 72–73(1973)—Du.

1190 Rausch, W.
Sulcorebutia steinbachii (Werdermann) Backeberg var. *horrida* Rausch. var. nov. *Kakt. And. Sukk.*, 24(9):193–194(1973)—Ge; illustration, coloured
4 illustration.

1191 Rausch, W.
Sulcorebutia vasqueziana Rausch. *Succulenta*, 52
4 (12):222(1973)—Du.

1192 Rausch, W.
Sulcorebutia verticillacantha var. *aureiflora* Rausch. *Succulenta*, 52(10):181,187–188(1973)—Du; coloured
4 illustration.

1193 Skarupke, E.
Sulcorebutia rauschii Frank. *Stachelpost*, 9(45):62
4 (1973)—Ge; illustration.

corroana (Card.) Brederoo & Donald in *Succulenta*,
4 52(10):192(1973): *Rebutia corroana.*
tuberculato-chrysantha (Card.) Brederoo & Donald *op. cit.* 193(1973): *Rebutia tuberculato-*
4 *chrysantha.*
steinbachii (Werdermann) Backeberg
var. **horrida** Rausch in *Kakt. And. Sukk.*, 24(9):
4 193(1973)—Bolivia.
vasqueziana Rausch
var. **albispina** Rausch in *Succulenta*, 52(12):222
4 (1973)—Bolivia.

TACINGA

1194 Buxbaum, F.
Gattung *Tacinga*. *Kakteen*, 55–56:B(1973)—
4 Illustrations.

THELOCACTUS

1195 Glass, C. *and* Foster, R.
Two new varieties of *Thelocactus conothele* (Reg. & Klein). Knuth. *Cact. Succ. J.* (U.S.A.), 44(2):47–50
4 (1972)—Illustrations, coloured illustration.

conothele (Reg. & Klein) Knuth
var. **argenteus** Glass & R. A. Foster in *Cact.*
4 *Succ. J.* (U.S.A.), 44(2):48(1972)—Mexico.
var. **aurantiacus** Glass & R. A. Foster *op. cit.* 48
4 (1972)—Mexico.

TOUMEYA

1196 Simon, W.
Toumeya papyracantha (Engelmann) Britton & Rose. *Kakt. And. Sukk.*, 24(2):38(1973)—Ge;
4 illustrations.

TRICHOCEREUS

1197 Donald, J. D.
Trichocereus huascha (Weber) Britt. & Rose.
4 *Ashingtonia*, 1(2):22(1973)—En; coloured illustration.

UEBELMANNIA

1198 Bleck, M.
Uebelmannia Buin. *Cact. Succ. J.* (U.S.A.), 45(3):
4 99–102(1973)—En; illustrations.

1199 Buining, A. F. H.
Uebelmannia flavispina Buining et Brederoo.
4 *Kakteen*, 54:CVIe(1973)—Ge; illustrations.

1200 Buining, A. F. H.
Uebelmannia meninensis Buining var. *meninensis*.
4 *Kakteen*, 55–56:CVIe(1973)—Ge; illustrations.

1201 Buining, A. F. H.
Uebelmannia pectinifera Buining var. *pectinifera*. *Kakteen*, 55–56:CVIe(1973)—Ge; illustrations,
4 coloured illustration.

1202 Buining, A. F. H. *and* Brederoo, A. J.
Uebelmannia flavispina Buining et Brederoo spec.
4 nova. *Succulenta*, 52(1):9–10(1973)—Du; illustration.

flavispina Buining & Brederoo in *Succulenta*, 52(1):
4 9(1973)—Brazil.
meninensis Buining
var. **rubra** Buining & Brederoo in *Kakteen*, 55–56:
4 CVIe(1973)—Brazil.

WEINGARTIA

1203 Backeberg, C.
About the genus *Weingartia* Werd. *Chileans*, 7(25):15–18(1973)—Illustrations. Translated by
4 H. Middleditch from *Cactus* (France), 20(85)1965.

1204 Bleck, M.
Weingartia Werd. *Cact. Succ. J.* (U.S.A.), 45(1):
3, 4 40–41(1973)—En; illustrations.

1205 Donald, J. D.
Purple flowered Weingartias. *Ashingtonia*, 1(1):1,
4 4–6(1973)—Illustrations, coloured illustrations.

1206 Middleditch, H.
4 The genus *Weingartia*. *Chileans*, 7(25):6–8(1973).

WIGGINSIA

1207 Vliet, D. J. van
Notocactus en *Wigginsia* één geslacht?: 3–5. *Succulenta*, 52(1): 2–7, (3): 49–52, (4): 66–70 (1973)—
4 Illustrations.

WILCOXIA

1208 Bourdoux, P.
Le genre *Wilcoxia* Br. & R.: 1–2. *Cactus* (Brussels),
4 4(4): 61–66, (5): 88–91 (1972)—Illustrations; key.

1209 Buxbaum, F.
Gattung *Wilcoxia*. *Kakteen*, 55–56: CIIa (1973)—
4 Illustrations, coloured illustration.

1210 Krainz, H.
Wilcoxia schmollii (Weingart) Backeb. *Kakteen*, 55–56: CVII (1973)—Ge; illustration, coloured
4 illustration.

WILMATTEA

1211 Croizat, L.
Nueva especie de *Wilmattea* (Cactaceae) en
4 Venezuela. *Pittieria*, no. 4: 39–41 (1972)—Illustration.

venezuelensis Croizat in *Pittieria*, 4: 39 (1972)—
4 Venezuela.

CALLITRICHACEAE

CALLITRICHE

1212 Favali, M. A. *and* Patrignani, G.
Ultrastructure of the stem of the aquatic plant *Callitriche stagnalis* L. *Österr. Bot. Zeitschr.*, 122(3):
1–4, 6, 7 167–175 (1973).

1213 Schotsman, H. D.
Note sur la répartition des Callitriches en Sologne et dans les régions limitrophes. *Bull. Centre Etudes*
1 *Rech. Sci. Biarritz*, 9(1): 19–52 (1972)—Map.

CAMPANULACEAE

1214 Dunbar, A.
A short report on the fine structure of some
1–7 Campanulaceae pollen. *Grana*, 13(1): 25–28 (1973).

ADENOPHORA

1215 Tutin, T. G.
Flora Europaea: notulae systematicae ad floram Europaeam spectantes: no. 14. Campanulaceae (226) *Adenophora lilifolia* (L.) Besser. *Bot. J. Linn. Soc.*,
1 67(3): 280 (1973).

liliifolia (L.) A. DC.
var. **pocsii** Soó in *Acta Bot. Acad. Sci. Hung.*,
1 18(1–2): 174 (1973)—Hungary.

CAMPANULA

1216 Bielawska, H.
Self fertilization in *Campanula rotundifolia* L. s.l. group. (Samozapłodnienie w grupie *Campanula rotundifolia* L. s.l.). *Acta Soc. Bot. Pol.*, 42(2): 253–264
1 (1973)—(Pol).

1217 Davis, P. H.
Campanula myrtifolia. *Curtis's Bot. Mag.*, 27(2): tab. 637 (1972 publ. 1973)—En; illustration, coloured
2 illustration.

1218 Fedorov, A. A.
Flora Europaea: notulae systematicae ad floram Europaeam spectantes: no. 14. Campanulaceae (227)
1, 2 *Campanula* L. *Bot. J. Linn. Soc.*, 67(3): 281 (1973).

1219 Geslot, A.
Contribution à l'étude cytotaxinomique de *Campanula rotundifolia* dans les Pyrénées françaises et espagnoles. *Phyton* (Austria), 15(1–2): 127–143 (1973)—
1 (En, Ge); chromosome numbers.

1220 Hubac, J. M.
Relations entre les dimensions du pollen et le niveau de polyploïdie dans le complexe du *Campanula rotundifolia* L. s.l. (Relations between pollen size and level of polyploidy in the *Campanula rotundifolia* L. s.l. complex). *Bull. Soc. Bot. France*, 119(5–6): 279–
1 285 (1972 publ. 1973)—(Ge, En).

1221 McAllister, H. A.
Campanula rotundifolia. *Glasgow Nat.*, 19(1): 66
1 (1973)—En. Local records.

affinis Roemer & Schultes
subsp. **bolosii** (Vayr.) Fedorov in *Bot. J. Linn. Soc.*,
1 67(3): 281 (1973): *C. bolosii*.
aizoon Boiss. & Spruner
subsp. **aizoides** (Zaffran) Fedorov *op. cit.* 281 (1973):
1 *C. aizoides*.
drabifolia Sibth.
subsp. **creutzburgii** (W. Greuter) Fedorov *op. cit.*
1 281 (1973): *C. creutzburgii*.
subsp. **pinatzii** (W. Greuter & Phitos) Fedorov
1 *op. cit.* 281 (1973): *C. pinatzii*.
glomerata L.
subsp. **subcapitata** (M. Popov) Fedorov *op. cit.*
1 281 (1973): *C. subcapitata*.
lusitanica L. ex Loefl.
subsp. **transtagana** (R. Fernandes) Fedorov *op. cit.*
1 281 (1973): *C. transtagana*.
patula L.
subsp. **costae** (Willk.) Fedorov *op. cit.* 281 (1973):
1 *C. costae*.
var. **vajdae** (Pénzes) Fedorov *op. cit.* 281 (1973):
1 *C. vajdae*.
persicifolia L.
var. **snjagovi** Velchev & P. Vasilev in *Izv. Bot.*
1 *Inst.* (Sofia), 24: 249 (1973)—Bulgaria.
subsp. **subpyrenaica** (Timb.-Lagr.) Fedorov in
1 *Bot. J. Linn. Soc.*, 67(3): 281 (1973): *C. subpyrenaica*.
rotundifolia L.
subsp. **legionensis** (Pau) Laínz in *Bol. Inst. Estud. Astur., Supl. Cienc.*, 16(Aport. Conoc. Fl.
1 Cántabro-Astur, 10: 35): 191 (1973): *C. legionensis*.
sibirica L.
subsp. **charkeviczii** (Fedorov) Fedorov in *Bot. J.*
1, 2 *Linn. Soc.*, 67(3): 281 (1973): *C. charkeviczii*.
subsp. **elatior** (Fomin) Fedorov *op. cit.* 281 (1973):
1, 2 *C. sibirica* f. *elatior*.
subsp. **talievii** (Juz.) Fedorov *op. cit.* 281 (1973):
1, 2 *C. talievii*.
subsp. **taurica** (Juz.) Fedorov *op. cit.* 281 (1973):
1, 2 *C. taurica*.
stevenii Bieb.
subsp. **altaica** (Ledeb.) Fedorov *op. cit.* 281 (1973):
1, 2 *C. altaica*.
subsp. **wolgensis** (Smirnov) Fedorov *op. cit.* 281
1 (1973): *C. wolgensis*.

CENTROPOGON

1222 Wilbur, R. L.
A new Costa Rican species of *Centropogon*: *C. irazuensis* (Campanulaceae-Lobelioideae). *Brittonia*,
3, 4 24(4): 420–424 (1972)—Key.

irazuensis Wilbur in *Brittonia*, 24(4):422(1972)—
4 Costa Rica.

CRATEROCAPSA Hilliard & B. L. Burtt in
5 *Notes Roy. Bot. Gard. Edinburgh*, 32(3):314(1973).
congesta Hilliard & B. L. Burtt *op. cit.* 320(1973)—
5 South Africa (Natal, Cape Prov.), Lesotho.
insizwae (Zahlbr.) Hilliard & B. L. Burtt *op. cit.*
5 323(1973): *Roella ? insizwae.*
montana (A. DC.) Hilliard & B. L. Burtt *op. cit.*
5 323(1973): *Wahlenbergia montana.*
tarsodes Hilliard & B. L. Burtt *op. cit.* 324(1973)—
South Africa (Natal, Transvaal, Orange Free
5 State), Rhodesia, Swaziland.

JASIONE

1223 Tutin, T. G.
Flora Europaea: notulae systematicae ad floram
Europaeam spectantes: no. 14. Notes on *Jasione* L.
1 in Europe. *Bot. J. Linn. Soc.*, 67(3):276–279(1973).

crispa (Pourret) Samp.
subsp. **amethystina** (Lag. & Rodr.) Tutin in *Bot.*
1 *J. Linn. Soc.*, 67(3):278(1973): *J. amethystina.*
subsp. **arvernensis** Tutin *op. cit.* 278(1973)—
1 France.
subsp. **cavanillesii** (C. Vicioso) Tutin *op. cit.* 278
1 (1973): *J. cavanillesii.*
subsp. **centralis** (Rivas-Martínez) Tutin *op. cit.*
1 278(1973): *J. humilis* subsp. *centralis.*
subsp. **maritima** (Duby) Tutin *op. cit.* 278(1973):
1 *J. montana* var. *maritima.*
subsp. **sessiliflora** (Boiss. & Reuter) Tutin *op. cit.*
1 278(1973): *J. sessiliflora.*
subsp. **tomentosa** (A. DC.) Tutin *op. cit.* 279
1 (1973): *J. humilis* var. *tomentosa.*
laevis Lam.
subsp. **orbiculata** (Griseb.) Tutin *op. cit.* 277
1 (1973): *J. orbiculata.*
subsp. **rosularis** (Boiss. & Reuter) Tutin *op. cit.*
1 277(1973): *J. rosularis.*
montana L.
monstr. **pedicellata** De Langhe in *Bull. Soc. Roy.*
1 *Bot. Belg.*, 106(1):71(1973)—Belgium.

MINDIUM

tchihatchewii (Fisch. & Mey.) Contandriopoulos in
Bull. Soc. Bot. France, 119(1–2):88(1972 publ. 1973),
2 without basionym ref.: *Michauxia tchihatchewii.*
thyrsoideum (Boiss. & Heldr.) Contandriopoulos
op. cit. 88(1972 publ. 1973), without basionym ref.:
2 *Michauxia thyrsoidea.*

PHYTEUMA

1224 Dirkse, G., Hoekstra, J. A. *and* Loode, W.
Carex brizoides L. en *Phyteuma spicatum* L. bij
1 Bentheim. *Gorteria*, 6(9):150–152(1973)—(Ge).

SYMPHYANDRA

1225 Forstner, W.
Die Ring-Glockenblume (*Symphyandra hofmannii*
Pantoczek) in der Steiermark. *Verh. Zool.-Bot. Ges.*
1 *Wien*, 112:92–93(1972)—Illustration.

CANNABACEAE

1226 Villiers, J.-F.
Cannabacées. Flore du Gabon, 22:55–58(1973)—
5 Illustrations.

CANNABIS

1227 Quimby, M. W., Doorenbos, N. J., Turner, C. E. *and*
Masoud, A.
Mississippi-grown marihuana—*Cannabis sativa*
cultivations and observed morphological variations.
6 *Econ. Bot.*, 27(1):117–127(1973)—Illustrations.

CANNACEAE

CANNA

1228 Richardson, J. W. *and* Smith, L. B.
Flora ilustrada catarinense 1: Canáceas. 1972. 39p.—
4 Illustrations, dot maps; keys. *Canna* only.

confusa J. W. Richardson & L. B. Smith in Fl. Ilustr.
Catar. 1 (Canác.):8(1972), nom. nov.: *C. paniculata*
4 sensu Petersen in Mart. Fl. Bras., 3(3):75(1890).
glauca L.
var. **angusta** J. W. Richardson *op. cit.* 24(1972)—
4 Argentina.

CAPPARACEAE

1229 Delaveau, P., Koudogbo, B. *and* Pousset, J.-L.
Alcaloïdes chez les Capparidaceae. *Phytochemistry*,
1–7 12(12):2893–2895(1973)—(En).

1230 Jafri, S. M. H.
Flora of West Pakistan: no. 34. Capparidaceae.
6 1973. 35p.—Illustrations, map.

1231 Legname, P. R.
Las Caparidaceas chaqueñas. Castelar, Ministerio de
Agricultura y Ganaderia de la Nacion, 1972. pp.
15–39. (Notas Preliminares para la Flora Chaqueña
(Formosa, Chaco y Santiago del Estero) no. 3)—
4 Illustrations, keys.

1232 Paiva, J. A. R.
Capparidaceae. *Anuár. Soc. Brot.*, 38:153–155(1972)
1 —Por; key.

1233 Wild, H. *and* Gonçalves, M. L.
Flora de Moçambique: 12. Capparaceae. 1973.
5 66p.—Illustrations; keys.

CADABA

farinosa Forssk.
subsp. **rariflora** Jafri in Fl. W. Pakistan, 34:13
6 (1973)—Pakistan.

CAPPARIS

1234 Iltis, H. H.
Capparis hypoleuca Presl: a synonym of *Solanum*
schlechtendalianum Walp. *Ann. Missouri Bot. Gard.*,
4 59(3):474(1972 publ. 1973).

1235 Raghavan, R. S.
A note on the typification of *Capparis roxburghii*
DC. (Capparaceae). *Blumea*, 20(2):356(1972 publ.
6 1973).

spinosa L.
var. **mucronifolia** (Boiss.) Hedge & Lam. ex
R. R. Stewart, Annot. Cat. Vasc. Pl. W.
6 Pakistan & Kashmir: 294(1972): *C. mucronifolia.*

CLEOME

heratensis Bunge & Bien. ex Boiss.
subsp. **pakistanica** Jafri *in* Fl. W. Pakistan, 34:25
6 (1973)—Pakistan.

MAERUA

juncea Pax
subsp. **crustata** (Wild) Wild in Fl. Moçamb.,
12(Capparac.):27(1973): *M. flagellaris* subsp.
5 *crustata.*

CAPRIFOLIACEAE

1236 D'Arcy, W. G.
Flora of Panama: part 9. Family 180. Caprifoliaceae.
Ann. Missouri Bot. Gard., 60(1):155–167(1973)—
4 Illustrations; keys.

ABELIA

1237 Golubkova, V. F.
O nekotorȳkh vidakh roda *Abelia* R. Br. (Caprifoliaceae Juss.). (De speciebus nonnullis generis
Abelia R. Br. (Caprifoliaceae Juss.)). *Novosti Sist.*
2 *Vȳssh. Rast.*, 10:241–248(1973)—Illustrations.

microphylla (W. W. Smith) Golubkova in *Novosti*
Sist. Vȳssh. Rast., 10:241(1973): *A. gracilenta* var.
2 *microphylla.*

LINNAEA

1238 Czarnota, K.
Stanowisko zimoziołu północnego *Linnaea borealis* i
barwinka pospolitego *Vinca minor* w Puszczy Białej
na Mazowszu. (The localities of the twin-flower and
the periwinkle in the Biała Primeval Forest in
Masovia). *Chrońmy Przyr. Ojczystą*, 29(2):69–70
1 (1973).

1239 Jakubowska-Gabara, J.
Nowe stanowisko *Linnaea borealis* w Polsce
środkowej. (A new locality of the twin-flower in
central Poland). *Chrońmy Przyr. Ojczystą*, 29(2):
1 66–68(1973)—Maps.

1240 Kaźmierski, J.
Stanowisko zimoziołu połnocnego *Linnaea*
borealis w województwie bydgoskim. (A locality of
the twin-flower, *Linnaea borealis*, in the province of
Bydgoszcz). *Chrońmy Przyr. Ojczystą*, 29(2):66
1 (1973).

LONICERA

1241 Barnes, W. J.
The autecology of the *Lonicera* × *bella* complex.
3 *Diss. Abstr. Int., B*, 33(6):2560(1972).

1242 Ferguson, D. K.
The identity of *Lonicera rhytidophylla*. *Notes Roy.*
2 *Bot. Gard. Edinburgh*, 32(3):415–416(1973).

1243 Májovský, J.
Neue und bemerkenswerte Arten in der Flora der
Slowakei: 3. Die Alpen-Heckenkirsche (*Lonicera*
alpigena L.) eine beachtenswerte Pflanzenart der
slowakischen Flora. *Acta Fac. Rerum Nat. Univ.*
1 *Comen., Bot.*, 20:93–118(1972).

1244 Mazurenko, M. T.
Osnovnȳe tsiklȳ trekh form rosta *Lonicera etrusca*
Santi. *Byull. Glavn. Bot. Sada* (Moscow), 90:50–53
1, 2 (1973)—Illustration.

1245 Möllerová, J.
Rozlišování některých druhů rodu *Lonicera*,
zvláště v nekvetoucím stavu. (Unterscheidungsmerkmale für einige *Lonicera*-Arten, besonders in
nicht blühendem Zustande). *Zprávy Cesk. Bot. Spol.*,
1 8(1):7–11(1973)—(Ge); illustration; key.

1246 Pringle, J. S.
Lonicera maackii (Caprifoliaceae) adventive in
Ontario. *Canad. Field.-Nat.*, 87(1):54–55(1973)—
2, 3 Illustration, map.

periclymenum L.
var. **parviflora** De Langhe in *Bull. Soc. Roy. Bot.*
1 *Belg.*, 106(1):71(1973)—Belgium.
f. **verticillata** De Langhe *op. cit.* 71(1973)—
1 Belgium.
pyrenaica L.
subsp. **majoricensis** (Gdgr.) Malagarriga, Las
Subesp. y Variac. Geogr.: 14(1973): *L.*
1 *majoricensis.*

TRIOSTEUM

1247 Duncan, T.
Three plant species new to Canada on Pelee
Island: *Triosteum angustifolium* L., *Valerianella*
umbilicata (Sull.) Wood, and *Valerianella intermedia*
Dyal. *Canad. Field-Nat.*, 87(3):261–265(1973)—Map,
3 dot maps.

1248 Lewis, W. H. *and* Frantz, P. R.
Tribal classification of *Triosteum* (Caprifoliaceae).
Rhodora, 75(801):120–121(1973)—Chromosome
3 numbers.

VIBURNUM

1249 Gibson, D. N.
Studies in American plants: 5. *Phytologia*, 25(3):
4 113–115(1973).

disjunctum Morton
var. **mendax** (Morton) D. N. Gibson in *Phytologia*,
4 25(3):114(1973): *V. mendax.*
jucundum Morton
var. **detractum** (Standl. & Steyerm.) D. N.
4 Gibson *op. cit.* 114(1973): *V. detractum.*
obtusatum D. N. Gibson *op. cit.* 114(1973)—
4 Mexico.

CARICACEAE

CARICA

1250 Prakash, N.
An unusual papaya from Kampong Kerinchi,
Kuala Lumpur. *Malayan Nat. J.*, 25(1):39–40(1972)—
6 Illustrations.

CARYOCARACEAE

1251 Araujo, P. A. de M. *and* Mattos Filho, A. de
Estrutura de madeiras de Caryocaraceae. *Arq.*
4 *Jard. Bot. Rio de Janeiro*, 19:5–48(1973).

CARYOCAR

1252 Barradas, M. M.
Morfologia do fruto e da semente de *Caryocar*
brasiliense (Piqui), em várias fases de desenvolvimento.
4 *Rev. Biol.*, 9(1–4):69–95(1973)—(En).

CARYOPHYLLACEAE

1253 Galiano, E. F. *and* Valdés, B.
Catálogo de las plantas vasculares de la provincia de
Sevilla: 4. Centrospermae: Caryophyllaceae.
1 *Lagascalia*, 3(1):71–79(1973)—(En).

1254 Ikonnikov, S. S.
Zametki o gvozdichnȳkh (Caryophyllaceae), 1. (Notae de Caryophyllaceis, 1). *Novosti Sist. Vȳssh.*
2, 3 *Rast.*, 10:136–142(1973)—Chromosome number.

1255 Wild, H., Hooper, S. S. *and* Diniz, M. A.
Flora de Moçambique: 19. Caryophyllaceae. 1973.
5 19p.—Keys.

AGROSTEMMA

1256 Thompson, P. A.
Effects of cultivation on the germination character of corn cockle (*Agrostemma githago* L.). *Ann. Bot.*
1 (U.K.), 36(149):133–154(1973)—Dot map.

ALLOCHRUSA

1257 Yukhananov, D. Kh.
Ob *Allochrusa paniculata* (Regel) Ovcz. et Czuk. subsp. *ferganensis* Jusup. ex Juch. (On *Allochrusa paniculata* (Regel) Ovcz. et Czuk. subsp. *ferganensis*
2 Jusup. ex Juch.). *Bot. Zhurn.*, 58(5):656–657(1973).

paniculata (Regel) Ovcz. & Czuk.
subsp. **ferganensis** Jusup. ex Yukhananov in *Bot.*
2 *Zhurn.*, 58(5):656(1973)—U.S.S.R. (Middle Asia).

ARENARIA

1258 Turner, B. F. H.
Allelopathic effects of *Petalostemon gattingeri* on the distribution of *Arenaria patula* in cedar glades. *Diss.*
3 *Abstr. Int., B*, 33(12):5777(1973).

bertolonii Fiori
f. **burnatii** (Briq.) Gamisans in *Candollea*, 28(1):60
1 (1973): *A. saxifraga* var. *burnatii*.
f. **morisii** (Briq.) Gamisans *op. cit.* 61(1973):
1 *A. saxifraga* var. *morisii*.
f. **salisii** (Briq.) Gamisans *op. cit.* 60(1973): *A.*
1 *saxifraga* var. *salisii*.
deflexa Decne.
var. **glabrata** Danin in *Notes Roy. Bot. Gard.*
2 *Edinburgh*, 32(2): 263(1973)—Sinai Peninsula.
grandiflora L.
subsp. **incrassata** (Lge.) Malagarriga, Las Subesp. y
1 Variac. Geogr.: 6(1973): *A. incrassata*.
subsp. **pseudoincrassata** Malagarriga *op. cit.* 6 (1973), nom. nov.: *A. incrassata* var. *glabrescens*
1 Lge.
hookeri Nutt.
subsp. **desertorum** (Maguire) W. A. Weber in *Southwest. Nat.*, 18(3):318(1973): *A. hookeri* var.
3 *desertorum*.
subsp. **pinetorum** (A. Nels.) W. A. Weber *op. cit.*
3 318(1973): *A. pinetorum*.
montana L.
subsp. **valentina** (Bss.) Malagarriga, Las Subesp.
1 y Variac. Geogr.: 6(1973): *A. valentina*.

CERASTIUM

1259 Möschl, W.
Über die Cerastien Österreichs. *Mitt. Naturwiss. Ver. Steiermark*, 103:141–169(1973)—Chromosome
1 numbers; key.

glomeratum Thuill.
subsp. **megacalyx** Kamelin in *Bot. Zhurn.*, 58(5):
2 628(1973)—U.S.S.R. (Middle Asia).
lucorum Schur
subsp. **macrocarpum** (Kot.) Laínz in *Bol. Inst. Estud. Astur., Supl. Cienc.*, 16(Aport. Conoc. Fl. Cántabro-Astur, 10:12):168(1973): *C. vulgatum*
1 subsp. *macrocarpum*.

DIANTHUS

1260 Fornaciari, G.
Una nuova stazione di *Dianthus glacialis* Haenke ssp. *glacialis* nel Parco Nazionale dello Stelvio, in Val di Cedech (Valfurva-Lombardia). (A new station of *Dianthus glacialis* Haenke ssp. *glacialis*, on the Stelvio National Park in Cedech Valley (Valfurva-Lombardy)). *Giorn. Bot. Ital.*, 107(5):243–248(1973)—
1 (En); dot map.

1261 Hecker, I. M.
Dianthus. Quart. Bull. Alp. Gard. Soc., 41(1):44–63 (1973)—En; illustrations. Popular horticultural
1–3, 5, 6 notes.

1262 Javeid, G. N.
A new *Dianthus* from Kashmir, *Dianthus minimus*
6 sp. nov. *Curr. Sci.*, 42(19):692–693(1973).

1263 Takahashi, H.
(On the scientific name of *Dianthus japonicus* f.
2 *albiflorus*). *J. Geobot.*, 20(1):13(1972)—Ja,En.

furcatus Balbis
subsp. **dissimilis** (Burnat) Pignatti in *Giorn. Bot.*
1 *Ital.*, 107(5):209(1973): *D. furcatus* var. *dissimilis*.
subsp. **lereschii** (Burnat) Pignatti *op. cit.* 209
1 (1973): *D. furcatus* var. *lereschii*.
glacialis Haenke
1 f. **reducta** Fornaciari *op. cit.* 246(1973)—Italy.
microlepis Boiss.
f. **alba** Delipavlov & S. Dimitrov in *Feddes Repert.*,
1 83(7–8):490(1972 publ. 1973)—Bulgaria.
minimus Javeid in *Curr. Sci.*, 42(19):692(1973)—
6 India.
sternbergii Sieber
subsp. **marsicus** (Ten.) Pignatti in *Giorn. Bot. Ital.*,
1 107(5):209(1973): *D. marsicus*.
subacaulis Vill.
subsp. **nivalis** (Wk.) Malagarriga, Las Subesp. y Variac. Geogr.: 6(1973): *D. brachyanthus* var.
1 *nivalis*.
sylvestris Wulfen
subsp. **garganicus** (Grande) Pignatti in *Giorn. Bot. Ital.*, 107(5):211(1973): *D. caryophyllus* subsp.
1 garganicus.

DICHODON

sect. **Perennia** S. Ikonnikov in *Novosti Sist. Vȳssh.*
1–3 *Rast.*, 10:142(1973).
argaeum (Boiss. & Bal.) S. Ikonnikov *op. cit.* 142
2 (1973): *Cerastium argaeum*.
dubium (Bast.) S. Ikonnikov *op. cit.* 141(1973):
1, 2 *Stellaria dubia*.
kotschyi (Boiss.) S. Ikonnikov *op. cit.* 142(1973):
2 *Cerastium kotschyi*.
persicum (Boiss.) S. Ikonnikov *op. cit.* 142(1973):
2 *Cerastium persicum*.
schischkinii (Grossh.) S. Ikonnikov *op. cit.* 142(1973):
2 *Cerastium schischkinii*.

DRYMARIA

1264 Jotani, Y.
(Notes on *Drymaria* from Taiwan). *J. Jap. Bot.*,
2 48(3):73(1973)—Ja.

EREMOGONE

sect. **Capillares** (McNeill) S. Ikonnikov in *Novosti Sist. Vȳssh. Rast.*, 10:138(1973): *Arenaria* sect.
2, 3 *Capillares*.
sect. **Glomeriflorae** (F. Williams) S. Ikonnikov *op.*
2 *cit.* 137(1973): *Arenaria* sect. *Glomeriflorae*.
sect. **Rigidae** (McNeill) S. Ikonnikov *op. cit.* 138
1, 2 (1973): *Arenaria* sect. *Rigidae*.

sect. **Sclerophyllae** (Boiss.) S. Ikonnikov *op. cit.*
2, 3 139(1973): *Arenaria* § 2 *Sclerophyllae.*
aberrans (M. E. Jones) S. Ikonnikov *op. cit.* 139
3 (1973): *Arenaria aberrans.*
acerosa (Boiss.) S. Ikonnikov *op. cit.* 139(1973):
2 *Arenaria acerosa.*
acicularis (F. Williams) S. Ikonnikov *op. cit.* 138
2 (1973): *Arenaria acicularis.*
aculeata (S. Wats.) S. Ikonnikov *op. cit.* 139(1973):
3 *Arenaria aculeata.*
angustisepala (McNeill) S. Ikonnikov *op. cit.* 138
2 (1973): *Arenaria angustisepala.*
androsacea (Grub.) S. Ikonnikov *op. cit.* 138(1973):
2 *Arenaria androsacea.*
acutisepala (Hausskn. ex F. Williams) S. Ikonnikov
2 *op. cit.* 139(1973): *Arenaria acutisepala.*
asiatica (Schischk.) S. Ikonnikov *op. cit.* 136(1973):
2 *Arenaria asiatica.*
blepharophylla (Boiss.) S. Ikonnikov *op. cit.* 136
2 (1973): *Arenaria blepharophylla.*
calcicola (Gilli) S. Ikonnikov *op. cit.* 139(1973):
2 *Arenaria calcicola.*
caricifolia (Boiss.) S. Ikonnikov *op. cit.* 138(1973):
2 *Arenaria caricifolia.*
congesta (Nutt. ex Torr. & Gray) S. Ikonnikov *op.*
3 *cit.* 139(1973): *Arenaria congesta.*
dianthoides (Smith) S. Ikonnikov *op. cit.* 138(1973):
2 *Arenaria dianthoides.*
drypidea (Boiss.) S. Ikonnikov *op. cit.* 140(1973):
2 *Arenaria drypidea.*
eastwoodiae (Rydb.) S. Ikonnikov *op. cit.* 139(1973),
3 as 'eastwoodii': *Arenaria eastwoodii.*
fendleri (A. Gray) S. Ikonnikov *op. cit.* 139(1973):
3 *Arenaria fendleri.*
ferganica (Schischk.) S. Ikonnikov *op. cit.* 140(1973):
2 *Arenaria ferganica.*
glaucescens (H. Winkl.) S. Ikonnikov *op. cit.* 140
2 (1973): *Arenaria glaucescens.*
griffithii (Boiss.) S. Ikonnikov *op. cit.* 140(1973):
2 *Arenaria griffithii.*
isaurica (Boiss.) S. Ikonnikov *op. cit.* 136(1973):
2 *Arenaria isaurica.*
kingii (S. Wats.) S. Ikonnikov *op. cit.* 140(1973):
3 *Stellaria kingii.*
koriniana (Fisch. ex Fenzl) S. Ikonnikov *op. cit.* 136
1, 2 (1973): *Arenaria koriniana.*
ledebouriana (Fenzl) S. Ikonnikov *op. cit.* 140(1973):
2 *Arenaria ledebouriana.*
litwinowii (Schischk.) S. Ikonnikov *op. cit.* 140
2 (1973): *Arenaria litwinowii.*
macradenia (S. Wats.) S. Ikonnikov *op. cit.* 140
3 (1973): *Arenaria macradenia.*
macrantha (Schischk.) S. Ikonnikov *op. cit.* 138
2 (1973): *Arenaria macrantha.*
meyeri (Fenzl) S. Ikonnikov *op. cit.* 139(1973):
2, 6 *Arenaria meyeri.*
micradenia (P. Smirn.) S. Ikonnikov *op. cit.* 137
1 (1973): *Arenaria micradenia.*
mongholica (Schischk.) S. Ikonnikov *op. cit.* 139
2 (1973): *Arenaria mongholica.*
multiflora (Gilli) S. Ikonnikov *op. cit.* 140(1973):
2 Arenaria multiflora.
paulsenii (H. Winkl.) S. Ikonnikov *op. cit.* 140(1973):
2 *Arenaria paulsenii.*
polaris (Schischk.) S. Ikonnikov *op. cit.* 137(1973):
1, 2 *Arenaria polaris.*
przewalskii (Maxim.) S. Ikonnikov *op. cit.* 139
2 (1973): *Arenaria przewalskii.*
pumicola (Coville & Leiberg) S. Ikonnikov *op. cit.*
3 140(1973): *Arenaria pumicola.*
saxatilis (L.) S. Ikonnikov *op. cit.* 137(1973):
1, 2 *Arenaria saxatilis.*
steveniana (Boiss.) S. Ikonnikov *op. cit.* 137(1973):
2 *Arenaria steveniana.*
szowitsii (Boiss.) S. Ikonnikov *op. cit.* 138(1973):
2 *Arenaria szowitsii.*
tetrasticha (Boiss.) S. Ikonnikov *op. cit.* 140(1973):
2 *Arenaria tetrasticha.*
tschuktschorum (Regel) S. Ikonnikov *op. cit.* 140
2 (1973): *Arenaria tschuktschorum.*
ursina (Robins.) S. Ikonnikov *op. cit.* 140(1973):
3 *Arenaria ursina.*

GASTROLYCHNIS

apetala (Fenzl) Tolm. & Kozh.
var. **nuda** Yurtsev in *Bot. Zhurn.*, 58(12):1747
2 (1973)—U.S.S.R. (Soviet Far East).

GYPSOPHILA

1265 McNeill, J.
Gypsophila and *Stellaria*: an unexpected problem in generic delimitation. *Notes Roy. Bot. Gard.*
2, 6 *Edinburgh*, 32(3):389–395(1973).

montserratii Fernández Casas in *Publ. Inst. Biol. Apl.*
1 *Barcelona*, 52:121(1972)—Spain.
perfoliata L.
subsp. **tomentosa** (L.) Malagarriga, Las Subesp. y
1 Variac. Geogr.: 6(1973): *G. tomentosa.*

HOLOSTEUM

1267 Baskin, J. M. *and* Baskin, C. C.
Studies on the ecological life cycle of *Holosteum umbellatum*. *Bull. Torrey Bot. Club*, 100(2):110–116
1–3 (1973).

IXOCA

1268 Soják, J.
Doplňky k nomenklatuře některých rodů (Phanerogamae). *Čas. Nár. Muz.* (Prague), 141(1–2):
1 61–63(1972).

alpestris (Jacq.) Soják in *Čas. Nár. Muz.* (Prague),
1 141(1–2):63(1972): *Silene alpestris.*

MINUARTIA

1269 Mackechnie, R.
Minuartia verna. *Glasgow Nat.*, 19(1):64(1973)—
1 En. Local records.

1270 Panov, P.
Diagnoses plantarum novarum ser. 1. *Dokl. Bolg.*
1 *Akad. Nauk*, 26(8):1057–1059(1973)—Illustrations.

1271 Panov, P.
Diagnoses plantarum novarum ser. 2. *Dokl. Bolg.*
1 *Akad. Nauk*, 26(9):1227–1230(1973).

ser. **Fragilis** Panov in *Dokl. Bolg. Akad. Nauk*,
1 26(8):1057(1973).
1 **diljanae** Panov *op. cit.* 1057(1973)—Bulgaria.
glomerata (MB) Degen
var. **glabra** Panov *op. cit.* 1059(1973)—U.S.S.R.
1 (Crimea).
laricifolia (L.) Sch. & Th.
subsp. **ophiolitica** Pignatti in *Giorn. Bot. Ital.*,
1 107(5):208(1973)—Italy.
lineata (C.A.M.) Bornm.
f. **foliosa** (Royle) R. R. Stewart, Annot. Cat. Vasc. Pl. W. Pakistan & Kashmir: 248(1972):
6 *Arenaria foliosa.*
f. **kashmirica** (Edgew.) R. R. Stewart *op. cit.* 248
6 (1972): *Arenaria kashmirica.*
mutabilis Schinz & Thell. ex Becherer
subsp. **balcanica** Panov in *Dokl. Bolg. Akad. Nauk*,
1 26(9):1227(1973)—Bulgaria.
1 var. **glabra** Panov *op. cit.* 1229(1973)—Bulgaria.
rumelica Panov in *Dokl. Bolg. Akad. Nauk*, 26(8):
1 1059(1973)—Bulgaria.

PETROCOPTIS

1272 Fernandez Casas, J.
Estudios sobre el género *Petrocoptis* A. Braun (Caryophyllaceae). *Cuad. Cienc. Biol.* (Granada),
1 2(1):43–45(1973)—(En); illustrations.

guarensis Fernandez Casas in *Cuad. Cienc. Biol.*
1 (Granada), 2(1):43(1973)—Spain.
pseudoviscosa Fernandez Casas *op. cit.* 44(1973)—
1 Spain.

POLYCARPAEA

repens (Forssk.) Asch. & Schweinf.
subsp. **tiranica** Danin in *Notes Roy. Bot. Gard.*
2 *Edinburgh*, 32(2):263(1973)—Arabia.

SAGINA

saginoides (L.) Karst.
subsp. **macrocarpa** (Rchb.) Soó in *Acta Bot. Acad. Sci. Hung.*, 18(1–2):177(1973): *Spergella macro-*
1 *carpa*.

SILENE

1273 Bari, E. A.
Cytological studies in the genus *Silene* L. *New*
1–3, 5 *Phytol.*, 72(4):833–838(1973)—Chromosome numbers.

1274 Chater, A. O.
A new species of *Silene* from south Spain.
1 *Lagascalia*, 3(2):219–222(1973)—(Sp); illustration.

1275 Melzheimer, V. *and* Damboldt, J.
Zur Morphologie und Cytologie tetraploider Sippen von *Silene vulgaris* (Caryophyllaceae). *Willdenowia*, 7(1):83–100(1973)—(En); dot map;
1 chromosome numbers.

1276 Mizushima, M.
A new species of *Silene* from northern Japan. *J.*
2 *Jap. Bot.*, 48(3):65–67(1973)—En (Ja); illustration.

1277 Panov, P.
Diagnoses plantarum novarum ser. 2. *Dokl. Bolg.*
1 *Akad. Nauk*, 26(9):1227–1230(1973).

1278 Smejkal, M.
K výskytu *Silene viridiflora* L. a kříženče *Silene nutans* (subsp. *nutans*) × *viridiflora* na Moravě. (Bemerkungen zum Vorkommen von *Silene viridiflora* L. und des Bastardes *Silene nutans* (subsp. *nutans*) × *viridiflora* in Mähren). *Zprávy Česk. Bot.*
1, 2 *Spol.*, 8(2):108–111(1973)—(Ge).

aomorensis Mizushima in *J. Jap. Bot.*, 48(3):65
2 (1973)—Japan.
colorata Poiret
subsp. **morisiana** (Béguinot & Ravano) Pignatti in *Giorn. Bot. Ital.*, 107(5):208(1973):
1 *S. morisiana*.
dioica (L.) Clairv.
var. foetida (Schur.) Guinochet & Vilmorin, Fl. France, 1:310(1973), without basionym ref.:
1 *Lychnis diurna* var. *subalpina foetida*.
× dubia (Hampe) Guinochet & Vilmorin *op. cit.* 310(1973), without basionym ref.: *Melandrium*
1 *dubium*.
nuratavica Kamelin in *Bot. Zhurn.*, 58(5):628
2 (1973)—U.S.S.R. (Middle Asia).
stockenii Chater in *Lagascalia*, 3(2):219(1973)—
1 Spain.
stojanovii Panov in *Dokl. Bolg. Akad. Nauk*, 26(9): 1229(1973), nom. nov.: *S. saxifraga* var. *pirinica*
1 Stoj.
vulgaris (Moench) Garke
subsp. **aetnensis** (Strobl) Pignatti in *Giorn. Bot.*
1 *Ital.*, 107(5):208(1973): *S. inflata* var. *aetnensis*.
wahlbergella Chawdhuri
subsp. **arctica** (Fr.) Hultén in *Bot. Notiser*, 126(4):
3 482(1973): *Wahlbergella apetala* β *arctica*.

SPERGULA

1279 Nezadal, W.
Über die Verbreitung von *Spergula arvensis* und *Euphorbia exigua* in Nordostbayern. *Göttinger Flor.*
1 *Rundbr.*, 7(3):54–57(1973)—Dot maps.

arvensis L.
subsp. **viscosa** (Lag.) Malagarriga, Las Subesp. y
1 Variac. Geogr.: 5 (1973): *S. viscosa*.

SPERGULARIA

1280 Ratter, J. A.
Cytogenetic studies in *Spergularia*: 7. Cryptic speciation in *S. media* (L.) C. Presl and *S. marina* (L.) Griseb. *Notes Roy. Bot. Gard. Edinburgh*, 32(2):
1, 2, 4 291–296(1973)—Chromosome numbers.

1281 Ratter, J. A.
Cytogenetic studies in *Spergularia*: 8. Barriers to the production of viable interspecific hybrids. *Notes*
1–4 *Roy. Bot. Gard. Edinburgh*, 32(2):297–301(1973).

STELLARIA

1282 Godeau, M.
Stellaria media (L.) Vill., *S. neglecta* Weihe, *S. pallida* (Dum.) Piré: observation des téguments séminaux au microscope électronique à balayage. *Compt. R. Hebd. Séances Acad. Sci., D* (Paris), 277(21):
1 2381–2384(1973).

1283 McNeill, J.
Gypsophila and *Stellaria*: an unexpected problem in generic delimitation. *Notes Roy. Bot. Gard. Edinburgh*,
2, 6 32(3):389–395(1973).

1284 Mukherjee, J. *and* Ghosh, K.
External and pollen morphological variations in the two populations of *Stellaria media* L. *Proc. Indian Sci.*
6 *Congr.*, 60(3):369(1973). Abstract only.

cherleriae (Fisch.) Williams
var. polyantha (Edgew. & Hk. f.) R. R. Stewart, Annot. Cat. Vasc. Pl. W. Pakistan & Kashmir: 254(1972), without basionym page or date:
6 (Basionym not stated).

CASUARINACEAE

CASUARINA

1285 Rao, A. N.
Anatomical studies on succulent cladodes in *Casuarina equisetifolia* Linn. *Proc. Indian Acad. Sci., B*,
2, 6 76(6):263–270(1972)—Illustrations.

1286 Sexton, R.
South Australian Casuarinas—sheathing teeth: a new key to the S.A. Casuarinas based entirely on the vegetative characters. *S. Austral. Nat.*, 47(4):77(1973)
7 —Illustrations.

CELASTRACEAE

1287 Villiers, J.-F.
Celastracées. Flore du Gabon, 22:3–13(1973)—
5 Illustrations; keys.

subfam. **Pottingerioïdeae** Airy Shaw in *Kew Bull.*,
2, 6 28(1):100(1973).

APATOPHYLLUM

1288 McGillivray, D. J.
Apatophyllum: a small Australian genus in the plant family Celastraceae. *Austral. Pl.*, 7(56):172(1973)—
7 Coloured illustration.

CATHA

1289 Getahun, A. *and* Krikorian, A. D.
Chat: coffee's rival from Harar, Ethiopia: 1. Botany, cultivation and use. *Econ. Bot.*, 27(4):353–377
5 (1973)—Illustrations, dot maps. *Catha edulis.*

CELASTRUS

glaber (Hatusima) Yamazaki in *J. Jap. Bot.*, 48(4):
2 128(1973): *C. kusanoi* var. *glaber.*

EUONYMUS

1290 Nath, J. *and* Clay, S. N.
Cytogenetic studies on some species of *Euonymus.*
1–3, 6 *Caryologia*, 25(4):417–427(1972 publ. 1973).

GYMINDA

orbicularis Borhidi & Muñiz in *Acta Bot. Acad. Sci.*
4 *Hung.*, 18(1–2):36(1973)—Cuba.

KOKOONA

1291 Jansen, W. T. *and* Baas, P.
Comparative leaf anatomy of *Kokoona* and *Lophopetalum* (Celastraceae). *Blumea*, 21(1):153–
6, 7 178(1973).

LOPHOPETALUM

1292 Jansen, W. T. *and* Baas, P.
Comparative leaf anatomy of *Kokoona* and *Lophopetalum* (Celastraceae). *Blumea*, 21(1):153–178
6, 7 (1973).

MAYTENUS

1293 Lander, N. S. *and* Johnson, L. A. S.
A new Australian species of *Maytenus. Contrib. New S. Wales Nation. Herb.*, 4(6):373–377(1973)—
7 Illustrations, dot map; key.

1294 Sands, M. J. S.
Maytenus chubutensis. Curtis's Bot. Mag., 27(2): tab.635(1972 publ. 1973)—En; illustration, coloured
4 illustration.

silvestris Lander & L. A. S. Johnson in *Contrib. New S. Wales Nation. Herb.*, 4(6):373(1973)—
7 Australia (New South Wales, Queensland).

PERROTTETIA

1295 Agostini, G.
Identity of *Perrottetia costaricensis* Lundell and *Perrottetia racemosa* Standley. *Phytologia*, 26(3):174
4 (1973).

POTTINGERIA

1296 Airy Shaw, H. K., Cutler, D. F. *and* Nilsson, S.
Pottingeria, its taxonomic position, anatomy and palynology. *Kew Bull.*, 28(1):97–104(1973)—
6 *Pottingeria* transferred from the Escalloniaceae.

acuminata Prain
var. **latifolia** Airy Shaw in *Kew Bull.*, 28(1):100
6 (1973)—Thailand.

PUTTERLICKIA

1297 Codd, L. E.
Notes on African plants: Celastraceae: author citation for *Putterlickia pyracantha. Bothalia*, 11(1–2):
5 115(1973).

CERATOPHYLLACEAE

CERATOPHYLLUM

1298 Imam, M., Chrtek, J. *and* Kosinová, J.
Novitates florae Aegyptiacae: 2. *Webbia*, 27(2): 461–466(1972 publ. 1973)—En(It). *Ceratophyllum*
2, 5 records from Egypt and Sudan.

1299 Lohammar, G.
Ceratophyllum demersum i Torne Lappmark. *Svensk Bot. Tidskr.*, 67(1):69–70(1973)—(En);
1 illustration.

1300 Turlier, M.-F.
Recherches sur l'organogenèse foliaire précoce: 1. Deux espèces-types pour les ramifications dichotome et latérale (*Ceratophyllum demersum* L. et *Myriophyllum spicatum* L.). *Rev. Gén. Bot.*, 79(941–943):297–308
1–7 (1972)—(En); illustrations.

CHENOPODIACEAE

1301 Bochantsev, V. P.
Vidy̆ podtriby̆ Kochieae Moq. (sem. Chenopodiaceae) Yuzhnoĭ i Yugo-Zapadnoĭ Afriki. (Species of the subtribe Kochieae Moq. (fam. Chenopodiaceae) from south and south-west Africa).
5 *Bot. Zhurn.*, 58(10):1462–1463(1973)—Keys.

ATRIPLEX

1302 Bassett, I. J. *and* Crompton, C. W.
The genus *Atriplex* (Chenopodiaceae) in Canada and Alaska: 3. Three hexaploid annuals: *A. subspicata*, *A. gmelinii*, and *A. alaskensis. Canad. J. Bot.*, 51(10): 1715–1723(1973)—(Fr); illustrations, dot maps;
3 chromosome numbers; key.

1303 Drysdale, F. R.
Variation of seed size in *Atriplex patula* var. *hastata.*
3 *Rhodora*, 75(801):106–110(1973).

1304 Gustafsson, M.
Evolutionary trends in the *Atriplex triangularis* group of Scandinavia: 1. Hybrid sterility and chromosomal differentiation. *Bot. Notiser*, 126(3):345–392
1 (1973)—Dot maps; chromosome numbers.

1305 Gustafsson, M.
Evolutionary trends in the *Atriplex triangularis* group of Scandinavia: 2. Spontaneous hybridization in relation to reproductive isolation. *Bot. Notiser*, 126(4):398–416(1973)—Illustrations, maps;
1 chromosome number.

BETA

1306 Buttler, K. P. *and* Schnedler, W.
Beta corolliflora, nicht *Beta trigyna* bei Blasbach (Kr. Wetzlar). *Hess. Flor. Briefe*, 262–264:58–61(1973)—
1 Illustrations; chromosome number.

1307 Jassem, M.
Endosperm development in diploid triploid and tetraploid seed of sugarbeet (*Beta vulgaris* L.). (Rozwój bielma w diploidalnych, triploidalnych i tetraploidalnych nasionach buraków cukrowych (*Beta vulgaris* L.)). *Genet. Pol.*, 14(3):295–303(1973)—
1, 2 (Pol,Ru).

1308 Keens, W. M.
Beta trigyna and *Linaria* × *dominii* in Berkshire.
1 *Watsonia*, 9(4):428(1973).

1309 Lundqvist, A., Østerbye, U., Larsen, K. *and* Linde-Laursen, I.
Complex self-incompatibility systems in *Ranunculus acris* L. and *Beta vulgaris* L. *Hereditas*,
1, 2 73(2):161–168(1973)—Chromosome numbers.

CERATOIDES

1310 Soják, J.
Doplňky k nomenklatuře některých rodů (Phanerogamae). *Čas. Nár. Muz.* (Prague), 141(1–2):
2 61–63(1972).

compacta (Lozin.) Soják in *Čas. Nár. Muz.*
2 (Prague), 141(1–2):63(1972): *Eurotia compacta.*
ewersmanniana (Stschegl. ex Borszcz.) Soják
2 *op. cit.* 63(1972): *Eurotia ewersmanniana.*

CHENOLEA

dinteri Bochantsev in *Bot. Zhurn.*, 58(10):1463
5 (1973)—South West Africa.

CHENOPODIUM

1311 Aellen, P.
Zum Formenkreis von *Chenopodium* L. sect. *Ambrina* (Spach) Benth. und Hook. und sect. *Nigrescentia* Aellen. *Acta Bot. Acad. Sci. Hung.*,
1–7 19(1–4):1–12(1973)—(En); illustrations; key.

1312 Asai, Y.
(Miscellaneous notes on the naturalized flora of Japan: 3). *J. Jap. Bot.*, 48(3):67–72(1973)—Ja(En);
2 illustrations. *Chenopodium leptophyllum.*

1313 Bornkamm, R.
Beiträge zur Ökologie von *Chenopodium botrys* L.: 5. Gehalt an organischem Stickstoff, Nitrat und Asche. *Verh. Bot. Ver. Prov. Brandenburg*, 108:56–63
1, 2 (1971).

1314 Bornkamm, R. *and* Sukopp, H.
Beiträge zur Ökologie von *Chenopodium botrys* L.: 6. Die ökologische Konstitution von *Chenopodium botrys. Verh. Bot. Ver. Prov. Brandenburg*, 108:64–74
1, 2 (1971).

1315 Crawford, D. J.
Morphology, flavonoid chemistry and chromosome number of the *Chenopodium neomexicanum* complex. *Madroño*, 22(4):185–195(1973)—
3, 4 Illustrations; chromosome numbers.

1316 Dapper, H.
Beiträge zur Ökologie von *Chenopodium botrys* L.: 2. Produktion an Ruderalstandorten. *Verh. Bot. Ver.*
1, 2 *Prov. Brandenburg*, 108:26–28(1971).

1317 Engstrand, L. *and* Gustafsson, M.
Drawings of Scandinavian plants: 83–85. *Chenopodium* L. *Bot. Notiser*, 126(1):1–6(1973)—Chromosome numbers. *Chenopodium glaucum, C.*
1 *rubrum, C. botryodes.*

1318 Engstrand, L. *and* Gustafsson, M.
Drawings of Scandinavian plants: 86–88. *Chenopodium* L. *Bot. Notiser*, 126(2):135–141(1973)—Chromosome numbers. *Chenopodium bonus henricus,*
1 *C. botrys, C. hybridum.*

1319 Engstrand, L. *and* Gustafsson, M.
Drawings of Scandinavian plants: 89–90. *Chenopodium* L. *Bot. Notiser*, 126(3):273–276(1973)—Illustrations; chromosome numbers. *Chenopodium*
1 *polyspermum, C. murale.*

1320 Engstrand, L. *and* Gustafsson, M.
Drawings of Scandinavian plants: 91–93. *Chenopodium* L. *Bot. Notiser*, 126(4):393–397(1973)—Illustrations; chromosome numbers. *Chenopodium*
1 *ficifolium, C. hircinum, C. vulvaria.*

1321 Jörgensen, M.
The genus *Chenopodium* in Norway. *Norw. J. Bot.*, 20(4):303–319(1973)—Illustrations, dot maps;
1 keys.

1322 Santo-Virzo, A. de
Beiträge zur Ökologie von *Chenopodium botrys* L.: 4. Wasserhaushalt. *Verh. Bot. Ver. Prov. Brandenburg*,
1, 2 108:37–55(1971).

1323 Uotila, P.
Chromosome counts on *Chenopodium* L. from S.E. Europe and S.W. Asia. *Ann. Bot. Fenn.*, 10(4):
1, 2, 6 337–440(1973).

1324 Zimmermann-Jaeger, S.
Beiträge zur Ökologie von *Chenopodium botrys* L.: 3. Substanzproduktion und Wuchsform in Abhängigkeit von Beleuchtungsstärke, Feuchtigkeit und Substrat. *Verh. Bot. Ver. Prov. Brandenburg*, 108:
1, 2 29–36(1971).

sect. **Nigrescentia** Aellen in *Acta Bot. Acad. Sci.*
4 *Hung.*, 19(1–4):3(1973).
album L.
var. **coronatum** Beaugé in *Bull. Centre Études Rech. Sci. Biarritz*, 8(4):696(1972 publ. 1973)—
1 France.
ambrosioides L.
var. **costei** Aellen in *Acta Bot. Acad. Sci. Hung.*,
1 19(1–4):11(1973)—France.
ficifolium Sm.
var. **coronatum** Beaugé in *Bull. Centre Études Rech. Sci. Biarritz*, 8(4):696(1972 publ. 1973)—
1 France.
minuatum Aellen in *Acta Bot. Acad. Sci. Hung.*,
4 19(1–4):8(1973)—Brazil.
sooanum Aellen *op. cit.* 5(1973), as 'sooianum'—
4 Argentina.
strictum Roth
subsp. **microphyllum** (Boenningh.) Jörgensen in *Norw. J. Bot.*, 20(4):317(1973): *C. album* γ
1 *microphyllum.*
zoellneri Aellen in *Acta Bot. Acad. Sci. Hung.*, 19(1–4):
4 5(1973)—Chile.

KRASCHENINNIKOVIA

1325 Aktaeva, M. K.
Kariologiya i morfologicheskiĭ analiz roda *Krascheninnikovia* Guldenst. (*Eurotia* Adans.). *Izv.*
1–4 *Akad. Nauk. Kaz. SSR, Biol.*, no. 6:4–6(1973)—(Ka).

MONOLEPIS

1326 Hakki, M. I.
Blütenmorphologische und embryologische Untersuchungen an Chenopodiaceae: *Monolepis*
trifida Schrad. *Bot. Jahrb.*, 93(4):481–497(1973)—
2 (En).

POLYCNEMUM

1327 Holzner, W.
Polycnemum heuffelii Láng. *Mitt. Bot. Arbeit.*
Oberösterr. Landesmus. Linz, 4(1–2):46(1972)—
1 Ge.

SALICORNIA

europaea L.
var. **brachystachya** (Meyer) Jacobsen in *Nation.*
Cact. Succ. J. (U.K.), 28(1):4(1973): *S.*
1 *brachystachya*.
prostrata Pall.
subsp. **simonkaiana** Soó in *Acta Bot. Acad. Sci.*
1 *Hung.*, 18(1–2):173(1973)—Hungary.

SALSOLA

1328 Aleksandrovskiĭ, E. S.
Biologiya tsveteniya i opӯleniya *Salsola paletzkiana* Litw. v yugo–zapadnӯkh Kӯzӯlkumakh. (Biology of flowering and pollination of *Salsola paletzkiana* Litw. in south-west Kyzylkums). *Bot. Zhurn.*, 58(8):118–
2 1126(1973)—(En); illustrations.

1329 Arnold, H. L.
An analysis of the taxonomic status of a new taxon in the genus *Salsola*. *Diss. Abstr. Int., B*, 33(7):2966
3 (1972).

1330 Bochantsev, V. P.
Novӯe vidӯ roda *Salsola* L. iz Yuzhnoĭ i Yugo-zapadnoĭ Afriki. (New species of the genus *Salsola* L. from South and South-West Africa). *Bot. Zhurn.*,
5 58(6):815–833(1973)—La.

1331 Kurbanov, Dzh.
Flora solyanok yugo-zapadnoĭ Turkmenii. (*Salsola* flora of south-western Turkmenistan). *Izv. Akad. Nauk Turkm. SSR, Biol. Nauk*, 4:23–28
2 (1973)—(En).

1332 Tölken, H. R.
A new species of *Salsola*. *J. S. Afr. Bot.*, 39(3):233–
5 234(1973)—(Af); illustration.

acocksii Bochantsev in *Bot. Zhurn.*, 58(6):815
5 (1973)—South West Africa.
albida Bochantsev *op. cit.* 816(1973)—South Africa
5 (Cape Prov.).
angolensis Bochantsev *op. cit.* 816(1973)—Angola,
5 South West Africa.
apiciflora Bochantsev *op. cit.* 817(1973)—South
5 Africa (Cape Prov.).
apterygea Bochantsev *op. cit.* 817(1973)—South
5 Africa (Cape Prov.).
araneosa Bochantsev *op. cit.* 818(1973)—South
5 West Africa.
aroabica Bochantsev *op. cit.* 818(1973)—South West
5 Africa.
atrata Bochantsev *op. cit.* 819(1973)—South Africa
5 (Cape Prov.).
ceresica Bochantsev *op. cit.* 820(1973)—South
5 Africa (Cape Prov.).
columnaris Bochantsev *op. cit.* 820(1973)—South
5 West Africa.
dealata Bochantsev *op. cit.* 820(1973)—South Africa
5 (Cape Prov.).
dinteri Bochantsev *op. cit.* 821(1973)—South West
5 Africa.
esterhuyseniae Bochantsev *op. cit.* 822(1973)—
5 South Africa (Cape Prov.).
etoshensis Bochantsev *op. cit.* 822(1973)—South
5 West Africa.
gemmifera Bochantsev *op. cit.* 823(1973)—South
5 West Africa.
giessii Bochantsev *op. cit.* 824(1973)—South West
5 Africa.
kalaharica Bochantsev *op. cit.* 824(1973)—
5 Botswana.
luederitzensis Bochantsev *op. cit.* 825(1973)—
5 South West Africa.
melanantha Bochantsev *op. cit.* 825(1973)—South
5 Africa (Cape Prov.).
microtricha Bochantsev *op. cit.* 826(1973)—South
5 Africa (Cape Prov.).
namaqualandica Bochantsev *op. cit.* 826(1973)—
5 South West Africa.
namibica Bochantsev *op. cit.* 827(1973)—South
5 West Africa.
patentipilosa Bochantsev *op. cit.* 827(1973)—South
5 Africa (Cape Prov.).
pearsonii Bochantsev *op. cit.* 828(1973)—South
5 Africa (Cape Prov.).
phillipsii Bochantsev *op. cit.* 829(1973)—South
5 Africa (Cape Prov.).
pillansii Bochantsev *op. cit.* 829(1973)—South
5 Africa (Cape Prov.).
seminuda Bochantsev *op. cit.* 830(1973)—South
5 West Africa.
sericata Bochantsev *op. cit.* 830(1973)—South
5 Africa (Cape Prov.).
squarrosula Bochantsev *op. cit.* 831(1973)—South
5 Africa (Cape Prov.).
tetramera Bochantsev *op. cit.* 831(1973)—South
5 Africa (Cape Prov.).
tuberculatiformis Bochantsev *op. cit.* 832(1973)—
5 South Africa (Cape Prov.).
verdoorniae Tölken in *J. S. Afr. Bot.*, 39(3):233
5 (1973)—South Africa (Cape Prov.).
warmbadica Bochantsev in *Bot. Zhurn.*, 58(6):833
5 (1973)—South West Africa.

SPINACIA

1333 Nair, P. K. K. *and* Kapoor, S. K.
Comparative pollen morphology of Japanese and Indian specimens of *Spinacia oleracea* Linn. *Curr. Sci.*,
2, 6 42(17):621–622(1973).

SUAEDA

1334 Dickson, J. H.
Suaeda maritima. *Glasgow Nat.*, 19(1):64(1973)—En.
1 Local records.

CHLORANTHACEAE

1335 Hatschbach, G. *and* Koczicki, C.
Clorantáceas do estado Paraná. *Bol. Mus. Bot. Mun.*
4 (Curitiba), no. 7. 5p.(1973)—Illustration, dot map.

1336 Reitz, R. *and* Klein, R. M.
Flórula da Ilha de Santa Catarina: Clorantáceas.
4 1972. 11p.—Illustrations.

HEDYOSMUM

1337 Burger, W. C.
Notes on the flora of Costa Rica: 2. *Hedyosmum* of the Chloranthaceae. *Phytologia*, 26(2):131–135
4 (1973)—Key.

costaricense C. E. Wood ex Burger in *Phytologia*,
4 26(2):132(1973)—Costa Rica.
4 **montanum** Burger *op. cit.* 133(1973)—Costa Rica.

CHRYSOBALANACEAE

1338 Prance, G. T.
New and interesting Chrysobalanaceae from
Amazonia. *Acta Amazonica*, 2(1):7–16(1972)—
4 Illustration, dot map.

ACIOA

edulis Prance in *Acta Amazonica*, 2(1):12(1972)—
4 Brazil.

COUEPIA

glabra Prance in *Acta Amazonica*, 2(1):10(1972)—
4 Brazil.

LICANIA

jimenezii Prance in *Acta Amazonica*, 2(1):7(1972)—
4 Surinam.

PARINARI

1339 Kostermans, A. J. G. H.
Some new taxa. *Bot. Tidsskr.*, 67(4):317–327
6 (1973)—Illustration.

euadenia Kostermans in *Bot. Tidsskr.*, 67(4):321
6 (1973)—Thailand.

CISTACEAE

CISTUS

1340 Poetsch, I. *and* Reznik, H.
Die Flavonoidmuster der Gattung *Cistus* L.: 1.
Untersuchung der Aglyca. (The flavonoids of the
genus *Cistus* L.: 1. The aglycones). *Ber. Deutsch. Bot.*
1 *Ges.*, 85(5–6):209–218(1972)—(En).

HELIANTHEMUM

1341 Conrad, M. *and* Deschatres, R.
Une plante corse absente des Flores de France:
l'*Helianthemum aegyptiacum* (L.) Miller. (On the
Helianthemum aegyptiacum in Corsica). *Bull. Soc. Bot.*
France, 119(7–8):503–508(1972 publ. 1973)—(En);
1 illustration.

1342 Uttal, L. J.
The genus *Helianthemum* in Virginia. *Castanea*,
3 38(2):196–199(1973).

croceum (Desf.) Pers.
subsp. **urrielense** (Laínz) Laínz in *Bol. Inst. Estud.*
Astur., *Supl. Cienc.*, 16(Aport. Conoc. Fl.
Cántabro-Astur, 10:20):176(1973): *H.*
1 *nummularium* subsp. *urrielense*.
marifolium (L.) Miller
subsp. **cinerascens** (Wk.) Malagarriga, Las
Subesp. y Variac. Geogr.: 12(1973): *H.*
1 *marifolium* var. *cinerascens*.
subsp. niveum (Wk.) Malagarriga *op. cit.* 12(1973),
1 without basionym date: *H. marifolium* var. *niveum*.
subsp. **tomentosum** (Wk.) Malagarriga *op. cit.*
1 12(1973): *H. marifolium* var. *tomentosum*.

HUDSONIA

1343 Skog, J. T. *and* Nickerson, N. H.
Variation and speciation in the genus *Hudsonia*.
Ann. Missouri Bot. Gard., 59(3):454–464(1972 publ.
3 1973).

ericoides L.
subsp. **andersonii** Nickerson & J. T. Skog in
Ann. Missouri Bot. Gard., 59(3):456(1972 publ.
1973)—Canada (Nova Scotia), U.S.A.
3 (Massachusetts, Maine).
subsp. **intermedia** (Peck) Nickerson & J. T. Skog
op. cit. 456(1972 publ. 1973): *H. tomentosa* var.
3 *intermedia*.
subsp. **montana** (Nutt.) Nickerson & J. T. Skog
3 *op. cit.* 456(1972 publ. 1973): *H. montana*.
subsp. **tomentosa** (Nutt.) Nickerson & J. T. Skog
3 *op. cit.* 456(1972 publ. 1973): *H. tomentosa*.

TUBERARIA

1344 Schotsman, H. D.
Deux races chromosomiques dans une population
de *Tuberaria guttata* (L.) Fourr. à Gran Canaria (Iles
Canaries). *Bull. Centre Etudes Rech. Sci. Biarritz*, 9(3):
1 235–241(1973)—Dot map; chromosome numbers.

CLETHRACEAE

1345 Harborne, J. B. *and* Williams, C. A.
A chemotaxonomic survey of flavonoids and simple
phenols in leaves of the Ericaceae. *Bot. J. Linn. Soc.*,
1–4, 6 66(1):37–54(1973).

CLUSIACEAE

1346 Adams, P.
Clusiaceae of the southeastern United States. *J.*
Elisha Mitchell Sci. Soc., 89(1–2):62–71(1973)—
3 Keys.

CALOPHYLLUM

1347 Kostermans, A. J. G. H.
Nouveautés pour la flore indochinoise. *Adansonia*,
6 13(3):333–340(1973)—(En); illustration.

1348 Seneviratne, E. W.
Variation in density in wood of six species of the
family Guttiferae. *Sri-Lanka Forest.*, 10(1–2):33–50
(1971). *Calophyllum bracteatum*, *C. calaba*, *C. soulattri*,
6 *C. walkeri*.

lineare Kostermans in *Adansonia*, 13(3):333(1973)—
6 Vietnam.

GARCINIA

1349 Seneviratne, E. W.
Variation in density in wood of six species of the
family Guttiferae. *Sri-Lanka Forest.*, 10(1–2):33–50
6 (1971). *Garcinia cambogia*.

MESUA

1350 Seneviratne, E. W.
Variation in density in wood of six species of the
family Guttiferae. *Sri-Lanka Forest.*, 10(1–2):33–50
6 (1971). *Mesua thwaitesii*.

COCHLOSPERMACEAE

1351 Keating, R. C.
The comparative morphology of the Cochlo-
spermaceae: 3. The flower and pollen. *Ann. Missouri*
4–7 *Bot. Gard.*, 59(2):282–296(1972 publ. 1973).

COMBRETACEAE

1352 Mehra, P. N. *and* Khosla, P. K.
Cytogenetical studies of East Himalayan Hamamelidaceae, Combretaceae and Myrtaceae. *Silvae Genet.*, 21(5):186–190(1972)—(Ge); chromosome
6 numbers.

1353 Mendes, E. J. *and* Exell, A. W.
Additiones et adnotationes florae Angolensi: 12. (Combretaceae). *Garcia de Orta, Sér. Bot.*, 1(1–2):
5 21–23(1973)—(En); illustrations; key.

1354 Wickens, G. E.
Flora of tropical East Africa: Combretaceae. London, Crown Agents, 1973. 99p.—Illustrations,
5 map; key.

COMBRETUM

1355 Stace, C. A.
The significance of the leaf epidermis in the taxonomy of the Combretaceae: 4. The genus *Combretum* in Asia. *Bot. J. Linn. Soc.*, 66(2):97–115
2 (1973).

TERMINALIA

1356 Coode, M. J. E.
Notes on *Terminalia* L. (Combretaceae) in Papuasia. *Contrib. Herb. Austral.*, no. 2. 33p.(1973)—Illustrations,
6 maps; keys.

avicapitis Coode in *Contrib. Herb. Austral.*, 2:3
6 (1973)—New Guinea (W. Irian).
calogemma Coode *op. cit.* 7(1973)—New Guinea
6 (Papua, W. Irian).
eddowesii Coode *op. cit.* 13(1973)—New Guinea
6 (Papua).
katikii Coode *op. cit.* 17(1973)—New Guinea (Terr.
6 New Guinea, W. Irian).
longespicata Sloot.
subsp. **sogerensis** (Bak.f.) Coode *op. cit.* 19(1973):
6 *T. sogerensis.*
menezesii Mendes & Exell in *Garcia de Orta, Sér.*
5 *Bot.*, 1(1–2):22(1973)—Angola.
microcarpa Decne.
subsp. **incana** Coode in *Contrib. Herb. Austral.*,
6 2:22(1973)—New Guinea (Papua, W. Irian).
morobensis Coode *op. cit.* 23(1973)—New Guinea
6 (Terr. New Guinea, W. Irian, Papua).

COMMELINACEAE

1357 Chimphamba, B. B.
Chromosome numbers in Malawian Commelinaceae.
5 *Bot. J. Linn. Soc.*, 66(4):303–306(1973).

DICHORISANDRA

1358 Steyermark, J. A.
Novedades venezolanas. *Acta Bot. Venezuel.*, 8(1–4): 111–119(1973)—Illustrations. *Dichorisandra*
4 *diederichsanae.*

diederichsanae Steyermark in *Acta Bot. Venezuel.*,
4 8(1–4):111(1973)—Venezuela.

GIBASIS

1359 Hunt, D. R.
Gibasis schiedeana. Curtis's Bot. Mag., 27(2):tab.636 (1972 publ. 1973)—En; illustration, coloured
4 illustration.

1360 Owens, S. J.
Self incompatibility in *Gibasis oaxacana* D. R. Hunt (Commelinaceae). *Incompat. Newsl.*, no. 2:43(1973)—
4 Chromosome numbers.

1361 Reveal, J. L. *and* Hess, W. J.
A new species of *Gibasis* (Commelinaceae) from northern Mexico. *Bull. Torrey Bot. Club*, 99(5):243–
4 245(1972 publ. 1973)—Illustration.

1362 Stant, M. Y.
Scanning electron microscopy of silica bodies and other epidermal features in *Gibasis* (*Tradescantia*)
4 leaf. *Bot. J. Linn. Soc.*, 66(3):233–243(1973).

sect. **Stenygrophylla** Reveal & W. J. Hess in *Bull.*
4 *Torrey Bot. Club*, 99(5):244(1972 publ. 1973).
heterophylla (Brandeg.) Reveal & W. J. Hess *op. cit.*
4 245(1972 publ. 1973): *Tradescantia heterophylla.*
matudae D. R. Hunt ex Stant in *Bot. J. Linn. Soc.*,
4 66(3):234(1973)—Mexico.
rhodantha (Torr.) Reveal & W. J. Hess in *Bull. Torrey Bot. Club*, 99(5):245(1972 publ. 1973):
4 *Tradescantia rhodantha.*
speciosa Reveal & W. J. Hess *op. cit.* 243(1972 publ.
4 1973)—Mexico.

MURDANNIA

1363 Cook, C. D. K.
New and noteworthy plants from the northern Italian ricefields. *Ber. Schweiz. Bot. Ges.*, 83(1):54–65
1 (1973). *Murdannia blumei* new to Europe.

RHOEO

1364 Lin, Y. J. *and* Paddock, E. F.
Ring-position and frequency of adjacent distribution of meiotic chromosomes in *Rhoeo spathacea*. *Amer. J.*
4 *Bot.*, 60(7):685–690(1973)—Chromosome number.

TRADESCANTIA

1365 Hunt, D. R.
Tradescantia andrieuxii. Curtis's Bot. Mag., 179(3): tab.646(1973)—En; illustration, coloured
4 illustration.

COMPOSITAE

1366 Adam, J. G.
Espèces nouvelles pour le Libéria et les Monts Nimba (Astéracées et Rubiacées). *Bull. Inst. Fond. Afr.*
5 *Noire, A*, 35(1):79–91(1973)—Illustrations.

1367 Bramwell, D. *and* Kunkel, G.
Notes on critical Compositae from Lanzarote. *Cuad. Bot. Canar.*, no. 18–19:49–53(1973)—(Sp);
1 illustrations.

1368 Cabrera, A. L.
Notas sobre tipos de compuestas sudamericanas en herbarios europeos: 4. (Notes on type specimens of South American Compositae in European herbaria: 4). *Bol. Soc. Argent. Bot.*, 15(1):113–125(1973)—(En);
4 illustrations.

1369 Clonts, J. A.
A revision of the genus *Elephantopus*, including *Orthopappus* and *Pseudelephantopus* (Compositae).
2–6 *Diss. Abstr. Int., B*, 33(4):1423(1972).

1370 Delay, J. *and* Petit, D.
Matériel chromatique singulier: 1. Composées. *Inform. Ann. Caryosyst. Cytogénét.*, no. 6:13–15 (1972 publ. 1973)—Illustrations; chromosome
1 numbers.

1371 Dostál, J.
Preliminary notes on the subtribe Centaureinae.
1–7 *Act Bot. Acad. Sci. Hung.*, 19(1–4):73–79(1973).

1372 Easterly, N. W.
The Compositae of the Oak Openings. *Ohio J.*
3 *Sci.*, 72(1):11–21(1972)—Map.

1373 Fay, J. J.
New species of Mexican Asteraceae. *Brittonia*,
4 25(2):192–199(1973)—Illustrations.

1374 Ferguson, I. K.
Flora Europaea: notulae systematicae ad floram
Europaeam spectantes: no. 14. Compositae (229)
1–3 *Arnica* L. *Bot. J. Linn. Soc.*, 67(3):282(1973).

1375 Fernandes, R. B.
Novitates et notulae systematicae et nomenclaturales
ad florae Lusitanicae, Azoricae et Madeirensis
1 pertinentes. *Anuár. Soc. Bot.*, 38:61–64(1972).

1376 Gommers, F. J.
Nematicidal principles in Compositae. *Meded.*
Landbouwhogesch. Wageningen, 73(17).71p.(1973)—
1–7 (Du).

1377 Grashoff, J. L., Bierner, M. W. *and* Northington, D. K.
Chromosome numbers in North and Central
American Compositae. *Brittonia*, 24(4):379–394
3, 4 (1973).

1378 Grierson, A. J. C.
Critical notes on the Compositae of Ceylon.
Ceylon J. Sci., Biol. Sci., 10(1):42–60(1972) (Critical
Notes on Ceylon Plants 5)—Illustrations; chromosome
6 number; keys.

1379 Gupta, P. K., Agarwal, D. K. *and* Srivastava, A. K.
Further cytological investigations in Indian
Compositae. *Cytologia*, 37(4):581–593(1972)—
6 Chromosome numbers.

1380 Herz, W.
Pseudoguaianolides in Compositae. *In* Nobel
Symposium 25: chemistry in botanical classification,
1–7 August 20–25, 1973. (1973). 1p. Abstract.

1381 Holub, J.
Some new nomenclatural combinations in
Centaureinae (Asteraceae). (Nové nomenklatorické
kombinace v podshluku Centaureinae čeledi
1 Asteraceae). *Preslia*, 45(2):142–146(1973)—(Cz).

1382 Husson, P.
Au sujet de la disposition des capitules chez quelques
Heliantheae. *Bull. Soc. Hist. Nat. Toulouse*, 109(1–2):
1–7 171–189(1973)—Illustrations.

1383 Kaul, V. *and* Singh, D.
Cytological studies in some members of Cichorieae
(Compositae). *Proc. Indian Sci. Congr.*, 59(3):366
6 (1972)—Chromosome numbers. Abstract only.

1384 Kitamura, S.
On some species of Compositae newly introduced
into Japan. *Acta Phytotax. Geobot.*, 25(4–6):189–191
2 (1973).

1385 Kuzmanov, B. *and* Kožuharov, S.
Karyological study of Bulgarian Compositae: 2.
Izv. Bot. Inst. (Sofia), 24:125–137(1973)—(Bu);
1 chromosome numbers.

1386 Leins, P.
Pollen systematische Studien an Inuleen: 2.
1, 2, 5 Filagininae. *Bot. Jahrb.*, 93(4):603–611(1973)—(En).

1387 Manilal, K. S.
Studies in the floral anatomy of Compositae: 2.
The tribe Asteroideae. *Proc. Nation. Acad. Sci. India*,
1–7 *B*, 41(3):210–220(1971 publ. 1972)—Illustrations.

1388 Nagl, W. *and* Ehrendorfer, F.
Chromosomengrösse und DNS-Gehalt bei drei
Anthemideen (Asteraceae-Anthemideae). (Chromo-
some size and DNA content in three species of
Asteraceae-Anthemideae). *Österr. Bot. Zeitschr.*,
1 121(3–4):165–169(1973)—(En).

1389 Robinson, H. *and* Brettell, R. D.
Tribal revisions in the Asteraceae: 3. A new tribe
4 Liabeae. *Phytologia*, 25(6):404–407(1973).

1390 Robinson, H. *and* Brettell, R. D.
Tribal revisions in the Asteraceae: 8. A new tribe,
Ursinieae. *Phytologia*, 26(3):76–85(1973)—Illustra-
5 tions; chromosome numbers.

1391 Robinson, H. *and* Brettell, R. D.
Tribal revisions in the Asteraceae: 11. A new tribe,
5 Eremothamneae. *Phytologia*, 26(3):163–166(1973).

1392 Roessler, H.
Nachträge zur Bearbeitung der Arctotideae-
Gorteriinae (Compositae). *Mitt. Bot. Staatssaml.*
5 *München*, 11:91–99(1973)—Dot map.

1393 Sheldon, J. C. *and* Burrows, F. M.
The dispersal effectiveness of the achene-pappus
units of selected Compositae in steady winds with
1–7 convection. *New Phytol.*, 72(3):665–675(1973).

1394 Singh, N. K.
Cytological investigations in some Compositae of
Bihar. *Proc. Indian Sci. Congr.*, 59(3):355(1972)—
6 Chromosome numbers. Abstract only.

1395 Smith, E. B.
An annotated list of the Compositae of Arkansas.
3 *Castanea*, 38(1):79–109(1973)—Map.

1396 Solbrig, O. T.
Cytology and cytogenetics of shrubs. *In* McKell,
C. M., Blaidell, J. R. *and* Goodin, J. R., *eds.* Wildland
shrubs and their biology and utilization: 127–137
1–7 (1972).

1397 Stuessy, T. F.
A systematic review of the subtribe Melampodiinae
(Compositae, Heliantheae.). *Contrib. Gray Herb.*,
2–4 no. 203:65–80(1973)—Chromosome numbers.

1398 Turner, B. L., Powell, A. M. *and* Watson, T. J.
Chromosome numbers in Mexican Asteraceae.
4 *Amer. J. Bot.*, 60(6):592–596(1973).

1399 Vuilleumier, B. S.
The genera of Lactuceae (Compositae) in the
southeastern United States. *J. Arnold Arbor.*, 54(1):
42–93(1973)—Illustrations; chromosome numbers;
3 key.

1400 Wunderlin, R. P.
New combinations in Compositae. *Ann. Missouri*
3 *Bot. Gard.*, 59(3):471–473(1972 publ. 1973).

subtribus **Crocidiinae** H. Robinson & Brettell in
6 *Phytologia*, 26(3):156(1973).
tribus **Eremothamneae** H. Robinson & Brettell
5 *op. cit.* 164(1973).
tribus **Liabeae** H. Robinson & Brettell in *Phytologia*,
4 25(6):405(1973).
tribus **Ursinieae** H. Robinson & Brettell in *Phytologia*,
5 26(2):83(1973).

ACANTHOSPERMUM

1401 Singh, V.
A new distributional record for *Acanthospermum*
4, 6 *australe* (Loefl.) Ktze. *Curr. Sci.*, 42(2):68–69(1973).

ACHILLEA

1402 Ehrendorfer, F.
New chromosome numbers and remarks on the *Achillea millefolium* polyploid complex in North America. *Österr. Bot. Zeitschr.*, 122(3):133–143
3 (1973)—(Ge).

ACHYROPHORUS

1403 Feráková, V. *and* Murín, A.
Cytotaxonomické a chorologické poznámky k druhu *Achyrophorus maculatus* (L.) Scop. (Cytotaxonomic and chorological notes on *Achyrophorus maculatus* (L.) Scop.). *Preslia*, 45(2):112–118(1973)—
1 (En); chromosome number.

ACOSTA

aplolepa (Moretti) Holub in *Preslia*, 45(2):142(1973):
1 *Centaurea aplolepa*.
attica (Nym.) Holub *op. cit.* 142(1973): *Centaurea*
1 *attica*.
cariensis (Boiss.) Holub *op. cit.* 142(1973): *Centaurea*
2 *cariensis*.
kochiana Holub *op. cit.* 142(1973), nom. nov.:
2 *Centaurea olympica* Wagenitz nom. illegit.
maculosa (Lam.) Rauschert in *Feddes Repert.*,
1 83(9–10):655(1972 publ. 1973): *Centaurea maculosa*.
polyclada (DC.) Holub in *Preslia*, 45(2):142(1973):
1, 2 *Centaurea polyclada*.
× **psammogena** (Gáyer) Rauschert in *Feddes Repert.*, 83(9–10):655(1972 publ. 1973): *Centaurea* ×
1 *psammogena*.
× **simonescui** (J. Wagn. & Prod.) Holub in *Preslia*, 45(2):142(25 April 1973), also Rauschert in *Feddes Repert.*, 83(9–10):655(1972 publ. 25 April
1 1973), as 'simionescui': *Centaurea* × *simonescui*.
spachii (Willk.) Holub in *Preslia*, 45(2):143(1973):
1 *Centaurea spachii*.
sublanata (DC.) Holub *op. cit.* 143(1973): *Centaurea*
1 *paniculata δ sublanata*.
tossiensis (Freyn & Sint.) Holub *op. cit.* 143(1973):
2 *Centaurea tossiensis*.

× **ACOSTITRAPA** Rauschert in *Feddes Repert.*,
1, 2 83(9–10):655(1972 publ. 1973). [*Acosta* × *Calcitrapa*].
hemiptera (Borbás) Rauschert *op. cit.* 655(1972 publ.
1 1973): *Centaurea hemiptera*.
pseudohemiptera (Wagner) Rauschert *op. cit.*
1 655(1972 publ. 1973): *Centaurea* × *pseudohemiptera*.
subdiffusa (Prodan) Rauschert *op. cit.* 655(1972 publ.
1 1973): *Centaurea* × *subdiffusa*.

ADENOSTYLES

1404 Tutin, T. G.
Flora Europaea: notulae systematicae ad floram Europaeam spectantes: no. 14. Compositae (228) *Adenostyles alliariae* subsp. *hybrida* (DC) Tutin.
1 *Bot. J. Linn. Soc.*, 67(3):282(1973).

alliariae Kern.
subsp. **hybrida** (DC.) Tutin in *Bot. J. Linn. Soc.*,
1 67(3):282(1973): *A. hybrida*.
briquetii Gamisans in *Candollea*, 28(1):75(1973)—
1 Corsica.

AJANIA

1405 Novozhilova, N. N.
Tsvetenie i opȳlenie édifikatora vȳsokogornȳkh pustȳn' vostochnogo Pamira ajanii Tibetskoĭ (*Ajania tibetica* (Hook. F. et Thoms.) Tzvel.). (*Ajania tibetica* flowering and pollition as an edification of the eastern Pamirs alpine desert). *Dokl. Akad. Nauk Tadzh. SSR*,
2 15(10):59–61(1972)—Ta; illustration.

ALEPIDOCLINE

1406 Robinson, H. *and* Brettell, R. D.
Tribal revisions in the Asteraceae: 4. The relationships of *Neurolaena*, *Schistocarpha* and
4 *Alepidocline*. *Phytologia*, 25(7):439–445(1973).

AMBROSIA

1407 Geissman, T. A., Lucas, A. J., Saitoh, T. *and* Payne, W. W.
Sesquiterpene lactones of *Ambrosia chamissonis*.
3, 4 *Biochem. Syst.*, 1(1):13–20(1973).

1408 Nakatani, N., Bohnstedt, C. *and* Mabry, T. J.
The origin of *Ambrosia chamissonis* in Chile.
3, 4 *Biochem. Syst.*, 1(3):129–132(1973)—Map.

1409 Payne, W. W., Geissman, T. A., Lucas, A. J. *and* Saitoh, T.
Chemosystematics and taxonomy of *Ambrosia chamissonis*. *Biochem. Syst.*, 1(1):21–33(1973)—
3, 4 Illustrations.

artemisiifolia L.
f. gracilissima D. Cîrţu & M. Cîrţu in *Stud. Comun., Bot.*,
1 5:116(1972), without type—Romania.

AMMOCYANUS (Boiss.) Dostál in *Acta Bot. Acad. Sci. Hung.*, 19(1–4):78(1973): *Centaurea* sect.
2 *Ammocyanus*.
arabicus Dostál in *Acta Bot. Acad. Sci. Hung.*, 19(1–4):
2 78(1973), nom. nov.: *Centaurea ammocyanus*.

ANTENNARIA

pulcherrima (Hook.) Greene
subsp. **anaphaloides** (Rydb.) W. A. Weber in
3 *Southwest. Nat.*, 18(3):317(1973): *A. anaphaloides*.

ANTHEMIS

triumfettii (L.) DC.
f. **flosculosa** (Briq. & Cavill.) R. Fernandes in *Anuár. Soc. Brot.*, 38:62(1972): *A. tinctoria* f.
1 *flosculosa*.

ARCTIUM

nemorosum Lej. & Court.
f. **mathei** Soó in *Acta Bot. Acad. Sci. Hung.*, 18(1–2):176(1973), nom. nov.: ? *A. pubentum* var.
1 *microcephalum* Máthé.

ARCTOTIS

1410 Watts, L. E.
The breeding of fertile *Venidio-Arctotis*. *J. Roy. Hort.*
0 *Soc.*, 98(8):359–360(1973).

ARGYROXIPHIUM

1411 Kobayashi, H. K.
Puttative generic hybrids of Haleakala's silver-sword and kupaoa (*Argyroxiphium sandwicense* × *Dubautia menziesii*) Compositae. *Pac. Sci.*, 27(2):
6 207–208(1973)—Illustration.

1412 Landgraf, L. K.
Mauna Kea and Mauna Loa silversword alive and perpetuating. *Bull. Pac. Trop. Bot. Gard.*, 3(4):64–66 (1973)—Illustrations, coloured illustration, dot map.
6 *Argyroxiphium sandwicense*, *A. kauense*.

ARNICA

angustifolia Vahl
subsp. **alpina** (L.) I. K. Ferguson in *Bot. J. Linn.*
1 *Soc.*, 67(3):282(1973): *A. montana* var. *alpina*.
subsp. **iljinii** (Maguire) I. K. Ferguson *op. cit.* 282
1 (1973): *A. alpina* subsp. *iljinii*.

ARTEMISIA

1413 Filatova, N. S.
Nomenklaturnȳe zametki o nekotorȳkh vidakh *Artemisia* iz podroda *Seriphidium* (Bess.) Rouy. (Nomenclature remarks on some *Artemisia* species from the subgenus *Seriphidium* (Bess.) Rouy). *Bot.*
1, 2 *Zhurn.*, 58(4):552–554(1973).

1414 Garrone, A., Lombard, V., Poldini, L., Rossetti, V., Sciortino, T. *and* Tourn, M. L.
Studio botanico e chemotassonomico delle *Artemisia vulgaris* e *verlotorum* della zone triestina. *Univ. Stud. Trieste Fac. Sci. Ist. Bot.*, no. 95. 22p.
1 (1973)—(Ge); chromosome numbers.

1415 Geissman, T. A.
Chemical constitution and botanical affinity in *Artemisia*. *In* Nobel Symposium 25: chemistry in botanical classification August 20–25, 1973. (1973).
1–6 1p. Abstract.

1416 Halligan, J. P.
Bare areas associated with shrub stands in grassland: the case of *Artemisia californica*. *BioScience*, 23(7):429–
3 432(1973).

1417 Kelsey, R. C., Morris, M. S., Bhadane, N. R. *and* Shafizadeh, F.
Sesquiterpene lactones of *Artemisia*: TLC analysis and taxonomic significance. *Phytochemistry*, 12(6):
3 1345–1350(1973).

1418 Korobkov, A. A.
Morfologo-anatomicheskie osobennosti semyanok polȳneĭ (*Artemisia* spp.) severo-vostoka SSSR. (The morpho-anatomical peculiarities of achenes of *Artemisia* spp. in the north-east of the U.S.S.R.). *Bot.*
1, 2 *Zhurn.*, 58(9):1302–1315(1973)—Key.

1419 Shah, G. L. *and* Parabia, M. H.
On the occurrence of *Artemisia scoparia* Waldst. &
6 Kit. in Gujarat. *Sci. Cult.*, 39(6):276–277(1973).

cretacea (Fiori) Pignatti in *Giorn. Bot. Ital.*, 107(5):
1 221(1973): *A. caerulescens* var. *cretacea*.

ASTER

1420 Bergh, L. M. J. van den *and* Hekking, W. H. A.
Zulte (*Aster tripolium* L.) langs de grote rivieren.
1 *Gorteria*, 6(8):138–139(1973)—Du(En).

1421 Hill, L. M.
Chemical, cytological and genetic considerations of the possible hybrid origin of *Aster blakei* (Porter) House. *Diss. Abstr. Int., B*, 33(8):3509–3510(1973)—
3 Chromosome number.

1422 Hill, L. M. *and* Rogers, O. M.
Chemical, cytological and genetic evidence for the hybrid origin of *Aster blakei* (Porter) House. *Rhodora*,
3 75(801):1–25(1973)—Illustrations.

1423 Lippert, W.
Revision der Gattung *Aster* in Afrika. *Mitt. Bot. Staatssamml. München*, 11:153–258(1973)—Illustra-
5 tions, map, dot maps; keys.

1424 Marchiori, S., Curti, L. *and* Lorenzoni, G. G.
Note su nuove stazioni di *Aster squamatus* Hieron. presso il Lago di Lesina (Foggia). (Notes about new stations of *Aster squamatus* Hieron. near Lesina Lake (Foggia)). *Inform. Bot. Ital.*, 5(1):77–78(1973)—
1 (En).

1425 Turner, B. L.
Transfer of *Aster gentryi* Standl. to the genus *Machaeranthera* sect. *Psilactis*. *Phytologia*, 25(2):57–58
4 (1972).

1426 Van Faasen, P. *and* Sterk, F. F.
Chromosome numbers in *Aster*. *Rhodora*, 75(801):
3 26–33(1973).

1427 Zdorov'eva, E. N. *and* Gorovoĭ, P. G.
K sistematike *Aster* L. dal'nego vostoka. *Byull.*
2 *Glavn. Bot. Sada* (Moscow), 90:35–39(1973).

ser. **Sibirica** Zdorov'eva in *Byull. Glavn. Bot. Sada*
2 (Moscow), 90:35(1973).
ageratoides Turcz.
f. **adustus** (Maxim.) Zdorov'eva *op. cit.* 38(1973):
2 *A. ageratoides* var. *adustus*.
f. **luxurifolius** (Tamamsch.) Zdorov'eva *op. cit.*
2 38(1973): *A. luxurifolius*.
f. **suputinicus** (Tamamsch.) Zdorov'eva *op. cit.* 38
2 (1973): *A. suputinicus*.
f. **sutschanensis** (Kom.) Zdorov'eva *op. cit.* 38
2 (1973): *A. sutschanensis*.
chimanimaniensis Lippert in *Mitt. Bot. Staatssamml.*
5 *München*, 11:206(1973)—Mozambique, Rhodesia.
comptonii Lippert *op. cit.* 225(1973)—Swaziland,
5 South Africa (Transvaal).
erucifolius (Thell.) Lippert *op. cit.* 246(1973):
5 *Felicia erucifolia*.
grauii Lippert *op. cit.* 236(1973)—South Africa
5 (Natal).
harveyanus O. Kuntze
subsp. **corymbosus** Lippert *op. cit.* 192(1973)—
5 Swaziland, South Africa (Transvaal).
subsp. **gracilis** *op. cit.* 195(1973)—Swaziland,
5 South Africa (Transvaal), Mozambique.
subsp. **nyikensis** Lippert *op. cit.* 198(1973)—
5 Zambia, Malawi, Rhodesia.
subsp. **robustus** Lippert *op. cit.* 201(1973)—
5 South Africa (Natal).
subsp. **xylophyllus** (Klatt) Lippert *op. cit.* 203
(1973): South Africa (Transvaal, Natal),
5 Swaziland.
lydenburgensis Lippert *op. cit.* 211(1973)—South
5 Africa (Transvaal, Orange Free State).
nubimontis Lippert *op. cit.* 239(1973)—South
5 Africa (Transvaal).
pseudobakeranus Lippert *op. cit.* 243(1973)—
5 Swaziland.

sibiricus L.
f. **subintegerrimus** (Trautv.) Zdorov'eva in *Byull. Glavn. Bot. Sada* (Moscow), 90:36(1973):
2 *A. sibiricus* var. *subintegerrimus*.

tansaniensis Lippert in *Mitt. Bot. Staatssamml. München*, 11:232(1973)—Malawi, Tanzania,
5 Zambia.

tripolium L.
subsp. **longicaulis** (Duf.) Malagarriga, Las Subesp.
1 y Variac. Geogr.: 14(1973): *A. longicaulis*.

zuluensis Lippert in *Mitt. Bot. Staatssamml. München*,
5 11:218(1973)—South Africa (Natal).

ATHANASIA

pinnatifida (Oliv.) Hilliard in *Notes Roy. Bot. Gard.*
5 *Edinburgh*, 32(3):330(1973): *Pentzia pinnatifida*.

villosa Hilliard *op. cit.* 330(1973)—South Africa
5 (Natal).

woodii (Thellung) Hilliard *op. cit.* 331(1973):
5 *Pentzia woodii*.

BACCHARIS

1428 Ariza Espinar, L.
Las especies de *Baccharis* (Compositae) de Argentina central. *Bol. Acad. Nac. Cienc.* (Cordoba), 50(1–4):
4 175–305(1973)—(En); illustrations, dot maps; key.

1429 Ariza Espinar, L.
Novedades sobre *Baccharis* (Compositae) de Argentina: 2. *Lorentzia*, no. 2:5–7(1973)—(En);
4 illustration.

1430 Ariza Espinar, L.
Novedades sobre *Baccharis* (Compositae) de Argentina: 3. *Kurtziana*, 7:187–193(1973)—(En);
4 illustrations.

1431 Rzedowski, J.
Tres adiciones al genero *Baccharis* (Compositae) en Mexico. *Brittonia*, 24(4):398–402(1972)—(En);
4 illustrations.

sect. **Racemosae** Ariza Espinar in *Bol. Acad. Nac.*
4 *Cienc.* (Cordoba), 50(1–4):183(1973).

cabrerae Ariza Espinar in *Kurtziana*, 7:187(1973)—
4 Argentina.

famatinensis Ariza Espinar in *Lorentzia*, 2:5(1973)—
4 Argentina.

flabellata Hooker & Arnott
var. **argentina** (Heer.) Ariza Espinar in *Bol. Acad. Nac. Cienc.* (Cordoba), 50(1–4):284(1973):
4 *B. argentina*.

matudae Rzedowski in *Brittonia*, 24(4):398(1972)—
4 Mexico.

niederleinii Heering
var. **potrerillana** Ariza Espinar in *Bol. Acad. Nac. Cienc.* (Cordoba), 50(1–4):240(1973)—
4 Argentina.

pyramidata (Rob. & Greenm.) Rzedowski in
4 *Brittonia*, 24(4):398(1972): *Bigelowia pyramidata*.

rodriguezii Ariza Espinar in *Kurtziana*, 7:190
4 (1973)—Argentina.

rufescens Sprengel
var. **subrufescens** (Heer.) Ariza Espinar in *Bol. Acad. Nac. Cienc.* (Cordoba), 50(1–4):288
4 (1973): *B. subrufescens*.

stenophylla Ariza Espinar *op. cit.* 241(1973)—
4 Argentina.

zamoranensis Rzedowski in *Brittonia*, 24(4):400
4 (1972)—Mexico.

BALDUINA

1432 Parker, E. S.
Biosystematic study of the genus *Balduina*
3 (Compositae). *Diss. Abstr. Int., B*, 33(9):4152(1973).

BALTIMORA

1433 Stuessey, T. F.
Revision of the genus *Baltimora* (Compositae, Heliantheae). *Fieldiana, Bot.*, 36(5):31–50(1973)—
4 Illustrations, dot maps.

geminata (Brandg.) Stuessy in *Fieldiana, Bot.*, 36(5):
4 42(1973): *Melampodium geminatum*.

BEDFORDIA

1434 Drury, D. G.
Nodes and leaf structure in the classification of some Australasian shrubby Senecioneae-Compositae. *New Zealand J. Bot.*, 11(3):525–554(1973)—Illustra-
7 tions; key.

1435 Gray, A. M.
The mainland *Bedfordia*. *Austral. Pl.*, 7(56):178–179
7 (1973). *B. arborescens*.

BERKHEYA

carlinopsis Welw. ex O. Hoffm.
subsp. **sylvicola** (S. Moore) Roessler in *Mitt. Bot.*
5 *Staatssamml. München*, 11:96(1973): *B. sylvicola*.

BIDENS

1436 Calduch, M.
Nota sobre *Bidens aurea* (Aiton) Sherff. *Lagascalia*,
1 3(1):59–60(1973)—(En).

1437 Dakshini, K. M. M. *and* Singh, P.
Bidens bipinnata (Asteraceae) in northern India. *Proc. Indian Sci. Congr.*, 58(3):436(1971)—Key.
6 Abstract only.

1438 Gillett, G. W.
Genetic relationships between *Bidens forbesii* and six species in the Hawaiian and the Marquesas
6 Islands. *Brittonia*, 25(1):10–14(1973).

aristosa (Michx.) Britt.
f. **fritcheyi** (Fern.) Wunderlin in *Ann. Missouri Bot. Gard.*, 59(3):471(1972 publ. 1973): *B.*
3 *aristosa* var. *fritcheyi*.
f. **involucrata** (Nutt.) Wunderlin *op. cit.* 472
3 (1972 publ. 1973): *Coreopsis involucrata*.
f. **mutica** (Gray) Wunderlin *op. cit.* 471(1972
3 publ. 1973): *Coreopsis aristosa* var. *mutica*.
var. **retrorsa** (Sherff) Wunderlin *op. cit.* 472(1972
3 publ. 1973): *B. polylepis* var. *retrorsa*.

BLENNOSPERMA

1439 Ornduff, R., Saleh, N. A. M. *and* Bohm, B. A.
The flavonoids and affinities of *Blennosperma* and
3, 4 *Crocidium* (Compositae). *Taxon*, 22(4):407–412(1973).

BLUMEA

1440 Rao, R. S. *and* Hemadri, K.
Blumea venkataramanii Rolla Rao et Hemadri—a new species of Asteraceae from Maharashtra State.
6 *Curr. Sci.*, 42(19):693–694(1973).

venkataramanii Rolla Rao & Hemadri in *Curr. Sci.*,
6 42(19):693(1973)—India.

BOMBYCILAENA

discolor (Pers.) Laínz in *Bol. Inst. Estud. Astur., Supl. Cienc.*, 16(Aport. Conoc. Fl. Cántabro-Astur, 10:
1 38):194(1973): *Micropus discolor*.

BRACHYGLOTTIS

1441 Drury, D. G.
Nodes and leaf structure in the classification of some Australasian shrubby Senecioneae-Compositae. *New Zealand J. Bot.*, 11(3):525–554(1973)—Illustrations;
7 key.

BUBONIUM

aquaticum (L.) Hill.
subsp. nanus (Bss.) Malagarriga, Las Subesp. y Variac. Geogr.: 14(1973), without basionym
1 ref.: (Basionym not stated).

CALCITRAPA

angusticeps (H. Lindb.) Holub in *Preslia*, 45(2):143
2 (1973): *Centaurea angusticeps.*
araneosa (Boiss.) Holub *op. cit.* 143(1973): *Centaurea*
2 *araneosa.*
calcitrapella (Bornm. & Dinsm.) Holub *op. cit.* 143
2 (1973): *Centaurea calcitrapella.*
carneola (Boiss.) Holub *op. cit.* 143(1973): *Centaurea*
2 *carneola.*
cilicica (Boiss. & Bal.) Holub *op. cit.* 143(1973):
2 *Centaurea cilicica.*
diluta (Ait.) Holub *op. cit.* 143(1973): *Centaurea*
1 *diluta.*
furfuracea (Coss. & Dur.) Holub *op. cit.* 143
1 (1973): *Centaurea furfuracea.*
maroccana (Ball) Holub *op. cit.* 143(1973): *Centaurea*
1 *maroccana.*
nicaensis (All.) Holub *op. cit.* 143(1973): *Centaurea*
1 *nicaensis.*
procurrens (Sieb.) Holub *op. cit.* 143(1973):
2 *Centaurea procurrens.*
pungens (Pomel) Holub *op. cit.* 143(1973): *Centaurea*
1 *pungens.*
sinaica (DC.) Holub *op. cit.* 143(1973): *Centaurea*
2 *sinaica.*

CALCITRAPOIDES

aspera (L.) Holub in *Folia Geobot. Phytotax.*, 8(2):
1 170(1973): *Centaurea aspera.*
bimorpha (Viviani) Holub *op. cit.* 170(1973):
1 *Centaurea bimorpha.*
gentilii (Br.-Bl. & Maire) Holub *op. cit.* 170(1973):
1 *Centaurea gentilii.*
glomerata (Vahl) Holub *op. cit.* 170(1973):
2 *Centaurea glomerata.*
lusitanica (Boiss. & Reut.) Holub *op. cit.* 170(1973):
1, 2 *Centaurea lusitanica.*
malacitana (Boiss.) Holub *op. cit.* 170(1973):
1, 2 *Centaurea malacitana.*
napifolia (L.) Holub *op. cit.* 170(1973): *Centaurea*
1 *napifolia.*
polyacantha (Willd.) Holub *op. cit.* 170(1973):
1 *Centaurea polyacantha.*
praecox (Oliver & Hiern) Holub *op. cit.* 170(1973):
5 *Centaurea praecox.*
rhizocephala (Oliver & Hiern) Holub *op. cit.* 170
5 (1973), as 'rhizocephalus': *Centaurea rhizocephala.*
seridis (L.) Holub *op. cit.* 170(1973): *Centaurea*
1 *seridis.*
sonchifolia (L.) Holub *op. cit.* 170(1973): *Centaurea*
1 *sonchifolia.*
sphaerocephala (L.) Holub *op. cit.* 170(1973):
1 *Centaurea sphaerocephala.*

CARDUUS

nutans L.
subsp. alpicola (Gillot) Malagarriga, Las Subesp. y Variac. Geogr.: 16(1973), without basionym
1 date: *C. alpicola.*

platypus Lge.
subsp. **granatensis** (WK.) Malagarriga *op. cit.*
1 16(1973): *C. granatensis.*

CELMISIA

1442 Gorer, R.
Plant awards 1972–73: *Celmisia incana* Hort. A.M.
7 *Quart. Bull. Alp. Gard. Soc.*, 41(4):328(1973).

1443 Merelie, G.
Sorting out Celmisias. *Quart. Bull. Alp. Gard. Soc.*,
7 41(2):173–180(1973)—Illustrations.

CENTAUREA

1444 Amico, A.
Ritrovamento a Brindisi di *Centaurea napifolia* L. (*Centaurea napifolia* L. in Apulia). *Inform. Bot. Ital.*,
1 5(1):85(1973)—(En).

1445 Gochu, D. I.
Nekotorȳe dannȳe o khromosomnȳkh chislakh vidov *Centaurea* L., proizrastayushchikh v Moldavii. (Some data on the chromosome numbers of species of *Centaurea* L. growing in Moldavia). *Bot. Zhurn.*,
1 58(1):77–78(1973).

1446 Gochu, D. I. *
O vidovom sostave vasil'kov Moldavskoi SSR. (Species composition of *Centaurea* in the Moldavian S.S.R.). *Izv. Akad. Nauk Mold. SSR, Biol. Khim. Nauk*, no. 2:15–18(1972). Abstr. in *Biol. Abstr.* 55(9):
1 47302(1973).

1447 Gochu, D. I.
Ob anatomii semyanok nekotorȳkh vidov *Centaurea* L. (On the anatomy of seeds of some species of the genus *Centaurea* L.). *Bot. Zhurn.*, 58(2):
1, 2 245–247(1973).

1448 Ninova, D.
Anatomichno ustroĭstvo na listnata druzhka pri vidovete ot rod *Centaurea* L. (metlichina) v Bulgariya. (La structure anatomique du petiole des espèces du genre *Centaurea* L. en Bulgarie). *Nauch. Trud. Plovdiv. Univ. 'Paisii Khilendarski'*, 10(3):155–165
1 (1972)—(Rus,Fr).

1449 Ninova, D.
Sravnitel'no-anatomicheskoe issledovanie vasil'kov Bolgarii: 1. Épiderma i ust'ichnȳĭ apparat. (Comparative-anatomical study of *Centaurea* spp. in Bulgaria: 1. The epiderm and the stomatic apparatus).
1 *Bot. Zhurn.*, 58(9):1357–1360(1973).

1450 Ninova, D.
Sravnitelno anatomichno izsledvane na listnata petura pri vidovete ot rod *Centaurea* L. (metlichina) v Bulgariya. (Recherche anatomique comparative des limbes foliaires des espèces du genre *Centaurea* L. en Bulgarie). *Nauch. Trud. Plovdiv. Univ. 'Paisii
1 Khilendarski'*, 11(1):123–132(1973)—(Rus,Fr).

1451 Phitos, D. *and* Kamari, G.
Cytologische Untersuchungen an einigen griechischen Blütenpflanzen (*Centaurea, Aethionema*). *Biol. Gallo-Hellen.*, 4(2):147–151(1973)—(Fr,En,Gre);
1 chromosome numbers.

1452 Plitmann, U.
Taxonomic studies in *Centaurea* sect. *Calcitrapa*: 1. New taxa in Palestine. *Israel J. Bot.*, 22(1):47–66
2 (1973)—Illustrations, dot maps; keys.

chrysolepis Vis.
f. **subcitrina** Gajić in *Glasn. Prir. Muz. Belgradu*,
1 27:34(1972)—Jugoslavia.
hyalolepis Boiss.
var. **brevispina** Plitmann in *Israel J. Bot.*, 22(1):
2 50(1973)—Israel, Jordan.
subsp. **negevensis** Plitmann *op. cit.* 50(1973)—
2 Israel.
iberica Trev.
var. **mesocentroides** Plitmann *op. cit.* 49(1973)—
2 Israel.
macroptilon Borb.
subsp. **oxylepis** (W. & Gr.) Soó in *Acta Bot. Acad. Sci. Hung.*, 18(1–2):176(1973): *C. jacea* subvar.
1 *oxylepis.*
montana L.
f. **semidecurrens** (Jordan) López-Fernández in
1 *Pirineos*, 105:30(1972): *C. semidecurrens.*
nervosa Willd.
f. **kojicii** Gajić in *Glasn. Prir. Muz. Belgradu*, 27:
1 34(1972)—Jugoslavia.
nigra L.
subsp. carpetana (Bss. R.) Malagarriga, Las Subesp. y Variac. Geogr.: 15(1973), without basionym
1 date: *C. carpetana.*
subsp. **debeauxii** (G.G.) Malagarriga *op. cit.* 15
1 (1973): *C. debeauxii.*
subsp. microptilon (Gren.) Malagarriga *op. cit.*
1 15(1973), without basionym date: *C. microptilon.*
subsp. nevadensis (Bss. R.) Malagarriga *op. cit.* 15
1 (1973), without basionym date: *C. nevadensis.*
subsp. **rivularis** (Brotero) Malagarriga *op. cit.* 15
1 (1973): *C. nigra* var. *pallida.*
pallescens Del.
var. **australis** Plitmann in *Israel J. Bot.*, 22(1):59
2 (1973)—Israel.
subsp. **calcitrapella** (Bornm. & Dinsm.) Plitmann
2 *op. cit.* 59(1973): *C. calcitrapella.*
subsp. **idumaea** Plitmann *op. cit.* 60(1973)—
2 Jordan.
subsp. **kurnubensis** (Eig) Plitmann *op. cit.* 60
2 (1973): *C. calcitrapella* subsp. *kurnubensis.*
procurrens Sieb. ex Spreng.
subsp. **sharonensis** Plitmann *op. cit.* 53(1973)—
2 Israel.
scabiosa L.
f. **dostalii** Soó in *Acta Bot. Acad. Sci. Hung.*, 18(1–2): 176(1973), nom. nov.: *C. scabiosa* subf.
1 *homoephylla* Dostál.
f. **subheterophylla** Soó *op. cit.* 177(1973), nom.
1 nov.: *C. scabiosa* subf. *variifolia* Dostál.

CENTIPEDA

1453 Dakshini, K. M. M., Prithipalsingh *and* Maheshwari, J. K.
Distributional records of *Soliva anthemifolia*, *Centipeda minima* and *Merremia emarginata* in and near
6 Delhi. *Curr. Sci.*, 42(11):396–397(1973).

CHAETOPAPPA

1454 Van Horn, G. S.
The taxonomic status of *Pentachaeta* and *Chaetopappa* with a revision of *Pentachaeta*. *Univ. Calif. Publ. Bot.*, 65. 41p.(1973)—Illustrations, dot map; chromosome
3, 4 numbers; keys.

CHARTOLEPIS

demirizii (Wagenitz) Holub in *Preslia*, 45(2):143
2 (1973): *Centaurea demirizii.*

CHEIROLOPHUS

argutus (Nees) Holub in *Preslia*, 45(2):143(1973):
1 *Centaurea arguta.*
benoistii (Humb.) Holub *op. cit.* 143(1973):
1 *Centaurea benoistii.*
canariensis (Willd.) Holub *op. cit.* 143(1973):
1 *Centaurea canariensis.*
tananicus (Maire) Holub *op. cit.* 143(1973):
1 *Centaurea tananica.*

CHONDRILLA

1455 Hull, V. J. *and* Groves, R. H.
Variation in *Chondrilla juncea* L. in southeastern Australia. *Austral. J. Bot.*, 21(1):113–135(1973)—
7 Illustrations, maps.

CHRYSANTHELLUM

tuberculatum (Hook. & Arn.) Cabrera in *Bol. Soc. Argent. Bot.*, 15(1):117(1973): *Adenospermum*
4 *tuberculatum.*

CHRYSANTHEMOIDES

1456 Skinner, G. C.
Cape trees no. 11: *Chrysanthemoides monilifera. East. Cape Nat.*, no. 49:9(1973)—Illustration.
5 Popular note.

CHRYSANTHEMUM

1457 Balbaa, S. I., Abdel-Kader, E. M., Abdel-Wahab, S. M., Zaki, A. Y. *and* El-Shamy, A. M.
A pharmacognostical study of *Chrysanthemum cinerariaefolium* Visiani growing in Egypt: part 1. Macro- and micromorphological study of the stem, the leaf, the rhizome and the root. *United Arab Republ. J. Pharm. Sci.*, 12(1):169–189(1972)—(Ara);
2 illustration.

1458 Balbaa, S. I., Abdel-Kader, E. M., Abdel-Wahab, S. M., Zaki, A. Y. *and* El-Shamy, A. M.
A pharmacognostical study of *Chrysanthemum cinerariaefolium* Visiani growing in Egypt: part 2. Macro- and micromorphology of the flower and fruit. *United Arab Republ. J. Pharm. Sci.*, 12(1):
2 191–207(1972)—(Ara).

1459 Ronald, W. G. *and* Ascher, P. D.
Self compatibility in *Chrysanthemum morifolium*
2 Ramat. *Incompat. Newsl.*, no. 3:27–28(1973).

1460 Schantz, M. V. *and* Forsén, K.
Chemotypen von *Chrysanthemum vulgare* (L.) Bernh. *In* Nobel Symposium 25: chemistry in botanical classification August 20–25, 1973. (1973). 2p.
1, 2 Abstract.

CHRYSOTHAMNUS

1461 Anderson, L. C.
Unique *Chrysothamnus* hybridization in Ash Meadows, Nevada. *Bull. Torrey Bot. Club*, 100(3):
3 171–177(1973).

CIRSIUM

1462 Anon.
Cirsium japonicum DC. × *C. norikurense* Nakai.
2 *Acta Phytotax. Geobot.*, 25(4–6):189(1973)—La,Ja.

1463 Greuter, W. *and* Dittrich, M.
Neuer Beitrag zur Kenntnis der Gattung *Lamyropsis* (Compositae): die Identität von *Cirsium microcephalum* Moris. (A new contribution to the knowledge of the genus *Lamyropsis* (Compositae): the identity of *Cirsium microcephalum* Moris). *Ann. Mus.*
1 *Goulandris*, 1:85–98(1973)—(En,Gre); illustrations.

1464 Moldenke, H. N.
Notes on new and noteworthy plants: 60.
3 *Phytologia*, 26(4):224–226(1973).

horridulum Michx.
f. **atropurpureum** Moldenke in *Phytologia*, 26(4):
3 224(1973)—U.S.A. (Alabama).
vulgare (Savi) Ten.
f. **erdneri** Soó in *Acta Bot. Acad. Sci. Hung.*, 18(1–2):
177(1973), nom. nov.: *C. vulgare* var.
1 *oligocephalum* Erdner.

COLOBIUM

hispidum (Roth) Holub in *Folia Geobot. Phytotax.*,
1 8(2):171(1973): *Thrincia hispida.*
maroccanum (Pers.) Holub *op. cit.* 171(1973):
1 *Thrincia maroccana.*
tingitanum (Boiss. & Reut.) Holub *op. cit.* 171
1 (1973): *Thrincia tingitana.*
tuberosum (L.) Holub *op. cit.* 171(1973): *Leontodon*
1, 2 *tuberosus.*

× **COLYMBACOSTA** Rauschert in *Feddes Repert.*,
83(9–10):656(1972 publ. 1973). [*Acosta* ×
1, 2 *Colymbada.*]

COLYMBADA

acicularis (Sibth. & Sm.) Holub in *Preslia*, 45(2):143
1, 2 (1973): *Centaurea acicularis.*
alpestris (Hegetschw.) Rauschert in *Feddes Repert.*,
1 83(9–10): 656(1972 publ. 1973): *Centaurea alpestris.*
araratica (Aznav.) Holub in *Preslia*, 45(2):143(1973):
2 *Centaurea araratica.*
chrysolepis (Vis.) Holub *op. cit.* 143(1973):
1, 2 *Centaurea chrysolepis.*
chrysantha (Wagenitz) Holub *op. cit.* 143(1973):
2 *Centaurea chrysantha.*
davisii (Wagenitz) Holub *op. cit.* 144(1973):
2 *Centaurea davisii.*
glehnii (Trautv.) Holub *op. cit.* 144(1973): *Centaurea*
1 *glehnii.*
litardierei (Jahand. & Maire) Holub *op. cit.* 144
1 (1973): *Centaurea litardierei.*
maireana (Emberger) Holub *op. cit.* 144(1973):
1 *Centaurea maireana.*
malinvaldiana (Battand.) Holub *op. cit.* 144(1973):
1 *Centaurea malinvaldiana.*
mannagettae (Podp.) Holub *op. cit.* 144(1973):
1 *Centaurea mannagettae.*
menteyrica (Chaix) Holub *op. cit.* 144(1973):
1 *Centaurea menteyrica.*
mixta (DC.) Holub *op. cit.* 144(1973): *Centaurea*
1 *mixta.*
nana (Desf.) Holub *op. cit.* 144(1973): *Centaurea*
1 *nana.*
nimrodis (Boiss. & Hausskn.) Holub *op. cit.* 144
2 (1973): *Centaurea nimrodis.*
pseudoscabiosa (Boiss. & Buhse) Holub *op. cit.* 144
1 (1973): *Centaurea pseudoscabiosa.*
psilacantha (Boiss. & Heldr.) Holub *op. cit.* 144
1, 2 (1973): *Centaurea psilacantha.*
rumelica (Boiss.) Holub *op. cit.* 144(1973): *Centaurea*
1 *rumelica.*
sadlerana (Janka) Holub *op. cit.* 144(1973), as
1 'sadleriana': *Centaurea sadleriana.*
scabiosa (L.) Rauschert in *Feddes Repert.*, 83(9–10):
1 656(1972 publ. 1973): *Centaurea scabiosa.*
scopulorum (Boiss. & Heldr.) Holub in *Preslia*,
1, 2 45(2):144(1973): *Centaurea scopulorum.*
scyria (Runemark) Holub *op. cit.* 144(1973):
1 *Centaurea scyria.*
sosnowskyi (Grossh.) Holub *op. cit.* 144(1973):
2 *Centaurea sosnowskyi.*
stereophylla (Bess.) Holub *op. cit.* 144(1973):
1, 2 *Centaurea stereophylla.*
× **sylvatica** (Pourr.) Holub *op. cit.* 144(1973):
1 *Centaurea* × *sylvatica.*
tauromenitana (Guss.) Holub *op. cit.* 144(1973):
1 *Centaurea tauromenitana.*
urvillei (DC.) Holub *op. cit.* 144(1973): *Centaurea*
1, 2 *urvillei.*

× **CONYGERON** Holub in *Folia Geobot. Phytotax.*,
8(2):156,176(28 Aug. 1973), nom. illegit. (see
1 *Conyzigeron*). [*Conyza* × *Erigeron.*]
huelsenii (Vatke) Holub in *Folia Geobot. Phytotax.*,
1 8(2):176(1973): *Erigeron huelsenii.*

CONYZA

1465 Cabrera, A. L.
Tres nuevas especies del genero *Conyza* (Compositae) del noroeste de la Argentina. *Bol. Soc. Argent. Bot.*, 14(4):347–353(1972)—(En);
4 illustrations.

1466 Marshall, J. B.
Conyza—taxa found in Britain. *Watsonia*, 9(4):
1 372–373(1973)—Key.

altoandina Cabrera in *Bol. Soc. Argent. Bot.*, 14(4):
4 347(1972)—Argentina.
floribunda Kunth
var. **subleiotheca** (Cuatr.) J. B. Marshall in
Watsonia, 9(4):372(1973): *C. bonariensis* f.
1, 4 *subleiotheca.*
glandulitecta Cabrera in *Bol. Soc. Argent. Bot.*,
4 14(4):349(1972)—Argentina.
4 **saltensis** Cabrera *op. cit.* 351(1972)—Argentina.

× **CONYZIGERON** Rauschert in *Feddes Repert.*,
83(9–10):656(1972 publ. 1973). [*Conyza* ×
1 *Erigeron.*]
huelsenii (Vatke) Rauschert in *Feddes Repert.*, 83
1 (9–10): 656(1972 publ. 1973): *Erigeron* × *huelsenii.*

COREOPSIS

1467 Kapelev, I. G.
Vidȳ koreopsisa kak efirnomaslichnȳe rasteniya. (The species of *Coreopsis* L. as essential-oil plants).
4–6 *Rast. Resursȳ*, 9(1):76–81(1973).

1468 Moldenke, H. N.
Notes on new and noteworthy plants: 60.
3 *Phytologia*, 26(4):224–226(1973).

1469 Smith, E. B.
A biosystematic study of *Coreopsis saxicola* (Compositae). *Brittonia*, 25(2):200–208(1973)—
3 Chromosome numbers.

grandiflora Hogg ex Sweet
var. **saxicola** (Alex.) E. B. Smith in *Brittonia*,
3 25(2):207(1973): *C. saxicola.*
lanceolata L.
var. **pumila** Moldenke in *Phytologia*, 26(4):224
3 (1973)—U.S.A. (Georgia).

CORRELLIA A. M. Powell in *Brittonia*, 25(2):116
4 (1973).

1470 Powell, A. M.
Correllia (Compositae: Peritylinae): a new monotypic genus from southern Chihuahua, Mexico.
4 *Brittonia*, 25(2):116–118(1973)—Illustrations.

montana A. M. Powell in *Brittonia*, 25(2):116(1973)—
4 Mexico.

COTULA

leariloba (DC.) Hilliard in *Notes Roy. Bot. Gard.*
5 *Edinburgh*, 32(3):332(1973): *Tanacetum linearilobum.*

membranifolia Hilliard *op. cit.* 332(1973)—South
5 Africa (Cape Prov., Natal).

paludosa Hilliard *op. cit.* 333(1973)—South Africa
5 (Cape Prov., Natal), Lesotho.

socialis Hilliard *op. cit.* 334(1973)—South Africa
5 (Cape Prov., Natal), Lesotho.

COUSINIA

1471 Huber-Morath, A.
Die anatolischen Arten der Gattung *Cousinia* Cass.
Ber. Schweiz. Bot. Ges., 82(3):223–268(1972)—
2 (En); key.

1472 Stanyukovich, K. V., Eremina, N. K. *and*
Stanyukovich, M. B.
Novȳe dannȳe o kolyuchetrav'yakh Tadzhikistana.
(New data on Tadjikistan prickly grass). *Dokl. Akad.*
2 *Nauk Tadzh. SSR*, 15(5):62–65(1972)—(Ta).

× paraatripurpurea Kamelin in *Bot. Zhurn.*, 58(5):
632(1973), without latin descr.—U.S.S.R. (Middle
2 Asia). (*C. microcarpa* × *C. radians*).

CRASSOCEPHALUM

1473 Henderson, R. J. F.
Crassocephalum crepidioides (Benth.) S. Moore in
Australia. *Proc. Roy. Soc. Queensl.*, 84:55–60(1972
publ. 1973)—Illustration, dot map; chromosome
7 numbers.

CREPIS

1474 Duvigneaud, J.
Complément à la distribution et à l'écologie de
Crepis pusilla (Sommier) Merxm. (Asteraceae).
1 *Lejeunia*, no. 71. 8p.(1973).

1475 Gerasimova-Havashina, E. N. *and* Gulyaev, V. A.
Nekotorȳe dannȳe ob ul'trastrukture kletok
zarodȳshevogo meshka *Crepis capillaris* (L.) Wallr.
posle opȳleniya. *Byull. Glavn. Bot. Sada* (Moscow),
1–3 89:14–20(1973)—Chromosome number.

1476 Houseman, F. *and* Leslie, A.
Field note: *Crepis mollis* at Buckden. *Naturalist*,
1 no. 927:151(1973).

1477 Kuvaev, V. B. *and* Demidova, L. S.
O novom vide skerdȳ (*Crepis* L.) s polyarnogo
Urala. (De generis *Crepis* L. specie nova e parte
polari montium Uralensium). *Novosti Sist. Vȳssh.*
1 *Rast.*, 10:275–276(1973).

albescens Kuvaev & L. S. Demidova in *Novosti Sist.*
1 *Vȳssh. Rast.*, 10:275(1973)—U.S.S.R. (Urals).

capillaris L.
f. **bischoffii** Soó in *Acta Bot. Acad. Sci. Hung.*,
18(1–2):177(1973), nom. nov.: *V. virens* var.
1 *diffusa* Bischoff.
f. **integra** (Bischoff) Soó *op. cit.* 177(1973):
1 *C. virens* f. *integra.*

froelichiana DC.
subsp. **dinarica** (Beck) Gutermann in *Österr. Bot.*
Zeitschr., 122(4):260(1973): *C. incarnata* var.
1 *dinarica.*

CROCIDIUM

1478 Ornduff, R., Saleh, N. A. M. *and*
Bohm, B. A.
The flavonoids and affinities of *Blennosperma* and
3, 4 *Crocidium* (Compositae). *Taxon*, 22(4):407–412(1973).

CYANUS

achtarovii (Urum.) Holub in *Preslia*, 45(2):144
1 (1973): *Centaurea achtarovii.*

canus (Sibth. & Sm.) Holub *op. cit.* 144(1973):
1, 2 *Centaurea cana.*

epirotus (Hal.) Holub *op. cit.* 144(1973): *Centaurea*
1 *epirota.*

germanicopolitanus (Bornm.) Holub *op. cit.* 144
2 (1973): *Centaurea germanicopolitana.*

lanigerus (DC.) Holub *op. cit.* 144(1973): *Centaurea*
2 *lanigera.*

lingulatus (Lag.) Holub *op. cit.* 144(1973): *Centaurea*
1 *lingulata.*

mentiens (Čerep.) Holub *op. cit.* 144(1973):
1 *Centaurea mentiens.*

pichleri (Boiss.) Holub *op. cit.* 144(1973): *Centaurea*
1, 2 *pichleri.*

pseudoaxillaris (Stef. & Georg.) Holub *op. cit.* 145
1 (1973): *Centaurea pseudoaxillaris.*

reuteranus (Boiss.) Holub *op. cit.* 145(1973), as
1, 2 'reuterianus': *Centaurea reuteriana.*

semi-decurrens (Jord.) Holub *op. cit.* 145(1973):
1, 2 *Centaurea semi-decurrens.*

thirkei (Schultz-Bip.) Holub *op. cit.* 145(1973):
1, 2 *Centaurea thirkei.*

CYNAROIDES (Boiss. ex Walp.) Dostál in *Acta*
Bot. Acad. Sci. Hung., 19(1–4):77(1973): *Centaurea*
1, 2 sect. *Cynaroides.*

regia (Boiss.) Dostál *op. cit.* 77(1973): *Centaurea*
1, 2 *regia.*

DACTYOTRICHIA Wild in *Garcia de Orta, Sér.*
5 *Bot.*, 1(1–2):67(1973).

1479 Wild, H.
A new genus of Compositae (Astereae) from the
Flora Zambesiaca area. *Garcia de Orta, Sér. Bot.*, 1
5 (1–2):67–68(1973)—(Por); illustrations.

robinsonii Wild in *Garcia de Orta, Sér. Bot.*, 1(1–2):
5 67(1973)—Zambia.

DIPLOSTEPHIUM

asplundii Cuatrecasas in *Ciencia*, 27(6):181(1972)—
4 Ecuador.

camargoanum Cuatrecasas *op. cit.* 180(1972)—
4 Colombia.

floribundum
subsp. **llanganatense** Cuatrecasas *op. cit.* 181
4 (1972)—Ecuador.

ritterbushii Cuatrecasas *op. cit.* 179(1972)—
4 Colombia.

4 **sagasteguii** Cuatrecasas *op. cit.* 182(1972)—Peru.

DITTRICHIA Greuter in *Exsicc. Genav.*, 4:71
1 (1973), nom. nov.: *Cupularia* Godron & Gren.

1480 Tutin, T. G.
Flora Europaea: notulae systematicae ad floram
Europaeam spectantes: no. 14. Compositae (230)
Dittrichia viscosa subsp. *revoluta* (Hoffmanns. & Link)
1 P. Silva & Tutin. *Bot. J. Linn. Soc.*, 67(3):282(1973).

graveolens (L.) Greuter in *Exsicc. Genav.*, 4:71
1 (1973): *Erigeron graveolens.*

viscosa (L.) Greuter *op. cit.* 71(1973): *Erigeron*
1 *viscosus.*
subsp. **revoluta** (Hoffmanns. & Link) P. Silva &
Tutin in *Bot. J. Linn. Soc.*, 67(3):282(1973):
1 *Inula revoluta.*

DORONICUM

1481 Edmondson, J. R.
Notes on *Doronicum* L. in S.W. Asia. *Notes Roy.*
2 *Bot. Gard. Edinburgh*, 32(2):255–258(1973).

1482 Gajić, M. R.
Nekoliko novih formi u flori Srbije. (Einige neue Formen in der Flora Serbiens). *Glasn. Prir. Muz.*
1 *Beogradu*, 27:31–35(1972)—(Ge); illustrations.

austriacum Jacq.
f. **longisepala** Gajić in *Glasn. Prir. Muz. Beogradu*,
1 27:32(1972)—Jugoslavia.
bithynicum Edmondson in *Notes Roy. Bot. Gard.*
2 *Edinburgh*, 32(2):257(1973)—Turkey.
subsp. **sparsipilosum** Edmondson *op. cit.* 258
2 (1973)—Turkey.
bracteatum Edmondson *op. cit.* 257(1973)—
2 Iraq.
hakkiaricum Edmondson *op. cit.* 255(1973)—
2 Turkey.
2 **tobeyi** Edmondson *op. cit.* 256(1973)—Turkey.

DUBAUTIA

1483 Kobayashi, H. K.
Putative generic hybrids of Haleakala's silversword and kupaoa (*Argyroxiphium sandwicense* × *Dubautia menziesii*) Compositae. *Pac. Sci.*, 27(2):207–208
6 (1973)—Illustration.

DYSSODIA

1484 Flyr, D.
Dyssodia tenuiloba (Compositae): new to Mississippi.
3 *Sida*, 5(2):130(1973)—Chromosome number.

ECHINACEA

1485 Cody, W. J. *and* Boivin, B.
Purple coneflower, *Echinacea purpurea*, in Ontario.
3 *Canad. Field.-Nat.*, 87(1):70(1973).

ELEPHANTOPUS

plurisetus (O. Hoffm.) Clonts in *Diss. Abstr. Int.*, *B*, 33(4):1423(1972), without basionym ref.:
5 (Basionym not stated).

EMILIA

1486 Fosberg, F. R.
Emilia (Compositae) in Ceylon. *Ceylon J. Sci.*, *Biol. Sci.*, 10(1):61–69(1972) (Critical Notes on
6 Ceylon Plant: 6)—Key.

1487 Olorode, O.
Identification of the genomic complements of *Emilia praetermissa* (Senecioneae—Compositae). *Amer. J. Bot.*, 60(1):55–60(1973)—Chromosome
5 numbers.

1488 Olorode, O. *and* Olorunfemi, A. E.
The hybrid origin of *Emilia praetermissa* (Senecioneae: Compositae). *Ann. Bot.* (U.K.), 36(149):
5 185–191(1973)—Chromosome numbers.

alstonii Fosberg in *Ceylon J. Sci., Biol. Sci.*, 10(1):66
6 (1972)—Sri Lanka.
6 **baldwinii** Fosberg *op. cit.* 68(1972)—Sri Lanka.
6 **exserta** Fosberg *op. cit.* 65(1972)—Sri Lanka.
6 **speeseae** Fosberg *op. cit.* 67(1972)—Sri Lanka.

ENCELIOPSIS

1489 Cronquist, A.
A new variety of *Enceliopsis nudicaulis* (Asteraceae) from southern Nevada. *Bull. Torrey Bot. Club*, 99(5):
3 246–247(1972)—Illustration.

nudicaulis (Gray) A. Nels.
var. **corrugata** Cronquist in *Bull. Torrey Bot. Club*,
3 99(5):246(1972 publ. 1973)—U.S.A. (Nevada).

ERIACHAENIUM

1490 Robinson, H. *and* Brettell, R. D.
Tribal revisions in the Asteraceae: 6. The relation-
4 ship of *Eriachaenium*. *Phytologia*, 26(2):71–72(1973).

ERIGERON

1491 Ling, Yong *and* Chen, Yi-ling
Notulae de genere *Erigeron* L. generibusque affinibus florae sinicae. *Acta Phytotax. Sin.*, 11(4):
2 399–430(1973)—Ch; illustrations; key.

1492 Spongberg, S. A.
A new Alaskan species of *Erigeron* L. (Compositae). *Rhodora*, 75(801):116–119(1973)—Illustrations;
3 chromosome numbers.

ser. **Breviscapi** Ling & Y. L. Chen in *Acta Phytotax.*
2 *Sin.*, 11(4):408(1973).
ser. **Himalajenses** Ling & Y. L. Chen *op. cit.* 418
2 (1973).
ser. **Multiradiati** Ling & Y. L. Chen *op. cit.* 410
2 (1973).
subsect. **Patentisquami** Ling & Y. L. Chen *op. cit.*
2 413(1973).
subsect. **Platyglossi** (Botsch.) Ling & Y. L. Chen
2 *op. cit.* 405(1973): *Erigeron* sect. *Platyglossae*.
subsect. **Siphonoglossi** (Botsch.) Ling & Y. L. Chen
2 *op. cit.* 421(1973): *Erigeron* sect. *Siphonoglossae*.
breviscapus (Vant.) Hand.-Mazz.
var. **alboradiatus** Ling & Y. L. Chen *op. cit.* 409
2 (1973)—China.
var. **tibeticus** Ling & Y. L. Chen *op. cit.* 409
2 (1973)—Tibet.
hultenii Spongberg in *Rhodora*, 75(801):116(1973)—
3 Alaska.
kiukiangensis Ling & Y. L. Chen in *Acta Phytotax.*
2 *Sin.*, 11(4):412(1973)—China.
kunshanensis Ling & Y. L. Chen *op. cit.* 417(1973)—
2 China.
leucoglossus Ling & Y. L. Chen *op. cit.* 411(1973)—
2 Tibet.
multifolius Hand.-Mazz.
var. **amplisquamus** Ling & Y. L. Chen *op. cit.*
2 416(1973)—China.
var. **pilanthus** Ling & Y. L. Chen *op. cit.* 415
2 (1973)—China.
multiradiatus (Wall.) Benth.
var. **glabrescens** Ling & Y. L. Chen *op. cit.* 411
2 (1973)—China.
var. **ovatifolius** Ling & Y. L. Chen *op. cit.* 411
2 (1973)—China.
var. **salicifolius** Chang ex Ling & Y. L. Chen
2 *op. cit.* 411(1973)—China.
porphyrolepis Ling & Y. L. Chen *op. cit.* 416(1973)—
2 Tibet.
pseudoseravschanicus Botsch.
f. **glabrescens** Ling & Y. L. Chen *op. cit.* 426
2 (1973)—China.
purpurascens Ling & Y. L. Chen *op. cit.* 419(1973)—
2 China.
taipeiensis Ling & Y. L. Chen *op. cit.* 406(1973)—
2 China.
tenuicaulis Ling & Y. L. Chen *op. cit.* 418(1973)—
2 China.

ERINACELLA (Rechinger f.) Dostál in *Acta Bot. Acad. Sci. Hung.*, 19(1–4):77(1973): *Centaurea* sect.
2 *Erinacella*.
rechingeri Dostál *op. cit.* 77(1973), nom. nov.:
2 *Centaurea erinacella* Rechinger f.

COMPOSITAE

ERIOPHYLLUM

1493 Mooring, J. S.
Chromosome counts in *Eriophyllum* and other Helenieae (Compositae). *Madroño*, 22(2):95–97
3 (1973).

ESPELETIA

1494 Rock, B. N.
Vegetative anatomy of *Espeletia* (Compositae).
4 *Diss. Abstr. Int., B*, 33(11):5175(1973).

garciae Cuatrecasas in *Ciencia*, 27(6):176(1972)—
4 Colombia.
4 **jabonensis** Cuatrecasas *op. cit.* 178(1972)—Venezuela.
4 **ruizii** Cuatrecasas *op. cit.* 176(1972)—Venezuela.
4 **ulotricha** Cuatrecasas *op. cit.* 177(1972)—Venezuela.

EUPATORIUM

carsticola Borhidi & Muñiz in *Acta Bot. Acad. Sci.*
4 *Hung.*, 18(1–2):45(1973)—Cuba.
× **polyneuron** (F. J. Herm.) Wunderlin in *Ann. Missouri Bot. Gard.*, 59(3):472(1972 publ. 1973):
3 *E. serotinum* var. *polyneuron*.

FELICIA

1495 Grau, J.
Revision der Gattung *Felicia* (Asteraceae). *Mitt. Bot. Staatssamml. München*, 9:195–705(1973)—Illustrations,
5 dot maps; key.

sect. **Dracontium** Grau in *Mitt. Bot. Staatssamml.*
5 *München*, 9:468(1973).
5 sect. **Lignofelicia** Grau *op. cit.* 275(1973).
5 sect. **Longistylus** Grau *op. cit.* 443(1973).
sect. **Neodetris** Grau *op. cit.* 476(1973), nom. nov.:
5 *Agathaea* sect. *Detris* (Adans.) DC.
abyssinica Sch. Bip. ex Rich.
subsp. **globularioides** (Mattf.) Grau *op. cit.* 407
5 (1973): *F. globularioides*.
subsp. **neghellensis** (Cufodontis) Grau *op. cit.*
5 404(1973): *F. neghellensis*.
aculeata Grau *op. cit.* 515(1973), nom. nov.: *Aster*
5 *hirtus* (DC.) Harv.
aethiopica (Burm. f.) Bolus & Wolley Dod ex Adams. & Salt.
subsp. **ecklonis** (Lees.) Grau *op. cit.* 508(1973):
5 *Aster ecklonis*.
5 **alba** Grau *op. cit.* 456(1973)—South West Africa.
amoena (Sch. Bip.) Levyns
subsp. **latifolia** Grau *op. cit.* 521(1973)—South
5 Africa (Cape Prov.).
subsp. **stricta** (DC.) Grau *op. cit.* 520(1973):
5 *Agathaea stricta*.
annectens (Harv.) Grau *op. cit.* 560(1973): *Aster*
5 *annectens*.
bellidioides Schlechter
subsp. **foliata** Grau *op. cit.* 542(1973)—South
5 Africa (Cape Prov.).
boehmii O. Hoffm.
subsp. **homochroma** (S. Moore) Grau *op. cit.* 450
5 (1973): *F. homochroma*.
brevifolia (DC.) Grau *op. cit.* 324(1973): *Agathaea*
5 *brevifolia*.
5 **caespitosa** Grau *op. cit.* 430(1973)—Lesotho.
canaliculata Grau *op. cit.* 303(1973)—South Africa
5 (Cape Prov.).
clavipilosa Grau *op. cit.* 363(1973)—South West
5 Africa.
subsp. **transvaalensis** Grau *op. cit.* 369(1973)—
5 Rhodesia, Botswana, South Africa (Transvaal).
comptonii Grau *op. cit.* 352(1973)—South Africa
5 (Cape Prov.).
cymbalariae (Aiton) Bolus & Wolley Dod ex Adams. & Salt.
subsp. **ionops** (Harv.) Grau *op. cit.* 491(1973):
5 *Aster cymbalariae* var. *ionops*.
cymbalarioides (DC.) Grau *op. cit.* 538(1973):
5 *Munichia cymbalarioides*.
dentata (Rich.) Dandy
5 subsp. **nubica** Grau *op. cit.* 437(1973)—Sudan.
denticulata Grau *op. cit.* 483(1973)—South Africa
5 (Cape Prov.).
desertii Schlechter ex Grau *op. cit.* 362(1973), as
5 'deserti'—South Africa (Cape Prov.).
5 **diffusa** (DC.) Grau *op. cit.* 534(1973): *Agathaea diffusa*.
subsp. **khamiesbergensis** Grau *op. cit.* 536(1973)—
5 South Africa (Cape Prov.).
ebracteata Grau *op. cit.* 512(1973)—South Africa
5 (Cape Prov.).
esterhuysenii Grau *op. cit.* 316(1973), as 'ester-
5 huyseniae'—South Africa (Cape Prov.).
filifolia (Vent.) Burtt-Davy
subsp. **bodkinii** (Compton) Grau *op. cit.* 294
5 (1973): *Aster bodkinii*.
subsp. **schaeferi** (Dinter) Grau *op. cit.* 292(1973):
5 *Aster schaeferi*.
subsp. **schlechteri** (Compton) Grau *op. cit.* 298
5 (1973): *Aster schlechteri*.
fruticosa (L.) Nichols.
subsp. brevipedunculata (Hutchinson) Grau *op. cit.* 278(1973), without basionym page: *Aster*
5 *brevipedunculatus*.
grantii (Oliv. & Hiern ex Oliv.) Grau *op. cit.* 466
5 (1973): *Erigeron grantii*.
heterophylla (Cass.) Grau *op. cit.* 558(1973):
5 *Charieis heterophylla*.
5 **hirta** (Thunb.) Grau *op. cit.* 371(1973): *Aster hirtus*.
hispida (DC.) Grau *op. cit.* 532(1973): *Agathaea*
5 *hispida*.
hyssopifolia (Berg.) Nees
subsp. **glabra** (DC.) Grau *op. cit.* 356(1973):
5 *F. angustifolia* var. *glabra*.
subsp. **polyphylla** (Harv.) Grau *op. cit.* 358(1973):
5 *Aster serrulatus* var. *polyphyllus*.
joubertinae Grau *op. cit.* 496(1973)—South Africa
5 (Cape Prov.).
subsp. **glabrescens** Grau *op. cit.* 497(1973)—
5 South Africa (Cape Prov.).
linifolia (Harv.) Grau *op. cit.* 510(1973): *Aster*
5 *linifolius*.
merxmuelleri Grau *op. cit.* 554(1973)—South
5 Africa (Cape Prov.).
microcephala Grau *op. cit.* 343(1973)—South
5 Africa (Cape Prov.).
minima (Hutchinson) Grau *op. cit.* 543(1973):
5 *Aster minimus*.
5 **monocephala** Grau *op. cit.* 475(1973)—Lesotho.
muricata (Thunb.) Nees
subsp. **cinerascens** Grau *op. cit.* 394(1973)—South West Africa, Botswana, South Africa (Cape
5 Prov.).
subsp. **strictifolia** Grau *op. cit.* 396(1973)—South
5 Africa (Transvaal).
nigrescens Grau *op. cit.* 350(1973)—South Africa
5 (Cape Prov.).
nordenstamii Grau *op. cit.* 336(1973)—South
5 Africa (Cape Prov.).
puberula Grau *op. cit.* 548(1973)—South Africa
5 (Cape Prov.).
oleosa Grau *op. cit.* 319(1973)—South Africa (Cape
5 Prov.).
quinquenervia (Klatt) Grau *op. cit.* 473(1973): *Aster*
5 *quinquenervius*.
serrata (Thunb.) Grau *op. cit.* 427(1973): *Aster*
5 *serratus*.
stenophylla Grau *op. cit.* 529(1973)—South Africa
5 (Cape Prov.).

tenella (L.) Nees
subsp. **cotuloides** (DC.) Grau *op. cit.* 415(1973):
5 *F. cotuloides.*
subsp. **longifolia** (DC.) Grau *op. cit.* 417(1973):
5 *F. tenella* var. *longifolia.*
subsp. **pusilla** (Harv.) Grau *op. cit.* 419(1973):
5 *Aster tenellus* var. *pusillus.*
tenera (DC.) Grau *op. cit.* 482(1973): *Agathaea*
5 *tenera.*
tsitsikamae Grau *op. cit.* 525(1973)—South Africa
5 (Cape Prov.).
uliginosa (Wood & Evans) Grau *op. cit.* 432(1973):
5 *Aster uliginosus.*
welwitschii (Hiern) Grau *op. cit.* 459(1973):
5 *Erigeron welwitschii.*
westiae (Fourc.) Grau *op. cit.* 335(1973), as 'westae':
5 *Aster westae.*
zeyheri (Lees.) Nees
subsp. **linifolia** (Harv.) Grau *op. cit.* 398(1973):
5 *Aster hyssopifolius* var. *linifolius.*

FERREYRELLA

1496 King, R. M. *and* Robinson, H.
Studies in the Eupatorieae (Asteraceae): 112. A new species of *Ferreyrella. Phytologia*, 26(3):167–169
4 (1973)—Illustration.

cuatrecasasii R. M. King & H. Robinson in
4 *Phytologia*, 26(3):167(1973)—Peru.

FILAGO

1497 Howell, J. T.
A new California record for *Filago vulgaris.*
3 *Madroño*, 22(3):151(1973).

GNAPHALIUM

1498 Vuppuluri, S. S. *and* Haridas, P.
Gnaphalium purpureum Linn. (Compositae): a new record for South India. *Curr. Sci.*, 42(10):368–369
6 (1973).

aecidiocephalum (Grierson) L. O. Williams in
4 *Phytologia*, 25(7):459(1973): *Anaphalis aecidiocephala.*
chiliastrum (Mattfeld) van Royen in *Bot. Notiser*,
6 126(4):418(1973): *Raoulia chiliastra.*
limicola Hilliard in *Notes Roy. Bot. Gard. Edinburgh*, 32(3):336(1973)—South Africa (Cape Prov.,
5 Natal), Lesotho.

GORTERIA

diffusa Thunb.
subsp. **calendulacea** (DC.) Roessler in *Mitt. Bot. Staatssamml. München*, 11:93(1973): *G.*
5 *calendulacea.*
subsp. **parviligulata** Roessler *op. cit.* 92(1973)—
5 South West Africa.

GROSSHEIMIA

polyphylla (Ledeb.) Holub in *Preslia*, 45(2):145
1, 2 (1973): *Centaurea polyphylla.*

GYNURA

micheliana J. G. Adam in *Bull. Inst. Fond. Afr.*
5 *Noire, A*, 35(1):81(1973)—Guinea, Mali.

HAPLOPAPPUS

1499 Gill, L. S. *and* Chinnappa, C. C.
A note on the karyology of *Haplopappus spinulosus* (Pursch) DC. ssp. *typicus* Hall. *Rhodora*, 75(802):
3, 4 220–224(1973)—Chromosome numbers.

1500 Jackson, R. C.
Chromosomal evolution in *Haplopappus gracilis*: a generic transposition race. *Evolution*, 27(2):243–256
3, 4 (1973)—Chromosome numbers.

HELENIUM

1501 Baskin, J. M. *and* Baskin, C. C.
Ecological life cycle of *Helenium amarum* in central Tennessee. *Bull. Torrey Bot. Club*, 100(2):117–124
3 (1973)—Illustration.

1502 Bierner, M. W.
Chemosystematic aspects of flavonoid distribution in twenty-two taxa of *Helenium. Biochem. Syst.*,
3 1(1):55–57(1973).

1503 Bierner, M. W.
Sesquiterpene lactones and the systematics of *Helenium quadridentatum* and *H. elegans. Biochem.*
3, 4 *Syst.*, 1(2):95–96(1973).

1504 Bierner, M. W.
Taxonomy of *Helenium* sect. *Tetrodus* and a conspectus of North American *Helenium* (Compositae). *Brittonia*, 24(4):331–355(1972)—Maps, dot maps;
3 chromosome numbers; keys.

3 sect. **Amarum** Bierner in *Brittonia*, 24(4):335(1972).

HELIANTHUS

1505 Georgieva-Todorova, Ĭ.
Tsitologicheskoe issledovanie nekotorȳkh diploidnȳkh (2n = 34) vidov *Helianthus*. (Cytological study of some diploid (2n = 34) species of *Helianthus*).
3 *Tsitol. Genet.*, 7(4):342–345(1973)—(En).

1506 Newcomb, W.
The development of the embryo sac of sunflower *Helianthus annuus* before fertilization. *Canad. J. Bot.*,
3, 4 51(5):863–878(1973)—(Fr).

1507 Newcomb, W.
The development of the embryo sac of sunflower *Helianthus annuus* after fertilization. *Canad. J. Bot.*,
3, 4 51(5):879–890(1973)—(Fr).

HELICHRYSUM

1508 Burtt, B. L. *and* Sunding, P.
Helichrysum monogynum, a new species from Lanzarote, Canary Islands. *Bot. Notiser*, 126(3):
1 340–344(1973)—Illustrations.

adenocarpum DC.
subsp. **ammophilum** Hilliard in *Notes Roy. Bot. Gard. Edinburgh*, 32(3):343(1973)—South Africa
5 (Natal), Mozambique.
basalticum Hilliard *op. cit.* 344(1973), nom. nov.:
5 *Gnaphalium alticola* Compton, as 'alticolum'.
chiliocephalum Hilliard & B. L. Burtt *op. cit.* 345
5 (1973), nom. nov.: *Chiliocephalum schimperi.*
cymosum (L.) D. Don
subsp. **calvum** Hilliard *op. cit.* 346(1973)—South
5 Africa (Natal, Cape Prov.), Lesotho.
dasymallum Hilliard *op. cit.* 347(1973), nom. nov.:
5 *H. lanatum* Harv.
glaciale Hilliard *op. cit.* 348(1973)—South Africa
5 (Cape Prov.), Lesotho.
griseolanatum Hilliard *op. cit.* 349(1973), nom. nov.:
5 *H. epapposum* var. *robustum* Bolus.
hyphocephalum Hilliard *op. cit.* 352(1973)—South
5 Africa (Natal).

lingulatum Hilliard *op. cit.* 352(1973)—South
5 Africa (Cape Prov., Natal), Lesotho.
monogynum B. L. Burtt & Sunding in *Bot. Notiser*,
1 126(3):340(1973)—Canary Is.
paleatum Hilliard in *Notes Roy. Bot. Gard. Edinburgh*,
5 32(3):354(1973)—South Africa (Natal).
pedunculatum Hilliard & B. L. Burtt *op. cit.* 356
5 (1973)—South Africa (Cape Prov.).
petiolare Hilliard & B. L. Burtt *op. cit.* 357(1973),
5 nom. nov.: *Gnaphalium tomentosum* Burm.
praecurrens Hilliard *op. cit.* 359(1973)—South
5 Africa (Natal), Lesotho.
sessilioides Hilliard *op. cit.* 360(1973), nom. nov.:
5 *H. aretioides* Thellung.
silvaticum Hilliard *op. cit.* 361(1973)—South Africa
5 (Natal), Mozambique.
subfalcatum Hilliard *op. cit.* 361(1973)—South
5 Africa (Natal), Lesotho.
vernum Hilliard *op. cit.* 363(1973)—South Africa
5 (Natal).

HELMINTHOTHECA

echioides (L.) Holub in *Folia Geobot. Phytotax.*,
1 8(2):176(1973): *Picris echioides*.

HEPTANTHUS

1509 Borhidi, A.
La taxonomia del género *Heptanthus* Griseb. *Ann. Univ. Budapest. Rolando Eötvos Nom., Biol.*, 14:85–90
4 (1972 publ. 1973)—Map, dot maps; key.

yumuriensis Borhidi in *Ann. Univ. Sci. Budapest. Rolando Eötvos Nom., Biol.*, 14:87(1972 publ. Feb. 1973), and in *Acta Bot. Acad. Sci. Hung.*, 18(1–2):
4 44(28 Feb. 1973)—Cuba.

HETEROTHECA

villosa (Pursh) Shinners
var. **camporum** (Greene) Wunderlin in *Ann. Missouri Bot. Gard.*, 59(3):471(1972 publ. 1973):
3 *Chrysopsis camporum*.

HIERACIUM

1510 Bräutigam, S.
Typen einiger infraspezifischer Taxa von *Hieracium laevigatum* Willd. im Herbarium des Botanischen Museums Berlin-Dahlem. *Willdenowia*, 7(1):105–111(1973) (Beiträge zur Kenntnis von *Hieracium*
1 *laevigatum* Willd.:1)—Illustrations.

1511 Delcourt, E.
Contribution a l'étude cytotaxinomique de *Hieracium pilosella* L. (Contribution to cytotaxinomic studies in *Hieracium pilosella*). *Bull. Soc. Bot. France*, 119(5–6):287–302(1972 publ. 1973)—(En); dot map;
1, 2 chromosome numbers.

1512 Karlsson, L.
Hieracier från Sareks Nationalpark. *Svensk Bot.*
1 *Tidskr.*, 67(3):225–237(1973)—(En).

1513 Óskarsson, I.
A new *Hieracium* species found in North Iceland (Eu-Hieracia—Sect. *Alpina*). *Acta Bot. Islandica*,
1 2:64–66(1973)—Illustration, dot map.

1514 Skalińska, M.
Further studies in facultative apomixis of *Hieracium aurantiacum* L. *Acta Biol. Cracov., Bot.*, 16(1):121–133
1 (1973)—Illustrations; chromosome numbers.

edentulum Oskarsson in *Acta Bot. Island.*, 2:64(1973),
1 without type—Iceland.
heterogynum (Froel.) Gutermann in *Österr. Bot.*
1 *Zeitschr.*, 122(4):262(1973): *Crepis heterogyna*.

HISPIDELLA

hispanica Barnad. ex Lam.
f. **tubuliflora** Pinto da Silva & Teles in *Agron.*
1 *Lusit.*, 34(3):195(1972 publ. 1973)—Portugal.

HYALOSERIS

1515 Ariza Espinar, L.
Revision del genero *Hyaloseris* (Compositae). *Kurtziana*, 7:195–211(1973)—(En); illustrations;
4 chromosome number; key.

HYMENOCLEA

1516 Peterson, K. M. *and* Payne, W. W.
The genus *Hymenoclea* (Compositae: Ambrosieae). *Brittonia*, 25(3):243–256(1973)—Illustrations, dot
3 maps; chromosome numbers; key.

salsola Torr. & Gray ex Gray
var. **fasciculata** (Nelson) K. M. Peterson & Payne
3 in *Brittonia*, 25(3):255(1973): *H. fasciculata*.

HYPOCHOERIS

1517 Asai, Y.
(Miscellaneous notes on the naturalized flora of Japan: 3). *J. Jap. Bot.*, 48(3):67–72(1973)—Ja(En);
2 illustrations. *Hypochoeris glabra*.

IFLOGA

rueppellii (Fresen.) Danin in *Notes Roy. Bot. Gard.*
2 *Edinburgh*, 32(2):269(1973): *Gnaphalium rüppellii*.

INULA

1518 Besold, B.
Pollenmorphologische Untersuchungen an Inuleen. (Angianthinae, Relhaniinae, Athrixiinae). Lehre, J. Cramer, 1971. 72p. (Dissertationes Botanicae,
5 Bd 14).

1519 Hrouda, L.
Československé druhy rodu *Inula* L. (Die Arten der Gattung *Inula* L, in der Tschechoslowakei). *Zprávy Česk. Bot. Spol.*, 8(2):85–95(1973)—(Ge);
1 illustrations; key.

1520 Tutin, T. G.
Flora Europaea: notulae systematicae ad floram Europaeam spectantes: no. 14. Compositae (231)
1 *Inula* L. *Bot. J. Linn. Soc.*, 67(3):282(1973).

candida (L.) Cass.
subsp. **decalvans** (Halácsy) P. W. Ball ex Tutin in *Bot. J. Linn. Soc.*, 67(3):282(1973): *I. candida* var.
1 *decalvans*.
cernua (Dalz.) R. R. Stewart, Annot. Cat. Vasc. Pl.
6 W. Pakistan & Kashmir: 753(1972): *Vicoa cernua*.
clarkei (Hk. f.) R. R. Stewart *op. cit.* 753(1972):
6 *I. obtusifolia* var. *clarkei*.
verbascifolia (Willd.) Hausskn.
subsp. **aschersoniana** (Janka) Tutin in *Bot. J. Linn.*
1 *Soc.*, 67(3):283(1973): *I. aschersoniana*.
subsp. **heterolepis** (Boiss.) Tutin *op. cit.* 283(1973):
1 *I. heterolepis*.
subsp. **methanaea** (Hausskn.) Tutin *op. cit.* 282
1 (1973): *I. methanaea*.
subsp. **parnassica** (Boiss. & Heldr.) Tutin *op. cit.*
1 283(1973): *I. parnassica*.

IOSTEPHANE

papposa Fay in *Brittonia*, 25(2):192(1973)—
4 Mexico.

ISCHNEA

1521 Robinson, H. *and* Brettell, R. D.
Tribal revisions in the Asteraceae: 9. The relationship of *Ischnea*. *Phytologia*, 26(3):153–158(1973)—
6 Keys.

brassii H. Robinson & Brettell in *Phytologia*, 26(3):
6 157(1973)—New Guinea (Terr. New Guinea).

ISOETOPSIS

1522 Robinson, H. *and* Brettell, R. D.
Tribal revisions in the Asteraceae: 7. The relationship of *Isoetopsis*. *Phytologia*, 26(3):73–75
7 (1973)—Illustrations.

JACEA

aterrima (Hayek) Holub in *Preslia*, 45(2):145(1973):
1 *Centaurea aterrima*.
× **austriacoides** (Woloszczak) Rauschert in *Feddes Repert.*, 83(9–10):656(1972 publ. 1973): *Centaurea* ×
1 *austriacoides*.
bosniaca (Murb.) Holub in *Preslia*, 45(2):145(1973):
1 *Centaurea pseudophrygia* subsp. *bosniaca*.
brevipappa (Boiss. & Reut.) Holub *op. cit.* 145
1 (1973): *Centaurea brevipappa*.
carpetana (Boiss. & Reut.) Holub *op. cit.* 145(1973):
1 *Centaurea carpetana*.
debeauxii (Godr. & Gren.) Holub *op. cit.* 145(1973):
1 *Centaurea debeauxii*.
degeniana (J. Wagn.) Holub *op. cit.* 145(1973):
1 *Centaurea degeniana*.
× **delicata** (Nyár.) Rauschert in *Feddes Repert.*, 83(9–10):656(1972 publ. 1973): *Centaurea* ×
1 *delicata*.
× **erdneri** (Wagner) Rauschert *op. cit.* 656(1972 publ.
1 1973): *Centaurea* × *erdneri*.
× **extranea** (Beck) Rauschert in *Feddes Repert.*, 83 (9–10):656(1972 publ. 25 April 1973), also Holub in
1 *Preslia*, 45(2):145(25 April 1973): *Centaurea* ×
1 *extranea*.
gaudinii (Boiss. & Reut.) Holub in *Preslia*, 45(2):145
1, 2 (1973): *Centaurea gaudinii*.
haynaldii (Hayek) Holub *op. cit.* 145(1973):
1 *Centaurea haynaldii*.
hyrcanica (Bornm.) Holub *op. cit.* 145(1973):
2 *Centaurea hyrcanica*.
jordaniana (Gren. & Godr.) Holub *op. cit.* 145
1 (1973): *Centaurea jordaniana*.
× **kuemmerlei** (Prodan & Wagner) Rauschert in *Feddes Repert.*, 83(9–10):656(1972 publ. 1973):
1 *Centaurea* × *kuemmerlei*.
melanocalathia (Czako) Holub in *Preslia*, 45(2):
1 145(1973): *Centaurea melanocalathia*.
microptilon (Godr.) Holub *op. cit.* 145(1973):
1 *Centaurea vulgaris* δ *microptilon*.
× **nanianii** (Prodan) Rauschert in *Feddes Repert.*, 83(9–10):656(1972 publ. 1973): *Centaurea* ×
1 *nanianii*.
neapolitana (Boiss.) Holub in *Preslia*, 45(2):145
1 (1973): *Centaurea neapolitana*.
procumbens (Balbis) Holub *op. cit.* 145(1973):
1, 2 *Centaurea procumbens*.
× **pseudopannonica** (Wagner) Rauschert in *Feddes Repert.*, 83(9–10):656(1972 publ. 1973): *Centaurea* ×
1 *pseudopannonica*.
ropalon (Pomel) Holub in *Preslia*, 45(2):145(1973):
1 *Centaurea ropalon*.
× **semiaustriaca** (Nyár.) Holub *op. cit.* 145(1973):
1 *Centaurea* × *semiaustriaca*.
× **similata** (Hausskn.) Rauschert in *Feddes Repert.*,
1 83(9–10):656(1972 publ. 1973): *Centaurea* × *similata*.
subcordata (Freyn & Sint.) Holub in *Preslia*, 45(2):
1 145(1973): *Centaurea subcordata*.

× **thaiszii** (Wagner) Rauschert in *Feddes Repert.*,
1 83(9–10):656(1972 publ. 1973): *Centaurea* × *thaiszii*.

× **JACEACOSTA** Rauschert in *Feddes Repert.*, 83(9–10):655(1972 publ. 1973). [*Acosta* ×
1, 2 *Jacea*.]
beckiana (F. Müllner) Rauschert in *Feddes Repert.*, 83(9–10):655(1972 publ. 1973): *Centaurea* ×
1 *beckiana*.
fortinata (Wagner) Rauschert *op. cit.* 655(1972 publ.
1 1973): *Centaurea* × *fortinata*.
juvenalis (Delile ex Godr.) Rauschert *op. cit.* 655
1 (1972 publ. 1973): *Centaurea* × *juvenalis*.
pseudomicrantha (Rech. fil.) Rauschert *op. cit.* 655
1 (1972 publ. 1973): *Centaurea* × *pseudomicrantha*.
teyberi (Hayek) Rauschert *op. cit.* 655(1972 publ.
1 1973): *Centaurea* × *teyberi*.

× **JACEITRAPA** Rauschert in *Feddes Repert.*,
1, 2 83(9–10):656(1972 publ. 1973). [*Calcitrapa* × *Jacea*.]
amphibola (Hausskn.) Rauschert *op. cit.* 656(1972
1 publ. 1973): *Centaurea* × *amphibola*.

JURINEA

1523 Akÿeva, M. *and* Vasilevskaya, V. K.
Anatomicheskoe stroenie nagolovatki derderovidnoĭ (*Jurinea derderoides* C. Winkl. v Karakumakh). (Anatomical composition of *Jurinea derderoides* C. Winkl. in Karakum). *Izv. Akad. Nauk Turkm. SSR,*
2 *Biol. Nauk*, 3:73–76(1973)—(En).

1524 Nemirova, E. S.
Novÿe vidÿ nagolovatok (*Jurinea* Cass.) severnogo Kavkaza. (Generis *Jurinea* Cass. species novae Caucasi borealis). *Novosti Sist. Vÿssh. Rast.*, 10:260–269(1973)
2 —Illustrations.

1525 Nemirova, E. S.
Novÿĭ vid *Jurinea* Cass. s Kavkaza. (New species of *Jurinea* Cass. from the Caucasus). *Bot. Zhurn.*, 58(7):
2 995-996(1973)—Illustration.

akinfievii Nemirova in *Novosti Sist. Vÿssh. Rast.*,
2 10:261(1973)—U.S.S.R. (Caucasus).
albovii Galushko & Nemirova *op. cit.* 326(1973)—
2 U.S.S.R. (Caucasus).
brachypappa Nemirova *op. cit.* 266(1973)—U.S.S.R.
2 (Caucasus).
galushkoi Nemirova *op. cit.* 264(1973)—U.S.S.R.
2 (Caucasus).
himalaica R. R. Stewart, Annot. Cat. Vasc. Pl. W. Pakistan & Kashmir: 757(1972), nom. nov.: *J.*
6 *macrocephala* Hook. f.
humilis DC.
subsp. bocconii (Gussone) Malagarriga, Las Subesp. y Variac. Geogr.: 16(1973), as 'bocconi'
1 and without basionym date: *J. bocconi*.
subsp. pyrenaica (G.G.) Malagarriga *op. cit.* 16
1 (1973), without basionym date: *J. pyrenaica*.
prokhanovii Nemirova in *Bot. Zhurn.*, 58(7):995
2 (1973)—U.S.S.R. (Caucasus).

KRIGIA

cespitosa (Raf.) Chambers in *J. Arnold Arbor.*, 54(1):
3 52(1973): *Serinia cespitosa*.

LACTUCA

viminea (L.) Presl.
subsp. chondrilliflora (Boreau) Malagarriga, Las Subesp. y Variac. Geogr.: 16(1973), without
1 basionym date: *L. chondrilliflora*.
subsp. ramosissima (All.) Malagarriga *op. cit.* 17 (1973), without basionym date: *Prenanthes*
1 *ramoisissima*.

LAMYROPSIS

1526 Greuter, W. *and* Dittrich, M.
Neuer Beitrag zur Kenntnis der Gattung *Lamyropsis* (Compositae): die Identität von *Cirsium microcephalum* Moris. (A new contribution to the knowledge of the genus *Lamyropsis* (Compositae): the identity of *Cirsium microcephalum* Moris). *Ann. Mus. Goulandris*,
1 1:85–98(1973)—(En,Gre); illustrations.

microcephala (Moris) Dittrich & Greuter in *Exsicc.*
1 *Genav.*, 3:47(1972): *Cirsium microcephalum.*

LASTHENIA

1527 Ornduff, R., Bohm, B. A. *and* Saleh, N. A. M.
Flavonoids of artificial interspecific hybrids in
4 *Lasthenia. Biochem. Syst.*, 1(3):147–151(1973).

LAUNAEA

1528 Banerjee, G. *and* Deshpande, B. D.
Foliar venation and leaf histology of certain members of Compositae: 1. Members with extreme open type of venation: (a) *Launaea asplenifolia* Hook. f.
1, 2, 5, 6 *Flora* (Germany), 162(6):529–532(1973).

1529 Sarkar, A. K., Dutta, N. *and* Chatterjee, U.
Confirmation on a new basic number for the genus *Launaea* Cass. *Sci. Cult.*, 39(2):95–96(1973)—
1, 2, 5, 6 Chromosome numbers.

LEONTODON

1530 Rousi, A.
Studies on the cytotaxonomy and mode of reproduction of *Leontodon* (Compositae). *Ann. Bot.*
1, 2 *Fenn.*, 10(3):201–215(1973)—Chromosome numbers.

1531 Serizawa, S.
(*Leontodon nudicaulis* naturalized in Japan). *J. Jap.*
2 *Bot.*, 48(7):214(1973)—Ja.

filii (Hochst. ex Seubert) Paiva & Ormonde in *Bol.*
1 *Soc. Brot.*, 46:447(1972 publ. 1973): *Microderis filii.*
rigens (Ait.) Paiva & Ormonde *op. cit.* 448(1972
1 publ. 1973): *Crepis rigens.*

LEONTOPODIUM

1532 Resmerită, I.
Cartarea speciei *Leontopodium aldinum* Cass. din Carpatii românești. *Stud. Circ. Biol., Bot.*, 25(5):385–
1 398(1973)–(Fr); map, dot map.

1533 Tutin, T. G.
Flora Europaea: notulae systematicae ad floram Europaeam spectantes: no. 14. Compositae (232) *Leontopodium* (Pers.) R.Br. *Bot. J. Linn. Soc.*, 67(3):
1 283(1973).

alpinum Cass.
subsp. **nivale** (Ten.) Tutin in *Bot. J. Linn. Soc.*,
1 67(3):283(1973): *Gnaphalium nivale.*

× **LEUCANTANACETUM** Rauschert in *Feddes Repert.*, 83(9–10):656(1972 publ. 1973).
1 (*Leucanthemum* × *Tanacetum.*)
rohlenae (Domin) Rauschert *op. cit.* 656(1972 publ.
1 1973): *Chrysanthemum* × *rohlenae.*

LEUZEA

1534 Holub, J.
Contribution to the taxonomy and nomenclature of *Leuzea* DC. and *Rhaponticum* auct. *Folia Geobot. Phytotax.*, 8(4):377–395(1973)—Chromosome
1, 2 numbers.

subgen. *Cestrinus* (Cass.) Holub in *Folia Geobot.*
1 *Phytotax.*, 8(4):391(1973): genus *Cestrinus.*
subgen. **Eleutherochaetum** (Soskov) Holub *op. cit.*
1 391(1973): *Rhaponticum* sect. *Eleutherochaetum.*
subgen. **Fornicium** (Cass.) Holub *op. cit.* 391(1973):
1, 2 genus *Fornicium.*
subgen. **Rhaponticellum** (Soskov) Holub *op. cit.* 391
2 (1973): *Rhaponticum* sect. *Rhaponticellum.*
1, 2 subgen. **Rhaponticina** Holub *op. cit.* 391(1973).
subgen. **Stemmacantha** (Cass.) Holub *op. cit.* 391
1 (1973): genus *Stemmacantha.*
1 **acaulis** (L.) Holub *op. cit.* 391(1973): *Cynara acaulis.*
annae-bentiae (Rech. f.) Holub *op. cit.* 391(1973):
2 *Rhaponticum annae-bentiae.*
aulieatensis (Iljin) Holub *op. cit.* 391(1973):
2 *Rhaponticum aulieatense.*
caulescens (Coss. & Bal.) Holub *op. cit.* 391(1973):
1 *Rhaponticum caulescens.*
centauroides (L.) Holub *op. cit.* 391(1973): *Cnicus*
1 *centauroides.*
froedinii (Rech. f.) Holub *op. cit.* 391(1973):
2 *Centaurea froedinii.*
imatongensis (Philipson) Holub *op. cit.* 391(1973):
5 *Centaurea imatongensis.*
insignis (Boiss.) Holub *op. cit.* 391(1973): *Phaeopappus*
2 *insignis.*
integrifolia (Winkler) Holub *op. cit.* 391(1973):
1 *Rhaponticum integrifolium.*
karatavica (Regel & Schmalh.) Holub *op. cit.* 391
2 (1973): *Rhaponticum karatavicum.*
lyrata (Winkler ex Iljin) Holub *op. cit.* 391(1973):
2 *Rhaponticum lyratum.*
namanganica (Iljin) Holub *op. cit.* 391(1973):
2 *Rhaponticum namanganicum.*
nana (Lipsky) Holub *op. cit.* 391(1973): *Rhaponticum*
2 *nanum.*
nitida (Fisch.) Holub *op. cit.* 392(1973): *Rhaponticum*
2 *nitidum.*
pulchra (Fisch. & C. A. Mey.) Holub *op. cit.* 392
2 (1973): *Rhaponticum pulchrum.*
rhapontica (L.) Holub *op. cit.* 392(1973): *Centaurea*
1 *rhapontica.*
subsp. **bicknellii** (Briq.) Holub *op. cit.* 392(1973):
1 *Centaurea rhapontica* var. *bicknellii.*
subsp. **heleniifolia** (Grenier & Godron) Holub
1 *op. cit.* 392(1973): *Rhaponticum heleniifolium.*
satzyperovii (Soskov) Holub *op. cit.* 392(1973):
2 *Rhaponticum satzyperovii.*
uniflora (L.) Holub *op. cit.* 392(1973): *Cnicus*
2 *uniflorus.*

LIATRIS

1535 Levin, D. A.
The age structure of a hybrid swarm in *Liatris*
3 (Compositae). *Evolution*, 27(3):532–535(1973).

LIGULARIA

amplectens (A. Gray) W. A. Weber in *Southwest.*
3 *Nat.*, 18(3):321(1973): *Senecio amplectens.*
bigelovii (A. Gray) W. A. Weber *op. cit.* 321(1973):
3 *Senecio bigelovii.*
subsp. **hallii** (A. Gray) W. A. Weber *op. cit.* 321
3 (1973): *Senecio bigelovii* var. *hallii.*
holmii (Greene) W. A. Weber *op. cit.* 321(1973):
3 *Senecio holmii.*
pudica (Greene) W. A. Weber *op. cit.* 321(1973):
3 *Senecio pudicus.*
soldanella (A. Gray) W. A. Weber *op. cit.* 322(1973):
3 *Senecio soldanella.*
taraxacoides (Greene) W. A. Weber *op. cit.* 322
3 (1973): *Senecio taraxacoides.*

MACHAERANTHERA

1536 Turner, B. L.
Machaeranthera restiformis (Asteraceae) a bizarre new gypsophile from northcentral Mexico. *Amer. J. Bot.*, 60(8):836–838(1973)—Illustrations; chromosome
4 numbers.

1537 Turner, B. L.
Transfer of *Aster gentryi* Standl. to the genus *Machaeranthera* sect. *Psilactis*. *Phytologia*, 25(2):57–58
4 (1972).

1538 Turner, B. L.
Two new gypsophilous species of *Machaeranthera* (Asteraceae-Astereae) from North-central Mexico. *Phytologia*, 26(2):116–120(1973)—Chromosome
4 numbers; key.

crutchfieldii B. L. Turner in *Phytologia*, 26(2):116
4 (1973)—Mexico.
gentryi (Standl.) Jackson ex B. L. Turner in *Phytologia*,
4 25(2):57(1972): *Aster gentryi*.
gypsophila B. L. Turner in *Phytologia*, 26(2):117
4 (1973)—Mexico.
restiformis B. L. Turner in *Amer. J. Bot.*, 60(8):836
4 (1973)—Mexico.

MATUDINA R. M. King & H. Robinson in
4 *Phytologia*, 26(3):171(1973).

1539 King, R. M. *and* Robinson, H.
Studies in the Eupatorieae (Asteraceae): 113. A new genus, *Matudina*. *Phytologia*, 26(3):170–173(1973)—
4 Illustrations.

corvii (McVaugh) R. M. King & H. Robinson in
4 *Phytologia*, 26(3):171(1973): *Eupatorium corvii*.

MELAMPODIUM

1540 Manilal, K. S.
Morphology of the capitulum of *Melampodium divaricatum* (Rich.) DC. *Curr. Sci.*, 42(16):578–580
4 (1973).

MELANTHERA

1541 Parks, J. C.
A revision of North American and Caribbean *Melanthera* (Compositae). *Rhodora*, 75(802):169–210 (1973)—Illustrations, dot map; chromosome
3 numbers; key.

aspera (Jacq.) Small
var. **glabriuscula** (O. Kuntze) J. C. Parks in *Rhodora*, 75(802):194(1973): *Amellus asper* var.
3, 4 *glabriusculus*.

MIKANIA

1542 Vaid, K. M.
A preliminary note on the identity of the contro-
2–4, 6 versial *Mikania*. *Indian Forester*, 99(1):19–22(1973).

1543 Van der Walt, J. J. A., Van der Schijff, H. P. *and* Schweickherdt, H. G.
Anomalous secondary growth in the stems of the lianes *Mikania cordata* (Burm. f.) Robins. (Compositae) and *Paullinia pinnata* Linn. (Sapindaceae). *Kirkia*,
4, 5 9(1):109–138(1973)—Illustrations.

NEUROLAENA

1544 Robinson, H. *and* Brettell, R. D.
Tribal revisions in the Asteraceae: 4. The relationships of *Neurolaena*, *Schistocarpha* and *Alepidocline*.
4 *Phytologia*, 25(7):439–445(1973).

NIDORELLA

tongensis Hilliard in *Notes Roy. Bot. Gard. Edinburgh*,
5 32(3):366(1973)—South Africa (Natal).

NOLLETIA

ruderalis Hilliard in *Notes Roy. Bot. Gard. Edinburgh*, 32(3):367(1973)—South Africa (Natal, Transvaal),
5 Swaziland.

NOTICASTRUM

1545 Cuatrecasas, J.
Miscellaneous notes on neotropical flora: 3.
4 *Phytologia*, 25(4):249–256(1973)—Illustrations.

acuminatum (DC.) Cuatrecasas in *Phytologia*, 25(4):
4 249(1973): *Aplopappus acuminatus*.
argentinense (Cabrera) Cuatrecasas *op. cit.* 250(1973):
4 *Aster argentinensis*.
calvatum (Baker) Cuatrecasas *op. cit.* 250(1973):
4 *Leucopsis calvata*.
decumbens (Baker) Cuatrecasas *op. cit.* 250(1973):
4 *Aster decumbens*.
gnaphalioides (Baker) Cuatrecasas *op. cit.* 250(1973):
4 *Leucopsis gnaphalioides*.
macrocephalum (Baker) Cuatrecasas *op. cit.* 251
4 (1973): *Leucopsis macrocephala*.
marginatum (H.B.K.) Cuatrecasas
f. **acaule** (Wedd.) Cuatrecasas *op. cit.* 253(1973), as
4 'acaulis': *Aster acaulis*.
montevidense (Sprengel) Cuatrecasas *op. cit.* 251
4 (1973): *Onoseris montevidensis*.
psammophilum (Klatt) Cuatrecasas *op. cit.* 251
4 (1973): *Aster psammophilus*.

ODONTOLOPHUS

trinervius (Steph. ex Willd.) Dostál in *Acta Bot. Acad.*
1, 2 *Sci. Hung.*, 19(1–4):78(1973): *Centaurea trinervia*.

OLEARIA

1546 Jeffrey, C.
Olearia ilicifolia. *Curtis's Bot. Mag.*, 179(4): tab.
7 654(1973)—En; illustration, coloured illustration.

1547 Jeffrey, C.
Olearia lacunosa. *Curtis's Bot. Mag.*, 179(3): tab.
7 645(1973)—En; illustration, coloured illustration.

OPHRYOSPORUS

1548 King, R. M. *and* Robinson, H.
Studies in the Eupatorieae (Asteraceae): 111. Additions to the genus *Ophryosporus*. *Phytologia*, 25(2):
4 65–67(1972).

anomalus R. M. King & H. Robinson in *Phytologia*,
4 25(2):66(1972), nom. nov.: *Piqueria cumingii*.
burkartii Cabrera in *Bol. Soc. Argent. Bot.*, 15(1):114
4 (1973)—Argentina.
densiflorus (Benth.) R. M. King & H. Robinson in
4 *Phytologia*, 25(2):66(1972): *Piqueria densiflora*.
floribundus (A. P. Decandolle) R. M. King &
4 H. Robinson *op. cit.* 66(1972): *Piqueria floribunda*.
galioides (A. P. Decandolle) R. M. King & H.
4 Robinson *op. cit.* 66(1972): *Piqueria galioides*.
hartwegii (B. L. Robinson) R. M. King & H.
4 Robinson *op. cit.* 66(1972): *Piqueria hartwegii*.
mathewsii (B. L. Robinson) R. M. King & H.
4 Robinson *op. cit.* 66(1972): *Piqueria mathewsii*.
peruvianus (Gmel.) R. M. King & H. Robinson
4 *op. cit.* 66(1972): *Flaveria peruviana*.
pinifolius (Phil.) R. M. King & H. Robinson *op. cit.*
4 66(1972): *Stevia pinifolia*.
pubescens (J. E. Smith) R. M. King & H. Robinson
4 *op. cit.* 66(1972): *Piqueria pubescens*.

ORITROPHIUM

figueirasii Cuatrecasas in *Ciencia*, 27(6): 183(1972)—
4 Venezuela.
4 **granatum** Cuatrecasas *op. cit.* 183(1972)—Venezuela.

OTHONNA

1549 Shields, M. E.
A garden under glass. *Bull. Afr. Succ. Pl. Soc.*,
5 8(5):167–170(1973)—Illustrations.

PALAEOCYANUS Dostál in *Acta Bot. Acad. Sci. Hung.*, 19(1–4):74(1973), nom. nov.: *Centaurea*
1 sect. *Palaeotypus* Nyman.
crassifolius (Bertol.) Dostál *op. cit.* 74(1973):
1 *Centaurea crassifolia.*

PARAPHYSIS (DC.) Dostál in *Acta Bot. Acad. Sci.*
2 *Hung.*, 19(1–4): 76(1973): *Amberboa* sect. *Paraphysis.*
decurrens (DC.) Dostál *op. cit.* 77(1973): *Amberboa*
2 *decurrens.*

PARTHENIUM

1550 Maheshwari, J. K. *and* Pandey, R. S.
Parthenium weed in Bihar State. *Curr. Sci.*, 42(20):
6 733(1973). *P. hysterophorus.*

PENTACHAETA

1551 Van Horn, G. S.
The taxonomic status of *Pentachaeta* and *Chaetopappa* with a revision of *Pentachaeta. Univ. Calif. Publ. Bot.*, 65. 41p.(1973)—Illustrations, dot map; chromosome
3, 4 numbers; keys.

exilis Gray
subsp. **aeolica** van Horn & Ornduff in *Univ. Calif.*
3 *Publ. Bot.*, 65:37(1973)—U.S.A. (California).

PEREZIA

1552 Simpson, B. B.
Contrasting modes of evolution in two groups of *Perezia* (Mutisieae; Compositae) of southern South America. *Taxon*, 22(5–6):525–536(1973)—Illustra-
4 tions, map, dot maps.

PERICOME

1553 Powell, A. M.
Taxonomy of *Pericome* (Compositae-Peritylinae). *Southwest. Nat.*, 18(3):335–339(1973)—Dot map;
3, 4 chromosome numbers; key.

PERITYLE

1554 Powell, A. M.
Taxonomy of *Perityle* section *Laphamia* (Compositae-Helenieae-Peritylinae). *Sida*, 5(2):61–128 (1973)—Illustrations, dot maps, chromosome
3 numbers; keys.

bisetosa (Torr. ex A. Gray) Shinners
var. **spathulata** A. M. Powell in *Sida*, 5(2):85
4 (1973)—Mexico.
coahuilensis A. M. Powell *op. cit.* 120(1973)—
4 Mexico.
cochisensis (Niles) A. M. Powell *op. cit.* 123(1973):
3 *Laphamia cochisensis.*
4 **gentryi** A. M. Powell *op. cit.* 125(1973)—Mexico.
gilensis (Jones) Macbride
var. **salensis** A. M. Powell *op. cit.* 104(1973)—
3 U.S.A. (Arizona).
megalocephala (Wats.) Macbride
var. **intricata** (Brandeg.) A. M. Powell *op. cit.* 94
3 (1973): *Laphamia intricata.*
var. **oligophylla** A. M. Powell *op. cit.* 91(1973)—
3 U.S.A. (California).

PETASITES

1555 Macpherson, P.
Petasites japonicus. Glasgow Nat., 19(1):67(1973)—
1 En. Local records.

PHILOGLOSSA

1556 Robinson, H. *and* Cuatrecasas, J.
Synopsis of the genus *Philoglossa* (Liabeae,
4 Asteraceae). *Phytologia*, 26(5):381–388(1973)—Key.

blakei H. Robinson & Cuatrecasas in *Phytologia*,
4 26(5):383(1973)—Peru.
mimuloides (Hieron.) H. Robinson & Cuatrecasas
4 *op. cit.* 384(1973): *Jaumea mimuloides.*
f. **sapida** (Bristol) H. Robinson & Cuatrecasas
4 *op. cit.* 385(1973): *P. peruviana* var. *sapida.*
f. **subintegrifolia** Hieron. ex H. Robinson &
4 Cuatrecasas *op. cit.* 385(1973)—Bolivia.

PICRADENIOPSIS

1557 Stuessy, T. F., Irving, R. S. *and* Ellison, W. L.
Hybridization and evolution in *Picradeniopsis* (Compositae). *Brittonia*, 25(1):40–56(1973)—
3 Illustrations, dot maps; chromosome numbers.

PICRIS

1558 Kitamura, S.
Picris hieracioides L. subsp. *japonica* Krӯlov var. *litoralis* Kitamura var. nov. *Acta Phytotax. Geobot.*,
2 25(4–6):188–189(1973)—Ja.

1559 Paiva, J. A. R. *and* Ormonde, J.
The species of *Picris* L. from Azores. *Bol. Soc. Brot.*,
1 46:447–448(1972 publ. 1973)—Illustrations.

hieracioides L.
var. **litoralis** Kitamura in *Acta Phytotax. Geobot.*,
2 25(4–6):188(1973)—Japan.

PILOSELLA

1560 Soják, J.
Přehled Československých plemen rodu *Pilosella* Hill. (Übersicht der tschechoslowakischen *Pilosella*-Unterarten.) *Čas. Nár. Mus.* (Prague), 141(1–2):41–61
1 (1972)—(Ge).

acrothyrsa (Näg. & Pet.) Soják
subsp. **fleischerana** (Zahn) Soják in *Čas. Nár. Muz.* (Prague), 141(1–2):57(1972), as 'fleischeriana':
1 *Hieracium acrothyrsum* subsp. *fleischerianum.*
subsp. **pseudoleptophyton** (Zahn) Soják *op. cit.* 57(1972): *Hieracium callimorphoides* subsp.
1 *pseudoleptophyton.*
albulana (Näg. & Pet.) Soják
subsp. **floccosa** (Näg. & Pet.) Soják *op. cit.* 51
1 (1972): *Hieracium florentinum* subsp. *floccosum.*
alpicola (Schleich. ex Steud. & Hochstett.) F. W. & C. H. Schultz
subsp. **furcotae** (Deg. & Zahn) Soják *op. cit.* 42
1 (1972): *Hieracium alpicola* subsp. *furcotae.*
subsp. **ullepitschii** (Błocki) Soják *op. cit.* 42(1972):
1 *Hieracium ullepitschii.*
anchusoides (Arv.-Touv.) Arv.-Touv.
subsp. **floccipedunculoides** (Zahn) Soják *op. cit.* 61(1972): *Hieracium anchusoides* subsp.
1 *floccipedunculoides.*
subsp. **polakiana** (Näg. & Pet.) Soják *op. cit.* 58
1 (1972): *Hieracium germanicum* subsp. *polakianum.*
apatelia (Näg. & Pet.) Soják
subsp. **altenbergensis** (Zahn) Soják *op. cit.*
1 46(1972): *Hieracium apatelium* subsp. *altenbergense.*
subsp. **aupaënsis** (Näg. & Pet.) Soják *op. cit.* 46
1 (1972): *Hieracium pilosellif lorum* subsp. *aupaënse.*

subsp. **bregensis** (Näg. & Pet.) Soják *op. cit.* 46
1 (1972): *Hieracium apatelium* subsp. *bregense.*
subsp. **carpathorum** (Zahn) Soják *op. cit.* 46(1972):
1 *Hieracium apatelium* subsp. *carpathorum.*
subsp. **pontimontana** (Zahn) Soják *op. cit.* 46
1 (1972): *Hieracium apatelium* subsp. *pontimontanum.*
subsp. **pratigena** (Näg. & Pet.) Soják *op. cit.* 46
1 (1972): *Hieracium apatelium* subsp. *pratigenum.*
subsp. **striifera** (Näg. & Pet.) Soják *op. cit.* 46
1 (1972): *Hieracium piloselliflorum* subsp. *striiferum.*
arvicola (Näg. & Pet.) Soják
subsp. **bohemica** (Vollm.) Soják *op. cit.* 57(1972):
1 *Hieracium arvicola* subsp. *bohemicum.*
subsp. **cincinnoclada** (Näg. & Pet.) Soják *op. cit.*
1 57(1972): *Hieracium arvicola* subsp. *cincinnocladum.*
subsp. **javorinensis** (Zahn) Soják *op. cit.* 57(1972):
1 *Hieracium arvicola* subsp. *javorinense.*
subsp. **molendiana** (Näg. & Pet.) Soják *op. cit.* 57
1 (1972): *Hieracium arvicola* subsp. *molendianum.*
subsp. **nosalica** (Näg. & Pet.) Soják *op. cit.* 57
1 (1972): *Hieracium arvicola* subsp. *nosalicum.*
subsp. **pseudocollina** (Näg. & Pet.) Soják *op. cit.*
1 57(1972): *Hieracium arvicola* subsp. *pseudocollinum.*
subsp. **remotiflora** (Näg. & Pet.) Soják *op. cit.*
1 57(1972): *Hieracium arvicola* subsp. *remotiflorum.*
subsp. **subhyperborea** (Peter) Soják *op. cit.* 57
(1972): *Hieracium hyperboreum* subsp. *subhyper-*
1 *boreum.*
atramentaria (Näg. & Pet.) Soják
subsp. **erubescens** (Näg. & Pet.) Soják *op. cit.* 56
1 (1972): *Hieracium floribundum* subsp. *erubescens.*
aurantiaca (L.) F. W. & C. H. Schultz
subsp. **auropurpurea** (Peter) Soják *op. cit.* 43(1972):
1 *Hieracium aurantiacum* subsp. *auropurpureum.*
subsp. **carpathicola** (Näg. & Pet.) Soják *op. cit.* 43
1 (1972): *Hieracium aurantiacum* subsp. *carpathicola.*
subsp. **melinoides** (Näg. & Pet.) Soják *op. cit.* 43
1 (1972): *Hieracium aurantiacum* subsp. *melinoides.*
subsp. **porphyranthes** (Näg. & Pet.) Soják *op. cit.*
43(1972): *Hieracium aurantiacum* subsp.
1 *porphyranthes.*
subsp. **porphyromelana** (Näg. & Pet.) Soják *op. cit.* 43(1972): *Hieracium aurantiacum* subsp.
1 *porphyromelanum.*
subsp. **subaurantica** (Näg. & Pet.) Soják *op. cit.* 43
(1972): *Hieracium aurantiacum* subsp. *sub-*
1 *aurantiacum.*
auriculina (Almq. ex Dahlst.) Soják
subsp. **firmisetiformis** (Zahn) Soják *op. cit.* 49
(1972): *Hieracium auriculinum* subsp. *firmiseti-*
1 *forme.*
subsp. **subteneriuscula** (Zahn) Soják *op. cit.* 49
1 (1972): *Hieracium subteneriusculum.*
auriculoides (Láng) F. Schultz
subsp. **arvensis** (Tausch ex Näg. & Pet.) Soják
op. cit. 50(1972): *Hieracium pannonicum* subsp.
1 *arvense.*
subsp. **asperrima** (Schur ex Näg. & Pet.) Soják
op. cit. 60(1972): *Hieracium pannonicum* subsp.
1 *asperrimum.*
subsp. **brachychaetophora** (Korb & Zahn) Soják
op. cit. 60(1972): *Hieracium auriculoides* subsp.
1 *brachychaetophorum.*
subsp. **echiocephala** (Näg. & Pet.) Soják *op. cit.*
60(1972): *Hieracium pannonicum* subsp.
1 *echiocephalum.*
subsp. **echiogenes** (Näg. & Pet.) Soják *op. cit.*
60(1972): *Hieracium pannonicum* subsp.
1 *echiogenes.*
subsp. **eumorpha** (Näg. & Pet.) Soják *op. cit.* 60
1 (1972): *Hieracium pannonicum* subsp. *eumorphum.*
subsp. **huljakii** (Zahn) Soják *op. cit.* 60(1972):
1 *Hieracium auriculoides* subsp. *huljakii.*

subsp. **lasiophora** (Näg. & Pet.) Soják *op. cit.* 60
1 (1972): *Hieracium pannonicum* subsp. *lasiophorum.*
subsp. **longiseta** (Näg. & Pet.) Soják *op. cit.* 60
1 (1972): *Hieracium pannonicum* subsp. *longisetum.*
subsp. **mira** (Näg. & Pet.) Soják *op. cit.* 60(1972):
1 *Hieracium pannonicum* subsp. *mirum.*
subsp. **parvicapitula** (Näg. & Pet.) Soják *op. cit.*
60(1972): *Hieracium pannonicum* subsp. *parvi-*
1 *capitulum.*
subsp. **polytila** (Näg. & Pet.) Soják *op. cit.* 60
1 (1972): *Hieracium pannonicum* subsp. *polytilum.*
subsp. **praecox** (Tausch ex Näg. & Pet.) Soják
op. cit. 60(1972): *Hieracium pannonicum* subsp.
1 *praecox.*
subsp. **pseudotanythrix** (Zahn) Soják *op. cit.* 59
(1972): *Hieracium auriculoides* subsp. *pseudotany-*
1 *thrix.*
subsp. **remota** (Näg. & Pet.) Soják *op. cit.* 59(1972):
1 *Hieracium pannonicum* subsp. *remotum.*
subsp. **sarmentosa** (Zahn) Soják *op. cit.* 59(1972):
1 *Hieracium auriculoides* subsp. *sarmentosum.*
subsp. **semiauriculoides** (Zahn) Soják *op. cit.* 59
(1972): *Hieracium auriculoides* subsp.
1 *semiauriculoides.*
subsp. **svevora** (Borb.) Soják *op. cit.* 59(1972):
1 *Hieracium megatrichum* var. *svevorum.*
subsp. **tanythrix** (Näg. & Pet.) Soják *op. cit.* 59
1 (1972): *Hieracium pannonicum* subsp. *tanythrix.*
subsp. **umbellosa** (Näg. & Pet.) Soják *op. cit.* 59
1 (1972): *Hieracium pannonicum* subsp. *umbellosum.*
bauhinii (Schult.) Arv.-Touv.
subsp. **adenoclada** (Rehm.) Soják *op. cit.* 53(1972):
1 *Hieracium magyaricum* subsp. *adenocladum.*
subsp. **albaecclesiensis** (Petrak) Soják *op. cit.* 54
1 (1972): *Hieracium bauhinii* subsp. *albaecclesiense.*
subsp. **anisoclada** (Rehm.) Soják *op. cit.* 54(1972):
1 *Hieracium magyaricum* subsp. *anisocladum.*
subsp. **arvorum** (Näg. & Pet.) Soják *op. cit.* 53
1 (1972): *Hieracium magyaricum* subsp. *arvorum.*
subsp. **chaunocyma** (Rehm.) Soják *op. cit.* 54(1972):
1 *Hieracium magyaricum* subsp. *chaunocymum.*
subsp. **cymantha** (Näg. & Pet.) Soják *op. cit.* 53
1 (1972): *Hieracium magyaricum* subsp. *cymanthum.*
subsp. **fastigiata** (Tausch ex Näg. & Pet.) Soják
op. cit. 53(1972): *Hieracium magyaricum* subsp.
1 *fastigiatum.*
subsp. **hispidissima** (Näg. & Pet.) Soják *op. cit.*
53(1972): *Hieracium magyaricum* subsp.
1 *hispidissimum.*
subsp. **holomnoon** (Näg. & Pet.) Soják *op. cit.* 54
1 (1972): *Hieracium magyaricum* subsp. *holomnoon.*
subsp. **hypermelana** (Rehm) Soják *op. cit.* 53
1 (1972): *Hieracium magyaricum* subsp. *hypermelanum.*
subsp. **macra** (Näg. & Pet.) Soják *op. cit.* 53(1972):
1 *Hieracium magyaricum* subsp. *macrum.*
subsp. **melachaeta** (Tausch) Soják *op. cit.* 54(1972):
1 *Hieracium melachaetum.*
subsp. **nematomastix** (Näg. & Pet.) Soják *op. cit.*
53(1972): *Hieracium magyaricum* subsp.
1 *nematomastix.*
subsp. **obscuribractea** (Näg. & Pet.) Soják *op. cit.* 53
(1972): *Hieracium magyaricum* subsp. *obscuri-*
1 *bracteatum.*
subsp. **plicata** (Tausch ex Näg. & Pet.) Soják *op. cit.* 54(1972): *Hieracium magyaricum* subsp.
1 *plicatum.*
subsp. **polyanthema** (Näg. & Pet.) Soják *op. cit.*
54(1972): *Hieracium magyaricum* subsp.
1 *polyanthemum.*
subsp. **pseudobauhinii** (Peter) Soják *op. cit.* 54
(1972): *Hieracium magyaricum* subsp.
1 *pseudobauhinii.*
subsp. **radiocaulis** (Tausch) Soják *op. cit.* 53(1972):
1 *Hieracium radiocaule.*

subsp. **rodnensis** (Näg. & Pet.) Soják *op. cit.* 53
1 (1972): *Hieracium magyaricum* subsp. *rodnense.*
subsp. **thaumasia** (Peter) Soják *op. cit.* 53(1972):
1 *Hieracium magyaricum* subsp. *thaumasium.*
subsp. **thaumasioides** (Peter) Soják *op. cit.* 53
1 (1972): *Hieracium magyaricum* subsp. *thaumasioides.*
subsp. **transgressa** (Näg. & Pet.) Soják *op. cit.* 54
1 (1972): *Hieracium magyaricum* subsp. *transgressum.*
subsp. **viscidula** (Tausch) Soják *op. cit.* 54(1972):
1 *Hieracium bauhinii* β *viscidulum.*
subsp. **weissiana** (Näg. & Pet.) Soják *op. cit.* 54
1 (1972): *Hieracium magyaricum* subsp. *weissianum.*
bifurca (M.B.) F. W. & C. H. Schultz
subsp. **breitenbrunnensis** (Rech. & Zahn) Soják
op. cit. 50(1972): *Hieracium bifurcum* subsp.
1 *breitenbrunnense.*
subsp. **comata** (Näg. & Pet.) Soják *op. cit.* 50
1 (1972): *Hieracium bifurcum* subsp. *comatum.*
subsp. **mesoschista** (Näg. & Pet.) Soják *op. cit.* 50
1 (1972): *Hieracium bifurcum* subsp. *mesoschistum.*
subsp. **pachyclada** (Näg. & Pet.) Soják *op. cit.* 50
1 (1972): *Hieracium bifurcum* subsp. *pachycladum.*
subsp. **praticola** (Tausch ex Näg. & Pet.) Soják *op. cit.*
1 50(1972): *Hieracium bifurcum* subsp. *praticola.*
subsp. **retzensis** (Zahn) Soják *op. cit.* 50(1972):
1 *Hieracium bifurcum* subsp. *retzense.*
subsp. **sterromastix** (Näg. & Pet.) Soják *op. cit.* 50
1 (1972): *Hieracium bifurcum* subsp. *sterromastix.*
subsp. **vindobonensis** (Näg. & Pet.) Soják *op. cit.*
1 50(1972): *Hieracium bifurcum* subsp. *vindobonense.*
blyttiana (Fries) F. W. & C. H. Schultz
subsp. **chrysochroa** (Peter) Soják *op. cit.* 43(1972):
1 *Hieracium chrysochroum.*
subsp. **detonatica** (Näg. & Pet.) Soják *op. cit.* 43
1 (1972): *Hieracium pyrrhanthes* subsp. *detonaticum.*
subsp. **laevisquama** (Näg. & Pet.) Soják *op. cit.*
43(1972): *Hieracium pyrrhanthes* subsp.
1 *laevisquamum.*
subsp. **latibractea** (Peter) Soják *op. cit.* 43(1972):
1 *Hieracium latibracteum.*
subsp. **pyrrhanthes** (Peter) Soják *op. cit.* 43(1972):
1 *Hieracium pyrrhanthes.*
brachiata (Bertol ex Lam. & DC.) F. W. & C. H.
Schultz
subsp. **amblyphylla** (Näg. & Pet.) Soják *op. cit.* 54
1 (1972): *Hieracium brachiatum* subsp. *amblyphyllum.*
subsp. **babaemontis** (Zahn) Soják *op. cit.* 54(1972):
1 *Hieracium brachiatum* subsp. *babaemontis.*
subsp. **bella** (Näg. & Pet.) Soják *op. cit.* 54(1972):
1 *Hieracium brachiatum* subsp. *bellum.*
subsp. **brachiatiformis** (Näg. & Pet.) Soják *op. cit.*
55(1972): *Hieracium brachiatum* subsp. *brachiati-*
1 *forme.*
subsp. **cineracea** (Näg. & Pet.) Soják *op. cit.* 54
1 (1972): *Hieracium brachiatum* subsp. *cineraceum.*
subsp. **crociflora** (Näg. & Pet.) Soják *op. cit.* 55
1 (1972): *Hieracium brachiatum* subsp. *crociflora.*
subsp. **flagelliflora** (Schur) Soják *op. cit.* 55(1972):
1 *Hieracium brachiatum* a. *flagelliflorum.*
subsp. **flava** (Näg. & Pet.) Soják *op. cit.* 55(1972):
1 *Hieracium brachiatum* subsp. *flavum.*
subsp. **gracillima** (Näg. & Pet.) Soják *op. cit.* 54
1 (1972): *Hieracium brachiatum* subsp. *gracillimum.*
subsp. **koscieliskoënsis** (Sag. & Schneid.) Soják
op. cit. 54(1972): *Hieracium brachiatum* var.
1 *koscieliskoënse.*
subsp. **leptocomoides** (Zahn) Soják *op. cit.* 55
1 (1972): *Hieracium brachiatum* subsp. *leptocomoides.*
subsp. **matrensis** (Näg. & Pet.) Soják *op. cit.* 55
1 (1972): *Hieracium brachiatum* subsp. *matrense.*
subsp. **micrantha** (Heuff. ex Fries) Soják *op. cit.* 55
1 (1972): *Hieracium brachiatum* subsp. *micranthum.*
subsp. **muranensis** (Rehm.) Soják *op. cit.* 55(1972):
1 *Hieracium brachiatum* subsp. *muranense.*
subsp. **nematocauloides** (Zahn) Soják *op. cit.* 55
(1972): *Hieracium brachiatum* subsp. *nemato-*
1 *cauloides.*
subsp. **nematocaulon** (Näg. & Pet.) Soják *op. cit.*
1 55(1972): *Hieracium brachiatum* subsp. *nematocaulon.*
subsp. **nigribractea** (Rehm.) Soják *op. cit.* 55(1972):
1 *Hieracium brachiatum* subsp. *nigribracteum.*
subsp. **pedunculata** (Wallr.) Soják *op. cit.* 55(1972):
1 *Hieracium pedunculatum.*
subsp. **pieniakensis** (Rehm.) Soják *op. cit.* 55(1972):
1 *Hieracium pieniakense.*
subsp. **pseudobrachiata** (Čelak.) Soják *op. cit.* 55
(1972): *Hieracium bifurcum* a (subsp.) *pseudo-*
1 *brachiatum.*
subsp. **sabranskyi** (Zahn) Soják *op. cit.* 54(1972):
1 *Hieracium brachiatum* subsp. *sabranskyi.*
subsp. **striatobrachiata** (Näg. & Pet.) Soják *op. cit.*
55(1972): *Hieracium brachiatum* γ (var.)
1 *striatobrachiatum.*
subsp. **subtilipes** (Zahn) Soják *op. cit.* 55(1972):
1 *Hieracium brachiatum* subsp. *subtilipes.*
subsp. **tapina** (Näg. & Pet.) Soják *op. cit.* 55(1972):
1 *Hieracium brachiatum* subsp. *tapinum.*
subsp. **valdestriata** (Näg. & Pet.) Soják *op. cit.* 54
1 (1972): *Hieracium brachiatum* subsp. *valdestriatum.*
subsp. **villarsii** (Näg. & Pet.) Soják *op. cit.* 54(1972):
1 *Hieracium brachiatum* subsp. *villarsii.*
subsp. **zittaviensis** (Zahn) Soják *op. cit.* 55(1972):
1 *Hieracium brachiatum* subsp. *zittaviense.*
caespitosa (Dumort.) Sell & C. West
subsp. **dissoluta** (Näg. & Pet.) Soják *op. cit.* 44
1 (1972): *Hieracium collinum* subsp. *dissolutum.*
subsp. **leptocaulon** (Näg. & Pet.) Soják *op. cit.* 44
1 (1972): *Hieracium collinum* subsp. *leptocaulon.*
subsp. **pseudosudetorum** (Zahn) Soják *op. cit.* 44
1 (1972): *Hieracium pratense* subsp. *pseudosudetorum.*
subsp. **redenicensis** (Zahn) Soják *op. cit.* 44(1972):
1 *Hieracium pratense* subsp. *redenicense.*
subsp. **stenocephala** (Näg. & Pet.) Soják *op. cit.* 44
1 (1972): *Hieracium collinum* subsp. *stenocephalum.*
subsp. **sudetorum** (Peter) Soják *op. cit.* 44(1972):
1 *Hieracium collinum* subsp. *sudetorum.*
subsp. **trichoderma** (Rehm.) Soják *op. cit.* 44(1972):
1 *Hieracium collinum* subsp. *trichoderma.*
subsp. **uechtritzii** (Näg. & Pet.) Soják *op. cit.* 44
1 (1972): *Hieracium collinum* subsp. *uechtritzii.*
callimorpha (Näg. & Pet.) Soják
subsp. **cryptoschista** (Zahn) Soják *op. cit.* 47(1972):
1 *Hieracium callomorphum* subsp. *cryptoschistum.*
subsp. **leipana** (Zahn) Soják *op. cit.* 46(1972):
1 *Hieracium leptophyton* subsp. *leipanum.*
subsp. **petrakiana** (Zahn) Soják *op. cit.* 47(1972):
1 *Hieracium callimorphum* subsp. *petrakianum.*
subsp. **proedlasiensis** (Zahn) Soják *op. cit.* 46(1972):
1 *Hieracium callimorphum* subsp. *proedlasiense.*
callimorphoides (Zahn) Soják
subsp. **eurotopolia** (Zahn) Soják *op. cit.* 47(1972):
1 *Hieracium callimorphoides* subsp. *eurotopolium.*
subsp. **pseudoschultesiiformis** (Zahn) Soják *op.*
cit. 47(1972): *Hieracium callimorphoides* subsp.
1 *pseudoschultesiiforme.*
calodon (Tausch ex Peter) Soják
subsp. **multiceps** (Näg. & Pet.) Soják *op. cit.* 59
1 (1972): *Hieracium calodon* subsp. *multiceps.*
subsp. **sphaleron** (Näg. & Pet.) Soják *op. cit.* 59
1 (1972): *Hieracium calodon* subsp. *sphaleron.*
subsp. **tenuiceps** (Näg. & Pet.) Soják *op. cit.* 59
1 (1972): *Hieracium calodon* subsp. *tenuiceps.*
calomastix (Näg. & Pet.) Soják
subsp. **acrosticta** (Näg. & Pet.) Soják *op. cit.* 56
1 (1972): *Hieracium calomastix* subsp. *acrostictum.*
cinerea (Tausch) F. W. & C. H. Schultz
subsp. **danubii** (Zahn) Soják *op. cit.* 50(1972):
1 *Hieracium bifurcum* subsp. *danubii.*

subsp. **langii** (Näg. & Pet.) Soják *op. cit.* 50(1972):
1 *Hieracium bifurcum* subsp. *langii*.
subsp. **tephraea** (Näg. & Pet.) Soják *op. cit.* 50
1 (1972): *Hieracium bifurcum* subsp. *tephraeum*.
subsp. **thayensis** (Näg. & Pet.) Soják *op. cit.* 50
1 (1972): *Hieracium bifurcum* subsp. *thayense*.
cochlearis (Norrl.) Weiss.
subsp. **pseudauricula** (Näg. & Pet.) Soják *op. cit.*
46(1972): *Hieracium floribundum* subsp.
1 *pseudauricula*.
subsp. **subauricula** (Näg. & Pet.) Soják *op. cit.* 61
1 (1972): *Hieracium floribundum* subsp. *subauricula*.
collina (Gochnat) Soják
subsp. **elegans** (Näg. & Pet.) Soják *op. cit.* 51(1972):
1 *Hieracium fallax* subsp. *elegans*.
subsp. **fallax** (Willd.) Soják *op. cit.* 51(1972):
1 *Hieracium fallax*.
subsp. **granensis** (Näg. & Pet.) Soják *op. cit.* 51
1 (1972): *Hieracium fallax* subsp. *granense*.
subsp. **znoymensis** (Näg. & Pet.) Soják *op. cit.* 51
1 (1972): *Hieracium fallax* subsp. *znoymense*.
cryptomastix (Näg. & Pet.) Soják
subsp. **empodista** (Näg. & Pet.) Soják *op. cit.* 52
1 (1972): *Hieracium magyaricum* subsp. *empodistum*.
subsp. **roxensis** (Sag. & Schneid.) Soják *op. cit.* 52
1 (1972): *Hieracium florentinum* var. *roxense*.
cymosa (L.) F. W. & C. H. Schultz
subsp. **chrysophaës** (Näg. & Pet.) Soják *op. cit.* 47
1 (1972): *Hieracium cymosum* subsp. *chrysophaës*.
subsp. **nestleri** (Villars, Lauth & Nestler) Soják *op.*
1 *cit.* 47(1972): *Hieracium nestleri*.
subsp. **regelii** (Näg. & Pet.) Soják *op. cit.* 47
1 (1972): *Hieracium cymosum* subsp. *regelii*.
decolorans (Fries) F. W. & C. H. Schultz
subsp. **subkajanensis** (Zahn) Soják *op. cit.* 43(1972):
1 *Hieracium aurantiacum* subsp. *subkajanense*.
densiflora (Tausch) Soják
subsp. **acrosciadia** (Näg. & Pet.) Soják *op. cit.* 58
1 (1972): *Hieracium pentaphylum* subsp. *acrosciadium*.
subsp. **actinophora** (Zahn) Soják *op. cit.* 58(1972):
1 *Hieracium tauschii* subsp. *actinophorum*.
subsp. **austrosaxonica** (Missb. & Zahn) Soják *op.*
cit. 58(1972): *Hieracium tauschii* subsp. *austro-*
1 *saxonicum*.
subsp. **bauhinifolia** (Näg. & Pet.) Soják *op. cit.* 58
1 (1972): *Hieracium pentaphylum* subsp. *bauhinifolium*.
subsp. **comocymosa** (Näg. & Pet.) Soják *op. cit.* 59
(1972): *Hieracium pentaphylum* subsp.
1 *comocymosum*.
subsp. **cymosiformis** (Näg. & Pet.) Soják *op. cit.*
58(1972): *Hieracium pentaphylum* subsp.
1 *cymosiforme*.
subsp. **manothyrsa** (Näg. & Pet.) Soják *op. cit.* 58
(1972): *Hieracium pentaphylum* subsp.
1 *manothyrsum*.
subsp. **melananthela** (Zahn) Soják *op. cit.* 58(1972):
1 *Hieracium tauschii* subsp. *melananthelum*.
subsp. **neilreichii** (Näg. & Pet.) Soják *op. cit.* 58
1 (1972): *Hieracium pentaphylum* subsp. *neilreichii*.
subsp. **ochrocephala** (Näg. & Pet.) Soják *op. cit.*
58(1972): *Hieracium pentaphylum* subsp.
1 *ochrocephalum*.
subsp. **pseudochrocephala** (Gerstlauer & Zahn)
Soják *op. cit.* 58(1972): *Hieracium tauschii* subsp.
1 *pseudochrocephalum*.
subsp. **pseudomagyariciformis** (Lengyel &
Zahn) Soják *op. cit.* 58(1972): *Hieracium tauschii*
1 subsp. *pseudomagyariciforme*.
subsp. **pseudo-tauschii** (Zahn) Soják *op. cit.* 58
1 (1972): *Hieracium tauschii* subsp. *pseudo-tauschii*.
subsp. **pycnomnoa** (Rehm.) Soják *op. cit.* 58 (1972):
1 *Hieracium umbelliferum* subsp. *pycnomnoum*.
subsp. **radiocauloides** (Zahn) Soják *op. cit.* 58
1 (1972): *Hieracium tauschii* subsp. *radiocauloides*.

subsp. **spaniopolia** (Zahn) Soják *op. cit.* 58(1972):
1 *Hieracium tauschii* subsp. *spaniopolium*.
subsp. **suebophila** (Zahn) Soják *op. cit.* 58(1972):
1 *Hieracium tauschii suebophilum*.
subsp. **umbellifera** (Näg. & Pet.) Soják *op. cit.* 58
1 (1972): *Hieracium pentaphylum* subsp. *umbelliferum*.
subsp. **xanthothyrsa** (Fest & Zahn) Soják *op. cit.*
58(1972): *Hieracium umbelliferum* subsp.
1 *xanthothyrsum*.
echioides (Lumn.) F. W. & C. H. Schultz
subsp. **freynii** (Näg. & Pet.) Soják *op. cit.* 49(1972):
1 *Hieracium echioides* subsp. *freynii*.
subsp. **macrocyma** (Näg. & Pet.) Soják *op. cit.* 49
1 (1972): *Hieracium echioides* subsp. *macrocymum*.
erythriophylla (Vukot.) Soják
subsp. **effusa** (Näg. & Pet.) Soják *op. cit.* 52(1972):
1 *Hieracium magyaricum* subsp. *effusum*.
subsp. **pseudosparsa** (Zahn) Soják *op. cit.* 52(1972):
1 *Hieracium bauhinii* subsp. *pseudosparsum*.
subsp. **rojowskii** (Rehm.) Soják *op. cit.* 52(1972):
1 *Hieracium magyaricum* subsp. *rojowskii*.
subsp. **scepusiensis** (Rehm.) Soják *op. cit.* 52(1972):
1 *Hieracium magyaricum* subsp. *scepusiense*.
euchaetia (Näg. & Pet.) Soják
subsp. **feichtingeri** (Zahn) Soják *op. cit.* 60(1972):
1 *Hieracium euchaetium* subsp. *feichtingeri*.
subsp. **longa** (Näg. & Pet.) Soják *op. cit.* 60(1972):
1 *Hieracium euchaetium* subsp. *longum*.
subsp. **padloschinensis** (Zahn) Soják *op. cit.* 60
1 (1972): *Hieracium euchaetium* subsp. *padloschinense*.
subsp. **pseudo-doelliana** (Zahn) Soják *op. cit.* 60
(1972): *Hieracium euchaetium* subsp. *pseudo-*
1 *doellianum*.
subsp. **tscherningii** (Zahn) Soják *op. cit.* 60(1972):
1 *Hieracium euchaetium* subsp. *tscherningii*.
fallacina (F. Schultz) C. H. Schultz
subsp. **alsatica** (Näg. & Pet.) Soják *op. cit.* 59
1 (1972): *Hieracium germanicum* subsp. *alsaticum*.
subsp. **fissicaulis** (Näg. & Pet.) Soják *op. cit.* 59
1 (1972): *Hieracium germanicum* β (var.) *fissicaule*.
subsp. **germanica** (Zahn) Soják *op. cit.* 59(1972):
1 *Hieracium fallacinum* subsp. *germanicum*.
subsp. **hyperdoxa** (Sag.) Soják *op. cit.* 59(1972):
1 *Hieracium hyperdoxum*.
subsp. **prantlii** (Näg. & Pet.) Soják *op. cit.* 59
1 (1972): *Hieracium germanicum* subsp. *prantlii*.
subsp. **vulsa** (Zahn) Soják *op. cit.* 59(1972):
1 *Hieracium fallacinum* subsp. *vulsum*.
fennica (Norrl.) Norrl.
subsp. **diatenta** (Näg. & Pet.) Soják *op. cit.* 46
(1972): *Hieracium spathophyllum* subsp.
1 *diatentum*.
subsp. **exorrhabda** (Näg. & Pet.) Soják *op. cit.* 45
(1972): *Hieracium spathophyllum* subsp.
1 *exorrhabdum*.
subsp. **fulvescens** (Näg. & Pet.) Soják *op. cit.* 45
1 (1972): *Hieracium spathophyllum* subsp. *fulvescens*.
subsp. **graciliglandula** (Zahn) Soják *op. cit.* 45
(1972): *Hieracium longiscapum* subsp. *gracili-*
1 *glandulum*.
subsp. **melanauricula** (Zahn) Soják *op. cit.* 45
(1972): *Hieracium spathophyllum* subsp.
1 *melanauricula*.
subsp. **polysarcon** (Rehm.) Soják *op. cit.* 45(1972):
1 *Hieracium spathophyllum* subsp. *polysarcon*.
subsp. **spathophylla** (Peter) Soják *op. cit.* 46(1972):
1 *Hieracium spathophyllum*.
flagellariformis (G. Schneider) Soják
subsp. **jaromirensis** (Zahn) Soják *op. cit.* 45
1 (1972): *Hieracium flagellariforme* subsp. *jaromirense*.
flagellaris (Willd.) Sell & C. West
subsp. **besnikensis** (Lengyel & Zahn) Soják *op. cit.*
1 44(1972): *Hieracium flagellare* subsp. *besnikense*.

subsp. **glatzensis** (Näg. & Pet.) Soják *op. cit.* 45
1 (1972): *Hieracium flagellare* subsp. *glatzense.*
subsp. **glaucochlora** (Rehm.) Soják *op. cit.* 44
1 (1972): *Hieracium flagellare* subsp. *glaucochlorum.*
subsp. **hahnii** (Zahn) Soják *op. cit.* 61(1972):
1 *Hieracium flagellare* subsp. *hahnii.*
subsp. **pertenuiglandula** (Lengyel & Zahn)
Soják *op. cit.* 44(1972): *Hieracium flagellare* subsp.
1 *pertenuiglandulum.*
subsp. **pratensifolia** (Zahn) Soják *op. cit.* 44(1972):
1 *Hieracium flagellare* subsp. *pratensifolium.*
subsp. **subsarmentosa** (Zahn) Soják *op. cit.*
44(1972): *Hieracium flagellare* subsp. *sub-*
1 *sarmentosum.*
subsp. **tatrensis** (Peter) Soják *op. cit.* 45(1972):
1 *Hieracium flagellare* subsp. *tatrense.*
floribunda (Wimm. & Grab.) Arv.-Touv.
subsp. **anomelana** (Näg. & Pet.) Soják *op. cit.* 45
1 (1972): *Hieracium floribundum* subsp. *anomelanum.*
subsp. **atrocrocea** (Peter) Soják *op. cit.* 45(1972):
1 *Hieracium floribundum* subsp. *atrocroceum.*
subsp. **hylaeophila** (Zahn & Petrak) Soják *op. cit.*
45(1972): *Hieracium floribundum* subsp. *hylaeo-*
1 *philum.*
subsp. **regiomontana** (Näg. & Pet.) Soják *op. cit.*
45(1972): *Hieracium floribundum* subsp.
1 *regiomontanum.*
subsp. **stenomacellata** (Zahn) Soják *op. cit.* 45
(1972): *Hieracium floribundum* subsp.
1 *stenomacellatum.*
subsp. **suecica** (Fries) Soják *op. cit.* 45(1972):
1 *Hieracium suecicum.*
subsp. **teplitzensis** (Näg. & Pet.) Soják *op. cit.* 45
1 (1972): *Hieracium floribundum* subsp. *teplitzense.*
glaucescens (Bess.) Soják
subsp. **adenocyma** (Näg. & Pet.) Soják *op. cit.* 53
1 (1972): *Hieracium magyaricum* subsp. *adenocymum.*
subsp. **adenocymigeriformis** (Lengyel & Zahn)
Soják *op. cit.* 53(1972): *Hieracium bauhinii* subsp.
1 *adenocymigeriforme.*
subsp. **amnoon** (Näg. & Pet.) Soják *op. cit.* 52
1 (1972): *Hieracium magyaricum* subsp. *amnoon.*
subsp. **besserana** (Spreng.) Soják *op. cit.* 52(1972),
1 as 'besseriana': *Hieracium besserianum.*
subsp. **branae** (Näg. & Pet.) Soják *op. cit.* 53
1 (1972): *Hieracium magyaricum* subsp. *branae.*
subsp. **decolor** (Näg. & Pet.) Soják *op. cit.* 52
1 (1972): *Hieracium magyaricum* subsp. *decolor.*
subsp. **doroghensis** (Näg. & Pet.) Soják *op. cit.* 52
1 (1972): *Hieracium magyaricum* subsp. *doroghense.*
subsp. **ferroviae** (Rehm.) Soják *op. cit.* 52(1972):
1 *Hieracium magyaricum* subsp. *ferroviae.*
subsp. **filifera** (Tausch) Soják *op. cit.* 52(1972):
1 *Hieracium filiferum.*
subsp. **heothina** (Näg. & Pet.) Soják *op. cit.* 53
1 (1972): *Hieracium magyaricum* subsp. *heothinum.*
subsp. **kerneri** (Näg. & Pet.) Soják *op. cit.* 53
1 (1972): *Hieracium magyaricum* subsp. *kerneri.*
subsp. **marginalis** (Näg. & Pet.) Soják *op. cit.* 53
1 (1972): *Hieracium magyaricum* subsp. *marginale.*
subsp. **megalocymophora** (Zahn) Soják *op. cit.* 52
(1972): *Hieracium bauhinii* subsp. *megalo-*
1 *cymophorum.*
subsp. **megalomastix** (Näg. & Pet.) Soják *op. cit.*
52(1972): *Hieracium magyaricum* subsp. *megalo-*
1 *mastix.*
subsp. **nigriseta** (Näg. & Pet.) Soják *op. cit.* 52
1 (1972): *Hieracium magyaricum* subsp. *nigrisetum.*
subsp. **pseudauriculoides** (Näg. & Pet.) Soják
op. cit. 53(1972): *Hieracium magyaricum* subsp.
1 *pseudauriculoides.*
subsp. **submagyarica** (Näg. & Pet.) Soják
op. cit. 52(1972): *Hieracium magyaricum* subsp.
1 *submagyaricum.*

subsp. **tephrops** (Näg. & Pet.) Soják *op. cit.* 52
1 (1972): *Hieracium magyaricum* subsp. *tephrops.*
glomerata (Froel.) Arv.-Touv.
subsp. **ambigua** (Ehrh. ex Näg. & Pet.) Soják
op. cit. 49(1972): *Hieracium glomeratum* subsp.
1 *ambiguum.*
subsp. **bialkana** (Zahn) Soják *op. cit.* 49(1972):
1 *Hieracium ambiguum* subsp. *bialkanum.*
subsp. **calomelana** (Rehm.) Soják *op. cit.* 49(1972):
1 *Hieracium glomeratum* subsp. *calomelanum.*
subsp. **cymigeriformis** (Näg. & Pet.) Soják
op. cit. 49(1972): *Hieracium glomeratum* subsp.
1 *cymigeriforme.*
subsp. **kunzii** (Zahn) Soják *op. cit.* 49(1972):
1 *Hieracium glomeratum* subsp. *kunzii.*
subsp. **lipovecianum** (Lengyel & Zahn) Soják
op. cit. 49(1972): *Hieracium ambiguum* subsp.
1 *lipovecianum.*
subsp. **nigrifontimontanum** (Zahn) Soják *op. cit.*
49(1972): *Hieracium ambiguum* subsp.
1 *nigrifontimontanum.*
subsp. **prolongata** (Näg. & Pet.) Soják *op. cit.*
1 49(1972): *Hieracium glomeratum* subsp. *prolongatum.*
subsp. **subambigua** (Näg. & Pet.) Soják *op. cit.*
49(1972): *Hieracium glomeratum* subsp.
1 *subambiguum.*
hoppeana (Schultes) F. W. & C. H. Schultz
subsp. **grundlii** (Näg. & Pet.) Soják *op. cit.* 41
1 (1972): *Hieracium hoppeanum* subsp. *grundlii.*
subsp. **testimonialis** (Nägeli) Soják *op. cit.* 41
1 (1972): *Hieracium hoppeanum* subsp. *testimoniale.*
iserana (Uechtr.) Soják
subsp. **anguinescens** (Zahn) Soják *op. cit.* 46
1 (1972): *Hieracium iseranum* subsp. *anguinescens.*
subsp. **atramentarioides** (Missbach & Zahn) Soják
op. cit. 46(1972): *Hieracium iseranum* subsp.
1 *atramentarioides.*
subsp. **confinium** (Näg. & Pet.) Soják *op. cit.* 46
1 (1972): *Hieracium nigriceps* subsp. *confinium.*
subsp. **fasciculata** (Näg. & Pet.) Soják *op. cit.* 61
1 (1972): *Hieracium nigriceps* subsp. *fasciculatum.*
subsp. **florida** (Näg. & Pet.) Soják *op. cit.* 46(1972):
1 *Hieracium nigriceps* subsp. *floridum.*
subsp. **luchavica** (Missb. & Zahn) Soják *op. cit.* 61
1 (1972): *Hieracium iseranum* var. *luchavicum.*
subsp. **pseudonigriceps** (Näg. & Pet.) Soják *op.*
cit. 46(1972): *Hieracium arvicola* subsp.
1 *pseudonigriceps.*
subsp. **schwarzerana** (Näg. & Pet.) Soják *op. cit.*
60(1972), as 'schwarzeriana': *Hieracium nigriceps*
1 subsp. *schwarzerianum.*
subsp. **seminigriceps** (Zahn) Soják *op. cit.* 46(1972):
1 *Hieracium iseranum* subsp. *seminigriceps.*
subsp. **subconfinium** (Zahn) Soják *op. cit.* 46(1972):
1 *Hieracium iseranum* subsp. *subconfinium.*
subsp. **subfasciculata** (Zahn) Soják *op. cit.* 46(1972):
1 *Hieracium iseranum* subsp. *subfasciculatum.*
subsp. **subnigriceps** (Zahn) Soják *op. cit.* 46(1972):
1 *Hieracium iseranum* subsp. *subnigriceps.*
kalksburgensis (Wiesb.) Soják
subsp. **almonica** (Näg. & Pet.) Soják *op. cit.* 48
1 (1972): *Hieracium canum* subsp. *almonicum.*
subsp. **anosciadia** (Näg. & Pet.) Soják *op. cit.* 47
1 (1972): *Hieracium canum* subsp. *anosciadium.*
subsp. **arenicola** (Näg. & Pet.) Soják *op. cit.* 48
1 (1972): *Hieracium canum* subsp. *arenicola.*
subsp. **catoschista** (Näg. & Pet.) Soják *op. cit.* 48
1 (1972): *Hieracium canum* subsp. *catoschistum.*
subsp. **charadraea** (Näg. & Pet.) Soják *op. cit.* 48
1 (1972): *Hieracium canum* subsp. *charadraeum.*
subsp. **cymosella** (Näg. & Pet.) Soják *op. cit.* 47
1 (1972): *Hieracium canum* subsp. *cymosella.*
subsp. **diffusa** (Näg. & Pet.) Soják *op. cit.* 48(1972):
1 *Hieracium canum* subsp. *diffusum.*

I subsp. **gracilis** (Tausch ex Näg. & Pet.) Soják *op. cit.* 48(1972): *Hieracium canum* subsp. *gracile*.
I subsp. **krausii** (Näg. & Pet.) Soják *op. cit.* 47(1972): *Hieracium canum* subsp. *krausii*.
I subsp. **lagara** (Näg. & Pet.) Soják *op. cit.* 48(1972): *Hieracium canum* subsp. *lagarum*.
I subsp. **laschii** (Zahn) Soják *op. cit.* 48(1972): *Hieracium laschii*.
I subsp. **lepteilema** (Näg. & Pet.) Soják *op. cit.* 48 (1972): *Hieracium canum* subsp. *lepteilema*.
I subsp. **leptobia** (Näg. & Pet.) Soják *op. cit.* 48 (1972): *Hieracium canum* subsp. *leptobium*.
I subsp. **melanopolia** (Näg. & Pet.) Soják *op. cit.* 48 (1972): *Hieracium canum* subsp. *melanopolium*.
I subsp. **melanoxantha** (Näg. & Pet.) Soják *op. cit.* 48(1972): *Hieracium canum* subsp. *melanoxanthum*.
I subsp. **pasta** (Näg. & Pet.) Soják *op. cit.* 47(1972): *Hieracium canum* subsp. *pastum*.
I subsp. **pleiantha** (Näg. & Pet.) Soják *op. cit.* 47 (1972): *Hieracium canum* subsp. *pleianthum*.
I subsp. **polianthes** (Näg. & Pet.) Soják *op. cit.* 48 (1972): *Hieracium canum* subsp. *polianthes*.
I subsp. **praelonga** (Näg. & Pet.) Soják *op. cit.* 47 (1972): *Hieracium canum* subsp. *praelongum*.
I subsp. **sublagara** (Oborny & Zahn) Soják *op. cit.* 48(1972): *Hieracium laschii* subsp. *sublagarum*.
I subsp. **virenticana** (Peter) Soják *op. cit.* 48(1972): *Hieracium virenticanum*.

koernickiana (Näg. & Pet.) Soják
I subsp. **denigrata** (Näg. & Pet.) Soják *op. cit.* 56 (1972): *Hieracium sulphureum* subsp. *denigratum*.
I subsp. **samoviae** (Näg. & Pet.) Soják *op. cit.* 56 (1972): *Hieracium sulphureum* subsp. *samoviae*.

lactucella (Wallr.) Sell & C. West
I subsp. **acutisquama** (Näg. & Pet.) Soják *op. cit.* 41(1972): *Hieracium auricula* subsp. *acutisquamum*.
I subsp. **amaureilema** (Näg. & Pet.) Soják *op. cit.* 41(1972): *Hieracium auricula* subsp. *amaureilema*.
I subsp. **auricula** Soják *op. cit.* 41(1972), nom. nov.: *Hieracium auricula* subsp. *auricula* Lam. & DC. ex Nägeli & Peter.
I subsp. **lampreilema** (Näg. & Pet.) Soják *op. cit.* 41 (1972): *Hieracium auricula* subsp. *lampreilema*.
I subsp. **littuanica** (Näg. & Pet.) Soják *op. cit.* 41 (1972): *Hieracium auricula* subsp. *littuanicum*.
I subsp. **magna** (Näg. & Pet.) Soják *op. cit.* 41(1972): *Hieracium auricula* subsp. *magnum*.
I subsp. **magnauricula** (Näg. & Pet.) Soják *op. cit.* 41(1972), as 'magnaricula': *Hieracium auricula* subsp. *magnauricula*.
I subsp. **melaneilema** (Peter) Soják *op. cit.* 41(1972): *Hieracium auricula* subsp. *melaneilema*.
I subsp. **melanocalathia** (Rehm.) Soják *op. cit.* 41 (1972): *Hieracium auricula* subsp. *melanocalathium*.
I subsp. **tricheilema** (Näg. & Pet.) Soják *op. cit.* 41 (1972): *Hieracium auricula* subsp. *tricheilema*.

macrostolona (G. Schneider) Soják
I subsp. **cernuiformis** (Näg. & Pet.) Soják *op. cit.* 45(1972): *Hieracium flagellare* subsp. *cernuiforme*.
I subsp. **piscinae** (Zahn) Soják *op. cit.* 45(1972): *Hieracium cernuiforme* subsp. *piscinae*.
I subsp. **polychaetobia** (Zahn) Soják *op. cit.* 45 (1972): *Hieracium cernuiforme* subsp. *polychaetobium*.

mollicaulis (Vukot.) Soják
I subsp. **anoclada** (Näg. & Pet.) Soják *op. cit.* 56 (1972): *Hieracium leptophyton* subsp. *anocladum*.
I subsp. **atriceps** (Näg. & Pet.) Soják *op. cit.* 56 (1972): *Hieracium leptophyton* subsp. *atriceps*.
I subsp. **bauhiniflora** (Näg. & Pet.) Soják *op. cit.* 56 (1972): *Hieracium leptophyton* subsp. *bauhiniflorum*.
I subsp. **bauhinifloriformis** (Lengyel & Zahn) Soják *op. cit.* 56(1972): *Hieracium leptophyton* subsp. *bauhinifloriforme*.
I subsp. **busocensis** (Nyárády & Zahn) Soják *op. cit.* 56(1972): *Hieracium leptophyton* subsp. *busocense*.
I subsp. **discolor** (Näg. & Pet.) Soják *op. cit.* 56 (1972): *Hieracium leptophyton* subsp. *discolor*.
I subsp. **leptophyton** Soják *op. cit.* 55(1972), nom. nov.: *Hieracium leptophyton* subsp. *leptophyton* Näg. & Pet.
I subsp. **leptosoma** (Näg. & Pet.) Soják *op. cit.* 56 (1972): *Hieracium leptophyton* subsp. *leptosoma*.
I subsp. **melanophila** (Rehm.) Soják *op. cit.* 55(1972): *Hieracium leptophyton* subsp. *melanophilum*.
I subsp. **orthochaeta** (Oborny & Zahn) Soják *op. cit.* 56(1972): *Hieracium leptophyton* subsp. *orthochaetum*.
I subsp. **pauciflora** (Näg. & Pet.) Soják *op. cit.* 56 (1972): *Hieracium leptophyton* subsp. *pauciflorum*.
I subsp. **sychnoschista** (Näg. & Pet.) Soják *op. cit.* 55 (1972): *Hieracium leptophyton* subsp. *sychnoschistum*.

obornyana (Näg. & Pet.) Soják
I subsp. **drumensis** (Korb & Zahn) Soják *op. cit.* 57 (1972): *Hieracium obornyanum* subsp. *drumense*.
I subsp. **polymastix** (Näg. & Pet.) Soják *op. cit.* 57 (1972): *Hieracium obornyanum* subsp. *polymastix*.
I subsp. **stricticaulis** (Näg. & Pet.) Soják *op. cit.* 57 (1972): *Hieracium obornyanum* subsp. *stricticaule*.
I subsp. **subtatrensis** (Rehm.) Soják *op. cit.* 57(1972): *Hieracium obornyanum* subsp. *subtatrense*.
I subsp. **superpratensis** (Korb & Zahn) Soják *op. cit.* 57(1972): *Hieracium obornyanum* subsp. *superpratense*.
I subsp. **valdeciliata** (Zahn) Soják *op. cit.* 57(1972): *Hieracium obornyanum* subsp. *valdeciliatum*.

obscura (Reichenb.) Soják
I subsp. **aquilonaris** (Näg. & Pet.) Soják *op. cit.* 51 (1972), as 'aguilonaris': *Hieracium florentium* β *aquilonare*.
I subsp. **basiphylla** (Peter) Soják *op. cit.* 51(1972): *Hieracium florentinum* subsp. *basiphyllum*.
I subsp. **berninae** (Griseb.) Soják *op. cit.* 51(1972): *Hieracium praealtum* β *berninae*.

onegensis (Norrl.) Norrl.
I subsp. **amauroderma** (Zahn) Soják *op. cit.* 44 (1972): *Hieracium pratense* subsp. *amaurodermum*.
I subsp. **glaucochroa** (Näg. & Pet.) Soják *op. cit.* 44 (1972): *Hieracium collinum* subsp. *glaucochroum*.
I subsp. **madara** (Näg. & Pet.) Soják *op. cit.* 44 (1972): *Hieracium collinum* subsp. *madarum*.

paragoga (Nag. & Pet.) Soják
I subsp. **biformis** (Rehm.) Soják *op. cit.* 56(1972): *Hieracium paragogum* subsp. *biforme*.
I subsp. **paragogiceps** (Zahn) Soják *op. cit.* 56(1972): *Hieracium paragogum* subsp. *paragogiceps*.
I subsp. **phalacrodes** (Zahn) Soják *op. cit.* 56(1972): *Hieracium paragogum* subsp. *phalacrodes*.

pilipes (Sael.) Norrl.
I subsp. **acrocoma** (Näg. & Pet.) Soják *op. cit.* 49 (1972): *Hieracium acrocomum*.
I subsp. **peckiana** (Näg. & Pet.) Soják *op. cit.* 60 (1972): *Hieracium acrocomum* subsp. *peckianum*.
I subsp. **schweidnitzensis** (Näg. & Pet.) Soják *op. cit.* 49(1972): *Hieracium acrocomum* subsp. *schweidnitzense*.

piloselliflora (Näg. & Pet.) Soják
I subsp. **paxii** (Näg. & Pet.) Soják *op. cit.* 46(1972): *Hieracium piloselliflorum* subsp. *paxii*.

pilosellina (F. Schultz) F. W. & C. H. Schultz
I subsp. **pseudoradians** (Deg. & Zahn) Soják *op. cit.* 59(1972): *Hieracium subgermaniciforme* subsp. *pseudoradians*.

piloselloides (Vill.) Soják
I subsp. **florentiniformis** (Näg. & Pet.) Soják *op. cit.* 52(1972): *Hieracium florentinum* subsp. *florentiniforme*.

praealta (Vill. ex Gochnat) F. W. & C. H. Schultz
subsp. **albidobractea** (Näg. & Pet.) Soják *op. cit.* 51(1972): *Hieracium florentinum* subsp.
I *albidobracteum.*
subsp. **hirsuticeps** (Näg. & Pet.) Soják *op. cit.* 51
I (1972): *Hieracium florentinum* subsp. *hirsuticeps.*
subsp. **ingens** (Näg. & Pet.) Soják *op. cit.* 51(1972):
I *Hieracium florentinum* subsp. *ingens.*
subsp. **stellata** (Tausch ex Zahn) Soják *op. cit.* 51
I (1972): *Hieracium florentinum* subsp. *stellatum.*

prussica (Näg. & Pet.) Soják
subsp. **eisbergensis** (Zahn) Soják *op. cit.* 44(1972):
I *Hieracium prussicum* subsp. *eisbergense.*
subsp. **elatior** (Rehm.) Soják *op. cit.* 44(1972):
I *Hieracium elatius.*
subsp. **scharlokiana** (Näg. & Pet.) Soják *op. cit.* 44(1972): *Hieracium prussicum* subsp.
I *scharlokianum.*

rothiana (Wallr.) F. W. & C. H. Schultz
subsp. **arenaria** (Tauscher ex Näg. & Pet.) Soják *op. cit.* 50(1972): *Hieracium setigerum* subsp.
I *arenarium.*
subsp. **carnuntii** (Vetter) Soják *op. cit.* 50(1972):
I *Hieracium rothianum* subsp. *carnuntii.*
subsp. **holopolia** (Peter) Soják *op. cit.* 50(1972):
I *Hieracium setigerum* subsp. *holopolium.*

roxolanica (Rehm.) Soják
subsp. **adenocymosa** (Nyár. & Zahn) Soják *op. cit.*
I 49(1972): *Hieracium rubellum* subsp. *adenocymosum.*
subsp. **rehmannii** (Näg. & Pet.) Soják *op. cit.* 49
I (1972): *Hieracium cruentum* subsp. *rehmannii.*
subsp. **rubricymigera** (Näg. & Pet.) Soják *op. cit.* 49(1972): *Hieracium cruentum* subsp. *rubri-*
I *cymigerum.*

sabina (Sebast.) F. W. & C. H. Schultz
subsp. **pseudosabina** (Näg. & Pet.) Soják *op. cit.* 47(1972): *Hieracium cymosum* γ (var.)
I *pseudosabinum.*

schultesii (F. Schultz) F. W. & C. H. Schultz
subsp. **acariaea** (Näg. & Pet.) Soják *op. cit.* 42
I (1972): *Hieracium auriculiforme* subsp. *acariaeum.*
subsp. **atra** (Näg. & Pet.) Soják *op. cit.* 42(1972):
I *Hieracium auriculiforme* subsp. *atrum.*
subsp. **austrobohemica** (Zahn) Soják *op. cit.* 42
I (1972): *Hieracium schultesii* subsp. *austrobohemicum.*
subsp. **erythrogramma** (Näg. & Pet.) Soják *op. cit.* 42(1972): *Hieracium auriculiforme* subsp.
I *erythrogrammum.*
subsp. **glaucoviridis** (Näg. & Pet.) Soják *op. cit.* 42
I (1972): *Hieracium auriculiforme* subsp. *glaucoviride.*
subsp. **holubyana** (Näg. & Pet.) Soják *op. cit.* 42 (1972), as 'hulubyana': *Hieracium auriculiforme*
I subsp. *holubyanum.*
subsp. **kiralymontis** (Zahn) Soják *op. cit.* 42(1972):
I *Hieracium schultesii* subsp. *kiralymontis.*
subsp. **megalophylla** (Näg. & Pet.) Soják *op. cit.* 42(1972): *Hieracium auriculiforme* subsp.
I *megalophyllum.*
subsp. **mendelii** (Peter) Soják *op. cit.* 42(1972):
I *Hieracium mendelii.*
subsp. **palatina** (Näg. & Pet.) Soják *op. cit.* 42
I (1972): *Hieracium auriculiforme* subsp. *palatinum.*
subsp. **parva** (Näg. & Pet.) Soják *op. cit.* 42(1972):
I *Hieracium furcatum* subsp. *parvum.*
subsp. **pictisquama** (Rehm.) Soják *op. cit.* 42
I (1972): *Hieracium auriculiforme* subsp. *pictisquamum.*
subsp. **poliomelanoides** (Zahn) Soják *op. cit.* 42
I (1972): *Hieracium schultesii* subsp. *poliomelanoides.*
subsp. **pseudauriculiformis** (Zahn) Soják *op. cit.* 42(1972): *Hieracium schultesii* subsp.
I *pseudauriculiforme.*
subsp. **pseudohandelii** (Lengyel & Zahn) Soják *op. cit.* 42(1972): *Hieracium schultesii* subsp.
I *pseudohandelii.*
subsp. **schultziorum** (Näg. & Pet.) Soják *op. cit.* 42(1972): *Hieracium auriculiforme* subsp.
I *schultziorum.*
subsp. **silvicola** (Näg. & Pet.) Soják *op. cit.* 42
I (1972): *Hieracium auriculiforme* subsp. *silvicola.*
subsp. **tricoronaemontis** (Zahn) Soják *op. cit.* 42
I (1972): *Hieracium schultesii* subsp. *tricoronaemontis.*

sciadophora (Näg. & Pet.) Soják
subsp. **atschavica** (Zahn) Soják *op. cit.* 48(1972):
I *Hieracium sciadophorum* subsp. *atschavicum.*
subsp. **digenes** (Näg. & Pet.) Soják *op. cit.* 48
I (1972): *Hieracium sciadophorum* subsp. *digenes.*
subsp. **humidicola** (Näg. & Pet.) Soják *op. cit.* 48
I (1972): *Hieracium sciadophorum* subsp. *humidicola.*
subsp. **ignota** (Näg. & Pet.) Soják *op. cit.* 48(1972):
I *Hieracium sciadophorum* subsp. *ignotum.*

stenosoma (Näg. & Pet.) Soják
subsp. **suprafloccosa** (Näg. & Pet.) Soják *op. cit.* 48 (1972): *Hieracium auriculiforme* subsp.
I *suprafloccosum.*

stoloniflora (W.K.) F.W. & C. H. Schultz
subsp. **oligocephala** (Schur ex Näg. & Pet.) Soják *op. cit.* 43(1972): *Hieracium stoloniflorum* subsp.
I *oligocephalum.*
subsp. **schuriana** (Näg. & Pet.) Soják *op. cit.* 43
I (1972): *Hieracium stoloniflorum* subsp. *schurianum.*

subcymiflora (Oborny & Zahn) Soják
subsp. **hyperdoxoides** (Zahn & Petrak) Soják *op.*
I *cit.* 59(1972): *Hieracium hyperdoxoides.*

subcymigera (Peter) Soják
subsp. **alethes** (Näg. & Pet.) Soják *op. cit.* 51(1972):
I *Hieracium florentinum* subsp. *alethes.*
subsp. **canipeduncula** (Näg. & Pet.) Soják *op. cit.* 51(1972): *Hieracium florentinum* subsp.
I *canipedunculum.*
subsp. **parciflocciformis** (Pal. & Zahn) Soják *op. cit.* 60(1972): *Hieracium piloselloides* subsp.
I *parciflocciforme.*
subsp. **pseudovistulina** (Oborny) Soják *op. cit.* 51 (1972): *Hieracium florentinum* subsp. *pseudovist-*
I *ulinum.*
subsp. **radiata** (Näg. & Pet.) Soják *op. cit.* 51(1972):
I *Hieracium florentinum* subsp. *radiatum.*
subsp. **subumbellosa** (Näg. & Pet.) Soják *op. cit.* 51(1972): *Hieracium florentinum* subsp.
I *subumbellosum.*

sulphurea (Doell) F.W. & C.H. Schultz
subsp. **glogaviensis** (Näg. & Pet.) Soják *op. cit.* 56
I (1972): *Hieracium sulphureum* subsp. *glogaviense.*
subsp. **pascuorum** (Näg. & Pet.) Soják *op. cit.* 56
I (1972): *Hieracium sulphureum* subsp. *pascuorum.*

tephroglauca (Näg. & Pet.) Soják
subsp. **niesslii** (Oborny) Soják *op. cit.* 50(1972):
I *Hieracium tephroglaucum* subsp. *niesslii.*

tubulascens Norrl.
subsp. **aestivae** (Gerstl. & Zahn) Soják *op. cit.* 49
I (1972): *Hieracium tubulascens* subsp. *aestivae.*

tubulata (Vollm.) Soják
subsp. **cymelliflora** (Zahn) Soják *op. cit.* 47(1972):
I *Hieracium spurium* subsp. *cymelliflorum.*
subsp. **pseudocymiflora** (Zahn) Soják *op. cit.* 61
I (1972): *Hieracium spurium* subsp. *pseudocymiflorum.*

vaillantii (Tausch) Soják
subsp. **cymigera** (Reichenb. f.) Soják *op. cit.* 47
I (1972): *Hieracium cymigerum.*
subsp. **pulverata** (Näg. & Pet.) Soják *op. cit.* 47
I (1972): *Hieracium cymosum* subsp. *pulveratum.*
subsp. **suomensis** (Norrl.) Soják *op. cit.* 47(1972):
I *P. suomensis.*

ziziana (Tausch) F. W. & C. H. Schultz
subsp. **affinis** (Näg. & Pet.) Soják *op. cit.* 57(1972):
I *Hieracium zizianum* subsp. *affine.*
subsp. **farinosa** (Näg. & Pet.) Soják *op. cit.* 57
I (1972): *Hieracium zizianum* subsp. *farinosum.*

subsp. **laevis** (Näg. & Pet.) Soják *op. cit.* 57(1972):
1 *Hieracium zizianum* subsp. *laeve.*
subsp. **leptophylla** (Näg. & Pet.) Soják *op. cit.* 58
1 (1972): *Hieracium zizianum* subsp. *leptophyllum.*
subsp. **petasodes** (Näg. & Pet.) Soják *op. cit.* 58
1 (1972): *Hieracium zizianum* subsp. *petasodes.*
subsp. **postdiluvialis** (Näg. & Pet.) Soják *op. cit.*
1 58(1972): *Hieracium zizianum* subsp. *postdiluviale.*

PITTOCAULON H. Robinson & Brettell in
4 *Phytologia,* 26(6): 451(1973).

1561 Robinson, H. *and* Brettell, R. D.
Studies in the Senecioneae (Asteraceae): 1. *a* new
4 genus, *Pittocaulon. Phytologia,* 26(6):451–453(1973).

bombycophole (Bullock) H. Robinson & Brettell in
4 *Phytologia,* 26(6):452(1973): *Senecio bombycopholis.*
filare (McVaugh) H. Robinson & Brettell *op. cit.* 452
4 (1973): *Senecio filaris.*
hintonii H. Robinson & Brettell *op. cit.* 452(1973)—
4 Mexico.
praecox (Cav.) H. Robinson & Brettell *op. cit.* 453
4 (1973): *Cineraria praecox.*
velatum (Greenm.) H. Robinson & Brettell *op. cit.*
4 453(1973): *Senecio velatus.*

PLAGIOCHEILUS

1562 Robinson, H. *and* Brettell, R. D.
Tribal revisions in the Asteraceae: 10. The relationship of *Plagiocheilus. Phytologia,* 26(3):159–162(1973)—
4 Key.

PSEPHELLUS

kemulariae Charadze in *Zam. Sist. Geogr. Rast.,* 29:
2 61(1972)—U.S.S.R. (Caucasus).

PSEUDOCONYZA

1563 Cuatrecasas, J.
Supplemental characterization of genus *Pseudoconyza* (Compositae, Inuleae-Plucheinae). *Phytologia,*
3, 4 26(6):410–412(1973)—Illustrations.

1564 D'Arcy, W. G.
A name change in *Pseudoconyza* (Compositae-
4, 5 Inuleae). *Phytologia,* 25(5):281(1973).

viscosa (Mill.) D'Arcy in *Phytologia,* 25(5):281(1973):
4 *Conyza viscosa.*
var. **lyrata** (H.B.K.) D'Arcy *op. cit.* 281(1973):
4 *Conyza lyrata.*

PSIADIA

ceylanica (Arn.) Grierson in *Ceylon J. Sci., Biol. Sci.,*
6 10(1):52(1972): *Solidago ceylanica.*

PSYCHROGETON

sect. **Conyzastrum** (Boiss.) Ling & Y. L. Chen in
Acta Phytotax. Sin., 11(4):427(1973): *Erigeron* sect.
2 *Conyzastrum.*
poncinsii (Franch.) Ling & Y. L. Chen *op. cit.* 427
2 (1973): *Aster poncinsii.*

PTILOSTEMON

1565 Greuter, W.
Monographie der Gattung *Ptilostemon* (Compositae).
Boissiera, 22:1–215(1973)—(En,Fr); illustrations, map,
1, 2 dot maps; chromosome numbers; key.

1 sect. **Candollea** Greuter in *Boissiera,* 22:132(1973).
2 sect. **Cassinia** Greuter *op. cit.* 99(1973).
sect. × **Cassion** Greuter *op. cit.* 160(1973). (*P.* sect.
1 *Cassinia* × *P.* sect. *Ptilostemon*).
1, 2 sect. **Echinocephale** Greuter *op. cit.* 128(1973).
1 sect. **Fontqueria** Greuter *op. cit.* 136(1973).
subgen. **Lamyra** (Cass.) Greuter *op. cit.* 149(1973):
1 *Cirsium* subgen. *Lamyra.*
1 sect. **Leptophyllon** Greuter *op. cit.* 126(1973).
sect. × **Platon** Greuter *op. cit.* 159(1973). (*P.* sect.
1 *Platyrhaphium* × *P.* sect. *Ptilostemon*).
1 sect. **Platyrhaphium** Greuter *op. cit.* 87(1973).
sect. × **Plinia** Greuter *op. cit.* 158(1973). (*P.* sect.
1 *Platyrhaphium* × *P.* sect. *Cassinia*).
1 sect. **Polyacantha** Greuter *op. cit.* 143(1973).
sect. × **Pterium** Greuter *op. cit.* 157(1973). (*P.* sect.
1 *Pterocaulos* × *P.* sect. *Platyrhaphium*).
1 sect. **Pterocaulos** Greuter *op. cit.* 77(1973).
abylensis (Maire) Greuter *op. cit.* 137(1973): *Cirsium*
1 *casabonae* subsp. *abylense.*
afer (Jacq.) Greuter
subsp. **eburneus** Greuter *op. cit.* 97(1973)—
2 Turkey.
chamaepeuce (L.) Less.
var. **camptolepis** (Boiss.) Greuter *op. cit.* 115
2 (1973): *Chamaepeuce alpini* var. *camptolepis.*
2 var. **cyprius** Greuter *op. cit.* 113(1973)—Cyprus.
var. **elegans** Greuter *op. cit.* 117(1973)—Karpathos,
1 Crete.
diacantha (Labill.) Greuter
subsp. **turcicus** Greuter *op. cit.* 103(1973)—Turkey,
2 Syria.
dyricola (Maire) Greuter *op. cit.* 142(1973): *Cirsium*
1 *casabonae* subsp. *dyricola.*
× **grandei** (Petrak) Greuter *op. cit.* 157(1973):
1 *Cirsium* × *grandei.*
leptophyllus (Pau & Font Quer) Greuter *op. cit.* 127
1 (1973): *Chamaepeuce leptophylla.*
× **pabotii** Greuter *op. cit.* 160(1973)—Syria. (*P.*
diacantha subsp. *turcicus* × *P. chamaepeuce* var.
1 *camptolepis*).
× **parisiensis** Greuter *op. cit.* 159(1973)—Syria. (*P.*
1 *afer* × *P. chamaepeuce* (an var. *camptolepis*?)).
rhiphaeus (Pau & Font Quer) Greuter *op. cit.* 138
1 (1973): *Chamaepeuce rhiphaea.*
var. **tetauensis** (Font Quer) Greuter *op. cit.* 141
1 (1973): *Cirsium rhiphaeum* var. *tetauense.*
× **tauricola** (Boiss. & Hausskn.) Greuter *op. cit.* 158
1 (1973): *Cirsium tauricola.*

PULICARIA

canariensis Bolle
subsp. **lanata** (FQ & Svent.) Bramwell & Kunkel
in *Cuad. Bot. Canar.,* 18–19:53(1973): *P.*
1 *canariensis* var. *lanata.*

PYRRHOPAPPUS

1566 Northington, D. K.
A new combination in *Pyrrhopappus* (Compositae:
3 Cichorieae). *Southwest. Nat.,* 18(3):343(1973).

1567 Northington, D. K.
Taxonomy of *Pyrrhopappus*: a cytotaxonomic and
chemosystematic study. *Diss. Abstr. Int., B,* 33(1):79
3 (1972).

multicaulis DC.
var. **geiseri** (Shinners) Northington in *South-*
3 *west. Nat.,* 18(3):343(1973): *P. geiseri.*

RAOULIA

1568 Simpson, M. J. A.
Cushion Raoulias of Canterbury. *Canterbury Bot.*
7 *Soc. J.,* no. 6:27–29(1973)—Illustrations.

1569 Simpson, M. J. A.
Raoulia australis Hook. f. on Banks Peninsula.
7 *Canterbury Bot. Soc. J.,* no. 6:4–6(1973).

1570 Van Royen, P.
Sertulum Papuanum: 18. Two superfluous genera in the New Guinean flora. *Bot. Notiser*, 126(4):417–425
6 (1973)—Illustrations; key.

REICHARDIA

famarae Bramwell & Kunkel in *Cuad. Bot. Canar.*,
1 18–19:49(1973)—Canary Is.

RHAPONTICUM

1571 Dittrich, M.
Proposal to conserve the generic name *Rhaponticum* Hill (1762) ("Rhapontica"), orth. mut. Lam. (1779).
1, 2, 5–7 *Taxon*, 22(2–3):314–315(1973).

1572 Holub, J.
Contribution to the taxonomy and nomenclature of *Leuzea* DC. and *Rhaponticum* auct. *Folia Geobot. Phytotax.*, 8(4):377–395(1973)—Chromosome
1, 2 numbers.

RIGIOPAPPUS

1573 Robinson, H. *and* Brettell, R. D.
Tribal revisions in the Asteraceae: 5. The relation-
3 ship of *Rigiopappus*. *Phytologia*, 26(2):69–70(1973).

RUDBECKIA

1574 Solntseva, M. P.
Semigamiya i oplodotvorenie u *Rudbeckia laciniata* L. (Semigamy and fertilization in *Rudbeckia laciniata* L.). *Bot. Zhurn.*, 58(9):1261–1277(1973)—(En);
3 chromosome numbers.

SABAZIA

4 **trifida** Fay in *Brittonia*, 25(2):197(1973)—Mexico.

SANTOLINA

etrusca (Lacaita) Marchi & D'Amato in *Inform. Bot.*
1 *Ital.*, 5(1):93(1973): *S. chamaecyparissus* var. *etrusca*.

SCHISTOCARPHA

1575 Robinson, H. *and* Brettell, R. D.
Tribal revisions in the Asteraceae: 4. The relationships of *Neurolaena*, *Schistocarpha* and *Alepidocline*.
4 *Phytologia*, 25(7):439–445(1973).

SCORZONERA

1576 Palamarchuk, A. S. *and* Palamarchuk, G. L.
Kozelets purpurovȳĭ-*Scorzonera purpurea* L. vo flore Belorussii. *Botanika* (Minsk), vȳp.15:193–196
1 (1973)—Illustration.

1577 Palamarchuk, A. S. *and* Palamarchuk, G. L.
Nove mistseznakhodzhennya skorzonerȳ purpurovoyi (*Scorzonera purpurea* L.) v Bilorusiyi. (A new habitat of *Scorzonera purpurea* L. in Byelorussia).
1 *Ukr. Bot. Zhurn.*, 30(1):117–118(1973)—Uk(En).

SENECIO

1578 Abbott, R.
Variability within populations of the self-fertilising species, *Senecio vulgaris* L. *Trans. Bot. Soc. Edinburgh*,
1, 2 42(1):120(1973).

1579 Boĭko, É. V.
O nekotorȳkh vidakh roda *Senecio* L. (De generis *Senecio* L. speciebus nonnullis). *Novosti Sist. Vȳssh.*
1–3 *Rast.*, 10:257–260(1973).

1580 Bruneryе, L.
Application de l'analyse factorielle des correspondances au traitement numérique de populations auvergnates du *Senecio helenitis* (L.) Cuf. *Rev. Gén. Bot.*,80(947–949):113–127(1973)—(En); chromo-
1 some numbers.

1581 Drury, D. G.
Nodes and leaf structure in the classification of some Australasian shrubby Senecioneae-Compositae. *New Zealand J. Bot.*, 11(3):525–554(1973)—Illustrations;
7 key.

1582 Habib, A.-A. M. *and* Metwally, A. M.
A new sesquiterpene keto-lactone from *Senecio*. *Pl.*
2 *Med.*, 23(1):88–93(1973). *S. aegyptiacus*.

1583 Ebel, F. *and* Kästner, A.
Morphologisch-anatomische Beobachtungen an einigen kubanischen Asteraceen-Sippen, insbesondere an *Senecio plumbeus* Griseb. s.l. und *Senecio rivalis* Greenm. (Morphological and anatomical observations in some Asteraceae species of Cuba, especially in *Senecio plumbeus* Griseb. s.l. and *Senecio rivalis* Greenm.). *Flora* (Germany), 162(3):191–205(1973)—
4 (En); illustrations.

1584 Khafagy, S. M., Saleh, M. R., Sabri, N. N. *and* Fathy, Z.
Pharmacognostical study of *Senecio vulgaris* L.: part 1. Macro- and micromorphology of the leaves and stems. *United Arab Republ. J. Pharm. Sci.*, 12(1):
1, 2 79–97(1972)—(Ara); illustrations.

1585 Khafagy, S. M., Saleh, M. R. I., Sabri, N. N. *and* Fathy, Z.
Pharmacognostical study of *Senecio vulgaris* L.: part 2. Macro- and micromorphology of the flower-heads and fruits. *United Arab. Republ. J. Pharm. Sci.*,
1, 2 12(1):99–128(1972)—(Ara); illustrations.

1586 Kowal, R. R. *and* Barkley, T. M.
Senecio anonymus Wood, an earlier name for *Senecio smallii* Britton. *Rhodora*, 75(802):211–219
3 (1973)—Illustration.

1587 Mabberley, D. L.
Evolution in the giant groundsels. *Kew Bull.*, 28(1):
5 61–96(1973)—Illustrations; keys.

1588 Popescu, A.
Consideratii asupra speciilor de *Senecio* L. din secția *Jacobaei* (Thunb.) Dumort. *Stud. Cerc. Biol., Bot.*,
1, 2 25(4):287–299(1973)—(En); illustrations; key.

1589 Prasher, R.
Senecio vulgaris var. *hibernicus*. *Glasgow Nat.*, 19(1):
1 66(1973)—En. Local records.

1590 Robinson, H. *and* Brettell, R. D.
A new species of *Senecio* from Costa Rica. *Phytologia*,
4 26(6):454(1973). *S. chirripoensis*.

1591 Sapaly, J.
Présence de *Senecio vimineus* (D.C.) Harvey dans le
1 département de la Gironde. *Monde Pl.*, no. 376:7(1973).

ser. **Barbaraeifolii** Popescu in *Stud. Cerc. Biol., Bot.*,
1 25(4):294(1973).
brassica R.E. & T.C.E. Fr.
subsp. **brassiciformis** (R.E. & T.C.E. Fr.) Mabberley in *Kew Bull.*, 28(1):89(1973): *S.*
5 *brassiciformis*.
brevilorus Hilliard in *Notes Roy. Bot. Gard.*
5 *Edinburgh*, 32(3):374(1973)—South Africa (Natal).
chirripoensis H. Robinson & Brettell in *Phytologia*,
4 26(6):454(1973)—Costa Rica.
doronicum L.
subsp. gerardii (G.G.) Malagarriga, Las Subesp. y Variac. Geogr.: 15(1973), as 'gerardi' and without
1 basionym date: *S. gerardi*.

helminthioides (Sch. Bip.) Hilliard in *Notes Roy. Bot. Gard. Edinburgh*, 32(3):379(1973): *S. quinquelobus* var. *helminthioides*.
5
johnstonii Oliver
subsp. **barbatipes** (Hedberg) Mabberley in *Kew*
5 *Bull.*, 28(1):86(1973): *S. barbatipes*.
subsp. **battiscombei** (R.E. & T.C.E. Fr.)
5 Mabberley *op. cit.* 86(1973): *S. battiscombei*.
subsp. **cheranganiensis** (Cotton & Blakelock)
5 Mabberley *op. cit.* 85(1973): *S. cheranganiensis*.
subsp. **cottonii** (Hutch. & Taylor) Mabberley *op.*
5 *cit.* 87(1973): *S. cottonii*.
subsp. **dalei** (Cotton & Blakelock) Mabberley *op.*
5 *cit.* 85(1973): *S. dalei*.
subsp. **elgonensis** (T.C.E. Fr.) Mabberley *op. cit.*
5 85(1973): *S. elgonensis*.
var. **friesiorum** (Mildbr.) Mabberley *op. cit.* 84
5 (1973): *S. friesiorum*.
subsp. **refractisquamatus** (De Wild.) Mabberley
5 *op. cit.* 82(1973): *S. refractisquamatus*.
lunulatus (Chiov.) Jacobsen in *Nation. Cact. Succ. J.*
5 (U.K.), 28(1):5(1973): *Monadenium lunulatum*.
macroglossoides Hilliard in *Notes Roy. Bot. Gard. Edinburgh*, 32(3):380(1973)—South Africa (Natal,
5 Cape Prov., Transvaal).
ngoyanus Hilliard *op. cit.* 382(1973)—South Africa
5 (Natal), Mozambique.
paludosus L.
var. **fukarekii** Gajić in *Glasn. Prir. Muz. Belgradu*,
1 27:34(1972)—Jugoslavia.
f. **gymnocarpus** Soó in *Acta Bot. Acad. Sci. Hung.*,
1 18(1–2):173(1973)—Hungary.
1 f. **hungaricus** Soó *op. cit.* 173(1973)—Hungary.
panduriformis Hilliard in *Notes Roy. Bot. Gard. Edinburgh*, 32(3):383(1973), nom. nov.: *S.*
5 *pandurifolius* Harv.
salisburgensis (Cufod.) Rauschert in *Feddes Repert.*, 83(9–10):654(1972 publ. 1973): *S. helenitis* subsp.
1 *salisburgensis*.
tanacetopsis Hilliard in *Notes Roy. Bot. Gard. Edinburgh*, 32(3):384(1973), nom. nov.: *S. tanace-*
5 *toides* Sond. ex Harv.
tournefortii Lap.
subsp. **aragonensis** (Wk.) Malagarriga, Las Subesp. y Variac. Geogr.: 15(1973): *S. tourne-*
1 *fortii* var. *aragonensis*.
subsp. **carpetanus** (Wk.) Malagarriga *op. cit.* 15
1 (1973): *S. tournefortii* var. *carpetanus*.
subsp. **granatensis** (Bss.) Malagarriga *op. cit.* 15 (1973), without basionym date: *S. tournefortii* var.
1 *granatensis*.

SERRATULA

nudicaulis (L.) DC.
subsp. **subinermis** (Cosson) Malagarriga, Las Subesp. y Variac. Geogr.: 16(1973): *S. nudicaulis*
1 var. *subinermis*.

3 **SHINNERSOSERIS** Tomb in *Sida*, 5(3):186(1973).

1592 Tomb, A. S.
Shinnersoseris gen. nov. (Compositae: Cichorieae). *Sida*, 5(3):183–189(1973)—En; illustration, dot map;
3 chromosome numbers.

rostrata (Gray) Tomb in *Sida*, 5(3):186(1973):
3 *Lygodesmia juncea* var. *rostrata*.

SIGESBECKIA

1593 Humbles, J. E.
Observations on the genus *Sigesbeckia* L. *Cienc. Nat.*, 13(1–2):2–19(1972)—(Sp); illustrations; chromosome
1–7 numbers.

SILPHIUM

1594 Settle, W. J. *and* Fisher, T. R.
A biosystematic study of *Silphium integrifolium Michaux* (Compositae). *Ohio J. Sci.*, 72(5):254–265
3 (1972)—Chromosome number.

SILYBUM

1595 Conacher, E. R. T.
Silybum marianum. *Glasgow Nat.*, 19(1):67(1973)—
1 (En). Local records.

eburneum Cosson & DR
subsp. **hispanicus** (Wk.) Malagarriga, Las Subesp. y Variac. Geogr.: 16(1973): *S. eburneum* var.
1 *hispanicus*.

SOLIDAGO

1596 Andreasen, M. L. *and* Eshbaugh, W. H.
Solidago albopilosa Braun, a little known goldenrod from Kentucky. *Castanea*, 38(2):117–132(1973)—
3 Illustration, map; chromosome numbers.

1597 McNeill, J.
Flora Europaea: notulae systematicae ad floram Europaeam spectantes: no. 14. Note on *Solidago* L. *Bot. J. Linn. Soc.*, 67(3):279–280(1973)—Chromosome
1–3 number.

1598 Morton, G. H.
Filiform creeping rhizomes in *Solidago juncea* (Compositae). *Sida*, 5(2):131–132(1973)—Chromo-
3 some number.

1599 Shul'ts, G. É.
Severoevropeĭskie taksonȳ iz rodstva *Solidago virgaurea* L. s.l. (sem. Asteraceae). (Taxa boreali-europaea ex affinitate *Solidago virgaurea* L. s.l. (Asteraceae)). *Novosti Sist. Vȳssh. Rast.*, 10:248–257
1, 2 (1973)—Illustrations.

gigantea Aiton
subsp. **serotina** (O. Kuntze) McNeill in *Bot. J. Linn. Soc.*, 67(3):280(1973): *Aster latissimifolius*
1, 3 var. *serotinus*.
lapponica With.
subsp. **stenophylla** Shul'ts in *Novosti Sist. Vȳssh. Rast.*, 10:249(1973), nom. nov.: *S. virgaurea* b.
1 *angustifolia* Meinsh.

SOLIVA

1600 Dakshini, K. M. M., Prithipalsingh *and* Maheshwari, J. K.
Distributional records of *Soliva anthemifolia*, *Centipeda minima* and *Merremia emarginata* in and near
4, 6 Delhi. *Curr. Sci.*, 42(11):396–397(1973).

SONCHUS

1601 Boulos, L.
The endemic species of *Sonchus* and related genera in the Macaronesian Islands. *Monogr. Biol. Canar.*,
1 no.4:60–63(1973).

1602 Boulos, L.
Révision systématique du genre *Sonchus* L. s.l.: 4. Sous-genre: 1. *Sonchus*. *Bot. Notiser*, 126(2):155–196 (1973)—Illustrations, dot maps; chromosome
1–3, 5–7 numbers.

1603 Harms, V. L.
A new record for *Sonchus uliginosus* Bieb. from interior Alaska. *Canad. Field.-Nat.*, 87(1):63(1973)—
3 Chromosome numbers.

asper (L.) Hill.
f. **adenotrichus** Soó in *Acta Bot. Acad. Sci. Hung.*,
1 18(1–2):173(1973)—Hungary.

STAEHELINA

petiolata (L.) Hilliard & B. L. Burtt in *Notes Roy. Bot. Gard. Edinburgh*, 32(3):384(1973): *Gnaphalium*
5 *petiolatum*.

STEPHANOMERIA

1604 Gottlieb, L. D.
Genetic differentiation, sympatric speciation and the origin of a diploid species of *Stephanomeria*. *Amer. J.*
3 *Bot.*, 60(6):545–553(1973).

1605 Gottlieb, L. D.
Levels of confidence in the analysis of hybridization in plants. *Ann. Missouri Bot. Gard.*, 59(3):435–446 (1972 publ. 1973)—Illustrations, dot maps;
3 chromosome numbers.

STEVIA

1606 Grashoff, J. L.
A systematic study of the north and central American species of *Stevia*. *Diss. Abstr. Int., B*, 33(9):
3, 4 4148(1973).

anadenotricha (Robins.) Grashoff in *Brittonia*, 24(4):
3 392(1972): *S. berlandieri* var. *anadenotricha*.
subpubescens Lag.
var. **intermedia** Grashoff *op. cit.* 392(1972)—
4 Mexico.

TAGETES

1607 Robinson, H.
Additions to the genus *Tagetes* (Helenieae,
4 Asteraceae). *Phytologia*, 26(5):378–380(1973).

iltisiana H. Robinson in *Phytologia*, 26(5):378(1973)—
4 Bolivia.
4 **moorei** H. Robinson *op. cit.* 379(1973)—Mexico.
praetermissa (Strother) H. Robinson *op. cit.* 380
4 (1973): *Vilobia praetermissa*.

TANACETUM

eriobasis (Rech. f.) R. R. Stewart, Annot. Cat. Vasc. Pl. W. Pakistan & Kashmir: 786(1972): *Chrysan-*
6 *themum eriobasis*.

TARAXACUM

1608 Arevshatyan, I. G.
Chisla khromosom i razmerȳ pȳl'tsevȳkh zeren vidov roda *Taraxacum* Weber, proizrastayushchikh v Armyanskoĭ SSR. (Numbers of chromosome and sizes of microspores of *Taraxacum* Weber in Armenia).
2 *Biol. Zhurn. Arm.*, 26(3):38–44(1973)—(Arm).

1609 Doll, R.
Neue *Taraxacum*-Arten aus der UdSSR. *Feddes Repert.*, 83(7–8):495–500(1972 publ. 1973)—(En);
2 illustrations.

1610 Doll, R.
Revision der sect. *Erythrosperma* Dahlst. emend. Lindb. f. der Gattung *Taraxacum* Zinn. *Feddes Repert.*, 83(9–10):673–740(1972 publ. 1973)—(En); dot maps;
1, 2, 6 chromosome numbers.

1161 Doll, R.
Revision der sect. *Erythrosperma* Dahlst. emend. Lindb. f. der Gattung *Taraxacum* Zinn. *Feddes Repert.*, 84(1–2):1–180(1973)—Illustrations, map;
1, 2 key.

1612 Doll, R.
Zwei neue Arten der Gattung *Taraxacum*. *Feddes Repert.*, 84(7–8):569–572(1973)—(En); illustrations;
2 chromosome numbers.

1613 Fridén, L.
En ny fyndort för Sankt Pers nycklar på Falbygden och anteckninger om Västergötlands maskrosor.
1 *Fauna Flora*, 68(3):101–105(1973)—Dot map.

1614 Hagendijk, A., Soest, J. L. van *and* Zevenbergen, H. A.
Neue Taraxacumarten der Niederlande: 2. *Acta Bot. Neerl.*, 22(6):616–636(1973)—(En); illustrations,
1 dot maps.

1615 Małecka, J.
Cyto-embryological studies in the genus *Taraxacum* section *Palustria* Dahlstedt. *Genet. Pol.*, 14(2):185–188
1 (1973)—(Pol,Rus); chromosome numbers.

1616 Małecka, J.
Further cyto-taxonomic studies in the genus *Taraxacum* section *Palustria* Dahlstedt. *Acta Biol. Cracov., Bot.*, 15(2):113–126(1972)—Illustrations,
1 map; chromosome numbers; key.

1617 Małecka, J.
Problems of the mode of reproduction in microspecies of *Taraxacum* section *Palustria* Dahlstedt. *Acta Biol. Cracov., Bot.*, 16(1):37–84(1973)—Chromosome
1 numbers.

1618 Øllgaard, H.
Nyfund af Maelkebøtter i Danmark. *Flora Fauna*,
1 79(3):79–83(1973)—Illustrations.

1619 Orlova, N. I.
Novȳĭvid oduvanchika (*Taraxacum* Wigg.) iz severo-vostochnoĭ Evropȳ. (Generis *Taraxacum* Wigg. species nova ex Europa boreali-orientali). *Novosti Sist. Vȳssh. Rast.*, 10:269–274(1973)—Illustrations,
1 dot map.

1620 Richards, A. J.
Dandelions. *In* Green, P. S. *ed.* Plants: wild and
1–4 cultivated: 172–178(1973).

1621 Richards, A. J.
The origin of *Taraxacum* agamospecies. *Bot. J. Linn. Soc.*, 66(3):189–211(1973)—Maps; chromosome
1–4, 6, 7 numbers.

1622 Singh, D. *and* Kaul, V.
Karyotype studies in the genus *Taraxacum* Weber. *Proc. Indian Sci. Congr.*, 60(3):317–318(1973)—
6 Chromosome numbers. Abstract only.

1623 Soest, J. L. van
Atlas of leaf shapes of *Taraxacum* species from the Netherlands. Leyden, Rijksherbarium, 1973. unpaged
1 —Illustrations.

subsect. **Obliqua** (Dahlst.) Doll in *Feddes Repert.*,
1 84(1–2):8(1973): *T.* sect. *Obliqua*.
arenicola Lindb. ex Doll *op. cit.* 79(1973), as
1 'arenicolum'—Finland.
austriacum v. Soest
var. **danubium** (Richards) Doll *op. cit.* 21(1973):
1 *T. danubium*.
var. **punctatum** (Richards) Doll *op. cit.* 22(1973):
1 *T. punctatum*.
austrinum Hagl.
f. **minor** Małecka in *Acta Biol. Cracov., Bot.*, 15(2):
1 118(1972)—Poland.

brabanticum Hagendijk, van Soest & Zevenbergen
1 in *Acta Bot. Neerl.*, 22(6):616(1973)—Netherlands.
brachyglossum (Dahlst.) Dahlst.
f. **aphanochroum** (Hagend., v. Soest & Zevenb.)
Doll in *Feddes Repert.*, 84(1–2):54(1973): *T.*
1 *aphanochroum.*
f. **argutum** (Dahlst. ex Druce) Doll *op. cit.* 54
1 (1973): *T. argutum.*
flevoense van Soest in *Acta Bot. Neerl.*, 22(6):620
1 (1973)—Netherlands.
fulvum Raunk.
f. **canulum** (Hagl.) Doll in *Feddes Repert.*, 84(1–2):
1 62(1973): *T. canulum.*
glaucinum Dahlst.
f. **glauciniforme** (Dahlst. ex Druce) Doll *op. cit.*
1 106(1973): *T. glauciniforme.*
hamatulum Hagendijk, van Soest & Zevenbergen in
Acta Bot. Neerl., 22(6):620(1973)—Netherlands,
1 Denmark, Belgium, France, Great Britain.
haneltii van Soest in *Feddes Repert.*, 84(7–8):569
2 (1973)—Mongolia.
kolymense Khokhryakov in *Byull. Glavn. Bot. Sada*
2 (Moscow), 88:47(1973)—U.S.S.R. (Soviet Far East).
limbatum Dahlst.
f. **egens** (Dahlst.) Doll in *Feddes Repert.*, 84(1–2):31
1 (1973): *T. obscurans* f. *egens.*
f. **psammophilum** (Hagl.) Doll *op. cit.* 31(1973):
1 *T. obscurans* f. *psammophilum.*
lucidiforme Hagendijk, van Soest & Zevenbergen
in *Acta Bot. Neerl.*, 22(6):622(1973)—Netherlands,
1 Belgium.
marginatum (Dahlst.) Dahlst.
f. **fennomarginatum** (Markl.) Doll in *Feddes*
1 *Repert.*, 84(1–2):102(1973): *T. fennomarginatum.*
pawlodarskum Doll in *Feddes Repert.*, 83(7–8):497
2 (1972 publ. 1973)—U.S.S.R. (Middle Asia).
perfiljevii Orlova in *Novosti Sist. Vÿssh. Rast.*, 10:269
1 (1973)—U.S.S.R. (European Russia).
ploegii Hagendijk, van Soest & Zevenbergen in *Acta*
1 *Bot. Neerl.*, 22(6):625(1973)—Netherlands.
polonicum Małecka & van Soest in *Acta Biol. Cracov.*,
1 *Bot.*, 15(2):119(1972)—Poland.
proximum (Dahlst.) Dahlst.
var. **detonsum** (Saltin & Olofsson) Doll in *Feddes*
1 *Repert.*, 84(1–2):75(1973): *T. detonsum.*
f. **proximiforme** (v. Soest) Doll *op. cit.* 75(1973):
1 *T. proximiforme.*
pseudo-tianschanicum Doll in *Feddes Repert.*, 83
(7–8):499(1972 publ. 1973)—U.S.S.R. (Middle
2 Asia).
1 **rostochiensis** Doll *op. cit.* 570(1973)—Germany.
simile Raunk.
f. **acutum** (Richards) Doll in *Feddes Repert.*, 84
1 (1–2):115(1973): *T. acutum.*
skalinskanum Małecka & van Soest in *Acta Biol.*
1 *Cracov., Bot.*, 15(2):120(1972)—Poland.
stenospermum Senn.
var. **asturiense** (v. Soest) Doll in *Feddes Repert.*,
1 84(1–2):132(1973): *T. asturiense.*
subarmatum Hagendijk, van Soest & Zevenbergen
in *Acta Bot. Neerl.*, 22(6):628(1973)—Netherlands,
1 Sweden, France.
subditivum Hagendijk, van Soest & Zevenbergen
1 *op. cit.* 630(1973)—Netherlands.
subericinum Hagendijk, van Soest & Zevenbergen
op. cit. 630(1973)—Netherlands, Germany, Belgium,
1 France.
taeniatum Hagl.
f. **alandicum** (Markl.) Doll in *Feddes Repert.*, 84
1 (1–2):104(1973): *T. alandicum.*
tarachodum Hagendijk, van Soest & Zevenbergen
1 in *Acta Bot. Neerl.*, 22(6):634(1973)—Netherlands.
ustamenum Doll in *Feddes Repert.*, 83(7–8):498(1972
2 publ. 1973)—U.S.S.R. (Middle Asia).
ziwaschum Doll *op. cit.* 495(1972 publ. 1973)—
1 U.S.S.R. (Ukraine).

TEPHROSERIS

sect. **Eriopappus** (Dum.) Holub in *Folia Geobot.*
Phytotax., 8(2):173(1973): *Cineraria* sect.
1–3 *Eriopappus.*
atropurpurea (Ledeb.) Holub *op. cit.* 173(1973):
1–3 *Cineraria atropurpurea.*
aucheri (DC.) Holub *op. cit.* 173(1973): *Senecio*
1, 2 *aucheri.*
balbisiana (DC.) Holub *op. cit.* 173(1973): *Senecio*
1 *balbisianus.*
elodes (DC.) Holub *op. cit.* 173(1973): *Senecio*
4 *elodes.*
flammea (DC.) Holub *op. cit.* 173(1973): *Senecio*
2 *flammeus.*
frigida (Richards.) Holub *op. cit.* 173(1973): *Cineraria*
1–3 *frigida.*
fuscata (Hayek) Holub *op. cit.* 173(1973): *Senecio*
1 *fuscatus.*
heldreichii (Boiss.) Holub *op. cit.* 173(1973): *Senecio*
1 *heldreichii.*
integrifolia (L.) Holub *op. cit.* 173(1973): *Othonna*
5 *integrifolia.*
1 **italica** Holub *op. cit.* 173(1973)—Italy.
jacutica (Šišk.) Holub *op. cit.* 173(1973): *Senecio*
2 *jacuticus.*
karjaginii (Sof.) Holub *op. cit.* 173(1973): *Senecio*
2 *karjaginii.*
kawakanii (Makino) Holub *op. cit.* 173(1973):
2 *Senecio kawakanii.*
kjellmanii (A. E. Porsild) Holub *op. cit.* 173(1973):
3 *Senecio kjellmanii.*
lanceolata (Lam.) Holub *op. cit.* 173(1973): *Cineraria*
1, 2 *lanceolata.*
subsp. **arvernensis** (Rouy) Holub *op. cit.* 173(1973):
1 *Senecio arvernensis.*
papposa (Reichenb.) Schur
subsp. **araneosa** (Griseb.) Holub *op. cit.* 173(1973):
1 *Cineraria crassifolia* var. *araneosa.*
subsp. **heuffelii** (Hoppe & Fürnr.) Holub *op. cit.*
1 174(1973): *Senecio heuffelii.*
subsp. **microrrhiza** (Schur) Holub *op. cit.* 174
1, 2 (1973): *T. microrrhiza.*
subsp. **rupicola** (Schur) Holub *op. cit.* 174(1973):
1 *T. rupicola.*
subsp. **wolffii** (Schur) Holub *op. cit.* 174(1973):
1, 2 *T. wolffii.*
pierotii (Miq.) Holub *op. cit.* 174(1973): *Senecio*
2 *pierotii.*
porphyrantha (Šišk.) Holub *op. cit.* 174(1973):
2 *Senecio porphyranthus.*
praticola (Šišk. & Serg.) Holub. *op. cit.* 174(1973):
2 *Senecio praticola.*
pricei (N. D. Simps.) Holub *op. cit.* 174(1973):
2 *Senecio pricei.*
pyrenaica (Godr. & Gren.) Holub *op. cit.* 174(1973):
1 *Senecio pyrenaicus.*
pyroglossa (Kar. & Kir.) Holub *op. cit.* 174(1973):
1, 2 *Senecio pyroglossus.*
serpentini (Gáyer) Holub *op. cit.* 174(1973): *Senecio*
1 *serpentini.*
stolonifera (Cuf.) Holub *op. cit.* 174(1973): *Senecio*
2 *stolonifera.*
subdentata (Bunge) Holub *op. cit.* 174(1973):
1, 2 *Cineraria subdentata.*
subfrigida (Komarov) Holub *op. cit.* 174(1973):
2 *Senecio subfrigidus.*
taitoensis (Hayata) Holub *op. cit.* 174(1973), as
2 'taitonensis': *Senecio taitonensis.*

takedana (Kitamura) Holub *op. cit.* 174(1973):
2 *Senecio takedanus.*
tenuifolia (Gaud.) Holub *op. cit.* 174(1973): *Cineraria*
1, 2 *tenuifolia.*
thyrsoidea (Host) Holub *op. cit.* 174(1973): *Cineraria*
1 *thyrsoidea.*
tundricola (Tolm.) Holub *op. cit.* 174(1973): *Senecio*
2, 3 *tundricola.*
turczaninovii (DC.) Holub *op. cit.* 174(1973):
2 *Senecio turczaninovii.*
wagneri (Degen.) Holub *op. cit.* 174(1973): *Senecio*
1 *wagneri.*
yukonensis (A. E. Porsild) Holub *op. cit.* 174(1973):
3 *Senecio yukonensis.*

TIAROCARPUS K. H. Rechinger in *Anz. Österr. Akad. Wiss., Mat.-Naturwiss.*, 108:5(1971 publ.
2 1972).
hymenostephanus K. H. Rechinger *op. cit.* 7(1971
2 publ. 1972)—Afghanistan.
neubaueri (K. H. Rechinger) K. H. Rechinger *op. cit.*
2 6(1971 publ. 1972): *Cousinia neubaueri.*
tragacanthoides (K. H. Rechinger & Gilli) K. H. Rechinger *op. cit.* 6(1971 publ. 1972): *Cousinia*
2 *tragacanthoides.*

TRAGOPOGON

1624 Chavdarov, I. P. *and* Georgiev, G. N.
Karyological study of *Tragopogon porrifolius* L. *Dokl. Bolg. Akad. Nauk*, 26(7):957–960(1973)—
1 Chromosome number.

TRAVERSIA

1625 Drury, D. G.
Nodes and leaf structure in the classification of some Australasian shrubby Senecioneae-Compositae. *New Zealand J. Bot.*, 11(3):525–554(1973)—Illustrations;
7 key.

TRICHOCLINE

boecheri Cabrera in *Bol. Soc. Argent. Bot.*, 15(1):122
4 (1973)—Argentina.

TRICHOSPIRA

1626 Robinson, H. *and* Brettell, R. D.
Tribal revisions in the Asteraceae: 2. The relation-
4 ship of *Trichospira. Phytologia*, 25(4):259–261(1973).

TRIPLEUROSPERMUM

1627 Hossain, A. B. M. E.
Materials for a flora of Turkey: 29. *Tripleurospermum* Schultz Bip. *Notes Roy. Bot. Gard. Edinburgh*,
1, 2 32(2):251–254(1973).

1628 Khandzhyan, N. S.
Novȳe dlya SSSR i Armenii vidȳ roda *Tripleurospermum* Sch. Bip. (New species of *Tripleurospermum* Sch. Bip. in the USSR and Armenia). *Biol. Zhurn. Arm.*, 26(3):86–87(1973)—(Arm). New distribution
2 records only for *T. grossheimii* and *T. tzvelevii.*

1629 Macpherson, G.
Tripleurospermum maritimum. Glasgow Nat., 19(1):
1 67(1973)—En. Local records.

1630 Óskarsson, I.
Nýtt af brigđi af baldursbrá (*Tripleurospermum maritimum* (L.) Koch ssp. *phaeocephalum* (Rupr.) Hämet-Ahti). (Eine neue Varietät der Geruchslosen Kamille). *Náttúrfrædingurinn*, 42(4):186–189(1973)—
1 (Ge); illustrations.

baytopianum E. Hossain in *Notes Roy. Bot. Gard.*
1 *Edinburgh*, 32(2):253(1973)—Turkey-in-Europe.
2 **corymbosum** E. Hossain *op. cit.* 253(1973)—Turkey.
rosellum (Boiss. & Orph.) Hayek
2 var. **album** E. Hossain *op. cit.* 254(1973)—Turkey.

TRIPOLIUM

pannonicum (Jacq.) Dobroč.
subsp. **maritimum** Holub in *Folia Geobot. Phytotax.*, 8(2):177(1973), nom. nov.: *Tripolium*
1, 2 *vulgare* var. α Nees.

TROMMSDORFIA

1631 Soják, J.
Doplňky k nomenklatuře některȳch rodů (Phanerogamae). *Čas. Nár. Muz.* (Prague), 141(1–2):
1, 2 61–63(1972).

crepidioides (Miyabe & Kudo) Soják in *Čas. Nár. Muz.*
1, 2 (Prague), 141(1–2):63(1972): *Picris crepidioides.*
uniflora (Vill.) Soják
subsp. **facchiniana** (Ambrosi) Soják *op. cit.* 63
1 (1972): *Hypochoeris facchiniana.*

VENIDIUM

1632 Watts, L. E.
The breeding of fertile *Venidio-Arctotis. J. Roy. Hort.*
0 *Soc.*, 98(8):359–360(1973).

VERBESINA

4 **costata** Fay in *Brittonia*, 25(2):192(1973)—Mexico.
4 **sousae** Fay *op. cit.* 195(1973)—Mexico.

VERNONIA

1633 Alencastro, F. M. M. R. de
Contribuição ao estudo da anatomia foliar das Vernonias do Brasil: 1. *Vernonia oppositifolia. Arq.*
4 *Jard. Bot. Rio de Janeiro*, 19:109–123(1973)—(En).

1634 Faust, W. Z.
A biosystematic study of the Interiores species group of the genus *Vernonia* (Compositae). *Brittonia*, 24(4):363–378(1972)—Illustrations, dot maps;
3 chromosome numbers; key.

1635 Faust, W. Z. *and* Jones, S. B.
The systematic value of trichome complements in a North American group of *Vernonia* (Compositae).
3 *Rhodora*, 75(804):517–528(1973)—Illustrations.

1636 Jones, S. B.
Revision of *Vernonia* section *Eremosis* (Compositae) in North America. *Brittonia*, 25(2):86–115(1973)—Illustrations, dot maps; chromosome numbers;
3, 4 key.

1637 Mitra, S. N.
Some new combinations in Indian plants. *Indian*
6 *Forester*, 99(2):100–101(1973).

baadii (McVaugh) S. B. Jones in *Brittonia*, 25(2): 113
4 (1973): *V. salicifolia* var. *baadii.*
baldwinii Torr.
subsp. **interior** (Small) Faust in *Brittonia*, 24(4):377
3 (1972): *V. interior.*
duncanii S. B. Jones in *Brittonia*, 25(2):108(1973)—
4 Mexico.
subsp. **bartlettii** S. B. Jones *op. cit.* 110(1973)—
4 Mexico.
gardneri Thw.
var. **brevior** Grierson in *Ceylon J. Sci., Biol. Sci.*,
6 10(1):43(1972)—Sri Lanka.
lankana Grierson *op. cit.* 43(1972), nom. nov.: *V.*
6 *scariosa* Arn.
var. **crassa** (Thw.) Grierson *op. cit.* 43(1972): *V.*
6 *scariosa* var. *crassa.*
macvaughii S. B. Jones in *Brittonia*, 25(2):105
4 (1973)—Mexico.
neocorymbosa Hilliard in *Notes Roy. Bot. Gard. Edinburgh*, 32(3):385(1973), nom. nov.: *Staehelina*
5 *corymbosa.*

neogleasoniana Cuatrecasas in *Ciencia*, 27(6):175
4 (1972)—Ecuador.
obtusa (Gleason) Blake
subsp. **parkeri** S. B. Jones in *Brittonia*, 25(2):108
4 (1973)—Mexico.
pectiniformis DC.
subsp. **puncticulata** (DC.) Grierson in *Ceylon J.*
6 *Sci., Biol. Sci.*, 10(1):45(1972): *V. puncticulata.*
praestans Ekm. ex Urb.
var. **cacuminis** Borhidi & Muñiz in *Acta Bot. Acad.*
4 *Sci. Hung.*, 18(1–2):48(1973)—Cuba.
pusilliflora Alain in *Phytologia*, 25(5):279(1973)—
4 Hispaniola (Dominican Republic).
pyramidale (D. Don) S. N. Mitra in *Indian Forester*,
6 99(2):100(1973): *Eupatorium pyramidale.*
triflosculosa H. B. K.
subsp. **palmeri** (Rose) S. B. Jones in *Brittonia*,
4 25(2):100(1973): *V. palmeri.*

VIGUIERA

1638 Mellinger, A. C.
Ecological life cycle of *Viguiera porteri* and factors
responsible for its endemism to granite outcrops of
Georgia and Alabama. *Diss. Abstr. Int., B*, 33(3):
3 1087(1972).

WAGENITZIA Dostál in *Acta Bot. Acad. Sci. Hung.*,
1, 2 19(1–4):76(1973).
lancifolia (Sieb. ex Spreng.) Dostál *op. cit.* 76(1973):
1, 2 *Centaurea lancifolia.*

XANTHISMA

1639 Semple, J. C.
The cytology, flavonoid chemistry and systematics
of the Texas sleepy daisy *Xanthisma texanum* DC.
3 (Asteraceae). *Diss. Abstr. Int., B*, 33(8):3513(1973).

XANTHIUM

1640 Amorin, J. L. *and* Orfila, E. N.
'Cepa caballo' *Xanthium spinosum* Linné (Composi-
tae), maleza utilizada en la medicina popular Argentins.
(*Xanthium spinosum* Linné (Compositae), weed used
in Argentine popular medicine). *Rev. Fac. Agron.
Univ. Nac. La Plata*, 48(2):155–169(1972 publ. 1973)—
4 (En); illustrations.

1641 Härdtl, H.
Über die Inhaltsstoffe einiger *Xanthium*-Arten.
1–7 *Phyton* (Austria), 15(1–2):1–25(1973).

1642 McMillan, C.
Photoperiod evidence in the introduction of
Xanthium strumarium to Tahiti and Hawaii from
Mexico. *Amer. J. Bot.*, 60(3):277–282(1973)—
4, 6 Illustration.

1643 McMillan, C.
Photoperiodic adaptation in *Xanthium strumarium*
populations introduced to Australia. *Canad. J. Bot.*,
1–7 51(1):221–229(1973)—Illustrations, dot map.

XEROLOMA

cylindracea (Sibth. & Sm.) Holub in *Folia Geobot.*
1, 2 *Phytotax.*, 8(2):177(1973): *Xeranthemum cylindraceum.*

XYLORHIZA

1644 Watson, T. J.
Systematics of *Xylorhiza* (Compositae-Astereae).
3, 4 *Diss. Abstr. Int., B*, 33(9):4155(1973).

CONNARACEAE

1645 Dickison, W. C.
Anatomical studies in the Connaraceae: 3. Leaf
anatomy. *J. Elisha Mitchell Sci. Soc.*, 89(1–2):121–138
4–6 (1973)—Illustrations.

CONVOLVULACEAE

1646 Austin, D. F.
The American Erycibeae (Convolvulaceae):
Maripa, Dicranostyles and *Lysiostyles*: 1. Systematics.
Ann. Missouri Bot. Gard., 60(2):306–412(1973)—
Illustrations, map, dot maps; chromosome numbers;
4 keys.

1647 Austin, D. F.
The American Erycibeae (Convolvulaceae) *Maripa,
Dicranostyles* and *Lysiostyles*: 2. Palynology. *Pollen
4 Spores*, 15(2):203–226(1973)—(Fr); illustrations.

1648 Falcão, J. I. de A.
Convolvulaceae de Minas Gerais. *Arq. Jard. Bot.
Rio de Janeiro*, 19:49–77(1973)—Illustrations;
4 keys.

1649 Falcão, J. I. de A. *
Convolvulaceae do Amazonas. *Acta Amazonica*,
1(1):15–20(1971 publ. 1972). Ref. in *Bull. Torrey Bot.
4 Club*, 99(6):324(1972).

1650 Hadač, E. *and* Chrtek, J.
Some further notes on the taxonomy and nomen-
clature of Cuscutaceae. *Folia Geobot. Phytotax.*, 8(2):
1–7 219–221(1973).

1651 Sampathkumar, R.
Cytology of some South-Indian Convolvulaceae.
Proc. Indian Sci. Congr., 58(3):474(1971)—Chromo-
6 some numbers. Abstract only.

1652 Wagner, H.
The chemistry of the glykolipid resins of the
Convolvulaceae family. *In* Nobel Symposium 25:
chemistry in botanical classification August 20–25,
1–7 1973. (1973). 1p. Abstract.

ARGYREIA

1653 Chao, J.-M. *and* DerMarderosian, A. H.
Identification of ergoline alkaloids in the genus
Argyreia and related genera and their chemotaxonomic
implications in the Convolvulaceae. *Phytochemistry*,
2–6 12(10):2435–2440(1973).

CALYSTEGIA

1654 Brummitt, R. K.
Calystegia—some British chromosome counts.
Watsonia, 9(4):369–370(1973)—Chromosome
1 numbers.

1655 Falcão, J. I. de A.
O gênero *Calystegia* R. Br. (Convolvulaceae) no
Brasil. *Arq. Jard. Bot. Rio de Janeiro*, 19:177–178
4 (1973)—Illustration.

1656 Dupont, P. *and* Dupont, S.
Recherches sur les *Calystegia* de Bretagne. *Bull.
Soc. Sci. Bretagne*, 46(3–4):209–214(1971 publ. 1972)—
1 Map.

1657 Skvortsov, A. K.
Calystegia inflata Sweet v Moskovskoĭ oblasti.
Byull. Glavn. Bot. Sada (Moscow), 90:24–27(1973)—
3 Illustration.

1658 Stace, C. A.
Calystegia—inheritance of the schizoflorous
1 character. *Watsonia*, 9(4):370–371(1973).

1659 Stace, C. A.
Chromosome numbers in British species of
Calystegia and *Convolvulus*. *Watsonia*, 9(4):363–367
1 (1973).

CONVOLVULUS

1660 Stace, C. A.
Chromosome numbers in British species of *Calystegia* and *Convolvulus*. *Watsonia*, 9(4):363–367
1 (1973).

CUSCUTA

1661 Bhan, A. K. *and* Kaul, M. L. H.
Host diversity and chromosomal polymorphism in *Cuscuta reflexa* Roxb. *Sci. Cult.*, 39(9):403–404(1973)—
6 Chromosome numbers.

1662 Bhattacharyya, P. K.
Stomata of the section *Collienche* (Cuscutaceae).
2, 6 *Sci. Cult.*, 39(2):91–92(1973).

1663 Hassawy, G. S.
Cuscuta species in Iraq: their hosts and seed germination. (*Cuscuta*-Arten und deren Wirte und Keimung in Irak). *In* European Weed Research Council Parasitic Weeds Research Group *and* Malta: Royal University. Symposium on parasitic weeds, 11–13 April 1973, Royal University of Malta, Malta:
2 280–288(1973)—(Ge,Fr).

1664 Stojanović, D. *and* Mijatović, K.
Distribution, biology and control of *Cuscuta* spp. in Yugoslavia. (Verbreitung, Biologie und Bekämpfung von *Cuscuta* spp. in Yugoslawien). *In* European Weed Research Council: Parasitic Weeds Research Group *and* Malta: Royal University. Symposium on parasitic weeds, 11–13 April 1973, Royal
1 University of Malta, Malta: 269–279(1973)—(Ge,Fr).

DACTYLOSTIGMA D. F. Austin in *Phytologia*,
5 25(7):426(1973).

1665 Austin, D. F.
Dactylostigma, a new genus of Convolvulaceae from Madagascar. *Phytologia*, 25(7):425–429(1973)—
5 Illustrations; key.

linearifolia D. F. Austin in *Phytologia*, 25(7):426
5 (1973)—Malagasy.

DICRANOSTYLES

subgen. **Kuhlmanniella** (Barroso) D. F. Austin in *Ann. Missouri Bot. Gard.*, 60(2):400(1973): genus
4 *Kuhlmanniella*.

ampla Ducke

var. **attenuata** D. F. Austin *op. cit.* 392(1973)—
4 Brazil.

var. **castanea** D. F. Austin *op. cit.* 391(1973)—
4 Brazil, Guyana.

globostigma D. F. Austin *op. cit.* 405(1973)—
4 Brazil.

EXOGONIUM

1666 Wasshausen, D. C. *and* Austin, D. F.
A *Ruellia* (Acanthaceae) species in *Exogonium* (Convolvulaceae). *Phytologia*, 25(7):433–437(1973)—
4 Illustrations.

GRAMMICA

acutiloba (Engelm.) Hadač & Chrtek in *Folia*
4 *Geobot. Phytotax.*, 8(2):220(1973): *Cuscuta acutiloba*.

californica (Choisy) Hadač & Chrtek *op. cit.* 220
3 (1973): *Cuscuta californica*.

cephalanthi (Engelm.) Hadač & Chrtek in *Folia Geobot. Phytotax.*, 8(2):220(28 Aug. 1973); also W. A. Weber in *Southwest. Nat.*, 18(3):319(5 Oct.
3, 4 1973): *Cuscuta cephalanthi*.

coryli (Engelm.) Hadač & Chrtek in *Folia Geobot. Phytotax.*, 8(2):220(28 Aug. 1973); also W. A. Weber in *Southwest. Nat.*, 18(3):319(5 Oct. 1973):
3 *Cuscuta coryli*.

cristata (Engelm.) Hadač & Chrtek in *Folia Geobot.*
4 *Phytotax.*, 8(2):220(1973): *Cuscuta cristata*.

cuspidata (Engelm. & Gray) Hadač & Chrtek in *Folia Geobot. Phytotax.*, 8(2):220(28 Aug. 1973); also W. A. Weber in *Southwest. Nat.*, 8(3):319
3 (5 Oct. 1973): *Cuscuta cuspidata*.

densiflora (Hook. fil.) Hadač & Chrtek in *Folia Geobot. Phytotax.*, 8(2):220(1973): *Cuscuta*
7 *densiflora*.

denticulata (Engelm.) W. A. Weber in *Southwest.*
3 *Nat.*, 18(3):319(1973): *Cuscuta denticulata*.

floribunda (Humb., Bonpl. & Kunth) Hadač & Chrtek in *Folia Geobot. Phytotax.*, 8(2):220(1973):
4 *Cuscuta floribunda*.

foetida (Humb., Bonpl. & Kunth) Hadač & Chrtek
4 *op. cit.* 220(1973): *Cuscuta foetida*.

gracillima (Engelm.) Hadač & Chrtek *op. cit.* 220
4 (1973): *Cuscuta gracillima*.

grandiflora (Humb., Bonpl. & Kunth) Hadač &
4 Chrtek *op. cit.* 220(1973): *Cuscuta grandiflora*.

indecora (Choisy) W. A. Weber in *Southwest. Nat.*,
3, 4 18(3):319(1973): *Cuscuta indecora*.

subsp. **neuropetala** (Engelm.) W. A. Weber *op. cit.*
3, 4 319(1973): *Cuscuta indecora* var. *neuropetala*.

jalapensis (Schlecht.) Hadač & Chrtek in *Folia*
4 *Geobot. Phytotax.*, 8(2):220(1973): *Cuscuta jalapensis*.

microstyla (Engelm.) Hadač & Chrtek *op. cit.* 220
4 (1973): *Cuscuta microstyla*.

obtusiflora (Humb., Bonpl. & Kunth) Hadač &
1–4, 7 Chrtek *op. cit.* 220(1973): *Cuscuta obtusiflora*.

occidentalis (Millspaugh) Hadač & Chrtek in *Folia Geobot. Phytotax.*, 8(2):220(28 Aug. 1973); also W. A. Weber in *Southwest. Nat.*, 18(3):319(5 Oct.
3 1973): *Cuscuta occidentalis*.

odontolepis (Engelm.) Hadač & Chrtek in *Folia Geobot. Phytotax.*, 8(2):220(1973): *Cuscuta*
4 *odontolepis*.

odorata (Ruiz & Pav.) Hadač & Chrtek *op. cit.* 220
4 (1973): *Cuscuta odorata*.

pentagona (Engelm.) W. A. Weber in *Southwest.*
3 *Nat.*, 18(3):319(1973): *Cuscuta pentagona*.

racemosa (Mart.) Hadač & Chrtek in *Folia Geobot.*
1–5, 7 *Phytotax.*, 8(2):220(1973): *Cuscuta racemosa*.

rostrata (Shuttl.) Hadač & Chrtek *op. cit.* 220(1973):
3 *Cuscuta rostrata*.

scandens (Brot.) Holub in *Preslia*, 45(4):359(1973):
1 *Cuscuta scandens*.

subsp. **australis** (R. Br.) Holub *op. cit.* 359(1973):
1 *Cuscuta australis*.

subsp. **cesattiana** (Bertol.) Holub *op. cit.* 359(1973):
1 *Cuscuta cesattiana*.

suksdorfii (Yuncker) Hadač & Chrtek in *Folia Geobot. Phytotax.*, 8(2):221(1973): *Cuscuta*
3 *suksdorfii*.

tinctoria (Mart.) Hadač & Chrtek *op. cit.* 221(1973):
4 *Cuscuta tinctoria*.

umbellata (Humb., Bonpl. & Kunth) Hadač & Chrtek in *Folia Geobot. Phytotax.*, 8(2):221(28 Aug. 1973); also W. A. Weber in *Southwest. Nat.*, 18(3):
3, 4 319(5 Oct. 1973): *Cuscuta umbellata*.

umbrosa (Hook.) W. A. Weber in *Southwest. Nat.*,
3 18(3):319(1973): *Cuscuta umbrosa*.

xanthochortos (Martius) Hadač & Chrtek in *Folia Geobot. Phytotax.*, 8(2):221(1973): *Cuscuta xantho-*
4 *chortos*.

IPOMOEA

1667 Kugler, H.
Zur Bestäubung von *Scaevola plumieri* (L.) Vahl und *Ipomoea pes-caprae* Sweet, zwei tropischen Strandpflanzen. (On the pollination of *Scaevola plumieri* (L.) Vahl and *Ipomoea pes-caprae* Sweet, two plants of tropical beaches). *Flora* (Germany), 162(4):381–391
2–7 (1973)—(En); illustrations.

1668 Matuda, E.
Una neuva *Ipomoea* de Michoacán. *Cact. Suc. Mex.*,
4 18(3):78–79(1973)—(En); illustrations. *I. albidiflora.*

1669 Rabson, A.
Ipomoea holubii Baker (Convolvulaceae). *Ashingtonia*, 1(3):29,31–32(1973)—En; coloured
5 illustration.

1670 Seago, J. L.
Developmental anatomy in roots of *Ipomoea purpurea*: 2. Initiation and development of secondary
4 roots. *Amer. J. Bot.*, 60(7):607–618(1973).

1671 Singh, N. P., Kulkarni, B. G. *and* Moorthy, S.
New plant records from Goa. *Curr. Sci.*, 42(13):
6 478(1973). *Ipomoea barlerioides.*

albidiflora Matuda in *Cact. Suc. Mex.*, 18(3):78
4 (1973)—Mexico.

MARIPA

subsect. **Melofructae** D. F. Austin in *Ann. Missouri*
4 *Bot. Gard.*, 60(2):377(1973).
sect. **Mouroucoa** (Aublet) D. F. Austin *op. cit.* 375
4 (1973): genus *Mouroucoa.*
4 subgen. **Ripama** D. F. Austin *op. cit.* 360(1973).
4 subsect. **Yacuascae** D. F. Austin *op. cit.* 383(1973).
janusiana D. F. Austin *op. cit.* 370(1973)—Brazil,
4 Peru.
pauciflora D. F. Austin *op. cit.* 384(1973)—Brazil,
4 Venezuela.
putumayana D. F. Austin *op. cit.* 385(1973)—
4 Colombia.
reticulata Ducke
var. **rugosa** (Ducke) D. F. Austin *op. cit.* 363(1973):
4 *M. rugosa.*

MERREMIA

1672 Dakshini, K. M. M., Prithipalsingh *and* Maheshwari, J. K.
Distributional records of *Soliva anthemifolia*, *Centipeda minima* and *Merremia emarginata* in and near
6 Delhi. *Curr. Sci.*, 42(11):396–397(1973).

MONOGYNELLA

exaltata (Engelm.) Hadač & Chrtek in *Folia Geobot.*
3 *Phytotax.*, 8(2):220(1973): *Cuscuta exaltata.*
timorensis (Decne.) Hadač & Chrtek *op. cit.* 220
6 (1973): *Cuscuta timorensis.*

CORIARIACEAE

1673 Ghafoor, A.
Flora of West Pakistan: no. 43. Coriariaceae. 1973.
6 3p.—Illustration, map.

CORNACEAE

1674 Patel, R. N.
Wood anatomy of the dicotyledons indigenous to New Zealand: 1. Cornaceae. *New Zealand J. Bot.*,
7 11(1):3–22(1973).

AFROCRANIA

1675 Bamps, P.
Distribution géographique d'*Afrocrania volkensii* (Harms) Hutch. (Cornaceae) de l'Afrique tropicale.
5 *Bull. Jard. Bot. Nation. Belg.*, 43(3–4):359–360(1973).

CORNUS

1676 Mignery, A. L.
Flowering dogwood: an American wood. *U.S. Dep. Agric. F.S.*, 232. 5p.(1973)—Illustrations. *Cornus*
3 *florida.*

CURTISIA

1677 Moll, E. J.
Curtisia dentata—a tree of the future. *Veld Flora*,
5 3(2):21,23–24(1973)—Coloured illustration, map.

MELANOPHYLLA

1678 Airy Shaw, H. K.
Melanophylla pachypoda Airy Shaw: correction.
5 *Kew Bull.*, 28(1):116(1973).

CORYLACEAE

CARPINUS

1679 Noack, H.
Das Porträt: *Carpinus betulus* L. *Kurzmitt. Deutsch.*
1 *Dendrol. Ges.*, no. 10:11–13(1973)—Illustration.

1680 Nožička, J.
Několik historických poznámek k výskytu habru (*Carpinus betulus* L.) v jižních Cechách. (Zur Geschichte des Vorkommens von *Carpinus betulus* L. in Südböhmen). *Zprávy Česk. Bot. Spol.*, 8(2):137–138
1 (1973).

CORYLUS

1681 Kasapligil, B.
A bibliography on *Corylus* (Betulaceae) with annotations. *Ann. Rep. North. Nut Growers Ass.*,
1–3, 6 63:107–162(1972)—Illustrations.

avellana L.
f. **peltata** De Langhe in *Bull. Soc. Roy. Bot. Belg.*,
1 106(1):70(1973)—Belgium.

OSTRYA

1682 Ceschi, I.
Nuovi ritrovamenti di carpinello (*Ostrya carpinifolia* Scop.) nel Canton Ticino. *Boll. Soc. Ticin. Sci. Nat.*,
1 61:25–32(1969–70 publ. 1971)—Map.

CRASSULACEAE

1683 Mandrik, V. Yu. *and* Golyshkin, L. V.
Émbriologicheskoe issledovanie nekotorykh vidov semeĭstva Crassulaceae. (Enbryological study of some species of the family Crassulaceae). *Bot. Zhurn.*, 58(2):
1–3 263–271(1973)—*Sempervivum, Rhodiola, Sedum* spp.

1684 Moran, R.
The stonecrop family: variations on a pattern.
1–5 *Environment Southwest*, 452:3–5(1973)—Illustrations.

1685 Tölken, R. H.
Crassulaceae. *Cact. Succ. J.* (U.S.A.), 45(6):293–295
1–7 (1973)—En; illustrations.

AEONIUM

1686 Voggenreiter, V.
Pflanzenverbreitungstypen auf La Palma. Geobotanisch-arealkundliche Untersuchungen: 1. *Aeonium* W. & B. *Cuad. Bot. Canar.*, 18–19:43–48
1 (1973)—(Sp,En); dot maps.

× anagensis Bramwell & Rowley in *Nation. Cact. Succ. J.* (U.K.), 28(1):5(1973), without type;
1 *A. lindleyi* × *tabulaeforme*—Canary Is.
× bramwellii Rowley *op. cit.* 5(1973), without type;
1 *A. canariense* × *cuneatum*—Canary Is.
× bravoanum Bramwell & Rowley *op. cit.* 5(1973), without type; *A. castellopaivae* × *viscatum*—
1 Canary Is.

× castelloplanum Bramwell & Rowley *op. cit.* 5
(1973), without type; *A. castellopaivae* × *subplanum*
1 —Canary Is.
× hawbicum Bramwell & Rowley *op. cit.* 5(1973),
1 without type; *A. haworthii* × *urbicum*—Canary Is.
× jacobsenii Bramwell & Rowley *op. cit.* 5(1973),
1 without type; A. *hierrense* × *palmense*—Canary Is.
× junionae Bramwell & Rowley *op. cit.* 5(1973),
1 without type; *A. ciliatum* × *palmense*—Canary Is.
× kunkelii Bramwell & Rowley *op. cit.* 5(1973),
1 without type; *A. ciliatum* × *holochrysum*—Canary Is.
× lambii Bramwell & Rowley *op. cit.* 5(1973),
1 without type; *A. palmense* × *valverdense*—Canary Is.
× lowei Bramwell & Rowley *op. cit.* 5(1973), without
1 type; *A. glandulosum* × *glutinosum*—Canary Is.
× ombriosum Bramwell & Rowley *op. cit.* 5(1973),
1 without type; *A. hierrense* × *valverdense*—Canary Is.
× rowleyi Bramwell ex Jacobsen & Rowley *op. cit.* 5
(1973), without type; *A. smithii* × *spathulatum*—
1 Canary Is.
× sanctisebastianii Bramwell & Rowley *op. cit.* 5
(1973), without type; *A. saundersii* × *subplanum*—
1 Canary Is.
× santosianum Bramwell & Rowley *op. cit.* 5(1973),
1 without type; *A. goochiae* × *palmense*—Canary Is.
× splendens Bramwell & Rowley *op. cit.* 5(1973),
1 without type; *A. ciliatum* × *nobile*—Canary Is.
× tabulicum Bramwell & Rowley *op. cit.* 5(1973),
1 without type; *A. tabulaeforme* × *urbicum*—Canary Is.
× tenensis Bramwell & Rowley *op. cit.* 5(1973),
1 without type; *A. ciliatum* × *haworthii*—Canary Is.
× teneriffae Bramwell & Rowley *op. cit.* 5(1973),
1 without type; *A. ciliatum* × *urbicum*—Canary Is.
× vegamorai Bramwell & Rowley *op. cit.* 5(1973),
1 without type; *A. subplanum* × *viscatum*—Canary Is.

AICHRYSON

× intermedium Bramwell & Rowley in *Nation. Cact.*
Succ. J. (U.K.), 28(1):5(1973), without type; *A.*
1 *laxum* × *punctatum*—Canary Is.

BRYOPHYLLUM

1687 Brossard, D.
Le bourgeonnement épiphylle chez le *Bryophyllum daigremontianum* Berger (Crassulacées): étude cytochimique, cytophotométrique et ultrastructurale. (The epiphyllous bud formation in *Bryophyllum daigremontianum* Berger: cytochemistry, cytophotometry and ultrastructure). *Ann. Sci. Nat., Bot. Biol.*
5 *Vég.*, 14(2):93–214(1973)—(En); illustrations.

COTYLEDON

1688 Herre, H.
Notes on South African succulents. *Cact. Succ. J.*
5 (U.S.A.), 45(5):238–239(1973)—Illustration.

1689 Nikolovski, T. *and* Matvejeva, J.
Cotyledon chloranthus (Heldr. et Sart.) Hal. kako nov vid vo florata na Makedonija. (*Cotyledon chloranthus* (Heldr. Sart.) Hal. as a new species in the flora from SRM). *Fragm. Balcan. Mus. Macedon. Sci. Nat.*, 8(20):
1 173–176(1971)—(En); illustration, map.

1690 Shields, M. E.
A garden under glass. *Bull. Afr. Succ. Pl. Soc.*, 7(6): 208–212, 8(1):3–7, (2):48–51,53,59–61, (3):85–89,
1, 2, 5 (4):126–128, 131–132(1973)—Illustrations.

CRASSULA

1691 Brown, J. R.
Crassulas in California gardens. *Cact. Succ. J.*
3 (U.S.A.), 45(1):36–38(1973)—Illustrations.

1692 Friedrich, H.-C.
Zur Cytotaxonomie der Gattung *Crassula*. *Garcia de Orta, Ser. Bot.*, 1(1–2):49–65(1973)—(Por,En);
1, 4, 5 chromosome numbers.

1693 [Rowley, G. D.]
From the President: tuberous Crassulas. *Bull. Afr.*
5 *Succ. Pl. Soc.*, 8(2):45–47(1973)—Illustrations.

1694 Smirnova, E. S.
Novyĭ taksonomicheskiĭ priznak dlya opredeleniya vidov roda *Crassula*. *Byull. Glavn. Bot. Sada* (Moscow),
1, 2, 5, 6 88:38–43(1973)—Key.

1695 Tölken, H. R.
Crassula streyi. *Flow. Pl. Afr.*, 42(3–4):pl.1672(1973)
5 —En; coloured illustration.

streyi Tölken in *Flow. Pl. Afr.*, 42(3–4): t. 1672
5 (1973)—South Africa (Natal).

DIAMORPHA

1696 Baskin, J. M. & Baskin, C. C.
Germination characteristics of *Diamorpha cymosa* seeds and an ecological interpretation. *Oecologia*,
3 10(1):17–28(1972).

ECHEVERIA

1697 Bleck, J.
Echeveria DeCandolle. *Cact. Succ. J.* (U.S.A.),
3, 4 45(4):190–193(1973)—En; illustrations.

1698 Carruthers, L.
Echeveria pulvinata Rose (*Cotyledon pulvinata* Hooker). *Bull. Afr. Succ. Pl. Soc.*, 7(6):216,218(1973)—
4 Illustration.

1699 Carruthers, L. *and* Ginns, R.
Echeverias: a guide to cultivation and identification. Edinburgh, Bartholomew, 1973. 110p.—Illustrations,
3, 4 coloured illustrations, map.

1700 Daniel, J. *and* Rowland, D.
An introduction to some cacti and succulent plants of New Mexico. *Cact. Succ. J. Gr. Brit.*, 35(3):53–54
3 (1973).

1701 Macdougall, T.
Comments on Walther's *Echeveria*. *Cact. Succ. J.*
4 (U.S.A.), 45(2):55(1973).

1702 Van Keppel, J. C.
Some notes on South American *Echeverias*. *Cact.*
4 *Succ. J.* (U.S.A.), 44(2):55–61(1972)—Illustrations.

× GREENONIUM

bramwellii Rowley in *Nation. Cact. Succ. J.* (U.K.), 28(1):7(1973), without type; *Greenovia dodrentalis* ×
1 *Aeonium* ? *spathulatum*—Canary Is.
rowleyi Bramwell ex Jacobsen & Rowley *op. cit.* 7 (1973), without type; *Greenovia dodrentalis* ×
1 *Aeonium haworthii*—Canary Is.

GREENOVIA

× aureozoon Bramwell & Rowley in *Nation. Cact. Succ. J.* (U.K.), 28(1):5(1973), without type;
1 *G. aizoon* × *aurea*—Canary Is.

JOVIBARBA

1703 Mitchell, P. J.
The *Sempervivum* and *Jovibarba* handbook. Burgess Hill, Sempervivum Society, 1973. [i]44p.—Coloured
1, 2 illustrations.

heuffelii (Schott) Á. & D. Löve
subsp. **patens** (Griseb. & Schenk) Holub in *Folia*
Geobot. Phytotax., 8(2):176(1973): *Sempervivum*
1 *patens.*
velenovskyi (Česmedžijev) Holub *op. cit.* 177(1973):
1 *Sempervivum velenovskyi.*

KALANCHOE

1704 Bleck, J.
Kalanchoe Adans. *Cact. Succ. J.* (U.S.A.), 45(2):
58–62(1973)—En; illustrations, coloured
2, 3, 5 illustration.

1705 Brown, J. R.
Interesting succulents. *Cact. Succ. J.* (U.S.A.), 45(1):
5 35(1973)—Illustration. *Kalanchoe rhombopilosa.*

1706 Graaf, A. de
Kalanchoe rhombopilosa. Succulenta, 52(12):227–228
5 (1973)—Du; illustration.

1707 Radcliffe-Smith, A.
Kalanchoe robusta. Curtis's Bot. Mag., 179(4):tab.657
5 (1973)—En; illustration, coloured illustration.

1708 Ungersböck, H.
Kalanchoe tomentosa Bak. *Stachelpost*, 9(43):14
4 (1973)—Ge; coloured illustration.

METEROSTACHYS

1709 Moran, R.
The genus *Meterostachys* Nakia (Crassulaceae).
Cact. Succ. J. (U.S.A.), 44(6):262–273(1972) –
Illustration, coloured illustration; chromosome
2, 3 numbers.

MONANTHES

× anagiflora Bramwell & Rowley in *Nation. Cact.*
Succ. J. (U.K.), 28(1):5(1973), without type;
1 *M. anagensis* × *laxiflora*—Canary Is.
× burchardii Bramwell & Rowley *op. cit.* 5(1973),
1 without type; *M. laxiflora* × *silensis*—Canary Is.
× hybrida Bramwell & Rowley *op. cit.* 5(1973),
without type; *M. brachycaulon* × *pallens*—
1 Canary Is.
× intermedia Bramwell & Rowley *op. cit.* 5(1973),
without type; *M. brachycaulon* × *polyphylla*—
1 Canary Is.
× polycaulis Bramwell & Rowley *op. cit.* 5(1973),
without type; *M. polyphylla* × *subcrassicaulis*—
1 Canary Is.
× pumila Bramwell & Rowley *op. cit.* 5(1973),
without type; *M. brachycaulon* × *silensis*—
1 Canary Is.
× silophylla Bramwell & Rowley *op. cit.* 5(1973),
without type; *M. silensis* × *polyphylla*—Canary
1 Is.
× sventenii Bramwell & Rowley *op. cit.* 5(1973),
1 without type; *M. laxiflora* × *pallens*—Canary Is.

PACHYPHYTUM

1710 Uhl, C. H. *and* Morgan, R.
The chromosomes of *Pachyphytum* (Crassulaceae).
Amer. J. Bot., 60(7):648–656(1973)—Chromosome
4 numbers.

PAGELLA

1711 Rowley, G. D.
Pagella archeri—a study in miniaturisation.
5 *Ashingtonia*, 1(3):29,30(1973)—Coloured illustration.

RHODIOLA

crassipes Boriss.
var. **cholaensis** (Praeger) Jacobsen in *Nation. Cact.*
Succ. J. (U.K.), 28(1):5(1973): *Sedum crassipes*
6 var. *cholaense.*
var. **cretinii** (Ham.) Jacobsen *op. cit.* 5(1973):
6 *Sedum cretinii.*
var. **stephanii** (Cham.) Jacobsen *op. cit.* 5(1973):
2 *Sedum stephanii.*
dumulosa (Franch.) Fu
f. **farreri** (Smith) Jacobsen *op. cit.* 5(1973): *Sedum*
2 *farreri.*
fastigiata (Hook. f. & Thoms.) Fu
var. **gelida** (Fisch. & Mey.) Jacobsen *op. cit.* 5
2 (1973): *Sedum gelidum.*
himalensis (Don) Fu
var. **bouvieri** (Ham.) Jacobsen *op. cit.* 5(1973):
6 *Sedum bouvieri.*
var. **ishidae** (Miy. & Kudo) Jacobsen *op. cit.* 5
2 (1973): *Sedum ishidae.*
kirilowii Regel
var. **rubra** (Praeger) Jacobsen *op. cit.* 5(1973):
2 *Sedum kirilowii.*
nuristanica (Kit.) Jacobsen *op. cit.* 5(1973): *Sedum*
2 *nuristanicum.*
primuloides (Franch.) Fu
var. **pachyclados** (Hemsl.) Jacobsen *op. cit.* 5
2 (1973): *Sedum pachyclados.*
rhodantha (Gray) Jacobsen *op. cit.* 5(1973): *Sedum*
3 *rhodanthum.*
rosea L.
var. **alaskana** (Rose) Jacobsen *op. cit.* 6(1973):
3 *R. alaskana.*
subsp. **atropurpurea** (Turcz.) Jacobsen *op. cit.* 5
2 (1973): *Sedum atropurpureum.*
subsp. **elongata** (Led.) Jacobsen *op. cit.* 5(1973):
2 *Sedum elongatum.*
subsp. **neomexicana** (Br.) Jacobsen *op. cit.* 5(1973):
3 *R. neomexicana.*
subsp. **polygama** (Rydb.) Jacobsen *op. cit.* 5(1973):
3 *Sedum polygamum.*
subsp. **roanensis** (Br.) Jacobsen *op. cit.* 6(1973):
3 *Sedum roanense.*
subsp. **tachiroi** (Franch. & Sav.) Jacobsen *op. cit.*
2 6(1973): *Sedum roseum* var. *tachiroi.*
2 **sinica** (Diels) Jacobsen *op. cit.* 6(1973): *Sedum sinicum.*
subopposita (Max.) Jacobsen *op. cit.* 6(1973): *Sedum*
2 *suboppositum.*
trifida (Hook. f. & Thoms.) Jacobsen *op. cit.* 6(1973):
6 *Sedum trifidum.*
tuberosa (Coss. & Let.) Jacobsen *op. cit.* 6(1973):
1 *Sedum tuberosum.*
yunnanensis (Franch.) Fu
var. **forrestii** (Ham.) Jacobsen *op. cit.* 6(1973):
2 *Sedum yunnanense* var. *forrestii.*
var. **henryi** (Diels) Jacobsen *op. cit.* 6(1973): *Sedum*
2 *henryi.*
var. **valerianoides** (Diels) Jacobsen *op. cit.* 6(1973):
2 *Sedum valerianoides.*

× **ROCHEASSULA** Rowley in *Nation. Cact. Succ.*
0 *J.* (U.K.), 28(1):7(1973). [*Crassula* × *Rochea.*]
langleyensis (Veitch) Rowley *op. cit.* 7(1973):
0 × Kalorochea langleyensis, nom. nud.

SEDUM

1712 Anon.
Sedum pilosum M. Bieb. *Sempervivum Soc. J.*,
2 1:7–9(1973)—En; coloured illustration.

1713 Barocio de las Fuentes, I.
Una especie nueva de *Sedum* (Crassulaceae)
encontrada cerca de Epazoyucan, Hidalgo (México).
Cact. Suc. Mex., 18(4):96–100(1973)—(En);
4 illustrations, coloured illustrations.

1714 Bleck, J.
Sedum L. *Cact. Succ. J.* (U.S.A.), 44(6):243–245
1–4 (1972)—En; illustrations.

CRASSULACEAE

1715 Ohba, H.
Notes on Himalayan *Sedum*: 1. *J. Jap. Bot.*, 48(11):
6 327–331(1973)—Illustrations.

1716 Tomida, M.
A new *Sedum* from Pref. Toyama. *J. Jap. Bot.*,
2 48(5):138–140(1973)—Ja,En; coloured illustrations.

batallae Barocio in *Cact. Suc. Mex.*, 18(4):96(1973)—
4 Mexico.
ettyuense Tomida in *J. Jap. Bot.*, 48(5):140(1973)—
2 Japan.
fui Rowley in *Nation. Cact. Succ. J.* (U.K.), 28(1):
2 6(1973), nom. nov.: *S. platyphyllum* Fu.
phariense Ohba in *J. Jap. Bot.*, 48(11):328(1973)—
6 Tibet.
reflexum L.
var. **aureum** (Wirtg.) Jacobsen in *Nation. Cact.*
Succ. J. (U.K.), 28(1):6(1973): *S. rupestre* var.
1 *aureum*.
thomsonianum Ohba in *J. Jap. Bot.*, 48(11):331
6 (1973), nom. nov.: *S. elongatum* Wall. ex Hook. f.

SEMPERVIVUM

1717 Anon.
Discussion of names in common use. *Sempervivum*,
no. 4:7–9(1973)—Coloured illustration. *Sempervivum*
2 *ingwersenii* Wale and *S. iranicum* Bornm. & Gauba.

1718 Anon.
Discussion of names in current use: *Sempervivum*
giuseppi Wale. *Sempervivum Soc. J.*, 2:9(1973)—
1 Coloured illustration.

1719 Anon.
Review of names. *Sempervivum Soc. J.*, 1:2–4(1973)
1, 2 —*Sempervivum albernellii* to *S.* '*Granat*'.

1720 Anon.
Sempervivum juvanii Strgar spec. nova. *Sempervivum*
1 *Soc. J.*, 2:4–5(1973)—En: chromosome number.

1721 Bellia, G. G. *and* Andrade, A. de
Studio sul genere *Sempervivum*: 3 parte. *Rea*,
1, 2 no.5:32–36(1972)—(En).

1722 Favarger, C. *and* Scherbatoff, M.
Sur une espèce intéressante des Alpes austro-occidentales: *Sempervivum calacreum* Jordan. *Candollea*,
28(2):219–235(1973)—(En,Ge); coloured illustration,
1 dot map, chromosome numbers.

1723 Gorer, R.
Plant awards 1972–73: *Sempervivum arachnoideum*
L. 'Minor'. A.M. *Quart. Bull. Alp. Gard. Soc.*, 41(4):
1 328–329(1973).

1724 Mitchell, P. J.
The *Sempervivum* and *Jovibarba* handbook. Burgess
Hill, Sempervivum Society, 1973. [i]44p.—Coloured
1, 2 illustrations.

1725 Rowley, G. D.
Sempervivum hybrids. *Sempervivum Soc. J.*, no.3:2–4
1, 2 (1973).

1726 Smith, M.
Sempervivum vincentei. *Sempervivum Soc. J.*, 2:2–4
1 (1973)—En; illustration.

marmoreum Gris.
f. **pallidiflorum** Soó in *Acta Bot. Acad. Sci. Hung.*,
18(1–2):175(1973), nom. nov.: *S. schlehani* var.
1 *genuinum* Domin.

tectorum L.
subsp. arvernense (Lecoq. & Lam.) Bellia & de
Andrade in *Rea*, 5:33(1972), without basionym
1 ref.: *S. arvernense*.
subsp. verlotii (Lam.) Bellia & de Andrade *op. cit.*
1 33(1972), without basionym ref.: *S. verlotii*.

SINOCRASSULA

indica Berg.
var. **longistyla** (Praeger) Jacobsen in *Nation. Cact.*
2 *Succ. J.* (U.K.), 28(1):6(1973): *Sedum longistylum*.

CRUCIFERAE

1727 Bramwell, D.
New species of Cruciferae from the Canary Islands.
Cuad. Bot. Canar., no. 17:19–26(1973)—Illustrations,
1 dot maps.

1728 Corbett, G. A.
New records for the West Virginia Cruciferae.
3 *Castanea*, 38(3):214–229(1973)—Dot maps.

1729 Dvořák, F.
The importance of the indumentum for the
investigation of evolutional relationship in the family
Brassicaceae. *Österr. Bot. Zeitschr.*, 121(3–4):155–164
1–7 (1973)—(Ge).

1730 Exell, A. W. *and* Gonçalves, M. L.
Flora de Moçambique: 11. Cruciferae. 1973. 11p.—
5 Illustrations; keys.

1731 Hadač, E. *and* Chrtek, J.
A contribution to the Brassicaceae of Iraq. *Acta*
Univ. Carol., Biol., no. 4:231–265(1971 publ. 1973)—
2 Illustrations, map, dot maps.

AETHIONEMA

1732 Phitos, D. *and* Kamari, G.
Cytologische Untersuchungen an einigen griechischen Blütenpflanzen (*Centaurea*, *Aethionema*). *Biol.*
Gallo-Hellen., 4(2):147–151(1973)—(Fr,En,Gre);
1 chromosome numbers.

1733 Phitos, D. *and* Snogerup. S.
A new species of *Aethionema* from Skiros, Greece.
Bot. Notiser, 126(2):142–145(1973)—Illustrations;
1 chromosome number.

retsina Phitos & Snogerup in *Bot. Notiser*, 126(2):
1 142(1973)—Greece.

ALYSSUM

1734 Avetisyan, V. E.
Alyssum persicum Boiss. vo flore Kavkaza. (*Alyssum*
persicum Boiss. in flora of the Caucasus). *Biol. Zhurn.*
2 *Arm.*, 26(5):82–84(1973)—(Arm).

1735 Contandriopoulos, J. *and* Afzal-Rafii, Z.
Contribution à l'étude cytotaxinomique des
Alyssum de Turquie. (Cytotaxinomic studies on
species of *Alyssum* from Turkey). *Ber. Schweiz. Bot.*
Ges., 83(1):14–29(1973)—(Ge,En); map; chromosome
2 number.

1736 Smejkal, M.
Československé druhy a poddruhy rodu *Alyssum* L.
(Tschechoslowakische Arten und Unterarten der
Gattung *Alyssum* L.). *Zprávy Cesk. Bot. Spol.*, 8(1):
1 1–7(1973)—(Ge); key.

ARABIS

1737 Rollins, R. C.
Purple-flowered *Arabis* of the Pacific coast of North America. *Contrib. Gray Herb.*, no.204:149–154
3 (1973)—Key.

1738 Titz, W.
Arabis borealis Andrz. ex Ledeb.—not in Ireland.
2 *Watsonia*, 9(4):369(1973).

brassica (Leers) Rauschert in *Feddes Repert.*, 83(9–10):
1 648(1972 publ. 1973): *Turritis brassica.*
serpentinicola Rollins in *Contrib. Gray Herb.*,
3 204:150(1973)—U.S.A. (California).

ATROPATENIA F. K. Meyer in *Feddes Repert.*,
2 84(5–6):467(1973).
rostrata (N. Busch) F. K. Meyer *op. cit.* 467(1973):
2 *Thlaspi rostratum.*
zangezurica (N. Tsvelev) F. K. Meyer *op. cit.* 467
2 (1973): *Thlaspi zangezuricum.*

BISCUTELLA

coronopifolia L.
subsp. **lusitanica** (Jordan) Malagarriga, Las
1 Subesp. y Variac. Geogr.: 7(1973): *B. lusitanica.*
subsp. **mediterranea** (Jordan) Malagarriga *op. cit.*
1 7(1973): *B. mediterranea.*
didyma L.
subsp. **applanata** (Mach.-Laur.) Hadač & Chrtek in *Acta Univ. Carol., Biol.*, 1971(4):238(1971 publ.
2 1973): *B. ciliata* var. *applanata.*
intermedia Gouan
subsp. **cuneata** (FQ) Malagarriga, Las Subesp. y
1 Variac. Geogr.: 7(1973): *B. cuneata.*
subsp. **glacialis** (Bss. R.) Malagarriga *op. cit.* 7
1 (1973): *B. glacialis.*
subsp. **gredensis** (Guinea) Malagarriga *op. cit.* 8
1 (1973): *B. gredensis.*
var. **macroclada** (Sennen ex Guinea) Malagarriga
1 *op. cit.* 8(1973), nom. illegit.: *B. macroclada.*
var. **scaposa** (Sennen) Malagarriga *op. cit.* 8(1973):
1 *B. scaposa.*
var. **subscaposa** (Sennen ex Guinea) Malagarriga
1 *op. cit.* 8(1973), nom. illegit.: *B. subscaposa.*
laevigata L.
var. **budensis** Soó in *Acta Bot. Acad. Sci. Hung.*,
1 18(1–2):172(1973)—Hungary.
subsp. **hungarica** Soó *op. cit.* 172(1973)—
1 Hungary.
sempervirens L.
subsp. **gibraltarica** (Wilmott ex Guinea) Malagarriga, Las Subesp. y Variac. Geogr.: 7
1 (1973), nom. illegit.: *B. gibraltarica.*
subsp. vicentina (Sampaio) Malagarriga *op. cit.* 6 (1973), without basionym ref.: (Basionym not
1 stated).
variegata Bss. R.
subsp. **foliosa** (Machatschki-Laurich) Malagarriga
1 *op. cit.* 7(1973): *B. foliosa.*
subsp. **latiorifolia** (Pau ex E. Guinea) Malagarriga
1 *op. cit.* 7(1973): *B. latiorifolia.*
subsp. **megacarpaea** (Bss. R.) Malagarriga *op. cit.*
1 7(1973): *B. megacarpaea.*

BRASSICA

1739 Durkee, A. B. *and* Harborne, J. B.
Flavonal glycosides in *Brassica* and *Sinapis*. *Phyto-*
1, 2 *chemistry*, 12(5):1085–1089(1973).

1740 Ferguson, L. F.
Brassica balearica. *Curtis's Bot. Mag.*, 179(3):tab.641
1 (1973)—En; illustration, coloured illustration.

1741 Poldini, L.
Brassica glabrescens, eine neue Art aus nordost-Italien. *Giorn. Bot. Ital.*, 107(4):181–189(1973)—
1 (En,It); illustrations, dot maps.

1742 Vaughan, J. G. *and* Gordon, E. I.
A taxonomic study of *Brassica juncea* using the techniques of electrophoresis, gas-liquid chromatography and serology. *Ann. Bot.* (U.K.), 36(149):167–
1–3 183(1973).

1743 Venkateswarlu, J. *and* Kamala, T.
Meiosis in double trisomic *Brassica campestris*. *Genetica*, 44(2):283–287(1973)—Chromosome
1, 2 numbers.

1744 Wellington, P. S. *and* Quartley, C. E.
A practical system for classifying, naming and identifying some cultivated Brassicas. *J. Nation. Inst. Agric. Bot.*, 12(3):413–432(1972)—Chromosome
0 numbers.

deserti Danin & Hedge in *Notes Roy. Bot. Gard.*
2 *Edinburgh*, 32(2):259(1973)—Sinai Peninsula.
glabrescens Poldini in *Giorn. Bot. Ital.*, 107(4):181
1 (1973)—Italy.

BRAYA

1745 Böcher, T. W.
Interspecific hybridization in *Braya* (Cruciferae). *Ann. Bot. Fenn.*, 10(1):57–65(1973)—Illustrations;
1–3 chromosome number.

CAKILE

subgen. **Eremocakile** (Pobed.) Hadač & Chrtek in *Acta Univ. Carol., Biol.*, 1971(4):239(1971 publ.
2 1973): *Cakile* sect. *Eremocakile.*

CALLOTHLASPI F. K. Meyer in *Feddes Repert.*,
2 84(5–6):457(1973).
abchasicum F. K. Meyer *op. cit.* 458(1973)—U.S.S.R.
2 (Caucasus).
antitauricum F. K. Meyer *op. cit.* 457(1973)—
2 Turkey.
lilacinum (Boissier & Huet du Pavillon) F. K. Meyer
2 *op. cit.* 457(1973): *Thlaspi lilacinum.*

CAPSELLA

1746 Yannitsaros, A.
Notes on the ecology and distribution of *Capsella grandiflora* (Fauche et Chaub.) Boiss. *Biol. Gallo-*
1 *Hellen.*, 4(2):163–168(1973)—(Fr,Gre); dot map; key.

CARDAMINE

1747 Almeida, M. T. de
Uma Crucífera nova para a flora de Portugal. *Anuár. Soc. Brot.*, 39:19–23(1973)—(Fr); illustrations,
1 dot map; key. *Cardamine parviflora.*

1748 Hart, T. W.
The generic relationships and biosystematics of *Cardamine bulbosa* (Muhl.) B. S. P. and *C. douglassii* Britt. *Diss. Abstr. Int., B*, 33(7):2968(1972)—
3 Chromosome numbers.

1749 Oesau, A.
Cardamine parviflora L. bei Lampertheim/Hessen.
1 *Hess. Flor. Briefe*, 256–258:18–22(1973)—Illustrations.

1750 Spasskaya, N. A.
O rasprostranenii *Cardamine rivularis* Schur. v Leningradskoĭ Oblasti. (On spreading *Cardamine rivularis* Schur. in Leningrad District). *Vestn. Leningrad. Univ., Biol.*, no. 3(1):131–135(1973)—
1 (En); chromosome numbers.

amara L.
f. **monochlamydea** Soó in *Acta Bot. Acad. Sci.*
1 *Hung.*, 18(1–2):173(1973)—Hungary.
hirsuta L.
1 f. **apetala** Soó *op. cit.* 172(1973)—Hungary.
raphanifolia Pourr.
subsp. **gallaecica** Laínz in *Bol. Inst. Estud. Astur., Supl. Cienc.*, 16(Aport. Conoc. Fl. Cántabro-
1 Astur, x:19): 175(1973)—Spain.

CAULOSTRAMINA Rollins in *Contrib. Gray*
3 *Herb.*, 204:155(1973).
jaegeri (Rollins) Rollins *op. cit.* 155(1973): *Thely-*
3 *podium jaegeri.*

CHORISPORA

1751 Májovský, J. *and* Feráková, V.
Chorispora tenella (Pall.) DC., nový druh Slovenskej flóry. (*Chorispora tenella* (Pall.) DC., a new species of Slovak flora). *Biológia, A*, (Czechoslovakia), 28(1):
1 73–76(1973)—(En,Rus); illustration.

COCHLEARIA

1752 Gill, J. J. B.
Cytogenetic studies in *Cochlearia* L. (Cruciferae). The origins of *C. officinalis* L. and *C. micacea* Marshall. *Genetica*, 44(2):217–233(1973)—Chromosome
1 numbers.

1753 Kakes, P.
The chromosome number of *Cochlearia pyrenaica* DC. near Moresnet (Belgium). *Acta Bot. Neerl.*, 22(3):
1 206–208(1973).

1754 López-Fernández, M. L.
Cochlearia aragonensis Coste & Soulié en la sierra de Satrústegui (Navarra). *Pirineos*, no. 109:31–34(1973)—
1 (Fr,En,Ge); illustration, map.

CRAMBE

1755 Kuznetsova, R. Ya.
Krambe abissinskaya. (*Crambe abyssinica*). *Trudȳ Prikl. Bot. Genet. Selek.*, 48(3):240–245(1972)—
1 (En).

1756 Mennema, J.
Zeekool (*Crambe maritima* L.) in Nederland. *Natura* (Netherlands), 70(1):1–4(1973)—(En); illustrations,
1 dot maps.

1757 Poldini, L.
Ad floram italicam notulae taxonomicae et geobotanicae: 8. *Crambe tataria* Sebeók. *Webbia*,
1, 2 28(1):31–36(1973)—It(Ge); illustration, dot maps.

scaberrima Webb ex Bramwell in *Cuad. Bot. Canar.*,
1 17:22(1973)—Canary Is.
sventenii Pettersson ex Bramwell & Sunding *op. cit.*
1 21(1973)—Canary Is.

DENTARIA

1758 Weber, A.
Stipularbildungen bei *Dentaria* und ihr Wert für die morphologische Deutung der Speicherschuppen als Phyllodien. *Österr. Bot. Zeitschr.*, 121(1–2):107–119
1 (1973)—(En).

integrifolia Nuttall
var. **cuneata** (Greene) J. T. Howell in Howitt & J. T. Howell, Suppl. Vasc. Pl. Monterey Co.,
3 Calif.; 13(1973): *Cardamine cuneata.*

DESCURAINIA

lemsii Bramwell in *Cuad. Bot. Canar.*, 17:24(1973)—
1 Canary Is.

DESIDERIA

1759 Suslova, T. A.
Novȳĭ vid roda *Desideria* Pamp. na Pamire. (Species nova generis *Desideria* Pamp. in Pamir). *Novosti Sist. Vȳssh. Rast.*, 10:160–165(1973)—
2 Illustration.

pamirica Suslova in *Novosti Sist. Vȳssh. Rast.*, 10:
2 163(1973)—U.S.S.R. (Middle Asia).

DIMITRIA Ravenna in *An. Mus. Hist. Nat.*
4 *Valparaíso*, 5:91(1972).

1760 Ravenna, P.
Dimitria, genero nuevo de Cruciferae de Patagonia (Argentina). *An. Mus. Hist. Nat. Valparaíso*, no. 5:
4 91–95(1972)—(En); illustrations.

onuridifolia Ravenna in *An. Mus. Hist. Nat.*
4 *Valparaíso*, 5:92(1972)—Argentina.

DIPLOTAXIS

crassifolia (Raf.) DC.
subsp. intricata (Wk.) Malagarriga, Las Subesp. y Variac. Geogr.: 10(1973), without basionym
1 date: *Pendulina intricata.*
subsp. lagascana (DC.) Malagarriga *op. cit.* 10
1 (1973), without basionym date: *D. lagascana.*

DRABA

cryophila Cuatrecasas in *Ciencia*, 27(6):172(1972)—
4 Colombia.
hispanica Bss.
subsp. dedeana (Bss.) Malagarriga, Las Subesp. y Variac. Geogr.: 9(1973), without basionym
1 date: *D. dedeana.*
subsp. **zapateri** (Wk.) Malagarriga *op. cit.* 9(1973),
1 as 'zapaterii': *D. zapateri.*
obscura Rollins, Genus *Lesquerella* in N. America, 10
4 (1973), nom. nov.: *Vesicaria frigida* Turcz.

EROPHILA

verna (L.) Chervall.
var. **hungarica** Borsos & Soó in *Acta Bot. Acad.*
1 *Sci. Hung.*, 18(1–2):173(1973)—Hungary.
var. **vertesensis** Borsos & Soó *op. cit.* 173(1973)—
1 Hungary.

ERYSIMUM

1761 Favarger, C.
Nouvelle contribution à l'étude cytologique du genre *Erysimum* L. *Ann. Sci. Univ. Besançon, Bot.*,
1 12:49–56(1972)—Dot maps; chromosome numbers.

1762 Polatschek, A.
Die Vertreter der Gattung *Erysimum* auf Kreta. (The taxa of *Erysimum* represented in Crete). *Ann. Mus. Goulandris*, 1:113–126(1973)—(En,Gre); dot maps;
1 chromosome numbers; key.

1763 Prima, V. M.
Novȳe vidȳ roda *Erysimum* L. s vostochnogo Kavkaza. (Generis *Erysimum* L. species novae e Caucaso orientalis). *Novosti Sist. Vȳssh. Rast.*, 10:156–
2 160(1973)—Illustration.

1764 Reveal, J. L.
Comments on two names in an early Utah flora. *Great Basin Nat.*, 32(4):221–222(1972)—*Erysimum*
3 *asperum.*

asperum (Nutt.) DC.
var. **amoenum** (Greene) Reveal in *Great Basin Nat.*,
3 32(4):222(1972): *Cheiranthus nivalis* var. *amoenus.*

babadagensis Prima in *Novosti Sist. Vÿssh. Rast.*, 10:
2 156(1973)—U.S.S.R. (Caucasus).
subnivale Prima *op. cit.* 157(1973)—U.S.S.R.
2 (Caucasus).

EUTREMA

heterophyllum (W. W. Smith) Hara in *J. Jap. Bot.*,
2, 6 48(4):97(1973): *Braya heterophylla.*

HEDINIA

1765 Velichkin, É. M.
O vzaimootnosheniyakh rodov *Smelowskia* C. A. Mey. i *Hedinia* Ostenf. (On interrelations between the genera *Smelowskia* C. A. Mey. and *Hedinia* Ostenf.)
2, 6 *Bot. Zhurn.*, 58(2):284–293(1973).

HESPERIS

1766 Dvořák, F.
Experimental taxonomic study of the species *Hesperis velenovskyi*. *Phyton* (Austria), 15(1–2):151–
1 154(1973)—(Ge); chromosome numbers.

1767 Dvořák, F.
Infrageneric classification of *Hesperis* L. *Feddes Repert.*, 84(4):259–271(1973)—(Ge); maps, dot
1, 2 maps.

1768 Dvořák, F.
A study of the species *Hesperis armena* and related species. *Phyton* (Austria), 15(1–2):107–121(1973)—
1, 2 (Ge); dot map.

subgen. **Cvelevia** Dvořák in *Feddes Repert.*, 84(4):
2 264(1973).
subgen. **Diaplictos** (Dvořák) Dvořák *op. cit.* 267
2 (1973): *Hesperis* sect. *Diaplictos.*
armena Boissier
var. **pisidica** (Huber-Morath) Dvořák in *Phyton*
2 (Austria), 15(1–2):109(1973): *H. pisidica.*
syriaca (DC.) Dvořák in *Feddes Repert.*, 84(4):267
2 (1973): *Cheiranthus ? syriacus.*
theophrasti Borbás
subsp. **graeca** (Dvořák) Dvořák in *Phyton* (Austria),
1, 2 15(1–2):119(1973): *H. graeca.*

IBERIS

pruitii Tineo
subsp. granatensis (Bss. R.) Malagarriga, Las Subesp. y Variac. Geogr.: 9(1973), without
1 basionym date: *I. granatensis.*
subsp. **hegelmaieri** (Wk.) Malagarriga *op. cit.* 9
1 (1973): *I. hegelmaieri.*
subsp. lagascana (DC.) Malagarriga *op. cit.* 9(1973),
1 without basionym date: *I. lagascana.*

IRANIA Hadač & Chrtek in *Acta Univ. Carol., Biol.*,
2 1971(4):248(1971 publ. 1973).
compacta (Rech. fil.) Hadač & Chrtek *op. cit.* 248
2 (1971 publ. 1973): *Fibigia compacta.*
membranacea (Rech. fil.) Hadač & Chrtek *op. cit.*
2 248(1971 publ. 1973): *Fibigia membranacea.*
multicaulis (Boiss. & Hohen.) Hadač & Chrtek *op.*
2 *cit.* 248(1971 publ. 1973): *Farsetia multicaulis.*
pendula (Boiss.) Hadač & Chrtek *op. cit.* 248(1971
2 publ. 1973): *Farsetia pendula.*
umbellata (Boiss.) Hadač & Chrtek *op. cit.* 248
2 (1971 publ. 1973): *Farsetia umbellata.*

ISATIS

subgen. **Sameraria** (Desv.) Hadač & Chrtek in *Acta Univ. Carol., Biol.*, 1971(4):250(1971 publ. 1973):
2, 6 genus *Sameraria.*
afghanica Hadač & Chrtek *op. cit.* 251(1971 publ.
2 1973), nom. nov.: *Sameraria stenophylla* Rechinger fil.
agnewii Hadač & Chrtek *op. cit.* 252(1971 publ.
2 1973)—Iraq.
besseri Trautv.
subsp. **tricostata** Hadač & Chrtek *op. cit.* 253
2 (1971 publ. 1973)—Iraq.
elegans (Boiss.) Hadač & Chrtek *op. cit.* 251(1971
2 publ. 1973): *Sameraria elegans.*
nummularia (Bornm.) Hadač & Chrtek *op. cit.* 251
2 (1971 publ. 1973): *Samararia nummularia.*
stylophora (Jaub. & Spach) Hadač & Chrtek *op. cit.*
2 251(1971 publ. 1973): *Tetrapterygium stylophorum.*

KONIGA

1769 Božilova, E. D.
Notes on the pollen grains of *Lepidotrichum* Vel. et Bornm. and its related genera. *Dokl. Bolg. Akad. Nauk*, 26(11):1521–1523(1973). Also *Koniga scardica* and
2 *Ptilotrichum elongatum.*

KOTSCHYELLA F. K. Meyer in *Feddes Repert.*,
2 84(5–6):457(1973).
cilicica (Schott & Kotschy ex Boiss.) F. K. Meyer
2 *op. cit.* 457(1973): *Carpoceras cilicicum.*
stenocarpa (Boissier) F. K. Meyer *op. cit.* 457(1973):
2 *Carpoceras stenocarpum.*

LEAVENWORTHIA

1770 Baskin, J. M. *and* Baskin, C. C.
Leavenworthia exigua var. *lutea* in Tennessee cedar
3 glades. *Castanea*, 38(2):195–196(1973).

LEPIDOTRICHUM

1771 Božilova, E. D.
Notes on the pollen grains of *Lepidotrichum* Vel. et Bornm. and its related genera. *Dokl. Bolg. Akad. Nauk*, 26(11):1521–1523(1973). Also *Koniga scardica* and
2 *Ptilotrichum elongatum.*

LESQUERELLA

1772 Rollins, R. C. *and* Shaw, E. A.
The genus *Lesquerella* (Cruciferae) in North America. Cambridge, Mass., Harvard University Press, 1973. x, 288p.—Illustrations, dot maps; chromosome
3, 4 numbers; keys.

1773 Rollins, R. C. *and* Solbrig, O. T.
Interspecific hybridization in *Lesquerella*. *Contrib.*
3 *Gray Herb.*, no. 203:3–48(1973)—Map.

alpina (Nutt.) Wats.
subsp. **condensata** (Nelson) Rollins & E. A. Shaw, Genus *Lesquerella* in N. America: 191(1973): *L.*
3 *condensata.*
subsp. **parvula** (Greene) Rollins & E. A. Shaw
3 *op. cit.* 189(1973): *L. parvula.*
arctica (Wormskj.) S. Wats.
subsp. **calderi** (Mull. & Pors.) Hultén in *Bot.*
3 *Notiser*, 126(4):488(1973): *L. calderi.*
arenosa (Richardson) Rydberg
var. **argillosa** Rollins & E. A. Shaw, Genus *Lesquerella* in N. America: 178(1973)—U.S.A.
3 (S. Dakota, Nebraska, Wyoming).
argentea (Schauer) Wats.
subsp. **lepidota** (Rose) Rollins & E. A. Shaw *op.*
4 *cit.* 159(1973): *Synthlipsis lepidota.*
argyraea (Gray) Wats.
subsp. **diffusa** (Rollins) Rollins & E. A. Shaw *op.*
4 *cit.* 83(1973): *L. diffusa.*
cordiformis (Rollins) Rollins & E. A. Shaw *op. cit.*
3 248(1973): *Physaria cordiformis.*
fremontii Rollins & E. A. Shaw *op. cit.* 228(1973)—
3 U.S.A. (Wyoming).
goddingii Rollins & E. A. Shaw *op. cit.* 164(1973)—
3 U.S.A. (New Mexico, Arizona).

gracilis (Hook.) Wats.
subsp. **nuttallii** (Torr. & Gray) Rollins & E. A.
3 Shaw *op. cit.* 53(1973): *Vesicaria nuttallii.*
inflata Rollins & E. A. Shaw *op. cit.* 150(1973)—
4 Mexico.
kingii Wats.
var. **cobrensis** Rollins & E. A. Shaw *op. cit.* 255
3 (1973)—U.S.A. (Nevada).
subsp. **diversifolia** (Greene) Rollins & E. A. Shaw
3 *op. cit.* 255 (1973)—*L. diversifolia.*
subsp. **latifolia** (Nelson) Rollins & E. A. Shaw
3 *op. cit.* 257(1973): *L. latifolia.*
occidentalis Wats.
subsp. **cinerascens** (Maguire & Holmgren)
Rollins & E. A. Shaw *op. cit.* 225(1973): *L.*
3 *occidentalis* var. *cinerascens.*
ovalifolia Rydberg
subsp. **alba** (Goodman) Rollins & E. A. Shaw *op.*
3 *cit.* 127(1973): *L. ovalifolia* var. *alba.*
purpurea (Gray) Wats.
subsp. **foliosa** (Rollins) Rollins & E. A. Shaw
3, 4 *op. cit.* 114(1973): *L. purpurea* var. *foliosa.*
thamnophila Rollins & E. A. Shaw *op. cit.* 86(1973)—
3 U.S.A. (Texas).
wyndii Rollins & E. A. Shaw *op. cit.* 154(1973)—
4 Mexico.

MALCOLMIA

1774 Dvořák, F.
Study of the characters of the genus *Malcolmia* R.
Br. in Aiton. *Feddes Repert.*, 84(4):315–325(1973)—
1, 2, 6 (Ge); chromosome numbers.

MASMENIA F. K. Meyer in *Feddes Repert.*, 84(5–6):
2 468(1973).
2 **crassiuscula** F. K. Meyer *op. cit.* 468(1973)—Turkey.
rosularis (Boissier & Balansa) F. K. Meyer *op. cit.*
2 468(1973): *Thlaspi rosulare.*

MATTHIOLA

longipetala (Vent.) DC.
subsp. **pendula** (Náběl.) Hadač & Chrtek in *Acta
Univ. Carol., Biol.*, 1971(4):257(1971 publ. 1973):
2 *Matthiola oxyceras* var. *pendula.*

MICROTHLASPI F. K. Meyer in *Feddes Repert.*,
1, 2 84(5–6):452(1973).
granatense (Boiss. & Reut.) F. K. Meyer *op. cit.* 453
1 (1973): *Thlaspi granatense.*
natolicum (Boissier) F.K.Meyer *op.cit.*452 (1973):
2 *Thlaspi natolicum.*
var. **drabiflorum** (Fenzl) F. K. Meyer *op. cit.* 453
2 (1973): *Thlaspi drabaeflorum.*
subsp. **gaillardotii** F. K. Meyer *op. cit.* 452(1973)—
2 Lebanon.
subsp. **longistylum** (G. E. Post) F. K. Meyer *op. cit.*
2 453(1973): *Thlaspi perfoliatum* var. *longistylum.*
var. **pentadactylonis** F. K. Meyer *op. cit.* 453
2 (1973)—Cyprus.
subsp. **sporadium** F. K. Meyer *op. cit.* 453(1973)—
2 Rhodes.
perfoliatum (L.) F. K. Meyer *op. cit.* 453(1973):
1, 2 *Thlaspi perfoliatum.*
umbellatum (Steven ex A. P. DC.) F. K. Meyer
2 *op. cit.* 453(1973): *Thlaspi umbellatum.*

NESODRABA

hyperborea (L.) Yurtsev in *Bot. Zhurn.*, 58(12):1748
1, 2 (1973): *Alyssum hyperboreum.*

NEUROTROPIS (A. P. DeCandolle) F. K. Meyer
in *Feddes Repert.*, 84(5–6):452(1973): *Thlaspi* sect.
2 *Neurotropis.*
orbiculata (Steven ex A. P. DeCandolle) F. K. Meyer
2 *op. cit.* 452(1973): *Thlaspi orbiculatum.*
platycarpa (Fischer & C. A. Meyer) F. K. Meyer
2 *op. cit.* 452(1973): *Thlaspi platycarpum.*
szowitsiana (Boissier) F. K. Meyer *op. cit.* 452(1973):
2 *Thlaspi szowitsianum.*

NOCCAEA

ser. **Alpestres** (Janchen) F. K. Meyer in *Feddes Repert.*,
1 84(5–6):463(1973): *Thlaspi* subsect. *Alpestria.*
2 ser. **Densiflores** F. K. Meyer *op. cit.* 465(1973).
2 ser. **Libanoticae** F. K. Meyer *op. cit.* 465(1973).
1, 2 ser. **Macranthae** F. K. Meyer *op. cit.* 464(1973).
1, 2 ser. **Minimae** F. K. Meyer *op. cit.* 459(1973).
1 ser. **Occitanicae** F. K. Meyer *op. cit.* 463(1973).
2 sect. **Papillifolia** F. K. Meyer *op. cit.* 460(1973).
1, 2 ser. **Praecoces** F. K. Meyer *op. cit.* 461(1973).
sect. **Pterotropis** (A. P. DeCandolle) F. K. Meyer
1, 2 *op. cit.* 461(1973): *Thlaspi* sect. *Pterotropis.*
1 ser. **Rivales** F. K. Meyer *op. cit.* 466(1973).
1 ser. **Stylosae** F. K. Meyer *op. cit.* 459(1973).
1 **albanica** F. K. Meyer *op. cit.* 461(1973)—Albania.
amani (G. E. Post) F. K. Meyer *op. cit.* 465(1973):
2 *Carpoceras amani.*
2 **angustifolia** F. K. Meyer *op. cit.* 465(1973)—Turkey.
annua (K. Koch) F. K. Meyer *op. cit.* 465(1973):
2 *Thlaspi annuum.*
aptera (Velenovsky) F. K. Meyer *op. cit.* 462(1973):
1 *Thlaspi apterum.*
arenaria (J. E. Duby) F. K. Meyer *op. cit.* 463(1973):
1 *Thlaspi alpestre* β *arenarium.*
banatica (v. Uechtritz) F. K. Meyer *op. cit.* 464(1973):
1 *Thlaspi banaticum.*
bellidifolia (Grisebach) F. K. Meyer *op. cit.* 460
1 (1973): *Thlaspi bellidifolium.*
1 **boeotica** F. K. Meyer *op. cit.* 462(1973)—Greece.
brachypetala (A. Jordan) F. K. Meyer *op. cit.* 464
1 (1973): *Thlaspi brachypetalum.*
subsp. **huteri** (Pernhoffer) F. K. Meyer *op. cit.* 464
1 (1973): *Thlaspi huteri.*
subsp. **tatrensis** (H. Zapałowicz) F. K. Meyer *op.*
1 *cit.* 464(1973): *Thlaspi tatrense.*
caerulescens (J. S. Presl & C. B. Presl) F. K. Meyer
1 *op. cit.* 463(1973): *Thlaspi caerulescens.*
subsp. **sylvestris** (A. Jordan) F. K. Meyer *op. cit.*
1 463(1973): *Thlaspi sylvestre.*
1 **cikaea** F. K. Meyer *op. cit.* 461(1973)—Albania.
corymbosa (J. Gay) F. K. Meyer *op. cit.* 460(1973):
1 *Hutchinsia corymbosa.*
crantzii F. K. Meyer *op. cit.* 461(1973), nom. nov.:
1 *Thlaspi alpinum* Crantz.
cretica (Degen & Javorka) F. K. Meyer *op. cit.* 466
1 (1973): *Thlaspi microphyllum* subsp. *creticum.*
cypria (J. Bornmüller) F. K. Meyer *op. cit.* 466(1973):
2 *Thlaspi cyprium.*
dacica (Heuffel) F. K. Meyer *op. cit.* 464(1973):
1 *Thlaspi dacicum.*
subsp. **montenegrina** F. K. Meyer *op. cit.* 464
1 (1973)—Yugoslavia.
densiflora (Boissier & Kotschy) F. K. Meyer *op. cit.*
2 465(1973): *Thlaspi densiflorum.*
eburneosa F. K. Meyer *op. cit.* 464(1973)—
1 Switzerland.
edinensium F. K. Meyer *op. cit.* 460(1973)—
2 Turkey.
epirota (Halacsy) F. K. Meyer *op. cit.* 459(1973):
1 *Thlaspi epirotum.*
1 **firmiensis** F. K. Meyer *op. cit.* 463(1973)—France.
goesingensis (Halacsy) F. K. Meyer *op. cit.* 462
1 (1973): *Thlaspi goesingense.*
graeca (A. Jordan) F. K. Meyer *op. cit.* 463(1973):
1 *Thlaspi graecum.*
× **gremliana** (Thellung) F. K. Meyer *op. cit.* 466
1 (1973): *Thlaspi* × *gremlianum.*
hausknechtii (Boissier) F. K. Meyer *op. cit.* 460
2 (1973): *Thlaspi hausknechtii.*

jankae (Kerner) F. K. Meyer *op. cit.* 462(1973):
1 *Thlaspi jankae.*
kovatsii (Heuffel) F. K. Meyer *op. cit.* 461(1973):
1 *Thlaspi kovatsii.*
2 **libanotica** F. K. Meyer *op. cit.* 466(1973)—Lebanon.
limosellifolia (Reuter ex Burnat) F. K. Meyer *op. cit.* 460(1973): *Thlaspi rotundifolium* var. *limo-*
1 *sellaefolium.*
lutescens (Velenovsky) F. K. Meyer *op. cit.* 462(1973):
1 *Thlaspilutescens.*
subsp. **drenovskii** F. K. Meyer *op. cit.* 462(1973)—
1 Bulgaria.
macrantha (Lipsky) F. K. Meyer *op. cit.* 464(1973):
1, 2 *Thlaspi praecox* (β) *macranthum.*
microphylla (Boissier & Orphanides) F. K. Meyer
1 *op. cit.* 460(1973): *Thlaspi microphyllum.*
microstyla (Boissier) F. K. Meyer *op. cit.* 466(1973):
2 *Thlaspi microstylum.*
minima (Arduino) F. K. Meyer *op. cit.* 459(1973):
1 *Thlaspi minimum.*
montana (L.) F. K. Meyer *op. cit.* 461(1973):
1 *Thlaspi montanum.*
nevadensis (Boissier & Reuter) F. K. Meyer *op. cit.*
1 460(1973): *Thlaspi nevadense.*
occitanica (A. Jordan) F. K. Meyer *op. cit.* 463(1973):
1 *Thlaspi occitanicum.*
subsp. **tallonis** (Sennen) F. K. Meyer *op. cit.* 463
1 (1973): *Thlaspi tallonis.*
ochroleuca (Boissier & Heldreich) F. K. Meyer *op.*
1, 2 *cit.* 461(1973): *Thlaspiochroleucum.*
papillosa (Boissier) F. K. Meyer *op. cit.* 460(1973):
2 *Thlaspiopapillosum.*
phrygia (J. Bornmüller) F. K. Meyer *op. cit.* 465
1, 2 (1973): *Thlaspi phrygium.*
praecox (Wulfen) F. K. Meyer *op. cit.* 462(1973):
1, 2 *Thlaspi praecox.*
pseudorivularis (Haussknecht ex Bornmüller)
1 F. K. Meyer *op. cit.* 466(1973): *Thlaspi pseudorivulare.*
rhodopensis F. K. Meyer *op. cit.* 462(1973)—
1 Bulgaria.
rivalis (C. B. Presl) F. K. Meyer *op. cit.* 466(1973):
1 *Thlaspi rivale.*
rotundifolia (L.) Moench
subsp. **grignensis** F. K. Meyer *op. cit.* 459(1973)—
1 Italy.
subsp. **intermedia** F. K. Meyer *op. cit.* 459(1973)—
1 France.
rubescens (Schott & Kotschy ex Boissier) F. K.
2 Meyer *op. cit.* 459(1973): *Aethionema rubescens.*
salisii (Brügger) F. K. Meyer *op. cit.* 464(1973):
1 *Thlaspi salisii.*
sarmatica F. K. Meyer *op. cit.* 464(1973)—U.S.S.R.
1 (European Russia).
sintenisii (Haussknecht ex J. Bornmüller) F. K. Meyer
2 *op. cit.* 458(1973): *Thlaspi sintenisii.*
stenoptera (Boissier & Reuter) F. K. Meyer *op. cit.*
1 461(1973): *Thlaspi stenopterum.*
sylvia (Gaudin) F. K. Meyer *op. cit.* 461(1973):
1 *Thlaspi sylvium.*
tatianae (E. Bordzilovski) F. K. Meyer *op. cit.*
2 465(1973): *Thlaspi tatianae.*
tenuis (Boissier & Buhse) F. K. Meyer *op. cit.* 459
2 (1973): *Iberidella tenuis.*
tymphaea (Haussknecht) F. K. Meyer *op. cit.* 462
1 (1973): *Thlaspi tymphaeum.*
valerianoides (K. H. Rechinger) F. K. Meyer *op. cit.*
2 460(1973): *Thlaspi valerianoides.*
versicolor (Stojanov & Kitanov) F. K. Meyer *op. cit.*
1 462(1973): *Thlaspi lutescens* f. *versicolor.*
violascens (Schott & Kotschy) F. K. Meyer *op. cit.*
2 465(1973): *Thlaspi violascens.*
virens (A. Jordan) F. K. Meyer *op. cit.* 463(1973):
1 *Thlaspi virens.*
viridisepala (Podpera) F. K. Meyer *op. cit.* 461(1973):
1 *Thlaspi praecox* var. *viridisepalum.*

NOCCIDIUM F. K. Meyer in *Feddes Repert.*,
2 84(5–6):456(1973).
hastulatum (A. P. DeCandolle) F. K. Meyer *op. cit.*
2 456(1973): *Hutchinsia hastulata.*
tuberculatum F. K. Meyer *op. cit.* 456(1973)—
2 Persia.

PHYSARIA

1775 Waite, S. B.
A taxonomic revision of *Physaria* (Cruciferae) in Utah. *Great Basin Nat.*, 33(1):31–36(1973)—Dot
3 maps; key.

PRINGLEA

1776 Delaveau, P., Hotellier, F., Guérin, H. P. *and* Courcelle, A.
Le chou des Kerguelen (*Pringlea antiscorbutica* H. Br.). *Pl. Méd. Phytothér.*, 7(3):208–214(1973)—(En);
7 illustrations.

1777 Young, S. B. *and* Schofield, E. K.
Palynological evidence for the Late Glacial occurrence of *Pringlea* and *Lyallia* on Kerguelen Islands. *Rhodora*, 75(802):239–247(1973)—
7 Illustration.

PROMICRANTHA Dvořák in *Folia Přírodověd. Fak. Univ. Purkyně Brně, Biol.*, 13(4):55(1972), without
2 type.
alaica (Korsh.) Dvořák *op. cit.* 59(1972), generic name
2 not validly publ.: *Matthiola albicaulis* var. *alaica.*
alyssifolia (DC.) Dvořák *op. cit.* 59(1972), generic
2 name not validly publ.: *Hesperis alyssifolia.*

PSEUDOCAMELINA

campylopoda (Bornm. & Gauba) Hadač & Chrtek in *Acta Univ. Carol., Biol.*, 1971(4):261(1971 publ.
2 1973): *Cochlearia campylopoda.*

PTILOTRICHUM

1778 Božilova, E. D.
Notes on the pollen grains of *Lepidotrichum* Vel. et Bornm. and its related genera. *Dokl. Bolg. Akad. Nauk*, 26(11):1521–1523(1973). Also *Koniga scardica* and
2 *Ptilotrichum elongatum.*

1779 Serve, L.
Sur le groupement à *Ptilotrichum spinosum* (L.) Boiss. du Malaza (Pyrénées-Orientales). *Bull. Soc. Bot. France*, 118(9):711–717(1971 publ. 1973)—Illustrations;
1 chromosome numbers.

RAPARIA F. K. Meyer in *Feddes Repert.*, 84(5–6):
1, 2 458(1973).
atlantica (Battandier) F. K. Meyer *op. cit.* 458(1973):
1 *Thlaspi atlanticum.*
bulbosa (Spruner) F. K. Meyer *op. cit.* 458(1973):
1, 2 *Thlaspi bulbosum.*

RHYNCHOSINAPIS

cheiranthos (Vill.) Dandy
subsp. cheiranthiflora (DC.) Malagarriga, Las Subesp. y Variac. Geogr.: 10(1973), without
1 basionym date: *Brassica cheiranthiflora.*
subsp. montana (DC.) Malagarriga *op. cit.* 10(1973),
1 without basionym date: *Brassica montana.*
subsp. **setigera** (Gay ex Lge.) Laínz in *Bol. Inst. Estud. Astur., Supl. Cienc.*, 16(Aport. Conoc. Fl.
1 Cántabro-Astur, x:20):176(1973): *Sinapis setigera.*
subsp. wrightii (O. E. Schulz) Malagarriga, Las Subesp. y Variac. Geogr.: 10(1973), without
1 basionym ref.: (Basionym not stated).

transtagana (P. Cout.) Pinto da Silva in *Agron. Lusit.*, 34(3):187(1972 publ. 1973): *Sinapis*
1 *longirostris* subsp. *transtagana.*

RORIPPA

1780 Jonsell, B.
Taxonomy and distribution of *Rorippa* (Cruciferae) in the southern U.S.S.R. *Svensk Bot. Tidskr.*, 67(3):
1, 2 281–302(1973)—Illustrations; key.

1781 Macpherson, P.
Rorippa spp. *Glasgow Nat.*, 19(1):63(1973)—En.
1 Local records.

1782 Resmeriţă, I.
Nouveaux taxons de la flore de Roumanie. *Rev. Roum. Biol., Bot.*, 17(5):311–312(1972)—
1 Illustration.

1783 Scannell, M. J. P.
Rorippa islandica (Oeder ex Murray) Borbás in Ireland. *Irish Nat. J.*, 17(10):348–349(1973)— European *Rorippa islandica* divided into *R. islandica*
1 and *R. palustris.*

1784 Zając, E. U. *and* Zając, A.
Nowe badania nad rozmieszczeniem gatunków synantropijnych w Americe Północnej. *Wiad. Bot.*,
3 17(3):163–165(1973)—Dot map. *Rorippa silvestris.*

barbaraeoides (Tsch.) Čel.
f. repens Resmeriţă in *Rev. Roum. Biol., Bot.*, 17(5):
1 312(1972), without type—Rumania.
silvestris (L.) Bess.
f. **olgae** Soó in *Acta Bot. Acad. Sci. Hung.*, 18(1–2):
1 173(1973)—Hungary.

SCHIMPERA

arabica Hochst. & Steud.
subsp. **persica** (Boiss.) Hadač & Chrtek in *Acta Univ. Carol., Biol.*, 1971(4):262(1971 publ. 1973):
2 *Schimpera persica.*

SINAPIS

1785 Durkee, A. B. *and* Harborne, J. B.
Flavonal glycosides in *Brassica* and *Sinapis. Phyto-*
1, 2 *chemistry*, 12(5):1085–1089(1973).

SMELOWSKIA

1786 Velichkin, É. M.
O vzaimootnosheniyakh rodov *Smelowskia* C. A. Mey. i *Hedinia Ostenf.* (On interrelations between the genera *Smelowskia* C. A. Mey. and *Hedinia Ostenf.*)
2, 3 *Bot. Zhurn.*, 58(2):284–293(1973).

SOLMS-LAUBACHIA

pumila (Kurz) Dvořák in *Folia Přírodověd. Fak. Univ.*
2 *Purkyně Brně, Biol.*, 13(4):24(1972): *Parrya pumila.*

STREPTOLOMA

1787 Bochantsev, V. P.
Vtoroĭ vid roda *Streptoloma* Bunge. (Generis *Streptolomatis* Bunge species secunda). *Novosti Sist.*
2 *Vÿssh. Rast.*, 10:155–156(1973).

sumbarense (Lipsky) Bochantsev in *Novosti Sist.*
2 *Vÿssh. Rast.*, 10:156(1973): *Sisymbrium sumbarense.*

THELYPODIOPSIS

ambigua (Wats.) Al-Shehbaz in *Contrib. Gray*
3 *Herb.*, 204:138(1973): *Thelypodium ambiguum.*
linearifolia (Gray) Al-Shehbaz *op. cit.* 140(1973),
3 without basionym ref.: (Basionym not stated.)

THELYPODIUM

1788 Al-Shehbaz, I. A.
The biosystematics of the genus *Thelypodium* (Cruciferae). *Contrib. Gray Herb.*, no. 204:3–148 (1973)—Illustrations, dot maps; chromosome
3, 4 numbers; keys.

1789 Rollins, R. C.
A reconsideration of *Thelypodium jaegeri. Contrib.*
3 *Gray Herb.*, no. 204:155–157(1973).

howellii Watson
subsp. **spectabilis** (Peck) Al-Shehbaz in *Contrib. Gray Herb.*, 204:116(1973): *T. howellii* var.
3 *spectabilis.*
integrifolium (Nuttall) Endlicher
subsp. **affine** (Greene) Al-Shehbaz *op. cit.* 110
3 (1973): *T. affine.*
subsp. **complanatum** Al-Shehbaz *op. cit.* 105 (1973)—U.S.A. (Oregon, California, Nevada,
3 Utah).
subsp. **gracilipes** (Robinson) Al-Shehbaz *op. cit.*
3 108(1973): *T. integrifolium* var. *gracilipes.*
subsp. **longicarpum** Al-Shehbaz *op. cit.* 111(1973)
3 —U.S.A. (Arizona).
laxiflorum Al-Shehbaz *op. cit.* 129(1973)—U.S.A.
3 (Colorado, Nevada, Utah).
rollinsii Al-Shehbaz *op. cit.* 97(1973)—U.S.A.
3 (Utah).
sagittatum (Nuttall) Endlicher
subsp. **ovalifolium** (Rydberg) Al-Shehbaz *op. cit.*
3 121(1973): *T. ovalifolium.*
wrightii Gray
subsp. **oklahomensis** Al-Shehbaz *op. cit.* 128
3 (1973)—U.S.A. (Oklahoma, Colorado).

THLASPI

1790 Dvořáková, M.
Thlaspi finitimum, eine neue Art aus dem Verwandtschaftskreis von *Thlaspi jankae. Folia Geobot.*
1 *Phytotax.*, 8(2):149–153(1973)—(En).

1791 Dvořáková, M.
Thlaspi pawlowskii sp. nova, eine neue Art aus den Ostkarpaten. (*Thlaspi pawlowskii* sp. nova, nový druh z Východních Karpat). *Preslia*, 45(4):315–319(1973)—
1 (Cz); illustrations.

1792 Lang, H.
Voralpen Hellerkraut (*Thlaspi alpestre* L.) in
1 Nordostoberfranken. *Nat. Mensch 1971*, 48–50(1972).

1793 Meyer, F. K.
Conspectus der '*Thlaspi*' Arten Europas, Africas und Vorderasiens. *Feddes Repert.*, 84(5–6):449–469
1, 2 (1973)—(En).

ser. **Grandifloria** F. K. Meyer in *Feddes Repert.*,
2 84(5–6):451(1973).
dacicum Heuff.
subsp. **banaticum** (Uechtr.) M. Dvořáková in
1 *Preslia*, 45(4):318(1973): *T. banaticum.*
finitimum M. Dvořáková in *Folia Geobot. Phytotax.*,
1 8(2):150(1973)—Czechoslovakia.
kochianum F. K. Meyer in *Feddes Repert.*, 84(5–6):
2 451(1973)—Turkey.
pawlowskii M. Dvořáková in *Preslia*, 45(4):315
1 (1973)—Rumania.

THLASPICERAS F. K. Meyer in *Feddes Repert.*,
2 84(5–6):453(1973).
2 ser. **Acornuta** F. K. Meyer *op. cit.* 455(1973).
2 ser. **Dolichocarpa** F. K. Meyer *op. cit.* 454(1973).
2 **bovis** F. K. Meyer *op. cit.* 454(1973)—Turkey.

cappadocicum (Boiss. & Balansa) F. K. Meyer *op. cit.*
2 455(1973): *Carpoceras cappadocicum.*
subsp. **microspermum** F. K. Meyer *op. cit.* 455
2 (1973)—Turkey.
capricornutum F. K. Meyer *op. cit.* 454(1973)—
2 Syria.
crassifolium Huber-Morath & F. K. Meyer *op. cit.*
2 454(1973)—Turkey.
dolichocarpum (Zohary) F. K. Meyer *op. cit.* 454
2 (1973): *Carpoceras dolichocarpum.*
eigii (Zohary) F. K. Meyer *op. cit.* 454(1973):
2 *Carpoceras eigii.*
subsp. **samuelssonii** F. K. Meyer *op. cit.* 454(1973)
2 —Syria.
elegans (Boissier) F. K. Meyer *op. cit.* 456(1973):
2 *Thlaspi elegans.*
huber-morathii F. K. Meyer *op. cit.* 455(1973)—
2 Turkey.
subsp. **divergens** F. K. Meyer *op. cit.* 455(1973)—
2 Turkey.
oxyceras (Boiss.) F. K. Meyer *op. cit.* 454(1973):
2 *Carpoceras oxyceras.*
2 **rechingeri** F. K. Meyer *op. cit.* 456(1973)—Syria.
triangulare F. K. Meyer *op. cit.* 455(1973)—
2 Turkey.

VANIA F. K. Meyer in *Feddes Repert.*, 84(5–6):467
2 (1973).
campylophylla F. K. Meyer *op. cit.* 467(1973)—
2 Turkey.
kurdica (I. C. Hedge) F. K. Meyer *op. cit.* 467(1973):
2 *Thlaspi kurdicum.*
2 **pulvinata** F. K. Meyer *op. cit.* 467(1973)—Turkey.

CTENOLOPHONACEAE

1794 Badré, F.
5 Ctenolophonaceae. Flore du Gabon, 21:55–58(1973).

CUCURBITACEAE

1795 Trivedi, R. N.
Cytological studies in four monotypic genera of the Cucurbitaceae. *J. Bihar Bot. Soc.*, 1(1–2):72(1972)—Chromosome numbers. Abstract; *Lagenaria, Benincasa,*
4–6 *Bryonopsis* and *Sechium.*

CITRULLUS

1796 Kaur, M., Dass, H. C. *and* Randhawa, G. S.
Systematic status of round melon (*Citrullus vulgaris* var. *fistulosus*) as studied by leaf phenolics. *Curr. Sci.*,
5 42(20):730–731(1973).

COCCINIA

1797 D'Cruz, R., Vyahalkar, G. R. *and* Ugale, S. D.
Cytogenetic studies in tetraploid *Coccinia indica* W. and A. *Caryologia*, 25(4):505–512(1972 publ.
6 1973)—Chromosome numbers.

1798 Getahøun, A.
Developmental anatomy and germination of seeds of anchoté, *Coccinia abyssinica* (W. & A.) Cogn. (Cucurbitaceae). *Bot. Notiser*, 126(4):437–449(1973)—
5 Illustrations.

CUCURBITA

1799 Bemis, W. P.
Triple haploids in *Cucurbita*. *J. Hered.*, 64(2):103–
3, 4 104(1973)—Illustration; chromosome number.

1800 Singh, D. *and* Dathan, A. S. R.
Structure and development of seed coat in Cucurbitaceae: 6. Seeds of *Cucurbita*. *Phytomorphology*,
4–6 22(1):29–45(1972 publ. 1973).

EUREIANDRA

1801 Jeffrey, C.
Eureiandra leucantha C. Jeffrey: new Cucurbitaceae from Zambia and Zaïre. *Bull. Jard. Bot. Nation. Belg.*,
5 43(3–4):425(1973).

leucantha C. Jeffrey in *Bull. Jard. Bot. Nation. Belg.*,
5 43(3–4): 425(1973)—Zambia, Zaïre.

IBERVILLEA

1802 Bourdoux, P.
Ibervillea sonorae. *Cactus* (Brussels), 5(1):10–11
4 (1973)—Fr.

LAGENARIA

1803 Heiser, C. B.
Variation in the bottle gourd. *In* Meggers, B. J., Ayensu, E. S. *and* Duckworth, W. D., *eds.* Tropical forest ecosystems in Africa and South America: 121–
2–6 128(1973)—*Lagenaria siceraria.*

LUFFA

1804 Roy, R. P. *and* Dutt, B.
Cytomorphological studies in induced polyploids of *Luffa acutangula* Roxb. *Nucleus*, 15(3):171–180
6 (1972)—Chromosome numbers.

MARAH

1805 Singh, D. *and* Dathan, A. S. R.
Structure and development of the seed coat in Cucurbitaceae: 8. Seeds of *Marah* Kell. *Bull. Torrey*
3 *Bot. Club*, 99(5):239–242(1972).

MELOTHRIA

1806 Singh, N. P., Kulkarni, B. G. *and* Moorthy, S.
New plant records from Goa. *Curr. Sci.*, 42(13):
6 478(1973)—*Melothria leucocarpa.*

MOMORDICA

1807 Dzevaltovskiĭ, A. K., Muratov, Z. M. *and* Zhalalov, G.
Apomixis u predstavitelya semeĭstva tȳkvennȳkh *Momordica charantia* L. *Dokl. Akad. Nauk Uzb. SSR*,
1–7 no.5:70–71(1973)—(Uz).

SECHIUM

1808 Sinha, B. M. B. *and* Thakur, G. K.
Chromosomal counts in *Thladiantha calcarata* (Wall.) Clarke and *Sechium edule* (Jacq.) Swartz. *Proc.*
6 *Indian Sci. Congr.*, 60(3):319(1973). Abstract only.

1809 Varghese, B. M.
Cytology of *Sechium edule* Sw. *Curr. Sci.*, 42(1):
4, 6 30(1973)—Chromosome numbers.

THLADIANTHA

1810 Sinha, B. M. B. *and* Thakur, G. K.
Chromosomal counts in *Thladiantha calcarata* (Wall.) Clarke and *Sechium edule* (Jacq.) Swartz. *Proc.*
6 *Indian Sci. Congr.*, 60(3):319(1973). Abstract only.

TRICHOSANTHES

1811 Roy, R. P., Singh, A. K. *and* Thakur, R.
Interspecific hybridisation in *Trichosanthes* Linn. *Proc. Indian Sci. Congr.*, 59(3):354–355(1972).
6 Abstract only.

1812 Thakur, G. K.
A natural tetraploid in the genus *Trichosanthes* from Bihar. *Proc. Indian Sci. Congr.*, 60(3):324(1973)—
6 Chromosome number. Abstract only.

CUNONIACEAE

WEINMANNIA

1813 Fabris, H. A.
Especies nuevas o criticas de la flora jujeña: 4. *Bol. Soc. Argent. Bot.*, 14(4):369–373(1972)—(En);
4 illustration. *Weinmannia boliviensis.*

parviflora Forst. f.
var. **glabrata** (F.Br.) Fosberg in *Micronesica*, 8(1–2):
6 45(1972): *W. marquesana* var. *glabrata.*
var. **marquesana** (F.Br.) Fosberg *op. cit.* 44(1972):
6 *W. marquesana.*
var. **myrsinites** Fosberg & Sachet *op. cit.* 45
6 (1972)—Marquesa Is.

CUSCUTACEAE

see

CONVOLVULACEAE

CYCLANTHACEAE

1814 Harling, G.
Flora of Ecuador: 216. Cyclanthaceae. *Op. Bot. Ser.*
4 *B.*, no. 1. 47p.(1973)—Keys.

ASPLUNDIA

aurantiaca Harling in *Op. Bot.*, B (Fl. Ecuador),
4 1:24(1973)—Ecuador.
4 **cayapensis** Harling *op. cit.* 22(1973)—Ecuador.
4 **domingensis** Harling *op. cit.* 25(1973)—Ecuador.
4 **lilacina** Harling *op. cit.* 20(1973)—Ecuador.
4 **lutea** Harling *op. cit.* 14(1973)—Ecuador.
4 **meraënsis** Harling *op. cit.* 13(1973)—Ecuador.
4 **sparrei** Harling *op. cit.* 21(1973)—Ecuador.

DICRANOPYGIUM

campii Harling in *Op. Bot.*, B (Fl. Ecuador), 1:36
4 (1973)—Ecuador.

SPHAERADENIA

sanctae-barbarae Harling in *Op. Bot.*, B (Fl. Ecuador),
4 1:42(1973)—Ecuador.

CYPERACEAE

1815 Cheadle, V. I. *and* Kosakai, H.
Vessels in the Cyperaceae. *Bot. Gaz.*, 133(3):214–
1–7 223(1972)—Illustrations.

1816 Hall, J. B.
The Cyperaceae within Nigeria—distribution and habitat. *Bot. J. Linn. Soc.*, 66(4):323–346(1973)—
5 Dot maps.

1817 Harvill, A. M.
Phytogeography of the Carices of Virginia.
3 *Rhodora*, 75(802):248–257(1973)—Dot map.

1818 Koyama, T.
New Cyperaceae from Nepal and adjoining Tibet. *Bot. Mag. Tokyo*, 86(1004):275–283(1973)—
2, 6 Illustrations.

1819 Rath, S. P., Choudhury, B. P. *and* Patnaik, S. N.
Flora of Bhubaneswar: 1. Cyperaceae. *Proc.*
6 *Indian Sci. Congr.*, 60(3):338–339(1973). Abstract only.

1820 Rath, S. P. *and* Patnaik, S. N.
Cytotaxonomic studies in Cyperaceae. *Proc. Indian Sci. Congr.*, 59(3):361(1972)—Chromosome numbers.
6 Abstract only.

1821 Raynal, J.
Notes cypérologiques: 19. Contribution à la classification de la sous-famille des Cyperoideae. *Adansonia*, 13(2):145–171(1973)—(En); illustrations;
1–7 keys.

1822 Raynal, J.
Répartition et évolution des modes de phytosynthèse chez les Cypéracées. *Compt. R. Hebd. Séances Acad. Sci., D* (Paris), 275(20):2231–2234
1–7 (1972)—Map.

1823 Saxena, H. O.
Contribution to the flora of Madhya Pradesh: 2. (Interesting records of Cyperaceae). *Indian Forester*,
6 99(8):505–507(1973).

1824 Siddiqui, M. O.
The sedges of Gorakhpur forests. *Proc. Indian Sci.*
6 *Congr.*, 60(3):340(1973). Abstract only.

tribus **ABILDGAARDIEAE** Lye in *Bot. Notiser*,
5 126(3):328(1973).

ABILDGAARDIA

1825 Lye, K. A.
Studies in African Cyperaceae: 8. The taxonomic position of *Abildgaardia* Vahl. and *Nemum* Hamilton.
5 *Bot. Notiser*, 126(3):325–329(1973)—Illustrations.

boeckelerana (Schweinf.) Lye in *Bot. Notiser*, 126(3):
5 327(1973), as 'boeckeleriana': *Scirpus boeckelerana.*

AFROTRILEPIS

1826 Haines, R. W. *and* Lye, K. A.
Studies in African Cyperaceae: 9. The morphology of *Coleochloa* Gilly and *Afrotrilepis* J. Rayn. *Bot. Notiser*,
5 126(3):330–339(1973)—Illustrations.

ALINIELLA J. Raynal in *Adansonia*, 13(2):157
5 (1973).
lipocarphioides (Kük.) Raynal *op. cit.* 157(1973):
5 *Ficinia lipocarphioides.*

ASCOLEPIS

ampullacea J. Raynal in *Adansonia*, 13(2):159(1973)—
5 Zambia.

BOLBOSCHOENUS

1827 Soják, J.
Doplňky k nomenklatuře některých rodů (Phanerogamae). *Čas. Nár. Muz.* (Prague), 141(1–2):
1–7 61–63(1972).

caldwellii (Cook) Soják in *Čas. Nár. Muz.* (Prague),
7 141(1–2):62(1972): *Scirpus caldwellii.*
fluviatilis (Torr.) Soják *op. cit.* 62(1972): *Scirpus*
2, 3, 7 *maritimus* var. *fluviatilis.*
maritimus (L.) Palla
subsp. **cymosus** (Reichenb.) Soják *op. cit.* 62
1 (1972): *Scirpus maritimus* var. *cymosus.*
medianus (Cook) Soják *op. cit.* 63(1972): *Scirpus*
7 *medianus.*
robustus (Pursh) Soják *op. cit.* 63(1972): *Scirpus*
3, 4 *robustus.*
schmidii (Raymond) Holub in *Folia Geobot. Phytotax.*, 8(2):176(1973): *Scirpus schmidii.*
1, 2

CAREX

1828 Alekseev, Yu. E.
Carex diluta Bieb. i rodstvennȳe eĭ vidȳ. (*Carex diluta* Bieb. and related species). *Byull. Mosk. Obshch.*
2 *Ispȳt. Prir., Biol.*, 78(5):141–145(1973).

1829 Alekseev, Yu. E.
Carex polyphylla Kar. et Kir. i blizkie ej vidy̆ iz vostochnoj evropy. (*Carex polyphylla* Kar. et Kir. and related species from Eastern Europe). *Vestn. Mosk.*
1 *Univ. Nauch. Zhurn.*, *6*, 28(5):66–71(1973)—(En).

1830 Alekseev, Yu. E.
Morfologiya pobegov i sistematicheskoe polozhenie *Carex alba* Scop. (The shoot morphology and taxonomic status of *Carex alba* Scop.). *Byull. Mosk. Obshch. Ispy̆t. Prir., Biol.*, 78(2):122–125(1973)—(En);
1, 2 illustrations.

1831 Anioł, J.
Nowe stanowisko *Carex dioica* L. var. *isogyna* Fr. na Śląsku. (A new locality of *Carex dioica* L. var. *isogyna* Fr. in Silesia). *Prace Bot.* (Wrocław), 16:3–5
1 (1973 publ. 1972)—Illustrations.

1832 Batoroeva, I. S.
Anatomicheskie osobennosti *Carex duriuscula* C. A. Mey. (Anatomical peculiarities of *Carex duriuscula*
2 C. A. Mey.). *Bot. Zhurn.*, 58(1):72–76(1973).

1833 Booth, E. M.
Carex appropinquata Schum. new to County
1 Carlow. *Irish Nat. J.*, 17(12):422(1973).

1834 Chevassus, P.
Le *Carex paupercula* Michx. var. *irrigua* (Wahlg.) Fernald en Haute-Savoie. *Trav. Lab. La Jaysinia*,
1 no.4:135–141(1972 publ. 1973)—(En).

1835 Dirkse, G., Hoekstra, J. A. *and* Loode, W.
Carex brizoides L. en *Phyteuma spicatum* L. bij
1 Bentheim. *Gorteria*, 6(9):150–152(1973)—(Ge).

1836 Egorova, T. V.
Zametki o vidakh roda *Carex* L. flory̆ SSSR. (Notulae de generis *Carex* L. speciebus florae URSS).
2 *Novosti Sist. Vy̆ssh. Rast.*, 10:98–104(1973).

1837 Faulkner, J. S.
Carex divulsa Stokes in Co. Armagh. *Irish Nat. J.*,
1 17(11):396(1973).

1838 Faulkner, J. S.
Experimental hybridization of north-west European species in *Carex* section *Acutae* (Cyperaceae).
1 *Bot. J. Linn. Soc.*, 67(3):233–253(1973)—Illustrations.

1839 Georgson, K.
Carex livida funnen i Halland. (*Carex livida* found
1 in Halland). *Svensk Bot. Tidskr.*, 67(1):66–67(1973).

1840 Greuter, W. *and* Charpin, A.
130. *Carex firma* Mygind ex Host. *In* Miège, J. *ed.*, Exsiccatorum genavensium a conservatorio botanico
1 distributorum: 4:57(1973). p.57.

1841 Gruber, M.
A propos de *Carex sempervirens* Vill. dans les
1 Pyrénées. *Monde Pl.*, no.376:4(1973).

1842 Kaule, G.
Zur Verbreitung von *Carex chordorrhiza* Ehrh. in Bayern. *Ber. Bayer. Bot. Ges.*, 44:183–186(1973)—
1 Dot map.

1843 Moravec, J.
Dvě nové lokality ostřice převislé—*Carex pendula* Huds.—v jižních Cechách. (Zwei neue Lokelitäten von *Carex pendula* Huds. in Südböhmen). *Zprávy*
1 *Česk. Bot. Spol.*, 8(2):114–115(1973)—(Ge).

1844 Novikov, V. S., Artemenko, V. I., Gubanov, I. A. *and* Tikhomirov, V. N.
O rasprostranenii *Carex bohemica* Schreb. v sredneĭ polose evropeĭskoĭ chasti SSSR. (On the distribution of *Carex bohemica* Schreb. in the central belt of the European part of the USSR). *Byull. Mosk. Obshch.*
1 *Ispy̆t. Prir., Biol.*, 78(2):143–145(1973)—Dot map.

1845 Proctor, H. G.
Carex aquatilis Wahlenb. in Upper Teesdale.
1 *Vasculum*, 57(2):17–24(1972)—Maps.

1846 Raymond, M.
La plus ancienne récolte québécoise de *Carex*
3 *heleonastes*. *Nat. Canad.*, 100(1):71–72(1973).

1847 Ribbons, B. W.
Carex elongata. *Glasgow Nat.*, 19(1):68–69(1973)—
1 En. Local record.

1848 Robertson, A. W. *and* Pollett, F. C.
Checklist of the genus *Carex* in Newfoundland. St. Johns, Newfoundland Forest Research Centre,
3 1973. 24p.

1849 Savelsbergh, E.
Die Verbreitung von *Carex binervis* Sm., *Carex laevigata* Sm. und *Carex pendula* Huds. (Cyperales: Cyperaceae) zwischen Roetgen Mulartshütte und
1 Schmithof. *Decheniana*, 125(1–2):49–53(1972).

1850 Shetler, S. G., *ed.*
A catalog of the genus *Carex* (Cyperaceae). *In* Shetler, S. G. *and others*. An introduction to the Botanical Type Specimen Register: 26–33(1973).
1–7 (Smithsonian Contrib. Bot., no.12).

1851 Steele, F. L. *and* Hodgdon, A. R.
Two interesting plants on Mt. Cardigan, Orange,
3 New Hampshire. *Rhodora*, 75(801):155–157(1973).

bigelowii Torr. ex Schwein.
subsp. **lugens** (H. T. Holm) Egorova in *Novosti*
2, 3 *Sist. Vy̆ssh. Rast.*, 10:104(1973): *C. lugens*.
subsp. **rigidioides** (Gorodk.) Egorova *op. cit.* 103
2 (1973): *C. rigida* subsp. *rigidioides*.
borii Nelmes
f. **lutea** R. R. Stewart, Annot. Cat. Vasc. Pl. W.
6 Pakistan & Kashmir: 74(1972)—Pakistan.
colchica J. Gay
subsp. **ligerica** (J. Gay) Egorova in *Novosti Sist.*
2 *Vy̆ssh. Rast.*, 10:104(1973): *C. ligerica*.
curaica Kunth
subsp. **pycnostachys** (Kar. & Kir.) Egorova *op. cit.*
2 104(1973): *C. pycnostachys*.
curta Gooden.
var. **maxima** (Kük.) R. R. Stewart, Annot. Cat. Vasc. Pl. W. Pakistan & Kashmir: 75(1972):
6 *C. canescens* var. *maxima*.
dichroa Freyn
subsp. **pamirensis** (Clarke) Egorova in *Novosti*
2 *Sist. Vy̆ssh. Rast.*, 10:103(1973): *C. pamirensis*.
dispalata Boott ex A. Gray
subsp. **laxiflorens** Koyama in *Bot. Mag. Tokyo*,
6 86(1004):275(1973)—Nepal.
flacca Schreber
subsp. claviformis (Hoppe) Malagarriga, Las Subesp. y Variac. Geogr.: 26(1973), without
1 basionym date: *C. claviformis*.
subsp. **praetutiana** (Parlat.) Holub in *Folia Geobot.*
1, 3 *Phytotax.*, 8(2):176(1973): *C. praetutiana*.
subsp. **serrulata** (Biv.) Malagarriga, Las Subesp. y
1 Variac. Geogr.: 26(1973): *C. serrulata*.

gracilis Curt.
subsp. **broterana** (Samp.) Laínz in *Bol. Inst. Estud. Astur., Supl. Cienc.*, 16(Aport. Conoc. Fl. Cántabro-Astur, x:44):200(1973), as 'broteriana':
1 *C. broteriana*.
hallerana Asso
subsp. **rorulenta** (Porta) Malagarriga, Las Subesp.
1 y Variac. Geogr.: 26(1973): *C. rorulenta*.
hemineuros Koyama in *Bot. Mag. Tokyo*, 86(1004):
6 276(1973)—Nepal, Tibet.
6 **himalaica** Koyama *op. cit.* 278(1973)—Nepal.
karoi (Freyn) Freyn
subsp. **selengensis** (Ivanova) Egorova in *Novosti*
2 *Sist. Vÿssh. Rast.*, 10:103(1973): *C. selengensis*.
leersiana Rauschert in *Feddes Repert.*, 83(9–10):660 (1972 publ. 1973), nom. nov.: *C. leersii* F. W.
1 Schultz.
mairii Cosson & Germain
subsp. nevadensis (Bss. R.) Malagarriga, Las Subesp. y Variac. Geogr.: 27(1973), without
1 basionym date: *C. nevadensis*.
ovalis Good.
var. **argyroglochin** (Hornem.) De Langhe & Lambinon in Nouv. Fl. Belg., Luxemb., N.
1 France etc.: 759(1973): *C. argyroglochin*.
parallela (Laest.) Sommerf.
subsp. **redowskiana** (C. A. Mey.) Egorova in *Novosti Sist. Vÿssh. Rast.*, 10:104(1973):
1, 2 *C. redowskiana*.
× **senayana** Soó in *Acta Bot. Acad. Sci. Hung.*, 18(1–2): 177(1973), nom. nov.: *C. haussknechtii* Senay.
1 (*C. cuprina* × *C. spicata*).
umbrosa Host
subsp. **huetiana** (Boiss.) Egorova in *Novosti Sist.*
2 *Vÿssh. Rast.*, 10:103(1973): *C. huetiana*.
subsp. **hypochlora** (Freyn) Egorova *op. cit.* 103
2 (1973): *C. hypochlora*.
vesicaria L.
subsp. **vesicata** (Meinsch.) Egorova *op. cit.* 103
2 (1973): *C. vesicata*.

COLEOCHLOA

1852 Haines, R. W. *and* Lye, K. A.
Studies in African Cyperaceae: 9. The morphology of *Coleochloa* Gilly and *Afrotrilepis* J. Rayn. *Bot. Notiser*,
5 126(3):330–339(1973)—Illustrations.

CYPERUS

1853 Johnson, L. A. S. *and* Evans, O. D.
Cyperus brevifolius and an allied species in eastern Australia. *Contrib. New S. Wales Nation. Herb.*,
7 4(6):378(1973)—Key.

1854 Marcks, B. G.
Population studies in North American *Cyperus* section *Laxiglumi* (Cyperaceae). *Diss. Abstr. Int., B*,
3 33(4):1425–1426(1972).

1855 Nagaraj, M. *and* Nijalingappa, B. H. M.
Embryological studies in *Cyperus alopecuroides* Rottb. *Proc. Indian Acad. Sci., B*, 77(6):252–263
5, 6 (1973)—Illustrations.

1856 Rath, S. P., Mohapatra, J. *and* Patnaik, S. N.
Cytotaxonomic studies in the genus *Cyperus* L. (s.l.). *Proc. Indian Sci. Congr.*, 60(3):315–316(1973)—
6 Chromosome numbers. Abstract only.

1857 Vorster, P.
Cyperus obtusiflorus var. *flavissimus*. *Flow. Pl. Afr.*,
5 42(3–4):pl. 1666(1973)—En; coloured illustration.

diaphanus Schrad. ex R. & S.
var. **gracilescens** (Kukenth.) Saxena in *Indian Forester*, 99(8):505(1973): *C. latespicatus* var.
6 *gracilescens*.

sphaeroideus L. A. S. Johnson & O. D. Evans in *Contrib. New S. Wales Nation. Herb.*, 4(6):378
7 (1973), nom. nov.: *Kyllinga intermedia* R. Br.

ELEOCHARIS

1858 Dunbar, A.
Pollen development in the *Eleocharis palustris* group (Cyperaceae): 1. Ultrastructure and ontogeny. *Bot.*
1–3, 6 *Notiser*, 126(2):197–254(1973).

1859 Soó, R. *and* Isépy, I.
Über einige Formenkreise der ungarischen und karpatischen Flora: 18. *Juncus bufonius*, *Eleocharis palustris* Agg. und *Agrostis canina*. *Ann. Univ. Budapest. Rolando Eötvos Nom., Biol.*, 14:169–177(1972 publ.
1 1973).

1860 Strandhede, S.-O.
Pollen development in the *Eleocharis palustris* group (Cyperaceae): 2. Cytokinesis and microspore
1–3, 6 degeneration. *Bot. Notiser*, 126(2):255–265(1973).

ERIOPHORUM

1861 Wein, R. W.
Eriophorum vaginatum L. *J. Ecol.*, 61(2):601–615 (1973) (Biological Flora of the British Isles)—
1 Map, dot map.

FIMBRISTYLIS

1862 Govindarajalu, E.
Studies in Cyperaceae: 8. Novelties in *Fimbristylis* (L.) Vahl. *Proc. Indian Acad. Sci., B*, 76(5):181–193
1–7 (1972)—Illustrations.

1863 Govindarajalu, E.
Studies in Cyperaceae: 9. Novelties in *Fimbristylis* (L.) Vahl. *Proc. Indian Acad. Sci., B*, 78(2):45–58
6 (1973)—Illustrations.

1864 Nijalingappa, B. H. M.
Cytological studies in *Fimbristylis dichotoma* (L.) Vahl. *Curr. Sci.*, 42(21):765–766(1973)—Chromosome
6 number.

rigidiuscula Govindarajalu in *Proc. Indian Acad. Sci.*,
6 *B*, 78(2):45(1973)—India.
6 **rugosa** Govindarajalu *op. cit.* 48(1973)—India.
semidisticha Govindarajalu *op. cit.* 51(1973)—
6 India.
6 **strigosa** Govindarajalu *op. cit.* 54(1973)—India.

HYPOLYTRUM

1865 Lorougnon, G.
Le vecteur pollinique chez les *Mapania* et les *Hypolytrum* Cypéracées du sous-bois des forêts tropicales ombrophiles. *Bull. Jard. Bot. Nation. Belg.*,
5 43(1–2):33–36(1973)—(En).

FUIRENA

1866 Imam, M., Chrtek, J. *and* Kosinová, J.
Novitates florae Aegyptiacae: 2. *Webbia*, 27(2): 461–466(1972 publ. 1973)—En(It). *Fuirena ciliaris*
2 recorded and compared with *F. pubescens*.

KOBRESIA

williamsii Koyama in *Bot. Mag. Tokyo*, 86(1004):
6 279(1973)—Nepal.

LIPOCARPHA

echinus J. Raynal in *Adansonia*, 13(2):159(1973)—
5 Zambia.

MAPANIA

1867 Lorougnon, G.
Le vecteur pollinique chez les *Mapania* et les *Hypolytrum* Cypéracées du sous-bois des forêts tropicales ombrophiles. *Bull. Jard. Bot. Nation. Belg.*,
5 43(1–2):33–36(1973)—(En).

1868 Thothathri, K., Banerjee, S. P. *and* Hazra, P. K.
Mapania cuspidata (Miq.) Uitt. var. *angustifolia* (Uitt.) Uitt.—an interesting Cyperaceae from Great Nicobar Island. *Indian Forester*, 98(12):708–709(1972)
6 —Illustration.

MARISCUS

malawicus J. Raynal in *Adansonia*, 13(2):159(1973)—
5 Malawi.

NEMUM

1869 Lye, K. A.
Studies in African Cyperaceae: 8. The taxonomic position of *Abildgaardia* Vahl. and *Nemum* Hamilton.
5 *Bot. Notiser*, 126(3):325–329(1973)—Illustrations.

bulbostyloides (S. Hooper) J. Raynal in *Adansonia*, 13(2):150(31 July 1973), as 'bulbostylidoides'; also Lye in *Bot. Notiser*, 126(3):328(9 November 1973):
5 *Scirpus bulbostyloides*.
equitans (Kükenth.) J. Raynal in *Adansonia*, 13(2): 150(31 July 1973); also Lye in *Bot. Notiser*, 126(3):
5 328(9 November 1973): *Scirpus equitans*.
megastachyum (Cherm.) J. Raynal in *Adansonia*,
5 13(2):150(1973): *Scirpus angolensis* var. *megastachyum*.

PTEROTHECA

1870 Mehra, P. N. *and* Mann, S. K.
Accessory chromosomes in *Pterotheca falconeri* H.F.
2, 6 *Nucleus*, 15(2):123–133(1972).

× RAYMONDIELLA Boivin in *Taxon*, 22(2–3):278 (1973), not in accordance with Art. H 7. [*Dulichium*
3 × *Rhynchospora*.]
weatherbiana (Fern.) Boivin *op. cit.* 278(1973), generic
3 name not validly publ.: *Cyperus weatherbianus*.

2, 5, 6 **RIKLIELLA** J. Raynal in *Adansonia*, 13(2):154(1973).
kernii (Raym.) J. Raynal *op. cit.* 155(1973): *Scirpus*
5 *kernii*.
rehmannii (Ridl.) J. Raynal *op. cit.* 155(1973):
5 *Scirpus rehmannii*.
squarrosa (L.) J. Raynal *op. cit.* 154(1973): *Scirpus*
2, 6 *squarrosus*.

SCHOENOPLECTUS

1871 Soják, J.
Doplňky k nomenklatuře některých rodů (Phanerogamae). *Čas. Nár. Muz.* (Prague), 141(1–2):
1–7 61–63(1972).

americanus (Pers.) Volkart
subsp. **longispicatus** (Britton) Soják in *Čas. Nár. Muz.* (Prague), 141(1–2):62(1972): *Scirpus*
3 *americanus* var. *longispicatus*.
chamissoi (Nees) Soják *op. cit.* 62(1972): *Malacochaete*
4 *riparius* β (var.) *chamissoi*.
deltarum (Schuyler) Soják *op. cit.* 62(1972): *Scirpus*
3 *deltarum*.
hondoensis (Ohwi) Soják *op. cit.* 62(1972): *Scirpus*
2 *hondoensis*.
juncoides (Roxb.) Palla
subsp. **hotarui** (Ohwi) Soják *op. cit.* 62(1972):
2 *Scirpus hotarui*.
subsp. **ohwianus** (T. Koyama) Soják *op. cit.* 62
2 (1972): *Scirpus ohwianus*.
subsp. **purshianus** (Fernald) Soják *op. cit.* 62
2, 3, 7 (1972): *Scirpus purshianus*.
subsp. **rockii** (Kükenth.) Soják *op. cit.* 62(1972):
6 *Scirpus rockii*.
lacustris (L.) Palla
subsp. **tenuiculmis** (Sheldon) Soják *op. cit.* 62
3 (1972): *Scirpus lacustris* var. *tenuiculmis*.
littoralis (Schrad.) Balla
subsp. **subulatus** (Vahl) Soják *op. cit.* 62(1972):
1–7 *Scirpus subulatus*.
monophyllus (Presl) Soják *op. cit.* 62(1972): *Scirpus*
1–4 *monophyllus*.
subsp. **conglomeratus** (H.B.K.) Soják *op. cit.* 62
1–3 (1972): *Scirpus conglomeratus*.
× **oblongus** (T. Koyama) Soják *op. cit.* 62(1972):
3 *Scirpus* × *oblongus*.
pseudotriqueter (Steud.) Soják *op. cit.* 62(1972):
3 *Scirpus pseudotriqueter*.
smithii (A. Gray) Soják *op. cit.* 62(1972): *Scirpus*
3 *smithii*.
subsp. **leiocarpus** (Kom.) Soják *op. cit.* 62(1972):
2, 3 *Scirpus supinus* var. *leiocarpus*.
subsp. **lateriflorus** (Gmel.) Soják *op. cit.* 62(1972):
1–7 *Scirpus lateriflorus*.
supinus (L.) Palla
subsp. **hallii** (A. Gray) Soják *op. cit.* 62(1972):
1, 3 *Scirpus hallii*.
subsp. **saximontanus** (Fernald) Soják *op. cit.* 62
3 (1972): *Scirpus saximontanus*.
triangulatus (Roxb.) Soják *op. cit.* 62(1972): *Scirpus*
2, 6 *triangulatus*.
validus (Vahl) Á. & D. Löve
subsp. **luxurians** (Miq.) Soják *op. cit.* 62(1972):
2 *Scirpus tabernaemontani* f. *luxurians*.

SCHOENUS

1872 Dizerbo, A. H.
Le *Schoenus nigricans* L. (Cypéracées) et son association dans la presqu'île de Crozon (Finistère). *Bull. Soc.*
1 *Sci. Bretagne*, 47(1–2):85–92(1972).

SCIRPOIDES

1873 Soják, J.
Doplňky k nomenklatuře některých rodů (Phanerogamae). *Čas. Nár. Muz.* (Prague), 141(1–2):
1–7 61–63(1972).

holoschoenus (L.) Soják
subsp. **australis** (Murr.) Soják in *Čas. Nár. Muz.*
1 (Prague), 141(1–2):61(1972): *Scirpus australis*.
subsp. **globifera** (L.f.) Soják *op. cit.* 61(1972):
1 *Scirpus globiferus*.
subsp. **thunbergii** (Schrad.) Soják *op. cit.* 61(1972):
5 *Isolepis thunbergii*.

SCIRPUS

1874 Levyns, M. R.
Scirpus rubicundus: a sedge with interesting seasonal
5 behaviour. *Veld Flora*, 3(1):14–16(1973)—Illustrations.

1875 Nakanishi, H.
(New localities of *Scirpus iseenis* Shimizu). *Hikobia*,
2 16(3–4):250(1973)—Ja.

subbisetosus Koyama in *Bot. Mag. Tokyo*, 86(1004):
6 281(1973)—Tibet.

SCLERIA

1876 Singh, N. P., Kulkarni, B. G. *and* Moorthy, S.
New plant records from Goa. *Curr. Sci.*, 42(13):
6 478(1973)—*Scleria africana*.

DATISCACEAE

1877 Davidson, C.
An anatomical and morphological study of Datiscaceae. *Aliso*, 8(1):49–110(1973)—Illustrations, dot
1–4, 6, 7 maps.

1878 Davidson, C.
An anatomical and morphological study of
1, 3, 4, 6 Datiscaceae. *Diss. Abstr. Int. B*, 33(12):5694(1973).

1879 Siddiqi, M. A.
Flora of West Pakistan: 37. Datiscaceae. 1973. 3p.—
6 Illustrations, map.

TETRAMELES *see* **TETRAMELACEAE**

DEGENERIACEAE

DEGENERIA

1880 Anon.
Degeneria, the living fossil. *Missouri Bot. Gard. Bull.*
6 62(6). 1p.(1973)—Illustration.

DIAPENSIACEAE

1881 Harborne, J. B. *and* Williams, C. A.
A chemotaxonomic survey of flavonoids and simple
phenols in leaves of the Ericaceae. *Bot. J. Linn. Soc.*,
1–3, 6 66(1):37–54(1973).

GALAX

1882 McVaugh, R.
Report of the Committee for Spermatophyta:
conservation of generic names: proposal 330 (6277)
Galax Sims (1804) vs. *Galax* Linnaeus (1753). *Taxon*,
3 22(1):156(1973).

DICHAPETALACEAE

1883 Breteler, F. J.
The African Dichapetalaceae: a taxonomical
revision. *Meded. Landbouwhogesch. Wageningen*, 73
5 (13). 143p. (1973)—Illustrations, dot maps; keys.

1884 Prance, G. T.
Flora neotropica monograph no.10.
Dichapetalaceae. New York, Hafner, 1972. 84p.—
4 Illustrations, dot maps; keys.

DICHAPETALUM

1885 Descoings, B.
Note sur les *Dichapetalum* (Dichapetalacées) de
Madagascar et description du *D. alaotrense* sp. nov.
(Note on the *Dichapetalum* (Dichapetalaceae) from
Madagascar and diagnosis of a new species, *D.
alaotrense*). *Bull. Soc. Bot. France*, 119(7–8):509–520
5 (1972 publ. 1973)—(En); key.

affine (Planch. ex Bth.) Breteler in *Meded. Land-
bouwhogesch. Wageningen*, 73(13):45(1973):
5 *Chailletia affinis*.
alaotrense Descoings in *Bull. Soc. Bot. France*, 119
5 (7–8):511(1972 publ. 1973)—Malagasy.
arachnoideum Breteler in *Meded. Landbouwhogesch.
Wageningen*, 73(13):61(1973)—Zaïre, Cameroun,
5 Gabon.
arenarium Breteler *op. cit.* 64(1973)—Tanzania,
5 Kenya.
barbatum Breteler *op. cit.* 76(1973)—Cameroun,
5 Gabon.
beilschmiedioides Breteler *op. cit.* 87(1973)—
5 Gabon.
bellum Breteler *op. cit.* 89(1973)—Gabon, Congo-
5 Brazzaville.
coelhoi Prance in Fl. Neotrop., 10:22(1972)—
4 Brazil.

donnell-smithii Engler
var. **chiapasense** (Standley) Prance *op. cit.* 28(1972):
4 *D. chiapasense*.
4 **froesii** Prance *op. cit.* 32(1972)—Brazil, Colombia.
madagascariense Poir.
var. **beniense** (Engl.) Breteler in *Meded. Land-
bouwhogesch. Wageningen*, 73(13):92(1973):
5 *D. beniense*.

STEPHANOPODIUM

angulatum (Little) Prance in Fl. Neotrop., 10:45
4 (1972): *Tapura angulata*.
organense (Rizzini) Prance *op. cit.* 49(1972): *S.
4 estrellense* var. *organense*.

TAPURA

amazonica Poeppig & Endlicher
var. **manausensis** Prance in Fl. Neotrop., 10:50
4 (1972)—Brazil.
cubensis (Poeppig & Endlicher) Grisebach
4 subsp. **minor** Prance *op. cit.* 63(1972)—Cuba.
peruviana Krause
var. **petioliflora** Prance *op. cit.* 54(1972)—Peru,
4 Ecuador.
tessmannii (Krause) Prance *op. cit.* 68(1972):
4 *Gonypetalum tessmannii*.

DILLENIACEAE

1886 Abedin, S.
Flora of West Pakistan: 42. Dilleniaceae. 1973. 4p.—
6 Illustration, map.

1887 Kubitzki, R.
Neue und bemerkenswerte neotropische Dillenia-
ceen. *Mitt. Bot. Staatssamml. München*, 9:707–720
4 (1973)—Illustrations.

1888 Wild, H. *and* Gonçalves, M. L.
5 Flora de Moçambique: 5. Dilleniaceae. 1973. 3p.

DAVILLA

strigosa Kubitzki in *Mitt. Bot. Staatssamml. München*,
4 9:718(1973)—Brazil.

DOLIOCARPUS

areolatus Kubitzki in *Mitt. Bot. Staatssamml. München*,
4 9:713(1973)—Brazil.
4 **gracilis** Kubitzki *op. cit.* 716(1973)—Brazil.
4 **lancifolius** Kubitzki *op. cit.* 715(1973)—Brazil.
4 **validus** Kubitzki *op. cit.* 715(1973)—Brazil.

DIONCOPHYLLACEAE

TRIPHYOPHYLLUM

1889 Kondo, K.
The chromosome numbers of *Striga asiatica* and
1, 5 *Triphyophyllum peltatum*. *Fyton*, 31(1):1–2(1973).

DIOSCOREACEAE

DIOSCOREA

1890 Miège, J.
La densité de nervation foliaire, caractère possible
d'identification et de séparation spécifiques chez les
Dioscoréacées. *Ann. Sci. Univ. Besançon, Bot.*, 12:
1–7 91–95(1972)—Illustrations.

1891 Prinsloo, H. D. A.
Dioscorea sylvatica—an elephant's foot. *Veld Flora*,
5 33(1):7–9(1973)—Illustrations.

1892 Siddiqi, M. A.
Flora of West Pakistan: 53. Dioscoreaceae. 1973.
6 6p.—Illustrations, map; key.

TESTUDINARIA

elephantipes Lindl.
f. **montana** (Burch.) Rowley in *Nation. Cact. Succ. J.* (U.K.), 28(1):6(1973): *T. montana.* 5
macrostachya (Benth.) Rowley *op. cit.* 6(1973): *Dioscorea macrostachya.* 4
sylvatica Knuth
var. **brevipes** (Burtt Davy) Rowley *op. cit.* 6 (1973): *Dioscorea brevipes.* 5
var. **multiflora** (Marl.) Rowley *op. cit.* 6(1973): *T. multiflora.* 5
var. **paniculata** (Duemm.) Rowley *op. cit.* 6(1973): *T. paniculata.* 5
var. **rehmannii** (Bak.) Rowley *op. cit.* 6(1973): *Dioscorea rehmannii.* 5

DIPSACACEAE

1893 Contandriopoulos, J. *and* Quézel, P.
Deux Dipsacées nouvelles de Grèce. *Biol. Gallo-Hellen.*, 5(1):131–137(1973)—(En,Gre); illustrations; chromosome numbers. *Cephalaria glaberrima*,
1 *Pterocephalus epiroticus.*

CEPHALARIA

mediterranea (Viv.) Szabó
subsp. **balearica** (Cosson) Malagarriga, Las Subesp. y Variac. Geogr.: 14(1973): *C. mediterranea* var. *balearica.* 1

KNAUTIA

dipsacifolia Kreutz.
subsp. **gracilis** (Szabó) Ehrendorfer in *Österr. Bot. Zeitschr.*, 122(4):264(1973): *K. gracilis.* 1
subsp. **sixtina** (Briq.) Ehrendorfer *op. cit.* 265 (1973): *K. sixtina.* 1
drymeia Heuff.
subsp. **centrifrons** (Borb.) Ehrendorfer *op. cit.* 263(1973): *K. centrifrons.* 1
subsp. **intermedia** (Pernh. & Wettst.) Ehrendorfer *op. cit.* 263(1973): *K. intermedia.* 1
subsp. **tergestina** (Beck) Ehrendorfer *op. cit.* 263 (1973): *K. sylvatica* var. *tergestina.* 1

SCABIOSA

× janchenii Soó in *Acta Bot. Acad. Sci. Hung.*, 18(1–2): 174(1973), without type—? Austria. (*S. columbaria* × *S. ochroleuca*). 1
ochroleuca L.
f. **baumgarteniana** Soó *op. cit.* 174(1973)—Rumania. 1

SUCCISELLA

inflexa (Kluk) Beck
f. **beckii** Soó in *Acta Bot. Acad. Sci. Hung.*, 18(1–2): 172(1973)—Hungary. 1

VIRGA

aspera (DC.) Holub in *Folia Geobot. Phytotax.*, 8(2): 175(1973): *Dipsacus asper.* 6
atrata (Hook. f. & Thoms.) Holub *op. cit.* 175 (1973): *Dipsacus atratus.* 6
inermis (Wall.) Holub *op. cit.* 175(1973): *Dipsacus inermis.* 6
leschenaultii (Coult.) Holub *op. cit.* 175(1973): *Dipsacus leschenaultii.* 6
stricta (D. Don) Holub *op. cit.* 175(1973): *Dipsacus strictus.* 6

DIPTEROCARPACEAE

1894 Ashton, P. S.
Precursor to a taxonomic revision of Ceylon Dipterocarpaceae. *Blumea*, 20(2):357–366(1972 publ.
6 1973)—Keys.

COTYLELOBIUM

lewisianum (Trim. ex Hook. f.) P. S. Ashton in *Blumea*, 20(2):358(1972 publ. 1973): *Stemonoporus lewisianus.* 6

HOPEA

brevipetiolaris (Thw.) P. S. Ashton in *Blumea*, 20(2):359(1972 publ. 1973): *Shorea brevipetiolaris.* 6

SHOREA

affinis (Thw.) P. S. Ashton in *Blumea*, 20(2):361 (1972 publ. 1973): *Doona affinis.* 6
congestiflora (Thw.) P. S. Ashton *op. cit.* 362(1972 publ. 1973): *Doona congestiflora.* 6
cordifolia (Thw.) P. S. Ashton *op. cit.* 362(1972 publ. 1973): *Doona cordifolia.* 6
disticha (Thw.) P. S. Ashton *op. cit.* 362(1972 publ. 1973): *Vateria disticha.* 6
gardneri (Thw.) P. S. Ashton *op. cit.* 362(1972 publ. 1973): *Doona gardneri.* 6
megistophylla P. S. Ashton *op. cit.* 362(1972 publ. 1973), nom. nov.: *Doona macrophylla* Thw. 6
ovalifolia (Thw.) P. S. Ashton *op. cit.* 363(1972 publ. 1973): *Doona ovalifolia.* 6
pallescens P. S. Ashton *op. cit.* 360(1972 publ. 1973)—Sri Lanka. 6
trapezifolia (Thw.) P. S. Ashton *op. cit.* 363(1972 publ. 1973): *Doona trapezifolia.* 6
zeylanica (Thw.) P. S. Ashton *op. cit.* 363(1972 publ. 1973): *Doona zeylanica.* 6

STEMONOPORUS

lancifolius (Thw.) P. S. Ashton in *Blumea*, 20(2):365 (1972 publ. 1973): *Monoporandra lancifolia.* 6

VATERIA

1895 Meijer, W.
Some taxonomic and nomenclatural notes on *Vateria copallifera* (Retz.) Alston. *Ceylon J. Sci., Biol. Sci.*, 10(1):76–78(1972) (Critical Notes on Ceylon
6 Plants 10)—Illustrations.

DONATIACEAE

DONATIA

1896 Philipson, W. R. *and* Philipson, M. N.
A comparison of the embryology of *Forstera* L. and *Donatia* J. R. & G. Forst. *New Zealand J. Bot.*, 11(3):
4, 7 449–459(1973).

DROSERACEAE

1897 Silva, A. R. Pinto da *and* Silva, Q. de J. G. Pinto da
Droseraceae. *Anuár. Soc. Brot.*, 38:147–151(1972)—
1 Por; key.

ALDROVANDA

1898 Sahashi, N. *and* Ikuse, M.
Pollen morphology of *Aldrovanda vesiculosa* L.
1, 2, 6, 7 *J. Jap. Bot.*, 48(12):374–379(1973).

DIONAEA

1899 Ewan, J.
Annals of 'The most wonderful plant in the world' [Darwin]. *In* Geck, E. *and* Pressler, G., *eds.* Festschrift für Claus Nissen zum siebzigsten Geburtstag 1971: 173–184(1973)—Illustrations, coloured illustration.
3 *Dionaea muscipula.*

DROSERACEAE

1900 Gennaro, J. F.
Kärbes huulheinal. (Fly in the sundew). *Eesti*
3 *Loodus*, 16(7):397–400(1973)—Illustrations.

1901 Williams, S. E.
The 'memory' of the Venus' flytrap. *Carniv. Pl.*
3 *Newsl.*, 2(2):23–25(1973).

DROSERA

1902 Hansen, C.
Note on *Drosera rotundifolia* L. in Greenland. *Bot.*
3 *Tidsskr.*, 67(4):342–343(1973)—Dot map.

1903 Kondo, K.
Chromosome numbers of five taxa. *Chromosome Inform. Serv.*, no.15:33–34(1973)—Chromosome
2, 7 numbers. *Drosera spathulata.*

1904 Kondo, K.
Chromosome numbers of some *Drosera* taxa.
J. Jap. Bot., 48(7):193–198(1973)—(Ja); illustration;
1–7 chromosome numbers.

1905 Križo, M.
Výskyt rosičky okrúhlolistej (*Drosera rotundifolia* L.) na Pol'ane v Slovenskom Stredohorí. (Occurrence of *Drosera rotundifolia* L. on the Polana in the Slovak midland range). *Biológia, A* (Czechoslovakia), 28(7):
1 595–597(1973)—(En,Rus); dot map.

1906 Kupriyanova, L. A.
Morfologicheskie issledovaniya pȳl'tsȳ *Drosera.* (Morphological studies of *Drosera* pollen). *Bot. Zhurn.*,
1–7 58(8):1155–1157(1973).

1907 Nash, R.
Tuberous Droseras in South Australia. *Carniv. Pl.*
7 *Newsl.*, 2(4):63–65(1973).

1908 Sivertsen, R.
Observations on tuberous Droseras. *Carniv. Pl.*
7 *Newsl.*, 2(4):62–65(1973).

1909 Whitehead, B.
Habitat of *Drosera peltata. Carniv. Pl. Newsl.*, 2(4):
7 61–62(1973).

1910 Ziemer, R. R.
In search of *Drosera anglica. Carniv. Pl. Newsl.*,
3 2(4):70–71(1973).

EBENACEAE

1911 Soukup, J.
Las Ebenáceas, Estiracáceas y Sapotáceas del Perú, sus generos y lista de especies (continuación). *Biota*,
4 9(74):265–271(1973).

DIOSPYROS

1912 Anon.
Persimmon, *Diospyros virginiana*: ebony family (Ebenaceae). *Morton Arbor. Quart.*, 9(1):14–15(1973)—
3 Illustration.

1913 Bouquet, A.
Plantes médicinales du Congo-Brazzaville: *Uvariopsis, Pauridiantha, Diospyros. Trav. Lab. Matière Méd.*
5 *Pharm. Galénique Fac. Pharm. Paris*, 57(1). 112p.(1973).

1914 Debray, M., Araud, O. *and* Paris, R. R.
Sur une Ébénacée endémique de Madagascar, le *Diospyros gracilipes* Hiern. *Pl. Méd. Phytothér.*, 7(1):
5 77–81(1973)—(En).

1915 Rathore, J. S.
Diospyros melanoxylon, a bread-winner tree of India. *Econ. Bot.*, 26(4):333–339(1973)—
6 Illustrations.

ELAEAGNACEAE

ELAEAGNUS

1916 Momiyama, Y. *and* Hara, H.
The genus *Elaeagnus* of eastern Himalaya. *J. Jap.*
6 *Bot.*, 48(9):257–267(1973)—Illustrations; key.

infundibularis Momiyama in *J. Jap. Bot.*, 48(9):262
6 (1973)—Nepal, India, Bhutan, Sikkim.
6 **kanaii** Momiyama *op. cit.* 264(1973)—Nepal.
6 **tricholepis** Momiyama *op. cit.* 262(1973)—Nepal.

HIPPOPHAË

1917 Filatov, I. I.
Zhiznennaya forma oblepikhi *Hippophaë rhamnoides* L. v Kazakhstane. (The life-form of *Hippophaë rhamnoides* L. in Kazakhstan). *Bot. Zhurn.*, 58(5):695–
2 697(1973)—Illustrations.

1918 Great Britain: Nature Conservancy
The management of sea buckthorn (*Hippophaë rhamnoides* L.) on selected sites in Great Britain: report of the *Hippophaë* Study Group (The Nature Conservancy), ed. by D. S. Ranwell. Norwich, Nature Conservancy, 1972.[vi]55p.—Illustrations, maps, dot
1 maps.

1919 Kapuściński, R.
Zbiór owoców rokitnika zwyczajnego *Hippophaë rhamnoides* na terenie wybrzeża w woj. gdańskim. (Collection of the fruit of the sea-buckthorn, *Hippophaë rhamnoides* on the Baltic shore in its section of the province of Gdańsk). *Chrońmy Przyr. Ojczystą*,
1 29(2):43–47(1973).

1920 Kondorskaya, V. R.
K teratologii oblepikhi. (To the teratology of the sea-buckthorn). *Byull. Mosk. Obshch. Ispȳt. Prir., Biol.*, 78(2):114–121(1973)—(En); illustrations.
1 *Hippophaë rhamnoides.*

ELATINACEAE

1921 Wild, H. *and* Diniz, M. A.
Flora de Moçambique: 22. Elatinaceae. 1973. 7p.—
5 Key.

ELATINE

1922 Lohammar, G.
Nordgränsen för *Elatine triandra. Svensk Bot. Tidskr.*,
1 67(3):306–307(1973)—(En); illustration.

EMPETRACEAE

EMPETRUM

1923 Bell, J. N. B. *and* Tallis, J. H.
Empetrum nigrum L. *J. Ecol.*, 61(1):289–305(1973) (Biological Flora of the British Isles)—Illustration,
1 dot map.

1924 Smith, R. I. L.
The occurrence of *Empetrum rubrum* Vahl ex Willd. on South Georgia. *Brit. Antarct. Surv. Bull.*, nos.33–34:
4 173–176(1973)—Illustrations.

EPACRIDACEAE

1925 Harborne, J. B. *and* Williams, C. A.
A chemotaxonomic survey of flavonoids and simple phenols in leaves of the Ericaceae. *Bot. J. Linn. Soc.*,
4, 6, 7 66(1):37–54(1973).

COLEANTHERA

1926 Jackes, B. R.
Coleanthera: the genus. *Austral. Pl.*, 7(56):190–191
7 (1973)—Illustrations; key.

ERICACEAE

1927 Harborne, J. B. *and* Williams, C. A.
A chemotaxonomic survey of flavonoids and simple phenols in leaves of the Ericaceae. *Bot. J. Linn. Soc.*, 66(1):37–54(1973)—Includes comparison with related
1–7 families.

1928 Hutira, M. *and* Moldovan, A.
Cîteva concluzii privind anatomia Ericaceelor. (Einige Schlussfolgerungen zur Anatomie der Ericaceae). *Bul. Şti., B* (Baia-Mare), 3:71–74(1971)—(Ge); key. Anatomical characters of *Vaccinium*,
1 *Oxycoccos*, *Calluna* and *Andromeda* spp.

1929 Muirhead, C. W.
Dwarf Ericaceae (other than *Rhododendron* collected by George Forrest. *J. Scott. Rock Gard. Club*, 13(3):
2, 6 246–252(1973).

1930 Oliver, E. G. H.
Notes on African plants: Ericaceae. *Bothalia*, 11(1–2):
5 115–116(1973)—Illustration.

1931 Visset, L.
Compléments pour la détermination au microscope electronique à balayage des Ericacées du Massif Amoricain. *Bull. Soc. Sci. Nat. Ouest France*, 70(4):
1 13–16(1972)—Illustrations.

1932 Visset, L.
Les pollens d'Ericacées du Massif Armoricain au microscope électronique à balayage. *Bull. Soc. Sci. Bretagne*, 46(3–4):193–204(1971 publ. 1972)—
1 Illustrations.

ARCTERICA

1933 Gorer, R.
Plant awards 1972–73: *Arcterica nana* P. C. Coville.
2 *Quart. Bull. Alp. Gard. Soc.*, 41(4):331–332(1973).

ARCTOSTAPHYLOS

1934 Davis, C. B.
'Bark striping' in *Arctostaphylos* (Ericaceae).
3 *Madroño*, 22(3):145–149(1973).

1935 Davis, C. B.
Comparative ecology of six members of the *Arctostaphylos andersonii* complex. *Diss. Abstr. Int., B*,
3 33(10):4766(1973).

1936 Denford, K. E.
Flavonoids of *Arctostaphylos uva-ursi* (Ericaceae). *Experientia*, 29(8):939(1973)—Chromosome
3 numbers.

1937 Resner, V.
Arctostaphylos alpina (L.) Sprengel v Belanských Tatrách. (*Arctostaphylos alpina* (L.) Sprengel in the Belanské Tatry mountains). *Biológia, A* (Czecho-
1 slovakia), 28(7):579–580(1973)—(En,Rus); illustration.

1938 Schmidt, M. G.
Western manzanitas. *Amer. Horticulturist*, 52(3):
3 26–32(1973)—Illustrations, coloured illustrations.

BRUCKENTHALIA

1939 Gay Wilson, D., Marchant, R. *and* Oldfield, F.
Fossil seeds of *Erica* from the Cromer Forest Bed Series. *New Phytol.*, 72(5):1235–1237(1973). Including
1 *Bruckenthalia*.

CALLUNA

1940 Aubert, G., Barbero, M. *and* Loisel, R.
Les Callunaies dans le sud-est de la France et le nord-ouest de l'Italie. *Bull. Soc. Bot. France*, 118(9):
1 679–699(1971 publ. 1973)—(En,It).

ERICA

1941 Baker, H. A.
New taxa in the genus *Erica*. *J.S. Afr. Bot.*, 39(3):
5 205–211(1973)—(Af).

1942 Gay Wilson, D., Marchant, R. *and* Oldfield, F.
Fossil seeds of *Erica* from the Cromer Forest Bed Series. *New Phytol.*, 72(5):1235–1237(1973). Including
1 *Bruckenthalia*.

1943 Frileux, P. N. *and* Wattez, J. R.
Les ultimes stations d'*Erica cinerea* L. subsistant dans l'ouest du nord de la France. (Nord et Pas-de-Calais; Picardie; Haute-Normandie): A. 1 partie. Etude chorologique. *Bull. Soc. Bot. France*, 118(9):813–821
1 (1971 publ. 1973)—(En); map.

1944 Oliver, E. G. H.
Erica vallis-aranearum. *Flow. Pl. Afr.*, 42(3–4):
5 pl.1680(1973)—En; coloured illustration.

1945 Scannell, M. J. P. *and* McClintock, D.
Double-flowered forms of *Erica mackaiana*.
1 *Watsonia*, 9(4):432(1973).

1946 Timbal, J.
Sur l'écologie d'*Erica multiflora* L. dans le nord-ouest de la Tunisie. (Ecological data about north western Tunisian stations of *Erica multiflora* L.). *Bull. Soc. Bot. France*, 119(5–6):325–333(1972 publ.
1 1973)—(En).

granulatifolia H. A. Baker in *J. S. Afr. Bot.*, 39(3):
5 206(1973)—South Africa (Cape Prov.).
kougabergensis H. A. Baker
var. **recurvifolia** H. A. Baker *op. cit.* 210(1973)—
5 South Africa (Cape Prov.).
krugeri E. G. H. Oliver in *Bothalia*, 11(1–2):115
5 (1973)—South Africa (Cape Province).
obconica H. A. Baker in *J. S. Afr. Bot.*, 39(3):205
5 (1973)—South Africa (Cape Prov.).
pustulata H. A. Baker *op. cit.* 207(1973)—South
5 Africa (Cape Prov.).
uysii H. A. Baker *op. cit.* 208(1973)—South Africa
5 (Cape Prov.).
vallis-aranearum E. G. H. Oliver in *Flow. Pl. Afr.*,
5 42(3–4):t.1680(1973)—South Africa (Cape Prov.).
vogelpoelii H. A. Baker in *J. S. Afr. Bot.*, 39(3):209
5 (1973)—South Africa (Cape Prov.).

GAULTHERIA

1947 Sealy, J. R.
Gaultheria stapfiana. *Curtis's Bot. Mag.*, 179(4):tab.
2, 6 651(1973)—En; illustration, coloured illustration.

KALMIA

1948 Hall, I. V., Jackson, L. P. *and* Everett, C. F.
The biology of Canadian weeds: 1. *Kalmia angustifolia* L. *Canad. J. Pl. Sci.*, 53(4):865–873(1973)—
3 Illustrations, dot map; chromosome number.

KALMIOPSIS

1949 Mulligan, B. O.
Kalmiopsis leachiana. Quart. Bull. Alp. Gard. Soc., 41(2):132–134(1973)—En. 3

LEDUM

1950 Korsmo, H.
Ledum palustre funnet i Eidsvoll, Akershus. *Blyttia*, 31(4):250(1973). 1

PHYLLODOCE

1951 Coker, P. D. *and* Coker, A. M.
Phyllodoce caerulea (L.) Bab. *J. Ecol.*, 61(3):901–913 (1973) (Biological Flora of the British Isles)—Illustrations, map, dot map; chromosome number. 1

ORNITHOSTAPHYLOS

1952 Morgan, R.
Ornithostaphylos (Ericaceae) in California. *Madroño*, 22(4):214(1973). 3

PIERIS

1953 Brown, E.
Pieris. Amer. Horticulturist, 52(1):20–23(1973)—En; illustrations, coloured illustrations. 2

RHODODENDRON

1954 Aleksandrova, M. S.
O redkikh rododendronakh Kavkaza. *Byull. Glavn. Bot. Sada* (Moscow), 87:36–40(1973)—Dot map. 2

1955 Breckle, S. W.
Alpenrosen im Hindukusch? *Jahrb. Ver. Schutze Alpenpfl. Tiere*, 37:140–146(1972)—Illustrations, dot map. *Rhododendron collettianum*, *R. afghanicum*. 2

1956 Cox, P. S.
Dwarf rhododendrons. London, Batsford, 1973. 296p.—Illustrations, coloured illustrations. 2, 6

1957 Craven, L. A.
Vireya rhododendrons: a history of their introduction and cultivation. *Rhododendron*, 12(3):9–15 (1973). 2, 6

1958 Ferrarini, E.
Rhododendron ferrugineum L. in fitocenosi relitte dell'Appennino settentrionale. (Relict communities of *Rhododendron ferrugineum* L. in northern Apennines). *Giorn. Bot. Ital.*, 107(3):143–156(1973)—(En); dot maps. 1

1959 Gibson, J. F. A.
Forrest rhododendrons suitable for the rock garden. *J. Scott. Rock Gard. Club*, 13(3):204–209(1973)—Illustrations. 2

1960 Heft, L.
Rhododendron yakusimanum and its hybrids. *Rhododendron*, 12(4):6–11(1973)—Illustration. 2

1961 Horton, J. H. *and* Pittillo, J. D.
An unusual specimen of *Rhododendron calendulaceum* (Michaux) Torrey (Ericaceae). *Castanea*, 38(2):204–205(1973)—Illustrations. 3

1962 Hunt, D. R.
Rhododendron concatenans. Curtis's Bot. Mag., 179(2): tab.634(1972 publ. 1973)—En; illustration, coloured illustration. 2

1963 Hunt, D. R.
Rhododendron sidereum. Curtis's Bot. Mag., 179(3): tab.638(1973)—En; illustration, coloured illustration. 2, 6

1964 Hunt, D. R.
Rhododendron wrightianum. Curtis's Bot. Mag., 179 (4):tab.653(1973)—En; illustration, coloured illustration. 6

1965 Lepiot, R.
A propos des azalées de l'Inde. *Jard. France*, no.10:12–14(1973)—Illustrations. Popular article. 2, 6

1966 Philipson, W. R. *and* Philipson, M. N.
A history of *Rhododendron* classification. *Notes Roy. Bot. Gard. Edinburgh*, 32(2):223–238(1973). 1–3, 6, 7

1967 Scoborio, R. W.
The Maddenii Series. *Rhododendron*, 12(2):13–14 (1973). 6

1968 Sleumer, H.
New species and noteworthy records of *Rhododendron* in Malesia (Ericaceae). *Blumea*, 21(2):357–376 (1973)—(Fr). 6

1969 Stanton, D. B.
Vireya Rhododendrons growing in Australia. *Rhododendron*, 12(1):8–10(1973). 7

1970 Sullivan, P.
Ten years of *Vireya* rhododendrons at Strybing Arboretum. *Rhododendron*, 12(3):2–7(1973)—Illustration. 2, 6

1971 Taylor, S.
Rhododendron macrophyllum D. Don ex G. Don: California rhododendron, California rosebay, West Coast rhododendron, Pacific rhododendron, Pacific rosebay. *Davidsonia*, 4(1):4–8(1973)—Illustrations. 3

blackii Sleumer in *Blumea*, 21(2):375(1973)—New Guinea (Terr. New Guinea). 6
buxoides Sleumer *op. cit.* 359(1973)—Borneo (Sarawak). 6
detzneranum Sleumer *op. cit.* 359(1973), as 'detzneriana'—New Guinea (Terr. New Guinea). 6
hartleyi Sleumer *op. cit.* 366(1973)—New Guinea (Terr. New Guinea). 6
hyacinthosmum Sleumer *op. cit.* 363(1973)—New Guinea (Papua). 6
micromalayanum Sleumer *op. cit.* 364(1973)—Borneo (Sarawak). 6
neobritannicum Sleumer *op. cit.* 361(1973)—New Britain. 6
praetervisum Sleumer *op. cit.* 376(1973)—Borneo (Sabah). 6
searleanum Sleumer *op. cit.* 367(1973)—New Guinea (Terr. New Guinea). 6
stevensianum Sleumer *op. cit.* 371(1973)—New Guinea (Terr. New Guinea). 6

SYMPHYSIA

floccosa (L. O. Wms.) Borhidi in *Acta Bot. Acad. Sci. Hung.*, 19(1–4):45(1973): *Hornemannia floccosa*. 4

VACCINIUM

1972 Nees, H. von, Pachaly, P. *and* Zymalkowski, F.
Zur Chemotaxonomie der Ericaceae: Isolierung und Identifizierung von Triterpenen und Sterinen aus *Vaccinium uliginosum*. (Chemotaxonomy of Ericaceae: isolation and identification of Triterpenes and steroids from *Vaccinium uliginosum*). *Pl. Med.*, 24(4):320–328(1973)—(En). 1–3

1973 Stewart, W. S. *and* Bannister, P.
Seasonal changes in carbohydrate content of three *Vaccinium* spp. with particular reference to *V. uliginosum* L. and its distribution in the British Isles.
1 *Flora* (Germany), 162(1–2):134–155(1973)—Maps.

pedris (Harshberger) Holub in *Folia Geobot. Phytotax.*,
3 8(2):177(1973): *V. uliginosum* var. *pedris*.

ERIOCAULACEAE

1974 Moldenke, H. N.
Additional notes on the Eriocaulaceae: 40–47. *Phytologia*, 25(2):68–69(1972), 25(3):121–166, (4):245–248, 26(1):15–47, (2):136–147, (3):179–202,
1–7 (4):226–262, (6):455–484(1973)—Map; key.

1975 Moldenke, H. N.
A fifth summary of the Verbenaceae, Avicenniaceae, Stilbaceae, Dicrastylidaceae, Symphoremaceae, Nyctanthaceae and Eriocaulaceae of the world as to valid taxa, geographic distribution, and synonymy:
1–6 supplement 2. *Phytologia*, 25(4):225–245(1973).

1976 Moldenke, H. N.
A fifth summary of the Verbenanceae, Avicenniaceae, Stilbaceae, Dicrastylidaceae, Symphoremaceae, Nyctanthaceae, and Eriocaulaceae of the world as to valid taxa, geographic distribution, and synonymy: supplement 3. *Phytologia*, 26(5):356–377
1–7 (1973).

ERIOCAULON

1977 Cook, C. D. K.
New and noteworthy plants from the northern Italian ricefields. *Ber. Schweiz. Bot. Ges.*, 83(1): 54–65(1973)—(Fr,Ge). *Eriocaulon cinereum* new to
1 Europe.

1978 Kulkarni, A. R. *and* Desai, M. H.
Eriocaulon tuberiferum—Kulkarni et Desai—a new species from Maharastra. *Proc. Indian Sci. Congr.*,
6 59(3):322–323(1972). Abstract only.

1979 Moldenke, H. N.
Notes on new and noteworthy plants: 58. *Phyto-*
4 *logia*, 25(7):430–432(1973).

1980 Satake, Y.
(A new species of *Eriocaulon* from Nepal). *J. Jap.*
6 *Bot.*, 48(10):314–317(1973)—Ja(En); illustrations.

reitzii Moldenke & L. B. Smith in *Phytologia*, 25(7):
4 430(1973)—Brazil.
sellowianum Kunth
var. **minor** Moldenke in *Phytologia*, 26(6):463
4 (1973)—Brazil.
var. **paranense** (Moldenke) Moldenke & L. B.
4 Smith in *Phytologia*, 26(1):37(1973): *E. paranense*.
staintonii Satake in *J. Jap. Bot.*, 48(10):316(1973)—
6 Nepal.
tuberiferum A. R. Kulkarni & Desai in *Proc. Indian Sci. Congr.*, 59(3):322(1972), without latin descr.
6 and without type—India.

LACHNOCAULON

1981 Moldenke, H. N.
Seven novelties from North and South America.
3 *Phytologia*, 25(3):117–120(1973).

engleri
var. **caulescens** Moldenke in *Phytologia*, 25(3):117
3 (1973)—U.S.A. (Alabama).

PAEPALANTHUS

1982 Moldenke, H. N.
Notes on new and noteworthy plants: 58.
4 *Phytologia*, 25(7):430–432(1973).

1983 Moldenke, H. N.
Notes on new and noteworthy plants: 61. *Phytologia*, 26(5):355–356(1973). *Paepalanthus*
4 *planifolius* var. *globulifer*.

catharinae
var. **hatschbachii** (Moldenke) Moldenke & L. B. Smith in *Phytologia*, 25(7):430(1973), as
4 'hatschbachi': *P.*
dianthoides
var. **langsdorffii** Moldenke in *Phytologia*, 26(4): 252(1973); *P. dianthoides* var. β Körn. in Mart. Fl. Bras., 3(1):339–340(1863), as 'langsdorffi'—
4 Brazil.
distichophyllus
var. **gardneri** Moldenke *op. cit.* 258(1973); *P. distichophyllus* var. β Körn. in Mart. Fl. Bras.,
4 3(1):349(1863)—Brazil.
hilairei
var. pohlianus Moldenke in *Phytologia*, 25(4):229, 241(1973), and in *Phytologia*, 25(7):430(1973),
4 without type—Brazil.
langsdorffii
var. **caracensis** Moldenke in *Phytologia*, 25(7):
4 430(1973)—Brazil.
leiseringii
var. **kleinii** Moldenke & L. B. Smith *op. cit.* 431
4 (1973)—Brazil.
planifolius
var. **globulifer** (Alv. Silv.) Moldenke & Smith in
4 *Phytologia*, 26(5):355(1973): *P. globulifer*.
polyanthus
f. villosus (Beauverd) Moldenke & L. B. Smith in *Phytologia*, 25(7):431(1973): *P. polyanthus* var.
4 *villosus*.

SYNGONANTHUS

1984 Moldenke, H. N.
Notes on new and noteworthy plants: 58.
4 *Phytologia*, 25(7):430–432(1973).

1985 Moldenke, H. N.
Notes on new and noteworthy plants: 59.
4 *Phytologia*, 26(3):177–179(1973).

1986 Moldenke, H. N.
Seven novelties from North and South America.
4 *Phytologia*, 25(3):117–120(1973).

appressus
var. **chapadensis** Moldenke in *Phytologia*, 25(3):
4 118(1973)—Brazil.
4 **bracteosus** Moldenke *op. cit.* 119(1973)—Brazil.
candidus
var. **bahiensis** Moldenke *op. cit.* 120(1973)—
4 Brazil.
chrysanthus
var. **castrensis** Moldenke & L. B. Smith in
4 *Phytologia*, 25(7):431(1973)—Brazil.
densiflorus
var. **glabrifolius** Moldenke in *Phytologia*, 25(3):
4 120(1973)—Brazil.
glandulosus
var. **epapillosus** Moldenke in *Phytologia*, 26(3):177
4 (1973)—Venezuela.
nitens
var. **viviparus** Moldenke in *Phytologia*, 25(4):223
4 (1973)—Brazil.
philcoxii Moldenke in *Phytologia*, 26(3):178(1973)—
4 Brazil.

prolifer
var. **parvus** Moldenke in *Phytologia*, 25(4):223
4 (1973)—Brazil.
ulei
var. **goyazensis** Moldenke *op. cit.* 223(1973)—
4 Brazil.
weddellii
var. **gracilis** Moldenke *op. cit.* 224(1973)—
4 Brazil.
xeranthemoides
f. **brevifolius** Moldenke in *Phytologia*, 26(3):178
4 (1973)—Venezuela.
var. **strigillosus** Moldenke in *Phytologia*, 25(4):224
4 (1973)—Brazil.
var. **tricostatus** (Gleason) Moldenke in *Phytologia*,
4 26(3):179(1973): *S. tricostatus.*

ERYTHROXYLACEAE

1987 Badré, F.
Erythroxylacées. Flore du Gabon, 21:49–54(1973)—
5 Illustrations, keys.

ERYTHROXYLUM

1988 Gentner, W. A.
The genus *Erythroxylum* in Columbia. *Cespedesia*,
4 1(4):481–560(1972)—Illustrations, maps; key.

ESCALLONIACEAE

1989 Patel, R. N.
Wood anatomy of the dicotyledons indigenous to New Zealand: 2. Escalloniaceae. *New Zealand J. Bot.*,
7 11(3):421–434(1973).

COROKIA

1990 Bhatnagar, A. K.
Morphological and embryological studies in
6, 7 *Corokia. Botanica* (Delhi), 23(4):149(1973).

1991 Bhatnagar, A. K. *and* Kapil, R. N.
Occurrence of stomata on the reproductive organs of *Corokia. Proc. Indian Sci. Congr.*, 60(3):289(1973).
6, 7 Abstract only.

ESCALLONIA

1992 Hunt, D. R.
Escallonia alpina. Curtis's Bot. Mag., 179(3):tab.642
4 (1973)—En; illustration, coloured illustration.

ITEA

1993 Petrov, S. *and* Drazheva-Stamatova, Ts.
Itea L. fossil pollen in tertiary sediments of Europe and North America. *Dokl. Bolg. Akad. Nauk*, 26(6): 811–814(1973). Including observations on present-
1–3, 6 day species.

POTTINGERIA

1994 Airy Shaw, H. K., Cutler, D. F. *and* Nilsson, S.
Pottingeria, its taxonomic position, anatomy and palynology. *Kew Bull.*, 28(1):97–104(1973)—
6 *Pottingeria* transferred to the Celastraceae.

EUPHORBIACEAE

1995 Croizat, L.
Les Euphorbiacées vues en elles-mêmes, et dans leurs rapports envers l'angiospermie en général. *Mem.*
1–7 *Soc. Brot.*, 23. 207p.(1973)—(En); illustrations, maps.

1996 Gill, B. S., Chawla, A. *and* Kanwal, R.
Cytology of Euphorbiaceae. *Proc. Indian Sci. Congr.*, 58(3):472–473(1971)—Chromosome numbers.
6 Abstract only.

1997 Hans, A. S.
Chromosomal conspectus of the Euphorbiaceae. *Taxon*, 22(5–6):591–636(1973)—Chromosome
1–7 numbers.

1998 Leach, L. C.
New and interesting taxa of the tribe Euphorbieae (Euphorbiaceae) from Portuguese Africa. *García de Orta, Ser. Bot.*, 1(1–2):31–41(1973)—(Por);
5 illustrations.

1999 Mukherjee, P. K.
Trends in specialization in the family Euphorbiaceae: 2. The female gametophyte. *Proc. Nation.*
1–7 *Acad. Sci., India, B*, 41(1):81–89(1971).

2000 Sharma, R.
Studies in the family Euphorbiaceae: 2. On the development of the cyathium. *Proc. Indian Sci. Congr.*,
1–7 60(3):297(1973). Abstract only.

2001 Webster, G. L. *and* Rupert, E. A.
Phylogenetic significance of pollen nuclear number in the Euphorbiaceae. *Evolution*, 27(3):524–531
1–7 (1973).

ACALYPHA

2002 Kahlem, G. *and* Champault, A.
Etude morphologique, caryologique, immunologique et enzymologique d'un hybride stérile d'*Acalypha* (Euphorbiacées). *Bull. Soc. Bot. France*, 118(9):621–626(1971 publ. 1973)—(En); illustrations;
2, 5 chromosome numbers.

2003 Radcliffe-Smith, A.
Allomorphic female flowers in the genus *Acalypha* (Euphorbiaceae). *Kew Bull.*, 28(3):525–529(1973)—
5 Illustrations; key.

2004 Radcliffe-Smith, A.
Notes on African Euphorbiaceae: 3. *Acalypha.*
5 *Kew Bull.*, 28(2):287–295(1973).

2005 Radcliffe-Smith, A.
Notes on African Euphorbiaceae: 4. *Kew Bull.*,
5 28(3):521–523(1973).

clutioides Radcliffe-Smith in *Kew Bull.*, 28(2):287
5 (1973)—Zambia.
fruticosa Forssk.
var. **eglandulosa** Radcliffe-Smith *op. cit.* 288(1973)
5 —Kenya, Sudan, Ethiopia, Uganda, Tanzania.

ANDROSTACHYS

2006 Airy Shaw, H. K.
Proposal to conserve the generic name 4336a *Androstachys* Prain (Euphorbiaceae) against *Androstachys* Grand Eury (gen. foss., Filicales). *Taxon*,
5 22(5–6):694(1973).

ANTIDESMA

2007 Airy Shaw, H. K.
New or noteworthy species of *Antidesma* (Stilaginaceae): 3. *Kew Bull.*, 28(2):269–281(1973).
6

cuspidatum Muell. Arg.
var. **orthocalyx** Airy Shaw in *Kew Bull.*, 28(2):274
6 (1973)—Sumatra.
hosei Pax & Hoffm.
var. **microcarpum** Airy Shaw *op. cit.* 271(1973)—
6 Borneo (Sarawak, Brunei, Sabah), Malaya.
katikii Airy Shaw *op. cit.* 278(1973)—New Guinea
6 (Terr. New Guinea).
leucopodum Miq.
var. **kinabaluense** Airy Shaw *op. cit.* 273(1973)—
6 Borneo (Sabah).
var. **platyphyllum** Airy Shaw *op. cit.* 273(1973)—
6 Borneo (Sarawak, Brunei).
montis-silam Airy Shaw *op. cit.* 269(1973)—Borneo
6 (Sabah).
neurocarpum Miq.
var. **angustatum** Airy Shaw *op. cit.* 270(1973)—
6 Borneo (Sarawak).
pachystachys Hook. f.
var. **palustre** Airy Shaw *op. cit.* 269(1973)—
6 Malaya.
plagiorrhynchum Airy Shaw *op. cit.* 272(1973)—
6 Sumatra.
vaccinioïdes Airy Shaw *op. cit.* 280(1973)—New
6 Guinea (Terr. New Guinea).

CEPHALOCROTON

2008 Radcliffe-Smith, A.
An account of the genus *Cephalocroton* Hochst. (Euphorbiaceae). *Kew Bull.*, 28(1):123–132(1973)—
5, 6 Key.

sect. **Cephalocrotonopsis** (Pax) Radcliffe-Smith in
5, 6 *Kew Bull.*, 28(1):131(1973): genus *Cephalocrotonopsis*.

CHAMAESYCE

jovetii (Huguet) Holub in *Folia Geobot. Phytotax.*,
1 8(2):176(1973): *Euphorbia jovetii*.

CICCA

2009 Rao, C. K.
Hermaphrodite flowers in Euphorbiaceae: *Cicca*
6 *acida* (L.) Merr. *Curr. Sci.*, 42(8):295(1973).

CROTON

2010 Croizat, L.
Una especie de *Croton* L. nueva para Venezuela: *Croton katoae* Croiz. *Pittieria*, no. 4:33–37(1972)—
4 Illustration.

2011 Leandri, J.
Deux espèces et une variété malgaches de *Croton*.
5 *Adansonia*, 13(3):295–297(1973)—(En); illustrations.

2012 Leandri, J.
Quelques *Croton* malgaches nouveaux. *Adansonia*,
5 13(2):173–176(1973)—(En); illustrations; key.

aymoniniorum Leandri in *Adansonia*, 13(2):175
5 (1973), as 'aymoninorum'—Malagasy.
bemarivensis Leandri in *Adansonia*, 13(4):423
5 (1973 publ. 1974)—Malagasy.
bergassae Leandri in *Adansonia*, 13(2):176(1973)—
5 Malagasy.
fianarantsoae Leandri in *Adansonia*, 13(3):295
5 (1973)—Malagasy.
var. **grandifolia** Leandri *op. cit.* 297(1973)—
5 Malagasy.
4 **katoae** Croizat in *Pittieria*, 4:33(1972)—Venezuela.
rakotoniainii Leandri in *Adansonia*, 13(3):295
5 (1973), as 'rakotonianii'—Malagasy.
rubricapitirupis Leandri in *Adansonia*, 13(2):173
5 (1973)—Malagasy.

ENDADENIUM Leach in *Garcia de Orta, Sér. Bot.*,
5 1(1–2):31(1973).
gossweileri (N. E. Br.) Leach *op. cit.* 32(1973):
5 *Monadenium gossweileri*.

EUPHORBIA

2013 Anon.
Euphorbia enormis N. E. Br. (Euphorbiaceae).
5 *Ashingtonia*, 1(3):27(1973)—En; coloured illustration.

2014 Bally, P. R. O.
Tropical African plants: 33. An amplified description of *Euphorbia quadrialata* Euphorbiaceae. *Kew Bull.*,
5 28(2):322–324(1973)—Illustrations.

2015 Bourdoux, P.
Euphorbia obesa Hook. *Cactus* (Brussels), 5(1):12–13
4 (1973)—Fr.

2016 Brewerton, D. V.
Euphorbias for you. *Nation. Cact. Succ. J.* (U.K.),
1, 2, 5 28(1):9(1973)—*E. echinus*, *E. fruticosa*.

2017 Dubovik, O. N., Krits'ka, L. I. *and* Moroz, I. I.
Novi vidomosti pro vȳdȳ rodu molochaĭ (*Euphorbia* L.) u flori Ukrayinȳ (New data on species of *Euphorbia* L. genus in the flora of the Ukraine). *Ukr. Bot. Zhurn.*, 30(1):80–88(1973)—Uk(En); dot
1 map.

2018 Dyer, R. A.
Euphorbia tortirama R. A. Dyer. *Ashingtonia*, 1(2): 15(1973)—En. Correction in *Ashingtonia*, 1(3):
5 35(1973).

2019 El Hadidi, M. N.
The genus *Euphorbia* L. In Egypt: 1. Section *Anisophyllum* Roep. *Bull. Jard. Bot. Nation. Belg.*, 43
2 (1–2):83–100(1973)—Illustrations; key.

2020 Geigy *Ltd*: Agricultural Chemicals Division
Euphorbiaceae: spurge family. *Geigy Weed Tables*,
1–7 60(1973)—Illustrations, coloured illustrations.

2021 Gilbert, M. G.
A new species of *Euphorbia* from the Ogaden, Ethiopia. *Kew Bull.*, 28(3):437–440(1973 publ. 1974)—
5 Illustrations.

2022 Herbst, D.
A new *Euphorbia* (Euphorbiaceae) from Hawaii. *Bull. Pac. Trop. Bot. Gard.*, 3(3):54–55(1973)—
6 Illustration.

2023 Herre, H.
Notes on South African succulents. *Cact. Succ. J.*
5 (U.S.A.), 45(5):238–239(1973)—Illustration.

2024 Jablonski, E.
Catalogus Euphorbiarum 1973: vol. 1–2. *Phytologia*,
1–7 25(6):329–349, 351–367, 369–397(1973)—Keys.

2025 Karrȳev, M. O. *and* Mamedov, K.
Khimicheskie i anatomicheskie issledovaniya *Euphorbia monostula* Prokh. (Chemical and anatomical investigation of *Euphorbia monostula* Prokh.). *Izv. Akad. Nauk Turkm. SSSR, Biol. Nauk*, 3:70–73
2 (1973)—(En); illustration.

2026 Leach, L. C.
Euphorbia species from the Flora Zambesiaca area: 10. *J. S. Afr. Bot.*, 39(1):3–22(1973)—(Af);
5 illustrations.

2027 Leach, L. C.
Euphorbia tirucalli L. its typification, synonymy and relationships, with notes 'Almeidina' and 'Cassoneira'.
5, 6 *Kirkia*, 9(1):69–86(1973)—Map.

2028 Nezadal, W.
Über die Verbreitung von *Spergula arvensis* und *Euphorbia exigua* in Nordostbayern. *Gottinger Flor.*
1 *Rundbr.*, 7(3):54–57(1973)—Dot maps.

2029 Noltee, F.
Euphorbia aeruginosa. *Succulenta*, 52(8):141,147–148
5 (1973)— Du; illustration, coloured illustration.

2030 Ponert, J.
A new species of *Euphorbia* from Iran. (Nový druh rodu *Euphorbia* z Íránu). *Preslia*, 45(4):361–363
2 (1973)—(Cz); illustrations. *E. sparsiglandulosa.*

2031 Rabson, A.
Euphorbia groenewaldii R. A. Dyer. *Ashingtonia*, 1(1):7,9(1973)—En; illustration, coloured
5 illustration.

2032 Raĭlyan, A. F.
Kriticheskiĭ obzor vidov *Euphorbia* L. Moldavii. (The critical review of the *Euphorbia* species of
1 Moldavia). *Bot. Zhurn.*, 58(7):1013–1023(1973)—Map.

2033 Raĭlyan, A. F.
O znachenii anatomicheskogo stroeniya semennoĭ kozhurȳ dlya sistematiki roda *Euphorbia* L. (On importance of anatomical structure of the seed-coat for the taxonomy of the genus *Euphorbia* L.). *Bot.*
1–3 *Zhurn.*, 58(2):239–245(1973).

2034 Raĭlyan, A. F. *
Novye dlya Moldavskoi SSR vidy roda *Euphorbia* L. (Species of *Euphorbia* L. new for the Moldavian S.S.R.) *Izv. Akad. Mold. SSR, Biol. Khim. Nauk*, no.2:80–81(1972). Abstr. in *Biol. Abstr.* 55(9):47301 (1973). *E. humifusa* Willd., *E. oblongifolia* C. Koch,
1 *E. volhynica* Bess ex Szaf, *E. bessarabica* Klok.

2035 Slavík, B.
Phytochorologische Charakteristik der Art *Euphorbia lucida* Waldst. et Kit. in bezug auf ihre Verbreitung in der Tschechoslowakei. (Fytochorologická charakteristika druhu *Euphorbia lucida* Waldst. et Kit. se zřetelem na rozšíření v Česko-
1 slovensku). *Preslia*, 45(2):119–131(1973)—(Cz); map.

2036 Stearn, W. T.
Mrs Robb and 'Mrs Robb's bonnett' (*Euphorbia*
2 *robbiae*). *J. Roy. Hort. Soc.*, 98(7):306–310(1973).

2037 Stillwell, M.
Les euphorbes. *Cactus* (Brussels), 4(3):52–57
1, 5, 6 (1972).

2038 Voggenreiter, V.
Euphorbia canariensis L. auf La Palma. Eine pflanzengeographische Kartierung. *Cuad. Bot. Canar.*,
1 18–19:55–64(1973)—(Sp,En); dot maps.

sect. **Conicocarpus** (Prochanov) Ponert in *Preslia*,
2 45(4):363(1973): *Tithymalus* sect. *Conicocarpus.*
ser. **Villosae** Raĭlyan in *Bot. Zhurn.*, 58(7):1016
1–3 (1973).
decliviticola Leach in *J. S. Afr. Bot.*, 39(1):13(1973)—
5 Mozambique.
esula L.
subsp. **nevadensis** (Bss. R.) Malagarriga, Las
1 Subesp. y Variac. Geogr.: 5(1973): *E. nevadensis.*
fanshawei Leach in *J. S. Afr. Bot.*, 39(1):8(1973)—
5 Zambia.
flavicoma DC.
subsp. **verrucosa** (Fiori) Pignatti in *Giorn. Bot. Ital.*, 107(5):220(1973): *E. epithymoides* var.
1 *verrucosa.*
gasparrinii Boiss.
subsp. **samnitica** (Fiori) Pignatti *op. cit.* 219(1973):
1 *E. epithymoides* var. *samnitica.*
granulata Forsk.
var. **turcomanica** (Boiss.) Hadidi in *Bull. Jard. Bot.*
2 *Nation. Belg.*, 43(1–2):93(1973): *E. turcomanica.*
klokoviana Raĭlyan in *Bot. Zhurn.*, 58(7):1019(1973),
1 nom. nov.: *E. bessarabica* Klok.
klokovii Dubovik in *Ukr. Bot. Zhurn.*, 30(1):84
1 (1973), nom. nov.: *E. pseudovillosa* Klokov.
knuthii Pax
subsp. **johnsonii** (N. E. Br.) Leach in *J. S. Afr. Bot.*,
5 39(1):20(1973): *E. johnsonii.*
linearibracteata Leach in *Garcia de Orta, Sér. Bot.*,
5 1(1–2):39(1973)—Angola.
lucida Waldst. & Kit.
f. **angustifolia** (Tausch) Slavík in *Preslia*, 45(2):120
1 (1973): *E. lucida* β *angustifolia.*
f. **leiocarpa** (Tausch) Slavík *op. cit.* 120(1973):
1 *E. lucida* β *leiocarpa.*
maresii Knoche
subsp. **balearica** (Wk.) Malagarriga, Las Subesp. y
1 Variac. Geogr.: 5(1973): *E. maresii* var. *balearica.*
mlanjeana Leach in *J. S. Afr. Bot.*, 39(1):3(1973)—
5 Malawi.
oschtenica Galushko in *Novosti Sist. Vȳssh. Rast.*,
2 10:325(1973)—U.S.S.R. (Caucasus).
pannonica Host.
f. **kitaibelii** Soó in *Acta Bot. Acad. Sci. Hung.*,
1 18(1–2):172(1973)—Hungary.
piscidermis M. G. Gilbert in *Kew Bull.*, 28(3):437
5 (1973 publ. 1974)—Ethiopia.
sojakii (Chrtek & Křísa) Dubovik in *Ukr. Bot. Zhurn.*,
1 30(1):80(1973): *E. austriaca* subsp. *sojakii.*
sparsiglandulosa Ponert in *Preslia*, 45(4):361
2 (1973)—Iran.
spinosa L.
subsp. **ligustica** (Fiori) Pignatti in *Giorn. Bot. Ital.*,
1 107(5):220(1973): *E. spinosa* f. *ligustica.*
triflora Schott, Nym. & K.
subsp. **kerneri** (Huter) L. Poldini ex Pignatti
1 *op. cit.* 220(1973): *E. kerneri.*

GLOCHIDION

2039 Bahadur, K. N., Gaur, R. C. *and* Sahni, K. C.
Glochidion montanum Thw. (Euphorbiaceae) a tree new to the flora of India. *Indian Forester*, 99(10):
6 623–627(1973)—Illustration; key.

HYMENOCARDIA

2040 Radcliffe-Smith, A.
Tropical African plants: 33. A new variety of *Hymenocardia acida* (Hymenocardiaceae). *Kew Bull.*,
5 28(2):324(1973).

acida Tul.
var. **mollis** (Pax) Radcliffe-Smith in *Kew Bull.*,
5 28(2):324(1973): *H. mollis.*

JATROPHA

2041 McClurkin, B.
Jatropha berlandieri Torrey. *Bull. Afr. Succ. Pl. Soc.*,
3, 4 8(3):104–105(1973)—En.

2042 Radcliffe-Smith, A.
Notes on African Euphorbiaceae: 2. *Jatropha*. *Kew*
5 *Bull.*, 28(2):283–286(1973)—Key.

2043 Radcliffe-Smith, A.
Notes on African Euphorbiaceae: 4. *Kew Bull.*,
5 28(3):521–523(1973 publ. 1974).

hildebrandtii Pax
var. **torrentis-lugardi** Radcliffe-Smith in *Kew*
5 *Bull.*, 28(2):283(1973)—Kenya.
microdonta Radcliffe-Smith *op. cit.* 284(1973)—
5 Tanzania.
oblanceolata A. Radcliffe-Smith in *Kew Bull.*,
5 28(3):522(1973 publ. 1974)—Kenya.
schweinfurthii Pax
subsp. **atrichocarpa** Radcliffe-Smith *op. cit.* 285
5 (1973)—Tanzania, Rhodesia.

LEUCOCROTON

moaënsis Borhidi & Muñiz in *Acta Bot. Acad. Sci.*
4 *Hung.*, 18(1–2):30(1973)—Cuba.

MANIHOT

2044 Rogers, D. J. *and* Appan, S. G.
Flora neotropica monograph no.13. *Manihot*, *Manihotoides* (Euphorbiaceae). New York, Hafner, 1973. 275p.—Illustrations, portraits, dot maps;
4 keys.

2045 Rogers, D. J. *and* Fleming, H. S.
A monograph of *Manihot esculenta* with an explanation of the taximetrics methods used. *Econ. Bot.*,
4 27(1):1–113(1973)—Illustrations.

2046 Umanah, E. E. *and* Hartmann, R. W.
Chromosome numbers and karyotypes of some *Manihot* species. *J. Amer. Soc. Hort. Sci.*, 98(3):272–
4 274(1973).

sect. **Anisophyllae** D. J. Rogers & Appan in Fl.
4 Neotrop., 13:105(1973).
sect. **Caerulescentes** D. J. Rogers & Appan *op. cit.*
4 238(1973).
sect. **Carthaginenses** D. J. Rogers & Appan *op. cit.*
4 112(1973).
sect. **Crotalariiformes** D. J. Rogers & Appan *op. cit.*
4 199(1973), as 'Crotalariaeformes'.
sect. **Foetidae** D. J. Rogers & Appan *op. cit.* 65
4 (1973).
sect. **Graciles** D. J. Rogers & Appan *op. cit.* 141
4 (1973).
sect. **Peruvianae** D. J. Rogers & Appan *op. cit.* 189
4 (1973).
sect. **Tripartitae** D. J. Rogers & Appan *op. cit.* 230
4 (1973).
sect. **Variifoliae** D. J. Rogers & Appan *op. cit.* 172
4 (1973).
alutacea D. J. Rogers & Appan *op. cit.* 130(1973)—
4 Brazil.
anomala Pohl
subsp. **cujabensis** (Mueller von Argau) D. J.
4 Rogers & Appan *op. cit.* 168(1973): *M. cujabensis.*
subsp. **glabrata** (Chodat & Hassler) D. J. Rogers &
4 Appan *op. cit.* 168(1973): *M. pubescens* f. *glabrata.*
subsp. **pavoniana** (Mueller von Argau) D. J.
4 Rogers & Appan *op. cit.* 170(1973): *M. pavoniana.*
subsp. **pubescens** (Pohl) D. J. Rogers & Appan
4 *op. cit.* 167(1973): *M. pubescens.*
caerulescens Pohl
subsp. **macrantha** (Pax & K. Hoffmann) D. J.
4 Rogers & Appan *op. cit.* 245(1973): *M. macrantha.*
subsp. **paraensis** (Mueller von Argau) D. J. Rogers
4 & Appan *op. cit.* 245(1973): *M. paraensis.*
condensata D. J. Rogers & Appan *op. cit.* 100(1973)—
4 Bolivia.

falcata D. J. Rogers & Appan *op. cit.* 125(1973)—
4 Brazil.
flemingiana D. J. Rogers & Appan *op. cit.* 143
4 (1973)—Brazil.
fruticulosa (Pax) D. J. Rogers & Appan *op. cit.*
4 149(1973): *M. triphylla* var. *fruticulosa.*
gracilis Pohl
subsp. **varians** (Pohl) D. J. Rogers & Appan *op. cit.*
4 160(1973): *M. varians.*
guaranitica Chodat & Hassler
subsp. **boliviana** (Pax & K. Hoffmann) D. J.
4 Rogers & Appan *op. cit.* 112(1973): *M. boliviana.*
irwinii D. J. Rogers & Appan *op. cit.* 137(1973)—
4 Brazil.
leptopoda (Mueller von Argau) D. J. Rogers &
4 Appan *op. cit.* 96(1973): *M. palmata* var. *leptopoda.*
maguireiana D. J. Rogers & Appan *op. cit.* 161
4 (1973)—Venezuela.
oaxacana D. J. Rogers & Appan *op. cit.* 46(1973)—
4 Mexico.
pentaphylla Pohl
subsp. **graminifolia** (Chodat & Hassler) D. J.
Rogers & Appan *op. cit.* 155(1973): *M.*
4 *graminifolia.*
subsp. **rigidula** (Mueller von Argau) D. J. Rogers
4 & Appan *op. cit.* 153(1973): *M. rigidula.*
subsp. **tenuifolia** (Pohl) D. J. Rogers & Appan
4 *op. cit.* 153(1973): *M. tenuifolia.*
quinquepartita Huber ex D. J. Rogers & Appan
4 *op. cit.* 196(1973)—Brazil.
rhomboidea Mueller von Argau
subsp. **microcarpa** (Mueller von Argau) D. J.
4 Rogers & Appan *op. cit.* 60(1973): *M. microcarpa.*
rubricaulis I. M. Johnston
subsp. **isoloba** (Standley) D. J. Rogers & Appan
4 *op. cit.* 46(1973): *M. isoloba.*
subspicata D. J. Rogers & Appan *op. cit.* 62(1973)—
4 Mexico.
surinamensis D. J. Rogers & Appan *op. cit.* 80
4 (1973)—Surinam, Venezuela, Guyana.
tomentosa Pohl
subsp. **araliifolia** (Pax) D. J. Rogers & Appan
op. cit. 215(1973), as 'araliaefolia': *Manihot*
4 *araliaefolia.*
tripartita (Sprengel) Mueller von Argau
subsp. **humilis** (Mueller von Argau) D. J. Rogers &
4 Appan *op. cit.* 236(1973): *M. humilis.*
subsp. **laciniosa** (Pohl) D. J. Rogers & Appan *op. cit.*
4 238(1973): *M. laciniosa.*
subsp. **vestita** (S. Moore) D. J. Rogers & Appan
4 *op. cit.* 236(1973): *M. tripartita* var. *vestita.*
tristis Mueller von Argau
subsp. **saxicola** (Lanjouw) D. J. Rogers & Appan
4 *op. cit.* 84(1973): *M. saxicola.*
subsp. **surumuensis** (Ule) D. J. Rogers & Appan
4 *op. cit.* 86(1973): *M. surumuensis.*
violacea Pohl
subsp. **recurvata** D. J. Rogers & Appan *op. cit.*
4 134(1973)—Brazil.
websteri D. J. Rogers & Appan *op. cit.* 72(1973),
4 as 'websterae'—Mexico.
xavantinensis D. J. Rogers & Appan *op. cit.* 124
4 (1973)—Brazil.

MANIHOTOIDES D. J. Rogers & Appan in Fl.
4 Neotrop., 13:247(1973).
pauciflora (T. S. Brandegee) D. J. Rogers & Appan
4 *op. cit.* 248(1973): *Manihot pauciflora.*

MERCURIALIS

2047 Geigy *Ltd*: Agricultural Chemicals Division
Euphorbiaceae: spurge family. *Geigy Weed Tables*,
1–7 61(1973)—Illustrations, coloured illustrations.

MICRANDRA

2048 Rodrigues, W. A. *
(*Micrandra scleroxylon*, a new Euphorbiaceous species of Brazilian Amazonia). *Acta Amazonica*, 1(3):3–8(1971)—Por(En). Abstr. in *Forest. Abstr.*
4 34(6):3284(1973).

MICRANTHEUM

2049 Anon.
The west coast *Micrantheum*. *Launceston Nat.*,
7 5(2):3(1971).

MONADENIUM

2050 Anon.
Monadenium coccineum (Euphorbiaceae). *Ashingtonia*,
5 1(3):33(1973)—En; coloured illustration.

cannellii Leach in *Garcia de Orta, Sér. Bot.*, 1(1–2):
5 35(1973)—Angola.
5 **torrei** Leach *op. cit.* 37(1973)—Mozambique.

PERA

2051 Rizzini, C. T.
Cassia spinigera et *Pera androgyna*, n.spp. (Leguminosae et Euphorbiaceae). *Rev. Brasil. Biol.*, 32(3):319–
4 322(1972)—Illustrations.

PHYLLANTHUS

2052 Rodrigues, W. A. *
Nôva *Phyllanthus* (Euphorbiaceae) da Amazônia Brasileira. *Acta Amazonica*, 1(2):17–18(1971 publ. 1972). Ref. in *Bull. Torrey Bot. Club*, 99(6):327
4 (1972).

PLATYGYNE

obovata Borhidi in *Acta Bot. Acad. Sci. Hung.*,
4 18(1–2):32(1973)—Cuba.
4 **triandra** Borhidi *op. cit.* 30(1973)—Cuba.

SAPIUM

2053 Volkart de Hualde, I.
Sapium glandulatum (Vell.) Pax, nueva especie para la flora Argentina. (*Sapium glandulatum* (Vell.) Pax, a new species for the Argentine flora). *Rev. Fac. Agron. Univ. Nac. La Plata*, 48(2):201–204(1972 publ.
4 1973)—(En); illustrations.

SINOPIMELODENDRON Tsiang in *Acta Bot.*
2 *Sin.*, 15(1):132(1973).

2054 Tsiang, Ying
(*Sinopimelodendron*, a new genus of Euphorbiaceae from Kwangsi). *Acta Bot. Sin.*, 15(1):131–135(1973)—
2 Ch, La; illustrations.

kwangsiense Tsiang in *Acta Bot. Sin.*, 15(1):133
2 (1973)—China.

TITHYMALUS

2055 Chrtek, J. *and* Křísa, B.
Eine neue ostkarpatische Art der Gattung *Tithymalus*
1 Gaertn. *Novit. Bot.*, 7–10(1972 publ. 1973).

dulcis (L.) Scop.
subsp. **purpuratus** (Thuill.) Holub in *Folia Geobot.*
1 *Phytotax.*, 8(2):174(1973): *Euphorbia purpurata.*
jasiewiczii Chrtek & Křísa in *Novit. Bot.*, 1972:
1 7(1972 publ. 1973)—Czechoslovakia.
sojakii (Chrtek & Křísa) Chrtek & Křísa in *Novit. Bot.*, 1972:9(1972 publ. 1 June 1973); also Holub in *Folia Geobot. Phytotax.*, 8(2):174(28 Aug. 1973), as
1 'sojaki': *Euphorbia austriaca* subsp. *sojakii.*

tauricola (Prochanov) Holub in *Folia Geobot. Phytotax.*, 8(2):175(1973), as 'tauricolus':
1 *Euphorbia tauricola.*
virgultosus (Klokov) Holub *op. cit.* 175(1973):
1 *Euphorbia virgultosa.*
volhynicus (Szafer, Kulcz. & Pawł.) Holub *op. cit.*
1 175(1973): *Euphorbia volhynica.*

TRAGIA

obovata (Borhidi) Borhidi in *Acta Bot. Acad. Sci.*
4 *Hung.*, 19(1–4):44(1973): *Platygyna obovata.*
triandra (Borhidi) Borhidi *op. cit.* 44(1973):
4 *Platygyna triandra.*

VERNICIA

2056 Radcliffe-Smith, A.
Typification of *Vernicia fordii* (Euphorbiaceae). *Kew*
2 *Bull.*, 28(2):296(1973).

EURYALACEAE
see
NYMPHAEACEAE

FABACEAE
see
LEGUMINOSAE

FAGACEAE

CASTANEA

2057 Browicz, K.
Kasztan jadalny (*Castanea sativa* Mill.) w Polsce. (The edible chestnut (*Castanea sativa* Mill.) in Poland). *Roczn. Sekc. Dendrol. Pol. Tow. Bot.*, 27:5–21(1973)—
1 (En); illustrations, dot map.

2058 Saucier, J. R.
American chestnut: an American wood (*Castanea dentata* Marsh. Borkh.). *U.S. Dep. Agric. F.S.*, no.230.
3 6p.(1973)—Illustrations, map.

FAGUS

2059 Grigoryan, R. A.
Novoe mestonakhozhdenie *Fagus orientalis* L. na yuzhnoĭ granitse malogo Kavkaza. (New location of *Fagus orientalis* L. on the south border of the minor
2 Caucasus). *Bot. Zhurn.*, 58(2):237–238(1973).

2060 Hanks, S. La B.
Palynotaxonomy of *Fagus* and *Nothofagus*. *Diss.*
1–4 *Abstr. Int., B*, 33(5):1959–1960(1972).

NOTHOFAGUS

2061 Hanks, S. La B.
Palynotaxonomy of *Fagus* and *Nothofagus*. *Diss.*
4, 6, 7 *Abstr. Int., B*, 33(5):1959–1960(1972).

2062 Howard, T. M. *and* Ashton, D. H.
The distribution of *Nothofagus cunninghamii* rainforest. *Proc. Roy. Soc. Victoria*, 86(1):47–75(1973)—
7 Dot maps.

QUERCUS

2063 Bărcă, C. V.
Noi stațiuni cu *Quercus frainetto* Ten. (gîrniță) în colinele Tutovei. (Nouvelles stations de *Quercus frainetto* Ten. en Colinele Tutovei-Moldavie). *Stud.*
1 *Comun., Bot.*, 751–754(1971)—(Fr).

2064 Becker, M.
Quelques observations morphologiques chez le
chêne sessile et chez le chêne pédoncule. *Bull. Soc.
Bot. France*, 119(3–4):231–236(1972 publ. 1973)—
1 (En). *Quercus sessiliflora, Q. pedunculata.*

2065 Cooper, G. A. *and* Watt, R. F.
Oak: an American wood. *U.S. Dep. Agric. F.S.*,
3 247. 9p.(1973)—Illustrations, maps.

2066 Cunliff, K. M.
Trees for tomorrow. *Trees S. Afr.*, 25(1):24–27
1 (1973)—Illustrations. *Quercus canariensis.*

2067 Globa-Mikhaĭlenko, D. A.
K formoobrazovaniyu v rode *Quercus*. *Byull.
Glavn. Bot. Sada* (Moscow), 88:35–38(1973)—
1 Illustrations.

2068 L'vov, P. L.
Dubovȳe lesa i redkoles'ya iz *Quercus pubescens*
Willd. v Dagestane. (Oak forests and redkolesjas of
Quercus pubescens Willd. in Daghestan). *Bot. Zhurn.*,
1 59(1):53–61(1973).

2069 Menitskiĭ, Yu. L.
Dubȳ (*Quercus* L.) podroda *Quercus* vostochnoĭ
Azii. (Conspectus specierum generis *Quercus* L.
(subgen. *Quercus*) Asiae orientalis). *Novosti Sist.
2 Vȳssh. Rast.*, 10:107–134(1973)—Key.

2070 Mukherjee, B. B.
Pollen morphological studies of *Quercus* species of
eastern Himalayas. *Proc. Indian Sci. Congr.*, 60(3):
6 368(1973). Abstract only.

2071 Sabato, S.
Considerazioni sul significato fitogeografico ed
ecologico di *Quercus coccifera* L. s.l. nel Salento
(Puglia.) *Webbia*, 27(2):517–549(1972 publ. 1973)—
1 (En); maps, dot map.

2072 Sanda, V., Popescu, A., Dihoru, G. *and* Roman, N.
Răspîndirea speciei *Quercus pendunculiflora* C. Koch
în România. *Stud. Cerc. Biol., Bot.*, 24(6):511–524
1 (1972)—(En); map, dot map.

2073 Stefanoff, B.
Neueres zur Herkunft und Taxonomie der
Quercus thracica Stef. et Ned. *Acta Bot. Acad. Sci. Hung.*,
1 19(1–4):343–357(1973)—(En); illustrations.

2074 Trede, R.
Uber die Nomenklatur einiger immergrüner und
halbimmergrüner Eichen des südlichen Mediterrange-
bietes. (Sur la nomenclature de quelques chênes à
feuilles caduques et sémicaduques de la région méditer-
ranéenne méridionale). *Mitt. Deutsch. Dendrol. Ges.*,
no.66:89–97(1973)—(Fr); illustration. *Quercus
1 lusitanica, Q. canariensis, Q. fruticosa, Q. infectoria.*

2075 Tucker, J. M. *and* Bogert, B.
Analysis of a progeny test of a hybrid oak, *Quercus
gambelii* × *Q. turbinella*. *Madroño*, 22(1):1–9(1973)—
3 Illustrations.

acutissima Carruth.
var. **chenii** (Nakai) Menitskiĭ in *Novosti Sist.
2 Vȳssh. Rast.*, 10:119(1973): *Q. chenii.*
subsp. **kingii** Menitskiĭ *op. cit.* 120(1973); nom.
nov.: *Q. acutissima* subsp. *roxburghii* (Endl.)
6 Camus.
dentata Thunb.
subsp. **yunnanensis** (Franch.) Menitskiĭ *op. cit.*
2 117(1973): *Q. yunnanensis.*
lanata Smith
subsp. **leiocarpa** (A. Camus) Menitskiĭ *op. cit.* 126
6 (1973): *Q. lanata* var. *leiocarpa.*
lusitanica Lamk.
subsp. alpestris (Bss.) Malagarriga, Las Subesp. y
Variac. Geogr.: 4(1973), without basionym
1 date: *Q. alpestris.*
mongolica Fisch. ex Ledeb.
subsp. **crispula** (Blume) Menitskiĭ in *Novosti Sist.
2 Vȳssh. Rast.*, 10:114(1973): *Q. crispula.*
phillyraeoides A. Gray
subsp. **fokiensis** (Nakai) Menitskiĭ *op. cit.* 123
2 (1973): *Q. fokiensis.*
var. **wrightii** (Nakai) Menitskiĭ *op. cit.* 123(1973):
2 *Q. wrightii.*
setulosa Hickel & A. Camus
subsp. **laotica** (A. Camus) Menitskiĭ *op. cit.* 128
6 (1973): *Q. laotica.*

FICOIDACEAE
see
AIZOACEAE

FLACOURTIACEAE

2076 Keating, R. C.
Pollen morphology and relationships of the Flacour-
tiaceae. *Ann. Missouri Bot. Gard.*, 60(2):273–305
2, 4–7 (1973).

2077 Wild, H. *and* Vidigal, M. P.
Flora de Moçambique: 16. Flacourtiaceae. 1973.
5 41p.—Illustration; key.

CALANTICA

2078 Sleumer, H.
Révision du genre *Calantica* Tul. (Flacourtiacées).
Adansonia, 12(4):539–544(1972 publ. 1973)—(En);
5 illustrations; key.

capuronii Sleumer in *Adansonia*, 12(4):543(1972 publ.
5 1973)—Malagasy.
chauvetiae Sleumer *op. cit.* 543 (1972 publ. 1973)—
5 Malagasy.

HOMALIUM

2079 Sleumer, H.
Révision du genre *Homalium* Jacq. (Flacourtiacées)
en Afrique (y compris Madagascar et les Mascareignes).
Bull. Jard. Bot. Nation. Belg., 43(3–4):239–328(1973)—
5 (En); keys.

sect. **Rhodonisa** Sleumer in *Bull. Jard. Bot. Nation.
5 Belg.*, 43(3–4):300(1973).
brachyrhachis Sleumer *op. cit.* 261(1973)—
5 Malagasy.
5 **capuronii** Sleumer *op. cit.* 319(1973)—Malagasy.
fulviflorum Sleumer *op. cit.* 310(1973)—Gabon,
5 Nigeria.
graciliflorum Sleumer *op. cit.* 321(1973)—
5 Malagasy.
5 **gracilipes** Sleumer *op. cit.* 269(1973)—Tanzania.
5 **intercedens** Sleumer *op. cit.* 300(1973)—Malagasy.
5 **myrtifolium** Sleumer *op. cit.* 258(1973)—Malagasy.
nudiflorum (DC.) Baill.
var. **ciliolatum** Sleumer *op. cit.* 299(1973)—
5 Malagasy.
perrieri Sleumer *op. cit.* 257(1973), nom. nov.:
5 *H. buxifolium* Perr.
5 **pulchrum** Sleumer *op. cit.* 320(1973)—Malagasy.
5 **retivenium** Sleumer *op. cit.* 264(1973)—Malagasy.

5 **rubriflorum** Sleumer *op. cit.* 302(1973)—Malagasy.
trigynum (Baker) Sleumer *op. cit.* 316(1973):
5 *Weinmannia trigyna.*

IDESIA

2080 Meikle, R. D.
Idesia polycarpa. Curtis's Bot. Mag., 179(4):tab.649
2 (1973)—En; illustration, coloured illustration.

2081 Meikle, R. D.
Proposal for the conservation of the generic name 5331 *Idesia* Maximowicz (1866) (Flacourtiaceae) against *Idesia* Scopoli (1777). *Taxon*, 22(5–6):694–
2 695(1973).

SCAPHOCALYX

2082 Heel, W. A. van
Flowers and fruits in Flacourtiaceae: 1. *Scaphocalyx spathacea* Ridl. *Blumea*, 21(2):259–279(1973)—
6 Illustrations.

SCOTTELLIA

2083 Sleumer, H.
A taxonomic revision of the genus *Scottellia* Oliv. (Flacourtiaceae). *Blumea*, 20(2):275–281(1972 publ.
5 1973)—Map; key.

FLAGELLARIACEAE

JOINVILLEA

2084 Degener, O. *and* Degener, I.
Flora Hawaiiensis: new illustrated flora of the Hawaiian Islands. Joinvilleaceae: *Joinvillea gaudi-*
6 *chaudiana.* 1973.[4]p.—Illustrations.

FOUQUIERIACEAE

2085 Henrickson, J.
World pollen and spore flora: 1. Angiospermae: Fouquieriaceae DC. Stockholm, Almqvist &
3, 4 Wiksell, 1973. 13p.—Chromosome numbers; key.

FRANKENIACEAE

FRANKENIA

2086 Roberts, R. H.
The sea heath in Anglesey. *B.S.B.I. Welsh Region*
1 *Bull.*, no.15:6(1971)—*Frankenia laevis.*

2087 Turner, B. L.
A new species of *Frankenia* (Frankeniaceae) from gypseous soil of north central Mexico. *Sida*, 5(2):
4 132–135(1973)—Illustration.

leverichii B. L. Turner in *Sida*, 5(2):132(1973)—
4 Mexico.

GENTIANACEAE

2088 Hall, K. S.
George Forrest's gentians and their hybrids. *J. Scott. Rock Gard. Club*, 13(3):179–183(1973)—
2 Illustration.

2089 Jossang, P., Carbonnier, J. *and* Molho, D.
Les xanthones des Gentianacées. Essai de taxinomie numérique au niveau moléculaire. *Trav. Lab. La Jaysinia*, no.4:143–167(1972 publ. 1973)—(En).
1–7 *Gentiana, Swertia, Frasera, Causcora.*

2090 Lebreton, P. *and* Dangy-Caye, M. P.
Contribution biochimique à l'étude taxonimique des Gentianacées. *Pl. Méd. Phytothér.*, 7(2):87–94
1–7 (1973)—(En).

2091 Taylor, A. W.
Gentians from the Himalayas. *Quart. Bull. Alp. Gard. Soc.*, 41(1):20–27(1973)—Illustrations. Popular
2, 6 horticultural notes.

2092 Weaver, R. E.
In search of tropical gentians. *Arnoldia*, 33(3):189–
3, 4 198(1973)—Illustrations.

CALATHIANA

angulosa (M. Bieb.) Holub in *Folia Geobot. Phytotax.*,
1, 2 8(2):169(1973): *Gentiana angulosa.*
bavarica (L.) Holub *op. cit.* 169(1973): *Gentiana*
1 *bavarica.*
brachyphylla (Vill.) Holub *op. cit.* 169(1973):
1 *Gentiana brachyphylla.*
nevadensis (Soltok.) Holub *op. cit.* 169(1973):
4 *Gentiana nevadensis.*
orbicularis (Schur) Holub *op. cit.* 169(1973):
1 *Gentiana orbicularis.*
pontica (Soltok.) Holub *op. cit.* 169(1973): *Gentiana*
2 *pontica.*
pumila (Jacq.) Holub *op. cit.* 169(1973): *Gentiana*
1 *pumila.*
rostanii (Reut.) Holub *op. cit.* 170(1973): *Gentiana*
1 *rostanii.*
schleicheri (Vacc.) Holub *op. cit.* 170(1973):
1, 2 *Gentiana verna* f. *schleicheri.*
tergestina (G. Beck) Holub *op. cit.* 170(1973):
1, 2 *Gentiana angulosa* f. *tergestina.*
terglouensis (Hacq.) Holub *op. cit.* 170(1973):
1 *Gentiana terglouensis.*
utriculosa (L.) Holub *op. cit.* 170(1973): *Gentiana*
1, 2 *utriculosa.*
1, 2 **verna** (L.) Holub *op. cit.* 170(1973): *Gentiana verna.*

CENTAURIUM

2093 Fijałkowski, D. *and* Taranowska, B.
Wstępne badania nad uprawą *Centaurium umbell-*
1 *atum* Gilib. *Wiad. Bot.*, 17(1):47–48(1973).

pulchellum (Sw.) Druce
var. caspica (Griseb.) R. R. Stewart, Annot. Cat. Vasc. Pl. W. Pakistan & Kashmir: 552(1972),
6 without basionym ref.: (Basionym not stated).

CIMINALIS

alpina (Vill.) Holub in *Folia Geobot. Phytotax.*, 8(2):
1, 2 171(1973): *Gentiana alpina.*
angustifolia (Vill.) Holub *op. cit.* 171(1973): *Gentiana*
1, 3 *angustifolia.*
clusii (Perr. & Song.) Holub *op. cit.* 171(1973):
1, 2 *Gentiana clusii.*
dinarica (G. Beck) Holub *op. cit.* 171(1973): *Gentiana*
1 *acaulis* var. *dinarica.*
ligustica (R. Vilmorin & Chopinet) Holub *op. cit.*
1 171(1973): *Gentiana ligustica.*
occidentalis (Jakowatz) Holub *op. cit.* 171(1973):
1–3 *Gentiana occidentalis.*

CURTIA

2094 Fabris, H. A.
Un nuevo genero de Gentianaceae para la Argentina: *Curtia* Cham. et Schlecht. *Bol. Soc.*
4 *Argent. Bot.*, 15(1):130–131(1973)—(En).

FAROA

2095 Taylor, P.
Gentianaceae: 190–207. *Distrib. Pl. Afr.*, 7.18p.
5 (1973)—maps, dot maps. *Faroa* spp. only.

2096 Taylor, P.
A revision of the genus *Faroa* Welwitsch. *Garcia de Orta, Sér. Bot.*, 1(1–2):69–82(1973)—(Por,Fr);
5 keys.

alata P. Taylor in *Garcia de Orta, Sér. Bot.*, 1(1–2):
5 75(1973)—Zambia.
5 **corniculata** P. Taylor *op. cit.* 76(1973)—Zambia.
5 **hutchinsonii** P. Taylor *op. cit.* 76(1973)—Zambia.
5 **minutiflora** P. Taylor *op. cit.* 79(1973)—Zambia.
5 **richardsiae** P. Taylor *op. cit.* 73(1973)—Tanzania.

FRASERA

2097 Beattie, A. J., Breedlove, D. E. *and* Ehrlich, P. R.
The ecology of the pollinators and predators of *Frasera speciosa*. *Ecology*, 54(1):81–91(1973)—
3, 4 Illustration.

GENTIANA

2098 Atlas rozšíření rostlin v jižní části Čech: 1. (Atlas der Pflanzenverbreitung im südlichen Teile Bohmens: 1). *Sborn. Jihocesk Muz. Cesk. Budejovicich Přír. Ved.*,
1 12 Suppl.3. 44p.(1972)—(Ge); dot maps.

2099 Carbonnier, J., Massias, M., Jarreau-Carbonnier, M.-C. *and* Molho, D.
Comparaison du point de vue chimiotaxinomique entre le *Gentiana ciliata* L. et quatre unités botaniques du type *Gentiana acaulis* L. Xanthones, triterpènes et taxinomie numérique. *Trav. Lab. La Jaysinia*, no.4:
1 169–194(1972 publ. 1973)—(En); illustrations.

2100 Kozlovskaya, N. V. *and* Simonovich, L. G.
Gorechavka vesennyaya—*Gentiana verna* L. vo flore Belorussii. *Botanika* (Minsk), vўp.15:191–192
1 (1973).

2101 Ohba, H.
A taxonomic note on *Gentiana nipponica* Maxim.
2 var. *robusta* Hara. *J. Jap. Bot.*, 48(12):373(1973).

2102 Orden, C. van *and* Ommen, F. van
Gentianen in Noord-Holland. *Natura* (Netherlands),
1 70(7):101–103(1973)—Illustrations.

2103 Pohl, K.
Enzianvorkommen im Lahn-Dillgebiet. *Hess. Flor.*
1 *Briefe*, 256–258:30–31(1973).

2104 Poldini, L.
Gentiana froelichii Jan anche nelle Alpi Giulie. (*Gentiana froelichii* Jan a new species for the Julian Alps). *Giorn. Bot. Ital.*, 107(1):29–36(1973)—(En,Ge);
1 maps.

2105 Slavík, B.
Gentiana asclepiadea L. v povodí Jizery. (*Gentiana asclepiadea* L. im Einzugsgebiet des Flusses Jizera (Nordböhmen)). *Zprávy Česk. Bot. Spol.*, 8(1):19–27
1 (1973)—(Ge); dot map.

pneumonanthe L.
subsp. depressa (Bss.) Malagarriga, Las Subesp. y Variac. Geogr.: 13(1973), without basionym
1 date: *G. pneumonanthe* var. *depressa*.
verna L.
subsp. pumila (Jacq.) Malagarriga *op. cit.* 13(1973),
1 without basionym date: *G. pumila*.

GENTIANELLA

longicarpa (Gilli) Holub in *Folia Geobot. Phytotax.*,
2 8(2):171(1973): *Gentiana longicarpa*.
minutissima (Boiss.) Holub *op. cit.* 171(1973):
2 *Gentiana minutissima*.
sibirica (Kuzn.) Holub *op. cit.* 171(1973): *Gentiana*
2 *aurea* var. *sibirica*.
stoliczka (Clarke) Holub *op. cit.* 171(1973): *Gentiana*
1, 2 *stoliczka*.

GENTIANOPSIS

stracheyi (Clarke) Holub in *Folia Geobot. Phytotax.*,
1 8(2):176(1973): *Gentiana detonsa* var. *stracheyi*.
stricta (Klotzsch) Holub *op. cit.* 176(1973): *Gentiana*
1, 3 *stricta*.

HOPPEA

2106 Kshetrapal, S.
Vascular anatomy of the node and flower of *Hoppea dichotoma* Willd. *Bot. Gaz.*, 134(1):1–4
6 (1973).

2107 Tiagi, Y. D. *and* Kshetrapal, S.
Vascular anatomy of the node and flower of *Hoppea dichotoma* Willd. *Proc. Indian Sci. Congr.*,
6 60(3):298(1973). Abstract only.

PURDIEANTHUS

splendens (Hook.) Cuatrecasas in *Ciencia*, 27(6):175
4 (1972): *Lisianthus splendens*.

SWERTIA

2108 Geesink, R.
A synopsis of the genus *Swertia* (Gent.) in Malesia.
6 *Blumea*, 21(1):179–183(1973)—Key.

2109 Natarajan, P. N. *and* Prasad, S.
Pharmacognostical studies on *Swertia alata*. *Proc.*
6 *Indian Acad. Sci., B*, 76(4):171–180(1972).

javanica Bl.
subsp. **coerulescens** (Zoll. & Mor.) Geesink in
6 *Blumea*, 21(1):181(1973): *Ophelia coerulescens*.
6 subsp. **confusa** Geesink *op. cit.* 181(1973)—Java.
subsp. **steenisii** (Geesink) Geesink *op. cit.* 181
6 (1973): *S. steenisii*.
subsp. **nana** Geesink *op. cit.* 181(1973)—Lesser
6 Sunda Is.
subsp. **sumatrensis** Geesink *op. cit.* 182(1973)—
6 Sumatra.
oxyphylla (Miq.) Gilg
var. **parvula** Geesink *op. cit.* 180(1973)—Lesser
6 Sunda Is.
piloglandulosa Geesink
subsp. **biovulata** Geesink *op. cit.* 182(1973)—
6 Sumatra.

GERANIACEAE

BALBISIA

2110 Xifreda, C. C.
Anatomia foliar de *Wendtia* y *Balbisia* (Geraniaceae).
4 *Kurtziana*, 7:213–232(1973)—(En); key to species.

ERODIUM

2111 Geigy *Ltd*: Agricultural Chemicals Division
Geraniaceae: Geranium family. *Geigy Weed Tables*,
1–3, 7 57(1972)—Illustrations, coloured illustrations.

2112 Guittonneau, G. G., Roux, M. *and* Uzan, J.
Essai d'application de l'analyse factorielle des correspondances à la subdivision du genre *Erodium* l'Hér. *Bull. Soc. Bot. France*, 118(9):719–731(1971
1–4, 7 publ. 1973)—(En).

2113 Mitsevski, K.
Erodium guicciardii Heldr. nov vid za florite na Makedonija i Jugoslavija. (*Erodium guicciardii* Heldr.—eine neue Art für die Flora von Mazedonien und Jugoslawien). *God. Zborn. Biol.*, 25:161–163(1973)—
1 (Ge); map.

2114 Montserrat, P.
Deux *Erodium* et un *Reseda* nouveaux de la province de Huesca, Espagne. *Bull. Soc. Bot. France*, 120(1–2):43–
1 48(1973)—(En); illustrations, chromosome numbers.

gaussenianum Montserrat in *Bull. Soc. Bot. France*,
1 120(1–2):43(1973)—Spain.
sanguis-christi Sennen
subsp. **durrieui** Montserrat *op. cit.* 45(1973)—
1 Spain.

GERANIUM

2115 Abbà, G.
Ad floram italicam notulae taxonomicae et geobotanicae: 11. *Geranium sibiricum* L. in Italia.
1 *Webbia*, 28(1):53–56(1973)—It(En); illustrations.

2116 Bate-Smith, E. C.
Chemotaxonomy of *Geranium*. *Bot. J. Linn. Soc.*,
1–7 67(4):347–359(1973)—Chromosome numbers.

2117 Geigy *Ltd*: Agricultural Chemicals Division
Geraniaceae: Geranium family. *Geigy Weed Tables*,
1–3, 7 58(1972)—Illustrations, coloured illustrations.

2118 Vivant, J.
L'ecologie du *Geranium endressi* Gay: à propos de la découverte d'une localité remarquable. *Bull. Soc. Bot.*
1 *France*, 118(9):823–825(1971 publ. 1973)—(En).

2119 Yeo, P. F.
The biology and systematics of *Geranium*, sections *Anemonifolia* Knuth and *Ruberta* Dum. *Bot. J. Linn. Soc.*, 67(4):285–346(1973)—Illustrations; chromosome
1 numbers; key.

2120 Yeo, P. F.
Geranium procurrens. *Curtis's Bot. Mag.*, 179(3):
6 tab.644(1973)—En; illustration, coloured illustration.

procurrens Yeo in *Curtis's Bot. Mag.*, 179(3):t.644
6 (1973)—Nepal, Sikkim.
tuberosum L.
subsp. **linearilobum** (DC) Malagarriga, Las Subesp. y Variac. Geogr.: 12(1973): *G.*
1 *linearilobum*.
subsp. **macrostylum** (Bss.) Malagarriga *op. cit.*
1 12(1973): *G. macrostylum*.
subsp. **malviflorum** (Bss. R.) Malagarriga *op. cit.*
1 11(1973): *G. malviflorum*.

PELARGONIUM

2121 Vorster, P.
Pelargonium pulchellum. *Flow. Pl. Afr.*, 42(3–4):
5 pl.1669(1973)—En; coloured illustration.

WENDTIA

2122 Xifreda, C. C.
Anatomia foliar de *Wendtia* y *Balbisia* (Geraniaceae).
4 *Kurtziana*, 7:213–232(1973)—(En); key to species.

GESNERIACEAE

2123 Skog, L. E.
A study of the tribe Gesneriaceae and a revision of part of *Gesneria* L. (Gesneriaceae-Gesnerioideae). *Diss. Abstr. Int., B*, 33(10):4694–4695(1973)—
4 Chromosome numbers.

2124 Wiehler, H.
Collecting Gesneriaceae in Ecuador and Panama:
4 part 2. *Gloxinian*, 23(4):21–24(1973)—Illustrations.

× **ACHIMENANTHA** H. E. Moore in *Baileya*,
4 19(1):35(1973). *Achimenes* × *Smithiantha*.
naegelioides (Van Houtte) H. E. Moore *op. cit.* 36
4 (1973): *Plectopoma* × *naegelioides*.

ALLOPLECTUS

2125 Steyermark, J. A.
Descripcion detallada de dos especies raras de Venezuela. *Pittieria*, no. 4:27–32(1972)—En(Sp);
4 illustrations.

2126 Teuscher, H.
Alloplectus teuscheri. *Gloxinian*, 23(3):13–14(1973)—
4 En; illustrations.

CAMPANEA

2127 Morton, C. V.
Lectotypification of *Isoloma* and *Campanea* (Gesn.).
4 *Taxon*, 22(2–3):317–318(1973).

CHIRITA

2128 Batcheller, F. N.
Chirita micromusa. *Gloxinian*, 23(2):3(1973)—En;
6 coloured illustration.

2129 Theobald, W. L. *and* Grupe, D. A.
A new name and combinations in the Gesneriaceae. *Ceylon J. Sci., Biol. Sci.*, 10(1):70–71(1972). (Critical
6 Notes on Ceylon Plants 7).

angusta (C. B. Cl.) Theobald & Grupe in *Ceylon J. Sci., Biol. Sci.*, 10(1):70(1972): *C. zeylanica* var.
6 *angusta*.
walkeri
subsp. **parviflora** (C. B. Cl.) Theobald & Grupe
6 *op. cit.* 70(1972): *C. walkeri* var. *parviflora*.

CODONANTHE

2130 Moore, H. E.
A synopsis of the genus *Codonanthe* Gesneriaceae. *Baileya*, 19(1):4–33(1973)—Illustrations; chromosome
4 numbers; key.

sect. **Codonanthopsis** (Mansfeld) H. E. Moore in
4 *Baileya*, 19(1):25(1973): genus *Codonanthopsis*.
4 **dissimulata** H. E. Moore *op. cit.* 25(1973)—Peru.
ulei (Mansfeld) H. E. Moore *op. cit.* 28(1973):
4 *Codonanthopsis ulei*.

COLUMNEA

2131 Morley, B.
Ecological factors of importance to *Columnea* taxonomy. *In* Heywood, V. H., *ed.* Taxonomy and
4 ecology: 265–281(1973).

2132 Morley, B.
Materials for a treatment of *Columnea* in Panama. *Ann. Missouri Bot. Gard.*, 60(2):449–460(1973)—
4 Illustrations.

consanguinea Hanst.
var. **adpressa** B. Morley in *Ann. Missouri Bot.*
4 *Gard.*, 60(2):451(1973)—Panama.
var. **darienensis** (Morton) B. Morley *op. cit.* 451
4 (1973): *C. darienensis.*
4 **cruenta** B. Morley *op. cit.* 458(1973)—Panama.
hirta Kl. & Hanst.
var. **mortonii** (Raymond) B. Morley *op. cit.* 455
4 (1973): *C. mortonii.*
4 **mira** B. Morley *op. cit.* 457(1973)—Panama.
4 **segregata** B. Morley *op. cit.* 459(1973)—Panama.
silvarum Morton
var. **butcheri** B. Morley *op. cit.* 457(1973)—
4 Panama.
tomentulosa Morton
var. **tulae** (Urban) B. Morley *op. cit.* 456(1973):
4 *C. tulae.*

CYRTANDRA

2133 Burtt, B. L. *and* Bokhari, M. H.
Studies in the Gesneriaceae of the Old World: 36. Foliar sclereids in New Guinea and Pacific *Cyrtandra*. *Notes Roy. Bot. Gard. Edinburgh*, 32(3):397–402
6 (1973).

2134 Gillett, G. W.
The genus *Cyrtandra* (Gesneriaceae) in Fiji—a
6 supplement. *Taxon*, 22(5–6):701(1973).

2135 Gillett, G. W.
The genus *Cyrtandra* (Gesneriaceae) in the South Pacific. *Univ. Calif. Publ. Bot.*, 66. 59p.(1973)—
2, 6 Illustrations, maps; key.

2136 Gillett, G. W.
The genus *Cyrtandra* in the Ryukyu and Caroline Islands. *J. Arnold Arbor.*, 54(1):105–110(1973)—
6 Key.

aurantiicarpa G. W. Gillett in *Univ. Calif. Publ.*
6 *Bot.*, 66:29(1973)—Samoa.
jonesii (F. Brown) G. W. Gillett *op. cit.* 55(1973),
6 without basionym page: *Cyrtandroidea jonesii.*
mooreaensis G. W. Gillett *op. cit.* 50(1973)—
6 Society Is.
natewaensis G. W. Gillett in *Taxon*, 22(5–6):701
6 (1973), nom. nov.: *C. attenuata* G. W. Gillett.
neo-hebridensis G. W. Gillett in *Univ. Calif. Publ.*
6 *Bot.*, 66:18(1973)—New Hebrides.
obovata G. W. Gillett *op. cit.* 17(1973)—New
6 Hebrides.
tohiveaensis G. W. Gillett *op. cit.* 41(1973)—
6 Society Is.

EPITHEMA

2137 Hallé, F. *and* Delmotte, A.
Croissance et floraison de la Gesnériacée africaine *Epithema tenue* C. B. Clarke. *Adansonia*, 13(3):273–
5 287(1973)—(En); illustrations, dot map.

GESNERIA

bonaoana Alain in *Phytologia*, 25(5):275(1973)—
4 Hispaniola (Dominican Republic).
sylvicola Alain *op. cit.* 276(1973)—Hispaniola
4 (Dominican Republic).

HABERLEA

2138 Stanev, S.
Kŭm istoriyata na Bŭlgarskite reliktni endemiti: 2. *Haberlea rhodopensis* Friv. (On the history of the Bulgarian relict endemics: 2. *Haberlea rhodopensis* Friv.). *Izv. Prir. Muz. Plovdiv*, 2:31–42(1972)—
1 (Rus, En); map.

ISOLOMA

2139 Morton, C. V.
Lectotypification of *Isoloma* and *Campanea* (Gesn.).
4 *Taxon*, 22(2–3):317–318(1973).

LYSIONOTUS

atropurpureus Hara in *J. Jap. Bot.*, 48(12):359
6 (1973)—India, Nepal.

NEGRIA

2140 Green, P. S.
Negria rhabdothamnoides. Curtis's Bot. Mag., 179(4):
7 tab.659(1973)—En; illustration, coloured illustration.

NEMATANTHUS

2141 Moore, H. E.
Comments on cultivated Gesneriaceae. *Baileya*,
0 19(1):35–41(1973)—Key to species of *Nematanthus.*

2142 Moore, H. E.
Nematanthus fritschii versus *Nematanthus mattosianus*
4 (Gesneriaceae). *Baileya*, 19(1):34(1973).

nervosus (Fritsch) H. E. Moore in *Baileya*, 19(1):
4 38(1973): *Hypocyrta nervosa.*
perianthomegus (Vellozo) H. E. Moore *op. cit.*
4 38(1973): *Orobanche perianthomega.*
radicans (Klotzsch & Hanstein) H. E. Moore *op. cit.*
4 38(1973): *Hypocyrta radicans.*
selloanus (Klotzsch & Hanstein) H. E. Moore *op. cit.*
4 38(1973): *Hypocyrta selloana.*
strigillosus (Martius) H. E. Moore *op. cit.* 38(1973):
4 *Hypocyrta strigillosa.*
wettsteinii (Fritsch) H. E. Moore *op. cit.* 38(1973):
4 *Hypocyrta wettsteinii.*

RAMONDA

2143 Velchev, V., Ĭordanov, D. *and* Ganchev, S.
Prouchvane na *Ramonda serbica* Panč v Bŭlgariya. (Untersuchung an *Ramonda serbica* Panč in Bulgarien). *Izv. Bot. Inst.* (Sofia), 24:139–167(1973)—(Rus,Ge);
1 coloured illustrations, map.

RHABDOTHAMNUS

2144 Lowry, J. B.
Rhabdothamnus solandri: some phytochemical
7 results. *New Zealand J. Bot.*, 11(3):555–560(1973).

RHYNCHOGLOSSUM

2145 Theobald, W. L. *and* Grupe, D. A.
A new name and combinations in the Gesneriaceae. *Ceylon J. Sci., Biol. Sci.*, 10(1):70–71(1972). (Critical
6 Notes on Ceylon Plants 7).

gardneri Theobald & Grupe in *Ceylon J. Sci., Biol.*
6 *Sci.*, 10(1):70(1972), nom. nov.: *Klugia zeylanica.*

RHYTIDOPHYLLUM

2146 Weber, A.
Über die 'Stipeln' von *Rhytidophyllum* Mart. (Gesneriaceae). *Österr. Bot. Zeitschr.*, 121(5):279–283
4 (1973)—(En); illustration.

mogoticola Borhidi & Muñiz in *Acta Bot. Acad. Sci.*
4 *Hung.*, 18(1–2):41(1973)—Cuba.

SAINTPAULIA

2147 Batcheller, F. N.
Gesneriads one by one—*Saintpaulia. Gloxinian*,
5 23(6):5–12(1973)—Illustrations.

SEEMANNIA

2148 Morley, B.
Seemannia latifolia. *J. Roy. Hort. Soc.*, 98(9):414
4 (1973)—En; coloured illustration.

SINNINGIA

cardinalis (Lehmann) H. E. Moore in *Baileya*, 19(1):
4 39(1973): *Gesneria cardinalis.*
claybergiana H. E. Moore *op. cit.* 39(1973), nom.
4 nov.: *Gesneria lindleyi* W. Hooker.
leucotricha (Hoehne) H. E. Moore *op. cit.* 39(1973):
4 *Rechsteineria leucotricha.*
macropoda (Sprague) H. E. Moore *op. cit.* 39(1973):
4 *Corytholoma macropodum.*
× **rosea** (H. E. Moore & R. G. Wilson) H. E. Moore
4 *op. cit.* 40(1973): × *Gloxinera rosea.*
tuberosa (Martius) H. E. Moore *op. cit.* 40(1973):
4 *Gesnera tuberosa.*
verticillata (Vellozo) H. E. Moore *op. cit.* 40(1973):
4 *Orobanche verticillata.*
warszewiczii (Bouché & Hanstein) H. E. Moore
4 *op. cit.* 40(1973): *Gesnera warszewiczii.*

TITANOTRICHUM

2149 Batcheller, F. N.
Gesneriads one by one: *Titanotrichum. Gloxinian*,
2 22(3):18–20(1973)—Illustration.

GOODENIACEAE

2150 Ali, S. I.
Flora of West Pakistan: 30. Goodeniaceae, 1972.
6 3p.—Illustrations, map.

2151 Reitz, R. *and* Klein, R. M.
Flórula da Ilha de Santa Catarina: Goodeniáceas.
4 1972. 10p.—Coloured illustration.

SCAEVOLA

2152 Gillett, G. W.
Hybridization and its taxonomic implications in the *Scaevola gaudichaudiana* complex of the Hawaiian Islands. *In* Kay, E. A., *ed.* A natural history of the Hawaiian Islands: 535–545(1972)—Illustrations, dot
6 maps.

2153 Kugler, H.
Zur Bestäubung von *Scaevola plumieri* (L.) Vahl und *Ipomoea pes-caprae* Sweet, zwei tropischen Strandpflanzen. (On the pollination of *Scaevola plumieri* (L.) Vahl and *Ipomoea pes-caprae*, two plants of tropical beaches). *Flora* (Germany), 162(4):381–391
3–7 (1973)—(En); illustrations.

GRAMINEAE

2154 Anton, A. M. *and* Astegiano, M. E.
Notas sobre la morfologia floral de Gramineas.
1–7 *Kurtziana*, 7:49–53(1973)—(En); illustrations.

2155 Blake, S. T.
Taxonomic and nomenclatural studies in the Gramineae: no.3. *Proc. Roy. Soc. Queensl.*, 84:61–70
7 (1972 publ. 1973). In Australia.

2156 Bor, N. L.
Bunches of grass. *J. Indian Bot. Soc.*, 50A:90–95
6 (1971?)—Illustrations.

2157 Bor, N. L.
Two new species of Gramineae from Asia. *Bot.*
6 *Tidsskr.*, 67(4):324–326(1973).

2158 Butzin, F.
Die Namen der supragenerischen Einheiten der Gramineae (Poaceae). (The names of the suprageneric units of Gramineae (Poaceae)). *Willdenowia*, 7(1):
1–7 113–168(1973)—(En).

2159 C., M.
New grasses added to the species list for Tasmania
7 February 1972. *Launceston Nat.*, 5(5):2(1972).

2160 Carolin, R. C., Jacobs, S. W. L. *and* Vesk, M.
The structure of the cells of mesophyll and parenchymatous bundle sheath of the Gramineae.
1–7 *Bot. J. Linn. Soc.*, 66(4):259–275(1973).

2161 Chopanov, P.
Opredelitel' zlakov Turkmenii. Ashkhabad,
2 Ȳlȳm, 1973. 133p.—Illustrations; keys.

2162 Chopanov, P. *and* Yurtsev, V. N.
Khromosomnȳe chisla nekotorȳkh zlakov Turkmenii. (Chromosome numbers of some grasses
2 of Turkmenia). *Bot. Zhurn.*, 58(2):301–302(1973).

2163 Clayton, W. D. *and* Richardson, F. R.
Studies in the Gramineae: 32. The tribe Zoysieae
2, 4–7 Miq. *Kew Bull.*, 28(1):37–48(1973)—Key.

2164 Clayton, W. D.
Studies in the Gramineae: 33. The awnless genera of
2, 4–7 Andropogoneae. *Kew Bull.*, 28(1):49–57(1973).

2165 Clayton, W. D.
Errata. (Studies in the Gramineae: 31. Andropogoneae.) *Kew Bull.*, 28(1):58(1973)—Corrections to
2, 4–7 *Kew Bull.*, 27(3):457–474(1972).

2166 Danilova, M. F. *and* Sokolovskaya, T. B.
Anatomiya prorostka nekotorȳkh vidov zlakov i vopros o prirode odnodol'nosti. (Seedling anatomy of some grass species and the problem of the origin of monocotyledonity). *Bot. Zhurn.*, 58(3):337–349
1–7 (1973)—(En).

2167 Delay, T. *and* Petit, D.
Chromosomes surnuméraires chez des Graminées. *Inform. Ann. Caryosyst. Cytogénét.*, no.6:1–4(1972
1, 2 publ. 1973)—Illustration; chromosome numbers.

2168 De Wet, J. M. J., Harlan, J. R. *and* Kurmarohita, B.
Origin and evolution of Guinea sorghums. *E. Afr. Agric. Forest. J.*, 38(2):114–119(1972)—
5 Illustrations, dot map.

2169 Dubovik, O. N.
Suchashȳĭ stan filohenetȳchnoyi sȳstematȳkȳ zlakiv. (Modern state of phylogenetic taxonomy of grasses). *Ukr. Bot. Zhurn.*, 30(5):545–561(1973)—
1–7 Uk(En).

2170 Exell, A. W. *and* Wild, H.
A statistical analysis of a sample of the Flora Zambesiaca: 2. Gramineae and Pteridophyta. *Kirkia*,
5 9(1):87–93(1973).

2171 Farragher, M. A.
Some recently recorded grass varieties. *Irish Nat. J.*,
1 17(10):345–347(1973)—Records for Ireland.

2172 Grant, C. A.
A scanning electron microscopy survey of some Maydeae pollen. *Grana*, 12(3):177–184(1972 publ.
4 1973).

2173 Grosser, D. *and* Liese, W.
Present status and problems of bamboo classification.
2, 4–6 *J. Arnold Arbor.*, 54(2):293–308(1973).

2174 Grosser, D. *and* Zamuco, G. I.
Anatomy of some bamboo species in the Philippines. *Philippine J. Sci.*, 100(1):57–73(1971 publ.
6 1973).

2175 Hartley, W.
Studies on the origin, evolution and distribution of the Gramineae: 5. The subfamily Festucoideae.
1–7 *Austral. J. Bot.*, 21(2):201–234(1973)—Dot maps.

2176 Hradecká, D.
Pozoruhodná lokalita rákosu obecného ze Slovenska. (Eine beachtenswerte Lokalität des gemeinen Schilfes in der Slowakei). *Zprávy Česk. Bot. Spol.*, 8(3):
1 188–192(1973).

2177 Hubbard, C. E.
Gräser: Beschreibung, Verbreitung, Verwendung. Stuttgart, E. Ulmer, 1973. 461p. (Uni-Taschenbücher
1 233)—Illustrations; chromosome numbers.

2178 Jain, S. K. *and* Doli Das
Notes on the nomenclature of some grasses. *Indian*
1–7 *Forester*, 99(9):570–578(1973)—Key.

2179 Johnson, B. L.
Polyploidy as a factor in the evolution and distribution of grasses. *In* Youngner, V. B. *and* McKell, C. M. *eds.* The biology and utilization of grasses: 18–35
1–7 (1972)—Chromosome numbers.

2180 Kammacher, P., Anoma, G., Adjanohoun, E. *and* Aké Assi, L.
Nombres chromosomiques de Graminées de Côte-d'Ivoire. *Candollea*, 28(2):191–217(1973)—
5 (En,Ge).

2181 Latting, J.
Differentiation in the grass inflorescence. *In* Youngner, V. B. *and* McKell, C. M. *eds.* The biology and utilization of grasses: 365–399(1972)—
1–7 Illustrations.

2182 Li, Y-H. C. *and* Phipps, J. B.
Studies in the Arundinelleae (Gramineae): 15. Taximetrics of leaf anatomy. *Canad. J. Bot.*, 51(3):
2, 4–7 657–680(1973)—(Fr).

2183 McClure, F. A.
Genera of bamboos native to the New World (Gramineae: Bambusoideae). *Smithsonian Contrib.*
3, 4 *Bot.*, no. 9. 148p.(1973)—Illustrations; keys.

2184 Martin, C. *and* Huber, M. L.
Zur Grasflora des Kanha-Nationalparkes. *Vierteljahrsschr. Naturforsch. Ges. Zürich*, 118(1):11–21
6 (1973)—Dot maps. Central India.

2185 Mohlenbrock, R. H.
Grasses: *Bromus* to *Paspalum*. Carbondale, etc., Southern Illinois University Press, 1972. xvi, 332p. (The Illustrated Flora of Illinois)—Illustrations, map,
3 dot maps; keys.

2186 Mohlenbrock, R. H.
Grasses: *Panicum* to *Danthonia*. Carbondale, etc., Southern Illinois University Press, 1973. xix, 378p. (The Illustrated Flora of Illinois)—Illustrations, map,
3 dot maps; keys.

2187 Myre, M.
Algumas Gramíneas novas ou pouco conhecidas para a província de Moçambique. *Bol. Soc. Brot.*,
5 46:345–353(1972).

2188 Nicora, E. G.
Gramineas nuevas para la flora Argentina. *Darwiniana*, 18(1–2):265–272(1973)—Illustrations;
4 chromosome numbers; keys.

2189 Nicora, E. G.
Novedades agrostológicas patagónicas. *Darwiniana*,
4 18(1–2):80–106(1973)—(En); illustrations; keys.

2190 Peshkova, G. A.
Zametki o zlakakh sredneĭ Sibiri. (Notae de Gramineis florae Sibiriae mediae). *Novosti Sist. Vȳssh.*
2 *Rast.*, 10:60–68(1973).

2191 Petrova, L. R.
Morfologiya reproduktivnȳkh organov nekotorȳkh vidov Bambusoideae (k filogenii podsemeĭstva). *Byull. Mosk. Obshch. Ispȳt. Prir., Biol.*,
2, 6, 7 78(4):113–123(1973)—Illustrations.

2192 Phipps, J. B.
Studies in the Arundinelleae (Gramineae): 16. Danthoniopsoids—the middle way. *Bol. Soc. Brot.*,
5 46:417–427(1972).

2193 Pohl, R. W.
Keys to the genera of grasses of Costa Rica. *Rev.*
4 *Biol. Trop.*, 20(2):189–219(1972)—En,Sp.

2194 Ponomarev, A. N., Bannikova, V. A. *and* Rusakova, M. B.
O sutochnoĭ periodichnosti opȳleniya u blizkikh allopaticheskikh taksonov zlakov. (On the daily periodicity of pollination in close allopatric taxa of
1, 2 grasses). *Bot. Zhurn.*, 58(4):484–492(1973)—(En).

2195 Probatova, N. S.
Novȳe i redkie vidȳ zlakov (Poaceae) s Dal'nego Vostoka. (Poacearum species novae et rariores ex Oriente Extremo). *Novosti Sist. Vȳssh. Rast.*, 10:68–79
2 (1973)—Chromosome numbers.

2196 Reddi, V. B. *and* Phipps, J. B.
Free amino acids as taxonomic characters in the tribe Arundinelleae (Gramineae). *Brittonia*, 24(4):
5–7 403–414(1972).

2197 Renvoize, S. A.
Studies in the Gramineae: 34. Chlorideae. *Kew Bull.*,
5 28(2):195–197(1973)—Illustration.

2198 Rost, T. L. *and* Lersten, N. R.
A synopsis and selected bibliography of grass caryopsis anatomy and fine structure. *Iowa State J.*
1–7 *Res.*, 48(1):47–87(1973).

2199 Sinha, R. R. P. *and* Jha, R. P.
Cytological studies in some grasses of Bihar. *Proc. Indian Sci. Congr.*, 59(3):352–353(1972)—Chromo-
6 some numbers. Abstract only.

2200 Stebbins, G. L.
The evolution of the grass family. *In* Youngner, V. B. *and* McKell, C. M. *eds.* The biology and
1–7 utilization of grasses: 1–17(1972).

2201 Sutton, D. D.
Leaf anatomy in the subfamily Eragrostoideae.
3 *Mich. Academician*, 5(3):373–383(1973).

2202 Tikhmenev, E. A. *and* Levkovskiĭ, V. P.
K antékologii zlakov arkticheskoĭ tundrȳ ostrova Vrangelya. (Anthecological studies of grasses in arctic tundras of Wrangel Island). *Bot. Zhurn.*, 58(10):
2 1474–1484(1973).

2203 Tothill, J. C. *and* Hacker, J. B.
The grasses of southeast Queensland. St. Lucia, Queensland University Press, 1973. x,300p.—
7 Illustrations, map; keys.

2204 Tran Van Nam
La ligule dorsale des Graminées. *Bull. Soc. Bot.*
1–7 *France*, 118(9):639–657(1971 publ. 1973)—(En).

2205 Tran Van Nam
Les 'oreillettes' des Graminées. (The 'auricles' of the Gramineae). *Bull. Soc. Bot. France*, 119(7–8):441–461
1–7 (1972 publ. 1973)—(En).

2206 Tran Van Nam
Sur la valeur morphologique des lemmes de Graminées. *Bull. Mus. Nation. Hist. Nat.* (Paris) 3e Sér., no.128 *Bot.*8:33–57(1973)—(En);
1–7 illustrations.

2207 Trouin, M.
Nombres chromosomiques de quelques Graminées du Soudan. *Adansonia*, 12(4):619–624(1972 publ.
5 1973)—(En).

2208 Tsvelev, N. N.
Obzor vidov tribȳ Triticeae Dum. semeĭstva zlakov (Poaceae) vo flore SSSR. (Conspectus specierum tribus Triticeae Dum. familiae Poaceae in flora URSS). *Novosti Sist. Vȳssh. Rast.*, 10:19–59(1973)—
2 Chromosome numbers.

2209 Tsvelev, N. N.
Zametki o zlakakh florȳ SSSR, 7. (Notae de Gramineis florae URSS, 7). *Novosti Sist. Vȳssh. Rast.*,
2 10:79–98(1973)—Chromosome numbers.

2210 Türpe, A. M.
Novedades en Gramineas para la flora Argentina. (News on Gramineae for the Argentine flora).
4 *Lilloa*, 33(13):287–291(1973)—(En); illustrations.

2211 Whyte, R. O.
The Gramineae, wild and cultivated, of monsoonal and equatorial Asia: 1. Southeast Asia. *Asian*
6 *Perspect.*, 15(2):127–151(1972).

2212 Youngner, V. B. *and* McKell, C. M. *eds.*
The biology and utilization of grasses. New York, etc., Academic Press, 1972. xx,426p.—Illustrations,
1–7 chromosome numbers.

ACIACHNE

2213 Astegiano, M. E.
Sobre la presencia del genero *Aciachne* (Gramineae) en la Argentina. *Kurtziana*, 7:43–47(1973)—(En);
4 illustrations.

AEGILOPS

2214 Coucoli, H. *and* Skorda, E. A.
Some aspects regarding behaviour of supernumerary chromosomes in *Aegilops speltoides* Tausch.
2 *Caryologia*, 26(2):171–185(1973).

2215 Gandilyan, P. A.
Dve novȳe nakhodki iz florȳ Armyanskoĭ SSR. (Two finds in Armenian flora). *Biol. Zhurn. Arm.*, 26(2):89–92(1973)—(Arm); illustrations; chromosome
2 numbers. *Aegilops tauschii.*

2216 Kruse, J.
Homologiebeziehungen bei den Grannenbildungen der Gattungen *Triticum* L. und *Aegilops* L. (Homology relations in awn growths of the genera *Triticum* L. and
2 *Aegilops* L. *Kulturpflanze*, 21:195–209(1973)—(En,Rus).

2217 Zeven, A. C.
Triticum lorentii (Hochst.) Zeven nov. comb.
2 *Taxon*, 22(2–3):321(1973)—En. *Aegilops biuncialis.*

cylindrica Host.
var. **aristulata** (Zhuk.) Tsvelev in *Novosti Sist. Vȳssh. Rast.*, 10:37(1973): *A. cylindrica* subsp.
2 *aristulata.*
var. **prokhanovii** Tsvelev *op. cit.* 37(1973)—
2 U.S.S.R. (Caucasus).
ovata L.
subsp. **biuncialis** (Vis.) Anghel & Beldie in Fl. Republ. Social. Român., 12:563(1972): *A.*
1 *biuncialis.*
tauschii Coss.
subsp. **strangulata** (Eig) Tsvelev in *Novosti Sist. Vȳssh. Rast.*, 10:37(1973): *A. squarrosa* subsp.
2 *strangulata.*

AEGOPOGON

2218 Türpe, A. M.
Anatomia foliar en *Aegopogon* (con especial referencia a la morfología *Aegopogon bryophilus* Doell) Eragrostoideae-Zoysieae. *Lilloa*, 33(12):261–280
4 (1973)—(En); illustration; key.

AGROPYRON

2219 Dewey, D. R.
Genome analysis of South American *Elymus patagonicus* and its hybrids with two North American and two Asian *Agropyron* species. *Bot. Gaz.*, 133(4):
2–4 436–443(1972)—Illustrations; chromosome numbers.

2220 Dihoru, G. *and* Negrean, G.
Nouveaux diagnèmes pour *Agropyron cristatum* et *A. brandzae*. *Rev. Roum. Biol., Bot.*, 18(2):61–71
1, 2 (1973)—Key.

2221 Dobrescu, C., Kovács, A. *and* Carlan, V.
O nouă specie în flora R. S. România: *Agropyron orientale* (L.) R. et Sch. *An Şti. Univ. Al. I. Cuza Iaşi, Biol.*, 19(1):141–144(1973)—(Fr); illustration;
1, 2 chromosome number.

2222 Taylor, R. J. *and* McCoy, G. A.
Proposed origin of tetraploid species of crested wheatgrass based on chromatographic and karyotypic analyses. *Amer. J. Bot.*, 60(6):576–583(1973)—
3 Chromosome numbers.

sect. **Australopyrum** Tsvelev in *Novosti Sist.*
7 *Vȳssh. Rast.*, 10:35(1973).
brandzae Panţu & Solacolu
subsp. ciliatum (G. Grinţ.) Dihoru & Negrean in *Rev. Roum. Biol., Bot.*, 18(2):66(1973), without
1, 2 basionym page: *A. bulbosum* var. *ciliatum.*
var. **nyaradyanum** Morariu in Fl. Republ. Social.
1 Român., 12:791(1972)—Romania.
brownei (Kunth) Tsvelev in *Novosti Sist. Vȳssh. Rast.*,
7 10:35(1973): *Triticum brownei.*

cristatum (L.) Beauv.
subsp. dunensis (G. Grinț.) Dihoru & Negrean in *Rev. Roum. Biol., Bot.*, 18(2):64(1973), without
1, 2 basionym page: *A. bulbosum* var. *dunensis*.
dasyanthum Ledeb.
var. **glabrum** (Pacz.) Tsvelev in *Novosti Sist. Vўssh. Rast.*, 10:34(1973): *A. dasyanthum* f.
1 *glabrum*.
elongatum (Host) P. Beauv.
subsp. **ruthenicum** (Gris.) Anghel & Morariu in Fl. Republ. Social. Român., 12:619(1972):
1 *Triticum rigidum* β *ruthenicum*.
intermedium (Host) P. Beauv.
f. **latronum** (Godr.) Anghel & Morariu *op. cit.* 614
1 (1972): *Triticum latronum*.
f. **majus** (Morariu) Morariu *op. cit.* 614(1972):
1 *A. intermedium* var. *major*.
junceum (L.) P. Beauv.
var. **bessarabicum** (Săvul. & Rayss.) Anghel
1 & Morariu *op. cit.* 616(1972): *A. bessarabicum*.
michnoi Roshev.
subsp. **nathaliae** (Sipl.) Tsvelev in *Novosti Sist.*
2 *Vўssh. Rast.*, 10:34(1973): *A. nathaliae*.
pectiniforme Roem. & Schult.
f. **barbatum** Nyár. ex Anghel & Morariu in Fl.
1 Republ. Social. Român., 12:791(1972)—Romania.
repens (L.) P. Beauv.
var. **arvense** (Schreb.) Anghel & Morariu *op. cit.*
1 609(1972): *Triticum arvense*.
tanaiticum Nevski
var. **glabriusculum** (Pidopl.) Tsvelev in *Novosti Sist. Vўssh. Rast.*, 10:33(1973): *A. dasyanthum* f.
1 *glabriusculum*.
trichophorum (Link) Richt.
f. **barbulatum** (Schur) Anghel & Morariu in Fl. Republ. Social. Român., 12:615(1972): *A.*
1 *barbulatum*.
var. **goiranicum** (Vis.) Anghel & Morariu *op. cit.*
1 615(1972): *A. goiranicum*.

AGROSTIS

2223 Holub, J.
Neue oder wenig bekannte Pflanzen der ungarischen Flora. *Ann. Univ. Budapest. Rolando Eötvos Nom., Biol.*,
1 14:91–104(1972 publ. 1973). *Agrostis coarctata*.

2224 Kerguélen, M.
Nos graminées à gazon: ray-grass, pâturins, *Agrostis*. *Rev. Hort.*, no.2319:37–40(1973)—
1–3 Illustrations. Horticultural article.

2225 Soó, R. and Isépy, I.
Über einige Formenkreise der ungarischen und karpatischen flora: 18. *Juncus bufonius, Eleocharis palustris* Agg. und *Agrostis canina*. *Ann. Univ. Budapest. Rolando Eötvos Nom., Biol.*, 14:169–177(1972 publ.
1 1973)—Key.

alpina Scop.
f. **aurata** (All.) Beldie in Fl. Republ. Social. Român.,
1 12:163(1972): *Avena aurata*.
subsp. minor (Bss.) Malagarriga, Las Subesp. y Variac. Geogr.: 28(1973), without basionym
1 date: *A. nevadensis* var. *minor*.
subsp. nevadensis Malagarriga *op. cit.* 28(1973),
1 without basionym date: *A. nevadensis*.
canina L.
var. **hybrida** (Gaud.) Beldie in Fl. Republ. Social.
1 Român., 12:158(1972): *A. hybrida*.
subsp. **pusilla** (Dum.) Malagarriga, Las Subesp. y
1 Variac. Geogr.: 28(1973): *A. pusilla*.
f. **varians** (Thuill.) Beldie in Fl. Republ. Social.
1 Român., 12:158(1972): *A. varians*.
f. **vinealis** (With.) Beldie *op. cit.* 158(1972):
1 *A. vinealis*.
coarctata Ehrh.
var. **nyaradyana** Beldie *op. cit.* 789(1972)—
1 Romania.
var. **subalpina** (Schur) Beldie *op. cit.* 161(1972):
1 *Trichodium subalpinum*.
mertensii Trin.
subsp. **borealis** (Hartm.) Tsvelev in *Novosti Sist.*
1–3 *Vўssh. Rast.*, 10:90(1973): *A. borealis*.
rupestris All.
f. **viridula** (Beck.) Beldie in Fl. Republ. Social.
1 Român., 12:162(1972): *A. rupestris* β *viridula*.
stolonifera L.
f. **flava** Beldie *op. cit.* 152(1972), nom. nov.: *A.*
1 *stolonifera* l. *flavida* Schur.
subsp. **gigantea** (Roth) Beldie *op. cit.* 152(1972):
1 *A. gigantea*.
1 f. **patula** (Gaud.) Beldie *op. cit.* 151(1972): *A. patula*.
f. **pontica** (Grec.) Beldie *op. cit.* 151(1972): *A. alba* γ
1 *pontica*.
var. **silvatica** (Host.) Beldie *op. cit.* 152(1972):
1 *A. silvatica*.
f. **subchlorantha** (Schur) Beldie *op. cit.* 152(1972):
1 *A. stolonifera* i. *subchlorantha*.
tenuis Sibth.
f. **aristata** (Tausch.) Beldie *op. cit.* 156(1972): *A.*
1 *vulgaris* var. *aristata*.
var. **hornungiana** (Schur) Beldie *op. cit.* 156
1 (1972): *A. hornungiana*.
f. **parviflora** (Schur) Beldie *op. cit.* 156(1972):
1 *A. vulgaris* l. *parviflora*.
var. **repens** (Schur) Beldie *op. cit.* 155(1972):
1 *A. vulgaris* i. *repens*.
var. **schurii** Beldie *op. cit.* 155(1972), nom. nov.:
1 *A. vulgaris* A.I.b. l. *arenicola* (Schur) A. & G.
f. **umbrosa** (Schur) Beldie *op. cit.* 156(1972):
1 *A. vulgaris* k. *umbrosa*.

AIRA

2226 Albers, F.
Zwei Ausgangskaryotypen bei der Gräser-Gattung *Aira* L. (Two basic karyotypes in the grass genus *Aira* L.). *Österr. Bot. Zeitschr.*, 121(3–4):251–254(1973)—
1–7 (En); chromosome numbers.

2227 Bor, N. L.
The typification of *Aira divaricata* Pourr. *Webbia*,
1 27(2):383–387(1972 publ. 1973)—(It).

elegantissima Schur
subsp. **hosteana** Holub in *Folia Geobot. Phytotax.*,
1, 2 8(2):176(1973), nom. nov.: *Aira capillaris* Host.

ALLOEOCHAETE

2228 Launert, E.
The genus *Alloeochaete* (Gramineae) in South tropical Africa. *Garcia de Orta, Sér. Bot.*, 1(1–2):91–92(1973)—
5 (Por); illustrations.

oreogena Launert in *Garcia de Orta, Sér. Bot.*, 1(1–2):91
5 (1973)—Malawi.

ALOPECURUS

pratensis L.
f. **brachyglossus** (Peterm.) Șerbănescu & E. I. Nyárády in Fl. Republ. Social. Român., 12:117
1 (1972): *A. brachyglossus*.
f. **glaucus** (Sond.) Șerbănescu & E. I. Nyárády
1 *op. cit.* 118(1972): *A. pratensis* var. *glaucus*.
f. **obscurus** (Griseb.) Șerbănescu & E. I. Nyárády
1 *op. cit.* 117(1972): *A. pratensis* β *obscurus*.

ventricosus Pers.
var. **exserens** (Gris.) Şerbănescu & E. I. Nyárády
1 *op. cit.* 114(1972): *A. ruthenicus* var. *exserens.*

AMPHIBROMUS

scabrivalvis (Trinius) Swallen
var. **indigestus** Nicora in *Darwiniana*, 18(1–2):101
4 (1973)—Argentina.

ANISANTHA

flabellata (Boiss.) Holub in *Folia Geobot. Phytotax.*,
2 8(2):176(1973): *Bromus flabellatus.*
haussknechtii (Boiss.) Holub *op. cit.* 176(1973):
2 *Bromus haussknechtii.*

ANTHOXANTHUM

2229 Valdés, B.
Revision de las especies anuales del genero
Anthoxanthum (Gramineae). *Lagascalia*, 3(1):99–141
(1973)—(En); illustrations, dot maps; chromosome
1 numbers; keys.

aristatum Boiss.
subsp. **macranthum** B. Valdés in *Lagascalia*, 3(1):
1 130(1973)—Spain, Italy, Morocco.
odoratum L.
f. **montanum** (A. & G.) A. Nyárády in Fl.
Republ. Social. Român., 12:101(1972): *A.*
1 *odoratum* var. *montanum.*
f. **tenerum** (A. & G.) A. Nyárády *op. cit.* 98
1 (1972): *A. odoratum* var. *tenerum.*

APERA

spica-venti (L.) P. Beauv.
f. **prorepens** (Schur) Şerbănescu in Fl. Republ.
Social. Român., 12:167(1972): *A. spica-venti* b.
1 *prorepens.*
f. **purpurea** (Schur) Şerbănescu *op. cit.* 167(1972):
1 *A. spica-venti* a. *purpurea.*

APOCLADA

arenicola McClure in *Smithsonian Contrib. Bot.*, 9:9
4 (1973)—Brazil.
cannavieira (Alvaro da Silveira) McClure *op. cit.*
4 12(1973): *Arundinaria cannavieira.*

× **ARCTODUPONTIA** Tsvelev in *Novosti Sist.*
1 *Vÿssh. Rast.*, 10:91(1973). *Arctophila* × *Dupontia.*
scleroclada (Rupr.) Tsvelev *op. cit.* 91(1973): *Poa*
1 *scleroclada.*

ARISTIDA

2230 DeLisle, D. G.
Chromosome numbers in the *Aristida purpurea*
complex (Gramineae). *Southwest. Nat.*, 18(1):79–83
3, 4 (1973).

adscencionis L.
subsp. coerulescens (Desf.) Malagarriga, Las
Subesp. y Variac. Geogr.: 27(1973), without
1, 2 basionym date: *A. coerulescens.*
ramosissima Engelm.
f. **uniaristata** (Gray) Mohlenbrock, Illus. Fl.
Illinois, Grasses, *Panicum* to *Danthonia*; 329
3 (1973): *A. ramosissima* var. *uniaristata.*

ARRHENATHERUM

elatius (L.) J. & C. Presl
f. **biflorum** (Schur) Buia in Fl. Republ. Social.
1 Român., 12:264(1972): *A. elatius* a. *biflorum.*
f. **pauciflorum** (Baenitz.) Buia *op. cit.* 264(1972):
1 *Avena elatior* d. *pauciflora.*
var. **subhirsutum** (Aschers.) Buia *op. cit.* 263
1 (1972): *Avena elatior* c. *subhirsuta.*
f. **versicolor** (Schur) Buia *op. cit.* 263(1972):
1 *A. elatius* c. *versicolor.*

ARTHRAXON

2231 Jain, S. K.
The genus *Arthraxon* P. Beauv. (Poaceae) in India.
Proc. Indian Sci. Congr., 58(3):436–437(1971).
6 Abstract only.

pusillus Bor in *J. Ind. Bot. Soc.*, 50A:92(1971 ?)—
2 Oman.

ARTHROSTYLIDIUM

simpliciusculum (Pilger) McClure in *Smithsonian*
4 *Contrib. Bot.*, 9:20(1973): *Arundinaria simpliciuscula.*

ARUNDINARIA

gigantea (Walter) Muhlenberg
subsp. **macrosperma** (Michaux) McClure in
Smithsonian Contrib. Bot., 9:28(1973): *A.*
3 *macrosperma.*
subsp. **tecta** (Walter) McClure *op. cit.* 26(1973):
3 *Arundo tecta.*

ATRACTANTHA McClure in *Smithsonian Contrib.*
4 *Bot.*, 9:42(1973).
4 **falcata** McClure *op. cit.* 48(1973)—Brazil.
4 **radiata** McClure *op. cit.* 50(1973)—Brazil.

AULONEMIA

amplissima (Nees) McClure in *Smithsonian Contrib.*
4 *Bot.*, 9:56(1973): *Arundinaria amplissima.*
aristulata (Doell) McClure *op. cit.* 56(1973):
4 *Arundinaria aristulata.*
deflexa (N. E. Brown) McClure *op. cit.* 56(1973):
4 *Arundinaria deflexa.*
effusa (Hackel) McClure *op. cit.* 56(1973):
4 *Arundinaria effusa.*
glaziovii (Hackel) McClure *op. cit.* 56(1973):
4 *Arundinaria glaziovii.*
goyazensis (Hackel) McClure *op. cit.* 56(1973):
4 *Arundinaria goyazensis.*
haenkei (Ruprecht) McClure *op. cit.* 56(1973):
4 *Arthrostylidium haenkei.*
herzogiana (Henrard) McClure *op. cit.* 56(1973):
4 *Arundinaria herzogiana.*
hirtula (Pilger) McClure *op. cit.* 56(1973): *Arundinaria*
4 *hirtula.*
humillima (Pilger) McClure *op. cit.* 58(1973):
4 *Arundinaria humillima.*
laxa (F. Maekawa) McClure *op. cit.* 58(1973):
4 *Matudacalamus laxus.*
parviflora (Presl) McClure *op. cit.* 58(1973): *Guadua*
4 *parviflora.*
patula (Pilger) McClure *op. cit.* 58(1973): *Arundinaria*
4 *patula.*
purpurata (McClure) McClure *op. cit.* 58(1973):
4 *Arthrostylidium purpuratum.*
ramosissima (Hackel) McClure *op. cit.* 58(1973):
4 *Arundinaria ramosissima.*
setigera (Hackel) McClure *op. cit.* 58(1973):
4 *Arundinaria setigera.*
sodiroana (Hackel) McClure *op. cit.* 61(1973):
4 *Arundinaria sodiroana.*
steyermarkii (McClure) McClure *op. cit.* 61(1973):
4 *Arthrostylidium steyermarkii.*
subpectinata (O. Kuntze) McClure *op. cit.* 61(1973):
4 *Arthrostylidium subpectinatum.*
trianae (Munro) McClure *op. cit.* 61(1973):
4 *Arundinaria trianae.*
viscosa (Hitchcock) McClure *op. cit.* 61(1973):
4 *Arundinaria viscosa.*

AVENA

2232 Baum, B. R., Rajhathy, T. *and* Sampson, D. R.
An important new diploid *Avena* species discovered on the Canary Islands. *Canad. J. Bot.*, 51(4):759–762
1 (1973)—(Fr); illustration, map; chromosome number.

2233 Thomas, H.
Ceirch tir âr a cheirch gwyllt. *Y Gwyddonydd*, 11(3):132–137(1973)—Illustration; chromosome
1–3 numbers.

canariensis Baum, Rajhathy & Sampson in *Canad. J.*
1 *Bot.*, 51(4):759(1973)—Canary Is.

AVENELLA

2234 Albers, F.
Cytosystematische Untersuchungen in der Subtribus Deschampsiineae Holub (Tribus Aveneae Nees.): 2. Die Gattungen *Vahlodea* Fr. und *Avenella* Koch. (Cytosystematics of the subtribus Deschampsiineae Holub (tribus Aveneae Nees.): 2. The genera *Vahlodea* Fr. and *Avenella* Koch. *Ber. Deutsch. Bot. Ges.*, 85
1–3 (5–6):279–285(1972)—(En); chromosome numbers.

AVENOCHLOA

2235 Gervais, C.
Nouvelles determination de nombres chromosomiques chez les avoines vivaces: 2. *Bull. Soc. Neuchâtel. Sci. Nat.*, 96:81–87(1973)—(En);
1 chromosome numbers.

2236 Holub, J.
Neue oder wenig bekannte Pflanzen der ungarischen Flora. *Ann. Univ. Budapest. Rolando Eötvos Nom., Biol.*, 14:91–104(1972 publ. 1973).
1 *Avenochloa adsurgens.*

adsurgens (Schur ex Simk.) Holub in *Ann. Univ. Sci. Budapest. Rolando Eötvos Nom., Biol.*, 14:94(1972
1 publ. 1973): *Avena adsurgens.*
pratensis (L.) Holub
subsp. **hirtifolia** (Podp.) Holub *op. cit.* 94(1972
1 publ. 1973): *Avenastrum pratense* var. *hirtifolium.*

AVENULA

cycladum (Rech. fil. & Scheffer) Greuter in *Ann. Mus.*
1 *Goulandris*, 1:75(1973): *Avenastrum cycladum.*

BAMBUSA

glomerata (Munro) McClure in *Smithsonian Contrib.*
4 *Bot.*, 9:66(1973): *Guadua glomerata.*
longifimbriata (E. G. Camus) McClure *op. cit.* 66
4 (1973): *Guadua longifimbriata.*
longifolia (Fournier) McClure *op. cit.* 66(1973):
4 *Arundinaria? longifolia.*
macrostachya (Ruprecht) McClure *op. cit.* 68
4 (1973): *Guadua macrostachya.*
philippinensis (Gamble) McClure *op. cit.* 68(1973):
6 *Guadua philippinensis.*
refracta (Munro) McClure *op. cit.* 68(1973): *Guadua*
4 *refracta.*
superba (Huber) McClure *op. cit.* 68(1973): *Guadua*
4 *superba.*
swalleniana McClure *op. cit.* 68(1973), nom. nov.:
4 *Guadua spinosa* (Swallen) McClure.
tessmannii (Pilger) McClure *op. cit.* 68(1973): *Guadua*
4 *tessmannii.*
tomentosa (Hackel & Lindman) McClure *op. cit.*
4 68(1973): *Guadua tomentosa.*
venezuelae (Munro) McClure *op. cit.* 68(1973):
4 *Guadua venezuelae.*
weberbaueri (Pilger) McClure *op. cit.* 68(1973):
4 *Guadua weberbaueri.*

BECKMANNIA

eruciformis (L.) Host
subsp. **borealis** Tsvelev in *Novosti Sist. Vŷssh. Rast.*,
1 10:81(1973)—U.S.S.R. (European Russia).

BELLARDIOCHLOA

2237 Resner, V.
Bellardiochloa violacea (Bell.) Chiov. v Belanskych Tatrách. (*Bellardiochloa violacea* (Bell.) Chiov. in the Belanské Tatry mountains). *Biológia, A* (Czecho-
1 slovakia), 28(7):575–577(1973)—(En,Rus); illustration.

violacea (Bell.) Chiov.
var. **popovii** Tsvelev in *Novosti Sist. Vŷssh. Rast.*,
1 10:87(1973)—U.S.S.R. (Ukraine).

BOTHRIOCHLOA

bladhii (Retz.) S. T. Blake
var. punctata (Roxb.) R. R. Stewart, Annot. Cat. Vasc. Pl. W. Pakistan & Kashmir: 106(1972),
6 without basionym date: *Andropogon punctatus.*

BOUTELOUA

2238 Nicholson, R. A.
Grama (*Bouteloua* Lag.) communities in a southern Arizona grassland. *Diss. Abstr. Int., B*, 33(8):3605
3 (1973).

BRACHIARIA

nilagirica Bor in *Bot. Tidsskr.*, 67(4):325(1973)—
6 India.

BRACHYPODIUM

pinnatum (L.) P. Beauv.
var. **ligulatum** (Buia) Anghel & Beldie in Fl. Republ. Social. Român., 12:343(1972): *B.*
1 *pinnatum* subsp. *ligulatum.*
f. **minus** (Schur) Anghel & Beldie *op. cit.* 344
1 (1972): *B. pinnatum* a. *minus.*
sylvaticum (Huds.) Beauv.
subsp. **pubescens** (Peterm.) Tsvelev in *Novosti Sist.*
1 *Vŷssh. Rast.*, 10:89(1973): *B. gracile* b. *pubescens.*

BREVIPODIUM

silvaticum (Huds.) Löve & Löve
var. **dumosum** (Vill.) Pinto da Silva & Teles in *Agron. Lusit.*, 34(3):180(1972 publ. 1973):
1 *Bromus dumosus.*

BRIZA

2239 Murray, B. C.
Cytogenetic studies in the genus *Briza* L.
1–4 (Gramineae). *Incompat. Newsl.*, no.1:20–21(1972).

2240 Murray, B. G. *and* Owens, S. J.
Interspecific incompatibility in the Umbelliferae and Gramineae. *Incompat. Newsl.*, no.2:15–18
1–4 (1973).

media L.
f. **pilosa** (Schur) Ghişa in Fl. Republ. Social.
1 Român., 12:433(1972): *B. pilosa.*
f. **repens** (Roth) Ghişa *op. cit.* 433(1972): *B. media* β
1 *repens.*

BROMELICA

cepacea (Philippi) Nicora in *Darwiniana*, 18(1–2):
4 270(1973): *Festuca cepacea.*

BROMOPSIS

angrenica (Drob.) Holub in *Folia Geobot. Phytotax.*,
2 8(2):167(1973): *Bromus angrenicus.*

4 **anomala** (Fourn.) Holub *op. cit.* 167(1973): *Bromus anomalus.*

3 **arctica** (Shear) Holub *op. cit.* 167(1973): *Bromus arcticus.*

1, 2 **aristata** (C. Koch) Holub *op. cit.* 167(1973): *Festuca aristata.*

2 **armena** (Boiss.) Holub *op. cit.* 167(1973): *Bromus armenus.*

1 **atlanticus** (Lindb.) Holub *op. cit.* 167(1973): *Bromus atlanticus.*

4 **attenuata** (Swallen) Holub *op. cit.* 167(1973): *Bromus attenuatus.*

1, 2 **benekenii** (Lange) Holub *op. cit.* 167(1973): *Schedonorus benekenii.*

2 **biebersteinii** (Roem. & Schult.) Holub *op. cit.* 167 (1973): *Bromus biebersteinii.*

2 **bikfayensis** (Camus & Gombault) Holub *op. cit.* 167(1973): *Bromus bikfayensis.*

3 **canadensis** (Michx.) Holub *op. cit.* 167(1973): *Bromus canadensis.*

1, 2 **cappadocica** (Boiss. & Bal.) Holub *op. cit.* 167(1973): *Bromus cappadocicus.*

1–3, 6 **ciliata** (L.) Holub *op. cit.* 167(1973): *Bromus ciliatus.*

1 **condensata** (Hackel) Holub *op. cit.* 167(1973): *Bromus condensatus.*

1 subsp. **australis** (Griseb.) Holub *op. cit.* 167(1973): *Bromus erectus* var. *australis.*

4 **densa** (Swallen) Holub *op. cit.* 167(1973): *Bromus densus.*

4 **dolichocarpa** (Wagnon) Holub *op. cit.* 167(1973): *Bromus dolichocarpus.*

4 **exaltata** (Bernh.) Holub *op. cit.* 167(1973): *Bromus exaltatus.*

2 **frigida** (Boiss. & Hausskn.) Holub *op. cit.* 167(1973): *Bromus frigidus.*

3 **frondosa** (Shear) Holub *op. cit.* 167(1973): *Bromus porteri* var. *frondosus.*

3 **grandis** (Shear) Holub *op. cit.* 167(1973): *Bromus orcuttianus* var. *grandis.*

1 **heterophylla** (Klokov) Holub *op. cit.* 167(1973): *Bromus heterophyllus.*

6 **himalaica** (Stapf) Holub *op. cit.* 167(1973): *Bromus himalaicus.*

2 **indurata** (Hausskn. & Bornm.) Holub *op. cit.* 167 (1973): *Bromus induratus.*

3 **kalmii** (A. Gray) Holub *op. cit.* 167(1973): *Bromus kalmii.*

1–3, 6 **inermis** (Leyss.) Holub *op. cit.* 167(1973): *Bromus inermis.*

2 **kopetdagensis** (Drob.) Holub *op. cit.* 168(1973): *Bromus kopetdagensis.*

2 **korotkyi** (Drob.) Holub *op. cit.* 168(1973): *Bromus korotkyi.*

1 **lacmonica** (Hausskn.) Holub *op. cit.* 168(1973): *Bromus lacmonicus.*

3 **laevipes** (Shear) Holub *op. cit.* 168(1973): *Bromus laevipes.*

3 **lanatipes** (Shear) Holub *op. cit.* 168(1973): *Bromus porteri* var. *lanatipes.*

1 **maroccana** (Pau & Font-Quer) Holub *op. cit.* 168 (1973), as 'marrocana': *Bromus marrocana.*

1 **microchaeta** (Font-Quer) Holub *op. cit.* 168(1973): *Bromus microchaetus.*

1 **moellendorfiana** (Aschers. & Graebn.) Holub *op. cit.* 168(1973): *Bromus erectus* subsp. *moellendorfianus.*

1 **moesiaca** (Velen.) Holub *op. cit.* 168(1973): *Bromus moesiacus.*

1 **monoclada** (Domin) Holub *op. cit.* 168(1973): *Bromus monocladus.*

3, 4 **mucroglumis** (Wagnon) Holub *op. cit.* 168(1973), as 'macroglumis': *Bromus macroglumis.*

3 **nottowayana** (Fernald) Holub *op. cit.* 168(1973): *Bromus nottowayanus.*

3 **orcuttiana** (Vasey) Holub *op. cit.* 168(1973): *Bromus orcuttianus.*

2 **ornans** (Komarov) Holub *op. cit.* 168(1973): *Bromus ornans.*

3 **pacifica** (Shear) Holub *op. cit.* 168(1973): *Bromus pacificus.*

2 **pamirica** (Drob.) Holub *op. cit.* 168(1973): *Bromus pamiricus.*

1 **pannonica** (Kumm. & Sendtn.) Holub *op. cit.* 168 (1973): *Bromus pannonicus.*

2 **paulsenii** (Hack.) Holub *op. cit.* 168(1973): *Bromus paulsenii.*

1 **permixta** (Lindb.) Holub *op. cit.* 168(1973): *Bromus erectus* subsp. *permixtus.*

1 **pindica** (Hausskn.) Holub *op. cit.* 168(1973): *Bromus pindicus.*

4 **pinetorum** (Swallen) Holub *op. cit.* 168(1973): *Bromus pinetorum.*

3 **porteri** (Coult.) Holub *op. cit.* 168(1973): *Bromus kalmii* var. *porteri.*

3 **pseudolaevipes** (Wagnon) Holub *op. cit.* 168(1973): *Bromus pseudolaevipes.*

3 **pubescens** (Willd.) Holub *op. cit.* 168(1973): *Bromus pubescens.*

3 **pumpelliana** (Scribn.) Holub *op. cit.* 168(1973): *Bromus pumpellianus.*

1–3, 6 **ramosa** (Huds.) Holub *op. cit.* 168(1973): *Bromus ramosus.*

2, 3 **richardsonii** (Link) Holub *op. cit.* 168(1973): *Bromus richardsonii.*

1, 2 **riparia** (Rehm.) Holub *op. cit.* 169(1973): *Bromus riparius.*

1, 2 subsp. **arenaria** (Heuffel) Holub *op. cit.* 169(1973): *Bromus erectus* δ *arenarius.*

1, 2 subsp. **barcensis** (Simonk.) Holub *op. cit.* 169 (1973): *Bromus barcensis.*

2 **sclerophylla** (Boiss.) Holub *op. cit.* 169(1973): *Bromus sclerophyllus.*

2 **stenostachya** (Boiss.) Holub *op. cit.* 169(1973): *Bromus stenostachyus.*

3 **suksdorfii** (Vasey) Holub *op. cit.* 169(1973): *Bromus suksdorfii.*

1, 2 **syriacus** (Boiss. & Bal.) Holub *op. cit.* 169(1973): *Bromus syriacus.*

4 **texensis** (Shear) Holub *op. cit.* 169(1973): *Bromus purgans* var. *texensis.*

1, 2 **tomentella** (Boiss.) Holub *op. cit.* 169(1973): *Bromus tomentellus.*

1 **transsilvanica** (Steudel) Holub *op. cit.* 169(1973): *Bromus transsilvanicus.*

2 **triniana** (Schult.) Holub *op. cit.* 169(1973): *Bromus triniana.*

2 **turkestanica** (Drob.) Holub *op. cit.* 169(1973): *Bromus turkestanicus.*

2 **tyttholepis** (Nevskij) Holub *op. cit.* 169(1973): *Zerna tyttholepis.*

2 **variegata** (M. Bieb.) Holub *op. cit.* 169(1973): *Bromus variegatus.*

1 **villosula** (Steudel) Holub *op. cit.* 169(1973): *Bromus villosulus.*

1–3 **vogulica** (Sočava) Holub *op. cit.* 169(1973): *Bromus vogulicus.*

3 **vulgaris** (Hook.) Holub *op. cit.* 169(1973): *Bromus purgans* var. *vulgaris.*

BROMUS

2241 Bakulina, É. V.
Nekotorȳe zakonomernosti ontogeneza pobega *Bromus benekenii* (Lange) Trin. (Some regularities of shoot ontogenesis in *Bromus benekenii* (Lange) Trin.).
1 *Bot. Zhurn.*, 58(1):64–71(1973).

2242 De Moraes, C. F.
Ecogenetic studies on variation and population structure in soft chess, *Bromus mollis* L. *Diss. Abstr.*
3 *Int., B*, 33(3):1086(1972).

2243 Grims, F.
Uber das ehemalige Vorkommen von *Bromus secalinus* L. im Sauwald. *Mitt. Bot. Arbeit. Oberösterr.*
1, 2 *Landesmus. Linz*, 4(1–2):51(1972).

2244 Grossman, A.
Was ist *Bromus unioloides*? *Göttinger Flor. Rundbr.*,
1–3 7(1):13–19(1973).

2245 Kožuharov, S. I. *and* Petrova, A. V.
Cytotaxonomic study of five rare Bulgarian grasses. *Dokl. Bolg. Akad. Nauk*, 26(6):829–832(1973)—
1 Chromosome numbers. *Bromus macrostachys*.

2246 Rivas Ponce, M. A.
Una nueva especie del genero *Bromus*. *Lagascalia*,
1 3(1):53–57(1973)—(En); illustrations.

2247 Smith, P. M.
Observations on some critical bromegrasses. *Watsonia*, 9(4):319–332(1973)—Illustrations, dot map;
1 chromosome numbers; key.

arvensis L.
f. **compactus** (A. & G.) Todor in Fl. Republ. Social. Român., 12:318(1972): *B. arvensis* var.
1 *compactus*.
f. **degener** (Schur) Todor *op. cit.* 318(1972): *B.*
1 *arvensis* e. *degener*.
f. **diffusus** (Schur) Todor *op. cit.* 318(1972): *B.*
1 *arvensis* c. *diffusus*.
f. **violaceus** (A. & G.) Todor *op. cit.* 318(1972):
1 *B. arvensis* subvar. *violaceus*.
barcensis Simk.
var. **romanicus** (F. Hermann) Todor *op. cit.* 311
1 (1972): *B. ramosus* var. *romanicus*.
hispanicus Rivas Ponce in *Lagascalia*, 3(1):53(1973)—
1 Spain.
inermis Leyss.
f. **pellitus** (Beck.) Todor in Fl. Republ. Social.
1 Român., 12:307(1972): *B. inermis* δ *pellitus*.
f. **villosus** (Mert. & Koch) Todor *op. cit.* 307
1 (1972): *Festuca inermis* β *villosa*.
japonicus Thunb.
var. **glabriglumis** Tsvelev in *Novosti Sist. Vўssh.*
1 *Rast.*, 10:88(1973)—U.S.S.R. (Crimea).
f. **tenuis** (Prod.) Todor in Fl. Republ. Social.
1 Român., 12:331(1972): *B. japonicus* var. *tenuis*.
var. **variegatus** (Schur) Todor *op. cit.* 331(1972):
1 *B. patulus* a. *variegatus*.
mollis L.
f. **contractus** (Lange) Todor *op. cit.* 325(1972):
1 *B. mollis* γ *contractus*.
1 f. **nanus** (Weig.) Todor *op. cit.* 325(1972): *B. nanus*.
f. **simplicissimus** (Ces.) Todor *op. cit.* 325(1972):
1 *B. simplicissimus*.
oxyodon Schrenk
var. **lanuginosus** Roshev ex Tsvelev in *Novosti Sist. Vўssh. Rast.*, 10:89(1973)—U.S.S.R.
2 (Middle Asia).
racemosus L.
f. **brachystachys** (Schur) Todor in Fl. Republ. Social. Român., 12:322(1972): *B. racemosus* a.
1 *brachystachys*.
f. **pratorum** (Schur) Todor *op. cit.* 322(1972):
1 *B. racemosus* b. *pratorum*.
secalinus L.
f. **divergens** (Rchb.) Todor *op. cit.* 317(1972):
1 *B. secalinus* var. *divergens*.
f. **velutinus** (Schrad.) Todor *op. cit.* 317(1972):
1, 2 *B. velutinus*.
squarrosus L.
var. **balsiensis** (Prod.) Todor *op. cit.* 327(1972):
1 *B. balsiensis*.
f. pauciflorus (Prod.) Todor *op. cit.* 327(1972),
1 without basionym ref.: (Basionym not stated).

CALAMAGROSTIS

2248 Balgooyen, T. G. *and* Moe, L. M.
Dispersal of grass fruits—an example of endornithochory. *Amer. Midl. Nat.*, 90(2):454–455(1973).
3 *Calamagrostis canadensis*.

2249 Heine, R. *
Calamagrostis neglecta [Ehrh.] Pal. Beauv. in der Oberlausitz. (*Calamagrostis neglecta* (E.) P.B. en Haute Lusace). *Abh. Ber. Naturkde-Mus. Görlitz*, 46(8):1–8(1971). Ref. in *Bull. Signal., Biol. Physiol.*
1 *Vég.*, 34(10):9616(1973).

2250 Kull, K.
Jäneskastiku (*Calamagrostis epigeios* (L.) Roth) varieeruvusest ja värdadest Eestis. (Variations and hybrids of *Calamagrostis epigeios* (L.) Roth in Estonia).
1 *Loodusuur. Seltsi Aastar.*, 61:97–106(1972)—(Rus,En).

2251 Langhe, J. E. de *and* Delvosalle, L.
Calamagrostis phragmitoides Hartman, Graminée indigène en Belgique. (*Calamagrostis phragmitoides* Hartman, in Belgien einheimische Gramineenart).
1, 2 *Bull. Soc. Roy. Bot. Belg.*, 106(3):61–64(1973)—(Ge).

2252 Leiris, H. de
Sur la répartition du *Calamagrostis villosa* dans les Alpes françaises. *Trav. Lab. La Jaysinia*, no.4:131–133
1 (1972 publ. 1973)—(En).

2253 Nieschalk, A. *and* Nieschalk, Ch.
Weitere Vorkommen von *Calamagrostis phragmitoides* Hartman (Purpurreitgras) in Nord-Hessen.
1 *Hess. Flor. Briefe*, 253–255:2–6(1973)—Illustrations.

2254 Tateoka, T.
Calamagrostis nana, a revived species of the family Gramineae. *Bull. Nation. Sci. Mus.* (Tokyo), 16(1): 45–59(1973)—Illustration, dot map; chromosome
1–3 numbers.

2255 Tateoka, T.
A taxonomic study of the genus *Calamagrostis* on Mount Yakeishidake. *Bot. Mag. Tokyo*, 86(1002):103–120(1973)—Illustrations, maps; chromosome
2 numbers.

arundinacea (L.) Roth
var. **alpina** (Schur) Şerbănescu & Beldie in Fl. Republ. Social. Român., 12:180(1972): *Deyeuxia*
1 *sylvatica* a. *alpina*.
epigeios (L.) Roth
f. **densiflora** (Ldb.) Şerbănescu & Beldie *op. cit.*
1 180(1972): *C. epigeios* var. *densiflora*.
nana Takeda
subsp. **hayachinensis** (Ohwi) Tateoka in *Bull. Nation. Sci. Mus.* (Tokyo), 16(1):57(1973): *C.*
2 *deschampsioides* var. *hayachinensis*.

CALDERONELLA Soderstrom & H. F. Decker in
4 *Ann. Missouri Bot. Gard.*, 60(2):427(1973).

2256 Soderstrom, T. R. *and* Decker, H. F.
Calderonella, a new genus of grasses, and its relationships to the centostecoid genera. *Ann. Missouri Bot. Gard.*, 60(2):427–441(1973)—Illustra-
4 tions, dot map; key.

sylvatica Soderstrom & H. F. Decker in *Ann.*
4 *Missouri Bot. Gard.*, 60(2):427(1973)—Panama.

CAPILLIPEDIUM

sulcatum Bor in *Bot. Tidsskr.*, 67(4):324(1973)—
6 Thailand.

CERATOCHLOA

arizonica (Shear) Holub in *Folia Geobot. Phytotax.*,
3 8(2):170(1973): *Bromus carinatus* var. *arizonicus*.
colorata (Steudel) Holub *op. cit.* 170(1973): *Bromus*
4 *coloratus*.
laciniata (Beal) Holub *op. cit.* 170(1973): *Bromus*
4 *laciniatus*.
mango (E. Desv.) Holub *op. cit.* 170(1973): *Bromus*
4 *mango*.
maritima (Piper) Holub *op. cit.* 170(1973): *Bromus*
3 *marginatus* subsp. *maritimus*.
pitensis (Humb., Bonpl. & Kunth) Holub *op. cit.*
4 171(1973): *Bromus pitensis*.
valdiviana (Philippi) Holub *op. cit.* 171(1973):
4 *Bromus valdivianus*.

CHAETIUM

2257 Montiel, M. B.
Determinación taxonómica de la especie *Chaetium bromoides* (Presl) Benth. basada en el estudio anatómico. *Rev. Biol. Trop.*, 20(1):45–79(1972)—
4 (En); illustrations; chromosome numbers.

CHASMOPODIUM

purpurascens (Robyns) Clayton in *Kew Bull.*, 28(1):
2, 5, 6 51(1973): *Rottboellia purpurascens*.

CHLORIS

2258 Shinners, L. H.
Chloris truncata R. Br. (Gramineae), an Australian wool-waste adventive in South Carolina. *Sida*, 5(3):
3, 7 182(1973).

ruahensis Renvoize in *Kew Bull.*, 28(2):195(1973)—
5 Tanzania.
5 **ferruginea** Renvoize *op. cit.* 195(1973)—Kenya.

CHRYSOPOGON

gryllus (L.) Trin.
var. aurea (Murr.) Ghişa in Fl. Republ. Social. Român., 12:57(1972), without basionym ref.:
1 (Basionym not stated).
subsp. **glabratus** (Trin.) Tsvelev in *Novosti Sist.*
2, 6 *Vȳssh. Rast.*, 10:79(1973): *C. glabratus*.

CHUSQUEA

perligulata (Pilger) McClure in *Smithsonian Contrib.*
4 *Bot.*, 9:75(1973): *Guadua ? perligulata*.

COLANTHELIA McClure & E. W. Smith in
4 *Smithsonian Contrib. Bot.*, 9:77(1973).
burchellii (Munro) McClure *op. cit.* 9:79(1973):
4 *Arthrostylidium burchellii*.
cingulata (McClure & L. B. Smith) McClure *op. cit.*
4 79(1973): *Aulonemia cingulata*.
distans (Trinius) McClure *op. cit.* 79(1973):
4 *Arundinaria distans*.
intermedia (McClure & L. B. Smith) McClure *op.*
4 *cit.* 79(1973): *Aulonemia intermedia*.
lanciflora (McClure & L. B. Smith) McClure *op. cit.*
4 79(1973): *Aulonemia lanciflora*.
macrostachya (Nees) McClure *op. cit.* 79(1973):
4 *Arundinaria macrostachya*.
rhizantha (Hackel) McClure *op. cit.* 79(1973):
4 *Arundinaria rhizantha*.

COLEANTHUS

2259 Nechaev, A. P. *and* Nechaev, A. A.
Coleanthus subtilis (Tratt.) Seidl. v priamurskoĭ chasti areala. (*Coleanthus subtilis* (Tratt.) Seidl. in the Amur part of its area). *Bot. Zhurn.*, 58(3):440–446
2 (1973)—Illustration.

CORTADERIA

2260 Connor, H. E.
Breeding systems in *Cortaderia* (Gramineae). *Evolution*, 27(4):663–678(1973)—Illustrations,
4, 6, 7 chromosome numbers.

pilosa (D'Urv.) Hackel
var. **minima** (Conert) Nicora in *Darwiniana*,
4 18(1–2):80(1973): *C. minima*.

CORYNEPHORUS

2261 Albers, F.
Cytosystematische Untersuchungen in der Subtribus Deschampsiineae Holub. (Tribus Aveneae Nees): 1. Zwei Arten der Gattung *Corynephorus* P.B. (Cytotaxonomický výzkum v subtribus Deschampsiineae Holub (tribus Aveneae Nees): 1. Dva druhy rodu *Corynephorus* P.B.). *Preslia*, 45(1):11–18
1 (1973)—(Cz); chromosome numbers.

2262 Bor, N. L.
The typification of *Aira divaricata* Pourr. *Webbia*,
1 27(2):383–387(1972 publ. 1973)—(It).

CYMBOPOGON

2263 Dutt, A. K. *and* Bradu, B. L.
Proliferation in *Cymbopogon* and its role in plant propagation. *Proc. Indian Acad. Sci.*, *B*, 77(5):181–185
6 (1973)—Illustrations.

2264 Gupta, B. K.
A note on the occurrence of natural hybrids in Indian *Cymbopogon*. *Pl. Sci.* (Lucknow), 3:120–121
6 (1971 ?publ. 1973).

CYNODON

dactylon (L.) Pers.
var. **septentrionalis** (A. & G.) Ravaruţ in Fl. Republ. Social. Român., 12:139(1972): *C.*
1 *dactylon* B. *septentrionalis*.

CYNOSURUS

cristatus L.
f. **tenuissimus** (Schur) Morariu in Fl. Republ. Social. Român., 12:341(1972): *C. cristatus* a.
1 *tenuissimus*.

DACTYLIS

2265 Egorova, V. N.
Zhiznennȳĭ tsikl sbornoĭ ezhĭ (*Dactylis glomerata* L.): soobshchenie 2. (The life cycle of orchard-grass (*Dactylis glomerata* L.): communication 2). *Byull. Mosk. Obshch. Ispȳt. Prir., Biol.*, 78(2):89–99(1973)—(En);
1, 2 illustrations.

2266 Miroslavov, E. A. *and* Zhigar, M. P.
Élektronnomikroskopicheskoe izuchenie naruzhnȳkh stenok épidermal'nȳkh kletok *Poa annua* L., *Dactylis glomerata* L. i *Leymus racemosus* (Lam.) Tzvel. (Electron microscopic study of the outer walls of epidermal cells in *Poa annua* L., *Dactylis glomerata* L. and *Leymus racemosus* (Lam.) Tzvel.) *Bot. Zhurn.*,
1, 2 58(10):1407–1412(1973)—(En).

2267 Williams, E. *and* Barclay, P. C.
Transmission of B-chromosomes in *Dactylis*. *New Zealand J. Bot.*, 10(4):573–584(1973)—Chromosome
1 numbers.

glomerata L.
f. **hispida** Ghişa in Fl. Republ. Social. Român., 12: 360(1972), nom. nov.: *D. glomerata* subsp. *hispida*
1 (Terrac.) Prod.
f. pubescens (Opiz) Ghişa *op. cit.* 360(1972), without
1 basionym ref.: *D. pubescens*.

DANTHONIA

2268 Baum, B. R. *and* Findlay, J. N.
Preliminary studies in the taxonomy of *Danthonia* in Canada. *Canad. J. Bot.*, 51(2):437–450(1973)—
3 (Fr); dot map; key.

chilensis Desvaux
var. **glabriflora** Nicora in *Darwiniana*, 18(1–2):82
4 (1973)—Argentina, Chile.
provincialis Lam. ex DC.
var. **major** Buia in Fl. Republ. Social. Român., 12:
1 790(1972)—Romania.
var. **microdanthonia** Nyár. ex Buia *op. cit.* 790
1 (1972)—Romania.

DANTHONIASTRUM

2269 Baum, B. R.
The genus *Danthoniastrum*, about its circumscription, past and present status, and some taxonomic principles. *Österr. Bot. Zeitschr.*, 122(1–2):51–57
1 (1973)—Illustrations.

DANTHONIOPSIS

sect. **Gazachloa** (J. B. Phipps) J. B. Phipps in *Bol.*
5 *Soc. Brot.*, 46:423(1972 publ. 1973): gen. *Gazachloa*.
sect. **Jacquesfelixia** (J. B. Phipps) J. B. Phipps *op. cit.*
5 423(1972 publ. 1973): gen. *Jacquesfelixia*.
sect. **Petrina** (J. B. Phipps) J. B. Phipps *op. cit.* 422
5 (1972 publ. 1973): gen. *Petrina*.
sect. **Xerodanthia** (J. B. Phipps) J. B. Phipps *op. cit.*
6 423(1972 publ. 1973): gen. *Xerodanthia*.
scopulorum (J. B. Phipps) J. B. Phipps *op. cit.* 423
5 (1972 publ. 1973): *Gazachloa scopulorum*.

DESCHAMPSIA

2270 McAllister, H. A.
Deschampsia aespitosa. *Glasgow Nat.*, 19(1):69 (1973)—En; chromosome numbers. A tetraploid
1 variant in Scotland.

caespitosa (L.) Pal. de Beauv.
var. **pulchra** (Nees & Mey.) Nicora in *Darwiniana*,
4 18(1–2):101(1973): *D. pulchra*.

DIGITARIA

2271 Veldkamp, J. F.
A revision of *Digitaria* Haller (Gramineae) in Malesia. Notes on Malesian grasses: 6. *Blumea*, 21(1): 1–80(1973)—Illustrations, dot maps; chromosome
6 numbers; key.

sect. Filiformes (Honda) Veldkamp in *Blumea*, 21(1): 16(1973), without basionym page: *Syntherisma* sect.
6 *Filiformes*.
abludens (R. & S.) Veldkamp *op. cit.* 53(1973):
6 *Panicum abludens*.
adscendens (H.B.K.) Henrard
subsp. chrysoblephara (Fig. & de Not.) R. R. Stewart, Annot. Cat. Vasc. Pl. W. Pakistan & Kashmir: 117(1972), name not accepted (see *D. ciliaris* var. *chrysoblephara* below): *D.*
6 *chrysoblephara*.
ciliaris (Retz.) Koel.
var. chrysoblephara (Fig. & de Not.) R. R. Stewart *op. cit.* 117(1972), name not accepted (see *D. adscendens* subsp. *chrysoblephara* above):
6 *D. chrysoblephara*.
var. criniformis (Hen.) R. R. Stewart *op. cit.* 117 (1972), without basionym ref.: (Basionym not
6 stated).
var. **fimbriata** (Link) Jain & Doli Das in *Indian*
6 *Forester*, 99(9):572(1973): *D. fimbriata*.
subsp. **marginata** (Link) Jain & Doli Das *op. cit.*
1–7 572(1973): *D. marginata*.
subsp. **nubica** (Stapf) Jain & Doli Das *op. cit.* 573
5 (1973): *D. marginata* var. *nubica*.
var. **pes-avis** (Buse) Jain & Doli Das *op. cit.* 572
6 (1973): *D. pes-avis*.
var. **sericea** (Honda) Jain & Doli Das *op. cit.* 573
2 (1973): *Syntherisma sericea*.
compacta (Roth. ex R. & S.) Veldkamp in *Blumea*,
6 21(1):71(1973): *Paspalum compactum*.
cruciata (Nees ex Steud.) A. Camus
var. **pectinata** Veldkamp *op. cit.* 72(1973)—India,
6 Vietnam.
fumida S. T. Blake in *Proc. Roy. Soc. Queensl.*, 84:
7 62(1973)—Australia (Queensland).
henrardii Veldkamp in *Blumea*, 21(1):55(1973)—
6 Philippine Is.
6 **jansenii** Veldkamp *op. cit.* 29(1973)—Lesser Sunda Is.
monodii Veldkamp *op. cit.* 57(1973)—New Guinea
6 (W. Irian).
porrecta S. T. Blake in *Proc. Roy. Soc. Queensl.*, 84:
7 63(1973)—Australia (Queensland).
setigera Roth ex R. & S.
var. **calliblepharata** (Henr.) Veldkamp in *Blumea*, 21(1):40(1973): *D. microbachne* subsp.
6 *calliblepharata*.
wallichiana (Steud.) Stapf
subsp. **remota** (Henr.) Veldkamp *op. cit.* 50(1973):
6 *D. remota*.

DINEBRA

2272 Phillips, S. M.
The genus *Dinebra* Jacq. (Gramineae). *Kew Bull.*, 28(3):411–418(1973 publ. 1974)—Illustrations, dot
5 map; keys.

polycarpha S. M. Phillips in *Kew Bull.*, 28(3):416
5 (1973 publ. 1974)—Uganda, Kenya, Tanzania.
retroflexa (Vahl) Panzer
var. **condensata** S. M. Phillips *op. cit.* 414(1973 publ. 1974)—Tanzania, Zaïre, Ethiopia, Uganda, Kenya, Mozambique, Zambia, Rhodesia, Botswana, Angola, Namibia, South Africa
5 (Transvaal, Natal, Cape Prov.).

DUPONTIA

fisheri R.Br.
subsp. **pelligera** (Rupr.) Tsvelev in *Novosti Sist.*
2 *Vÿssh. Rast.*, 10:91(1973): *Poa pelligera*.

ECHINOCHLOA

2273 Duwensee, H. A.
Zum Vorkommen von *Echinochloa crus-galli* (L.) P.B. in und bei Bad Soden und Salmünster. *Hess. Flor.*
1 *Briefe*, 253–255:14–15(1973).

2274 Gasquez, J. *and* Compoint, J.-P.
Premières données taxonomiques sur *Echinochloa crus galli* (L.) P.B. et espèces voisines par électrophorèse en tube sur gel de polyacrylamide. *Compt. R. Hebd. Séances Acad. Sci., D* (Paris), 277(10):837–840
1 (1973).

2275 Hyde-Wyatt, B.
Barnyard grass: a potential threat to vegetable crops and lucerne. *J. Agric. Tasman.*, 44(1):52–56
1–7 (1973)—Illustration. *Echinochloa crus-galli*.

crus-galli (L.) P. Beauv.
f. **aristata** (Vasing.) Morariu in Fl. Republ. Social.
1 Român., 12:89(1972): *E. macrocarpa* var. *aristata*.
f. **breviseta** (Döll) Morariu *op. cit.* 86(1972):
1 *Panicum crus-galli* B. *brevisetum*.
f. **echinata** (Willd.) Morariu *op. cit.* 86(1972):
1 *Panicum echinatum*.
var. **macrocarpa** (Vasing.) Morariu *op. cit.* 86
1, 2 (1972): *E. macrocarpa*.
f. **mutica** (Vasing.) Morariu *op. cit.* 89(1972):
1, 2 *E. macrocarpa* var. *mutica*.
f. **submutica** (Neilr.) Morariu *op. cit.* 89(1972):
1 *E. crus-galli* var. α *submutica*.

ECTROSIA

lasioclada (Merr.) S. T. Blake in *Proc. Roy. Soc.*
7 *Queensl.*, 84:65(1973): *Eragrostis lasioclada*.

ELEUSINE

2276 Chennaveeraiah, M. S. *and* Hiremath, S. C.
Morphology, cytology and fertility in *Eleusine tristachya*, *E. floccifolia* and their hybrid. *Proc. Indian Sci. Congr.*, 60(3):306(1973)—Chromosome number.
4, 5 Abstract only.

2277 Seetharam, A., Nayar, K. M. D. *and* Sreekantaradhya, R.
Accessory chromosomes in *Eleusine coracana*. *Sci.*
5, 6 *Cult.*, 38(9):416(1972).

× ELYHORDEUM

jordalii (Meld.) Tsvelev in *Novosti Sist. Vȳssh. Rast.*,
3 10:58(1973): × *Agrohordeum jordalii*.

× ELYMOTRYGIA

kirghizica Tsvelev in *Novosti Sist. Vȳssh. Rast.*, 10: 57(1973)—U.S.S.R. (Middle Asia). (*Elymus*
2 *schrenkianus* × *Elytrigia batalinii*).

ELYMUS

2278 Dewey, D. R.
Genome analysis of South American *Elymus patagonicus* and its hybrids with two North American and two Asian *Agropyron* species. *Bot. Gaz.*, 133(4):
2–4 436–443(1972)—Illustrations; chromosome numbers.

2279 Jensen, E. R.
A taxonomic study of single spikelet *Elymus* species of western United States. *Diss. Abstr. Int., B*, 33(9):
3 4149(1973)—Chromosome numbers.

sect. **Anthosachne** (Steud.) Tsvelev in *Novosti Sist.*
2, 6, 7 *Vȳssh. Rast.*, 10:25(1973): genus *Anthosachne*.
sect. **Clinelymopsis** (Nevski) Tsvelev *op. cit.* 25
2 (1973): *Roegneria* sect. *Clinelymopsis*.
arenarius L.
subsp. **sabulosus** (M.B.) Beldie in Fl. Republ.
1 Social. Român., 12:579(1972): *E. sabulosus*.
australasicus (Steud.) Tsvelev in *Novosti Sist. Vȳssh.*
7 *Rast.*, 10:25(1973): *Anthosachne australasica*.
confusus (Roshev.) Tsvelev
var. **pruinosus** (Roshev.) Tsvelev *op. cit.* 27(1973):
2 *Agropyron confusum* var. *pruinosum*.
var. **pubiflorus** (Roshev.) Tsvelev *op. cit.* 27
2 (1973): *Agropyron confusum* var. *pubiflorum*.
dentatus (Hook. f.) Tsvelev
subsp. **lachnophyllus** (Ovcz. & Sidor.)
2 Tsvelev *op. cit.* 21(1973): *Roegneria lachnophylla*.
subsp. **ugamicus** (Drob.) Tsvelev *op. cit.* 21(1973):
2 *Agropyron ugamicum*.
elongatus (Host) Greuter in *Ann. Mus. Goulandris*, 1:
1 73(1973): *Triticum elongatum*.
fedtschenkoi Tsvelev in *Novosti Sist. Vȳssh. Rast.*,
2 10:21(1973), nom. nov.: *Agropyron curvatum* Nevski.
fibrosus (Schrenk) Tsvelev
subsp. **subfibrosus** (Tsvelev) Tsvelev *op. cit.* 25
2 (1973): *Roegneria subfibrosa*.
hystrix L.
var. **bigeloviana** (Fern.) Mohlenbrock, Illustr. Fl. Illinois, Grasses, *Bromus* to *Paspalum*: 206
3 (1972): *Asperella hystrix* var. *bigeloviana*.
kronokensis (Kom.) Tsvelev
subsp. **borealis** (Turcz.) Tsvelev in *Novosti Sist.*
1, 2 *Vȳssh. Rast.*, 10:24(1973): *Triticum boreale*.
subsp. **subalpinus** (L. Neum.) Tsvelev *op. cit.*. 24
1 (1973): *Triticum violaceum* f. *subalpinum*.
lenensis (M. Pop.) Tsvelev *op. cit.* 24(1973): *Agropyron*
2 *lenense*.
longe-aristatus (Boiss.) Tsvelev
subsp. **flexuosissimus** (Nevski) Tsvelev *op. cit.* 26
2 (1973): *Agropyron flexuosissimum*.
subsp. **litvinovii** Tsvelev *op. cit.* 26(1973)—
2 U.S.S.R. (Middle Asia).
macrourus (Turcz.) Tsvelev
subsp. **neplianus** (V. Vassil.) Tsvelev *op. cit.* 25
2 (1973): *Roegneria nepliana*.
mutabilis (Drob.) Tsvelev
subsp. **barbulata** Nevski ex Tsvelev *op. cit.* 22
2 (1973)—U.S.S.R. (Caucasus).
subsp. **praecaespitosus** (Nevski) Tsvelev *op. cit.* 22
2 (1973): *Agropyron praecaespitosum*.
subsp. **transbaicalensis** (Nevski) Tsvelev *op. cit.* 22
2 (1973): *Agropyron transbaicalensis*.
pallidissimus (M. Pop.) Peshkova *op. cit.* 67(1973):
2 *Agropyron pallidissimum*.
sajanensis (Nevski) Tsvelev
subsp. **hyperarticus** (Polun.) Tsvelev *op. cit.* 24
2, 3 (1973): *Agropyron violaceum* var. *hyperarticum*.
subsp. **villosus** (V. Vassil.) Tsvelev *op. cit.* 24
2 (1973): *Roegneria villosa*.
trachycaulus (Link) Gould ex Shinners
subsp. **kamczadalorum** (Nevski) Tsvelev *op. cit.*
2 24(1973): *Roegneria kamczadalorum*.
subsp. **major** (Vasey) Tselev *op. cit.* 24(1973):
3 *Agropyron violaceum* var. *majus*.
subsp. **novae-angliae** (Scribn.) Tsvelev *op. cit.* 23
2, 3 (1973): *Agropyron novae-angliae*.
uralensis (Nevski) Tsvelev
subsp. **komarovii** (Nevski) Tsvelev *op. cit.* 22
2 (1973): *Agropyron komarovii*.
subsp. **prokudinii** (Sered.) Tsvelev *op. cit.* 22
2 (1973): *Roegneria prokudinii*.
subsp. **tianschanicus** (Drob.) Tsvelev *op. cit.* 22
2 (1973): *Agropyron tianschanicum*.

ELYTRIGIA

2280 Larsen, E. *and* Pedersen, A.
Påbegyndt spredning af *Elytrigia junceiformis* × *pungens*, Strand-Kvik × Stiv Kvik, i Storstrøms Amt.
1 *Flora Fauna*, 79(2):47–51(1973)—(Ge); dot map.

2281 Roos, A. *and* Viljasoo, L.
Uus orashein meie liivarandadel. (A new couch-grass on our beaches). *Eesti Loodus*, 16(3):139–143 (1973)—(Rus,En); illustrations, maps; chromosome
1 numbers. *Elytrigia repens*.

2282 Viljasoo, L. *and* Roos, A.
Elytrigia juncea (L.) Nevski on diploid. (*Elytrigia juncea* (L.) Nevski is a diploid). *Eesti NSV Tead. Akad. Toim., Biol.*, 22(1):29–31(1973)—Es(Rus,En);
1 chromosome numbers.

sect. **Caesitosae** (Rouy) Tsvelev in *Novosti Sist.*
1, 2 *Výssh. Rast.*, 10:28(1973): *Agropyron* sect. *Caespitosa.*
sect. **Junceae** (Prat) Tsvelev *op. cit.* 32(1973):
1, 2 *Agropyron* sect. *Juncea.*
× **acuta** (DC.) Tsvelev *op. cit.* 32(1973): *Triticum*
1 *acutum.*
batalinii (Krasn.) Nevski
subsp. **alaica** (Drob.) Tsvelev *op. cit.* 28(1973):
2 *Agropyron alaicum.*
caespitosa (C. Koch) Nevski
subsp. **nodosa** (Nevski) Tsvelev *op. cit.* 30(1973):
1 *Agropyron nodosum.*
subsp. **sinuata** (Nevski) Tsvelev *op. cit.* 30(1973):
2 *Agropyron sinuatum.*
divaricata (Boiss. & Bal.) Nevski
subsp. **attenuatiglumis** (Nevski) Tsvelev *op. cit.*
2 28(1973): *Agropyron attenuatiglume.*
farcta (Viviani) Holub in *Folia Geobot. Phytotax.*,
1 8(2):171(1973): *Triticum farctum.*
geniculata (Trin.) Nevski
subsp. **ferganensis** (Drob.) Tsvelev in *Novosti Sist.*
2 *Výssh. Rast.*, 10:29(1973): *Agropyron ferganense.*
subsp. **pruinifera** (Nevski) Tsvelev *op. cit.* 29
1, 2 (1973): *Agropyron pruiniferum.*
subsp. **scythica** (Nevski) Tsvelev *op. cit.* 29(1973):
1 *Agropyron scythicum.*
gentryi (Meld.) Tsvelev *op. cit.* 30(1973): *Agropyron*
2 *gentryi.*
intermedia (Host) Nevski
subsp. **pulcherrima** (Grossh.) Tsvelev *op. cit.* 31
2 (1973): *Agropyron pulcherrimum.*
subsp. **trichophora** (Link) Tsvelev *op. cit.* 31
1, 2 (1973): *Triticum trichophorum.*
juncea (L.) Nevski
subsp. **bessarabica** (Savul. & Rayss) Tsvelev *op.*
1 *cit.* 32(1973): *Agropyron bessarabicum.*
kasteki (M. Pop.) Tsvelev *op. cit.* 31(1973): *Agropyron*
2 *kasteki.*
pontica (Podp.) Holub in *Folia Geobot. Phytotax.*,
1 8(2):171(1973): *Triticum ponticum.*
pubiflora (Steud.) Tsvelev in *Novosti Sist. Výssh.*
4 *Rast.*, 10:33(1973): *Triticum pubiflorum.*
repens (L.) Nevski
subsp. **elongatiformis** (Drob.) Tsvelev *op. cit.* 31
2 (1973): *Agropyron elongatiforme.*
subsp. **pseudocaesia** (Pacz.) Tsvelev *op. cit.* 31
1 (1973): *Agropyron repens* var. *pseudocaesium.*
scirpea (C. Presl) Holub in *Folia Geobot. Phytotax.*,
1 8(2):171(1973): *Agropyron scirpeum.*
stipifolia (Czern. ex Nevski) Nevski
subsp. **armena** (Nevski) Tsvelev in *Novosti Sist.*
2 *Výssh. Rast.*, 10:29(1973): *Agropyron armenum.*
var. **cretacea** (Klok. & Prokud.) Tsvelev *op. cit.* 29
1 (1973): *Agropyron cretaceum.*
tauri (Boiss. & Bal.) Tsvelev *op. cit.* 30(1973):
2 *Agropyron tauri.*
subsp. **pertenuis** (C. A. Mey.) Tsvelev *op. cit.* 30
2 (1973): *Triticum intermedium* var. *pertenue.*

ERAGROSTIS

2283 Ceynowa-Giełdon, M.
Eragrostis pilosa P. Beauv. nad Wisłą kolo Włocławka. (*Eragrostis pilosa* P. Beauv. an der Wisła bei Włocławek (Zentralpolen)). *Fragm. Flor. Geobot.*, 19(1):3–4
1 (1973)—(Ge); illustration.

2284 Hudson, J. H.
Of *Eragrostis* and range extensions; 1970 botanical
3 remarks. *Blue Jay*, 29(2):96–98(1971)—Key.

2285 Kabuye, C. H. S.
A new species of *Eragrostis* (Gramineae) from East
5 Africa. *Kew Bull.*, 28(3):530(1973 publ. 1974).

2286 Koch, S. D.
A re-evaluation of the cycle of *Eragrostis tracyi* (Gramineae, Eragrostoideae) and its taxonomic implications. *J. Elisha Mitchell Sci. Soc.*, 88(4):211–217
3 (1972)—Chromosome number.

abrumpens Kabuye in *Kew Bull.*, 28(3):530(1973
5 publ. 1974)—Kenya.

EREMOPYRUM

bonaepartis (Spreng.) Nevski
var. **turkestanicum** (Gand.) Tsvelev in *Novosti Sist. Výssh. Rast.*, 10:36(1973): *Agropyron*
2 *turkestanicum.*

ERIONEURON

pilosum (Buckl.) Nash
var. **argentinum** (O. Kuntze) Nicora in *Darwiniana*,
4 18(1–2):104(1973): *Sieglingia argentina.*
var. **mendocinum** (Parodi) Nicora *op. cit.* 104
4 (1973): *Tridens pilosa* var. *mendocina.*

FESTUCA

2287 Alekseev, E. B.
Obsyanitsy gruppy Intravaginales Hack. sektsii *Festuca* (*Ovinae* Fr.) na Kavkaze. (Fescues of the group Intravaginales Hack. of the section *Festuca* (*Ovinae* Fr.) in the Caucasus). *Byull. Mosk. Obshch. Ispýt. Prir., Biol.*, 78(3):94–110(1973)—(En); chromosome
2 numbers; keys.

2288 Alekseev, E. B.
Uzkolistnye ovsyanitsy izvestnyakov severnogo Urala. (Narrow leaved fescues of calcareous rocks of the north Urals). *Byull. Mosk. Obshch. Ispýt. Prir.,*
1, 2 *Biol.*, 78(5):143–145(1973).

2289 Auquier, P.
La distribution de *Festuca pallens* Host en Belgique et dans les régions limitrophes. *Nat. Mosana*, 25(4):
1 114–116(1972 publ. 1973)—Map.

2290 Auquier, P.
Une fétuque nouvelle de Bretagne: *Festuca huonii.*
1 *Candollea*, 28(1):15–19(1973)—(En,Ge); illustrations.

2291 Auquier, P.
Qu'est-ce que le *Festuca caesia* Sm. (Poaceae) ?
1 *Lejeunia*, no.70.12p.(1973)—Illustrations.

2292 Auquier, P. *and* Rammeloo, J.
Nombres chromosomiques dans le genre *Festuca* en Belgique et les régions limitrophes. *Bull. Soc. Roy.*
1 *Bot. Belg.*, 106(2):317–328(1973)—(Du,En).

2293 Baudière, A., Geslot, A., Chiglione, C. *and* Negre, R.
La pelouse à *Festuca eskia* en Pyrénées centrales et orientales: esquisse taxinomique et écologique. *Acta*
1 *Bot. Acad. Sci. Hung.*, 19(1–4):23–35(1973)—(En).

2294 Bidault, M.
A propos de la systématique des *Festuca* du groupe *ovina* L. s.l. de France. *Ann. Sci. Univ. Besançon Bot.*,
1–3 12:119–123(1972)—Key.

2295 Gallinet, J. P.
Remarques sur l'étude chimiotaxonomique du *Festuca glauca* var. *macrophylla* (St Y.) Bid. *Ann. Sci.*
1 *Univ. Besançon, Bot.*, 10:35–41(1971).

2296 Gasquez, J.
Apport de quelques méthodes modernes de la taxonomie à l'étude de *Festuca ovina* L. ssp. *macrophylla* (St Y.) Bid. *Ann. Sci. Univ. Besançon, Bot.*, 10:43–116
1 (1971)—Illustrations; chromosome numbers.

2297 Gymer, P. T. *and* Whittington, W. J.
Hybrids between *Lolium perenne* and *Festuca pratensis*: 2. Comparative morphology. *New Phytol.*,
1, 2 72(4):861–865(1973)—Chromosome numbers.

2298 Malik, C. P. *and* Tripathi, R. C.
Cytological studies in the genus *Festuca*. *Chromosome Inform. Serv.*, no.15:25–27(1973)—Chromosome
1–7 numbers.

2299 Markgraf-Dannenberg, I.
Eine neue *Festuca*-Art aus China. *Acta Bot. Acad.*
2 *Sci. Hung.*, 19(1–4):207–209(1973)—Illustrations.

2300 Mayor, M., Andres, J. *and* Martinez, G. *
Hibridos intergenericos en gramineas en la peninsula iberica (*Festuca* × *Lolium*). (Hybrides intergénériques des Graminées de la Péninsule Ibérique (*F.* × *L.*)). *An. Fac. Veter. Leon*, 17(17):69–80(1971)—(Fr,En). Ref. in *Bull. Signal., Biol. Physiol. Vég.*,
1 34(10):9606(1973).

2301 Siplivinskiĭ, V. N.
Viviparnȳe ovsyanitsȳ (*Festuca* L.) vnearkticheskoĭ Sibiri. (Generis *Festuca* L. species viviparae e Sibiria extraarctica). *Novosti Sist. Vȳssh. Rast.*, 10:11–19(1973)
2 —Chromosome numbers.

2302 Soó, R.
Zeitgemässe Taxonomie der *Festuca ovina*-Gruppe. *Acta Bot. Acad. Sci. Hung.*, 18(3–4):363–377(1973)—
1 (En); key.

2303 Trist, P. J. O.
Festuca glauca Lam. and its var. *caesia* (Sm.) K.
1 Richt. *Watsonia*, 9(3):257–262, (4):432(1973).

2304 Wilde-Duyfjes, B. E. E. de
Over de twee ondersoorten van *Festuca rubra* L. in Nederland. *Gorteria*, 6(8):128–131(1973)—(En);
1 chromosome number.

subsect. **Extravaginales** (Hack.) Beldie in Fl. Republ. Social Român., 12:510(1972): *Festuca* sect.
1 *Ovinae* 2. *Extravaginales* vel *Mixtae*.
alexeenkoi Alekseev in *Byull. Mosk. Obshch. Ispȳt.*
2 *Prir., Biol.*, 78(3):104(1973)—U.S.S.R. (Caucasus).
arenicola (Prodan) Soó in *Acta Bot. Acad. Sci. Hung.*,
1 18(3–4):365(1973): *F. pallens* subsp. *arenicola*.
arundinacea Schreb.
f. **baltica** (A. & G.) Beldie in Fl. Republ. Social.
1 Român., 12:502(1972): *F. arundinacea* var. *baltica*.
f. **obtusiflora** (Schur) Beldie *op. cit.* 505(1972):
1 *F. arundinacea* a. *obtusiflora*.
azgarica Alekseev in *Byull. Mosk. Obshch. Ispȳt.*
2 *Prir., Biol.*, 78(3):107(1973)—U.S.S.R. (Caucasus).
callieri (Hack.) Mgf.
f. **hirta** (Nyár.) Beldie in Fl. Republ. Social.
1 Român., 12:541(1972): *F. pallens* f. *hirta*.
castilloniana Türpe in *Lilloa*, 33(13):285(1973)—
4 Argentina.
cinerea Vill.
subsp. **arenicola** (Prod.) Beldie in Fl. Republ. Social. Român., 12:537(1972): *F. pallens* subsp.
1 *arenicola*.
subsp. **curvula** (Gaud.) Beldie *op. cit.* 534(1972):
1 *F. curvula*.
f. **depauperata** (Hack.) Beldie *op. cit.* 537(1972):
1 *F. ovina* f. *depauperata*.
f. **longifolia** (Viv.) Beldie *op. cit.* 534(1972):
1 *F. longifolia*.
var. **mamaiae** (Prod.) Beldie *op. cit.* 538(1972):
1 *F. pallens* subsp. *mamaiae*.

f. **stenostachya** (Hack.) Beldie *op. cit.* 537(1972):
1 *F. ovina* f. *stenostachya*.
circummediterranea Patzke in *Österr. Bot. Zeitschr.*, 122(4):261(1973), nom. nov.: *F. ovina* subsp. *laevis*
1, 2 var. *genuina* Hackel.
dalmatica (Hack.) K. Richt.
subsp. **panciciana** (Hack.) Beldie in Fl. Republ. Social. Român., 12:549(1972): *F. ovina* var.
1 *panciciana*.
var. **pseudodalmatica** (Kraj.) Beldie *op. cit.* 548
1 (1972): *F. pseudodalmatica*.
f. **pseudorupicola** (Thaisz) Beldie *op. cit.* 549
1 (1972): *F. dalmatica* subvar. *pseudorupicola*.
drymeia Mert. & Koch
f. **lucorum** (Schur) Beldie *op. cit.* 476(1972):
1 *F. lucorum*.
f. **timpae** Morariu ex Beldie *op. cit.* 790(1972)—
1 Romania.
elbrusica Alekseev in *Byull. Mosk. Obshch. Ispȳt.*
2 *Prir., Biol.*, 78(3):107(1973)—U.S.S.R. (Caucasus).
gigantea (L.) Vill.
f. **uliginosa** (Schur) Beldie in Fl. Republ. Social. Român., 12:509(1972): *F. gigantea* b.
1 *uliginosa*.
heldreichii (Hack.) Alekseev in *Byull. Mosk. Obshch. Ispȳt. Prir., Biol.*, 78(3): 105(1973): *F. ovina* subvar.
2 *heldreichii*.
heteropachys (St. Yves) Patzke ex Auquier in Nouv. Fl. Belg., Luxemb., N. France etc.: 759
1 (1973): *F. ovina* subvar. *heteropachys*.
heterophylla Lam.
f. **leiophylla** (Hack.) Beldie in Fl. Republ. Social. Român., 12:513(1972): *F. rubra* subvar.
1 *leiophylla*.
huonii Auquier in *Candollea*, 28(1):16(1973)—
1 France.
inguschetica Alekseev in *Byull. Mosk. Obshch. Ispȳt.*
2 *Prir., Biol.*, 78(3):106(1973)—U.S.S.R. (Caucasus).
jacutica Drob.
subsp. **pobedimovae** Tsvelev in *Novosti Sist. Vȳssh. Rast.*, 10:91(1973), as 'pobedimoviae'—
2 Kuril Is.
kansuensis Markgraf-Dannenberg in *Acta Bot. Acad.*
2 *Sci. Hung.*, 19(1–4):207(1973)—China.
karsiana Alekseev in *Byull. Mosk. Obshch. Ispȳt.*
2 *Prir., Biol.*, 78(3):108(1973)—U.S.S.R. (Caucasus).
ovina L.
f. **nitida** Beldie in Fl. Republ. Social. Român., 12:
1 790(1972)—Romania.
pallens Host
subsp. **pannonica** (Wulf. ex Host) Soó in *Acta Bot.*
1 *Acad. Sci. Hung.*, 18(3–4):371(1973): *F. pannonica*.
pohleana Alekseev in *Byull. Mosk. Obshch. Ispȳt. Prir., Biol.*, 78(5):144(1973)—U.S.S.R. (European
1 Russia).
porcii Hack.
f. **hirsuta** (A. Nyár.) Beldie in Fl. Republ. Social.
1 Român., 12:524(1972): *F. porcii* var. *hirsuta*.
pratensis Huds.
subsp. **apennina** (De Not.) Beldie *op. cit.* 501(1972):
1 *F. apennina*.
primae Alekseev in *Byull. Mosk. Obshch. Ispȳt.*
2 *Prir., Biol.*, 78(3):106(1973)—U.S.S.R. (Caucasus).
pseudovina Hack. ex Wiesbaur
var. **salina** (Kern. ex Hack.) Soó in *Acta Bot.*
1 *Acad. Sci. Hung.*, 18(3–4):373(1973): *F. salina*.
pumila Vill.
var. **hackeliana** (Zapał.) Beldie in Fl. Republ.
1 Social. Român., 12:492(1972): *F. hackeliana*.
rubra L.
subsp. **daghestanica** Tsvelev in *Novosti Sist.*
2 *Vȳssh. Rast.*, 10:92(1973)—U.S.S.R. (Caucasus.)

f. **grandiflora** (Hack.) Beldie in Fl. Republ. Social. Român., 12:522(1972): *F. rubra* subvar.
1 *grandiflora.*
f. **pilosa** Beldie *op. cit.* 522(1972), nom. nov.: *F.*
1 *rubra* subvar. *barbata* Hack.
f. **subheterophylla** (Nyár.) Beldie *op. cit.* 523
1 (1972): *F. rubra* subf. *subheterophylla.*
var. **uliginosa** (Schur) Beldie *op. cit.* 522(1972):
1 *F. rubra* d. *uliginosa.*
rupicola Heuff.
f. **almasului** (Nyár. & Şerb.) Beldie *op. cit.* 552
1 (1972): *F. sulcata* f. *almasului.*
f. **colorata** (Schur) Beldie *op. cit.* 555(1972):
1 *F. colorata.*
var. **coziae** (Nyár.) Beldie *op. cit.* 552(1972):
1 *F. sulcata* var. *coziae.*
1 var. **dacica** Beldie *op. cit.* 790(1972)—Romania.
f. **glaberrima** (Schur) Beldie *op. cit.* 555(1972):
1 *F. glaberrima.*
f. **hirsuta** (Host) Beldie *op. cit.* 551(1972): *F.*
1 *hirsuta.*
f. **hispida** (Hack.) Beldie *op. cit.* 551(1972): *F. ovina*
1 subvar. *hispida.*
1 f. **minor** Beldie *op. cit.* 791(1972)—Romania.
subsp. **pachyphylla** (Deg. ex Nyár.) Beldie *op. cit.*
1 555(1972): *F. pachyphylla.*
var. **retezatensis** Beldie *op. cit.* 791(1972)—
1 Romania.
var. **rodnensis** Beldie *op. cit.* 791(1972)—
1 Romania.
subsp. **saxatilis** (Schur) Beldie *op. cit.* 552(1972):
1 *F. saxatilis.*
f. **strictiflora** (Nyár.) Beldie *op. cit.* 551(1972):
1 *F. sulcata* f. *strictiflora.*
f. **submutica** Beldie *op. cit.* 555(1972), nom. nov.:
1 *F.* × *subspicata* Nyár.
f. **supinoides** (Nyár.) Beldie *op. cit.* 555(1972):
1 *F.* × *supinoides.*
scoparia (Kern. & Hack.) Nym.
subsp. **lutea** (Hack.) Beldie *op. cit.* 496(1972):
1 *F. varia* var. *lutea.*
trachyphylla (Hack.) Kraj.
var. **multinervis** (Stohr) Stohr ex Auquier in Nouv. Fl. Belg., Luxemb., N. France etc.:
1 759(1973): *F. duvalii* var. *multinervis.*
uralensis (Tsvelev) Alekseev in *Byull. Mosk. Obshch. Ispȳt. Prir., Biol.*, 78(5):143(1973): *F. callieri* subsp.
1 *uralensis.*
vaginata W. & K.
subsp. **buiae** (Prod.) Beldie in Fl. Republ. Social.
1 Român., 12:539(1972): *F. glauca* subsp. *buiae.*
1 var. **dubia** Beldie *op. cit.* 790(1972)—Romania.
subvar. **mucronata** (Hack.) Soó in *Acta Bot. Acad. Sci. Hung.*, 18(3–4):371(1973): *F. ovina* subvar.
1 *mucronata.*
valesiaca Schleich. ex Gaud.
var. **asperrima** (Hack.) Alekseev in *Byull. Mosk. Obshch. Ispȳt. Prir., Biol.*, 78(3):109(1973):
2 *F. sulcata* var. *asperrima.*
var. **banatica** (Deg.) Beldie in Fl. Republ. Social.
1 Român., 12:546(1972): *F. valesiaca* f. *banatica.*
subsp. **brunnescens** (Tsvelev) Alekseev in *Byull. Mosk. Obshch. Ispȳt. Prir., Biol.*, 78(3):109(1973):
2 *F. rupicola* subsp. *brunnescens.*
f. **craiovensis** (Buia & Nyár.) Beldie in Fl. Republ. Social Român., 12:547(1972):
1 *F. craiovensis.*
var. **hirsuta** (Link) Alekseev in *Byull. Mosk. Obshch. Ispȳt. Prir., Biol.*, 78(3):109(1973):
2 *F. ovina* var. *hirsuta.*
f. **hirta** (Nyár.) Beldie in Fl. Republ. Social.
1 Român., 12:546(1972): *F. valesiaca* subf. *hirta.*
f. **hirtula** (Nyár.) Beldie *op. cit.* 547(1972):
1 *F. valesiaca* subf. *hirtula.*
f. **longifolia** (Ráv. ex Nyár.) Beldie *op. cit.* 547
1 (1972): *F. valesiaca* subf. *longifolia.*
f. **viridis** (Nyár.) Beldie *op. cit.* 547(1972): *F.*
1 *valesiaca* subf. *viridis.*
var. **wagneri** (Degen., Thaisz & Flatt) Beldie *op.*
1 *cit.* 547(1972): *F. sulcata* subvar. *wagneri.*
violacea Gaud.
f. **aureo-flava** (Schur) Beldie *op. cit.* 514(1972):
1 *F. flaccida* a. *aureoflava.*
var. **flaccida** (Schur) Beldie *op. cit.* 514(1972):
1 *F. flaccida.*
var. **subracemosa** (Schur) Beldie *op. cit.* 514(1972):
1 *F. subracemosa.*
wagneri Degen., Thaisz & Flatt
var. **hungarica** (Soó) Soó in *Acta Bot. Acad. Sci.*
1 *Hung.*, 18(1–2):174(1973): *F. stricta* var. *hungarica.*
xanthina Roem. & Schult.
var. **klasterskyi** (Kraj.) Beldie in Fl. Republ. Social. Român., 12:495(1972): *F. xanthina* subvar.
1 *klasterskyi.*

GAUDINIA

2305 Scannell, M. J. P.
An early record for *Gaudinia fragilis* (L.) Beauv.
1 *Irish Nat. J.*, 17(12):425(1973).

2306 Wacker, F.
Sesshaftwerden eines fremden Grases. Ährenhaber (*Gaudinia fragilis* (L.) P.B.) auf Naturwiesen am Unterlauf der Schutter. *Mitt. Bad. Landesver. Naturk.*
1 *Naturschutz*, 10(3):543–546(1971 publ. 1972).

GLYCERIA

2307 Berger, E. *and* Simon, C.
Zur Verbreitung von *Glyceria declinata* Bréb., insbesondere in der Region der Basler Flora. *Bauhinia*,
1 5(1):28–32(1973).

2308 Holub, J.
Neue oder wenig bekannte Pflanzen der ungarischen Flora. *Ann. Univ. Budapest. Rolando Eötvos Nom.*,
1 *Biol.*, 14:91–104(1972 publ. 1973). *Glyceria declinata.*

2309 Sokolovskaya, A. P. *and* Probatova, N. S.
Chisla khromosom dal'nevostochnȳkh vidov *Glyceria* R. Br. (The chromosome numbers of the far eastern species of *Glyceria* R. Br.). *Bot. Zhurn.*, 58(9):
2 1342–1347(1973)

fluitans (L.) R.Br.
var. **declinata** (Bréb.) Ghişa in Fl. Republ. Social.
1 Român., 12:441(1972): *G. declinata.*
plicata Fries
var. **triticea** (M. T. Lange) Ghişa *op. cit.* 445(1972):
1 *G. plicata* B. *triticea.*

HAYNALDIA

villosa (L.) Schur
var. **glabrata** (Borb.) Anghel in Fl. Republ. Social. Român., 12:643(1972): *Dasypyrum* sive
1 *Pseudosecale villosum* var. *glabratum.*

HELEOCHLOA

2310 Etienne, C. E.
A new grass for California. *Madroño*, 22(1):30
3 (1973)—*Heleochloa alopecuroides.*

HELICTOTRICHON

2311 Gervais, C.
Nouvelles determinations de nombres chromosomiques chez les avoines vivaces: 2. *Bull. Soc. Neuchâtel. Sci. Nat.*, 96:81–87(1973)—(En); map;
1 chromosome numbers.

decorum (Janka) Henrard
f. **heterophyllum** (Schur) Morariu & Beldie in
Fl. Republ. Social. Român., 12:279(1972):
1 *Heuffelia convoluta* b. *heterophylla.*
f. **hirsutum** (Schur) Morariu & Beldie *op. cit.*
1 279(1972): *Heuffelia convoluta* a. *hirsuta.*
filifolium (Lag.) Henrard
subsp. **cantabricum** (Lag.) Paunero in *Bol. Inst. Estud. Astur., Supl. Cienc.*, 16 (Aport. Conoc. Fl.
Cántabro-Astur, x:47):203(1973): *Avena*
1 *cantabrica.*
montanum (Vill.) Henr.
subsp. velutinum (Bss.) Malagarriga, Las Subesp. y
Variac. Geogr.: 28(1973), without basionym date:
1 *Avena filifolia* var. *velutina.*
pratense (L.) Pilger
var. **subdecurrens** (Borb.) Morariu & Beldie in
Fl. Republ. Social. Român., 12:283(1972):
1 *Avena pratensis* subsp. *subdecurrens.*
pubescens (Huds.) Pilger
var. **alpinum** (Gaud.) Morariu & Beldie *op. cit.*
1 280(1972): *Avena pubescens* b. *alpina.*
f. **pallida** (Schur) Morariu & Beldie *op. cit.* 281
1 (1972): *Heuffelia pubescens* c. *pallida.*

HEMARTHRIA

2312 Oakes, A. J.
Hemarthria collection from South Africa. *Turrialba*,
5 23(1):37–40(1973)—(Sp); dot map.

2313 Schank, S. C. *
Chromosome numbers in eleven new *Hemarthria*
(Limpograss) introductions. *Crop Science*, 12(4):550–
551(1972). Abstr. in *Ref. Zhurn. Biol.*, 12(1):v490
4, 5 (1973).

HENRARDIA

persica (Boiss.) C. E. Hubb.
subsp. **glabriglumis** (Nevski) Chopanov, Opred.
Zlakov Turkmenii: 34(1973): *Pholiurus*
2 *glabriglumis.*

HIEROCHLOË

2314 Duckert-Henriod, M.-M.
Hierochloë odorata dans le Jura. *Bull. Soc. Neuchâtel.*
1 *Sci. Nat.*, 96:33–35(1973)—Dot maps.

2315 Weimarck, G.
Male meiosis in some amphimictic and apomictic
Hierochloë (Gramineae). *Bot. Notiser*, 126(1):7–36
1–3 (1973)—Chromosome numbers.

2316 Zotov, V. D.
Hierochloe R. Br. (Gramineae) in New Zealand. *New Zealand J. Bot.*, 11(3): 561–579(1973)— Illustrations,
7 dot maps.

cuprea Zotov in *New Zealand J. Bot.*, 11(3):571
7 (1973)—New Zealand.
7 **equiseta** Zotov *op. cit.* 568(1973)—New Zealand.
7 **fusca** Zotov *op. cit.* 576(1973)—New Zealand.
glabra Trin.
subsp. **sachalinensis** (Printz) Tsvelev in *Novosti Sist. Vȳssh. Rast.*, 10:83(1973): *H. odorata* var.
2 *sachalinensis.*
subsp. **sibirica** (Tsvelev) Tsvelev *op. cit.* 83(1973):
2 *H. odorata* subsp. *sibirica.*
odorata (L.) Beauv.
subsp. **hirta** (Schrank) Tsvelev *op. cit.* 81(1973):
1, 2 *Savastana hirta.*
recurvata (Hackel ex Cheeseman) Zotov in *New Zealand J. Bot.*, 11(3):566(1973): *H. fraseri* var.
7 *recurvata.*

×**zinserlingii** Tsvelev in *Novosti Sist. Vȳssh. Rast.*,
10:83(1973)—U.S.S.R. (European Russia). (*H.*
1 *odorata* subsp. *hirta* × *H. alpina* subsp. *alpina*).

HOLCUS

mollis L.
subsp. reuteri (Bss.) Malagarriga, Las Subesp. y
Variac. Geogr.: 28(1973), without basionym
1 date: *H. reuteri.*

HOMOZEUGOS

2317 Clayton, W. D.
A new species of *Homozeugos* (Gramineae). *Garcia de Orta, Sér. Bot.*, 1(1–2):11–12(1973)—(Por);
5 key.

katakton Clayton in *Garcia de Orta, Sér. Bot.*, 1(1–2):
5 11(1973)—Angola, Zambia.

HORDELYMUS

2318 Danielsson, B.
Ny lokal för *Hordelymus europaeus* i Småland. (A
new locality for *Hordelymus europaeus* (L.) Harz in
Småland, SE. Sweden). *Svensk Bot. Tidskr.*, 67(2):
1 210(1973).

asper (Simk.) Beldie in Fl. Republ. Social. Român.,
1 12:583(1972): *Cuviera caput-medusae* var. *aspera.*

HORDEUM

2319 Anioł, A. *and* Nowacki, E.
Serological relationships within *Hordeum* genus.
Genet. Pol., 14(3):255–267(1973)—(Pol,Rus);
1–3 illustrations.

2320 Gandilyan, P. A.
Dve novȳe nakhodki iz florȳ Armyanskoĭ SSR.
(Two finds in Armenian flora). *Biol. Zhurn. Arm.*,
26(2):89–92(1973)—(Arm); illustrations; chromosome
2 numbers. *Hordeum murinum.*

2321 Hunziker, J. H., Naranjo, C. A. *and* Zeiger, E.
Las relaciones evolutivas entre *Hordeum compressum*
y otras especies diploides americanas afines. *Kurtziana*,
1–4 7:7–26(1973)—(En); chromosome numbers.

2322 Kruse, A.
Hordeum × *Triticum* hybrids. *Hereditas*, 73(1):157–
1, 2 161(1973)—Chromosome numbers.

distichon L.
convar. **nudum** (L.) Tsvelev in *Novosti Sist. Vȳssh. Rast.*, 10:55(1973): *H. distichon* var.
0 *nudum.*
convar. **zeocrithon** (L.) Tsvelev *op. cit.* 55(1973):
0 *H. zeocrithon.*
marinum Huds.
subsp. **gussoneanum** (Parl.) Anghel & Velican in
Fl. Republ. Social. Român., 12:591(1972): *H.*
1 *gussoneanum.*
vulgare L.
convar. **revelatum** (Koern.) Tsvelev in *Novosti Sist. Vȳssh. Rast.*, 10:55(1973): *H. vulgare* var.
0 *revelatum.*

ICHNANTHUS

2323 Rogers, K. E.
Studies in *Ichnanthus* (Gramineae): 3. Additional
new taxa in section *Foveolata. Phytologia*, 26(2):57–66
4 (1973).

breviacuminatus K. E. Rogers in *Phytologia*, 26(2):
4 57(1973)—Brazil.

4 **exilis** K. E. Rogers *op. cit.* 58(1973)—Brazil.
papillatus K. E. Rogers *op. cit.* 59(1973)—Ecuador,
4 Guyana, Brazil.
parodii K. E. Rogers
var. **villosissimus** K. E. Rogers *op. cit.* 60(1973)—
4 Brazil.
4 **petraeus** K. E. Rogers *op. cit.* 61(1973)—Peru.
sabulosus K. E. Rogers *op. cit.* 61(1973)—
4 Venezuela.
sucrensis K. E. Rogers *op. cit.* 63(1973)—
4 Venezuela.
4 **swallenii** K. E. Rogers *op. cit.* 64(1973)—Brazil.
4 **tipuaniensis** K. E. Rogers *op. cit.* 65(1973)—Bolivia.

ISACHNE

2324 Singh, N. P. *and* Deshpande, U. R.
Report of an endemic Ceylonese grass from India.
Indian Forester, 99(11):674–675(1973)—Illustration.
6 *Isachne globosa* var. *effusa*.

ISCHAEMUM

2325 Oke, J. G.
Karyotypic variations, meiotic behaviour and phylogenetic inter-relationships in some species of *Ischaemum* Linn. and *Sehima* Forsk. in parts of western India. *Proc. Indian Acad. Sci.*, *B*, 78(2):80–88(1973)—
6 Chromosome numbers.

KOELERIA

2326 Ujhelyi, J.
Data to the systematics of the sections *Bulbosae* and *Caespitosae* of the genus *Koeleria*, II. *Ann. Hist.-Nat.*
1 *Mus. Nation. Hung.*, 65:83–108(1973)—Illustrations.

2327 Ujhelyi, J. *
Evolutionary problems of the European Koelerias.
Symp. Biol. Hung., 12:163–176(1972)—Ref. in *Bull.*
1 *Signal., Biol. Physiol. Vég.*, 34(1):151(1973).

ser. **Tuzsoniae** Ujhelyi in *Ann. Hist.-Nat. Mus.*
1 *Nation. Hung.*, 65:90(1973).
macrantha (Ldb.) J. A. & J. H. Schult.
var. **arenicola** (Domin) Ghişa in Fl. Republ. Social. Român., 12:238(1972): *K. gracilis* var.
1 *arenicola*.
f. **aristulata** (Domin) Ghişa *op. cit.* 237(1972):
1 *K. gracilis* f. *aristulata*.
var. **borbasii** (Domin) Ghişa *op. cit.* 237(1972):
1 *K. gracilis* var. *borbasii*.
var. **colorata** (Heuff.) Ghişa *op. cit.* 237(1972):
1 *K. cristata* δ *colorata*.
var. **eminens** (Csató) Ghişa *op. cit.* 237(1972):
1 *K. eminens*.
var. **fenzliana** (Schur) Ghişa *op. cit.* 236(1972):
1 *K. fenzliana*.
var. **flaccida** (Domin) Ghişa *op. cit.* 237(1972):
1 *K. gracilis* var. *flaccida*.
f. **glabrescens** (Domin) Ghişa *op. cit.* 237(1972):
1 *K. gracilis* f. *glabrescens*.
f. **lanata** (Domin) Ghişa *op. cit.* 237(1972): *K.*
1 *gracilis* f. *lanata*.
var. **montana** (Hausm.) Ghişa *op. cit.* 236(1972):
1 *K. cristata* var. *montana*.
var. **pseudocristata** (Domin) Ghişa *op. cit.* 237
1 (1972): *K. cristata* var. *pseudocristata*.
var. **rigidula** (Simk.) Ghişa *op. cit.* 236(1972):
1 *K. rigidula*.
var. **rohlenae** (Domin.) Ghişa *op. cit.* 236(1972):
1 *K. gracilis* var. *rohlenae*.
tuzsonii Ujhelyi in *Ann. Hist.-Nat. Mus. Nation.*
1 *Hung.*, 65:92(1973)—Bulgaria.
var. **albanica** (Dom.) Ujhelyi *op. cit.* 95(1973):
1 *K. splendens* var. *albanica*.

LAGURUS

2328 Misri, B.
On the occurrence of *Lagurus ovatus* Linn. in India.
6 *Curr. Sci.*, 42(12):439(1973).

LASIACIS

2329 Davidse, G.
A systematic study of the genus *Lasiacis* (Gramineae, Panicoideae). *Diss. Abstr. Int.*, *B*, 33(10):4689(1973)—
4 Chromosome numbers.

LASIURUS

2330 Mehta, K. *and* Bhandari, M. M.
On the chromosome study of *Lasiurus* species from Indian desert. *Chromosome Inform. Serv.*, no.15:
6 22–23(1973)—Chromosome numbers.

LEPTOCHLOA

acuminata (Nash) Mohlenbrock, Illus. Fl. Illinois, Grasses, *Panicum* to *Danthonia*: 293 (1973):
3 *Diplachne acuminata*.

×**LEYMOSTACHYS** Tsvelev in *Novosti Sist.*
2 *Vўssh. Rast.*, 10:59(1973). *Leymus* × *Psathyrostachys*.
korovinii Tsvelev *op. cit.* 59(1973)—U.S.S.R. (Middle Asia). (*Leymus ? racemosus* × *Psathyro-*
2 *stachys juncea*).

×LEYMOTRIGIA

stricta (Dethard.) Tsvelev in *Novosti Sist. Vўssh.*
1 *Rast.*, 10:58(1973): *Triticum strictum*.
wiluica (Drob.) Tsvelev *op. cit.* 58(1973): *Agropyron*
2 *wiluicum*.

LEYMUS

2331 Miroslavov, E. A. *and* Zhigar, M. P.
Élektronnomikroskopicheskoe izuchenie naruzhnўkh stenok épidermal'nўkh kletok *Poa annua* L., *Dactylis glomerata* L. i *Leymus racemosus* (Lam.) Tzvel. (Electron microscopic study of the outer walls of epidermal cells in *Poa annua* L., *Dactylis glomerata* L. and *Leymus racemosus* (Lam.) Tzvel.) *Bot. Zhurn.*,
1, 2 58(10):1407–1412(1973)—(En).

alaicus (Korsh.) Tsvelev
subsp. **karataviensis** (Roshev.) Tsvelev in *Novosti Sist. Vўssh. Rast.*, 10:50(1973): *Elymus*
2 *karataviense*.
subsp. **petraeus** (Nevski) Tsvelev *op. cit.* 50(1973):
2 *Aneurolepidium petraeum*.
×**jenisseiensis** (Turcz.) Tsvelev *op. cit.* 51(1973):
2 *Elymus jenisseiensis*.
karelinii (Turcz.) Chopanov, Opred. Zlakov Turkmenii: 28(1973): *Elymus karelinii*, combination already made by Tsvelev in *Novosti Sist. Vўssh.*
2 *Rast.*, 9:59(1972).
var. **kirghisorum** (Drob.) Tsvelev in *Novosti*
2 *Sist. Vўssh. Rast.*, 10:49(1973): *Elymus kirghisorum*.
nikitinii (Czopan.) Tsvelev in *Novosti Sist. Vўssh. Rast.*, 10:50 (30 March 1973), also Chopanov, Opred. Zlakov Turkmenii: 28 (5 July 1973):
2 *Elymus nikitinii*.
×**ramosoides** Kolmak. ex Tsvelev in *Novosti Sist. Vўssh. Rast.*, 10:51(1973)—U.S.S.R. (European
1 Russia). (*L. paboanus* × *L. ramosus*).
secalinus (Georgi) Tsvelev
subsp. **ovatus** (Trin.) Tsvelev *op. cit.* 49(1973):
2 *Elymus ovatus*.

LOLIUM

2332 Beddows, A. R.
Lolium multiflorum Lam. *J. Ecol.*, 61(2):587–600
(1973) (Biological Flora of the British Isles)—
1 Illustrations, dot map.

2333 Gymer, P. T. *and* Whittington, W. J.
Hybrids between *Lolium perenne* and *Festuca*
pratensis: 2. Comparative morphology. *New Phytol.*,
1, 2 72(4):861–865(1973)—Chromosome numbers.

2334 Kerguélen, M.
Un *Lolium* nouveau pour la flore française *Lolium*
parabolicae Sennen ex Sampaio. *Lejeunia*, no. 65. 7p.
1 (1972)—Illustrations.

multiflorum Lam.
f. **submuticum** (Mutel.) Anghel & Beldie in Fl.
Republ. Social. Român., 12:575(1972): *L.*
1 *multiflorum submuticum*.
remotum Schrank
f. **asperum** (Roth) Anghel & Beldie *op. cit.* 572
1 (1972): *L. asperum*.

LOUDETIA

2335 Lubke, R. A. *and* Phipps, J. B.
Taximetrics of *Loudetia* (Gramineae) based on leaf
anatomy. *Canad. J. Bot.*, 51(11):2127–2146(1973)—
5 (Fr); illustrations.

MACLUROLYRA C. E. Calderón & Soderstrom
4 in *Smithsonian Contrib. Bot.*, 11:6(1973).

2336 Calderón, C. E. *and* Soderstrom, T. R.
Morphological and anatomical considerations of
the grass subfamily Bambusoideae based on the new
genus *Maclurolyra*. *Smithsonian Contrib. Bot.*, no. 11.
4 [iv]55p.(1973)—Illustrations, map.

tecta C. E. Calderón & Soderstrom in *Smithsonian*
4 *Contrib. Bot.*, 11:6(1973)—Panama.

MELICA

2337 Hempel, W.
Die systematische Stellung von *Melica altissima* L.
und des *Melica minuta* L.-*Melica ramosa* Vill.-Kom-
plexes (*Melica* L. subgen. *Altimelica* Hempel.)
(Vorarbeiten zu einer Revision der Gattung *Melica*
L.-3). *Feddes Repert.*, 84(7–8):533–568(1973)—(En,
4 Fr,Sp,Rus); illustrations, dot maps.

2338 Kalheber, H.
Zur Verbreitung von *Melica ciliata* L. und *Melica*
transsilvanica Schur im mittleren Lahngebiet. *Hess.*
1 *Flor. Briefe*, 253–255:10–11(1973).

subgen. **Altimelica** Hempel in *Feddes Repert.*, 84
1, 2 (7–8):539(1973).
subsect. **Chondrachyrum** (Nees) Hempel *op. cit.*
4 541(1973): gen. *Chondrachyrum*.
sect. **Husnotchloa** Tsvelev in *Novosti Sist. Výssh.*
Rast., 10:84(1973), nom. nov.: *Melica* subsect.
1, 3 *Uniflorae* Hempel.
sect. **Macrae** Hempel in *Feddes Repert.*, 84(7–8):541
4 (1973).
4 sect. **Melicella** Hempel *op. cit.* 541(1973).
4 subsect. **Melicula** Hempel *op. cit.* 544(1973).
argyrea Hack.
f. **pilosa** (Papp) Hempel *op. cit.* 553(1973): *M.*
4 *argyrea* var. *pilosa*.
brasiliana Ard.
var. **aurantiaca** (Desr. ex Lam.) Hempel *op. cit.*
4 554(1973): *M. aurantiaca*.
f. **tandilensis** (Ktze.) Hempel *op. cit.* 554(1973):
4 *M. aurantiaca* var. *tandilensis*.
f. **truncata** (Papp) Hempel *op. cit.* 554(1973):
4 *M. hyalina* var. *truncata*.
ciliata L.
f. **aequiglumis** (Papp) C. Papp & Beldie in Fl.
Republ. Social. Român., 12:243(1972): *M.*
1 *ciliata* var. *aequiglumis*.
f. **flavescens** (Schur) C. Papp & Beldie *op. cit.* 243
1 (1972): *M. ciliata* b. *flavescens*.
f. **glabrata** (Čel.) C. Papp & Beldie *op. cit.* 244
1 (1972): *M. ciliata* var. *glabrata*.
1 f. **latifolia** C. Papp *op. cit.* 789(1972)—Romania.
var. **lobata** (Schur) C. Papp & Beldie *op. cit.* 244
1 (1972): *M. lobata*.
f. **violacea** (Papp) C. Papp & Beldie *op. cit.* 243
1 (1972): *M. ciliata* var. *violacea*.
laxiflora Cav.
f. **hirta** (Ph.) Hempel in *Feddes Repert.*, 84(7–8):
4 555(1973): *M. hirta*.
nutans L.
f. **fissurae** (Nyár.) C. Papp & Beldie in Fl. Republ.
Social. Român., 12:248(1972): *M. picta* f.
1 *fissurae*.
picta K. Koch
1 f. **pilosa** C. Papp *op. cit.* 789(1972)—Romania.
scabra H.B.K.
f. **pubescens** (Papp) Hempel in *Feddes Repert.*,
4 84(7–8):556(1973): *M. scabra* var. *pubescens*.
transsilvanica Schur
subsp. **atropatana** (Schischk.) Tsvelev in *Novosti*
2 *Sist. Výssh. Rast.*, 10:85(1973): *M. atropatana*.
violacea Cav.
f. **mollis** (Papp) Hempel in *Feddes Repert.*, 84(7–8):
4 557(1973): *M. alata* var. *mollis*.

MEROSTACHYS

polyantha McClure in *Smithsonian Contrib. Bot.*, 9:
4 91(1973)—Brazil.

MILIUM

effusum L.
f. **lerchenfeldianum** (Schur) Morariu in Fl.
Republ. Social. Român., 12:185(1972): *M.*
1 *transsilvanicum* a. *lerchenfeldianum*.

MOLINIA

caerulea (L.) Mnch.
var. **altissima** (Link.) Dobrescu in Fl. Republ.
1 Social. Român., 12:353(1972): *M. altissima*.

NARDURUS

2339 Kožuharov, S. I. *and* Petrova, A. V.
Cytotaxonomic study of five rare Bulgarian
grasses. *Dokl. Bolg. Akad. Nauk*, 26(6):829–832
(1973)—Chromosome numbers. *Nardurus maritimus*,
1 *N. tenellus*.

maritimus (L.) Murb.
subsp. **krausei** (Regel) Tsvelev in *Novosti Sist.*
2 *Výssh. Rast.*, 10:87(1973): *Festuca krausei*.
var. **villosus** (Maire) Tsvelev *op. cit.* 87(1973):
1 *N. maritimus* subvar. *villosus*.

NASELLA

2340 Tasmania: Department of Agriculture
Weeds and weed control: serrated tussock (*Nasella*
trichotoma), declared dangerous weed. *Tasman. J.*
4, 7 *Agric.*, 44(4):241–243(1973)—Illustration.

ORYZA

2341 Krasnook, N. P., Erÿgin, I. S. *and* Petibskaya, V. S. *
Zakonomernosti formirovaniya zernovki risa.
Biologiya, 6(6):836–843(1971)—(En). Abstr. in *Ref.*
2, 6 *Zhurn. Biol.*, 3(1):v378(1973). *Oryza sativa.*

2342 Mehta, R. N. *and* Oza, G. M.
Botanical identity of carbonized cereal grains from
6 Nagara. *Curr. Sci.*, 42(5):179(1973).

PANICUM

2343 Adam, J.-G.
Un *Panicum* américain colonise l'Ouest Africain
Panicum laxum Sw. *J. Agric. Trop. Bot. Appl.*, 19(1–3):
4, 5 59–62(1972 publ. 1973)—Illustration.

2344 Anderson, C. A.
Cytology and histology of vine mesquite grass
(*Panicum obtusum* HBK). *Diss. Abstr. Int.*, *B*, 33(9):
4 4145(1973).

2345 Le Clerch, J.
Introduction d'une nouvelle Graminée en France:
Panicum dichotomiflorum Michx. *Bull. Soc. Bot. France*,
1 120(5–6):223–226(1973)—(En).

2346 Muniyamma, M. *and* Narayan, K. N.
Cytoembryological studies in *Panicum paludosum*
Roxb. *Proc. Indian Sci. Congr.*, 59(3):359(1972)—
6 Chromosome number. Abstract only.

2347 Nickell, G. L.
The physiological ecology of upland and lowland
Panicum virgatum. *Diss. Abstr. Int.*, *B*, 33(8):3512
3 (1973).

2348 Palmer, P. G.
A biosystematic study of the *Virgata* group of
Panicum (Gramineae) in the United States. *Diss.*
3 *Abstr. Int.*, *B*, 33(12):5697(1973).

rigidulum Nees
var. **condensum** (Nash) Mohlenbrock, Illus. Fl.
Illinois, Grasses, *Panicum* to *Danthonia*: 71
3 (1973): *P. condensum.*

PASPALIDIUM

disjunctum S. T. Blake in *Proc. Roy. Soc. Queensl.*,
84: 65(1973)—Australia (Queensland), New
7 Guinea (Papua, Terr. New Guinea).
gausum S. T. Blake *op. cit.* 68(1973)—Australia
7 (Queensland, New South Wales).
scabrifolium S. T. Blake *op. cit.* 69(1973)—
7 Australia (Queensland).

PASPALUM

2349 Murata, G.
[Short note on *Paspalum notatum* and *P. minus*.].
6 *Acta Phytotax. Geobot.*, 25(4–6):185(1973)—Ja.

sect. **Diplostachys** (Steud.) Tsvelev in *Novosti*
Sist. Vÿssh. Rast., 10:79(1973): *Panicum* sect.
1–7 *Diplostachys.*
canarae (Steud.) Veldkamp in *Blumea*, 21(1):72
6 (1973): *Panicum canarae.*
var. **fimbriatum** (Bor) Veldkamp *op. cit.* 72
6 (1973): *Panicum canarae* var. *fimbriatum.*

PENNISETUM

2350 Bono, M.
Contribution à la morphosystématique des
Pennisetum annuels cultivés pour leur grain en Afrique
occidentale francophone. (Contribution to the
morpho-systematics of the annual *Pennisetum* grown
for their grains in francophone West Africa). *Agron.*
Trop., 28(3):229–355(1973)—(En,Sp); illustrations,
5 maps, dot maps; key.

PHALAROIDES

arundinacea (L.) Rausch.
subsp. **caesia** (Nees) Tsvelev in *Novosti Sist.*
5 *Vÿssh. Rast.*, 10:80(1973): *Phalaris caesia.*
subsp. **japonica** (Steud.) Tsvelev *op. cit.* 80(1973):
2 *Phalaris japonica.*
var. **picta** (L.) Tsvelev *op. cit.* 80(1973): *Phalaris*
1–3 *arundinacea* var. *picta.*

PHLEUM

alpinum L.
subsp. **commutatum** (Gaud.) Malagarriga, Las
Subesp. y Variac. Geogr.: 27(1973): *P.*
1 *commutatum.*
f. **subcapitatum** (Schur) Şerbănescu & E. I.
Nyárády in Fl. Republ. Social. Român., 12:124
1 (1972): *P. alpinum* b. *subcapitatum.*
phleoides (L.) Karsten
var. **interruptum** (Zabel) Şerbănescu & E. I.
Nyárády *op. cit.* 125(1972): *P. boehmeri* var.
1 *interruptum.*
pratense L.
subsp. **bertolonii** (DC.) Şerbănescu & E. I.
1 Nyárády *op. cit.* 123(1972): *P. bertolonii.*
subsp. **brachystachyum** (Salis) Gamisans in
Candollea, 28(1):48(1973): *P. pratense* var.
1, 4 *brachystachyum.*
f. **elongatum** (Schur) Şerbănescu & E. I. Nyárády
in Fl. Republ. Social. Român., 12:123(1972):
1 *P. pratense* c. *elongatum.*
f. **fallax** (C. J. Klinggr.) Şerbănescu & E. I.
Nyárády *op. cit.* 124(1972): *P. pratense* c.
1 *fallax.*
subsp. **microstachyum** (Ruiz) Malagarriga, Las
Subesp. y Variac. Geogr.: 27(1973): *P. micro-*
1 *stachyum.*
var. **stoloniferum** (Host) Şerbănescu & E. I.
Nyárády in Fl. Republ. Social. Român., 12:123
1 (1972): *P. stoloniferum.*

PHRAGMITES

2351 Pazourek, J.
The density of stomata in leaves of two ecotypes of
Phragmites communis Trin. in southern Bohemia.
(Hustota průduchů u listů dvou ekotypů rákosu
Phragmites communis (Trin.) v jižních Čechách).
1–7 *Preslia*, 45(3):242–249(1973)—(Cz).

2352 Raicu, P., Staicu, S., Stoian, V. *and* Roman, T.
The *Phragmites communis* Trin. chromosome
complement in the Danube delta. *Hydrobiologia*,
1 39(1):83–89(1972).

2353 Wallentinus, H.-G., Gustafsson, K. *and*
Söderström, B.
Bladvassen, *Phragmites communis* Trin., i Brunns-
viken, Stockholm 1971. (The reed, *Phragmites*
communis Trin., in the Brunnsviken, Stockholm
1971). *Svensk Bot. Tidskr.*, 67(2):81–96(1973)—(En);
1 dot map.

australis (Cavanilles) Trin. ex Steud.
var. **stenophylla** (Boissier) Jain & Doli Das in
Indian Forester, 99(9):575(1973): *P. communis*
2, 6 var. *stenophylla.*
f. **subuniflora** (DC.) Dobrescu in Fl. Republ.
Social. Român., 12:212(1972): *Arundo phragmites*
1 var. *subuniflora.*

PLEIONEURA (C. E. Hubbard) J. B. Phipps in
Bol. Soc. Brot., 46:418(1972 publ. 1973): *Loudetia*
5 sect. *Pleioneura*.
sect. **Rattraya** (J. B. Phipps) J. B. Phipps *op. cit.* 420
5 (1972 publ. 1973): gen. *Rattraya*.
anomala (C. E. Hubbard & Schweick.) J. B. Phipps
5 *op. cit.* 420(1972 publ. 1973): *Loudetia anomala*.
petiolata (J. B. Phipps) J. B. Phipps *op. cit.* 421 (1972
5 publ. 1973): *Rattraya petiolata*.
ramosa (Stapf) J. B. Phipps *op. cit.* 419(1972 publ.
5 1973): *Trichopteryx ramosa*.
simulans (C. E. Hubbard) J. B. Phipps *op. cit.* 420
5 (1972 publ. 1973): *Loudetia simulans*.

POA

2354 Ellis, W. M.
The breeding system and variation in populations
of *Poa annua* L. *Evolution*, 27(4):656–662(1973)—
1–7 Chromosome numbers.

2355 Miroslavov, E. A. *and* Zhigar, M. P.
Élektronnomikroskopicheskoe izuchenie naruzh-
nȳkh stenok épidermal'nȳkh kletok *Poa annua* L.,
Dactylis glomerata L. i *Leymus racemosus* (Lam.)
Tzvel. (Electron microscopic study of the outer walls
of epidermal cells in *Poa annua* L., *Dactylis glomerata* L.
and *Leymus racemosus* (Lam.) Tzvel.) *Bot. Zhurn.*,
1, 2 58(10):1407–1412(1973)—(En).

2356 Probatova, N. S.
O sistematicheskom polozhenii *Poa irkutica*
Roshev. (De *Poa irkutica* Roshev. nota). *Novosti*
2 *Sist. Vȳssh. Rast.*, 10:9–11(1973).

2357 Sokolovskaya, A. P. *and* Probatova, N. S.
Kariosistematicheskoe issledovanie dal'nevo
stochnȳkh vidov *Poa* L.: 2. (Karyosystematic investi-
gations of the Far East species of *Poa* L.). *Bot. Zhurn.*,
2 58(1):89–96(1973)—Chromosome numbers.

2358 Tateoka, T.
A taxonomic study of the *Poa macrocalyx* complex,
with particular reference to the populations of eastern
Hokkaido. *Bot. Mag. Tokyo*, 86(1003):213–228
2, 3 (1973)—Illustrations, maps; chromosome numbers.

abbreviata R.Br.
subsp. **jordalii** (Ports.) Hultén in *Bot. Notiser*,
3 126(4):468(1973): *P. jordalii*.
atropidiformis Hackel
var. **patagonica** (Parodi) Nicora in *Darwiniana*,
4 18(1–2):97(1973): *Dissanthelium patagonicum*.
bactriana Roshev.
subsp. **drobovii** Tsvelev in *Novosti Sist. Vȳssh.*
2 *Rast.*, 10:96(1973)—U.S.S.R. (Middle Asia).
subsp. **glabriflora** (Roshev. ex Ovcz.) Tsvelev
2 *op. cit.* 96(1973): *P. glabriflora*.
var. **vivipara** Tsvelev *op. cit.* 96(1973)—U.S.S.R.
2 (Middle Asia).
subsp. **zaprjagajevii** (Ovcz.) Tsvelev *op. cit.* 96
2 (1973): *P. zaprjagajevii*.
badensis Haenke
var. **imeretica** (Somm. & Levier) Tsvelev *op. cit.*
2 93(1973): *P. imeretica*.
bulbosa L.
subsp. **nevskii** (Roshev ex Ovcz.) Tsvelev *op. cit.*
2 95(1973): *P. nevskii*.
subsp. **sinaica** (Steud.) Tsvelev *op. cit.* 95(1973):
2 *P. sinaica*.
delicatula Tsvelev *op. cit.* 94(1973)—U.S.S.R.
2 (Caucasus).

deylii Chrtek & Jirásek
f. **breazensis** (Nyár.) Ghişa & Beldie in Fl.
Republ. Social. Român., 12:399(1972): *P.* ×
1 *breazensis*.
f. **pietrosuana** (Zapał.) Ghişa & Beldie *op. cit.*
1 399(1972): *P. cenisia* b. *pietrosuana*.
var. **retezatensis** (A. Nyár.) Ghişa & Beldie *op. cit.*
1 399(1972): *P. granitica* subsp. *retezatensis*.
f. **subgranitica** (Nyár.) Ghişa & Beldie *op. cit.*
1 399(1972): *P. granitica* var. *subgranitica*.
kamczatensis Probatova in *Novosti Sist. Vȳssh.*
2 *Rast.*, 10:70(1973)—U.S.S.R. (Soviet Far East).
laxa Haenke
f. **racemosa** (Schur) Ghişa & Beldie in Fl. Republ.
Social. Român., 12:391(1972): *P. supina* a.
1 *racemosa*.
media Schur
f. **ovata** (Nyár.) Ghişa & Beldie *op. cit.* 380(1972):
1 *P. media* var. *ovata*.
molineri Balbis
var. **subtilis** (Schur) Ghişa & Beldie *op. cit.* 387
1 (1972): *P. subtilis*.
nemoralis L.
f. **glaucantha** (Gaud.) Ghişa & Beldie *op. cit.* 404
1 (1972): *P. glaucantha*.
sichotensis Probatova in *Novosti Sist. Vȳssh. Rast.*,
2 10:68(1973)—U.S.S.R. (Siberia).
skvortzovii Probatova *op. cit.* 72(1973), nom. nov.:
2 *P. pseudonemoralis* Skvortz.
timoleontis Heldr. ex Boiss.
var. **dshilgensis** (Roshev) Tsvelev *op. cit.* 94
2 (1973): *P. dshilgensis*.
tremula Schur
f. **pendula** (Tuzson) Ghişa & Beldie in Fl. Republ.
Social. Român., 12:392(1972): *P. minor* f.
1 *pendula*.
trivialis L.
f. **humilis** (Schur) Ghişa & Beldie *op. cit.* 421
1 (1972): *P. trivialis* b. *humilis*.
var. **multiflora** (Rchb.) Ghişa & Beldie *op. cit.*
1 417(1972): *P. trivialis multiflora*.
f. **prorepens** (Schur) Ghişa & Beldie *op. cit.* 418
1 (1972): *P. trivialis* a. *prorepens*.
violacea Bell.
f. **flavescens** (Mert. & Koch) Ghişa & Beldie
1 *op. cit.* 368(1972): *Festuca rhaetica* β *flavescens*.

PSATHYROSTACHYS

2359 Ponert, J.
Eine neue Art und eine neue Sektion in der Gattung
Psathyrostachys aus Iran. *Feddes Repert.*, 83(7–8):507–
2 509(1972 publ. 1973)—(En); illustration.

sect. **Elborsia** Ponert in *Feddes Repert.*, 83(7–8):509
2 (1972 publ. 1973).
scabriphylla Ponert *op. cit.* 507(1972 publ. 1973)—
2 Iran.

PSILURUS

2360 Kožuharov, S. I. *and* Petrova, A. V.
Cytotaxonomic study of five rare Bulgarian grasses.
Dokl. Bolg. Akad. Nauk, 26(6):829–832(1973)—
1 Chromosome numbers. *Psilurus aristatus*.

aristatus (L.) Duval-Jouve
f. **hirtellus** (Simk.) Csűrös in Fl. Republ. Social.
1 Român., 12:567(1972): *P. hirtellus*.

PTILAGROSTIS

2361 Peshkova, G. A.
O rode *Ptilagrostis* Griseb. v vostochnoĭ Sibiri. (De
genere *Ptilagrostis* Griseb. in Sibiria orientalis).
2 *Novosti Sist. Vȳssh. Rast.*, 10:4–9(1973).

PUCCINELLIA

2362 Seybold, S.
Der Salzschwaden (*Puccinellia distans* (Jacq.) Parl.)
an Bundesstrassen und Autobahnen. *Göttinger Flor.*
1 *Rundbr.*, 7(4):70–73(1973)—Illustration, dot map.

sect. **Pseudocolpodium** Tsvelev in *Novosti Sist.*
1, 2 *Vÿssh. Rast.*, 10:85(1973).
× **beringensis** Tsvelev *op. cit.* 86(1973)—U.S.S.R.
2 (Soviet Far East). (*P. wrightii* × *P. tenella*).
transsilvanica (Schur) Jáv.
var. **arenaria** (Schur) Morariu in Fl. Republ.
Social. Român., 12:450(1972): *A. intermedia* b.
1 *arenaria*.
var. **pseudomaritima** (Schur) Morariu *op. cit.*
1 450(1972): *Atropis intermedia* a. *pseudomaritima*.
wrightii (Scribn. & Merr.) Tsvelev
subsp. **colpodioides** (Tsvelev) Tsvelev in *Novosti*
2 *Sist. Vÿssh. Rast.*, 10:86(1973): *P. colpodioides*.

RHIPIDOCLADUM McClure in *Smithsonian*
4 *Contrib. Bot.*, 9:101(1973).
ampliflorum (McClure) McClure *op. cit.* 105(1973):
4 *Arthrostylidium ampliflorum*.
angustiflorum (Stapf) McClure *op. cit.* 105(1973):
4 *Arthrostylidium angustiflorum*.
bartlettii (McClure) McClure *op. cit.* 105(1973):
4 *Arthrostylidium bartlettii*.
geminatum (McClure) McClure *op. cit.* 105(1973):
4 *Arthrostylidium geminatum*.
harmonicum (Parodi) McClure *op. cit.* 105(1973):
4 *Arthrostylidium harmonicum*.
maxonii (Hitchcock) McClure *op. cit.* 105(1973):
4 *Arthrostylidium maxonii*.
parviflorum (Trinius) McClure *op. cit.* 105(1973):
4 *Arundinaria parviflora*.
pittieri (Hackel) McClure *op. cit.* 105(1973):
4 *Arthrostylidium pittieri*.
prestoei (Munro) McClure *op. cit.* 106(1973):
4 *Arthrostylidium prestoei*.
racemiflorum (Steudel) McClure *op. cit.* 106(1973):
4 *Arthrostylidium ? racemiflorum*.
verticillatum (Nees) McClure *op. cit.* 106(1973):
4 *Arundinaria verticillata*.

RYTIDOSPERMA

glabra (Phil.) Nicora in *Darwiniana*, 18(1–2):87
4 (1973): *Danthonia glabra*.
picta (Nees & Meyen) Nicora *op. cit.* 91(1973):
4 *Danthonia picta*.
var. **bimucronata** Nicora *op. cit.* 91(1973)—
4 Argentina.
4 **sorianoi** Nicora *op. cit.* 89(1973)—Argentina.
violacea (Desvaux) Nicora *op. cit.* 91(1973):
4 *Danthonia violacea*.
virescens (Desv.) Nicora *op. cit.* 93(1973): *Danthonia*
4 *virescens*.
var. **parvispicula** Nicora *op. cit.* 95(1973)—
4 Argentina.
var. **patagonica** (Spegazzini) Nicora *op. cit.* 93
4 (1973): *Danthonia picta* var. *patagonica*.

SACCHARUM

2363 Bhandari, M.
A note on the occurrence of *Saccharum rufipilum*
Steud. in the Dehra Dun Siwaliks. *Indian Forester*,
6 99(4):241–242(1973).

2364 Nair, M. K.
Cytogenetics of *Saccharum*: 3. Karyotype analysis
and meiosis in *S. spontaneum*. *Nucleus*, 15(2):107–117
1, 2, 5, 6 (1972).

2365 Nair, M. K.
Cytogenetics of *Saccharum officinarum* L., *Saccharum*
spontaneum L. and *S. officinarum* × *S. spontaneum*
hybrids: 1. Chromosome mosaics. *Cytologia*, 37(4):
1, 2, 6 565–573(1972)—Chromosome numbers.

SASA

2366 Suzuki, S.
The range of the genus *Sasa* Makino et Shibata (6).
Hikobia, 16(3–4):260–264(1973)—Ja(En); dot maps;
2 key.

SASAELLA

2367 Hatusima, S.
A new bamboo from Kyusyu. *J. Geobot.*, 20(2):
2 36–37(1972)—En, Ja; illustration.

marunoi Hatusima in *J. Geobot.*, 20(2):37(1972)—
2 Japan.

SCOLOCHLOA

2368 Smith, A. L.
Life cycle of the marsh grass, *Scolochloa festucacea*.
Canad. J. Bot., 51(9):1661–1668(1973)—(Fr);
3 illustration, dot map.

SECALE

2369 Kandelaki, G. V. *and* Meliya, N. S.
O skreshchivaemosti *Tr. araraticum* Jakubz. i
Tr. timopheevi Zhuk. s gornoĭ rozh'yu *Secale*
montanum Guss. (Concerning the crossability of *Tr.*
araraticum Jukub. and *Tr. timopheevi* Zhuk. with
mountain rye). *Soobshch. Akad. Nauk Gruz., SSR*,
1, 2 70(1):177–180(1973)—Rus(Geo,En).

2370 Loxton, A. E.
Wild rye. (Wilderog, *Secale africanum* Stapf).
5 *Veld Flora*, 3(4):75(1973)—En; map.

× **derzhavinii** Tsvelev in *Novosti Sist. Vÿssh. Rast.*,
10:47(1973)—Cultivation. (*S. montanum* × *S.*
0 *cereale*).
montanum Guss.
subsp. **anatolicum** (Boiss.) Tsvelev *op. cit.* 46
2 (1973): *S. anatolicum*.
subsp. **chaldicum** (An. Fed.) Tsvelev *op. cit.* 46
2 (1973): *S. chaldicum*.
subsp. **kuprijanovii** (Grossh.) Tsvelev *op. cit.* 46
2 (1973): *S. kuprijanovii*.
segetale (Zhuk.) Roshev.
var. **afghanicum** (Vav.) Tsvelev *op. cit.* 46(1973):
2 *S. cereale* var. *afghanicum*.
subsp. **dighoricum** (Vav.) Tsvelev *op. cit.* 46
2 (1973): *S. cereale* subsp. *dighoricum*.

SEHIMA

2371 Oke, J. G.
Karyotypic variations, meiotic behaviour and
phylogenetic interrelationships in some species of
Ischaemum Linn. and *Sehima* Forsk. in parts of western
India. *Proc. Indian Acad. Sci., B*, 78(2):80–88(1973)—
6 Chromosome numbers.

SESLERIA

coerulans Friv.
subsp. **bielzii** (Schur) Gergely & Beldie in Fl.
1 Republ. Social. Român., 12:223(1972): *S. bielzii*.
1 f. **solacolui** Gergely *op. cit.* 789(1972)—Romania.
heuflerana Schur
f. **praelonga** (Schur) Gergely & Beldie *op. cit.*
1 224(1972): *S. heuflerana* a. *praelonga*.

rigida Heuff.
f. **aestivalis** (Schur) Gergely & Beldie *op. cit.* 220
1 (1972): *S. haynaldiana* b. *aestivalis.*
f. **gracilis** (Schur) Gergely & Beldie *op. cit.* 220
1 (1972): *S. gracilis.*
subsp. **haynaldiana** (Schur) Gergely & Beldie
1 *op. cit.* 219(1972): *S. haynaldiana.*
f. **leucostachys** (Schur) Gergely & Beldie *op. cit.*
1 219(1972): *S. haynaldiana* d. *leucostachys.*
uliginosa Opiz
f. **subscabrida** (Simk.) Gergely & Beldie *op. cit.*
1 225(1972): *S. barcensis* var. *subscabrida.*

SETARIA

2372 Fujita, Y.
(Some considerations on the vernacular names of *Setaria italica* Beauv. in south-eastern Asia). *J. Jap. Bot.*,
2, 6 48(6):187–188(1973)—Ja(En).

2373 Gregg, W. P.
Ecology of the annual grass *Setaria lutescens* on oldfields of the Pennsylvania Piedmont. *Proc. Acad.*
3 *Nat. Sci. Philadelphia*, 124:135–196(1973).

2374 Gupta, P. K. *and* Yashvir
Abnormal meiosis in hexaploid *Setaria verticillata. Phyton* (Austria), 15(1–2):31–36(1973)—(Ge);
6 chromosome numbers.

decipiens C. Schimp.
var. **schulteissii** (A. Schwarz) Morariu in Fl. Republ. Social. Român., 12:78(1972): *S. ambigua*
1 f. *schulteisii.*
verticillata (L.) P. Beauv.
f. **arenosa** (Schur) Morariu *op. cit.* 77(1972):
1 *S. verticillata* a. *arenosa.*
viridis (L.) P. Beauv.
f. **arenosa** (Schur) Morariu *op. cit.* 78(1972):
1 *S. viridis* b. *arenosa.*

SIEGLINGIA

2375 Kožuharov, S. I. *and* Petrova, A. V.
Cytotaxonomic study of five rare Bulgarian grasses. *Dokl. Bolg. Akad. Nauk*, 26(6):829–832(1973)—
1 Chromosome numbers. *Sieglingia decumbens.*

decumbens (L.) Bernh.
f. **gigantea** Buia in Fl. Republ. Social. Român., 12:
1 790(1972)—Romania.

SORGHUM

2376 Ivanoyukovich, L. K. *and* Yakushevskiĭ, E. S.
Anatomicheskoe stroenie lista nekotorȳkh vidov sorgo (*Sorghum* Moench subgen. *Sorghum*). (Anatomical structure of leaf in some species of sorgho (*Sorghum* Moench subgenus *Sorghum*)). *Bot. Zhurn.*, 58(7):
1–7 1028–1037(1973).

sect. **Blumenbachia** (Koel.) Tsvelev in *Novosti Sist.*
1, 2, 6 *Vȳssh. Rast.*, 10:79(1973): genus *Blumenbachia.*
bicolor (L.) Moench
var. **caffrorum** (Retz.) Mohlenbrock, Illus. Fl. Illinois, Grasses, *Panicum* to *Danthonia*: 192
3 (1973): *Panicum caffrorum.*
var. **cernuum** (Ard.) Ghişa in Fl. Republ. Social.
1, 2, 6 Român., 12:61(1972): *Holcus cernuus.*
var. **drummondii** (Nees) Mohlenbrock, Illus. Fl. Illinois, Grasses, *Panicum* to *Danthonia*: 192
3 (1973): *Andropogon drummondii.*
var. **saccharatum** (L.) Mohlenbrock *op. cit.* 192
3 (1973): *Holcus saccharatus.*

SPARTINA

2377 Miège, J.
La naissance d'une espèce ou l'histoire d'un fils indigne. *Mus. Genève*, 131:17–22(1973)—Illustrations; chromosome numbers. *Spartina* × *townsendii* and
1 *S. anglica.*

SPOROBOLUS

2378 Baskin, J. M. *and* Baskin, C. C.
Observation of the ecology of *Sporobolus vaginiflorus* in cedar glades. *Castanea*, 38(1):25–35(1973)—
3 Illustrations.

2379 Jovet, P. *and* Guédès, M.
Validation of names in *Sporobolus. Taxon*, 22(1):163
1, 3–6 (1973).

STIPA

2380 Caro, J. A. *and* Sánchez, E.
Las especies de *Stipa* (Gramineae) del subgenero *Jarava. Kurtziana*, 7:61–116(1973)—(En); illustra-
4 tions, key.

2381 Karavaev, M. N. *and* Skryabin, S. Z.
Fragmentȳ tȳrsova—kovȳl'nȳkh stepeĭ so *Stipa decipiens* P. Smirn. v Yakutii. (Fragments of steppes with *Stipa decipiens* P. Smirn. in Yakutia). *Bot. Zhurn.*,
2 58(8):1131–1142(1973)—Dot map.

2382 Maze, J. *and* Bohm, L. R.
Comparative embryology of *Stipa elmeri* (Gramineae). *Canad. J. Bot.*, 51(1):235–247(1973)—
3 (Fr); illustrations.

2383 Roig, F. A.
Novedades en *Stipa* (Gramineae) de la Patagonia. *Bol. Soc. Argent. Bot.*, 14(4):311–318(1972)—(En);
4 illustrations.

2384 Vivant, J.
Rumex cantabricus Rech. et *Stipa parviflora* Desf., phanerogames méconnues de la flore française. (*Rumex cantabricus* and *Stipa parviflora*, unknown phanerogams of the French flora). *Bull. Soc. Bot. France*, 119(5–6):
1 335–338 (1972 publ. 1973)—(En).

breviseta Caro & E. Sánchez in *Kurtziana*, 7:100
4 (1973)—Argentina.
capillata L.
f. **orthopogon** (A. & G.) Morariu in Fl. Republ. Social. Român., 12:197(1972): *S. capillata* A.
1 *orthopogon.*
var. **rumelica** (Velen.) Morariu *op. cit.* 197(1972):
1 *S. capillata* var. β *rumelica.*
f. **ulopogon** (A. & G.) Morariu *op. cit.* 197
1 (1972): *S. capillata* B. *ulopogon.*
chubutensis Speg.
var. **hirsutissima** F. A. Roig in *Bol. Soc. Argent.*
4 *Bot.*, 14(4):316(1972)—Argentina.
4 **maeviae** F. A. Roig *op. cit.* 314(1972)—Argentina.
parviflora Desf.
subsp. contorta (Lge) Malagarriga, Las Subesp. y Variac. Geogr.: 27(1973), without basionym
1 date: *S. parviflora* var. *contorta.*
pennata L.
subsp. **sabulosa** (Pacz.) Tsvelev in *Novosti Sist.*
1 *Vȳssh. Rast.*, 10:80(1973): *S. pennata* f. *sabulosa.*
poeppigiana T. & R.
var. **pubescens** F. A. Roig in *Bol. Soc. Argent. Bot.*,
4 14(4):318(1972)—Argentina.
pugionata Caro & E. Sánchez in *Kurtziana*, 7:109
4 (1973)—Argentina.

subaristata (Matthei) Caro & E. Sánchez *op. cit.* 96
4 (1973): *S. leptostachya* var. *subaristata*.
subplumosa Hicken ex F. A. Roig in *Bol. Soc. Argent.*
4 *Bot.*, 14(4):311(1972)—Argentina.

SWALLENOCHLOA McClure in *Smithsonian*
4 *Contrib. Bot.*, 9:106(1973).
depauperata (Pilger) McClure *op. cit.* 112(1973):
4 *Chusquea depauperata*.
spicata (Munro) McClure *op. cit.* 112(1973):
4 *Chusquea spicata*.
subtessellata (Hitchcock) McClure *op. cit.* 113(1973):
4 *Chusquea subtessellata*.
tessellata (Munro) McClure *op. cit.* 113(1973):
4 *Chusquea tessellata*.
weberbaueri (Pilger) McClure *op. cit.* 113(1973):
4 *Chusquea weberbaueri*.

TRIDENS

2385 Crooks, P. *and* Kucera, C. L.
Tridens × *oklahomensis* (*T. flavus* × *T. strictus*), an
interspecific sterile hybrid in the Eragrosteae
(Gramineae). *Amer. J. Bot.*, 60(3):262–267(1973)—
3 Illustration; chromosome numbers.

TRIPSACUM

2386 Gray, J. R. *and* Newell, C. A.
Fatty acid variation in seeds of *Tripsacum dactyloides*.
Phytochemistry, 12(8):1993–1994(1973)—Chromosome
3, 4 numbers.

TRISETUM

2387 Chrtek, J.
Bemerkungen zum *Trisetum flavescens* Komplex im
Pindus Gebirge. *Acta Univ. Carol., Biol.*, no.4:227–230
1 (1971 publ. 1973).

flavescens (L.) P. Beauv.
f. **lutescens** (Rchb.) Buia & Morariu in Fl. Republ.
Social Român., 12:292(1972): *Avena flavescens*
1 var. *lutescens*.
gracile (Moris) Boiss.
var. **morisianum** Gamisans in *Candollea*, 28(1):
1 54(1973)—Corsica.
tenue (Hack.) Chrtek in *Acta Univ. Carol., Biol.*,
1971(4):228(1971 publ. 1973): *T. flavescens* var.
1 *tenue*.

× TRITICALE

2388 McVaugh, R.
Report of the Committee for Spermatophyta:
conservation of generic names: Proposal 322 (407 ×
408) × *Triticale* Müntzing (1936) vs. × *Triticosecale*
1, 2 Wittmack (1899). *Taxon*, 22(1):155(1973).

× TRITICOSECALE

2389 McVaugh, R.
Report of the Committee for Spermatophyta:
conservation of generic names: Proposal 322 (407 ×
408) × *Triticale* Müntzing (1936) vs. × *Triticosecale*
1, 2 Wittmack (1899). *Taxon*, 22(1):155(1973).

TRITICUM

2390 Barash, S. I.
Staromestnȳe pshenitsȳ zakavkaz'ya. (Ancient local
wheats of the Transcaucasia). *Trudȳ Prikl. Bot. Genet.*
2 *Selek.*, 50(1):294–312(1973)—(En); map.

2391 Dennell, R. W.
The phylogenesis of *Triticum dicoccum*: a recon-
2 sideration. *Econ. Bot.*, 27(3):329–331(1973).

2392 Gorgidze, A. D.
Osnovnȳe puti proiskhozhdeniya i stanovleniya
initsial'nȳkh vidov kul'turnoĭ pshenitsȳ. (The basic
sources of the origin and formation of initial species of
cultural wheat). *Soobshch. Akad. Nauk Gruz. SSR*,
69(2):425–428(1973)—(Geo,En); chromosome
1, 2 numbers.

2393 Hillel, J., Simchen, G. *and* Feldman, M. W.
Mating systems and population structure in two
closely related species of the wheat group: 2.
Environmental factors and population structure.
1, 2 *Heredity*, 30(1):73–83(1973).

2394 Kandelaki, G. V. *and* Meliya, N. S.
O skreshchivaemosti *Tr. araraticum* Jakubz. i *Tr.*
timopheevi Zhuk. s gornoĭ rozh'yu *Secale montanum*
Guss. (Concerning the crossability of *Tr. araraticum*
Jukub. and *Tr. timopheevi* Zhuk. with mountain rye).
Soobshch. Akad. Nauk Gruz. SSR, 70(1):177–180
1, 2 (1973)—Rus(Geo,En).

2395 Kruse, A.
Hordeum × *Triticum* hybrids. *Hereditas*, 73(1):157–
1, 2 161(1973)—Chromosome numbers.

2396 Kruse, J.
Homologiebeziehungen bei den Grannenbildungen
der Gättungen *Triticum* L. und *Aegilops* L. (Homology
relations in awn growths of the genera *Triticum* L. and
Aegilops L. *Kulturpflanze*, 21:195–209(1973)—
1, 2 (En,Rus).

2397 Maan, S. S.
Cytoplasmic and cytogenetic relationships among
tetraploid *Triticum* species. *Euphytica*, 22(2):287–300
2 (1973).

2398 Nigmatullin, F. G.
Novaya raznovidnost' sharozernoĭ pshenitsȳ. (New
variety of roundfruited wheat (*Tr. sphaerococcum*
Perc.)). *Dokl. Akad. Nauk Tadzh. SSR*, 15(10):58
2 (1972)—(Ta).

2399 Nigmatullin, F. G.
Novȳe raznovidnosti myagkoĭ pshenitsȳ. (New
varieties of common wheat (*Tr. aestivum* L.)). *Dokl.*
2 *Akad. Nauk Tadzh. SSR*, 15(12):47–49(1972)—(Ta).

2400 Stankov, I. Ts. *and* Tsikov, D. K.
Two new botanical varieties of *Triticum sphaero-*
coccum Perc. *Dokl. Bolg. Akad. Nauk*, 26(6):807–810
1 (1973)—Chromosome number.

2401 Udachin, R. A.
K poznaniyu vida *Triticum turanicum* Jakubz. (To
the knowledge of *Triticum turanicum*). *Trudȳ Prikl.*
2 *Bot. Genet. Selek.*, 50(1):35–43(1973)—(En); key.

2402 Waines, J. G. *and* Kimber, G.
Satellite number and size in *Triticum monococcum* L.
and the evolution of the polyploid wheats. *Canad. J.*
Genet. Cytol., 15(1):117–122(1973)—Chromosome
1, 2 numbers.

2403 Zeven, A. C.
Triticum lorentii (Hochst.) Zeven nov. comb. *Taxon*,
2 22(2–3):321(1973)—En. *Aegilops biuncialis*.

aestivum L.
var. **erythrospermum** (Körn.) Velican in Fl.
Republ. Social. Român., 12:636(1972): *T.*
1 *vulgare* var. *erythrospermum*.
var. **ferrugineum** (Alef.) Velican *op. cit.* 636
1 (1972): *T. vulgare* Var.-Gruppe *ferrugineum*.

subsp. **hadropyrum** (Flaksb.) Tsvelev in *Novosti Sist. Vȳssh. Rast.*, 10:43(1973): *T. vulgare* subsp.
0 *hadropyrum.*
subsp. **inflatum** (Kudr.) Tsvelev *op. cit.* 43(1973):
0 *T. inflatum.*
var. **japschorvi** Nigmatullin in *Dokl. Akad. Nauk Tadzh. SSR*, 15(12):47(1972)—U.S.S.R.
2 (Middle Asia).
var. **lutescens** (Alef.) Velican in Fl. Republ. Social. Român., 12:636(1972): *T. vulgare* Var.-
1 Gruppe *lutescens.*
var. **milturum** (Alef.) Velican *op. cit.* 636(1972):
1 *T. vulgare* Var.-Gruppe *milturum.*
var. **ruchczianum** Nigmatullin in *Dokl. Akad. Nauk Tadzh. SSR*, 15(12):47(1972)—U.S.S.R.
2 (Middle Asia).
dicoccon (Schrank) Schuebl.
subsp. **subspontaneum** Tsvelev in *Novosti Sist.*
2 *Vȳssh. Rast.*, 10:41(1973)—U.S.S.R. (Caucasus).
subsp. **volgense** (Flaksb.) Tsvelev *op. cit.* 41(1973):
1 *T. dicoccon* f. *volgense.*
lorentii (Hochst.) Zeven in *Taxon*, 22(2–3):321
1, 2 (1973): *Aegilops lorentii.*
sphaerococcum Pers.
var. **ischkaschimicum** Nigmatullin in *Dokl. Akad. Nauk Tadzh. SSR*, 15(10):58(1972)—U.S.S.R.
2 (Middle Asia).
var. kostoffii I. T. Stankov & Tsikov in *Dokl. Bolg. Akad. Nauk*, 26(6):810(1973), without type—
1 Bulgaria in cultivation?
var. popovii I. T. Stankov & Tsikov *op. cit.* 807
1 (1973), without type—Bulgaria in cultivation?

× **TRITITRIGIA** Tsvelev in *Novosti Sist. Vȳssh.*
1 *Rast.*, 10:59(1973). *Triticum* × *Elytrigia.*
cziczinii Tsvelev *op. cit.* 59(1973)—U.S.S.R. (European Russia). (*Elytrigia intermedia* × *Triticum*
1 *aestivum*).

UNIOLA

2404 Yates, H. O. *and* Channell, R. B.
Nomenclatural history and taxonomic status of the Linnaean genus *Uniola* (Gramineae). *J. Tenn. Acad.*
3 *Sci.*, 48(1):12–15(1973).

VAHLODEA

2405 Albers, F.
Cytosystematische Untersuchungen in der Subtribus Deschampsiineae Holub (Tribus Aveneae Nees.): 2. Die Gattungen *Vahlodea* Fr. und *Avenella* Koch. (Cytosystematics of the subtribus Deschampsiineae Holub (tribus Aveneae Nees.): 2. The genera *Vahlodea* Fr. and *Avenella* Koch. *Ber. Deutsch. Bot. Ges.*, 85(5–6):279–285(1972)—(En); chromosome
1–3 numbers.

VULPIA

2406 Cotton, R. *and* Stace, C. A.
Vulpia sicula and the genus *Vulpia*. *Watsonia*, 9(4):
1 426(1973).

2407 Stace, C. A. *and* Cotton, R.
Vulpia sicula and the genus *Vulpia*. *Nature New Biol.*,
1 244(134):128(1973)—Chromosome number.

YUSHANIA

subgen. **Otatea** McClure & E. W. Smith in
4 *Smithsonian Contrib. Bot.*, 9:116(1973).
acuminata (Munro) McClure *op. cit.* 119(1973):
4 *Arundinaria acuminata.*
aztecorum McClure & E. W. Smith *op. cit.* 116
4 (1973)—Mexico.

ZERNA

2408 Tubli, E.
Ohtetu püsikluste (*Zerna inermis* Lindm.) varieeruvusest Eesti NSV-s. (Varieties of *Zerna inermis* Lindm. in the Estonian SSR). *Loodusuur. Seltsi Aastar.*, 61:107–115(1972)—(Rus,En); illustration,
1, 2 map, dot map.

ZOYSIA

seslerioides (Bal.) Clayton & F. R. Richardson in
6 *Kew Bull.*, 28(1):41(1973): *Brousemichea seslerioides.*

GROSSULARIACEAE

RIBES

2409 Arrigoni, P. V., Corrias, B., Corrias, S. D., Nardi, E. *and* Valsecchi, F.
Ad floram italicam notulae taxonomicae et geobotanicae: 12. Nuove stazioni di *Hypericum annulatum* Moris e *Ribes multiflorum* Kit. ssp. *sandalioticum* Arrig. in Sardegna. *Webbia*, 28(2):423–425(1973)—(En);
1 map.

2410 Semenikhina, L. V.
Mikrosporogenez i morfologiya pȳl'tsȳ u vidov smorodinȳ *Ribes janczewskii* A. Pojark. i *Ribes villosum* Wall. *Izv. Akad. Nauk Tadzh. SSR, Biol. Nauk*, 1(46):81–83(1972)—(Ta); chromosome
2, 6 numbers.

uva-crispa L.
subsp. **hunyadensis** (Simk.) Soó in *Acta Bot. Acad.*
1 *Sci. Hung.*, 18(1–2): *R. grossularia* var. *hunyadensis.*

GUNNERACEAE

GUNNERA

2411 Badillo, V. M. *and* Steyermark, J. A.
Una nueva especie de *Gunnera*. *Acta Bot. Venezuel.*,
4 8(1–4):311–315(1973)—Illustrations.

pittierana Badillo & Steyermark in *Acta Bot.*
4 *Venezuel.*, 8(1–4):311(1973)—Venezuela.

GUTTIFERAE

see

CLUSIACEAE, HYPERICACEAE

and

QUIINACEAE

HALORAGACEAE

LAUREMBERGIA

2412 Obermeyer, A. A.
Notes on African plants: Halorrhagidaceae. A note on *Laurembergia repens*. *Bothalia*, 11(1–2):116–117
5 (1973).

repens Berg.
subsp. **brachypoda** (Hiern) Obermeyer in *Bothalia*, 11(1–2):117(1973): *Serpicula repens* var.
5 *brachypoda.*
walkeri (Ohwi) Ohwi in *J. Geobot.*, 20(3):59
2 (1972): *Haloragis walkeri.*

MYRIOPHYLLUM

2413 Moravec, J.
Příspěvek k rozšíření stolístku střídavokvětého—*Myriophyllum alterniflorum* DC.—v jižních Čechách. (Beitrag zur Verbreitung von *Myriophyllum alterniflorum* DC. in Südböhmen). *Zprávy Česk. Bot. Spol.*,
1 8(1):16–19(1973)—(Ge).

2414 Nijalingappa, B. H. M.
Cytological studies in *Myriophyllum oliganthum* (Haloragaceae). *Curr. Sci.*, 42(22):802–803(1973)—
6 Chromosome number.

2415 Turlier, M.-F.
Recherches sur l'organogenèse foliaire précoce: 1. Deux espèces-types pour les ramifications dichotome et latérale (*Ceratophyllum demersum* L. et *Myriophyllum spicatum* L.). *Rev. Gén. Bot.*, 79(941–943):297–308
1–7 (1972)—(En); illustrations.

2416 Verdcourt, B.
A new combination in *Myriophyllum* (Haloragaceae).
4, 5 *Kew Bull.*, 28(1):36(1973).

aquaticum (Vellozo) Verdcourt in *Kew Bull.*, 28(1):
4, 5 36(1973): *Enydria aquatica.*

HAMAMELIDACEAE

2417 Mehra, P. N. *and* Khosla, P. K.
Cytogenetical studies of East Himalayan Hamamelidaceae, Combretaceae and Myrtaceae. *Silvae Genet.*,
6 21(5):186–190(1972)—(Ge); chromosome numbers.

2418 Melikyan, A. P.
Anatomiya semennoĭ kozhurȳ i sistematika semeĭstva Hamamelidaceae. (Anatomy of seed coat and systematics of the Hamamelidaceae family). *Biol.*
2–7 *Zhurn. Arm.*, 26(3):104–105(1973).

2419 Melikyan, A. P.
Tipȳ semennoĭ kozhurȳ Hamamelidaceae i blizkikh semeĭst v svyazi s ikh sistematicheskimi vzaimootnosheniyami. (Seed-coat types of Hamamelidaceae and allied families in relation to their systematics).
2–7 *Bot. Zhurn.*, 58(3):350–359(1973)—(En).

2420 Rao, T. A. *and* Bhupal, O. P.
Typology of foliar sclereids in various taxa of Hamamelidaceae. *Proc. Indian Sci. Congr.*, 60(3):296–
2–7 297(1973). Abstract only.

SYCOPSIS

2421 Carder, V. A.
Sycopsis sinensis. *Curtis's Bot. Mag.*, 179(4):tab.655
2 (1973)—En; illustration, coloured illustration.

HECTORELLACEAE

LYALLIA

2422 Young, S. B. *and* Schofield, E. K.
Palynological evidence for the Late Glacial occurrence of *Pringlea* and *Lyallia* on Kerguelen Islands. *Rhodora*, 75(802):239–247(1973)—
7 Illustration.

HELICONIACEAE
see
MUSACEAE

HIPPOCASTANACEAE

AESCULUS

2423 Hsiao, J.-Y. *and* Li, H.-L.
Chromatographic studies on the red horse-chestnut (*Aesculus* × *carnea*) and its putative parent species.
1, 3 *Brittonia*, 25(1):57–63(1973).

2424 Kosaev, M. N.
Introduktsiya konskogo kashtana obȳknovennogo. (Tsentralnȳĭ Botanicheskiĭ Sad AN KazSSR). *Izv.*
1–3 *Akad. Nauk. Kaz. SSR., Biol.*, 5:15–19(1973)—(Ka).

HIPPURIDACEAE

HIPPURIS

2425 Jeune, B. *and* Cusset, G.
A propos de la ramification de l'*Hippuris vulgaris* L. *Bull. Soc. Bot. France*, 118(9):627–638(1971 publ.
1, 2 1973)—(En); illustrations.

HUMIRIACEAE

2426 Badré, F.
Humiriacées. Flore du Gabon, 21:45–48(1973)—
5 Illustration.

2427 Narayana, L. L. *and* Rao, D.
Contributions to the floral anatomy of the Humiriaceae: 2. *J. Jap. Bot.*, 48(5):143–146(1973)—
4, 5 (Ja).

2428 Narayana, L. L. *and* Rao, D.
Contributions to the floral anatomy of Humiriaceae: 3. *J. Jap. Bot.*, 48(8):242–246(1973)—(Ja);
4, 5 illustration.

SACOGLOTTIS

trichogyna Cuatrecasas in *Ciencia*, 27(6):171
4 (1972)—Costa Rica.

HYDRANGEACEAE

2429 McClintock, E.
Climbing hydrangeas. *Calif. Hort. J.*, 34(4):141–
2, 4, 6 145(1973)—Illustrations.

DEUTZIA

2430 Saxena, N. P.
Studies in the family Saxifragaceae: 6. Structure and development of gametophyte in *Deutzia corymbosa* R. Br. and *Dichroa febrifuga* Lour. *Ber. Schweiz. Bot.*
6 *Ges.*, 81:91–96(1972)—(Ge).

2431 Zaikonnikova, T. I.
Novȳĭ vid roda *Deutzia* Thunb. iz Butana. (Generis *Deutzia* species nova e districtu Bhutan).
6 *Novosti Sist. Vȳssh. Rast.*, 10:166–167(1973).

bhutanensis Zaikonnikova in *Novosti Sist. Vȳssh.*
6 *Rast.*, 10:166(1973)—Bhutan.

DICHROA

2432 Saxena, N. P.
Studies in the family Saxifragaceae:6. Structure and development of gametophyte in *Deutzia corymbosa* R. Br. and *Dichroa febrifuga* Lour. *Ber. Schweiz. Bot.*
6 *Ges.*, 81:91–96(1972)—(Ge).

PHILADELPHUS

2433 Steinwendtner, R.
Philadelphus coronarius L. bei Leonstein an der
Steyr. (8151/1). *Mitt. Bot. Arbeit. Oberösterr.*
1 Landesmus. Linz, 4(1–2):85–86(1972).

HYDROCHARITACEAE

2434 Den Hartog, C.
Flora of Panama: part 2. Family 5a. Hydrocharita-
ceae. *Ann. Missouri Bot. Gard.*, 60(1):7–15(1973)—
4 Illustrations; keys.

BLYXA

2435 Den Hartog, C.
Blyxa quadricostata nov. sp., a new Hydrocharitacea
from Thailand. *Blumea*, 21(1):185–187(1973)—
6 Illustration.

quadricostata den Hartog in *Blumea*, 21(1):185
6 (1973)—Thailand.

ELODEA

2436 Hamashima, S.
(*Elodea nuttallii* is naturalized in Tokai Provinces).
2 *J. Geobot.*, 20(1):16(1972)—Ja.

2437 Hongve, D.
Vasspest, *Elodea canadensis* Michx. i Jevnaker
(*Elodea canadensis* Michx. at Jevnaker, S.E. Norway).
1 *Blyttia*, 31(1):17–18(1973)—(En).

2438 Jonsson, P. N.
Två växtfynd fran Lits socken, Jämtland. (Two
records of vascular plants from Lit parish, Jämtland).
Svensk Bot. Tidskr., 67(1):67(1973)—*Elodea*
1 *canadensis*.

HALOPHILA

2439 Sachet, M.-H. *and* Fosberg, F. R.
Remarks on *Halophila* (Hydrocharitaceae). *Taxon*,
2, 5–7 22(4):439–443(1973).

HYDRILLA

2440 Misra, M. P.
Cytological studies of *Hydrilla verticillata* Presl.
from Bihar & Bengal. *J. Bihar Bot. Soc.*, 1(1–2):72
6 (1972)—Chromosome numbers. Abstract.

2441 Speta, F.
Hydrilla verticillata. *Mitt. Bot. Arbeit. Oberösterr.*
1 *Landesmus. Linz*, 4(1–2):46(1972)—Ge.

LIMNOBIUM

2442 Bruggen, H. W. E. van
Eine Namensänderung! *Inform. ZAG Wasser-*
3 *pflanzen*, no.3: 6(1973). *Limnobium laevigatum*.

OTTELIA

2443 Chaudhuri, J. B.
Cytology of two members of Hydrocharitaceae.
Proc. Indian Sci. Congr., 59(3):346(1972). Abstract
6 only; *Ottelia alismoides*.

THALASSIA

2444 Jagels, R.
Studies of a marine grass, *Thalassia testudinum*: 1.
Ultrastructure of the osmoregulatory leaf cells. *Amer.*
3, 4 *J. Bot.*, 60(10):1003–1009(1973).

VALLISNERIA

2445 Anon.
Vallisneria spiralis Linné. *Inform. ZAG Wasser-*
1 *pflanzen*, no.3, Beilage. [2]p.(1973)—Ge; illustration.

2446 Chaudhuri, J. B.
Cytology of two members of Hydrocharitaceae.
Proc. Indian Sci. Congr., 59(3):346(1972). Abstract
6 only; *Vallisneria spiralis*.

2447 Harris, S. *and* Lording, T. A.
Distribution of *Vallisneria spiralis* L. in the River Lea
Navigation Canal (Essex-Hertfordshire border).
1 *Watsonia*, 9(3):253–256(1973)—Map.

HYDROPHYLLACEAE

2448 Flaster, B. *and* Peixoto, A. L.
Flora ilustrada catarinense: Hidrofiláceas. 1972.
4 12p.—Illustrations, dot map.

PHACELIA

2449 Atwood, N. D.
A revision of the *Phacelia Crenulatae* group
(Hydrophyllaceae) for North America. *Diss. Abstr.*
3 *Int.*, *B*, 33(2):596(1972).

2450 Atwood, N. D.
Two new species of *Phacelia* (Hydrophyllaceae).
Phytologia, 26(6):437–438(1973). *P. argillacea*, *P.*
3 *mammillarensis*.

2451 Mulligan, B. O.
Plant portraits: *Phacelia sericea* A. Gray. *Quart. Bull.*
Alp. Gard. Soc., 41(3):201,204–205(1973)—
3 Illustration.

argillacea Atwood in *Phytologia*, 26(6):437(1973)—
3 U.S.A. (Utah).
mammillarensis Atwood *op. cit.* 437(1973)—U.S.A.
3 (Utah).

HYDROSTACHYACEAE

2452 Cusset, C.
Révision des Hydrostachyaceae. *Adansonia*, 13(1):
5 75–119(1973)—(En); illustrations; keys.

HYDROSTACHYS

distichophylla A. Juss.
var. **hildebrandtii** (Engl.) C. Cusset in *Adansonia*,
5 13(1):84(1973): *H. hildebrandtii*.
5 **lufirensis** C. Cusset *op. cit.* 102(1973)—Zaïre.
lukungensis (Hauman) C. Cusset *op. cit.* 117(1973):
5 *H. myriophylla* var. *lukungensis*.
trifaria Perr.
var. **multisquamata** (Perr.) C. Cusset *op. cit.* 110
5 (1973): *H. multisquamata*.

HYMENOCARDIACEAE
see
EUPHORBIACEAE

HYPERICACEAE

HYPERICUM

2453 Arrigoni, P. V., Corrias, B., Corrias, S. D., Nardi, E.
and Valsecchi, F.
Ad floram italicam notulae taxonomicae et geo-
botanicae: 12. Nuove stazioni di *Hypericum*
annulatum Moris e *Ribes multiflorum* Kit. ssp.
sandalioticum Arrig. in Sardegna. *Webbia*, 28(2):
1 423–425(1973)—(En); map.

2454 Geigy *Ltd*: Agricultural Chemicals Division
Hypericaceae: St John's wort family. *Geigy Weed*
1 *Tables*, 63(1973)—Illustrations, coloured illustration.

2455 Healy, A. J.
Weedy St. John's worts (*Hypericum* spp.) in New Zealand. *In* Proceedings of the 25th New Zealand Weed and Pest Control Conference: 180–190(1972)—
7 Illustrations; key.

2456 Reynaud, C.
Contribution à l'étude cytotaxinomique du genre *Hypericum* L. en Turquie: 1. *Bull. Soc. Bot. France*, 120(5–6):201–216(1973)—(En); map; chromosome
2 numbers.

2457 Robson, N. K. B.
Flora of West Pakistan: no. 32. Guttiferae. 1973.
6 12p.—Illustrations, map.

2458 Robson, N. K. B.
Hypericum. *In* Green, P. S. *ed*. Plants: wild and
1, 2, 5 cultivated: 79–82(1973)—En.

2459 Robson, N. K. B.
Hypericum sect. *Brathys* in Africa and Madagascar. *Garcia de Orta, Sér. Bot.*, 1(1–2):83–90(1973)—(Por);
5 dot map; key.

2460 Robson, N. K. B.
Identification lists of Malaysian specimens: 46. *Hypericum*. Leyden, Foundation Flora Malesiana,
6 1972. pp.801–811.

2461 Robson, N. K. B.
Notes on Malesian species of *Hypericum* (Guttiferae) Florae Malesianae Praecursores: 52. *Blumea*, 20(2):
2–7 251–274(1972 publ. 1973)—Maps; keys.

2462 Rodríguez-Jiménez, C.
Distribution géographique du genre *Hypericum* L. section *Brathys* (Mutis) sous section *Spachium* Keller. *Compte R. Séances Soc. Biogéogr.*, nos.425–433:
1–7 87–96(1973)—Maps, dot maps.

2463 Webb, D. A. *and* Halliday, G.
The distribution, habitat and status of *Hypericum canadense* L. in Ireland. *Watsonia*, 9(4):333–344
1 (1973)—Map; chromosome number.

sect. **Pulogensia** N. Robson in *Blumea*, 20(2):259
2, 6 (1972 publ. 1973).
sect. **Takasagoya** (Y. Kimura) N. Robson *op. cit.*
2, 6 252(1972 publ. 1973): *Takasagoya*.
beccarii N. Robson *op. cit.* 260(1972 publ. 1973)—
6 Sumatra.
subsp. **steenisii** N. Robson *op. cit.* 261(1972 publ.
6 1973)—Sumatra.
bifurcatum N. Robson *op. cit.* 256(1972 publ. 1973)—
6 New Guinea (Terr. New Guinea).
choisianum Wall. ex N. K. B. Robson in Fl. W.
6 Pakistan, 32(Guttiferae): 6(1973)—India.
geminiflorum Hemsley
var. **simplicistylum** (Hayata) N. Robson in *Blumea*, 20(2):254(1972 publ. 1973): *H. simpli-*
2, 6 *cistylum*.
nakamurai (Masamune) N. Robson *op. cit.* 253
2 (1972 publ. 1973): *Takasagoya nakamurai*.
perrieri N. Robson in *Garcia de Orta, Sér. Bot.*,
5 1(1–2):87(1973)—Malagasy.
var. **rupestre** (R. Keller) N. Robson *op. cit.* 89
5 (1973): *H. japonicum* var. *rupestre*.
petiolulatum Hook. f. & Thoms. ex Dyer
subsp. **yunnanense** (Franchet) N. Robson in *Blumea*, 20(2):262(1972 publ. 1973): *H.*
2 *yunnanense*.
sewense N. Robson *op. cit.* 254(1972 publ. 1973)—
6 New Guinea (Terr. New Guinea).

HYPOXIDACEAE

2464 Thompson, M. F.
Anatomical and cytological studies in the Hypoxidaceae. *Forum Bot.*, 11(7):47(1973). Notes on the work
2–7 of Miss M. F. Thompson for the degree of M.Sc.

ICACINACEAE

2465 Lobreau-Callen, D.
Les caractères du pollen et la répartition géographique des Icacinaceae. *Compte R. Séances Soc. Biogéogr.*, nos. 425–433: 4–17(1972 publ. 1973)—
2, 4–7 Maps.

2466 Lobreau-Callen, D.
Pollen des Icacinaceae: 1. Atlas (1). *Pollen Spores*,
2, 4–7 14(4):345–388(1972 publ. 1973)—Fr(Ge).

2467 Lobreau-Callen, D.
Le pollen des Icacinaceae: 2. Observations en microscopie électronique, corrélations, conclusions.
2, 4–7 *Pollen Spores*, 15(1):47–89(1973)—(Ge,Fr).

2468 Staveren, M. G. G. van *and* Baas, P.
Epidermal leaf characters of the Malesian Icacina-
6, 7 ceae. *Acta Bot. Neerl.*, 22(4):329–359(1973).

2469 Villiers, J.-F.
Icacinacées. Flore du Cameroun, 15:3–100(1973)—Illustrations; keys. Also published as Flore du Gabon:
5 20(1973).

2470 Villiers, J. F.
Icacinacées. Flore du Gabon, 20: 3–100(1973)—Illustrations; keys. Also published as Flore du
5 Cameroun: 15(1973).

DESMOSTACHYS

oblongifolia (Engler) Villiers in Fl. Cameroun, 15: 48(Jan. 30 1973), and in Fl. Gabon, 20: 48(Jan. 30
5 1973): *Alsodeiopsis oblongifolia*.
tenuifolius Oliver
var. **angustifolius** Pellegrin ex Villiers in Fl. Cameroun, 15:46(Jan. 30 1973), and in Fl.
5 Gabon, 20:46(Jan. 30 1973)—Gabon.

IODES

klaineana Pierre
var. **tomentosa** Villiers in Fl. Cameroun, 15:12 (Jan. 30 1973), and in Fl. Gabon, 20:12(Jan. 30
5 1973)—Cameroun, Gabon.

LASIANTHERA

africana Palisot de Beauvois
var. **microphylla** Pellegrin ex Villiers in Fl. Cameroun, 15:18(Jan. 30 1973), and in Fl.
5 Gabon, 20:18(Jan. 30 1973)—Gabon.

LEPTAULUS

zenkeri Engler
var. **longipedicellata** Villiers in Fl. Cameroun, 15: 62(Jan. 30 1973), and in Fl. Gabon, 20:62(Jan. 30
5 1973)—Gabon.

PYRENACANTHA

cordicula Villiers in Fl. Cameroun, 15:86 (Jan. 30 1973), and in Fl. Gabon, 20:86 (Jan. 30 1973), nom.
5 nov.: *P. cordata* Villiers.

RHAPHIOSTYLIS

ferruginea Engler
var. **villosa** (Pellegrin) Villiers in Fl. Cameroun, 15: 40(Jan. 30 1973), and in Fl. Gabon, 20:40(Jan. 30
5 1973): *R. fusca* var. *villosa*.

ILLECEBRACEAE

2471 Ghafoor, A.
Flora of West Pakistan: no.47. Illecebraceae, 1973.
6 12p.—Illustrations, map; keys.

2472 Wild, H. *and* Diniz, M. A.
Flora de Moçambique: 20. Illecebraceae. 1973. 7p.—
5 Chromosome number; keys.

HERNIARIA

2473 Trpin, D.
Zur Kenntnis und Bewertung von *Herniaria rotundifolia* Visiani. *Feddes Repert.*, 84(4):295–301
1 (1973)—(En).

glabra L.
subsp. **rotundifolia** (Visiani) Trpin in *Feddes*
1 *Repert.*, 84(4):300(1973): *H. rotundifolia*.

PARONYCHIA

kapela (Hacq.) Kerner
subsp. aretioides (Pourret) Malagarriga, Las Subesp. y Variac. Geogr.: 5(1973), without
1 basionym ref.: *Illecebrum aretioides*.

SCLERANTHUS

annuus L.
subsp. **aetnensis** (Strobl) Pignatti in *Giorn. Bot.*
1 *Ital.*, 107(5):207(1973): *S. aetnensis*.
marginatus Guss.
subsp. **dichotomus** (Schur) Pignatti *op. cit.* 207
1 (1973): *S. dichotomus*.
polycarpos L.
subsp. **collinus** (Hornung) Pignatti *op. cit.* 208
1 (1973): *S. annuus* subsp. *collinus*.
subsp. **ruscinonensis** (Gillot & Coste) Pignatti *op.*
1 *cit.* 208(1973): *S. candolleanus* subvar. *ruscinonensis*.

IRIDACEAE

2474 Banerji, M. *and* Chaudhuri, M.
Further studies on chromosomes of some Orchidaceae and Iridaceae from the temperate Himalayas. *Proc. Indian Sci. Congr.*, 59(3):347(1972)—
6 Chromosome numbers. Abstract only.

2475 Banerjee, M. *and* Sharma, A. K.
A cytotaxonomical analysis of several genera of the family Iridaceae. *Pl. Sci.* (Lucknow), 3:14–29(1971 ?
1–7 publ. 1973).

2476 Jakimow-Barras, M.
Les polysaccharides des graines de quelques Liliacees et Iridacees. *Phytochemistry*, 12(6):1331–1339
1–3 (1973)—(En).

2477 Marais, W.
Notes on African Iridaceae. *Kew Bull.*, 28(2):311–
5 317(1973)—Key.

2478 Rodionenko, G. I.
Irises—some history, the present and the future. *Iris Year Book 1972*, 81–92(1973)—Chromosome
2 number.

CHAMELUM

2479 Ariza Espinar, L.
Nota aclaratoria sobre el tipo de *Chamelum*
4 *badenbenderi* (Iridaceae). *Kurtziana*, 7:266(1973).

CROCUS

2480 Brickell, C. D. *and* Mathew, B.
Three recent discoveries in *Crocus*. *J. Roy. Hort. Soc.*, 98(8):360–366(1973)—Coloured illustrations; chromosome numbers. *C. almehensis*, *C. robertianus*,
1, 2 *C. antalyensis*.

2481 Brighton, C. A., Mathew, B. *and* Marchant, C. J.
Chromosome counts in the genus *Crocus* (Iridaceae).
1, 2 *Kew Bull.*, 28(3):451–464(1973 publ. 1974).

2482 Frey, A.
Występowanie szafrana spiskiego *Crocus scepusiensis* (Rehm. et Woł.) Borb. na nizinie Sandomierskiej. (The occurrence of *Crocus scepusiensis* (Rehm. et Woł.) Borb. in the Sandomierz lowland). *Ochr.*
1 *Přzyr.*, 37:285–294(1972)—(En); dot map.

2483 Mathew, B.
Crocus antalyensis. *Curtis's Bot. Mag.*, 179(2):tab. 629(1972 publ. 1973)—En; illustration, coloured
2 illustration.

2484 Mathew, B.
Crocus candidus Clarke. *J. Roy. Hort. Soc.*, 98(5):222
2 224(1973)—En; coloured illustration.

2485 Mathew, B.
Crocus olivieri. *Curtis's Bot. Mag.*, 179(3):tab.639
1, 2 (1973)—En; illustration, coloured illustration.

2486 Mathew, B.
Plant awards 1972–73: *Crocus cyprius* Kotschy A. M.
2 *Quart. Bull. Alp. Gard. Soc.*, 41(4):326(1973).

2487 Mathew, B.
Plant awards 1972–73: *Crocus michelsonii* B. Fedtsch. P. C. *Quart. Bull. Alp. Gard. Soc.*, 41(4):330–
2 331(1973).

2488 Mathew, B.
Plant awards 1972–73: *Crocus scardicus* Kosanin P. C.
1 *Quart. Bull. Alp. Gard. Soc.*, 41(4):330(1973).

2489 Pacyna, A. *and* Piękoś, H.
Rozmieszczenie szafrana spiskiego *Crocus scepusiensis* w Karpatach Polskich i na Nizinie Sandomierskiej. (Distribution of *Crocus scepusiensis* in the West Carpathians and on the Sandomierz Lowland). *Chrońmy Przyr. Ojczystą*, 29(3):59–62
1 (1973).

2490 Simon, T. *and* Molnár, A.
A *Crocus heuffelianus* Herb. új észak-alföldi
termöhelye. (Der neue Standort von *Crocus heuf-*
felianus Herb. im nördlichen Alföld). *Bot. Közlem.*,
59(3):193–195(1972 publ. 1973)—(Ge);
1 illustrations.

2491 Šopova, M.
The cytology of one more *Crocus* species from
Macedonia. (Citološka stulija na ušte eden *Crocus* od
florata na Makedonija). *Fragm. Balcan. Mus. Macedon.*
Sci. Nat., 8(22):189–195(1972)—(Se); chromosome
1 numbers. *C. chrysanthus* var. *citrinus.*

almehensis C. D. Brickell & B. F. Mathew in *J. Roy.*
2 *Hort. Soc.*, 98(8):360(1973)—Iran.
olivieri Gay
subsp. **balansae** (Gay ex Bak.) B. F. Mathew in
Curtis's Bot. Mag., 179(3): sub t. 639(1973):
6 *C. balansae.*
robertianus C. D. Brickell in *J. Roy. Hort. Soc.*,
1 98(8):362(1973)—Greece.

DIETES

2492 Goldblatt, P.
Proposal for the conservation of the generic name
1265a *Dietes* Salisb. (1812) against *Naron* Medikus
5, 7 (1790) (Iridaceae). *Taxon*, 22(4):504–505(1973).

GLADIOLUS

2493 Barnard, T. T.
Painted ladies. *Veld Flora*, 3(3):41–44(1973)—
5 Coloured illustration.

2494 Hamilton, A. P.
The Corne Flagge—a neolithic plant? *Watsonia*,
9(4):427–428(1973)—Chromosome number;
5 *Gladiolus italicus.*

aequinoctialis Herbert
var. **divina** (Vaupel) Marais in *Kew Bull.*, 28(2):
5 312(1973): *Acidanthera divina.*
var. **tomentosus** Marais *op. cit.* 312(1973)—Sierra
5 Leone.
callianthus Marais *op. cit.* 312(1973), nom. nov.:
5 *Acidanthera bicolor* Hochst.
chevalieranus Marais *op. cit.* 313(1973), nom. nov.:
5 *Dortania amoena* A. Chev.
curtifolius Marais *op. cit.* 313(1973), nom. nov.:
5 *Acidanthera goetzei* Harms.
gunnisii (Rendle) Marais *op. cit.* 313(1973):
5 *Acidanthera gunnisii.*
iroënsis (A. Chev.) Marais *op. cit.* 313(1973):
5 *Acidanthera iroënsis.*
5 **leonensis** Marais *op. cit.* 315(1973)—Sierra Leone.
praecostatus Marais *op. cit.* 314(1973)—Guinea,
5 Ivory Coast.
ukambanensis (Baker) Marais *op. cit.* 313(1973):
5 *Acidanthera ukambanensis.*
var. **alatus** Marais *op. cit.* 314(1973)—Kenya,
5 Ethiopia, Somali Republic, Tanzania.

HOMERIA

2495 Goldblatt, P.
Biosystematic and taxonomic studies in *Homeria*
(Iridaceae). *J. S. Afr. Bot.*, 39(2):133–140(1973)—
5 (Af); illustrations; chromosome numbers.

IRIDODICTYUM

2496 Furse, P.
Iris reticulata section or the genus *Iridodictyum. Iris*
Year Book 1972, 103–111(1973)—Semi-popular notes;
2 with appendix by C. Grey-Wilson.

IRIS

2497 Blanchard, D. K. S.
Some brief notes on the occurrence of *Iris spuria* in
England and Western Europe. *Iris Year Book 1973*,
1 140–142(1973).

2498 Brummitt, M.
Some apogon irises for the garden. *J. Roy. Hort.*
1–3 *Soc.*, 98(5):210–214(1973)—Illustrations.

2499 Furse, P.
Iris reticulata section or the genus *Iridodictyum. Iris*
Year Book 1972, 103–111(1973)—Semi-popular notes;
2 with appendix by C. Grey-Wilson.

2500 Gorer, P.
Plant awards 1972–73: *Iris kolpakowskiana* Regel
P. C. *Quart. Bull. Alp. Gard. Soc.*, 41(4):329–330
2 (1973).

2501 Grey-Wilson, C.
Irises seen in Iran and Afghanistan during 1971.
2 *Iris Year Book 1973*, 105–120(1973)—Keys.

2502 Ingram, C.
A new *Iris* from Morocco. *Iris Year Book, 1973*,
1 96–97(1973)—Illustration.

2503 Kollár, A.
Rozšírenie *Iris sibirica* L. na východnom Slovensku.
(Extension of *Iris sibirica* L. in eastern Slovakia).
Biológia, A (Czechoslovakia), 28(7):581–583(1973)—
1 (En,Rus); dot map.

2504 Le Cocq, C.
La megasporogénèse chez l'*Iris pseudacorus* L. 1.
Etude cytologique qualitative. *Rev. Cytol. Biol. Vég.*,
1, 2 35(1–2):41–164(1972)—Illustrations.

2505 Luscombe, E. G. B.
Iris pumila Linnaeus: descriptive notes of its
botanical characteristics including morphology,
variability, the geographical range of its natural
distribution and the like: a résumé of available data:
1, 2 part 1. *Iris Year Book, 1972*, 149–157(1973).

2506 Luscombe, E. G. B.
Iris pumila Linnaeus: descriptive notes of its
botanical characteristics including morphology,
variability, the geographical range of its natural
distribution and the like: a résumé of available data.
Part 2. *Iris Year Book 1973*, 71–79(1973)—
1, 2 Chromosome numbers.

2507 Luscombe, E. G. B.
Iris species identified. *Iris Year Book 1973*, 153–157
1, 2 (1973)—Illustration.

2508 Mangini, E. *and* Maleci, L. B.
Indagine cariologica su alcune *Iris* barbate nane a 40
cromosomi (*Iris chamaeiris* Bertol. s.l.). *Caryologia*,
1 26(1):133–155(1973)—(En).

2509 Mathew, B.
Iris pamphylica. Curtis's Bot. Mag., 179(3):tab.648
2 (1973)—En; illustration, coloured illustration.

2510 Mathew, B.
Plant awards 1972–73: *Iris afghanica* Wendelbo
P. C. *Quart. Bull. Alp. Gard. Soc.*, 41(4):332,337
2 (1973).

2511 Mathew, B.
Plant awards 1972–73: *Iris fosterana* Aitch. & Bak. A. M. *Quart. Bull. Alp. Gard. Soc.*, 41(4):325–326
2 (1973).

2512 Mathew, B.
Plant awards 1972–73: *Iris helena* K. Koch A. M.
2 *Quart. Bull. Alp. Gard. Soc.*, 41(4):327–328(1973).

2513 Mathew, B.
Plant awards 1972–73: *Iris histrioides* var. *sophenensis* Foster A. M. *Quart. Bull. Alp. Gard. Soc.*, 41(4):
2 326–327(1973).

2514 Mathew, B.
Plant awards 1972–73: ? *Iris vvedensky* ? *Iris rosenbachiana* Regel. *Quart. Bull. Alp. Gard. Soc.*,
2 41(4):338(1973).

2515 Soboleva, L.
Spuria irises in Turkmenia. *Iris Year Book 1972*,
2 93–97(1973).

2516 Vogel, S.
Iris fulva Ker-Gawl., a humming bird flower. *Iris*
1, 3–5 *Year Book 1972*, 115–122(1973).

fontanesii Godr.
var. **mellorii** C. Ingram in *Iris Year Book 1973*:
1 97(1973)—Morocco.
heweri Grey-Wilson & Matthew *op. cit.* 106,117 (1973), without latin descr. & without type—
2 Afghanistan.
hymenospatha Matthew & Wendelbo ex Grey-Wilson *op. cit.* 108(1973), without type & without
2 latin descr.—Iran.
kuschkensis Grey-Wilson & Matthew *op. cit.* 107,117
2 (1973), without latin descr.—Afghanistan.
spuria L.
var. **notha** (M. Bieb.) R. R. Stewart, Annot. Cat. Vasc. Pl. W. Pakistan & Kashmir: 65(1972):
6 *I. notha.*
wendelboi Grey-Wilson & Matthew in *Iris Year Book 1973*: 111,120(1973), without latin descr.
2 & without type—Afghanistan.

LAPEIROUSIA

2517 Goldblatt, P.
Lapeirousia oreogena. *Flow. Pl. Afr.*, 42(3–4):pl.1670
5 (1973)—En; coloured illustration.

MORAEA

2518 Goldblatt, P.
Contributions to the knowledge of *Moraea* (Iridaceae) in the summer rainfall region of South Africa. *Ann. Missouri Bot. Gard.*, 60(2):204–259 (1973)—Illustrations, map, dot maps; chromosome
5 numbers; keys.

albicuspa Goldblatt in *Ann. Missouri Bot. Gard.*,
5 60(2):230(1973)—South Africa (Cape Prov.).
algoensis Goldblatt *op. cit.* 226(1973)—South Africa
5 (Cape Prov.).
alpina Goldblatt *op. cit.* 225(1973)—South Africa
5 (Natal), Lesotho.
alticola Goldblatt *op. cit.* 255(1973)—South Africa
5 (Natal, Cape Prov.), Lesotho.
ardesiaca Goldblatt *op. cit.* 243(1973)—South
5 Africa (Natal).
bellendenii N. E. Brown
subsp. **cormifera** Goldblatt *op. cit.* 236(1973)—
5 South Africa (Cape Prov.).
carnea Goldblatt *op. cit.* 242(1973)—South Africa
5 (Natal).
dracomontana Goldblatt *op. cit.* 231(1973)—South
5 Africa (Cape Prov.).
galpinii (Bak.) N. E. Br.
subsp. **robusta** Goldblatt *op. cit.* 248(1973)—South
5 Africa (Transvaal, Cape Prov.).
graminicola Oberm.
subsp. **notata** Goldblatt *op. cit.* 245(1973)—South
5 Africa (Cape Prov.).
hiemalis Goldblatt *op. cit.* 248(1973)—South Africa
5 (Natal).
inclinata Goldblatt *op. cit.* 222(1973)—South Africa
5 (Natal, Cape Prov.).
moggii N. E. Br.
subsp. **albescens** Goldblatt *op. cit.* 239(1973)—
5 South Africa (Natal).
pubiflora N. E. Br.
subsp. **brevistyla** Goldblatt *op. cit.* 229(1973)—
5 South Africa (Natal, Cape Prov.).
reticulata Goldblatt *op. cit.* 249(1973)—South
5 Africa (Cape Prov.).
spathulata (L. f.) Klatt
subsp. **autumnalis** Goldblatt *op. cit.* 254(1973)—
5 South Africa (Cape Prov.).
subsp. **saxosa** Goldblatt *op. cit.* 254(1973)—South
5 Africa (Transvaal).
subsp. **transvaalensis** Goldblatt *op. cit.* 253(1973)—
5 South Africa (Transvaal), Swaziland.
unibracteata Goldblatt *op. cit.* 241(1973)—South
5 Africa (Natal).

NARON

2519 Goldblatt, P.
Proposal for the conservation of the generic name 1265a *Dietes* Salisb. (1812) against *Naron* Medikus
5, 7 (1790) (Iridaceae). *Taxon*, 22(4):504–505(1973).

ONA Ravenna in *An. Mus. Hist. Nat. Valparaíso*, 5:
4 98(1972).

2520 Ravenna, P.
Ona, genero nuevo de Iridaceae nativo de las provincias de Magallanes (Chile) y Tierra del Fuego (Argentina). *An. Mus. Hist. Nat. Valparaíso*, no.5:
4 97–103(1972)—(En); illustration.

obscura (Cav.) Ravenna in *An. Mus. Hist. Nat.*
4 *Valparaíso*, 5:99(1972): *Galaxia obscura.*

ROMULEA

2521 De Vos, M. P.
Note on *Romulea parviflora* Ecklon. *J. S. Afr. Bot.*,
5 39(2):111(1973).

2522 Steyn, E.
'N embriologiese ondersoek van *Romulea rosea* Eckl. var. *reflexa* Beg.: 1. Die bou, ontwikkeling en bevrugting van die saadknop. (The embryology of *Romulea rosea* Eckl. var. *reflexa* Beg.: 1. The structure, development and fertilisation of the ovule). *J. S. Afr.*
1, 5 *Bot.*, 39(2):113–121(1973)—(En).

2523 Steyn, E.
'N embriologiese ondersoek van *Romulea rosea* Eckl. var *reflexa* Beg.: 2. Die ontwikkeling van die endosperm, embrio en caadhuid. *J. S. Afr. Bot.*, 39(3):
1, 5 235–243(1973)—(En).

SISYRINCHIUM

2524 Henderson, D. M.
A biosystematic study of the genus *Sisyrinchium* (Iridaceae) in the Pacific Northwest. *Diss. Abstr. Int.*,
3 *B*, 33(8):3508–3509(1973).

ITEACEAE
see
ESCALLONIACEAE

IXONANTHACEAE

2525 Badré, F.
Ixonanthaceae. Flore du Gabon, 21:55–58(1973)—
5 Illustrations; key.

JOINVILLEACEAE
see
FLAGELLARIACEAE

JUGLANDACEAE

2526 Stone, D. E.
New World Juglandaceae: 3. A new perspective of the tropical members with winged fruits. *Ann. Missouri Bot. Gard.*, 59(2):297–322(1972 publ. 1973)—
4 Illustrations; chromosome number; keys.

2527 Sushko, M. M.
Man' chzhurskiĭ orekh na Dal'nem Vostoke. (Manchurian walnut in the Far East). *Trudȳ Prikl.*
2 *Bot. Genet. Selek.*, 49(1):145–150(1973)—En.

CARYA

2528 Clark, A.
Pecan: an American wood. *U.S. Dep. Agric. F.S.*,
3 249. 8p.(1973)—Illustrations, maps.

2529 Dennis, W. M.
A new record water hickory for South Carolina.
3 *Castanea*, 28(2):205(1973). *Carya aquatica.*

2530 Gibbon, E. L.
The taxonomic and ecologic distinctness of *Carya ovata* (Mill.) K. Koch variety *ovata* and *C. ovata* variety *australis* (Ashe) Little. *Diss. Abstr. Int., B*,
3 33(12):5695(1973).

2531 Manning, W. E.
The northern limit of the distribution of the mockernut hickory. *Mich. Bot.*, 12(4):203–209(1973)
3 Dot map. *Carya tomentosa.*

2532 Manning, W. E.
The northern limits of the distributions of hickories in New England. *Rhodora*, 75(801):34–51(1973)—
3 Dot maps; key.

2533 Phillips, D. R.
Hickory: an American wood. *U.S. Dep. Agric.*,
3, 4 *F.S.*, 241. 7p.(1973)—Illustrations, maps.

JUGLANS

2534 Howell, J. T.
A lectotype for the Hinds walnut. *Madrõno*, 22(3):
3 144(1973).

2535 Landt, E. F. *and* Phares, R. E.
Black walnut: an American wood. *U.S. Dep. Agric. F.S.*, no.270. 7p.(1973)—Illustration, map. *Juglans*
3 *nigra.*

OREOMUNNEA

mexicana (Standley) Leroy
subsp. **costaricensis** Stone in *Ann. Missouri Bot.*
4 *Gard.*, 59(2):320(1972 publ. 1973)—Costa Rica.

JUNCACEAE

2536 Fredga, A., Bendz, G. *and* Apell, A.
Red pigments in the Juncaceae family. *In* Nobel Symposium 25: chemistry in botanical classification
1–7 August 20–25, 1973. (1973). 1p. Abstract.

2537 Lisowski, S., Malaisse, F. *and* Symoens, J. J.
Flore d'Afrique centrale (Zaïre – Rwanda – Burundi): Spermatophytes. Juncaceae. 1973.[ii]13p.—
5 Illustrations; keys.

2538 Ross-Craig, S.
Drawings of British plants: being illustrations of the species of flowering plants growing naturally in the British Isles: part 30. Juncaceae, Typhaceae, Sparganiaceae, Araceae. London, Bell, 1973. 41pl.
1 Juncaceae pl. 1–32.

JUNCUS

2539 Benoit, P. M.
Juncus foliosus Desf.—in Wales. *Watsonia*, 9(4):
1 373(1973).

2540 Conacher, E. R. T. *and* Ribbons, B. W.
Juncus filiformis. Glasgow Nat., 19(1):67–68(1973)—
1 En. Local records.

2541 Cope, T. A. *and* Stace, C. A.
The segregates of *Juncus bufonius* Agg. *Watsonia*,
1 9(4):426(1973).

2542 Eichwald, K.
Nakhozhdenie sitnika tonkogo v Estonskoĭ SSR. (*Juncus tenuis* Willd. in Estonia). *Flor. Märkmed*,
1 1(5):310–311(1973).

2543 Londo, G.
Enige ervaringen met *Juncus tenageia* L.f. in de leemputten bij Udenhout (N.-Br.). *Gorteria*, 6(12):
1 198–202(1973)—(En).

2544 Ploeg, D. T. E. van der *and* Zimmerman Azn, A.
De draadrus (*Juncus filiformis* L.) in Friesland.
1 *Gorteria*, 6(7):123–124(1973)—(En).

2545 Scannell, M. J. P.
Juncus planifolius R. Br. in Ireland. *Irish Nat. J.*,
1 17(9):308–309(1973)—En.

2546 Schölch, H. F.
Juncus sphaerocarpus Nees im Rhein-Main-Gebiet. *Hess. Flor. Briefe*, 259–261:41–48(1973)—
1 Illustration.

2547 Soó, R. *and* Isépy, I.
Über einige Formenkreise der ungarischen und karpatischen Flora: 18. *Juncus bufonius, Eleocharis palustris* Agg. und *Agrostis canina. Ann. Univ.*
1 *Budapest. Rolando Eötvos Nom., Biol.*, 14:169–177(1972).

2548 Yaroshevskaya, A. S.
Mikrosporogenez u *Juncus* i *Luzula*. (The microsporogenesis of *Juncus* and *Luzula*). *Vestn. Mosk. Univ.*
1–7 *Nauch. Zhurn.*, 6, 27(6):83–85(1972)—(En).

LUZULA

2549 Dvořák, J. *and* Vorel, J.
Über den Fund einer neuen Hybride *Luzula forsteri* × *luzulina* (*Luzula* × *romanica* J. Dvořák et Vorel hybr. nova). (O nálezu nového křížence *Luzula forsteri* × *luzulina* (*Luzula* × *romanica* J. Dvořák et Vorel hybr. nova)). *Preslia*, 45(1):87–89
1 (1973)—(Cz).

2550 Hämet-Ahti, L.
Notes on the *Luzula arcuata* and *L. parviflora* groups in eastern Asia and Alaska. *Ann. Bot. Fenn.*, 10(2):
2, 3 123–130(1973)—Illustrations, dot map.

2551 Synnott, D.
Luzula × *borreri* in Ireland. *Irish Nat. J.*, 17(10):327–
1 333(1973).

2552 Yaroshevskaya, A. S.
Mikrosporogenez u *Juncus* i *Luzula*. (The microsporogenesis of *Juncus* and *Luzula*). *Vestn. Mosk. Univ.*
1–7 *Nauch. Zhurn.*, 6, 27(6):83–85(1972)—(En).

× **romanica** Dvořák & Vorel in *Preslia*, 45(1):87
1 (1973)—Rumania. (*L. forsteri* × *L. luzulina*).

JUNCAGINACEAE

TRIGLOCHIN

2553 Delay, J.
Triglochin maritimum L.: étude ultrastructurale des caryocinèses. *Inform. Ann. Caryosyst. Cytogénét.*, no.
1–3 6:25–32(1972 publ. 1973).

2554 Serbănescu-Jitariu, G.
Betrachtungen über Gynäzeum, Frucht und Keimung bei *Triglochin maritimum* L. und *Triglochin palustre* L. *Rev. Roum. Biol., Bot.*, 18(1):9–20(1973)—
1–3 Illustrations.

KRAMERIACEAE

2555 Robertson, K. R.
The Krameriaceae in the south-eastern United States. *J. Arnold Arbor.*, 54(2):322–327(1973)—
3 Chromosome numbers.

LABIATAE

2556 Arustamova, D. M.
Tomillyarȳ Armyanskoĭ SSR. (Tomillares of the Armenian SSR). *Byull. Mosk. Obshch. Ispȳt. Prir.,*
2 *Biol.*, 78(5):121–129(1973)—(En).

2557 Morton, J. K.
A cytological study of the British Labiatae (excluding *Mentha*). *Watsonia*, 9(3):239–246(1973)—
1 Chromosome numbers.

2558 Pal, S.
Cytotaxonomy of Labiatae. *Proc. Indian Sci. Congr.*, 58(3):475(1971)—Chromosome numbers. Abstract
6 only.

2559 Pereira, C. *and* Pereira, E.
Flora do Estado do Paraná fam. Labiatae. *Arq. Jard. Bot. Rio de Janeiro*, 19:79–107(1973)—Illustrations;
4 keys.

2560 Singh, U. P.
Cytology of some members of the Labiatae of Bihar. *J. Bihar Bot. Soc.*, 1(1–2):68–69(1972)—
6 Chromosome numbers. Abstract.

2561 Standley, P. C. *and* Williams, L. O.
Flora of Guatemala: part 9, number 3. Labiatae. *Fieldiana, Bot.*, 24(9 nos. 3 4):237–317(1973)—
4 Illustrations; keys.

ACINOS

arvensis (Lam.) Dandy
subsp. **eglandulosus** (Klok.) Tsvelev in *Novosti*
1 *Sist. Vȳssh. Rast.*, 10:365(1973): *A. eglandulosus.*
pseudacinos (Lacaita) Fenaroli in *Webbia*, 28(2):373 (1973), without basionym ref.: *Calamintha*
1 *pseudacinos.*

BALLOTA

2562 Vanecek, J.
Je známa zlutokvetá *Ballota nigra* L. subsp. *nigra*? (Ist die gelbblütige *Ballota nigra* L. subsp. *nigra*
1 bekannt?). *Zprávy Cesk. Bot. Spol.*, 8(3):209(1973).

hispanica (L.) Pau
subsp. **bullata** (Pomel) Malagarriga, Las Subesp. y
1 Variac. Geogr.: 23(1973): *B. bullata.*
subsp. hispida (Bth.) Malagarriga *op. cit.* 23(1973),
1 without basionym date: *B. hirsuta* var. *hispida.*
subsp. mollissima (Bth.) Malagarriga *op. cit.* 23
1 (1973), without basionym date: *B. mollissima.*

BETONICA

alopecuros L.
subsp. **godronii** (Rouy) Laínz in *Bol. Inst. Estud. Astur., Supl. Cienc.*, 16(Aport. Conoc. Fl. Cántabro-Astur, x:33): 189(1973): *Stachys*
1 *alopecuros* r. *godronii.*

CALAMINTHA

officinalis Mönch
subsp. **subnuda** (W. & K. Pl.) Soó in *Acta Bot. Acad. Sci. Hung.*, 18(1–2):176(1973): *Melissa*
1 *subnuda.*

CORIDOTHYMUS

2563 Philianos, S. M. *and* Véléri-Triandaphyllou, J.
Sur les constituants du thym capité (*Coridothymus capitatus* Reichb. f.) du Bassin de l'Attique: 1. *Biol.*
1 *Gallo-Hellen.*, 4(2):153–162(1973)—(En, Gre).

DRACOCEPHALUM

2564 Měsíček, J. *and* Soják, J.
Karyological and taxonomic observations on *Dracocephalum foetidum* Bunge and *Koenigia islandica* L. *Folia Geobot. Phytotax.*, 8(1):105–112(1973)—
1, 2 Chromosome numbers.

EPLINGIA L. O. Williams in *Fieldiana, Bot.*, 36(3):
4 18(1973).

2565 Williams, L. O.
Eplingia, a new genus of the Labiatae from Mexico.
4 *Fieldiana*, Bot., 36(3):17–20(1973)—Illustration.

saxicola L. O. Williams in *Fieldiana, Bot.*, 36(3):18
4 (1973)—Mexico.

ERIOPE

2566 Harley, R. M.
A Cuban *Hyptis* transferred from *Eriope*. Notes on
4 New World Labiatae: 2. *Kew Bull.*, 28(1):24(1973).

2567 Harley, R. M.
Eriope horridula, a member of the Verbenaceae. Notes on New World Labiatae: 1. *Kew Bull.*, 28(1):
4 121–122(1973).

GALEOPSIS

2568 Duvigneaud, J.
Une forme curieuse et nouvelle de *Galeopsis*
1 *segetum*. *Bull. Soc. Roy. Bot. Belg.*, 106(2):367(1973).

segetum Necker
f. **glabra** J. Duvigneaud in *Bull. Soc. Roy. Bot. Belg.*,
1 106(2):367(1973)—Luxembourg, Belgium.

GLECHOMA

hederaceum L.
subsp. sardoum (Béguinot) Soó in *Acta Bot. Acad. Sci. Hung.*, 18(1–2):176(1973), without basionym
1 date: *G. sardoum*.

HAUMANIASTRUM

2569 Lisowski, S., Malaisse, F. *and* Symoens, J. J.
Un nouvel *Haumaniastrum* à port cespiteux des hauts plateaux du Shaba (Zaïre). *Bull. Jard. Bot.*
5 *Nation. Belg.*, 43(3–4):329–332(1973)—Illustration.

dilunguense Lisowski, Malaisse & Symoens in *Bull.*
5 *Jard. Bot. Nation. Belg.*, 43(3–4):330(1973)—Zaïre.

HYPTIS

2570 Harley, R. M.
A Cuban *Hyptis* transferred from *Eriope*. Notes on New World Labiatae: 2. *Kew Bull.*, 28(1):24
4 (1973).

2571 Wulff, R.
Intrapopulation variation in the germination of seeds in *Hyptis suaveolens*. *Ecology*, 54(3):646–649
4 (1973).

trichopes (Epling) Harley in *Kew Bull.*, 28(1):24
4 (1973): *Eriope trichopes*.

LAGOCHILUS

olgae Kamelin in *Bot. Zhurn.*, 58(5):631(1973)—
2 U.S.S.R. (Middle Asia).

LAMIASTRUM

flavidum (F. Herm.) Ehrendorfer in *Österr. Bot.*
1 *Zeitschr.*, 122(4):266(1973): *Lamium flavidum*.
montanum (Pers.) Ehrendorfer *op. cit.* 266(1973):
1 *Pollichia montana*.

LAMIUM

2572 Conolly, A. P.
Lamium album in Wales. *Watsonia*, 9(4):425 426
1 (1973).

LEUCOPHAË

2573 Kunkel, G.
Notes on the genus *Leucophaë* Webb & Berth. (Lamiaceae) in the Canary Islands. *Cuad. Bot. Canar.*,
1 18–19:65–74(1973)—(Sp); illustrations.

cabrerae (Ceb. & Ort.) Kunkel in *Cuad. Bot. Canar.*,
1 18–19:67(1973): *Sideritis cabrerae*.
cystosiphon (Svent.) Kunkel *op. cit.* 68(1973):
1 *Sideritis cystosiphon*.
dendro-chahorra (Bolle) Christ
var. **albida** (Pitard) Kunkel *op. cit.* 68(1973):
1 *Sideritis massoniana* var. *albida*.
gomerae (De Noé ex Bolle) Kunkel *op. cit.* 72
1 (1973): *Sideritis gomerae*.
kueglerana (Bornm.) Kunkel *op. cit.* 72(1973), as
1 'kuegleriana': *Sideritis kuegleriana*.
lotsyi Pitard
var. **mascaënsis** (Svent.) Kunkel *op. cit.* 72(1973):
1 *Sideritis lotzyi* var. *mascaënsis*.
nutans (Svent.) Kunkel *op. cit.* 73(1973): *Sideritis*
1 *nutans*.
1 **sventenii** Kunkel *op. cit.* 69(1973)—Canary Is.

LYCOPUS

2574 Zając, E. U. *and* Zając, A.
Nowe badania nad rozmieszczeniem gatunków synantropijnych w Americe Północnej. *Wiad. Bot.*,
3 17(3):163–165(1973)—Dot map. *Lycopus europaeus*.

MENTHA

2575 Gill, L. S., Lawrence, B. M. *and* Morton, J. K.
Variation in *Mentha arvensis* L. (Labiatae): 1. The North American populations. *Bot. J. Linn. Soc.*,
3 67(3):213–232(1973)—Illustrations.

2576 Harley, R. M.
Mints. *In* Green, P. S. *ed.* Plants: wild and
1, 2 cultivated: 126–128(1973)—Chromosome number.

× *piperita* L.
nm. **nepetoides** (Lej.) Lebeau in Nouv. Fl. Belg., Luxemb., N. France etc.: 759(1973): *M.* ×
1 *nepetoides*.
spicata L.
subsp. **glabrata** (Lej. & Court.) Lebeau *op. cit.*
1 759(1973): *M. viridis* var. *glabrata*.

MICROMERIA

fruticosa (L.) Druce
var. **italica** (Huter) Fenaroli in *Webbia*, 28(2):375
1 (1973): *M. marifolia* β *italica*.
serbaliana Danin & Hedge in *Notes Roy. Bot. Gard.*
2 *Edinburgh*, 32(2):261(1973)—*Sinai Peninsula*.

MONARDA

2577 Egler, F.
The hybrid nature of *Monarda media* Willd.
3 *Castanea*, 38(3):209–214(1973).

NEPETA

2578 Valsecchi, F. *and* Corrias, S. D.
Notizie ecologiche, cariologiche e sistematiche su *Nepeta foliosa* Moris. (Some ecological, caryological and systematic data on *Nepeta foliosa* Moris). *Giorn. Bot. Ital.*, 107(4):173–180(1973)—(En); illustrations,
1 dot map; chromosome numbers.

murcica Guir. ex Willk.
var. **cuneata** Segura Zubizarreta in *Pirineos*, 109:
1 43(1973)—Spain
subsp. **toranzii** Segura Zubizarreta *op. cit.* 42
1 (1973)—Spain.
numantina Segura Zubizarreta *op. cit.* 42(1973)—
1 Spain.
royleana R. R. Stewart, Annot. Cat. Vasc. Pl. W. Pakistan & Kashmir: 625(1972), nom. nov.:
6 *N. salviaefolia* Royle ex Benth.

OCIMUM

2579 Figueiredo, R. de C. L.
Sobre a anatomia dos órgãos vegetativos de *Ocimum nudicaule* Benth. (Labiatae). *An. Acad. Brasil.*
4 *Cienc.*, 44(3–4):549–570(1972)—(En); illustrations.

ORIGANUM

vulgare L.
f. **chlorescens** Simk. ex Soó & Borhidi in *Acta Bot. Acad. Sci. Hung.*, 18(1–2):173(1973)—
1 Hungary.

OTOSTEGIA

2580 Sebald, O.
Die Gattung *Otostegia* Bentham (Labiatae) in Africa und auf der arabischen Halbinsel. *Stuttgarter Beitr. Naturk.*, *A*, no. 263:1–84(1973)—(En);
2, 5 illustrations, dot maps; keys.

subsect. **Cuneifolia** Sebald in *Stuttgarter Beitr.*
5 *Naturk., Biol.*, 263:32(1973).
subsect. **Cyclodonta** (Chiovenda) Sebald *op. cit.* 35
5 (1973): *Otostegia* sect. *Cyclodonta.*
subsect. **Heterostegia** (Kudr.) Sebald *op. cit.* 71
5 (1973): *Otostegia* sect. *Heterostegia.*
5 sect. **Mucrophyllon** Sebald *op. cit.* 75(1973).
5 subsect. **Somalae** Sebald *op. cit.* 29(1973).
fruticosa (Forsk.) Schweinfurth ex Penzig
subsp. **schimperi** (Benth.) Sebald *op. cit.* 66(1973):
5 *Ballota schimperi.*
hildebrandtii (Vatke & Kurtz) Sebald *op. cit.* 56
5 (1973): *Ballota hildebrandtii.*
migiurtiana Sebald *op. cit.* 34(1973)—Somali
5 Republic.
modesta S. Moore
var. **buranensis** Sebald *op. cit.* 53(1973)—Somali
5 Republic.
somala (Patzak) Sebald *op. cit.* 29(1973): *Ballota*
5 *somala.*
tomentosa A. Richard
subsp. **ambigens** (Chiovenda) Sebald *op. cit.* 42
5 (1973): *O. ambigens.*
subsp. **steudneri** (Schweinf.) Sebald *op. cit.* 40
5 (1973): *O. steudneri.*

PLECTRANTHUS

2581 Lye, K. A.
Studies in African plants: 2. A new species of *Plectranthus* (Labiatae) from Uganda. *Norw. J. Bot.*,
5 20(1):57–59(1973)—Illustrations.

ugandensis Lye in *Norw. J. Bot.*, 20(1):57(1973)—
5 Uganda.

POGOGYNE

2582 Howell, J. T.
3 A new *Pogogyne. Four Seasons*, 4(3):22(1973).

clareana J. T. Howell in *Four Seasons*, 4(3):22
3 (1973)—U.S.A. (California).

RABDOSIA

2583 Anon.
Chromosome number of *Rabdosia umbrosa* (Maxim.)
2 *Acta Phytotax. Geobot.*, 25(4–6):153(1973).

2584 Codd, L. E.
Notes on African plants: Labiatae. The genus *Rabdosia* in South Africa. *Bothalia*, 11(1–2):117
5 (1973).

2585 Murata, G.
[Short nomenclatural notes on *Rabdosia*.] *Acta*
2, 6, 7 *Phytotax. Geobot.*, 25(4–6):185(1973)—La.

calycina (Benth.) Codd in *Bothalia*, 11(1–2):117
5 (1973): *Plectranthus calycinus.*
×**inamii** (Murata) Murata in *Acta Phytotax. Geobot.*,
2 25(4–6):185(1973): *Isodon* × *inamii.*

ROSMARINUS

eriocalix Jordan & Fourr.
subsp. **tomentosus** (Huber-Morath & Maire) Fernandez Casas in *Cuad. Cienc. Biol.* (Granada),
1 2(1):40(1973): *R. tomentosus.*
officinalis L.
subsp. **palaui** (O. Bolòs & R. Mol.) Malagarriga, Las Subesp. y Variac. Geogr.: 23(1973):
1 *R. officinalis* var. *palaui.*

SALVIA

2586 Afzal-Rafii, Z.
Contribution à l'étude cytotaxinomique des *Salvia* de Turquie: 2. Bull. Soc. Bot. France, 119(3–4):167–175(1972 publ. 1973)—(En); chromosome
1, 2 numbers.

2587 Moldenke, H. N.
Notes on new and noteworthy plants: 60. *Phytologia*,
3 26(4):224–226(1973).

2588 Morley, B.
Four interesting *Salvia* species. *J. Roy. Hort. Soc.*,
4 98(12):550(1973)—Coloured illustrations.

2589 Viţalariu, G. *and* Leucov, M.
Salvia reflexa Hornem. o nouă plantă adventivă în România. (*Salvia reflexa* Hornem., une nouvelle plante adventive de la Roumanie). *Stud. Comun., Bot.*,
1 689–693(1971)—(Fr); illustration.

decumbens Alain in *Phytologia*, 25(5):273(1973)—
4 Hispaniola (Dominican Republic).
holwayi Blake
var. **puberulenta** L. O. Williams in *Fieldiana, Bot.*,
4 24(9) nos. 3–4:286(1973). Guatemala.
lyrata
f. **purpureorubra** Moldenke in *Phytologia*, 26(4):
3 225(1973)—U.S.A. (S. Carolina).

SATUREJA

2590 Hérisset, A., Jolivet, J., Zoll, A. *and* Chaumont, J. P.
A propos des falsifications de la sarriette des jardins (*Satureia hortensis* L.). *Pl. Méd. Phytothér.*,
1 7(2):121–134(1973).

2591 San Martin, R., Granger, R., Adzet, T., Passet, J. *and* Teulade-Arbousset, G.
Le polymorphisme chimique chez deux Labiées mediterranéennes: *Satureia montana* L. et *Satureia obovata* Lag. (stations naturelles espagnoles). *Pl. Méd.*
1 *Phytothér.*, 7(2):95–103(1973).

SIDERITIS

2592 Mendoza-Heuer, I.
La morfologia de la *Sideritis* en Macaronesia y su grado de diferenciación. *Monogr. Biol. Canar.*, no. 4:
1 64–70(1973)—Illustrations; chromosome numbers.

hirsuta L.
subsp. **hirtula** (Brot.) Pinto da Silva in *Agron. Lusit.*, 34(3):192(1972 publ. 1973): *S. hirsuta* raç.
1 *hirtula.*

STACHYS

2593 Fernandez, A. M. F. de *and* Sarmiento, M. N. R. de
Contribucion al conocimiento del genera *Stachys* en la provincia de Tucuman con especial referencia a su taxonomia. *Lilloa*, 33(10):175–204(1973)—(En);
4 chromosome numbers.

2594 Hinojo, M. E. C. de *and* Fernandez, A. M. F. de
Contribucion al conocimiento del genero *Stachys* en la provincia de Tucuman: 3. Diferencias fenotipicas entre dos poblaciones geneticamente similares de la especie *Stachys petiolosa* Bryq. *Lilloa*, 33(17):311–318
4 (1973)—(En).

2595 Sarmiento, M. N. R. de
Contribucion al conocimiento del genero *Stachys* en la provincia de Tucuman: 2. Estudio anatómico de ejemplares de *Stachys petiolosa* Briq. en zonas de altura.
4 *Lilloa*, 33(16):305–310(1973)—(En).

TEUCRIUM

2596 Meusel, H.
Teucrium subspinosum Pourret et Willd., ein Dornpolster-Endemit der Balearen. *Acta Bot. Acad. Sci. Hung.*, 19(1–4):211–222(1973)—(En);
1 illustrations.

chamaedrys L.
subsp. multiflorum (L.) Malagarriga, Las Subesp. y Variac. Geogr.: 25(1973), without basionym
1 date: *T. multiflorum.*
marum L.
subsp. subspinosum (Pourret) Malagarriga *op. cit.*
1 24(1973), without basionym date: *T. subspinosum.*
polium L.
subsp. achaemenis (Schreb.) Malagarriga *op. cit.* 24
1 (1973), without basionym date: *T. achaemenis.*
subsp. **carthaginense** (Lge.) Malagarriga *op. cit.* 24
1 (1973): *T. carthaginense.*
subsp. lusitanicum (Schreb.) Malagarriga *op. cit.*
1 24(1973), without basionym date: *T. lusitanicum.*
subsp. pulverulentum (Cosson) Malagarriga *op. cit.* 24(1973), without basionym date: *T. pulveru-*
1 *lentum.*
subsp. verticillatum (Cav.) Malagarriga *op. cit.* 24
1 (1973), without basionym date: *T. verticillatum.*
subsp. **vincentinum** (Rouy) Malagarriga *op. cit.* 24
1 (1973): *T. vincentinum.*
pyrenaicum L.
subsp. cinereum (Bss.) Malagarriga *op. cit.* 23(1973),
1 without basionym date: *T. cinereum.*
subsp. rotundifolia (Schreb.) Malagarriga *op. cit.* 23
1 (1973), without basionym ref.: *T. rotundifolia.*

THYMUS

2597 Gogina, E. E.
O nasledovanii zhenskoĭ dvudomnosti u *Thymus loevyanus* Opiz. *Byull. Glavn. Bot. Sada* (Moscow),
1 88:54–59(1973).

2598 Granger, R. *and* Passet, J.
Thymus vulgaris spontane de France: races chimiques et chemotaxonomie. *Phytochemistry*, 12(7):1683–1691
1 (1973)—(En); dot map.

2599 Granger, R., Passet, J., Teulade-Arbousset, G. *and* Auriol, P.
Types chimiques de *Thymus nitens* Lamotte, endémique cévénol. *Pl. Méd. Phytothér.*, 7(3):225–
1 233(1973).

2600 Hérisset, A., Jolivet, J. *and* Rey, P.
Différenciation de quelques huiles essentielles présentant une constitution voisine (en particulier par l'examen de leurs spectres UV,IR et Raman) 8. Essences de divers *Thymus*. *Pl. Méd. Phytothér.*, 7(1):
1 37–47(1973).

2601 Jalas, J.
Thymus subsect. *Pseudomarginati* in the Himalayas and adjoining western mountain ranges, and in Caucasia. *Ann. Bot. Fenn.*, 10(2):104–122(1973)—
2, 6 Illustrations, dot map.

2602 Menitskiĭ, Yu. L.
Nadvidovye̅ taksony̅ roda *Thymus* L. (Labiatae): 1. (Supraspecific taxa in the genus *Thymus* L. (Labiatae):
1, 2 1). *Bot. Zhurn.*, 58(6):794–805(1973).

2603 Menitskiĭ, Yu. L.
Nadvidovye̅ taksony̅ roda *Thymus* L. (Labiatae): 2. (Supraspecific taxa in the genus *Thymus* L. (Labiatae): 2). *Bot. Zhurn.*, 58(7):988–994(1973)—Chromosome
1, 2 numbers.

2604 Oberdorfer, E.
Thymian, Quendel, *Thymus* L. *Göttinger Flor. Rundbr.*, 7(4):67–69(1973)—Ge; chromosome
1 numbers; key.

2605 Schmidt, P.
Übersicht über die mitteleuropäischen Arten der Gattung *Thymus* L. *Feddes Repert.*, 83(9–10):663–671
1 (1972 publ. 1973)—(En).

subsect. **Micantes** (Velen.) Menitskiĭ in *Bot. Zhurn.*,
1 58(6):800(1973): *Thymus* sect. *Micantes.*
1 sect. **Saturejoides** Menitskiĭ *op. cit.* 798(1973).
1, 2 sect. **Serrulatus** Menitskiĭ *op. cit.* 799(1973).
1 sect. **Thymbropsis** Minitskiĭ *op. cit.* 798(1973).
subsect. **Verticillati** (Klok. & Shost.) Menitskiĭ in *Bot. Zhurn.*, 58(7):991(1973): *T.* stat. indefin.
1, 2 *Verticillati.*
caucasicus Willd. ex Ronn.
subsp. **grossheimii** (Ronn.) Jalas in *Ann. Bot. Fenn.*,
2 10(2):107(1973): *T. grossheimii.*
cephalotus L.
subsp. ciliatus (Desf.) Malagarriga, Las Subesp. y Variac. Geogr.: 22(1973), without
1 basionym date: *Thymbra ciliata.*
subsp. longiflorus (Bss.) Malagarriga *op. cit.* 22
1 (1973), without basionym date: *T. longiflorus.*
glabrescens Willd.
subsp. **degenianus** (Lyka) Soó in *Acta Bot. Acad. Sci. Hung.*, 18(1–2):172(1973): *T. serpyllum* subsp.
1 *degenianus.*
hispanicus Poiret
subsp. granatensis (BSS.) Malagarriga, Las Subesp. y Variac. Geogr.: 21(1973), without basionym
1 date: *T. granatensis.*
subsp. serpylloides (Bory) Malagarriga *op. cit.* 22
1 (1973), without basionym date: *T. serpylloides.*
linearis Bentham
subsp. **hedgei** Jalas in *Ann. Bot. Fenn.*, 10(2):117
6 (1973)—Pakistan.
praecox Opiz
subsp. **britannicus** (Ronn.) Holub in *Preslia*, 45(4):
1 359(1973): *T. britannicus.*
subsp. **widderi** (Mach.) P. Schmidt in *Feddes*
1 *Repert.*, 83(9–10):669(1972 publ. 1973): *T. widderi.*
pulcherrimus Schur.
subsp. **sudeticus** (Lyka) P. Schmidt *op. cit.* 669 (1972 publ. 1973): *T. serpyllum* subsp.
1 *sudeticus.*
pulegioides L.
subsp. **carniolicus** (Borb.) P. Schmidt *op. cit.* 664
1 (1972 publ. 1973): *T. carniolicus.*
var. **parvifolius** (Borb.) P. Schmidt *op. cit.* 664
1 (1972 publ. 1973): *T. chamaedrys* var. *parvifolius.*
serpyllum L.
var. **carniolicus** (Borbás) Malagarriga, Las Subesp. y Variac. Geogr.: 21(1973): *T.*
1 *carniolicus.*
subsp. confertus (G.G.) Malagarriga *op. cit.* 21 (1973), without basionym date: *T. serpyllum* var.
1 *confertus.*
subsp. **froelichianus** (Opiz) Malagarriga *op. cit.*
1 21(1973): *T. froelichianus.*
subsp. piperella (L.) Malagarriga *op. cit.* 21(1973),
1 without basionym date: *T. piperella.*
subsp. pseudoserpyllum (Nègre) Malagarriga *op. cit.* 21(1973), without basionym page:
1 *T. pseudoserpyllum* nom. illegit.

subsp. **richardii** (P.) Malagarriga *op. cit.* 21(1973):
1 *T. richardii.*
× *sooi* Lyka
nm. **nyirsegensis** Soó in *Acta Bot. Acad. Sci.*
1 *Hung.*, 18(1–2):172(1973)—Hungary.

TRICHOSTEMA

2606 Montgomery, F. H. *and* Morton, J. K.
Trichostema dichotomum L. (Labiatae) new to
3 Canada. *Canad. Field.-Nat.*, 87(1):63–64(1973).

ZIZIPHORA

2607 Taarna, K.
Chromosome counts on *Ziziphora* L. Labiatae. *Ann.*
2 *Bot. Fenn.*, 10(4):341(1973).

LACISTEMATACEAE

LOZANIA

2608 Agostini, G.
El género *Lozania mutis* (Lacistemaceae). *Acta Bot.*
Venezuel., 8(1–4):167–175(1973)—Illustrations, map;
4 keys.

2609 Agostini, G.
Lozania mutisiana (Lacistemataceae) especie nueva
4 para Venezuela. *Pittieria*, no.4:49–50(1972)—(En).

2610 Agostini, G.
Revaluacion de *Lozania pittieri* (Lacistemataceae).
4 *Pittieria*, no. 4:43–47(1972)—(En); illustration.

LAMIACEAE
see
LABIATAE

LAURACEAE

ALSEODAPHNE

2611 Kostermans, A. J. G. H.
A synopsis of *Alseodaphne* Nees (Lauraceae).
Candollea, 28(1):93–136(1973)—(Fr,Ge);
2, 6 illustrations.

albifrons Kostermans in *Candollea*, 28(1):99(1973)—
6 Malaya, Borneo (Sabah, Sarawak, Kalimantan).
birmanica Kostermans *op. cit.* 102(1973)—
6 Thailand.
cavaleriei (Lév.) Kostermans *op. cit.* 104(1973):
2 *Machilus cavaleriei.*
6 **dura** Kostermans *op. cit.* 104(1973)—Malaya.
foxiana (Gamble) Kostermans *op. cit.* 107(1973):
6 *Beilschmiedia foxiana.*
garciniicarpa Kostermans *op. cit.* 107(1973), as
6 'garciniaecarpa'—Malaya.
glauciflora Kostermans *op. cit.* 108(1973)—Sumatra,
6 Borneo (Kalimantan).
gracilis Kostermans *op. cit.* 109(1973)—Locality not
6 stated.
6 **habrotricha** Kostermans *op. cit.* 110(1973)—India.
6 **himalayana** Kostermans *op. cit.* 111(1973)—Nepal.
intermedia Kostermans *op. cit.* 113(1973)—Malaya,
2, 6 Singapore, Sumatra.
kochummenii Kostermans *op. cit.* 115(1973)—
6 Malaya.
lanuginosa Kostermans *op. cit.* 116(1973)—
6 Vietnam.
macrantha Kostermans *op. cit.* 117(1973)—
6 Malaya.
micrantha Kostermans *op. cit.* 118(1973)—
6 Malaya.
nigrescens (Gamble) Kostermans *op. cit.* 119(1973):
6 *Dehaasia nigrescens.*
rhododendropsis Kostermans *op. cit.* 126(1973)—
6 Vietnam.
6 **rubriflora** Kostermans *op. cit.* 127(1973)—Burma.
6 **siamensis** Kostermans *op. cit.* 129(1973)—Thailand.
suboppositifolia Kostermans *op. cit.* 130(1973)—
6 Vietnam.
sulcata Kostermans *op. cit.* 131(1973)—Borneo
6 (Sabah).
tomentosa Kostermans *op. cit.* 132(1973)—Borneo
6 (Sarawak).
6 **utilis** Kostermans *op. cit.* 133(1973)—Vietnam.
2 **yunnanensis** Kostermans *op. cit.* 133(1973)—China.

ANIBA

2612 Morais, A. Alpande de *and others*
Óleos essenciais de espécies do gênero *Aniba*. *Acta*
4 *Amazonica*, 2(1):41–44(1972)—(En).

BEILSCHMIEDIA

2613 Fouilloy, R.
Trois arbres nouveaux d'Afrique tropicale (Laur. –
Myrist. – Sapind.). *Adansonia*, (4):545–555(1972 publ.
5 1973)—Illustrations. *Beilschmiedia descoingsii.*

2614 Kostermans, A. J. G. H.
Nouveautés pour la flore indochinoise. *Adansonia*,
6 13(3):333–340(1973)—(En); illustrations.

2615 Ross, J. H.
Notes on African plants: Lauraceae, a new generic
record for South Africa and a new species. *Bothalia*,
5 11(1–2):118(1973). *Beilschmiedia natalensis.*

descoingsii Fouilloy in *Adansonia*, 12(4):545(1972
5 publ. 1973)—Congo-Brazzaville.
natalensis J. H. Ross in *Bothalia*, 11(1–2):118(1973)—
5 South Africa (Natal).
vidalii Kostermans in *Adansonia*, 13(3):336(1973)—
6 Vietnam.

CINNADENIA Kostermans in *Adansonia*, 13(2):223
2, 6 (1973).

2616 Kostermans, A. J. G. H.
Cinnadenia Kosterm. genus novum Lauracearum.
2, 6 *Adansonia*, 13(2):223–227(1973)—En; illustrations.

paniculata (Hooker f.) Kostermans in *Adansonia*,
2, 6 13(2):223(1973): *Dodecadenia paniculata.*

CINNAMOMUM

2617 Kostermans, A. J. G. H.
A forgotten Ceylonese cinnamon-tree (*Cinna-*
momum capparu-conde Bl.). *Ceylon J. Sci., Biol. Sci.*,
6 10(2):119–121(1973).

DEHAASIA

2618 Kostermans, A. J. G. H.
A synopsis of the genus *Dehaasia* Bl. (Lauraceae).
6 *Bot. Jahrb.*, 93(3):424–480(1973)—(Ge); illustrations.

annamensis Kostermans in *Bot. Jahrb.*, 93(3):427
6 (1973)—Vietnam.
brachybotrys (Merr.) Kostermans *op. cit.* 429(1973):
6 *Beilschmiedia brachybotrys.*
candolleana (Meisn.) Kostermans *op. cit.* 431(1973):
6 *Dictyodaphne candolleana.*
6 **celebica** Kostermans *op. cit.* 432(1973)—Celebes.

corynantha Kostermans *op. cit.* 433(1973)—Borneo
6 (Brunei, Sabah, Sarawak, Kalimantan).
gigantocarpa Kostermans *op. cit.* 439(1973)—
6 Borneo (Sabah).
hainanensis Kostermans *op. cit.* 439(1973)—
2 Hainan.
6 **kerrii** Kostermans *op. cit.* 444(1973)—Thailand.
kwangtungensis Kostermans *op. cit.* 461(1973)—
2 China.
longipetiolata Kostermans *op. cit.* 462(1973)—
6 Thailand, Malaya.
palembanica Kostermans *op. cit.* 463(1973)—
6 Sumatra, ? Borneo (Kalimantan).
polyneura (Miq.) Kostermans *op. cit.* 466(1973):
6 *Alseodaphne polyneura.*
sumatrana Kostermans *op. cit.* 470(1973)—
6 Sumatra.
sumbaensis Kostermans *op. cit.* 471(1973)—Lesser
6 Sunda Is.
teijsmannii Kostermans *op. cit.* 471(1973)—Lesser
6 Sunda Is.
tomentosa (Bl.) Kostermans *op. cit.* 473(1973):
6 *Cyanodaphne tomentosa.*
turfosa Kostermans *op. cit.* 474(1973)—Borneo
6 (Brunei, Sarawak, Kalimantan).
velutinosa Kostermans *op. cit.* 476(1973)—Borneo
6 (Kalimantan, Sarawak).

LINDERA

2619 Kostermans, A. J. G. H.
Nouveautés pour la flore indochinoise. *Adansonia,*
6 13(3):333–340(1973)—(En); illustrations.

2620 Minami, A.
(*Lindera strychnifolia* Vill. found on the roadside at Tokuyama City, Yamaguchi Prefect). *J. Geobot.,*
2 20(3):64(1972)—Ja.

hamiltonii Kostermans in *Adansonia,* 13(3):337 (1973), nom. nov.: *L. reticulata* (Ham. ex Nees)
6 Hook f.
6 **spicata** Kostermans *op. cit.* 339(1973)—Vietnam.

OCOTEA

2621 Skinner, G. C.
Cape trees no. 12: *Ocotea bullata* common names: black stinkwood, stinkhout. *East. Cape Nat.,* no. 50:
5 24–25(1973)—Illustration.

PERSEA

2622 Bergh, B. O., Scora, R. W. *and* Storey, W. B.
A comparison of leaf terpenes in *Persea* subgenus
4 *Persea. Bot. Gaz.,* 134(2):130–134(1973).

PHOEBE

2623 Kostermans, A. J. G. H.
Nouveautés pour la flore indochinoise. *Adansonia,*
6 13(3):333–340(1973)—(En); illustrations.

poilanei Kostermans in *Adansonia,* 13(3):340(1973)—
6 Vietnam.

SASSAFRAS

2624 Ghent, A. W.
Gravity and the distribution of leaf shape in the trees of *Sassafras albidum. New Phytol.,* 72(5):1141–
3 1158(1973)—Illustration.

TEMMODAPHNE Kostermans in *Nat. Hist. Bull. Siam Soc.,* 24(3–4):479(January 1973), and in *Bot.*
6 *Tidsskr.,* 67(4):319(9 August 1973).

2625 Kostermans, A. J. G. H.
Some new taxa. *Bot. Tidsskr.,* 67(4):317–323(1973)
6 —Illustration.

thailandica Kostermans in *Nat. Hist. Bull. Siam Soc.,* 24(3–4):479 (January 1973), and in *Bot. Tidsskr.,*
6 67(4):319 (9 August 1973)—Thailand.

LECYTHIDACEAE

2626 Muller, J.
Pollen morphological evidence for subdivision and affinities of Lecythidaceae. *Blumea,* 20(2):350–355
4–7 (1972 publ. 1973).

BARRINGTONIA

2627 Muller, J.
Pollen morphology of *Barringtonia calyptrocalyx*
6 K. Sch. (Lecythidaceae). *Grana,* 13(1):29–44(1973).

LEDOCARPACEAE

2628 Hunziker, A. T. *and* Ariza Espinar, L.
Aporte a la rehabilitacion de Ledocarpaceae, familia monotipica. *Kurtziana,* 7:233–240(1973)—
4 (En).

BALBISIA

sect. **Wendtia** (Meyen) A. T. Hunziker & Ariza
4 Espinar in *Kurtziana,* 7:238(1973): gen. *Wendtia.*
aphanifolia (Gris.) A. T. Hunziker & Ariza
4 Espinar *op. cit.* 238(1973): *Wendtia aphanifolia.*
calycina (Gris.) A. T. Hunziker & Ariza Espinar *op.*
4 *cit.* 238(1973): *Viviania calycina.*
gracilis (Meyen) A. T. Hunziker & Ariza Espinar
4 *op. cit.* 238(1973): *Wendtia gracilis.*

LEEACEAE

2629 Descoings, B.
Vitacées et Leeacées de Madagascar: localités nouvelles et corrigenda. *Bull. Soc. Bot. France,* 118(9):
5 745–748(1971 publ. 1973).

LEEA

2630 Yeim Yok Siv *and* Paris, R. R.
Sur quelques plantes à flavonoïdes du Cambodge appartenant aux genres *Cananga, Colona, Grewia, Leea* et *Melastoma. Pl. Méd. Phytothér.,* 6(4):299–305 (1972)—also publ. in *Trav. Lab. Matière Méd. Pharm.*
5 *Galénique Fac. Pharm. Paris,* 57(2)(1973).

LEGUMINOSAE

2631 Baudet, J. C.
Les legumineuses vivrières au Burundi. *Rev. Univ.*
5 *Burundi,* 1(2):50–63(1973 ?).

2632 Cowan, R. S.
Studies of tropical American Leguminosae: 4.
4 *Proc. Biol. Soc. Wash.,* 86(39):447–460(1973)—Keys.

2633 Hijwegen, T.
Autonomous and induced pterocarpanoid formation in the Leguminosae. *Phytochemistry,* 12(2):375–380
1–7 (1973).

2634 Langenheim, J. H.
Leguminous resin-producing trees in Africa and South America. *In* Meggers, B. J., Ayensu, E. S. *and* Duckworth, W. D., *eds*. Tropical forest ecosystems in
4, 5 Africa and South America: 89–104(1973).

2635 Schreiber, A.
Nachtrag 1 zu den Familien 58–60 (Leguminosae) im *Prodromus einer Flora von Südwestafrica. Mitt. Bot.*
5 *Staatssamml. München*, 11:115–152(1973).

2636 Sinha, S. S. N. *and* Prasad, R.
Meiotic studies in some Indian Leguminosae. *J. Bihar Bot. Soc.*, 1(1–2):45–47(1972)—Chromosome numbers. *Vicia sativa, Indigofera tinctoria, Tephrosia purpurea, Clitoria ternatea, Crotalaria sericea, Cassia*
6 *occidentalis.*

2637 Solbrig, O. T.
Cytology and cytogenetics of shrubs. *In* McKell, C. M., Blaidell, J. R. *and* Goodin, J. R., *eds*. Wildland shrubs and their biology and utilization: 127–137
1–7 (1972)—Chromosome numbers.

Subfam. CAESALPINOIDEAE

2638 Liberato, M. C.
Flora da Guiné Portuguesa: Caesalpinaceae. 1973.
5 46p.—Illustrations; chromosome numbers; keys.

2639 Schmitz, A.
Contribution palynologique à la taxonomie des Bauhinieae (Caesalpiniaceae). *Bull. Jard. Bot. Nation.*
4–6 *Belg.*, 43(3–4):369–423(1973).

2640 Watson, R. *and* Fowden, L.
Amino acids of *Caesalpinia tinctoria* and some allied species. *Phytochemistry*, 12(3):617–622(1973)—
2–7 Covers 12 genera.

ADENOLOBUS

rufescens (Lam.) A. Schmitz in *Bull. Jard. Bot.*
1, 5 *Nation. Belg.*, 43(3–4):399(1973): *Bauhinia rufescens.*

ARAPATIELLA Rizzini & A. Mattos in *Rev.*
4 *Brasil. Biol.*, 32(3):323(1972).

2641 Rizzini, C. T. *and* Mattos Filho, A. de
Sobre *Arapatiella trepocarpa* n.g. & sp. (Leguminosae, Caesalpinioideae). *Rev. Brasil. Biol.*, 32(3):323–
4 333(1972)—(En); illustrations.

psilophylla (Harms) R. S. Cowan in *Proc. Biol. Soc.*
4 *Wash.*, 86(39):447(1973): *Tachigalia psilophylla.*
trepocarpa Rizzini & A. Mattos in *Rev. Brasil. Biol.*,
4 32(3):324(1972)—Brazil.

BAUHINIA

2642 Larsen, K. *and* Larsen, S. S.
The genus *Bauhinia* in Thailand. *Nat. Hist. Bull.*
6 *Siam Soc.*, 25(1–2):1–22(1973)—Key.

2643 Wunderlin, R. P.
Seven new combinations in *Bauhinia. Ann. Missouri*
4 *Bot. Gard.*, 60(2):570–571(1973).

aculeata L.
subsp. **grandiflora** (Juss.) Wunderlin in *Ann.*
4 *Missouri Bot. Gard.*, 60(2):570(1973): *B. grandiflora.*
beguinotii Cufod.
var. **gorgonae** (Killip ex Cowan) Wunderlin *op.*
4 *cit.* 570(1973): *B. gorgonae.*
bidentata Jack.
subsp. **bicornuta** (Miq.) K. & S. Larsen in *Nat. Hist. Bull. Siam Soc.*, 25(1–2):8(1973): *Phanera*
6 *bicornuta.*
forficata Link
var. **platypetala** (Burch. ex Benth.) Wunderlin in *Ann. Missouri Bot. Gard.*, 60(2):571(1973):
4 *B. platypetala.*
glabrifolia (Benth.) Baker
subsp. **sericea** (Lace) K. & S. Larsen in *Nat. Hist.*
6 *Bull. Siam Soc.*, 25(1–2):10(1973): *B. sericea.*
glauca (Wall. ex Benth.) Benth.
subsp. **tenuiflora** (Watt ex C. B. Clarke) K. & S.
6 Larsen *op. cit.* 10(1973): *B. tenuiflora.*
guianensis Aubl.
var. **kunthiana** (Vog.) Wunderlin in *Ann.*
4 *Missouri Bot. Gard.*, 60(2):571(1973): *B. kunthiana.*
harmsiana Hoss.
var. **media** (Craib) K. & S. Larsen in *Nat. Hist.*
6 *Bull. Siam Soc.*, 25(1–2):11(1973): *B. media.*
hymeniifolia Triana ex Hemsl.
var. **stuebeliana** (Harms) Wunderlin in *Ann.*
4 *Missouri Bot. Gard.*, 60(2):571(1973): *B. stuebeliana.*
longicuspis Spruce ex Benth.
var. **paraensis** (Ducke) Wunderlin *op. cit.* 571
4 (1973): *B. holophylla* var. *paraensis.*
pottsii G. Don
var. **decipiens** (Craib) K. & S. Larsen in *Nat. Hist.*
6 *Bull. Siam Soc.*, 25(1–2):15(1973): *B. decipiens.*
scandens Linn.
var. **horsfieldii** (Watt ex Prain) K. & S. Larsen
6 *op. cit.* 17(1973): *B. anguina* var. *horsfieldii.*
stenoloba (Britt. & Killip) Wunderlin in *Ann. Missouri Bot. Gard.*, 60(2):571(1973): *Schnella*
4 *stenoloba.*

BINARIA

angulosa (Vog.) A. Schmitz in *Bull. Jard. Bot. Nation.*
4 *Belg.*, 43(3–4):405(1973): *Bauhinia angulosa.*
var. **densiflora** (Mart. ex Benth.) A. Schmitz *op.*
4 *cit.* 405(1973): *Bauhinia angulosa* var. *densiflora.*
confertiflora (Benth.) A. Schmitz *op. cit.* 406(1973):
4 *Bauhinia confertiflora.*
coronata (Benth.) A. Schmitz *op. cit.* 403(1973):
4 *Bauhinia coronata.*
herrerae (Britt. & Rose) A. Schmitz *op. cit.* 403
4 (1973): *Schnella herrerae.*
hondurensis (Standley) A. Schmitz *op. cit.* 403
4 (1973): *Bauhinia hondurensis.*
hymeniifolia (Triana ex Hemsl.) A. Schmitz *op. cit.* 404(1973), as 'hymenaeifolia': *Bauhinia hymenaei-*
4 *folia.*
kleiniana (Burk.) A. Schmitz *op. cit.* 404(1973):
4 *Bauhinia kleiniana.*
kunthiana (Vog.) A. Schmitz *op. cit.* 406(1973):
4 *Bauhinia kunthiana.*
leiopetala (Benth.) A. Schmitz *op. cit.* 406(1973):
4 *Bauhinia leiopetala.*
longipetala (Benth.) A. Schmitz *op. cit.* 404(1973):
4 *Schnella longipetala.*
maximilianii (Benth.) A. Schmitz *op. cit.* 404(1973),
4 as 'maximiliani': *Bauhinia maximiliani.*
nigra (Britt. & Killip) A. Schmitz *op. cit.* 404(1973):
4 *Schnella nigra.*
platycalyx (Benth.) A. Schmitz *op. cit.* 406(1973):
4 *Bauhinia platycalyx.*
punctata (Burch.) A. Schmitz *op. cit.* 406(1973):
4 *Bauhinia punctata.*
riedeliana (Bong.) A. Schmitz *op. cit.* 406(1973):
4 *Bauhinia riedeliana.*
rubiginosa (Bong.) A. Schmitz *op. cit.* 404(1973):
4 *Bauhinia rubiginosa.*
rutilans (Spruce ex Benth.) A. Schmitz *op. cit.* 405
4 (1973): *Bauhinia rutilans.*

splendens (H.B.K.) A. Schmitz *op. cit.* 406(1973):
4 *Bauhinia splendens.*
var. **latifolia** (Mart. ex Benth.) A. Schmitz *op. cit.*
4 406(1973): *Bauhinia splendens* var. *latifolia.*
sprucei (Benth.) A. Schmitz *op. cit.* 405(1973):
4 *Bauhinia sprucei.*
var. **acuminata** (Mart. ex Benth.) A. Schmitz *op.*
4 *cit.* 405(1973): *Bauhinia sprucei* var. *acuminata.*
uruguayensis (Benth.) A. Schmitz *op. cit.* 406(1973):
4 *Bauhinia uruguayensis.*
vestita (Macbr.) A. Schmitz *op. cit.* 405(1973):
4 *Bauhinia vestita.*

CAESALPINIA

2644 Fosberg, F. R.
Caesalpinia major a legitimate name. *Taxon*, 22(1):
4–6 162–163(1973).

CASSIA

2645 Datta, S. *and* Datta, K. B.
Karyotype studies in some ligneous taxa of *Cassia* L.
Proc. Indian Sci. Congr., 60(3):307(1973)—Chromosome number. Abstract only. *C. alata*, *C. fistula*,
1–7 *C. siamea.*

2646 Rizzini, C. T.
Cassia spinigera et *Pera androgyna*, n.spp. (Leguminosae et Euphorbiaceae). *Rev. Brasil. Biol.*, 32(3):
4 319–322(1972)—Illustrations.

2647 Tandon, S. L. *and* Bhat, R. N.
Cytotaxonomy of *Cassia*. *Proc. Indian Sci. Congr.*, 58(3):470(1971)—Chromosome numbers. Abstract
6 only.

CERCIS

hebecarpa (Bornm.) Ponert in *Feddes Repert.*, 83 (9–10): 618(1972 publ. 1973): *C. siliquastrum* var.
2 *hebecarpa.*
× **yaltirikii** Ponert *op. cit.* 618(1972 publ. 1973), as 'yaltiriki'.—Turkey. (*C. hebecarpa* × *C.*
2 *siliquastrum*).

DIMORPHANDRA

2648 Ragonese, A. M.
Systematic anatomical characters of the leaves of *Dimorphandra* and *Mora* (Leguminosae: Caesal-
4 pinioideae). *Bot. J. Linn. Soc.*, 67(3):255–274(1973).

HYMENAEA

2649 Lee, Y. T. *and* Langenheim, J. H.
New taxa from Brazil and Guyana in the genus *Hymenaea* (Leguminosae, Caesalpinioideae). *J. Arnold*
4 *Arbor.*, 54(1):94–104(1973)—Illustrations, dot map.

aurea Y.-T. Lee & Langenheim in *J. Arnold Arbor.*,
4 54(1):96(1973)—Brazil.
oblongifolia
var. **davisii** (Sandwith) Y.-T. Lee & Langenheim
4 *op. cit.* 102(1973): *H. davisii.*
var. **latifolia** Y.-T. Lee & Langenheim *op. cit.* 98
4 (1973)—Brazil.
var. **palustris** (Ducke) Y.-T. Lee & Langenheim
4 *op. cit.* 101(1973): *H. palustris.*

LASIOBEMA

dolichobotrys (Merr.) A. Schmitz in *Bull. Jard. Bot. Nation. Belg.*, 43(3–4):377(1973): *Bauhinia*
6 *dolichobotrys.*

MACROLOBIUM

conjunctum R. S. Cowan in *Proc. Biol. Soc. Wash.*,
4 86(39):450(1973)—Brazil.
4 **dressleri** R. S. Cowan *op. cit.* 451(1973)—Panama.
4 **extensum** R. S. Cowan *op. cit.* 452(1973)—Colombia.
4 **rigidum** R. S. Cowan *op. cit.* 453(1973)—Brazil.

MORA

2650 Ragonese, A. M.
Systematic anatomical characters of the leaves of *Dimorphandra* and *Mora* (Leguminosae: Caesal-
4 pinioideae). *Bot. J. Linn. Soc.*, 67(3):255–274(1973).

PAULETIA

acuminata (L.) A. Schmitz in *Bull. Jard. Bot. Nation.*
2, 5, 6 *Belg.*, 43(3–4):387(1973): *Bauhinia acuminata.*
acuruana (Moricand) A. Schmitz *op. cit.* 387(1973):
4 *Bauhinia acuruana.*
affinis (Vog.) A. Schmitz *op. cit.* 388(1973):
4 *Bauhinia affinis.*
angulicaulis (Harms) A. Schmitz *op. cit.* 388(1973):
4 *Bauhinia angulicaulis.*
bauhinioides (Mart.) A. Schmitz *op. cit.* 388(1973):
4 *Perlebia bauhinioides.*
bicuspidata (Benth.) A. Schmitz *op. cit.* 394(1973):
4 *Bauhinia bicuspidata.*
bongardii (Steud.) A. Schmitz *op. cit.* 388(1973):
4 *Bauhinia bongardii.*
bowkeri (Harv.) A. Schmitz *op. cit.* 394(1973):
5 *Bauhinia bowkeri.*
breviloba (Benth.) A. Schmitz *op. cit.* 389(1973):
4 *Bauhinia breviloba.*
burchellii (Benth.) A. Schmitz *op. cit.* 394(1973):
4 *Bauhinia burchellii.*
candicans (Benth.) A. Schmitz *op. cit.* 389(1973):
4 *Bauhinia candicans.*
corniculata (Benth.) A. Schmitz *op. cit.* 394(1973):
4 *Bauhinia corniculata.*
cupulata (Benth.) A. Schmitz *op. cit.* 394(1973):
4 *Bauhinia cupulata.*
curvula (Benth.) A. Schmitz *op. cit.* 389(1973):
4 *Bauhinia curvula.*
depauperata (Glaz.) A. Schmitz *op. cit.* 389(1973):
4 *Bauhinia depauperata.*
dubia (G. Don) A. Schmitz *op. cit.* 394(1973):
1 *Bauhinia dubia.*
dumosa (Benth.) A. Schmitz *op. cit.* 394(1973):
4 *Bauhinia dumosa.*
ellenbeckii (Harms) A. Schmitz *op. cit.* 394(1973):
5 *Bauhinia ellenbeckii.*
forficata (Link) A. Schmitz *op. cit.* 389(1973):
4 *Bauhinia forficata.*
var. **latifolia** (Mart.) A. Schmitz *op. cit.* 390(1973):
4 *Bauhinia forficata* var. *latifolia.*
var. **longiflora** (D. Dietr.) A. Schmitz *op. cit.* 390
4 (1973): *Bauhinia longiflora.*
gardneri (Benth.) A. Schmitz *op. cit.* 395(1973):
4 *Bauhinia gardneri.*
geminata (Vog.) A. Schmitz *op. cit.* 395(1973):
4 *Bauhinia geminata.*
glaziovii (Taub.) A. Schmitz *op. cit.* 390(1973):
4 *Bauhinia glaziovii.*
hirsuta (Weinm.) A. Schmitz *op. cit.* 395(1973):
2, 6 *Bauhinia hirsuta.*
integerrima (Mart.) A. Schmitz *op. cit.* 395(1973):
4 *Bauhinia integerrima.*
kalantha (Harms) A. Schmitz *op. cit.* 391(1973):
5 *Bauhinia kalantha.*
loesenerana (Harms) A. Schmitz *op. cit.* 395(1973), as
5 'loeseneriana': *Bauhinia loeseneriana.*
longicuspis (Spruce ex Benth.) A. Schmitz *op. cit.*
4 395(1973): *Bauhinia longicuspis.*

macrostachya (Benth.) A. Schmitz *op. cit.* 391
4 (1973): *Bauhinia macrostachya.*
membranacea (Benth.) A. Schmitz *op. cit.* 395(1973):
4 *Bauhinia membranacea.*
nitida (Benth.) A. Schmitz *op. cit.* 395(1973):
4 *Bauhinia nitida.*
obtusata (Vog.) A. Schmitz *op. cit.* 391(1973):
4 *Bauhinia obtusata.*
odoratissima (Moricand) A. Schmitz *op. cit.* 395
4 (1973): *Bauhinia odoratissima.*
platypetala (Burch.) A. Schmitz *op. cit.* 396(1973):
4 *Bauhinia platypetala.*
platyphylla (Benth.) A. Schmitz *op. cit.* 396(1973):
4 *Bauhinia platyphylla.*
pruinosa (Vog.) A. Schmitz *op. cit.* 396(1973):
4 *Bauhinia pruinosa.*
pulchella (Benth.) A. Schmitz *op. cit.* 391(1973):
4 *Bauhinia pulchella.*
rufa (Steud.) Bong.
var. **cordata** (Vog.) A. Schmitz *op. cit.* 392(1973):
4 *Bauhinia cordata.*
var. **dodecandra** (Steud.) A. Schmitz *op. cit.* 392
4 (1973): *Bauhinia dodecandra.*
var. **intermedia** (Vog.) A. Schmitz *op. cit.* 392
4 (1973): *Bauhinia intermedia.*
selerana (Harms) A. Schmitz *op. cit.* 392(1973), as
4 'seleriana': *Bauhinia seleriana.*
subclavata (Benth.) A. Schmitz *op. cit.* 396(1973):
4 *Bauhinia subclavata.*
taitensis (Taub.) A. Schmitz *op. cit.* 392(1973):
5 *Bauhinia taitensis.*
tarapotensis (Benth.) A. Schmitz *op. cit.* 396(1973):
4 *Bauhinia tarapotensis.*
tenella (Benth.) A. Schmitz *op. cit.* 396(1973):
4 *Bauhinia tenella.*
tomentosa (L.) A. Schmitz *op. cit.* 393(1973):
5, 6 *Bauhinia tomentosa.*
ungulata (L.) A. Schmitz *op. cit.* 393(1973):
4 *Bauhinia ungulata.*
viridescens (Desv.) A. Schmitz *op. cit.* 396(1973):
6 *Bauhinia viridescens.*
var. **baviensis** (Drake del Cast.) A. Schmitz *op. cit.*
2, 6 396(1973): *Bauhinia baviensis.*
viscidula (Harms) A. Schmitz *op. cit.* 393(1973):
4 *Bauhinia viscidula.*

PELTOGYNE

2651 Areia, C. Alves de *and* Marquete, O.
Pontuações guarnecidas em duas especies de *Peltogyne* Vog. *P. gracilipes* Ducke e *P. lecointei* Ducke subfamília Caesalpinoideae (Leguminosae). *Arq. Jard.*
4 *Bot. Rio de Janeiro*, 19:179–181(1973).

2652 Mattos Filho, A. de
Anatomia do lenho de *Peltogyne recifensis* Ducke. *Arq. Jard. Bot. Rio de Janeiro*, 19:125–131(1973)—
4 Key.

PERLEBIA

blakeana (Dunn.) A. Schmitz in *Bull. Jard. Bot.*
2 *Nation. Belg.*, 43(3–4):387(1973): *Bauhinia blakeana.*
exellii (Torre & Hillc.) A. Schmitz *op. cit.* 387
5 (1973): *Bauhinia exellii.*
galpinii (N.E. Br.) A. Schmitz *op. cit.* 382(1973):
5 *Bauhinia galpinii.*
macrantha (Oliv.) A. Schmitz *op. cit.* 384(1973):
5 *Bauhinia macrantha.*
subsp. **serpae** (Fic. & Hiern) A. Schmitz *op. cit.*
5 384(1973): *Bauhinia serpae.*
mendoncae (Toore & Hillc.) A. Schmitz *op. cit.* 384
5 (1973): *Bauhinia mendoncae.*
natalensis (Oliv. ex Hook.) A. Schmitz *op. cit.* 385
5 (1973): *Bauhinia natalensis.*
nervosa (Wall. ex Prain) A. Schmitz *op. cit.* 382
6 (1973): *Bauhinia nervosa.*
petersiana (Bolle) A. Schmitz *op. cit.* 385(1973):
5 *Bauhinia petersiana.*
platypetala (Benth. ex Burch.) A. Schmitz *op. cit.*
4 385(1973): *Bauhinia platypetala.*
pottsii (G. Don) A. Schmitz *op. cit.* 383(1973):
6 *Bauhinia pottsii.*
var. **elongata** (Korth.) A. Schmitz *op. cit.* 383
6 (1973): *Bauhinia elongata.*
var. **subsessilis** (Craib) A. Schmitz *op. cit.* 384
6 (1973): *Bauhinia subsessilis.*
purpurea (L.) A. Schmitz *op. cit.* 382(1973):
2, 5, 6 *Bauhinia purpurea.*
var. **corneri** (de Wit) A. Schmitz *op. cit.* 383
6 (1973): *Bauhinia purpurea* var. *corneri.*
var. **violacea** (de Wit) A. Schmitz *op. cit.* 383
6 (1973): *Bauhinia purpurea* var. *violacea.*
urbaniana (Schinz) A. Schmitz *op. cit.* 386(1973):
5 *Bauhinia urbaniana.*
variegata (L.) A. Schmitz *op. cit.* 386(1973):
2, 5, 6 *Bauhinia variegata.*
variegata (L.) A. Schmitz
var. **alboflava** (de Wit) A. Schmitz *op. cit.* 386
6 (1973): *Bauhinia variegata* var. *alboflava.*

PHANERA

tauranensis (Gagn.) A. Schmitz in *Bull. Jard. Bot.*
2, 6 *Nation. Belg.*, 43(3–4):381(1973): *Bauhinia tauranensis.*

RECORDOXYLON

irwinii R. S. Cowan in *Proc. Biol. Soc. Wash.*, 86(39):
4 453(1973)—Brazil.

SCHNELLA

macrostachya Raddi
var. **bahiensis** (Bong.) A. Schmitz in *Bull. Jard. Bot. Nation. Belg.*, 43(3–4):402(1973): *Bauhinia*
4 *bahiensis.*
var. **brachystachya** (Mart. ex Benth.) A. Schmitz *op. cit.* 402(1973): *Bauhinia raddiana* var. *brachy-*
4 *stachya.*
spicata (Vog.) A. Schmitz *op. cit.* 402(1973):
4 *Bauhinia spicata.*

SCHOTIA

2653 Codd, L. E.
Schotia afra var. *angustifolia. Flow. Pl. Afr.*, 42(3–4):
5 pl.1665(1973)—En; coloured illustration.

SWARTZIA

2654 Cowan, R. S.
A new *Swartzia* from Suriname. *Phytologia*, 26(4):
4 279–280(1973).

calva R. S. Cowan in *Proc. Biol. Soc. Wash.*, 86(39):
4 454(1973)—Peru.
flaemingii Raddi
var. **cognata** R. S. Cowan *op. cit.* 456(1973)—
4 Brazil.
4 var. **polita** R. S. Cowan *op. cit.* 455(1973)—Brazil.
jimenezii R. S. Cowan *op. cit.* 456(1973)—
4 Venezuela.
4 **micrantha** R. S. Cowan *op. cit.* 457(1973)—Brazil.
4 **peremarginata** R. S. Cowan *op. cit.* 457(1973)—
4 Brazil.
rediviva R. S. Cowan in *Phytologia*, 26(4):279
4 (1973)—Surinam.
tillettii R. S. Cowan in *Proc. Biol. Soc. Wash.*, 86(39):
4 458(1973)—Guyana.

SWEETIA

2655 Yakovlev, G. P.
Zametki po sistematike rodov *Sweetia* Spreng., *Machaerium* Pers., *Angylocalyx* Taub., *Fedorovia* Yakovl., *Placolobium* Miq., *Ormosia* Jacks. (Fabaceae) (De generibus *Sweetia* Spreng., *Machaerium* Pers., *Angylocalyx* Taub., *Fedorovia* Yakovl., *Placolobium* Miq. et *Ormosia* Jacks. (Fabaceae) notulae systematicae). *Novosti Sist. Vÿssh. Rast.*, 10:190–196
2, 6 (1973).

TOURNAYA A. Schmitz in *Bull. Jard. Bot. Nation.*
5 *Belg.*, 43(3–4):397(1973).
gossweileri (Bak. f.) A. Schmitz *op. cit.* 398(1973):
5 *Bauhinia gossweileri*.

TRACHYLOBIUM

2656 Martin, S. S., Langenheim, J. H. *and* Zavarin, E.
Compositional variation of leaf pocket sesquiterpenes in *Trachylobium verrucosum*. *Biochem. Syst.*, 1(1):
5 35–37(1973).

Subfam. MIMOSOIDEAE

2657 Ali, S. I.
Flora of West Pakistan: no.36. Mimosaceae, 1973.
6 41p.—Illustrations, map; keys.

2658 Caccavari de Filice, M. A.
Granos de polen de Leguminosas de la Argentina: 2. Subfam. 'Mimosoideae'; tribu 'Adenanthereae'. *Rev. Mus. Argent. Cienc. Nat. 'Bernardino Rivadavia'*,
4 *Bot.*, 4(3):281–320(1972).

2659 Isely, D.
Leguminosae of the United States: 1. Subfamily Mimosoideae. *Mem. New York Bot. Gard.*, 25(1):
3 1–152(1973)—Dot maps; chromosome numbers; keys.

2660 Krauss, G.-J. *and* Reinbothe, H.
Die freien Aminosäuren in Samen von Mimosaceae.
1–7 *Phytochemistry*, 12(1):125–142(1973)—(En).

2661 Liberato, M. C.
Flora da Guiné Portuguesa: Mimosaceae. 1972.
5 38p.—Keys.

2662 Liberato, M. C.
Flora de S. Tomé e Príncipe: Mimosaceae. 1973.
5 29p.—Keys.

ACACIA

2663 Cheema, M. S. Z. A. *and* Qadir, S. A.
Autecology of *Acacia senegal* (L.) Willd. *Vegetatio*,
6 27(1–3):131–162(1973)—Illustrations.

2664 Datta, M. K.
Cytotaxonomy of Leguminosae. *Proc. Indian Sci. Congr.*, 58(3):471–472(1971)—Chromosome numbers.
6 Abstract only.

2665 Lothian, N.
Acacia gilesiana F. Muell.: its occurrence and likely distribution in South Australia. *S. Austral. Nat.*, 47(4):
7 69–73(1973)—Illustrations, map.

2666 McMillan, C.
Prostrateness in *Acacia farnesiana* from the western coast of Mexico under uniform environmental conditions. *Madroño*, 22(3):140–144(1973)—
4 Illustrations.

2667 Pedley, L.
Taxonomy of the *Acacia aneura* complex. *Trop.*
7 *Grasslands*, 7(1):3–8(1973)—Key.

2668 Pettigrew, C. J. *and* Watson, L.
On the identification of sterile *Acacias*, and the feasibility of establishing an automatic key-generating
7 system. *Austral. J. Bot.*, 21 (1): 141-150 (1973)—Key.

2669 Ross, J. H.
Notes on *Acacia* species in Southern Africa: 3. *Bothalia*, 11(1–2):127–131(1973)—Illustrations, dot
5 map; keys.

2670 Ross, J. H.
Towards a classification of the African Acacias. *Bothalia*, 11(1–2):107–113(1973)—Chromosome
5 numbers.

2671 Sinha, S. C. *
Floral morphology of Acacias. *Carib. J. Sci.*, 11: 137–153(1971 publ. 1972). Ref. in *Bull. Torrey Bot.*
4–7 *Club*, 99(6):332(1972).

2672 Vassal, J.
Apport des recherches ontogéniques et séminologiques à l'étude morphologique, taxonomique et phylogénique du genre *Acacia*. *Bull. Soc. Hist. Nat. Toulouse*, 108(1–2):125–247(1972)—(En); map; chromosome number; keys to subgenera, sections
2–7 and species.

2673 Vassal, J.
Contribution à l'étude morphologique des graines d'*Acacia*. *Bull. Soc. Hist. Nat. Toulouse*, 107(1–2):
2–7 191–246(1971)—Illustrations.

2674 Vassal, J.
Contribution à l'étude de la morphologie des plantules d'*Acacia*. Acacias américains, Acacias asiatiques. *Bull. Soc. Hist. Nat. Toulouse*, 107(1–2):
2–7 157–190(1971)—Illustrations.

2675 Vassal, J.
Contribution à l'étude de la morphologie des plantules d'*Acacia*. Acacias insulaires des océans Indien et Pacifique: Australie, Formose, Iles Maurice et Hawaii: 2me partie. Les Acacias bipennés: Pulchellae, Botryocephalae, Gummiferae. *Bull. Soc. Hist. Nat.*
2–7 *Toulouse*, 107(1–2):247–278(1971)—Illustrations.

ANADENANTHERA

2676 Hunziker, A. T.
El 'cebil' (*Anadenanthera colubrina* var *cebil*) en la
4 Provincia de Córdoba. *Kurtziana*, 7:265(1973).

CALLIANDRA

2677 Chen, Y.-Y.
Studies on the development of polyad grains of *Calliandra haematocephala* Hassk. with electron
4–6 microscopic technique. *Taiwania*, 18(1):18–28(1973).

2678 Datta, M. K.
Cytotaxonomy of Leguminosae. *Proc. Indian Sci. Congr.*, 58(3):471–472(1971)—Chromosome numbers.
6 Abstract only.

DICHROSTACHYS

2679 Datta, M. K.
Cytotaxonomy of Leguminosae. *Proc. Indian Sci. Congr.*, 58(3):471–472(1971)—Chromosome numbers.
6 Abstract only.

ENTADA

2680 Ross, J. H.
Notes on *Entada* in South Africa. *Bothalia*, 11(1–2):
5 125–126(1973)—Dot maps.

arenaria Schinz
subsp. **microcarpa** (Brenan) J. H. Ross in *Bothalia*,
5 11(1–2):126(1973): *E. nana* subsp. *microcarpa*.

INGA

2681 Baretta-Kuipers, T.
Some aspects of wood-anatomical research in the genus *Inga* (Mimosaceae) from the Guianas and especially Suriname. *Acta Bot. Neerl.*, 22(3):193–205
4 (1973)—Key.

PARKIA

2682 Loureiro, A. A. *and* Silva, M. Freitas da
Contribuição ao estudo dendrológico de 5 Parkias (Leguminosae) da Amazônia. *Acta Amazonica*, 2(2):
4 71–86(1972)—(En).

PROSOPIS

2683 Werker, E., Dafni, A. *and* Negbi, M.
Variability in *Prosopis farcata* in Israel: anatomical features of the seed. *Bot. J. Linn. Soc.*, 66(3):223–232
2 (1973).

STRYPHNODENDRON

2684 Martins, E. M. Occhioni
Nova espécie de *Stryphnodendron* para a flora Amazônica. *Leandra*, 2(2):121–123(1972)—
4 Illustration.

2685 Martins, E. M. Occhioni
Sôbre a nomenclatura científica do 'Barbatimão' do Brasil. (Leguminosae – Mimosoideae). *Leandra*, 2(3): 79–81(1972). *Stryphnodendron barbadetimam* (Vell.)
4 Mart.

2686 Martins, E. M. Occhioni *and* Martins, A. Guedes
Stryphnodendron Mart. (Leg. Mim.) as espécies da Amazônia Brasileira. *Leandra*, 2(2):11–40(1972)—
4 Illustrations, map; key.

occhionianum E. M. O. Martins in *Leandra*, 2(2):121
4 (1972)—Brazil.

Subfam. PAPILIONOIDEAE

2687 Baudet, J. C.
Intérêt taxonomique des caractères épidermiques dans le complexe *Phaseolus-Vigna*. (Taxonomic interest of the epidermal characters in the *Phaseolus-Vigna* complex). *Bull. Soc. Roy. Bot. Belg.*, 106(3):
5, 6 53–59(1973)—(En).

2688 Breedlove, D. E. *and* Ehrlich, P. R.
Co-evolution: patterns of legume predation by a lycaenid butterfly. *Oecologia*, 10(2):99–104(1972)—
3 *Lupinus*, *Thermopsis* and *Vicia*.

2689 D'Orey, J. *and* Liberato, M. C.
Flora da Guiné Portuguesa: Papilionaceae 1° apênd.
5 fenologia (apontamento). 1972. 27p.—(En,Fr,Ge).

2690 Faugeras, G., Valdes-Bermejo, E. *and* Paris, E.
Alcaloïdes et polyphénols des Légumineuses: 27. Distribution des alcaloïdes chez diverses Génistées d'Espagne et du Portugal appartenant aux genres *Cytisus*, *Genista*, *Echinospartium*, *Stauracanthus* et
1 *Adenocarpus*. *Pl. Méd. Phytothér.*, 7(1):68–76(1973).

2691 Forissier, R.
Recherches cytotaxonomiques préliminaires sur les genres *Lembotropis*, *Cytisus*, *Chamaecytisus*, *Genista* et *Chamaespartium*. *Bull. Soc. Neuchâtel. Sci. Nat.*, 96:
1 51–65(1973)—(Ge,En); chromosome numbers.

2692 Freire de Carvalho, L. d'A. *and* Valente, M. da C.
Contribuição ao estudo da nervação foliar das Leguminosas dos Cerrados: 3. Faboideae. *Arq. Jard. Bot. Rio de Janeiro*, 19:227–247(1973)—(En);
4 illustrations; keys.

2693 Lawrence, J. M.
Homoserine in seedlings of the tribe Vicieae of the Leguminosae. *Phytochemistry*, 12(9):2207–2209
1–3 (1973).

2694 Liberato, M. C.
Flora de S. Tome e Príncipe: Papilionaceae. 1972.
5 90p.—Keys.

2695 Ohashi, H.
Euchresteae, a new tribe of the family Leguminosae. *J. Jap. Bot.*, 48(8):225–234(1973)—En,Ja; illustration;
2, 6 chromosome numbers.

2696 Rakova, M. V.
Tverdȳe semena v predelakh osobi u dikorastushchikh bobovȳkh. (Hard seeds within the individual in wild-growing legumes). *Bot. Zhurn.*, 58(3):
1–3 436–439(1973).

2697 Shah, G. L. *and* Kothari, M. J.
On the structure of stomata and hairs and its bearing on the systematics in the tribe Vicieae (Papilionaceae)
1–7 *Flora* (Germany), 62(6):533–548(1973).

2698 Subramaniam, D.
Karyotype analysis in some wild leguminous plants. *Proc. Indian Sci. Congr.*, 59(3):358–359(1972)—Chromosome numbers. Abstract only; *Crotalaria*,
6 *Canavalia* and *Psoralia* spp.

2699 Yakovlev, G. P. *
(Notes sur la systématique des genres de la tribu des Sophorae Spreng et Millettiae Hutch.). *Nauch. Dokl. Vȳssh. Shkolȳ.*, *Biol. Nauki*, no.1:57–59(1973)—Rus. Ref. in *Bull. Signal.*, *Biol. Physiol. Vég.*, 34(10):8864
1–7 (1973).

2700 Zelenov, A. N. *
(Chromosome groups of genera in the Fabaceae.). *Izv. Timiryazev. Selskokhoz. Akad.* no.5:53–62(1972)
1–7 —Rus(En). Abstr. in *Herb. Abstr.* 43(6):1629(1973).

tribus **EUCHRESTEAE** Ohashi in *J. Jap. Bot.*,
2, 6 48(8):229(1973).

ADENOCARPUS

2701 Sañudo, A.
Variabilídad cromosómica de las genisteas de la flora española en relación con su ecología. Nota previa sobre los géneros *Cytisus* L. y *Adenocarpus* DC. *Cuad. Cienc. Biol.* (Granada), 2(1):37–38(1973)—(En);
1 chromosome numbers.

2702 Sañudo, A.
Variabilídad cromosómica de las Genisteas de la flora española en relación con su ecología: 1. Número y comportamiento de los cromosomas durante la meiosis: C. Géneros *Chronanthus* (DC.) C. Koch, *Adenocarpus* DC. y *Erinacea* Adanson. *Lagascalia*, 3(2):
1 205–210(1973)—Chromosome numbers.

AESCHYNOMENE

2703 Verdcourt, B.
Studies in the Leguminosae-Papilionoideae-Hedysareae (sensu lato) for the 'Flora Zambesiaca': 3. *Kew Bull.*, 28(3):429–431(1973 publ. 1974)—
5 Illustrations.

mediocris Verdcourt in *Kew Bull.*, 28(3):429(1973
5 publ. 1974)—Rhodesia, Zambia.

ALHAGI

mannifera Desv.
var. **assyriaca** (Nábělek) Ponert in *Feddes Repert.*, 83(9–10):644(1972 publ. 1973): *A. maurorum* var.
2 *assyriacum*.

ALYSICARPUS

2704 Maurya, A. N. *and* Ambasht, R. S.
Significance of seed dimorphism in *Alysicarpus*
5–7 *monilifer* DC. *J. Ecol.*, 61(1):213–217(1973).

ANARTHROPHYLLUM

2705 Sorarú, S. B.
Tres especies nuevas de Leguminosas: *Anarthrophyllum pedicellatum*, *A. capitatum* y *A. strigulipetalum*.
4 *Darwiniana*, 18(1–2):37–43(1973)—Illustrations.

capitatum Sorarú in *Darwiniana*, 18(1–2):40
4 (1973)—Argentina.
4 **pedicellatum** Sorarú *op. cit.* 37(1973)—Argentina.
4 **strigulipetalum** Sorarú *op. cit.* 41(1973)—Argentina.

ANATROPOSTYLIA (Plitmann) Kupicha in *Notes Roy. Bot. Gard. Edinburgh*, 32(2):247(1973): *Vicia*
2 sect. *Anatropostylia*.

2706 Kupicha, F. K.
Studies in the Vicieae: 1. The new genus *Anatropostylia*. *Notes Roy. Bot. Gard. Edinburgh*, 32(2):247–250
2 (1973).

koeieana (Rech. fil.) Kupicha in *Notes Roy. Bot. Gard.*
2 *Edinburgh*, 32(2):247(1973): *Vicia koeieana*.

ANGYLOCALYX

2707 Yakovlev, G. P.
Zametki po sistematike rodov *Sweetia* Spreng., *Machaerium* Pers., *Angylocalyx* Taub., *Fedorovia* Yakovl., *Placolobium* Miq., *Ormosia* Jacks. (Fabaceae). (De generibus *Sweetia* Spreng., *Machaerium* Pers., *Angylocalyx* Taub., *Fedorovia* Yakovl., *Placolobium* Miq. et *Ormosia* Jacks. (Fabaceae) notulae systematicae). *Novosti Sist. Vȳssh. Rast.*, 10:190–196
2, 6 (1973).

ANTHYLLIS

2708 Couderc, H.
Nouvelle contribution à l'étude chimiotaxinomique des espèces françaises du genre *Anthyllis*. *Compt. R. Hebd. Séances Acad. Sci., D* (Paris), 277(21):2377–
1 2380(1973).

2709 Ritter, J.
Sur la présence, dans le Vercors méridional, des cytotypes diploïde et tétraploïde d'*Anthyllis montana* L. et du cytotype diploïde de *Ranunculus pyrenaeus* L. *Compt. R. Séances Soc. Biol. Filiales*, 166(12):1622–
1 1625(1972 publ. 1973)—Map.

ARACHIS

2710 Mukherji, C.
Cytological investigation of certain species of *Arachis*. *Proc. Indian Sci. Congr.*, 60(3):313(1973)—
4 Chromosome numbers. Abstract only.

2711 Raman, V. S.
Genome relationships in *Arachis*. *Oléagineux*, 28(3):
4 137–140(1973).

2712 Raman, V. S. *and* Sree Rangasamy, S. R. *
A consideration of the phylogenetic relationships of species of *Arachis*. (Considération sur les affinités phylogénétique des espèces de *A.*). *Madras Agric. J.*, 59(5):262–271(1972)—Abstr. in *Bull. Signal.*, *Biol.*
4 *Physiol. Vég.*, 34(1):157(1973).

2713 Sigafus, R. E.
Peanuts: a limited review of some worldwide laboratory and field research. Lexington, University of Kentucky College of Agriculture, 1973. 32p.—
4 Chromosome number.

ASTRAGALUS

2714 Ali, S. I.
Contributions to the genus *Astragalus* (Leguminosae) from Pakistan: 6. Further additions and corrections.
6 *Kew Bull.*, 28(2):297–304(1973)—Illustrations.

2715 Barneby, R. C.
A range extension and a new synonym for *Astragalus polaris* Benth. *Rhodora*, 75(802):304–305
2, 3 (1973).

2716 Ishchenko, L. E.
Zhiznennȳe formȳ astragalov Turkmenii. *Byull.*
2 *Glavn. Bot. Sada* (Moscow), 87:61–66(1973).

2717 Ledingham, G. F. *and* Pepper, B. M.
Chromosome numbers of some South American species of *Astragalus*. *Kurtziana*, 7:27–37(1973)—
4 (Sp).

2718 Mitsevski, K.
Nekolku nepoznati i retki *Astragalus*- i *Oxytropis*-vidovi za florata na Makedonija. (Einige unbekannte und seltene *Astragalus*- und *Oxytropis*-Arten der Flora von Mazedonien). *God. Zborn.*, *Biol.*, 25:157–160
1 (1973)—(Ge); dot map. *Astragalus gremlii*, *A. australis*.

2719 Pavlova, N. S. *and* Basargin, D. D.
Novȳĭ vid astragala s dal'nego vostoka. *Byull. Glavn. Bot. Sada* (Moscow), 90:42–44(1973)—
1 Illustration; chromosome numbers.

2720 Podlech, D.
Neue und bemerkenswerte *Astragalus*-Arten aus Afghanistan. *Mitt. Bot. Staatssamml. München*, 11:259–321(1973) (Beiträge zur Flora von Afghanistan 6)—
2 Illustrations.

2721 Vestergaard, P.
An edaphic approach to the Danish distribution of *Astragalus danicus*. *Bot. Tidsskr.*, 68(2):109–121(1973)—
1, 2 Illustrations, dot map; chromosome numbers.

adraskanensis Podlech in *Mitt. Bot. Staatssamml.*
2 *München*, 11:287(1973)—Afghanistan.
aduncus Willd.
subsp. **chaborasicus** (Boiss. & Hausskn.) Ponert in *Feddes Repert.*, 83(9–10):630(1972 publ. 1973):
2 *A. chaborasicus*.

f. **kotschyanus** (Boiss.) Ponert *op. cit.* 630(1972
2 publ. 1973): *A. kotschyanus.*
amoenus Fenzl
subsp. **squalidus** (Boiss. & Noë) Ponert *op. cit.* 631
2 (1972 publ. 1973): *A. squalidus.*
ancistrocarpus Boiss. & Hausskn.
subsp. **gudrunensis** (Boiss. & Hausskn.) Ponert
2 *op. cit.* 632(1972 publ. 1973): *A. gudrunensis.*
andersianus Podlech in *Mitt. Bot. Staatssamml.*
2 *München*, 11:286(1973)—Afghanistan.
angustiflorus C. Koch
subsp. **filamentosus** (Bunge) Ponert in *Feddes*
Repert., 83(9–10):622(1972 publ. 1973): *A.*
2 *filamentosus.*
subsp. **fraxinella** (Bunge) Ponert *op. cit.* 622(1972
2 publ. 1973): *A. fraxinella.*
subsp. **hellenicus** (Boiss.) Ponert *op. cit.* 622(1972
1, 2 publ. 1973): *A. hellenicus.*
angustifolius Lam.
subsp. **violaceus** (Boiss.) Ponert *op. cit.* 632(1972
2 publ. 1973): *A. angustifolius* var. *violaceus.*
antilihani Bunge
subsp. **goeznensis** (Eig) Ponert *op. cit.* 620(1972
2 publ. 1973): *A. goeznensis.*
aqrabatensis Podlech in *Mitt. Bot. Staatssamml.*
2 *München*, 11:289(1973)—Afghanistan.
asper Wulf.
f. **karpatii** Soó in *Acta Bot. Acad. Sci. Hung.*,
1 18(1–2):172(1973)—Hungary.
bakakrianus Ponert in *Feddes Repert.*, 83(9–10):
624(1972 publ. 1973), nom. nov.: *A. andrachne*
2 var. *sintenisianus* Širj.
bamianicus Podlech in *Mitt. Bot. Staatssamml.*
2 *München*, 11:291(1973)—Afghanistan.
barbatus Lam.
subsp. **nanus** (DC.) Ponert in *Feddes Repert.*,
2 83(9–10):620(1972 publ. 1973): *A. nanus.*
bashgalensis Podlech in *Mitt. Bot. Staatssamml.*
2 *München*, 11:265(1973)—Afghanistan.
bicolor Lam.
subsp. **karputanus** (Boiss. & Noë) Ponert in
Feddes Repert., 83(9–10):631(1972 publ. 1973):
2 *A. karputanus.*
subsp. **owerinii** (Bunge) Ponert *op. cit.* 631(1972
2 publ. 1973): *A. owerinii.*
brevicalyx (Eig) Ponert *op. cit.* 624(1972 publ. 1973):
1, 2 *A. cephalotes* var. *brevicalyx.*
calcicola Podlech in *Mitt. Bot. Staatssamml. München*,
2 11:279(1973), as 'calcicolus'—Afghanistan.
cariensis Boiss.
subsp. **ehrendorferi** Ponert in *Feddes Repert.*,
2 83(9–10):632(1972 publ. 1973)—Turkey.
chloroxanthinus (Freyn & Bornm.) Ponert *op. cit.*
631(1972 publ. 1973), as 'chlorxanthinus': *A.*
2 *squalidus* var. *chloroxanthinus.*
christianus L.
subsp. **cilicius** (Boiss.) Ponert *op. cit.* 621(1972
2 publ. 1973): *A. cilicius.*
subsp. **suboccidentalis** Ponert *op. cit.* 621(1972
2 publ. 1973)—Turkey.
compactus Lam.
subsp. **albifolius** (Freyn & Sint.) Ponert *op. cit.*
2 624(1972 publ. 1973): *A. albifolius.*
subsp. **isparticus** Ponert *op. cit.* 624(1972 publ.
2 1973)—Turkey.
subsp. **micropterus** (Fischer) Ponert *op. cit.* 624
2 (1972 publ. 1973): *A. micropterus.*
connectens Podlech in *Mitt. Bot. Staatssamml.*
2 *München*, 11:293(1973)—Afghanistan.
cretaceus Boiss. & Kotschy
var. **globosa** Ponert in *Feddes Repert.*, 83(9–10):
2 620(1972 publ. 1973)—Turkey.
creticus Lam.
subsp. **samius** (Širj. & Rech.) Ponert *op. cit.* 622
2 (1972 publ. 1973): *A. creticus* var. *samius.*
cylindraceus DC.
subsp. **melinus** (Boiss.) Ponert *op. cit.* 631(1972
2 publ. 1973): *A. melinus.*
dasysemius (Chamb. & Matthews) Ponert *op. cit.*
2 629(1972 publ. 1973): *A. maximus* var. *dasysemius.*
declinatus Willd.
subsp. **pinetorum** (Boiss.) Ponert *op. cit.* 622(1972
2 publ. 1973): *A. pinetorum.*
decurrens Boiss.
subsp. **dichroanthus** (Freyn & Sint.) Ponert
2 *op. cit.* 629(1972 publ. 1973): *A. dichroanthus.*
delanensis Širj. & Rech. f.
subsp. **baytopianus** (Chamb. & Matthews)
Ponert *op. cit.* 622(1972 publ. 1973): *A.*
2 *baytopianus.*
delicatulus Podlech in *Mitt. Bot. Staatssamml.*
2 *München*, 11:295(1973)—Afghanistan.
2 **dieterlei** Podlech *op. cit.* 299(1973)—Afghanistan.
diphtherites Fenzl
subsp. **karataschensis** (Širj.) Ponert in *Feddes*
Repert., 83(9–10):624(1972 publ. 1973): *A.*
2 *karataschensis.*
subsp. **leucomallus** (Bornm. & Širj.) Ponert *op.*
2 *cit.* 624(1972 publ. 1973): *A. leucomallus.*
dipsaceus Bunge
var. **germanicopolitanus** (Bornm.) Ponert *op. cit.*
2 629(1972 publ. 1973): *A. germanicopolitanus.*
edinburghensis Ponert *op. cit.* 632(1972 publ.
2 1973)—Turkey.
eriocephalus Willd.
subsp. **bracteolatus** Ponert *op. cit.* 622(1972
2 publ. 1973)—Turkey.
faryabensis Podlech in *Mitt. Bot. Staatssamml.*
2 *München*, 11:301(1973)—Afghanistan.
flabellatus Podlech *op. cit.* 272(1973)—
2 Afghanistan.
fumosus Boriss.
f. **subelatior** Ponert in *Feddes Repert.*, 83(9–10):
2 630(1972 publ. 1973)—Turkey.
ghoratensis Podlech in *Mitt. Bot. Staatssamml.*
2 *München*, 11:259(1973)—Afghanistan.
ghorbandicus Podlech *op. cit.* 302(1973)—
2 Afghanistan.
gummifer Lab.
subsp. **tournefortii** (Boiss.) Ponert in *Feddes*
Repert., 83(9–10):622(1972 publ. 1973): *A.*
2 *tournefortii.*
guttatus Banks & Sol.
subsp. **latileguminis** Ponert *op. cit.* 620(1972 publ.
2 1973)—Turkey.
hauarensis Boiss.
var. **brachycarpus** (Širj. & Rech. f.) Ali in *Kew*
Bull., 28(2):299(1973): *A. gyzensis* var. *brachy-*
6 *carpus.*
hirsutus Vahl
subsp. **alindanus** (Boiss.) Ponert in *Feddes Repert.*,
2 83(9–10):631(1972 publ. 1973): *A. alindanus.*
subsp. **paraglobosus** Ponert *op. cit.* 631(1972
2 publ. 1973)—Turkey.
hyalolepis Bunge
subsp. **angustibracteata** Ponert *op. cit.* 631(1972
2 publ. 1973)—Turkey.
imitans Podlech in *Mitt. Bot. Staatssamml. München*,
2 11:266(1973)—Afghanistan.
incertus Ledeb.
subsp. **sparsevillosus** Ponert in *Feddes Repert.*,
2 83(9–10):630(1972 publ. 1973)—Turkey.
inflaticarpus Ponert *op. cit.* 621(1972 publ. 1973)—
2 Turkey.
irmingardis Podlech in *Mitt. Bot. Staatssamml.*
2 *München*, 11:282(1973)—Afghanistan.

karamasicus Boiss. & Bal.
subsp. **ulaschensis** (Huber-Morath & Reese) Ponert in *Feddes Repert.*, 83(9–10):630(1972 publ.
2 1973): *A. ulaschensis.*
khojakensis Ali in *Kew Bull.*, 28(2):300(1973)—
6 Pakistan.
lagurus Willd.
subsp. **hirticalyx** (Boiss. & Kotschy) Ponert in *Feddes Repert.*, 83(9–10):628(1972 publ. 1973):
2 *A. hirticalyx.*
subsp. **persicus** (DC.) Ponert *op. cit.* 629(1972
2 publ. 1973): *A. lagopoides* var. *persicus.*
subsp. **sivasianus** Ponert *op. cit.* 629(1972 publ.
2 1973)—Turkey.
lamarckii Boiss.
subsp. **bombycalyx** (Eig) Ponert *op. cit.* 624(1972
2 publ. 1973): *A. bombycalyx.*
subsp. **eginensis** (Freyn & Sint.) Ponert *op. cit.* 624
2 (1972 publ. 1973): *A. eginensis.*
subsp. **nimrudensis** (Širj.) Ponert *op. cit.* 624
2 (1972 publ. 1973): *A. nimrudensis.*
subsp. **tinctus** (Freyn & Sint.) Ponert *op. cit.* 624
2 (1972 publ. 1973): *A. tinctus.*
latifolius Lam.
subsp. **candolleanus** (Boiss.) Ponert *op. cit.* 631
2 (1972 publ. 1973): *A. candolleanus.*
leporinus Boiss.
f. **hirsutus** (Post) Ponert *op. cit.* 622(1972 publ.
2 1973): *A. brachystachys* var. *hirsutus.*
lineatus Lam.
f. **jildisianus** (Bornm.) Ponert *op. cit.* 629(1972
2 publ. 1973): *A. jildisianus.*
subsp. **longidens** (Freyn) Ponert *op. cit.* 629(1972
2 publ. 1973): *A. longidens.*
lowarensis Ali in *Kew Bull.*, 28(2):299(1973)—
6 Pakistan.
macrourus Fisch. & Mey.
subsp. **adpressipilosus** Ponert in *Feddes Repert.*,
2 83(9–10):620(1972 publ. 1973)—Turkey.
subsp. **brevilobus** Ponert *op. cit.* 620(1972 publ.
2 1973)—Turkey.
subsp. **pseudotauricolus** Ponert *op. cit.* 620(1972
2 publ. 1973)—Turkey.
maymanensis Podlech in *Mitt. Bot. Staatssamml.*
2 *München*, 11:305(1973)—Afghanistan.
nitens Boiss. & Heldr.
subsp. **viridis** (Bunge) Ponert in *Feddes Repert.*,
2 83(9–10):632(1972 publ. 1973): *A. viridis.*
onobrychis L.
subsp. **goktschaicus** (Grossh.) Ponert *op. cit.* 630
2 (1972 publ. 1973):*A. goktschaicus.*
oocephalus Boiss.
var. **stachyophorus** (Hub.-Mor. & Chamb.) Ponert *op. cit.* 629(1972 publ. 1973): *A. oocephalus*
2 subsp. *stachyophorus.*
paktiensis Podlech in *Mitt. Bot. Staatssamml. München*,
2 11:314(1973)—Afghanistan.
paragriseus Ponert in *Feddes Repert.*, 83(9–10):
2 620(1972 publ. 1973)—Turkey.
pennatus Bunge
subsp. **idae** (Širj.) Ponert *op. cit.* 623(1972 publ.
2 1973): *A. idae.*
pilosicarpa Ponert *op. cit.* 621(1972 publ. 1973)—
2 Turkey.
pisidicus Boiss. & Heldr.
subsp. **rugosicarpus** Ponert *op. cit.* 621(1972 publ.
2 1973)—Turkey.
plumosus Willd.
subsp. **akardaghicus** (Eig) Ponert *op. cit.* 623
2 (1972 publ. 1973): *A. akardaghicus.*
var. **dubius** (Širj.) Ponert *op. cit.* 623(1972 publ.
2 1973): *A. krugeanus* var. *dubius.*
var. **karagoelldaghensis** (Širj.) Ponert *op. cit.* 623 (1972 publ. 1973): *A. krugeanus* var. *karagoell-*
2 *daghensis.*
subsp krugeanus (Freyn & Bornm.) Ponert *op. cit.* 623(1972 publ. 1973), without basionym page:
2 *A. krugeanus.*
var. **kurdomerensis** (Širj.) Ponert *op. cit.* 623
2 (1972 publ. 1973): *A. krugeanus* var. *kurdomerensis.*
subsp. **nitens** (Freyn & Bornm.) Ponert *op. cit.* 623
2 (1972 publ. 1973): *A. krugeanus* var. *nitens.*
var. **sipikordaghensis** (Širj.) Ponert *op. cit.* 623 (1972 publ. 1973): *A. krugeanus* var. *sipikor-*
2 *daghensis.*
× **pseudoparnassi** Ponert *op. cit.* 628(1972 publ.
2 1973)—Turkey. (*A. parnassi* × *A. thracicus*).
psoraloides Lam.
subsp. **canescens** (DC.) Ponert *op. cit.* 630(1972
2 publ. 1973): *A. canescens.*
subsp. **mukusiensis** (Rech. f.) Ponert *op. cit.* 630
2 (1972 publ. 1973): *A. mukusiensis.*
ptilodes Boiss.
subsp. **paraptilodes** Ponert *op. cit.* 628(1972 publ.
2 1973)—Turkey.
pycnocephalus Fischer
subsp. **munzurica** Ponert *op. cit.* 622(1972 publ.
2 1973)—Turkey.
subsp. **seytunensis** (Bunge) Ponert *op. cit.* 622
2 (1972 publ. 1973): *A. seytunensis.*
rassoulii Podlech in *Mitt. Bot. Statssamml. München*,
2 11:307(1973)—Afghanistan.
robustus Bunge
subsp. **subrobustus** (Boriss.) Ponert in *Feddes Repert.*, 83(9–10):632(1972 publ. 1973): *A.*
2 *subrobustus.*
saganlugensis Trautv.
var. **sapozhnikovii** (Šiškin) Ponert *op. cit.* 620
2 (1972 publ. 1973): *A. sapozhnikovii.*
salangensis Podlech in *Mitt. Bot. Staatssamml.*
2 *München*, 11:309(1973)—Afghanistan.
schizopterus Boiss.
subsp. **glaucophyllus** (Bunge) Ponert in *Feddes Repert.*, 83(9–10):631(1972 publ. 1973): *A.*
2 *glaucophyllus.*
scorpiurus Bunge
var. **glaber** Ali in *Kew Bull.*, 28(2):297(1973)—
6 Pakistan.
sempervirens Lam.
subsp. **alpinus** Pignatti in *Giorn. Bot. Ital.*, 107(5):
1 215(1973)—Italy.
1 subsp. **gussonei** Pignatti *op. cit.* 216(1973)—Italy.
setulosus Boiss. & Bal.
subsp. **pseudostrigillosus** Ponert in *Feddes Repert.*,
2 83(9–10):630(1972 publ. 1973)—Turkey.
shebarensis Podlech in *Mitt. Bot. Stattssamml.*
2 *München*, 11:274(1973)—Afghanistan.
sommieri Freyn
subsp. **acmophylloides** (Grossh.) Ponert in *Feddes Repert.*, 83 (10):625(1972 publ. 1973): *A.*
2 *acmophylloides.*
steineranus Podlech in *Mitt. Bot. Staatssamml.*
2 *München*, 11:318(1973)—Afghanistan.
stevenianus DC.
subsp. **gevasicus** Ponert in *Feddes Repert.*, 83(9–10):
2 629(1972 publ. 1973)—Turkey.
subsp. **jodostachys** (Boiss. & Buhse) Ponert *op. cit.*
2 629(1972 publ. 1973): *A. jodostachys.*
strictifolius Boiss.
subsp. **kutepovii** (Širj.) Ponert *op. cit.* 623(1972
2 publ. 1973): *A. strictifolius* var. *kutepovii.*
subsp. **stenonychioides** (Freyn & Bornm.) Ponert *op. cit.* 623(1972 publ. 1973): *A.*
2 *stenonychioides.*
subsp. **zoharyi** Ponert *op. cit.* 623(1972 publ. 1973)
2 —Turkey.

subuliferus Boiss.
subsp. **wallii** (Eig) Ponert *op. cit.* 621(1972 publ.
2 1973): *A. wallii.*
sultani Ali in *Kew Bull.*, 28(2):301(1973)—
6 Pakistan.
surugensis Boiss. & Hausskn.
subsp. **zederbaueri** (Stadlmann) Ponert in *Feddes Repert.*, 83(9–10):629(1972 publ. 1973): *A.*
2 *zederbaueri.*
takharensis Podlech in *Mitt. Bot. Staatssamml.*
2 *München*, 11:313(1973)—Afghanistan.
tashqurghanicus K. H. Rechinger & Podlech *op. cit.*
2 269(1973)—Afghanistan.
tauricolus Boiss.
subsp. **austro-aegaeus** (Rech. f.) Ponert in *Feddes Repert.*, 83(9–10):620(1972 publ. 1973): *A.*
1, 2 *austro-aegaeus.*
thracicus Gris.
subsp. **afyonicus** Ponert *op. cit.* 628(1972 publ.
2 1973)—Turkey.
subsp. **baytopianus** Ponert *op. cit.* 628(1972 publ.
2 1973)—Turkey.
subsp. **trojanus** (Stev.) Ponert *op. cit.* 628(1972
1, 2 publ. 1973): *A. trojanus.*
tmoleus Boiss.
subsp. **bounacanthus** (Boiss.) Ponert *op. cit.* 628
2 (1972 publ. 1973): *A. bounacanthus.*
toachius Woron.
subsp. **erzincanicus** Ponert *op. cit.* 632(1972 publ.
2 1973)—Turkey.
tumninensis N. S. Pavlova & Basargin in *Byull. Glavn. Bot. Sada* (Moscow), 90:42(1973)—U.S.S.R.
2 (Soviet Far East).
urgunensis Podlech in *Mitt. Bot. Staatssamml.*
2 *München*, 11:263(1973)—Afghanistan.
wiedermannianus Fischer
subsp. **brachypterus** (Fischer) Ponert in *Feddes Repert.*, 83(9–10):628(1972 publ. 1973): *A.*
2 *brachypterus.*
xerophilus Ledeb.
subsp. **caudiculosus** (Boiss. & Huet) Ponert *op. cit.*
2 631(1972 publ. 1973): *A. caudiculosus.*
xylobasis Freyn & Bornm.
f. **angustus** (Freyn & Sint.) Ponert *op. cit.* 630
2 (1972 publ. 1973): *A. xylorrhizus* var. *angustus.*

ATYLOSIA

2722 Hepper, F. N.
Tropical African plants: 33. Another synonym of *Atylosia scarabaeoides.* (Leguminosae-Papilionoideae).
5 *Kew Bull.*, 28(2):319(1973).

CARAGANA

2723 Kordyum, E. L. *and* Nedukha, O. M.
Dva tȳpȳ dilennya khloroplastiv u chereshkakh lȳstkiv akatsiyi zhovtoyi. (Two types of chloroplast division in leaf petioles of *Caragana arborescens* Lam.).
2 *Ukr. Bot. Zhurn.*, 30(6):787–790(1973)—Uk(En).

CHAMAECYTISUS

2724 Forissier, R.
Recherches cytotaxonomiques préliminaires sur les genres *Lembotropis*, *Cytisus*, *Chamaecytisus*, *Genista* et *Chamaespartium. Bull. Soc. Neuchâtel. Sci. Nat.*, 96:51–
1 65(1973)—(Ge,En); chromosome numbers.

2725 Petrowicz, M.
Badania nad zmiennościąe kologia i rozmnażaniem *Chamaecytisus albus* Rothm. (Studies on the variability, ecology and propagation of *Chamaecytisus albus* Rothm.). *Ann. Univ. Mariae Curie-Skłodowska, C*, 28:
1 245–264(1973)—(Rus,En); illustrations, map.

austriacus (L.) Link
subsp. **pygmaeus** (Willd.) Ponert in *Feddes Repert.*,
1, 2 83(9–10):619(1972 publ. 1973): *Cytisus pygmaeus.*
subsp. **tommasinii** (Visiani) Ponert *op. cit.* 618
1, 2 (1972 publ. 1973): *Cytisus tommasinii.*
drepanolobus (Boiss.) Rothm.
subsp. **gibbsii** Ponert *op. cit.* 618(1972 publ. 1973),
2 as 'gibbsi'—Turkey.
hirsutus (L.) Link
subsp. **hirsutissimus** (C. Koch) Ponert *op. cit.* 619
1, 2 (1972 publ. 1973): *Cytisus hirsutissimus.*
subsp. **leucotrichus** (Schur.) Ponert *op. cit.* 619
1, 2 (1972 publ. 1973): *Cytisus leucotrichus.*
subsp. **polytrichus** (Bieb.) Ponert *op. cit.* 619
1, 2 (1972 publ. 1973): *Cytisus polytrichus.*

CHAMAESPARTIUM

2726 Forissier, R.
Recherches cytotaxonomiques préliminaires sur les genres *Lembotropis*, *Cytisus*, *Chamaecytisus*, *Genista* et *Chamaespartium. Bull. Soc. Neuchâtel. Sci. Nat.*, 96:
1 51–65(1973)—(Ge,En); chromosome numbers.

2727 Soják, J.
Doplňky k nomenklatuře některých rodů (Phanerogamae). *Čas. Nár. Muz.* (Prague), 141(1–2):
1, 2 61–63(1972).

delphinensis (Verlot) Soják in *Čas. Nar. Muz.*
1 (Prague), 141(1–2):63(1972): *Genista delphinensis.*
nevadense Esteve Chueca & Varo Alcala in *Trab.*
1 *Dep. Bot. Univ. Granada*, 1:3(1972)—Spain.
sagittale (L.) Gibbs
subsp. **undulatum** (Ern) Soják in *Čas. Nár. Muz.* (Prague), 141(1–2):63(1972): *Genistella sagittalis*
1 subsp. *undulata.*
tridentatum (L.) Gibbs
subsp. **lasianthum** (Spach) Soják *op. cit.* 63(1972):
1, 2 *Genista lasiantha.*
subsp. **stenopterum** (Spach) Soják *op. cit.* 63
1, 2 (1972): *Genista stenoptera.*

CHRONANTHUS

2728 Sañudo, A.
Variabilidad cromosimica de las Genisteas de la flora española en relación con su ecología: 1. Número y comportamiento de los cromosomas durante la meiosis: C. Géneros *Chronanthus* (DC.) C. Koch, *Adenocarpus* DC. y *Erinacea* Adanson. *Lagascalia*, 3(2):
1 205–210(1973)—Chromosome numbers.

CICER

2729 Prima, V. M.
Cicer minutum Boiss. et Hohen. s vostochnogo Kavkaza (gora Nesin-dag). (*Cicer minutum* Boiss. et Hohen. e Caucaso orientali (mons Nesin-dag)).
2 *Novost. Sist. Vȳssh. Rast.*, 10:189–90(1973).

pimpinellifolium Jaub. & Spach
subsp. **minutum** (Boiss. & Hohen.) Ponert in *Feddes Repert.*, 83(9–10):633(1972 publ. 1973):
2 *C. minutum.*
pinnatifidum Jaub. & Spach
subsp. **judaicum** (Boiss.) Ponert *op. cit.* 633(1972
2 publ. 1973): *C. judaicum.*

COLUTEA

2730 Timbal, J.
Le baguenaudier (*Colutea arborescens* L.) en Lorraine. *Bull. Soc. Bot. France*, 118(9):837–840(1971 publ. 1973)
1 —(En).

arborescens L.
subsp. **atlantica** (Browicz) Ponert in *Feddes Repert.*, 83(9–10):619(1972 publ. 1973): *Colutea*
1, 2 *atlantica.*
subsp. **cilicica** (Boissier & Balansa) Ponert *op. cit.*
1, 2 620(1972 publ. 1973): *C. cilicica.*

COROTHAMNUS

acutangulus (Jaub. & Spach.) Ponert in *Feddes Repert.*, 83(9–10):619 (1972 publ. 1973): *Cytisus*
2 *acutangulus.*
balansae (Boiss.) Ponert *op. cit.* 619(1972 publ. 1973):
2 *Sarothamnus balansae.*
multiflorus (Sweet) Ponert *op. cit.* 619(1972 publ.
1, 2 1973): *Cytisus multiflorus.*
purgans (L.) Ponert *op. cit.* 619(1972 publ. 1973):
1, 2 *Genista purgans.*

COURSETIA

2731 Rudd, V. E.
Studies in *Coursetia* (Leguminosae): 1. *Phytologia*,
4 25(4):257–258(1973).

guatemalensis Rudd in *Phytologia*, 25(4):257
4 (1973)—Guatemala.
4 **hintonii** Rudd *op. cit.* 258(1973)—Mexico.

CROTALARIA

2732 Bisby, F. A.
The role of taximetrics on angiosperm taxonomy: 1. Empirical comparisons of methods using *Crotalaria* L.
5 *New Phytol.*, 72(3):699–726(1973).

2733 Bisby, F. A. *and* Polhill, R. M.
The role of taximetrics in angiosperm taxonomy: 2. Parallel taximetric and orthodox studies in
5 *Crotalaria* L. *New Phytol.*, 72(3):727–742(1973).

2734 Chennaveeraiah, M. S. *and* Patil, B. C.
Chromosome number in the genus *Crotalaria* Linn. *Proc. Indian Sci. Congr.*, 59(3): 351–352 (1972).
6 Abstract only.

2735 Nadkarni, R. S.
Karyotypic study in *Crotalaria linifolia* Linn. f. suppl. *Chromosome Inform. Serv.*, no.14.27–29
2, 6, 7 (1973).

2736 Sinha, R. P. *and* Gupta, A. K.
Cytotaxonomic studies in some *Crotalaria* species. *Proc. Indian Sci. Congr.*, 60(3):319(1973)—Chromosome numbers. Abstract only; *C. sericea, C.*
6 *mysorensis.*

2737 Windler, D. R.
Field and garden studies in *Crotalaria sagittalis* L. and related species. *Phytologia*, 26(5):289–354(1973)—
3–7 Illustrations.

subsect. **Glaucae** (Benth.) Bisby & Polhill in *New*
5 *Phytol.*, 72(3):741(1973): *Crotalaria* group *Glaucae.*
subsect. **Incanae** (Benth.) Bisby & Polhill *op. cit.*
5 741(1973): *Crotalaria* group *Incanae.*
subsect. **Macrostachyae** (Benth.) Bisby & Polhill
5 *op. cit.* 741(1973): *Crotalaria* group *Macrostachyae.*
subsect. **Tetralobocalyx** (Harms) Bisby & Polhill
5 *op. cit.* 741(1973): *Crotalaria* sect. *Tetralobocalyx.*

CYTISANTHUS

2738 Leute, G. H.
Der Kugelginster *Cytisanthus radiatus* (L.) Lang, eine floristische Kostbarkeit am Weissensee.
1 *Carinthia II*, 82:221–223(1972).

sericopetalus (Buchegger) Holub in *Folia Geobot. Phytotax.*, 8(2):176(1973): *Genista radiata* var.
1 *sericopetala.*

CYTISUS

2739 Faugeras G., Paris, R. *and* Valdes-Bermejo, E.
Alcaloïdes et polyphénols des Légumineuses: 26. Sur les alcaloïdes du *Sarothamnus patens* (L.) Webb (*Cytisus striatus* (Hill) Rothm.). (Alkaloids and polyphenols of Leguminosae: 26. The alkaloids of *Sarothamnus patens* (L.) Webb (*Cytisus striatus* (Hill) Rothm.)). *Ann. Pharm. Franc.*, 30(7–8):527–533(1972)—(En). Also publ. in *Trav. Lab. Matière Méd. Pharm.*
1 *Galénique Fac. Pharm. Paris*, 57(2)(1973).

2740 Forissier, R.
Recherches cytotaxonomiques préliminaires sur les genres *Lembotropis, Cytisus, Chamaecytisus, Genista* et *Chamaespartium. Bull. Soc. Neuchâtel. Sci. Nat.*, 96:51–
1 65(1973)—(Ge,En); chromosome numbers.

2741 Keeley, J. E.
Cytisus monspessulanus L., established in southern
3 California. *Madroño*, 22(1):30(1973).

2742 Sañudo, A.
Variabilídad cromosómica de las genisteas de la flora española en relación con su ecología. Nota previa sobre los géneros *Cytisus* L. y *Adenocarpus* DC. *Cuad. Cienc. Biol.* (Granada), 2(1):37–38(1973)—(En);
1 chromosome numbers.

DALEA

2743 Barneby, R. C.
New names in *Dalea* and *Marina* (Fabaceae tribus
4 Amorpheae). *Phytologia*, 26(1):1–2(1973).

aenigma Barneby in *Phytologia*, 26(1):1(1973)—
4 Mexico.
foliolosa (Aiton) Barneby *op. cit.* 1(1973): *Psoralea*
4 *foliolosa.*
var. **citrina** (Rydberg) Barneby *op. cit.* 1(1973):
4 *Parosela citrina.*
leucostachya A. Gray
var. **eysenhardtioides** (Hemsley) Barneby *op. cit.*
4 1(1973): *D. eysenhardtioides.*
obovatifolia Ortega
var. **uncifera** (Schlechtendahl & Chamisso)
4 Barneby *op. cit.* 1(1973): *D. uncifera.*
versicolor Zuccarini
var. **involuta** (Rydberg) Barneby *op. cit.* 1(1973):
4 *Parosela involuta.*
var. **sessilis** (A. Gray) Barneby *op. cit.* 1(1973):
3 *D. wislizeni* var. *sessilis.*

DENDROLOBIUM

lanceolatum (Dunn) Schindler
var. **microcarpum** Ohashi in *Ginkgoana*, 1:65
6 (1973)—Thailand.
rugosum (Prain) Schindler
var. **moniliferum** Ohashi *op. cit.* 74(1973), nom.
6 nov.: *Desmodium wallichii* Prain.
triangulare (Retzius) Schindler
subsp. **cephalotoides** (Craib) Ohashi *op. cit.* 80
6 (1973): *Desmodium cephalotoides.*
6 f. **villosum** Ohashi *op. cit.* 81(1973)—Burma.
umbellatum (L.) Benth.
f. **hirsutum** (DC.) Ohashi *op. cit.* 84(1973):
6 *Desmodium umbellatum* var. *hirsutum.*

DESMODIUM

2744 Ohashi, H.
The Asiatic species of *Desmodium* and its allied genera (Leguminosae). *Ginkgoana*, no.1.[ii]318p.
2, 5–7 (1973)—Illustrations, dot maps; keys.

sect. **Angustistipulosa** Ohashi in *Ginkgoana*, 1:97
2, 6 (1973).
ser. **Calcarata** (Schindler) Ohashi *op. cit.* 130(1973):
6 *D.* subsect. *Calcaratae.*
subgen. **Hanslia** (Schindler) Ohashi *op. cit.* 112(1973):
6 genus *Hanslia.*
sect. **Monarthrocarpus** (Merr.) Ohashi *op. cit.* 125
6 (1973): genus *Monarthrocarpus.*
subgen. **Ougeinia** (Benth.) Ohashi *op. cit.* 116(1973):
6 genus *Ougeinia.*
sect. **Oxytes** (Schindler) Ohashi *op. cit.* 227(1973):
6, 7 *Desmodium* subgen. *Oxytes.*
subgen. **Podocarpium** (Benth.) Ohashi *op. cit.* 120
6 (1973): *D.* sect. *Podocarpium.*
2, 6, 7 sect. **Renifolia** Ohashi *op. cit.* 229(1973).
2, 6 sect. **Repanda** Ohashi *op. cit.* 155(1973).
barbatum (L.) Benth.
subsp. **saulierei** (Schindler) Ohashi *op. cit.* 203
6 (1973): *Nicolsonia barbata* var. *saulierei.*
benthamii Ohashi *op. cit.* 249(1973), nom. nov.:
6 *D. brachystachyum* Grah. ex Benth.
concinnum DC.
var. **retusum** (D. Don) Ohashi *op. cit.* 182(1973):
6 *Hedysarum retusum.*
duclouxii Pampanini
var. **henryi** (Schindler) Ohashi *op. cit.* 159(1973):
2, 6 *D. henryi.*
ferrugineum Wall. ex Thwaites
subsp. **wynaadense** (Bedd ex Gamble) Ohashi *op.*
6 *cit.* 206(1973): *D. wynaadense.*
heterocarpon (L.) DC.
subsp. **angustifolium** (Craib) Ohashi *op. cit.* 212
2, 6 (1973): *D. polycarpum* var. *angustifolium.*
var. **birmanicum** (Watt ex Prain) Ohashi *op. cit.*
6 213(1973): *D. birmanicum.*
laxiflorum DC.
subsp. **lacei** (Schindler) Ohashi *op. cit.* 103(1973):
6 *D. lacei.*
laxum DC.
subsp. **laterale** (Schindler) Ohashi *op. cit.* 137
2, 6 (1973): *D. laterale.*
subsp. **leptopus** (A. Gray ex Benth.) Ohashi *op.*
2, 6 *cit.* 141(1973): *D. leptopus.*
nemorosum F. v. Muell. ex Benth.
var. **whitfordii** (Schindler) Ohashi *op. cit.* 220
6, 7 (1973): *D. nemorosum* subvar. *whitfordii.*
oojeinense (Roxb.) Ohashi *op. cit.* 117(1973):
6 *Dalbergia oojeinensis.*
renifolium (L.) Schindler
var. **oblatum** (Baker ex Kurz) Ohashi *op. cit.* 233
2, 6 (1973): *D. oblatum.*
rubrum (Lour.) DC.
var. **macrocarpum** Ohashi *op. cit.* 222(1973)—
6 Vietnam.
velutinum (Willd.) DC.
subsp. **longibracteatum** (Schindler) Ohashi *op.*
2, 6 *cit.* 194(1973): *D. longibracteatum.*
williamsii Ohashi *op. cit.* 163(1973)—Tibet, Nepal,
2, 6 Assam, Bhutan, China.
subsp. **magnibracteatum** Ohashi *op. cit.* 166
2 (1973)—China.

DICERMA

biarticulatum (L.) DC.
subsp. **australiense** (Schindler) Ohashi in *Ginkgoana*,
6, 7 1:255(1973): *D. biarticulatum* var. *australiense.*
subsp. **hispidum** (Schindler) Ohashi *op. cit.* 257
6 (1973): *D. hispidum.*

DOLICHOS

2745 Dardenne, G. A., Thonart, P., Otoul, E. *and* Marechal, R.
Etude chimiotaxonomique dans les genres *Macrotyloma*, *Dolichos* et *Pseudovigna*. *Phytochemistry*,
5 12(8):1983–1992(1973)—(En).

DORYCNIUM

pentaphyllum Scop.
subsp. **amani** (Zohary) Ponert in *Feddes Repert.*,
2 83(9–10):640(1972 publ. 1973): *D. amani.*

EBENUS

bourgaei Boiss.
subsp. **cappadocica** (Hausskn. & Siehe ex Bornm. Ponert in *Feddes Repert.*, 83(9–10):644(1972 publ.
2 1973): *E. cappadocica.*
laguroides Boiss.
subsp. **cilicica** (Boiss.) Ponert *op. cit.* 644(1972
2 publ. 1973): *E. montbretii* var. *cilicica.*
subsp. **pisidica** (Huber-Morath & Reese) Ponert
2 *op. cit.* 644(1972 publ. 1973): *E. pisidica.*
longipes Boiss. & Bal.
subsp. **argentea** (Siehe ex Bornm.) Ponert *op. cit.*
2 644(1972 publ. 1973): *E. argentea.*
plumosa Boiss. & Bal.
subsp. **speciosa** (Boiss. & Bal.) Ponert *op. cit.* 644
2 (1972 publ. 1973): *E. plumosa.*
reesei Huber-Morath
subsp. **minor** (Huber-Morath) Ponert *op. cit.* 644
2 (1972 publ. 1973): *E. reesei* var. *minor.*

ECHINOSOPHORA

2746 Yakovlev, G. P.
O rode *Keyserlingia* Bunge (vklyuchaya *Echinosophora* Nakai). (On the genus *Keyserlingia* Bunge (including *Echinosophora* Nakai)). *Bot. Zhurn.*, 58(6):
2, 6 833–838(1973)—Keys.

ERINACEA

2747 Sañudo, A.
Variabilidad cromosimica de las Genisteas de la flora española en relación con su ecología: 1. Número y comportamiento de los cromosomas durante la meiosis: C. Géneros *Chronanthus* (DC.) C. Koch, *Adenocarpus* DC. y *Erinacea* Adanson. *Lagascalia*,
1 3(2):205–210(1973)—Chromosome numbers.

ERIOSEMA

2748 Baudet, J. C.
Notes sur quelques espèces africaines du genre *Eriosema* (DC) Desv. (Papilionaceae). *Bull. Jard. Bot. Nation. Belg.*, 43(1–2):17–32(1973)—Map, dot
5 maps.

2749 Baudet, J. C.
Papilionaceae: 179–189. *Distrib. Pl. Afr.*, 7. 11p.
5 (1973)—Dot maps. *Eriosema* spp. only.

2750 Ramcharan, E. K., McAndrews, J. H. *and* Grear, J. W.
Pollen morphology of *Rhynchosia* and *Eriosema* (Fabaceae). *Canad. J. Bot.*, 51(12):2395–2399
3, 4 (1973).

ERYTHRINA

2751 Krukoff, B. A. *and* Barneby, R. C.
Notes on the species of *Erythrina*: 6. *Phytologia*, 25(1):1–31(1972). Includes synopsis of all known
2–7 species.

2752 Krukoff, B. A. *and* Barneby, R. C.
Notes on the species of *Erythrina*: 7. *Phytologia*, 27(2):108–141(1973)—Key to the subspecies of
4 *E. lanata.*

berenices Krukoff & Barneby in *Phytologia*, 27(2):
4 120(1973)—Mexico.
caribaea Krukoff & Barneby in *Phytologia*, 25(1):
4 9(1972)—Mexico.
herbacea L.
subsp. **nigrorosea** Krukoff & Barneby *op. cit.* 6
4 (1972)—Mexico.
lanata Rose
subsp. **occidentalis** (Standley) Krukoff & Barneby
4 in *Phytologia*, 27(2):117(1973): *E. occidentalis.*
pudica Krukoff & Barneby *op. cit.* 114(1973)—
4 Mexico.
salviiflora Krukoff & Barneby in *Phytologia*, 25(1):
4 14(1972)—Guatemala.
tuxtlana Krukoff & Barneby *op. cit.* 11(1972)—
4 Mexico.

EYSENHARDTIA

2753 Lang, J. M.
Eysenhardtia (Leguminosae): taxonomic revision and relationships. *Diss. Abstr. Int., B*, 33(10):4692
3, 4 (1973).

FEDOROVIA

2754 Yakovlev, G. P.
Zametki po sistematike rodov *Sweetia* Spreng., *Machaerium* Pers., *Angylocalyx* Taub., *Fedorovia* Yakovl., *Placolobium* Miq., *Ormosia* Jacks. (Fabaceae). (De generibus *Sweetia* Spreng., *Machaerium* Pers., *Angylocalyx* Taub., *Fedorovia* Yakovl., *Placolobium* Miq. et *Ormosia* Jacks. (Fabaceae) notulae systemati-
2, 6 cae). *Novosti Sist. Vÿssh. Rast.*, 10:190–196(1973).

sect. **Pinnatae** Yakovlev in *Novosti Sist. Vÿssh. Rast.*,
2, 6 10:196(1973).
sect. **Striatae** (Prain) Yakovlev *op. cit.* 196(1973):
2 *Ormosia* subsect. *Striatae.*
inopinata (Prain) Yakovlev *op. cit.* 195(1973):
2 *Ormosia inopinata.*
penangensis (Ridley) Yakovlev *op. cit.* 195(1973):
6 *Ormosia penangensis.*
pinnata (Lour.) Yakovlev *op. cit.* 196(1973):
2, 6 *Cynometra pinnata.*
striata (Dunn) Yakovlev *op. cit.* 196(1973): *Ormosia*
2 *striata.*

GENISTA

2755 Forissier, R.
Recherches cytotaxonomiques préliminaires sur les genres *Lembotropis*, *Cytisus*, *Chamaecytisus*, *Genista* et *Chamaespartium*. *Bull. Soc. Neuchâtel. Sci. Nat.*, 96:51–
1 65(1973)—(Ge,En); chromosome numbers.

2756 Murin, A. *and* Neischlová, E.
Chromosomenzahlen einige Taxa aus dem Verwandtschaftskreis von *Genista tinctoria* L. (Chromosome numbers for some taxa of *Genista tinctoria* complex). *Biológia, A* (Czechoslovakia), 28(7):493–497(1973)—(En,Cz,Rus); chromosome
1 numbers.

lobelii DC.
var. **lobelioides** Gamisans in *Candollea*, 28(1):68
1 (1973)—Corsica.
lydia Boiss.
subsp. **antalayaensis** Ponert in *Feddes Repert.*,
2 83(9–10):619(1972 publ. 1973)—Turkey.
subsp. **antiochia** (Boiss.) Ponert *op. cit.* 619(1972
2 publ. 1973): *G. antiochia.*
subsp. **rumelica** (Velenovský) Ponert *op. cit.* 619
1, 2 (1972 publ. 1973): *G. rumelica.*
salzmannii DC.
subsp. **parnassica** (Halácsy) Ponert *op. cit.* 619
1, 2 (1972 publ. 1973): *G. parnassica.*
tinctoria L.
subsp. **insubrica** (Brügger) Pignatti in *Giorn.*
1 *Bot. Ital.*, 107(5):214(1973): *G. insubrica.*
subsp. **scariosa** (Viv.) Pignatti *op. cit.* 214(1973):
1 *G. scariosa.*

GLYCYRRHIZA

aspera Pall.
subsp. **iconica** (Hub.-Mor.) Ponert in *Feddes Repert.*, 83(9–10):633(1972 publ. 1973): *G.*
2 *iconica.*
glabra L.
subsp. **glandulifera** (Waldst. & Kit.) Ponert *op.*
1, 2 *cit.* 633(1972 publ. 1973): *G. glandulifera.*

GOMPHOLOBIUM

2757 Lebler, B. A.
Wedge peas of south-eastern Queensland. *Queensl.*
7 *Agric. J.*, 99(5):239–242(1973)—Illustrations.

HARPALYCE

2758 Irwin, H. S. *and* Arroyo, M. T. K.
New endemic species of *Harpalyce* (Leguminosae: Brongniartieae) from south-central Brazil with a key to the Brazilian species. *Brittonia*, 25(1):21–25(1973)—
4 Illustrations.

parvifolia H. S. Irwin & Arroyo in *Brittonia*, 25(1):
4 21(1973)—Brazil.
robusta H. S. Irwin & Arroyo *op. cit.* 23(1973)—
4 Brazil.

HEDYSARUM

pogonocarpum Boiss.
subsp. **pannosum** (Boiss.) Ponert in *Feddes Repert.*, 83(9–10):642 (1972 publ. 1973): *Onobrychis*
2 *pannosum.*
varium Willd.
subsp. **leucocladum** (Boiss.) Ponert *op. cit.* 641
2 (1972 publ. 1973): *H. leucocladum.*
subsp. **nitidum** (Willd.) Ponert *op. cit.* 641(1972
2 publ. 1973): *H. nitidum.*
subsp. **pestalozzae** (Boiss.) Ponert *op. cit.* 641
2 (1972 publ. 1973): *H. pestalozzae.*
subsp. **pluriphylla** Ponert *op. cit.* 641(1972 publ.
2 1973)—Turkey.
subsp. **syriacum** (Boiss.) Ponert *op. cit.* 641(1972
2 publ. 1973): *H. syriacum.*

HIPPOCREPIS

2759 Fearn, G. M.
Hippocrepis comosa L. *J. Ecol.*, 61(3):915–926(1973) (Biological Flora of the British Isles)—Illustrations,
1 dot maps; chromosome numbers.

HYMENOCARPOS

circinnatus (L.) Savi
subsp. **nummularius** (DC.) Ponert in *Feddes Repert.*, 83(9–10):641(1972 publ. 1973): *Medicago*
1, 2 *nummularia.*

INDIGOFERA

2760 Oomman, C. I.
Studies in the Papilionaceae: 1. Male and female gametophytes of *Indigofera tinctoria* Linn. *Proc. Nation. Acad. Sci., India, B*, 41(3):275–280(1971 publ.
2, 4–6 1972)—Illustrations.

KEYSERLINGIA

2761 Yakovlev, G. P.
O rode *Keyserlingia* Bunge (vklyuchaya *Echino-*
sophora Nakai). (On the genus *Keyserlingia* Bunge
(including *Echinosophora* Nakai)). *Bot. Zhurn.*, 58(6):
2, 6 833–838(1973)—Keys.

hortensis (Boiss. & Buhse) Yakovlev in *Bot. Zhurn.*,
2 58(6):837(1973): *Edwardsia hortensis.*
var. **hydaspidis** (Baker) Yakovlev *op. cit.* 838
6 (1973): *Sophora mollis* var. *hydaspidis.*
koreënsis (Nakai) Yakovlev *op. cit.* 838(1973):
2 *Sophora koreënsis.*
mollis (Royle) Boiss.
var. **ferruginea** (Yakovlev) Yakovlev *op. cit.* 837
6 (1973): *Sophora interrupta* var. *ferruginea.*

LATHYRUS

2762 Bässler, M.
Revision der eurasiatischen Arten von *Lathyrus* L.
Sect. *Orobus* (L.) Gren. et Godr. *Feddes Repert.*, 84
(5–6):329–447(1973)—(En); illustrations, dot map;
1–3 chromosome numbers; key.

2763 Geigy *Ltd*: Agricultural Chemicals Division
Leguminosae: legumes. *Geigy Weed Tables*, 54
(1972)—Illustrations, coloured illustrations.
1–5 *Lathyrus.*

2764 Gruber, M.
A propos d'une sous-espèce pyrénéenne méconnue
du *Lathyrus laevigatus* (Waldst. et Kit.) Gren. *Bull. Soc.*
Bot. France, 119(3–4):237–242(1972 publ. 1973)—
1 (En); illustrations.

2765 Moldenke, H. N.
Notes on new and noteworthy plants: 61.
Phytologia, 26(5):355–356(1973). *Lathyrus latifolius* f.
3 *albiflorus.*

2766 Stork, A. L.
Lathyrus sphaericus discovered on Mön, Denmark.
Svensk Bot. Tidskr., 67(4):456–457(1973)—Dot
1 maps.

brachypterus Čelak.
subsp. **hausskncchtii** (Širj.) Ponert in *Feddes*
Repert., 83(9–10):635(1972 publ. 1973): *L.*
2 *haussknechtii.*
cyaneus (Stev.) Koch
subsp. **digitatus** (Bieb.) Ponert *op. cit.* 635(1972
1, 2 publ. 1973): *Orobus digitatus.*
subsp. **pinnatus** (P. H. Davis) Ponert *op. cit.* 635
2 (1972 publ. 1973): *L. cyaneus* var. *pinnatus.*
emodi (Wall. ex Fritsch) Fritsch
var. **majalis** (Fritsch) Bässler in *Feddes Repert.*,
6 84(5–6):359(1973): *Orobus emodi* var. *majalis.*
gorgoni Parl.
subsp. **lineatus** (Post) Ponert in *Feddes Repert.*,
83(9–10):635(1972 publ. 1973): *L. cicera* var.
1, 2 *lineatus.*
subsp. **tiriopolitanus** (Davidoff) Ponert *op. cit.*
2 635(1972 publ. 1973): *L. tiriopolitanus.*
japonicus Willd.
var. **acutifolius** (Bab.) Bässler in *Feddes Repert.*,
1 84(5–6):410(1973): *L. maritimus* β *acutifolius.*
laevigatus (Waldst. & Kit.) Gren.
subsp. **grandifolius** (Boiss. ex Fritsch) Gruber in
Bull. Soc. Bot. France, 119(3–4):237(1972 publ.
1 1973): *Orobus occidentalis* var. *grandifolius.*
latifolius
f. **albiflorus** Moldenke in *Phytologia*, 26(5):355
3 (1973)—U.S.A. (Pennsylvania).

linifolius (Reichard) Bässler
lus. **albiflorus** (Alefeld) Bässler in *Feddes Repert.*,
84(5–6):435(1973): *Orobus tuberosus* var.
1 *albiflorus.*
f. **emarginatus** (Hertzsch) Bässler *op. cit.* 436
1 (1973): *L. montanus* f. *emarginatus.*
f. **pubescens** (Beck) Bässler *op. cit.* 436(1973):
1 *L. montanus* var. *pubescens.*
f. **pyrenaicus** (L.) Bässler *op. cit.* 436(1973):
1 *Orobus pyrenaicus.*
lus. **variegatus** (Rouy) Bässler *op. cit.* 435(1973):
1, 2 *L. macrorrhizus* subvar. *variegatus.*
lycicus Boiss.
subsp. **phaselitanus** (Huber-Morath & Davis)
Ponert in *Feddes Repert.*, 83(9–10):635(1972 publ.
2 1973): *L. phaselitanus.*
nivalis Hand.-Mazz.
subsp. **atropatanus** (Grossh.) Ponert *op. cit.* 635
2 (1972 publ. 1973): *Orobus atropatanus.*
occidentalis (Fischer & Meyer) Fritsch
f. **cirrhosus** (Rouy) Bässler in *Feddes Repert.*, 84
1 (5–6):367(1973): *L. gmelinii* β *cirrhosus.*
subsp. **grandifolius** (Fritsch) Bässler *op. cit.* 368
1 (1973): *L. grandifolius.*
f. **lanceolatus** (Rouy) Bässler *op. cit.* 367(1973): *L.*
1 *occidentalis* β *lanceolatus.*
f. **parvifolius** (Fritsch) Bässler *op. cit.* 366(1973):
1 *Orobus occidentalis* var. *parvifolius.*
subsp. **scopolii** (Fritsch) Bässler *op. cit.* 367(1973):
1 *L. scopolii.*
f. velutinus (St. Lager) Bässler *op. cit.* 367(1973),
1 without basionym ref.: (Basionym not stated.)
palustris L.
f. **angustifolius** (Coutinho) Bässler *op. cit.* 422
1 (1973): *L. nudicaulis* f. *angustifolius.*
f. **latifolius** (Lambertye) Bässler *op. cit.* 420
1, 2 (1973): *L. palustris* var. *latifolius.*
f. **linearifolius** (Ser.) Bässler *op. cit.* 420(1973):
1, 2 *L. palustris* β *linearifolius.*
rotundifolius Willd.
subsp. **undulatus** (Boiss.) Ponert in *Feddes Repert.*,
1, 2 83(9–10):635(1972 publ. 1973): *L. undulatus.*
spathulatus Čelak.
subsp. **elongatus** (Bornm.) Ponert *op. cit.* 635
(1972 publ. 1973): *Orobus sessilifolius* subsp.
2 *elongatus.*
sylvestris L.
subsp. **latifolius** (L.) Ponert *op. cit.* 635(1972 publ.
1, 2 1973): *L. latifolius.*
venetus (Mill.) Wohlf.
lus. **albescens** (Rohlena) Bässler in *Feddes Repert.*,
1 84(5–6):392(1973): *L. venetus* f. *albescens.*
vernus (L.) Bernh.
f. **macranthus** (Rohlena) Bässler *op. cit.* 382(1973):
1 *L. vernus* var. *macranthus.*
lus. **variegatus** (Schuster) Bässler *op. cit.* 382(1973):
1 *L. vernus* f. *variegatus.*

LEMBOTROPIS

2767 Forissier, R.
Recherches cytotaxonomiques préliminaires sur les
genres *Lembotropis*, *Cytisus*, *Chamaecytisus*, *Genista* et
Chamaespartium. *Bull. Soc. Neuchâtel. Sci. Nat.*, 96:51–
1 65(1973)—(Ge,En); chromosome numbers.

2768 Geideman, T. S. *and* Istratii, A. I. *
(Présence de *Lembotropis nigricans* (L.) G. en
Moldavie). *Okhrana Prirody Moldav.*, no.10:50–53
(1972)—Rus. Ref. in *Bull. Signal.*, *Biol. Physiol. Vég.*,
1 34(10):9611(1973).

LENS

2769 Sinha, S. S. H. *and* Acharia, S. S.
Karyotype analysis in some varieties of *Lens culinaris*. *J. Bihar Bot. Soc.*, 1(1–2):70–71(1972).
2 Abstract.

2770 Zohary, D.
The wild progenitor and the place of origin of the cultivated lentil: *Lens culinaris*. *Econ. Bot.*, 26(4):326–
2 332(1973)—Illustration.

culinaris Medikus
subsp. **orientalis** (Boiss.) Ponert in *Feddes Repert.*, 83(9–10):634(1972 publ. 1973): *Ervum*
1, 2 *orientalis.*

LOTUS

2771 Borsos, O.
Cytophotometric studies on the DNA contents of diploid *Lotus* species. *Acta Bot. Acad. Sci. Hung.*,
1 18(1–2):49–58(1973)—Chromosome numbers.

2772 Breitfeld, C.
Lotus maculatus, eine bisher unbeschriebene Art von Tenerife. *Cuad. Bot. Canar.*, 17:27–31(1973)—
1 Illustration.

2773 Cheng, R. I-J. *and* Grant, W. F.
Species relationships in the *Lotus corniculatus* group as determined by karyotype and cytophotometric analyses. *Canad. J. Genet. Cytol.*, 15(1):101–115
1, 2 (1973)—Chromosome numbers.

2774 Chrtková-Žertová, A.
A monographic study of *Lotus corniculatus* L.: 1. Central and Northern Europe. *Rozpr. Česk. Akad. Věd, Mat. Přr.*, 83(4).94p.(1973)—Illustrations;
1 key.

2775 Laínz, M.
Lotus subbiflorus Lag. et *L. castellanus* Boiss. & Reut.: feliciter, binomina in consueto usu retinenda.
1 *Candollea*, 28(1):13–14(1973)—(Fr,En,Ge).

2776 Kunkel, M. A.
On the rediscovery of *Lotus kunkelii*. *Cuad. Bot.*
1 *Canar.*, 17:33–34(1973).

angustissimus L.
subsp. **palustris** (Willd.) Ponert in *Feddes Repert.*,
1, 2 83(9–10):640(1972 publ. 1973): *L. palustris.*
corniculatus L.
subsp. **brachyodon** (Boiss.) Ponert *op. cit.* 640
1, 2 (1972 publ. 1973): *L. corniculatus* var. *brachyodon.*
subsp. **preslii** (Tenore) Ponert *op. cit.* 640(1972
1, 2 publ. 1973): *L. preslii.*
gebelia Vent.
subsp. **hirsutissimus** (Ledeb.) Ponert *op. cit.* 641
1, 2 (1972 publ. 1973): *L. corniculatus* var. *hirsutissimus.*
subsp. **libanoticus** (Boiss.) Ponert *op. cit.* 640
2 (1972 publ. 1973): *L. libanoticus.*
hakkariensis Ponert *op. cit.* 641(1972 publ. 1973)—
2 Turkey.
maculatus Breitfeld in *Cuad. Bot. Canar.*, 17:27
1 (1973)—Canary Is.
malatayicus Ponert in *Feddes Repert.*, 83(9–10):641
2 (1972 publ. 1973)—Turkey.
peregrinus L.
subsp. **carmeli** (Boiss.) Ponert *op. cit.* 640(1972
2 publ. 1973): *L. carmeli.*

LUPINUS

2777 Cox, B. J.
A chemosystematic comparison of the *Lupinus lepidus–L. caespitosus* complex. *Bull. Torrey Bot. Club*,
3 100(1):12–17(1973).

2778 Cox, B. J.
Protein relationship among the perennial caespitose lupines. *Bull. Torrey Bot. Club*, 100(3):153–158(1973)—
3 *Lupinus* in North America.

2779 Cox, B. J.
The subspecies of *Lupinus culbertsonii* and *L. cusickii* (Leguminosae). *Madroño*, 22(4):169–177(1973)—
3 Illustrations, dot maps; chromosome numbers; key.

2780 Dunn, D. B. *and* Cox, B. J.
A new subspecies of *Lupinus burkei*. *Madroño*, 22(1):
3 25–29(1973).

2781 Elliott, M. H.
Reproductive barriers in the micranthi group of *Lupinus* L. and patterns of differentiation in the species *Lupinus nanus* Dougl. *Diss. Abstr. Int.*, *B*, 33(10):
3 4689(1973).

2782 Scogin, R.
Leucine amonopeptidase polymorphism in the genus *Lupinus* (Leguminosae). *Bot. Gaz.*, 134(1):
1–6 73–76(1973)—Chromosome numbers.

albus
subsp. **termis** (Forssk.) Ponert in *Feddes Repert.*,
2 83(9–10):619(1972 publ. 1973): *L. termis.*
burkei Wats.
subsp. **caeruleomontanus** D. B. Dunn & B. J. Cox
3 in *Madroño*, 22(1):26(1973)·—U.S.A. (Oregon).
culbertsonii Greene
subsp. **hypolasius** (Greene) B. J. Cox in *Madroño*,
3 22(4): 171 (1973): *L. hypolosius*
cusickii S. Wats.
subsp. **abortivus** (Greene) B. J. Cox *op. cit.* 175
3 (1973): *L. abortivus.*
subsp. **brachypodus** (Piper) B. J. Cox *op. cit.* 176
3 (1973): *L. brachypodus.*

MACHAERIUM

2783 Rudd, V. E.
New taxa and combinations in *Machaerium*
4 (Leguminosae): 3. *Phytologia*, 25(6):398–403(1973).

2784 Rudd, V. E.
New taxa and combinations in *Machaerium*
4 (Leguminosae): 4. *Phytologia*, 23(3):100–102(1973).

2785 Yakovlev, G. P.
Zametki po sistematike rodov *Sweetia* Spreng., *Machaerium* Pers., *Angylocalyx* Taub., *Fedorovia* Yakovl., *Placolobium* Miq., *Ormosia* Jacks. (Fabaceae). (De generibus *Sweetia* Spreng., *Machaerium* Pers., *Angylocalyx* Taub., *Fedorovia* Yakovl., *Placolobium* Miq. et *Ormosia* Jacks. (Fabaceae) notulae systematicae).
2, 6 *Novosti Sist. Vўssh. Rast.*, 10:190–196(1973).

acutifolium Vogel
var. **enneandrum** (Hoehne) Rudd in *Phytologia*,
4 25(6):399(1973): *M. enneandrum.*
var. **pseudacutifolium** (Pittier) Rudd *op. cit.* 399
4 (1973): *M. pseudacutifolium.*
atratum (Mohlenb.) Yakovlev in *Novosti Sist. Vўssh.*
4 *Rast.*, 10:191(1973): *Sweetia atrata.*
brasiliense Vogel
var. **erianthum** (Bentham) Rudd in *Phytologia*,
4 25(6):399(1973): *M. erianthum.*

4 **conzattii** Rudd *op. cit.* 399(1973)—Mexico.
costulatum Rudd in *Phytologia*, 26(2):101(1973)—
4 Brazil.
cuzcoense Rudd in *Phytologia*, 25(6):403(1973)—
4 Peru.
4 **fluminense** Rudd *op. cit.* 400(1973)—Brazil.
4 **froesii** Rudd *op. cit.* 400(1973)—Brazil.
4 **guanaiense** Rusby ex Rudd in *Phytologia*, 26(2):
4 101(1973)—Bolivia.
4 **hatschbachii** Rudd *op. cit.* 100(1973)—Brazil.
huanucoense Rudd in *Phytologia*, 25(6):401(1973)—
4 Peru.
mutisii Killip ex Rudd *op. cit.* 401(1973)—
4 Colombia.
nyctitans (Vellozo) Bentham
var. **gardneri** (Bentham) Rudd in *Phytologia*,
4 26(2):100(1973): *M. gardneri*.
parvifolium Killip ex Rudd in *Phytologia*, 25(6):402
4 (1973)—Colombia.
penduliflorum Rudd in *Phytologia*, 26(2):102
4 (1973)—Peru, Bolivia.
subrhombiforme Rudd in *Phytologia*, 25(6):402
4 (1973)—Bolivia.
4 **uribei** Rudd *op. cit.* 403(1973)—Colombia.

MACROTYLOMA

2786 Dardenne, G. A., Thonart, P., Otoul, E. *and* Marechal, R.
Etude chimiotaxonomique dans les genres *Macrotyloma, Dolichos* et *Pseudovigna*. *Phytochemistry*,
5–7 12(8):1983–1992(1973)—(En).

MARINA

2787 Barneby, R. C.
New names in *Dalea* and *Marina* (Fabaceae tribus
4 Amorpheae). *Phytologia*, 26(1):1–2(1973).

neglecta (B. L. Robinson) Barneby in *Phytologia*,
4 26(1):2(1973): *Dalea neglecta*.
var. **elongata** (Rose) Barneby *op. cit.* 2(1973):
4 *Parosela elongata*.
nutans (Cavanilles) Barneby *op. cit.* 2(1973):
4 *Psoralea nutans*.

MEDICAGO

2788 Ignasiak, T. *and* Lesins, K.
Carotenoids in petals of some perennial *Medicago*
1, 2, 5 species. *Biochem. Syst.*, 1(2):97–100(1972).

2789 Kałasa-Balicka, M.
Development of microsporangia in di- and tetraploid *Trifolium, Melilotus, Medicago* and *Trigonella* species and forms. (Rozwoj mikrosporangiów u di- i tetraploidalnych gatunków i form z rodzaju *Trifolium, Melilotus, Medicago, Trigonella.*) *Genet. Pol.*, 14(4): 389–395(1973)—(Pol,Rus); chromosome
1–7 numbers.

2790 Kałasa-Balicka, M.
Microsporogenesis in di- and tetraploid species of *Melilotus, Trifolium, Medicago* and *Trigonella* genera. (Mikrosporogeneza di- i tetraploidalnych gatunków z rodzaju *Melilotus, Trifolium, Medicago* and *Trigonella.*).
1–7 *Genet. Pol.*, 14(3):305–320(1973)—(Pol,Rus).

2791 Kizene, B. P. *and* Lapinskene, N. A.
Medicago falcata L. i ee soobshchestva na kholmakh Baltiĭskoĭ gryadȳ Litovskoĭ SSR. (*Medicago falcata* L. and its communities on the hills of the Baltic ridge in the Lithuanian SSR.). *Bot. Zhurn.*, 58(8):1172–1184
1 (1973).

2792 Lesins, K. *and* Singh, S. M.
Relationships of taxa in the genus *Medicago* as revealed by hybridization: 7. *M. tarnata* complex. *Canad. J. Genet. Cytol.*, 15(2):321–325(1973)—
1, 2 Illustrations.

2793 Mitsevski, K.
Prilog kon taksonomijata i horologijata na *Medicago prostrata* Jacq.-agg. vo Makedonija. (Beitrag zur Taxonomie und Chorologie des *Medicago prostrata*-agg. in Mazedonien). *God. Zborn., Biol.*, 25:137–147
1 (1973)—(Ge); illustrations, dot map.

2794 Terziĭsku, D. P. *and* Poddubnaya-Arnol'di, V. A.
Tsitoémbriologicheskoe issledovanie lyutsernȳ rodopskoĭ (*Medicago rhodopaea* Vel.). *Byull. Glavn.*
1 *Bot. Sada* (Moscow), 89:20–29(1973).

2795 Vollrath, H.
Medicago sativa in Mitteleuropa angebaut und verwildert? *Göttinger Flor. Rundbr.*, 7(1):9–13
1 (1973).

glutinosa Bieb.
subsp. **papillosa** (Boiss.) Ponert in *Feddes Repert.*,
2 83(9–10):639(1972 publ. 1973): *M. papillosa*.
intertexta (L.) Miller
subsp. **ciliaris** (L.) Ponert *op. cit.* 640(1972 publ.
1, 2 1973): *M. polymorpha* var. *ciliaris*.
minima (L.) Bart.
subsp. **brevispina** (Benth.) Ponert *op. cit.* 639
1, 2 (1972 publ. 1973): *M. minima* var. *brevispina*.
polymorpha L.
subsp. **hispida** (Gaertn.) Ponert *op. cit.* 639(1972
1, 2 publ. 1973): *M. hispida*.
prostrata Jacq.
f. **glandulosa** Mitsevski in *God. Zborn., Biol.*, 25:
1 141(1973)—Jugoslavia.
var. **matkae** Mitsevski *op. cit.* 144(1973)—
1 Jugoslavia.
1 f. **pilosa** Mitsevski *op. cit.* 143(1973)—Jugoslavia.
subsp. **pseudorupestris** (Hayek) Mitsevski *op. cit.*
1 140(1973): *M. pseudorupestris*.
var. **topolkae** Mitsevski *op. cit.* 144(1973)—
1 Jugoslavia.
subsp. **vukovicii** Mitsevski *op. cit.* 142(1973)—
1 Jugoslavia.
rigidula (L.) All.
subsp. **agrestis** (Burniat) Ponert in *Feddes Repert.*, 83(9–10):640(1972 publ. 1973): *M. rigidula* var.
1, 2 *agrestis*.
subsp. **cinerascens** (Jord.) Ponert *op. cit.* 639
2 (1972 publ. 1973): *M. cinerascens*.
subsp. **constricta** (Durieu) Ponert *op. cit.* 640
2 (1972 publ. 1973): *M. constricta*.
var. **submitis** (Boiss.) Ponert *op. cit.* 640(1972 publ.
1, 2 1973): *M. gerardii* var. *submitis*.
subsp. **zoharyi** Ponert *op. cit.* 640(1972 publ.
2 1973)—Turkey.
rotata Boiss.
subsp. **blancheana** (Boiss.) Ponert *op. cit.* 639
1, 2 (1972 publ. 1973): *M. blancheana*.
subsp. **bonarotiana** (Arc.) Ponert *op. cit.* 639
2 (1972 publ. 1973): *M. bonarotiana*.
subsp. **eliezeri** (Eig) Ponert *op. cit.* 639(1972 publ.
2 1973): *M. rotata* var. *eliezeri*.
truncatula Gaertn.
subsp. **littoralis** (Rodhe ex Lois.) Ponert *op. cit.*
1, 2 639(1972 publ. 1973): *M. littoralis*.
subsp. **longiaculeata** (Urb.) Ponert *op. cit.* 639
2 (1972 publ. 1973): *M. truncatula* var. *longiaculeata*.
turbinata (L.) All.
subsp. **spinulosa** (DC.) Ponert *op. cit.* 640(1972
2 publ. 1973): *M. spinulosa*.

MELILOTUS

2796 Kałasa-Balicka, M.
Development of microsporangia in di- and tetraploid *Trifolium, Melilotus, Medicago* and *Trigonella* species and forms. (Rozwoj mikrosporangiów u di-tetraploidalnych gatunków i form z rodzaju *Trifolium, Melilotus, Medicago, Trigonella.*) *Genet. Pol.*, 14(4):
1–7 389–395(1973)—(Pol,Rus); chromosome numbers.

2797 Kałasa-Balicka, M.
Microsporogenesis in di- and tetraploid species of *Melilotus, Trifolium, Medicago* and *Trigonella* genera. (Mikrosporogeneza di- i tetraploidalnych gatunków z rodzaju *Melilotus, Trifolium, Medicago* and *Trigonella*).
1–7 *Genet. Pol.*, 14(3):305–320(1973)—(Pol,Rus).

2798 Sano, Y. *and* Kita, F.
Studies of interspecific hybrids in the genus *Melilotus* from the plant breeding standpoint: 8. Cytological studies of the interspecific hybrids *Melilotus alba* × *M. suaveolens, M. polonica* × *M. suaveolens,* and *M. hirsuta* × *M. suaveolens. Mem. Fac.*
1, 2 *Agric. Hokkaido Univ.*, 8(4):411–419(1973)—(En).

2799 Sano, Y. *and* Kita, F.
Studies of interspecific hybrids in the genus *Melilotus* from the plant breeding standpoint: 9. Cytological studies of the interspecific hybrids. *Melilotus taurica* × *M. alba, M. taurica* × *M. wolgica,* and *M. dentata* × *M. taurica. Mem. Fac. Agric.*
1, 2 *Hokkaido Univ.*, 9(1):1–9(1973)—(En).

MOGHANIA

2800 Singh, H. B. *and* Arora, R. K.
Soh-phlong, *Moghania vestita*: a leguminous root crop of India. *Econ. Bot.*, 27(3):332–338(1973)—
6 Illustrations, dot maps.

MUCUNA

2801 Ishikura, N. *and* Shibata, M.
Paper chromatographic survey of anthocyanins in Leguminosae: 2. On the flower pigments of *Mucuna*
2 *sempervirens. Bot. Mag. Tokyo*, 86(1001):1–4(1973).

ONOBRYCHIS

2802 Rechinger, K. H.
Onobrychis aliacmonia, eine neue griechisch-mazedonische Art aus der Sektion *Hymenobrychis.* (*Onobrychis aliacmonia*, a new species of sect. *Hymenobrychis* from Greek Macedonia). *Ann. Mus. Goulandris*, 1:127–131(1973)—(En,Gre);
1 illustration.

2803 Vlasov, M. I.
Polimorfizm éspartseta kurinskogo (*Onobrychis cyri* Grossh.) (The polymorphism of *Onobrychis cyri* Grossh.). *Trudy̆ Prikl. Bot. Genet. Selek.*, 49(1):81–84
2 (1973)—(En).

alba (W. & K.) Desv.
subsp. **tenoreana** (Lacaita) Pignatti in *Giorn. Bot.*
1 *Ital.*, 107(5):219(1973): *O. tenoreana.*
aliacmonia K. H. Rechinger in *Ann. Mus. Goulandris,*
1 1:128(1973)—Greece.
brnoensis Ponert in *Feddes Repert.*, 83(9–10):642 (1972 publ. 1973), nom. nov.: *O. megataphros* var.
2 *podperae* Širj. & Nabělek.
davisii Ponert *op. cit.* 642(1972 publ. 1973), as
2 'davisi'—Turkey.
hakkariensis Ponert *op. cit.* 643(1972 publ. 1973)—
2 Turkey.
huetiana Boiss.
subsp. **bornmuelleri** (Freyn.) Ponert *op. cit.* 643
2 (1972 publ. 1973): *O. bornmuelleri.*
laxiflora Baker
var. **schugnanica** (B. Fedtsch.) Ali in R. R. Stewart, Annot. Cat. Vasc. Pl. W. Pakistan &
6 Kashmir: 414(1972): *O. schugnanica.*
megataphros Boiss.
subsp. **major** (Boiss.) Ponert in *Feddes Repert.*, 83(9–10):642(1972 publ. 1973): *O. balansae* var.
2 *major.*
ornata (Willd.) Desv.
subsp. **atropatana** (Boiss.) Ponert *op. cit.* 643
2 (1972 publ. 1973): *O. atropatana.*
subsp. **vanensis** (Širj.) Ponert *op. cit.* 643(1972
2 publ. 1973): *O. atropatana* var. *vanensis.*
oxyodonta Boiss.
subsp. **armena** (Boiss. & Huet) Ponert *op. cit.* 642
2 (1972 publ. 1973): *O. armena.*
subsp. **bithynica** (Širj.) Ponert *op. cit.* 642(1972
2 publ. 1973): *O. bithynica.*
paucijuga Bornm.
subsp. **densipilosa** Ponert *op. cit.* 643(1972 publ.
2 1973)—Turkey.
subsp. **subglabra** Ponert *op. cit.* 643(1972 publ.
2 1973)—Turkey.
reesei Ponert *op. cit.* 642(1972 publ. 1973)—
2 Turkey.
sulphurea Boiss. & Ball
subsp. **pallida** (Boiss. & Ky.) Ponert *op. cit.* 642
2 (1972 publ. 1973): *O. pallida.*
tournefortii Desv.
subsp. **cappadocica** (Boiss.) Ponert *op. cit.* 643
2 (1972 publ. 1973): *O. cappadocica.*
f. **concolor** (Širj.) Ponert *op. cit.* 643(1972 publ.
2 1973): *O. cappadocica* f. *concolor.*
subsp. **nitida** (Boiss.) Ponert *op. cit.* 643(1972 publ.
2 1973): *O. nitida.*
vanensis (Hedge) Ponert *op. cit.* 642(1972 publ.
2 1973): *O. sulphurea* var. *vanensis.*
viciifolia Scop.
subsp. **altissima** (Grossh.) Ponert *op. cit.* 642(1972
2 publ. 1973): *O. altissima.*

ONONIS

2804 Gutterman, Y. *and* Heydecker, W.
Studies in the surfaces of desert plant seeds: 1. Effect of day length upon maturation of the seed-coat of *Ononis sicula* Guss. *Ann. Bot.* (U.K.), 37(153):
1, 2 1049–1050(1973).

2805 Villar-Pérez, L.
Comunidades de *Ononis fruticosa* en la parte subcantábrica de Aragón y Navarra. *Pirineos*, no. 105:
1 61–68(1972).

aragonensis Asso
subsp. reuteri (Bss.) Malagarriga, Las Subesp. y Variac. Geogr.: 11(1973), without basionym
1 date: *O. reuteri.*
natrix L.
subsp. pyrenaica (Wk. & Costa) Malagarriga *op. cit.*
1 11(1973), without basionym date: *O. pyrenaica.*
pusilla L.
subsp. saxicola (Bss. R.) Malagarriga *op. cit.* 11
1 (1973), without basionym date: *O. saxicola.*
striata Gouan
subsp. **cephalotes** (Bss.) Malagarriga *op. cit.* 11
1 (1973): *O. cephalotes.*

ORMOSIA

2806 Yakovlev, G. P.
Zametki po sistematike rodov *Sweetia* Spreng., *Machaerium* Pers., *Angylocalyx* Taub., *Fedorovia* Yakovl., *Placolobium* Miq., *Ormosia* Jacks. (Fabaceae). (De generibus *Sweetia* Spreng., *Machaerium* Pers., *Angylocalyx* Taub., *Fedorovia* Yakovl., *Placolobium* Miq. et *Ormosia* Jacks. (Fabaceae) notulae systematicae).
2, 6 *Novosti Sist. Vÿssh. Rast.*, 10:190–196(1973).

OXYTROPIS

2807 Ali, S. I.
New taxa and combinations in the genus *Oxytropis* (Leguminosae) from Pakistan. *Kew Bull.*, 28(2):305–
6 309(1973)—Illustrations.

2808 Looman, J.
Biological flora of the Canadian Prairie Provinces: 1. *Oxytropis besseyi* (Rydb.) Blank. *Canad. J. Pl. Sci.*, 53(3):677–687(1973)—(Fr); illustrations, map, dot
3 map.

2809 Mitsevski, K.
Nekolku nepoznati i retki *Astragalus*- i *Oxytropis*-vidovi za florata na Makedonija. (Einige unbekannte und seltene *Astragalus*- und *Oxytropis*-Arten der Flora von Mazedonien). *God. Zborn., Biol.*, 25:157–160 (1973)—(Ge); dot map. *Oxytropis lapponica*, *O.*
1 *purpurea*.

birirensis Ali in *Kew Bull.*, 28(2):306(1973)—
6 Pakistan.
hypoglottoides (Bak.) Ali *op. cit.* 306(1973):
2, 6 *Astragalus hypoglottoides*.
immersa (Baker ex Aitchison) Bunge & B. Fedtsch.
var. **jinaliensis** Ali *op. cit.* 308(1973)—
6 Pakistan.
luteo-caerulea (Bak.) Ali *op. cit.* 306(1973)—
2, 6 *A. luteo-caeruleus*.
sikaramensis (Širj. & Rech. f.) Ali *op. cit.* 305(1973):
2, 6 *A. sikaramensis*.
6 **staintoniana** Ali *op. cit.* 307(1973)—Pakistan.

PETALOSTEMON

2810 Baskin, J. M. *and* Baskin, C. C.
The past and present geographical distribution of *Petalostemon foliosus* and notes on its ecology. *Rhodora*,
3 75(801):132–140(1973)—Map.

2811 Turner, B. F. H.
Allelopathic effects of *Petalostemon gattingeri* on the distribution of *Arenaria patula* in cedar glades. *Diss.*
3 *Abstr. Int., B*, 33(12):5777(1973).

PHASEOLUS

2812 Arora, R. K., Chandel, K. P. S. *and* Joshi, B. S.
Morphological diversity in *Phaseolus sublobatus*
6 Roxb. *Curr. Sci.*, 42(10):359–361(1973)—Illustrations.

2813 Salgare, S. A.
On the early embryogeny of *Phaseolus aconitifolius*.
2, 6 *Sci. Cult.*, 39(7):315–316(1973).

2814 Smartt, J., Kloz, J. *and* Klozová, E.
Segregation of serological characters in the cross *Phaseolus coccineus* L. × *Ph. vulgaris* L. *Biol. Pl.*, 15(6):
1–7 369–373(1973).

PHYLLODIUM

subgen. **Prainia** Ohashi in *Ginkgoana*, 1:266
6 (1973).
kurzianum (O. Kuntze) Ohashi *op. cit.* 272 (1973):
2, 6 *Meibomia kurziana*.

PISUM

2815 Ben-Ze'ev, N. *and* Zohary, D.
Species relationships in the genus *Pisum* L. *Israel J. Bot.*, 22(2):73–91(1973)—Illustrations;
1, 2 chromosome number.

pumilio (Meikle) Greuter in *Ann. Mus. Goulandris*,
1, 2 1:42(1973): *P. sativum* var. *pumilio*.
sativum L.
subsp. **brevipedunculatum** (Davis & Meikle) Ponert in *Feddes Repert.*, 83(9–10):636(1972 publ.
1, 2 1973): *P. sativum* var. *brevipedunculatum*.
subsp. **pumilio** (Meikle) Ponert *op. cit.* 636(1972
1, 2 publ. 1973): *P. sativum* var. *pumilio*.

PLACOLOBIUM

2816 Yakovlev, G. P.
Zametki po sistematike rodov *Sweetia* Spreng., *Machaerium* Pers., *Angylocalyx* Taub., *Fedorovia* Yakovl., *Placolobium* Miq., *Ormosia* Jacks. (Fabaceae). (De generibus *Sweetia* Spreng., *Machaerium* Pers., *Angylocalyx* Taub., *Fedorovia* Yakovl., *Placolobium* Miq. et *Ormosia* Jacks. (Fabaceae) notulae systemati-
2, 6 cae). *Novosti Sist. Vÿssh. Rast.*, 10:190–196(1973).

cambodianum (Gagnep.) Yakovlev in *Novosti Sist.*
6 *Vÿssh. Rast.*, 10:192(1973): *Ormosia cambodiana*.
crassivalve (Gagnep.) Yakovlev *op. cit.* 192(1973):
2, 6 *Ormosia crassivalvis*.
gracile (Prain) Yakovlev *op. cit.* 192(1973): *Ormosia*
6 *gracilis*.
paniculatum (Merr.) Yakovlev *op. cit.* 192(1973):
6 *Ormosia paniculata*.
robustum (Roxb.) Yakovlev *op. cit.* 191(1973):
6 *Sophora robusta*.
tavoyanum (Prain) Yakovlev *op. cit.* 192(1973):
6 *Ormosia tavoyana*.
travancoricum (Bedd.) Yakovlev *op. cit.* 192(1973):
6 *Ormosia travancorica*.
watsonii (C. E. C. Fisher) Yakovlev *op. cit.* 192
6 (1973): *Ormosia watsonia*.

POECILANTHE

ulei (Harms) Arroyo & Rudd in *Phytologia*, 25(6):
4 398(1973): *Machaerium* ? *ulei*.

PSEUDOVIGNA

2817 Dardenne, G. A., Thonart, P., Otoul, E. *and* Marechal, R.
Etude chimiotaxonomique dans les genres *Macrotyloma*, *Dolichos* et *Pseudovigna*. *Phytochemistry*,
5 12(8):1983–1992(1973)—(En).

PSORALEA

2818 Toft, C. A. *and* Welsh, S. L.
A revision of the *Psoralea lanceolata* complex (Leguminosae). *Great Basin Nat.*, 32(2):76–87(1972)—
3 Illustrations, map, dot map; key.

RHYNCHOSIA

2819 Rameharan, E. K., McAndrews, J. H. *and* Grear, J. W.
Pollen morphology of *Rhynchosia* and *Eriosema* (Fabaceae). *Canad. J. Bot.*, 51(12):2395–2399
3, 4 (1973).

SAROTHAMNUS

2820 Faugeras, G., Paris, R. *and* Valdes-Bermejo, E.
Alcaloïdes et polyphénols des Légumineuses: 26. Sur les alcaloïdes du *Sarothamnus patens* (L.) Webb (*Cytisus striatus* (Hill) Rothm.). (Alkaloids and polyphenols of Leguminosae: 26. The alkaloids of

Sarothamnus patens (L.) Webb (*Cytisus striatus* (Hill) Rothm.)). *Ann. Pharm. Franc.*, 30(7–8):527–533(1972) —(En). Also publ. in *Trav. Lab. Matière Méd. Pharm.*
1 *Galénique Fac. Pharm. Paris*, 57(2)(1973).

SESBANIA

2821 Datta, R. M. *and* Bagchi, S.
On the nature of chromosome in meiotic mechanism of a natural tetraploid *Sesbania benthamiana* Domin. *Broteria, Ciênc. Nat.*, 42(1–2):31–37
7 (1973).

2822 Salgare, S. A.
On the megagametophyte of *Sesbania aculeata*
2, 4–6 Poir. *Sci. Cult.*, 39(7):309–311(1973).

SOPHORA

2823 Markham, K. R.
Flavonoids and the chemotaxonomy of three *Sophora* species. *Phytochemistry*, 12(5):1091–1094
4, 7 (1973).

alopecuroides
subsp. **tomentosa** (Boissier) Ponert in *Feddes Repert.*, 83(9–10):618(1972 publ. 1973): *Goebelia*
2 *alopecuroides* var. *tomentosa*.

STRONGYLODON

2824 Polhill, R. M.
Strongylodon macrobotrys. *Curtis's Bot. Mag.*, 179(2): tab.627(1972 publ. 1973)—En; illustration, coloured
6 illustration.

TADEHAGI Ohashi in *Ginkgoana*, 1:280(1973),
2, 6, 7 nom. nov.: *Pteroloma* Desvaux ex Benth.
subgen. **Godefroya** (Schindler) Ohashi *op. cit.* 284
6 (1973): *Droogmansia* subgen. *Godefroya*.
6 subgen. **Kerria** Ohashi *op. cit.* 287(1973).
godefroyanum (O. Kuntze) Ohashi *op. cit.* 284
6 (1973): *Meibomia godefroyana*.
kerrii (Schindler) Ohashi *op. cit.* 287(1973):
6 *Pteroloma kerrii*.
triquetrum (L.) Ohashi *op. cit.* 290(1973): *Hedysarum*
2, 6 *triquetrum*.
subsp. **alatum** (DC.) Ohashi *op. cit.* 292(1973):
6 *Desmodium alatum*.
subsp. **auriculatum** (DC.) Ohashi *op. cit.* 293
6 (1973): *Desmodium auriculatum*.
subsp. **pseudotriquetrum** (DC.) Ohashi *op. cit.*
2, 6 295(1973): *Desmodium pseudotriquetrum*.
subsp. **rodgeri** (Schindler) Ohashi *op. cit.* 295
6 (1973): *Pteroloma rodgeri*.

TEPHROSIA

2825 Lebrun, J.-P.
Tephrosia paucijuga Harms (Leguminosae) new to the flora of West Africa. *Kew Bull.*, 28(1):140
5 (1973).

TETRAGONOLOBUS

2826 Cherevko, M. V.
Pro nove mistseznakhodzhennya Tetrahonolobusa struchkovoho [*Tetragonolobus siliquosus* (L.) Roth.] na Prўkarpatti. (On new occurrence of *Tetragonolobus siliquosus* (L.) Roth. in the Forecarpathians). *Ukr. Bot.*
1 *Zhurn.*, 30(3):387(1973)—Uk(En).

purpureus Moench.
subsp. **palaestinus** (Boiss.) Ponert in *Feddes Repert.*, 83(9–10):641(1972 publ. 1973): *Lotus*
2 *palaestinus*.

THERMOPSIS

2827 Scogin, R.
Isoenzyme polymorphism in the genus *Thermopsis*.
3 *Biochem. Syst.*, 1(2):79–81(1973).

TRIFOLIUM

2828 Géhu, J.-M.
Trifolium occidentale D. E. Coombe: espèce nouvelle pour le littoral du nord Portugal. *Agron. Lusit.*, 34(3): 197–204(1972 publ. 1973)—(Por); chromosome
1 numbers.

2829 Geigy *Ltd*: Agricultural Chemicals Division
Leguminosae: legumes. *Geigy Weed Tables*, 56
1–5 (1972)—Illustrations, coloured illustrations. *Trifolium*.

2830 Gillett, J. M., Bassett, I. J. *and* Crompton, C. W.
Pollen morphology and its relationship to the taxonomy of North American *Trifolium* species. *Pollen Spores*, 15(1):91–108(1973)—(Fr); chromosome
3 numbers.

2831 Gillett, J. M. *and* Cochrane, T. S.
Preliminary reports on the flora of Wisconsin: no.63. The genus *Trifolium*—the clovers. *Trans. Wis. Acad. Sci. Arts Lett.*, 61:59–74(1973)—Illustrations, dot
3 maps; chromosome numbers; key.

2832 González Bernáldez, F., Sánchez de Rivera, A. M. *and* Angulo, M. D.
Estudios cromosomicos en el genero *Trifolium*: 4. *Lagascalia*, 3(2):195–203(1973)—(En); chromosome
1 numbers.

2833 Kałasa-Balicka, M.
Development of microsporangia in di- and tetraploid *Trifolium*, *Melilotus*, *Medicago* and *Trigonella* species and forms. (Rozwoj mikrosporangiów u di- i tetraploidalnych gatunków i form z rodzaju *Trifolium*, *Melilotus*, *Medicago*, *Trigonella*.) *Genet. Pol.*, 14(4):389–395(1973)—(Pol,Rus);
1–7 chromosome numbers.

2834 Kałasa-Balicka, M.
Microsporogenesis in di- and tetraploid species of *Melilotus*, *Trifolium*, *Medicago* and *Trigonella* genera. (Mikrosporogeneza di- i tetraploidalnych gatunków z rodzaju *Melilotus*, *Trifolium*, *Medicago* and *Trigonella*.).
1–7 *Genet. Pol.*, 14(3):305–320(1973)—(Pol,Rus).

2835 Kolev, I. D. *and* Sedmakova-Louchanska, E. *
A karyological study of *Trifolium arvense* L., *T. preslianum* Boiss. and *T. smyrnaeum* Boiss. in Bulgaria. *Compt. R. Acad. Agric. Georgi Dimitrov*, 5(1):39–43(1972)—Chromosome numbers. Abstr. in
1 *Herb. Abstr.*, 43(1):27(1973).

2836 Kolev, I. D. *and* Sedmakova-Louchanska, E. *
A karyological study of *Trifolium arvense* L., *T. preslianum* Boiss. and *T. smyrnaeum* Boiss. in Bulgaria. (Une étude caryologique de *T. a.* L., *T. p.* B. et *T. s.* B. en Bulgarie). *C.R. Acad. Agric. Georgi Dimitrov*, 5(1):39–43(1972). Abstr. in *Bull. Signal.*,
1 *Biol. Physiol. Vég.*, 34(1):136(1973).

2837 Putiyevsky, E. *and* Katznelson, J.
Cytogenetic studies in *Trifolium* spp. related to berseem: 1. Intra- and interspecific hybrid seed
1, 2 formation. *Theor. Appl. Genet.*, 43(8):351–358(1973).

2838 Toma, C.
Contribuţii la studiul morfologic şi histoanatomical stipelelor de la speciile de *Trifolium* L. din România. *An. Şti. Univ. Al. I. Cuza Iaşi, Biol.*, 19(2):317–337
1 (1973)—(Fr); illustrations.

alexandrinum L.
subsp. **berytheum** (Boiss. & Bl.) Ponert in *Feddes Repert.*, 83(9–10):638(1972 publ. 1973): *T.*
2 *berytheum.*
subsp. **vavilovii** (Eig) Ponert *op. cit.* 638(1972
2 publ. 1973): *T. vavilovii.*
angustifolium L.
subsp. **desvauxii** (Boiss. & Bl.) Ponert *op. cit.* 637
1, 2 (1972 publ. 1973): *T. desvauxii.*
subsp. **gibellianum** Pignatti in *Giorn. Bot. Ital.*, 107(5):217(1973), nom. nov.: *T. intermedium*
1 Guss.
subsp. **intermedium** (Guss.) Ponert in *Feddes Repert.*, 83(9–10):637(1972 publ. 1973):
1, 2 *T. intermedium.*
subsp. **pamphylicum** (Boiss. & Heldr.) Ponert
2 *op. cit.* 638(1972 publ. 1973): *T. pamphylicum.*
subsp. **purpureum** (Lois.) Ponert *op. cit.* 637(1972
1, 2 publ. 1973): *T. purpureum.*
apertum Bobrov
subsp. **kilaeum** (Zoh. & Lern.) Ponert *op. cit.* 638
1, 2 (1972 publ. 1973): *T. apertum* var. *kilaeum.*
arvense L.
subsp. **gracile** (Thuill.) Ponert *op. cit.* 637(1972
1, 2 publ. 1973): *T. gracile.*
f. **komlodiae** Soó in *Acta Bot. Acad. Sci. Hung.*,
1 18(1–2):174(1973)—Hungary.
echinatum Bieb.
subsp. **carmelii** (Boiss.) Ponert in *Feddes Repert.*, 83(9–10):638(1972 publ. 1973), as 'carmeli':
2 *T. carmelii.*
subsp. **serrulatum** (Hossain) Ponert *op. cit.* 638
2 (1972 publ. 1973): *T. carmelii* var. *serrulatum.*
fragiferum L.
subsp. **pulchellum** (Lange) Ponert *op. cit.* 636
1, 2 (1972 publ. 1973): *T. fragiferum* var. *pulchellum.*
glanduliferum Boiss.
subsp. **nervulosum** (Boiss. & Heldr.) Ponert *op.*
1, 2 *cit.* 637(1972 publ. 1973): *T. nervulosum.*
medium L.
subsp. **eriocalycinum** (Hausskn.) Ponert *op. cit.* 637(1972 publ. 1973): *T. medium* var.
1, 2 *eriocalycinum.*
michelianum Savi
subsp. **balansae** (Boiss.) Ponert *op. cit.* 638(1972
1, 2 publ. 1973): *T. balansae.*
montanum L.
subsp. **rupestre** (Ten.) Pignatti in *Giorn. Bot. Ital.*,
1 107(5):217(1973): *T. rupestre.*
noricum Wulfen
subsp. **praetutianum** (Guss.) Pignatti *op. cit.* 218
1 (1973): *T. praetutianum.*
physodes Stev. ex Bieb.
f. **sclerorrhizum** (Boiss.) Ponert in *Feddes Repert.*,
1, 2 83(9–10):636(1972 publ. 1973): *T. sclerorrhizum.*
pratense L.
subsp. **expansum** (Waldst. & Kit.) Ponert *op. cit.*
1, 2 637(1972 publ. 1973): *T. expansum.*
subsp. **sativum** (Schreb.) Ponert *op. cit.* 637(1972
1, 2 publ. 1973): *T. pratense* var. *sativum.*
subsp. **semipurpureum** (Strobl) Pignatti in *Giorn. Bot. Ital.*, 107(5):218(1973): *T. pratense* var.
1 *semipurpureum.*
repens L.
subsp. **giganteum** (Lag.-Foss.) Ponert in *Feddes Repert.*, 83(9–10):636(1972 publ. 1973): *T.*
2 *repens* var. *giganteum.*
subsp. **macrorrhizum** (Boiss.) Ponert *op. cit.* 636
2 (1972 publ. 1973): *T. macrorrhizum.*
resupinatum L.
subsp. **suaveolens** (Willd.) Ponert *op. cit.* 636
1, 2 (1972 publ. 1973): *T. suaveolens.*

roussaeanum Boiss.
subsp. **dichroanthum** (Boiss.) Ponert *op. cit.* 638
2 (1972 publ. 1973): *T. dichroanthum.*
rytidosemium Boiss. & Hoh.
subsp. **rivulare** (Boiss. & Bal.) Ponert *op. cit.* 636
2 (1972 publ. 1973): *T. rivulare.*
scabrum L.
subsp. **lucanicum** (Gasparr.) Pignatti in *Giorn.*
1 *Bot. Ital.*, 107(5):217(1973): *T. lucanicum.*
stellatum L.
subsp. **adpressum** (Turrill) Ponert in *Feddes Repert.*, 83(9–10):637(1972 publ. 1973): *T. stellatum* var.
2 *adpressum.*
subsp. **longiflorum** (Gris.) Ponert *op. cit.* 637
2 (1972 publ. 1973): *T. stellatum* var. *longiflorum.*
tomentosum L.
subsp. **bullatum** (Boiss. & Hausskn.) Ponert *op. cit.*
2 636(1972 publ. 1973): *T. bullatum.*
subsp. macrosphaerum (Zohary) Ponert *op. cit.* 637(1972 publ. 1973), without basionym page:
2 *T. bullatum* var. *macrosphaerum.*
trichocephalum Bieb.
subsp. **parantzemae** (Aznav.) Ponert *op. cit.* 637
2 (1972 publ. 1973): *T. parantzemae.*
vesiculosum Savi
subsp. **mutabile** (Portenschl.) Ponert *op. cit.* 638
1, 2 (1972 publ. 1973): *T. mutabile.*

TRIGONELLA

2839 Kałasa-Balicka, M.
Development of microsporangia in di- and tetraploid *Trifolium*, *Melilotus*, *Medicago* and *Trigonella* species and forms. (Rozwoj mikrosporangiów u di- i tetraploidalnych gatunków i form z rodzaju *Trifolium*, *Melilotus*, *Medicago*, *Trigonella*.) *Genet. Pol.*, 14(4): 389–395(1973)—(Pol,Rus); chromosome
1–7 numbers.

2840 Kałasa-Balicka, M.
Microsporogenesis in di- and tetraploid species of *Melilotus*, *Trifolium*, *Medicago* and *Trigonella* genera. (Mikrosporogeneza di- i tetraploidalnych gatunków z rodzaju *Melilotus*, *Trifolium*, *Medicago* and *Trigonella*.).
1–7 *Genet. Pol.*, 14(3):305–320(1973)—(Pol,Rus).

2841 Singh, A.
Studies on the interspecific hybrids of *Trigonella corniculata* L., *T. hamosa* L. and *T. cretica* L. *Genetica*,
6 44(2):264–269(1973)—Illustrations.

2842 Singh, A. *and* Singh, D.
Meiotic study of an autotriploid *Trigonella foenum-graecum*. *Chromosome Inform. Serv.*, no.14:12–14(1973)
1, 2, 6 —Chromosome numbers.

halophila Boiss.
subsp. **arenicola** (Hub.-Mor.) Ponert in *Feddes Repert.*, 83(9–10):639(1972 publ. 1973):
2 *T. arenicola.*
sprunerana Boiss.
subsp. **hierosolymitana** (Boiss.) Ponert *op. cit.*
2 638(1972 publ. 1973): *T. hierosolymitana.*
subsp. **mesopotamica** (Hub.-Mor.) Ponert *op. cit.*
2 638(1972 publ. 1973): *T. mesopotamica.*
subsp. **sibthorpii** (Boiss.) Ponert *op. cit.* 638(1972
2 publ. 1973): *T. sibthorpii.*

ULEX

2843 Conacher, E. R. T.
Ulex minor. *Glasgow Nat.*, 19(1):65(1973)—En.
1 Local records.

VICIA

2844 Chrtková-Žertová, A.
Cytotaxonomic study of the *Vicia cracca* complex: 1. Czechoslovak taxa. *Folia Geobot. Phytotax.*, 8(1): 67–93(1973)—(Ge); illustrations, dot maps;
1 chromosome numbers.

2845 Chrtková-Žertová, A.
The cytotaxonomic study of the *Vicia cracca* complex: 2. The taxa of the southern part of northern Europe. *Folia Geobot. Phytotax.*, 8(3):249–253(1973)—
1 (Ge); chromosome numbers.

2846 Činčura, F.
Angaben über die Zahl und die Morphologie der Chromosomen von *Vicia dumetorum* L. und deren Vergleich mit *Vicia pisiformis* L. (Dannȳe o chisle i morfologii khromosom u *Vicia dumetorum* L. i ikh sopostavlenie s vidom *Vicia pisiformis* L.). *Acta Fac. Rerum Nat. Univ. Comen., Bot.*, 21:103–107(1973)—
1 (Cz,Rus); chromosome numbers.

2847 Cubero, J. I.
Evolutionary trends in *Vicia faba* L. *Theor. Appl.*
1, 2, 6 *Genet.*, 43(2):59–65(1973)—Dot map.

2848 Fouquet, B.
Merkmalsvariabilität bei *Vicia faba* L. 2. Tabellarische Erfassung der Merkmalsvariabilität von Kulturpflanzen. (Variability of characters within *Vicia faba* L. 2. Tabular documentation of the character variability of cultivated plants). *Kulturpflanze*,
1, 2 21:57–60(1973)—(En,Rus).

2849 Geigy *Ltd*: Agricultural Chemicals Division
Leguminosae: legumes. *Geigy Weed Tables*, 55 (1972)—Illustrations, coloured illustrations.
1–5 *Vicia*.

2850 Gonyan, G. G.
Geografiya i ékologiya mnogoletnikh vik Armenii. (Geographical distribution and ecology of perennial vetch in Armenia). *Biol. Zhurn. Arm.*, 26(4):89–98
2 (1973)—(Arm); dot maps.

2851 Hanelt, P.
Merkmalsvariabilität bei *Vicia faba* L.: 1. Künstliche morphologische Systeme bei Kulturpflanzen-Arten. (Variability of characters within *Vicia faba* L.: 1. Artificial morphological classification of cultivated plants). *Kulturpflanze*, 21:55–57,59–60(1973)—
1, 2 (En,Rus).

2852 Killick, H. J.
Where has *Vicia sativa* gone? *Watsonia*, 9(4):
1 428–429(1973)—Key.

2853 Lassetter, J. S.
A biosystematic study of the *Vicia ludoviciana* complex (Leguminosae). *Diss. Abstr. Int., B*, 33(10):
3 4692(1973)—Chromosome number.

2854 Mettin, D. *and* Hanelt, P.
Über Speziationsvorgänge in der Gattung *Vicia* L. (On speciation processes in the genus *Vicia* L.). *Kulturpflanze*, 21:25–54(1973)—(En,Rus); chromo-
1–4 some numbers.

2855 Negrean, G. *and* Dihoru, G.
Cleistogamie subterană la *Vicia amphicarpa* (L. nomen) Dorthes. *Stud. Cerc. Biol., Bot.*, 24(5):377–
1 380(1972)—(Fr); illustration.

2856 Plitmann, U.
Biological flora of Israel: 4. *Vicia sativa* subsp. *amphicarpa* (Dorth.) Aschers. & Graebn. *Israel J. Bot.*, 22(3):178–194(1973)—Illustrations, dot maps;
1, 2 chromosome numbers.

2857 Rousi, A.
Cytotaxonomical studies on *Vicia cracca* L. and *V. tenuifolia* Roth 3. The relation between karyotype and morphology. *Ann. Bot. Fenn.*, 10(2):89–96
1, 2 (1973)—Illustration; chromosome numbers.

2858 Schäfer, H. I.
Zur Taxonomie der *Vicia narbonensis*-Gruppe. (Taxonomy of the *Vicia narbonensis* group). *Kulturpflanze*, 21:211–273(1973)—(En,Rus); illustrations,
1, 2 dot maps; key.

2859 Toma, C., Rugină, R. *and* Diaconu, R.
Contributions à l'étude morphologique et histo-anatomique des stipules chez des espèces de *Vicia* L. de flore de la Roumanie. *An. Şti. Univ. Al. I. Cuza*
1 *Iaşi, Biol.*, 19(1):120–121(1973)—Illustrations.

galilaea Plitm. & Zoh.
var. **faboidea** (Plitm. & Zoh.) Schäfer in *Kultur-*
2 *pflanze*, 21:265(1973): *V. galilaea* subsp. *faboidea*.
hyrcanica Fisch. & Mey.
subsp. **assyriaca** (Boiss.) Ponert in *Feddes Repert.*,
2 83(9–10):633(1972 publ. 1973): *V. assyriaca*.
subsp. **megalodonta** (Rech. fil.) Ponert *op. cit.* 633
2 (1972 publ. 1973): *V. noeana* var. *megalodonta*.
subsp. **noeana** (Reuter ex Boiss.) Ponert *op. cit.*
2 633(1972 publ. 1973): *V. noeana*.
johannis Tamamsch.
var. **ecirrhosa** (M. Pop.) Schäfer in *Kulturpflanze*, 21:
1, 2 261(1973): *V. narbonensis* var. *ecirrhosa*.
var. **procumbens** Schäfer *op. cit.* 261(1973)—
0 Cultivation.
lunata (Boiss. & Bal.) Boiss.
subsp. **grandiflora** (Plitm.) Ponert in *Feddes Repert.*, 83(9–10):633(1972 publ. 1973): *Vicia lunata* var.
2 *grandiflora*.
narbonensis L.
subsp. **faboidea** (Plitm. & Zoh.) Ponert *op. cit.* 634(1972 publ. 1973): *V. galilaea* subsp.
2 *faboidea*.
subsp. **galilaea** (Plitm. & Zoh.) Ponert *op. cit.*
2 634(1972 publ. 1973): *V. galilaea*.
var. **jordanica** Schäfer in *Kulturpflanze*, 21:250
0 (1973)—Cultivation.
var. **salmonea** (Mout.) Schäfer *op. cit.* 253(1973):
1, 2 *V. serratifolia* subsp. *salmonea*.
pannonica Crantz
subsp. **striata** (Bieb.) Ponert in *Feddes Repert.*, 83 (9–10):634(1972 publ. 1973), combination already made in Nyman, Conspectus, 209(1878):
1, 2 *V. striata*.
peregrina L.
subsp. **aintabensis** (Boiss. & Hausskn. ex Boiss.) Ponert *op. cit.* 634(1972 publ. 1973): *V.*
2 *aintabensis*.
subsp. **megalosperma** (Bieb.) Ponert *op. cit.* 634
1, 2 (1972 publ. 1973): *V. megalosperma*.
subsp. **michauxii** (Sprengel) Ponert *op. cit.* 634
2 (1972 publ. 1973): *V. michauxii*.
subsp. **persepolitana** (Boiss.) Ponert *op. cit.* 634
2 (1972 publ. 1973): *V. persepolitana*.
sepium L.
subsp. **eriocalyx** (Čelak.) Holub in *Folia Geobot. Phytotax.*, 8(2):177(1973): *V. sepium* var.
1, 2, 6 *eriocalyx*.

sericocarpa Fenzl
subsp. **mardinica** Ponert in *Feddes Repert.*, 83
2 (9–10):634(1972 publ. 1973)—Turkey.
subsp. **microphylla** (Boiss.) Ponert *op. cit.* 634
2 (1972 publ. 1973): *V. sericocarpa* var. *microphylla.*
subsp. **mollis** (Boiss. & Hausskn. ex Boiss.)
2 Ponert *op. cit.* 634(1972 publ. 1973): *V. mollis.*
subsp. **urfae** Ponert *op. cit.* 634(1972 publ. 1973)—
2 Turkey.
tetrasperma (L.) Schreb.
subsp. **pubescens** (DC.) Ponert *op. cit.* 633(1972
1, 2 publ. 1973): *Ervum pubescens.*
villosa Roth
subsp. **semicretica** Ponert *op. cit.* 633(1972 publ.
2 1973)—Turkey.

VIGNA

2860 Swindell, R. E., Watt, E. E. *and* Evans, G. M.
A natural tetraploid mungbean of suspected amphidiploid origin. *J. Hered.*, 64(2):107(1973)—
5, 6 Chromosome number. *Vigna radiata.*

WISTERIA

2861 McClintock, E.
The generic name *Wisteria. Calif. Hort. J.*, 34(4):
2, 3 159(1973)—Illustration.

ZORNIA

2862 Chakrabarti, A. *and* Jain, S. K.
On the genus *Zornia* Gmel. (Fabaceae) in India and
6 Ceylon. *Curr. Sci.*, 42(21):744–745(1973)—Key.

LEMNACEAE

2863 Hepper, F. N.
Flora of tropical East Africa: Lemnaceae. London,
5 Crown Agents, 1973. 9p.—Illustrations; keys.

2864 Ivanova, I. E.
K sistematike semeĭstva Lemnaceae S. Gray. (On the taxonomy of Lemnaceae S. Gray). *Bot. Zhurn.*,
1–7 58(10):1413–1423(1973)—(En); illustrations.

2865 Plas, F. van der
Identification list of Malaysian specimens: 45. Lemnaceae. Leyden, Foundation Flora Malesiana,
6 1972. pp. 797–800.

2866 Sloover, J. L., de
Lemnaceae du Rwanda. *Bull. Jard. Bot. Nation.*
5 *Belg.*, 43(3–4):361–368(1973)—Illustrations.

LEMNA

2867 Lange, L. de *and* Pieterse, A. H.
A comparative study of the morphology of *Lemna gibba* L. and *Lemna minor* L. *Acta Bot. Neerl.*,
1–7 22(5):510–517(1973).

rwandensis de Sloover in *Bull. Jard. Bot. Nation.*
5 *Belg.*, 43(3–4):366(1973)—Rwanda.

SPIRODELA

2868 Berriman, D. J.
Tropical African plants: 33. The first record of *Spirodela punctata* (Lemnaceae) from Rhodesia. *Kew*
5 *Bull.*, 28(2):325–326(1973).

WOLFFIA

2869 Burmeister, W.
Ein neuer Fundort von *Wolffia arrhiza* (L.) Horkel ex Wimmer. *Inform. ZAG Wasserpflanzen*, no.4:2
1–7 (1973)—Illustration.

2870 Lipkin, Y.
Note: *Wolffia arrhiza* (L.) Hork. ex Wimmer on the
2 Golan Plateau. *Israel J. Bot.*, 22(3):175–177(1973).

2871 Weber-Oldecop, D. W.
Anmerkung zur Entdeckungsgeschichte von *Wolffia arrhiza* in Niedersachsen. *Göttinger Flor.*
1 *Rundbr.*, 7(2):29–30(1973).

LENTIBULARIACEAE

2872 Komiya, S.
New subdivision of the Lentibulariaceae. *J. Jap. Bot.*,
3, 4 48(5):147–153(1973)—(Ja).

PINGUICULA

2873 Kondo, K.
The chromosome numbers of four species of carnivorous plants. *Phyton* (Argentina), 31(2):93–94
3 (1973). *Pinguicula planifolia.*

2874 Moldenke, H. N.
Notes on new and noteworthy plants: 60.
3 *Phytologia*, 26(4):224–226(1973).

2875 Yoshimura, K.
Studies on the chromosome number and karyotype of *Pinguicula ramosa* Miyoshi (Lentibulariaceae). *J.*
2 *Jap. Bot.*, 48(10):289–294(1973)—(Ja).

2876 Ziemer, R. R.
Some field observations of *Darlingtonia* and
3 *Pinguicula. Carniv. Pl. Newsl.*, 2(2):25–27(1973).

pumila
f. **alba** Moldenke in *Phytologia*, 26(4):225(1973)—
3 U.S.A. (Florida).

UTRICULARIA

2877 Anon.
[*Utricularia*]. *Carniv. Pl. Newsl.*, 2(3):34(1973)—
4 En; illustrations; chromosome numbers.

2878 Kondo, K.
The chromosome number of *Utricularia denticulata* Benjamin. *Ann. Missouri Bot. Gard.*, 59(3):474–476
4 (1972 publ. 1973).

2879 Kondo, K.
Chromosome numbers of five taxa. *Chromosome Inform. Serv.*, no.15:33–34(1973). Chromosome
3 numbers. *Utricularia vulgasis.*

2880 Kondo, K.
The chromosome numbers of four species of carnivorous plants. *Phyton* (Argentina), 31(2):93–94
2, 4 (1973). *Utricularia racemosa, U. tricolor.*

2881 Kondo, K.
Key to the North American species of *Utricularia* (Bladderworts). *Carniv. Pl. Newsl.*, 2(4):66–69
3 (1973)—Illustrations.

2882 Kondo, K.
Some taxonomic problems in the Lentibulariaceae, especially *Utricularia*. Systematic seminar, Duke University, Durham, North Carolina. November 1,
3 1972. *Carniv. Pl. Newsl.*, 2(1):12(1973).

2883 Ceska, A. *and* Bell, M. A. M.
Utricularia (Lentibulariaceae) in the Pacific northwest. *Madroño*, 22(2):74–84(1973)—Illustrations,
3, 4 dot maps; key.

LEPIDOBOTRYACEAE

2884 Badré, F.
Lepidobotryaceae. Flore du Gabon, 21:40–42
5 (1973)—Illustration.

LILIACEAE

2885 Brandham, P. E.
The chromosomes of the Liliaceae: 3. New cases of interchange hybridity in the Aloineae. *Kew Bull.*, 28(3):341–348(1973 publ. 1974)—Chromosome
5 numbers.

2886 Chandra, S.
Segregation of Xanthorrhoeaceae from Liliaceae. *Proc. Indian Sci. Congr.*, 58(3):456(1971). Abstract
1–7 only.

2887 Harland, S.
Classification problems of Haworthias. *Bull. Afr.*
5 *Succ. Pl. Soc.*, 8(1):24–27(1973).

2888 Holitscher, O.
Průhonický sortiment tulipámi. (The Průhonice assortment of tulips). *Acta Pruhoniciana*, 27:17–48
1, 2 (1972)—(En,Rus,Ge). Includes garden varieties.

2889 Jakimow-Barras, N.
Les polysaccharides des graines de quelques Liliacées et Iridacées. *Phytochemistry*, 12(6):1331–1339
1, 2, 6 (1973)—(En).

2890 Knox Finlay, M. W.
Liliaceae. *J. Scott. Rock Gard. Club*, 13(3):240–244
2 (1973)—Illustrations. Found by George Forrest.

2891 Neves, J. Barros
Contribution à la connaissance cytotaxinomique des spermatophyta du Portugal: 8. Liliaceae. *Bol. Soc.*
1 *Brot.*, 47:157–212(1973)—Chromosome numbers.

2892 Obermeyer, A. A.
Notes on African plants: Liliaceae. *Aloe, Chamaealoe, Haworthia, Astroloba, Poellnitzia* and *Chortolirion.*
5 *Bothalia*, 11(1–2):119(1973).

2893 Raicu, P., Radu, M., Bogdan, D. *and* Kirillova, M.
Cercetări citogenetice la cîteva specii de plante floricole bulboase. (Cytogenetic researches of some bulbous ornamental plants). *An. Univ. Bucureşti, Biol. Veg.*, 20:49–60(1971)—(Rus,En); chromosome numbers. *Hyacinthus orientalis, Tulipa gesneriana,*
1 *Lilium candidum, L. regale.*

2894 Ravenna, P.
Estudios sobre Alliaceae. *Mus. Nac. Hist. Nat. Not.*
4 *Mens.*, (Chile), (200):3–6(1972)—Illustrations.

2895 Sharma, A. K., *comp.*
Chromosome census of the plant kingdom: 1. Monocotyledons: part 2. Liliales, Liliaceae tribes I
1–7 and II. *Nucleus*, 15(3). 12p.(1972).

2896 Sterling, C.
Comparative morphology of the carpel in the Liliaceae: Wurmbaeae. *Bot. J. Linn. Soc.*, 66(1):
5, 7 75–82(1973).

2897 Wendelbo, P.
Contributions to the Flora of Iraq: 12. New species and new combinations in the Liliaceae. *Kew Bull.*,
2 28(1):29–35(1973)—Illustrations, dot map.

AGAPANTHUS

2898 Hunt, D. R.
Agapanthus caulescens ssp. *gracilis*. *Curtis's Bot. Mag.*, 179(2):tab.632(1972 publ. 1973)—En; illustration,
5 coloured illustration.

2899 Rourke, J. P.
Agapanthus walshii. *Flow. Pl. Afr.*, 42(3–4): pl.1675
5 (1973)—En; coloured illustration.

ALBUCA

2900 Gledhill, D. *and* Oyewole, S. O.
The taxonomy of *Albuca* in West Africa. *Bol. Soc. Brot.*, 46:93–106(1972 publ. 1973)—Map; chromosome
5 number; key.

2901 Oyewole, S. O.
Cytological and cytogenetic studies in the genus *Albuca* L., in West Africa. *Bol. Soc. Brot.*, 46:149–170
5 (1972)—Chromosome numbers.

fibrotunicata Gledhill & Oyewole in *Bol. Soc. Brot.*,
5 46:98(1972 publ. 1973)—Nigeria.

ALETRIS

2902 Naruhashi, N.
A new series of *Aletris* with an interesting corm. *Acta Phytotax. Geobot.*, 25(4–6):131–135(1973)—
2 (Ja); illustrations.

2903 Sullivan, V. I.
Biosystematics of *Aletris lutea* Small, *Aletris obovata* Nash, and natural hybrids (Liliaceae). *Brittonia*, 25(3):
2, 3, 5 294–303(1973)—Dot map; chromosome numbers.

makiyataroi Naruhashi in *Acta Phytotax. Geobot.*,
2 25(4–6):131(1973)—Japan.

ALLIUM

2904 Cheshmedzhiev, I. V.
K tsitosistematike nekotorӯkh bolgarskikh vidov *Allium* L. (To the cytotaxonomy of some Bulgarian *Allium* L. species). *Bot. Zhurn.*, 58(6):864–875
1 (1973)—Chromosome numbers.

2905 Dietrich, J.
Genre *Allium*—espèces de collection (suite). *Inform. Ann. Caryosyst. Cytogénét.*, no.6:17–20(1972 publ.
2 1973)—Chromosome numbers.

2906 Fulvio, T. E. di
Sobre el gineceo de *Allium* y *Nothoscordum*. *Kurt-*
1–6 *ziana*, 7:241–253(1973)—(En).

2907 Gohil, R. N. *and* Koul, A. K.
Some adaptive genetic-evolutionary processes accompanying polyploidy in the Indian Alliums. *Bot. Notiser*, 126(4):426–432(1973)—Illustration;
6 chromosome numbers.

2908 Kollmann, F.
Allium paniculatum, another case of proximally localized chiasmata. *Israel J. Bot.*, 22(2):113–115
1, 2 (1973)—Chromosome numbers.

2909 Kollmann, F.
Karyology of some species of *Allium* section *Molium* in Israel. *Israel J. Bot.*, 22(2):92–112(1973)—
1, 2, 6 Chromosome numbers.

2910 Kuusk, V.
Nurmlauk Eestis. (*Allium vineale* L. in Estonia).
1 *Flor. Märkmed.* 1(5):304–305(1973)—Dot map.

2911 McNeal, D. W. *and* Ownbey, M.
Bulb morphology in some western north American
3 species of *Allium*. *Madroño*, 22(1):10–24(1973).

2912 Murín, A. *and* Feráková, V.
Karyotypy niektorých kultúrnych druhov rodu
Allium L. (Karyotypes of some cultivated species of
the genus *Allium* L.). *Biológia, A*, (Czechoslovakia),
28(1):65–71(1973)—(En,Rus). *A. cepa*, *A. sativum*, *A.*
1, 2 *schoenoprasum*.

2913 Nagai, Y.
(Reproductive biology of *Allium monanthum*
Maxim). *J. Geobot.*, 20(4):84–91(1972)—Ja(En);
2 illustrations.

2914 Panov, P.
Diagnoses plantarum novarum ser. 2. *Dokl. Bolg.*
1 *Akad. Nauk*, 26(9):1227–1230(1973).

2915 Tahbaz, F.
L'*Allium* 'Tarée irani' du groupe *ampeloprasum* L.
cultivé en Iran, region de Téhéran. *Bull. Soc. Bot.*
France, 118(9):753–761(1971 publ. 1973)—(En);
2 Illustrations.

2916 Wilde-Duyfjes, B. E. E. de
Typification of 23 *Allium* species described by
Linnaeus and possibly occurring in Africa. *Taxon*,
1, 2, 5 22(1):57–91(1973)—Illustrations.

akaka Gmelin
subsp. **haemanthoides** (Boiss. & Reut. Regel)
Wendelbo in *Kew Bull.*, 28(1):29(1973): *A.*
2 *haemanthoides*.
subsp. **shelkovnikovii** (Grossh.) Wendelbo *op. cit.*
2 29(1973): *A. shelkovnikovii*.
rilaense Panov in *Dokl. Bolg. Akad. Nauk*, 26(9):
1 1227(1973)—Bulgaria.
vineale L.
lusus **virescens** Soó in *Acta Bot. Acad. Sci. Hung.*,
18(1–2):173(1973), nom. nov.: *A. vineale* var.
1 *virens* A. & Gr.
vinicolor Wendelbo in *Kew Bull.*, 28(1):29(1973)—
2 Iraq.

ALOE

2917 Anon.
Aloe niebuhriana Lavranos (Liliaceae). *Ashingtonia*,
2 1(3):34(1973)—En; coloured illustration.

2918 Brandham, P. E.
The production of new hybrids in the Aloineae.
Nation. Cact. Succ. J. (U.K.), 28(1):16–19(1973)—
5 Illustrations.

2919 Christian, H. B.
Notes on African Aloes. *Aloe, Cact. Succ. Soc.*
Rhodesia Quart. Newsl., no.20:15–18(1973).
Continuation of reprint of notes written by the late
H. Basil Christian and published in *The Rhodesia*
5 *Agric. J.*, 1933.

2920 Giess, W.
Notes on African plants. A new species of *Aloe*
from South West Africa. *Bothalia*, 11(1–2):120–122
5 (1973)—Illustrations. *A. dewinteri*.

2921 Lavranos, J. J.
Aloe chlorantha: a new species from the south
western Karoo (South Africa). *J. S. Afr. Bot.*, 39(1):
5 85–90(1973)—(Af); illustrations.

2922 Lavranos, J. J.
A new variety of *Aloe* from Namaqualand. *J. S. Afr.*
5 *Bot.*, 39(1):41–43(1973)—(Af); illustration.

2923 Lavranos, J. J.
Three new Aloes from Somalia. *Cact. Succ. J.*
5 (U.S.A.), 45(3):114–119(1973)—Illustrations.

2924 Newton, L. E.
Tropical African plants: 33. Notes on West
African *Aloë* (Liliaceae). *Kew Bull.*, 28(2):324
5 (1973).

2925 Roy, S. C.
Chromosome studies in the genus *Aloe*. *Pl. Sci.*
5, 6 (Lucknow), 3:30–35(1971 ?publ. 1973).

2926 Wetselaar, A.
Aloë speciosa Bak. *Succulenta*, 52(5):91–93(1973)—
2, 5 Du; illustrations.

2927 Wright, I.
The *Aloe chabaudii*. *East. Cape Nat.*, no. 48:25–26
5 (1973).

bargalensis Lavranos in *Cact. Succ. J.* (U.S.A.), 45(3):
5 116(1973)—Somali Republic.
chlorantha Lavranos in *J. S. Afr. Bot.*, 39(1):87
5 (1973)—South Africa (Cape Prov.).
dewinteri Giess in *Bothalia*, 11(1–2):120(1973)—
5 Namibia.
heliderana Lavranos in *Cact. Succ. J.* (U.S.A.), 45(3):
5 114(1973)—Somali Republic.
krapohliana Marloth
var. **dumoulinii** Lavranos in *J. S. Afr. Bot.*, 39(1):
5 41(1973)—South Africa (Cape Prov.).
pulchra Lavranos in *Cact. Succ. J.* (U.S.A.), 45(3):118
5 (1973)—Somali Republic.
subacutissima Rowley in *Nation. Cact. Succ. J.*
(U.K.), 28(1):6(1973), nom. nov.: *A. intermedia*
5 Reyn.

× **ALOLIRION** Rowley in *Nation. Cact. Succ. J.*
0 (U.K.), 28(1):7(1973). (*Aloë* × *Chortolirion*).

× **ALWORTHIA** Rowley in *Nation. Cact. Succ. J.*
0 (U.K.), 28(1):7(1973). (*Aloë* × *Haworthia*).

ANDROCYMBIUM

2929 Sterling, C.
Comparative morphology of the carpel in the
Liliaceae: Colchiceae (*Androcymbium*). *Bot. J. Linn. Soc.*,
1, 2, 5 67(2):149–156(1973)—Chromosome numbers.

ANTHERICUM

2930 Jakuczun, B.
Pajęcznica liliowata *Anthericum liliago* w Wolińskim
Parku Narodowym. (The antheric, *Anthericum*
liliago, in the Wolin National Park). *Chrońmy Przyr.*
1 *Ojczystą*, 29(1):61–62(1973)—Illustration.

liliago L.
subsp. australe (Wk.) Malagarriga, Las Subesp. y
Variac. Geogr.: 25(1973), without basionym date:
1 *A. liliago* var. *australe*.
subsp. baeticum (Bss.) Malagarriga *op. cit.* 25
1 (1973), without basionym date: *A. baeticum*.

APHYLLANTHES

2931 Abbà, G.
Una nuova stazione per il Piemonte dell' *Aphyl-*
lanthes monspeliensis L. (A new station in Piedmont of
the *Aphyllanthes monspeliensis* L.). *Inform. Bot. Ital.*,
1 5(1):87–88(1973)—(En).

ASPARAGUS

2932 Rivière, S.
Les activités méristématiques durant l'ontogenèse d'une plantule de Monocotylédone à germination hypogée : l'*Asparagus officinalis* L. (Liliacées). *Compt. R. Hebd. Séances Acad. Sci., D* (Paris), 277(3):293–
1, 2 296(1973)—Illustrations.

2933 Subramaniam, D.
Cytological investigations in *Haemanthus coccineus* L., *Asparagus racemosus* Willd. and *Yucca gloriosa* Linn. *Proc. Indian Sci. Congr.*, 60(3):323–324(1973)—
5 Chromosome number. Abstract only.

ASPHODELUS

2934 Britton, R. S.
Onion weed: good management can help. *J. Agric. S. Austral.*, 76(4):100–104(1973)—Illustrations.
1, 2 *Asphodelus fistulosus.*

ASTROLOBA

2935 Majumdar, S. K. *and* Riley, H. P.
Chromosome morphology in *Astroloba* and
5 *Poellnitzia*. *Nucleus*, 15(2):85–91(1972).

× ASTROWORTHIA

bicarinata (Haw.) Rowley in *Nation. Cact. Succ. J.*
5 (U.K.), 28(1):7(1973): *Apicra bicarinata.*
nm. **skinneri** (Bgr.) Rowley *op. cit.* 7(1973):
5 *Apicra skinneri.*

BELLEVALIA

2936 Bothmer, R. von *and* Bentzer, B.
Chromosome morphology in *Bellevalia dubia* subsp. *boissieri* and *B. trifoliata* (Liliaceae) from Greece. *Ann. Mus. Goulandris*, 1:11–14(1973)—(Gre);
1 chromosome numbers.

parva Wendelbo in *Kew Bull.*, 28(1):32(1973)—
2 Iraq.

BRODIAEA

hooveri (Niehaus) Traub in *Pl. Life*, 29:47(1973):
3 *B. elegans* subsp. *hooveri.*

BULBINE

frutescens (L.) Willd.
var. **incurva** (Thunb.) Rowley in *Nation. Cact. Succ. J.* (U.K.), 28(1):6(1973): *Anthericum*
5 *incurvum.*
var. **rostrata** (Jacq.) Rowley *op. cit.* 6(1973):
5 *Anthericum rostratum.*

CAESIA

2937 Obermeyer, A. A.
Notes on African plants: *Caesia* and *Nanolirion*.
5 *Bothalia*, 11(1–2):122–123(1973).

capensis (H. Bol.) Obermeyer in *Bothalia*, 11(1–2):
5 122(1973): *Herpolirion capense.*

CALOCHORTUS

2938 Hill, A. J.
A distinctive new *Calochortus* (Liliaceae) from Marin County, California. *Madroño*, 22(2):100–104(1973)—
3 Illustrations.

tiburonensis A. J. Hill in *Madroño*, 22(2):100(1973)—
3 U.S.A. (California).

CHAMAEALOE

2939 Brandham, P. E.
The production of new hybrids in the Aloineae. *Nation. Cact. Succ. J.* (U.K.), 28(1):16–19(1973)—
5 Illustrations.

× **CHAMAELEPTALOË** Rowley in *Nation. Cact. Succ. J.* (U.K.), 28(1):7(1973). (*Chamaealoë* ×
0 *Leptaloë*).

CHLOROPHYTUM

2940 Charoenphol, C. C.
Studies in the genus *Chlorophytum* of Thailand.
6 *Thai Forest Bull., Bot.*, no.7:67–69(1973)—Key.

intermedium Craib
var. **pubescens** Charoenphol in *Thai Forest Bull.,*
6 *Bot.*, 7:69(1973)—Thailand.
var. **subundulatum** (K. Larsen) Charoenphol *op.*
6 *cit.* 69(1973): *C. subundulatum.*
orchidastrum Lindl.
var. **simplex** (Craib) Charoenphol *op. cit.* 68
6 (1973): *C. simplex.*

CLINTONIA

2941 Gunther, P. P.
The phenodynamics of *Clintonia borealis* (Ait.) Raf. and *C. umbellulata* (Michx.) Morong (Liliaceae) as related to their general distributions. *Diss. Abstr. Int., B,*
2, 3 33(10):4690–4691(1973).

2942 Utech, F. *and* Thien, L. B.
Chromosome numbers in *Clintonia borealis* (Liliaceae). *Mich. Bot.*, 12(2):122–128(1973)—Dot
2, 3 maps.

COLCHICUM

2943 Bailey, D.
Planhigion meddyginiaethol o gefn gwlad Cymru: 4. Saffrwm y gweunydd (*Colchicum autumnale*). *Y*
1 *Gwyddonydd*, 11(2):83–84(1973)—Illustration.

2944 Sterling, C.
Comparative morphology of the carpel in the Liliaceae: Colchiceae (*Colchicum*). *Bot. J. Linn. Soc.*,
1 66(3):213–221(1973).

EUCOMIS

2945 Reyneke, W. F.
Pynappellelies in Suid-Afrika. *Veld Flora*, 3(3):58–
5 59(1973)—Illustrations.

DRACAENA

2946 Kunkel, G.
Novedades en la flora de las islas orientales, con una nota sobre *Dracaena draco* en Gran Canaria. *Cuad.*
1 *Bot. Canar.*, no.17:35–37(1973)—(En).

2947 Sheriff, A. *and* Sadashiva Singh, B. K.
Cytological studies in a wild Indian species of *Dracaena*. *Curr. Sci.*, 42(9):324–325(1973)—
6 Chromosome numbers.

DRIMIOPSIS

2948 Mahalakshmi, N. *and* Sheriff, A.
Karyomorphology of *Drimiopsis kirkii*. *Proc. Indian*
5 *Sci. Congr.*, 58(3):469–470(1971). Abstract only.

FRITILLARIA

2949 Dvorakovskaya, V. M.
O prorastanii semyan *Fritillaria lutea* Mill. i *F. ruthenica* Wikstr. *Byull. Glavn. Bot. Sada* (Moscow),
2 89:82–83(1973).

2950 Gramuglio, G. *and* Arena, M.
Nuove stazioni di *Fritillaria messanensis* Raf. e considerazioni ecologiche desunte da preliminari osservazioni anatomiche. *Webbia*, 28(2):411–416
1 (1973)—(En); illustration, dot map.

2951 Gulino, A. *and* Gulino, G.
La presenza di *Fritillaria meleagris* L. e di *Fritillaria tenella* M. Bieb in Val Maira (Alpi Cozie). *Rea*,
1 no.5:36–38(1972)—(En).

2952 Mathew, B.
Fritillaria reuteri. *Curtis's Bot. Mag.*, 179(4):tab.658
2 (1973)—En; illustration, coloured illustration.

2953 Mathew, B.
Plant portraits: *Fritillaria bucharica* Regel. *Quart. Bull. Alp. Gard. Soc.*, 41(3):206,208(1973)—
2 Illustration.

2954 Richard, J.-L.
A propos de la sociologie de la fritillaire pintade (*Fritillaria meleagris* L.) dans le Jura. *Bull. Soc.*
1 *Neuchâtel. Sci. Nat.*, 96:5–15(1973)—(Ge,En).

2955 Rotov, R. A.
Ryabchik shakhmatnyĭ (*Fritillaria meleagris* L.) v Moskovskoĭ oblasti. *Byull. Glavn. Bot. Sada* (Moscow),
1 90:27–29(1973).

messanensis Raf.
subsp. boissieri (Costa) Malagarriga, Las Subesp. y Variac. Geogr.: 25(1973), without basionym
1 date: *F. boissieri*.
subsp. hispanica (Bss.R.) Malagarriga *op. cit.* 25
1 (1973), without basionym date: *F. hispanica*.
subsp. lusitanica (Wicstr.) Malagarriga *op. cit.* 25
1 (1973), without basionym date: *F. lusitanica*.

GAGEA

2956 Jansen, M. T. *and* Mennema, J.
Gagea villosa (Bieb.) Duby in Nederland. *Gorteria*,
1 6(8):125–128(1973)—(En); dot map.

villosa (M.B.) Duby
subsp. **hervieri** (Degen ex Hervier) A. Löve & Kjellqvist in *Lagascalia*, 3(2):166(1973): *G.*
1 *arvensis* subsp. *hervieri*.

× GASTROLEA

beguinii (Radl.) Walth.
nm. **chludowii** (Radl.) Rowley in *Nation. Cact.*
0 *Succ. J.* (U.K.), 28(1):7(1973): *Aloë chludowii*.

HAWORTHIA

2957 Bayer, M. B.
Natural variation and species delimitation in *Haworthia* Duval: part 3. *Haworthia reinwardtii* and *Haworthia coarctata* Haw. *Nation. Cact. Succ. J.* (U.K.),
5 28(3):80–86(1973)—Illustration, map.

2958 Bayer, M. B.
Two new species of *Haworthia* (Liliaceae). *J. S. Afr.*
5 *Bot.*, 39(3):249–254(1973)—(Af); illustrations.

2959 Harland, S.
Classification problems of Haworthias. *Bull. Afr.*
5 *Succ. Pl. Soc.*, 8(2):50,52,54–59(1973).

2960 Majumdar, S. K. *and* Riley, H. P.
Chromosome numbers, pollen fertility and pollen size in *Haworthia* species and hybrids. *Canad. J. Bot.*,
5 51(10):1753–1759(1973)—(Fr).

coarctata Haworth
subsp. **adelaidensis** (VP) Bayer in *Nation. Cact. Succ. J.* (U.K.), 28(3):86(1973): *H. reinwardtii* var.
5 *adelaidensis*.
var. **greenii** (Baker) Bayer *op. cit.* 86(1973):
5 *H. greenii*.
var. **tenuis** (Smith) Bayer *op. cit.* 86(1973):
5 *H. reinwardtii* var. *tenuis*.
pulchella Bayer in *J. S. Afr. Bot.*, 39(3):252(1973)—
5 South Africa (Cape Prov.).
serrata Bayer *op. cit.* 249(1973)—South Africa (Cape
5 Prov.).

HYACINTHELLA

2961 Mathew, B.
A new *Hyacinthella* (Liliaceae) from Turkey. *Kew Bull.*, 28(3):517–519(1973 publ. 1974)—
2 Illustrations.

sect. **Alrawia** Wendelbo in *Kew Bull.*, 28(1):35
2 (1973).
bellii (Baker) Wendelbo *op. cit.* 35(1973): *Scilla*
2 *bellii*.
2 **nutans** Wendelbo *op. cit.* 33(1973)—Iraq.
siirtensis B. F. Mathew in *Kew Bull.*, 28(3):517
2 (1973 publ. 1974)—Turkey.

IPHIGENIA

2962 Arekal, G. D. *and* Ramaswamy, S. N.
A new species of *Iphigenia*. *Proc. Indian Sci. Congr.*,
6 58(3):435(1971). Abstract only.

2963 Sheriff, A. *and* Rao, U. G.
Karyotype studies in the genus *Iphigenia* and the problem of speciation. *Proc. Indian Sci. Congr.*, 59(3):
6 357(1972)—Chromosome numbers. Abstract only.

mysorensis Arekal & S. N. Ramaswamy in *Proc. Indian Sci. Congr.*, 58(3):435(1971), without latin
6 descr.—India.

KNIPHOFIA

2964 Codd, L. E.
Kniphofia tabularis. *Flow. Pl. Afr.*, 42(3–4):pl.1661
5 (1973)—En; coloured illustration.

2965 Marais, W.
A revision of the tropical species of *Kniphofia* (Liliaceae). *Kew Bull.*, 28(3):465–483(1973 publ.
5 1974)—Key.

2966 Morley, B.
Kniphofia schimperi, a correction. *J. Roy. Hort. Soc.*,
5 98(11):504(1973).

nana Marais in *Kew Bull.*, 28(3):478(1973 publ.
5 1974)—Zaïre.
princeae (Berger) Marais *op. cit.* 470(1973 publ.
5 1974): *Aloe princeae*.
thomsonii Baker
var. **snowdenii** (C. H. Wright) Marais *op. cit.* 468
5 (1973 publ. 1974): *K. snowdenii*.

LEDEBOURIA

2967 Jessop, J. P.
Studies on the South African bulbous Liliaceae: 4. A new species of *Ledebouria* from South West Africa.
5 *J. S. Afr. Bot.*, 39(1):45–48(1973)—(Af); illustrations.

scabrida J. P. Jessop in *J S. Afr. Bot.*,
5 39(1): 45 (1973)—South West Africa.

LEOPOLDIA

2968 Bentzer, B.
Taxonomy, variation and evolution in representatives of *Leopoldia* Parl. (Liliaceae) in the southern and central Aegean. *Bot. Notiser*, 126(1):69–132(1973) —Illustrations, coloured illustrations, map, dot maps;
1, 2 chromosome numbers; key.

2969 Garbari, F.
Le specie del genere *Leopoldia* Parl. (Liliaceae) in Italia. *Webbia*, 28(1):57–80(1973)—(En); dot maps;
1, 2 chromosome numbers; key.

cycladica (Davis & Stuart) Garbari
subsp. **subsessilis** Bentzer in *Bot. Notiser*, 126(1):
1 82(1973)—Greece.
dionysica (Rech. f.) Bentzer *op. cit.* 82(1973):
1 *Muscari dionysicum.*

× **LEPTALOINELLA** Rowley in *Nation. Cact. Succ. J.* (U.K.), 28(1):7(1973). (*Aloinella* ×
0 *Leptaloë*).

LEUCOCORYNE

2970 Zoellner, O.
El genero *Leucocoryne*. *An. Mus. Hist. Nat. Valparaíso*, no.5:9–83(1972)—Illustrations, dot maps;
4 chromosome number; key.

conferta Zoellner in *An. Mus. Hist. Nat. Valparaíso*,
4 5:51(1972), without type—Chile.
coquimbensis F. Phil.
var. alba Zoellner *op. cit.* 48(1972), without type—
4 Chile.

LILIUM

2971 Baranova, M. V.
O formirovanii tsvetkov v steblevȳkh lukovichkakh *Lilium*. (On the formation of flowers in stem bulblets of *Lilium*). *Bot. Zhurn.*, 58(6):890–893(1973)—Illustrations. *L. tigrinum*, *L. bulbiferum* and *L.*
1, 2 *sargentiae*.

2972 Dijkstra, S. J.
Lilium bulbiferum L. in Limburg. *Gorteria*, 6(12):203
1 (1973)—Du.

2973 Noda, S.
Karyotype and meiosis in *Lilium hansonii*. *Chromo-*
2 *some Inform. Serv.*, no.15:7–9(1973).

2974 Weeda, E. J.
En recente vondst van *Lilium bulbiferum* L. subsp.
1 *croceum* (Chaix) Arcang. *Gorteria*, 6(8):140(1973).

MASSONIA

2975 Merxmüller, H. *and* Roessler, H.
Eine bisher verkannte Liliacee der Gattung *Massonia* in Südwestafrika. *Mitt. Bot. Staatssamml. München*,
5 11:83–85(1973)—Illustrations.

MERENDERA

bulbocodium Ram.
f. **alba** Q. G. Pinto da Silva in *Agron. Lusit.*, 34(3):
1 182(1972 publ. 1973)—Portugal.
subsp. filifolia (Camb.) Malagarriga, Las Subesp. y Variac. Geogr.: 26(1973), without basionym date:
1 *M. filifolia.*

MILLIGANIA

2976 Curtis, W. M.
A new Tasmanian species of *Milligania* (family Liliaceae). *Rec. Queen Victoria Mus.* (Tasmania),
7 no.45. 2p.(1972). *M. lindoniana.*

NANOLIRION

2977 Obermeyer, A. A.
Notes on African plants: *Caesia* and *Nanolirion*.
5 *Bothalia*, 11(1–2):122–123(1973).

NOLINA

2978 Zedler, P. H. *and* Keeley, J. E.
A second location for *Nolina interrata* Gentry
3 (Agavaceae). *Madroño*, 22(4):214(1973).

NOTHOSCORDUM

2979 Crosa, O.
Estudios cariologicos en el genero *Nothoscordum* (Liliaceae). *Bol. Fac. Agron.* (Montevideo), no.122. 8p.
4 (1972)—Chromosome numbers.

2980 Fulvio, T. E. di
La semilla y la plántula de *Nothoscordum arenarium*
4 (Liliaceae). *Kurtziana*, 7:267–268(1973).

2981 Fulvio, T. E. di
Sobre el gineceo de *Allium* y *Nothoscordum*.
1–6 *Kurtziana*, 7:241–253(1973)—(En).

2982 Guaglianone, E. R.
Nothoscordum andinum especie de Liliaceae nueva para la flora argentina. *Darwiniana*, 18(1–2):31–36
4 (1973)—Illustrations.

2983 Traub, H. J.
Nothoscordum mahuii sp. nov. *Pl. Life*, 29(1–4):34
4 (1973)—En.

entrerianum Ravenna in *Mus. Nac. Hist. Nat. Not.*
4 *Mens.* (Chile), 200:3(1972)—Argentina.
mahui Traub in *Pl. Life*, 29:34(1973), as 'mahuii'—
4 Chile.
nublense Ravenna in *Mus. Nac. Hist. Nat. Not Mens.*
4 (Chile), 200:4(1972)—Chile.
4 **serenense** Ravenna *op. cit.* 3(1972)—Chile.

OPHIOPOGON

2984 Karagyozova, M. D.
Anatomy of the leaf blade of *Ophiopogon jaburan* Lodd. var. *variegatus*. *Dokl. Bolg. Akad. Nauk*, 26(5):
2 703–705(1973).

ORNITHOGALUM

2985 Czapik, R.
Cytoembryology of experimental hybrids between two related species of *Ornithogalum* L. *Acta Biol. Cracov., Bot.*, 15(2):165–176(1972)—Chromosome
1, 2 numbers. *O. umbellatum* and *O. gussonei*.

2986 Żabińska, D.
Embryo sac and endosperm development in some representatives of the group *Ornithogalum gussonei* from Poland. *Acta Biol. Cracov., Bot.*, 15(2):177–190 (1972)—Chromosome numbers. *O. umbellatum*, *O.*
1 *gussonei*, *O. nutans* and *O. boucheanum*.

nutans L.
var. multiflora Voliotis in *Feddes Repert.*, 84(4):328
1 (1973), without latin descr. or type—Greece.

PARIS

2987 Cyunel, E.
An anatomical analysis of *Paris quadrifolia* L. *Acta*
1 *Biol. Cracov., Bot.*, 15(2):205–208(1972).

× **POELINERIA** Rowley in *Nation. Cact. Succ. J.*
0 (U.K.), 28(1):7(1973). (*Gasteria* × *Poellnitzia*).

POELLNITZIA

2988 Majumdar, S. K. *and* Riley, H. P.
Chromosome morphology in *Astroloba* and
5 *Poellnitzia. Nucleus*, 15(2):85–91(1972).

POLYGONATUM

2989 Abramova, L. I.
Polygonatum laoticum Gagn. i ego mesto v sisteme roda *Polygonatum* Mill. (*Polygonatum laoticum* Gagn. and its place in the system of the genus *Polygonatum* Mill.). *Bot. Zhurn.*, 58(12):1794–1799(1973)—
6 Illustrations; chromosome numbers.

2990 Levering, D. F.
An autecological study of *Polygonatum pubescens* (Willd.) Pursh and *Polygonatum biflorum* (Walt.) Ell. emphasizing inter- and intra-specific photosynthetic and respiratory rate variation. *Diss. Abstr. Int., B*,
3 33(4):1425(1972).

PSEUDOMUSCARI

2991 Garbari, F.
Note sul genere *Pseudomuscari* (Liliaceae). *Webbia*, 27(2):369–381(1972 publ. 1973)—(En); coloured
1, 2 illustrations; chromosome numbers.

RUSCUS

2992 Bolman, J.
Muizendoorn (*Ruscus aculeatus* L.), een plant met bessen op de bladeren. *Natura* (Netherlands), 70(3):
1 37–41(1973)—Illustrations.

2993 Hillman, E. M. *and* Warren, A.
Survey of *Ruscus aculeatus* on Bookham Common: the first two years. *London Nat.*, no.52:93–103(1973)—
1 Illustrations, map, dot map.

SANSEVIERIA

2994 Guédès, M.
L'ovule de *Sansevieria trifasciata* Prain (Liliacées): sa vascularisation et sa signification morphologique. *Compt. R. Hebd. Séances Acad. Sci., D* (Paris), 276(6):
5 953–956(1973).

2995 Marais, W.
(350): proposal to amend the entry of 1100 *Sansevieria* in the list of Nomina Generica Conser-
2, 5, 6 vanda. *Taxon*, 22(1):158–159(1973).

2996 Wijnands, D. O.
Typification and nomenclature of two species of *Sansevieria* (Agavaceae). *Taxon*, 22(1):109–114
5, 6 (1973)—Illustrations.

SCILLA

2997 Hillesheim-Kimmel, U., Karafiat, H. *and* Marquardt, F.
Kartierung des Zweiblättrigen Blausterns (*Scilla bifolia* L.) im Naturschutzgebiet 'Kuhkopf-Knoblochsaue'. *Hess. Flor. Briefe*, 259–261:34–37
1 (1973)—Illustration, map.

2998 Komar, G. A.
Arilloidȳ nekotorȳkh vidov *Scilla*. (Arilloids of some *Scilla* species). *Bot. Zhurn.*, 58(10):1503–1508
1, 2 (1973).

2999 Satyesh, C. R. *
Chromosome study in different species of *Scilla*. *Rev. Roum. Embryol. Cytol.*, 8(1):29–35(1971)—Chromosome numbers. Abstr. in *Ref. Zhurn. Biol.*,
1, 2, 5 3(1):v474(1972).

SMILAX

aspera L.
subsp. balearica (Wk.) Malagarriga, Las Subesp. y Variac. Geogr.: 26(1973), without basionym
1 date: *S. aspera* var. *balearica*.
subsp. mauritanica (Desf.) Malagarriga *op. cit.* 26
1 (1973), without basionym date: *S. mauritanica*.
china L.
f. **variegata** Takahashi in *J. Geobot.*, 20(2):44(1972)—
2 Japan.

STREPTOPUS

parasimplex Hara & Ohashi in *J. Jap. Bot.*, 48(4):101
6 (1973)—Nepal, Sikkim.

TOFIELDIA

calyculata (L.) Wahlenb.
subsp. glacialis (Gaud.) Malagarriga, Las Subesp. y Variac. Geogr.: 26(1973), without basionym
1 date: *T. glacialis*.

TRACHYANDRA

3000 Roessler, H.
Über *Trachyandra* (Liliaceae) in Südwestafrika. *Mitt. Bot. Staatssamml. München*, 11:87–90(1973)—
5 Dot map.

ensifolia (Sölch) Roessler in *Mitt. Bot. Staatssamml.*
5 *München*, 11:88(1973): *Anthericum ensifolium*.
laxa (N. E. Br.) Oberm.
var. **rigida** (Suesseng.) Roessler *op. cit.* 90(1973):
5 *Anthericum arvense* var. *rigidum*.

TRILLIUM

3001 De Vries, B.
Nodding Trilliums in eastern Saskatchewan. *Blue*
3 *Jay*, 31(4):214–217(1973). Site locations.

3002 Fukuda, I.
The spontaneous chromosome mutation in *Trillium ovatum*. *Chromosome Inform. Serv.*, no.12.
3 4–5(1971).

3003 Munz, P. A. *and* Thorne, R. F.
A new northern California *Trillium*. *Aliso*, 8(1):
3 15–17(1973)—Illustration.

3004 Murrell, J. T. *and* Channell, R. B.
Fragrance analyses of *Trillium luteum* and *Trillium cuneatum* (Liliaceae). *J. Tenn. Acad. Sci.*, 48(3):101–
3 103(1973)—Dot map.

ovatum Pursh
subsp. **oettingeri** Munz & Thorne in *Aliso*, 8(1):
3 15(1973)—U.S.A. (California).

TULIPA

3005 Abramova, S. N.
Semennoe razmnozhenie dikorastushchikh tyul'panov Turkmenii. *Byull. Glavn. Bot. Sada*
2 (Moscow), 89:90–93(1973)—Illustration.

3006 Abramova, S. N. *and* Zakalyabina, L. G.
Novȳi vid tyul'pana iz Turkmenii. (Generis *Tulipae* L. species nova e Turcomania). *Novosti Sist.*
2 *Vȳssh. Rast.*, 10:104–106(1973).

3007 Ivanova, I. A.
Vnutrisemennoĭ rost zarodȳsha u tyul'panov. *Byull.*
2 *Glavn. Bot. Sada* (Moscow), 90:70–73(1973).

botschantzevae S. N. Abramova & Zakalyabina in *Novosti Sist. Vȳssh. Rast.*, 10:104(1973)—U.S.S.R.
2 (Middle Asia).

URGINEA

3008 Boraiah, A. *and* Fathima, T.
Cytotaxonomy of a new species of *Urginea*. *Proc. Indian Sci. Congr.*, 59(3):359–360(1972)—Chromo-
6 some number. Abstract only.

3009 Naik, V. N.
A natural tetraploid of *Urginea coromandeliana* Hk. f. *Curr. Sci.*, 42(12):439–440(1973)—
6 Chromosome number.

govindappensis A. Boraiah & Fathima in *Proc. Indian Sci. Congr.*, 59(3):359(1972), without latin
6 descr.—India.

VERATRUM

3010 Cyunel, E.
Pharmacobotanical investigations of species of *Veratrum* L. in Poland: part 1. Anatomical analysis of *Veratrum nigrum* L. *Acta Biol. Cracov., Bot.*, 16(1):
1, 2 23–28(1973).

3011 Sokołowska-Kulczycka, A.
Development of bisporic embryo sacs in *Veratrum lobelianum* Bernh. *Acta Biol. Cracov., Bot.*, 16(1):
1, 2 85–98(1973)—Chromosome number.

ZIGADENUS

elegans Pursh
subsp. **glaucus** (Nutt.) Hultén in *Bot. Notiser*,
3 126(4):473(1973): *Melanthium glaucum*.

LIMNANTHACEAE

FLOERKEA

3012 Gauthier, B. *and* Rousseau, C.
L'écologie du *Floerkea proserpinacoides* Willd. à l'Ile aux Grues, Montmagny, (Québec). *Nat. Canad.*,
3 100(4):371–383(1973)—(En); illustrations, map.

LIMNANTHES

3013 Arroyo, M. T. K.
Chiasma frequency evidence on the evolution of autogamy in *Limnanthes floccosa* (Limnanthaceae).
3 *Evolution*, 27(4):679–688(1973)—Illustrations, map.

3014 Arroyo, M. T. K.
A taximetric study of infraspecific variation in autogamous *Limnanthes floccosa* (Limnanthaceae).
3 *Brittonia*, 25(2):177–191(1973)—Dot map.

floccosa Howell
subsp. **bellingerana** (M. E. Peck) Arroyo in *Brittonia*, 25(2):187(1973), as 'bellingeriana': *L.*
3 *bellingeriana*.
subsp. **californica** Arroyo *op. cit.* 187(1973)—
3 U.S.A. (California).
subsp. **grandiflora** Arroyo *op. cit.* 188(1973)—
3 U.S.A. (Oregon).
subsp. **pumila** (Howell) Arroyo *op. cit.* 187(1973):
3 *L. pumila*.

LINACEAE

3015 Badré, F.
Linaceae. Flore du Gabon, 21:23–39(1973)—
5 Illustrations; key.

3016 Heimsch, C. *and* Tschabold, E. E.
Xylem studies in the Linaceae. *Bot. Gaz.*, 133(3):
1–7 242–253(1972)—Illustrations.

3017 Narayana, L. L. *and* Rao, D.
Contributions to floral anatomy of Linaceae: 5.
1–7 *J. Jap. Bot.*, 48(7):205–208(1973)—(Ja).

LINUM

3018 D'Alascio Deschamps, R.
Organisation du sac embryonnaire du *Linum catharticum* L., espèce récoltée en station naturelle; étude ultrastructurale. *Bull. Soc. Bot. France*, 120(5–6):
1, 2 189–200(1973)—(En).

3019 Dubey, D. K. *and* Kumar, S.
Cross-relationship between two *Linum* species bearing different basic chromosome numbers. *Indian*
1 *J. Agric. Sci.*, 43(1):18–20(1973).

3020 Dulberger, R.
Distyly in *Linum pubescens* and *L. mucronatum*. *Bot. J.*
1, 2 *Linn. Soc.*, 66(2):117–126(1973)—Illustrations.

3021 Lobana, K. S., Lal, R. *and* Gupta, M. L.
Morphology of pachytene chromosomes in *Linum grandiflorum* Desf. *Nucleus*, 15(3):167–170(1972)—
1 Chromosome number.

3022 Petrova, A. V.
Taksonomicheskoe izuchenie dikorastushchikh predstaviteleĭ roda *Linum* L. (len) v Bolgarii. (Taxonomic study of the wild species of the genus *Linum* L. (Flax) in Bulgaria). *Izv. Bot. Inst.* (Sofia),
1 23:70–76(1973)—(Rus,En).

3023 Petrova, A. V.
Taksonomichno prouchvane na divorastyashtite predstaviteli ot rod *Linum* L. (len) v Bŭlgariya: 2. Palinologichna i kariolichna kharakteristika. (Taxonomic study of the wild species from genus *Linum* (flax) in Bulgaria: 2. (Palynological and karyological characteristics)). *Izv. Bot. Inst.* (Sofia),
1 24:5–29(1973)—(Rus,En); chromosome numbers.

3024 Rogers, C. M.
The taxonomic significance of the fatty acid content
1–7 of seeds of *Linum*. *Brittonia*, 24(4):415–419(1972).

3025 Seetharam, A. *and* Srinivasachar, D.
Cytomorphological studies in the genus *Linum*.
1–7 *Cytologia*, 37(4):661–671(1972).

subsect. **Adenolinum** (Reichenb.) A. V. Petrova in *Izv. Bot. Inst.* (Sofia), 23:51(1973): gen.
1 *Adenolinum*.
subsect. **Dichrolinum** (Planch.) A. V. Petrova *op. cit.*
1 60(1973): *Linum* ser. *Dichrolinum*.
alpinum Jacq.
subsp. **pirinicum** A. V. Petrova *op. cit.* 56(1973)—
1 Bulgaria.

capitatum Kit. ex Schult.
var. **laxiflorum** A. V. Petrova *op. cit.* 31(1973)—
1 Bulgaria.
f. **minimum** A. V. Petrova *op. cit.* 31(1973)—
1 Bulgaria.
flavum L.
subsp. **sparsiflorum** (Stoj.) A. V. Petrova *op. cit.* 37
1 (1973): *L. flavum* var. *sparsiflorum.*
hirsutum L.
var. **longipedunculatum** A. V. Petrova *op. cit.*
1 44(1973)—Bulgaria.
var. **montanum** A. V. Petrova *op. cit.* 43(1973)—
1 Bulgaria.
var. **racemosa** A. V. Petrova *op. cit.* 43(1973)—
1 Bulgaria.
var. **slavjankae** A. V. Petrova *op. cit.* 43(1973)—
1 Bulgaria.
nodiflorum L.
f. **micranthum** A. V. Petrova *op. cit.* 39(1973)—
1 Bulgaria.
pallasianum Schult.
subsp. **borzeanum** (Nyar.) A. V. Petrova *op. cit.*
1 28(1973): *L. borzeanum.*
tauricum Willd.
subsp. **bulgaricum** (Podp.) A. V. Petrova *op. cit.*
1 23(1973): *L. bulgaricum.*
subsp. **linearifolium** (Lindem.) A. V. Petrova
1 *op. cit.* 23(1973): *L. flavum* var. *linearifolium.*
subsp. **serbicum** (Podp.) A. V. Petrova *op. cit.* 23
1 (1973): *L. serbicum.*
thracicum (Griseb.) Degen
subsp. **multiflorum** A. V. Petrova *op. cit.* 35
1 (1973)—Bulgaria.

LOASACEAE

3026 Leins, P. *and* Winhard, W.
Entwicklungsgeschichtliche Studien an Loasaceen-Blüten. (Developmental studies on flowers of Loasaceae). *Österr. Bot. Zeitschr.*, 122(3):145–165
4 (1973)—(En).

LOGANIACEAE

3027 Angenot, L., Bisset, N. G. *and* Denoël, A.
Alcaloïdes des Loganiacées. Sur l'isolement de l'hormone a partir du *Strychnos usambarensis* Gilg. *Pl.*
5 *Méd. Phytothér.*, 7(1):33–36(1973).

3028 Ariza Espinar, L.
Notas sobre Buddlejaceae argentinas. *Kurtziana*,
4 7:153–163(1973)—(En).

3029 Melhem, T. S. *and* Takahashi, M. C.
Pollen morphological studies—Loganiaceae. *Hoehnea*, 2:11–18(1972)—(Sp); illustrations. Species
4 from Brazil.

3030 Soukup, J.
Las Simplocáceas, Oleáceas, Loganiáceas y Desfontaniáceas del Perú, sus géneros y lista de
4 especies. *Biota*, 9(74):272–282(1973)—Keys.

3031 Tiagi, Y. D.
Nodal and floral anatomy of Loganiaceae. *Proc.*
2–7 *Indian Sci. Congr.*, 58(3):512(1971). Abstract only.

ANDROYA

3032 Leeuwenburg, A. J. M.
The Loganiaceae of Africa: 10. *Androya* H. Perr. *Acta Bot. Neerl.*, 22(4):456–459(1973)—
5 Illustration.

ANTHOCLEISTA

3033 Guédès, M.
Morphologie de la feuille d'*Anthocleista* (Loganiacées). *Phyton* (Austria), 15(1–2):45–55(1973)—(Ge);
5 illustrations.

3034 Leeuwenberg, A. J. M.
The Loganiaceae of Africa: 11. *Anthocleista* 2.
5 *Acta Bot. Neerl.*, 22(5):597–598(1973).

longifolia (Lamark) Boiteau in *Adansonia*, 13(2):250
5 (1973): *Plumeria longifolia.*

ANTONIA

3035 Paula, J. E. de
Estudo anatômico e palinológico de *Antônia ovata* Pohl. (Loganiaceae). *Acta Amazonica*, 2(2):55–70
4 (1972)—(En).

BUDDLEJA

3036 Grootendorst, H. J.
2–6 *Buddleia. Dendroflora*, no.8:38–42(1972)—Du.

STRYCHNOS

3037 Angenot, L. *and* Denoël, A.
Découvertes récentes dans la chimie des alcaloïdes du *Strychnos usambarensis* Gilg. *Pl. Méd. Phytothér.*,
5 7(4):284–292(1973).

3038 Bissett, N. G., Leenhouts, P. W., Leeuwenberg, A. J. M., Philcox, D., Tirel-Roudet, C. *and* Vidal, J. E.
The Asian species of *Strychnos*: part 2. Typification, miscellaneous notes, synoptic key and sectional
2, 6–7 classification. *Lloydia*, 36(2):179–201(1973).

3039 Krukoff, B. A.
Supplementary notes on the American species of
4 *Strychnos*: 12. *Phytologia*, 25(1):49–54(1972).

3040 Krukoff, B. A. *and* Barneby, R. C.
Supplementary notes on the American species of
4 *Strychnos*: 13. *Phytologia*, 27(2):97–107(1973).

alvimiana Krukoff & Barneby in *Phytologia*, 27(2):
4 105(1973)—Brazil.
cayennensis Krukoff & Barneby *op. cit.* 101(1973)—
4 French Guiana.
recognita Krukoff & Barneby *op. cit.* 103(1973)—
4 Brazil.
schunkei Krukoff & Barneby in *Phytologia*, 25(1):
4 53(1972)—Peru.

LORANTHACEAE

3041 Abdulla, P.
Flora of West Pakistan: no. 35. Loranthaceae. 1973.
6 9p.—Illustrations, map.

3042 Johri, B. M. *and* Bhatnagar, S. P.
Loranthaceae. New Delhi, Council of Scientific and Industrial Research, 1972. [viii] 155p. (Botanical Monograph no.8)—Illustrations, coloured
1–7 illustration, maps, dot maps.

3043 Perring, F. H.
Mistletoe. *In* Green, P. S. *ed.* Plants: wild and cultivated: 139–145(1973)—Dot maps. Distribution
1 in the British Isles.

3044 Vohra, S. C. A.
Morphological and embryological studies in
1–7 Viscaceae. *Botanica* (Delhi), 23(4):145–148(1973).

ALLOHEMIA

3045 Kuijt, J.
Proposal for the conservation of the generic name *Oryctanthus* Eichler (1868) (Loranthaceae), against *Glutago* Poiret (1821) and *Allohemia* Rafinesque
4 (1838). *Taxon*, 22(4):505–506(1973).

ARCEUTHOBIUM

3046 Hawksworth, F. G.
Dwarf mistletoes (*Arceuthobium*) of coniferous forests of the world. (Die auf Koniferen vorkommenden Zwergmisteln (*Arceuthobium* ssp.)). *In* European Weed Research Council: Parasitic Weeds Research Group *and* Malta: Royal University. Symposium on parasitic weeds, 11–13 April 1973, Royal University of Malta, Malta: 231–235(1973)—
1–6 (Ge,Fr).

3047 Mijušković, M.
Arceuthobium oxycedri (DC) M.B. (Loranthaceae), ranije nezabilježena parazitna biljka u flori Crne Gore. (*Arceuthobium oxycedri* (DC) M. B. (Loranthaceae) plante semiparasite non signalée auparavant au Montenegro). *Glasn. Republ. Zavoda Zašt. Prir. Prirodnj. Muz. Titogradu*, no.5:25–28(1972 publ.
1 1973)—(Fr); map.

6 **CECARRIA** Barlow in *Brittonia*, 25(1):34(1973).
obtusifolia (Merr.) Barlow *op. cit.* 34(1973):
6 *Phrygilanthus obtusifolius*.

DENDROPHTHORA

3048 Kuijt, J.
Proposal for the conservation of the generic name *Dendrophthora* Eichler (1868) (Viscaceae), against *Ozarthris* Rafinesque (1836(1938)). *Taxon*, 22(4):
4 506(1973).

GLUTAGO

3049 Kuijt, J.
Proposal for the conservation of the generic name *Oryctanthus* Eichler (1868) (Loranthaceae), against *Glutago* Poiret (1821) and *Allohemia* Rafinesque (1838).
4 *Taxon*, 22(4):505–506(1973).

HYPHEAR

3050 Chao, J.-M.
Genus *Hyphear* of Taiwan. *Taiwania*, 18(2):169–
2 172(1973)—Illustrations; key.

kaoi J.-M. Chao in *Taiwania*, 18(2):169(1973)—
2 Taiwan.

LORANTHUS

3051 Bhaumik, G. H.
Meiotic studies in the genus *Loranthus*. *Proc. Indian Sci. Congr.*, 59(3):346–347(1972)—Chromosome numbers. Abstract only; *L. ligustrinus*, *L. longiflorus*,
6 *L. scurrula*.

ORYCTANTHUS

3052 Kuijt, J.
Proposal for the conservation of the generic name *Oryctanthus* Eichler (1868) (Loranthaceae), against *Glutago* Poiret (1821) and *Allohemia* Rafinesque
4 (1838). *Taxon*, 22(4):505–506(1973).

3053 Rizzini, C. T.
Duas Lorantáceas novas. *Leandra*, 2(3):73–77
4 (1972)—Illustration. *Oryctanthus chlamydatus*.

chlamydatus Rizzini in *Leandra*, 2(3):73(1972)—
4 Venezuela.

PHRYGILANTHUS

3054 Barlow, B. A. *and* Wiens, D.
The classification of the generic segregates of *Phrygilanthus* (=*Notanthera*) of the Loranthaceae. *Brittonia*, 25(1):26–39(1973)—Chromosome
3, 4, 6, 7 numbers; key.

PHTHIRUSA

3055 Kuijt, J. *and* Weberling, F.
The flower of *Phthirusa pyrifolia* (Loranthaceae). *Ber. Deutsch. Bot. Ges.*, 85(7–9):467–480(1973)—(Ge);
4 illustrations.

PSITTACANTHUS

3056 Kuijt, J.
Further evidence for the systematic position of *Psittacanthus sonorae* (Loranthaceae). *Madroño*, 22(4):
4 177–185(1973)—Illustrations.

palmeri (Watson) Barlow & Wiens in *Brittonia*,
4 25(1):39(1973): *Loranthus palmeri*.

STRUTHANTHUS

3057 Rizzini, C. T.
Duas Lorantáceas novas. *Leandra*, 2(3):73–77
4 (1972)—Illustration. *Struthanthus maricensis*.

maricensis Rizzini in *Leandra*, 2(3):76(1972), without
4 type—Locality not stated.
panamensis (Rizzini) Barlow & Wiens in *Brittonia*,
4 25(1):39(1973): *Phrygilanthus panamensis*.

TRISTERIX

aphyllus (Miers ex DC.) Tiegh. ex Barlow & Wiens
4 in *Brittonia*, 25(1):39(1973): *Loranthus aphyllus*.
grandiflorus (R. & P.) Barlow & Wiens *op. cit.*
4 39(1973): *Loranthus grandiflorus*.

VISCUM

3058 Bailey, D.
Planhigion meddyginiaethol o gefn gwlad Cymru: 3. Yr uchelwydd (*Viscum album*). *Y Gwyd-*
1 *donydd*, 11(1):15–16(1973)—Illustration.

3059 Harms, H.
Die Mistel und ihre Verbreitung in Ostwestfalen. *Mitt. Deutsch. Dendrol. Ges.*, no.66:69–88(1973)—
1 Illustrations, map. *Viscum album*.

3060 Houfek, J.
Rozšíření jmelí (*Viscum* L.) v. Československu Pokračování. (Verbreitung von Mistel (*Viscum* L.) in der Tschechoslowakei: Fortsetzung). *Zprávy Česk.*
1 *Bot. Spol.*, 8(3):210–214(1973).

3061 Perring, F. H.
Mistletoe: a survey. *North. Gard.*, 27(5):138–140,
1 145(1973). *Viscum album* in the British Isles.

3062 Preywisch, K.
Zur Ökologie der Laubholzmistel (*Viscum album* L. ssp. *album*) im Oberen Weserbergland. *Decheniana*,
1 125(1–2):103–109(1972)—Maps.

LYTHRACEAE

3063 Guha, S.
Cytotaxonomical studies on the family Lythraceae.
Proc. Indian Sci. Congr., 59(3):344–345(1972)—
6 Chromosome numbers. Abstract only.

3064 Huynh, K.-L.
Étude de l'arrangement du pollen dans la tétrade
chez les angiospermes sur la base de données cyto-
logiques: 6. Lythraceae et Bombacaceae. *Bol. Soc.*
1–7 *Brot.*, 46:171–180(1972)—(En).

LYTHRUM

3065 Levin, D. A. *and* Kerster, H. W.
Assortative pollination for stature in *Lythrum*
1–3 *salicaria*. *Evolution*, 27(1):144–152(1973).

ROTALA

3066 Cook, C. D. K.
New and noteworthy plants from the northern
Italian ricefields. *Ber. Schweiz. Bot. Ges.*, 83(1):
54–65(1973). *Rotala densiflora* and *R. ramosior* new to
1 Europe.

MAGNOLIACEAE

3067 Agababian, V. S.
Pollen morphology of the family Magnoliaceae.
2–4, 6 *Grana*, 12(3):166–176(1972 publ. 1973).

3068 Jelitto, C. R.
Vielgestaltige Magnolien: 1–2. *Palmengarten*,
37(1):21–24,37(2):51–53(1973)—Illustrations.
2, 3 Popular article.

LIRIODENDRON

3069 Vick, C. B.
Yellow-poplar: an American wood. *U.S. Dep.*
Agric. F.S., no.272. 7p.(1973)—Illustrations, map.
3 *Liriodendron tulipifera* L.

MAGNOLIA

3070 Dandy, J. E.
Magnolia hypoleuca. *Baileya*, 19(1):44(1973)—
2 En.

3071 Murray, E.
Magnolia species descriptions. *Kalmia*, no.5. 17p.
2–4, 6 (1973).

MALPIGHIACEAE

3072 Badré, F.
Malpighiaceae. Flore du Gabon, 21:3–18(1973)—
5 Illustrations; keys.

3073 Mallick, R.
Taxonomic revision of the family Malpighiaceae in
India. *Proc. Indian Sci. Congr.*, 60(3):337–338(1973)—
6 Chromosome numbers.

3074 Nasir, Y.
Flora of West Pakistan: no.45. Malpighiaceae.
6 1973. 4p.—Illustrations, map.

BYRSONIMA

3075 Lewis, W. H.
A new species of *Byrsonima* (Malpighiaceae) from
Panama. *Brittonia*, 25(3):304–306(1973)—
4 Illustrations.

dressleri W. H. Lewis in *Brittonia*, 25(3):304(1973)—
4 Panama.

TRIASPIS

3076 Launert, E.
A new species of *Triaspis* (Malpighiaceae) named
in honour of G. Le Testu. *Bol. Soc. Brot.*, 46:385–
5 387(1972 publ. 1973)—Illustration.

letestuana Launert in *Bol. Soc. Brot.*, 46:386(1972
5 publ. 1973)—Gabon.

MALVACEAE

3077 Heslop-Harrison, J., Heslop-Harrison, Y., Knox, R. B.
and Howlett, B.
Pollen-wall proteins: 'gametophytic' and 'sporo-
phytic' fractions in the pollen walls of the Malvaceae.
1–7 *Ann. Bot.* (U.K.), 39(151):403–412(1973).

ABELMOSCHUS

3078 Arora, R. K. *and* Singh, H. B.
Abelmoschus tuberculatus and its relation to *A.*
esculentus (Malvaceae). *Baileya*, 19(2):90(1973)—
6 Chromosome numbers.

3079 Kundu, B. C. *and* Biswas, C.
Anatomical characters for distinguishing the genera
Abelmoschus and *Hibiscus*. *Proc. Indian Sci. Congr.*,
5–7 60(3):295(1973). Abstract only.

ALTHAEA

officinalis L.
subsp. **pseudoarmeniaca** (Polgár) Kárpáti ex Soó
in *Acta Bot. Acad. Sci. Hung.*, 18(1–2):176(1973):
1 *A. officinalis* var. *pseudoarmeniaca*.

BAKERIDESIA

3080 Bates, D. M.
A revision of *Bakeridesia* Hochreutiner subgenus
Bakeridesia (Malvaceae). *Gentes Herb.* (Ithaca),
10(5):425–491(1973)—Illustrations, dot maps;
3, 4 chromosome numbers; key.

sect. **Minores** D. M. Bates in *Gentes Herb.* (Ithaca),
4 10(5):466 (1973).
4 sect. **Tubicalyces** D. M. Bates *op. cit.* 449(1973).
bakerana (Rose) D. M. Bates *op. cit.* 473(1973), as
4 'bakeriana': *Abutilon bakerana*.
exalata D. M. Bates *op. cit.* 449(1973)—Guatemala,
4 El Salvador, Nicaragua.
gaumeri (Standley) D. M. Bates *op. cit.* 480(1973):
4 *Abutilon gaumeri*.
4 **gloriosa** D. M. Bates *op. cit.* 452(1973)—Mexico.
integerrima (Hooker) D. M. Bates *op. cit.* 467(1973):
4 *Sida integerrima*.
4 **molinae** D. M. Bates *op. cit.* 466(1973)—Honduras.
nelsonii (Rose) D. M. Bates *op. cit.* 455(1973):
4 *Abutilon nelsonii*.
pittieri (Donn. Smith) D. M. Bates *op. cit.* 478
4 (1973): *Abutilon pittieri*.
subcordata (Hochreutiner) D. M. Bates *op. cit.* 476
4 (1973): *Robinsonella subcordata*.
vulcanicola (Standley) D. M. Bates *op. cit.* 457(1973):
4 *Abutilon vulcanicola*.
yucatana (Standley) D. M. Bates *op. cit.* 452(1973):
4 *Abutilon yucatanum*.

GOSSYPIUM

3081 Fryxell, P. A.
The correct name of 'kidney cotton'. *Baileya*, 19(2):
4 91–92(1973).

3082 Kavathekar, A. K.
Origin and distribution of cotton. *Botanica* (Delhi),
2, 4–7 23(2–3):43–47(1973)—Chromosome number.

3083 Narayanan, S. S. *and* Sree Rangasamy, S. R.
Differences in anther number in reciprocal hybrids
between cultivated Asiatic cottons and *Gossypium*
anomalum. *J. Hered.*, 64(5):307–308(1973). Includes
2, 5, 6 notes on phylogeny.

3084 Omar, M. A. *and* Abdel'-Bari, A. A.
Tsitologicheskoe izuchenie nekotorȳkh mezh-
vidovȳkh gibridov roda *Gossypium* L. (Cytological
study of some interspecific hybrids in the genus
Gossypium L.). *Genetika*, 9(7):163–166(1973)—(En);
2–7 chromosome numbers.

barbadense L.
var. **braziliense** (Rafinesque) Fryxell in *Baileya*,
4 19(2):91(1973): *G. guyanense* var. *braziliense*.

HERISSANTIA

3085 Fryxell, P. A.
New species and other notes in the Malvaceae.
Brittonia, 25(2):77–85(1973)—Illustrations, dot
4 map.

dressleri Fryxell in *Brittonia*, 25(2):77(1973)—
4 Mexico.

HIBISCADELPHUS

3086 Bishop, L. E. *and* Herbst, D.
A new *Hibiscadelphus* (Malvaceae) from Kauai.
Brittonia, 25(3):290–293(1973)—Illustrations, map;
6 key.

3087 Stewart, M.
New species found on Kauai. *Bull. Pac. Trop. Bot.*
6 *Gard.*, 3(4):71(1973).

distans L. E. Bishop & Herbst in *Brittonia*, 25(3):290
6 (1973)—Hawaiian Is.

HIBISCUS

3088 Fryxell, P. A.
New species and other notes in the Malvaceae.
4 *Brittonia*, 25(2):77–85(1973)—Illustrations, dot map.

3089 Kundu, B. C. *and* Biswas, C.
Anatomical characters for distinguishing the genera
Abelmoschus and *Hibiscus*. *Proc. Indian Sci. Cong.*,
3–7 60(3):295(1973). Abstract only.

3090 Lebler, B. A.
Native *Hibiscus*. *Queensl. Agric. J.*, 99(3):128–134
7 (1973)—Illustrations.

3091 Rusanov, F. N.
Mezhvidovaya gibridizatsiya u vidov gibiskusa i
yukki kak primer akklimatizatsii. (Interspecific
hybridization of *Hibiscus* and *Yucca* species as an example
of the acclimatisation). *In* U.S.S.R.: Akademiya Nauk
SSSR: Sovet Botanicheskikh Sadov: Uspekhi
introduktsii rasteniĭ: 224–231(1973)—Coloured
2–7 illustrations.

3092 Wise, D. A.
Genome relationship of *Hibiscus mutabilis*. *J. Hered.*,
64(5):285–288(1973)—Illustrations; chromosome
2, 3 numbers.

longifilus Fryxell in *Brittonia*, 25(2):78(1973)—
4 Mexico.
4 **wilsonii** Fryxell *op. cit.* 80(1973)—Brazil.

LEBRONNECIA

3093 Sachet, M.-H.
The discovery of *Lebronnecia kokioides*. *Bull. Pac.*
6 *Trop. Bot. Gard.*, 3(3):41–43(1973)—Illustrations.

MALVA

3094 Geigy *Ltd*: Agricultural Chemicals Division
Malvaceae: mallow family. *Geigy Weed Tables*,
1–3 62(1973)—Illustration, coloured illustration.

MALVASTRUM

3095 Hall, J. B.
Malvastrum corchorifolium (Desr.) Britton in Ghana.
5 *Nigerian Field*, 38(4):189–192(1973)—Illustration.

PAVONIA

3096 Fryxell, P. A.
New species and other notes in the Malvaceae.
4 *Brittonia*, 25(2):77–85(1973)—Illustrations, dot map.

4 **gentryi** Fryxell in *Brittonia*, 25(2):81(1973)—Mexico.

ROBINSONELLA

3097 Fryxell, P. A.
A revision of *Robinsonella* Rose and E. G. Baker
(Malvaceae). *Gentes Herb.* (Ithaca), 11(1):1–26
4 (1973)—Illustrations, dot map; key.

sect. **Grayella** Fryxell in *Gentes Herb.* (Ithaca),
4 11(1):8 (1973).
4 sect. **Mirandella** Fryxell *op. cit.* 16(1973).
4 **brevituba** Fryxell *op. cit.* 16(1973)—Mexico.
4 **densiflora** Fryxell *op. cit.* 22(1973)—Salvador.
4 **hintonii** Fryxell *op. cit.* 21(1973)—Mexico.
lindeniana (Turcz.) Rose & E. G. Baker
subsp. **divergens** (Rose & E. G. Baker) Fryxell
4 *op. cit.* 12(1973): *R. divergens*.
4 **macvaughii** Fryxell *op. cit.* 24(1973)—Mexico.
pilosa Rose
subsp. **septentrionalis** Fryxell *op. cit.* 21(1973)—
4 Mexico.
4 **pilosissima** Fryxell *op. cit.* 19(1973)—Mexico.
pleiopoda (J. Donnell Smith) Fryxell *op. cit.* 15
4 (1973): *Abutilon pleiopodum*.
4 **samaricarpa** Fryxell *op. cit.* 13(1973)—Mexico.
speciosa Fryxell *op. cit.* 18(1973)—Guatemala,
4 Mexico, Salvador.

SIDA

3098 Hatusima, S.
(*Sida subspicata* F. v. M., a plant recently naturalized
in Japan). *J. Geobot.*, 20(4):102–103(1972)—Ja;
2 illustration.

3099 Savchenko, M. I. *and* Dmitrashko, P. I.
Stroenie semeni u *Sida hermaphrodita* Rusby. (Seed
structure in *Sida hermaphrodita* Rusby). *Bot. Zhurn.*,
3 58(4):570–576(1973).

MARANTACEAE

3100 Kennedy, H.
Notes on Central American Marantaceae: 1. New species and records from Panama and Costa Rica. *Ann. Missouri Bot. Gard.*, 60(2):413–426(1973)—
4 Illustrations.

CALATHEA

dressleri H. Kennedy in *Ann. Missouri Bot. Gard.*,
4 60(2):414(1973)—Panama.
robiniae H. Kennedy *op. cit.* 419(1973), as 'robinae'—
4 Costa Rica.
4 **venusta** H. Kennedy *op. cit.* 416(1973)—Panama.
4 **vinosa** H. Kennedy *op. cit.* 422(1973)—Costa Rica.

HYLACANTHE

panamaensis (Standley) H. Kennedy in *Ann. Missouri Bot. Gard.*, 60(2):426(1973): *Myrosma*
4 *panamaensis.*

MONOTAGMA

3101 Steyermark, J. A.
Novedades venezolanas. *Acta Bot. Venezuel.*, 8(1–4): 111–119(1973)—Illustrations. *Monotagma*
4 *yapacanensis.*

yapacanensis Steyermark & Bunting in *Acta Bot.*
4 *Venezuel.*, 8(1–4):114(1973)—Venezuela.

MELASTOMATACEAE

3102 Fernandes, A. *and* Fernandes, R. B.
Melastomataceae africanae novae vel minus cognitae: 6. *Bol. Soc. Brot.*, 46:63–80(1972)—
5 Illustrations.

3103 Uribe Uribe, L.
Catalogo ilustrado de las plantas de Cundinamarca: vol. 5. Passifloraceae, Begoniaceae, Melastomataceae.
4 1972. 165p. Melastomataceae p.61–159.

3104 Vinha, S. Guimarães da
Novas Melastomataceae encontradas na região sul da Bahia. *Bradea*, 1(30):319–326(1973)—(En);
4 illustrations.

3105 Wurdack, J. J.
Certamen Melastomataceis: 22. *Phytologia*, 26(6):
4 397–409(1973).

AXINANDRA

3106 Meijer, W.
The genus *Axinandra*—Melastomataceae: a missing link in Myrtales? *Ceylon J. Sci., Biol. Sci.*, 10(1):72–74 (1972) (Critical notes on Ceylon Plants 8)—
6 Illustrations.

CLIDEMIA

ablusa Wurdack in *Phytologia*, 26(6):402(1973)—
4 Ecuador.
4 **asplundii** Wurdack *op. cit.* 404(1973)—Ecuador.
4 **caudata** Wurdack *op. cit.* 403(1973)—Ecuador.
4 **imparilis** Wurdack *op. cit.* 401(1973)—Ecuador.
4 **ostentata** Wurdack *op. cit.* 400(1973)—Guyana.

CREAGHIELLA

3107 Nayar, M. P.
A review of the genus *Creaghiella* Stapf (Melastomataceae). *Proc. Indian Sci. Congr.*, 60(3):338(1973).
6 Abstract only.

philippinensis (Merr.) Nayar in *Proc. Indian Sci. Congr.*, 60(3):338(1973), without basionym ref.:
6 *Anerincleistus philippinensis.*

DISSOTIS

melleri Hook. f.
var. **greenwayi** (A. Fernandes & R. Fernandes) A. Fernandes & R. Fernandes in *Bol. Soc. Brot.*,
5 46:68(1972): *D. greenwayi.*
senegambiensis (Guill. & Perr.) Triana
var. **alpestris** (Taub.) A. Fernandes & R. Fernandes
5 *op. cit.* 69(1972): *D. alpestris.*
f. **irvingiana** (Hook.) A. Fernandes & R. Fernandes
5 *op. cit.* 69(1972): *D. irvingiana.*

GRAVESIELLA

speciosa A. Fernandes & R. Fernandes in *Bol. Soc.*
5 *Brot.*, 46:71(1972)—Burundi.

HAPLOPHYLLOPHORA (Brenan) A. Fernandes & R. Fernandes in *Bol. Soc. Brot.*, 46:70 (1972), as 'Haplophyllophorus': *Amphiblemma* sect.
5 *Haplophyllophora.*
acaulis (Cogn.) A. Fernandes & R. Fernandes *op. cit.*
5 70(1972): *Amphiblemma acaule.*
var. **brevipes** (Brenan) A. Fernandes & R. Fernandes *op. cit.* 70(1972): *Amphiblemma acaule*
5 var. *brevipes.*
seretii (De Wild.) A. Fernandes & R. Fernandes *op.*
5 *cit.* 71(1972): *Cincinnobotrys seretii.*

HETEROCENTRON

3108 Whiffin, T.
Analysis of a hybrid swarm between *Heterocentron elegans* and *H. glandulosum* (Melastomataceae).
4 *Taxon*, 22(4):413–423(1973)—Illustrations, dot map.

3109 Whiffin, T. P.
A systematic study of the genus *Heterocentron* (Melastomataceae). *Diss. Abstr. Int., B*, 33(9):4156
4 (1973).

MELASTOMA

3110 Yeim Yok Siv *and* Paris, R. R.
Sur quelques plantes à flavonoïdes du Cambodge appartenant aux genres *Cananga, Colona, Grewia, Leea* et *Melastoma. Pl. Méd. Phytothér.*, 6(4):299–305 (1972)—also publ. in *Trav. Lab. Matière Méd. Pharm.*
5 *Galénique Fac. Pharm. Paris*, 57(2) (1973).

MEMECYLON

insulare A. Fernandes & R. Fernandes in *Bol. Soc.*
5 *Brot.*, 46:65(1972)—Mozambique.
sansibaricum Taub.
var. **buchananii** (Gilg) A. Fernandes & R.
5 Fernandes *op. cit.* 66(1972): *M. buchananii.*
sousae A. Fernandes & R. Fernandes *op. cit.* 67(1972)
5 —Mozambique.
torrei A. Fernandes & R. Fernandes *op. cit.* 63(1972)—
5 Mozambique.

MICONIA

3111 Wurdack, J. J.
Certamen Melastomataceis: 21. *Phytologia*, 26(1):
4 3–11(1973).

bolivarensis Wurdack in *Phytologia*, 26(1):8(1973)—
4 Ecuador.
centrodesmoides Wurdack *op. cit.* 4(1973)—
4 Colombia, Ecuador.
cercophora Wurdack
subsp. **canelosana** Wurdack in *Phytologia*, 26(6):
4 397(1973)—Ecuador.

dapsiliflora Wurdack in *Phytologia*, 26(1):7(1973)—
4 Ecuador.
4 **explicita** Wurdack *op. cit.* 3(1973)—Ecuador.
harlingii Wurdack *op. cit.* 10(1973)—Ecuador,
4 Colombia.
lugonis Wurdack in *Phytologia*, 26(6):397(1973)—
4 Ecuador.
oligantha Wurdack in *Phytologia*, 26(1):5(1973)—
4 Ecuador.
pastazana Wurdack in *Phytologia*, 26(6):398(1973)—
4 Ecuador.
rimbachii Wurdack in *Phytologia*, 26(1):10(1973)—
4 Ecuador.
4 **seposita** Wurdack *op. cit.* 6(1973)—Ecuador.
trichopoda (Gleason) Wurdack *op. cit.* 4(1973):
4 *Clidemia trichopoda.*

OSSAEA

asplundii Wurdack in *Phytologia*, 26(6):405(1973)—
4 Ecuador.
capitata Guimarães da Vinha in *Bradea*, 1(30):322
4 (1973)—Brazil.
laxivenula Wurdack in *Phytologia*, 26(6):407(1973)—
4 Colombia, Ecuador.
quadrisulca (Naud.) Wurdack *op. cit.* 408(1973):
4 *Clidemia quadrisulca.*
sessilifolia (Triana) Wurdack *op. cit.* 407(1973):
4 *Oxymeris sessilifolia.*

POIKILOGYNE

3112 Nayar, M. P.
Three new species of *Poikilogyne* (Melastomataceae) from New Guinea. *J. Jap. Bot.*, 48(8):238–241
6 (1973)—(Ja); illustrations.

hooglandii Nayar in *J. Jap. Bot.*, 48(8):239(1973)—
6 New Guinea (Terr. New Guinea).
neglecta Nayar *op. cit.* 238(1973)—New Guinea
6 (W. Irian).
paucifolia Nayar *op. cit.* 241(1973)—New Guinea
6 (Papua).

ROUSSEAUXIA

3113 Jacques-Félix, H.
Contribution à l'étude du genre *Rousseauxia* (Melast.). *Adansonia*, 13(2):177–193(1973)—(En);
5 illustrations; key.

andringitrensis (H. Perr.) Jacques-Félix in *Adansonia*,
5 13(2):188(1973): *Osbeckia andringitrensis.*
aurata (H. Perr.) Jacques-Félix *op. cit.* 182(1973):
5 *Osbeckia aurata.*
chrysophylla (Descr.) DC.
var. **setulosa** (H. Perr.) Jacques-Félix *op. cit.* 181
5 (1973): *Osbeckia chrysophylla* var. *setulosa.*
cistoides Jacques-Félix *op. cit.* 185(1973)—
5 Malagasy.
dionychoides (Cogn.) Jacques-Félix *op. cit.* 182
5 (1973): *Osbeckia dionychoides.*
subsp. **fulva** Jacques-Félix *op. cit.* 183(1973)—
5 Malagasy.
5 **glauca** Jacques-Félix *op. cit.* 184(1973)—Malagasy.
5 **gracilis** Jacques-Félix *op. cit.* 187(1973)—Malagasy.
humbertii (H. Perr.) Jacques-Félix *op. cit.* 188(1973):
5 *Osbeckia humbertii.*
madagascariensis (Cogn.) Jacques-Félix *op. cit.*
5 185(1973): *Osbeckia madagascariensis.*
mandrarensis (H. Perr.) Jacques-Félix *op. cit.* 186
5 (1973): *Osbeckia mandrarensis.*
marojejensis Jacques-Félix *op. cit.* 188(1973)—
5 Malagasy.
var. **hirsuta** Jacques-Félix *op. cit.* 190(1973)—
5 Malagasy.
subsp. **lepidota** Jacques-Félix *op. cit.* 189(1973)—
5 Malagasy.
subsp. **sericea** Jacques-Félix *op. cit.* 191(1973)—
5 Malagasy.
minimifolia (Jum. & Perr.) Jacques-Félix *op. cit.* 188
5 (1973): *Osbeckia minimifolia.*
tamatavensis (H. Perr.) Jacques-Félix *op. cit.* 185
5 (1973): *Osbeckia tamatavensis.*

TIBOUCHINA

3114 Teixeira, A. H.
Arvore do papel—*Tibouchina papyrus* (Pohl.) Toledo. *Orquídea*, 30(3):104–110(1973)—
4 Illustrations.

paulo-alvinii Guimarães da Vinha in *Bradea*, 1(30):
4 320(1973)—Brazil.
stipulacea Guimarães da Vinha *op. cit.* 321(1973), as
4 'stipulaceae'—Brazil.

TOCOCA

chuivensis Wurdack
var. **ecuadorensis** Wurdack in *Phytologia*, 26(6): 399
4 (1973)—Ecuador, Colombia.

TRIURANTHERA

3115 Nayar, M. P.
Notes on the genus *Triuranthera* Back. (Melastomataceae). *J. Jap. Bot.*, 48(2):49–52(1973)—(Ja);
6 illustration; key.

dispar (Cogn.) Nayar in *J. Jap. Bot.*, 48(2):50(1973):
6 *Medinilla dispar.*

MELIACEAE

3116 Grijpma, P. *and* Styles, B. T., *comps*
Bibliografia selectiva sobre Meliaceas. Turrialba, Centro Interamericano de Documentación e Información Agrícola 1973. 143p. (IICA Biblio-
1–7 grafías, no. 14).

3117 Styles, B. T.
The flower biology of the Meliaceae and its bearing on tree breeding. *Silvae Genet.*, 21(5):175–182(1972)—
4–7 (Ge); illustrations.

APHANAMIXIS

3118 Ghosh, R. B.
Studies in the embryology of the family Meliaceae: 4. Fertilisation, endosperm and embryogeny of *Aphanamixis polystachya* (Wall.) Parker (syn. *Amoora rohituka* W. & A.): a medicinal plant with a discussion on its taxonomic status and horticulture. *An. Estac. Exp. Aula Dei*, 11(3–4):396–403(1972)—
6 (Sp); illustrations.

CEDRELA

3119 Cunliff, K. M.
Trees for tomorrow. *Trees S. Afr.*, 24(4):107–111
6 (1973)—Illustration. *Cedrela toona.*

MELIANTHACEAE

BERSAMA

3120 Jackson, B. P. *and* Jethwa, K. R.
Morphology and anatomy of the leaves of *Bersama abyssinica* Fresen. from Kenya and Uganda. *Bot. J.*
5 *Linn. Soc.*, 66(3):245–257(1973)—Illustration.

MENISPERMACEAE

3121 Krukoff, B. A. *and* Barneby, R. C.
Supplementary notes on American Menisper-
4 maceae: 9. *Phytologia*, 25(1):32–48(1972).

3122 Troupin, G. *and* Gonçalves, M. L.
Flora de Moçambique: 7. Menispermaceae. 1973.
5 27p.—Illustrations; keys.

ABUTA

3123 Albuquerque, B.W. P. de
Contribuição ao estudo da nervação foliar de plantas da flora amazônica: 3. Cinco espécies do gênero *Abuta* (Menispermaceae). *Acta Amazonica*,
4 2(1):21–28(1972)—(En).

fluminum Krukoff & Barneby in *Phytologia*, 25(1):
4 38(1972)—Peru, Ecuador.

TELITOXICUM

negroense (Krukoff & Moldenke) Krukoff in *Phytologia*, 25(1):37(1972), without basionym ref.:
4 *Abuta negroensis*.

TILIACORA

3124 Kundu, B. C. *and* Guha, S.
The genus *Tiliacora* in India. *Proc. Indian Sci.*
6 *Congr.*, 60(3):337(1973). Abstract only.

MENYANTHACEAE

3125 Hatschbach, G. *and* Imaguire, N.
Meniantaceas do estado do Paraná. *Bol. Mus. Bot. Mun.* (Curitiba), no. 9. 5p.(1973)—Illustrations, dot
4 map.

3126 Nilsson, S. *and* Ornduff, R.
World pollen and spore flora: 2. Angiospermae: Menyanthaceae Dum. Stockholm, Almqvist &
1–7 Wiksell, 1973. 20p.—Chromosome numbers.

NYMPHOIDES

3127 Kröhnke, G.
Die Verbreitung der Seekanne (*Nymphoides peltata*, Gentianaceae) in Niedersachsen unter besonderer Berücksichtigung ihrer Soziologie im Bremer Becken. *Nat. Kultur Jagd*, 25(4):90–97(1972)—
1 Illustration, dot maps.

MESEMBRYANTHEMACEAE
see
AIZOACEAE

MIMOSACEAE
see
LEGUMINOSAE

MOLLUGINACEAE
see
AIZOACEAE

MONIMIACEAE

SIPARUNA

3128 Araujo, P. A. de M. *and* Mattos Filho, A. de
Estrutura das madeiras brasileiras de angiospermas dicotiledôneas: 10. Monimiaceae (*Siparuna bifida* (Poepp. & Endl.) A. DC). *Brasil Florest.*, 4(14):41–45
4 (1973)—(En).

MONTINIACEAE

GREVEA

3129 Verdcourt, B.
Notes on the genus *Grevea* Bail. (Montiniaceae).
5 *Kew Bull.*, 28(1):147–150(1973).

eggelingii Milne-Redh.
subsp. **echinocarpa** (Mendes) Verdcourt in *Kew Bull.*, 28(1):149(1973): *G. eggelingii* var.
5 *echinocarpa*.
madagascariensis Baill.
subsp. **keniensis** Verdcourt *op. cit.* 147(1973)—
5 Kenya.
subsp. **sublevis** Verdcourt *op. cit.* 148(1973)—
5 Malagasy.

MORACEAE

3130 Burger, W. C.
Notes on the flora of Costa Rica: 3. New species in
4 the Moraceae. *Phytologia*, 26(6):421–434(1973).

COUSSAPOA

glaberrima Burger in *Phytologia*, 26(6):422(1973)—
4 Nicaragua, Costa Rica.

DORSTENIA

3131 Carauta, J. P. P., Valente, M. da C. *and* Sucre B., D.
Dorstenia grazielae (Moraceae)—Espécie nova do Estado do Rio de Janeiro. *Atas Soc. Biol. Rio de Janeiro*, 16(2–3):59–61(1973)—Illustration, coloured
4 illustration.

3132 Carauta, J. P. P., Valente, M. da C. *and* Sucre B., D.
Flora do Estado da Guanabara. *Dorstenia* L. (Moraceae) do Parque Nacional da Tijuca. *Anais Soc. Bot. Brasil*, 149–164(1972)—(En); illustrations, map;
4 key.

3133 Granville, J. J. de
Notes sur la biologie florale de quelques espèces du genre *Dorstenia* (Moracées). *Cah. O.R.S.T.O.M.*,
4–6 *Biol.*, no.15:61–97(1971)—(En).

3134 Lavranos, J. J.
Notes on the succulent flora of north-east Africa and southern Arabia. *Cact. Succ. J.* (U.S.A.), 44(6): 257–261(1972)—Illustrations, coloured illustration.
2, 5 *Dorstenia gypsophila*.

grazielae Carauta, Valente & Sucre in *Atas Soc.*
4 *Biol. Rio de Janeiro*, 16(2–3):59(1973)—Brazil.
gypsophila Lavranos in *Cact. Succ. J.* (U.S.A.),
4 44(5):257(1972)—Somali Republic.
ramosa (Desvaux) Carauta, Valente & Sucre in *Anais Soc. Bot. Brasil*, 1972:158(1972): *Sychinium*
4 *ramosum*.

FICUS

3135 Corner, E. J. H.
Identification list of Malaysian specimens: 43. *Ficus* (Moraceae) from India, Burma, Thailand, Indochina, China, Korea, Japan, Ryukyu, Formosa and Hainan. Leyden, Foundation Flora Malesiana,
2, 6 1972. pp.735–784.

3136 Corner, E. J. H.
New taxa of *Ficus* (Moraceae). *Blumea*, 20(2):427–
6 432(1972 publ. 1973)—Illustrations.

3137 D'Arcy, W. G.
New names for neotropical plants. *Phytologia*,
4 25(3):116(1973).

3138 Nagata, K. M. *and* Hirano, R. T.
A checklist of *Ficus* (Moraceae) in the Harold L. Lyon Arboretum. Honolulu, University of Hawaii Foundation, and Harold L. Lyon Arboretum Fund,
2, 4–7 1972. 12p.—Illustrations.

brevibracteata Burger in *Phytologia*, 26(6):423
4 (1973)—Costa Rica, Honduras.
citrifolia Mill.
var. **brevifolia** (Nutt.) D'Arcy in *Phytologia*, 25(3):
4 116(1973): *F. brevifolia*.
crassivenosa Burger in *Phytologia*, 26(6):424(1973)—
4 Costa Rica.
4 **laterisyce** Burger *op. cit.* 426(1973)—Costa Rica.
morazaniana Burger *op. cit.* 427(1973)—Guatemala,
4 Mexico, Costa Rica.
neriifolia J. E. Sm.
var. **nemoralis** (Wall. ex Miq.) Corner ex R. R. Stewart, Annot. Cat. Vasc. Pl. W. Pakistan &
6 Kashmir: 193(1972): *F. nemoralis*.
sarawakensis Corner in *Blumea*, 20(2):427(1972
6 publ. 1973)—Borneo (Sarawak).
spadicea Corner *op. cit.* 429(1972 publ. 1973)—New
6 Guinea (Papua).
turrialbana Burger in *Phytologia*, 26(6):429(1973)—
4 Costa Rica, Mexico.

HELIANTHOSTYLIS

3139 Romero-Castañeda, R.
Apuntes botanicos: 4. *Mutisia*, no.38. 18p.(1972)—
4 (En); illustrations. *Helianthostylis suerpo*, pp.1–7.

4 **suerpo** Romero in *Mutisia*, 38:3(1972)—Colombia.

MACLURA

3140 Burton, J. D.
Osage-orange: an American wood. *U.S. Dep. Agric. F.S.*, no.248. 7p.(1973)—Illustration, map.
3 *Maclura pomifera* Raf. Schneid.

3141 Kaastra, R. C.
Description and taxonomic position of *Maclura brasiliensis* (Moraceae). *Acta Bot. Neerl.*, 22(1):69–74
4 (1973)—Illustrations.

MORUS

3142 Gzȳryan, M. S. *and* Papyan, S. S.
Stroenie chereshka i zhilkovanie lista u nekotorȳkh sortov shelkovitsȳ. (The structure of petiole and leaf venation in mulberry). *Biol. Zhurn. Arm.*, 26(6):44–51
1, 2 (1973)—(Arm).

3143 Hotta, T.
Supplement to the knowledge of the systematics of *Morus* in Quelpaert. *Acta Phytotax. Geobot.*, 25(4–6):
6 186–187(1973)—Key.

bombycis Koidz.
var. **saishuensis** Hotta in *Acta Phytotax. Geobot.*,
2 25(4–6):186(1973)—Korea.

POUROUMA

umbellifera Burger in *Phytologia*, 26(6):430(1973)—
4 Costa Rica.

TROPHIS

involucrata Burger in *Phytologia*, 26(6):432(1973)—
4 Costa Rica.

MORINGACEAE

3144 Qaiser, M.
Flora of West Pakistan: no.38. Moringaceae. 1973.
6 4p.—Illustrations, map.

MUSACEAE

HELICONIA

3145 Barreiros, H. de Souza
Espécies críticas de *Heliconia* L.: 1. Ecologiae morfologia (Heliconiaceae). *Rev. Brasil. Biol.*, 32(3):
4 297–302(1972)—Illustration; key.

3146 Barreiros, H. de Souza *
Lista das espécies de *Heliconia* L. (Musaceae) de inflorescência pêndula. *Atas Soc. Biol. Rio de Janeiro*, 15(2):55–58(1972). Abstr. in *Ref. Zhurn. Biol.*, 2(1):
4 v507(1973).

3147 Linhart, Y. B.
Ecological and behavioral determinants of pollen dispersal in hummingbird-pollinated *Heliconia*. *Amer.*
4 *Nat.*, 107(956):511–523(1973).

MYOPORACEAE

3148 Sahay, S. K.
Trend of apertural evolution in pollen of some members of Myoporaceae. *Proc. Indian Sci. Congr.*,
4–7 60(3):369(1973). Abstract only.

EREMOPHILA

3149 Ey, T. M. *and* Barlow, B. A.
Distribution of chromosome races in the *Eremophila glabra* complex. *Search*, 3(9):337–338
7 (1972)—Dot maps.

MYRICACEAE

3150 Halim, A. F. *and* Collins, R. P.
Essential oil analysis of the Myricaceae of the eastern United States. *Phytochemistry*, 12(5):1077–1083
3 (1973).

3151 Silva, A. R. Pinto da
Myricaceae. *Anuár. Soc. Brot.*, 38:101–108(1972)—
1 Por; illustrations, dot maps; key.

MYRICA

3152 Abbe, E. C.
The inflorescence and flower in male *Myrica esculenta* var. *farquhariana*. *Bot. Gaz.*, 133(3):206–213
6 (1972)—Illustrations.

rothmalerana Pinto da Silva in *Anuár. Soc. Brot.*, 38:
1 106(1972), as 'rothmaleriana'—Portugal.

MYRISTICACEAE

3153 Hatschbach, G.
Miristicáceas do estado do Paraná. *Bol. Mus. Bot. Mun.* (Curitiba), no.4. 5p.(1972)—Illustration, dot
4 map.

3154 Reitz, R. *and* Klein, R. M.
Flórula da Ilha de Santa Catarina: Miristicáceas.
4 1972. 16p.—Illustrations; key.

BROCHONEURA

3155 Capuron, R.
Observations sur les Myristicacées de Madagascar: les genres *Brochoneura* Warb. et *Mauloutchia* Warb. *Adansonia*, 13(2):203–221(1973)—(En); illustrations;
5 key.

COELOCARYON

3156 Fouilloy, R.
Trois arbres nouveaux d'Afrique tropicale (Laur.-Myrist.-Sapind.). *Adansonia*, 12(4):545–555(1972 publ. 1973)—Illustrations; key to 4 spp. *Coelocaryon*
5 *sphaerocarpum*.

sphaerocarpum Fouilloy in *Adansonia*, 12(4):548 1972
5 publ. 1973)—Guinea, Ivory Coast.

MAULOUTCHIA

3157 Capuron, R.
Observations sur les Myristicacées de Madagascar: les genres *Brochoneura* Warb. et *Mauloutchia* Warb. *Adansonia*, 13(2):203–221(1973)—(En); illustrations;
5 key.

chapelieri (Baill.) Warb.
f. **ecristata** Capuron in *Adansonia*, 13(2):211(1973)
5 —Malagasy.
5 var. **media** Capuron *op. cit.* 211(1973)—Malagasy.
f. **sambiranensis** Capuron *op. cit.* 212(1973)—
5 Malagasy.
5 **coriacea** Capuron *op. cit.* 220(1973)—Malagasy.
5 **heckelii** Capuron *op. cit.* 217(1973)—Malagasy.
humblotii (H. Perr.) Capuron *op. cit.* 212(1973):
5 *Brochoneura humblotii*.
5 **parvifolia** Capuron *op. cit.* 214(1973)—Malagasy.
rarabe (H. Perr.) Capuron *op. cit.* 213(1973):
5 *Brochoneura rarabe*.

VIROLA

3158 Honda, M. *
Madeiras 'Ucuúba': 1. *Virola divergens* Ducke e *V. multinervia* Ducke. *Acta Amazonica*, 1(2):79–83 (1971 publ. 1972). Ref. in *Bull. Torrey Bot. Club*,
4 99(6):331(1972).

3159 Rodrigues, W. A.
A ucuuba de várzea e suas aplicações. *Acta Amazonica*, 2(2):29–48(1972)—(En); illustrations;
4 key. *Virola surinamensis*.

MYRSINACEAE

3160 Smith, A. C.
Studies of Pacific island plants: 25. The Myrsinaceae of the Fijian region. *J. Arnold Arbor.*, 54(1):1–41,(2):
6 228–292(1973)—Illustrations; keys.

CORRELLIANA D'Arcy in *Ann. Missouri Bot.*
4 *Gard.*, 60(2):442(1973).

3161 D'Arcy, W. G.
Correlliana (Myrsinaceae), a new palmoid genus of the tropical rain forest. *Ann. Missouri Bot. Gard.*,
4 60(2):442–448(1973)—Illustrations.

chamaephyta (Diels) D'Arcy in *Ann. Missouri Bot.*
4 *Gard.*, 60(2):445(1973): *Weigeltia chamaephyta*.
multiflora (A. C. Smith) D'Arcy *op. cit.* 445(1973):
4 *Weigeltia multiflora*.
schlimii (Hook. f.) D'Arcy *op. cit.* 443(1973):
4 *Comomyrsine schlimii*.
spectabilis (Standl.) D'Arcy *op. cit.* 443(1973):
4 *Ardisia spectabilis*.

DISCOCALYX

amplifolia A. C. Smith in *J. Arnold Arbor.*, 54(2):
6 269(1973)—Fiji.
6 **crinita** A. C. Smith *op. cit.* 272(1973)—Fiji.

RAPANEA

amischocarpa A. C. Smith in *J. Arnold Arbor.*, 54(2):
6 280(1973)—New Hebrides, Santa Cruz Is.
6 **hadrocarpa** A. C. Smith *op. cit.* 290(1973)—Fiji.
6 **longipes** A. C. Smith *op. cit.* 287(1973)—Samoa.
6 **polyantha** A. C. Smith *op. cit.* 289(1973)—Fiji.

TAPEINOSPERMA

divaricatum (Gillespie) A. C. Smith in *J. Arnold*
6 *Arbor.*, 54(2):241(1973): *Discocalyx divaricata*.
multiflorum (Gillespie) A. C. Smith *op. cit.* 252
6 (1973): *Discocalyx multiflora*.

MYRTACEAE

3162 Borhidi, A., Acuña, J. *and* Muñiz, O.
New Plants in Cuba: 3. *Acta Bot. Acad. Sci. Hung.*, 19(1–4):37–45(1973)—Illustrations. Myrtaceae
4 pp.37–41.

3163 Dawson, J. W.
Pacific capsular Myrtaceae: 4. The *Metrosideros* complex: *Xanthostemon*, *Nani*, *Pleurocalyptus*, *Purpureostemon*. *Blumea*, 20(2):315–322(1972 publ.
6, 7 1973)—Illustrations.

3164 Legrand, C. D.
Analisis de un trabajo de Mirtaceas: 'E Kausel – Mirtáceas sudamericanas nuevas o criticas'. *Bradea*,
4 1(28):309–312(1973).

3165 McVaugh, R.
Notes on West Indian Myrtaceae. *J. Arnold Arbor.*,
4 54(2):309–314(1973)—Key.

3166 Mehra, P. N. *and* Khosla, P. K.
Cytogenetical studies of East Himalayan Hamamelidaceae, Combretaceae and Myrtaceae. *Silvae Genet.*,
6 21(5):186–190(1972)—(Ge); chromosome numbers.

3167 Singh Raikwar, S. K. *and* Roy, S. K.
Relationships of Myrtaceae. *Proc. Indian Sci. Congr.*,
1–7 58(3):476(1971). Abstract only.

ABDULMAJIDIA

3168 Whitmore, T. C.
New Malaysian genus. *Commonw. Forest. Rev.*, 52(4):289(1973). *Abdulmajidia*: short nomenclatural
6 note.

BAECKEA

3169 Lebler, B. A.
Baeckeas. *Queensl. Agric. J.*, 99(10):519–527(1973)—
6, 7 En; illustrations.

CALLISTEMON

3170 Lothian, T. R. N.
Callistemon or 'bottle brushes'. *J. Roy. Hort. Soc.*, 98(10):453–457(1973)—Coloured illustrations.
7 Popular account of some cultivated spp.

CALYCORECTES

3171 Legrand, C. D. *and* Klein, R. M.
Flora ilustrada catarinense: Mirtaceas: 7. *Calycorectes* Berg. 1972: 553–569.—Illustrations, dot
4 maps; key.

EUCALPYPTUS

3172 Cunliff, K. M.
Trees for tomorrow. *Trees S. Afr.*, 24(4):107–111
7 (1973)—Illustrations. *Eucalyptus melliodora.*

3173 Hall, N. *and* Brooker, I.
Albany blackbutt *Eucalyptus staeri* (Maiden) Kessel et C. A. Gardn. *Austral. Forest. Timber Bur. Forest Tree Ser.*, no. 73. 3p.(1973)—Illustrations,
7 map.

3174 Hall, N. *and* Brooker, I.
Alpine cider gum *Eucalyptus archeri* Maiden et Blakely. *Austral. Forest. Timber Bur. Forest Tree Ser.*,
7 no.54. 3p.(1972)—Illustrations, map.

3175 Hall, N. *and* Brooker, I.
Baeuerlen's gum *Eucalyptus baeuerlenii* F. Muell. *Austral. Forest. Timber Bur. Forest Tree Ser.*, no.92.
7 3p.(1973)—Illustrations, map.

3176 Hall, N. *and* Brooker, I.
Barlee box *Eucalyptus lucasii* Blakely. *Austral. Forest. Timber Bur. Forest Tree Ser.*, no.74. 3p.
7 (1973)—Illustrations, map.

3177 Hall, N. *and* Brooker, I.
Bogong gum *Eucalyptus chapmaniana* A. K. Cameron. *Austral. Forest. Timber Bur. Forest Tree Ser.*,
7 no.86. 3p.(1973)—Illustrations, map.

3178 Hall, N. *and* Brooker, I.
Broad-leaved carbeen *Eucalyptus confertiflora* F. Muell. *Austral. Forest. Timber Bur. Forest Tree Ser.*,
7 no. 62. 3p.(1973)—Illustrations, map.

3179 Hall, N. *and* Brooker, I.
Bull mallee *Eucalyptus behriana* F. Muell. *Austral. Forest. Timber Bur. Forest Tree Ser.*, no.60.3p.(1973)—
7 Illustrations, map.

3180 Hall, N. *and* Brooker, I.
Bullich *Eucalyptus megacarpa* F. Muell. *Austral. Forest. Timber Bur. Forest Tree Ser.*, no.84. 3p.(1973)—
7 Illustrations, map.

3181 Hall, N. *and* Brooker, I.
Camden white gum *Eucalyptus benthamii* Maiden et Cambage. *Austral. Forest. Timber Bur. Forest Tree Ser.*,
7 no.57. 3p.(1973)—Illustrations, map.

3182 Hall, N. *and* Brooker, I.
Darling Range ghost gum *Eucalpytus laeliae* F. D. Podger et G. M. Chippendale. *Austral. Forest. Timber Bur. Forest Tree Ser.*, no.72. 3p.(1973)—Illustrations,
7 map.

3183 Hall, N. *and* Brooker, I.
Desert gum *Eucalyptus gongylocarpa* Blakely. *Austral. Forest. Timber Bur. Forest Tree Ser.*, no.81.
7 3p.(1973)—Illustrations, map.

3184 Hall, N. *and* Brooker, I.
Heart-leaved silver gum *Eucalyptus cordata* Labill. *Austral. Forest. Timber Bur. Forest Tree Ser.*, no.53.
7 3p.(1972)—Illustrations, map.

3185 Hall, N. *and* Brooker, I.
Gippsland mallee *Eucalyptus kitsoniana* (Luehm.) Maiden. *Austral. Forest. Timber Bur. Forest Tree Ser.*,
7 no.78. 3p.(1973)—Illustrations, map.

3186 Hall, N. *and* Brooker, I.
Gippsland manna gum *Eucalyptus pryoriana* L. Johnson. *Austral. Forest. Timber Bur. Forest Tree Ser.*,
7 no.89. 3p.(1973)—Illustrations, map.

3187 Hall, N. *and* Brooker, I.
Grampians gum *Eucalyptus alpina* Lindl. *Austral. Forest. Timber Bur. Forest Tree Ser.*, no.79. 3p.(1973)—
7 Illustrations, map.

3188 Hall, N. *and* Brooker, I.
Inland bloodwood *Eucalyptus terminalis* F. Muell. *Austral. Forest. Timber Bur. Forest Tree Ser.*, no.63.
7 3p.(1973)—Illustrations, map.

3189 Hall, N. *and* Brooker, I.
Kamarooka mallee *Eucalyptus froggattii* Blakely. *Austral. Forest. Timber Bur. Forest Tree Ser.*, no.80.
7 3p. (1973)—Illustrations, map.

3190 Hall, N. *and* Brooker, I.
Kopi gum *Eucalyptus striaticalyx* W. V. Fitzg. *Austral. Forest. Timber Bur. Forest Tree Ser.*, no.83. 3p.
7 (1973)—Illustrations, map.

3191 Hall, N. *and* Brooker, I.
Large-leaved cabbage gum *Eucalyptus grandifolia* R. Br. ex. Benth. *Austral. Forest. Timber Bur. Forest Tree*
7 *Ser.*, no.61. 3p.(1973)—Illustrations, map.

3192 Hall, N. *and* Brooker, I.
Mealy stringybark *Eucalyptus cephalocarpa* Blakely. *Austral. Forest. Timber Bur. Forest Tree Ser.*, no.58.
7 3p.(1973)—Illustrations, map.

3193 Hall, N. *and* Brooker, I.
Morrisby's gum *Eucalyptus morrisbyi* R. G. Brett. *Austral. Forest. Timber Bur. Forest Tree Ser.*, no.52. 3p.
7 (1972)—Illustrations, map.

3194 Hall, N. *and* Brooker, I.
Mount Buffalo gum *Eucalyptus mitchelliana* Cambage. *Austral. Forest. Timber Bur. Forest Tree Ser.*,
7 no.90. 3p.(1973)—Illustrations, map.

3195 Hall, N. *and* Brooker, I.
Mountain marri *Eucalyptus haematoxylon* Maiden. *Austral. Forest. Timber Bur. Forest Tree Ser.*, no.71.
7 3p.(1973)—Illustrations, map.

3196 Hall, N. *and* Brooker, I.
Mountain swamp gum *Eucalyptus camphora* R. T. Bak. *Austral. Forest. Timber Bur. Forest Tree Ser.*,
7 no.56. 3p.(1973)—Illustrations, map.

3197 Hall, N. *and* Brooker, I.
Omeo gum *Eucalyptus neglecta* Maiden. *Austral. Forest. Timber Bur. Forest Tree Ser.*, no.88. 3p.(1973)—
7 Illustrations, map.

3198 Hall, N. *and* Brooker, I.
Powderbark wandoo *Eucalyptus accedens* W. V. Fitzg. *Austral. Forest. Timber Bur. Forest Tree Ser.*,
7 no.82. 3p.(1973)—Illustrations, map.

3199 Hall, N. *and* Brooker, I.
Pricklybark *Eucalyptus todtiana* F. Muell. *Austral. Forest. Timber Bur. Forest Tree Ser.*, no.85. 3p.(1973)—
7 Illustrations, map.

3200 Hall, N. *and* Brooker, I.
Salmonbark wandoo *Eucalyptus lane-poolei* Maiden. *Austral. Forest. Timber Bur. Forest Tree Ser.*, no.75. 3p.
7 (1973)—Illustrations, map.

3201 Hall, N. *and* Brooker, I.
Shiny-barked gum *Eucalyptus pachycalyx* Maiden et Blakely. *Austral. Forest. Timber Bur. Forest Tree Ser.*,
7 no.64. 3p.(1973)—Illustrations, map.

3202 Hall, N. *and* Brooker, I.
Small-leaved gum *Eucalyptus parvifolia* Cambage. *Austral. Forest. Timber Bur. Forest Tree Ser.*, no.91. 3p.
7 (1973)—Illustrations, map.

3203 Hall, N. *and* Brooker, I.
Snow gum *Eucalyptus niphophila* Maiden et Blakely. *Austral. Forest. Timber Bur. Forest Tree Ser.*, no. 59.
7 3p.(1973)—Illustrations, map.

3204 Hall, N. *and* Brooker, I.
Swamp peppermint *Eucalyptus rodwayi* R. T. Bak. et G. G. Sm. *Austral. Forest. Timber Bur. Forest Tree*
7 *Ser.*, no. 51. 3p.(1972)—Illustrations, map.

3205 Hall, N. *and* Brooker, I.
Thin-leaved stringybark *Eucalyptus eugenioides* Sieb. ex Spreng. *Austral. Forest. Timber Bur. Forest*
7 *Tree Ser.*, no.93. 3p.(1973)—Illustrations, map.

3206 Hall, N. *and* Brooker, I.
Umbrawarra gum *Eucalyptus umbrawarrensis* Maiden. *Austral. Forest. Timber Bur. Forest Tree Ser.*,
7 no.65. 3p.(1973)—Illustrations, map.

3207 Hall, N. *and* Brooker, I.
Victorian eurabbie *Eucalyptus pseudoglobulus* Naudin ex Maiden. *Austral. Forest. Timber Bur. Forest Tree*
7 *Ser.*, no.77. 3p.(1973)—Illustrations, map.

3208 Hall, N. *and* Brooker, I.
Victorian silver gum *Eucalyptus crenulata* Blakely et de Beuzeville. *Austral. Forest. Timber Bur. Forest*
7 *Tree Ser.*, no.87. 3p.(1973)—Illustrations, map.

3209 Hall, N. *and* Brooker, I.
White peppermint *Eucalyptus pulchella* Desf. syn. *E. linearis* Dehnh. *Austral. Forest. Timber Bur. Forest*
7 *Tree Ser.*, no.55. 3p.(1972)—Illustrations, map.

3210 Hall, N. *and* Brooker, I.
Yarra gum *Eucalyptus yarraensis* Maiden et Cambage. *Austral. Forest. Timber Bur. Forest Tree Ser.*, no.76. 3p.
7 (1973)—Illustrations, map.

3211 Hall, N. *and* Gray, A. M.
Silver peppermint *Eucalyptus tenuiramis* Miq. *Austral. Forest. Timber. Bur. Forest Tree Ser.*, no.70.
7 3p.(1973)—Illustrations, map.

3212 Hall, N. *and* Gray, A. M.
Tasmanian alpine yellow gum *Eucalyptus subcrenulata* Maiden et Blakely. *Austral. Forest. Timber Bur. Forest Tree Ser.*, no.68. 3p.(1973)—Illustrations,
7 map.

3213 Hall, N. *and* Gray, A. M.
Tasmanian yellow gum *Eucalyptus johnstonii* Maiden. *Austral. Forest. Timber Bur. Forest Tree Ser.*,
7 no.69. 3p.(1973)—Illustrations, map.

3214 Hall, N. *and* Gray, A. M.
Urn gum *Eucalyptus urnigera* Hook. f. *Austral. Forest. Timber Bur. Forest Tree Ser.*, no.66. 3p.(1973)—
7 Illustrations, map.

3215 Hall, N. *and* Gray, A. M.
Varnished gum *Eucalyptus vernicosa* Hook f. *Austral. Forest. Timber Bur. Forest Tree Ser.*, no.67. 3p.(1973)—
7 Illustrations, map.

3216 Johnson, L. A. S. *and* Blaxell, D. F.
Current research on *Eucalyptus*. *Austral. Pl.*, 7(56):
7 174-177(1973)—Illustrations.

3217 Johnson, L. A. S. *and* Blaxell, D. F.
New taxa and combinations in *Eucalyptus*: 2. *Contrib. New S. Wales Nation. Herb.*, 4(6):379–383
7 (1973).

3218 Kirkpatrick, J. B., Simmons, D. *and* Parsons, R. F.
The relationship of some populations involving *Eucalyptus cypellocarpa* and *E. globulus* to the problem of phantom hybrids. *New Phytol.*, 72(4):867–876
7 (1973)—Maps.

3219 Parsons, R. F.
Disjunctions in the distribution of *Eucalyptus* species between western Victoria and the Mount Lofty—Flinders Ranges area, South Australia. *Proc.*
7 *Roy. Soc. Victoria*, 86(1):29–34(1973)—Maps.

3220 Sheikh, M. I.
Identification of *Eucalyptus tereticornis*. *Pakistan J.*
6, 7 *Forest.*, 22(3):389(1972)—Illustration.

3221 Southwell, I. A.
Variation in the leaf oil of *Eucalyptus punctata*.
7 *Phytochemistry*, 12(6):1341–1343(1973).

andrewsii Maiden
subsp. **campanulata** (R. T. Baker & H. G. Smith) L. A. S. Johnson & Blaxell in *Contrib. New S. Wales Nation. Herb.*, 4(6):381(1973): *E.*
7 *campanulata*.
gregsoniana L. A. S. Johnson & Blaxell *op. cit.* 380
7 (1973), nom. nov.: *E. pauciflora* var. *nana* Blakely.
macrorhyncha F. Muell. ex Benth.
subsp. **cannonii** (R. T. Baker) L. A. S. Johnson &
7 Blaxell *op. cit.* 379(1973): *E. cannonii*.
paliformis L. A. S. Johnson & Blaxell *op. cit.* 382
7 (1973)—Australia (New South Wales).
pauciflora Sieber ex Spreng.
subsp. **debeuzevillei** (Maiden) L. A. S. Johnson &
7 Blaxell *op. cit.* 379(1973): *E. debeuzevillei*.
subsp. **niphophila** (Maiden & Blakely) L.A.S.
7 Johnson & Blaxell *op. cit.* 379(1973): *E. niphophila*.
piperita Sm.
subsp. **urceolaris** (Maiden & Blakely) L. A. S.
7 Johnson & Blaxell *op. cit.* 381(1973): *E. urceolaris*.
radiata Sieber ex DC.
subsp. **robertsonii** (Blakely) L. A. S. Johnson &
7 Blaxell *op. cit.* 380(1973): *E. robertsonii*.
sclerophylla (Blakely) L. A. S. Johnson & Blaxell
7 *op. cit.* 381(1973): *E. haemastoma* var. *sclerophylla*.

EUGENIA

alainii Borhidi in *Acta Bot. Acad. Sci. Hung.*, 19(1–4):
4 39(1973)—Cuba.
chacueyana Alain in *Phytologia*, 25(5):267(1973)—
4 Hispaniola (Dominican Republic).
hodgei McVaugh in *J. Arnold Arbor.*, 54(2):310
4 (1973)—Windward Is.
leonis Borhidi & Muñiz in *Acta Bot. Acad. Sci. Hung.*,
4 19(1–4):39(1973)—Cuba.
pubicalyx Alain in *Phytologia*, 25(5):268(1973)—
4 Hispaniola (Dominican Republic).
sooana Borhidi in *Acta Bot. Acad. Sci. Hung.*, 19(1–4):
4 37(1973), as 'sooiana'—Cuba.
subspinulosa Borhidi & Muñiz *op. cit.* 38(1973)—
4 Cuba.
yumana Alain in *Phytologia*, 25(5):268(1973)—
4 Hispaniola (Dominican Republic).

HOTTEA

neibensis Alain in *Phytologia*, 25(5):269(1973)—
4 Hispaniola (Dominican Republic).

LINDSAYOMYRTUS B. P. M. Hyland & van
6, 7 Steenis in *Blumea*, 21(2):190(1973).

3222 Hyland, B. P. M. *and* Steenis, C. G. G. J. van
The generic identity of *Xanthostemon brachyandrus* C. T. White. *Lindsayomyrtus* novum genus
6, 7 (Myrtaceae). *Blumea*, 21(2):189–192(1973).

brachyandrus (C. T. White) B. P. M. Hyland & van Steenis in *Blumea*, 21(2):190(1973): *Xanthostemon*
6, 7 *brachyandrus*.

MELALEUCA

3223 Prakash, N.
A study of the embryology of *Melaleuca*. *Malaysian*
6, 7 *J. Sci.*, 2(A):81–85(1973)—Illustrations.

3224 Verdon, D.
Melaleuca species at Canberra Botanic Gardens: part 1. *Austral. Pl.*, 7(53):2–12,25–44(1972)—Coloured
7 illustrations.

METROSIDEROS

3225 Corn, C. A. *and* Hiesey, W. M.
Altitudinal variation in Hawaiian *Metrosideros*. *Amer. J. Bot.*, 60(10):991–1002(1973)—Illustrations,
6 maps.

3226 Dawson, J. W.
Pacific capsular Myrtaceae: 5. The *Metrosideros* complex: *M. elegans* group. *Blumea*, 20(2):323–326
6 (1972 publ. 1973)—Illustrations.

3227 Dawson, J. W.
Pacific capsular Myrtaceae: 6. The *Metrosideros* complex: *M. perforata* and the *M. operculata* group. *Blumea*, 20(2):327–329(1972 publ. 1973)—
7 Illustrations.

3228 Porter, J. R.
The growth and phenology of *Metrosideros* in
6 Hawaii. *Diss. Abstr. Int., B*, 33(9):4153(1973).

3229 Smith, A. C.
Studies of Pacific island plants: 26. *Metrosideros collina* (Myrtaceae) and its relatives in the southern Pacific. *Amer. J. Bot.*, 60(5):479–490(1973)—
6 Illustrations; keys.

ochrantha A. C. Smith in *Amer. J. Bot.*, 60(5):490
6 (1973)—Fiji.

MOORIA

3230 Dawson, J. W.
Pacific capsular Myrtaceae: 7. *Mooria*. *Blumea*,
6 20(2):331–334(1972 publ. 1973)—Illustrations.

MYRCIA

antillana McVaugh in *J. Arnold Arbor.*, 54(2):311
4 (1973)—Guadeloupe Is., Windward Is.

MYRTUS

3231 Ahmed, M. *and* Vardar, Y.
Distribution and plasticity of *Myrtus communis*. *Phyton* (Austria), 15(1–2):145–150(1973)—(Ge);
1–7 illustrations, dot maps.

PSIDIUM

3232 Contin, L. F.
Contribuição ao estudo anatômico do *Psidium hatschbachii* Legrand. (Contribution to the anatomical study of *Psidium hatschbachii* Legrand). *Acta Biol. Parana.*, 1(3–4):27–32(1972)—(En,Fr);
4 illustrations.

3233 Sree Rangasamy, S. R. *and* Das, L. D. V.
Seedlessness in triploid guava (*Psidium guajava* L.): embryological studies. *Canad. J. Genet. Cytol.*, 15(2):
2–7 331–334(1973)—Chromosome numbers.

gracilipes Alain in *Phytologia*, 25(5):269(1973)—
4 Hispaniola (Dominican Republic).
longipes (Berg) McVaugh in *J. Arnold Arbor.*, 54(2):
4 312(1973): *Eugenia longipes*.
var. **orbiculare** (Berg) McVaugh *op. cit.* 314
4 (1973): *Eugenia orbicularis*.
nannophyllum Alain in *Phytologia*, 25(5):270
4 (1973)—Hispaniola (Dominican Republic).
sessilifolium Alain *op. cit.* 270(1973)—Hispaniola
4 (Dominican Republic).

SYZYGIUM

3234 Hartley, T. G. *and* Perry, L. M.
A provisional key and enumeration of species of *Syzygium* (Myrtaceae) from Papuasia. *J. Arnold*
6 *Arbor.*, 54(2):160–227(1973)—Key.

3235 Mitra, S. N.
Some new combinations in Indian plants. *Indian*
6 *Forester*, 99(2):100–101(1973).

amplum T. G. Hartley & Perry in *J. Arnold Arbor.*, 54(2):192(1973)—New Guinea (Terr. New
6 Guinea, Papua).
burepense T. G. Hartley & Perry *op. cit.* 184(1973)—
6 New Guinea (Terr. New Guinea).
busuense T. G. Hartley & Perry *op. cit.* 206(1973)—
6 New Guinea (Terr. New Guinea).
carrii T. G. Hartley & Perry *op. cit.* 209(1973)—New
6 Guinea (W. Irian).
cauliflorum T. G. Hartley & Perry *op. cit.* 182
6 (1973)—New Guinea (Papua).
coalitum (Greves) T. G. Hartley & Perry *op. cit.*
6 196(1973): *Eugenia coalita*.
cumini (L.) Skeels
var. caryophyllifolia (Du.) R. R. Stewart, Annot. Cat. Vasc. Pl. W. Pakistan & Kashmir: 504 (1972), without basionym date: *Eugenia*
6 *jambolana* var. *caryophyllifolia*.
diospyrifolium (Wall. ex Duthie) S. N. Mitra in
6 *Indian Forester*, 99(2):100(1973): *Eugenia diospyrifolia*.
erythropetalum T. G. Hartley & Perry in *J. Arnold Arbor.*, 54(2):186(1973)—New Guinea (Terr. New
6 Guinea).
fibrosum (F. M. Bailey) T. G. Hartley & Perry *op.*
6, 7 *cit.* 201(1973): *Eugenia fibrosa*.
flavidum T. G. Hartley & Perry *op. cit.* 183(1973)—
6 New Guinea (Terr. New Guinea).
gratum (Wt.) S. N. Mitra in *Indian Forester*, 99(2):
6 100(1973): *Eugenia grata*.
insulare T. G. Hartley & Perry in *J. Arnold Arbor.*, 54(2):183(1973)—New Britain, Bougainville,
6 British Solomon Is.
madangense T. G. Hartley & Perry *op. cit.* 188
6 (1973)—New Guinea (Terr. New Guinea).
normanbiense T. G. Hartley & Perry *op. cit.* 189
6 (1973)—New Guinea (Papua).
pluviatile T. G. Hartley & Perry *op. cit.* 199(1973)—
6 New Guinea (W. Irian, Terr. New Guinea, Papua).

robbinsii T. G. Hartley & Perry *op. cit.* 209(1973)—
6 New Guinea (Terr. New Guinea).
sambogense T. G. Hartley & Perry *op. cit.* 207
6 (1973)—New Guinea (Papua).
suborbiculare (Benth.) T. G. Hartley & Perry *op. cit.*
6, 7 189(1973): *Eugenia suborbicularis.*
sylvicola T. G. Hartley & Perry *op. cit.* 198(1973)—
6 New Guinea (Terr. New Guinea, Papua).
thornei T. G. Hartley & Perry *op. cit.* 206(1973)—
6 New Guinea (Terr. New Guinea, Papua).
tierneyanum (F. Muell.) T. G. Hartley & Perry *op.*
6 *cit.* 200(1973): *Eugenia tierneyana.*
validinerve T. G. Hartley & Perry *op. cit.* 193
6 (1973)—New Guinea (Terr. New Guinea).
variabile T. G. Hartley & Perry *op. cit.* 195(1973)—
6 New Guinea (Terr. New Guinea).
waikaiunense T. G. Hartley & Perry *op. cit.* 194
6 (1973)—New Guinea (Papua).
womersleyi T. G. Hartley & Perry *op. cit.* 193
6 (1973)—New Guinea (Terr. New Guinea, Papua).

TEPUALIA

3236 Dawson, J. W.
Pacific capsular Myrtaceae: 8. *Tepualia. Blumea,*
4 20(2):335–337(1972 publ. 1973)—Illustrations.

XANTHOSTEMON

3237 Hyland, B. P. M. *and* Steenis, C. G. G. J. van
The generic identity of *Xanthostemon brachyandrus*
C. T. White. *Lindsayomyrtus* novum genus (Myrtaceae).
6 *Blumea,* 21(2):189–192(1973).

NAJADACEAE

3238 Wentz, W. A. *and* Haynes, R. R.
Flora of Panama: part 2. Family 3b. Najadaceae.
Ann. Missouri Bot. Gard., 60(1):1–5(1973)—
4 Illustrations; key.

NAJAS

3239 Clemedson, C.-J.
Najas flexilis funnen i Södermanland. *Svensk Bot.*
1 *Tidskr.,* 67(3):303–306(1973)—(En); illustration.

3240 Dorofeev, P. I.
K sistematike iskopaemȳkh nayad podroda
Caulinia (Willd.) Aschers. (On the taxonomy of fossil
naiads of the subgenus *Caulinia* (Willd.) Aschers.).
1–7 *Bot. Zhurn.,* 58(3):385–394(1973).

3241 Hsu, C.-C.
Some noteworthy plants found in Taiwan.
Taiwania, 18(1):62–72(1973)—Illustrations; key.
2 *Najas,* pp.67–71.

3242 Hunziker, A. T. *and* Subils, R.
Sobre una nueva especie de *Najas* del centro de
Argentina: *N. arcana* nov. sp. *Lorentzia,* no.2:9–13
4 (1973)—(En); illustration.

3243 Zając, E. U. *and* Zając, A.
Nowe badania nad rozmieszczeniem gatunków
synantropijnych w Americe Północnej. *Wiad. Bot.,*
3 17(3):163–165(1973)—Dot map. *Najas minor.*

arcana Subils & A. T. Hunziker in *Lorentzia,* 2:9
4 (1973)—Argentina.

NECTAROPETALACEAE
see
OLACACEAE

NEPENTHACEAE

NEPENTHES

3244 Kondo, K.
The chromosome number of *Nepenthes* × *mixta.*
2, 5–7 *J. Jap. Bot.,* 48(6):189–190(1973).

3245 Lim Ah-Lan *and* Prakash, N.
Life history of *Nepenthes gracilis. Malaysian J. Sci.,*
5–7 2(A):45–53(1973)—Illustrations.

3246 Stewart, M.
Nepenthes hunting on Mt. Kinabalu. *Bull. Pac. Trop.*
6 *Bot. Gard.,* 1(4):8–11(1971)—Illustrations.

3247 Toyoda, Y.
Nepenthes and I—Mt. Kinabalu (Borneo Malaysia)
6 trip. *Carniv. Pl. Newsl.,* 1(4):62–63(1973).

NYCTAGINACEAE

3248 Fernandes, R. B.
Nyctaginaceae. *Anuár. Soc. Brot.,* 38:117–118
1 (1972)—Por.

MIRABILIS

3249 Cruden, R. W.
Reproductive biology of weedy and cultivated
Mirabilis (*Nyctaginaceae*). *Amer. J. Bot.,* 60(8):802–809
3, 4 (1973)—Illustrations; chromosome numbers.

NYMPHAEACEAE

3250 Dubina, D. V.
Poshȳrennya fitotsenolohiya ta produktȳvnist
latattyevȳkh u vodoskhovȳshchakh Dniprovs'koho
kaskadu. (Distribution, phytocenology and reserves
of Nymphaeaceae in reservoirs of the Dnieper
Cascade). *Ukr. Bot. Zhurn.,* 30(6):694–701(1973)—
1 Uk(En); maps.

3251 Mendonça, F. A.
Flora de Moçambique: 8. Nymphaeaceae. 1973.
5 7p.—Key.

EURYALE

3252 Gupta, P. P. *and* Roy, S. K.
Cytology of *Euryale ferox* Salisb. *Proc. Indian Sci.*
Congr., 59(3):364–365(1972)—Chromosome number.
6 Abstract only.

3253 Gupta, P. P. *and* Roy, S. K.
Karyotype of *Euryale ferox* Salisb. *Curr. Sci.,* 42(3):
6 106(1973).

NELUMBO

3254 Shilov, M. P.
K ekologii nekotorȳkh redkikh vodnȳkh rasteniĭ
na severnom predele ikh rasprostraneniya. (To
the ecology of some rare aquatic plants on the northern
limits of their area). *Bot. Zhurn.,* 58(10):1554–1559
1, 2 (1973). *Nelumbo comarovii.*

NUPHAR

3255 Wicks, J. C.
Some morphological aids in distinguishing *Nuphar*
microphyllum from similar aquatics. *Rhodora,* 75
(801):65–74(1973)—Illustrations. The species in New
3 Hampshire.

NYMPHAEA

alba L.

f. **csepelensis** Soó in *Acta Bot. Acad. Sci. Hung.*,
1 18(1–2):172(1973)—Hungary.

NYSSACEAE

NYSSA

3256 Mohana Rao, P. R.
Embryology of *Nyssa sylvatica*, and systematic consideration of the family Nyssaceae. *Phyto-*
2–4, 6 *morphology*, 22(1):8–21(1972 publ. 1973).

OCHNACEAE

3257 Kanis, A.
Ochnaceae. Flore du Cambodge, du Laos et du
6 Viêt-nam fasc. 14:3–16(1973)—Illustrations; key.

3258 Robson, N. K. B. *and* Garcia, J. G.
Flora de Moçambique: 40. Ochnaceae. 1973. 29p.—
5 Keys.

OCTOKNEMACEAE
see
OLACACEAE

OLACACEAE

3259 Badré, F.
Nectaropetalaceae. Flore du Gabon, 21:19–22
5 (1973)—Illustration; key.

3260 Villiers, J.-F.
Octoknemaceae. Flore du Cameroun, 15:181–192 (1973)—Illustrations; keys. Also published as Flore du
5 Gabon, 20(1973).

3261 Villiers, J.-F.
Oktoknemaceae. Flore du Gabon, 20: 181–192 (1973)—Illustrations; keys. Also published as Flore du
5 Cameroun, 15(1973).

3262 Villiers, J.-F.
Olacaceae. Flore du Cameroun, 15:101–162(1973)—Illustrations; keys. Also published as Flore du Gabon,
5 20(1973).

3263 Villiers, J.-F.
Olacaceae. Flore du Gabon, 20: 101–162(1973)—Illustrations; keys. Also published as Flore du
5 Cameroun, 15(1973).

ANACOLOSA

3264 Hallé, N.
Crochets de lianes du Gabon: *Ancistrocladus* et *Anacolosa* (Ancistrocladacées et Olacacées). *Adansonia*,
5 13(3):299–302(1973)—(En); illustrations.

STROMBOSIA

pustulata Oliver

var. **lucida** (Léonard) Villiers in Fl. Cameroun, 15:139(Jan. 30, 1973); and in Fl. Gabon, 20:139
5 (Jan. 30, 1973): *S. glaucescens* var. *lucida*.

OLEACEAE

3265 Liben, L.
Flore d'Afrique Centrale (Zaïre-Rwanda-Burundi): Spermatophytes. Oleaceae. 1972. 36p.—Illustrations;
5 keys.

3266 Soukup, J.
Las Simplocáceas, Oleáceas, Loganiáceas y Desfontaniáceas del Perú, sus géneros y lista de especies.
4 *Biota*, 9(74):272–282(1973)—Keys.

FORSYTHIA

3267 Green, P. S.
Forsythia cv. 'Karl Sax'. *Curtis's Bot. Mag.*, 179(4): tab.652(1973)—En; illustration, coloured illustration;
0 chromosome numbers. Its origin and biology.

3268 Hara, H.
A new species of *Forsythia* from Is. Shôdoshima.
2 *J. Jap. Bot.*, 48(8):256(1973)—(Ja).

togashii Hara in *J. Jap. Bot.*, 48(8):256(1973)—
2 Japan.

FRAXINUS

3269 Resmeriţă, I.
Nouveaux taxons de la flore de Roumanie. *Rev. Roum. Biol., Bot.*, 17(5):311–312(1972)—
1 Illustration.

3270 Stewart, H. A. *and* Krajicek, J. E.
Ash: an American wood. *U.S. Dep. Agric. F.S.*,
3 216. 7p.(1973)—Illustrations, maps.

3271 Taylor, S. M. O.
Ecological and genetic isolation of *Fraxinus americana* and *Fraxinus pennsylvanica*. *Diss. Abstr.*
3 *Int., B*, 33(11):5176(1973).

angustifolia Vahl

subsp. **pannonica** Soó & Simon in *Acta Bot. Acad. Sci. Hung.*, 18(1–2):174(1973)—Hungary,
1 Rumania.

ornus L.

f. microphylla Resmeriţă in *Rev. Roum. Biol.*,
1 *Bot.*, 17(5):312(1972), without type—Rumania.

LINOCIERA

3272 Liben, L.
Distribution géographique de deux *Linociera* (Oleaceae) d'Afrique centrale. *Bull. Jard. Bot. Nation.*
5 *Belg.*, 43(3–4):357–358(1973).

OLINIACEAE

3273 Liben, L.
Flore d'Afrique Centrale (Zaïre-Rwanda-Burundi): Spermatophytes. Oliniaceae. 1973. 5p.—
5 Illustration.

OLINIA

3274 Liben, L.
Réhabilitation de *Olinia rochetiana* A. Juss. (Oliniacée). *Bull. Jard. Bot. Nation. Belg.*, 43(2–3):235–
5 236(1973).

ONAGRACEAE

3275 Bizzarri, M. P. *and* Raven, P. H.
Adumbratio florae Aethiopicae: 23. Onagraceae. *Webbia*, 27(2):467–504(1972 publ. 1973)—En(It); illustrations, dot maps; chromosome numbers.
5 *Oenothera*, *Epilobium*, *Ludwigia*.

3276 Dement, W. A. *and* Raven, P. H.
Distribution of the chalcone, isosalipurposide in the Onagraceae. *Phytochemistry*, 12(4):807–808
1–7 (1973).

3277 Eyde, R. H. *and* Morgan, J. T.
Floral structure and evolution in Lopezieae (Onagraceae). *Amer. J. Bot.*, 60(8):771–787(1973)—
4 Illustrations; chromosome number.

3278 Plitmann, U., Raven, P. H. *and* Breedlove, D. E.
The systematics of Lopezieae (Onagraceae). *Ann. Missouri Bot. Gard.*, 60(2):478–563(1973)—Illustra-
4 tions, dot maps; chromosome numbers; keys.

3279 Vu Van Cuong *and* Vidal, J. E.
Onagraceae. Flore du Cambodge, du Laos et du Viêt-nam, fasc. 14:17–39(1973)—Illustrations;
6 keys.

CAMISSONIA

3280 Linsley, E. G., Macswain, J. W. *and* Raven, P. H.
Comparative behavior of bees and Onagraceae: 5. *Camissonia* and *Oenothera* bees of Cismontane California and Baja California. *Univ. Calif. Publ. Bot.*,
3, 4 71.[ii]68p.(1973)—Illustrations, dot maps.

CLARKIA

3281 Gottlieb, L. D.
Enzyme differentiation and phylogeny in *Clarkia franciscana*, *C. rubicunda* and *C. amoena*. *Evolution*,
4 27(2):205–214(1973).

3282 Lewis, H.
The origin of diploid neospecies in *Clarkia*. *Amer. Nat.*, 107(954):116–170(1973)—Chromosome
3, 4 numbers.

3283 Macswain, J. W., Raven, P. H. *and* Thorp, R. W.
Comparative behavior of bees and Onagraceae: 4. *Clarkia* bees of the western United States. *Univ.*
3 *Calif. Publ. Entomol.*, 70.[ii]80p.(1973)—Maps.

3284 Stevens, F. A.
Clarkia unguiculata, adaptation of a population to a Mediterranean climate. *Diss. Abstr. Int.*, *B*, 33(12):
3, 4 5776–5777(1973).

3285 Tin-Wa, M., Vasek, F. C. *and* Scora, R. W.
Phytochemical investigations of three species of
3 *Clarkia*. *Pl. Med.*, 24(4):397–401(1973).

DIPLANDRA

3286 McVaugh, R.
Report of the Committee for Spermatophyta: conservation of generic names: proposal 325(5826) *Diplandra* Hooker and Arnott (1838) vs. *Diplandra*
4 Bertero (1829). *Taxon*, 22(1):155–156(1973).

EPILOBIUM

3287 Einarsson, E.
Eyrarós (*Epilobium latifolium* L.). *Náttúrúfraeðingurinn*, 43(1–2):1–11(1973)—(En); coloured
1–3 illustration.

3288 Geigy *Ltd*: Agricultural Chemicals Division
Oenotheraceae: evening primrose family (= Onagraceae). *Geigy Weed Tables*, 65(1973)—
1–7 Illustrations, coloured illustrations.

3289 Holub, J.
Neue oder wenig bekannte Pflanzen der ungarischen Flora. *Ann. Univ. Budapest. Rolando Eötvos Nom., Biol.*, 14:91–104(1972 publ. 1973). *Epilobium*
1 *adenocaulon*.

3290 Hovda, J. T.
De amerikanske artene av slekta *Epilobium* L. s.str. som er funnet i Norge. (The American species of the genus *Epilobium* L. s.str. found in Norway). *Blyttia*,
1 31(1):19–28(1973)—(En); dot maps.

3291 Nylander, C.-E.
Fynd av *Epilobium adenocaulon* i Blekinge (*Epilobium adenocaulon* found in Blekinge). *Svensk Bot. Tidskr.*,
1 67(1):68(1973).

3292 Zając, E. U. and Zając, A.
Nowe badania nad rozmieszczeniem gatunków synantropijnych w Americe Północnej. *Wiad. Bot.*, 17(3):163–165(1973)—Dot map. *Epilobium*
3 *hirsutum*.

FUCHSIA

3293 Saunders, E.
Wagtails Book of Fuchsias: vol.3. Godalming, E. R. M. Saunders, 1973. 112p.—Illustrations, coloured
0 illustration. English bred species.

LOPEZIA

3294 Plitmann, U., Raven, P. H. *and* Breedlove, D. E.
New taxa and recombinations in *Lopezia* (Onagraceae). *Ann. Missouri Bot. Gard.*, 59(2):279–
4 281(1972 publ. 1973).

sect. **Diplandra** (Hook. & Arn.) Plitmann, Raven & Breedlove in *Ann. Missouri Bot. Gard.*, 60(2):498
4 (1973): genus *Diplandra*.
sect. **Jehlia** (Rose) Plitmann, Raven & Breedlove
4 *op. cit.* 507(1973): genus *Jehlia*.
sect. **Nannolopezia** Plitmann, Raven & Breedlove
4 *op. cit.* 552(1973).
sect. **Pelozia** (Rose) Plitmann, Raven & Breedlove
4 *op. cit.* 554(1973): genus *Pelozia*.
sect. **Riesenbachia** (Presl) Plitmann, Raven &
4 Breedlove *op. cit.* 501(1973): genus *Riesenbachia*.
ciliatula Plitmann, Raven & Breedlove in *Ann. Missouri*
4 *Bot. Gard.*, 59(2): 279 (1972 publ. 1973)—Mexico.
gentryi (Munz) Plitmann, Raven & Breedlove *op. cit.*
4 281(1972 publ. 1973): *L. laciniata* subsp. *gentryi*.
grandiflora Zuccar.
subsp. **macrophylla** (Benth.) Plitmann, Raven & Breedlove *op. cit.* 281(1972 publ. 1973): *L.*
4 *macrophylla*.
laciniata (Rose) Jones
subsp. **ovata** Plitmann, Raven & Breedlove *op. cit.*
4 280(1972 publ. 1973)—Mexico.
lopezioides (Hook. & Arn.) Plitmann, Raven & Breedlove *op. cit.* 281(1972 publ. 1973): *Diplandra*
4 *lopezioides*.
miniata Lag. ex DC.
subsp. **paniculata** (Seemann) Plitmann, Raven & Breedlove in *Ann. Missouri Bot. Gard.*, 60(2): 523
4 (1973): *L. paniculata*.
nuevo-leonis Plitmann, Raven & Breedlove in *Ann. Missouri Bot. Gard.*, 59(2): 280(1972 publ. 1973)—
4 Mexico.
racemosa Cav.
subsp. **moelchenensis** Plitmann, Breedlove & Raven in *Ann. Missouri Bot. Gard.*, 60(2): 547
4 (1973)—Mexico.

riesenbachia Plitmann, Raven & Breedlove in *Ann. Missouri Bot. Gard*,. 59 (2): 281(1972 publ. 1973),
4 nom. nov.: *Riesenbachia racemosa* Presl.
semeiandra Plitmann, Raven & Breedlove *op. cit.* 281(1972 publ. 1973), nom. nov.: *Semeiandra*
4 *grandiflora* Hook. & Arn.

LUDWIGIA

3295 Mühlberg, H.
Wo ist *Ludwigia palustris* beheimatet? *Inform. ZAG*
1–3 *Wasserpflanzen*, no.3 :3(1973).

OENOTHERA

3296 Andrade, A. de
Oenothera × *teframila* hort. (*O. tetragona* Roth. var. *fraseri* (Pursh.) Munz. × *O. pumila* L.). *Rea*, no.5:
0 28–30(1972)—It(En); illustration.

3297 Asai, Y.
(Notes on some species of *Oenothera* in Japan: 2). *J. Jap. Bot.*, 48(6):182–186(1973)—Ja(En);
2 illustrations.

3298 Linsley, E. G., Macswain, J. W. *and* Raven, P. H.
Comparative behavior of bees and Onagraceae: 5. *Camissonia* and *Oenothera* bees of Cismontane California and Baja California. *Univ. Calif. Publ.*
3, 4 *Bot.*, 71.[ii]68p.(1973)—Illustrations, dot maps.

3299 Pyšek, A.
Oenothera hoelscheri Renner ex Rostański var. *albinervis* Rostanski – nový druh československé květeny. (*Oenothera hoelscheri* Renner ex Rostański var. *albinervis* Rostanski – eine neue Art der tchecho-slowakischer Flora). *Zprávy Česk. Bot. Spol.*, 8(2):
1 111–114(1973)—(Ge).

campestris Greene
var. **obispoensis** (Raven) J. T. Howell in Howitt & J. T. Howell, Suppl. Vasc. Pl. Monterey Co., Calif.: 38(1973): *Camissonia campestris* subsp.
3 *obispoensis*.
fruticetorum J. T. Howell in Howitt & J. T. Howell *op. cit.* 39(1973), nom. nov.: *Camissonia*
3 *intermedia* Raven.
hardhamiae (Raven) J. T. Howell in Howitt & J. T.
3 Howell *op. cit.* 39(1973): *Camissonia hardhamiae*.
luciae (Raven) J. T. Howell in Howitt & J. T.
3 Howell *op. cit.* 39(1973): *Camissonia luciae*.
× teframila de Andrade in *Rea*, 5:29(1972), without type—Cultivation. (*O. tetragona* var. *fraseri* ×
0 *O. pumila*).

OPILIACEAE

3300 Villiers, J.-F.
Opiliaceae. Flore du Cameroun, 15:168–180 (1973)—Illustrations; keys. Also published as Flore
5 du Gabon: 20(1973).

3301 Villiers, J.-F.
Opiliaceae. Flore du Gabon, 20: 168–180(1973)—Illustrations; keys. Also published as Flore du
5 Cameroun: 15(1973).

ORCHIDACEAE

3302 Anon.
Nuevas especies reportadas para Mexico, desde la obra de L. O. Williams. *Orquidea* (Mexico), 2(5):
4 127–130(1972)—(En).

3303 Allison, C. M.
In the open air. *J. Orchid Soc. Gr. Brit.*, 22(4):97–98
1 (1973). Orchids in Argyll.

3304 Arditti, J. *and* Shechter, Y.
Orchid nomenclature: scientific names of orchids and their use. *Paphioped. World*, 2(4):79–82
1–7 (1972).

3305 Baker, R. K.
Foliar anatomy of the Laeliinae (Orchidaceae).
4 *Diss. Abstr. Int.*, *B*, 33(8):3506(1973).

3306 Banerji, M. *and* Chaudhuri, M.
Further studies on chromosomes of some Orchidaceae and Iridaceae from the temperate Himalayas. *Proc. Indian Sci. Congr.*, 59(3):347(1972)—
6 Chromosome numbers. Abstract only.

3307 Benzing, D.
The monocotyledons: their evolution and comparative biology: 1. Mineral nutrition and related phenomena in Bromeliaceae and Orchidaceae.
1–7 *Quart. Rev. Biol.*, 48(2):277–290(1973).

3308 Brieger, F. G.
Sobre a distribuição fitogeográfica das orquídeas brasileiras. *Orquídea*, 30(1):4–10, (2):76–80(1973)—
4 (En).

3309 Ceballos, A.
Las orquideas españolas. *Bol. Estac. Central Ecol.*, 2(3):29–39(1973)—(En); coloured illustrations;
1 chromosome numbers.

3310 Choon, Y. B.
Some Penang orchids. *Bull. Orchid Soc. S. E. Asia*,
6 no. 6:5–9(1973).

3311 Chow Cheng
Formosan orchids. Taichung, Chow Cheng,
2 [1973 ?] 102p.—Coloured illustrations.

3312 Claydon, A.
Some observations of native orchids of the Dorrigo.
7 *Orchadian*, 4(8):71–72(1973).

3313 Comber, J.
Some Indonesian species. *Malayan Orchid Rev.*, 10(4):18–28(1971)—Illustrations, coloured
6 illustrations.

3314 Dodson, C. H.
Significado de los estudios sobre la polinizacion de las orquideas. *Orquidea* (Mexico), 2(4):99–108
1–7 (1972)—(En).

3315 Dressler, R. L.
Dos nuevas orquideas del Mexico occidental. *Orquidea* (Mexico), 2(2):40–44(1972)—(En);
4 illustration.

3316 Dungs, F. *and* Pabst, G. F. J.
Orchidaceae Brasiliensis. *Orquídea*, 29(5):215–218, (6):274–286(1967 publ. 1972), 30(1):35–38, (2):65–70
4 (1973).

3317 Dunsterville, G. C. K.
Algumas orquídeas das mais atlas regiões do Brasil.
4 *Orquídea*, 30(1):16–31(1973)—Illustrations, maps.

3318 Dunsterville, G. C. K.
Orchids of Cerro Autana, Venezuela. *Amer. Orchid Soc. Bull.*, 42(5):388–401(1973)—Illustrations,
4 coloured illustrations, map.

3319 Duveen, D. I.
Further easy orchid hunting in Brazil. *Orchid Rev.*,
4 81(959):145–150(1973)—Illustrations, portraits.

3320 Echavarria R., A.
Las orquídeas, una familia de plantas que se
1–7 extingue. *Orquideologia*, 8(2):123–128(1973).

3321 Escobar R., R.
Revista 'Orquideologia' informe: especies nuevas.
4 *Orquideologia*, 7(4):254–264(1972).

3322 Esser, G.
Einige saugewählte Orchideenstandorte in
Paraguay und ihre Ökologie. *Orchidee*, 24(2):51–56
4 (1973)—Illustrations, map.

3323 Frei, J. K.
Orchid ecology in a cloud forest in the mountains
of Oaxaca, Mexico. *Amer. Orchid Soc. Bull.*, 42(4):
4 307–314(1973)—Maps.

3324 Garay, L. A.
Orquideas colombianas nuevas o críticas: decena
4 VIII. *Orquideologia*, 7(4):197–202(1972).

3325 Garay, L. A.
Orquideas colombianas nuevas o críticas: decena
IX. *Orquideologia*, 8(1):52–57,41–44(1973)—Coloured
4 illustrations.

3326 Garay, L. A.
Orquideas colombianas nuevas o críticas: decena X.
Orquideologia, 8(2):95–102(1973)—(En); coloured
4 illustrations.

3327 Garay, L. A.
Studies in the American orchids: 8. *Bradea*, 1(27):
301–308(1973)—Illustration; *Epidendrum*, *Laelia*,
Orleanesia, *Bulbophyllum*, *Zygopetalum*, *Pabstia*,
4 *Notylia*.

3328 Garay, L. A.
Studies in American Orchids: 9. *Bot. Mus. Leafl.
Harvard Univ.*, 23(7):299–304(1973)—Illustration.
Nomenclatural notes on *Scaphyglottis*, *Neokoehleria*,
4 *Rusbyella*, *Oncidium*, *Anthosiphon* and *Pterostemma*.

3329 Gölz, P. *and* Reinhard, H. R.
Biostatistische Untersuchungen an europäischen
Orchideen. (Biostatistical studies on European
orchids). *Ber. Schweiz. Bot. Ges.*, 83(2):93–105(1973)—
1 En.

3330 Gumprecht, R.
Kurzmitteilungen über bemerkenswerte Orchideen
aus Europa und dem Mittelmeergebiet. *Orchidee*,
1 24(6):256–257(1973)—Illustration.

3331 Gumprecht, R. *and* Uhlherr, P.
Erdorchideen in Australien unter besonderer
Berücksichtigung der Blütenbiologie. *Orchidee*,
24(5):195–200(1973)—Illustration, coloured
7 illustrations.

3332 Hagsater, E.
Nuevas orquideas de Mexico. *Orquidea* (Mexico),
4 2(3):68–69(1972)—(En).

3333 Halls, W.
Orchids in the south of France. *Orchid Rev.*, 81
(963):271–274(1973)—Illustrations. Popular
1 article.

3334 Hilgers, J. H. M.
De achteruitgang van de Orchidaceae in Limburg:
13. *Natuurhist. Maandbl.*, 62(1):4–7(1973)—
1 Illustrations.

3335 Hu, S. Y.
The Orchidaceae of China: 4. *Quart. J. Taiwan*
2 *Mus.*, 25(3–4):199–230(1972)—Illustrations.

3336 Hu, S. Y.
The Orchidaceae of China (5). *Quart. J. Taiwan*
2 *Mus.*, 26(1–2):131–165(1973)—Illustrations.

3337 Jones, H. G.
Notes on some natural hybrid orchids of tropical
America and the West Indies. *Bol. Soc. Brot.*, 46:239–
4 252(1972).

3338 Lawler, L. J. *and* Slaytor, M.
Uses of Australian orchids by Aborigines and early
settlers. *Orchid Rev.*, 81(959):160–164(1973)—
Reprinted from the *Medical Journal of Australia*,
7 1970.

3339 Leinig, M.
Contribuição à lista geral das Orquidáceas
4 Paranaenses: 7. *Orquídea*, 30(2):82(1973).

3340 Marchant, C. J.
Terrestrial orchids of Europe. *Davidsonia*, 4(1):
1 1–3(1973)—Illustrations.

3341 Masamune, G. *and* Satomi, N.
(Distribution maps of Orchidaceae in Japan: 43–46).
2 *J. Geobot.*, 20(1):1, (2):31, (3):55, (4):83(1972)—Ja.

3342 Moir, W. W. G.
Oncidiums and the Oncidiinae. *Orchid Rev.*,
3, 4 81(964):297–299(1973).

3343 Moir, W. W. G.
Some lesser known Oncidiinae. *Amer. Orchid Soc.*
4 *Bull.*, 42(4):292–299(1973)—Illustrations.

3344 Nackejima, C.
Preliminary notes on the noteworthy Orchidaceae
from Formosa, Ryukyus, Bonin Islands and Southern
Japan. *Biol. Mag. Okinawa*, 9:34–45(1972)—Ja;
illustrations; chromosome numbers. With their
2 distributions.

3345 Nieschalk, A. *and* Nieschalk, C.
Beiträge zur Orchideenflora Spaniens. *Orchidee*,
1 24(4):163–168,(5):211–216(1973).

3346 Nový, P.
K rozšíření vstavačovitých rostlin na Tachovsku.
(Zur Verbreitung der Orchideenpflanzen im Kreis
von Tachau). *Zprávy Česk. Bot. Spol.*, 8(3):167–170
1 (1973).

3347 Ospina H., M.
Los antipolinizadores. *Orquidea* (Mexico), 2(6):
4 163–167(1972)—(En); illustrations.

3348 Pabst, G. F. J.
As orquideas de Minas Gerais. *Oréades*, 3(5):5–11
4 (1972)—Illustrations; key.

3349 Pabst, G. F. J.
Additamenta ad orchideologiam brasiliensem: 14.
4 *Bradea*, 1(31):327–338(1973)—Por,En; illustrations.

3350 Pabst, G. F. J.
Additamenta ad orchideologiam brasiliensem: 15.
4 *Bradea*, 1(36):361–370(1973)—Por,En; illustrations.

3351 Pabst, G. F. J.
Orquídeas do Parque Nacional do Tocantins.
4 *Orquídea*, 30(3):112–116(1973).

3352 Pabst, G. F. J. *and* Dungs, F.
Ein Blick in die Werkstatt der Natur. *Orchidee*,
4 24(6):233–234(1973)—Coloured illustrations.

3353 Pocock, M. R.
Ground orchids of Australia. Brisbane, Jacaranda
7 Press, 1972. xii,99p.—Coloured illustrations.

3354 Porcher, R. D.
New county records for Orchidaceae from South Carolina coastal plain. *J. Elisha Mitchell Sci. Soc.*,
3 88(4):218–219(1972).

3355 Pradhan, U. C.
Orchideen des östlichen Himalaya. *Orchidee*, 24(5):
6 201–204(1973).

3356 Querl, H. F.
Standortbedingungen epiphytischer Orchideen in Queensland. *Orchidee*, 24(4):150–153(1973)—
7 Illustrations.

3357 Reinikka, M. A.
A chronological guide to descriptive and taxonomic orchid literature. *In* Reinikka, M. A. A history of the
1–7 orchid: 97–108(1972).

3358 Reinikka, M. A.
A history of the Orchid. Coral Gables, Florida, University of Miami Press, 1972. xx,316p.—
1–7 Illustrations, portraits.

3359 Russenberger, H.
Orchideen des Randens. *Neujahrsbl. Naturforsch. Ges. Schaffhausen*, 25. 86p.(1973 publ. 1972)—
1 Illustrations, coloured illustrations.

3360 Sanford, W. W. *and* Adanlowo, I.
Velamen and exodermis characters of West African epiphytic orchids in relation to taxonomic grouping and habit tolerance. *Bot. J. Linn. Soc.*,
5 66(4):307–321(1973).

3361 Schelpe, E. A.
Hunting orchids in the Natal Drakensberg. *S. Afr.*
5 *Orchid J.*, Dec.:10–11(1973)—Illustration.

3362 Schlechter, R.
Die Orchideen: ihre Beschreibung, Kultur und Züchtung. 3. völlig. neubearb. Aufl. Lfg. 4. Berlin, etc., P. Parey, 1973. pp.193–256.—Illustrations,
1–7 coloured illustration; keys.

3363 Seidenfaden, G.
Contributions to the orchid flora of Thailand. *Bot. Tidsskr.*, 68(1):41–95(1973)—Illustrations, coloured
6 illustrations; keys.

3364 Seidenfaden, G.
An enumeration of Laotian orchids. *Bull. Mus. Nation. Hist. Nat.* (Paris), 3e sér., no.71 *Bot.* 5:101–152
6 (1972 publ. 1973)—(Fr); illustrations, map.

3365 Senghas, K.
Einige neue Namen in der Unterfamilie
3, 4 Orchidoideae. *Orchidee*, 24(3):97–98(1973).

3366 Small, P. F. *and* Barclay, F. H.
New county records of Tennessee orchids.
3 *Castanea*, 38(2):189–195(1973)—Map.

3367 Stewart, J.
A second checklist of the orchids of Kenya. *Amer. Orchid Soc. Bull.*, 42(6):525–531(1973)—Illustrations,
5 map.

3368 Tolmatschen, A. I.
Neue Daten über die Orchideen-Verbreitung in äussersten Norden Eurasiens. *Acta Bot. Acad. Sci.*
1, 2 *Hung.*, 19(1–4):375–378(1973)—(En).

3369 Wells, G.
Orchid collecting in southern Queensland.
7 *Orchadian*, 4(7):58–60(1973).

3370 Würmli, M.
Zur Kenntnis der Orchideenflora des Königreiches Bhutan (Osthimalaya). (Studies on the orchids of Bhutan (Eastern Himalaya)). *Ber. Schweiz. Bot. Ges.*,
6 83(1):30–38(1973)—(En); map.

gen.-ser. CRASPEDOSTIGMATATA Schltr. ex
4 Senghas in *Orchidee*, 24(3):97(1973), without type.
gen.-ser. STELEOSTIGMATATA Schltr. ex
4 Senghas *op. cit.* 97(1973), without type.

ACAMPE

3371 Balganur, B. M.
Cyto-embryological studies in the orchid *Acampe wightiana* Lindl. *Proc. Indian Sci. Congr.*, 59(3):352
6 (1972)—Chromosome number. Abstract only.

ACANTHEPHIPPIUM

3372 Pradhan, G. M.
Indian orchids: *Acanthephippium sylhetense*. *Amer. Orchid Soc. Bull.*, 42(10):879–880(1973)—
6 Illustration.

3373 Pradhan, G. M.
The north Indian species of *Acanthephippium*. *Orchid Dig.*, 37(4):133–134(1973)—Coloured
6 illustrations, map.

ACERAS

3374 Mennema, J.
De poppenorchis, *Aceras anthropophorum* (L.) Ait.f., in Nederland. *Natuurhist. Maandbl.*, 62(9):115–
1 116(1973)—Dot map.

× **AËRIDITIS** hort. in *Orchid Rev.*, 81(957): centre
0 page pull-out p.7(1973). (*Aërides* × *Doritis*).

ANACAMPTIS

3375 Bruggen, H. W. E. van
Een explosie van *Anacamptis pyramidalis* (L.) Rich.
1 te Wijk aan Zee. *Gorteria*, 6(12):203(1973).

ANGRAECUM

3376 Garay, L. A.
Systematics of the genus *Angraecum* (Orchidaceae). *Kew Bull.*, 28(3):495–516(1973 publ. 1974)—
5 Key.

3377 Senghas, K.
Angraecum bosseri, ein neuer 'Stern von Madagaskar'.
5 *Orchidee*, 24(5):191–193(1973)—(En); illustrations.

sect. **Acaulia** Garay in *Kew Bull.*, 28(3):498 (1973
5 publ. 1974).
sect. **Angraecoides** (Cordem.) Garay *op. cit.* 500
5 (1973 publ. 1974): *Mystacidium* sect. *Angraecoides.*
sect. **Dolabrifolia** (Pfitz.) Garay *op. cit.* 499(1973
5 publ. 1974): *Mystacidium* sect. *Dolabrifolia.*
5 sect. **Filangis** Garay *op. cit.* 500(1973 publ. 1974).
sect. **Gomphocentrum** (Benth.) Garay *op. cit.* 501
5 (1973 publ. 1974): *Mystacidium* sect. *Gomphocentrum.*
5 sect. **Lemurangis** Garay *op. cit.* 501(1973 publ. 1974).
sect. **Nana** (Cordem.) Garay *op. cit.* 502(1973 publ.
5 1974): *Mystacidium* sect. *Nana.*
bosseri Senghas in *Orchidee*, 24(5):193(1973), nom.
nov.: *A. sesquipedale* var. *angustifolium* Bosser &
5 Morat.

ANTHOGONIUM

3378 Teuscher, H.
Collector's item: *Anthogonium gracile. Amer. Orchid*
6 *Soc. Bull.*, 42(4):323–325(1973)—Illustrations.

ANTHOSIPHON

pseudobulbosus (C. Schweinf.) Garay in *Bot. Mus.*
Leafl. Harvard Univ., 23(7):303(1973): *Cryptocentrum*
4 *pseudobulbosum.*

APOSTASIA

3379 Rao, V. S.
The floral anatomy of the rare Apostasias: *Apostasia*
species. *Proc. Indian Acad. Sci., B*, 78(1):17–21(1973)—
6, 7 Illustrations.

ARTORIMA

3380 Hagsater, E.
Artorima erubescens, quien la puede cultivar ?
(*Artorima erubescens*: who can grow it ?) *Orquidea*
4 (Mexico), 3(1):6–12(1973)—(En); illustration.

BARKERIA

3381 Halbinger, F.
Barkeria barkeriola. Orquidea (Mexico), 3(8):
4 229–233(1973)—Sp(En); illustrations.

3382 Halbinger, F.
Barkeria chinensis. Orquidea (Mexico), 3(1):13–20
4 (1973)—Sp(En); illustrations.

3383 Halbinger, F.
Barkeria halbingeri Thien. *Orquidea* (Mexico),
3(4):99–109(1973)—Sp(En); illustrations, coloured
4 illustration.

3384 Halbinger, F.
Barkeria melanocaulon Richard & Galeotti.
(*Barkeria melancaulon* Richard and Galeotti). *Orquidea*
4 (Mexico), 3(5):131–137(1973)—Sp(En); illustration.

3385 Halbinger, F.
Barkeria naevosa. Orquidea (Mexico), 3(3):77–83
4 (1973)—Sp(En); illustration, coloured illustration.

3386 Halbinger, F.
Barkeria palmeri. Orquidea (Mexico), 3(2):37–44
4 (1973)—Sp(En); illustrations.

3387 Halbinger, F.
Barkeria skinneri. Orquidea (Mexico), 3(6):163–168
4 (1973)—Sp(En); illustrations.

3388 Halbinger, F.
The genus *Barkeria. Amer. Orchid Soc. Bull.*, 42(7): 620–
4 626(1973)—Coloured illustrations; key to species.

3389 Halbinger, F.
Historia y estudios preliminares sobre el genero
Barkeria. Orquidea (Mexico), 2(7):177–191(1972)—
4 (En); illustrations.

3390 Thien, L. B.
Barkeria halbingeri—a new species of orchid from
Oaxaca, Mexico. *Amer. Orchid Soc. Bull.*, 42(5):421–
4 422(1973)—Illustration, coloured illustration.

halbingeri Thien in *Amer. Orchid Soc. Bull.*, 42(5):421
4 (1973)—Mexico.

BIFRENARIA

3391 Dunsterville, G. C. K. *and* Dunsterville, E.
Bifrenaria maguirei, a beautiful rarity of Venezuela's
Amazonas highlands. *Orchid Dig.*, 37(5):187–188
4 (1973)—Coloured illustration, map.

BLETIA

3392 Dressler, R. L.
Notas sobre el genero *Bletia. Orquidea* (Mexico),
4 1(7):9–21(1971)—Illustrations; keys.

3393 Jones, H. G.
Three additions to the flora of Barbados. *J.*
Barbados Mus. Hist. Soc., 34(3):113–116(1973).
4 *Bletia florida.*

3394 Peña de Sousa, M.
Nuevo status de *Crybe* en el genero *Bletia. Orquidea*
4 (Mexico), 2(7):192–198(1972)—Illustrations; keys.

subgen. **Crybe** (Lindl.) Peña de Sousa in *Orquidea*
4 (Mexico), 2(7):194(1972): gen. *Crybe.*

BLETILLA

3395 Leuchs, O.

Bletilla striata—die Japanorchidee. *Palmengarten,*
2 37(1):18(1973)—Illustration.

BRACHYPEZA

laotica (Seidenfaden) Seidenfaden in *Bot. Tidsskr.*,
6 68(1): 66(1973): *Pteroceras laotica.*

× **BRADEARA** hort. in *Orchid Rev.*, 81(961):
centre page pull-out p.7(1973). (*Comparettia* ×
0 *Gomesa* × *Rodriguezia*).

BRASSAVOLA

3396 Jones, H. G.
The genus *Brassavola* as an example of infrageneric
evolution in the Orchidaceae. *Adansonia*, 13(2):235–
3, 4 239(1973)—Key.

3397 Jones, H. G.
Synopsis of Middle American *Brassavola. Amer.*
3, 4 *Midl. Nat.*, 89(2):499–503(1973)—Key.

BRASSIA

3398 Teuscher, H.
Brassia bidens, B. neglecta and *B. wageneri. Amer.*
4 *Orchid Soc. Bull.*, 42(2):125–129(1973)—Illustrations.

3399 Teuscher, H.
The two sections of the genus *Brassia. Amer. Orchid*
Soc. Bull., 42(12):1089–1094(1973)—Illustrations;
4 key.

escobariana Garay in *Orquideologia*, 8(1):57(1973)—
4 Colombia.

× BRASSOCATTLEYA

araguaiensis (Pabst) H. G. Jones in *Bol. Soc. Brot.*,
4 46:248(1972 publ. 1973): *Cattleya araguaiensis*.

BROMHEADIA

3400 Laverack, P.
Two rare terrestrial orchids from Cape York
Peninsula. *Orchadian*, 4(6):51(1972)—*Bromheadia*
7 *venusta.*

BULBOPHYLLUM

3401 Anon.
A correction. *Amer. Orchid Soc. Bull.*, 42(7):619
(1973). *Bulbophyllum seidenfadeni* corrected to *B.*
6 *seidenfadenii*.

3402 Hallé, N.
Observations nouvelles sur quatre *Bulbophyllum*,
orchidées des serres du Muséum, trois espèces de
Nouvelle-Calédonie, une de Birmanie. *Bull. Mus.*
Nation. Hist. Nat., (Paris) 3e Sér., no.127 *Bot.* 7:21–30
6 (1973)—Illustrations.

3403 Kerr, A. D.
A new species – *Bulbophyllum seidenfadeni*. *Amer.*
2 *Orchid Soc. Bull.*, 42(5):411–412(1973)—Illustrations.

3404 Lee, T. C. *and* Young-Saye, J. L.
Hong Kong orchid species. *Malayan Orchid Rev.*,
10(4):15–17(1971)—Illustration. *Bulbophyllum*
2 *watsonianum*.

3405 Pabst, G. F. J.
New or critical orchids from Brazil: 6. *Bulbophyllum*
sanderianum Rolfe. *Orchid Rev.*, 964:288–289(1973)—
4 Illustration.

albociliatum (Liu & Su) Seidenfaden in *Dansk Bot.*
6 *Ark.*, 29(1):89(1973): *Cirrhopetalum albociliatum*.
bootanoides (Guill.) Seidenfaden *op. cit.* 60(1973):
6 *Cirrhopetalum bootanoides*.
capillipes (Guillaumin) N. Hallé in *Bull. Mus.*
Nation. Hist. Nat. (Paris), 3e sér., no. 127, *Bot.*
6 7: 23(1973): *Cirrhopetalum capillipes*.
chondriophorum (Gagnep.) Seidenfaden in *Dansk*
Bot. Ark., 29(1):53(1973): *Cirrhopetalum*
6 *chondriophorum*.
dalatensis Seidenfaden *op. cit.* 64(1973), nom. nov.:
6 *Cirrhopetalum sigaldii* Guill.
dyeranum (King & Pantl.) Seidenfaden *op. cit.* 175
6 (1973), as 'dyerianum': *Cirrhopetalum dyeranum*.
electrinum Seidenfaden *op. cit.* 77(1973), nom. nov.:
6 *Cirrhopetalum melinanthum* Schlechter.
farreri (W. W. Smith) Seidenfaden *op. cit.* 212(1973):
6 *Cirrhopetalum farreri*.
fibratum (Gagnep.) Seidenfaden *op. cit.* 194(1973):
6 *Cirrhopetalum fibratum*.
fischeri Seidenfaden *op. cit.* 202(1973), nom. nov.:
6 *Cirrhopetalum gamblei*.
flaviflorum (Liu & Su) Seidenfaden in *Bull. Mus.*
Nation. Hist. Nat. (Paris), 3e Sér., no. 71, *Bot.* 5:
6 109(1972 publ. 1973): *Cirrhopetalum flaviflorum*.
forestii Seidenfaden in *Dansk Bot. Ark.*, 29(1):120
(1973), nom. nov.: *Cirrhopetalum aemulum* W. W.
6 Smith.
formosanum (Rolfe) Seidenfaden *op. cit.* 172(1973):
6 *Cirrhopetalum formosanum*.
hirundinis (Gagnep.) Seidenfaden *op. cit.* 76(1973):
6 *Cirrhopetalum hirundinis*.
insulsoides Seidenfaden *op. cit.* 89(1973), nom. nov.:
6 *B. racemosum* Hayata.
insulsum (Gagnep.) Seidenfaden *op. cit.* 141(1973):
6 *Cirrhopetalum insulsum*.
leratiae (Kränzlin) Seidenfaden *op. cit.* 227(1973):
6 *Cirrhopetalum leratiae*.
longidens (Rolf) Seidenfaden *op. cit.* 57(1973):
6 *Cirrhopetalum longidens*.
nanobulbon Seidenfaden *op. cit.* 87(1973), nom. nov.:
6 *Cirrhopetalum microbulbon* Ridley.
oreogenes (W. W. Smith) Seidenfaden *op. cit.* 212
6 (1973): *Cirrhopetalum oreogenes*.
pabstii Garay in *Bradea*, 1(27):305(1973), nom. nov.:
4 *B. fractiflexum* Pabst.
pachybulbum (Schltr.) Seidenfaden in *Dansk Bot. Ark.*
6 29(1):221(1973): *Cirrhopetalum pachybulbum*.
pectenveneris (Gagnep.) Seidenfaden *op. cit.* 37
6 (1973): *Cirrhopetalum pectenveneris*.
peyeranum (Kränzlin) Seidenfaden *op. cit.* 151
6 (1973), as 'peyerianum': *Cirrhopetalum peyeranum*.
polliculosum Seidenfaden in *Bot. Tidsskr.*, 68(1):
6 60(1973)—Thailand.
sanguineopunctatum Seidenfaden & A. D. Kerr in
6 *Dansk Bot. Ark.*, 29(1):39(1973)—Laos.
seidenfadenii A. D. Kerr in *Amer. Orchid Soc. Bull.*,
42(5):411(1973), as 'seidenfadeni' and without type
6 —Laos.
serratotruncatum Seidenfaden in *Dansk Bot. Ark.*,
29(1):50(1973), nom. nov.: *Cirrhopetalum ochraceum*
6 Ridl.
taiwanense (Fukuyama) Seidenfaden *op. cit.* 89
6 (1973): *Cirrhopetalum taiwanense*.
unciniferum Seidenfaden in *Bot. Tidsskr.*, 68(1):58
6 (1973)—Thailand.

CALANTHE

3406 Pirie, D. H.
The Calanthes. *S. Afr. Orchid J.*, Mar.:9–10(1973)—
2, 4–7 Illustrations.

discolor Lindl.
f. **kanashiroi** (Fukuyama) Nackejima in *Biol. Mag.*
2 *Okinawa*, 9:36(1972): *C. discolor* var. *kanashiroi*.
patsinensis S. Y. Hu in *Chung Chi Jour.*, 11:13(1972);
cf. S. Y. Hu in *Quart. J. Taiwan Mus.*, 25(3–4):
2 205(1972)—Hong Kong.

CALOCHILUS

3407 Cady, L.
Notes on the pollination of *Calochilus campestris*
7 R. Br. *Orchadian*, 4(6):52–53(1972).

CALOPOGON

3408 Thien, L. B.
Isolating mechanisms in the genus *Calopogon*. *Amer.*
Orchid Soc. Bull., 42(9):794–797(1973)—Illustrations;
3, 4 chromosome numbers.

CAPANEMIA

3409 Pabst, G. F. J.
El género *Capanemia*, Rodr. *Orquideologia*, 7(4):
4 215–228,237–242(1972)—Illustrations; key.

fluminensis Pabst in *Orquideologia*, 7(4):223(1972)—
4 Brazil.
4 **riograndensis** Pabst *op. cit.* 240(1973)—Brazil.

CATASETUM

3410 Couret, P.
Research on the 'red' *Catasetum pileatum* Rchb. F.
in Venezuela – a preliminary report. *Amer. Orchid*
Soc. Bull., 42(11):992–998(1973)—Coloured
4 illustrations.

3411 Teuscher, H.
Collector's item: *Catasetum bicolor*. *Amer. Orchid*
4 *Soc. Bull.*, 42(1):29–33(1973)—Illustrations.

3412 Teuscher, H.
Collector's item: *Catasetum tenebrosum*. *Amer. Orchid Soc. Bull.*, 42(5):406–408(1973)—
4 Illustrations.

3413 Teuscher, H.
Catasetum integerrimum, *C. oerstedii* and *C. macrocarpum*. *Amer. Orchid Soc. Bull.*, 42(3):216 219
4 (1973)—Illustrations.

CATTLEYA

3414 Bawden, L. E.
Cattleya velutina. *J. Orchid Soc. Gr. Brit.*, 22(4):96
4 (1973)—En; illustration.

3415 Dunsterville, G. C. K. *and* Dunsterville, E.
Habitat notes on *Cattleya lawrenceana* Rchb. f. *Orchid Dig.*, 37(4):147–150(1973)—Illustration,
4 coloured illustrations, map.

3416 Fatio, J.-P.
Formas de *Cattleya loddigesii*. *Orquídea*, 30(2):83
4 (1973).

3417 Fowlie, J. A.
Cattleya lawrenceana Rchb. f. *Orchid Dig.*, 37(4):
4 145–147(1973)—En.

3418 Fowlie, J. A.
Cattleya schroederae Rchb. f. *Orchid Dig.*, 37(5):
4 183–186(1973)—Coloured illustrations, map.

3419 Hartmann, W. L.
El genero *Cattleya* algo sobre su historia y su
4 cultivo. *Orquidea* (Mexico), 1(11):5–9(1971).

3420 Machado, E. de Freitas
Cattleya warneri var. *amethystina*. *Orquídea*, 30(1):
4 32(1973)—Por; coloured illustration.

3421 Rodriguez, R.
Las Cattleyas. *Orquidea* (Mexico), 1(2):7–10
4 (1971).

3422 Rodriguez, R.
Las Cattleyas: 2a parte Cattleyas bifoliadas.
4 *Orquidea* (Mexico), 1(3):3–6(1971).

3423 Stort, M. N. S.
Natural pollination in *Cattleya loddigesii* Lindl.
4 *Amer. Orchid Soc. Bull.*, 42(7):606–608(1973).

CAULARTHRON

3424 Jones, H. G.
Three additions to the flora of Barbados. *J. Barbados Mus. Hist. Soc.*, 34(3):113–116(1973)—
4 *Caularthron amazonicum*.

CEPHALANTHEROPSIS

gracilis (Lindl.) S. Y. Hu in *Quart. J. Taiwan Mus.*,
6 25(3–4):213(1972): *Calanthe gracilis*.
venusta (Schlechter) S. Y. Hu *op. cit.* 213(1972):
2 *Calanthe venusta*.

CERATOSTYLIS

3425 Taylor, P.
Ceratostylis retisquama. *Curtis's Bot. Mag.*, 179(3): tab. 640(1973)—En; illustration, coloured
6 illustration.

CHAUBARDIA

klugii (C. Schweinf.) Garay in *Orquideologia*, 8(1):34
4 (1973): *Zygopetalum klugii*.

× **CHEWARA** hort. in *Orchid Rev.*, 81(957): centre page pull-out p.7(1973). (*Aërides* ×
0 *Renanthera* × *Rhynchostylis*).

CHILOGLOTTIS

3426 Beek, M.
Chiloglottis cornuta Hook. f. in South Australia.
7 *Orchadian*, 4(7):63(1973).

CHYTROGLOSSA

3427 Duveen, D. I.
Chytroglossa aurata Reichb. f. A little known but attractive and desirable orchid. *Orchid Dig.*, 37(6):
4 207–208(1973)—Coloured illustrations.

CIRRHOPETALUM

3428 Seidenfaden, G.
Notes on *Cirrhopetalum* Lindl. *Dansk. Bot. Ark.*,
6 29(1):1–260(1973)—Illustrations; keys.

CISCHWEINFIA

3429 Teuscher, H.
Cischweinfia dasyandra (*Trichopilia dasyandra*). *Amer. Orchid Soc. Bull.*, 42(11):988–990(1973)—
4 En; illustrations; key.

colombiana Garay in *Orquideologia*, 8(1):54(1973)—
4 Colombia.
platychila Garay in *Orquideologia*, 7(4):201(1972)—
4 Colombia.

COELOGLOSSUM

viride (L.) Hartm.
f. integrum Soó in *Acta Bot. Acad. Sci. Hung.*, 18 (1–2):173(1973), without full ref. and without
1 type—Switzerland.

COELOGYNE

3430 Ferguson Beall, G.
Apuntes sobre la *Coelogyne cristata*. *Orquidea*
2, 6 (Mexico), 2(6):168–171(1972)—(En); illustration.

leungiana S. Y. Hu in *Quart. J. Taiwan Mus.*, 25
2 (3–4):223(1972)—Hong Kong.

CORYANTHES

3431 Horich, C. K.
Coryanthes hunteriana, eine neuentdeckung für Costa Rica. *Orchidee*, 24(1):3–5(1973)—(En);
4 illustration, coloured illustration.

CORYBAS

3432 Jones, D. L.
A new *Corybas* species from Victoria and N.S.W.
7 *Victorian Nat.*, 90(4):96–99(1973)—Illustrations.

hispidus D. L. Jones in *Victorian Nat.*, 90(4):96
7 (1973)—Australia (Victoria, New South Wales).

CRANICHIS

3433 Weber, D.
Deux Orchidacées nouvelles pour la flore des Iles Galapagos. *Bull. Soc. Neuchâtel. Sci. Nat.*, 96:18–32 (1973)—(En,Ge); illustration, maps. *Cranichis*
4 *lichenophila*.

lichenophila D. Weber in *Bull. Soc. Neuchâtel. Sci.*
4 Nat., 96:18(1973)—Galapagos Is.

CRYBE

3434 Peña de Sousa, M.
Nuevo status de *Crybe* en el genero *Bletia*. *Orquidea*
4 (Mexico), 2(7):192–198(1972)—Illustrations; keys.

CYMBIDIUM

3435 Barrow, L. R.
Cymbidium tigrinum. *Cymbidium Soc. News*, 28(1):
6 3–5(1973)—En; illustrations.

3436 Maslen, R.
Cymbidium canaliculatum in New South Wales.
7 *Orchadian*, 4(6):51(1972).

3437 Pradhan, G. M.
Notes on *Cymbidium devonianum*. *Orchid Dig.*,
6 37(2):40–42(1973)—Map.

szechuanensis S. Y. Hu in *Quart. J. Taiwan Mus.*,
26(1–2):134(1973), nom. nov.: *C. aphyllum* Ames
2 & Schlechter.

CYNORKIS

3438 O'Connor, M. J.
Cynorkis compacta. *S. Afr. Orchid J.*, June:8–11
5 (1973)—En; illustrations.

CYPRIPEDIUM

3439 Castaño, G.
Bitacora en la vida del *Cypripedium irapeanum*.
Orquidea (Mexico), 2(8):209–216(1972)—(En);
4 illustration.

3440 Eberle, G.
Prächtiger Frauenschuh (*Cypripedium calceolus* L.).
Jahrb. Ver. Schutze Alpenpfl. Tiere, 37:7–18(1972)—
1 Illustrations, coloured illustration.

3441 Hackenberg, D.
Plant portraits: *Cypripedium montanum* Dougl. ex
Lindl. *Quart. Bull. Alp. Gard. Soc.*, 41(2):159,163–
3 164(1973)—Illustration.

3442 Hagsater, E.
Cypripedium irapeanum. *Orquidea* (Mexico), 1(10):
4 5–8(1971)—Sp.

3443 Harms, V. L.
New record for the yellow lady's slipper orchid,
Cypripedium calceolus L. subsp. *parviflorum* (Salisb.)
3 Hult. from Alaska. *Rhodora*, 75(803):491(1973).

3444 Pirie, M.
So the lady's slipper is not lost in England after all.
S. Afr. Orchid J., Mar.:6(1973)—Illustrations.
1 *Cypripedium calceolus*.

3445 Pontes, M.
Mas sobre el *Cypripedium irapeanum*. *Orquidea*
4 (Mexico), 1(12):25–28(1971 publ. 1972).

3446 Schrenk, W. J.
Die Orchideen der nordöstlichen Vereinigten
Staaten: Teil 1. *Cypripedium* L. (The orchids of the
north-eastern United States). *Orchidee*, 24(3):108–
3 111(1973)—(En); coloured illustrations.

× DACTYLOGLOSSUM

3447 Reinhardt, R.
× *Dactyloglossum dominianum* (Camus) Soó aus der
polnischen Hohen Tatra. *Orchidee*, 24(3):118–119
1 (1973).

conigerum (Norman) Rauschert in *Feddes Repert.*,
83(9–10):658(1972 publ. 1973): *Coeloglossum*
1 *conigerum*.

DACTYLORHIZA

3448 Beisenherz, W.
Zur Verbreitung von *Dactylorhiza cruenta* und
D. pseudocordigeria in Skandinavien. *Orchidee*, 24(3):
1 119–120(1973)—Coloured illustration.

3449 Procházka, F. *and* Potucek, O.
Dactylorhiza incarnata var. *haematodes* (Reichenb. f.)
Soó – nový taxón československé flóry. (*Dactylorhiza*
incarnata var. *haematodes* (Reichenb. f.) Soó – ein
neues Taxon der tschechoslowakischen Flora).
1 *Preslia*, 45(1):90–93(1973)—(Ge).

3450 Scannell, M. J. P.
Dactylorhiza × *kellerana* P. F. Hunt in Westmeath
1 and east Mayo. *Irish Nat. J.*, 17(12):426(1973).

3451 Scannell, M. J. P.
Dactylorhiza traunsteineri (Sauter) Soó in east Cork,
mid Cork, Offaly, Meath, Leitrim and east Mayo.
1 *Irish Nat. J.*, 17(12):426(1973).

3452 Sundermann, H. *and* Wattke, R.
Neue Beiträge zur Zytotaxonomie der Erdor-
chideen (*Dactylorhiza*, *Orchis* und *Serapias*).
Orchidee, 24(3):116–118(1973)—Illustrations;
1, 2 chromosome numbers.

maurusii (Emberg. & Maire) Holub in *Folia Geobot.*
1 *Phytotax.*, 8(2):176(1973): *Orchis maurusii*.
× vermeuleniana Soó in *Acta Bot. Acad. Sci. Hung.*,
18(1–2):174(1973), without type.—Locality
1 unknown. (*D. majalis* × *D. maculata*).

DENDROBIUM

3453 Anon.
Dendrobium johnsoniae F. v. Muell. *Austral. Orchid*
6 *Rev.*, 38(1):25(1973)—En; illustration.

3454 Hartmann, W. L.
El genero *Dendrobium* y su cultivo. (The *Dendrobium*
and its culture). *Orquidea* (Mexico), 2(10):283–289
4 (1972 publ. 1973)—(En); illustration.

3455 Herklots, G. A. C.
Nepalese and Indian orchids: *Dendrobium* – section
Nigrohirsutae. *Orchid Rev.*, 81(960):186–189(1973)—
6 Illustrations. *Dendrobium formosum*, *D. infundibulum*.

3456 Herklots, G. A. C.
Nepalese and Indian orchids. *Orchid Rev.*, 81
(961):224,xi(1973)—Illustration. *Dendrobium*
6 *longicornu*.

3457 Herklots, G. A. C.
Nepalese and Indian Orchids. *Orchid Rev.*, 81
(962):238–239(1973)—Illustration. *Dendrobium*
6 *williamsonii*.

3458 Herklots, G. A. C.
Nepalese and Indian Orchids. *Orchid Rev.*, 81
(963):264–265(1973)—Illustration. *Dendrobium*
6 *chrysanthum*.

3459 Herklots, G. A. C.
Nepalese and Indian orchids. *Orchid Rev.*, 81(964):
290–292(1973)—Illustrations. *Dendrobium amoenum*,
6 *D. aureum*.

3460 Herklots, G. A. C.
Nepalese and Indian Orchids. *Orchid. Rev.*, 81 (965):319–322(1973)—Illustrations. *Dendrobium*
6 *candidum, D. crepidatum.*

3461 Herklots, G. A. C.
Nepalese and Indian Orchids. *Orchid Rev.*, 81 (966):357–360(1973)—Illustrations. *Dendrobium*
6 *denneanum* and *D. aphyllum.*

3462 Laverack, P.
Two rare terrestrial orchids from Cape York Peninsula. *Orchadian*, 4(6):51(1972)—*Dendrobium*
7 *lobbii.*

3463 Pradhan, G. M.
Dendrobium candidum. Amer. Orchid Soc. Bull.,
6 42(1):39(1973)—En; illustration.

lindleyi Steudel
var. **majus** (Rolfe) S. Y. Hu in *Quart. J. Taiwan*
2 *Mus.*, 26(1–2):151(1973): *D. alboviride* var. *majus.*

DISA

3464 Hall, A. V.
Disa maculata. Flow. Pl. Afr., 42(3–4):pl.1678
5 (1973)—En; coloured illustration.

3465 Hofmann, H.
Krabbenspinnen auf *Disa racemosa* L. f. *Orchidee*,
5 24(3):113–114(1973)—Coloured illustrations.

3466 Stewart, J.
Disa woodii. S. Afr. Orchid J., Mar.:7–8(1973)—
5 En.

DORITIS

3467 Sheehan, T. *and* Sheehan, M.
Orchid genera, illustrated: 32. *Amer. Orchid Soc.*
6 *Bull.*, 42(3):248–249(1973)—Coloured illustration.

ELEORCHIS

3468 Murata, G.
[Short note on *Eleorchis japonica.*] *Acta Phytotax.*
2 *Geobot.*, 25(4–6):185(1973)—Ja.

ELLEANTHUS

3469 Dressler, R. L.
Elleanthus capitatus – a name that must be changed, or is it? *Amer. Orchid Soc. Bull.*, 42(5):419–420
4 (1973)—Coloured illustration.

ELYTHRANTHERA

3470 George, A., Gonzales, C., Strauss, M. S. *and* Arditti, J.
Chemotaxonomic and ecological implications of anthocyanins in *Elythranthera. Biochem. Syst.*, 1(1):
7 45–49(1973).

ENCYCLIA

3471 Dressler, R. L.
Una reconsideracion del genero *Encyclia. Orquidea*
4 (Mexico), 1(11):10–27(1971)—Illustrations.

3472 Hagsater, E.
Encyclia ghiesbreghtiana. Orquidea (Mexico), 2(5):
4 113–115(1972)—Sp(En).

3473 Hagsater, E.
Una especie recientemente descrita del noroeste de Mexico: *Encyclia kennedyi* (Fowl. and Withner) comb. nov. (A recently described species for Mexico: *Encyclia kennedyi* (Fowl. & Withner) comb. nov.). *Orquidea* (Mexico), 3(3):70–76(1973)—(En);
4 illustration.

kennedyi (Fowl. & Withner) Hagsater in *Orquidea*
4 (Mexico), 3(3):70(1973): *Epidendrum kennedyi.*

EPIDENDRUM

3474 Hagsater, E.
Epidendrum eximium. Orquidea (Mexico), 2(9):243–
4 244,246(1972 publ. 1973)—Sp(En).

3475 Hagsater, E.
Epidendrum ibaguense. Orquidea (Mexico), 1(4):
4 11–15(1971)—Sp; illustration.

3476 Hagsater, E.
Epidendrum longicaule: una especie jalisciense poco conocida. *Orquidea* (Mexico), 2(11):325–328(1972)—
4 (En); illustration.

3477 Hagsater, E.
Epidendrum matudae comentarios a una especie nueva. *Orquidea* (Mexico), 2(11):305–308(1973)—
4 (En); coloured illustration.

3478 Kennedy, G.
About the occurrence and discovery of the showy new species of *Encyclia* from west Mexico. *Orchid Dig.*, 37(2):44–46(1973)—Illustration, coloured illustrations, map. An account of the discovery of
4 *Epidendrum kennedyi.*

3479 Pollard, G. E.
Comentarios sobre del *Epidendrum microcharis. Orquidea* (Mexico), 2(2):33–36(1972)—Illustrations.
4 Adapted from *Orchidata*, 8(1):26–28(1968).

3480 Pollard, G. E.
En busca de Epidendrums Mexicanos. *Orquidea*
4 (Mexico), 1(5):11–19(1971).

3481 Withner, C. L.
Epidendrum rowleyi una nueva especie para honrar a un naturalista. *Orquidea* (Mexico), 1(12):5–8(1971 publ.
4 1972)—Illustrations.

3482 Withner, C. L.
A new *Epidendrum* of the *Encyclia* section from Mexico, *Epidendrum kennedyi*, sp. nov. *Orchid Dig.*,
4 37(2):43(1973).

kautskyi Pabst in *Bradea*, 1(36):364(1973)—
4 Brazil.
kennedyi Fowlie & Withner in *Orchid Dig.*, 37(2):
4 43(1973)—Mexico.

EPIGENEIUM

sanseiense (Hayata) Nackejima in *Biol. Mag. Okinawa*,
2 9:37(1972): *Dendrobium sanseiense.*

EPIPACTIS

3483 Corbineau, R.
Quelques Orchidacées de l'ouest de la France *Epipactis helleborine* (L) Crantz. *Bull. Soc. Sci. Nat.*
1 *Ouest France*, 71(2):40–42(1973).

3484 Hermjakob, G.
Eine für Griechenland neue, autogame *Epipactis*-Art: *E. leptochila* (Godf.) Godf. (Orchidaceae). (An autogamous species of *Epipactis* new to Greece: *E. leptochila* (Godf.) Godf. (Orchidaceae)). *Ann. Mus.*
1 *Goulandris*, 1:99–104(1973)—(En,Gre); illustrations.

3485 Holub, J.
Neue oder wenig bekannte Pflanzen der ungarischen Flora. *Ann. Univ. Budapest. Rolando Eötvos Nom., Biol.*, 14:91–104(1972 publ. 1973). *Epipactis*
1 *leptochila* and *E. muelleri*.

3486 Mitra, S. N.
Some new combinations in Indian plants. *Indian*
6 *Forester*, 99(2):100–101(1973).

3487 Renz, J.
Epipactis-Arten des Kaspischen Waldgebietes. Vorarbeiten zu K. H. Rechinger *Flora Iranica*, no. 19. *Orchidee*, 24(6):251–255(1973)—(En); illustrations,
2 coloured illustration; key.

helleborine (L.) Cr.
f. **gracilis** (Dageförde) Pаucă & Morariu in Fl. Republ. Social. Român., 12:761(1972): *E.*
1 *latifolia gracilis*.
var. **herbacea** (Lindl.) S. N. Mitra in *Indian*
6 *Forester*, 99(2):101(1973): *E. herbacea*.
var. **intrusa** (Lindl.) S. N. Mitra *op. cit.* 101(1973):
6 *E. intrusa*.
var. **thomsonii** (Hook. f.) R. R. Stewart, Annot. Cat. Vasc. Pl. W. Pakistan & Kashmir: 68(1972), without basionym date; also S. N. Mitra in *Indian Forester*, 99(2):101(1973), as 'thomsoni':
6 *E. latifolia* var. *thomsoni*.
persica (Soó) Renz in *Orchidee*, 24(6):251,253(1973),
2 without basionym ref.: (Basionym not stated).
2 **rechingeri** Renz *op. cit.* 253(1973)—Iran.

EPIPOGIUM

3488 Holmes, F. M. *and* Holmes, W. B. K.
Some notes on *Epipogium roseum* (D. Don.) Lindl.
2, 5–7 *Orchadian*, 4(8):67–70(1973).

ERIA

3489 Pradhan, G. M.
Eria convallarioides. *Amer. Orchid Soc. Bull.*, 42(4):
6 327–328(1973)—En; illustration.

ERIOPSIS

3490 Charpentier, E. V.
Orquideas de Costa Rica: la *Eriopsis wercklei* Schltr. (Orchids of Costa Rica: *Eriopsis wercklei* Schltr.). *Orquidea* (Mexico), 3(5):141–149(1973)—
4 Sp(En); illustration.

EUANTHE

3491 Senghas, K.
Zum Titelbild Reichenbachia. *Orchidee*, 24(2):45
6 (1973)—Coloured illustration. *Euanthe sanderiana*.

EULOPHIA

3492 Ghosh, R. B.
Eulophia graminea Ldl. an ornamental terrestrial orchid in the Indian Botanic Garden, Calcutta.
6 *Planto*, 2(1):12–13(1972).

EULOPHIDIUM

3493 Senghas, K.
Eulophidium rauhii, ein pflanzengeographisch bedeutsamer Neufund aus Madagaskar. *Orchidee*,
5 24(2):57–61(1973)—Illustrations.

rauhii Senghas in *Orchidee*, 24(2):61(1973)—
5 Malagasy.

EULOPHIELLA

3494 Taylor, P.
Eulophiella elisabethae. *Curtis's Bot. Mag.*, 179(4): tab.656(1973)—En; illustration, coloured
5 illustration.

GASTROCHILUS

3495 Pradhan, G. M.
Indian orchids: *Gastrochilus acutifolium*. *Amer.*
6 *Orchid Soc. Bull.*, 42(8):694–695(1973)—Illustrations.

3496 Sheehan, T. *and* Sheehan, M.
Orchid genera, illustrated: 34. *Amer. Orchid Soc.*
2, 6 *Bull.*, 42(7):632–633(1973)—Coloured illustration.

× **GOFFARA** hort. in *Orchid Rev.*, 81(961):centre page pull-out p.8(1973). (*Luisia* × *Rhynchostylis* ×
0 *Vanda*).

GOVENIA

3497 Dressler, R. L.
Notas sobre el genero *Govenia* en Mexico. (Notes on the genus *Govenia* in Mexico). *Orquidea* (Mexico), 2(6):143–162(1972)—(En); illustrations, dot maps;
4 keys.

3498 Dressler, R. L. *and* Hagsater, E.
Una *Govenia* nueva del estado de Jalisco: *Govenia tequilana*. (A new *Govenia* from the State of Jalisco: *Govenia tequilana*). *Orquidea* (Mexico), 3(6):175–183
4 (1973)—(En); illustrations, coloured illustrations.

tequilana Dressler & Hagsater in *Orquidea* (Mexico),
4 3(6):175(1973)—Mexico.

GRAMMATOPHYLLUM

3499 Ragan, M. *and* Ragan, J.
Grammatophyllum elegans: a fabulous species. *Amer. Orchid Soc. Bull.*, 42(11):981–982(1973)—Illustration,
6 coloured illustration.

3500 Sheehan, T. *and* Sheehan, M.
Orchid genera, illustrated: 36. *Amer. Orchid Soc. Bull.*, 42(11):1000–1001(1973)—Coloured
6 illustrations.

HABENARIA

3501 Bowden, J. K.
Habenaria englerana. *Curtis's Bot. Mag.*, 179(2): tab.631.(1972 publ. 1973)—En; illustration, coloured
5 illustration.

3502 Weber, D.
Deux Orchidacées nouvelles pour la flore des Iles Galapagos. *Bull. Soc. Neuchâtel. Sci. Nat.*, 96:18–32 (1973)—(En,Ge); illustration, maps. *Habenaria distans*
4 var. *jamaicensis*.

drepanopetala Pabst in *Bradea*, 1(36):361(1973)—
4 Brazil.

× **HAGERARA** hort. in *Orchid Rev.*, 81(957): centre page pull-out p.9(1973). (*Doritis* ×
0 *Phalaenopsis* × *Vanda*).

HERSCHELIA

3503 Hall, A. V.
Herschelia charpentierana. *Flow. Pl. Afr.*, 42(3–4):
5 pl.1673(1973)—En; coloured illustration.

3504 Hall, A. V.
Herschelia hians. *Flow. Pl. Afr.*, 42(3–4):pl.1674
5 (1973)—En; coloured illustration.

hians (L. f.) A. V. Hall in *Flow. Pl. Afr.*, 42(3–4):
5 t. 1674(1973): *Satyrium hians*.

HIMANTOGLOSSUM

3505 Sundermann, H.
Himantoglossum (*Loroglossum*) *hircinum-caprinum-calcaratum-affine*. *Acta Bot. Acad. Sci. Hung.*, 19(1–4): 367–374(1973)—Ge(En); illustrations; chromosome
1, 2 numbers.

hircinum (L.) Spreng.
var. **obtusum** De Langhe in *Bull. Soc. Roy. Bot.*
1 *Belg.*, 106(1):72(1973)—Belgium.

× **JIMENEZARA** hort. in *Orchid Rev.*, 81(964): centre page pull-out p.8(1973). (*Broughtonia* ×
0 *Laelia* × *Laeliopsis*).

× **JOANNARA** hort. in *Orchid Rev.*, 81(964): centre page pull-out p.8(1973). (*Renanthera* ×
0 *Rhynchostylis* × *Vanda*).

KOELLENSTEINIA

florida (Rchb. f.) Garay in *Orquideologia*, 8(1):23
4 (1973): *Promenaea florida*.
lineata (Barb. Rodr.) Garay *op. cit.* 23(1973):
4 *Cyrtopodium lineatum*.

LAELIA

3506 Fowlie, J. A.
Three confused *Laelia* species from Brazil and their known varieties: *Laelia pumila* (Hook.) Rchb. f.; *Laelia praestans* Rchb. f.; and *Laelia dayana* Rchb. f. with notes on a missing species, *Laelia sincorana* Schltr. *Orchid Dig.*, 37(2):63–71(1973)—Illustrations,
4 coloured illustrations, map; key.

3507 Pontes, M.
Descripcion de la *Laelia majalis* en la obra original de Llave y Lexarza. *Orquidea* (Mexico), 2(10):278–
4 282(1972 publ. 1973)—Illustration.

4 **bradei** Pabst in *Bradea*, 1(31):332(1973)—Brazil.
4 **ghillanyi** Pabst *op. cit.* 332(1973)—Brazil.
4 **itambana** Pabst *op. cit.* 333(1973)—Brazil.
4 **liliputana** Pabst *op. cit.* 334(1973)—Brazil.
4 **macrobulbosa** Pabst *op. cit.* 335(1973)—Brazil.

LEPANTHES

3508 Hespenheide, H. A.
New *Lepanthes* from Jamaica and Mexico. *Amer. Orchid Soc. Bull.*, 42(10):870–874(1973)—
4 Illustrations.

3509 Hespenheide, H. A.
A revision of the West Indian species of *Lepanthes* (Orchidaceae): 3. Cuba. *Brittonia*, 25(3):257–283
4 (1973)—Illustrations; key.

acunae Hespenheide in *Brittonia*, 25(3):263(1973)—
4 Cuba.
blepharophylla (Griseb.) Hespenheide *op. cit.* 260
4 (1973): *Pleurothallis blepharophylla*.
byfieldii H. A. Hespenheide in *Amer. Orchid Soc.*
4 *Bull.*, 42(10):871(1973)—Jamaica.
cubensis Hespenheide in *Brittonia*, 25(3): 269(1973)—
4 Cuba.
4 **dressleri** Hespenheide *op. cit.* 268(1973)—Cuba.
4 **grisebachiana** Hespenheide *op. cit.* 272(1973)—Cuba.
4 **obliquiloba** Hespenheide *op. cit.* 273(1973)—Cuba.
4 **occidentalis** Hespenheide *op. cit.* 276(1973)—Cuba.
pollardii H. A. Hespenheide in *Amer. Orchid Soc.*
4 *Bull.*, 42(10):872(1973)—Mexico.

LIPARIS

3510 Gehu, J.-M. *and* Wattez, J.-R.
Liparis loeselii (L.) Rich. dans le nord de la France; ses stations anciennes et son maintien actuel. *Bull. Soc. Bot. France*, 118(9):801–811(1971 publ. 1973)—
1 (En); map.

3511 Rodway, G.
The fen orchid – *Liparis loeselii*. *Orchid Rev.*, 81
1 (965):312–314(1973)—Illustrations.

LISTERA

3512 Rodway, F. G.
Listera cordata. *Glasgow Nat.*, 19(1):68(1973)—En.
1 Local record.

cordata (L.) R.Br.
f. **trifolia** (A. & G.) Pauca & Ştefureac in Fl. Republ. Social. Român., 12:734(1972): *L.*
1 *cordata* lus. *trifolia*.
ovata (L.) R.Br.
f. trifoliata (Cariot) Pauca & Ştefureac *op. cit.* 733
1 (1972), without basionym ref.: *L. trifoliata*.

LUISIA

curtisii Seidenfaden in *Bot. Tidsskr.*, 68(1):83(1973),
6 nom. nov.: *L. tristis* auct. non Hooker f.

× **LUTHERARA** hort. in *Orchid Rev.*, 81(957): centre page pull-out p. 9(1973). (*Phalaenopsis* ×
0 *Renanthera* × *Rhynchostylis*).

LYCASTE

3513 Hagsater, E.
Lycaste skinneri. *Orquidea* (Mexico), 12(2):37–39
4 (1972)—Sp(En).

MALAXIS

3514 Nackejima, C.
(*Malaxis matsudai* (Yamamoto) Hatusima, new for the Ryukyus). *J. Geobot.*, 20(3):65–69(1972)—Ja;
2 illustrations.

cogniauxiana (Schltr.) Pabst in *Orquídea*, 30(3):
4 133(1973): *Microstylis cogniauxiana*.
matsudai (Yamamoto) Hatusima ex Nackejima in
2 *J. Geobot.*, 20(3):65(1972): *Microstylis matsudai*.
pabstii (Schltr.) Pabst in *Orquídea*, 30(3):133(1973):
4 *Microstylis pabstii*.
roohutuensis (Fukuyama) Nackejima in *Biol. Mag.*
2, 6 *Okinawa*, 9:41(1972): *Microstylis roohutuensis*.

MASDEVALLIA

3515 Fiske, M. D.
Masdevallia deformis. *Amer. Orchid Soc. Bull.*, 42(9):
4 783–784(1973)—En; coloured illustration.

3516 Maekawa, F. *and* Hashimoto, T.
A new *Masdevallia* species from Bolivia. *J. Jap.*
4 *Bot.*, 48(1):27–29(1973)—(Ja); illustrations.

3517 Pérez, A. F. *and* Sarmiento, F. *
Sinopsis sobre el genero *Masdevallia* en Colombia. *Rev. Fac. Nac. Agron. Medellin*, 27:31–42(1972). Ref.
4 in *Bull. Torrey Bot. Club*, 99(6):326(1972).

3518 Richardson, D.
Masdevallia elephanticeps Reichb. f. *Amer. Orchid Soc. Bull.*, 42(9):772–773(1973)—En; coloured
4 illustration. Popular article.

3519 Richardson, D.
Some members of the *Coccineae* section of *Masdevallia*. *Amer. Orchid Soc. Bull.*, 42(3):238–243
4 (1973)—Coloured illustrations; key.

espirito-santensis Pabst in *Bradea*, 1(31):329(1973)—
4 Brazil.
4 **kautskyi** Pabst *op. cit.* 330(1973)—Brazil.

MAXILLARIA

3520 Peña de Sousa, M.
Lo variable en *Maxillaria variabilis* Batem. ex Lindl. *Orquidea* (Mexico), 2(5):118–122(1972)—(En);
4 illustrations.

MENDONCELLA

acuminata (C. Schweinf.) Garay in *Orquideologia*,
4 8(1):28(1973): *Zygopetalum acuminatum*.
antioquiana (Krzl.) Garay *op. cit.* 28(1973):
4 *Batemania antioquiana*.
burkei (Rchb. f.) Garay *op. cit.* 28(1973):
4 *Zygopetalum burkei*.
ciliata (Morel) Garay *op. cit.* 28(1973): *Zygopetalum*
4 *ciliatum*.
4 **colombiana** Garay *op. cit.* 28(1973)—Colombia.
marginata (Garay) Garay *op. cit.* 32(1973):
4 *Zygopetalum marginatum*.
prainiana (Rolfe) Garay *op. cit.* 32(1973): *Zygopetalum*
4 *prainianum*.

MESOSPINIDIUM

3521 Garay, L. A.
Die Gattung *Mesospinidium*. *Orchidee*, 24(5):183–190(1973)—Illustrations, coloured illustrations;
4 key.

ecuadorense Garay in *Orchidee*, 24(5):186(1973)—
4 Ecudaor.
endresii (Krzl.) Garay *op. cit.* 187 (1973): *Solenidium*
4 *endresii*.
4 **lehmannii** Garay *op. cit.* 189(1973)—Colombia.
4 **panamense** Garay *op. cit.* 185(1973)—Panama.
4 **peruvianum** Garay *op. cit.* 187(1973)—Peru.

MICROCOELIA

3522 Stewart, J.
African orchids: more about Microcoelias. *Amer. Orchid Soc. Bull.*, 42(9):801–805(1973)—Coloured
5 illustrations.

MICROSTYLIS

3523 Pabst, G. F. J.
Redescoberta de *Microstylis pabstii* Schltr. *Orquídea*,
4 30(3):133–134(1973)—Sp(En); illustrations.

MILTONIA

3524 Moir, W. W. G.
Brazilian Miltonias and *Miltonia warscewiczii*. *Orchid Dig.*, 37(6):200–205(1973)—Illustrations,
4 coloured illustrations.

3525 Moir, W. W. G.
The peculiar Miltonias. *Orchid Dig.*, 37(5):160–162 (1973)—Coloured illustrations; chromosome
4 numbers.

3526 Horich, C. K.
Ein Albino von *Odontoglossum confusum* (*Miltonia schroederiana*). *Orchidee*, 24(5):193–195(1973)—(En);
4 coloured illustrations.

MORMODES

3527 Hagsater, E.
Nueva especie vistosa de *Mormodes* de Mexico: *Mormodes* (seccion *Coryodes*) *sanguineoclaustrum* Fowl.
4 *Orquidea* (Mexico), 1(6):14–20(1971)—Illustration.

3528 Kennedy, D. E.
Una narracion del redescubrimiento de *Mormodes luxatum* (Lindl.). *Orquidea* (Mexico), 2(10):273–277
4 (1972 publ, 1973)—(En); coloured illustration, map.

MYSTACIDIUM

3529 Lester, R.
Mystacidium capense (Schltr.) subtribe Sarcanthinea.
5 *Orchidata*, 13(1):9–10(1973)—Illustration.

NEOESCOBARIA Garay in *Orquideologia*, 7(4):194
4 (1972).
brevis (Rolfe) Garay *op. cit.* 194(1972): *Trichopilia*
4 *brevis*.
callichroma (Rchb.f.) Garay *op. cit.* 196(1972):
4 *Trichopilia callichroma*.

NEOGARDNERIA Schlechter ex Garay in
4 *Orquideologia*, 8(1):32(1973).
murrayana (Gardn. ex Hook.) Schlechter ex Garay
4 *op. cit.* 33(1973): *Zygopetalum murrayanum*.

NEOKOEHLERIA

corydaloides (Krzl.) Garay in *Bot. Mus. Leafl. Harvard Univ.*, 23(7):300(1973): *Rodriguezia*
4 *corydaloides*.

NERVILIA

3530 Trouslot, M. F.
Biologie et tubérisation de quelques *Nervilia* de Côte-d'Ivoire. *Compt. R. Hebd. Séances Acad. Sci.*,
5 *D* (Paris), 276(4):533–535(1973)—Illustrations.

OBERONIA

kanburiensis Seidenfaden in *Bot. Tidsskr.*, 68(1):47
6 (1973)—Thailand.

ODONTOGLOSSUM

3531 Fowlie, J. A.
A prediction come true . . . the discovery and rediscovery of *Odontoglossum harryanum*. *Orchid Dig.*, 37(6):223–226(1973)—Coloured illustrations, map. Based in part on and reprinted from F. Boyle's *The*
4 *Woodlands Orchids*.

3532 Halbinger, F.
El genero *Odontoglossum* en Mexico. *Orquidea*
4 (Mexico), 1(5):3–10(1971).

3533 Halbinger, F.
Odontoglossum londesboroughianum. *Orquidea*
4 (Mexico), 2(1):5–8(1972)—Sp(En).

3534 Halbinger, F. *and* Hagsater, E.
The rediscovery of *Odontoglossum londesboroughianum* Reichenbach filius, *Gard. Chron.* n.s. 6. 772 (1876). *Orchid Dig.*, 37(3):84–85(1973)—Coloured illustrations. Reprinted from *Orquidea*, 2(1)1972.
4 English version prepared by the authors.

3535 Hands, G.
Odontoglossum odoratum. *Orchid Rev.*, 81(957):98–
4 100(1973)—En.; illustrations.

3536 Hartmann, W. L.
Odontoglossum citrosmum. *Orquidea* (Mexico),
4 1(9):5–7(1971)—Sp.

3537 Horich, C. K.
Ein Albino von *Odontoglossum confusum* (*Miltonia schroederiana*). *Orchidee*, 24(5):193–195(1973)—(En);
4 coloured illustrations.

ONCIDIUM

3538 Charanasri, U., Kamemoto, H. *and* Takeshita, M.
Chromosome numbers in the genus *Oncidium* and some allied genera. *Amer. Orchid Soc. Bull.*, 42(6):518–
3, 4 524(1973)—Illustrations.

3539 Duveen, D. I.
The checkered career of *Oncidium spilopterum*. *Orchid Dig.*, 37(5):165–166(1973)—Coloured
4 illustration, map.

3540 Duveen, D. I.
The rediscovery of *Oncidium spilopterum*. *Orchid Rev.*, 81(960):177–179(1973)—Coloured
4 illustrations.

3541 Hagsater, E.
Una especie nueva del estado de Michoacan, Mexico: *Oncidium oviedomotae*. (A new species from the state of Michoacan, Mexico: *Oncidium oviedomotae*). *Orquidea* (Mexico), 3(8):234–243(1973)—(En);
4 illustrations, coloured illustration.

3542 Hagsater, E.
Una especie poco conocida: *Oncidium stelligerum* Reichb. F. var. *ernesti* Williams. *Orquidea* (Mexico),
4 2(8):217–221(1972)—(En); coloured illustration.

3543 Moir, W. W. G.
Synonymous? *Hawaii Orchid J.*, 2(1):19–28(1973)—
3, 4 Illustrations.

3544 Osment, W.
Oncidium moirianum, a new species. *Amer. Orchid*
4 *Soc. Bull.*, 42(4):305–306(1973)—Illustrations.

3545 Sinotô, Y.
Oncidium hybrids having odd 2n numbers. *Chromosome Inform. Serv.*, no.12:3(1971)—Chromosome
3, 4 numbers.

3546 Teuscher, H.
Collector's item: *Oncidium barbatum*, *Onc. micropogon* and *Onc. suscephalum*. *Amer. Orchid Soc. Bull.*,
4 42(8):696–700(1973)—Illustrations.

3547 Williams, N. H. *and* Dressler, R. L.
Oncidium described by Jacquin and the typification of *Oncidium*. *Taxon*, 22(2–3):221–227(1973)—
4 Illustrations; chromosome numbers.

guianense (Aubl.) Garay in *Bot. Mus. Leafl. Harvard*
4 *Univ.*, 23(7):300(1973): *Ophrys guianensis*.
oviedomotae Hagsater in *Orquidea* (Mexico), 3(8):
4 234(1973)—Mexico.
stachyi Garay in *Bot. Mus. Leafl. Harvard Univ.*,
4 23(7):301(1973)—Bolivia.

OPHRYS

3548 Canehl, G.
Über Vorkommen von *Ophrys mammosa* var. *helenae* im nördlichen Epirusgebiet. *Orchidee*, 24(6):
1 258–259(1973)—Coloured illustrations.

3549 Danesch, O. *and* Danesch, E.
Nachtrag zu: *Ophrys fuciflora* (Crantz) Moensch ssp. *gracilis* Büel et Danesch ssp. nov. *Orchidee*, 24(1):36
1 (1973).

3550 Danesch, O. *and* Danesch, E.
Zum Problem der Bastardierung innerhalb der Gattung *Ophrys*, eine Entgegnung. *Orchidee*, 24(1):
1 33–36(1973).

3551 Forster, J. *and* Peisl, P.
Des fleurs pour les mâles. *Mus. Genève*, 132:8–10
1 (1973)—Illustrations. Pollination of *Ophrys*.

3552 Krey, W.-D.
Ein bisher nicht bekannter *Ophrys*-Bastard auf
2 Rhodos. *Orchidee*, 24(6):257–258(1973).

3553 Renz, J.
Orchidaceae aus dem südwestlichen Iran – *Ophrys luristanica* Renz spec. nov. Vorarbeiten zu K. H. Rechinger *Flora Iranica*, Nr. 18. *Orchidee*, 24(2):46–50
2 (1973)—(En); illustrations, coloured illustrations.

3554 Soó, R.
Supplement to species and subspecies of the genus *Ophrys*. *Acta Bot. Acad. Sci. Hung.*, 18(3–4):379–384
1 (1973).

3555 Sundermann, H.
Ergänzung zu *Ophrys luristanica* Renz spec. nov. *Orchidee*, 24(2):50–51(1973)—Illustrations, coloured
2 illustration.

carmeli Fleischm. & Bornm.
subsp. **orientalis** (Renz) Soó in *Acta Bot. Acad. Sci. Hung.*, 18(3–4):381(1973): *O. cornuta* subsp.
2 *orientalis*.
× *chatenieri* Rouy
nm. **bakonyensis** Soó in *Acta Bot. Acad. Sci. Hung.*, 18(1–2):173(1973)—Hungary. (*O. fuciflora* var.
1 *cornigera* × *O. sphegodes* var. *fucifera*).
exaltata Ten.
subsp. **sundermannii** Soó *op. cit.* 173(1973)—
1 Italy.
fuciflora (Crantz) Moench
subsp. **gracilis** Büel, O. Danesch & E. Danesch in
1 *Orchidee*, 24(1):36(1973)—Italy.
subsp. **lacaitae** (Lojacono) Soó in *Acta Bot. Acad.*
1 *Sci. Hung.*, 18(3–4):382(1973): *O. lacaitae*.
subsp. **maxima** (Fleischm.) Soó *op. cit.* 382,384
1, 2 (1973): *O. fuciflora* var. *maxima*.
f. **tallosii** Soó in *Acta Bot. Acad. Sci. Hung.*, 18(1–2):
1 173(1973)—Hungary.
1 f. **triloboviridis** Soó *op. cit.* 173(1973)—Hungary.
2 **luristanica** Renz in *Orchidee*, 24(2):47(1973)—Iran.
lutea Cav.
var. **fleischmannii** (Hay.) Soó in *Acta Bot. Acad.*
1 *Sci. Hung.*, 18(3–4):379(1973): *O. fleischmannii*.
mammosa Desf.
subsp. **caucasica** (Woronow) Soó *op. cit.* 383
2 (1973): *O. caucasica*.
subsp. helenae (Renz) Soó *op. cit.* 380,383(1973),
1 without basionym ref.: *O. helenae*.
subsp. **vierhapperi** (Soó) Soó *op. cit.* 380(1973):
2 *O. aranifera* var. *vierhapperi*.

sphegodes Mill.
subsp. × **panormitana** (Tod.) Soó *op. cit.* 384
1 (1973): *Arachnites fuciflora* var. *panormitana.*
tommasinii Vis.
subsp. litigiosa (Camus) Soó *op. cit.* 383(1973),
1 without basionym page: *O. litigiosa.*

× ORCHIACERAS
bivonae (Tod.) Soó
nm. **welwitschii** (Rchb. f.) Pinto da Silva in
Agron. Lusit., 34(3):183(1972 publ. 1973):
1 *Orchis welwitschii.*

ORCHIS

3556 Corrias, B. *and* Villa, R.
Ad floram italicam notulae taxonomicae et geobotanicae: 10. *Orchis purpurea* Huds. nuovo reperto per la flora sarda. *Webbia*, 28(1):49–51(1973)—
1 (En); chromosome numbers.

3557 Klein, E.
Orchis sancta L. × *Orchis coriophora* L. ssp. *fragrans* (Poll.) Camus–die erste intragenerische Hybride der Sektion *Coriophorae* Parlat. *Orchidee*, 24(5):209–211
1 (1973)—(En); illustration, coloured illustrations.

3558 Procházka, F.
Vstavač zvrhlý (*Orchis* × *hybrida* (Lindl.) Boenningh.) v. Čechách. (Der Bastard *Orchis* × *hybrida* (Lindl.) Boenningh in Böhmen). *Acta Mus. Reginaehrad.*, *A*, 12:69–71(1971 publ. 1972)—
1 (Ge,Rus).

3559 Sundermann, H. *and* Wattke, R.
Neue Beiträge zur Zytotaxonomie der Erdorchideen (*Dactylorhiza*, *Orchis* und *Serapias*). *Orchidee*, 24(3):
1, 2 116–118(1973)—Illustrations; chromosome numbers.

3560 Teschner, W.
Vergleichende Beobachtungen an *Orchis patens* Desf. in Algerien und Ligurien. *Orchidee*, 24(1):25–
1 29(1973)—(En); coloured illustration.

3561 Vöth, W.
Orchis italica var. *purpurea*. *Orchidee*, 24(6):259–
1 260(1973)—Ge; coloured illustration.

cordigera Fries
f. **macrobracteata** (Schur) Pauca in Fl. Republ. Social. Român., 12:701(1972): *O. latifolia* var.
1 *macrobracteata.*
f. **rivularis** (Heuff.) Paucă *op. cit.* 702(1972):
1 *O. rivularis.*
var. **siculorum** (Soó) Paucă *op. cit.* 701(1972):
1 *O. cordigera* subsp. *siculorum.*
× *gennarii* Rchb. f.
nm. popii Soó in *Acta Bot. Acad. Sci. Hung.*, 18(1–2):177(1973), without type—Rumania.
1 (*O. papilonacea* × *O. morio* var. *caucasica*).
incarnata L.
var. **haematodes** (Rchb. f.) Paucă in Fl. Republ. Social. Român., 12:692(1972): *O. incarnata* lus.
1 *haematodes.*
× kallithea E. Klein in *Orchidee*, 24(5):210(1973), without type—Rhodes. (*O. sancta* × *O. coriophora*
1 subsp. *fragrans*).
maculata L.
var. **heuffelii** (Borsos & Soó) Paucă in Fl. Republ. Social. Român., 12:698(1972): *Dactylorchis*
1 *maculata* var. *heuffelii.*
var. **hunyadensis** (Borsos & Soó) Paucă *op. cit.*
1 698(1972): *Dactylorchis maculata* var. *hunyadensis.*
var. **olgae** (Soó) Pauca *op. cit.* 698(1972):
1 *Dactylorhiza maculata* var. *olgae.*

var. **schurii** (Klinge) Paucă *op. cit.* 697(1972):
1 *O. angustifolia* f. *schurii.*
× *maculatiformis* Rouy
nm. **claudiopolitana** (Simk.) Paucă *op. cit.* 707
1 (1972): *O. ambigua* var. *claudiopolitana.*
morio L.
f. **velutina** (Schur) Paucă *op. cit.* 671(1972):
1 *O. morio* a. *velutina.*
patens Desf.
subsp. **spitzelii** (Sauter) A. Löve & Kjellqvist in
1 *Lagascalia*, 3(2):176(1973): *O. spitzelii.*
ustulata L.
f. **leopoliensis** (Zapał.) Paucă in Fl. Republ. Social. Român., 12:674(1972): *O. ustulata* var.
1 *leopoliensis.*

× ORCHISERAPIAS
maura Schrenk in *Senckenberg. Biol.*, 52(6):453
1 (1971)—France.

ORLEANESIA
mineirosensis Garay in *Bradea*, 1(27):303(1973)—
4 Brazil.

ORTHOPENTHEA

3562 Hall, A. V.
Orthopenthea rosea. *Flow. Pl. Afr.*, 42(3–4):pl.1679
5 (1973)—En; coloured illustration.

× **OTOPABSTIA** Garay in *Bradea*, 1(27):308
4 (1973). (*Otostylis* × *Pabstia*).

PABSTIA Garay in *Bradea*, 1(27):306(1973), nom.
4 nov.: *Colax* Lindl. (1843).
jugosa (Lindl.) Garay *op. cit.* 307(1973): *Maxillaria*
4 *jugosa.*
var. **rufina** (Rchb. f.) Garay *op. cit.* 307(1973):
4 *Colax jugosus* var. *rufinus.*
var. **viridis** (Godefr.) Garay *op. cit.* 307(1973):
4 *Colax jugosus* var. *viridis.*
modestior (Rchb. f.) Garay *op. cit.* 307(1973):
4 *Colax modestior.*
placanthera (Hook.) Garay *op. cit.* 307(1973):
4 *Maxillaria placanthera.*
triptera (Rolfe) Garay *op. cit.* 308(1973): *Colax*
4 *tripterus.*
viridis (Lindl.) Garay *op. cit.* 308(1973): *Maxillaria*
4 *viridis.*
var. **parvifolia** (Hoehne) Garay *op. cit.* 308(1973):
4 *Colax viridis* var. *parviflorus.*

× **PALMERARA** hort. in *Orchid Rev.*, 81(959): centre page pull-out p. x(1973). (*Batemannia* ×
0 *Otostylis* × *Zygosepalum*).

PAPHIOPEDILUM

3563 Fowlie, J. A.
Malaya revisited, part 2. *Paphiopedilum johorensis* and its habitat on the plateau of Gunong Panti. *Orchid Dig.*, 37(5):191–194(1973)—Coloured
6 illustrations.

3564 Fowlie, J. A.
Malaya revisited: part 3. *Paphiopedilum niveum* in the Langkawi Islands. *Orchid Dig.*, 37(6):230–234(1973)—
6 Illustrations, coloured illustrations, map, dot maps.

3565 Fowlie, J. A.
Notes on two new color varieties of *Paphiopedilum chamberlainianum* of the *primulinum* form. *Orchid Dig.*, 37(3):102–105(1973)—Illustration, coloured
6 illustrations.

3566 Han, S.
The Burmese *Paphiopedilum*. *Paphioped. World*,
6 2(3):53–56(1972).

3567 Liem Khe Wie
The dreamlike reality of discovering the new yellow
Paphiopedilum. *Orchid Rev.*, 81(961):220–222(1973)—
6 Coloured illustrations.

3568 Pontes, M.
El genero *Paphiopedilum*. *Orquidea* (Mexico),
4 2(11):309–311(1973)—Illustrations.

3569 Pontes, M. *and* Castaño, G.
El genero *Paphiopedilum*. *Orquidea* (Mexico),
2(8):226–228, (9):240–242(1972 publ. 1973)—
4 Illustrations. *P. insigne* and *P. barbatum*.

3570 Pradhan, G. M.
Paphiopedilum fairieanum. *Paphioped. World*, 2(1):
6 12–14(1972)—En; illustrations.

3571 Pradhan, G. M.
Paphiopedilum hirsutissimum. *Paphioped. World*,
6 2(4):84–85(1972)—En.

3572 Romero-Price, G.
A review of the Philippine species of *Paphiopedilum*.
Orchid Dig., 37(4):120–125(1973)—Coloured
6 illustrations, map; key.

3573 Sheehan, T. *and* Sheehan, M.
Orchid genera, illustrated: 35. *Amer. Orchid Soc.*
6 *Bull.*, 42(9):808–809(1973)—Coloured illustrations.

3574 Taylor, P.
[*Paphiopedilum*]. *Orchid Rev.*, 81(965):329(1973)—
6 En. Short note.

3575 Wood, M. W. *and* Taylor, P.
Paphiopedilum primulinum M. W. Wood and P.
Taylor, sp. nov. *Orchid Rev.*, 81(961):220–221
6 (1973)—En; coloured illustrations.

chamberlainianum (O'Brien) Pfitz.
var. flavescens Fowlie in *Orchid Dig.*, 37(3):
104(1973), rank misplaced, see Arts. 5 & 33—
6 Sumatra.
var. flavum Fowlie *op. cit.* 104(1973), rank
6 misplaced, see Arts. 5 & 33—Sumatra.
f. **primulinum** (Wood & Taylor) Fowlie *op. cit.*
6 104(1973): *P. primulinum*.
primulinum M. W. Wood & P. Taylor in *Orchid*
6 *Rev.*, 81(955):x(1973)—Sumatra.

PARAPHALAENOPSIS

3576 Sweet, H. R.
The genus *Paraphalaenopsis*. *Orchid Dig.*, 37(1):
6 11–14(1973)—Coloured illustrations, maps.

PARODIA

neglectoides F. H. Brandt in *Stachelpost*, 9(47):
4 129(1973)—Bolivia.

PECTEILIS

sagarikii Seidenfaden in *Bot. Tidsskr.*, 68(1):46
6 (1973)—Thailand.

PERISTERIA

cochlearis Garay in *Orquideologia*, 7(4):199(1972)—
4 Colombia.

PHAIUS

Crinita (Gagnep.) Seidenfaden in *Bull. Mus. Nation.*
Hist. Nat. (Paris), 3e Sér., no.71, *Bot.* 5:141(1972
6 publ. 1973): *Calanthe crinita*.

PHALAENOPSIS

3577 Anon.
Phalaenopsis violacea. *Bull. Orchid Soc. S.E. Asia*,
no.10:7–9(1973)—En. Reprinted from 7th Annual
6 Orchid Show Programme of Perak Orchid Society.

3578 Sweet, H. R.
Observations on the genus *Phalaenopsis*: 12. *P.*
fuscata and its related species. *Orchid Dig.*, 37(3):
6 107–108(1973)—Illustrations, coloured illustrations.

3579 Sweet, H. R.
Observations on the genus *Phalaenopsis*: 13.
Phalaenopsis parishii var. *lobbii* Rchb. f. *Orchid Dig.*,
37(5):167–168(1973)—Illustrations, coloured
6 illustration.

PHRAGMIPEDIUM

3580 Kuhn, J. *and* Kuhn, L.
Phragmipedium pearcei. *Amer. Orchid Soc. Bull.*,
4 42(8):710–711(1973)—En.

PITYPHYLLUM

3581 Sweet, H. R.
Orquídeas Andinas poco conocidas: 6. *Pity-*
phyllum Schltr. *Orquideologia*, 7(4):203–214(1972)—
4 Illustrations; key.

pinoides H. R. Sweet in *Orquideologia*, 7(4):214
4 (1972)—Ecuador.

PLATANTHERA

bifolia (L.) L. C. Rich.
f. **trifoliata** (Thielens) Paucă & A. Nyárády in
Fl. Republ. Social. Român., 12:728(1972):
1 *P. solstitialis* var. *trifoliata*.
brevifolia (Greene) Senghas in *Orchidee*, 24(3):97
4 (1973): *Habenaria brevifolia*.
minor (Miq.) Reichenbach f.
var. **mikurensis** Takahashi in *J. Geobot.*, 20(1):
2 15(1972)—Japan.

PLEIONE

3582 Esslemont, H.
Pleione forrestii. *J. Scott. Rock Gard. Club*, 13(3):
2 184(1973)—En; coloured illustration.

3583 Ghòse, B. N.
Pleione uniflora. *Orchid Rev.*, 81(965):317–318
6 (1973)—En.

PLEUROTHALLIS

4 **dutrai** Pabst in *Bradea*, 1(36):362(1973)—Brazil.
4 **fluminensis** Pabst *op. cit.* 363(1973)—Brazil.
4 **garayi** Pabst *op. cit.* 364(1973)—Brazil.
4 **kautskyi** Pabst in *Bradea*, 1(31):330(1973)—Brazil.

POLYCYCNIS

3584 Teuscher, H.
Collector's item: *Polycycnis barbata* and *Polycycnis*
gratiosa. *Amer. Orchid Soc. Bull.*, 42(7):601–605(1973)—
4 Illustrations.

POLYSTACHYA

3585 Duckitt, F.
Polystachya tessellata. *S. Afr. Orchid J.*, June:12
5 (1973)—En.

3586 Stewart, J.
An outstanding new species of *Polystachya* from Tanzania. *Amer. Orchid Soc. Bull.*, 42(7):590–593
5 (1973)—Coloured illustrations.

porphyrochila J. Stewart in *Amer. Orchid Soc. Bull.*,
5 42(7):592(1973)—Tanzania.

× **POMATISIA** hort. in *Orchid Rev.*, 81(964): centre page pull-out p.9(1973). (*Luisia* ×
0 *Pomatocalpa*).

PONTHIEVA

3587 Dulieu, D.
Étude morphologique de la surface pollinique de *Ponthieva maculata* Lindl. (Orchidaceae) en microscopie électronique à balayage. *Adansonia*, 13(2):
3, 4 229–234(1973).

PORROGLOSSUM

3588 Sweet, H. R.
An orchid genus new to Peru. *Amer. Orchid Soc.*
4 *Bull.*, 42(3):221–223(1973)—Illustrations.

peruvianum H. R. Sweet in *Amer. Orchid Soc. Bull.*,
4 42(3):221(1973)—Peru.

PSEUDOLAELIA

cipoensis Pabst in *Bradea*, 1(36):365(1973)—
4 Brazil.
4 **irwiniana** Pabst *op. cit.* 366(1973)—Brazil.

PSEUDORCHIS

3589 Soják, J.
Doplňky k nomenklatuře některých rodů (Phanerogamae). *Cas. Nár. Muz.* (Prague), 141(1–2):
3 61–63(1972).

straminea (Fernald) Soják in *Čas. Nár. Muz.*
3 (Prague), 141(1–2):63(1972): *Habenaria straminea.*

PTEROSTEMMA

brasiliensis Pabst in *Bradea*, 1(36):368(1973)—
4 Brazil.
frigidum (Dodson & Dressler) Garay in *Bot. Mus. Leafl. Harvard Univ.*, 23(7):304(1973): *Cypholoron*
4 *frigida.*

PTEROSTYLIS

3590 [Gee, B.]
Exciting re-discovery. *Launceston Nat.*, 5(3):1–2
7 (1971)—*Pterostylis cucullata.*

RENANTHERA

3591 Sheehan, T. *and* Sheehan, M.
Orchid genera, illustrated: 32. *Amer. Orchid Soc.*
2, 6 *Bull.*, 42(5):436–437(1973)—Coloured illustration.

RHYNCHOGYNA Seidenfaden & Garay in
6 *Bot. Tidsskr.*, 68(1):88(1973).
luisifolia (Ridl.) Seidenfaden & Garay *op. cit.* 90
6 (1973): *Saccolabium luisifolium.*
saccata Seidenfaden & Garay *op. cit.* 91(1973)—
6 Thailand.

RHYNCHOLAELIA

3592 Hartmann, W. L.
Rhyncholaelia digbyana (Benth). *Orquidea* (Mexico),
4 1(2):10–16(1971)—Sp; illustration.

ROBIQUETIA

3593 Kerr, A. D.
Orchids of southeast Asia: the genus *Robiquetia. Amer. Orchid Soc. Bull.*, 42(1):51–52(1973)—
6 Illustration, coloured illustration.

× **RODRIGLOSSUM** hort. in *Orchid Rev.*, 81 (958): centre page pull-out p.x(1973). (*Odonto-*
0 *glossum* × *Rodriguezia*).

RODRIGUEZIA

3594 Braga, P. I. S.
Rodriguezia sucrei Braga (Orchidaceae) espécie nova da flora fluminense. *Bradea*, 1(35):357–360(1973)—
4 Illustrations.

3595 Mejía R., A.
Comentarios sobre el género *Rodriguezia* en Colombia. *Orquideologia*, 8(2):104–116(1973)—
4 (En); coloured illustrations.

sucrei P. I. S. Braga in *Bradea*, 1(35):357(1973)—
4 Brazil.

ROGRIGUEZIELLA

3596 Berman, J. D.
An interesting discovery. *J. Orchid Soc. Gr. Brit.*,
4 22(3):71(1973).

3597 Berman, J. D.
An interesting discovery: *Rodrigueziella gomeziodes.*
4 *Orchid Rev.*, 81(956):56(1973)—Illustration.

gomeziodes (Rodriguez) Berman in *Orchid Rev.*,
4 80(956):56(1973): *Theodorea gomeziodes.*
jucunda (Rchb. f.) Garay in *Orchidee*, 24(5):190
4 (1973): *Mesospinidium jucundum.*

SARCOCHILUS

3598 Dockrill, A. W.
Sarcochilus, the genus. *Austral. Pl.*, 7(53):13–18
7 (1972)—Coloured illustrations.

SARCOGLOTTIS

magdalenensis (Brade & Pabst) Pabst in *Bradea*,
4 1(31):329(1973): *Pelexia magdalenensis.*

SARCOGLYPHIS

pensilis (Ridl.) Seidenfaden in *Bot. Tidsskr.*, 68(1):87
6 (1973): *Sarcanthus pensilis.*

× **SARTYLIS** hort. in *Orchid Rev.*, 81(964): centre page pull-out p.10(1973). (*Rhynchostylis* ×
0 *Sarcochilus*).

SATYRIUM

3599 Anon.
Cover photo: *Satyrium carneum*, R.Br. *S. Afr.*
5 *Orchid J.*, June: 5(1973)—Illustration.

3600 Shechter, Y. *and* Arditti, J.
Orchid nomenclature and phytochemistry: caproic acid and the derivation of the name *Satyrium*: a curious relationship. *Orchid Rev.*, 81(962):249–251
1–7 (1973).

SAUNDERSIA

3601 Pabst, G. F. J.
Saundersia mirabilis (uma orquídea vista muito
4 raramente). *Orquídea*, 30(2):60(1973)—Illustrations.

SCAPHYGLOTTIS

minuta (Rich. & Gal.) Garay in *Bot. Mus. Leafl.*
4 *Harvard Univ.*, 23(7):299(1973): *Polystachya minuta.*

SCHLIMIA

3602 Sweet, H. R.
Orquídeas andinas poco conocidas: 7. *Schlimia*
Planchon & Linden ex Lindley. *Orquideologia*, 8(1):
3–14, 35–36(1973)—(En); illustrations, coloured
4 illustrations; key.

garayana H. R. Sweet in *Orquideologia*, 8(1):8
4 (1973)—Ecuador.

SCHOMBURGKIA

3603 Jones, H. G.
The genus *Schomburgkia*: a study in the history and
bibliography of plant taxonomy. *Taxon*, 22(2–3):
4 229–239(1973).

3604 Jones, H. G.
Notas sobre dos especies interesantes de
Schomburgkia. (Notes on two interesting species of
Schomburgkia). *Orquidea* (Mexico), 3(2):45–57(1973)—
4 (En); illustration, coloured illustration.

brysiana Lem.
var. **intermedia** H. G. Jones in *Bol. Soc. Brot.*, 46:
4 244(1972 publ. 1973)—Cayman Is.

SCUTICARIA

irwiniana Pabst in *Bradea*, 1(31):336(1973)—
4 Brazil.
4 **itirapinensis** Pabst *op. cit.* 337(1973)—Brazil.

SERAPIAS

3605 Ramin, I. von
Serapias olbia bei Nizza. *Orchidee*, 24(3):120–121
1 (1973)—Illustration.

3606 Sundermann, H. *and* Wattke, R.
Neue Beiträge zur Zytotaxonomie der Erdorchideen
(*Dactylorhiza*, *Orchis* und *Serapias*). *Orchidee*, 24(3):
116–118(1973)—Illustrations; chromosome
1, 2 numbers.

SIGMATOSTALIX

cuculligera (Schltr.) Garay in *Orquideologia*, 7(4):202
4 (1972): *Roezliella cuculligera.*

SOLENIDIUM

3607 Sweet, H. R.
Orquídeas Andinas poco conocidas: 8. *Solenidium*
Lindley. *Orquideologia*, 8(2):87–93(1973)—Sp,En;
4 illustration, coloured illustrations.

SOPHRONITELLA

3608 Lester, R.
Sophronitella violacea (Schltr.). *Amer. Orchid Soc.*
Bull., 42(8):676–677(1973)—En; coloured
4 illustration.

SPIRANTHES

3609 Grazi, G. F.
Orchidee spontanee liguri: *Spiranthes autumnalis*
1 Rich. *Rea*, no.5:23–25(1972)—(En).

3610 Hagsater, E.
Spiranthes aurantiaca. Orquidea (Mexico), 1(8):
4 5–6(1971)—Sp.

3611 Millner, P.
Spiranthes spiralis (L.) Chevall a Mecsek-hegységben.
(*Spiranthes spiralis* (L.) Chevall im Mecsek-Gebirge
(Südungarn)). *Bot. Közlem.*, 59(3):172(1972 publ.
1 1973).

3612 Sheviak, C. J.
A new *Spiranthes* from the grasslands of central
North America. *Bot. Mus. Leafl. Harvard Univ.*,
3 23(7):285–297(1973)—Illustration.

3613 Williams, L. O.
A new *Spiranthes* (Orchidaceae) from Guatemala.
Phytologia, 26(6):435–436(1973)—Illustration. *S.*
4 *umbraticola.*

magnicamporum Sheviak in *Bot. Mus. Leafl. Harvard*
3 *Univ.*, 23(7):287(1973)—U.S.A. (Illinois).
petenensis L. O. Williams in *Phytologia*, 25(7):460
4 (1973)—Guatemala.
umbraticola L. O. Williams in *Phytologia*, 26(6):
4 435(1973)—Guatemala.

× **STACYARA** hort. in *Orchid Rev.*, 81(958):
centre page pull-out p.x(1973). (*Cattleya* ×
0 *Epidendrum* × *Sophronitis*).

STANHOPEA

3614 Charpentier, E. V.
La sistematica de *Stanhopea*: un caos. *Orquidea*
4 (Mexico), 2(9):259–267(1972 publ. 1973)—(En);
illustration.

3615 Teuscher, H.
Collector's item: *Stanhopea connata* Klotzsch
(1854). *Amer. Orchid Soc. Bull.*, 42(10):896–899
4 (1973)—Illustrations.

3616 Teuscher, H.
Collector's item: *Stanhopea eburnea* and *Stanhopea*
saccata. Amer. Orchid Soc. Bull., 42(6):510–515(1973)—
4 Illustrations.

3617 Teuscher, H.
Collector's item: *Stanhopea insignis. Amer. Orchid*
4 *Soc. Bull.*, 42(9):790–793(1973)—Illustrations.

STIGMATODACTYLUS

3618 Murata, G.
[Short note on *Stigmatodactylus sikokianus* Maxim.]
2 *Acta Phytotax. Geobot.*, 25(4–6):187(1973)—Ja(En).

SYMPHYGLOSSUM

bowmannii (Rchb. f.) Garay in *Orchidee*, 24(5):
4 190(1973), as 'bowmanni': *Mesospinidium bowmanni.*

TRAUNSTEINERA

globosa (L.) Rchb. f.
f. **major** (Schur) Beldie in Fl. Republ. Social.
1 Român., 12:708(1972): *Orchis globosa* a. *major.*

TRICHOPILIA

3619 Garay, L. A.
Notas sobre la alianza *Trichopilia. Orquideologia*,
4 7(4):191–196(1972)—Illustration; key.

3620 Sheehan, T. *and* Sheehan, M.
Orchid genera, illustrated: 31. *Amer. Orchid Soc.*
4 *Bull.*, 42(1):56–57(1973)—Coloured illustration.

3621 Teuscher, H.
Cischweinfia dasyandra (*Trichopilia dasyandra*).
Amer. Orchid Soc. Bull., 42(11):988–990(1973)—En;
4 illustrations; key.

TULOTIS

flava (L.) Senghas in *Orchidee*, 24(3):97(1973):
3 *Orchis flava.*
sonoharae (Masamune) Nackejima in *Biol. Mag.*
2 *Okinawa*, 9:43(1972): *Platanthera sonoharai.*

VANDA

3622 Dennis, G. F. G.
A recent finding of *Vanda hindsii*. *Austral. Orchid*
6 *Rev.*, 38(1):28–29(1973).

3623 Fowlie, J. A.
The quest for *Vanda sanderiana*. *Orchid Dig.*, 37(4):
6 151–153(1973)—Coloured illustration, map.

3624 Halbinger, F.
Vanda coerulea. *Orquidea* (Mexico), 1(8):9–11
6 (1971)—Sp.

3625 Pirie, D.
Strap leaf Vandas. *S. Afr. Orchid J.*, Sept.: 7–9
6 (1973)—Illustrations.

3626 Pradhan, G. M.
The habitat and growing conditions of *Vanda coerulea* in nature with cultural hints gleaned therefrom. *Orchid Dig.*, 37(3):87–91(1973)—Coloured
6 illustration, map.

3627 Pradhan, G. M.
A note on *Vanda coerulescens*. *Orchid Dig.*, 37(3):
6 91–92(1973).

VANDOPSIS

3628 Millar, A.
Vandopsis muelleri—we found it! *Austral. Orchid*
6 *Rev.*, 38(1):22–23(1973).

WARREELLA

patula Garay in *Orquideologia*, 8(1):52(1973)—
4 Colombia.

WARREOPSIS Garay in *Orquideologia*, 8(1):51
4 (1973).
colorata (Linden & Rchb. f.) Garay *op. cit.* 51
4 (1973): *Zygopetalum coloratum.*
pardina (Rchb. f.) Garay *op. cit.* 51(1973): *Zygo-*
4 *petalum pardinum.*
parviflora (L. O. Williams) Garay *op. cit.* 51(1973):
4 *Zygopetalum parviflorum.*

× **WILKINSARA** hort. in *Orchid Rev.*, 81(958):
centre page pull-out p.x(1973). (*Ascocentrum* ×
0 *Vandopsis*).

× **ZYGOPABSTIA** Garay in *Bradea*, 1(27):308
4 (1973). (*Pabstia* × *Zygopetalum*).

ZYGOPETALUM

3629 Garay, L. A.
El complejo *Zygopetalum*. *Orquideologia*, 8(1):
15–34,37–39(1973)—(En); illustrations, coloured illustrations; keys to species of the component genera.
4 Correction in *Orquideologia*, 8(2):103(1973).

3630 [Self, E. W.]
Zygopetalum. *J. Orchid Soc. Gr. Britain*, 22(1):16–17(1973)—En; illustration, coloured illustration;
4 *Z. intermedium.*

pedicellatum (Thunb.) Garay in *Bradea*, 1(27):305
4 (1973): *Cymbidium pedicellatum.*

ZYGOSEPALUM

angustilabium (C. Schweinf.) Garay in *Orquideologia*,
4 8(1):34(1973): *Zygopetalum tatei* var. *angustilabium.*

OROBANCHACEAE

3631 Mądalski, J.
Atlas flory Polskiej i ziem ościennych. (Florae Polonicae terrarumque adiacentium iconographia): tom 17 zeszyt 1. Scrophulariaceae (pars 4), Orobanchaceae. Warszawa, etc., Wydawnictwo Naukowe,
1 1973. 77p.—Illustrations.

CISTANCHE

3632 Fernández Casas, J. *and* Laínz, M.
Sobre *Cistanche phelypaea* (L.) P. Cout. *Anuár.*
1 *Soc. Brot.*, 39.129–130(1973).

LATHRAEA

3633 Anchisi, E.
Une plante méconnue: la clandestine (*Lathraea*
1, 2, 6 *squamaria*). *Rea*, no.5:38–41(1972)—Illustrations.

3634 Kuijt, J.
Biologische und morphologische Hinweise zur systematischen Stellung von *Lathraea*. *Beitr. Biol. Pfl.*,
1, 2 49(1):137–146(1973)—(En).

OROBANCHE

3635 Asai, Y.
(Miscellaneous notes on the naturalized flora of Japan: 3). *J. Jap. Bot.*, 48(3):67–72(1973)—Ja(En);
2 illustrations. *Orobanche minor* var. *flava.*

3636 Heckard, L. R.
A taxonomic re-interpretation of the *Orobanche californica* complex. *Madroño*, 22(2):41–70(1973)—
3, 4 Illustrations, dot maps; keys.

3637 Hepper, F. N.
Problems in naming *Orobanche* and *Striga*. (Probleme bei der Benennung von *Orobanche* und *Striga* Zusammenfassung). *In* European Weed Research Council: Parasitic Weeds Research Group *and* Malta: Royal University. Symposium on parasitic weeds, 11–13 April 1973, Royal University of Malta, Malta: 9–17(1973)—(Ge,Fr); illustrations;
1 key.

3638 Kropáč, Z.
Weedy *Orobanche* spp. of Czechoslovakia and the range of their parasitism. (Les espèces nuisibles d' *Orobanche* en Tchécoslovaquie et l'importance de leur parasitisme). *In* European Weed Research Council: Parasitic Weeds Research Group *and* Malta: Royal University. Symposium on parasitic weeds, 11–13 April 1973, Royal University of Malta, Malta:
1 35–43(1973)—(Fr, Ge).

3639 Mijatović, K. *and* Stojanović, D.
Distribution of *Orobanche* spp. on the agricultural crops in Yugoslavia. (Distribution de l'orobanche (*Orobanche* spp.) sur les plantes cultivées en Yugoslavie). *In* European Weeed Research Council: Parasitic Weeds Research Group *and* Malta: Royal University. Symposium on parasitic weeds, 11–13 April 1973, Royal University of Malta, Malta:28–34(1973)—
1 (Fr,Ge).

3640 Piehl, M. A. *and* Piehl, P. N.
Orobanche in Tennessee. *J. Tenn. Acad. Sci.*, 48(1):
3 39(1973)—Illustration. *O. ludoviciana.*

3641 Saadi Gharib, M.
Biological and economical aspects of the broom-rapes (*Orobanche* spp.) in northern Iraq. (Biologische und wirtschaftliche Aspekte von *Orobanche* spp. im nördlichen Irak). *In* European Weed Research Council: Parasitic Weeds Research Group *and* Malta: Royal University. Symposium on parasitic weeds, 11–13 April 1973, Royal University of Malta, Malta: 44–47
2 (1973)—(Ge,Fr).

3642 Swart, J.
Een bijzondere plant op een bijzondere plaats. *Natura* (Netherlands), 69(4):63–64(1972)—*Orobanche*
1 *purpurea.*

3643 Ungurean, L. *and* Şerbănescu-Jitariu, G.
Some morpho-anatomical aspects of *Orobanche racemosa* L. (Einige morpho-anatomische Aspekte von *Orobanche racemosa* L.). *In* European Weed Research Council: Parasitic Weeds Research Group *and* Malta: Royal University. Symposium on parasitic weeds, 11–13 April 1973, Royal University of Malta,
1 Malta: 132–139(1973)—(Ge,Fr).

sect. **Nothaphyllon** (A. Gray) Heckard in *Madroño,*
3 22(2):41(1973): *Aphyllon* sect. *Nothaphyllon.*
californica Chamisso & Schlechtendal
subsp. **condensa** Heckard *op. cit.* 59(1973)—U.S.A.
3 (California).
subsp. **feudgei** (Munz) Heckard *op. cit.* 62(1973):
3, 4 *O. grayana* var. *feudgei.*
subsp. **grandis** Heckard *op. cit.* 60(1973)—U.S.A.
3 (California).
subsp. **grayana** (Beck) Heckard *op. cit.* 54(1973):
3 *O. grayana.*
subsp. **jepsonii** (Munz) Heckard *op. cit.* 57(1973):
3 *O. grayana* var. *jepsonii.*
parishii (Jepson) Heckard *op. cit.* 66(1973): *O.*
3, 4 *californica* var. *parishii.*
subsp. **brachyloba** Heckard *op. cit.* 68(1973)—
3, 4 U.S.A. (California), Mexico.
reticulata Wimm. & Grab.
subsp. **pallidiflora** (Wimm. & Grab.) Tsvelev in *Novosti Sist. Vўssh. Rast.,* 10:362(1973): *O.*
1 *pallidiflora.*
vallicola (Jepson) Heckard in *Madroño,* 22(2):64
3 (1973): *O. comosa* var. *vallicola.*

PHACELLANTHUS

3644 Terëkhin, É. S.
Émbriologiya Orobanchaceae: 2. K émbriologii i ékologii *Phacellanthus tubiflorus* Sieb. et Zucc. (The embryology of Orobanchaceae: 2. On the embryology and ecology of *Phacellanthus tubiflorus* Sieb. et Zucc.).
2 *Bot. Zhurn.,* 58(12):1775–1788(1973)—Illustration.

OXALIDACEAE

OXALIS

3645 Bleck, M.
Oxalis gigantea Barn. *Cact. Succ. J.* (U.S.A.), 44(5):
4 204–208(1972)—En; illustrations.

3646 Denton, M. F.
A monograph of *Oxalis*, section *Ionoxalis* (Oxalidaceae) in North America. *Publ. Mus. Mich. State Univ., Biol.,* 4(10):455–615(1973)—Illustrations,
3 dot maps; keys.

3647 Geigy *Ltd*: Agricultural Chemicals Division
Oxalidaceae: wood sorrel family. *Giegy Weed Tables,* 59(1972)—Illustrations, coloured
1–7 illustration.

3648 Mulgura de Romero, M. E.
Sinopsis de las especies de *Oxalis* L. de la Mesopotamia Argentina. *Darwiniana,* 18(1–2):44–69
4 (1973)—(En); illustrations; key.

3649 Ornduff, R.
Oxalis dines, a new species from the Western Cape. *J. S. Afr. Bot.,* 39(3):201–203(1973)—(Af); illustra-
5 tions; chromosome numbers.

dines Ornduff in *J. S. Afr. Bot.,* 39 (3): 201(1973),
5 nom. nov.: *O. disticha* var. β *alba* Salter.
incisa Denton in *Publ. Mus. Mich. State Univ.,*
4 *Biol.,* 4(10):550(1973)—Mexico.
sellowiana Zuccarini
var. **alba** Múlgura in *Darwiniana,* 18(1–2):51
4 (1973)—Argentina.
tetraphylla Cav.
var. **guerreroensis** Denton in *Publ. Mus. Mich.*
4 *State Univ., Biol.,* 4(10):592(1973)—Mexico.
4 var. **mexicana** Denton *op. cit.* 594(1973)—Mexico.

XANTHOXALIS

amara (A. St.-Hilaire) Holub in *Folia Geobot.*
4 *Phytotax.,* 8(2):175(1973): *Oxalis amara.*
aureoflava (Steudel) Holub *op. cit.* 175(1973): *Oxalis*
4 *aureoflava.*
breviramulosa (Rusby) Holub *op. cit.* 175(1973):
4 *Oxalis breviramulosa.*
chrysantha (Progel) Holub *op. cit.* 175(1973):
4 *Oxalis chrysantha.*
cordobensis (R. Knuth) Holub *op. cit.* 175(1973):
4 *Oxalis cordobensis.*
dumetorum (Barn.) Holub *op. cit.* 175(1973):
4 *Oxalis dumetorum.*
filiformis (Humb., Bonpl. & Kunth) Holub *op. cit.*
4 175(1973): *Oxalis filiformis.*
laxa (Hook. & Arn.) Holub *op. cit.* 175(1973):
4 *Oxalis laxa.*
lineata (Gill.) Holub *op. cit.* 175(1973): *Oxalis*
4 *lineata.*
lotoides (Humb., Bonpl. & Kunth) Holub *op. cit.*
4 175(1973): *Oxalis lotoides.*
lyonii (Pursh) Holub *op. cit.* 175(1973): *Oxalis*
4 *lyonii.*
medicaginea (Humb., Bonpl. & Kunth) Holub
4 *op. cit.* 175(1973): *Oxalis medicaginea.*
micrantha (Colla) Holub *op. cit.* 175(1973): *Oxalis*
4 *micrantha.*
ortgiesii (Regel) Holub *op. cit.* 175(1973): *Oxalis*
4 *ortgiesii.*
parvifolia (DC.) Holub *op. cit.* 175(1973): *Oxalis*
4 *parvifolia.*
phaeotricha (Diels) Holub *op. cit.* 175(1973): *Oxalis*
4 *phaeotricha.*
polyantha (Walpers) Holub *op. cit.* 175(1973):
4 *Oxalis polyantha.*
refracta (A. St.-Hilaire) Holub *op. cit.* 176(1973):
4 *Oxalis refracta.*
rosea (Jacq.) Holub *op. cit.* 176(1973): *Oxalis*
4 *rosea.*
spiralis (G. Don) Holub *op. cit.* 176(1973): *Oxalis*
4 *spiralis.*
sternbergii (Zucc.) Holub *op. cit.* 176(1973):
4 *Oxalis sternbergii.*
thelyoxys (Focke) Holub *op. cit.* 176(1973): *Oxalis*
4 *thelyoxys.*
trinidadensis (R. Knuth) Holub *op. cit.* 176(1973):
4 *Oxalis trinidadensis.*
tuberosa (Molina) Holub *op. cit.* 176(1973): *Oxalis*
4 *tuberosa.*
valdiviensis (Barn.) Holub *op. cit.* 176(1973):
4 *Oxalis valdiviensis.*

PAEONIACEAE

3650 Almeida, M. T. de
Paeoniaceae. *Anuár. Soc. Brot.*, 38:145–146(1972)—
1 Por.

PAEONIA

3651 Dadasheva, Sh. G.
Anatomicheskaya kharakteristika pionov Azerbaĭdzhana. *Izv. Akad. Nauk Az. SSR, Biol.*
2 *Nauk*, no.3:24–25(1972)—(Az).

3652 Filipescu, G.
Cercetări anatomice comparative la speciile *Paeonia peregrina* Mill. var. *romanica* (Brandza) A. Nyár, *P. tenuifolia* L. si *P. triternata* Pall. *An. Sti.*
1 *Univ. "Al. I. Cuza" Iaşi, Biol.*, 18(2):431–440(1972).

3653 Uspenskaya, M. S.
Razvitie seyantsev nekotorȳkh vidov roda *Paeonia* L. (The development of seedlings of some species of the genus *Paeonia*). *Vestn. Mosk. Univ. Nauch. Zhurn.*, 6, 27(6):89–93(1972)—(En);
1–3 illustration.

PALMAE

3654 Bernardi, L.
Les palmiers: 1. Les palmes du triomphe? *Mus.*
1–7 *Genève*, 138:13–17(1973)—Illustrations.

3655 Bernardi, L.
Les palmiers: 2. Les palmes des martyres. *Mus.*
1–7 *Genève*, 139:2–7(1973)—Illustrations, map.

3656 Davis, T. A. *and* Basu, S. K.
Structure of the inflorescence in Palmae. *Proc.*
1–7 *Indian Sci. Congr.*, 60(3):290–291(1973). Abstract only.

3657 Essig, F. B.
Pollination in some New Guinea palms. *Principes*,
6 17(3):75–83(1973)—Illustrations.

3658 Glassman, S. F.
A revision of B. E. Dahlgren's *Index of American Palms*. Lehre, J. Cramer, 1972. vi,294p. (Phanero-
4 gamarum Monographiae, tomus 6).

3659 Moore, H. E.
The major groups of Palms and their distribution. *Gentes Herb.* (Ithaca), 11(2):27–141(1973)—
1–7 Illustrations, maps; chromosome numbers; key.

3660 Moore, H. E.
Palms in the tropical forest ecosystems of Africa and South America. *In* Meggers, B. J., Ayensu, E. S. *and* Duckworth, W. D., *eds.* Tropical forest ecosystems in Africa and South America: 63–88(1973)—Maps,
4, 5 dot maps; chromosome number.

3661 Murray, S. G.
The formation of endocarp in palm fruits. *Principes*,
1–7 17(3):91–102(1973)—Illustrations.

3662 Uhl, N. W. *and* Moore, H. E.
The protection of pollen and ovules in palms.
1–7 *Principes*, 17(4):111–149(1973)—Illustrations.

3663 Whitmore, T. C.
Palms of Malaya. London, etc., Oxford University
6 Press, 1973. xvi,132p.—Illustrations, maps; keys.

3664 Williams, C. A., Harborne, J. B. *and* Clifford, H. T.
Negatively charged flavones and tricin as chemosystematic markers in the Palmae. *Phytochemistry*,
1–7 12(10):2417–2430(1973).

BACTRIS

dahlgreniana Glassman in Phanerogam. Monogr., 6 (Rev. B. E. Dahlgren's Ind. Amer. Palms): 34
4 (1972), nom. nov.: *Guilielma microcarpa.*

BORASSUS

3665 Dassanayake, M. D. *and* Sivakadachchan, B.
Germination and seedling structure of *Borassus flabellifer* L. *Ceylon J. Sci., Biol. Sci.*, 10(2):159–166
6 (1973).

BUTIA

3666 Silberbauer-Gottsberger, I.
Blüten- und Fruchtbiologie von *Butia leiospatha* (Arecaceae). (Flower and fruit biology of *Butia leiospatha* (Arecaceae)). *Österr. Bot. Zeitschr.*, 121(3–4):
4 171–185(1973)—(En); illustrations.

CATOBLASTUS

inconstans (Dugand) Glassman in Phanerogam. Monogr., 6 (Rev. B. E. Dahlgren's Ind. Amer.
4 Palms): 63(1972): *Catostigma inconstans.*

CHRYSALIDOCARPUS

3667 Fisher, J. B.
Unusual branch development in the palm, *Chrysalidocarpus. Bot. J. Linn. Soc.*, 66(1):83–95(1973)—
5 Illustrations.

COPERNICIA

3668 Karpowiczowa, L.
Mikołaj Kopernik a botanika. *Wiad. Bot.*, 17(3):
4 169–172(1973)—Illustration, portrait.

CORYPHA

3669 Meijer, W.
Endangered plant life. *Biol. Conserv.*, 5(3):163–
6, 7 167(1973)—Illustrations. *Corypha elata.*

LIVISTONA

3670 Moore, H. E.
Livistona brassii, *L. crustacea* and *L. muelleri.*
6, 7 *Principes*, 17(2):60(1973).

METROXYLON

3671 Whitmore, T. C.
On the Solomons' sago palm. *Principes*, 17(2):
6 46–48(1973)—Illustrations. *Metroxylon sagu.*

NEOPHLOGA

3672 Guillaumet, J.-L.
Une nouvelle espèce de palmier nain de Madagascar. *Adansonia*, 13(3):341–349(1973)—(En);
5 illustrations; key.

brevicaulis Guillaumet in *Adansonia*, 13(3):343
5 (1973)—Malagasy.

ORANIA

3673 Meijer, W.
Endangered plant life. *Biol. Conserv.*, 5(3):163–
6, 7 167(1973)—Illustrations.

PIGAFETTA

3674 Dransfield, J.
Pigafetta filaris in Sibolangit. *Principes*, 17(3):105–
6 107(1973)—Illustrations.

SERENOA

3675 Fisher, J. B. *and* Tomlinson, P. B.
Branch and inflorescence production in saw palmetto
(*Serenoa repens*). *Principes*, 17(1):10–19(1973)—
3 Illustrations.

TRITHRINAX

3676 Maldonado de Magnano, S.
Trithrinax campestris (Palmae): inflorescencia y flor
con especial referencia al gineceo. *Kurtziana*, 7:137–
4 152(1973)—(En); illustration.

PANDACEAE

3677 Villiers, J.-F.
Pandaceae. Flore du Gabon, 22:14–30(1973)—
5 Illustrations; keys.

PANDANACEAE

3678 Stone, B. C.
A reconsideration of the evolutionary status of the
family Pandanaceae and its significance in mono-
cotyledon phylogeny. *Quart. Rev. Biol.*, 47(1):34–45
4–7 (1972).

FREYCINETIA

3679 Stone, B. C.
Materials for a monograph of *Freycinetia* Gaudich:
14. On the relation between *F. banksii* A. Cunn. of
New Zealand and *F. baueriana* Endl. of Norfolk
Island, with notes on the structure of the seeds. *New*
7 *Zealand J. Bot.*, 11(2):241–246(1973).

baueriana Endl.
subsp. **banksii** (A. Cunn.) Stone in *New Zealand*
7 *J. Bot.*, 11(2):242(1973): *F. banksii*.

PANDANUS

3680 Cheah, C. H. *and* Stone, B. C.
Chromosome studies of the genus *Pandanus*
Pandanaceae. *Bot. Jahrb.*, 93(4):498–529(1973)
5–7 (Studies in Malesian Pandanaceae 7)—(Ge).

3681 Chu, S. P. *and* Stone, B. C.
Morphological studies in Pandanaceae: 5. A further
survey of foliar anatomy in the genus *Pandanus*: 1.
5, 6 *J. Jap. Bot.*, 48(2):55–64(1973).

3682 Chu, S. P. *and* Stone, B. C.
Morphological studies in Pandanaceae: 5. A
further survey of foliar anatomy in the genus
2, 5–7 *Pandanus*: 2. *J. Jap. Bot.*, 48(3):74–81(1973)—(Ja).

3683 Gillett, J. B.
Pandanus embuensis (St. John ined.) an interesting
screw pine in need of preservation. *EANHS Bull.*,
5, 6 Oct.:142(1973).

3684 Guillaumet, J. L.
Formes et développement des *Pandanus* malgaches.
5 *Webbia*, 28(2):495–519(1973)—(En,It); illustrations.

3685 St. John, H.
Revision of the genus *Pandanus* Stockman: 35.
Additional *Pandanus* species from New Guinea. *Pac.*
6 *Sci.*, 27(1):44–91(1973)—Illustrations; keys.

3686 Stone, B. C.
The genus *Pandanus* in the Solomon Islands with
notes on adjacent regions: part 2. *Malaysian J. Sci.*,
6 2A:59–80(1973)—Illustrations; keys.

3687 Stone, B. C.
A synopsis of the African species of *Pandanus*. *Ann.*
Missouri Bot. Gard., 60(2):260–272(1973)—
5 Illustrations.

3688 Stone, B. C.
The systematic position of *Pandanus obovatus* and
P. calcis, Thailand Pandanaceae. *Malayan Nat. J.*,
6 26(1–2):53–54(1973)—Illustration.

3689 Stone, B. C. *and* Guillaumet, J.-L.
Un nouveau *Pandanus* (Pandanacées) sub-aquatique
de Madagascar. *Adansonia*, 12(4):525–530(1972 publ.
5 1973)—(En); illustrations.

sect. **Galeatistigma** St. John in *Pac. Sci.*, 27(1):52
6 (1973).
sect. **Heterostigma** (Gaudichaud) Stone in *Ann.*
Missouri Bot. Gard., 60(2):270(1973): genus
5 *Heterostigma*.
sect. **Liniobtutus** St. John in *Pac. Sci.*, 27(1):56
6 (1973).
sect. Megastigma Stone in *Malaysian J. Sci.*, 2(A):
6 59(1973), without latin descr.
acicularis St. John in *Pac. Sci.*, 27(1):44(1973)—
6 New Guinea (Terr. New Guinea).
aecuatus St. John *op. cit.* 47(1973)—New Guinea
6 (Papua).
alatus St. John *op. cit.* 83(1973)—Louisiade
6 Archipelago.
antaresensis St. John *op. cit.* 58(1973)—New Guinea
6 (W. Irian).
aprilensis St. John *op. cit.* 62(1973)—New Guinea
6 (W. Irian).
biciliatus St. John *op. cit.* 64(1973)—Louisiade
6 Archipelago.
biformatus St. John *op. cit.* 67(1973)—New Guinea
6 (Papua).
calostigma Martelli
var. **tenaruensis** Stone in *Malaysian J. Sci.*, 2(A):
6 62(1973)—British Solomon Is.
clausus St. John in *Pac. Sci.*, 27(1):69(1973)—New
6 Guinea (Terr. New Guinea).
concavus St. John *op. cit.* 71(1973)—New Guinea
6 (W. Irian).
congregatus St. John *op. cit.* 49(1973)—New Guinea
6 (W. Irian).
daymanensis St. John *op. cit.* 79(1973)—New Guinea
6 (Papua).
galeatus St. John *op. cit.* 74(1973)—Louisiade
6 Archipelago.
hooglandii St. John *op. cit.* 56(1973)—New Guinea
6 (Papua).
luteus St. John *op. cit.* 77(1973)—Louisiade
6 Archipelago.
marinus St. John *op. cit.* 87(1973)—New Guinea
6 (Terr. New Guinea).
navigatorum Martelli
var. **elbertii** Stone in *Malaysian J. Sci.*, 2(A):64
6 (1973)—British Solomon Is.
novohibernicus Martelli
var. **contractus** Stone *op. cit.* 65(1973)—British
6 Solomon Is.
var. **leptomeris** Stone *op. cit.* 66(1973)—British
6 Solomon Is.
var. **praeacutus** Stone *op. cit.* 66(1973)—
6 Bougainville Is.
var. **pseudoturritus** Stone *op. cit.* 67(1973)—
6 British Solomon Is.

pedunculatus R. Br.
var. **insularis** Stone *op. cit.* 67(1973)—British
6 Solomon Is.
var. **malagunensis** Stone *op. cit.* 68(1973)—New
6 Britain.
var. **rendovensis** Stone *op. cit.* 68(1973)—British
6 Solomon Is.
pentagonos St. John in *Pac. Sci.*, 27(1):50(1973)—
6 New Guinea (Terr. New Guinea).
peyrierasii Stone & Guillaumet in *Adansonia*, 12(4):
5 525(1972 publ. 1973)—Malagasy.
polyacris Martelli
var. **malaitensis** Stone in *Malaysian J. Sci.*, 2(A):
6 69(1973)—British Solomon Is.
var. **pseudo-linnaei** (Martelli) Stone *op. cit.* 70
6 (1973): *P. odoratissimus* var. *pseudo-linnaei*.
schoddei St. John in Stone *op. cit.* 59(1973), without
6 latin descr.—Bougainville Is.
vogelensis St. John in *Pac. Sci.*, 27(1):89(1973)—
6 New Guinea (Papua).

PAPAVERACEAE

3690 Exell, A. W. *and* Gonçalves, M. L.
Flora de Moçambique: 9. Papaveraceae. 1973. 5p.—
5 Chromosome numbers.

3691 Koul, A. K., Gohil, R. N., Sharma, M. C. *and* Kapoor, H. K.
Genetic stocks of Kashmir Papavers: 1. Plant morphology, chromosome numbers and meioses. *Nucleus*, 15(2):117–123(1972)—Chromosome
6 numbers.

ARGEMONE

3692 Malik, C. P. *and* Grover, I. S.
The genus *Argemone*: 1. Cytogenetic relationships between diploid species. *Genetica*, 44(1):55–70(1973)—
3, 4 Chromosome numbers.

3693 Malik, C. P. *and* Grover, I. S.
The genus *Argemone*: 2. Cytogenetic relationships of *A. ochroleuca* ssp. *ochroleuca* (2n = 56) and some diploid (2n = 28) *Argemone* species. *Theor. Appl.*
3, 4 *Genet.*, 43(7):329–334(1973)—Illustrations.

3694 Stermitz, F. R., Kim, D. K. *and* Larson, K. A.
Alkaloids of *Argemone albiflora*, *A. brevicornuta* and *A. turnerae*. *Phytochemistry*, 12(6):1355–1357
3, 4 (1973).

3695 Stermitz, F. R., Ito, R. J., Workman, S. M. *and* Klein, W. M.
Alkaloids of *Argemone fruticosa* and *A. echinata*.
4 *Phytochemistry*, 12(2):381–382(1973).

BOCCONIA

3696 Vent, W.
Beiträge zur Kenntnis der Sippenstruktur der Gattungen *Bocconia* L. und *Macleaya* R. Br. (Papaveraceae). *Acta Bot. Acad. Sci. Hung.*, 19(1–4):385–
4 391(1973)—(En); chromosome numbers.

CORYDALIS

3697 Khokhryakov, A. P.
Novaya sektsiya roda *Corydalis* Vent. (Generis *Corydalis* Vent. sectio nova). *Novosti Sist. Vȳssh.*
2 *Rast.*, 10:152–155(1973).

3698 Mathew, B.
Plant awards 1972–73: *Corydalis caucasica* DC. P.C.
2 *Quart. Bull. Alp. Gard. Soc.*, 41(4):331(1973).

sect. **Raphanituber** Khokhryakov in *Novosti Sist.*
2 *Vȳssh. Rast.*, 10:155(1973).

DICENTRA

3699 Davis, T. A. W.
Dicentra formosa. *B.S.B.I. News*, 2(2):34(1973)—
1 En.

3700 Gorer, R.
Plant awards 1972–73: *Dicentra cucullaria* (L.) DC.
3 A.M. *Quart. Bull. Alp. Gard. Soc.*, 41(4):327(1973).

3701 Khánh, T. C.
Beitrag zur Kenntnis der Sippenstruktur des Genus *Dicentra* Bernh. subgenus *Dactylicapnos* (Wall.) K. R. Stern (Fumariaceae). *Feddes Repert.*, 83(7–8):511–
2, 6 575(1972 publ. 1973)—(En); illustrations, dot maps.

sect. **Minicalcara** Khánh in *Feddes Repert.*, 83(7–8):
6 566(1972 publ. 1973).
ventii Khánh *op. cit.* 540(1972 publ. 1973)—
6 India.

MACLEAYA

3702 Vent, W.
Beiträge zur Kenntnis der Sippenstruktur der Gattungen *Bocconia* L. und *Macleaya* R. Br. (Papaveraceae). *Acta Bot. Acad. Sci. Hung.*, 19(1–4):385–
2 391(1973)—(En); chromosome numbers.

MECONOPSIS

3703 Taylor, *Sir* G.
Meconopsis collections of George Forrest. *J. Scott.*
2 *Rock Gard. Club*, 13(3):244–246(1973)—Illustrations.

PAPAVER

3704 Arthur, A. E., Gale, J. S. *and* Lawrence, M. J.
Variation in wild populations of *Papaver dubium*:
1 7. Germination time. *Heredity*, 30(2):189–197(1973).

3705 Arthur, A. E., Rana, M. S., Gale, J. S., Humphreys, M. O. *and* Lawrence, M. J.
Variation in wild populations of *Papaver dubium*: 6. Dominance relationships for genes contvolling
1 metrical characters. *Heredity*, 30(2):177–187(1973).

3706 Duke, T. A., Gunn, C. R., Leppik, E. E., Reed, C. F., Solt, M. L. *and* Terrell, E. E.
Annotated bibliography on opium and oriental poppies and related species. [Washington, D.C.],
1–7 U.S.D.A., 1973. 349p. (ARS-NE-28).

3707 Kiger, R. W.
Sectional nomenclature in *Papaver* L. *Taxon*,
1–3 22(5–6):579–582(1973).

3708 Kitagawa, M.
(Notulae fractae ob floram Asiae orientalis: 30). *J. Jap. Bot.*, 48(8):235–237(1973)—*Papaver*
2 *borealisinense*.

3709 Malik, C. P. *and* Grover, I. S.
Cytogenetic studies in *Papaver*: 2. Hybrids among species with 7 as the haploid chromosome number.
1–7 *Caryologia*, 26(1):13–25(1973).

3710 Mary, T. N. *and* Malik, C. P.
Cytogenetic studies in *Papaver*: 3. Induced polyploids in some *Papaver* species. *Chromosome Inform.*
1–7 *Serv.*, no.15. 27–29(1973)—Chromosome numbers.

3711 Mary, T. N. *and* Malik, C. P.
Cytogenetic studies in *Papaver*: 4. Trisomic in garden poppy. *Chromosome Inform. Serv.*, no.15.
1–7 34–35(1973)—Chromosome numbers.

3712 Phillipson, J. D., Sariyar, G. *and* Baytop, T.
Alkaloids from *Papaver fugax* of Turkish origin.
2 *Phytochemistry*, 12(10):2431–2434(1973).

3713 Preininger, V. *and* Tošnarová, V.
Isolierung und Chemie der Alkaloide der Gattung *Papaver*: 58. Isolierung und Identifizierung der Alkaloide aus *Papaver urbanianum* Fedde. *Pl. Med.*,
1 23(3):233–238(1973)—(En).

3714 Reckin, J.
A contribution to the cytology of *Papaver gracile* Auch. including proposals for a revision of the section *Mecones*. *Caryologia*, 26(2):245–251(1973)—
2 Illustration; chromosome numbers.

3715 Tolmachev, A. I. *and* Petrovskiĭ, V. V.
Novȳe vidȳ *Papaver* s kraĭnego severo-vostoka Azii. (New species of *Papaver* from the extreme north-east Asia). *Bot. Zhurn.*, 58(8):1127–1130(1973)—
2 Illustrations.

borealisinense Kitagawa in *J. Jap. Bot.*, 48(8):237
2 (1973), nom. nov.: *P. alpinum* lusus *chinense* Regel.
degenii (Urumoff & Javorka) Pignatti in *Giorn. Bot.*
1 *Ital.*, 107(5):211(1973): *P. pyrenaicus* subsp. *degenii*.
gorodkovii Tolmachev & Petrovskiĭ in *Bot. Zhurn.*,
2 58(8):1128(1973)—U.S.S.R. (Siberia).
lapeyrousianum Gutermann in *Österr. Bot. Zeitschr.*,
1 122(4):268(1973), nom. nov.: *P. suaveolens* Lapeyr.
paucistaminum Tolmachev & Petrovskiĭ in *Bot.*
2 *Zhurn.*, 58(8):1129(1973)—U.S.S.R. (Siberia).
popovii Siplivinskiĭ in *Novosti Sist. Vȳssh. Rast.*,
2 10:360(1973)—U.S.S.R. (Siberia).
tatricum (Nyár.) Ehrendorfer in *Österr. Bot. Zeitschr.*,
1 122(4):268(1973): *P. alpinum* subsp. *tatricum*.
uschakovii Tolmachev & Petrovskiĭ in *Bot. Zhurn.*,
2 58(8):1128(1973)—U.S.S.R. (Siberia).

PISTOLOCHIA

afganica (Gilli) Holub in *Folia Geobot. Phytotax.*,
2 8(2):171(1973): *Corydalis afganica*.
alekseenkoana (Busch) Holub *op. cit.* 171(1973):
2 *Corydalis alekseenkoana*.
angustifolia (M. Bieb.) Holub *op. cit.* 172(1973):
2 *Fumaria angustifolia*.
blanda (Schott) Holub *op. cit.* 172(1973): *Corydalis*
1 *blanda*.
boissieri (Prain) Holub *op. cit.* 172(1973): *Corydalis*
2 *boissieri*.
× **budensis** (Vajda) Holub *op. cit.* 172(1973):
1 *Corydalis* × *budensis*.
cabulica (Gilli) Holub *op. cit.* 172(1973): *Corydalis*
2 *cabulica*.
decumbens (Thunb.) Holub *op. cit.* 172(1973):
2 *Fumaria decumbens*.
feddeana (Poellnitz) Holub *op. cit.* 172(1973):
2 *Corydalis feddeana*.
griffithii (Boiss.) Holub *op. cit.* 172(1973): *Corydalis*
2, 6 *griffithii*.
hallaisanensis (Leveillé) Holub *op. cit.* 172(1973):
2 *Corydalis hallaisanensis*.
× **hausmannii** (Klebelsb.) Holub *op. cit.* 172(1973):
0 *Corydalis* × *hausmannii*.
nm. **campylochila** (Teyber) Holub *op. cit.* 172
0 (1973): *Corydalis* × *campylochila*.
nm. **samuelsonii** (Fedde) Holub *op. cit.* 172(1973):
0 *Corydalis* × *samuelsonii*.

kiautschouensis (Poellnitz) Holub *op. cit.* 172 (1973), as 'kiautschowensis': *Corydalis kiaut-*
2 *schowensis*.
kurdica (Cullen & Davis) Holub *op. cit.* 172(1973):
2, 6 *Corydalis rutifolia* subsp. *kurdica*.
majorii (Poellnitz) Holub *op. cit.* 172(1973), as
1 'majori': *Corydalis majori*.
marschalliana (Pall.) Holub *op. cit.* 172(1973):
1, 2 *Corydalis marschalliana*.
nariniana (Andr. Fedorov) Holub *op. cit.* 172(1973):
2 *Corydalis nariniana*.
nivalis (Boiss. & Huet.) Holub *op. cit.* 172(1973):
2, 3, 6 *Corydalis nivalis*.
oppositiflora (DC.) Holub *op. cit.* 172(1973):
2, 6 *Corydalis oppositiflora*.
papilligera (Ohwi) Holub *op. cit.* 172(1973):
2 *Corydalis papilligera*.
slivenensis (Velen.) Holub *op. cit.* 172(1973):
2 *Corydalis slivenensis*.
solida (L.) Bernh.
subsp. **densiflora** (J.S. & C.B. Presl) Holub *op. cit.*
1 172(1973): *Corydalis densiflora*.
tarkiensis (Prochanov) Holub *op. cit.* 172(1973), as
2 'tarkensis': *Corydalis tarkensis*.
turtschaninovii (Bess.) Holub *op. cit.* 172(1973), as
2 'turtschaninowii': *Corydalis turtschaninowii*.
verticillaris (DC.) Holub *op. cit.* 172(1973):
2, 6 *Corydalis verticillaris*.
wettsteinii (Adamović) Holub *op. cit.* 172(1973):
1 *Corydalis wettsteinii*.
× **zahlbruckneri** (Scheffer) Holub *op. cit.* 173(1973):
2 *Corydalis* × *zahlbruckneri*.

PARNASSIACEAE

3716 Siddiqui, M. A.
Flora of West Pakistan: no.31. Parnassiaceae. 1973.
6 5p.—Illustrations, map.

PASSIFLORACEAE

3717 Uribe Uribe, L.
Catalogo ilustrado de las plantas de Cundinamarca: vol. 5. Passifloraceae, Begoniaceae, Melastomataceae.
4 1972. 165p.—Illustrations. Passifloraceae p.11–14.

ADENIA

3718 Glass, C. *and* Foster, R.
The succulent passion flowers. *Cact. Succ. J.* (U.S.A.), 45(2):74–84(1973)—Illustrations, coloured
2, 5–7 illustration.

BASANANTHE

3719 Wilde, W. J. J. O. de
Revision of *Basananthe*, formerly *Tryphostemma* (Passifloraceae). *Blumea*, 21(2):327–356(1973)—
5 Illustrations; key.

apetala (Baker f.) de Wilde in *Blumea*, 21(2):338
5 (1973): *Tryphostemma apetala*.
baumii (Harms) de Wilde *op. cit.* 350(1973):
5 *Tryphostemma baumii*.
var. **caerulescens** (A. & R. Fernandes) de Wilde
5 *op. cit.* 351(1973): *Tryphostemma caerulescens*.
berberoides (Chiov.) de Wilde *op. cit.* 355(1973):
5 *Carania berberoides*.
gossweileri (Hutch. & Pearce) de Wilde *op. cit.*
5 346(1973): *Tryphostemma gossweileri*.
hanningtoniana (Mast.) de Wilde *op. cit.* 342
5 (1973): *Tryphostemma hanningtonianum*.
5 **hederae** de Wilde *op. cit.* 348(1973)—Tanzania.
5 **hispidula** de Wilde *op. cit.* 347(1973)—Zaïre.

kottoensis de Wilde *op. cit.* 349(1973)—Central
5 African Republic.
lanceolata (Eng.) de Wilde *op. cit.* 345(1973):
5 *Tryphostemma lanceolatum.*
papillosa (A. & R. Fernandes) de Wilde *op. cit.* 351
5 (1973): *Tryphostemma papillosum.*
parvifolia (Baker f.) de Wilde *op. cit.* 340(1973):
5 *Tryphostemma parvifolium.*
pedata (Baker f.) de Wilde *op. cit.* 333(1973):
5 *Tryphostemma pedatum.*
phaulantha (Dandy) de Wilde *op. cit.* 341(1973):
5 *Tryphostemma phaulanthum.*
polygaloides (Hutch. & Pearce) de Wilde *op. cit.*
5 338(1973): *Tryphostemma polygaloides.*
pseudostipulata de Wilde *op. cit.* 337(1973)—
5 Zaïre, Rhodesia.
5 **pubiflora** de Wilde *op. cit.* 347(1973)—Tanzania.
reticulata (Baker f.) de Wilde *op. cit.* 350(1973):
5 *Tryphostemma reticulatum.*
sandersonii (Harvey) de Wilde *op. cit.* 339(1973):
5 *Tryphostemma sandersonii.*
scabrifolia (Dandy) de Wilde *op. cit.* 353(1973):
5 *Tryphostemma scabrifolium.*
5 **spinosa** de Wilde *op. cit.* 354(1973)—Kenya.
triloba (Bolus) de Wilde *op. cit.* 335(1973):
5 *Tryphostemma trilobum.*
zanzibarica (Mast.) de Wilde *op. cit.* 342(1973):
5 *Tryphostemma zanzibaricum.*

PASSIFLORA

3720 Beal, P. R.
Cytology of the native Australian and several exotic *Passiflora* species: 2. Chromosome morphology. *Queensl. J. Agric. Anim. Sci.*, 30(1):17–18(1973)—
7 Chromosome numbers.

3721 Beal, P. R.
Cytology of the native Australian and several exotic *Passiflora* species: 3. Morphology of satellited chromosomes. *Queensl. J. Agric. Anim. Sci.*, 30(1):
7 19–24(1973)—Chromosome numbers.

3722 Dathan, A. S. R. *and* Singh, D.
Development and structure of seed in *Tacsonia* Juss. and *Passiflora* L. *Proc. Indian Acad. Sci., B*, 77(1):
2–7 5–18(1973)—Illustrations.

3723 Marshall, E. D.
Where are the plants of yesteryear ?: the Passifloras. *Calif. Hort. J.*, 34(4):146–153(1973)—Illustra-
1–7 tions.

3724 Sacco, J. da Costa
Contribuiçao ao estudo das Passifloraceae do Brasil 4—*Passiflora castellanosii* Sacco n.sp. *Bradea*,
4 1(32):345–348(1973)—(En); illustration.

3725 Sacco, J. da Costa
Uma nova especie de *Passiflora* da Bolívia. *Bradea*,
4 1(33):349–352(1973)—Illustrations.

castellanosii Sacco in *Bradea*, 1(32):346(1973)—
4 Brazil.
pilosicorona Sacco in *Bradea*, 1(33):350(1973)—
4 Bolivia.

TRYPHOSTEMMA *see* BASANANTHE

PEDALIACEAE

3726 Abedin, S.
Flora of West Pakistan: no.33. Pedaliaceae, 1973.
6 4p.—Illustrations, map.

SESAMUM

3727 Ilyas, M.
Famous plants: sesame. *Botanica* (Delhi), 23(2–3):
2, 5, 6 64–66(1973).

3728 Kavathekar, A. K., Ganapathy, P. S. *and* Paliwall, G. S.
Observations on the morphology and anatomy of floral appendages showing phyllody in *Sesamum mulayanum*. *Phyton* (Austria), 15(1–2):155–163(1973)—
6 (Ge); illustrations.

PENTADIPLANDRACEAE

3729 Villiers, J.-F.
Pentadiplandraceae. Flore du Cameroun, 15:163–167(1973)—Illustration; key. Also published as Flore
5 du Gabon, 20(1973).

3730 Villiers, J.-F.
Pentadiplandraceae. Flore du Gabon, 20:163–167 (1973)—Illustration; key. Also published as Flore du
5 Cameroun, 15(1973).

PENTADIPLANDRA

brazzeana Baillon
var. **valida** Pellegrin ex Villiers in Fl. Cameroun, 15:167 (Jan. 30, 1973); and in Fl. Gabon, 20:167
5 (Jan. 30, 1973)—Cameroun, Gabon.

PHILADELPHACEAE
see
HYDRANGEACEAE

PHRYMACEAE

3731 Qaiser, M.
Flora of West Pakistan: no.46. Phrymaceae, 1973.
6 3p.—Illustration, map.

PHRYMA

3732 Kundu, B. C. *and* Bose, S.
Taxonomic position of the genus *Phryma*. *Proc.*
2, 3 *Indian Sci. Congr.*, 58(3):437(1971). Abstract only.

PHYTOLACCACEAE

3733 Hatschbach, G. *and* Guimarães, O.
Fitolacáceas do estado do Paraná. *Bol. Mus. Bot. Mun.* (Curitiba), no.8. 34p.(1973)—Illustrations, dot
4 maps; keys.

PIPERACEAE

3734 Agostini, G.
Florula de Macarao: Piperaceae. fasc. 44. Caracas,
4 Inst. Pedagógico, 1973. 15p.—Illustrations; key.

3735 Howard, R. A.
Notes on the Piperaceae of the Lesser Antilles. *J.*
4 *Arnold Arbor.*, 54(3):377–411(1973).

PEPEROMIA

3736 Düll, R.
Die *Peperomia*-Arten Afrikas. *Bot. Jahrb.*, 93(1):
5 56–129(1973)—Illustrations; key.

3737 Rauh, W. *and* Hutchison, P. C.
Peperomia columella a new succulent, dwarf species from northern Peru. *Cact. Succ. J.* (U.S.A.), 45(4): 4 152–156(1973)—Illustrations.

3738 Steyermark, J. A.
Novedades venezolanas. *Acta Bot. Venezuel.*, 8(1–4):111–119(1973)—Illustrations. *Peperomia* 4 *buntingii.*

abyssinica Miquel
var. **stuhlmannii** (C.DC.) Düll in *Bot. Jahrb.*, 5 93(1):97(1973): *P. stuhlmannii.*
blanda (Jacquin) Humboldt, Bonpland & Kunth
var. **leptostachya** (Hooker & Arnott) Düll *op. cit.* 5, 7 110(1973): *P. leptostachya.*
buntingii Steyermark in *Acta Bot. Venezuel.*, 8(1–4): 4 116(1973)—Venezuela.
columella Rauh & P. C. Hutchison in *Cact. Succ.* 4 *J.* (U.S.A.), 45(4):152(1973)—Peru.
fernandopoiana C.DC.
var. **butaquensis** (de Wildeman) Düll in *Bot.* 5 *Jahrb.*, 93(1):116(1973): *P. butaquensis.*
pallida (Forst. f.) A. Dietr.
var. **anderssonii** (Yuncker) Fosberg & Sachet in 6 *Micronesica*, 8(1–2):43(1972): *P. anderssonii.*
var. **mannii** (Hook. f.) Düll *op. cit.* 90(1973): 5 *P. mannii.*
retusa (L. f.) A. Dietrich
var. **bachmannii** (DC.) Düll in *Bot. Jahrb.*, 93(1): 5 90(1973): *P. bachmannii.*

PIPER

3739 Burger, W. C.
Evolutionary trends in the Central American species of *Piper* (Piperaceae). *Brittonia*, 24(4):356–362 4 (1972)—Illustrations.

3740 Dhar, K. L. *and* Raina, M. L.
Further chemical studies of *Piper peepuloides*. *Pl.* 1 *Med.*, 23(3):295–297(1973).

3741 Yuncker, T. G.
The Piperaceae of Brazil: 1. *Piper*—group 1,2,3,4. 4 *Hoehnea*, 2:21–366(1972)—Illustrations; keys.

PITTOSPORACEAE

3742 Cufodontis, G. *and* Faria, M.T.
5 Flora de Moçambique: 17. Pittosporaceae. 1973. 5p.

3743 Faria, M. T.
Revisão das Pittosporaceae de Moçambique. (Revision of Pittosporaceae in Mozambique). *Agron.* 5 *Moçamb.*, 7(2): 97–103(1973)— (En, Fr); dot map.

PLANTAGINACEAE

3744 Bassett, J.
The plantains of Canada. *Canada Dep. Agric. Monogr.*, no.7. 47p.(1973)—Illustrations, dot maps; 3 chromosome numbers; keys.

3745 Fernandes, A. *and* França, F.
Contribution à la connaissance cytotaxinomique des spermatophyta du Portugal: 6. Plantaginaceae. *Bol.* 1 *Soc. Brot.*, 46:465–501(1972)—Chromosome numbers.

PLANTAGO

3746 Arnold, S. M.
Interactions of light and temperature on the germination of *Plantago maritima* L. *New Phytol.*, 1 72(3):583–593(1973).

3747 Balbaa, S. I., Karawya, M. S. *and* Afifi, M. S.
A pharmacognostical study of the seeds of certain *Plantago* species growing in Egypt. *United Arab Republ. J. Pharm. Sci.*, 12(1):35–52(1971)— 2 (Ara).

3748 Cartier, D.
Etude biosystématique de quelques espèces du genre *Plantago* (Tourn.) L. (section *Coronopus* DC.): 3. Apports de croisements expérimentaux chez les *P. alpina* L. et *P. serpentina* All. Conclusions sur le complexe *P. alpina–P. serpentina*. *Rev. Gén. Bot.*, 1 80(944–946):5–40(1973)—(En); chromosome numbers.

3749 Cartier, D.
Etude biosystématique de quelques espèces du genre *Plantago* (Tourn.) L.: 4. La section *Oreades* Decne. *Rev. Gén. Bot.*, 80(947–949):129–149(1973)—(En); 1, 2 illustrations, maps; chromosome numbers.

3750 Delay, J.
Genre *Plantago* (suite). *Inform. Ann. Caryosyst. Cytogenet.*, no. 6:5–8(1972 publ. 1973)—Illustrations; 1 chromosome numbers.

3751 Fernandes, A. *and* Franca, F.
La meiose chez *Plantago serraria* L. *Genet. Iber.*, 1 24(1–2):31–66(1972)—Chromosome numbers.

3752 Kalda, A. *and* Roos, A.
Eesti floora *Plantago* ja *Filipendula* liikide kromosoomide arv. (Chromosome numbers in *Plantago* and *Filipendula* species in the flora of Estonia). *Flor. Märkmed*, 1(5):312–314(1973)—Chromosome 1 numbers.

3753 Kazanfarova, V. K. *and* Gasanov, A. M.
Morfologo-anatomicheskaya kharakteristika variatsiĭ i formȳ *Plantago coronopus* L. *Izv. Akad. Nauk* 1, 2 *Az. SSR*, *Biol. Nauk*, no.4:110–113(1972)—(Az).

3754 Kazanfarova, V. K. *and* Gasanov, A. M.
Sravnitel'naya morfologo-anatomicheskaya kharakteristika *Plantago media* L. i *Plantago stepposa* Kupr. *Izv. Akad. Nauk Az. SSR*, *Biol. Nauk*, no.1: 1–3 7–11(1973)—(Az).

3755 Ross, M. D.
Inheritance of self-incompatibility in *Plantago* 1, 2 *lanceolata*. *Heredity*, 30(2):169–176(1973).

3756 Vasilevskaya, V. K., Baranov, M. P. *and* Borisovskaya, G. M.
Stroenie rozetochnogo rasteniya *Plantago major* L. v pervȳĭ god zhizni. (The structure of rosette plant *Plantago major* L. during the first year of its life). *Bot.* 1, 2 *Zhurn.*, 58(1):33–42(1973)—(En).

radicata Hoffgg. & Lk.
f. **gigantea** Pinto da Silva in *Agron. Lusit.*, 34(3): 1 194(1972 publ. 1973)—Portugal.

PSYLLIUM

chamaepsyllium (Zohary) Holub in *Folia Geobot.* 2 *Phytotax.*, 8(2):173(1973): *Plantago chamaepsyllium.*
maris-mortui (Eig) Holub *op. cit.* 173(1973): 2 *Plantago maris-mortui.*
sarcophyllum (Zohary) Holub *op. cit.* 173(1973): 2 *Plantago sarcophylla.*
sinaicum (Barnéoud) Holub *op. cit.* 173(1973): 2 *Plantago psyllium* var. *sinaica.*

PLATANACEAE

PLATANUS

3757 Hsiao, J. Y.
Biochemical systematic and numerical taxonomic studies of the genus *Platanus*. *Diss. Abstr. Int., B,*
1–4, 6 33(4):1424–1425(1972).

3758 Hsiao, J.-Y.
A numerical taxonomic study of the genus *Platanus* based on morphological and phenolic characters.
1–4, 6 *Amer. J. Bot.*, 60(7):678–684(1973).

3759 McAlpine, R. G. *and* Applefield, M.
American sycamore: an American wood. *U.S. Dept. Agric. F.S.*, 267. 7p.(1973)—Illustrations, map.
3, 4 *Platanus occidentalis.*

3760 Vasil'eva, L. I. *and* Vasil'chenko, I. T.
Imeetsya li na territorii Sovetskogo Soyuza dikorastushchii platan (*Platanus orientalis* L.)? (Are there wild growing plane trees (*Platanus orientalis* L. on the territory of the Soviet Union?). *Byull. Mosk.*
2 *Obshch. Ispyt. Prir., Biol.*, 78(5):146–148(1973).

PLUMBAGINACEAE

ACANTHOLIMON

3761 Tripod, R. *and* Auw, P. von
Un genre asiatique peu répandu: l'*Acantholimon*.
1, 2, 6 *Mus. Genève*, 134:9–11(1973)—Illustrations.

ARMERIA

3762 Donadille, P.
Au sujet des *Armeria* de la Cordillère Cantabrique.
1 *Monde Pl.*, no.375:5(1973).

3763 Donadille, P.
Révision du genre *Armeria* Willd. en Afrique du Nord: 1. *A. mauritanica* Wallroth: nomenclature et taxonomie. (Revision of the genus *Armeria* Willd. in North Africa: 1. *A. mauritanica* Wallroth. Nomenclature and taxonomy). *Bull. Soc. Bot. France*,
1 119(7–8):487–501(1972 publ. 1973)—(En).

3764 Donadille, P.
Révision du genre *Armeria* Willd. en Afrique du Nord: 2. *A. gaditana* Boiss., *A. tingitana* Boiss. et Reut.
1 *Bull. Soc. Bot. France*, 120(3–4):141–152(1973)—(En).

3765 Lefevre, C., Gorenflot, R. *and* Roux, M.
Traitment numérique de populations d'*Armeria maritima* (Mill.) Willd. d'Europe occidentale et centrale. *Rev. Gén. Bot.*, 79(941–943):341–356(1972)—
1 (En).

3766 Silva, A. R. Pinto da
O género *Armeria* Willd. em Portugal. *Anuár. Soc.*
1 *Brot.*, 38:159–180(1972)—Key.

belgenciensis Donadille ex Guinochet & Vilmorin, Fl. France, 1:338(1973), nom. nov., without
1 replaced synonym ref.: *A. filicaulis* Rouy.
daveaui (P. Cout.) Pinto da Silva in *Anuár. Soc. Brot.*,
1 38:170(1972): *A. alliacea* var. *daveaui.*
maghrebensis Donadille in *Bull. Soc. Bot. France*, 119(7–8):491(1972 publ. 1973), nom. nov., without replaced synonym page: *A. mauritanica* sensu Boiss. in De Candolle, Prod. Syst., 12:674–677(1848),
1 non Wallr.
var. ebracteolata Donadille *op. cit.* 492(1972 publ.
1 1973), specific name invalid—Algeria.
var. mamorensis Donadille *op. cit.* 493(1972 publ.
1 1973), specific name invalid—Morocco.
var. minor (Blatt.) Donadille *op. cit.* 492(1972 publ. 1973), without basionym page and specific name
1 invalid: (Basionym not stated).
var. simplex (Pomel) Donadille *op. cit.* 492(1972 publ. 1973), without basionym page and specific
1 name invalid: (Basionym not stated).
var. soloitana (Maire) Donadille *op. cit.* 493(1972 publ. 1973) specific, name invalid: (Basionym not
1 stated).
tingitana B. & R.
var. chamaeropicola (Pau) Donadille in *Bull. Soc. Bot. France*, 120(3–4):149(1973), without
1 basionym ref.: *A. gaditana* var. *chamaeropicola.*

LIMONIUM

3767 Bokhari, M. H.
Variation and taxonomic importance of anatomical characters in *Limonium*. *Notes Roy. Bot. Gard.*
1, 2 *Edinburgh*, 32(2):275–290(1973).

duriusculum (Gir.) Fourr.
subsp. bianorii (Sen. & Pau) Malagarriga, Las Subesp. y Variac. Geogr.: 13(1973), without
1 basionym page or date: *Statice bianorii.*
serotinum (Rchb.) Pignatti in *Giorn. Bot. Ital.*,
1 107(5):220(1973): *Statice serotina.*

POACEAE
see
GRAMINEAE

PODOSTEMACEAE

3768 Cusset, C.
Podostemaceae. Flore du Cambodge, du Laos et du
6 Viêt-nam, fasc. 14:65–74(1973)—Illustrations; keys.

3769 Cusset, C.
Les Podostemaceae de Madagascar. *Adansonia*, 12(4):557–568(1972 publ. 1973)—(En); illustrations;
5 keys.

3770 Cusset, C.
Tristichaceae. Flore du Cambodge, du Laos et du Viêt-nam, fasc. 14:75–79(1973)—Illustration;
6 key.

CLADOPUS

sect. **Griffithella** (Tulasne) C. Cusset in Fl. Cambodge, Laos & Viêtn., 14:73(1973): *Mniopsis* sect.
6 *Griffithella.*
6 **fallax** C. Cusset *op. cit.* 72(1973)—Vietnam.
hookerana (Tul.) C. Cusset *op. cit.* 71(1973), as
6 'hookeriana': *Mniopsis hookeriana.*
pierrei (Lecomte) C. Cusset *op. cit.* 73(1973):
6 *Mniopsis pierrei.*
willisiana (Warm.) C. Cusset *op. cit.* 72(1973):
6 *Griffithella willisiana.*

DALZELLIA

carinata (Lecomte) C. Cusset in Fl. Cambodge, Laos
6 & Viêtn., 14:78(1973): *Terniola carinata.*
diversifolia (Lecomte) C. Cusset *op. cit.* 78(1973):
6 *Terniola diversifolia.*

ENDOCAULOS C. Cusset in *Adansonia*, 12(4):560
5 (1972 publ. 1973).
mangorense (Perr.) C. Cusset in *Adansonia*, 12(4):
5 560(1972 publ. 1973): *Spaerothylax mangorensis.*

PALEODICRAEIA C. Cusset in *Adansonia*, 12(4):
5 562(1972 publ. 1973).
imbricata (Tul.) C. Cusset in *Adansonia*, 12(4):563
5 (1972 publ. 1973): *Dicraeia imbricata.*

STONESIA

3771 Cusset, C.
Contribution à l'étude des Podostemaceae: 3. Le genre *Stonesia. Adansonia*, 13(3):307–312(1973)—
5 Illustrations; key.

taylorii C. Cusset in *Adansonia*, 13(3):308(1973)—
5 Guinea.

THELETHYLAX C. Cusset in *Adansonia*, 12(4):
5 564(1972 publ. 1973).
isalensis (Perr.) C. Cusset in *Adansonia*, 12(4):567
5 (1972 publ. 1973): *Dicraeia isalensis.*
minutiflora (Tul.) C. Cusset *op. cit.* 566(1972 publ.
5 1973): *Dicraeia minutiflora.*

POLEMONIACEAE

3772 Zarubina, G. M.
O zhiznesposobnosti pÿl'tsÿ dvukh vidov semeĭstva Polemoniaceae. *Byull. Glavn. Bot. Sada*
1–4 (Moscow), 88:67–68(1973).

COLLOMIA

3773 Pignatti Wikus, E.
Collomia linearis Nutt. Una Polemoniacea nordamericana avventizía nel Bellunese. (*Collomia linearis* Nutt. (Polemoniaceae) naturalized near Belluno, northern Italy). *Giorn. Bot. Ital.*, 107(1):
1, 3 43–46(1973)—(En); illustrations.

PHLOX

3774 Moldenke, H. N.
Notes on new and noteworthy plants: 60.
3 *Phytologia*, 26(4):224–226(1973).

3775 Zarubina, G. M.
Morfogenez floksa Sibirskogo. *Byull. Glavn. Bot.*
2 *Sada* (Moscow), 87:72–75(1973)—Illustration.

drummondii
f. **albiflora** Moldenke in *Phytologia*, 26(4):224
3 (1973)—U.S.A. (Florida).
nivalis
f. **rubella** Moldenke *op. cit.* 225(1973)—U.S.A.
3 (Florida).
sibirica L.
subsp. **pulvinata** (Wherry) W. A. Weber in *Southwest. Nat.*, 18(3):322(1973): *P. caespitosa*
2, 3 subsp. *pulvinata.*

POLEMONIUM

3776 Ito, K.
(*Polemonium* in Hokkaido, the Kuriles and Sakhalin: 1). *J. Geobot.*, 20(4):91–93(1972)—Ja;
2 key.

3777 Środoń, A.
Polemonium caeruleum L.—rozmieszczenie współczesne oraz wysterpowanie w plejstocenie Polski. (*Polemonium caeruleum* L.—present distribution and occurrence during the Pleistocene in Poland.) *Fragm.*
1 *Flor. Geobot.*, 19(1):9–21(1973)—(En); dot maps.

POLYGALACEAE

MONNINA

3778 Ferreyra, R.
Dos especies nuevas de *Monnina* para el Peru (Polygalaceae). *Publ. Mus. Hist. Nat. Javier Prado, B,*
4 no.26. 14p.(1973)—(En); illustrations.

ebracteata Ferreyra in *Publ. Mus. Hist. Nat. Javier*
4 *Prado, B*, 26:8(1973)—Peru.
4 **valcareliana** Ferreyra *op. cit.* 6(1973)—Peru.

POLYGALA

3779 Burtt, B. L.
Polygala arvensis, chinensis and *glomerata. Notes Roy.*
2, 6 *Bot. Gard. Edinburgh*, 32(3):403–404(1973).

3780 Dar, M. I.
Flora of West Pakistan: no.52. Polygalaceae. 1973.
6 11p.—Illustrations, map. *Polygala.*

3781 Falcão, W. Fraga de Almeida, Alencastro, F. M. M. R. de *and* Correia, I. de Lima
Notas sôbre a anatomia e morfologia da espécie *Polygala paniculata* L. *Arq. Jard. Bot. Rio de Janeiro*,
4 19:281–294(1973)—Illustrations.

3782 Kulkarni, B. G. *and* Singh, N. P.
On the distribution of *Polygala jacobii* Chandr.
6 *Curr. Sci.*, 42(10):359(1973)—Illustration.

3783 Saulmon, J. G.
The genus *Polygala* in the southeastern United States. *Proc. W. Virg. Acad. Sci.*, 43:9–13(1971)—
3 Key.

amarella Cr.
var. **salsa** Degen ex Soó in *Acta Bot. Acad. Sci.*
1 *Hung.*, 18(1–2):173(1973)—Hungary.

POLYGONACEAE

ACETOSA

nebroides (Campdera) Holub in *Folia Geobot.*
1 *Phytotax.*, 8(2):176(1973): *Rumex nebroides.*

CALLIGONUM

3784 Soskov, Yu. D.
Novÿĭ vid roda *Calligonum* L. iz Aravii. (Generis *Calligonum* L. species nova ex Arabia). *Novosti Sist.*
2 *Vÿssh. Rast.*, 10:134–135(1973).

3785 Soskov, Yu. D.
Rod *Calligonum* L. v peredneĭ Azii. (*Calligonum* L. in Front Asia). *Izv. Akad. Nauk. Turkm. SSR, Biol.*
2 *Nauk*, (1):20–25(1973)—(En); key.

3786 Soskov, Yu. D., Ivanov, A. I. *and* Kozulya, I. E.
Izmenchivost (*Calligonum aphyllum* (Pall.) Guerke v kul'ture. (Variability of *Calligonum aphyllum* (Pall.) Guerke in culture). *Bot. Zhurn.*, 58(6):784–792(1973)—
2 (En).

3787 Valovich, E. M. *and* Soskov, Yu. D.
Khromosomnÿe chisla vidov roda *Calligonum* L. (The chromosome numbers of species of the genus
1, 2 *Calligonum* L.). *Bot. Zhurn.*, 58(4):554–561(1973).

arabicum Yu. D. Soskov in *Novosti Sist. Vÿssh. Rast.*,
2 10:134(1973)—Arab Emirates.

ERIOGONUM

3788 Reveal, J. L.
Eriogonum (Polygonaceae) of Utah. *Phytologia*,
3 25(4):169–217(1973)—Keys.

3789 Reveal, J. L.
Five new varieties of *Eriogonum* (Polygonaceae).
3 *Great Basin Nat.*, 32(2):113–116(1972)—Key.

3790 Reveal, J. L.
A new subfruticose *Eriogonum* (Polygonaceae) from western Colorado. *Great Basin Nat.*, 33(2):120-
3 122(1973).

brevicaule Nutt.
var. **cottamii** (S. Stokes) Reveal in *Great Basin Nat.*, 32(2):113(1972): *E. tenellum* subsp.
3 *cottamii*.
var. **wasatchense** (M. E. Jones) Reveal *op. cit.* 113
3 (1972): *E. wasatchense*
deflexum Torr.
var. **nevadense** Reveal in *Phytologia*, 25(4):206
3 (1973)—U.S.A. (Nevada).
3 **grayi** Reveal *op. cit.* 193(1973)—U.S.A. (Utah).
heracleoides Nutt.
var. **leucophaeum** Reveal *op. cit.* 200(1973)—
3 U.S.A. (Washington).
howellianum Reveal *op. cit.* 204(1973)—U.S.A.
3 (Utah).
jamesii Benth.
var. **rupicola** Reveal *op. cit.* 202(1973)—U.S.A.
3 (Utah).
var. **wootonii** Reveal *op. cit.* 201(1973)—U.S.A.
3 (New Mexico).
3 **nanum** Reveal *op. cit.* 194(1973)—U.S.A. (Utah).
ovalifolium Nutt.
var. **caelestinum** Reveal in *Great Basin Nat.*,
3 32(2):115(1972)—U.S.A. (Nevada).
pelinophilum Reveal *op. cit.* 120(1973)—U.S.A.
3 (Colorado).
shockleyi S. Wats.
var. **longilobum** (M. E. Jones) Reveal *op. cit.* 114
3 (1972): *E. longilobum*.
umbellatum Torr.
var. **devestivum** Reveal *op. cit.* 115(1972)—U.S.A.
3 (Idaho).

FAGOPYRUM

3791 Krotov, A. S. *and* Dranenko, E. T.
Amfidiploid grechikhi *Fagopyrum giganteum* Krot. sp. nova. *Byull. Vses. Ord. Lenina Inst. Rast. N.I.*
1, 2 *Vavilova*, 30:41–44(1973)—Illustration.

× **giganteum** Krotov in *Byull. Vses. Ord. Lenina Inst. Rast. N.I. Vavilova*, 30:44(1973)—Cultivation.
0 (*F. tataricum* × *F. cymosum*).

KOENIGIA

3792 Měsíček, J. *and* Soják, J.
Karyological and taxonomic observations on *Dracocephalum foetidum* Bunge and *Koenigia islandica* L. *Folia Geobot. Phytotax.*, 8(1):105–112(1973)—
1–3 Chromosome numbers.

cyanandra (Diels) Měsíček & Soják in *Folia Geobot.*
2 *Phytotax.*, 8(1):110(1973): *Polygonum cyanandrum*.
forrestii (Diels) Měsíček & Soják *op. cit.* 110(1973):
2 *Polygonum forrestii*.
hubertii (Lingelsh.) Měsiček & Soják *op. cit.* 110
2 (1973): *Polygonum hubertii*.
nummularifolia (Meisn.) Měsíček & Soják *op. cit.*
6 110(1973): *Polygonum nummularifolium*.

PERSICARIA

mitis (Schrank) Holub in *Folia Geobot. Phytotox.*,
1 8(2):177(1973): *Polygonum mite*.

POLYGONUM

3793 Brooks, G. M. *and* Mertens, T. R.
A biosystematic study of *Polygonum ramosissimum* and *Polygonum tenue*. *Proc. Indiana Acad. Sci.*, 81:277-
3 283(1971 publ. 1972).

3794 Dalci, M.
The taxonomy of the section *Persicaria* (Tourn.) L. in the genus *Polygonum* (Tourn.) L. (Polygonaceae) in the United States east of the Rocky Mountains.
1–3 *Diss. Abstr. Int., B*, 33(6):2503(1972).

3795 Engell, K.
A preliminary morphological, cytological, and embryological investigation in *Polygonum viviparum*. *Bot. Tidsskr.*, 67(4):305–316(1973)—Illustrations;
1, 3, 6 chromosome numbers.

3796 Kitagawa, M.
(Two new *Polygonum* from Japan). *J. Geobot.*, 20(2):
2 32–35(1972)—Ja; illustrations.

3797 Kongar, É. T.
Polygonum middendorfii—novyĭ vid s nizov'ev Amura. (*Polygonum middendorfii*—a new species from
2 the lower Amur). *Bot. Zhurn.*, 58(3):432–433(1973).

3798 Rao, B. A. *and* Mukherjee, P. K.
A note on the embryology of *Polygonum glabrum*
4–6 Willd. *Curr. Sci.*, 42(7):250–252(1973)—Illustrations.

3799 Visser, A. de
Polygonum polystachyum Wall. ex Meissn. op
1 Walcheren. *Gorteria*, 6(12):203–204(1973).

middendorfii Kongar in *Bot. Zhurn.*, 58(3):432
2 (1973)—U.S.S.R. (Siberia).
pteranthum Kitagawa in *J. Geobot.*, 20(2):32(1972)—
2 Japan.
rottboellioides Jaub. & Spach
var. tibeticum (Hk. f.) R. R. Stewart, Annot. Cat. Vasc. Pl. W. Pakistan & Kashmir: 210(1972), without basionym date: *P. tubulosum* var.
6 *tibeticum*.
ryukyuense Kitagawa in *J. Geobot.*, 20(2):34(1972)—
2 Ryukyu Is.

RHEUM

3800 Nordhagen, R.
Über ein spontanes Vorkommen von *Rheum rhaponticum* L. in Aurland im inneren Sognefjord-Gebiet, Norwegen, sowie über das Vorkommen der Art in Bulgarien. Oslo, etc., Universitetsforlaget, 1973. 90p. (Norske Videnskaps-Akademi I Oslo 1. Mat.-Naturv. Klasse. Ny Serie no. 31)—Illustrations,
2 dot maps; chromosome numbers.

RUMEX

3801 Harris, W.
Leaf form and panicle height variability in *Rumex*
1–7 *acetosella* L. *New Zealand J. Bot.*, 11(1):115–144(1973).

3802 Kenneth, A. G.
Rumex maritimus. *Glasgow Nat.*, 19(1):65–66
1 (1973)—En. Local records.

3803 Kopecký, K.
Je šťovík alpinský (*Rumex alpinus* L.) v Orlických horách původní? (Ist der Alpenampfer (*Rumex alpinus* L.) im Adlergebirge ursprünglich?). *Preslia*,
1 45(2):132–139(1973)—(Ge); dot map.

3804 Lakušić, R.
Rumex balcanicus Rech. fil. novi tercijerni relikt u flori Crne Gore. (*Rumex balcanicus* Rech. fil. neues tertiers Relikt in der Flora von Montenegro). *Glasn. Republ. Zavoda Zašt. Prir. Prirodnj. Muz. Titogradu*,
1 no.5:29–32(1972 publ. 1973)—Dot map.

3805 Perez-Moreau, R. L. *and* Crespo, S.
Notas sobre el genero *Rumex* (Polygonaceae). *Bol. Soc. Argent. Bot.*, 14(4):325–329(1972)—(En);
4 illustrations.

3806 Reichert, H.
Morphologisch-taxonomische Beobachtungen an einer Mischpopulation der beiden Ampferarten *Rumex obtusifolius* und *Rumex crispus*. *Faun.-Flor.*
1 *Notizen Saarland*, 4(3–5):17–23(1971).

3807 Rychlewski, J. *and* Zarzycki, K.
Analysis of natural populations of *Rumex acetosa* L. and *R. thyrsiflorus* Fing. in respect of the sex ratio. *Genet. Pol.*, 14(2):189–191(1973)—(Pol,Rus);
1 chromosome numbers.

3808 Valta, A.
Rumex confertus Willd. × *obtusifolius* L. and *R. crispus* L. × *obtusifolius* L. found in Finland. *Acta Bot.*
1 *Fenn.*, 10(1):68–69(1973).

3809 Vivant, J.
Rumex cantabricus Rech. et *Stipa pariflora* Desf., phanérogames méconnues de la flore française. (*Rumex cantabricus* and *Stipa parviflora*, unknown phanerogams of the French flora). *Bull. Soc. Bot.*
1 *France*, 119(5–6):335–338(1972 publ. 1973)—(En).

beringensis Yurtsev & Petrovskiĭ in *Bot. Zhurn.*,
2 58(12):1745(1973)—U.S.S.R. (Soviet Far East).
krausii Yurtsev & Petrovskiĭ *op. cit.* 1745(1973)—
2 U.S.S.R. (Soviet Far East).
scutatus L.
subsp. **induratus** (Bss. R.) Malagarriga, Las
1 Subesp. y Variac. Geogr.: 5(1973): *R. induratus*.

PONTEDERIACEAE

EICHHORNIA

3810 Agostini, G.
El genero *Eichhornia* (Pontederiaceae) en Venezuela. *Acta Bot. Venezuel.*, 9(1–4):303–310(1974 publ. 1973)—
4 Illustration; key.

3811 Schöpfel, H.
Eichhornia diversifolia. *Inform. ZAG Wasserpflanzen*,
4 no.3:7(1973)—Ge.

HETERANTHERA

3812 Agostini, G.
El genero *Heteranthera* (Pontederiaceae). *Acta Bot. Venezuel.*, 9(1–4):295–301(1974 publ. 1973)—
4 Illustration; keys.

PONTEDERIA

3813 Lowden, R. M.
Revision of the genus *Pontederia* L. *Rhodora*, 75 (803):426–487(1973)—Illustrations, dot maps;
3, 4 chromosome numbers; key.

subovata (Seub.) Lowden in *Rhodora*, 75(803):478
4 (1973): *Eichhornia subovata*.

PORTULACACEAE

3814 Ghafoor, A.
Flora of West Pakistan: no.51. Portulacaceae,
6 1973. 8p.—Illustration, map; key.

3815 Wild, H. *and* Diniz, M. A.
Flora de Moçambique: 21. Portulacaceae. 1973.
5 15p.—Illustrations; chromosome numbers; keys.

LEWISIA

3816 Fosback, J.
Two Oregon plant records. *Madroño*, 22(4):213
3 (1973). *Lewisia cotyledon* var. *howellii*.

3817 MacBryde, B.
A *Lewisia* native in Mexico and Guatemala. *J. Scott. Rock. Gard. Club*, 13(4):294–296(1973)—
4 Chromosome numbers; key.

3818 McNeill, J.
Lewisia triphylla (S. Watson) Robinson and *Spraguea umbellata* Torrey, new species for Canada.
3 *Syesis*, 6:179–181(1973).

megarhiza (Hemsley) B. MacBryde in *J. Scott. Rock Gard. Club*, 13(4):295(1973): *Calandrinia*
4 *megarhiza*.

MONTIA

3819 Wedeck, H.
Über Vorkommen und soziologische Bindung von *Montia verna* Neck. im Raum Zülpich (Spermatophyta, Portulacaceae). *Decheniana*, 125(1–2):141–145
1 (1972).

PORTULACA

3820 Anon.
Portulaca pilosa Linn. (Portulaceae). *Ashingtonia*,
4 1(3):36(1973)—En; coloured illustration.

PORTULACARIA

3821 Moran, R.
Spekboom blooms. *Cact. Succ. J.* (U.S.A.), 45(2): 71–73(1973)—Illustrations, coloured illustration.
5 *Portulacaria afra*.

3822 Oakes, A. J.
Portulacaria afra Jacq.: a potential browse plant.
5 *Econ. Bot.*, 27(4):413–416(1973)—Illustrations.

SPRAGUEA

3823 McNeill, J.
Lewisia triphylla (S. Watson) Robinson and *Spraguea umbellata* Torrey, new species for Canada.
3 *Syesis*, 6:179–181(1973).

TALINUM

3824 Black, C. B. *and* Murdy, W. H.
The evolutionary origin of *Talinum teretifolium* Pursh. *Bot. Gaz.*, 133(4):405–410(1972)—Map;
3 chromosome numbers.

3825 Khan, A. M.
Development of ovule and female gametophyte in *Talinum patens* Willd. *Curr. Sci.*, 42(10):364–365
6 (1973).

POTAMOGETONACEAE

POTAMOGETON

3826 Denny, P. *and* Lye, K. A.
The *Potamogeton schweinfurthii* complex in Uganda.
5 *Kew Bull.*, 28(1):117–120(1973).

3827 Haynes, R. R. *and* Reveal, J. L.
A re-evaluation of *Potamogeton fibrillosus* fern (Potamogetonaceae). *Rhodora*, 75(801):75–77
3 (1973).

3828 Hose, R. d'
Potamogeton berchtoldii Fieber, nieuw voor de Belgische flora. (*Potamogeton berchtoldii* Fieber, espèce nouvelle pour la flore belge). *Bull. Soc. Roy. Bot.*
1 *Belg.*, 106(3):65–68(1973)—(Fr,Ge).

3829 Myhre, A.
Potamogeton lucens funnet i Sør-Trøndelag. (*Potamogeton lucens* found in Sør-Trøndelag, central
1 Norway). *Blyttia*, 31(1):35–37(1973)—(En); dot map.

3830 Naqshi, A. R. *and* Javeid, G. N.
Two new plant records for Kashmir valley. *Curr.*
6 *Sci.*, 42(2):70(1973). *Potamogeton lucens, P. crispus.*

3831 Posluszny, U. *and* Sattler, R.
Floral development of *Potamogeton densus*. *Canad.*
1, 2 *J. Bot.*, 51(3):647–656(1973)—(Fr).

3832 Serbănescu-Jitariu, G.
La morphologie du gynécée chez certaines espèces du genre *Potamogeton*. *Bull. Soc. Hist. Nat. Afr. Nord*,
2–7 63(3–4):3–7(1972 publl. 1973).

× **bunyonyiensis** Denny & Lye in *Kew Bull.*, 28(1):120(1973)—Uganda. (*P. schweinfurthii* ×
5 *P. thunbergii*).
foliosus Raf.
var. **fibrillosus** (Fern.) R. R. Haynes & Reveal
3 in *Rhodora*, 75(801):76(1973): *P. fibrillosus*.

PRIMULACEAE

ANDROSACE

3833 Gorer, R.
Plant awards 1972–73: *Androsace muscoidea* Duby. var. *longiscapa* P. C. *Quart. Bull. Alp. Gard. Soc.*,
2 41(4):332(1973).

rioxana Segura Zubizarreta in *Pirineos*, 109:47
1 (1973)—Spain.

CYCLAMEN

3834 Saunders, D. E.
Cyclamen: a gardener's guide to the genus: rev. and amended by R. D. Meikle and C. Grey-Wilson. [London] Alpine Garden Society, 1973. [ii]49p.—Illustrations; key. First published in 1959 in the
1–2 *Alpine Gard. Soc. Bull.*

repandum S. S.
subsp. balearicum (WK.) Malagarriga, Las Subesp. y Variac. Geogr.: 13(1973), without basionym
1 date: *C. balearicum*.

LYSIMACHIA

3835 Marks, G. E. *and* Barber, S.
The reproductive biology of *Lysimachia nummularis*. *Ann. Rep. John Innes Inst.*, 63:62–63(1972 publ. 1973)
1, 2 —Chromosome numbers.

PRIMULA

3836 Bullard, E. R. *and* Shearer, H.
Primula scotica project. *Orkney Field Club Bull.*,
1 no.1:9–10(1973)—Ecological study.

3837 Eaton, R. D.
The evolution of seed incompatibility in *Primula*.
1–3 *New Phytol.*, 72(4):855–860(1973).

3838 Crosland, J. D.
Primula forrestii. *J. Scott. Rock Gard. Club*, 13(3):
2 238–239(1973)—Illustrations.

3839 Grey-Wilson, C.
Primula edelbergii. *Curtis's Bot. Mag.*, 179(2):tab. 628(1972 publ. 1973)—En; illustration, coloured
2 illustration.

3840 Happy, C. *photogr.*
Primula saxatilis. *Quart. Amer. Primrose Soc.*, 31(4):
2, 6 1,6(1973)—En; illustration.

3841 Knight, G. H.
The primrose paradox. *Ann. Rep. Warwick Nat. Hist. Soc.*, 19:16–20(1973). Pollination studies in
1 *Primula*.

3842 Kress, A.
Über *Primula-villosa-pedemontana*-Hybriden und ihre Stammsippen. *Ber. Bayer. Bot. Ges.*, 44:187–200
1 (1973)—Maps; chromosome numbers.

3843 Livingstone, D. *and* Masterton, R. S.
Primulas. *J. Scott. Rock Gard. Club*, 13(3):196–204
2 (1973)—Illustrations. Found by George Forrest.

3844 Magomedmirzaev, M. M.
O zavisimosti kharakteristiki populyatsiĭ ot puteĭ evolyutsii razmernykh priznakov (na primere dvukh vidov pervotsveta). (On the relationship between population properties and the pathways of the evolution of meristic characters (two species of *Primula* L. taken as an example)). *Zhurn. Obshch.*
1, 2 *Biol.*, 34(3):388–395(1973)—(En).

3845 Magomedmirzaev, M. M. *and* Khabibov, A. D.
O geneticheskom dimorfizme v prirodnÿkh populyatsiyakh pervotsveta. (On genetical dimorphism in natural populations of *Primula macrocalyx* Bge and *P. ruprechtii* Kusn.). *Zhurn.*
2 *Obshch. Biol.*, 34(4):483–487(1973)—(En).

3846 Meadows, G. E. M.
The yellow primulas of the Cantabrian mountains.
1 *Quart. Bull. Alp. Gard. Soc.*, 41(2):181–184(1973).

3847 Ricciardi, M.
Nuove stazioni di *Primula palinuri* Petagna lungo la costa tirrenica meridionale. *Webbia*, 28(2):417–421
1 (1973)—(En).

3848 Sarkar, A. K.
Cytotaxonomic studies on some members of Primulaccac from the eastern Himalayas. *Proc. Indian Sci. Congr.*, 60(3):316(1973)—Chromosome
6 numbers. Abstract only.

3849 Ştefureac, T. I. *and* Peicea, I.
Corologia speciei *Primula minimal* L. în România. *Stud. Cerc. Biol., Bot.*, 24(4):339–347(1972)—(Ge);
1 dot map.

veris L.
subsp. **suaveolens** (Bertol.) Gutermann & Ehrendorfer in *Österr. Bot. Zeitschr.*, 122(4):268
1 (1973): *P. suaveolens.*

SOLDANELLA

3850 Jeslík, R.
Zbarvení kořenu druhů rodu *Soldanella* L. jako určovací znak. (Die Farbe der Wurzeln der Arten der Gattung *Soldanella* L. als ein Bestimmungsmerkmal).
1 *Zprávy Česk. Bot. Spol.*, 8(1):14–16(1973)—(Ge).

TRIENTALIS

3851 Anderson, R. C. *and* Loucks, O. L.
Aspects of the biology of *Trientalis borealis* Raf. *Ecology*, 54(4):798–808(1973)—Illustrations, map,
3 dot maps.

3852 Ştefureac, T. I.
A new relict in the flora of Bukovinian ridges. *Rev. Roum. Biol., Bot.*, 18(1):21–26(1973)—Illustration.
1 *Trientalis europaea.*

PROTEACEAE

3853 Lamont, B.
Proteaceae: the ecology of the Australian Proteaceae with implications for their cultivation. *Austral. Pl.*, 7(54):47–56(1973)—Coloured
7 illustrations.

BANKSIA

3854 Ferguson, I. K.
Banksia coccinea. Curtis's Bot. Mag., 179(2):tab.630 (1972 publ. 1973)—En; illustration, coloured
7 illustration.

3855 Gee, B.
Banksia integrifolia. Launceston Nat., 5(2):1–2
7 (1971)—En.

3856 Lebler, B. A.
Banksias of south-eastern Queensland. *Queensl.*
7 *Agric. J.*, 99(7):345–354(1973)—Illustrations.

3857 Mullins, B.
Australia's Banksias. *Austral. Nat. Hist.*, 17(7):213–
7 216(1972)—Illustrations. Popular article.

FAUREA

3858 Rourke, J. P.
Faurea, a possible ancestor of the *Protea*? *Veld*
5 *Flora*, 3(2):28–29(1973)—Illustration.

HAKEA

3859 Lamont, B.
Hakea rubriflora (Proteaceae) a new species from Western Australia. *J. Roy. Soc. West. Austral.*, 55(4):
7 97–99(1972)—Illustrations.

3860 Maconochie, J. R.
Studies on some species of *Hakea* (Proteaceae). *Trans. Roy. Soc. S. Austral.*, 97(2):127–133(1973)—
7 Illustrations, dot maps; key.

coriacea Maconochie in *Trans. Roy. Soc. S. Austral.*,
7 97(2):132(1973)—Australia (W. Australia).
minyma Maconochie *op. cit.* 130(1973)—Australia
7 (S. Australia).
rubriflora Lamont in *J. Roy. Soc. West. Austral.*,
7 55(4):97(1972)—Australia (W. Australia).
standleyensis Maconochie in *Trans. Roy. Soc. S.*
7 *Austral.*, 97(2):127(1973)—Australia (N. Terr.).

HELICIA

3861 Na Songkhla, B.
Proteaceae. *Thai Forest Bull., Bot.*, no.7:49–66
2, 6 (1973)—En; illustrations, map; keys.

LEUCADENDRON

3862 Rourke, J. P.
The use of the name *Leucadendron aureum* Burm. f.
5 *J. S. Afr. Bot.*, 39(4):273–276(1973)—(Af); illustrations.

LOMATIA

3863 Moir, M. *and* Thomson, R. H.
Naphthaquinones in *Lomatia* species. *Phytochemistry*,
4, 7 12(6):1351–1353(1973).

PROTEA

3864 Beard, J. S.
The correct application of the name *Protea crinita*.
5 *Kirkia*, 9(1):86(1973)—Illustrations.

3865 Rourke, J. P.
Protea rubropilosa. Flow. Pl. Afr., 42(3–4):pl. 1677
5 (1973)—En; coloured illustration.

neocrinita Beard in *Kirkia*, 9(1):86(1973), nom. nov.:
5 *P. crinita* Beard.

ROUPALA

3866 Gentry, A. H.
A new species of Proteaceae from Panama. *Ann.*
4 *Missouri Bot. Gard.*, 60(2):571–572(1973)—Key.

percoriacea A. H. Gentry in *Ann. Missouri Bot. Gard.*,
4 60(2):571(1973)—Panama.

TELOPEA

3867 Boden, R. W. *and* Powell, R. H.
A Waratah hybrid. *Austral. Pl.*, 7(56):168–170, 204(1973)—Coloured illustrations. *Telopea mon-*
7 *gaensis* × *speciosissima.*

PUNICACEAE

PUNICA

3868 Shilkina, I. A.
K anatomii drevesinȳ roda *Punica* L. (On the xylem anatomy of the genus *Punica*). *Bot. Zhurn.*, 58(11):
5, 6 1628–1630(1973).

PYROLACEAE

3869 Gambaryan, P. P.
Grushankovȳe (Pyrolaceae)—novoe semeĭstvo dlya florȳ Armenii. (Pyrolaceae—a new family in flora of Armenia). *Biol. Zhurn. Arm.*, 26(5):85–86(1973)—
2 (Arm).

3870 Harborne, J. B. *and* Williams, C. A.
A chemotaxonomic survey of flavonoids and simple phenols in leaves of the Ericaceae. *Bot. J. Linn. Soc.*,
1–3, 6 66(1):37–54(1973).

CHIMAPHILA

3871 Hodgdon, A. R. *and* Eastman, L. M.
Chimaphila maculata (L.) Pursh in Maine and New
3 Hampshire. *Rhodora*, 75(801):162–165(1973).

ORTHILIA

kareliniana (Skvorcov) Holub in *Folia Geobot.*
2 *Phytotax.*, 8(2):177(1973): *Ramischia kareliniana.*

PYROLA

3872 Haber, E.
A biosystematic study of the eastern North American species of the genus *Pyrola. Diss. Abstr.*
3 *Int., B*, 33(10):4691(1973)—Chromosome numbers.

QUIINACEAE

3873 Hatschbach, G.
Quiinaceas do estado do Paraná. *Bol. Mus. Bot. Mun.*
4 (Curitiba), no.6.4p.(1972)—Illustration, dot map.

RAFFLESIACEAE

3874 Gentry, A. H.
Flora of Panama: part 4. Family 50a: Rafflesiaceae. *Ann. Missouri Bot. Gard.*, 60(1):17–21(1973)—
4 Illustrations; key.

3875 Hansen, B.
Rafflesiaceae. Flore du Cambodge, du Laos et du
6 Viêt-nam, fasc. 14:59–64(1973)—Illustration; keys.

3876 Paiva, J. A. R.
Rafflesiaceae. *Anuár. Soc. Brot.*, 38:141–143(1972)—
1 Por; key.

3877 Vattimo, I. de
Flora ilustrada catarinense: Raflesiáceas. 1972.
4 16p.—Illustrations, dot map; key.

RAFFLESIA

3878 Bastin, J.
Dr Joseph Arnold and the discovery of *Rafflesia arnoldi* in west Sumatra in 1818. *J. Soc. Bibliogr. Nat.*
6 *Hist.*, 6(5):305–372(1973)—Illustration, portraits.

3879 Meijer, W.
Endangered plant life. *Biol. Conserv.*, 5(3):163–
6 167(1973)—Illustrations. *Rafflesia arnoldii.*

RANUNCULACEAE

3880 Exell, A. W., Milne-Redhead, E. *and* Gonçalves, M. L.
Flora de Moçambique: 4. Ranunculaceae. 1973.
5 17p.—Illustrations; key.

3881 Kürbs, S.
Vergleichend-entwicklungsgeschichtliche Studien an Ranunculaceen-Fiederblättern: 1. *Bot. Jahrb.*, 93(1):130–167(1973)—(En); illustrations. *Thalictrum,*
1–4, 7 *Aquilegia, Anemone, Caltha, Ranunculus* and *Nigella.*

3882 Kürbs, S.
Vergleichend-entwicklungsgeschichtliche Studien an Ranunculaceen-Fiederblättern: 2. *Bot. Jahrb.*, 93(3):325–371(1973)—(En). *Aconitum, Trollius,*
1–7 *Eranthis, Anemone, Ranunculus, Helleborus.*

3883 Salisbury, *Sir* E.
The organization of the ranunculaceous flower wih especial regard to the correlated variations of its constituent members. *Proc. Roy. Soc., B*,183 (1072):
1–7 205–225(1973).

ACONITUM

3884 Descimon, H.
Sur la présence d'*Aconitum paniculatum* LMK (Rénonculacées) dans les Hautes-Pyrénées. *Bull. Trim.*
1 *Soc. Hist. Nat. Amis Mus. Autun*, no.63:28–29(1972).

3885 Filipescu, G.
Anatomia organelor vegetative la *Aconitum romanicum* Wol. unitate taxonomică endemică. *An. Şti. Univ. Al. I. Cuza Iaşi, Biol.*, 19(2):311–315
1 (1973)—(Fr); illustrations.

3886 Kartashova, N. N. *and* Malakhova, L. A.
Kariologicheskiĭ analiz spontannogo gibrida mezhdu dvumya vidami *Aconitum*. (Karyological analysis of spontaneous interspecific hybrid of two *Aconitum* species). *Genetika*, 9(7):167–168(1973)—
1–3 (En).

3887 Lauener, L. A.
Aconitum of the Himalaya: 2. *Notes Roy. Bot. Gard.*
6 *Edinburgh*, 32(3):411–414(1973)—Illustrations; key.

deltoideum Lauener in *Notes Roy. Bot. Gard.*
6 *Edinburgh*, 32(3):411(1973)—Nepal.

ADONIS

3888 Petrov, S. *and* Borissova-Ivanova, O.
Pollen morphology of the Bulgarian representatives of the family Ranunculaceae: *Adonis* L. *Dokl.*
1 *Bolg. Akad. Nauk*, 26(7):961–964(1973).

3889 Poshkurlat, A. P. *and* Ryabova, N. N.
Osobennosti anatomo-morfologicheskogo stroeniya goritsveta zolotistogo. *Byull. Glavn. Bot. Sada* (Moscow),90:54–59(1973)—Illustration.
2 *Adonis vernalis.*

ANEMONASTRUM Holub in *Folia Geobot. Phytotax.*, 8(2):158(1973), nom. nov.: *Homalocarpus*
1–3, 6 (DC.) Schur.

brevipedunculatum (Juz.) Holub *op. cit.* 165
2 (1973): *Anemone brevipedunculata.*

chinense (Kitagawa) Holub *op. cit.* 165(1973):
2 *Anemone narcissiflora* var. *chinense.*

chosenicola (Ohwi) Holub *op. cit.* 165(1973), as
2 'chosenicolum': *Anemone chosenicola.*

crinitum (Juz.) Holub *op. cit.* 165(1973): *Anemone*
2 *crinita.*

demissum (Hook. f. & Thoms.) Holub *op. cit.* 165
6 (1973): *Anemone demissa.*

elongatum (D. Don) Holub *op. cit.* 165(1973):
6 *Anemone elongata.*

fasciculatum (L.) Holub *op. cit.* 165(1973): *Anemone*
1, 2 *fasciculata.*

imbricatum (Maxim.) Holub *op. cit.* 165(1973):
6 *Anemone imbricata.*

impexum (Juz.) Holub *op. cit.* 165(1973): *Anemone*
2 *impexa.*

narcissiflorum (L.) Holub *op. cit.* 165(1973):
1–3 *Anemone narcissiflora.*

polyanthes (D. Don) Holub *op. cit.* 165(1973):
6 *Anemone polyanthes.*

protractum (Ulbrich) Holub *op. cit.* 165(1973):
1, 2 *Anemone narcissiflora* var. *protracta.*

schantungensis (Hand.-Mazz.) Holub *op. cit.* 165
2 (1973): *Anemone schantungensis.*

sibiricum (L.) Holub *op. cit.* 165(1973): *Anemone*
2 *sibirica.*

sikokianum (Mak.) Holub *op. cit.* 165(1973):
2 *Anemone narcissiflora* var. *sikokiana.*

tetrasepalum (Royle) Holub *op. cit.* 165(1973):
6 *Anemone tetrasepala.*

zephyrum (A. Nels.) Holub *op. cit.* 165(1973):
3 *Anemone zephyra.*

ANEMONE

3890 Atlas rozšíření rostlin v jižní části Čech: 1. (Atlas der Pflanzenverbreitung im südlichen Teile Bohmens: 1). *Sborn. Jihočesk. Muz. Česk. Budějovicich Přír. Ved.*,
1 12 *Suppl.* 3. 44p.(1972)—(Ge); dot maps.

3891 Chupov, V. S.
Immunoélektroforeticheskie issledovaniya v rodakh *Anemone* L., *Pulsatilla* Mill. i *Hepatica* Mill. (Immunoelectrophoretical investigations in genera *Anemone* L., *Pulsatilla* Mill. and *Hepatica* Mill.). *Bot. Zhurn.*, 58(2):
1–3 189–199(1973)—(En); illustrations.

3892 Makushenko, L. M.
Chisla khromosom nekotorȳkh vidov sem. Ranunculaceae Juss. (Chromosome numbers of some species in Ranunculaceae Juss.). *Bot Zhurn.*, 58(7):
1 1023–1026(1973).

fuscopurpurea Hara in *J. Jap. Bot.*, 48(12):353
6 (1973)—Nepal.

ANEMONOIDES

altaica (C. A. Mey.) Holub in *Folia Geobot. Phytotax.*,
1, 2 8(2):165(1973): *Anemone altaica.*
americana (L.) Holub *op. cit.* 165(1973): *Anemone*
3 *americana.*
amurensis (Korž.) Holub *op. cit.* 165(1973): *Anemone*
2 *nemorosa* subsp. *amurensis.*
apennina (L.) Holub *op. cit.* 165(1973), as 'appenina':
1, 2 *Anemone appenina.*
baicalensis (Turcz.) Holub *op. cit.* 166(1973): *Anemone*
2 *baicalensis.*
blanda (Schott & Kotschy) Holub *op. cit.* 166(1973):
1, 2 *Anemone blanda.*
caerulea (DC.) Holub *op. cit.* 166(1973): *Anemone*
2 *caerulea.*
caucasica (Rupr.) Holub *op. cit.* 166(1973): *Anemone*
2 *caucasica.*
debilis (Turcz.) Holub *op. cit.* 166(1973): *Anemone*
2 *debilis.*
delavayi (Franch.) Holub *op. cit.* 166(1973): *Anemone*
2 *delavayi.*
fischerana (DC.) Holub *op. cit.* 166(1973), as
2 'fischeriana': *Anemone fischeriana.*
flaccida (F. Schmidt) Holub *op. cit.* 166(1973):
2 *Anemone flaccida.*
griffithii (Hook. & Thoms.) Holub *op. cit.* 166
6 (1973): *Anemone griffithii.*
jenisejensis (Korž.) Holub *op. cit.* 166(1973):
2 *Anemone ranunculoides* subsp. *jenisejensis.*
keiskeana (Maxim.) Holub *op. cit.* 166(1973):
2 *Anemone keiskeana.*
maximowiczii (Juz.) Holub *op. cit.* 166(1973):
2 *Anemone maximowiczii.*
nemorosa (L.) Holub *op. cit.* 166(1973): *Anemone*
1–3 *nemorosa.*
nikoënsis (Maxim.) Holub *op. cit.* 166(1973):
2 *Anemone nikoënsis.*
piperi (Rydb.) Holub *op. cit.* 166(1973): *Anemone*
3 *piperi.*
prattii (Ulbrich) Holub *op. cit.* 166(1973): *Anemone*
2 *prattii.*
pseudo-altaica (Hara) Holub *op. cit.* 166(1973):
2 *Anemone pseudo-altaica.*
quinquefolia (L.) Holub *op. cit.* 166(1973): *Anemone*
1–3 *quinquefolia.*
raddeana (Regel) Holub *op. cit.* 166(1973): *Anemone*
2 *raddeana.*
ranunculoides (L.) Holub *op. cit.* 166(1973):
1, 2 *Anemone ranunculoides.*
reflexa (Willd.) Holub *op. cit.* 166(1973): *Anemone*
1, 2 *reflexa.*
rossii (S. Moore) Holub *op. cit.* 166(1973): *Anemone*
2 *rossii.*
stolonifera (Maxim.) Holub *op. cit.* 166(1973):
2 *Anemone stolonifera.*
trifolia (L.) Holub *op. cit.* 166(1973): *Anemone*
1, 3 *trifolia.*
udensis (Trautv. & Mey.) Holub *op. cit.* 166(1973):
2 *Anemone udensis.*
ulbrichiana (Diels) Holub *op. cit.* 166(1973):
2 *Anemone ulbrichiana.*
umbrosa (C. A. Mey.) Holub *op. cit.* 166(1973):
2 *Anemone umbrosa.*
uralensis (DC.) Holub *op. cit.* 166(1973): *Anemone*
1, 2 *uralensis.*

AQUILEGIA

vulgaris L.
var. **sooi** É. Kovács ex Soó in *Acta Bot. Acad. Sci.*
1 *Hung.*, 18(1–2):174(1973)—Hungary.

ATRAGENE

koreana (Kom.) Zvirgdz, Maurin' & Tsinovskis, Skriver. Dendrar.: 50(1972), without basionym
2 ref.: *Clematis koreana.*

BATRACHIUM

3893 Filipescu, G.
Cercetări anatomíce comparative la speciile *Batrachium aquatile* (L.) Dumort., *Batrachium divaricatum* (Schrank) Schur., *Batrachium trichophyllum* (Chaix) Bossche. *An. Şti. Univ. Al. I. Cuza Iaşi, Biol.*,
1 19(1):123–134(1973)—(Fr); illustrations.

× cookii Soó in *Acta Bot. Acad. Sci. Hung.*, 18(1–2): 175(1973), nom. nud.: Locality unknown. (*B.*
1 *circinatum* × *B. trichophyllum*).
× glueckii Soó *op. cit.* 175(1973), as 'glückii', nom. nov., without type and name not accepted: *B.*
1 *lambertii* (Félix) Soó. (*B. aquatile* × *B. baudotii*).
× segrettii (Félix) Soó *op. cit.* 175(1973), as 'segretti': *Ranunculus segretti* nom. nud. (*B. baudotii* × *B.*
1 *trichophyllum*).
× **virzionense** (Félix) Soó *op. cit.* 175(1973):
1 *Ranunculus virzionensis.* (*B. aquatile* × *B. radians*).

CALLIANTHEMUM

3894 Decamps, O.
Développement ontogénique de *Callianthemum* (Rénonculacées). *Bull. Soc. Hist. Nat. Toulouse*, 107
1 (1–2):79–88(1971)—Illustrations.

CALTHA

3895 Moore, D. M. *and* Goodall, R. N. P.
Interspecific hybridization in Fuegian *Caltha* L. *Bol. Soc. Argent. Bot.*, 15(1):72–76(1973)—(Sp);
4 illustrations; chromosome number.

3896 Morozova, N. S.
Sravnitel'no-anatomicheskaya kharakteristika semennoĭ kozhurȳ nekotorȳkh vidov *Caltha*. (Comparative anatomical characteristics of spermoderm of some species of *Caltha*). *Bot. Zhurn.*, 58(4):
1–3 523–528(1973)—Chromosome numbers.

3897 Smit, P. G.
A revision of *Caltha* (Ranunculaceae). *Blumea*, 21(1):119–150(1973)—Illustrations, maps, dot maps;
1–4, 6, 7 chromosome numbers; keys.

leptosepala DC.
subsp. **howellii** (Huth) P. G. Smit in *Blumea*,
3 21(1):143(1973): *C. leptosepala* var. *howellii*.
palustris L.
f. **arrabonensis** Soó & Kovács-Láng in *Acta Bot.*
1 *Acad. Sci. Hung.*, 18(1–2):174(1973)—Hungary.
var. **borbasii** Soó *op. cit.* 176(1973), nom. nov.:
1 *C. palustris* var. *truncata* Beck.
f. **czarnohorensis** (Zapał.) Soó & Kovács-Láng
1 *op. cit.* 176(1973): *C. laeta* var. *czarnohorensis*.
f. **platycarpa** Soó & Kovács-Láng *op. cit.* 174
1 (1973)—Hungary.
f. **retyezatensis** Soó & Kovács-Láng *op. cit.* 174
1 (1973)—Rumania.
f. **simonkaiana** Soó & Kovács-Láng *op. cit.* 174
1 (1973)—Rumania.
f. **transsilvanica** Soó & Kovács-Láng *op. cit.* 174
1 (1973)—Rumania.

CLEMATIS

3898 Beskaravaĭnaya, M. A.
Klematisȳ i ikh vosiriimchivost k muchinstoĭ rose.
1–3, 5, 6 *Byull. Glavn. Bot. Sada* (Moscow), 89:94–97(1973).

3899 Pringle, J. S.
The cultivated taxa of *Clematis* sect. *Atragene* (Ranunculaceae). *Baileya*, 19(2):49–89(1973)—
3 Illustrations, maps; chromosome numbers; keys.

3900 Shimizu, T.
A new species of *Clematis* (Ranunculaceae) from Taiwan. *Taiwania*, 18(2):173–175(1973)—Illustrations;
2 key.

3901 Starek, J.
Clematis alpina and its hybrids. *Quart. Bull. Alp.*
1 *Gard. Soc.*, 41(4):283–287(1973).

sasakii Shimizu in *Taiwania*, 18(2):173(1973)—
2 Taiwan.

CONSOLIDA

3902 Trifonova, V. I.
Biosistematicheskoe issledovanie kavkazskikh vidov *Consolida* S. F. Gray. (A biosystematic study of Caucasian species of *Consolida* S. F. Gray). *Bot. Zhurn.*, 58(4):505–518(1973)—(En); dot maps; chromosome
2 number.

DELPHINIUM

3903 Badis, H.
General description of a mature ovule and embryo sac, early stages of development of the endosperm and of embryo development in some species of *Delphinium* L. genera. *Genet. Pol.*, 14(1):29–36(1973)—(Pol,
1–3 Rus).

3904 Billore, K. V.
A new variety of *Delphinium malabaricum* (Huth.) Munz from the Western Ghats. *Indian Forester*, 99(7):
6 436–439(1973)—Illustrations.

3905 Kondrat'eva-Mel'vil, E. A.
Razvitie strukturȳ odnoletnego dvudol'nogo rasteniya s monopodial'nȳm i simpodial'nȳm narastaniem. (Development of structure of an annual dicotyledonous plant with monopodial and sympodial growth). *Bot. Zhurn.*, 58(8):1157–1172(1973)—
1 Illustrations. *Delphinium ajacis*.

3906 Makushenko, L. M.
Chisla kromosom nekotorykh vidov sem. Ranunculaceae Juss. (Chromosome numbers of some species of Ranunculaceae Juss.). *Bot. Zhurn.*, 58(7):
1 1023–1026(1973).

3907 Malyutin, N. I.
Filogeniya i sistematika roda *Delphinium* L. (The phylogeny and the taxonomy of the *Delphinium* L. genus). *Bot. Zhurn.*, 58(12):1710–1722(1973)—(En);
1, 2 chromosome numbers.

connectens Pakhomova in Opred. Rast. Sred. Azii, 3:
2 239(1972)—U.S.S.R. (Middle Asia).
keminense Pakhomova *op. cit.* 240(1972)—U.S.S.R.
2 (Middle Asia).
malabaricum (Huth) Munz
var. **ghaticum** Billore in *Indian Forester*, 99(7):436
6 (1973)—India.
nikitinae Pakhomova in Opred. Rast. Sred. Azii, 3:
2 240(1972), as 'nicitinae'—U.S.S.R. (Middle Asia).
popovii Pakhomova *op. cit.* 238(1972)—U.S.S.R.
2 (Middle Asia).
raikovae Pakhomova *op. cit.* 241(1972)—U.S.S.R.
2 (Middle Asia).
vvedenskyi Pakhomova *op. cit.* 238(1972)—U.S.S.R.
2 (Middle Asia).

FICARIA

verna Huds.
f. **divergens** (F. Schultz) Soó in *Acta Bot. Acad. Sci. Hung.*, 18(1–2):176(1973): ? *Ranunculus*
1 *vernus* f. *divergens*.
f. dolichopetala (Zapał.) Soó *op. cit.* 176(1973), without basionym date: *Ranunculus ficaria* f.
1 *dolichopetalus*.
f. elongata (Zapał.) Soó *op. cit.* 176(1973), without
1 basionym date: *Ranunculus ficaria* f. *elongatus*.
f. podolica (Zapał.) Soó *op. cit.* 176(1973), without
1 basionym date: *Ranunculus ficaria* f. *podolicus*.
f. polypetala (Zapał.) Soó *op. cit.* 176(1973), without basionym date: *Ranunculus ficaria* f.
1 *polypetalus*.

HELLEBORUS

3908 Vazúr, M.
Helleborus purpurascens Waldst. et Kit. v Čerhovskom Pohorí. *Biológia, A* (Czechoslovakia), 28(4):
1 307(1973)—Cz.

HEPATICA

3909 Chupov, V. S.
Immunoélektroforeticheskie issledovaniya v rodakh *Anemone* L., *Pulsatilla* Mill. i *Hepatica* Mill. (Immuno-electrophoretical investigations in genera *Anemone* L., *Pulsatilla* Mill. and *Hepatica* Mill.). *Bot.*
1, 2 *Zhurn.*, 58(2):189–199(1973)—(En); illustrations.

3910 Gorer, R.
Plant awards 1972–73: *Hepatica transsilvanica* Fuss. Flore Plena A. M. *Quart. Bull. Alp. Gard. Soc.*, 41(4):
1 327(1973).

LACCOPETALUM

3911 Miranda, A. L.
Contribucion al mejor conocimiento de la histologia floral de tres species de Ranunculaceas utilizadas en medicina popular. *Bol. Soc. Bot. Libertad*,
4 3(2):119–136(1971)—(En); illustrations.

MYOSURUS

3912 Cadbury, C. J.
Mousetail, *Myosurus minimus* L. at the Ouse
1 Washes. *Nat. Cambridge.*, no.16:37–39(1973).

NIGELLA

3913 Bhandari, N. N. *and* Kishori, R.
Development of tapetal membrane and Ubisch-granules in *Nigella damascena*—a histochemical approach. *Beitr. Biol. Pfl.*, 49(1):59–72(1973)—
1, 2 (Ge).

media Pakhomova in Opred. Rast. Sred. Azii, 3:237
2 (1972)—U.S.S.R. (Middle Asia).

PARAQUILEGIA

scabrifolia Pakhomova in Opred. Rast. Sred. Azii,
2 3:237(1972)—U.S.S.R. (Middle Asia).

PULSATILLA

3914 Atlas rozšíření rostlin v jižní části Čech: 1. (Atlas der Pflanzenverbreitung im südlichen Teile Bohmens: 1). *Sborn. Jihočesk. Muz. Česk. Budejovicich Prir. Ved.*,
1 12 *Suppl.* 3. 44p.(1972)—(Ge); dot maps.

3915 Chupov, V. S.
Immunoélektroforeticheskie issledovaniya v rodakh *Anemone* L., *Pulsatilla* Mill. i *Hepatica* Mill. (Immuno-electrophoretical investigations in genera *Anemone* L., *Pulsatilla* Mill. and *Hepatica* Mill.). *Bot.*
1, 2 *Zhurn.*, 58(2):189–199(1973)—(En); illustrations.

3916 Marvet, A.
Kõik karused lilled pole karukellad. (The genus *Geum rivale* and the species *Pulsatilla*). *Eesti Loodus*,
1, 2 16(4):224–226(1973).

3917 Voroshilov, V. N. *and* Khokhryakov, A. P.
Novyĭ vid prostrela iz okrestnosteĭ Magadana. *Byull. Glavn. Bot. Sada* (Moscow), 90:40–41(1973).
2 *Pulsatilla magadanensis.*

australis (Heuff.) Soó in *Acta Bot. Acad. Sci. Hung.*, 18(1–2):175(1973): *Anemone pulsatilla* var.
1 *australis.*
grandis Wender.
f. **borbasiana** Soó *op. cit.* 175(1973), nom. nov.:
1 *P. grandis* var. *angustisecta* Neilr.
f. **pseudoslavica** Soó *op. cit.* 175(1973)—
1 Hungary.
magadanensis Khokhryakov & Voroshilov in *Byull. Glavn. Bot. Sada* (Moscow), 90:40(1973)—
2 U.S.S.R. (Soviet Far East).

RANUNCULUS

3918 Atlas rozšíření rostlin v jižní části Čech: 1. (Atlas der Pflanzenverbreitung im südlichen Teile Bohmens: 1). *Sborn. Jihočesk. Muz. Česk. Budějovicich Přír. Ved.*,
1 12 *Suppl.* 3. 44p.(1972)—(Ge); dot maps.

3919 Banerji, M. L.
Morphological studies on petal venation of *Ranunculus diffusus* DC. *Proc. Indian Sci. Congr.*,
6 60(3):288(1973). Abstract only.

3920 Băra, I. I. *and* Ghiorghiţă, G. I.
Analiza statistică a unor caractere de diagnoză la specii ale genului *Ranunculus*. (L'analyse statistique des quelques caractères de diagnose de espèces du genre *Ranunculus*). *Stud. Comun., Bot.*, 889–896(1971)—(Fr). *R. auricomus*, *R. cassubicus*, *R. binatus*, *R.*
1 *flabellifolius.*

3921 Barykina, R. P. *and* Pustovojtova, V. I.
Morfologo-anatomicheskoe issledovanie *Ranunculus repens* L. i *R. reptans* L. v protsesse ikh individual'nogo razvitiya. (Anatomical-morphological study of *Ranunculus repens* L. and *R. reptans* L. in the ontogenesis). *Vestn. Mosk. Univ. Nauch. Zhurn.*, *6*, 28(6):
1–3 28–39(1973)—(En); illustrations.

3922 Coles, S. M.
Ranunculus bulbosus L. in Europe. *Watsonia*, 9(3):
1 207–228(1973)—En; illustrations, dot maps.

3923 Eliasson, U.
Korrigerad korrigering ? (A corrected correction ?). *Svensk Bot. Tidskr.*, 67(1):67–68(1973)—*Ranunculus*
1 *trichophyllus.*

3924 Ginzo, H. D. *and* Lovell, P. H.
Aspects of the comparative physiology of *Ranunculus bulbosus* L. and *Ranunculus repens* L.: 1. Response to nitrogen. *Ann. Bot.* (U.K.),37(152):
1, 2 753–764(1973).

3925 Ginzo, H. D. *and* Lovell, P. H.
Aspects of the comparative physiology of *Ranunculus bulbosus* L. and *Ranunculus repens* L.: 2. Carbon dioxide assimilation and distribution of photosynthates. *Ann. Bot.* (U.K.), 37(152):765–776
1, 2 (1973)—Illustrations.

3926 Jonsson, P. N.
Två växtfynd från Lits socken, Jämtland. (Two records of vascular plants from Lit parish, Jämtland). *Svensk Bot. Tidskr.*, 67(1):67(1973)—*Ranunculus*
1 *lingua.*

3927 Kuusk, V. *and* Puusepp, V.
Villtulika (*Ranunculus lanuginosus* L.) leidudest Eesti NSV-s. (*Ranunculus lanuginosus* L. in the Estonian
1 SSR). *Flor. Märkmed*, 1(5):301–303(1973).

3928 Lundqvist, A., Østerbye, U., Larsen, K. *and* Linde-Laursen, I.
Complex self-incompatibility systems in *Ranunculus acris* L. and *Beta vulgaris* L. *Hereditas*, 73(2):161–168
1–3 (1973)—Chromosome numbers.

3929 Makushenko, L. M.
Chisla khromosom nekotorykh vidov sem. Ranunculaceae Juss. (Chromosome numbers of some species of Ranunculaceae Juss.). *Bot. Zhurn.*, 58(7):
1 1023–1026(1973).

3930 Mason, R.
On two native Ranunculi. *Canterbury Bot. Soc. J.*,
7 no.6:2–4(1973).

3931 Miranda, A. L.
Contribucion al mejor conocimiento de la histologia floral de tres species de Ranunculaceas utilizadas en medicina popular. *Bol. Soc. Bot. Libertad*, 3(2):119–
4 136(1971)—(En); illustrations.

3932 Nogler, G. A.
Genetik der Aposporie bei *Ranunculus auricomus*: 2. Endospermzytologie. *Ber. Schweiz. Bot. Ges.*, 82(1):
1, 2 54–63(1972)—(En); chromosome numbers.

3933 Pogan, E. *and* Wcisło, H.
Studies in *Ranunculus ficaria* L.: 1. Karyological analysis of *R. ficaria* L. ssp. *bulbifer* (Marsden-Jones) Lawalrée and *R. ficaria* L. ssp. *calthifolius* (Rchb.) Arc. *Acta Biol. Cracov., Bot.*, 16(1):135–143(1973)—Dot
1, 2 map; chromosome numbers.

3934 Popova, M.
Tsitotaksonomichno prouchvane v rod *Ranunculus* L. (Cytotaxonomic investigation in genus *Ranunculus* L.). *Izv. Bot. Inst.* (Sofia), 24:233–239(1973)—(Rus,
1–7 En); chromosome numbers.

3935 Proskuryakova, G. M.
Ob izmenchivosti nekotorȳkh vidov lyutika podroda *Ranunculastrum* DC. *Byull. Glavn. Bot. Sada*
1, 2, 5, 6 (Moscow), 90:29–34(1973)—Illustrations.

3936 Ritter, J.
Sur la présence, dans le Vercors méridional, des cytotypes diploïde et tétraploïde d'*Anthyllis montana* L. et du cytotype diploïde de *Ranunculus pyrenaeus* L. *C. R. Séanc. Soc. Biol. Filiales*, 166(12):1622–1625
1 (1972 publ. 1973)—Map.

3937 Smith, J. M. B.
Natural history of the Kinabalu buttercup. *Malayan Nat. J.*, 25(2):90–100(1972)—Illustration.
6 *Ranunculus lowii*.

3938 Turała, K.
Embryological studies in a tetraploid hybrid of *Ranunculus* subgenus *Batrachium* from the Nowy Targ Basin (Poland). *Acta Biol. Cracov., Bot.*, 15(2):191–203
1 (1972)—Chromosome numbers.

3939 Turała, K.
Uwagi o polskich gatunkach rodzaju *Ranunculus* podrodzaju *Batrachium* (DC.) A. Gray z kluczem do ich oznaczania. (Some remarks on the taxonomy of the Polish species of genus *Ranunculus* subgenus *Batrachium* (DC.) A. Gray with a key for their identification). *Fragm. Flor. Geobot.*, 19(3):291–304
1 (1973)—(En); illustrations; chromosome numbers.

convexiusculus Kovalevskya in Opred. Rast. Sred.
2 Azii, 3:242(1972)—U.S.S.R. (Middle Asia).
flavidus (Hand. Mazz.) R. R. Stewart, Annot. Cat. Vasc. Pl. W. Pakistan & Kashmir: 270(1972):
2, 6 *Batrachium flavidum*.
michaelis Kovalevskya in Opred. Rast. Sred. Azii,
2 3:242(1972)—U.S.S.R. (Middle Asia).
ovczinnikovii Kovalevskya *op. cit.* 218(1972), nom.
2 nov.: *R. stenopetalus* Ovcz.

THALICTRUM

3940 Bochantseva, V. V.
O *Thalictrum leptophyllum* F. Nyl. (On *Thalictrum leptophyllum* F. Nyl.). *Bot. Zhurn.*, 58(7):1026–
1 1027(1973)—Rus.

3941 Bochantseva, V. V.
Thalictrum minus L. s. lat. Evropeĭskoĭ chasti SSSR. (*Thalictrum minus* L. s. lat. of the U.S.S.R. European
1 part). *Bot. Zhurn.*, 58(11):1641–1644(1973)—Key.

3942 Schönfelder, P.
Thalictrum simplex L.—auch in Südniedersachsen. *Göttinger Flor. Rundbr.*, 7(3):51–54(1973)—Dot map;
1 key.

3943 Shi, Yi-chen
A study on the pollen morphology of *Thalictrum* L. *Acta Bot. Sin.*, 15(2):159–162(1973). Chinese
2 species.

macrophyllum V. V. Bochantseva in *Bot. Zhurn.*, 58(11):1643(1973), as 'macrophylliym'—Finland, U.S.S.R. (European Russia, Caucasus, Siberia,
1 Soviet Far East, Middle Asia).

TRAUTVETTERIA

3944 Tobe, H.
Embryo sac development in *Trautvetteria japonica* Sieb. et Zucc. (Ranunculaceae). *Sci. Rep. Tôhoku*
2, 3 *Univ., IV*, 36(4):255–259(1973).

TROLLIUS

3945 Igoshina, K. N.
O yuzhnoural'skikh vidakh roda *Trollius* L. (De generis *Trollius* L. speciebus austrouralensibus).
1, 2 *Novosti Sist. Vȳssh. Rast.*, 10:142–152(1973).

bargusinensis Siplivinskiĭ in *Novosti Sist. Vȳssh. Rast.*,
2 10:358(1973)—U.S.S.R. (Siberia).
micranthus (Winkl. & Kom.) Pakhomova in Opred. Rast. Sred. Azii, 3:145(1972): *Hegemone*
2 *lilacina* var. *micrantha*.

RESEDACEAE

RESEDA

3946 Montserrat, P.
Deux *Erodium* et un *Reseda* nouveaux de la province de Huesca, Espagne. *Bull. Soc. Bot. France*, 120(1–2):
1 43–48(1973)—(En).

suffruticosa Leofling
subsp. **baetica** (Mueller-Arg.) Malagarriga, Las
1 Subesp. Variac. Geogr.: 10(1973): *R. baetica*.
subsp. **macrostachya** (Lge) Malagarriga *op. cit.*
1 10(1973): *R. macrostachya*.
vivantii Montserrat in *Bull. Soc. Bot. France*, 120(1–2):
1 47(1973)—Spain.

RESTIONACEAE

EMPODISMA L. A. S. Johnson & D. F. Cutler in
7 *Kew Bull.*, 28(3):383(1973 publ. 1974).

3947 Johnson, L. A. S. *and* Cutler, D. F.
Empodisma: a new genus of Australasian Restionaceae. *Kew Bull.*, 28(3):381–385(1973 publ. 1974)—
7 Chromosome numbers.

gracillimum (F. Muell.) L. A. S. Johnson & D. F. Cutler in *Kew Bull.*, 28(3):383(1973 publ. 1974):
7 *Calorophus gracillimus*.
minus (Hook. f.) L. A. S. Johnson & D. F. Cutler
7 *op. cit.* 383(1973 publ. 1974): *Calorophus minor*.

RHAMNACEAE

3948 Drummond, R. B. *and* Torre, A. R.
Flora de Moçambique: 49. Rhamnaceae. 1973.
5 23p.—Keys.

3949 Johnston, M. C. *and* Soares, M. A. de Freitas
Flora ilustrada catarinense: Ramnáceas. 1972. 50p.—
4 Illustrations, dot maps; keys.

CEANOTHUS

× **humboldtensis** Roof in *Four Seasons*, 4(3):21 (1973)—U.S.A. (California). (*C. pumilus* ×
3 *C. ramulosus*).

COLUBRINA

cubensis (Jacq.) Brogn.
subsp. **ekmanii** (M. C. Johnst.) Borhidi in *Acta Bot. Acad. Sci. Hung.*, 19(1–4):44(1973), without
4 basionym page: *C. cubensis* var. *ekmanii*.

subsp. **floridana** (M. C. Johnst.) Borhidi *op. cit.*
4 44(1973): *C. cubensis* var. *floridana.*
glandulosa Perkins
subsp. **antillana** (M. C. Johnst.) Borhidi *op. cit.*
4 44(1973): *C. rufa* var. *antillana.*
subsp. **nipensis** (M. C. Johnst.) Borhidi *op. cit.* 44
4 (1973): *C. nipensis.*
subsp. **reitzii** (M. C. Johnst.) Borhidi *op. cit.* 43
4 (1973): *C. rufa* var. *reitzii.*
retusa (Pittier) Cowan
subsp. **latifolia** (Reissek) Borhidi *op. cit.* 44(1973):
4 *Cormonema spinosum* var. (beta) *latifolium.*
spinosa Donnell-Smith
subsp. **mexicana** (Rose) Borhidi *op. cit.* 44(1973):
4 *Cormonema mexicanum.*

GOUANIA

3950 St. John, H.
A new living species of *Gouania* (Rhamnaceae) on Oahu, Hawaiian Islands. *Pac. Sci.*, 27(3):269–273
6 (1973) (Hawaiian Plant Studies: 40)—Illustrations.

gagnei St. John in *Pac. Sci.*, 27(3):269(1973)—
6 Hawaiian Is.

KENTROTHAMNUS

3951 Johnston, M. C.
Revision of *Kentrothamnus* (Rhamnaceae). *J. Arnold*
4 *Arbor.*, 54(4):471–473(1973).

weddellianus (Miers) M. C. Johnston in *J. Arnold*
4 *Arbor.*, 54(4):472(1973): *Trevoa weddelliana.*

RHAMNELLA

3952 Yamazaki, T.
On *Rhamnella* and its allied genera. *J. Jap. Bot.*,
2 48(1):30–32(1973)—Ja; illustration.

RHAMNIDIUM

potrerilloanum Borhidi & Muñiz in *Acta Bot.*
4 *Acad. Sci. Hung.*, 18(1–2):37(1973)—Cuba.

RHAMNUS

lycioides L.
subsp. **bourgeanus** (Gdgr.) Malagarriga, Las
1 Subesp. y Variac. Geogr.: 12(1973): *R. bourgeanus.*
utilis Decne.
var. **chingshuiensis** (T. Shimizu) Yamazaki in
2 *J. Jap. Bot.*, 48(5):141(1973): *R. chingshuiensis.*

SCUTIA

3953 Valente, M. da C., Freire de Carvalho, L. d'A., Sucre B., D. *and* Vianna, M. C.
Flora do Estado da Guanabara. Rhamnaceae: 2. Taxonomia e anatomia de *Scutia arenicola* (Casaretto) Reissek. *An. Acad. Brasil. Cienc.*, 44(1):99–117(1972)—
4 (En); illustrations.

VENTILAGO

3954 Oza, G. M.
Critical notes on the identity and nomenclature of *Ventilago calpicarpa* (Kurz) Oza. *Indian Forester*, 99(8):
6 516–517(1973).

ZIZIPHUS

mauritiana Lam.
var. fruticosa (Haines) R. R. Stewart, Annot. Cat. Vasc. Pl. W. Pakistan & Kashmir: 469 (1972), without basionym ref.: *Z. jujuba* var.
6 *fruticosa.*
var. spontanea (Edgew.) R. R. Stewart *op. cit.* 469 (1972), without basionym ref.: *Z. jujuba* var.
6 *spontanea.*

RHIZOPHORACEAE

3955 Araujo, P. A. de M. *and* Mattos Filho, A. de
Estrutura das madeiras de Rhizophoraceae: 2.
4 *Arq. Jard. Bot. Rio de Janeiro*, 19:133–148(1973).

ANISOPHYLLEA

3956 Pires, J. M. *and* Rodrigues, W. A. *
Notas sôbre os gêneros *Polygonanthus* e *Anisophyllea. Acta Amazonica*, 1(2):7–15(1971 publ. 1972). Ref. in
4 *Bull. Torrey Bot. Club*, 99(6):326(1972).

CERIOPS

3957 Steinke, T. D. *and* Ward, C. J.
Ceriops tagal (Perr.) C. B. Robinson at Kosi Bay.
5 *J. S. Afr. Bot.*, 39(3):245–247(1973)—(Af); dot map.

POLYGONANTHUS

3958 Pires, J. M. *and* Rodrigues, W. A. *
Notas sôbre os gêneros *Polygonanthus* e *Anisophyllea. Acta Amazonica*, 1(2):7–15(1971 publ. 1972). Ref. in
4 *Bull. Torrey Bot. Club*, 99(6):326(1972).

RHIZOPHORA

3959 De Fonseka, R. N. *and* Balasubramaniam, S.
A double-seedling condition in *Rhizophora apiculata* Bl. *Ceylon J. Sci., Biol. Sci.*, 10(1):81–83
6 (1972).

ROSACEAE

3960 Aitchinson, R. R. *and* Pankhurst, R. J.
Identification of blackberries and their allies in Cambridgeshire. *Nat. Cambridge.*, no.16:31–36
1 (1973)—Key.

3961 Challice, J. S.
Phenolic compounds of the subfamily of Pomoideae: a chemotaxonomic survey. *Phytochemistry*, 12(5):
1–4, 6 1095–1101(1973).

3962 Ĭordanov, D. *ed.*
Flora na Narodna Republika Bŭlgariya: 5. (Flora Republicae Bŭlgariya: 5.) Sofiya, Bŭlgarskata Akad.
1 Nauk, 1973. 442p.—Illustrations, map. Rosaceae.

3963 [Journet, P.]
Les *Crataegus*, *Sorbus* et genres voisins, végétaux robustes et décoratifs. *Jard. France*, no.10:21–24(1973)
1 —Illustrations. Popular article.

3964 Kalkman, C.
The Malesian species of the subfamily Maloideae (Rosaceae). *Blumea*, 21(2):413–442(1973)—Illustra-
2, 6 tions; keys.

3965 Kania, W.
Entwicklungsgeschichtliche Untersuchungen an Rosaceenblüten. *Bot. Jahrb.*, 93(2):175–246(1973)—
1–7 (En); illustrations.

3966 Ponomarenko, V. V.
Dikorastushchie yabloni vostochnoĭ Sibiri i dal'nego vostoka SSSR. (Wild growing apple trees in east Siberia and the far east). *Trudȳ Prikl. Bot. Genet.*
2 *Selek.*, 49(1):85–94(1973)—Illustrations.

subfam. **AGRIMONIOIDEAE** (S. F. Gray) Skalický in *Novit. Bot.*, 1972:11(1972 publ. 1973):
1–5 fam. Agrimoniaceae.
tribus **ALCHEMILLEAE** Skalický in *Novit. Bot.*,
1–3, 5 1972:13(1972 publ. 1973).

ACAENA

3967 Yeo, P. F.
Acaena. In Green, P. S. *ed.* Plants: wild and culti-
4 vated: 51–55(1973)—En.

3968 Yeo, P. F.
The species of *Acaena* with spherical heads cultivated and naturalized in the British Isles. *In* Green, P. S. *ed.* Plants: wild and cultivated: 193–221(1973)—
1, 4, 7 Illustrations; key.

AGRIMONIA

3969 Jain, D. K. *and* Singh, V.
Structure and development of hairs in *Agrimonia eupatorium* L. (family Rosaceae). *Curr. Sci.*, 42(12):
1–3 434–436(1973).

3970 Skalický, V.
Amerikanische Arten der Gattung *Agrimonia* L. ser. *Tuberosae* ser. nova. *Folia Geobot. Phytotax.*, 8(1):
3 95–104(1973)—(En); maps; key.

3971 Skalický, V.
Miscellanea taxonomica et nomenclatorica ad generis *Agrimonia* L. investigationem pertinentia. *Novit. Bot.*,
1–5 11–18(1972 publ. 1973).

subser. **Rostellatae** Skalický in *Folia Geobot.*
3 *Phytotax.*, 8(1):97(1973).
3 ser. **Tuberosae** Skalický *op. cit.* 95(1973).
eupatoria L.
f. **rosulata** Skalický in *Novit. Bot.*, 1972:15(1972
1 publ. 1973)—Czechoslovakia.
procera Wallr.
f. **glandulosa** (Simonkai) Skalický *op. cit.* 17
1 (1972 publ. 1973): *A. odorata* var. *glandulosa.*

ALCHEMILLA

3972 Lonsing, A.
Alchemilla glabra Ney genfind im Mühlvierel. *Mitt. Bot. Arbeit. Oberösterr. Landesmus.*, 5(2):216–217
1 (1973)—Map.

3973 Plocek, A.
Alchemilla suavis, a new species from The Western Carpathians. *Österr. Bot. Zeitschr.*, 122(3):195–198
1 (1973)—(Ge).

3974 Ponert, J.
Generis *Alchemilla* novitates turcicae kurdistanicaeque. (Neue Arten in der Gattung *Alchemilla* aus der Türkei und Kurdistan). *Feddes Repert.*, 83(7–8):
2 501–505(1972 publ. 1973—(Ge,En).

3975 Walters, S. M.
Alchemilla. In Green, P. S. *ed.* Plants: wild and
1–3, 7 cultivated: 48–50—(1973)En.

cavillieri (Burnat) Pignatti in *Giorn. Bot. Ital.*,
1 107(5):212(1973): *A. vulgaris* var. *cavillieri.*
hungarica Soó in *Acta Bot. Acad. Sci. Hung.*, 18(1–2):
1 172(1973)—Hungary.
paracompactilis Ponert in *Feddes Repert.*, 83(7–8):
2 502(1972 publ. 1973)—Turkey.
parapersica Ponert *op. cit.* 502(1972 publ. 1973)—
2 Turkey.
pawlowskii Asenov in Fl. Nar. Republ. Bŭlgariya, 5:
1 414(1973)—Bulgaria.
rivularis Ponert in *Feddes Repert.*, 83(7–8):503
2 (1972 publ. 1973)—Turkey.
sordidicalyx Ponert *op. cit.* 504(1972 publ. 1973)—
2 Turkey.
suavis Plocek in *Österr. Bot. Zeitschr.*, 122(3):195
1 (1973)—Czechoslovakia.

AMELANCHIER

alnifolia (Nutt.) Nutt.
subsp. **florida** (Lindl.) Hultén in *Bot. Notiser*,
3 126(4):496(1973): *A. florida.*
ovalis Medicus
subsp. **cretica** (Willd.) Pignatti in *Giorn. Bot. Ital.*,
1 107(5):213(1973): *Pyrus cretica.*

ARONIA

3976 Hardin, J. W.
The enigmatic chokeberries (*Aronia*, Rosaceae). *Bull. Torrey Bot. Club*, 100(3):178–184(1973)—
3 Illustration, maps.

CERASUS

3977 Browicz, K.
Distribution of woody Rosaceae in W. Asia: 12. Miscellaneous notes. *Arbor. Kórnickie*, 18:23–33 (1973)—(Pol,Rus); illustrations, dot maps. *Cerasus*
1, 2 *angustifolia.*

3978 Gracza, P. *and* Rácz, Z.
A hypanthium szervezödésének néhány kérdése a *Cerasus avium* Mönch-ben. (Some problems of the hypanthium organization of *Cerasus avium* Mönch).
1, 2 *Bot. Közlem.*, 60(2):135(1973)—(En).

3979 Sobko, V. G.
Novyĭ vyd vyshni (*Cerasus klokovii* Sobko sp. nov.) z hranitnykh vidslonen pivdennoho Buhu. (A new species of cherry (*Cerasus klokovii* Sobko sp. nov.) from the granite outcroppings of the south Bug area). *Ukr. Bot. Zhurn.*, 30(5):663–665(1973)—
1 Uk(En).

avium (L.) Mönch.
var. **albida** (Ehrh.) Soó in *Acta Bot. Acad. Sci. Hung.*, 18(1–2):175(1973): *Prunus varia* var.
1 *albida.*
var. **duracina** (L.) Vŭlev in Fl. Nar. Republ. Bŭlgariya, 5:396(1973): *Prunus avium* var.
1 *duracina.*
var. **yuliana** (L.) Vŭlev *op. cit.* 395(1973): *Prunus*
1 *avium* var. *yuliana.*
klokovii Sobko in *Ukr. Bot. Zhurn.*, 30(5):664
1 (1973)—U.S.S.R. (Ukraine).
sargentii (Rehd.) Tsinovskis in Zvirgdz, Maurin' & Tsinovskis, Skriver. Dendrar.: 104(1972), without
2 basionym ref: (Basionym not stated).

CERCOCARPUS

3980 Mortenson, T. H.
Ecological variation in the leaf anatomy of selected species of *Cercocarpus. Aliso*, 8(1):19–48(1973)—
3, 4 Illustrations, dot maps.

COTONEASTER

3981 Hurusawa, I.
Taxonomische Untersuchung der Gattung *Cotoneaster* (Rosaceen) auf karpologischer Grundlage. *J. Fac. Sci. Univ. Tokyo, Bot.*, 11(5–7):195–242
1, 2, 6 (1973)—Illustrations; chromosome numbers; key.

3982 Klotz, G.
Neue oder kritische *Cotoneaster*-Arten: 6. *Wiss. Zeitschr. Friedrich-Schiller Univ. Jena, Mat.-Naturwiss.*,
2, 6 21(5–6):969–977(1972)—Illustrations.

3983 Klotz, G.
Neue oder kritische *Cotoneaster*-Arten: 7. *Wiss. Zeitschr. Friedrich-Schiller Univ. Jena, Mat.-Naturwiss.*,
2, 6 21(5–6):979–1021(1972)—Illustrations.

3984 Terletskiĭ, V. K.
Geograficheskaya izmenchivost' ritmiki razvitiya
kizil'nikov *Cotoneaster* Medic. (The geographic
variability in developmental rhythms of *Cotoneaster*
1, 2 Medic.). *Bot. Zhurn.*, 58(9):1325–1331(1973).

sect. **Adpressi** (Hurusawa) Hurusawa in *J. Fac. Univ.*
2, 6 *Tokyo, Bot.*, 11(5–7):214(1973): *C.* subsect. *Adpressi.*
2, 6 subsect. **Buxifolii** Hurusawa *op. cit.* 225(1973).
ser. **Dielsiani** Klotz in *Wiss. Zeitschr. Friedrich-*
Schiller Univ. Jena, Mat.-Naturwiss., 21(5–6):991
2 (1972).
ser. **Divaricati** (Hurusawa) Hurusawa in *J. Fac. Sci.*
Univ. Tokyo, Bot., 11(5–7):218(1973): *C.* subsect.
2 *Divaricati.*
2 ser. **Glaucophylli** Hurusawa *op. cit.* 240(1973).
ser. **Horizontales** (Hurusawa) Klotz in *Wiss.*
Zeitschr. Friedrich-Schiller Univ. Jena, Mat.-
Naturwiss., 21(5–6):1011(1972); also Hurusawa in *J.*
Fac. Sci. Univ. Tokyo, Bot., 11(5–7):216(1973): *C.*
2 subsect. *Horizontales.*
ser. **Humiles** Hurusawa in *J. Fac. Sci. Univ. Tokyo,*
2, 6 *Bot.*, 11(5–7):205(1973).
2 ser. **Hupehenses** Hurusawa *op. cit.* 232(1973).
2 ser. **Ignavi** Hurusawa *op. cit.* 208(1973).
2 ser. **Morrisonenses** Hurusawa *op. cit.* 226(1973).
subsect. **Nitidi** Hurusawa *op. cit.* 218(1973), nom.
2 nov.: *C.* ser. *Distichi* Yü.
ser. **Orientales** Klotz in *Wiss. Zeitschr. Friedrich-*
Schiller Univ. Jena, Mat.-Naturwiss., 21(5–6):996
2, 6 (1972).
2, 6 ser. **Sanguinei** Klotz *op. cit.* 969(1972).
2, 6 ser. **Strigosi** Klotz *op. cit.* 998(1972).
2, 6 ser. **Verruculosi** Klotz *op. cit.* 1004(1972).
2 ser. **Yunnanenses** Klotz *op. cit.* 992(1972).
subsect. **Zabelioidei** (Flinck & Hylmö) Klotz *op. cit.*
2 982(1972): *C.* ser. *Zabelioides.*
subsect. Zabelioides (Flinck & Hylmö) Hurusawa in
J. Fac. Sci. Univ. Tokyo, Bot., 11(5–7):205(1973),
1, 2 without basionym ref.: *C.* ser. *Zabelioides.*
assamensis Klotz in *Wiss. Zeitschr. Friedrich-Schiller*
Univ. Jena, Mat.-Naturwiss., 21(5–6):996(1972)—
6 Assam.
6 **bakeri** Klotz *op. cit.* 970(1972)—Tibet, Bhutan.
6 **bis-ramianus** Klotz *op. cit.* 994(1972)—Nepal.
0 **boisianus** Klotz *op. cit.* 987(1972)—Cultivation.
6 **burmanicus** Klotz *op. cit.* 1004(1972)—Burma.
2 **difficilis** Klotz *op. cit.* 1017(1972)—China.
2 **dissimilis** Klotz *op. cit.* 984(1972)—China.
2, 6 **dokeriensis** Klotz *op. cit.* 974(1972)—China, Tibet.
2 **farreri** Klotz *op. cit.* 1006(1972)—Burma.
giraldii Flinck & Hylmö ex Klotz *op. cit.* 1018
2 (1972)—China.
globosus (Hurusawa) Klotz *op. cit.* 1017(1972):
2 *C. dielsianus* var. *globosus.*
2 **hurusawaianus** Klotz *op. cit.* 984(1972)—China.
2 **improvisus** Klotz *op. cit.* 1005(1972)—China.
2 **kansuensis** Klotz *op. cit.* 1001(1972)—China.
kerstanii Klotz *op. cit.* 1010(1972), holotype not
6 correctly indicated—Pakistan.
2 **kongboensis** Klotz *op. cit.* 997(1972)—China.
laetevirens (Rehd. & Wils.) Klotz *op. cit.* 985
2 (1972): *C. acutifolia* var. *laetevirens.*
2 **langei** Klotz *op. cit.* 1000(1972)—China.
2 **mingkwongensis** Klotz *op. cit.* 989(1972)—China.
2 **notabilis** Klotz *op. cit.* 972(1972)—China.
2 **salwinensis** Klotz *op. cit.* 1003(1972)—China.
2 **schantungensis** Klotz *op. cit.* 1018(1972)—China.
2 **sichuanensis** Klotz *op. cit.* 1013(1972)—China.
soulieanus Klotz *op. cit.* 1014(1972), nom. nov.:
2 *C. disticha* var. *tongolensis* C. Schneider.
6 **staintonii** Klotz *op. cit.* 971(1972)—Nepal.
2 **strigosus** Klotz *op. cit.* 999(1972)—China.
2 **subalpinus** Klotz *op. cit.* 1003(1972)—China.
2 **taoensis** Klotz *op. cit.* 1010(1972)—China.
2 **vilmorinianus** Klotz *op. cit.* 992(1972)—China.

CRATAEGUS

3985 Browicz, K.
Distribution of woody Rosaceae in W. Asia: 11.
Crataegus aronia (L.) Bosc. ex DC. *Arbor. Kórnickie,*
1, 2 18:5–21(1973)—(Pol,Rus); illustrations, dot map.

3986 Hrabětová-Uhrová, A.
Ergänzungsbeitrag zur Taxonomie der Weissdorne
(*Crataegus* L.) in der Tschechoslowakei. (Doplněk k
taxonomii hlohů (*Crataegus* L.) v Československu).
1 *Preslia*, 45(2):108–111(1973).

3987 Hrabětová-Uhrová, A.
Hloh (*Crataegus* L.) v Dolním Pomoraví. (Weissdorn
(*Crataegus* L.) im Niedermarchgebiet). *Preslia*, 45(1):
1 31–36(1973)—(Ge); Illustrations.

3988 Hrabětová-Uhrová, A.
Stručný přehled hlohů v ČSSR. (Kurze Uebersicht
der Crataegen in der ČSSR). *Zprávy Česk. Bot. Spol.*,
1, 2 8(2):115–117(1973)—Key.

3989 O'Mahony, A.
Crataegus laevigata (Poir.) DC.—a first definite
record for Ireland. *Irish. Nat. J.*, 17(10):354–355
1 (1973).

3990 Shilova, N. V.
Razvitie listovogo sleda i stroenie steblya i neko-
torÿkh vidov *Crataegus* L. (Development of leaf trace
and stem structure in some species of the genus
1, 2 *Crataegus* L.). *Bot. Zhurn.*, 58(2):177–188(1973)—(En).

austromoravica Hrabětová-Uhrová in *Preslia*,
1 45(1):35(1973)—Czechoslovakia.
bohemica Hrabětová-Uhrová in *Preslia*, 45(2):109
1 (1973)—Czechoslovakia.
calciphila Hrabětová-Uhrová
var. **hadensis** Hrabětová-Uhrová *op. cit.* 110(1973)
1 —Czechoslovakia.
lepida Hrabětová-Uhrová *op. cit.* 108(1973)—
1 Czechoslovakia.
macrocarpa Hegetschw.
var. **belanensis** Hrabětová-Uhrová *op. cit.* 110
1 (1973)—Czechoslovakia.
var. **cebinensis** Hrabětová-Uhrová *op. cit.* 110
1 (1973)—Czechoslovakia.
mikulcicensis Hrabětová-Uhrová in *Preslia*, 45(1):
1 32(1973)—Czechoslovakia.
monogyna Jacq.
var. **tauscheri** (Gandoger) Soó in *Acta Bot. Acad.*
Sci. Hung., 18(1–2):175(1973): (Basionym not
1 stated).
oxyacantha L.
subsp. **nemorensis** Hrabětová-Uhrová in *Preslia*,
1 45(1):33(1973)—Czechoslovakia.
silesiaca Hrabětová-Uhrová in *Preslia*, 45(2):109
1 (1973)—Czechoslovakia.
× uhrovae Soó in *Acta Bot. Acad. Sci. Hung.*, 18(1–2):
175(1973), without type—Locality unknown. (*C.*
1 *curvisepala* × *C. oxyacantha*).

EPICOSTORUS

3991 McVaugh, R.
Report of the Committee for Spermatophyta:
conservation of generic names: proposal 329(3316)
Physocarpus (Cambessèdes) Maximowicz (1879), nom.
conserv., vs. *Physocarpum* (A. P. De Candolle)
Berchtold and Presl (1823), and *Epicostorus* Rafinesque
3 (1832). *Taxon*, 22(1):156(1973).

FILIPENDULA

3992 Kalda, A. *and* Roos, A.
Eesti floora *Plantago* ja *Filipendula* liikide kromosoomide arv. (Chromosome numbers in *Plantago* and *Filipendula* species in the flora of Estonia). *Flor. Märkmed*, 1(5):312–314(1973)—Chromosome
1 numbers.

FRAGARIA

3993 Staudt, G.
Fragaria iturupensis, eine neue Erdbeerart aus Ostasien. (*Fragaria iturupensis*, a new strawberry species from East Asia). *Willdenowia*, 7(1):101–104 (1973)—(En); illustrations, dot map; chromosome
2 numbers.

iturupensis Staudt in *Willdenowia*, 7(1):102(1973)—
2 Kuril Is.

GEUM

3994 Marvet, A.
Kõik karused lilled pole karukellad. (The genus *Geum rivale* and the species *Pulsatilla*). *Eesti Loodus*, 16(4):224–226(1973)—Illustration, coloured
1–3 illustration.

LAUROCERASUS

officinalis M. J. Roemer
var. **sipkaensis** (Späth) Vŭlev in Fl. Nar. Republ. Bŭlgariya, 5:412(1973): *Prunus laurocerasus* var.
1 *sipkaensis*.

MALUS

3995 Argimbaeva, R. Zh. *and* Bÿkov, B. A.
O vnutrividovÿkh formakh yabloni siversa. *Izv. Akad. Nauk. Kaz. SSR, Biol.*, 6:1–4(1973)—(Ka);
2 illustrations. *Malus sieversii*.

3996 Huckins, C. A.
A revision of the sections of the genus *Malus* Miller.
1–3 *Diss. Abstr. Int., B*, 33(3):1031(1972).

3997 Ponomarenko, V. V.
Botaniko-geograficheskie i taksonomicheskie dannÿe o *Malus sachalinensis* (Kom.) Juz. (Botanic-geographical and taxonomical data on *Malus sachalinensis* (Kom.) Juz.). *Bot. Zhurn.*, 58(11):1630–
2 1641(1973)—Illustrations.

manshurica (Max.) Kom.
subsp. **sachalinensis** (Juz.) Ponomarenko in *Bot.*
2 *Zhurn.*, 58(11):1640(1973): *M. sachalinensis*.
praecox (Pall.) Borkh.
var. **paradisiaca** (L.) Vŭlev in Fl. Nar. Republ.
1 Bŭlgariya, 5:347(1973): *Pyrus malus paradisiaca*.
silvestris (L.) Mill.
subsp. **paradisiaca** (L.) Soó in *Acta Bot. Acad. Sci. Hung.*, 18(1–2):175(1973): *Pyrus malus* var.
1 *paradisiaca*.

MICROMELES

corymbifera (Miq.) Kalkman in *Blumea*, 21(2):437
2, 6 (1973): *Vaccinium ? corymbiferum*.

PADELLUS Vasil'chenko in *Novosti Sist. Vÿssh.*
1, 2 *Rast.*, 10:185(1973).
mahaleb (L.) Vasil'chenko *op. cit.* 185(1973): *Prunus*
1, 2 *mahaleb*.

PADUS

3998 Browicz, K.
Distribution of woody Rosaceae in W. Asia: 12. Miscellaneous notes. *Arbor. Kórnickie*, 18:23–33 (1973)—(Pol,Rus); illustrations, dot maps. *Padus*
1, 2 *avium*.

avium Mill.
subsp. **pubescens** (Rgl. & Tilling) Browicz in *Arbor. Kórnickie*, 18:23(1973): *Prunus padus* var.
2 *pubescens*.
virginiana (L.) Mill.
var. demissa (T. & G.) Zvirgdz in Zvirgdz, Maurin' & Tsinovskis, Skriver. Dendrar.:62
3 (1972), without basionym ref.: *P. demissa*.

PETROPHYTUM

3999 De Mezey, A. A.
Plant portraits: *Petrophytum hendersonii* Rydberg. *Quart. Bull. Alp. Gard. Soc.*, 41(2):160–161(1973)—
3 Illustration.

PHOTINIA

nussia (D. Don) Kalkman in *Blumea*, 21(2):429(1973):
2, 6 *Pyrus nussia*.
serratifolia (Desf.) Kalkman *op. cit.* 424(1973):
6 *Crataegus serratifolia*.

PHYSOCARPUS

4000 McVaugh, R.
Report of the Committee for Spermatophyta: conservation of generic names: proposal 329(3316) *Physocarpus* (Cambessèdes) Maximowicz (1879), nom. conserv., vs. *Physocarpum* (A. P. De Candolle) Berchtold and Presl (1823), and *Epicostorus* Rafinesque
3 (1832). *Taxon*, 22(1):156(1973).

POTENTILLA

4001 Murata, G.
[Short note on *Potentilla fragarioides* L. var. *togashii* (Ohwi)]. *Acta Phytotax. Geobot.*, 25(4–6):185(1973)—
2 Ja.

4002 Pignatti, S. *and* Perticarari, S.
Su *Potentilla tommasiniana* F. Schultz ed altre '*Potentilla*' italiane a peli stellati. *Atti Mus. Civ. Stor. Nat. Trieste*, 28(1):211–220(1973)—Key. Repr. in
1, 2 *Univ. Stud. Trieste Fac. Sci. Inst. Bot.*, [94]93.

4003 Richards, A. J.
An upland race of *Potentilla erecta* (L.) Räusch in the British Isles. *Watsonia*, 9(4):301–317(1973)—
1 Illustrations, dot maps; chromosome number.

caulescens L.
subsp. **alchimilloides** (Lap.) Malagarriga, Las Subesp. y Variac. Geogr.: 11(1973): *P. alchimil-*
1 *loides*.
subsp. **nivalis** (Lap.) Malagarriga *op. cit.* 11(1973):
1 *P. nivalis*.
cinerea Chaix ex Vill.
var. **tommasiniana** (Schultz) Markova in Fl. Nar.
1 Republ. Bŭlgariya, 5:263(1973): *P. tommasiniana*.
erecta (L.) Räuschel
subsp. **strictissima** (Zimmeter) A. J. Richards in
1 *Watsonia*, 9(4):312(1973): *P. strictissima*.
impolita Wahlbg.
f. **borbasiana** Soó in *Acta Bot. Acad. Sci. Hung.*,
1 18(1–2):173(1973)—Hungary.
inclinata Vill.
var. **incisoserrata** (T. Wolf) Markova in Fl. Nar. Republ. Bŭlgariya, 5:231(1973): *P. canescens* var.
1 *incisoserrata*.

var. **laciniosa** (Mert. & Koch) Markova *op. cit.*
1 231(1973): *P. canescens* var. *laciniosa.*
f. **lanuginosa** (T. Wolf) Markova *op. cit.* 231
1 (1973): *P. canescens* f. *lanuginosa.*
f. **polydonta** (T. Wolf) Markova *op. cit.* 231
1 (1973): *P. canescens* f. *polydonta.*
neglecta Baumg.
var. **dissecta** (Wallr.) Markova *op. cit.* 228(1973):
1 *P. argentea* var. *dissecta.*
var. **pseudocalabra** (T. Wolf) Markova *op. cit.*
1 227(1973): *P. argentea* var. *pseudocalabra.*
obscura Willd.
var. **fallacina** (Blocki) Markova *op. cit.* 238(1973):
1 *P. fallacina.*
pilosa Willd.
var. **balcanica** (T. Wolf) Markova *op. cit.* 237
1 (1973): *P. recta* var. *balcanica.*
pirotensis (Borb.) Markova *op. cit.* 250(1973):
1 *P. taurica* var. *pirotensis.*
sulphurea Lam.
var. **achtarovii** Markova *op. cit.* 414(1973)—
1 Bulgaria.

PRUNUS

4004 Baiashvili, E. I.
(Caryological study of the plum (tkemali) *Prunus divaricata* Ldb). *Soobshch. Akad. Nauk Gruz. SSR*, 69(1):193–195(1973)—Geo(Rus,En); chromosome
2 number.

4005 Kawasaki, T.
(A new hybrid of *Prunus* from Mt. Ibuki, Prov. Ohmi.). *J. Jap. Bot.*, 48(11):332–336(1973)—Ja;
2 illustrations.

4006 Rozanov, B. S. *and* Nikodimova, E. N.
Razvitie generativnȳkh pochek slivȳ v Tadzhikistane. *Byull. Glavn. Bot. Sada* (Moscow), 90:45–49
2 (1973).

4007 Seymour, F. C.
The range of *Prunus americana* in New England.
3 *Phytologia*, 23(3):97–99(1973).

4008 Surányi, D.
Sexual correlation in self-compatible and self-incompatible varieties of some *Prunus. Acta Bot. Acad.*
1 *Sci. Hung.*, 18(1–2):179–185(1973).

4009 Vasil'chenko, I. T.
K voprosu o sistematicheskom polozhenii magalebki (*Prunus mahaleb* L.). (De positione systematica *Prunus mahaleb* L.). *Novosti Sist. Vȳssh.*
1, 2 *Rast.*, 10:180–188(1973)—Illustrations, map.

× **pseudaffinis** Kawasaki in *J. Jap. Bot.*, 48(11):336 (1973)—Japan. (*P. jamasakura* × *P. incisa* var.
2 *kinkiensis*).
rechingeri (Browicz) R. R. Stewart, Annot. Cat. Vasc. Pl. W. Pakistan & Kashmir: 363(1972), without
6 basionym date: *Cerasus rechingeri.*

PYRUS

4010 Browicz, K.
Distribution of woody Rosaceae in W. Asia: 12. Miscellaneous notes. *Arbor. Kórnickie*, 18:23–33 (1973)—(Pol,Rus); illustrations, dot maps. *Pyrus*
1, 2 *boissieriana.*

4011 Challice, J. S.
Phenolic compounds of the genus *Pyrus*: 6. Distribution of phenols amongst the various tissues of
1, 2 *Pyrus* stem. *J. Sci. Food Agric.*, 24:285–293(1973).

4012 Challice, J. S. *and* Westwood, M. N.
Numerical taxonomic studies of the genus *Pyrus* using both chemical and botanical characters. *Bot. J.*
1, 2 *Linn. Soc.*, 67(2):121–148(1973).

4013 Kosȳkh, V. M.
Lokholistnaya grusha *Pyrus elaeagnifolia* Pall. v. Krȳmu. (*Pyrus elaeagnifolia* Pall. in Crimea). *Bot. Zhurn.*, 58(9):1348–1357(1973)—Illustrations;
1 chromosome number.

4014 Mulkidzhanyan, Ya. I.
Novȳe materialȳ k systematike grush Peredneĭ Azii. (New data on the pears of the Near East). *Byull. Mosk. Obshch. Ispȳt. Prir., Biol.*, 78(2):145–146
1 (1973).

bourgaeana Dcne.
f. **triloba** (P. Silva & Fontes) Pinto da Silva in *Agron. Lusit.*, 34(3):188(1972 publ. 1973): *P.*
1 *communis* f. *triloba.*
elaeagnifolia Pall.
subsp. **bulgarica** (Khutath. & Sachok.) Vŭlev in Fl. Nar. Republ. Bŭlgariya, 5:338(1973): *P.*
1 *bulgarica.*

RHAPHIOLEPIS

philippinensis (Vidal) Kalkman in *Blumea*, 21(2):
6 434(1973): *Eriobotrya philippinensis.*

ROSA

4015 Dubovik, O. N.
Karlikovȳe shipovniki i ikh proiskhozhdenie. (Dwarf forms of the genus *Rosa* and their origin). *Byull. Mosk. Obshch. Ispȳt. Prir., Biol.*, 78(5):135–142
1, 2 (1973)—(En).

4016 Gadzhieva, G. G.
Novȳe shipovniki (*Rosa* L.) iz sektsii *Caninae* Crep. *Izv. Akad. Nauk Az. SSR, Biol. Nauk*, no.3:
2 14–17(1973)—(Az); illustrations.

4017 Gauckler, K.
Französische Rose und Gebirgs-Rose in Nordbayern. (*Rosa gallica* et *Rosa pendulina* in Bavaria septentrionale). *Nat. Mensch 1972*,: 7–9(1973)—
1 Illustrations, maps, dot maps.

4018 Heiserer, L. *and* Weinmeister, B.
Rosa pomifera J. Herrm. var. *resinosa* (Sternb.) R. Keller = Harz-Rose. *Mitt. Bot. Arbeit. Oberösterr.*
1 *Landesmus. Linz*, 4(1–2):11–12(1972).

4019 Kalkman, C.
The genus *Rosa* in Malesia. *Blumea*, 21(2):281–291
6 (1973)—Illustrations; key.

4020 Končalová, M. N. *and* Jičínská, D.
Početní zastoupení druhů rodu *Rosa* v okolí Sitna. (The frequency of the *Rosa* species from the region of the mountains Stiavnické pohőří (Slovakia)). *Zprávy*
1 *Česk. Bot. Spol.*, 8(2):127–129 (1973)—(En).

4021 Noack, H.
Wildrosen. *Kurzmitt. Deutsch. Dendrol. Ges.*,
1, 2 no.11:22–24(1973).

4022 Ponert, Ĭ.
Dva novȳkh vida shipovnika (*Rosa* L.) iz severnogo Irana. (Generis *Rosa* L. duae species novae ex Irano boreali). *Novosti Sist. Vȳssh. Rast.*, 10:175–180(1973)—
2 Illustration.

4023 Starikova, V. V.
Morfologo-anatomicheskaya kharakteristika
oreshkov nekotorȳkh vidov *Rosa*. (Morphological-
anatomical characteristics of the nutlets in some *Rosa*
1–3, 6 species). *Bot. Zhurn.*, 58(6):893–898(1973).

4024 Werger, J. *and* Burton, R. E.
Roses: a bibliography of botanical, horticultural
and other works related to the genus *Rosa*. Metuchen,
1–3 Scarecrow Press, 1972. vi,169p.

abutalybovii Gadzhieva in *Izv. Akad. Nauk Az.*
SSR, Biol. Nauk, 1973(3):16(1973)—U.S.S.R.
2 (Caucasus).
iljinii Chrshan. ex Gadzhieva *op. cit.* 14(1973), as
2 'iljini'—U.S.S.R. (Caucasus).
parasicula Ponert in *Novosti Sist. Vȳssh. Rast.*, 10:
2 178(1973)—Iran.
× schwertschlageri Soó in *Acta Bot. Acad. Sci. Hung.*,
18(1–2):176(1973), without type—Germany. (*R.*
1 *canina* × *R. rubiginosa*).
steppofruticetorum Ponert in *Novosti Sist. Vȳssh.*
2 *Rast.*, 10:176(1973)—Iran.

RUBUS

4025 Edees, E. S.
Notes on British *Rubi*: 1. *Watsonia*, 9(3):247–251
1 (1973).

4026 Gilli, A.
Die Hybride *Rubus bifrons* Vest × *R. hirtus*
1 W. K. in Kärnten. *Carinthia II*, 82:197–199(1972).

4027 Oredsson, A.
Frequency mapping of blackberry species (*Rubus* L.
subgen. *Rubus*) in Sweden: 1. Method and pre-
liminary results. *Bot. Notiser*, 126(1):37–68(1973)—
1 Maps.

4028 Piotrowska, H. *and* Żukowski, W.
Malina moroszka *Rubus chamaemorus* w Słowińskim
Parku Narodowym. (The cloudberry, *Rubus*
chamaemorus in the Słowiński National Park). *Chrońmy*
Przyr. Ojczystą, 29(1):63–67(1973)—Illustrations,
1 map.

4029 Shealy, H. E. *and* Herr, J. M.
Carpelloid stamens in *Rubus trivialis* Michx. *Bot.*
Gaz., 134(1):77–87(1973)—Illustrations. From a
3 South Carolinan clone.

4030 Sin'kova, G. M.
O sibirskoĭ maline—*Rubus sibiricus* (Kom.)
Sinjkova (De specie *Rubus sibiricus* (Kom.) Sinjkova).
Novosti Sist. Vȳssh. Rast., 10:172–175(1973)—
2 Chromosome numbers.

4031 Sokolova, N. P.
Morfogenez vegetativnȳkh organov i zhiznennȳe
formȳ v rode *Rubus* L. (Morphogenesis of vegetative
organs and life forms in the genus *Rubus* L.). *Byull.*
Mosk. Obshch. Ispȳt. Prir., Biol., 78(1):84–99(1973)—
1–3 (En).

4032 Weber, H. E.
Die gattung *Rubus* L. (Rosaceae) im nordwestlichen
Europa vom Nordwestdeutschen Tiefland bis
Skandinavien mit besonderer Berücksichtigung
Schleswig-Holsteins. Lehre, J. Cramer, 1972. vii,
504p. (Phanerogamarum Monographiae, tomus 7)—
Illustrations, map, dot maps; chromosome numbers;
1 keys.

adspersus Weihe ex H. E. Weber in *Österr. Bot.*
Zeitschr., 122(4):278(1973), nom. nov.: *R.*
1 *carpinifolius* Weihe.
aequiserrulatus H. E. Weber in *Mitt. Arbeitsgem.*
Floristik Schleswig-Holstein und Hamburg, 20:107
(1972): cf. H. E. Weber in *Phanerogam. Monogr.*,
1 7:359(1972 publ. 1973)—Germany.
albisequens H. E. Weber in *Phanerogam. Monogr.*,
1 7:279(1972 publ. 1973)—Germany.
anoplocladus Sudre
var. **appendiculatus** (Hruby) Markova in Fl. Nar.
Republ. Bŭlgariya, 5:105(1973): *R. hirtus* f.
1 *appendiculatus*.
1 var. **pirinica** Markova *op. cit.* 414(1973)—Bulgaria.
var. **rilaensis** (Hruby) Markova *op. cit.* 105(1973):
1 *R. hirtus* f. *rilaensis*.
canescens DC.
var. **rhodopaeus** (Vel.) Markova *op. cit.* 53(1973):
1 *R. tomentosus* var. *rhodopaeus*.
christiansenorum H. E. Weber in *Phanerogam.*
1 *Monogr.*, 7:322(1972 publ. 1973)—Germany.
condensatus Mueller
var. **achtarovii** Markova in Fl. Nar. Republ.
1 Bŭlgariya, 5:413(1973)—Bulgaria.
correctispinosus H. E. Weber in *Phanerogam.*
1 *Monogr.*, 7:141(1972 publ. 1973)—Germany.
crassus Holuby
var. **orbicularis** (Hruby) Markova in Fl. Nar.
Republ. Bŭlgariya, 5:102(1973): *R. hirtus* var.
1 *orbicularis*.
drejeriformis (K. Friderichsen) H. E. Weber in
Phanerogam. Monogr., 7:252(1972 publ. 1973):
1 *R. mucronatus* var. *drejeriformis*.
eideranus (K. Frid.) H. E. Weber *op. cit.* 307
1 (1972 publ. 1973): *R. thyrsiflorus* var. *eideranus*.
erichsenii H. E. Weber *op. cit.* 186(1972 publ. 1973)—
1 Germany.
finitimus Sudre
var. **gracilis** (Hruby) Markova in Fl. Nar. Republ.
Bŭlgariya, 5:78(1973): *R. tereticaulis* f.
1 *gracilis*.
var. **inaequiserratus** (Hruby) Markova *op. cit.*
1 81(1973): *R. tereticaulis* f. *inaequiserratus*.
hirsutior (Fitschen) H. E. Weber in *Phanerogam.*
Monogr., 7:306(1972 publ. 1973): *R. pallidus* subsp.
1 *hirsutior*.
incultus Wirtg. ex Focke
var. **grossidens** (Hruby) Markova in Fl. Nar.
Republ. Bŭlgariya, 5:93(1973): *R. rivularis* f.
1 *grossidens*.
insulariopsis H. E. Weber in *Phanerogam. Monogr.*,
1 7:214(1972 publ. 1973)—Germany.
koehleri Weihe & Nees
var. **trifoliatus** Markova in Fl. Nar. Republ.
1 Bŭlgariya, 5:413(1973)—Bulgaria.
lloydianus Genev.
var. **setosoglandulosus** (Wirtg. ex Focke)
Markova *op. cit.* 57(1973): *R. tomentosus* var.
1 *setosoglandulosus*.
macrostachys Mueller
var. **balcanica** Markova *op. cit.* 413(1973)—
1 Bulgaria.
marianus (Krause) H. E. Weber in *Phanerogam.*
Monogr., 7:276(1972 publ. 1973): *R. infestus* var.
1 *marianus*.
melanoxylon Mueller
var. **obovatus** Markova in Fl. Nar. Republ.
1 Bulgariya, 5:413(1973)—Bulgaria.
milfordensis Edees in *Watsonia*, 9(3):250(1973)—
1 Great Britain.
minutiflorus Mueller
var. **crassiusculus** (Hruby) Markova in Fl. Nar.
Republ. Bŭlgariya, 5:109(1973): *R. hirtus* var.
1 *crassiusculus*.

miostilus Boulay
var. **rhodopaensis** (Hruby) Markova *op. cit.* 74
1 (1973): *R. hirtus* var. *rhodopaensis.*
nessensis W. Hall
subsp. **scissoides** H. E. Weber in *Phanerogam.*
1 *Monogr.*, 7:108(1972 publ. 1973)—Germany.
noltei H. E. Weber *op. cit.* 192(1972 publ. 1973)—
1 Germany.
nuptialis H. E. Weber *op. cit.* 262(1972 publ. 1973)—
1 Germany.
oblongoobovatus Markova in Fl. Nar. Republ.
1 Bŭlgariya, 5:414(1973)—Bulgaria.
pogonantherus H. E. Weber in *Phanerogam. Monogr.*, 7:367(1972 publ. 1973), nom. nov.: *R. divergens*
1 L. M. Neuman.
sanguineus Friv.
var. **alnifolius** Markova in Fl. Nar. Republ.
1 Bŭlgariya, 5:413(1973)—Bulgaria.
var. **cyclophyllus** (Hruby) Markova *op. cit.* 46
1 (1973): *R. sanctus* f. *cyclophyllus.*
var. **obtriangulatus** Markova *op. cit.* 413(1973)—
1 Bulgaria.
var. **truncatus** Markova *op. cit.* 413(1973)—
1 Bulgaria.
schleicheri Weihe
var. **vitosensis** Markova *op. cit.* 414(1973)—
1 Bulgaria.
selmeri Lindeberg
var. **argyriophyllus** (Ranke) H. E. Weber in *Phanerogam. Monogr.*, 7:208(1972 publ. 1973):
1 *R. villicaulis* var. *argyriophyllus.*
sibiricus (Kom.) Sin'kova in *Novosti Sist. Vȳssh.*
2 *Rast.*, 10:173(1973): *R. idaeus* subsp. *sibiricus.*
var. **fontaneus** (Sin'kova) Sin'kova *op. cit.* 174
2 (1973): *R. sachalinensis* var. *fontaneus.*
var. **setosus** Sin'kova *op. cit.* 174(1973)—U.S.S.R.
2 (Siberia).
stormanicus H. E. Weber in *Phanerogam. Monogr.*,
1 7:310(1972 publ. 1973)—Germany.
subcalvatus (K. Frid.) H. E. Weber *op. cit.* 332(1972
1 publ. 1973): *R. apricus* var. *subcalvatus.*
yabei Lév. & Van.
var. **shikokianus** Ohwi & Inobe in *J. Geobot.*,
2 20(3):60(1972)—Japan.

SANGUISORBA

4033 Asai, Y.
On a new form of *Sanguisorba officinalis* Linn. *J. Jap.*
2 *Bot.*, 48(1):19–20(1973)—Ja(En); illustration.

4034 Becherer, A.
Nochmals *Sanguisorba dodecandra* Moretti im
1 Veltlin. *Ber. Schweiz. Bot. Ges.*, 83(1):75–76(1973).

4035 Mulligan, G. A. *and* Cody, W. J.
A clarification of the chromosome number situation in *Sanguisorba canadensis. Canad. J. Bot.*,
3 51(11):2075–2077(1973)—(Fr).

SIBBALDIA

cuneata Kunze
var. micrantha (Hk. f.) R. R. Stewart, Annot. Cat. Vasc. Pl. W. Pakistan & Kashmir: 369(1972), without basionym date: *Potentilla sibbaldii* var.
6 *micrantha.*

SORBUS

4036 Berggren, G.
Seed characters for taxonomic distinction between *Sorbus teodori* and *Sorbus meinichii. Svensk Bot. Tidskr.*,
1 67(4):321–332(1973)—Illustrations.

4037 Gabriélyan, É. Ts. *and* Tumanchan, S. A.
Anatomicheskoe issledovanie uzlov i chereshkov roda *Sorbus* L. (Nodal and petiolar anatomy of *Sorbus* L.). *Biol. Zhurn. Arm.*, 26(1):57–71(1973)—
1–3 (Arm).

4038 Gostyńska-Jakuszewska, M.
Nowe stanowisko jarząbu szwedzkiego *Sorbus intermedia* nad Zatoką Gdańska. (A new locality of the white beam *Sorbus intermedia*, at the Bay of Gdańsk).
1 *Chrońmy Przyr. Ojczystą*, 29(2):70–72(1973)—Map.

4039 [Taylor, S.]
Sorbus sitchensis Roemer: Western mountain-ash, Pacific mountain-ash, Sitka mountain-ash, rowan, California mountain-ash. *Davidsonia*, 4(3):30–33
3 (1973)—Illustrations.

4040 Zaikonnikova, T. I.
Sorbus migarica Zinserl. na Kavkaze. (*Sorbus migarica* Zinserl. in Caucaso). *Novosti Sist. Vȳssh. Rast.*,
2 10:167–172(1973)—Illustration, dot map.

umbellata (Desf.) Fritsch
subsp. **meridionalis** (Guss.) Vŭlev in Fl. Nar. Republ. Bŭlgariya, 5:365(1973):
1 *Pyrus meridionalis.*

SPIRAEA

4041 Jørgensen, P. M.
Forvillede arter og hybrider av slekten *Spiraea* i Norge. (Garden-escaped *Spiraea* species and hybrids in Norway). *Blyttia*, 31(1):29–33(1973)—(En);
1 illustration.

4042 Uttal, L. J.
An infraspecific nomenclatorial correction in
1–4, 6 *Spiraea. Castanea*, 38(2):205(1973).

4043 Yamanaka, T.
Notes on *Spiraea ogawae* Nakai. *J. Jap. Bot.*, 48(3):
2 94–96(1973)—(Ja); illustrations.

nipponica Maxim.
var. **ogawae** (Nakai) Yamanaka in *J. Jap. Bot.*,
2 48(3):95(1973): *S. ogawae.*

RUBIACEAE

4044 Adam, J. G.
Espèces nouvelles pour le Libéria et les Monts Nimba (Astéracées et Rubiacées). *Bull. Inst. Fond. Afr.*
5 *Noire, A*, 35(1):79–91(1973)—Illustrations.

4045 Anderson, W. R.
A morphological hypothesis for the origin of heterostyly in the Rubiaceae. *Taxon*, 22(5–6):537–542
1–7 (1973)—Illustrations.

4046 Majumdar, L.
Cytological analysis of some members of the tribe Hedyotideae (fam. Rubiaceae). *Proc. Indian Sci. Congr.*, 60(3):311(1973)—Chromosome numbers.
6 Abstract only.

4047 Majumdar, L.
Karyomorphology of Rubiaceae. *Proc. Indian Sci. Congr.*, 58(3):475–476(1971)—Chromosome
6 numbers. Abstract only.

4048 Sahay, S. K.
Eurypalyny in Rubiaceae. *Proc. Indian Sci. Congr.*,
1–7 58(3):455–456(1971). Abstract only.

4049 Soukup, J.
Las Rubiaceas del Peru, sus generos y lista de especies. *Biota*, 9(74):315–328,(75):329–346(1973)—
4 Key.

4050 Steyermark, J. A.
Novedades Rubiaceae de Venezuela. *Acta Bot.*
4 *Venezuel.*, 8(1–4):247–253(1973).

4051 Williams, L. O.
Tropical American plants: 14. *Phytologia*, 26(6):
4 487–493(1973)—Illustrations.

ADINA

4052 McVaugh, R.
Report of the Committee for Spermatophyta: conservation of generic names: proposal 308 (8226) *Adina* Salisbury (1807) vs. *Nauclea* Linnaeus (1762).
2, 5, 6 *Taxon*, 22(1):155(1973).

ALLENANTHUS

hondurensis
var. **parvifolia** L. O. Williams in *Phytologia*,
4 25(7):461(1973)—Mexico.

AMARACARPUS

vaccinioides (C. H. Wright) van Royen in *Bot.*
6 *Notiser*, 126(4):423(1973): *Dolianthus vaccinioides.*

ASPERULA

4053 Kucowa, I.
O geograficznym rozmieszczeniu trzech gatunków marzanek (*Asperula* L.) w Polsce i na obszarach sąsiednich. (The geographical distribution of three species of *Asperula* L. in Poland and the adjacent territories). *Fragm. Flor. Geobot.*, 19(1):29–42(1973)—
1 (En); dot maps. *A. cynanchica, A. tinctoria, A. glauca.*

purpurea (L.) Ehrendorfer in *Österr. Bot. Zeitschr.*,
1 122(4):260(1973): *Galium purpureum.*

BORRERIA

4054 Hsu, C.-C.
Some noteworthy plants found in Taiwan. *Taiwania*, 18(1):62–72(1973)—Illustration; key.
2 *Borreria*, pp.65–66.

gracilis L. O. Williams in *Phytologia*, 26(6):487
4 (1973)—Costa Rica.

BREONIA

chinensis (Lam.) Capuron in *Adansonia*, 13(4):472
5 (1973 publ. 1974): *Cephalanthus chinensis.*

CANEPHORA

4055 Paris, R. R. *and* Debray, M.
Sur une Rubiacée malgache: le *Canephora madagascariensis*. Présence de C-flavonoïdes dans les feuilles. *Pl. Méd. Phytothér.*, 7(2):135–139(1973)—
5 (En).

CAPTAINCOOKIA N. Hallé in *Adansonia*, 13(2):
6 197(1973).

4056 Hallé, N.
Captaincookia, genre nouveau monotypique néocalédonien de Rubiaceae-Ixoreae. *Adansonia*,
6 13(2):195–202(1973)—Illustrations.

margaretae N. Hallé in *Adansonia*, 13(2):197
6 (1973)—New Caledonia.

CARPHALEA

4057 Verdcourt, B.
A revision of the African species of *Carphalea* Juss. (= *Dirichletia* Klotzsch) (Rubiaceae). *Kew Bull.*,
5 28(3):423–428(1973 publ. 1974)—Key.

sect. **Dirichletia** (Klotzsch) Verdcourt in *Kew Bull.*,
5 28(3):424(1973 publ. 1974): gen. *Dirichletia.*
glaucescens (Hiern) Verdcourt *op. cit.* 424(1973
5 publ. 1974): *Dirichletia glaucescens.*
subsp. **angustifolia** Verdcourt *op. cit.* 425(1973
5 publ. 1974)—Tanzania.
obovata (Balf. f.) Verdcourt *op. cit.* 427(1973 publ.
5 1974): *Dirichletia obovata.*
pubescens (Klotzsch) Verdcourt *op. cit.* 426(1973
5 publ. 1974): *Dirichletia pubescens.*

CEPHAELIS

4058 Fisher, H. H.
Origin and uses of Ipecac. *Econ. Bot.*, 27(2):231–
4 234(1973)—Illustrations. *Cephaelis ipecacuanha.*

micheliae J. G. Adam in *Bull. Inst. Fond. Afr. Noire, A*,
5 35(1):83(1973)—Liberia.

CHIMARRHIS

perijaënsis Steyermark in *Acta Bot. Venezuel.*,
4 8(1–4):248(1973)—Venezuela.

CHIONE

allenii L. O. Williams in *Phytologia*, 25(7):461(1973)
4 —Costa Rica.
tetramera Alain in *Phytologia*, 25(5):277(1973)—
4 Hispaniola (Dominican Republic).

CINCHONA

henleana Karst.
var. **trichophylla** Steyermark in *Acta Bot.*
4 *Venezuel.*, 8(1–4):247(1973)—Venezuela.

COCCOCYPSELUM

hirsutum
var. **glabrum** (Bartling ex DC.) L. O. Williams in
4 *Phytologia*, 25(7):462(1973): *C. glabrum.*

COUSSAREA

imitans L. O. Williams in *Phytologia*, 26(6):488
4 (1973)—Guatemala, British Honduras.
terepaimensis Steyermark in *Acta Bot. Venezuel.*,
4 8(1–4):250(1973)—Venezuela.

CRUCIATA

4059 Surova, T. D.
Khromosomnȳe chisla nekotorȳkh vidov roda *Galium* L. i roda *Cruciata* Mill. severo-zapada evropeĭskoĭ chasti SSSR. (Chromosome numbers of some species of genera *Galium* L. and *Cruciata* Mill. of the north-west European part of the U.S.S.R.).
1 *Bot. Zhurn.*, 58(12):1819–1824(1973).

pedemontana (Bell.) Ehrend.
var. **vestita** (Rouy) Soó in *Acta Bot. Acad. Sci. Hung.*, 18(1–2):176(1973): *Galium pedemontanum*
1 γ *vestitum.*

COFFEA

4060 Hall, W. van
Coffea (fam. der Rubiaceae). *Wachendorffia*, 1(2):
4–6 21–22(1973)—Du; illustration.

RUBIACEAE

4061 Payne, R. C., Oliveira, A. R. *and* Fairbrothers, D. E.
Disc electrophoretic investigation of *Coffea*
arabica and *C. canephora*: general protein and malate
dehydrogenase of mature seeds. *Biochem. Syst.*, 1(1):
4 59–61(1973).

CRATERISPERMUM

4062 Verdcourt, B.
The identity of the common East African species of
Craterispermum Benth. (Rubiaceae) with some other
notes on the genus. *Kew Bull.*, 28(3):433–436(1973
5 publ. 1974)—Key.

brieyi De Wild.
var. **inquisitorium** (Wernh.) Verdcourt in *Kew*
5 *Bull.*, 28(3):436(1973 publ. 1974): *C. inquisitorium*.

DEPPEA

anisophylla L. O. Williams in *Phytologia*, 26(6):
4 490(1973)—Guatemala.

DOLIANTHUS

4063 Van Royen, P.
Sertulum Papuanum: 18. Two superfluous genera
in the New Guinean flora. *Bot. Notiser*, 126(4):417–
6 425(1973).

ELAEAGIA

4064 Steyermark, J. A.
Novedades de los Andes Venezolanos. *Pittieria*,
4 no.4:9–25(1972)—Sp,En; illustrations.

ruiz-teranii Steyermark in *Pittieria*, 4:15(1972)—
4 Venezuela.

ERITHALIS

fruticosa L.
subsp. **odorifera** (Jacq.) Steyermark in *Acta Bot.*
4 *Venezuel.*, 8(1–4):248(1973): *E. odorifera*.

EXOSTEMA

caribaeum (Jacq.) Roem. & Schult.
var. **pubescens** Borhidi & Muñiz in *Acta Bot. Acad.*
4 *Sci. Hung.*, 18(1–2):44(1973)—Cuba.

FARAMEA

standleyana L. O. Williams in *Phytologia*, 26(6):
4 490(1973)—Guatemala.

GAILLONIA

4065 Linchevskiĭ, I. A.
O rode *Gaillonia* A. Rich. ex DC. (Rubiaceae).
(De genere *Gaillonia* A. Rich. ex DC. (Rubiaceae)).
2 *Novosti Sist. Vȳssh. Rast.*, 10:224–236(1973)—Key.

GALIUM

4066 Arrigoni, P. V.
Su due *Galium* endemici di Sardegna: *Galium*
schmidii nova sp. e *Galium glaucophyllum* E. Schmid.
Webbia, 27(2):505–516(1972 publ. 1973)—(En);
1 illustrations, dot map.

4067 Bacigalupo, N. M.
Aclaración sobre la verdadera patria y la sinonimia
de *Galium latoramosum* Clos. *Darwiniana*, 18(1–2):
4 279–280(1973).

4068 Bacigalupo, N. M. *and* Ehrendorfer, F.
Una nueva especie de *Galium* sect. *Lophogalium* de
los Andes argentinos. *Darwiniana*, 18(1–2):27–30
4 (1973)—Illustrations; chromosome numbers.

4069 Dempster, L. T.
The fleshy-fruited Galiums (Rubiaceae) of Mexico
and Central America. *Brittonia*, 25(1):15–20(1973)—
4 Illustrations, dot map; chromosome numbers.

4070 Dempster, L. T.
A new species of *Galium* (Rubiaceae) from
Guatemala. *Phytologia*, 26(1):12–14(1973)—
4 Illustrations.

4071 Dempster, L. T.
The polygamous species of the genus *Galium*
(Rubiaceae), section *Lophogalium*, of Mexico and
south-western United States. *Univ. Calif. Publ. Bot.*,
64. 36p.(1973)—Illustrations, dot maps; chromosome
3, 4 numbers; keys.

4072 Dempster, L. T.
Transfer of *Relbunium sphagnophilum* (Rubiaceae)
and *R. mazocarpum* to *Galium*. *Phytologia*, 26(4):
4 221–223(1973)—Illustrations, dot map.

4073 Ehrendorfer, F. *and* Verdcourt, B.
Two new species of *Galium* (Rubiaceae—Rubieae)
from tropical East Africa. *Kew Bull.*, 28(3):485–489
5 (1973 publ. 1974)—Illustrations.

4074 Kliphuis, E.
Cytotaxonomic notes on some *Galium* species.
Galium boreale L.: 1. *Proc. Kon. Nederl. Akad.*
Wetensch., *B*, 76(4):359–372(1973)—Illustrations;
1–3 chromosome numbers.

4075 Kliphuis, E.
Cytotaxonomic notes on some *Galium* species.
Galium boreale L.: 2. *Proc. Kon. Nederl. Akad.*
Wetensch., *C*, 76(5):449–464(1973)—Illustrations;
1–3 chromosome numbers.

4076 Lawson, C. A.
The genus *Galium* (Rubiaceae) in Oklahoma.
Southwest. Nat., 18(1):45–72(1973)—Illustrations, dot
3 maps; keys.

4077 Pobedimova, E. G.
O dvukh vidakh roda *Galium* L., ryada *Borealia*
Pobed. (De speciebus duobus e generis *Galium* L.
serie *Borealia* Pobed.). *Novosti Sist. Vȳssh. Rast.*,
2 10:237–240(1973)—Illustrations, dot map.

4078 Surova, T. D.
Khromosomnȳe chisla nekotorȳkh vidov roda
Galium L. i roda *Cruciata* Mill. severo-zapada
evropeĭskoĭ chasti SSSR. (Chromosome numbers of
some species of genera *Galium* L. and *Cruciata* Mill.
of the north-west European part of the U.S.S.R.).
1 *Bot. Zhurn.*, 58(12):1819–1824(1973).

4079 Verdcourt, B.
Galium glaciale Krause (Rubiaceae) and its allies in
eastern Africa. *Kew Bull.*, 28(3):491–494(1973 publ.
5 1974)—Illustrations; key.

4080 Verdcourt, B.
The identity of *Galium aparinoides* Forssk.
2, 5 (Rubiaceae). *Kew Bull.*, 28(1):59–60(1973).

4081 Vicherek, J.
Galium exoletum Klokov, nová rostlina česko-
slovenské květeny. (*Galium exoletum* Klokov, eine
neue Pflanze der tschechoslowakischen Flora).
1 *Preslia*, 45(1):37–40(1973)—(Ge).

aculeolatum Dempster in *Univ. Calif. Publ. Bot.*,
4 64:27(1973)—Mexico.
boreale L.
var. incurvatum (Urschler) Kliphuis in *Proc. Kon. Nederl. Akad. Wetensch.*, *C*, 76(5):459(1973), without basionym ref.: *G. septentrionale* var.
2, 3 *incurvatum.*
var. septentrionale (Roemer & Schultes) Kliphuis *op. cit.* 459(1973), without basionym ref.: *G.*
2, 3 *septentrionale.*
brenanii Ehrendorfer & Verdcourt in *Kew Bull.*,
5 28(3):485(1973 publ. 1974)—Tanzania, Kenya.
canescens H. B. & K.
f. **alstonii** Steyermark in *Acta Bot. Venezuel.*,
4 8(1–4):253(1973)—Venezuela.
catalinense Gray
subsp. **acrispum** Dempster in *Univ. Calif. Publ.*
3 *Bot.*, 64:11(1973)—U.S.A. (California).
coronadoense Dempster *op. cit.* 31(1973)—
4 Mexico.
4 **diabolense** Dempster *op. cit.* 28(1973)—Mexico.
firmum Tausch
subsp. **balearicum** (Knoche) Malagarriga, Las Subesp. y Variac. Geogr.: 13(1973): *G. firmum*
1 var. *balearicum.*
fuscum Mart. & Gal.
subsp. **altiplanicum** Dempster in *Univ. Calif.*
4 *Publ. Bot.*, 64:16(1973)—Mexico.
subsp. **guerrericum** Dempster *op. cit.* 19(1973)—
4 Mexico.
subsp. **hypadenium** (Schauer) Dempster *op. cit.*
4 16(1973): *G. hypadenium.*
glaciale Krause
var. **satimmae** Verdcourt in *Kew Bull.*, 28(3):493
5 (1973 publ. 1974)—Kenya.
glaucum L.
var. **hirsutum** (Wallr.) Soó in *Acta Bot. Acad. Sci. Hung.*, 18(1–2):176(1973): *Asperula galioides* var.
1 *hirsuta.*
gracilicaule Ehrendorfer & N. M. Bacigalupo in
4 *Darwiniana*, 18(1–2):27(1973)—Argentina.
humifusum M. B.
subsp. **besseranum** (Klokow) Soó in *Acta Bot. Acad. Sci. Hung.*, 18(1–2):176(1973), as
1 'besserianum': *Asperula besseriana.*
subsp. **cincinnatum** (Klokow) Soó *op. cit.* 176
1 (1973): *Asperula cincinnata.*
kenyanum Verdcourt in *Kew Bull.*, 28(3):493
5 (1973 publ. 1974)—Kenya, Tanzania.
langei (P. Cout.) Pinto da Silva in *Agron. Lusit.*, 34(3):194(1972 publ. 1973): *G. uliginosum* subsp.
1 *langei.*
quichense Dempster in *Phytologia*, 26(1):12(1973)—
4 Guatemala.
rubrum L.
subsp. **balearicum** (Briq.) Malagarriga, Las Subesp. y Variac. Geogr.: 14(1973): *G.*
1 *balearicum.*
schelkownikowianum (Bordz.) Holub in *Folia Geobot. Phytotax.*, 8(2):176(1973): *Asperula rivalis*
1, 2 var. *schelkownikowiana.*
schmidii Arrigoni in *Webbia*, 27(2):507(1972 publ.
1 1973)—Sardinia.
sphagnophilum (Greenm.) Dempster in *Phytologia*,
4 26(4):221(1973): *Relbunium sphagnophilum.*
var. **mazocarpum** (Greenm.) Dempster *op. cit.*
4 222(1973): *Relbunium mazocarpum.*
tanganyikense Ehrendorfer & Verdcourt in *Kew*
5 *Bull.*, 28(3):488(1973 publ. 1974)—Tanzania.
volcanense Dempster in *Univ. Calif. Publ. Bot.*, 64:
4 27(1973)—Mexico.

GEOPHILA

4082 Verdcourt, B.
Tropical African plants: 33. Notes on *Geophila*, *Psychotria* and *Rubia* (Rubiaceae). *Kew Bull.*, 28(2):
5 320–321(1973)—Illustrations.

4083 Williams, L. O.
Geophila (Rubiaceae) in North America. *Phytologia*,
3 26(4):263–264(1973)—Key.

HEDYOTIS

4084 Lewis, W. H.
Hedyotis correllii (Rubiaceae): a new Texas species.
3 *Brittonia*, 24(4):395–397(1972)—Illustrations.

4085 Rao, R. S.
The genus *Hedyotis* Linn. in Maharastra State.
Proc. Indian Sci. Congr., 60(3):338(1973). Abstract
6 only.

4086 Rao, R. S. *and* Hemadri, K.
The genus *Hedyotis* Linn. in Maharastra State.
6 *Indian Forester*, 99(6):372–379(1973)—Key.

anamalayana (Gamble) Rolla Rao & Hemadri in *Indian Forester*, 99(6):378(1973): *Oldenlandia*
6 *anamalayana.*
bourdillonii (Gamble) Rolla Rao & Hemadri *op. cit.* 378(1973), without basionym page: *Oldenlandia*
6 *bourdillonii.*
correllii W. H. Lewis in *Brittonia*, 24(4):395(1972)—
3 U.S.A. (Texas).
maheshwarii (Sant. & Merch.) Rolla Rao & Hemadri in *Indian Forester*, 99(6):376(1 June 1973), and in *Proc. Indian Sci. Congr.*, 60(3):338 (no exact date
6 given, 1973): *Oldenlandia maheshwarii.*
nagporensis (Brace ex Haines) Rolla Rao & Hemadri in *Indian Forester*, 99(6):377 (1 June 1973), and in *Proc. Indian Sci. Congr.*, 60(3):338 (no exact date
6 given, 1973): *Oldenlandia nagporensis.*
prainiana (W. A. Talbot) Rolla Rao & Hemadri in
6 *Indian Forester*, 99(6):378(1973): *Anotis prainiana.*
ramarowii (Gamble) Rolla Rao & Hemadri *op. cit.*
6 378(1973): *Oldenlandia ramarowii.*
retorsa (Boiss.) Rolla Rao & Hemadri *op. cit.* 378
6 (1973): *Oldenlandia retorsa.*
shuteri (Hook.f.) Rolla Rao & Hemadri *op. cit.* 378
6 (1973): *Oldenlandia shuteri.*
stocksii (Hook.f. & Thom.) Rolla Rao & Hemadri
6 *op. cit.* 378(1973): *Oldenlandia stocksii.*
stricta (L.) Rolla Rao & Hemadri *op. cit.* 378(1973):
6 *Oldenlandia stricta.*
villosastripulata (Gamble) Rolla Rao & Hemadri *op. cit.* 379(1973), without basionym page:
6 *Oldenlandia villosastripulata.*
wynaadensis (Gamble) Rolla Rao & Hemadri *op. cit.*
6 379(1973): *Oldenlandia wynaadensis.*

HOFFMANNIA

4087 Steyermark, J. A.
Novedades de los Andes Venezolanos. *Pittieria*,
4 no.4:9–25(1972)—Sp,En; illustrations.

4088 Williams, L. O.
Hoffmannias from Mexico and Central America.
4 *Fieldiana, Bot.*, 36(6):51–60(1973).
aroensis Steyermark in *Pittieria*, 4:17(1972)—
4 Venezuela.
4 **bernardii** Steyermark *op. cit.* 20(1972)—Venezuela.
var. **aracalensis** Steyermark *op. cit.* 20(1972)—
4 Venezuela.

bullata L. O. Williams in *Fieldiana, Bot.*, 36(6):52
(1973)—Mexico, Guatemala, British Honduras,
4 Honduras, Costa Rica.
larensis Steyermark in *Pittieria*, 4:20(1972)—
4 Venezuela.
liesnerana L. O. Williams in *Fieldiana, Bot.*, 36(6):
4 54(1973), as 'liesneriana'—Costa Rica.
montana L. O. Williams *op. cit.* 55(1973)—
4 Guatemala, Mexico.
nicotanifolia (Mart. & Gal.) L. O. Williams *op. cit.*
4 56(1973): *Psychotria nicotanifolia.*
4 **oreogena** L. O. Williams *op. cit.* 57(1973)—Honduras.
oreophila L. O. Williams *op. cit.* 57(1973)—
4 Nicaragua.
pauciflora
subsp. **venezuelensis** Steyermark in *Pittieria*, 4:25
4 (1972)—Venezuela.
sessilifolia L. O. Williams in *Phytologia*, 25(7):463
4 (1973)—Guatemala.
stenocarpa Steyermark in *Pittieria*, 4:21(1972)—
4 Venezuela.
terepaimensis Steyermark *op. cit.* 24(1972)—
4 Venezuela.
teruae L. O. Williams & Molina in *Fieldiana, Bot.*,
4 36(6):59(1973)—Guatemala.
triosteoides
subsp. **costensis** (Standl. & Steyermark)
4 Steyermark in *Pittieria*, 4:25(1972): *H. costensis.*

IXORA

micheliana J. G. Adam in *Bull. Inst. Fond. Afr.*
5 *Noire, A*, 38(1):85(1973)—Liberia.

KOTCHUBAEA

4089 Steyermark, J. A.
Una nueva especie de *Kotchubaea* (Rubiaceae) de la
Guayana venezolana. *Acta Bot. Venezuel.*, 9(1–4):
4 287–290(1974 publ. 1973)—Illustration; key.

morilloi Steyermark in *Acta Bot. Venezuel.*, 9(1–4):
4 287(1974 publ. 1973), preprint—Venezuela.

KOHAUTIA

4090 Sharma, V. S.
Kohautia nagporensis (Brace ex Haines) Sant. et
Merch.—an imperfectly known species from Upper
Gangetic Plain. *Indian Forester*, 99(3):159–162(1973)—
6 Illustration.

LEROYA

4091 Cavaco, A.
A new species of *Leroya* (Rubiaceae—Vanguerieae)
from Madagascar. *Kew Bull.*, 28(3):349–350(1973
5 publ. 1974).

richardiae Cavaco in *Kew Bull.*, 28(3):349(1973 publ.
5 1974)—Malagasy.

MASCHALODESME

4092 Ridsdale, C. E., Bakhuizen van den Brink, R. C. *and*
Koek-Noorman, J.
Notes on New Guinea Rubiaceae: *Versteegia* and
Maschalodesme. *Blumea*, 20(2):339–348(1972 publ.
6 1973)—Keys.

versteegii (Val.) Ridsdale in *Blumea*, 20(2):344(1972
6 publ. 1973): *Tricalysia ? versteegia.*

MITRATHECA

4093 Verdcourt, B.
The identity of *Mitratheca richardsonioides*
(Rubiaceae). *Kew Bull.*, 28(3):419–421(1973 publ.
5 1974). Synonym of *Oldenlandia* L.

NAUCLEA

4094 McVaugh, R.
Report of the Committee for Spermatophyta:
conservation of generic names: proposal 307 (8231)
Nauclea Merrill 1915 vs. *Nauclea* Linnaeus (1762).
2, 5, 6 *Taxon*, 22(1):154(1973).

4095 McVaugh, R.
Report of the Committee for Spermatophyta:
conservation of generic names: proposal 308 (8226)
Adina Salisbury (1807) vs. *Nauclea* Linnaeus (1762).
2, 5, 6 *Taxon*, 22(1):155(1973).

NEOGAILLONIA Linchevskiĭ in *Novosti Sist.*
Vȳssh. Rast., 10:226(1973), nom. nov.: *Gaillonia*
2, 5, 6 A. Rich. ex DC.
2 sect. **Bucharia** Linchevskiĭ *op. cit.* 230(1973).
2, 5, 6 sect. **Eriantha** Linchevskiĭ *op. cit.* 229(1973).
2 sect. **Surchania** Linchevskiĭ *op. cit.* 232(1973).
afghanica (Ehrendf.) Linchevskiĭ *op. cit.* 230(1973):
2 *Gaillonia afghanica.*
asperuliformis (Linchevskiĭ) Linchevskiĭ *op. cit.*
2, 6 230(1973): *Gaillonia asperuliformis.*
botschantzevii Linchevskiĭ *op. cit.* 232(1973)—
2 U.S.S.R. (Middle Asia).
bruguierei (A. Rich. ex DC.) Linchevskiĭ *op. cit.*
2 226(1973), as 'bruguierii': *Gaillonia bruguierii.*
bucharica (B. Fedtsh. & Desjat.) Linchevskiĭ *op. cit.*
2 230(1973): *Gaillonia bucharica.*
dubia (Aitch. & Hemsl.) Linchevskiĭ *op. cit.* 230
2 (1973): *Gaillonia dubia.*
eriantha (Jaub. & Spach) Linchevskiĭ *op. cit.* 229
2 (1973): *Gaillonia eriantha.*
iljinii Linchevskiĭ *op. cit.* 227(1973)—U.S.S.R.
2 (Middle Asia).
incana (Jaub. & Spach) Linchevskiĭ *op. cit.* 227(1973):
2 *Gaillonia incana.*
inopinata (Linchevskiĭ) Linchevskiĭ *op. cit.* 232(1973):
2 *Gaillonia inopinata.*
kerstanii (Ehrendf.) Linchevskiĭ *op. cit.* 230(1973):
2 *Gaillonia kerstanii.*
macrantha (Blatt. & Hallb.) Linchevskiĭ *op. cit.* 230
6 (1973): *Gaillonia macrantha.*
mestscherjakovii Lincheveskiĭ *op. cit.* 230(1973)—
2 U.S.S.R. (Middle Asia).
olivieri (A. Rich. ex DC.) Linchevskiĭ *op. cit.* 226
2 (1973): *Gaillonia olivieri.*
puberula (Balf. f.) Linchevskiĭ *op. cit.* 229(1973):
2, 5 *Gaillonia puberula.*
pulchella (Podlech) Linchevskiĭ *op. cit.* 230(1973):
2 *Gaillonia pulchella.*
putorioides (A. R. Smith) Linchevskiĭ *op. cit.* 230
5 (1973): *Jaubertia putorioides.*
szovitsii (DC.) Linchevskiĭ *op. cit.* 226(1973):
2 *Gaillonia szowitzii.*
thymoides (Balf. f.) Linchevskiĭ *op. cit.* 230(1973):
2, 5 *Gaillonia thymoides.*
tinctoria (Balf. f.) Linchevskiĭ *op. cit.* 230(1973):
5 *Gaillonia tinctoria.*
trichophylla (M. Pop. ex Tscherneva) Linchevskiĭ
2 *op. cit.* 231(1973): *Gaillonia trichophylla.*
vassilczenkoi Linchevskiĭ *op.cit.* 228(1973)—U.S.S.R.
2 (Middle Asia).

OCTODON

4096 Andronova, N. N.
K émbriologii *Octodon setosum* Hiern. (Rubiaceae).
(To the embryology of *Octodon setosum* Hiern.
5 (Rubiaceae)). *Bot. Zhurn.*, 58(9):1371–1377(1973).

OLDENLANDIA

4097 Shivaramaiah, G. *and* Rajan, S. S.
A contribution to the embryology of *Oldenlandia umbellata* Linn. *Proc. Indian Acad. Sci., B*, 77(1):19–24
5, 6 (1973)—Illustrations.

4098 Verdcourt, B.
The identity of *Mitratheca richardsonioides* (Rubiaceae). *Kew Bull.*, 28(3):419–421(1973 publ.
5 1974). Synonym of *Oldenlandia* L.

chiovendae (Bremek.) Verdcourt in *Kew Bull.*, 28(3):421(1973 publ. 1974), as 'chiovendii':
5 *Eionitis chiovendii*.
richardsonioides (K. Schum.) Verdcourt *op. cit.*
5 420(1973 publ. 1974): *Mitratheca richardsonioides*.
var. **gracilis** Verdcourt *op. cit.* 421(1973 publ.
5 1974)—Somali Republic/Kenya border.
var. **hirtella** (Chiov.) Verdcourt *op. cit.* 421(1973
5 publ. 1974): *O. psammophila* f. *hirtella*.
var. **laxiflora** (Bremek.) Verdcourt *op. cit.* 420 (1973 publ. 1974): *Eionitis psammophila* var.
5 *laxiflora*.

OTIOPHORA

4099 Verdcourt, B.
A description of two new species of *Otiophora* Zucc. (Rubiaceae—Anthospermeae) and a revised key to the genus. *Garcia de Orta, Ser. Bot.*, 1(1–2):
5 25–29(1973)—(Por,Fr); illustrations; key.

angustifolia Verdcourt in *Garcia de Orta, Sér. Bot.*,
5 1(1–2):27(1973)—Zambia.
lanceolata Verdcourt *op. cit.* 28(1973)—Mozambique,
5 Rhodesia.
pauciflora Bak.
subsp. **burttii** (Milne-Redh.) Verdcourt *op. cit.*
5 29(1973): *O. burttii*.

PAURIDIANTHA

4100 Bouquet, A.
Plantes médicinales du Congo-Brazzaville, *Uvariopsis, Pauridiantha, Diospyros. Trav. Lab. Matière Méd. Pharm. Galénique Fac. Pharm. Paris*,
5 57(1). 112p.(1973).

camposii (G. Taylor) Exell in *Bull. Brit. Mus. (Nat.*
5 *Hist.), Bot.*, 4(8):375(1973): *Urophyllum camposii*.

PAVETTA

micheliana J. G. Adam in *Bull. Inst. Fond. Afr. Noire,*
5 *A*, 35(1):87(1973)—Liberia, Guinea, Ivory Coast.

PSEUDOGAILLONIA Linchevskiĭ in *Novosti Sist.*
2, 6 *Vȳssh. Rast.*, 10:235(1973).
hymenostephana (Jaub. & Spach) Linchevskiĭ *op.*
2, 6 *cit.* 236(1973): *Gaillonia hymenostephana*.

PSYCHOTRIA

4101 Petit, E.
Rubiaceae: 123–154. *Distrib. Pl. Afr.*, 5. 32p.
5 (1973)—Dot maps.

4102 Petit, E.
Rubiaceae: 155–178. *Distrib. Pl. Afr.*, 6. 24p.(1973)—
5 Dot maps.

4103 Verdcourt, B.
Tropical African plants: 33. Notes on *Geophila, Psychotria* and *Rubia* (Rubiaceae). *Kew Bull.*, 28(2):320–
5 321(1973)—Illustrations.

hierniana Exell in *Bull. Brit. Mus. (Nat. Hist.), Bot.*, 4(8):376(1973), nom. nov.: *Grumilea macrocarpa*
5 Hiern.
kirkii Hiern
var. **hirtella** (Oliv.) Verdcourt in *Kew Bull.*, 28(2):
5 321(1973): *P. hirtella*.
micheliana J. G. Adam in *Bull. Inst. Fond. Afr.*
5 *Noire, A*, 35(1):89(1973)—Liberia.
nervosa
var. **rufescens** (H.B.K.) L. O. Williams in
4 *Phytologia*, 26(6):493(1973): *P. rufescens*.
perijaënsis Steyermark in *Acta Bot. Venezuel.*, 8(1–4):
4 250(1973)—Venezuela.

PTEROGAILLONIA Linchevskiĭ in *Novosti Sist.*
2 *Vȳssh. Rast.*, 10:233(1973).
2 sect. **Bicornia** Linchevskiĭ *op. cit.* 235(1973).
calycoptera (Decne.) Linchevskiĭ *op. cit.* 233(1973):
2 *Spermacoce calycoptera*.
crucianelloides (Jaub. & Spach) Linchevskiĭ *op. cit.*
2 235(1973): *Gaillonia crucianelloides*.
stscherbinovskii Linchevskiĭ *op. cit.* 234(1973)—
2 Iran.

RACHICALLIS

4104 Nickerson, N. H. *and* Tripp, J. W.
Floral dimorphism in *Rachicallis americana* (Jacq.) Hitch. (saltwater-bush). *Rhodora*, 75(801):111–113
3 (1973).

RELBUNIUM

4105 Dempster, L. T.
Transfer of *Relbunium sphagnophilum* (Rubiaceae) and *R. mazocarpum* to *Galium*. *Phytologia*, 26(4):221–
4 223(1973)—Illustrations, dot map.

RICHARDIA

4106 Bahadur, K. N. *and* Naithani, H. B.
New distributional records of *Richardia scabra* Linn. in India. *Indian Forester*, 99(7):449–453(1973)—
6 Illustrations.

RONDELETIA

4107 Williams, L.O.
Tropical American plants: 14. *Rondeletiae* Centro-
4 Americanae. *Phytologia*, 26(2):127–130(1973).

myriantha Standl. & Steyerm.
var. **armentalis** L. O. Williams in *Phytologia*,
4 26(2):127(1973)—Guatemala.
4 **rubens** L. O. Williams *op. cit.* 128(1973)—Guatemala.
silvicola L. O. Williams *op. cit.* 128(1973)—
4 Honduras, Guatemala.
urophylla Standl. & L. O. Williams *op. cit.* 129
4 (1973)—Costa Rica.

RUBIA

4108 Dempster, L. T.
Note on the type of *Rubia acuminata* Martens and
4 Galeotti. *Brittonia*, 25(3):313(1973).

4109 Verdcourt, B.
Tropical African plants: 33. Notes on *Geophila, Psychotria* and *Rubia* (Rubiaceae). *Kew Bull.*, 28(2):
5 320–321(1973)—Illustrations.

4110 Voggenreiter, V.
Bermerkungen zur Pflanzengeographie von *Rubia fruticosa* ssp. *melanocarpa* auf Gran Canaria und Tenerife. *Cuad. Bot. Canar.*, 18–19:33–37(1973)—
1 (Sp,En); dot maps.

SCOLOSANTHUS

subsessilis Alain in *Phytologia*, 25(5):278(1973)—
4 Hispaniola (Dominican Republic).

SERISSA

4111 Kimura, K., Shimada, H., Nomura, S., Hisada, Y. *and* Tanaka, T.
(Variation of the flower of *Serissa japonica* (Thunb.) Thunb. (Rubiaceae)). *J. Jap. Bot.*, 48(4):117–121
2 (1973)—Ja,(En); illustration.

4112 Rao, V. S. *and* Gupte, K.
An interesting feature of the stipule bundles of *Serissa foetida*. *Curr. Sci.*, 42(16):580(1973)—
6 Illustration.

SIMIRA

lezamae Steyermark in *Acta Bot. Venezuel.*, 8(1–4):
4 249(1973)—Venezuela.

SOMMERA

4113 Williams, L. O.
Sommera (Rubiaceae) in North America. *Phytologia*,
3, 4 26(2):121–126(1973).

dunlapii L. O. Williams in *Phytologia*, 26(2):122
4 (1973)—Panama.
montana L. O. Williams *op. cit.* 124(1973)—
4 Honduras.
rivularis L. O. Williams *op. cit.* 125(1973)—Costa
4 Rica.
4 **zygocalyx** L. O. Williams *op. cit.* 126(1973)—Mexico.

SPERMACOCE

4114 Amaratunga, K. L. D.
Spermacoce latifolia Aublet, a new record for Sri Lanka. *Ceylon J. Sci., Biol. Sci.*, 10(2):155–156(1973)—
6 Illustration.

perijaënsis Steyermark in *Acta Bot. Venezuel.*, 8(1–4):
4 252(1973)—Venezuela.

VERSTEEGIA

4115 Ridsdale, C. E., Bakhuizen van den Brink, R. C. *and* Koek-Noorman, J.
Notes on New Guinea Rubiaceae: *Versteegia* and *Maschalodesme*. *Blumea*, 20(2):339–348(1972 publ.
6 1973)—Keys.

puberula Ridsdale in *Blumea*, 20(2):340(1972 publ.
6 1973)—New Guinea (W. Irian).
solomonensis Ridsdale *op. cit.* 340(1972 publ. 1973)—
6 British Solomon Is., Bougainville Is.

RUPPIACEAE

4116 Ba, L. T., Mestre, J.-C. *and* Guignard, J.-L.
Embryogénie des Ruppiacées. Développement de l'embryon chez le *Ruppia maritima* L. *Compt. R. Hebd. Séances Acad. Sci., D* (Paris), 276(5):737–740
1–3 (1973)—Illustrations.

RUTACEAE

4117 Albuquerque, B. W. P. de *and* Honda, M. *
Nota prévia sobre Rutaceae nova da Amazonia. *Acta Amazonica*, 1(1):24(1971 publ. 1972). Ref. in *Bull.*
4 *Torrey Bot. Club*, 99(6):324(1972).

4118 Albuquerque, B. W. P. de *and* Honda, M.
Rutaceae nova de Amazônia. *Acta Amazonica*,
4 2(2):49–54(1972)—(En); illustrations.

4119 Andon, T. M. *
(Particularités du développement de *Ruta graveolens* en Moldavie). *Izv. Akad. Nauk Moldav. SSR, Biol. Khim. Nauk*, no.1:9–13(1973)—Rus. Ref. in *Bull. Signal., Biol. Physiol. Vég.*, 34(10):9526
1 (1973).

4120 Borhidí, A., Acuña, J. *and* Muñiz, O.
New plants in Cuba: 3. *Acta Bot. Acad. Sci. Hung.*, 19(1–4):37–45(1973)—Illustrations. Rutaceae pp.
4 42–45.

4121 Calabrese, F.
Considerazioni sulla classificazione botanica delle Aurantioideae (Rutaceae). *Webbia*, 28(1):161–187
1, 2, 6 (1973)—(En).

4122 Guédès, M.
Carpel morphology and axis-sharing in syncarpy in some Rutaceae, with further comments on 'New
1–4, 6 Morphology'. *Bot. J. Linn. Soc.*, 66(1):55–74(1973).

4123 Mendonça, F. A. *and* Sousa, E. P.
Contribuição para o conhecimento das Rutaceae de Moçambique. *Garcia de Orta, Ser. Bot.*, 1(1–2):
5 95–96(1973)—(En).

4124 Mendonça, F. A. *and* Sousa, E. P.
Flora de Moçambique: 37. Rutaceae. 1973. 29p.—
5 Keys.

4125 Prance, G. T.
Flora neotropica monograph no.11. Rhabdodendraceae. New York, Hafner, 1972. 23p.—Illustrations,
4 portrait, dot maps; key.

4126 Waterman, P. G.
Alkaloids and triterpenes from the African
5 Toddaliodeae. *Biochem. Syst.*, 1(3):153–161(1973).

AGATHOSMA

4127 Williams, I.
Studies in the genera of the Diosmeae (Rutaceae): 1. *Agathosma rosmarinifolia* (Bartl.) Williams comb. nov. *J. S. Afr. Bot.*, 39(2):179–183(1973)—(Af); illustra-
5 tions, map, dot map.

rosmarinifolia (Bartl.) I. J. M. Williams in *J. S. Afr. Bot.*, 39(2):181(1973): *Acmadenia*
5 *rosmarinifolia*.

AMYRIS

stromatophylla P. Wils.
subsp. **moaënsis** Borhidi & Muñiz in *Acta Bot.*
4 *Acad. Sci. Hung.*, 18(1–2):30(1973)—Cuba.
subsp. **yumuriensis** Borhidi & Muñiz *op. cit.* 29
4 (1973)—Cuba.

BONINIA

4128 Yamazaki, T.
(Generic relationships of the genus *Boninia* (Rutaceae)). *J. Jap. Bot.*, 48(6):190–192(1973)—Ja,En;
2 illustrations.

BORONIA

4129 Lebler, B. A.
A long-lost *Boronia*. *Queensl. Agric. J.*, 99(12):618–
7 620(1973)—Illustration.

CITRUS

4130 Hérisset, A., Jolivet, J., Rey, P. *and* Lavault, M.
Différenciation de quelques huiles essentielles présentant une constitution voisine (en particulier par l'examen de leurs spectres UV, IR et Raman): 10. Essences de divers *Citrus*. *Pl. Méd. Phytothér.*,
2, 6 7(4):306–318(1973).

4131 Stone, B. C., Lowry, J. B., Scora, R. W. *and* Jong, K.
Citrus halimii: a new species from Malaya and Peninsular Thailand. *Biotropica*, 5(2):102–110(1973)—
6 Illustrations, dot map; chromosome number.

halimii Stone in *Biotropica*, 5(2):106(1973)—Malaya,
6 Thailand.
rangpuriensis Traub in *Pl. Life*, 29:142(1973)–
3 Cultivation in U.S.A. (California).

EUCHAETIS

4132 Williams, I.
Studies in the genera of the Diosmeae (Rutaceae): 2. *Euchaetis glabra* Williams sp. nov. *J. S. Afr. Bot.*,
5 39(3):229–232(1973)—(Af); dot map.

glabra I. J. M. Williams in *J. S. Afr. Bot.*, 39(3):229
5 (1973)—South Africa (Cape Prov.).

FAGARA

4133 Albuquerque, B. W. P. de *
Contribuição ao estudo da nervação foliar de plantas da flora Amazonica: 2. *Fagara prancei* Albuq. (Rutaceae). *Acta Amazonica*, 1(1):11–13(1971 publ. 1972). Ref. in *Bull. Torrey Bot. Club*, 99(6):330
4 (1972).

4134 Fish, F. *and* Waterman, P. G.
Chemosystematics in the Rutaceae: 2. The chemosystematics of the *Zanthoxylum/Fagara* complex.
2, 4–7 *Taxon*, 22(2–3):177–203(1973).

4135 Killick, D. J. B.
Fagara capensis. *Flow. Pl. Afr.*, 42(3–4):pl.1671
5 (1973)—En; illustration, coloured illustration.

HAPLOPHYLLUM

4136 Bacon, L.
Plant portraits: *Haplophyllum boissieranum* Vis. et Pancić, syn. *H. albanicum* (Bald.) Bornm. *Quart. Bull.*
1 *Alp. Gard. Soc.*, 41(2):160–162(1973)—Illustration.

KODALYODENDRON Borhidi & Acuña in
4 *Acta Bot. Acad. Sci. Hung.*, 19(1–4):42(1973).
cubensis Borhidi & Acuña *op. cit.* 43(1973)—
4 Cuba.

PHELLODENDRON

4137 Starshova, N. P. *and* Solntseva, M. P.
K kharakteristike zarodȳshevogo meshka *Phellodendron amurense* Rupr. (To the characteristic of the embryo sac of *Phellodendron amurense* Rupr.). *Bot.*
2 *Zhurn.*, 58(11):1644–1654(1973).

RAVENIA

4138 Ghosh, R. B.
Some preliminary observations on the anatomy of an ornamental plant *Ravenia spectabilis* Engl. as an aid to its identity. *Broteria, Sci. Nat.*, 41(3–4):139–147
4 (1972).

SPATHELIA

4139 Adams, C. D., Taylor, D. R. *and* Warner, J. M.
N-methylflindersine from *Spathelia sorbifolia*.
4 *Phytochemistry*, 12(6):1359–1360(1973).

TECLEA

4140 Exell, A. W. *and* Mendonça, F. A.
Duas espécies novas de *Teclea* (Rutaceae). *Garcia de Orta, Ser. Bot.*, 1(1–2):93–94(1973)—(En);
5 illustrations.

macedoi Exell & Mendonça in *Garcia de Orta, Sér.*
5 *Bot.*, 1(1–2):93(1973)—Mozambique.
myrei Exell & Mendonça *op. cit.* 93(1973)—
5 Mozambique.

ZANTHOXYLUM

4141 Fish, F. *and* Waterman, P. G.
Chemosystematics in the Rutaceae: 2. The chemosystematics of the *Zanthoxylum/Fagara* complex.
2, 4–7 *Taxon*, 22(2–3):177–203(1973).

SABIACEAE

MELIOSMA

4142 Hatusima, S.
(*Meliosma oldhami* Maxim. and its alliance). *J.*
2, 6 *Geobot.*, 20(3):56–58(1972)—Ja,En.

SALICACEAE

4143 Golȳsheva, M. D.
Gipoderma v embrional'nom i vzroslom liste i znachenie étogo priznaka pri reshenii nekotorȳkh voprosov évolyutsii i sistematiki v Sem. Salicaceae i drugikh taksonakh. (The hypodermis of embryonal and adult leaves and the importance of this character in some problems of evolution and taxonomy in the family Salicaceae and other taxa). *Byull. Mosk. Obshch. Ispȳt. Prir., Biol.*, 78(5):93–106(1973)—
1–7 (En).

4144 Mehra, P. N. *and* Sareen, T. S.
Cytology of West Himalayan Betulaceae and Salicaceae. *J. Arnold Arbor.*, 54(3):412–418(1973)—
6 Chromosome numbers.

CHOSENIA

4145 Golȳsheva, M. D.
O nekotorȳkh chertakh stroeniya i formirovaniya lista *Chosenia arbutifolia* (Pall.) A. Skvorts. (On some features of structure and formation of leaves in *Chosenia arbutifolia* (Pall.) A. Skvorts.). *Bot. Zhurn.*,
1–3 58(12):1764–1774(1973).

POPULUS

4146 Allegri, E.
Sul riconoscimento delle species e varietà dei pioppi indigeni in Italia. (Identification of indigenous poplar species and varieties in Italy). *Ann. Inst. Sper. Selvic*, 2:3–62(1971 publ. 1973)—(Fr,En);
1 illustrations.

4147 Jones, J. R. *and* Markstrom, D. C.
Aspen: an American wood. *U.S. Dep. Agric. F.S.*, 217. 8p.(1973)—Illustrations, maps. *Populus tremuloides*
3 Michx. and *P. grandidentata* Michx.

4148 Milne-Redhead, E.
The Black Poplar Survey. *B.S.B.I. News*, 2(2):
1 32–33(1973)—*Populus nigra.*

4149 Milne-Redhead, E.
Rescuing the black poplar in Britain. *Biol. Conserv.*,
1 5(3):232–233(1973). *Populus nigra.*

4150 Müller, R. *and* Sauer, E.
Urteilsgrundlagen für die trichocarpa Pappel:
Versuch einer Analyse auch der bisherigen Pappel-
wirtschaft. Strassenhaus, Wirtschafts-und Forstverlag
Euting KG, 1972. 150p.—Illustrations. Review in *Allg.*
3 *Forst-Zeitschr.* 28(19):461(1973); *Populus trichocarpa.*

4151 Ronald, W. G., Lenz, L. M. *and* Cumming, W. A.
Biosystematics of the genus *Populus* L.: 1.
Distribution and morphology of native Manitoba
species and variants. *Canad. J. Bot.*, 51(12):2431–
3 2442(1973)—Illustrations, dot map; keys.

4152 Ronald, W. G., Lenz, L. M. *and* Cumming, W. A.
Biosystematics of the genus *Populus* L.: 2. Chemo-
taxonomy, phenology, fertility and segregates of
native Manitoba species and variants. *Canad. J. Bot.*,
3 51(12):2443–2450(1973)—Illustrations.

4153 Usbekistan: Akademiya Nauk Usbekskoĭ SSR:
Botanicheskiĭ Sada
Dendrologiya Usbekistana: vol. 3: Topol', by
A. U. Usmanov. Tashkent, Nauka, 1971. 262p.—
1–3 Illustrations, map.

SALIX

4154 Brayshaw, T. C.
The sandbar willow on Vancouver Island. *Syesis*,
3 6:147–152(1973)—Illustrations. *Salix sessilifolia.*

4155 Bridle, P., Stott, K. G. *and* Timberlake, C. F.
Anthocyanins in *Salix* species: a new anthocyanin in
Salix purpurea bark. *Phytochemistry*, 12(5):1103–1106
1, 2 (1973).

4156 Kimura, A.
Salicis nova species ex regione Okutonensi in
Japonia. *J. Jap. Bot.*, 48(11):321–326(1973)—(Ja);
2 illustrations.

4157 McCleod, K. W. *and* McPherson, J. K.
Factors limiting the distribution of *Salix nigra*. *Bull.*
Torrey Bot. Club, 100(2):102–110(1973)—
3 Illustrations.

4158 Malyutina, E. T.
O prichinakh izmeneniya pola u *Salix triandra* f.
quadrivalvis. *Byull. Glavn. Bot. Sada* (Moscow),
1 88:59–61(1973)—Illustrations.

4159 Rejewski, M.
Nowe stanowiska wierzby borówkolistnej
(*Salix myrtilloides* L.) w powiecie bydgoskim. (Neue
Standorte der Heidelbeer-Weide (*Salix myrtilloides*
L.) im Kreis Bydgoszcz). *Zesz. Nauk. Uniw. Mikol.*
Kopernika Toruniu, Biol., 30:41–44(1972)—(Ge);
1 maps.

4160 Tikhmenev, E. A.
K antékologii nekotorȳkh vidov *Salix* v arkti-
cheskoĭ tundre o. Vrangelya. (On the anthecology of
some *Salix* species in arctic tundra (Wrangel Island)).
2 *Bot. Zhurn.*, 58(8):1209–1216(1973).

cascadensis Cockerell
var. **thompsonii** Brayshaw in *Syesis*, 6:147(1973)—
3 Canada (British Columbia).
hukaoana A. Kimura in *J. Jap. Bot.*, 48(11):321
2 (1973)—Japan.
sessilifolia Nuttall
var. **vancouverensis** Brayshaw in *Syesis*, 6:147
3 (1973)—Canada (British Columbia).

SALVADORACEAE

4161 Diniz, M. A.
Notas sobre as Salvadoraceae de Moçambique.
Rev. Ciênc. Biol. (Lourenço Marques), 5:67–77
5 (1972)—(En); keys.

4162 Qureshi, S.
Flora of West Pakistan: no.29. Salvadoraceae,
6 1972. 4p.—Illustrations, map.

SALVADORA

4163 Evenari, M. *and* Gutterman, Y.
Some notes on *Salvadora persica* L. in Sinai and its
use as a toothbrush. *Flora* (Germany), 162(1–2):
2 118–125(1973)—(Ge); illustrations, dot map.

SAMBUCACEAE

SAMBUCUS

4164 Conrad, P. W. *and* McDonaugh, W. T.
Growth and reproduction of red elderberry on
subalpine rangeland in Utah. *Northwest Sci.* (U.S.A.),
46(2):140–148(1972)—Illustrations. *Sambucus*
3 *racemosa.*

4165 Helebrant, L.
Laciniate Kulturvarietäten der Gattung *Sambucus.*
1–3 *Mitt. Deutsch. Dendrol. Ges.*, no.66:173–175(1973).

4166 Singh, G.
On Indian *Sambucus ebulus* (sensu C. B. Clarke).
Indian Forester, 99(4):235–237(1973). Nomenclatural
6 note.

SANTALACEAE

4167 Lawalrée, A.
Santalaceae. Flore du Gabon, 21:63–68(1973)—
5 Illustrations; key.

THESIUM

4168 Delay, J. *and* Petit, D.
Matériel chromatique singulier: 2. Une
Santalacée. *Inform. Ann. Caryosyst. Cytogénét.*, no.6:
1 16(1972 publ. 1973)—Chromosome numbers.

SAPINDACEAE

4169 Abdulla, D.
Flora of West Pakistan: no.39. Sapindaceae. 1973.
6 10p.—Illustrations, map; keys.

4170 Fernandes, A.
Flora de Moambique: 51. Sapindaceae. 1973.
5 48p.—Keys.

4171 Fouilloy, R. *and* Hallé, N.
Sapindaceae. Flore du Cameroun, 16. 202p.
(1973)—Illustrations; keys. Also published as Flore
5 du Gabon, 23.(1973).

4172 Fouilloy, R. *and* Hallé, N.
Sapindaceae. Flore du Gabon, 23. 202p.(1973)—
Illustrations; keys. Also published as Flore du
5 Cameroun, 16.(1973).

4173 Sousa, E. P.
Contribuição para o conhecimento das Sapindaceae
de Moçambique. *Garcia de Orta, Sér. Bot.*, 1(1–2):
5 99–100(1973)—(En).

ALLOPHYLUS

4174 Exell, A. W. *and* Mendonça, F. A.
A new species of *Allophylus* (Sapindaceae). *Garcia*
5 *de Orta, Sér. Bot.*, 1(1–2):97(1973)—(Por).

hallaei Fouilloy in Fl. Cameroun, 16:32 (Dec. 10
1973); and in Fl. Gabon, 23:32(Dec. 10 1973)—
5 Gabon.
torrei Exell & Mendonça in *Garcia de Orta, Sér. Bot.*,
5 1(1–2):97(1973)—Mozambique.

CHYTRANTHUS

4175 Hallé, N.
Chytranthus prieurianus Baillon (Sapind.) et le
problème de son origine. *Bull. Mus. Nation. Hist. Nat.*,
(Paris) 3e Sér., no.127 *Bot.* 7: 11–19(1973)—(En);
5 illustrations.

sect. **Chytrolepis** N. Hallé in Fl. Cameroun, 16:85
(Dec. 10 1973); and in Fl. Gabon, 23:85(Dec. 10
5 1973).
prieurianus Baillon
subsp. **longiflorus** (Verdcourt) N. Hallé in *Bull.*
Mus. Nation. Hist. Nat. (Paris), 3e sér., no. 127,
5 *Bot.* 7:19(1973): *C. sacleuxii* subsp. *longiflorus*.

CROSSONEPHELIS

4176 Fouilloy, R.
Trois arbres nouveaux d'Afrique tropicale (Laur.-
Myrist.-Sapind.) *Adansonia*, 12(4):545–555(1972 publ.
1973)—Illustrations; key to 5 spp. *Crossonephelis*
5 *adami*.

4177 Leenhouts, P. W.
A revision of *Crossonephelis* (Sapindaceae). *Blumea*,
5, 6 21(1):91–103(1973)—Key.

4178 Muller, J.
Pollen morphology of the genus *Crossonephelis*
5, 6 (Sapindaceae). *Blumea*, 21(1):105–117(1973).

adamii Fouilloy in *Adansonia*, 12(4):551(1972 publ.
5 1973), as 'adami'—Liberia.
africanus (Radlk.) Leenhouts in *Blumea*, 21(1):95
5 (1973): *Melanodiscus africanus*.
palawanicus (Radlk.) Leenhouts *op. cit.* 97(1973):
6 *Lepisanthes palawanica*.
penangensis (Ridl.) Leenhouts *op. cit.* 98(1973):
6 *Tristira penangensis*.
philippinensis (Radlk.) Leenhouts *op. cit.* 100(1973):
6 *Hedyachras philippinensis*.
thorelii (Pierre) Leenhouts *op. cit.* 101(1973):
6 *Cnemidiscus thorelii*.
unijugatus (Pellegr.) Leenhouts *op. cit.* 102(1973):
5 *Melanodiscus unijugatus*.

DIMOCARPUS

4179 Leenhouts, P. W.
A new species of *Dimocarpus* (Sapindaceae) from
Australia. *Blumea*, 21(2):377–380(1973)—Illustrations,
7 dot map.

australianus Leenhouts in *Blumea*, 21(2):377(1973)—
7 Australia (Queensland).

LYCHNODISCUS

brevibracteatus Fouilloy in Fl. Cameroun, 16:166
(Dec. 10 1973); and in Fl. Gabon, 23:166(Dec. 10
5 1973)—Gabon.

PAULLINIA

4180 Van der Walt, J. J. A., Van der Schijff, H. P. *and*
Schweickerdt, H. G.
Anomalous secondary growth in the stems of the
lianes *Mikania cordata* (Burm. f.) Robins. (Compositae)
and *Paullinia pinnata* Linn. (Sapindaceae). *Kirkia*,
4 9(1):109–138(1973)—Illustrations.

THINOUIA

4181 Fabris, H. A.
Especies nuevas o criticas de la flora jujeña: 4. *Bol.*
Soc. Argent. Bot., 14(4):369–373(1972)—(En);
4 illustration. *Thinouia repanda*.

THOUINIA

elliptica Radlk.
var. **rotundata** Borhidi & Muñiz in *Acta Bot.*
4 *Acad. Sci. Hung.*, 18(1–2):37(1973)—Cuba.

SAPOTACEAE

4182 Aubréville, A.
Géophylétique florale des Sapotacées. *Adansonia*,
1–7 13(3):255–271(1973).

4183 Aubréville, A.
Les Sapotacées de l'Ile Maurice. *Adansonia*, 13(2):
5 135–143(1973)—Illustrations; keys.

4184 Soukup, J.
Las Ebenáceas, Estiracáceas y Sapotáceas del
Perú, sus géneros y lista de especies (continuación).
4 *Biota*, 9(74):265–271(1973).

CHRYSOPHYLLUM

collinum Little in *Phytologia*, 25(2):60(1972)—
4 Ecuador.

SIDEROXYLON

annithomae Aubréville in *Adansonia*, 13(2):138
5 (1973)—Réunion.
sessiliflorum (Poir.) Capuron ex Aubréville *op. cit.*
5 142(1973): *Achras sessiliflora*.

SARCOLAENACEAE

PENTACHLAENA

4185 Capuron, R.
Un *Pentachlaena* (Sarcolaenacées) nouveau.
Adansonia, 13(3):289–293(1973)—Illustration.
5 Introduction by J.-F. Leroy.

orientalis Capuron in *Adansonia*, 13(3):289(1973)—
5 Malagasy.

SARRACENIACEAE

4186 Mazrimas, J. A.
Darlingtonia vs. *Chrysamphora*. *Carniv. Pl. Newsl.*,
3 2(3):47(1973)—En.

DARLINGTONIA

4187 Ziemer, R. R.
Some field observations of *Darlingtonia* and
3 *Pinguicula. Carniv. Pl. Newsl.*, 2(2):25–27(1973).

HELIAMPHORA

4188 Kondo, K.
The chromosome numbers of four species of
carnivorous plants. *Phyton* (Argentina), 31(2):93–94
4 (1973). *Heliamphora minor.*

SARRACENIA

4189 Cody, W. J. *and* Talbot, S. S.
The pitcher plant *Sarracenia purpurea* L. in the
Northwestern part of its range. *Canad. Field-Nat.*,
3 87(3):318–320(1973)—Dot map.

4190 Marmelstein, A. D.
Field observations on *Sarracenia. Carniv. Pl. Newsl.*,
3 2(1):5–6(1973).

SAXIFRAGACEAE

4191 Kloper, K.
Florale Morphogenese und Taxonomie der
Saxifragaceae sensu lato. *Feddes Repert.*, 84(7–8):
1–7 475–516(1973)—(En); chromosome numbers.

4192 Pastre, A. *and* Pons, A.
Quelques aspects de la systématique des
Saxifragacées à la lumière des données de la
palynologie. *Pollen Spores*, 15(1):117–133(1973)—
1–6 (En).

4193 Rampi, A.
Adumbratio Florae Aethiopicae: 25. Saxifragaceae.
Webbia, 28(2):521–542(1973)—Fr(It,En); illustrations,
5 dot maps; key.

ASTILBE

4194 Murray-Lyon, D. M.
Plant portraits: *Astilbe* × *crispa* 'Perkeo'. *Quart.
Bull. Alp. Gard. Soc.*, 41(2):158,163(1973)—
1 Illustration.

BERGENIA

4195 Korobova-Semenchenko, L. V.
K embriologii semeĭstva kamnelomkovȳkh: 5.
Mikrosporogenez i razvitie muzhskogo gametofita
u *Bergenia crassifolia* (L.) Fritsch. (The embryology of
Saxifragaceae: 5. Microsporogenesis and male
gametophyte development in *Bergenia crassifolia*).
Vestn. Mosk. Univ. Nauch. Zhurn., *6*, 28(3):58–63
2 (1973)—(En).

CHRYSOSPLENIUM

saxatile Khokhryakov in *Byull. Glavn. Bot. Sada*
(Moscow), 88:45(1973)—U.S.S.R. (Soviet Far
2 East).

DEUTZIA *see* **HYDRANGEACEAE**

DICHROA *see* **HYDRANGEACEAE**

MITELLA

4196 Savile, D. B. O.
Vegetative distinctions in Canadian species of
Mitella and *Tiarella. Canad. Field-Nat.*, 87(4):460–
3 462(1973)—Illustration.

4197 Wakabayashi, M.
(A note on the genus *Mitella* of Japan). *Acta
Phytotax. Geobot.*, 25(4–6):136–153(1973)—Ja(En);
2 illustrations, dot maps; chromosome numbers; key.

SAXIFRAGA

4198 Krause, D. L. *and* Beamish, K. I.
Notes on *Saxifraga occidentalis* and closely related
species in British Columbia. *Syesis*, 6:105–113(1973)—
3 Dot map; chromosome numbers; key.

4199 Simon, C.
Saxifraga pickeringi nov. spec. aus Madeira.
1 *Bocagiana*, no.33. 4p.(1973)—Illustrations.

4200 Teeri, J. A.
Microenvironmental adaptations of local popula-
tions of *Saxifraga oppositifolia* in the high Arctic.
3 *Diss. Abstr. Int., B*, 33(11):5251(1973).

4201 Tiffney, W. N.
A new station for *Saxifraga rivularis* L. in the White
Mountains, New Hampshire. *Rhodora*, 75(801):
3 153–155(1973)—Illustration.

4202 Wakabayashi, M.
On *Saxifraga* sect. *Diptera* of Japan, with description
of a new species. *Acta Phytotax. Geobot.*, 25(4–6):
154–169(1973)—(Ja); illustrations; chromosome
2 numbers; key.

acerifolia Wakabayashi & Satomi in *Acta Phytotax.*
2 *Geobot.*, 25(4–6):167(1973)—Japan.
calycina Sternb.
subsp. **unalaschcensis** (Sternb.) Hultén in *Bot.*
3 *Notiser*, 126(4):494(1973): *S. unalaschcensis.*
nelsoniana D. Don
subsp. **cascadensis** (Calder & Savile) Hultén *op.*
3 *cit.* 494(1973): *S. punctata* subsp. *cascadensis.*
subsp. carlottae (Calder & Savile) Hultén *op. cit.*
494(1973), without basionym page and as
3 'charlottae': *S. punctata* subsp. *carlottae.*
subsp. **insularis** (Hultén) Hultén *op. cit.* 494(1973):
3 *S. punctata* subsp. *insularis.*
subsp. **pacifica** (Hultén) Hultén *op. cit.* 494(1973):
3 *S. punctata* subsp. *pacifica.*
subsp. **porsildiana** (Calder & Savile) Hultén *op.*
3 *cit.* 494(1973): *S. punctata* subsp. *porsildiana.*
subsp. **reniformis** (Ohwi) Hultén *op. cit.* 494
3 (1973): *S. reniformis.*
oppositifolia L.
subsp. **glandulisepala** Hultén *op. cit.* 492(1973)—
3 Canada (British Columbia).
pickeringii C. Simon in *Bocagiana*, 33:4(1973), as
1 'pickeringi'—Madeira.
reflexa Hook.
subsp. **occidentalis** (S. Wats.) Hultén in *Bot.*
3 *Notiser*, 126(4):495(1973): *S. occidentalis.*

TELLIMA

4203 Ratnaparkhi, S.
Co-existence of monosporic and bisporic embryo
sacs in *Tellima grandiflora. Phyton* (Austria), 15(1–2):
3 123–126(1973)—(Ge).

TIARELLA

4204 Savile, D. B. O.
Vegetative distinctions in Canadian species of
Mitella and *Tiarella. Canad. Field-Nat.*, 87(4):460–
3 462(1973)—Illustration.

SCHEUCHZERIACEAE

SCHEUCHZERIA

4205 Murata, G.
[Short note on *Scheuchzeria palustris* L.] *Acta*
1–3 *Phytotax. Geobot.*, 25(4–6):187(1973)—Ja(En).

SCROPHULARIACEAE

4206 Klaren, C. H.
Compounds of benzoic acid in hemiparasitic
Scrophulariaceae. *Acta Bot. Neerl.*, 22(4):452–455
1–3 (1973).

4207 Kondo, K.
The chromosome numbers of four species in the
Scrophulariaceae from the United States. *Chromo-*
3 *some Inform. Serv.*, no.14:25–26(1973).

4208 Mądalski, J.
Atlas flory Polskiej i ziem ościennych. (Florae
Polonicae terrarumque adiacentium iconographia):
tom 17, zeszyt 1. Scrophulariaceae (pars 4), Oroban-
chaceae. Warszawa, etc., Wydawnictwo Naukowe,
1 1973. 77p.—Illustrations.

4209 Musselman, L. J.
On the anatomy of the haustoria of parasitic
Scrophulariaceae. (Uber die Morphologie des
Haustoriums schmartotzender Scrophulariaceae). *In*
European Weed Research Council: Parasitic Weeds
Research Group *and* Malta: Royal University.
Symposium on parasitic weeds, 11–13 April 1973,
Royal University of Malta, Malta: 140–148(1973)—
1–7 (Ge,Fr).

4210 Standley, P. C. *and* Williams, L. O.
Flora of Guatemala: part 9, no.4. Scrophulariaceae.
Fieldiana, Bot., 24(9) nos.3–4:319–418(1973)—
4 Illustrations; keys.

4211 Varghese, T. M.
Structure of the anther in some Scrophulariaceae. *J.*
1–7 *Palynol.*, 8:84–88(1972 publ. 1973)—Illustrations.

ALECTRA

4212 Vijayaraghavan, M. R. *and* Ratnaparkhi, S.
Some aspects of embryology of *Alectra thomsoni*.
6 *Phytomorphology*, 22(1):1–8(1972 publ. 1973).

ANARRHINUM

4213 Rigual, A.
Un *Anarrhinum* nuevo para la flora europea
(*Anarrhinum fruticosum* Desf. subsp. *fruticosum*). *An.*
1 *Inst. Bot. A. J. Cavanilles*, 30:89–97(1973)—Illustrations.

bellidifolium (L.) Desf.
subsp. laxiflorum (Bss.) Malagarriga, Las Subesp. y
Variac. Geogr.: 17(1973), without basionym
1 date: *A. laxiflorum*.

ANTIRRHINUM

4214 Harkiss, K. J.
Studies in the Scrophulariaceae: part 7. Structure of
the aerial parts of *Antirrhinum orontium* L. (*Misopates*
orontium (L.) Raf.). *Pl. Med.*, 23(2):182–190(1973)—
1, 5, 6 Illustrations.

majus L.
subsp. ambiguum (Lge.) Malagarriga, Las Subesp.
y Variac. Geogr.: 19(1973), without basionym
1 date: *A. ambiguum*.
subsp. barrelieri (Bor.) Malagarriga *op. cit.* 18
1 (1973), without basionym page: *A. barrelieri*.
subsp. charidemi (Lge.) Malagarriga *op. cit.* 19
1 (1973), without basionym date: *A. charidemi*.
subsp. glutinosum (Bss. R.) Malagarriga *op. cit.* 19
1 (1973), without basionym date: *A. glutinosum*.
subsp. hispanicum (Chav.) Malagarriga *op. cit.*
1 19(1973), without basionym date: *A. hispanicum*.
subsp. latifolium (DC.) Malagarriga *op. cit.* 19
1 (1973), without basionym date: *A. latifolium*.
subsp. meonanthum (H. Lk.) Malagarriga *op. cit.*
1 19(1973), without basionym date: *A. meonanthum*.
subsp. molle (L.) Malagarriga *op. cit.* 19(1973),
1 without basionym date: *A. molle*.
subsp. rupestre (Bss. R.) Malagarriga *op. cit.* 19
1 (1973), without basionym date: *A. rupestre*.
subsp. sempervirens (Lap.) Malagarriga *op. cit.* 19
1 (1973), without basionym date: *A. sempervirens*.
subsp. **valentinum** (FQ) Malagarriga *op. cit.* 19
1 (1973): *A. valentinum*.

molle L.
subsp. **lopesianum** (Rothm.) Pinto da Silva in
Agron. Lusit., 34(3):193(1972 publ. 1973): *A.*
1 *lopesianum*.

ARAGOA

4215 Romero-Castañeda, R.
Apuntes botanicos: 4. *Mutisia*, no.38. 18p.(1972)—
4 (En); illustrations. *Aragoa perez-arbelaeziana* pp.8–11.

perez-arbelaeziana Romero in *Mutisia*, 38:8(1972)—
4 Colombia.

BACOPA

4216 Shah, J. D.
Dispersal of *Bacopa monnieri* (L.) Penn. *Proc. Indian*
2, 5–7 *Sci. Congr.*, 60(3):398(1973). Abstract only.

CASTILLEJA

4217 Dobbins, D. R.
A study on the haustorium of *Castilleja* (Scrophu-
lariaceae). (Etude sur l'haustorium de *Castilleja*
(Scrophulariaceae)). *In* European Weed Research
Council: Parasitic Weeds Research Group *and* Malta:
Royal University, Symposium on parasitic weeds,
11–13 April 1973, Royal University of Malta, Malta:
1, 4 117–131(1973)—(Fr,Ge); Illustration.

4218 Dobbins, D. R. *and* Kuijt, J.
Studies on the haustorium of *Castilleja* (Scrophu-
lariaceae). 2. The endophyte. *Canad. J. Bot.*, 51(5):
2–4 923–931(1973)—(Fr).

4219 Duffield, W. J.
Pollination ecology of *Castilleja* in Mount Rainier
3 National Park. *Ohio J. Sci.*, 72(2):110–114(1972).

4220 Holmgren, N. H.
Five new species of *Castilleja* (Scrophulariaceae)
from the Inter-mountain Region. *Bull. Torrey Bot.*
3 *Club*, 100(2):83–93(1973)—Illustrations; keys.

aquariensis N. H. Holmgren in *Bull. Torrey Bot.*
3 *Club*, 100(2):87(1973)—U.S.A. (Utah).
christii N. H. Holmgren *op. cit.* 91(1973)—U.S.A.
3 (Idaho).
kaibabensis N. H. Holmgren *op. cit.* 89(1973)—
3 U.S.A. (Arizona).
revealii N. H. Holmgren *op. cit.* 87(1973)—U.S.A.
3 (Utah).
salsuginosa N. H. Holmgren *op. cit.* 83(1973)—
3 U.S.A. (Nevada).

SCROPHULARIACEAE

CELSIA

4222 Huber-Morath, A.
Verbascum L. s.l. (incl. *Celsia* L. et *Staurophragma*
1, 2 Fisch. & Mey.). *Bauhinia*, 5(1):7–16(1973).

CHAENORRHINUM

4223 Fernandes, R. Batarda
Une correction corrigée. *Anuár. Soc. Brot.*, 39:125–
1 127(1973). *Chaenorrhinum origanifolium.*

cadevallii (O. Bolós & Vigo) Holub in *Preslia*,
1 45(4):359(1973): *Linaria origanifolia* subsp. *cadevallii.*
origanifolium (L.) Lge.
subsp. capitatum (Wk.) Malagarriga, Las Subesp. y Variac. Geogr.: 18(1973), without basionym
1 date: *C. crassifolium* var. *capitatum.*
subsp. crassifolium (Cav.) Malagarriga *op. cit.* 18 (1973), without basionym date: *Antirrhinum*
1 *crassifolium.*
subsp. robustum (Loscos) Malagarriga *op. cit.* 18
1 (1973), without basionym date: *C. robustum.*
subsp. **segoviense** (Reuter) Malagarriga *op. cit.*
1 18(1973): *Linaria segoviense.*
subsp. **serpyllifolium** (Lge.) Malagarriga *op. cit.*
1 18(1973): *C. serpyllifolium.*
rubrifolium (Rob. & Cast.) Lge.
subsp. **bianorii** (Knoche) Malagarriga *op. cit.* 18
1 (1973): *C. rubrifolium* var. *bianorii.*

CONOBEA

4224 Morais, A. Alpande de *and others*
Óleos essenciais da Amazônia contendo Timol. *Acta Amazonica*, 2(1):45–46(1972)—(En). *Conobea*
4 *scoparioides.*

CORDYLANTHUS

4225 Chuang, T. I. *and* Heckard, L. R.
Taxonomy of *Cordylanthus* subgenus *Hemistegia* (Scrophulariaceae). *Brittonia*, 25(2):135–158(1973)—
3, 4 Illustrations, dot maps; chromosome numbers; keys.

subgen. **Hemistegia** (Gray) Chuang & Heckard in *Brittonia*, 25(2):144(1973): *Cordylanthus* sect.
3 *Hemistegia.*
maritimus Nutt. ex Benth.
subsp. **canescens** (A. Gray) Chuang & Heckard
3 *op. cit.* 149(1973): *C. canescens.*
subsp. **palustris** (Behr) Chuang & Heckard *op. cit.*
3 148(1973): *Chloropyron palustre.*
mollis A. Gray
subsp. **hispidus** (Pennell) Chuang & Heckard *op.*
3 *cit.* 152(1973): *C. hispidus.*

DIGITALIS

4226 Bailey, D.
Planhigion meddyginiaethol o gefn gwlad Cymru: 2. Bysedd y Cwn (*Digitalis purpurea*). *Y Gwyddonydd*,
1 10(4):159–161(1972)—Illustration.

4227 Craffmann, F.
Bemerkenswerke Pflanzenfunde in südlichen Dillkreis. *Hess. Flor. Briefe*, 262–264:52–54(1973)—
1 Illustrations. *Digitalis lutea* and *D. lanata.*

4228 Kotukov, G. N.
Mezhvidovaya gibridizatsiya v rode naperstyanki *Digitalis.* (Interspecific hybridization in the genus *Digitalis*). *Tsitol. Genet.*, 7(4):361–364(1973)—(En);
1 chromosome numbers.

grandiflora Mill.
f. **peterfii** Soó in *Acta Bot. Acad. Sci. Hung.*, 18
1 (1–2):174(1973)—Rumania.

purpurea L.
subsp. mariana (Bss.) Malagarriga, Las Subesp. y Variac. Geogr.: 19(1973), without basionym
1 date: *D. mariana.*
subsp. nevadensis (Kze.) Malagarriga *op. cit.* 19
1 (1973), without basionym date: *D. nevadensis.*

ERINUS

4229 Steinwendtner, R.
Erinus alpinus L. neu für Oberösterreich. *Mitt. Bot.*
1 *Arbeit. Oberösterr. Landesmus. Linz*, 4(1–2):45(1972).

EUPHRASIA

4230 Harkiss, K. J. *and* Timmins, P.
Studies in the Scrophulariaceae: part 8. Phytochemical investigation of *Euphrasia officinalis. Pl. Med.*,
1 23(4):342–347(1973).

4231 Jalas, J. *and* Kukkonen, I.
Typification of the taxa of *Euphrasia* (Scrophulariaceae) described by Finnish botanists. *Ann. Bot. Fenn.*, 10(1):27–42(1973)—Illustrations. Includes
1 biographical notes on the authors of the new taxa.

4232 Scannell, M. J. P.
Euphrasia frigida Pugsl. on the Galtees. *Irish Nat. J.*,
1 17(11):396(1973).

4233 Swann, E. L.
Norfolk *Euphrasiae. Watsonia*, 9(4):345–356
1 (1973).

4234 Yeo, P. F.
Catalogue of the late Dr E. F. Warburg's
1 *Euphrasia* collection. *B.S.B.I. News*, 2(1):16(1973).

GLOSSOSTIGMA

4235 Bruggen, H. W. E. van
Glossostigma elatinoides, een nieuwe bodembedekker. *Aquarium* (Netherlands), 43(7):156–157(1973)—
7 Illustrations.

HEBE

4236 Green, P. S.
Hebe. In Green, P. S. *ed.* Plants: wild and cultivated:
1, 4, 7 164–166(1973)—(En).

4237 Green, P. S.
Hebe × *franciscana* (Eastwood) Souster, not *H.* × *lewisii* naturalized in Britain. *Watsonia*, 9(4):371–372
1 (1973).

4238 Moore, L. B.
Protogyny in *Hebe* 'Paniculatae'. *New Zealand J.*
7 *Bot.*, 11(1):173–176(1973)—Illustration.

KICKXIA

macilenta (Decne.) Danin in *Notes Roy. Bot. Gard.*
2 *Edinburgh*, 32(2):265(1973): *Linaria macilenta.*

LAMOUROUXIA

4239 Reveal, J. L. *and* Hess, W. J.
A new combination in *Lamourouxia* (Scrophu-
4 lariaceae). *Phytologia*, 25(3):105–106(1973).

longiflora Benth.
var. **hyssopifolia** (A. Gray) Reveal & W. J. Hess
4 in *Phytologia*, 25(3):106(1973): *L. hyssopifolia.*

LINARIA

4240 Keens, W. M.
Beta trigyna and *Linaria* × *dominii* in Berkshire.
1 *Watsonia*, 9(4):428(1973).

anticaria Boiss. & Reuter
subsp. **rossmassleri** (Willk.) Rivas Goday in
Trab. Dep. Bot. Fisiol. Veg. Univ. Madrid, 6:25
1 (1973), as 'rossmasslerii': *L. rossmasslerii*.
supina (L.) Desf.
subsp. aeruginea (Loscos & Pardo) Malagarriga,
Las Subesp. y Variac. Geogr.: 17(1973), without
1 basionym date: *L. aeruginea*.
subsp. maritima (DC.) Malagarriga *op. cit.* 17
1 (1973), without basionym date: *L. maritima*.
subsp. melanantha (Bss. R.) Malagarriga *op. cit.* 17
1 (1973), without basionym date: *L. melanantha*.
subsp. nevadensis (Bss. R.) Malagarriga *op. cit.*
1 17(1973), without basionym date: *L. nevadensis*.
subsp. tristis (L.) Malagarriga *op. cit.* 17(1973),
1 without basionym date: *Antirrhinum tristis*.

MELAMPYRUM

4241 Greilhuber, J.
Über die Entwicklung des Embryosacks von
Melampyrum und *Parentucellia latifolia* (Scrophulariaceae, Pedicularieae). *Österr. Bot. Zeitschr.*, 121
1 (1–2):81–97(1973)—(En); chromosome numbers.

bihariense Kern.
lusus **devanum** Soó in *Acta Bot. Acad. Sci. Hung.*,
1 18(1–2):175(1973)—Rumania.

MIMULUS

4242 Alam, M. T. *and* Vickery, R. K.
Crossing relationships in the *Mimulus glabratus*
heteroploid complex. *Amer. Midl. Nat.*, 90(2):449–
3, 4 454(1973).

4243 De Vries, B.
Mimulus glabratus HBK. var. *fremontii* (Benth.)
Grant. An obscure species from the eastern Qu'-
Appelle Valley, Saskatchewan. *Blue Jay*, 29(1):39–40
3 (1971)—Illustration.

4244 Kiang, Y. T.
Floral structure, hybridization and evolutionary
relationship of two species of *Mimulus*. *Rhodora*,
75(802):225–238(1973)—Illustration. *M. guttatus*
1, 3 and *M. nasutus*.

4245 Vickery, R. K.
Mimulus wiensii (Scrophulariaceae) a new species
from western Mexico. *Madroño*, 22(4):161–168(1973)—
4 Illustration; chromosome numbers.

wiensii R. K. Vickery in *Madroño*, 22(4):161(1973)—
4 Mexico.

ODONTITES

tenuifolia
subsp. **albarracinensis** (Laínz) Laínz in *Candollea*,
28(2):187(1973): *Odontitella virgata* subsp.
1 *albarracinensis*.

PARENTUCELLIA

4246 Greilhuber, J.
Über die Entwicklung des Embryosacks von
Melampyrum und *Parentucellia latifolia* (Scrophulariaceae, Pedicularieae). *Österr. Bot. Zeitschr.*, 121(1–2):
1 81–97(1973)—(En); chromosome numbers.

PEDICULARIS

4247 Balls, E. K.
Plant portraits: *Pedicularis groenlandica* Retz. *Quart.
Bull. Alp. Gard. Soc.*, 41(3):202–204(1973)—
3 Illustration.

4248 Borg, S. J. ter *and* Koeman-Kwak, M. M.
Eén-en tweejarige *Pedicularis palustris* L. op de
buitendijkse waarden bij Makkum. *Gorteria*, 6(8):
1 139–140(1973).

4249 Koeman-Kwak, M.
The pollination of *Pedicularis palustris* by nectar
thieves (short tongued bumblebees). *Acta Bot. Neerl.*,
1, 2 22(6):608–615(1973)—Illustrations.

4250 Macior, L. W.
The pollination ecology of *Pedicularis* on Mount
Rainer. *Amer. J. Bot.*, 60(9):863–871(1973)—
3 Illustrations.

lanata Cham. & Schlecht.
subsp. **adamsii** (Hultén) Hultén in *Bot. Notiser*,
3 126(4):504(1973): *P. adamsii*.
subsp. **dasyantha** (Hadač) Hultén *op. cit.* 504
3 (1973): *P. dasyantha*.
subsp. **pallasii** (Vved.) Hultén *op. cit.* 504(1973):
3 *P. pallasii*.

PSEUDOLYSIMACHION

vandasii (Rohl.) Holub in *Folia Geobot. Phytotax.*,
1 8(2):177(1973): *Veronica vandasii*.

RHINANTHUS

gracilis Schur
subsp. **stojanovii** Soó in *Acta Bot. Acad. Sci. Hung.*,
1 18(1–2):174(1973), as 'stojanovi'—Bulgaria.

SCROPHULARIA

crithmifolia Bss.
subsp. **alpina** (Bss.) Malagarriga, Las Subesp. y
Variac. Geogr.: 17(1973): *S. crithmifolia* var.
1 *alpina*.

STAUROPHRAGMA

4251 Huber-Morath, A.
Verbascum L. s.l. (incl. *Celsia* L. et *Staurophragma*
1, 2 Fisch. & Mey.). *Bauhinia*, 5(1):7–16(1973).

STRIGA

4252 Hepper, F. N.
Problems in naming *Orobanche* and *Striga*.
(Probleme bei der Benennung von *Orobanche* und
Striga Zusammenfassung). *In* European Weed
Research Council: Parasitic Weeds Research Group
and Malta: Royal University. Symposium on
parasitic weeds, 11–13 April 1973, Royal University
of Malta, Malta: 9–17(1973)—(Ge,Fr); illustrations,
1–7 key.

4253 Kondo, K.
The chromosome numbers of *Striga asiatica* and
6 *Triphyophyllum peltatum*. *Fyton*, 31(1):1–2(1973).

4254 Musselman, L. J.
Preliminary studies on the anatomy of seedling
roots and haustoria of *Striga lutea* Lour. (Scrophulariaceae). (Recherches préliminaires sur l'anatomie
des racines des plantules et du suçoir chez *Striga lutea*
Lour. (Scrofulariacées)). *In* European Weed Research
Council: Parasitic Weeds Research Group *and* Malta:
Royal University. Symposium on parasitic weeds,
11–13 April 1973, Royal University of Malta, Malta:
5–7 149–153(1973)—(Ge,Fr).

TETRANEMA

4255 Weber, A.
Der Blütenstand von *Tetranema mexicanum* Benth. (Scrophulariaceae) und seine Beziehungen zu jenem der Gesneriaceen. *Mitt. Bot. Arbeit. Oberösterr.*
4 *Landesmus. Linz*, 4(1–2):27–43(1972)—Illustrations.

TOZZIA

4256 Nikiticheva, Z. I. *and* Terekhin, É. S.
K émbriologii *Tozzia alpina* L. (Scrophulariaceae). (To the embryology of *Tozzia alpina* L. (Scrophulariaceae)). *Bot. Zhurn.*, 58(5):680–691(1973)—
1 Illustrations.

VERBASCUM

4257 Hansen, A. *and* Hansen, C.
Verbascum osbeckii L. and the *Triguera* species (Solanaceae): a revision. *Lagascalia*, 3(2):183–193
1 (1973)—Illustrations.

4258 Huber-Morath, A.
Verbascum L. s.l. (incl. *Celsia* L. et *Staurophragma*
1, 2, 5, 6 Fisch. & Mey.). *Bauhinia*, 5(1):7–16(1973).

akdarense (Murb.) Huber-Morath in *Bauhinia*, 5(1):
2 10(1973): *Celsia akdarensis*.
arbusculum (A. Rich.) Huber-Morath *op. cit.* 10
5 (1973): *Celsia arbuscula*.
aucheri (Boiss.) Huber-Morath *op. cit.* 10(1973):
2 *Celsia aucheri*.
ballii (Battand.) Huber-Morath *op. cit.* 10(1973):
1 *Celsia ballii*.
battandieri (Murb.) Huber-Morath *op. cit.* 10
1 (1973): *Celsia battandieri*.
bornmuelleranum Huber-Morath *op. cit.* 10 (1973), nom. nov., as 'bornmuellerianum': *Celsia*
2 *bornmuelleri*.
bottae (Deflers) Huber-Morath *op. cit.* 11(1973):
2 *Celsia bottae*.
bourgeauanum Huber-Morath *op. cit.* 11(1973),
2 nom. nov.: *V. lycium* (Boiss.) O. Ktze.
brevipedicellatum (Engler) Huber-Morath *op. cit.*
5 11(1973): *Celsia brevipedicellata*.
capitis-viridis Huber-Morath *op. cit.* 11(1973), nom.
5 nov.: *Celsia insularis* Murb.
carmanicum (Bornm.) Huber-Morath *op. cit.* 11
2 (1973): *Celsia carmanica*.
chiovendae Huber-Morath *op. cit.* 11(1973), nom.
5 nov.: *Celsia micrantha* Chiovenda.
commixtum (Murb.) Huber-Morath *op. cit.* 11
1 (1973): *Celsia commixta*.
coromandelianum (Vahl) Huber-Morath *op. cit.*
2, 6 11(1973): *Celsia coromandeliana*.
cystolithicum (Pettersson) Huber-Morath *op. cit.*
5 12(1973): *Celsia cystolithica*.
densifolium (Hook. f.) Huber-Morath *op. cit.* 12
5 (1973): *Celsia densifolia*.
deserticola (Murb.) Huber-Morath *op. cit.* 12
2 (1973): *Celsia deserticola*.
disjectum (Murb.) Huber-Morath *op. cit.* 12(1973):
2 *Celsia disjecta*.
dudleyanum (Huber-Morath) Huber-Morath *op. cit.*
2 12(1973): *Celsia dudleyana*.
farsistanicum (Murb.) Huber-Morath *op. cit.* 12
2 (1973): *Celsia farsistanica*.
faurei (Murb.) Huber-Morath *op. cit.* 12(1973):
1 *Celsia faurei*.
flabellifolium (Huber-Morath) Huber-Morath
2 *op. cit.* 12(1973): *Celsia flabellifolia*.
gabrielae (Bornm.) Huber-Morath *op. cit.* 12(1973):
6 *Celsia gabrielae*.
× **gabrielianae** (Huber-Morath) Huber-Morath
2 *op. cit.* 13(1973): × *Celsioverbascum gabrielianae*.
glanduliferum (Post) Huber-Morath *op. cit.* 13
2 (1973): *Celsia glandulifera*.
haussknechtianum Huber-Morath *op. cit.* 13(1973),
2 nom. nov.: *V. persicum* Hausskn. ex Bornm.
keniense (Murb.) Huber-Morath *op. cit.* 13(1973):
5 *Celsia keniensis*.
ledermannii ((Schlecht.) Murb.) Huber-Morath
5 *op. cit.* 13(1973): *Celsia ledermannii*.
longirostre (Murb.) Huber-Morath *op. cit.* 13
1 (1973): *Celsia longirostris*.
lychnitis L.
f. **rochelianum** Soó in *Acta Bot. Acad. Sci. Hung.*,
1 18(1–2):172(1973)—? Hungary or Rumania.
lyprocarpum (Murb.) Huber-Morath in *Bauhinia*,
2 5(1):13(1973): *Celsia lyprocarpa*.
mairei (Murb.) Huber-Morath *op. cit.* 13(1973):
1 *Celsia mairei*.
maroccanum (Ball) Huber-Morath *op. cit.* 13
1 (1973): *Celsia maroccana*.
megricum (Tsvelev) Huber-Morath *op. cit.* 14
2 (1973): *Celsia megrica*.
melhanense (Murb.) Huber-Morath *op. cit.* 14
2 (1973): *Celsia melhanensis*.
× **mirabile** (Rech. f. & Huber-Morath) Huber-Morath *op. cit.* 14(1973): × *Celsioverbascum*
1 *mirabile*.
natolicum (Fisch. & Mey.) Huber-Morath *op. cit.*
2 14(1973): *Staurophragma natolicum*.
nudicaule (Wydl.) Huber-Morath *op. cit.* 14(1973):
2 *Scrophularia nudicaulis*.
pubescens (Skan) Huber-Morath *op. cit.* 14(1973):
5 *Celsia pedunculosa* var. *pubescens*.
rhiphaeum (Murb.) Huber-Morath *op. cit.* 14
1 (1973): *Celsia rhiphaea*.
rupicola (Hayek & Siehe) Huber-Morath *op. cit.*
2 14(1973): *Celsia rupicola*.
scabridum (Skan) Huber-Morath *op. cit.* 15(1973):
5 *Celsia scabrida*.
scrophulariifolium (Hochst.) Huber-Morath *op. cit.*
5 15(1973): *Celsia scrophulariifolia*.
sedgwickianum (Schimp.) Huber-Morath *op. cit.*
5 15(1973): *Celsia sedgwickiana*.
serpenticola Huber-Morath *op. cit.* 15(1973), nom.
2 nov.: *Celsia serpentinica* Huber-Morath.
serratifolium (Huber-Morath) Huber-Morath *op. cit.*
2 15(1973): *Celsia serratifolia*.
sorgerae (Huber-Morath) Huber-Morath *op. cit.*
2 15(1973): *Celsia sorgerae*.
spodiotrichum (Huber-Morath) Huber-Morath
2 *op. cit.* 15(1973): *Celsia spodiotricha*.
straussii (Bornm.) Huber-Morath *op. cit.* 15(1973):
2 *Celsia straussii*.
tenuicaule (Murb.) Huber-Morath *op. cit.* 15(1973):
2 *Celsia tenuicaulis*.
tibesticum (Quézel) Huber-Morath *op. cit.* 15
2 (1973): *Celsia tibestica*.
trapifolium (Stapf) Huber-Morath *op. cit.* 16(1973):
2 *Celsia trapaefolia*.
valerianifolium (A. Rich.) Huber-Morath *op. cit.*
5 16(1973): *Celsia valerianifolia*.
zaianense (Murb.) Huber-Morath *op. cit.* 16(1973):
1 *Celsia zaianensis*.

VERONICA

4259 Berg, R. L., Kalinin, O. M. *and* Kolosova, L. D.
Sopostavlenie vnutrividovoĭ i mezhvidovoĭ izmenchivosti u veronik (rod *Veronica*). (Comparison of intraspecies variability in speedwells (*Veronica*)).
1, 2 *Zhurn. Obshch. Biol.*, 34(2):216–226(1973)—(En).

4260 Blažková, D.
Poznámky k rozšíření *Veronica filiformis* Smith v Československu. (Beitrag zur Verbreitung von *Veronica filiformis* Smith in der Tschechoslowakei).
1 *Zprávy Česk. Bot. Spol.*, 8(1):11–13(1973)—(Ge).

4261 Elenevskiĭ, A. G.
Proiskhozhdenie i razvitie roda *Veronica* L. (Origin and development of the genus *Veronica* L.). *Byull. Mosk. Obshch. Ispȳt. Prir., Biol.*, 78(1):70–83(1973)—(En); chromosome numbers. Including *Hebe*, *Parahebe*,
1–7 *Pygmaea* etc.

4262 Fischer, M.
Zur Cytotaxonomie von *Veronica chamaedrys* L. agg.: 2. Subsp. *micans* M. Fischer, subsp. nova, eine weitere diploide Sippe. *Österr. Bot. Zeitschr.*, 121 (1–2):73–79(1973)—(En); illustrations; chromosome
1 numbers.

4263 Grayer-Barkmeijer, R. J.
A chemosystematic study of *Veronica*: iridoid
1–7 glucosides. *Biochem. Syst.*, 1(2):101–110(1973).

4264 Jongh, S. E. de *and* Kern, J. H.
De variabiliteit van *Veronica hederifolia* L. in Nederland aanvullende gegevens. *Gorteria*, 6(12):
1 202–203(1973)—(En); chromosome numbers.

4265 Resmeriţă, I.
Nouveaux taxons de la flore de Roumanie. *Rev.*
1 *Roum. Biol. Bot.*, 17(5):311–312(1972)—Illustration.

chamaedrys L.
subsp. **micans** M. Fischer in *Österr. Bot. Zeitschr.*,
1 121(1–2):73(1973)—Austria.
scheereri (J. P. Brandt) Holub in *Folia Geobot. Phytotax.*, 8(2):177(1973): *V. prostrata* subsp.
1 *scheereri*.
× tzesnae Resmeriţă in *Rev. Roum. Biol., Bot.*, 17(5): 311(1972), without type—Rumania. (*V. crassifolia* ×
1 *V. spicata*).
vindobonensis (M. Fischer) M. Fischer in *Österr. Bot. Zeitschr.*, 122(4):287(1973): *V. chamaedrys*
1, 2 subsp. *vindobonensis*.

SELAGINACEAE

GLOBULARIA

4266 Murray-Lyon, D. M.
Plant portraits: *Globularia incanescens*. *Quart. Bull.*
1 *Alp. Gard. Soc.*, 41(3):205–207(1973)—Illustration.

SIMAROUBACEAE

4267 Porter, D. M.
Flora of Panama: part 6. Family 90. Simaroubaceae. *Ann. Missouri Bot. Gard.*, 60(1):23–39(1973)—
4 Illustrations; keys.

PICRAMNIA

4268 Porter, D. M.
Three new species of *Picramnia* (Simaroubaceae) from Central America. *J. Arnold Arbor.*, 54(2):
4 315–321(1973)—Illustrations.

allenii D. M. Porter in *J. Arnold Arbor.*, 54(2):315
4 (1973)—Panama, Costa Rica.
4 **cooperi** D. M. Porter *op. cit.* 317(1973)—Panama.
4 **dwyeri** D. M. Porter *op. cit.* 319(1973)—Panama.

SIMMONDSIACEAE
see
BUXACEAE

SOLANACEAE

4269 D'Arcy, W. G.
New names for neotropical plants. *Phytologia*,
4 25(3):116(1973).

4270 Gentry, J. L.
Studies in Mexican and Central American Solanaceae. *Phytologia*, 26(4):265–278(1973)—
4 Illustrations.

4271 Hsiao, Pei-ken, Hsia, Kuang-cheng *and* Ho, Li-yi
The occurrence of some important tropane alkaloids in Chinese solanaceous plants. *Acta Bot. Sin.*, 15(2):
2 194–202(1973)—Ch(En).

4272 Lavie, D.
Applying chemistry to genetics in certain Solanaceae. *In* Nobel Symposium 25: chemistry in botanical classification, August 20–25, 1973. (1973).
1–7 1p. Abstract.

ACNISTUS

4273 Mora, D. E.
Desarrollo y receimiento del ápice vegetativo en *Acnistus arborescens*. 1. Estructura, cambios plastocrónicos y primeros estadios en el desarrollo de la hoja. *Rev. Biol. Trop.*, 19(1–2):171–189(1971)—(En);
4 illustrations.

ATHENAEA

orinocensis (Kunth) D'Arcy in *Phytologia*, 25(3):
4 116(1973): *Physalis orinocensis*.

BRUGMANSIA

4274 Lockwood, T. E.
Generic recognition of *Brugmansia*. *Bull. Mus. Leafl. Harvard Univ.*, 23(6):273–284(1973)—
4 Illustration. Separated from *Datura*.

BRUNFELSIA

4275 Plowman, T.
Four new Brunfelsias from north-western South America. *Bot. Mus. Leafl. Harvard Univ.*, 23(6): 245–272(1973)—Illustrations, dot map; chromosome
4 number; keys.

chiricaspi Plowman in *Bot. Mus. Leafl. Harvard*
4 *Univ.*, 23(6):255(1973)—Colombia.
chocoensis Plowman *op. cit.* 245(1973)—Colombia,
4 Panama.
grandiflora D. Don
subsp. **schultesii** Plowman *op. cit.* 259(1973)—Colombia, Venezuela, Ecuador, Peru, Bolivia,
4 Brazil.
macrocarpa Plowman *op. cit.* 251(1973)—Ecuador,
4 Colombia.

CAPPARIS

4276 Iltis, H. H.
Capparis hypoleuca Presl: a synonym of *Solanum schlechtendalianum* Walp. *Ann. Missouri Bot. Gard.*,
4 59(3):474(1972 publ. 1973).

SOLANACEAE

CAPSICUM

4277 D'Arcy, W. G. *and* Eshbaugh, W. H.
The name for the common bird pepper. *Phytologia*,
1–7 25(6):350(1973).

4278 Ferrari, J-P. *and* Aillaud, G.
Bibliographie du genre *Capsicum*. *J. Agric. Trop.*
4 *Bot. Appl.*, 18(11):385–480(1971 publ. 1972).

annuum
var. **aviculare** (Dierb.) D'Arcy & Eshbaugh in *Phytologia*, 25(6):350(1973): *C. indicum* var.
4 *aviculare.*

CESTRUM

acutifolium Alain in *Phytologia*, 25(5):274(1973)—
4 Hispaniola (Dominican Republic).
lewisii D'Arcy in *Phytologia*, 25(3):116(1973), nom.
4 nov.: *C. acuminatum* Francey.
mortonianum J. L. Gentry in *Phytologia*, 26(4):267
4 (1973)—Guatemala.

CHAMAESARACHA

4279 Averett, J. E.
Biosystematic study of *Chamaesaracha* (Solanaceae). *Rhodora*, 75(803):325–365(1973)—Illustrations, dot
3, 4 maps; chromosome numbers; keys.

CYPHOMANDRA

stuckertii (Bitt.) D'Arcy in *Ann. Missouri Bot. Gard.*,
4 59(2):269(1972 publ. 1973): *Solanum stuckertii.*

DATURA

4280 Kondrat'eva-Mel'vil, E. A.
Razvitie strukturȳ odnoletnego dvudol'nogo rasteniya s monopodial'nȳm i simpodial'nȳm narastaniem. (Development of structure of an annual dicotyledonous plant with monopodial and sympodial growth). *Bot. Zhurn.*, 58(8):1157–1172
6 (1973)—Illustrations. *Datura fastuosa.*

4281 Lockwood, T. E.
Generic recognition of *Brugmansia*. *Bull. Mus. Leafl. Harvard Univ.*, 23(6):273–284(1973)—
4 Illustration. Separated from *Datura.*

4282 Ngô Tru'c Nhã *and* Danert, S.
Über die Fruchtentwicklung und über zentrische Leitbündel in der Gattung *Datura* (Solanaceae). (About development of fruits and centric vascular bundles in the genus *Datura* (Solanaceae)). *Kultur-*
1–7 *pflanze*, 21:119–193(1973)—(En,Rus); illustrations.

4283 Sharma, M.
Studies in the flower of *Datura stramonium* Linn. in relation to bee-botany. *J. Palynol.*, 8:17–21(1972 publ.
1–7 1973)—Illustrations.

HYOSCYAMUS

4284 Sibri, N. N., El-Masry, S. *and* Khafagy, S. M.
Pharmacognostical study of *Hyoscyamus desertorum*: part 1. Morphology and histology of the flower. *Pl.*
2 *Med.*, 23(1):10–22(1973).

LITHOPHYTUM

4285 D'Arcy, W. G. *and* Keating, R. C.
The affinities of *Lithophytum*: a transfer from Solanaceae to Verbenaceae. *Brittonia*, 25(3):213–225
4 (1973)—Illustrations.

LYCIANTHES

4286 McVaugh, R.
Report of the Committee for Spermatophyta: conservation of generic names: proposal 328 (sub 7407) *Lycianthes* (Dunal) Hassler (1917) vs. *Otilix* Rafinesque (1828) and *Parascopolia* Baillon (1888).
4–6 *Taxon*, 22(1):156(1973).

armentalis J. L. Gentry in *Phytologia*, 26(7):269
4 (1973)—Mexico, British Honduras, Guatemala.
connata J. L. Gentry in *Phytologia*, 26(4):271(1973)—
4 Guatemala, Mexico.
cuchumatanensis J. L. Gentry *op. cit.* 273(1973)—
4 Guatemala.
escuintlensis (Coult.) D'Arcy in *Phytologia*, 25(3):
4 116(1973): *Brachistus escuintlensis.*
gongylodes J. L. Gentry in *Phytologia*, 26(4):274
4 (1973)—Guatemala.
limitanea (Standl.) J. L. Gentry *op. cit.* 275(1973):
4 *Solanum limitaneum.*

LYCOPERSICON

4287 Nettancourt, D. de, Devreux, M., Laneri, U., Pacini, M., Cresti, M. *and* Sarfatti, G.
Ultrastructural aspects of unilateral interspecific incompatibility between *Lycopersicum peruvianum* and *Lycopersicum esculentum*. *Incompat. Newsl.*, no.2:
4 28(1973).

4288 Nettancourt, D. de *and* Capaccio, E.
Polyploidy and aneuploidy in the self-compatible species *Lycopersicum peruvianum* Mill. *Incompat.*
4 *Newsl.*, no.2:42(1973)—Chromosome numbers.

MARKEA

megalandra (Dun.) D'Arcy in *Phytologia*, 25(3):
4 116(1973): *Metternickia megalandra.*

NICOTIANA

4289 Fredman, K. I. *and* Fredman, H. B.
Tobacco—a historical view. *Herbarist*, no.39:33–41
3, 4 (1973).

4290 Luneva, M. Z. *and* Poddubnaya-Arnol'di, V. A.
Sravnitel'noe morfologicheskoe i tsitoémbriologicheskoe issledovanie diploidnoĭ i avtotetraploidnoĭ form *Nicotiana glauca* Grah. *Byull. Glavn. Bot. Sada*
4, 7 (Moscow), 89:36–45(1973)—Chromosome numbers.

OTILIX

4291 McVaugh, R.
Report of the Committee for Spermatophyta: conservation of generic names: proposal 328 (sub 7407) *Lycianthes* (Dunal) Hassler (1917) vs. *Otilix* Rafinesque (1828) and *Parascopolia* Baillon (1888).
4–6 *Taxon*, 22(1):156(1973).

PARASCOPOLIA

4292 McVaugh, R.
Report of the Committee for Spermatophyta: conservation of generic names: proposal 328 (sub 7407) *Lycianthes* (Dunal) Hassler (1917) vs. *Otilix* Rafinesque (1828) and *Parascopolia* Baillon (1888).
4–6 *Taxon*, 22(1):156(1973).

SCOPOLIA

4293 Krȳlova, I. L.
O morfologicheskoĭ strukture vidov roda *Scopolia* Jacq. emend. Link v svyazi s évolyutseĭ roda. (On the morphological structure of *Scopolia* species in connection with the evolution of the genus). *Byull. Mosk. Obshch. Ispȳt. Prir., Biol.*, 78(5):107–112(1973)—
1, 2, 6 (En); illustration.

4294 Lyubchenko, V. M.
Nove mistsezrostannya skopoliyi karnioliyis'koyi (*Scopolia carniolica* Jacq.) na Ukrayini. (On a new habitat of *Scopolia carniolica* Jacq. in the Ukraine).
1 *Ukr. Bot. Zhurn.*, 30(1):116–117(1973)—Uk(En).

SOLANUM

4295 Baráth, Z. Máthé, I., *jr and* Máthé, I.
A *Solanum* genusz *Morella* szekciójának vizsgálatairól: 1. Adatok a *Solanum nigrum* L. morfológiájához. (Investigations of the *Morella* section of *Solanum* genus: 1. Data to the morphology of *Solanum nigrum* L.). *Bot. Közlem.*, 60(3):191–198
1, 2 (1973)—(En); illustrations.

4296 D'Arcy, W. G.
Solanaceae studies: 2. Typification of subdivisions of *Solanum*. *Ann. Missouri Bot. Gard.*, 59(2):262–
4 278(1972 publ. 1973).

4297 Gerasimenko, I. I.
Dve novÿe raznovidnosti *Solanum vescum* F. Muell. kak ishkhodnÿĭ material dlya selektsii. (Two new varieties of *Solanum vescum* F. Müll. as the initial material for breeding). *Rast. Resursÿ*, 9(4):420–425 (1973)—Chromosome number. *S. vescum* var.
7 *kibalcziczii*, *S. vescum* var. *davidii*.

4298 Hossain, M.
Observations on stylar heteromorphism in *Solanum torvum* Sw. (Solanaceae). *Bot. J. Linn. Soc.*, 66(4):291–
5, 7 301(1973)—Illustrations.

4299 Khan, A. H., Rao, G. R. *and* Khan, R.
Interrelationship within the *Solanum nigrum* complex. *Proc. Indian Sci. Congr.*, 59(3):364(1972)—
6 Chromosome numbers. Abstract only.

4300 Mäki-Tanila, A.
Eurooppalaisen perunan evoluutio. (The evolution of the European potato). *Luonnon Tutkija*, 77(2):
1, 4 38–43(1973)—(En); chromosome numbers.

4301 Ochoa, C.
Nueva especie de papa silvestre diploide para el Peru. *Bol. Soc. Argent. Bot.*, 14(4):330–334(1972)—(En); illustrations; chromosome number. *Solanum*
4 *ariduphilum*.

sect. **Aquartia** (Jacq.) D'Arcy in *Ann. Missouri Bot.*
4 *Gard.*, 59(2):266(1972 publ. 1973): genus *Aquartia*.
subgen. **Brevantherum** (Seithe) D'Arcy *op. cit.*
4 267(1972 publ. 1973): *Solanum* sect. *Brevantherum*.
sect. **Extensum** D'Arcy *op. cit.* 268(1972 publ.
4 1973).
ser. **Juglandifolia** D'Arcy *op. cit.* 269(1972 publ.
4 1973).
sect. **Lasiocarpa** D'Arcy *op. cit.* 270(1972 publ.
2 1973): *Solanum* ** *Lasiocarpa* Dun.
subgen. **Potatoe** (G. Don) D'Arcy *op. cit.* 272(1972
3, 4 publ. 1973): *Solanum* subsect. *Potatoe*.
sect. **Regmandra** Ugent ex D'Arcy *op. cit.* 273
4 (1972 publ. 1973): *Solanum* **Regmandra* Dun.
ariduphilum Ochoa in *Bol. Soc. Argent. Bot.*, 14(4):
4 330(1972)—Peru.
cobanense J. L. Gentry in *Phytologia*, 26(4):276
4 (1973), nom. nov.: *Cyphomandra aculeata* Donn.-Sm.
dulcamara L.
var. **pusztarum** Soó in *Acta Bot. Acad. Sci. Hung.*,
1 18(1–2):172(1973)—Hungary.
molinarum J. L. Gentry in *Phytologia*, 26(4):276
4 (1973)—Honduras, Guatemala.
quitoënse
f. **septentrionale** (Schultes & Cuatrec.) D'Arcy
in *Phytologia*, 25(3):116(1973): *S. quitoënse* var.
4 *septentrionale*.

TRIANAEA

4302 Romero-Castañeda, R.
Apuntes botanicos: 4. *Mutisia*, no.38. 18p.(1972)—
4 (En); illustrations. *Trianaea neovisae* pp.11–17.

neovisae Romero in *Mutisia*, 38:11(1972)—
4 Colombia.

TRIGUERA

4303 Hansen, A. *and* Hansen, C.
Verbascum osbeckii L. and the *Triguera* species (Solanaceae): a revision. *Lagascalia*, 3(2):183–193
1 (1973)—Illustrations.

WITHERINGIA

dichotoma (Rusby) D'Arcy in *Phytologia*, 25(3):
4 116(1973): *Cyphomandra dichotoma*.
physocalycia (Donn.-Sm.) J. L. Gentry in *Phytologia*,
4 26(4):278(1973): *Brachistus physocalycius*.

SPARGANIACEAE

4304 Ross-Craig, S.
Drawings of British plants: being illustrations of the species of flowering plants growing naturally in the British Isles: part 30. Juncaceae, Typhaceae, Sparganiaceae, Araceae. London, Bell, 1973. 41pl.
1 Sparganiaceae pl. 33–34.

SPARGANIUM

4305 Asplund, I.
Embryological studies in the genus *Sparganium*. *Svensk Bot. Tidskr.*, 67(2):177–200(1973)—
1–3, 6, 7 Chromosome numbers.

4306 Harms, V. L.
Taxonomic studies of North American *Sparganium*: 1. *S. hyperboreum* and *S. minimum*. *Canad. J. Bot.*, 51(9):1629–1641(1973)—(Fr); illustrations, dot maps;
3 chromosome number; key.

SPHAEROSEPALACEAE

DIALYCERAS

4307 Leroy, J.-F.
Recherches sur la spéciation et l'endémisme dans la flore malgache: 3. Note sur le genre *Dialyceras* R. Cap. (Sphaerosépalacées). *Adansonia*, 13(1):37–53
5 (1973)—Illustrations, dot map.

coriaceum (R. Cap.) Leroy in *Adansonia*, 13(1):39
5 (1973): *D. parvifolium* var. *coriaceum*.
discolor (R. Cap.) Leroy *op. cit.* 39(1973): *D. parvi-*
5 *folium* f. *discolor*.

RHOPALOCARPUS

4308 Bosser, J.
Sur trois *Rhopalocarpus* de Madagascar. *Adansonia*,
5 13(1):55–62(1973)—(En); illustrations.

SPHENOCLEACEAE

4309 Jafri, S. M. H.
Flora of West Pakistan: no.25. Sphenocleaceae.
6 1972. 3p.—Illustration, map.

4310 Thulin, M.
Flore d'Afrique centrale (Zaïre-Rwanda-Burundi): Spermatophytes. Sphenocleaceae. 1973. [ii]5p.—
5 Illustrations.

STAPHYLEACEAE

STAPHYLEA

4311 Gostyńska-Jakuszewska, M. *and* Wojterska, H.
Cechy diagnostyczne owoców i nasion niektórych gatunków z rodzaju *Staphylea* L. (Diagnostic characters of the fruits and seeds of some species from the genus *Staphylea* L.). *Roczn. Sekc. Dendrol. Pol.*
1–3, 6 *Tow. Bot.*, 27:45–67(1973)—(En).

STERCULIACEAE

BRACHYCHITON

4312 Prentice, J. R.
Two spectacular trees in Australia. *J. Roy. Hort. Soc.*, 98(6):270–271(1973)—Coloured illustration.
7 *Brachychiton acerifolius.*

COLA

4313 Jacob, V. J.
Overcoming self-incompatibility in *Cola nitida* through differential pollination. *Incompat. Newsl.*,
5 no.3:62–64(1973).

4314 Jacob, V. J.
Self-incompatibility mechanism in *Cola nitida.*
5 *Incompat. Newsl.*, no.3: 60–61(1973).

HERITIERA

4315 Kostermans, A. J. G. H.
Nouveautés pour la flore indochinoise. *Adansonia*,
6 13(3):333–340(1973)—(En).

cochinchinensis (Pierre) Kostermans in *Adansonia*,
6 13(3):335(1973): *Tarrietia cochinchinensis.*

HERRANIA

4316 Wingate, J. C.
Anatomy of the seed and seedling of *Herrania nycterodendron. Diss. Abstr. Int., B*, 33(12):5699–5700
4 (1973).

MEGATRITHECA

4317 Arbo, M. M.
Los nectarios foliares de *Megatritheca* (Sterculiaceae).
5 *Darwiniana*, 18(1–2):272–276(1973)—(En).

WALTHERIA

4318 Kohler, E.
Über einen bemerkenswerten Pollendimorphismus in der Gattung *Waltheria* L. (Vorläufige Mitteilung über die heterostyle *W. viscosissima* St. Hil.). *Grana*,
4 13(1):57–64(1973)—(En); illustrations.

XYLOSTERCULIA Kostermans in *Bot. Tidsskr.*,
4 67(4):317(1973).

4319 Kostermans, A. J. G. H.
Some new taxa. *Bot. Tidsskr.*, 67(4):317–323(1973)
4 —Illustration.

pilosa (Ducke) Kostermans in *Bot. Tidsskr.*, 67(4):
4 317(1973): *Sterculia pilosa.*
rugosa (R.Br.) Kostermans *op. cit.* 317(1973):
4 *Sterculia rugosa.*

STILAGINACEAE
see
EUPHORBIACEAE

STYLIDIACEAE

FORSTERA

4320 Philipson, W. R. *and* Philipson, M. N.
A comparison of the embryology of *Forstera* L. and *Donatia* J. R. & G. Forst. *New Zealand J. Bot.*,
7 11(3):449–459(1973).

STYRACACEAE

4321 Silva, M. Freitas da *
Nota prévia sobre Styracaceae novas da Amazônia. *Acta Amazonica*, 1(1):23(1971 publ. 1972). Ref. in
4 *Bull. Torrey Bot. Club*, 99(6):325(1972).

4322 Soukup, J.
Las Ebenáceas, Estiracáceas y Sapotáceas del Perú, sus generos y lista de especies (continuación). *Biota*,
4 9(74):265–271(1973).

HALESIA

4323 Anon.
Silver bell, *Halesia carolina:* storax family (Styracaceae). *Morton Arbor. Quart.*, 9(2):28–29
3 (1973)—Illustrations.

STYRAX

4324 Silva, M. Freitas da
Uma nova *Styrax* (Styracaceae) na Amazônia. *Acta*
4 *Amazonica*, 2(1):17–19(1972)—(En); illustrations.

4325 Steyermark, J. A.
Novedades de los Andes Venezolanos. *Pittieria*, no.4:9–25(1972)—Sp,En; illustration. *Styrax*
4 *andinus.*

andinus Steyermark in *Pittieria*, 4:12(1972)—
4 Venezuela.
discolor M. F. Silva in *Acta Amazonica*, 2(1):17
4 (1972)—Brazil.

SYMPLOCACEAE

4326 Soukup, J.
Las Simplocáceas, Oleáceas, Loganiáceas y Desfontaniáceas del Perú, sus géneros y lista de especies.
4 *Biota*, 9(74):272–282(1973)—Keys.

TACCACEAE

4327 Drenth, E.
Identification lists of Malaysian specimens: 44. Taccaceae. Leyden, Foundation Flora Malesiana,
6 1972:785–796.

4328 Drenth, E.
A revision of the family Taccaceae. *Blumea*, 20(2): 367–406(1972 publ. 1973)—Illustrations, dot maps;
4–7 keys.

TACCA

bibracteata Drenth in *Blumea*, 20(2):395(1972 publ.
6 1973)—Borneo (Sarawak).
ebeltajae Drenth *op. cit.* 401(1972 publ. 1973)—
6 British Solomon Is.
plantaginea (Hance) Drenth *op. cit.* 391(1972 publ.
2, 6 1973): *Schizocapsa plantaginea.*

TETRAGONIACEAE
see
AIZOACEAE

TETRAMELACEAE

TETRAMELES

4329 Hyland, B. P. M.
Tetrameles nudiflora R. Brown (Datiscaceae): a new record for Australia. *Blumea*, 20(2):338(1972 publ.
7 1973).

THEACEAE

CAMELLIA

4330 Ackerman, W. L.
Aneuploidy in the *Camellia*. *J. Hered.*, 64(4):197–
2 202(1973)—Illustration; chromosome numbers.

4331 Bezbaruah, H. P. *and* Gogoi, S. C.
An interspecific hybrid between tea (*Camellia sinensis* L.) and *C. japonica* L. *Proc. Indian Acad. Sci., B*, 76(5):219–220(1972)—Illustrations; chromosome
2 numbers.

4332 Savige, T. J.
Camellia irrawadiensis. *Camellia News*, no.50:13–15
6 (1973)—En; illustration.

4333 Sealy, J. R.
Camellia pitardii var. *yunnanica*. *Curtis's Bot. Mag.*, 179(2):tab.633(1972 publ. 1973)—En; illustration,
2 coloured illustration.

4334 Tateishi, S.
(On the tissue of the leaves of *Camellia sasanqua*, *Camellia sinensis* and their presumed hybrids). *J. Jap.*
2, 6 *Bot.*, 48(7):199–204(1973)—Ja(En).

4335 Valder, P. G.
Life cycle of the *Camellia*. *Camellia News*, no.51:
2, 6 17–20(1973).

4336 Waterhouse, E. G.
Chinese accounts of the Yunnan reticulatas. *Int.*
2 *Camellia J.*, 1(5):14–22(1973).

4337 Woodroof, W. E., *ed.*
Camellia nomenclature. 13th rev. ed. Lynwood, South California Camellia Society, 1972. 187p.—
2, 6 Coloured illustrations.

4338 Yoshioka, M.
Looking for the *Camellia* in South Vietnam. *Camellia News*, no.49: 8–13(1973)—Map. Translated from the *Yokohama Camellia Soc. J.*, no.3, 1973 by
6 H. Haga.

GORDONIA

4339 Kondo, K.
Chromosome numbers of five taxa. *Chromosome Inform. Serv.*, no.15:33–34(1973)—Chromosome
2 numbers. *Gordonia chrysandra* × *G. axillaris.*

KIELMEYERA

rupestris A. P. Duarte in *Arq. Jard. Bot. Rio de*
4 *Janeiro*, 19:217(1973)—Brazil.

PYRENARIA

4340 Keng, H.
Nomenclatural note on Wu P'i Ch'a or the black-barked tea plant. *Taiwania*, 18(1):55–56(1973).
2, 6 *Pyrenaria shingkoensis.*

TERNSTROEMIA

glandulosa Alain in *Phytologia*, 25(5):266(1973)—
4 Hispaniola (Dominican Republic).

THELIGONACEAE

4341 Almeida, M. T. de
Theligonaceae. *Anuár. Soc. Brot.*, 38:157(1972)—
1 Por.

THYMELAEACEAE

4342 Aymonin, G. G.
Chorologie régionale et chorologie générale intérêt pour l'étude du polymorphisme: un exemple chez les Thyméléacées méditerranéennes. *Bull. Soc.*
1, 2 *Bot. France*, 118(9):827–833(1971 publ. 1973)—(En).

4343 Rao, T. A. *and* Bhupal, O. P.
Typology of foliar sclereids in a few taxa of the Thymelaeaceae. *Curr. Sci.*, 41(24):884–886(1972)—
1–7 Illustrations.

DAPHNE

4344 Mathew, B.
Plant awards 1972–73: *Daphne arbuscula* Cel. F.C.C.
1 *Quart. Bull. Alp. Gard. Soc.*, 41(4):325(1973).

4345 Mattanovich, H.
Ein neues Vorkommen von *Daphne laureola* L.
1 in den Karawanken. *Carinthia II*, 82:225–226(1972).

4346 Müller, K.-W.
Gemeiner Seidelbast (*Daphne mezereum*). *Palmengarten*, 37(1):20–21(1973)—Illustrations. Popular
1 article.

DIRCA

4347 Dale Thomas, R.
Dirca palustris (Thymelaeaceae): new to Louisiana.
3 *Sida*, 5(2):132(1973).

GNIDIA

4348 Röer, H. F.
Gnidia polystachya found in the Canary Islands.
1 *Cuad. Bot. Canar.*, 18–19:15–16(1973)—Illustration.

GONYSTYLUS

4349 Airy Shaw, H. K.
Two new taxa in *Gonystylus* Teijsm. & Binnend.
6 (Thymelaeaceae). *Kew Bull.*, 28(2):267–268(1973).

affinis Radlk.
var. **elegans** Airy Shaw in *Kew Bull.*, 28(2):268
6 (1973)—Borneo (Sarawak).
eximius Airy-Shaw *op. cit.* 267(1973)—Borneo
6 (Sarawak).

THYMELAEA

myrtifolia (Poir.) D. A. Webb.
subsp. montana Malagarriga, Las Subesp. y Variac.
1 Geogr.: 5(1973), without type—Spain ?

WIKSTROEMIA

4350 Degener, O. *and* Degener, I.
Flora Hawaiiensis: new illustrated flora of the Hawaiian Islands. Thymelaeaceae: *Wikstroemia.*
6 1973. [4]p.—Illustrations.

TILIACEAE

4351 Yeim Yok Siv *and* Paris, R. R.
Sur quelques plantes à flavonoïdes du Cambodge appartenant aux genres *Cananga*, *Colona*, *Grewia*, *Leea* et *Melastoma*. *Pl. Méd. Phytothér.*, 6(4):299–305 (1972)—also publ. in *Trav. Lab. Matière Méd. Pharm.*
5 *Galénique Fac. Pharm. Paris*, 57(2) (1973).

DIPLODISCUS

scortechinii (King) Ashton ex Kochummen in
6 Tree Fl. Malaya, 2:396(1973): *Pentace scortechinii.*

GREWIA

hirsuta (Korth.) Kochummen in Tree Fl. Malaya, 2:
6 397(1973): *Omphacarpus hirsutus.*

SPARRMANNIA

4352 Killick, D. J. B.
Sparrmannia africana. Flow. Pl. Afr., 42(3–4):pl.1663
5 (1973)—En; coloured illustration.

TILIA

4353 Hatusima, S.
A new linden from Japan. *J. Geobot.*, 20(1):2–4
2 (1972)—Ja,En; illustrations.

4354 Kallio, E. *and* Godman, R. M.
American basswood: an American wood. *U.S. Dep. Agric. F.S.*, 219.8p.(1973)—Illustrations, map.
3 *Tilia americana.*

chugokuensis Hatusima in *J. Geobot.*, 20(1):2
2 (1972)—Japan.

TRAPACEAE

4355 Vu Van Cuong, *and* Vidal, J. E.
Trapaceae. Flore du Cambodge, du Laos et du
6 Viêt-nam, fasc. 14:40–47(1973)—Illustrations; key.

TRAPA

4356 Shilov, M. P.
K ekologii nekotorȳkh redkikh vodnȳkh rasteniĭ na severnom predele ikh rasprostraneniya. (To the ecology of some rare aquatic plants on the northern limits of their area). *Bot. Zhurn.*, 58(10):1554–1559
1, 2 (1973).

4357 Vasil'ev, V. N.
Novȳe vidȳ vodyanogo orekha (*Trapa* L.). (Generis *Trapa* L. species novae). *Novosti Sist. Vȳssh.*
2 *Rast.*, 10:197–211(1973)—Illustrations.

alatyrica Spryg. ex Vasil'ev in *Novosti Sist. Vȳssh.*
1 *Rast.*, 10:207(1973)—U.S.S.R. (European Russia).
caspica Vasil'ev *op. cit.* 203(1973)—U.S.S.R.
1 (European Russia).
2 **fastigiata** Vasil'ev *op. cit.* 205(1973)—China.
kazakorum Vasil'ev *op. cit.* 201(1973)—U.S.S.R.
1 (European Russia).
1 **krauselii** Vasil'ev *op. cit.* 210(1973)—Poland.
metschorica Vasil'ev *op. cit.* 208(1973)—U.S.S.R.
1 (Ukraine).
okensis Vasil'ev *op. cit.* 201(1973)—U.S.S.R.
1 (European Russia).
pseudorossica Vasil'ev *op. cit.* 199(1973)—U.S.S.R.
1 (European Russia).
sajanensis Vasil'ev *op. cit.* 202(1973)—U.S.S.R.
2 (Middle Asia).
turbinata Vasil'ev *op. cit.* 207(1973)—U.S.S.R.
1 (European Russia).
ucrainica Vasil'ev *op. cit.* 209(1973)—U.S.S.R.
1 (Ukraine).
uralensis Vasil'ev *op. cit.* 200(1973)—U.S.S.R.
1 (Urals).
wolgensis Vasil'ev *op. cit.* 197(1973)—U.S.S.R.
1 (European Russia).

TRITICHACEAE

see
PODOSTEMACEAE

TROPAEOLACEAE

4358 Sparre, B.
Flora ilustrada catarinense: Tropeoláceas. 1972.
4 26p.—Illustrations, dot maps; keys.

4359 Sparre, B.
Flora of Ecuador: 89. Tropaeolaceae. *Op. Bot. Ser.*
4 *B.*, no.2.30p.(1973)—Illustrations; keys.

TROPAEOLUM

4360 Fernandez, J.
Sobre la dispersion meridional de *Tropaeolum tuberosum* R.P. *Bol. Soc. Argent. Bot.*, 15(1):106–112
4 (1973)—Illustrations, map.

asplundii Sparre in *Op. Bot., B* (Fl. Ecuador), 2:26
4 (1973)—Ecuador.
4 **brideanum** Sparre *op. cit.* 28(1973)—Ecuador.
carchense Killip ex Sparre *op. cit.* 20(1973)—
4 Ecuador.
fintelmannii Wagener ex Schlchtd.
var. **olmosense** (Mansf.) Sparre *op. cit.* 16(1973):
4 *T. olmosense.*
4 **harlingii** Sparre *op. cit.* 15(1973)—Ecuador.
4 **leonis** Sparre *op. cit.* 17(1973)—Ecuador.
4 **magnificum** Sparre *op. cit.* 6(1973)—Ecuador.
penthaphyllum Lam.
subsp. **megapetalum** (Buch.) Sparre in Fl. Ilustr. Catar., 1(Tropeolác.): 8(1972): *T. penthaphyllum*
4 var. *megapetalum.*
pubescens H. B. K.
var. **coma-beatae** Sparre in *Op. Bot., B* (Fl.
4 Ecuador), 2:20(1973)—Ecuador.
sanctae-catharinae Sparre in Fl. Ilustr. Catar.,
4 1(Tropeolác.):13(1972)—Brazil.
tuberosum R. & P.
subsp. **silvestre** Sparre in *Op. Bot., B* (Fl. Ecuador),
4 2:9(1973)—Ecuador.
warmingianum Rohrb.
var. **glaziovii** (Buch.) Sparre in Fl. Ilustr. Catar.,
4 1(Tropeolác.): 22(1972): *T. glaziovii.*
subsp. **occidentale** Sparre *op. cit.* 22(1972)—
4 Bolivia.

TURNERACEAE

STAPFIELLA

4361 Verdcourt, B.
Tropical African plants: 33. A new variety of *Stapfiella lucida* (Turneraceae). *Kew Bull.*, 28(2):320
5 (1973).

lucida Robyns
var. **pubescens** Verdcourt in *Kew Bull.*, 28(2):320
5 (1973)—Burundi.

TURNERA

4362 Moura, C. A. F. de
A new species of *Turnera* from Pará, Brazil. *Acta Bot. Neerl.*, 22(2):159–163(1973)—(Por);
4 illustrations.

4363 Steyermark, J. A.
Descripcion detallada de dos especies raras de Venezuela. *Pittieria*, no.4:27–32(1972)—En(Sp);
4 illustration.

tapajoensis Moura in *Acta Bot. Neerl.*, 22(2):159
4 (1973)—Brazil.

TYPHACEAE

4364 Ross-Craig, S.
Drawings of British plants: being illustrations of the species of flowering plants growing naturally in the British Isles: part 30. Juncaceae, Typhaceae, Sparganiaceae, Araceae. London, Bell, 1973. 41pl.
1 Typhaceae pl.35–38.

TYPHA

4365 Čvančara, A. *and* Šourková, M.
Poznámky k rozšíření a variabilitě *Typha laxmannii* Lepech. v Československu. (Bemerkungen zur Verbreitung und Variabilität von *Typha laxmannii* Lepech. in der Tschechoslowakei). *Preslia*, 45(3):
1 265–275(1973)—(Ge); illustrations, dot map; key.

4366 Geissert, F.
Typha laxmannii Lepechin (= *T. stenophylla* Fischer & Meyer) in der Pfalz. *Mitt. Bad. Landesver. Naturk. Naturschutz*, 16(3):547–548(1971 publ. 1972)—
1 Illustration.

4367 Lee, D. W. *and* Fairbrothers, D. E.
Enzyme differences between adjacent hybrid and parent populations of *Typha*. *Bull. Torrey Bot. Club*,
3 100(1):3–11(1973). Local plant populations.

4368 McNaughton, S. J.
Comparative photosynthesis of Quebec and California ecotypes of *Typha latifolia*. *Ecology*, 54(6):
3 1260–1270(1973).

4369 Pilous, Z. *and* Houfek, J.
Typha laxmanni Lepech. u Chotína (okres Komárno) neroste. (*Typha laxmanni* Lepech. wächst nicht bei Chotín (Bezirk Komárno, südslowaki)).
1 *Zprávy Česk. Bot. Spol.*, 8(1):64–65(1973)—(Ge).

4370 Tutayuk, V. Kh. *and* Arazov, B. M.
(Anatomicheskoe stroenie vegetativnȳkh organov rogoza uzkolistogo (*Typha angustifolia*)). *Izv. Akad.*
1–3, 5 *Nauk Az. SSR, Biol. Nauk*, no.4:3–11(1973)—Az(Rus).

ULMACEAE

4371 Diglio, A. P. L.
Las ulmaceas chaqueñas. Castelar, Ministerio Agricultura y Ganaderia Nacion, 1972. 9p. (Notas preliminares para la flora chaqueña (Formosa, Chaco
4 y Santiago del Estero) no. 3)—Illustrations.

4372 Phuphathanaphong, L.
Ulmaceae. *Thai Forest Bull., Bot.*, no.7:39–48
2, 6 (1973)—En; illustration; keys.

APHANANTHE

yunnanensis (H. H. Hu) Phuphathanaphong in *Thai Forest Bull., Bot.*, 7:41(1973): *Gironniera*
2, 6 *yunnanensis*.

CELTIS

4373 Eroshkov, N. A.
Morfologicheskoe opisanie vskhodov karkasa introduktsirovannogo v Turkmeniyu. (Morphological description of seedlings of hackberry introduced into Turkmenistan). *Izv. Akad. Nauk Turkm. SSR, Biol.*
2 *Nauk*, 5:42–47(1973)—(En); key.

4374 Poli, E., Maugeri, G. *and* D'Urso, A.
La *Celtis tournefortii* Lam. sull'Etna. (The Etnean *Celtis tournefortii* Lam.). *Inform. Bot. Ital.*, 5(2):157–
1 158(1973)—(En).

4375 Smalley, G. W.
Hackberry: an American wood. *U.S. Dep. Agric.*
3 *F.S.*, 238. 7p.(1973)—Illustrations, maps.

ULMUS

4376 Bate-Smith, E. C. *and* Richens, R. H.
Flavonoid chemistry and taxonomy in *Ulmus*.
1–7 *Biochem. Syst.*, 1(3):141–146(1973).

4377 Chen, P. Y. S. *and* Schlesinger, R. C.
Elm: an American wood. *U.S. Dep. Agric. F.S.*,
3 233. 7p.(1973)—Illustrations, maps.

4378 Grudzinskaya, I. A.
O sistematicheskom polozhenii i areale sektsii *Chaetoptelea* roda *Ulmus*. (On taxonomic position and area of the section *Chaetoptelea*, genus *Ulmus*.). *Bot.*
3, 4, 6 *Zhurn.*, 59(1):61–66(1973)—Map.

minor Mill.
var. **suberosa** (Mönch.) Soó in *Acta Bot. Acad. Sci.*
1 *Hung.*, 18(1–2):177(1973): *U. suberosa*.

UMBELLIFERAE

4379 Hedge, I. C.
Umbelliferae in 1672 and 1972. *Notes Roy. Bot.*
1–7 *Gard. Edinburgh*, 32(2):151–158(1973).

4380 Hedge, I. C. *and* Lamond, J. M.
A review of the tribe Echinophoreae (Umbelliferae). *Notes Roy. Bot. Gard. Edinburgh*, 32(2):167–188 (1973)—Illustrations, map, dot maps; keys.
Pycnocycla, Dicyclophora, Echinophora, Anisosciadium,
1, 2, 5, 6 *Ergocarpon, Thecocarpus*.

4381 Heywood, V. H.
The Umbelliferae of Macaronesia. *Cuad. Bot.*
1 *Canar.*, no.18–19:17–23(1973).

4382 Hore, A.
New chromosome reports and cytotaxonomy of Umbelliferae. *Proc. Indian Sci. Congr.*, 58(3):473
6 (1971)—Chromosome numbers. Abstract only.

4383 Lloyd, D. G.
Sex ratios in sexually dimorphic Umbelliferae.
1–7 *Heredity*, 31(2):239–249(1973).

4384 Mathias, M. E. *and* Constance, L.
New and reconsidered Mexican Umbelliferae.
Contrib. Univ. Mich. Herb., 11(1):1–24(1973)—
4 Illustrations; chromosome numbers; key.

4385 Mathias, M. E., Constance, L. *and* Araujo, D.
Flora ilustrada catarinense: Umbelíferas. 1972.
4 205p.—Illustrations, dot maps; keys.

4386 Queirós, M.
Contribuição ao conhecimento citotaxonómico das Spermatophyta de Portugal: 7. Umbelliferae. *Anuár. Soc. Brot.*, 38:293–314(1972)—(Fr); chromo-
1 some numbers.

4387 Sobolevskaya, K. A., Tyurina, E. V. *and* Gus'kova, I. N.
Efirnomaslichnȳe rasteniya sem. zontichnȳkh vo flore zapadnoĭ Sibiri. (The umbelliferous essential oil bearing plants in the flora of West Siberia).
2 *Rast. Resursȳ*, 9(1):68–72(1973).

4388 Thorne, R. F.
Inclusion of the Apiaceae (Umbelliferae) in the Araliaceae. *Notes Roy. Bot. Gard. Edinburgh*, 32(2):
1–7 161–165(1973).

ACRONEMA

4389 Babu, C. R.
Acronema johrianum Babu (Apiaceae): a new species from the Himalayas of Nepal. *Brittonia*,
6 25(2):159–161(1973)—Illustration.

johrianum Babu in *Brittonia*, 25(2):159(1973)—
6 Nepal.

AGROCHARIS

4390 Heywood, V. H.
The taxonomic position of *Agrocharis* Hochst. and allied genera. *Notes Roy. Bot. Gard. Edinburgh*, 32(2):
5 211–215(1973)—Illustrations.

ANGELICA

4391 Kitagawa, M.
(Notulae fractae ob floram Asiae orientalis: 30). *J. Jap. Bot.*, 48(8):235–237(1973)—Ja. *Angelica*
2 *furcijuga*, *A. ubatakensis*.

4392 Méndez, J., Rubido, J. *and* Mato, M. C.
Coumarins of *Angelica pachycarpa*. *Experientia*,
1 29(3):371–372(1973)—(Sp).

4393 Okada, M.
(On the flower-bud control of medicinal angelica 'Obuka-toki' (*Angelica acutiloba* Kit.)). *J. Jap. Bot.*,
2 48(11):344(1973)—Ja.

4394 Okada, M.
(On a peculiar technique of medicinal angelica 'Obuka-toki' (*Angelica acutiloba* Kit.)). *J. Jap. Bot.*,
2 48(11):326(1973)—Ja.

4395 Weinert, E.
Die taxonomische Stellung und das Areal von *Angelica archangelica* L. und *A. lucida* L. *Feddes Repert.*,
1–4 84(4):303–313(1973)—(En); maps.

archangelica L.
var. **decurrens** (Ledeb.) Weinert in *Feddes Repert.*,
1, 2, 6 84(4):309(1973): *Archangelica decurrens*.
f. **himalaica** (Clarke) Weinert *op. cit.* 310(1973):
6 *Archangelica officinalis* var. *himalaica*.
f. **komarovii** (Schischk.) Weinert *op. cit.* 310
2 (1973): *Archangelica komarovii*.
f. **tschimganica** (Korov.) Weinert *op. cit.* 309
2 (1973): *Archangelica decurrens* var. *tschimganica*.

lucida L.
var. **angustifolia** (Fr. Schmidt ex Maxim.) Weinert *op. cit.* 307(1973): *Physolophium*
1, 2 *saxatile* var. *angustifolia*.
var. **trichocarpa** (Hara) Weinert *op. cit.* 307
2 (1973): *A. trichocarpa*.

ubatakensis (Makino) Kitagawa
var. **valida** Kitagawa in *J. Jap. Bot.*, 48(8):236
2 (1973)—Japan.

ANTHRISCUS

caucalis M.B.
var. **gymnocarpa** (Moris) Soó in *Acta Bot. Acad. Sci. Hung.*, 18(1–2):175(1973): (Basionym not
1 stated).

APIUM

4396 Longevialle, M.
Sur la présence à Paris de deux Ombellifères étrangères à la flore française: *Hydrocotyle sibthorpioides* Lam. et *Apium ammi* Urb., adventices envahissantes au Jardin des Plantes. *Bull. Soc. Bot.*
1 *France*, 118(9):835–836(1971 publ. 1973)—(En).

ARRACACIA

ebracteata (Rose) Mathias & Constance in *Contrib. Univ. Mich. Herb.*, 11(1):12(1973): *Coaxana*
4 *ebracteata*.
4 **filipes** Mathias & Constance *op. cit.* 6(1973)—Mexico.
macvaughii Mathias & Constance *op. cit.* 8(1973)—
4 Mexico.
papillosa Mathias & Constance *op. cit.* 10(1973)—
4 Mexico.

BUNIUM

4397 Engstrand, L.
Generic delimitation of *Bunium*, *Conopodium* and *Geocaryum* (Umbelliferae). *Bot. Notiser*, 126(2):146–
1, 2 154(1973).

4398 Silvestre, S.
Estudio taxonómico de los géneros *Conopodium* Koch y *Bunium* L. en la Península Ibérica: 2. Parte sistemática. *Lagascalia*, 3(1):3–48(1973)—(En); dot
1 maps; chromosome numbers; keys.

BUPLEURUM

4399 Linchevskiĭ, I. A.
Zametki o rode *Bupleurum* L. (Apiaceae), 1. (Notulae de genere *Bupleuro* L. (Apiaceae), 1).
2 *Novosti Sist. Vȳssh. Rast.*, 10:222–224(1973).

4400 Šourková, M.
Bupleurum × *sitenskýi* Šourková (*B. falcatum* L. × *B. longifolium* L.)—ein neuer Bastard innerhalb der Gattung *Bupleurum*. *Novit. Bot.*, 19–27(1972 publ.
1 1973)—Illustrations; chromosome number.

falcatum L.
subsp. **dilatatum** (Schur) Soó in *Acta Bot. Acad. Sci. Hung.*, 18(1–2):177(1973): *B. falcatum* var.
1 *dilatatum*.
feranense Linchevskiĭ in *Novosti Sist. Vȳssh. Rast.*,
2 10:222(1973)—U.S.S.R. (Middle Asia).

longicaule Wall. ex DC.
var. **hazarica** (Nasir) Nasir in R. R. Stewart, Annot. Cat. Vasc. Pl. W. Pakistan & Kashmir:
6 513(1972): *B. hazarica.*
× **sitenskyi** Šourková in *Novit. Bot.*, 1972:21(1972 publ. 1973)—Czechoslovakia. (*B. falcatum* × *B.*
1 *longifolium*).

CARUM

4401 Henderson, A. A. R.
Carum verticillatum. Glasgow Nat., 19(1):65(1973)—
1 En. Local records.

CACHRYS

4402 Gruenberg-Fertig, I., Heyn, C. C. *and* Herrnstadt, I.
Typification of *Cachrys libanotis* L. (Umbellif.).
1 *Taxon*, 22(4):425–434(1973)—Illustrations.

CEPHALOPODUM Korovin in *Izv. Akad. Nauk*
2 *Tadzh. SSR, Biol. Nauk*, 1(50):17(1973).

4403 Korovin, E. P.
Novy͡e taksony͡ semeĭstva zontichny͡kh iz Pamiro-Alaya (soobshchenie 1). *Izv. Akad. Nauk Tadzh. SSR,*
2 *Biol. Nauk*, 1(50):14–19(1973)—(Ta); illustration.

badachshanicum Korovin in *Izv. Akad. Nauk Tadzh. SSR, Biol. Nauk*, 1(50):17(1973)—U.S.S.R.
2 (Middle Asia).

CHAEROPHYLLUM

hirsutum L.
subsp. **magellense** (Ten.) Pignatti in *Giorn. Bot.*
1 *Ital.*, 107(5):220(1973): *C. magellense.*

CNIDIUM

4404 Kitagawa, M.
(Notulae fractae ob floram Asiae orientalis: 30). *J. Jap. Bot.*, 48(8):235–237(1973)—Ja. *Cnidium*
2 *monnieri.*

monnieri (L.) Cusson
var. **formosanum** (Yabe) Kitagawa in *J. Jap. Bot.*,
2 48(8):237(1973): *C. formosanum.*

COAXANA

bambusioides Mathias & Constance in *Contrib. Univ.*
4 *Mich. Herb.*, 11(1):13(1973)—Mexico.

CONOPODIUM

4405 Engstrand, L.
Generic delimitation of *Bunium*, *Conopodium* and *Geocaryum* (Umbelliferae). *Bot. Notiser*, 126(2):
1, 2 146–154(1973).

4406 Silvestre, S.
Estudio taxonómico de los géneros *Conopodium* Koch y *Bunium* L. en la Península Ibérica: 2. Parte sistemática. *Lagascalia*, 3(1):3–48(1973)—(En); dot
1 maps; chromosome numbers; keys.

glaberrimum (Desf.) Engstrand in *Bot. Notiser*,
1 126(2):153(1973): *Scandix glaberrima.*
majus (Gouan) Loret
subsp. **ramosum** (Costa) Silvestre in *Lagascalia*,
1 2(2):151(1972): *C. ramosum.*

DAUCUS

4407 Cerceau-Larrival, M. T. *and* Roland-Heydacker, F.
Ultrastructure du pollen de *Daucus carrota* L. en microscopies à balayage et à transmission. *Compt. R. Hebd. Séances Acad. Sci., D* (Paris), 275(21):2331–
1, 2 2333(1972).

4408 Murray, B. G. *and* Owens, S. J.
Interspecific incompatibility in the Umbelliferae and Gramineae. *Incompat. Newsl.*, no.2:15–18(1973).
1–3 Chromosome numbers. *Daucus.*

DIPLOLOPHIUM

4409 Cannon, J. F. M.
Studies in tropical African Umbelliferae: 1. *Frommia ceratophylloides* and the *Diplolophium buchananii* complex. *Notes Roy. Bot. Gard. Edinburgh*, 32(2):195–
5 201(1973)—Illustrations, dot map.

buchananii (Benth. ex Oliv.) Norman
subsp. **swynnertonii** (Bak. f.) Cannon in *Notes Roy. Bot. Gard. Edinburgh*, 32(2):200(1973):
5 *Physotrichia swynnertonii.*

DONNELLSMITHIA

ampulliformis Mathias & Constance in *Contrib.*
4 *Univ. Mich. Herb.*, 11(1):2(1973)—Mexico.
juncea (Humb. & Bonpl. ex Spreng.) Mathias &
4 Constance *op. cit.* 4(1973): *Peucedanum junceum.*
var. **purpurea** (Coult. & Rose) Mathias &
4 Constance *op. cit.* 5(1973): *Museniopsis purpurea.*
pinnatisecta (Riley) Mathias & Constance *op. cit.*
4 1(1973): *Museniopsis biennis* var. *pinnatisecta.*
ternata (S. Wats.) Mathias & Constance *op. cit.* 5
4 (1973): *Eulophus ternatus.*

ECHINOPHORA

subgen. **Chrysophora** (DC.) Hedge & J. M. Lamond in *Notes Roy. Bot. Gard. Edinburgh*, 32(2):
1, 2, 6 180(1973): *Echinophora* sect. *Chrysophora.*
subgen. **Lamprosciadium** Hedge & J. M. Lamond
2 *op. cit.* 181(1973).
subgen. **Trichophora** Hedge & J. M. Lamond *op. cit.*
2 179(1973).

ERYNGIUM

jaliscense Mathias & Constance in *Contrib. Univ.*
4 *Mich. Herb.*, 11(1):23(1973)—Mexico.

FROMMIA

4410 Cannon, J. F. M.
Studies in tropical African Umbelliferae: 1. *Frommia ceratophylloides* and the *Diplolophium buchananii* complex. *Notes Roy. Bot. Gard. Edinburgh*,
5 32(2):195–201(1973)—Illustrations, dot map.

GEOCARYUM

4411 Engstrand, L.
Generic delimitation of *Bunium*, *Conopodium* and *Geocaryum* (Umbelliferae). *Bot. Notiser*, 126(2):146–
1, 2 154(1973).

GYNOPHYGE Gilli in *Feddes Repert.*, 84(3):181
5 (1973).

4412 Gilli, A.
Gynophyge eine neue Umbelliferengattung. *Feddes*
5 *Repert.*, 84(3):181–184(1973)—(En); key.

tansaniensis Gilli in *Feddes Repert.*, 84(3):181(1973)—
5 Tanzania.

HERACLEUM

4413 Basargin, D. D.
O nekotory͡kh morfologicheskikh anomaliyakh u borshchevikov (*Heracleum moellendorffii* i *H. dulce*) Dal'nego Vostoka. (On some morphological anomalies in *Heracleum moellendorffii* and *H. dulce* of
2 Far East). *Bot. Zhurn.*, 58(5):697–700(1973).

4414 McClintock, D.
Heracleum sphondylium × *H. mantegazzianum*.
1, 2 *Watsonia*, 9(4):429–430(1973)—En.

sphondylium L.
subsp. **flavescens** (Willd.) Soó in *Acta Bot. Acad.*
1 *Sci. Hung.*, 18(1–2):175(1973): *H. flavescens*.

HYDROCOTYLE

4415 Longevialle, M.
Sur la présence à Paris de deux Ombellifères étrangères à la flore française: *Hydrocotyle sibthorpioides* Lam. et *Apium ammi* Urb., adventices envahissantes au Jardin des Plantes. *Bull. Soc. Bot. France*,
1 118(9):835–836(1971 publ. 1973)—(En).

chiangdaoensis Murata in *Acta Phytotax. Geobot.*,
6 25(4–6):97(1973)—Thailand.

IMPERATORIA

4416 Kopecký, K.
Die Beziehung zwischen Siedlungsgeschichte und Verbreitung von *Imperatoria ostruthium* im Gebirge Orlické hory (Adlergebirge, Nordostböhmen). *Folia Geobot. Phytotax.*, 8(3):241–248(1973)—(En,Ge);
1 dot maps.

LEIBERGIA

4417 Mathias, M. E. *and* Constance, L.
Another look at *Leibergia* Coult. & Rose (Umbelli-
3 ferae). *Madroño*, 22(2):71–73(1973)—Illustrations.

LEVISTICUM

4418 McVaugh, R.
Report of the Committee for Spermatophyta: conservation of generic names. proposal 278(6083)
1–3 *Levisticum* Hill. *Taxon*, 22(1):154(1973).

LIBANOTIS

4419 Bȳlova, A. M. *and* Grosheva, N. P.
Morfologicheskaya i fiziologobiokhimicheskaya kharakteristika vozrastnȳkh sostoyanii u *Libanotis intermedia* Rupr. (Morphological and physiologo-biochemical characteristics of age states of *Libanotis intermedia* Rupr.). *Bot. Zhurn.*, 58(10):1508–1515
1, 2 (1973)—Illustrations.

4420 Pimenov, M. G. *and* Sdobnina, L. I.
O nekotorȳkh vidakh *Libanotis* (Hall.) Zinn, opisannȳkh iz Kazakhstana. (On some species of *Libanotis* (Hall.) Zinn, described for Kazakhstan).
2 *Bot. Zhurn.*, 58(10):1485–1492(1973)—Dot maps.

pyrenaica (L.) Bourg.
subsp. **athamanthoides** (Spr.) Soó in *Acta Bot. Acad. Sci. Hung.*, 18(1–2):175(1973): *Ligusticum*
1 *pyrenaica*.

LIGUSTICUM

stewartii (Hiroe) Nasir in R. R. Stewart, Annot. Cat. Vasc. Pl. W. Pakistan & Kashmir: 520(1972):
6 *Selinum stewartii*.

MANDENOVIA Alava in *Notes Roy. Bot. Gard.*
2 *Edinburgh*, 32(2):191(1973).

4421 Alava, R.
A new genus of Umbelliferae (Tordyliinae) from Caucasia. *Notes Roy. Bot. Gard. Edinburgh*, 32(2):
2 189–193(1973)—Illustrations; chromosome number.

komarovii (Manden.) Alava in *Notes Roy. Bot. Gard.*
2 *Edinburgh*, 32(2):191(1973): *Tordylium komarovi*.
f. **crenatiloba** Alava *op. cit.* 193(1973)—U.S.S.R.
2 (Caucasus).

MEUM

4422 Reichert, H.
Verbreitung und Soziologie der Bärwurz (*Meum athamanticum* Jacq.) im Hunsrück. *Decheniana*, 125
1 (1–2):15–22(1972)—(En); illustration.

athamanticum Jacq.
subsp. **nevadense** (Bss.) Malagarriga, Las Subesp. y
1 Variac. Geogr.: 12(1973): *M. nevadense*.

MUTELLINA

mutellinoides (Crantz) Holub in *Folia Geobot.*
1 *Phytotax.*, 8(2):177(1973): *Laserpitium mutellinoides*.

ODONTITES

linkii Boiss.
subsp. **cretica** (Boiss.) Greuter in *Ann. Mus*
1 *Goulandris*, 1:51(1973): *O. cretica*.

OENANTHE

4423 Bertová, L.
Taxonómia druhov rodov *Phellandrium* L. a *Oenanthe* L. na Slovensku. *Biol. Práce*, 19(4).76p. (1973)—(Rus,Ge); illustrations, dot maps; chromo-
1 some numbers; keys.

4424 Wattez, J. R., Vignon, F. *and* Douchet, M.
L'*Oenanthe* safranée, une espèce nouvelle pour la flore de la Somme. *Rev. Féd. Franç. Soc. Sci. Nat.*,
1 12(52):1–4(1973).

javanica (Bl.) DC.
subsp. **linearis** (Wall. ex DC.) Murata in *Acta*
6 *Phytotax. Geobot.*, 25(4–6):103(1973): *O. linearis*.
subsp. **stolonifera** (DC.) Murata *op. cit.* 105(1973):
6 *O. stolonifera*.
silaifolia M. Bieb.
f. **grandisecta** (E. Simon) Bertová in *Biol. Práce*,
1 19(4):34(1973): *O. biebersteinii* f. *grandisecta*.
subsp. **hungarica** (E. Simon) Bertová *op. cit.* 37
1 (1973): *O. biebersteinii* f. *hungarica*.
f. **lanceolata** (E. Simon) Bertová *op. cit.* 34(1973):
1 *O. biebersteinii* var. *lanceolata*.
subsp. **media** (Griseb.) Bertová *op. cit.* 38(1973):
1 *O. media*.
subsp. **montana** (E. Simon) Bertová *op. cit.* 39
1 (1973): *O. biebersteinii* f. *montana*.

PACHYPLEURUM

4425 Pimenov, M. G. *and* Soobnina, L. I.
O sistematicheskom polozhenii *Pachypleurum mucronatum* (Schrenk) Schischk. (On the taxonomic status of *Pachypleurum mucronatum* (Schrenk) Schischk. *Byull. Mosk. Obshch. Ispȳt. Prir., Biol.*,
1, 2 78(5):132–140(1973)—(En); dot map.

PARALIGUSTICUM Tikhomirov in *Byull. Mosk.*
2 *Obshch. Ispȳt. Prir., Biol.*, 78(1):107(1973).

4426 Tikhomirov, V. N.
Paraligusticum V. Tichom. gen. nov.—novȳĭ rod iz semeĭstva Umbelliferae. (*Paraligusticum* V. Tichom.—genus novum Umbelliferarum). *Byull. Mosk.*
2 *Obshch. Ispȳt. Prir., Biol.*, 78(1):100–108(1973).

discolor (Ledeb.) Tikhomirov in *Byull. Mosk.*
Obshch. Ispȳt. Prir., Biol., 78(1):107(1973):
2 *Ligusticum discolor*.

PAULIA Korovin in *Izv. Akad. Nauk Tadzh. SSR, Biol. Nauk*, 1(50):14(1973). 2

4427 Korovin, E. P.
Novȳe taksonȳ semeĭstva zontichnȳkh iz Pamiro-Alaya (soobshchenie 1). *Izv. Akad. Nauk Tadzh. SSR, Biol. Nauk*, 1(50):14–19(1973)—(Ta);
2 illustration.

ovczinnikovii Korovin in *Izv. Akad. Nauk Tadzh. SSR, Biol. Nauk*, 1(50):14(1973)—U.S.S.R.
2 (Middle Asia).

PETROSELINUM

crispum (Mill.) A. W. Hill
subsp. **tuberosum** (Bernh. ex Rchb.) Soó in *Acta Bot. Acad. Sci. Hung.*, 18(1–2):177(1973): *P.*
1 *tuberosum.*

PEUCEDANUM

4428 Cannon, J. F. M.
Studies in tropical African Umbelliferae: 2. The genus *Peucedanum* in the *Flora Zambesiaca* area, with the description of a new species. *Garcia de Orta, Sér.*
5 *Bot.*, 1(1–2):45–48(1973)—(Por); illustration; key.

4429 Kowal, T. *and* Wojterska, H.
Morfologiczne i anatomiczne cechy diagnostyczne owoców wybranych gatúnkow rodzaju *Peucedanum* L. (The morphological and anatomical diagnostic features of fruits of the selected species of the genus *Peucedanum* L.). *Práce Kom. Biol., Poznań, Tow. Przy. Nauk, Mat.-Przyrod.*, 35(7):33–67(1973)—
1, 2, 5 Illustrations; key.

4430 Tyurina, E. V.
Anatomicheskoe stroenie chereshka u nekotorȳkh vidov *Peucedanum. Byull. Glavn. Bot. Sada* (Moscow),
1, 2, 5, 6 89:55–60(1973).

rhodesicum Cannon in *Garcia de Orta, Sér. Bot.*,
5 1(1–2):47(1973)—Rhodesia.

PHELLANDRIUM

4431 Bertová, L.
Taxonómia druhov rodov *Phellandrium* L. a *Oenanthe* L. na Slovensku. *Biol. Práce*, 19(4). 76p. (1973)—(Rus,Ge); illustrations, dot maps;
1 chromosome numbers; keys.

aquaticum L.
var. **microcarpum** (Beck) Bertová in *Biol. Práce*, 19(4):23(1973): *Oenanthe phellandrium* var.
1 *microcarpum.*

PHLYCTIDOCARPA

4432 Theobald, W. L. *and* Cannon, J. F. M.
A survey of *Phlyctidocarpa* (Umbelliferae) using the light and scanning electron microscope. *Notes Roy. Bot. Gard. Edinburgh*, 32(2):203–210(1973)—
5 Illustrations.

PIMPINELLA

saxifraga L.
var. **procera** (Weide) Soó in *Acta Bot. Acad. Sci. Hung.*, 18(1–2):176(1973): *P. saxifraga* subsp.
1 *procera.*
tragium Vill.
subsp. **balearica** (Knoche) Malagarriga, Las Subesp. y Variac. Geogr.: 13(1973): *P. tragium*
1 var. *balearica.*

PRANGOS

4433 Korovin, E. P.
Novȳe taksonȳ semeĭstva zontichnȳkh iz Pamiro-Alaya (soobshchenie 1). *Izv. Akad. Nauk Tadzh. SSR, Biol. Nauk*, 1(50):14–19(1973)—(Ta);
2 illustration.

hissarica Korovin in *Izv. Akad. Nauk Tadzh. SSR,*
2 *Biol. Nauk*, 1(50):18(1973)—U.S.S.R. (Middle Asia).

PRIONOSCIADIUM

bellii Mathias & Constance in *Contrib. Univ. Mich.*
4 *Herb.*, 11(1):15(1973)—Mexico.
lilacinum Mathias & Constance *op. cit.* 17(1973)—
4 Mexico.

RHODOSCIADIUM

macvaughiae Mathias & Constance in *Contrib. Univ.*
4 *Mich. Herb.*, 11(1):21(1973)—Mexico.
rzedowskii Mathias & Constance *op. cit.* 19(1973)—
4 Mexico.

SCALIGERIA

stewartianum (Nasir) Nasir in R. R. Stewart, Annot. Cat. Vasc. Pl. W. Pakistan & Kashmir: 525
6 (1972): *Bunium stewartianum.*
vvedenskyi Kamelin in *Bot. Zhurn.*, 58(5):630
2 (1973)—U.S.S.R. (Middle Asia).

SCANDIX

4434 Vinogradova, V. M.
O novom dlya SSSR vide roda *Scandik* L. (De generis *Scandix* L. specie pro flora URSS nova). *Novosti Sist. Vȳssh. Rast.*, 10:218–221(1973)—
1, 2 Illustrations.

SCHRENKIA

transitoria (K.-Pol.) Kamelin in *Bot. Zhurn.*, 58(5):
2 630(1973): *Anidrum transitorium.*

SESELI

4435 Burduja, C.
Seseli hippomarathrum Jacq. în flore României. (*Seseli hippomarathrum* Jacq. dans la flore de Roumanie). *Ocrot. Nat.*, 17(2):197–201(1973)—
1 Illustration.

4436 Markova, M. *and* Robeva, P.
Karyological study of three Bulgarian taxa of the genus *Seseli* L. *Izv. Bot. Inst.* (Sofia), 24:199–213
1 (1973)—(Bu); chromosome numbers.

sect. **Condensatae** Pimenov & Sdobnina in *Byull.*
2 *Mosk. Obshch. Ispȳt. Prir., Biol.*, 78(4):139(1973).
sect. **Sclerorhiza** Pimenov & Sdobnina in *Bot.*
2 *Zhurn.*, 58(10):1488(1973).
libanotis (L.) Koch
f. **callayi** (J. Duvigneaud) J. Duvigneaud in Nouv. Fl. Belg., Luxemb., N. France etc.: 759(1973):
1 *Libanotis pyrenaica* f. *callayi.*
marginatum (Korov.) Pimenov & Sdobnina in *Bot. Zhurn.*, 58(10):1490(1973): *Libanotis*
2 *marginata.*
mironovii (Korov.) Pimenov & Sdobnina *op. cit.*
2 1489(1973): *Libanotis mironovii.*
montanum L.
subsp. nanum (Duf.) Malagarriga, Las Subesp. y Variac. Geogr.:12(1973), without basionym date:
1 *S. nanum.*
mucronatum (Schrenk) Pimenov & Sdobnina in *Byull. Mosk. Obshch. Ispȳt. Prir., Biol.*, 78(4):139
2 (1973): *Neogaya mucronata.*
nevskii (Korov.) Pimenov & Sdobnina in *Bot.*
2 *Zhurn.*, 58(10):1489(1973): *Libanotis nevskii.*

TAUSCHIA

tenuissima (Geyer ex Hook.) Mathias & Constance in
3 *Madroño*, 22(2):73(1973): *Peucedanum tenuissimum*.

THECOCARPUS

carvifolius (Boiss.) Hedge & J. M. Lamond in
Notes Roy. Bot. Gard. Edinburgh, 32(2):182(1973):
2 *Echinophora carvifolia*.

TORDYLIUM

4437 Malato-Beliz, J. *and* Amaral Franco, J. do
Uma nova umbelífera para a flora portuguesa. (A
new umbellifer for the Portuguese flora). *Bol. Soc.
Brot.*, 46:429–436(1972)—(En); illustrations, dot map.
1 *Tordylium apulum*.

URTICACEAE

4438 Lopes, A. Queiroz
Urticaceae. *Anuár. Soc. Brot.*, 38:109–115(1972)—
1 Por; keys.

PARIETARIA

4439 Chrtek, J.
Taxonomic observations on some species of the
genus *Parietaria*. *Folia Geobot. Phytotax.*, 8(4):425–
1 427(1973)—Key.

VAHLIACEAE

VAHLIA

4440 Rangaswami Ayyangar, K.
Cytology of *Vahlia oldenlandioides*. *Proc. Indian Sci.
Congr.*, 58(3):469(1971)—Chromosome numbers.
2, 5, 6 Abstract only.

VALERIANACEAE

4441 Kao, M-T. *and* De Vol, C. E.
The Valerianaceae of Taiwan. *Taiwania*, 18(2):
146–159(1973)—Illustrations; keys to genera and
2 species.

VALERIANA

4442 Keller, L.
Contribution à la cytogéographie de *Valeriana
officinalis* L. en Suisse (note préliminaire). *Bull. Soc.
Neuchâtel. Sci. Nat.*, 96:67–79(1973)—(Ge,En);
1 dot map; chromosome numbers.

officinalis L.
var. **sarkanyi** Soó in *Acta Bot. Acad. Sci. Hung.*,
1 18(1–2):172(1973), as 'Sárkányii'—Hungary.
sambucifolia Mikan
var. **calvescens** (E. Walther) Soó *op. cit.* 172(1973):
1 *V. procumbens* var. *calvescens*.
tripteris
1 var. **carpatica** Soó *op. cit.* 172(1973)—Rumania.

VALERIANELLA

4443 Duncan, T.
Three plant species new to Canada on Pelee Island:
Triosteum angustifolium L., *Valerianella umbilicata*
(Sull.) Wood, and *Valerianella intermedia* Dyal. *Canad.
3 Field-Nat.*, 87(3):261–265(1973)—Map, dot maps.

4444 Greger, H. *and* Ernet, D.
Flavonoid-Muster, Systematik und Evolution bei
Valerianella. *Phytochemistry*, 12(7):1693–1699(1973)—
1–3 (En).

eriocarpa Desv.
subsp. **truncata** (Betcke) Malagarriga, Las Subesp.
1 y Variac. Geogr.: 14(1973): *V. truncata*.

VELLOZIACEAE

4445 Ayensu, E. S.
Phytogeography and evolution of the Velloziaceae.
In Meggers, B. J., Ayensu, E. S. *and* Duckworth,
W. D., *eds.* Tropical forest ecosystems in Africa and
South America: 105–119(1973)—Illustrations, map;
4, 5 key to genera.

4446 Ayensu, E. S.
Studies on pollen morphology in the Velloziaceae.
2, 4, 5 *Proc. Biol. Soc. Wash.*, 85(40):469–480(1972).

4447 Coetzee, H., Van der Schijff, H. P. *and* Steyn, E.
External morphology of the species of the South
African Velloziaceae including a key based on external
morphological characteristics. (Uitwendige
morfologie van die spesies van Velloziaceae in Suid-
Afrika en'n sleutel gebaseer op uitwendige morfolo-
giese kenmerke). *Dinteria*, no.9:3–21(1973)—En,Af;
5 illustrations; key.

BARBACENIOPSIS

4448 Noher de Halac, R. I.
Una nueva especie de *Barbaceniopsis* (Velloziaceae).
4 *Kurtziana*, 7:261–264(1973)—(En).

humahuaquensis Noher in *Kurtziana*, 7:262(1973)—
4 Argentina.

XEROPHYTA

4449 Coetzee, H. *and* Van der Schijff, H. P.
New combinations in *Xerophyta* Juss. (*Vellozia*
5 Vand. pro parte). *J. S. Afr. Bot.*, 39(3):269(1973).

retinervis Baker
var. **equisetoides** (Baker) Coetzee in *Dinteria*, 9:4
(April 1973), without basionym ref., and in
J. S. Afr. Bot., 39(3):269 (July 1973): *X.
5 equisetoides*.
var. **wentzeliana** (Harms) Coetzee in *Dinteria*, 9:
4 (April 1973), without basionym ref., and in
J. S. Afr. Bot., 39(3):269 (July 1973): *Barbacenia
5 wentzeliana*.

VERBENACEAE

4450 Chaudhri, I.
Chromosome studies of Verbenaceae. *Proc. Indian
6 Sci. Congr.*, 58(3):474–475(1971). Abstract only.

4451 Moldenke, H. N.
A fifth summary of the Verbenaceae, Avicenniaceae,
Stilbaceae, Dicrastylidaceae, Symphoremaceae,
Nyctanthaceae and Eriocaulaceae of the world as to
valid taxa, geographic distribution, and synonymy:
1–7 Supplement 2. *Phytologia*, 25(4):225–245(1973).

4452 Moldenke, H. N.
A fifth summary of the Verbenaceae, Avicenniaceae,
Stilbaceae, Dicrastylidaceae, Symphoremaceae,
Nyctanthaceae, and Eriocaulaceae of the world as to
valid taxa, geographic distribution and synonymy:
1–7 Supplement 3. *Phytologia*, 26(5):356–377(1973).

4453 Moldenke, H. N.
Flora of Panama: part 9. Family 168: Verbenaceae. *Ann. Missouri Bot. Gard.*, 60(1):41–148(1973)—
4 Illustrations; chromosome number; keys.

4454 Moldenke, H. N.
Notes on new and noteworthy plants: 56. *Phytologia*, 25(4):220–224(1973). *Aegiphila, Kalaharia,*
4 *Lantana, Stachytarpheta.*

4455 Moldenke, H. N.
Notes on new and noteworthy plants: 57. *Phytologia*, 25(6):368(1973). *Aegiphila, Citharexylum,*
4, 6 *Clerodendrum, Verbena.*

4456 Moldenke, H. N.
Notes on new and noteworthy plants: 58. *Phytologia*, 25(7):430–432(1973). *Aegiphila, Verbena,*
4, 6 *Vitex.*

4457 Moldenke, H. N.
Notes on new and noteworthy plants: 59. *Phytologia*, 26(3):177–179(1973). *Caryopteris, Duranta,*
3, 4, 6 *Lantana.*

4458 Moldenke, H. N.
Notes on new and noteworthy plants: 61. *Phytologia*, 26(5):355–356(1973). *Hymenopyramis,*
6 *Teijsmanniodendron.*

tribus **Lithophytoideae** D'Arcy & Keating in
4 *Brittonia*, 25(3):224(1973).

AEGIPHILA

4459 Moldenke, H. N.
Additional notes on the genus *Aegiphila*: 18.
4 *Phytologia*, 25(5):282–325(1973).

4460 Moldenke, H. N.
Additional notes on the genus *Aegiphila*: 19.
4 *Phytologia*, 25(6):408–417(1973).

bogotensis
f. **ternata** Moldenke in *Phytologia*, 25(4):220(1973)
4 —Colombia.
cordata Poepp.
var. **villosissima** (Moldenke) Moldenke in
4 *Phytologia*, 25(5):298(1973): *A. villosissima.*
martinicensis
f. **barbadensis** (Moldenke) Moldenke in *Phytologia*,
4 25(6):368(1973): *A. barbadensis.*
mollis
var. **surfaceana** (Moldenke) Moldenke in
4 *Phytologia*, 25(7):430(1973)—*A. surfaceana.*

CARYOPTERIS

nepalensis
var. **parvifolia** Moldenke in *Phytologia*, 26(3):177
6 (1973)—Nepal.

CITHAREXYLUM

4461 Tomlinson, P. B.
Dioecism in *Citharexylum* (Verbenaceae): an
3, 4 addendum. *J. Arnold Arbor.*, 54(1):120(1973).

schottii
var. **pubescens** Moldenke in *Phytologia*, 25(6):368
4 (1973)—Guatemala.

CLERODENDRUM

4462 Tanaka, H.
(Moth pollination of *Clerodendron trichotomum*). *J. Jap. Bot.*, 48(7):209–214(1973)—Ja(En);
2, 6 illustrations.

philippinum
f. **subfertile** Moldenke in *Phytologia*, 25(6):368
6 (1973)—Sumatra.

DURANTA

repens
var. **lopez-palacii** Moldenke in *Phytologia*, 26(3):
4 177(1973)—Venezuela.

HYMENOPYRAMIS

parvifolia
var. **nitida** Moldenke in *Phytologia*, 26(5):355
6 (1973)—Thailand.

KALAHARIA

uncinata
var. **parviflora** ((Schinz) Gürke) Moldenke in *Phytologia*, 25(4):220(1973): *Clerodendron*
5 *spinescens* var. *parviflora.*

LANTANA

4463 Moldenke, H. N.
Seven novelties from North and South America.
4 *Phytologia*, 25(3):117–120(1973).

glandulosissima
f. **albiflora** Moldenke in *Phytologia*, 26(3):177
4 (1973)—Guatemala.
minasensis
var. **puberulenta** Moldenke in *Phytologia*, 25(4):
4 220(1973)—Brazil.
moritziana
f. **parvifolia** Moldenke in *Phytologia*, 25(3):117
4 (1973)—Venezuela.

LIPPIA

4464 Harley, R. M.
Eriope horridula, a member of the Verbenaceae. Notes on New World Labiatae: 1. *Kew Bull.*, 28(1):
4 121–122(1973).

4465 Morais, A. Alpande de *and others*
Óleos essenciais da Amazônia contendo Timol.
4 *Acta Amazonica*, 2(1):45–46(1972). *Lippia origanoides.*

LITHOPHYTUM

4466 D'Arcy, W. G. *and* Keating, R. C.
The affinities of *Lithophytum*: a transfer from Solanaceae to Verbenaceae. *Brittonia*, 25(3):213–225
4 (1973)—Illustrations.

PARODIANTHUS

4467 Troncoso, N. S.
Una nueva especie de *Parodianthus* (Verbenaceae) *P. capillaris* nov. spec. *Darwiniana*, 18(1–2):19–26
4 (1973)—(En); illustrations.

capillaris Troncoso in *Darwiniana*, 18(1–2):19
4 (1973)—Argentina.

PHYLA

4468 Estes, J. R. *and* Brown, L. S.
Entomophilous, intrafloral pollination in *Phyla incisa*. *Amer. J. Bot.*, 60(3):228–230(1973)—
3 Illustrations.

STACHYTARPHETA

4469 Moldenke, H. N.
Seven novelties from North and South America.
4 *Phytologia*, 25(3):117–120(1973).

chapadensis Moldenke in *Phytologia*, 25(3):117
4 (1973)—Brazil.
diamantinensis Moldenke in *Phytologia*, 25(4):220
4 (1973)—Brazil.
lacunosa
var. **angustifolia** Moldenke *op. cit.* 221(1973)—
4 Brazil.
var. **cordifolia** Moldenke *op. cit.* 221(1973)—
4 Brazil.
var. **ovatifolia** Moldenke *op. cit.* 222(1973)—
4 Brazil.
matogrossensis Moldenke *op. cit.* 222(1973)—
4 Brazil.
viscidula
var. **brevipilosa** Moldenke *op. cit.* 222(1973)—
4 Brazil.

TEIJSMANNIODENDRON

bintuluense Moldenke in *Phytologia*, 26(5):355
6 (1973)—Borneo (Sarawak).

VERBENA

4470 Moldenke, H. N.
Notes on new and noteworthy plants—62.
4 *Phytologia*, 26(6):409(1973). *Verbena rectiloba.*

hasslerana
var. **glandulosa** Moldenke in *Phytologia*, 25(6):
4 368(1973)—Brazil.
pedicellata Moldenke in *Phytologia*, 25(7):431
4 (1973)—Brazil.
rectiloba Moldenke in *Phytologia*, 26(6):409(1973)—
4 Brazil.

VITEX

4471 Cunliff, K. M.
Trees for tomorrow. *Trees S. Afr.*, 25(1):24–27
5 (1973)—Illustrations. *Vitex zeyheri.*

negundo
var. **trifoliolata** Moldenke in *Phytologia*, 25(7):
6 432(1973)—Pakistan.

VIOLACEAE

4472 Robson, N. K. B. *and* Vidigal, M. P.
Flora de Moçambique: 13. Violaceae. 1973. 19p.—
5 Keys.

4473 Taylor, F. H.
The secondary xylem of the Violaceae: a comparative study. *Bot. Gaz.*, 133(3):230–242(1972)—
1–7 Illustrations.

VIOLA

4474 Anon.
Viola pilosa Blume. *Acta Phytotax. Geobot.*, 25(4–6):
6 118(1973)—Ja.

4475 Chatterjee, A. *and* Sharma, A. K.
Cytological studies on Indian representatives of the genus *Viola. J. Genet.*, 61(1):52–63(1973)—Chromo-
6 some numbers.

4476 Dathan, A. S. R.
Structure and development of seed coat in *Viola* L. spp. *Proc. Indian Sci. Congr.*, 60(3):290(1973).
1, 2, 6 Abstract only.

4477 Diklić, N. *and* Nikolić, V.
Novo natazište vrste *Viola grisebachiana* Vis. u Srbiji. (Neue Fundstelle der Art *Viola grisebachiana* Vis. in Serbien). *Glasn. Prir. Muz. Beogradu*, 27:
1 213–217(1972)—(Ge); illustration, map.

4478 Dizerbo, A. H.
Les *Viola* de la section *Nomimium* du Massif Armoricain. *Bull. Soc. Sci. Bretagne*, 47(3–4):137–139
1 (1972 publ. 1973)—Chromosome numbers.

4479 Dizerbo, A. *
Les *Viola* de la section *Nomimium* du Massif Armoricain. [France], 1972. 241p. (Thèse Doct. Sci. Nat. Univ. Rouen, 1972). Ref. in *Bull. Signal., Biol.*
1 *Physiol. Vég.*, 34(1):421(1973).

4480 Geigy *Ltd*: Agricultural Chemicals Division
Violaceae: violet family. *Geigy Weed Tables*,
1–3 64(1973)—Illustrations, coloured illustrations.

4481 Harms, V. L.
Violets of Saskatchewan. *Blue Jay*, 31(2):105–112
3 (1973)—Illustrations, dot maps; key.

4482 Moldenke, H. N.
Notes on new and noteworthy plants: 60.
3 *Phytologia*, 26(4):224–226(1973).

4483 Morariu, I.
Două specii de *Viola* controversate în flora României. (Deux espèces de *Viola* controversé dans la flore de Roumanie). *Stud. Comun., Bot.*, 95–102(1972)
1 —(Fr). *V. palustris, V. epipsila.*

4484 Skalický, V.
Viola cornuta L., nová adventivni violka v ČSSR. (*Viola cornuta* L., ein neues adventives Veilchen in der Tschechoslowakei). *Preslia*, 45(2):147–150(1973)—
1 (Ge).

bhutanica Hara in *J. Jap. Bot.*, 48(12):356(1973)—
6 Bhutan.
rafinesquii
f. **caerulea** Moldenke in *Phytologia*, 26(4):225
3 (1973)—U.S.A. (S. Carolina, N. Carolina).
f. **minor** Moldenke *op. cit.* 226(1973)—U.S.A.
3 (N. Carolina).

VISCACEAE
see
LORANTHACEAE

VITACEAE

4485 Descoings, B.
Vitacées et Leeacées de Madagascar: localités nouvelles et corrigenda. *Bull. Soc. Bot. France*, 118(9):
5 745–748(1971 publ. 1973).

CISSUS

4486 Croat, T. B.
A new species of *Cissus* (Vitaceae) for Central and South America. *Ann. Missouri Bot. Gard.*, 60(2):564–
4 567(1973).

pseudosicyoides Croat in *Ann. Missouri Bot. Gard.*, 60(2):564(1973)—Panama, Colombia, Costa
4 Rica.

VITIS

4487 Bertova, L. *and* Berta, J.
Vitis vulpina L. zdivočený druh flóry Slovenska.
(*Vitis vulpina* L.: a run-wild species of Slovak flora).
Biológia, A (Czechoslovakia), 28(10):841–844(1973)—
1, 3 (En,Rus); key.

VOCHYSIACEAE

VOCHYSIA

4488 Maria, J. *and* Ferreira, M. Brandão
Estudos sobre Vochysiaceae: 3. Contribuição para
conhecimento de *Vochysia rufa* (Spr.) Mart. *Oréades*,
4 3(5):38–45(1972)—(En); illustrations.

4489 Maria, J. *and* Ferreira, M. Brandão
Estudos sobre Vochysiaceae: 4. Contribuição ao o
conhecimento de *Vochysia thyrsoideae* Pohl. *Oréades*,
4 3(5):30–37(1972)—(En); illustrations.

4490 Rodrigues, W. A. *
Nova espécie de *Vochysia* (*V. rufescens*) de Amazonia
Brasileira. *Acta Amazonica*, 1(1):7–10(1971 publ.
1972). Ref. in *Bull. Torrey Bot. Club*, 99(6):327
4 (1972).

WINTERACEAE

4491 Bongers, J. M.
Epidermal leaf characters of the Winteraceae.
4–7 *Blumea*, 21(2):381–411(1973)—Illustrations.

XANTHOPHYLLACEAE

XANTHOPHYLLUM

4492 Dickison, W. C.
Nodal and leaf anatomy of *Xanthophyllum* (Poly-
6, 7 galaceae). *Bot. J. Linn. Soc.*, 67(2):103–115(1973).

4493 McVaugh, R.
Report of the Committee for Spermatophyta:
conservation of generic names: proposal 324 (4281)
Xanthophyllum Roxburgh (1820), nom. conserv., vs.
2, 6, 7 *Eystathes* Loureiro (1790). *Taxon*, 22(1):155(1973).

4494 Meijden, R. van der
Twenty-four new names in *Xanthophyllum*
(Polygalaceae). *Bot. J. Linn. Soc.*, 67(2):117–120
6, 7 (1973).

subgen. **Brunophyllum** van der Meijden in *Bot. J.*
6 *Linn. Soc.*, 67(2):117(1973).
sect. **Eystathes** (Lour.) van der Meijden *op. cit.* 117
6 (1973): gen. *Eystathes*.
subgen. **Macintyria** (F. v. Mueller) van der Meijden
6 *op. cit.* 117(1973): gen. *Macintyria*.
brevipes van der Meijden *op. cit.* 117(1973)—
6 Borneo.
ceraceifolium van der Meijden *op. cit.* 117(1973)—
6 Borneo.
chartaceum van der Meijden *op. cit.* 118(1973)—
6 Malaya.
discolor Chodat
subsp. **macranthum** Chod. ex van der Meijden
6 *op. cit.* 118(1973)—Philippine Is.
ferrugineum van der Meijden *op. cit.* 118(1973)—
6 Borneo.
hainanense Hu
var. **occidentale** van der Meijden *op. cit.* 118
6 (1973)—Burma.
kalimantanum van der Meijden *op. cit.* 118(1973)—
6 Borneo.
laevis van der Meijden *op. cit.* 118(1973)—
6 Sumatra.
malayanum van der Meijden *op. cit.* 118(1973)—
6 Malaya.
neglectum van der Meijden *op. cit.* 119(1973)—
6 Borneo.
6 **ngii** van der Meijden *op. cit.* 119(1973)—Sumatra.
nigricans van der Meijden *op. cit.* 119(1973)—
6 Borneo.
papuanum Whitmore ex van der Meijden *op. cit.*
6 119(1973)—New Guinea.
parvifolium van der Meijden *op. cit.* 119(1973)—
6 Borneo.
pauciflorum van der Meijden *op. cit.* 119(1973)—
6 Borneo.
poilanei van der Meijden *op. cit.* 120(1973)—
6 Vietnam.
resupinatum van der Meijden *op. cit.* 120(1973)—
6 Borneo.
subcoriaceum (Chodat) van der Meijden *op. cit.*
6 120(1973): *X. ellipticum* var. *subcoriaceum*.
tardicrescens van der Meijden *op. cit.* 120(1973)—
6 Borneo.
tenuipetalum van der Meijden *op. cit.* 120(1973)—
6 New Guinea.
vitellinum (Bl.) D. Dietr.
var. **clovis** Steen. ex van der Meijden *op. cit.* 120
6 (1973)—Borneo.

XANTHORRHOEACEAE

4495 Chandra, S.
Segregation of Xanthorrhoeaceae from Liliaceae.
6, 7 *Proc. Indian Sci. Congr.*, 58(3):456(1971). Abstract only.

ZANNICHELLIACEAE

CYMODOCEA

4496 Van Tonder, E.
Die blomplant wat in die see groei, *Cymodocea*
5 *ciliata*. *Veld Flora*, 3(4):76–77(1973)—Illustration.

ZINGIBERACEAE

4497 Kumar, V.
Course of vascular differentiation in the axis of
certain Zingiberaceae. *Flora* (Germany), 162(5):
2–7 420–425(1973).

4498 Kumar, V.
Morphological studies in the roots of Zingiberaceae.
2–7 *Flora* (Germany), 162(5):467–478(1973).

AFRAMOMUM

4499 Lock, J. M. *and* Hall, J. B.
Three new species of *Aframomum* (Zingiberaceae)
from Ghana, with notes on spherical geocarpic fruits
in the genus. *Kew Bull.*, 28(3):441–449(1973 publ.
5 1974)—Illustrations; key.

atewae Lock & J. B. Hall in *Kew Bull.*, 28(3):441
5 (1973 publ. 1974)—Ghana.
cordifolium Lock & J. B. Hall *op. cit.* 447(1973 publ.
5 1974)—Ghana, Sierra Leone, Liberia, Ivory Coast.
geocarpum Lock & J. B. Hall *op. cit.* 445(1973 publ.
1974)—Ghana, Guinea, Sierra Leone, Liberia,
5 Ivory Coast.

ALPINIA

4500 Raynal, A. *and* Raynal, J.
Un nouvel *Alpinia* L. (Zingiberaceae) des Nouvelles-
Hébrides. *Adansonia*, 13(1):63–69(1973)—(En);
6 illustrations.

nidus-vespae A. Raynal & J. Raynal in *Adansonia*,
6 13(1):63(1973)—New Hebrides.

CAULOKAEMPFERIA

4501 Larsen, K.
Studies in Zingiberaceae: 6. *Bot. Tidsskr.*, 68(2):
6 157–159(1973)—Illustrations.

saksuwaniae K. Larsen in *Bot. Tidsskr.*, 68(2):157
6 (1973)—Thailand.
6 **thailandica** K. Larsen *op. cit.* 157(1973)—Thailand.

GLOBBA

4502 Lim Siew-Ngo
Cytogenetics and taxonomy of the genus *Globba*
L. (Zingiberaceae) in Malaya: 5. Introgressive
hybridization in hexaploids. *Bot. J. Linn. Soc.*, 66(2):
6 143–156(1973)—Illustrations; chromosome numbers.

ROSCOEA

4503 S., R. M.
What the label doesn't tell. *Gard. July, Roy. Bot.*
1 *Gard. Edinburgh*, 4–5(1973)—Illustration.

ZOSTERACEAE

ZOSTERA

4504 Phillips, R. C.
Ecological life history of *Zostera marina* L. (eelgrass)
in Puget Sound, Washington. *Diss. Abstr. Int.*, *B*,
3 33(8):3512–3513(1973).

4505 Yamashita, T.
Über die Embryo- und Wurzelentwicklung bei
Zostera japonica Aschers. et Graebn. *J. Fac. Sci. Univ.*
2 *Tokyo, Bot.*, 11(5–7):175–193(1973)—(En).

ZYGOPHYLLACEAE

FAGONIA

4506 El Hadidi, M. N.
Revision of *Fagonia* species (Zygophyllaceae) with
tri- to unifoliolate and simple leaves. *Österr. Bot.*
1, 2, 5, 6 *Zeitschr.*, 121(5):269–278(1973)—Illustrations; keys.

socotrana (Balf. f.) Schweinf.
var. **luntii** (Baker) Hadidi in *Österr. Bot. Zeitschr.*,
2 121(5):276(1973): *F. luntii.*

LARREA

4507 Hunziker, J. H., Palacios, R. A., De Valesi, A. G. *and*
Poggio, L.
Species disjunctions in *Larrea*: evidence from
morphology, cytogenetics, phenolic compounds and
seed albumins. *Ann. Missouri Bot. Gard.*, 59(2):224–
233(1972 publ. 1973)—Illustrations, map, dot map;
3, 4 chromosome numbers.

PEGANUM

4508 Sulaĭmanov, É. S.
Novoe v biologii garmalȳ obȳknovennoĭ. *Uzb.*
Biol. Zhurn., no.3:76–77(1973)—Rus. *Peganum*
1, 2 *harmala.*

TRIBULUS

4509 Porter, D. M.
Tribulus cistoides in Panama. *Ann. Missouri Bot.*
4 *Gard.*, 60(2):570(1973).

4510 Sahibzada, M. A. *and* Khan, A. G.
Starch-filled nodular hypertrophies in *Tribulus*
1–7 *terrestris*. *Pakistan J. Bot.*, 5(1):87–88(1973).

ZYGOPHYLLUM

4511 Crookston, R. K. *and* Moss, D. N.
C-4 and C-3 carboxylation characteristics in the
genus *Zygophyllum* (Zygophyllaceae). *Ann. Missouri*
1, 2, 5, 6 *Bot. Gard.*, 59(3):465–470(1972 publ. 1973).

Phytogeography

4512 Baker, H. G.
Evolutionary relationships between flowering
plants and animals in American and African tropical
forests. *In* Meggers, B. J., Ayensu, E. S. *and*
Duckworth, W. D., *eds*. Tropical forest ecosystems in
4, 5 Africa and South America: 145–159(1973).

4513 Bernard, A.
Note sur la phénologie des Arrhénathéraies de la
plaine d'Alsace. *Compt. R. Hebd. Séances Acad. Sci.*, *D*
1 (Paris), 275(23):2667–2670(1972).

4514 Cox, C. B., Healey, I. N. *and* Moore, P. D.
Biogeography: an ecological and evolutionary
approach. Oxford, Blackwell, 1973. viii,184p.—
1–7 Illustrations, maps.

4515 Croizat, L.
La panbiogeografia in breve. *Webbia*, 28(1):189–
1–7 226(1973)—Maps.

4516 Eilart, J.
Peipsi osa taimede sisserännus. (The role of Lake
Peipsi in plant immigration.) *Eesti Loodus*, 16(12):
1 722–725(1973)—(En,Rus); illustrations, map.

4517 Hedberg, O.
Adaptive evolution in a tropical-alpine environ-
ment. *In* Heywood, V. H., *ed*. Taxonomy and
5 ecology: 71–92(1973)—Illustrations, dot maps.

4518 Hughes, N. F., *ed*.
Organisms and continents through time: methods
of assessing relationships between past and present
biologic distributions and the positions of continents:
a symposium volume of 23 papers. London, Palaeon-
tological Association, 1973. 334p. (Special Papers in
1–7 Palaeontology, no.12)—Maps, dot maps.

4519 International Union for the Protection of Nature
A working system for classification of world
1–7 vegetation. *IUCN Occ. Pap.*, no.5. [ii]23p.(1973).

4520 Jäger, E.
Areal- und Florenkunde (floristische Geobotanik).
1–7 *Fortschr. Bot.*, 35:307–320(1973).

4521 Meggers, B. J., Ayensu, E. S. *and* Duckworth, W. D.,
eds
Tropical forest ecosystems in Africa and South
America: a comparative review. Washington,
Smithsonian Institution Press, 1973. viii,350p.—
Illustrations, maps, dot maps; chromosome numbers.
4, 5 Introduction by J. P. M. Brenan.

4522 Mokhamed, A. Z. *and* Shmidt, V. M.
O sistematicheskoj strukture flor stran yuzhnogo sredizemnomor'ya: 3. Sravnenie struktury flor otdel'nykh rajonov. (On the systematical structure of flora of south mediterranean countries: 3. Comparison of the structure of flora of separate regions). *Vestn.*
1, 2 *Leningrad. Univ., Biol.*, 21(4):44–53(1973)—(En).

4523 Moor, M.
Das Corydalido-Aceretum, ein Beitrag zur Systematik der Ahornwälder. (The Corydalido-Aceretum, a contribution to the taxonomy of maple forests). *Ber. Schweiz. Bot. Ges.*, 83(2):106–132
1–3 (1973)—(En).

4524 Raven, P. H.
An introduction to continental drift. *Austral. Nat.*
5–7 *Hist.*, 17(8):245–248(1972).

4525 Raven, P. H.
Plant species disjunctions: a summary. *Ann. Missouri Bot. Gard.*, 59(2):234–246(1972 publ.
1–7 1973).

4526 Smith, A. C.
Angiosperm evolution and the relationship of the floras of Africa and America. *In* Meggers, B. J., Ayensu, E. S. *and* Duckworth, W. C., *eds*. Tropical forest ecosystems in Africa and South America: 49–61
4, 5 (1973)—Maps.

4527 Solbrig, O. T.
The floristic disjunctions between the 'Monte' in Argentina and the 'Sonoran Desert' in Mexico and the United States. *Ann. Missouri Bot. Gard.*, 59(2):218–
3, 4 223(1972 publ. 1973).

4528 Terborgh, J.
On the notion of favorableness in plant ecology.
1–7 *Amer. Nat.*, 107(956): 481–501(1973).

4529 Thorne, R. F.
Floristic relationships between tropical Africa and tropical America. *In* Meggers, B. J., Ayensu, E. S. *and* Duckworth, W. D., *eds*. Tropical forest ecosystems in
4, 5 Africa and South America: 27–47(1973).

4530 Thorne, R. F.
Major disjunctions in the geographic ranges of seed plants. *Quart. Rev. Biol.*, 47(4):365–411(1972)—
1–7 Maps.

4531 Turner, B. L.
Chemosystematic data: their use in the study of disjunctions. *Ann. Missouri Bot. Gard.*, 59(2):152–164
1–7 (1972 publ. 1973)—Map; chromosome numbers.

4532 Steenis, C. G. G. J. van
Woodiness in island floras. *Taiwania*, 18(1):45–
1–7 48(1973).

4533 Voronov, A. G.
Osobennosti florȳ i rastitel' nosti ostrovov mirovogo okeana. (Peculiarities of the insular vegetation of the world ocean). *Byull. Mosk. Obshch.*
1–7 *Ispȳt. Prir., Biol.*, 78(4):54–66(1973)—(En).

4534 Wagner, W. H.
Disjunctions in homosporous vascular plants. *Ann. Missouri Bot. Gard.*, 59(2):203–217(1972 publ. 1973)—
1–7 Chromosome numbers.

4535 Walter, H. *and* Flohn, H.
Die Vegetation der Erde in öko-physiologischer Betrachtung: Bd 1. Die tropischen und subtropischen Zonen. Dritte Aufl. Jena, G. Fischer, 1973. 743p.—
1–7 Illustrations, maps, dot maps.

4536 Watts, D. *
Principles of biogeography. New York, McGraw-Hill, 1971. xvi, 401p.—Ref. in *Ref. Zhurn. Biol.*,
1–7 2(1):v406(1972).

4537 Whitehead, D. R.
Approaches to disjunct populations: the contribution of palynology. *Ann. Missouri Bot. Gard.*,
1–7 59(2):125–137(1972 publ. 1973).

4538 Whitmore, T. C.
Plate tectonics and some aspects of Pacific plant geography. *New Phytol.*, 72(5):1185–1190(1973)—
6, 7 Map.

4539 Wood, C. E.
Morphology and phytogeography: the classical approach to the study of disjunctions. *Ann. Missouri Bot. Gard.*, 59(2):107–124(1972 publ. 1973)—Maps,
1–7 dot maps.

Floristics

EUROPE

4540 Ciba-Geigy *Ltd*
Unkrautgemeinschaften Europas... (Weed communities of Europe)... Basle, Ciba-Geigy Ltd, 1971. var. paging.—Ge,Fr,En,It,Sp,Por,Rus; coloured
1 illustrations of 'weed' seedlings.

4541 Defilipps, R. A.
Index to taxa cited in *A survey of taxonomic and floristic research in Europe since 1945, Webbia* 18,
1 1963. *Webbia*, 27(2):299–368(1972 publ. 1973).

4542 Dobrescu, C. *and* Kovács, A.
Pe marginea operei *Flora Europaea*, vol. 3. *An. Şti. Univ. Al. I. Cuza Iaşi, Biol.*, 19(2):459 464(1973). Comparative nomenclature of *Flora Europaea* and
1 *Flora R.S. România.*

4543 Eberle, G.
Polsterpflanzen: Ausblick aus den Alpen auf eine weitverbreitete Wuchsform. *Jahrb. Ver. Schutze*
1 *Alpenpfl. Tiere*, 38:61–70(1973)—Illustrations.

4544 Fekete, G. *and* Priszter, S.
A Balkáni flóra és vegetáció elsó nemzetközi szimpóziuma, Várna, 1973. Június 7–10. (Das I. International Symposium der Balkanischen Flora und Vegetation in Warna, 7–10.6.1973). *Bot. Közlem.*,
1 60(3):211–212(1973).

4545 Heywood, V. H., *ed.*
Flora Europaea: notulae systematicae ad floram Europaeam spectantes: no.14. *Bot. J. Linn. Soc.*,
1 67(3):275–283(1973).

4546 Holub, J.
Brief notes on the third volume of *Flora Europaea*. (Stručne poznámky k tretímu svazku díla *Flora*
1 *Europaea*). *Preslia*, 45(4):355–360 (1973)—(Cz).

4547 Jalas, J. *and* Suominen, J., *eds.*
Atlas florae europaeae: distribution of vascular plants in Europe: 2. Gymnospermae (Pinaceae to Ephedraceae). Helsinki, Committee Mapping Flora Europe and Societas Biologica Fennica Vanamo,
1 1973. 40p.

4548 Küpfer, P.
Sur quelques problèmes de spéciation dans les montagnes du bassin occidental de la Méditerranée. *Bull. Soc. Bot. France*, 118(9):667–678(1971 publ.
1 1973)—(En); chromosome numbers.

4549 Laínz, M.
In Floram Europaeam animadversiones: 2. *Candollea*, 28(2):181–190(1973)—(Fr,En,Ge); dot
1 map.

4550 Polunin, O. *and* Smythies, B. E.
Flowers of south-west Europe: a field guide. London, etc., Oxford University Press, 1973. xvi, 480p.—Illustrations, coloured illustrations, maps;
1 keys.

4551 Schönfelder, P.
Die Flora Europaea-Kartierung. *Gottinger Flor.*
1 *Rundbr.*, 7(1):20–23(1973)—Dot maps.

4552 Sojàk, J.
Nomenklatorické poznámky (Phanerogamae). (Nomenclatural notes (Phanerogamae)). *Čas. Nár.*
1 *Muz.* (Prague), 140(3–4):127–134(1972).

AUSTRIA

4553 Grims, F.
Die Flora des Sauwaldes und der umgrenzenden Täler von Pram, Inn und Donau. *Jahrb. Oberösterr.*
1 *Musealver.*, 117(1):335–376(1972)—Dot maps.

4554 Gutte, P.
Die Vegetation der Aschehalde Rositz bei Altenburg. *Abh. Ber. Naturk. Mus. 'Mauritianum' Altenburg*,
1 7(1):7–16(1971).

4555 Holzner, W.
Die Ackerunkrautvegetation Niederösterreichs. *Mitt. Bot. Arbeit. Oberösterr. Landesmus. Linz*, 5(1):
1 1–157(1973)—Map.

4556 Holzner, W. *
Niederösterreichs Ackervegetation als Umweltzeiger. *Bodenkultur*, 22(4):397–414(1971)—(En). Abstr.
1 in *Ref. Zhurn. Biol.*, 3(1):v422(1972).

4557 Jelem, H.
Über die Schwarzföhre (*Pinus nigra*) in Niederösterreich. *Allg. Forstzeitung* (Vienna), 84(5):123–125
1 (1973)—Map.

4558 Kusel-Fetzmann, E. *and* Lew, H.
Die Makrophytenvegetation des Klopeiner Sees (Kärnten). *Verh. Zool.-Bot. Ges. Wien*, 112:94–99
1 (1972).

4559 Melzer, H.
Floristische neuigkeiten aus Kärnten. *Carinthia II*,
1 82:201–220(1972).

4560 Melzner, H.
Neues zur Flora von Steiermark: 14. *Mitt.*
1 *Naturwiss. Ver. Steiermark*, 102:101–115(1972).

4561 Niklfeld, H.
Der Niederösterreichische Alpenostrand: ein Glazialrefugium montaner Pflanzensippen. *Jahrb. Ver.*
1 *Schutze Alpenfl. Tiere*, 37:42–94(1972)—Dot maps.

4562 Traxler, G.
Die burgenländischen Pflanzenstandorte bei Carolus Clusius. *Burgenländ. Heimatbl.*, 35(2):49–59
1 (1973).

4563 Traxler, G.
Floristische Neuigkeiten aus dem Burgenland: 7.
1 *Burgenländ. Heimatbl.*, 35(4):163–171(1973).

4564 Wagner, R.
Flora von Eisenerz und Umgebung. *Mitt. Abteil. Bot. Landesmus. 'Joanneum' Graz*, 43–44. 259p.
1 (1973).

4565 Zimmermann, A.
Hochgebirgspflanzen in der Laubwaldstufe (2): die niederösterreichisch-steirischen Kalkvoralpen und das Grazer Bergland. *Alpengarten*, 16(2):2–7
1 (1973)—Illustrations.

BALEARIC ISLANDS

4566 Finschow, G., Guerau de Arellano Tur, C. *and* Kuhbier, H.
Contribución al estudio de la flora de las Pitiusas.
1 *Eivissa*, no.1. 3a época:24–26(1972)—(Ge,En).

BELGIUM

4567 Hose, R. d' *and* Langhe, J.-E. de
Nieuwe groeiplaatsen van zeldzame planten in België. (New localities of rare plants in Belgium). *Bull. Soc. Roy. Bot. Belg.*, 106(2):273–277(1973)—
1 (Fr).

4568 Lambinon, J., Duvigneaud, J. *and* Lebeau, J.
La publication du volume 3 de '*Flora Europaea*' et ses incidences sur la floristique en Belgique et dans les régions limitrophes. *Nat. Mosana*, 26(1–2):8–20
1 (1973).

4569 Langhe, J.-E. de
Formes aberrantes de plantes indigènes en Belgique. (Abweichende Formen von in Belgien einheimischen Pflanzenarten). *Bull. Soc. Roy. Bot. Belg.*, 106(3):
1 69–73(1973)—(Ge).

4570 Langhe, J.-E. de, Delvosalle, L., Duvigneaud, J., Lambinon, J. *and* Berghen, C. vanden
Nouvelle flore de la Belgique, du Grand-Duché de Luxembourg, du Nord de la France et des régions voisines (ptéridophytes et spermatophytes). Bruxelles, Jardin Botanique, 1973. xcviii, 821p.—
1 Illustrations, coloured illustrations, map; keys.

4571 Parent, G. H.
Notes chorologiques et écologiques sur la flore de la province de Luxembourg. *Lejeunia*, no.68. 88p.
1 (1973).

BULGARIA

4572 Delipavlov, D. *and* Dimitrov, S.
Beitrag zum Studium der Flora Bulgariens. *Feddes*
1 *Repert.*, 83(7–8): 489–493(1972 publ. 1973)—(En).

4573 Ĭordanov, D., *ed.*
Flora na Narodna Republika Bŭlgariya: 5. (Flora Reipublicae Popularis Bulgaricae: 5). Sofiya, Bŭlgarskata Akad. Nauk, 1973. 442p.—Illustrations,
1 map.

4574 Iordanov, D. *and* Kochev, Khr.
Materiali i kritichni belezhki za florata na Bŭlgariya (Materials and critical notes for the flora in Bulgaria). *Izv. Bot. Inst.* (Sofia), 24:241–245(1973)—(Rus,
1 En).

4575 Kochev, Khr.
Geobotanichni i fitoklimatichni izsledvaniya na kestenovite gori v raĭona na Berkovitsa (Zapadna Stara planina) (Geobotanical and phytoclimatic investigations of the chestnut-forests in the district of Berkovitza (western Balkan mountains)). *Izv. Bot.*
1 *Inst.* (Sofia), 24:31–73(1973)—(Rus,En); map.

4576 Lubenov, J.
Sur le problème posé par les mauvaises herbes parasites en Bulgarie. (The problem of parasitic weeds in Bulgaria). *In* European Weed Research Council: Parasitic Weeds Research Group *and* Malta: Royal University. Symposium on parasitic weeds, 11–13 April 1973, Royal University of Malta, Malta: 18–22
1 (1973)—(En,Ge).

4577 Panov, P.
Novi materiali i floristichni belezhki. (New materials and floristic notes). *Izv. Bot. Inst.* (Sofia),
1 23:235–236(1973)—Bu(Rus,En).

4578 Panov, P. *and* Panova, L.
Novi materiali i kriticheski belezhki po florata na Bŭlgariya. (New materials and critical notes on Bulgarian flora). *Izv. Bot. Inst.* (Sofia), 23:237–239
1 (1973)—Bu(Rus,En).

4579 Stanev, S.
Floristichni materiali ot iztochna Sredna Gora s kritichni belezhki. (Floristic material from eastern Sredna Gora mountain with critical notes: 1). *Iz.v Bot. Inst.* (Sofia), 23:219–227(1973)—Bu(Rus,En);
1 dot maps.

4580 Stanev, S. *and* Ninova, D.
Pregled na anatomosistematichnite izledvaniya vŭrkhu visshite rasteniya v Bŭlgariya (1887–1969).
1 *Izv. Bot. Inst.* (Sofia), 23:207–218(1973)—Bu(Rus,En).

4581 Vasilev, P.
Novi i redki rasteniya za florata na planinata Golo Bŭrdo. (New and rare plants for the flora of the Golo Bardo Mountain). *Izv. Bot. Inst.* (Sofia), 23:229–233
1 (1973)—Bu(Rus,En); dot map.

4582 Velchev, V.
Fitotsenologichno prouchvane na byalata mura (*Pinus peuce* Griseb.) v Sredna Stara planina. (Phytozönologische Untersuchungen des *Pinus peuce* Griseb. in der mittleren Stara Planina). *Izv. Bot. Inst.* (Sofia), 24:169–198(1973)—(Rus,Ge);
1 illustrations, coloured illustrations, map.

4583 Velchev, V., Vasilev, P. *and* Yanev, Ya.
Novi rasteniya i khorologichni danni za florata na visshite rasteniya v Bŭlgariya. (New plants and horological data about higher flora in Bulgaria). *Izv.*
1 *Bot. Inst.* (Sofia), 24:242–250(1973)—(Rus,En).

CORSICA

4584 Gamisans, J.
Contribution à l'étude de la flore de la Corse: 5. *Candollea*, 28(1):39–82(1973)—(En,Ge); illustrations;
1 chromosome numbers.

4585 Lanza, B.
The natural history of the Cerbicale islands (south-eastern Corsica) with particular reference to their herpetofauna. (Storia naturale delle Isole Cerbicale (Corsica sudorientale) con particolore riferimento alla loro erpetifauna). *Natura* (Italy), 63(4):345–407
1 (1972)—(It); illustrations.

4586 Staav, R.
Biologiska intryk från en resa på Korsika våren 1973. *Fauna Flora*, 68(5):193–200(1973)—(En);
1 illustrations, map.

CRETE

4587 Greuter, W.
Additions to the flora of Crete, 1938–1972. *Ann.*
1 *Mus. Goulandris*, 1:15–83(1973)—(Gre).

CZECHOSLOVAKIA

4588 Atlas rozšíření rostlin v jižní části Čech: 1. (Atlas der Pflanzenverbreitung im südlichen Teile Böhmens: 1). *Sborn. Jihocesk. Muz. Česk. Budějovicich Přír. Ved.*,
1 12 *Suppl.* 3. 44p.(1972)—(Ge); dot maps.

4589 Brno: Universita J. E. Purkyně
Schedae ad floram exsiccatam reipublicae socialisticae Čechoslovacae . . . Centuria xvi (No. 1501–1600).
1 Brno, Univ. J. E. Purkyne, 1972. 24p.

4590 Bureš, L.
Příspěvek k poznání květeny Doupovských hor. (Beitrag zur Flora des Gebirges Doupovské hory (Duppauer Berge)). *Zprávy Česk. Bot. Spol.*, 8(1):
1 29–34(1973).

4591 Chrtková-Žertová, A. *and* Chrtek, J.
Poznámky ke květeně Kysucké vrchoviny: 2. (Anmerkungen zur Flora der Höhe Kysucká vrchovina (Slowakei) 2.). *Zprávy Česk. Bot. Spol.*,
1 8(1):49–56(1973).

4592 Chrtková-Žertová, A. *and* Chrtek, J.
Poznámky ke květeně Kysucké vrchoviny: 3. (Bemerkungen zur Flora der Höhe Kysucká vrchovina in der Slowakei: 3). *Zprávy Česk. Bot.*
1 *Spol.*, 8(2):121–126(1973).

4593 Deyl, Č.
Příspěvek ke květeně širšího okolí Olomouce. (Beitrag zur Flora der weiteren Umgebung von Olomouc (Olmütz)). *Zprávy Česk. Bot. Spol.*, 8(1):
1 40–48(1973).

4594 Deyl, Č.
Příspěvek ke květeně širšího okolí Olomouce (dokoncenční). (Beitrag zur Flora der weiteren Umgebung von Olomouc (Olmütz) (Schluss)).
1 *Zprávy Česk. Bot. Spol.*, 8(2):96–108(1973).

4595 Deyl, M.
Poznámky ke květeně Blatenska. (Notes on the flora of the Blatná region). *Čas. Nár. Muz.* (Prague),
1 141(1–2):111–113(1972).

4596 Eliáš, P.
Potentillo (Argenteae)-Absinthietum im Nördlichen Teil des Tafellands Žitavská Tabula (Südwest Slowakei). (Potentillo (Argenteae)-Absinthietum in the northern part of Žitava flat land (southwest Slovakia)). *Biológia A*, (Czechoslovakia), 28(1):23–37
1 (1973)—(En,Cz,Rus).

4597 Fajmonová, E. *and* Šimeková, J.
Beitrag zur phytocoenologischen Klassifikation der Kalkstein-Buchenwälder in den Westkarpaten. (K tsitotsenologicheskoĭ klassifikatsii i vestnyakovȳkh vukovȳkh porosleĭ v zapadnȳkh Karpatakh). *Acta Fac. Rerum Nat. Univ. Comen., Bot.*, 21:31–49(1973)—
1 (Cz,Rus).

4598 Feltl, J.
Floristické poměry mezihoří Polička Litomyšl. (Floristische Verhältnisse im Zwischengebirge Polička-Litomyšl). *Zprávy Česk. Bot. Spol.*, 8(3):170–
1 171(1973).

4599 Fiedler, J.
Nové floristické nálezy z východních Čech. (Neue floristische Funde in Ostböhmen). *Zprávy*
1 *Česk. Bot. Spol.*, 8(3):179–181(1973).

4600 Hendrych, R., *ed.*
Phytokartogramme einiger Gebirgspflanzenarten der Tschechischen Sozialistichen Republik (ČSR). *Acta Univ. Carol., Biol.*, no.4:267–306(1971 publ.
1 1973)—Dot maps.

4601 Houfek, J.
Krátká floristická sdělení a výsledky floristické akce v Čechách. (Kurze floristische Mitteilungen und Ergebnisse der floristischen Durchforschung Böhmens). *Zprávy Česk. Bot. Spol.*, 8(2):148–156,
1 (3):222–225(1973).

4602 Hrabětová-Uhrová, A.
Adventivní rostliny na Ivančicku (okr. Brno). (Adventivpflanzen in der Umgebung der Stadt Ivančice (Bez. Brno)). *Zprávy Česk. Bot. Spol.*, 8(1):
1 35–39(1973).

4603 Kovár, P.
Příspěvek k vegetaci v údolí Trebovky u Ceské Trebové: 1. (Beitrag zur Vegetati on im Tal des Baches Trebovka bei Ceská Trebová: 1). *Zprávy*
1 *Česk. Bot. Spol.*, 8(2):130–136(1973)—Map.

4604 Kubíková, J. *and* Slavíková, J.
Zajímavá lokalita slatinné vegetace u Pavčíné Lehoty (okr. Liptovsky Mikuláš). (Eine beachtenswerte Lokalität mit Niedermoorvegetation bei der Gemeinde Pavčina Lehota). *Zprávy Česk. Bot. Spol.*, 8(1):56–58
1 (1973)—(Ge).

4605 Lhotská, M.
Příspěvek ke květeně dolního Posázaví: 3. (Beitrag zur Flora des Gebietes am unteren Sazava-
1 fluss). *Zprávy Česk. Bot. Spol.*, 8(3):172–179(1973).

4606 Majeríková, J.
Floristický materiál ke květeně okolí Horské Kvildy na šumavě, 1. Rostliny cévnaté. (Floristisches Material zur Flora der Umgebung von Horská Kvilda im Böhmerwald, 1. Gefässpflanzen). *Zprávy*
1 *Česk. Bot. Spol.*, 8(3):181–183(1973).

4607 Miadok, D.
Kurzgefasster Überblick über die Vegetation des Alluviums des Baches Banský Potok. (Kratkiĭ obzor vegetatsii allyuviya Banskogo ruch'ya). *Acta Fac. Rerum Nat. Univ. Comen., Bot.*, 21:51–80(1973)—
1 (Cz,Rus).

4608 Miadok, D.
Příspěvek k flóre Lysca. (Beitrag zur Flora des Berges Lysec (Südslowakii)). *Zprávy Česk. Bot. Spol.*,
1 8(1):59–60(1973).

4609 Myczkowski, S.
Native tree species of the Tatra National Park. *Int. Dendrol. Soc. Year Book 1972*: 70–80(1973)—
1 Illustrations.

4610 Šachl, J.
Nové a vzácné druhy flóry Kriváňské Malé Fatry. (New and rare plant species in the Kriváň region of the Malá Fatra). *Biológia, A* (Czechoslovakia), 27
1 (10):805–810(1972)—(En,Rus).

4611 Skalický, V.
Publikování krátkých floristických sdělení podle pokynu floristické sekce Československé botanické společnosti. (Das Publizieren kurzer floristischer Mitteilungen nach Anweisungen der floristischen Sektion der Tschechoslowakischen Botanischen Gesellschaft). *Zprávy Česk. Bot. Spol.*, 8(3):226–228
1 (1973).

4612 Slavík, B.
Rozšíření některých druhů rostlin v ČSR: 3. (Verbreitung einiger Pflanzenarten in der ČSR: 3). *Zprávy Česk. Bot. Spol.*, 8(3):157–167(1973)—(Ge);
1 dot maps.

4613 Šomšák, L.
Vegetationsverhältnisse des Zipser Teiles des slowakischen Erzgebirges: slovenské Rudohorie 3. (Verbreitetste Waldgesellschaften). (Usloviya vegetatsii Spishskoĭ chasti Slovatskogo Rudogor'ya: 3. (Preobladayushie soobshchestva lesnȳkh porosleĭ)). *Acta Fac. Rerum Nat. Univ. Comen., Bot.*, 21:1–29
1 (1973)—(Cz,Rus).

4614 Svobodová, Z.
Příspěvek k adventivní květeně Slovenska. (Beitrag zur adventiven Flora der Slowakei).
1 *Zprávy Česk. Bot. Spol.*, 8(1):60–68(1973).

4615 Vaněčková, L.
Příspěvek k fytogeografickým poměrum údolí Křetínky. (Beitrag zu den phytogeographischen Verhältnissen im Tal des Flusses Kretínka). *Zprávy*
1 *Geogr. Úst., ČSAV*, 9(5):16–23(1972)—(Ge,Rus); map.

4616 Vološčuk, I. *and* Vnuk, J.
Výskyt niektorých pozoruhodnejších druhov v juhozápadnej časti Vel'kej Fatry. (Occurrence of some remarkable species in the southwest part of Velka Fatra mountains). *Biológia, A*, (Czechoslovakia),
1 28(7):585–589(1973)—(En,Rus); dot maps.

4617 Zavřel, H.
Několik vzácnějších zplanělých rostlin z okresu Kroměřížského. (Einige seltenere verwilderte Pflanzen im Bezirk Kroměříž). *Zprávy Česk. Bot.*
1 *Spol.*, 8(3):184–185(1973).

4618 Zavřel, H.
Rozšíření jmelí bílého (*Viscum album* L.s.l.) na Kroměřížsku. (Verbreitung von *Viscum album* L.s.l. in der Umgebung von Kroměříž (Mittelmähren)).
1 *Zprávy Česk. Bot. Spol.*, 8(1):76–78(1973).

DENMARK

4619 Pinborg, U.
Murruden på ny sjaellandsk lokalitet. *Bot. Tidsskr.*, 67(4):337–341(1973)—(En). The flora on the wall of a
1 mediaeval ruin.

4620 Worsøe, E.
Øster Lovnkaer: en tidligere dansk løveng og dens
1 flora. *Flora Fauna*, 79(1):1–10(1973)—Map.

ENGLAND *see* GREAT BRITAIN

FAROES

4621 Hensen, K.
Vertical vegetation zones and vertical distribution types in the Faeroes. *Bot. Tidsskr.*, 67(1–2):33–63
1 (1972).

FINLAND

4622 Kaakinen, E.
Studies on herb-rich forest vegetation in southern Kainuu, northern Finland. *Aquilo, Bot.*, 11:23–42
1 (1972)—Map.

FRANCE

4623 Barbe, J.
Aperçu sur la flore et la végétation de la forêt de Levier. *Bull. Féd. Soc. Hist. Nat. Franche-Comté*,
1 74(2):33–41(1972 publ. 1973)—Illustrations.

4624 Barbero, M., Loisel, R. *and* Poirion, L.
L'originalité biogéographique des préalpes de
1 Grasse. *Riviera Sci.*, 59(4):81–92(1972)—Map.

4625 Bernard, C. *and* Fabre, G.
Sur cinq phanérogames nouvelles pour l'Aveyron. *Monde Pl.*, no.375:2(1972). *Iris sibirica*, *Claytonia perfoliata*, *Myosurus minimus*, *Cnidium silaifolium* and
1 *Parentucellia latifolia*.

4626 Bouby, H.
Intérêt botanique et protection du Site de Brimborion (Hauts-de-Seine). *Rev. Féd. Franç. Soc.*
1 *Sci. Nat.*, 12(50):53–58(1973).

4627 Bournérias, M.
Flore et végétation du massif forestier de Rambouillet (Yvelines). *Cah. Nat.*, 28(2):17–58(1972 publ.
1 1973).

4628 Corillion, R.
La flore médicinale du massif armoricain: remarques phytogéographiques et taxonomiques. *Pl. Méd.*
1 *Phytothér.*, 7(3):200–207(1973)—Dot maps.

4629 Corillion, R.
Observations sur les végétations des sables du lit mineur de la Loire en Anjou—Basse Loire. *Bull.*
1 *Mayenne-Sci.*, 1970–1971:143–175(1972).

4630 Couderc, J.-M.
Biogéographie de la rive ouest de l'étang des Vigneaux. *Rev. Féd. Franç. Soc. Sci. Nat.*, 11(49):
1 91–95(1972).

4631 Couderc, J.-M., Guédès, M. *and* Le Goff, P.
Plantes nouvelles du lit des rivières Tourangelles. *Cah. Nat.*, 27(2):33–44(1971 publ. 1972). Plants
1 originating from various distributions.

4632 Dendaletche, C. *
Ecologie et peuplement végétal des Pyrénées occidentales. Essai d'écologie montagnarde: thèse. Nantes, C.N.R.S., 1973. [488p.]. Ref. in *Bull. Signal.*,
1 *Biol. Physiol. Vég.*, 34(10):9627(1973).

4633 Dhien, R.
Contribution à la connaissance de la flore Nivernaise. *Cah. Nat.*, 27(2):44–45(1971 publ. 1972)—
1 La Nièvre.

4634 Dhien, R.
Flore des fouilles d'Alésia (Côte d'Or). *Cah. Nat.*,
1 28(4):99(1972 publ. 1973).

4635 Dhien, R.
Un peuplement initial de végétation. *Bull. Trim.*
1 *Soc. Hist. Nat. Amis Mus. Autun*, no.63:4–6(1972).

4636 Dupont, P.
La cartographie de la flore française et le contexte européen. *Bull. Soc. Bot. France*, 118(9):765–782(1971
1 publ. 1973)—(En,Ge).

4637 Dupont, P.
Une oeuvre collective à réaliser: la cartographie floristique de la Loire-Atlantique. *Bull. Soc. Sci. Nat.*
1 *Ouest France*, 70(4):1–4(1972)—Map.

4638 Gaussen, H.
1 Catalogue flore. *Monde Pl.*, no.375:6–7(1972).

4639 Gaussen, H.
1 Catalogue flore. *Monde Pl.*, no.376:8(1973).

4640 Guinochet, M. *and* Vilmorin, R. de
Flore de France: fasc. 1. Paris, C.N.R.S., 1973.
1 366p.—Illustrations; keys.

4641 Hesse, J. *and* Henry, C.
Sites historiques et naturels à préserver dans la vallée de la Cisse. *Sci. Nat.*, no.115:27–31(1973)—
1 Illustrations, map.

4642 Jovet, P.
Observations et remarques sur la flore et la végétation du Pays Basque français. *Bull. Centre Etudes Rech.*
1 *Sci. Biarritz*, 8(4):645–656(1971 publ. 1972).

4643 Jovet, P. *and* Vilmorin, R. de
Flore descriptive et illustrée de la France par l'abbé H. Coste. Premier supplément. Paris, A. Blanchard,
1 1972 [1973]. xi,86p.—Keys.

4644 Lebreton, P.
Sur les endémiques du Mercantour. *Bull. Mens. Soc.*
1 *Linn. Lyon*, 41(9):173–175(1972).

4645 Litzler, P.
Loisirs botaniques dolois. *Bull. Féd. Soc. Hist. Nat. Franche-Comté*, 74(3):1–2(1972). List of species
1 introduced to France.

4646 Millet, B.
Vue d'ensemble sur la végétation du lac de Remoray. *Bull. Féd. Soc. Hist. Nat. Franche-Comté*, 73(3):55–57
1 (1971)—Map.

4647 Perney, J.-P.
Quelques nouvelles stations de plantes rares en Haute-Saône. *Ann. Sci. Univ. Besançon, Bot.*, 12:115–
1 117(1972).

4648 Petit, D.
Les végétaux thermophiles peu communs de la région minière du Nord et du Pas de Calais.
1 *Monde Pl.*, no.375:5(1973).

4649 Poirion, L.
Le Marguareis: une cime importante des Alpes Maritimes. *Sci. Nat.*, no.118:28–32(1973)—
1 Illustrations.

4650 Prost, J.-F.
Les étangs du Bas-Jura. *Monde Pl.*, no.376:1–3
1 (1973).

4651 Prost, J. F.
Excursion botanique au Crêt de la Neige et au Reculet. *Bull. Féd. Soc. Hist. Nat. Franche-Comté*,
1 73(1):13–14(1971).

4652 Rameau, J. C.
Etude phytosociologique des pelouses de l'Oxfordien Haut-Marnais. *Ann. Sci. Univ.*
1 *Besançon, Bot.*, 10:139–170(1971)—Map.

4653 Richard, J.-L.
Un facteur écologique nouveau dans le Jura: le chamois. *Bull. Soc. Neuchâtel. Sci. Nat.*, 95:89–95 (1972)—(Ge,En); effect of the chamois on the
1 vegetation.

4654 Ritter, J.
Sur les relations floristiques entre le Vercors meridional les Alps et les Pyrénées. *Caryologia*, 26(2):
1 187–191(1973)—(En); chromosome numbers.

4655 Rosskopf, G.
Pflanzengesellschaften der Talmoore an der Schwarzen und Weissen Laber in Oberpfälzer Jura. *Denkschr. Regensburg. Bot. Ges.*, 28(1):3–115(1971)—
1 Map.

4656 Royer, J. M.
Reconnaissance phytosociologique en Basse-Bourgogne: 2. À propos des inules de la vallée de la Cure. *Bull. Trim. Soc. Hist. Nat. Amis Mus. Autun*,
1 no.62:19–21(1972).

4657 Royer, J. M. *and* Rameau, J. C.
A propos de la végétation des éboulis de la montagne Chatillonnaise et du pays de Champlitte. *Ann. Sci.*
1 *Univ. Besançon, Bot.*, 10:131–139(1971)—Map.

4658 Tronchet, A.
Promenade botanique dans l'étage alpin du Valais et des Grisons. *Bull. Féd. Soc. Hist. Nat. Franche-*
1 *Comté*, 74(1):8–15(1972 publ. 1973).

4659 Vivant, J.
Plantes vasculaires intéressantes récoltées aux Pyrénées occidentales françaises (suite). *Monde Pl.*,
1 no.375:3–4(1972).

4660 Vivien, J.
Herborisations regionales au cours de l'année 1972. *Bull. Ass. Nat. Vallée Loing Massif Fontainebleau*,
1 49(3–4):31–33(1973).

4661 Vivien, J.
Plantes interessantes rencontrées dans le massif de Fontainebleau au cours de l'année 1971. *Bull. Ass. Nat. Vallée Loing Fontainebleau*, 48(3–4):37–38
1 (1972).

GERMANY

4662 Adler, R.
Die Pflanzenwelt des Osnabrücker Raumes. *Jahrb.*
1 *Ver. Schutze Alpenfl. Tiere*, 38:104–108(1973).

4663 Cramer, H.
Zur Flora des Augsberger Ostens—Botanische Beobachtungen im Landkreis Aichach. *Ber. Naturwiss.*
1 *Ver. Schwaben*, 77(1–2):28–33(1973).

4664 Dörr, E.
Flora des Allgäus: 7. Teil. Polygonaceae—Papaveraceae. *Ber. Bayer. Bot. Ges.*, 44:143–181
1 (1973).

4665 Dörr, E.
Floristische Notizen zur Allgäu-Flora 1973. *Mitt. Naturwiss. Arbeitskr. Kempten (Allgäu)*, 17(3):2–15
1 (1973).

4666 Dörr, E.
Zur Allgäu-Flora: Arbeitsergebnisse für 1972. *Mitt. Naturwiss. Arbeitskr. Kempten (Allgäu)*, 17(1):
1 41–58(1973).

4667 Duwensee, H. A.
Einige Funde im Oberharz bei Goslar. *Göttinger*
1 *Flor. Rundbr.*, 7(4):73–75(1973).

4668 Eschelmüller, A.
Grünten-Report 1972: eine Phanerogamenliste als Beitrag zur floristischen Landesaufnahme in Südbayern. *Mitt. Naturwiss. Arbeitskr. Kempten*
1 *(Allgäu)*, 17(1):1–40(1973).

4669 Gauckler, K.
Einstrahlungen der Alpenflora im Bayerischen Wald und Oberpfälzer Wald. *Jahrb. Ver. Schutze Alpenpfl. Tiere*, 37:25–41(1972)—Illustrations, map,
1 dot maps.

4671 Merxmüller, H.
Neue Übersicht der im rechtsrheinischen Bayern einheimischen Farne und Blütenpflanzen. *Ber.*
1 *Bayer. Bot. Ges.*, 44:221–238(1973).

4672 Rauschert, S.
Zur Nomenklatur der Farn- und Blütenpflanzen Deutschlands (3). *Feddes Repert.*, 83(9–10):645–662
1 (1972 publ. 1973).

4673 Reichert, H.
Neue Angaben über die Verbreitung von 60 Pflanzenarten im südwestlichen Hunsrück. *Decheniana*, 125(1–2):147–154(1972)—(En). 30 mosses, 30 vascular
1 plants.

4674 Runge, F.
Ein artenreicher Trockenrasen bei Battenberg.
1 *Hess. Flor. Briefe*, 256–258:29–30(1973).

4675 Schmidt, E.
Die Wolladventivpflanzen der Kämmerei Döhren/ Hannover. *Göttinger Flor. Rundbr.*, 7(4):75–80(1973)—
1 Illustrations.

4676 Schnedler, W.
Floristische Kartierung im Messtischblatt 5317 (Rodheim-Bieber). *Göttinger Flor. Rundbr.*, 7(2):
1 30–41(1973).

4677 Steinwendtner, R.
Bemerkenswerte Funde 1971. *Mitt. Bot. Arbeit.*
1 *Oberösterr. Landesmus. Linz*, 4(1–2):107(1972).

4678 Sukopp, H. *
Wandel von Flora und Vegetation in Mitteleuropa unter dem Einfluss des Menschen. *Ber. Landwirt.*, 50(1):112–139(1972)—Ref. in *Curr. Adv. Pl. Sci.*,
1 1(5):6095(1972).

4679 Wagenitz, G.
Zur Flora von Baltrum. *Göttinger Flor.*
1 *Rundbr.*, 7(1):3–9(1973).

4680 Walter, E.
Beiträge zur Flora von Bamberg aus der nördlichen Frankenalb: 2. *Ber. Naturforsch. Ges. Bamberg*, 47:
1 36–41(1972 publ. 1973).

4681 Wedeck, H.
Unkrautgesellschaften der Hackfruchtkulturen in Osthessen. *Philippia*, 1(4):194–212(1972)—(En);
I maps.

4682 Wittenberger, W.
Der Oberhorstweiher bei Offenbach am Main.
I *Hess. Flor. Briefe*, 253–255:12–13(1973).

GREAT BRITAIN

4683 Airy Shaw, H. K., *ed.*
Additions to the wild fauna and flora of the Royal Botanic Gardens, Kew: 29. Miscellaneous records,
I 1960–72. *Kew Bull.*, 28(3):387–409(1973).

4684 Anon.
Butlers Grove Reserve: being an initial report on the wildlife in the Reserve (4th July, 1972). *Basildon Nat. Hist. Soc. Bull.*, no.32:5–71(1972)—Plants
I pp.6–7.

4685 Anon.
Flora of Breconshire recording scheme. *B.S.B.I.*
I *Welsh Region Bull.*, 19:2–5(1973).

4686 Anon.
The new Warwickshire flora. *Proc. Birmingham Nat. Hist. Soc.*, 22(3):161–175(1973). Reprints of reviews of the book (publ. 1971) and later additions
I to the flora.

4687 Anon.
I Plant records. *Watsonia*, 9(3):273–278(1973).

4688 Anon.
Welsh botanical records. *Ann. Report Nation. Mus.*
I *Wales*, 117(1972–1973 publ. 1973).

4689 Birks, H. J. B.
Past and present vegetation of the Isle of Skye: a palaeoecological study. Cambridge, University Press,
I 1973. xii,415p.—Maps.

4690 Birks, H. J. B. *and* Deacon, J.
A numerical analysis of the past and present flora of the British Isles. *New Phytol.*, 72(4):877–902(1973)—
I Map, dot maps.

4691 Brummitt, J. M.
Plants of the Denbighshire coast. *B.S.B.I. Welsh*
I *Region Bull.*, 15:3–6(1971)—Map.

4692 Bullard, E. R.
Lagoons and 'oyces' in Orkney. *Orkney Field Club*
I *Bull.*, no.3:7–8(1972).

4693 Conolly, A. P.
A note on the flora of Welsh castles and abbeys.
I *B.S.B.I. Welsh Region Bull.*, 17:3–6(1972).

4694 Crackles, E.
Plant records: East Riding. *Ann. Rep. York. Nat.*
I *Union 1972*, 111:28(1973).

4695 Davis, T. A. W., *comp.*
Field notes: plants. *Nat. Wales*, 13(3):197–207
I (1973).

4696 Ellis, E. A. *and* Swann, E. L.
Albinism, doubling, pelory and cleistogamy in Norfolk plants. *Trans. Norfolk Norwich Nat. Soc.*,
I 22(6):401–405[1973].

4697 Falk, P.
The Dunsmore flora. *Rep. Rugby Sch. Nat. Hist.*
I *Soc.*, 16–18(1972–1973 publ. 1973).

4698 Gilbert, O. L., Holligan, P. M. *and* Holligan, M. S.
The flora of North Rona, 1972. *Trans. Bot. Soc.*
I *Edinburgh*, 42(1):43–68(1973)—Map.

4699 Green, P. S. *ed.*
Plants: wild and cultivated. A Conference on Horticulture and Field Botany. The Roy. Hort. Soc. and the Bot. Soc. of the British Isles 2nd and 3rd September 1972. London, The Botanical Society of the British Isles, 1973. 231p. (B.S.B.I. Conference Reports no. 13)—Illustrations, dot maps; chromosome
I number.

4700 Green, R.
Some British alpines. *Quart. Bull. Alp. Gard. Soc.*,
I 41(3):214–225,227(1973)—Illustration.

4701 Hardy, E.
The naturalist in Lakeland. Newton Abbot, David & Charles, 1973. 192p. (The Regional Naturalist)—
I Illustrations, map.

4702 Hart, C.
British trees in colour. London, Joseph, 1973. 63p.—Illustrations, coloured illustrations. Illustrations by
I C. Raymond.

4703 Hillman, E. M.
Changes in the distribution of trees and shrubs on Bookham Common, 1953–1971. *London Nat.*, no.51:
I 45–54(1971 publ. 1972)—Maps.

4704 Hird, L. P.
I Holy Island plants. *Vasculum*, 58(4):54(1973).

4705 Hird, L. P.
Plants at Craster, Northumberland. *Vasculum*, 58(2):
I 12(1973).

4706 Holmes, N. T. H., Lloyd, E. J. H., Potts, M. *and* Whitton, B. A.
Plants of the river Tyne and future water transfer
I scheme. *Vasculum*, 57(2):56–78(1972)—Map.

4707 Houseman, F.
Plant records: aliens and casuals. *Ann. Rep. York.*
I *Nat. Union 1972*, 111:31–32(1973).

4708 Kennedy, M. E. *and* Holland, P. C.
Plant mapping scheme. *London Nat.*, no.52:29–30
I (1973).

4709 Kenneth, A. G.
Dubh Loch, Glen Shira. *Glasgow Nat.*, 19(1):69–70
I (1973). Marsh plants.

4710 Libbey, R. P. *and* Swann, E. L.
Additions to the Norfolk flora. *Trans. Norfolk*
I *Norwich Nat. Soc.*, 22(6):393–399[1973]—Keys.

4711 Lousley, J. E.
Botanical records for 1971 and 1972. *London Nat.*,
I no.52:116–120(1973).

4712 Mackechnie, R. *and* Ribbons, B. W.
Ailsa Craig. *Glasgow Nat.*, 19(1):70(1973)—En.
I Plant records.

4713 Maddams, W. F. *and* Maddams, B.
Comments on the succulent flora of Tresco. *Cact.*
1 *Succ. J. Gr. Brit.*, 34(3):55–57(1973)—Illustrations.

4714 Marcan, B.
Botanical records. *Middle-Thames Nat.*, no.25:
1 31–33(1972 publ. 1973).

4715 Mason, J. L.
Bird-seed aliens: 1. *Wild Flower Mag.*, 368:11–17
1 (1973)—Illustrations.

4716 Medd, T. F.
Plant records: North Riding. *Ann. Rep. York. Nat.*
1 *Union 1972*, 111:29–30(1973).

4717 Mercer, I.
District ecological records: 1. The South Hams.
J. Devon Trust Nat. Conserv., 5(2):84–90(1973)—
1 Maps.

4718 Murgatroyd, F.
Botanical excursion to canal banks near Hebden
Bridge and Luddendenfoot, July 8th, 1973. *Naturalist*,
1 no.927:152(1973).

4719 Murgatroyd, F.
Plant records: West Riding. *Ann. Rep. York. Nat.*
1 *Union 1972*, 111:30–31(1973).

4720 [National Trust for Scotland]
Ben Lawers and its alpine flowers. [Edinburgh],
National Trust for Scotland, 1972. 87p.—Illustrations,
1 map. First published in 1955.

4721 Pankhurst, R. J.
Vascular plant records. *Nat. Cambridge.*, no.16:
1 41–42(1973).

4722 Plumb, W. J.
Some observations on the flora of the margins of
the A12. *Suffolk Nat. Hist.*, 16(3):162–172(1973).
1 S.W. of Ipswich.

4723 Ross-Craig, S.
Drawings of British plants: being illustrations of the
species of flowering plants growing naturally in the
British Isles: part 30. Juncaceae, Typhaceae,
1 Sparganiaceae, Araceae. London, Bell, 1973. 41pl.

4724 Side, A. G.
Vegetational changes at Egypt Bay, Kent, following
the exclusion of tidal water 1964–71. *Trans. Kent*
1 *Field Club*, 5(1):60–73(1973).

4725 Sillar, F. C. *and* Meyler, R. M.
Skye. Newton Abbot, David & Charles, 1973.
240p. (The Islands Series)—Illustrations, maps.
1 Plants pp.41–48.

4726 Slack, A. A. P.
Lochside vegetation. *Inverness Surv. 1970–74*
1 *Newsl.*, no.2:8–25(1973)—Map; key.

4727 Smith, P.
District ecological records: 2. Teignbridge. *J. Devon*
Trust Nat. Conserv., 5(3):131–137(1973)—Illustration,
1 maps.

4728 Swan, G. A. *and* Swan, M.
Flowering plants and ferns. *Vasculum*, 58(2):15
1 (1973). Plant records for the North of England.

4729 Swan, G. A. *and* Swan, M.
Records: flowering plants and ferns. *Vasculum*,
1 58(1):6–8(1973).

4730 Swanborough, J., *comp.*
Wiltshire plant notes: 33. *Wiltshire Archaeol. Nat.*
1 *Hist. Mag.*, *A*, 68:29–31(1973).

4731 Tutin, T. G.
Weeds of a Leicester garden. *Watsonia*, 9(3):263–
1 267(1973).

4732 Wells, T. C. E.
Botanical notes. *Ann. Rep. Huntingdon. Fauna Flora*
1 *Soc.*, (24):3–10(1973).

4733 Williams, J. T.
A preliminary survey of salt-tolerant vegetation
near Droitwich. *Proc. Birmingham Nat. Hist. Soc.*,
1 22(3):185–190(1973)—Map.

GREECE

4734 Économidou, E.
Contribution à l'étude de la flore et la phyto-
géographie de l'Ile de Skopelos: phytogéographie des
sporades du Nord: 2. *Biol. Gallo-Hellen.*, 5(1):
1 77–121(1973)—(En,Gre).

4735 Économidou, E.
Étude et cartographie de la végétation halophile
littorale de l'Ile de Skiathos. *Biol. Gallo-Hellen.*,
1 4(2):115–137(1973)—(En,Gre); maps.

4736 Quézel, P.
Contribution à l'étude de la végétation du
Vardoussia. *Biol. Gallo-Hellen.*, 5(1):139–166(1973)—
1 (En,Gre).

4737 Voliotis, D.
Contribution to the knowledge of the Greek flora:
new taxa of the mountainous region Cholomon
Greek Macedonia. *Feddes Repert.*, 84(4):327–328
1 (1973)—(Ge).

4738 Zahariadi, C.
Quelques taxons rares ou nouvellement découverts
de la flore de la Grèce. (Some rare or newly dis-
covered taxa of the Greek flora). *Ann. Mus. Goulandris*,
1 1:165–183(1973)—(En,Gre); illustrations, dot map.

HUNGARY

4739 Csürös, S.
Geobotanische Forschungen im Casin (Kászon)-
Becken (Ostkarpaten). *Acta Bot. Acad. Sci. Hung.*,
19(1–4):55–71(1973)—(En); illustrations.
1 Ecological.

4740 Holub, J.
Neue oder wenig bekannte Pflanzen der ungarischen
Flora. *Ann. Univ. Budapest. Rolando Eötvos Nom., Biol.*,
1 14:91–104(1972 publ. 1973).

4741 Horvát, A. O.
Die Vegetation des Mecsekgebirges und seiner
Umgebung. Budapest, Akad. Kiadó, 1972. 376p.—
1 Illustrations, coloured illustrations, maps.

4742 Soó, R.
A Magyar flóra és vegetáció rendszertani-
növényföldrajzi kézikönyve: 5. (Synopsis systematico-
geobotanica florae vegetationisque Hungariae: 5).
1 Budapest, Akad. Kiadó, 1973. 724p.

4743 Soó, R.
Nomina a nobis 'non rite' publicata. *Acta Bot.*
I *Acad. Sci. Hung.*, 18(1–2):171–177(1973)—(En,Ge).

IRELAND

4744 Carvill, P. H. *and* Curtis, T. G.
New plant records for County Wicklow. *Irish*
I *Nat. J.*, 17(11):386–388(1973).

4745 Farrell, L.
New plant records for Co. Monaghan. *Irish Nat. J.*,
I 17(10):355(1973).

4746 Scannell, M. J. P.
Recent additions to the Irish flora. *Watsonia*, 9(4):
I 431–432(1973).

4747 Scannell, M. J. P. *and* Synnott, D.
Census catalogue of the flora of Ireland: a concise
conspectus of the Irish Pteridophyta, Gymnospermae
and Angiospermae including all the native plants and
established aliens known to occur in Ireland with the
distribution of each species. Dublin, Stationery
I Office, 1972. xvi,127p.—Map.

ITALY

4748 Becherer, A.
Kleine Beiträge zur Flora des Aostatales. *Ber.*
I *Schweiz. Bot. Ges.*, 83(1):39–41(1973).

4749 Becherer, A.
Zur Flora der nördlichen Lombardei. *Bauhinia*,
I 5(1):3–6(1973).

4750 Brilli-Cattarini, A. J. B. *and* Sialm, R.
Segnalazione di pianta nuove, inedite, o notevoli
per la regione marchigiana: 3. (A report about new
or interesting plants in the Marche region: 3). *Giorn.*
I *Bot. Ital.*, 107(2):59–73(1973).

4751 Cook, C. D. K.
New and noteworthy plants from northern Italian
ricefields. *Ber. Schweiz. Bot. Ges.*, 83(1):54–65
I (1973)—(Ge,Fr).

4752 Favarger, C.
Cytotaxonomie de quelques orophytes des
Abruzzes. *Acta Bot. Acad. Sci. Hung.*, 19(1–4):81–92
I (1973)—(En); chromosome numbers.

4753 Fenaroli, L.
Florae Garganicae prodromus: pars tertia. *Webbia*,
I 28(2):323–410(1973)—It; illustrations, map.

4754 Ferrarini, E.
Carta della vegetazione delle Alpi Apuane e zone
limitrofe. *Webbia*, 27(2):551–582(1972 publ. 1973)—
I (En); illustrations, maps.

4755 Marcello, A.
La flora urbica de Venezia: notizia preliminare.
Atti Ist. Veneto Sci. Lett. Arti, Sci. Mat. Nat., 130:425–
I 432(1972).

4756 Marcello, A.
Notizia di reperti floristici e del loro ritmo antesico
presso le grotte di Costozza (Vicenza). *Atti Ist. Veneto*
Sci. Lett. Arti, Sci. Mat. Nat., 129:371–387(1971)—
I (En).

4757 Pignatti, S.
Note critiche sulla flora d'Italia: 1. Appunti
miscellanei. (Critical notes on the flora of Italy: 1.
Miscellaneous remarks). *Giorn. Bot. Ital.*, 107(5):207–
I 221(1973)—(En).

4758 Sanmarchi, A.
Der Wald von Somadida—Dolomitische Alpen.
Jahrb. Ver. Schutze Alpenfl. Tiere, 38:119–125(1973)—
I Illustrations, map.

JUGOSLAVIA *and* TRIESTE

4759 Birks, H. J. B. *and* Walters, S. M.
The flora and vegetation of Barno Jezero, Durmitor,
Montenegro. (Flora i vegetacija Barno Jezera na
Durmitoru Crna Gora). *Glasn. Republ. Zavoda Zašt.*
Prir. Prirodnj. Muz. Titogradu, no.5:5–23(1972 publ.
I 1973)—(Se).

4760 Blečić, V. *and* Krasnići, F.
Zajednica endemičnog šibljaka forzicije i krstušca
(Polygalo-forsythietum Europaeae Blečić i Krasnići)
u Jugozapadnoj Srbiji. (Gesellschaft des endemischen
Šibljaks (Polygalo-Forsythietum europaeae Blečić
und Krasn.) in südwestlichem Serbien). *Glasn.*
Republ. Zavoda Zašt. Prir. Prirodnj. Muz. Titogradu,
I no.4:35–40(1971 publ. 1972)—(Ge).

4761 Diklić, N. *and* Nikolić, V.
Zur Flora und Vegetation des jugoslawischen
Anteiles des neuen Djerdap-Stauseegebietes an der
Donau. *Jahrb. Ver. Schutze Alpenpfl. Tiere*, 38:109–117
I (1973)—Illustrations, maps.

4762 Jovanović, B.
Fitocenoze crnoga bora (*Pinus nigra* Arn.) na
Kopaoniku. (Phytocenosen der Schwarzkiefer
(*Pinus nigra* Arn.) am Kopaonik). *Glasn. Prirodnj.*
I *Muz. Beogradu, B*, 27:11–29(1972)—(Ge).

4763 Lakušić, R.
Specifičnosti flore i vegetacije Crnogorskih
Kanjona. (Die Besonderheiten der Flora und die
Vegetation des Montenegrinisches Kanons). *Glasn.*
Republ. Zavoda Zašt. Prir. Prirodnj. Muz. Titogradu,
I no.4:157–169(1971 publ. 1972)—(Ge).

4764 Lausi, D.
Notizie sulla vegetazione del Triestino. *Inform. Bot.*
I *Ital.*, 3(3):172–174(1971 publ. 1972).

4765 Lausi, D. *and* Poldini, L.
Schizzo botanico della Val Rosandra. *Inform. Bot.*
I *Ital.*, 3(3):181–185(1971 publ. 1972).

4766 Lausi, D. *and* Poldini, L.
Vegetazione carsica attorno ad Aurisina-Cave.
I *Inform. Bot. Ital.*, 3(3):174(1971 publ. 1972).

4767 Lausi, D. *and* Poldini, L.
Vegetazione delle Doline e delle pendici meridionali
del Monte Cocusso. *Inform. Bot. Ital.*, 3(3):175(1971
I publ. 1972).

4768 Kelly, J.
Some notes on Triglav. *Quart. Bull. Alp. Gard.*
I *Soc.*, 41(3):209–211(1973).

4769 Poldini, L.
Appunti sulla flora del Monte Nanos (m1261).
I *Inform. Bot. Ital.*, 3(3):176–180(1971 publ. 1972).

4770 Poldini, L.
Notizie sulla flora del Triestino. *Inform. Bot. Ital.*,
1 3(3):169–172(1971 publ. 1972).

4771 Wraber, T.
Arbutus unedo L. in *Myrtus communis* L. v Slovenski Istri. (*Arbutus unedo* L. and *Myrtus communis* L. in Slovenian Istria). *Biol. Vestn.*, 20:127–133(1972)—
1 (En,Ge).

LUXEMBOURG

4772 Langhe, J. E. de, Delvosalle, L., Duvigneaud, J., Lambinon, J. *and* Berghen, C. vanden
Nouvelle flore de la Belgique, du Grand-Duché de Luxembourg, du Nord de la France et des régions voisines (ptéridophytes et spermatophytes). Bruxelles, Jardin Botanique, 1973. xcviii,821p.—Illustrations,
1 coloured illustrations, map; keys.

MALTA

4773 Lanfranco, E.
Additions and corrections to the Maltese flora.
1 *Maltese Nat.*, no.4:8–10(1973).

NETHERLANDS

4774 Beeftink, W. G., Daane, M. C. *and* Munck, W. de
Tien jaar botanisch-œcologische verkenningen langs het Veerse Meer. (Ten years of phyto-ecological research along the borders of the Veerse Meer). *Nat.*
1 *Landschap*, 25(2):50–63(1971)—(En).

4775 Groot, C.
Planten op een fabrieksterrein. *Natura* (Netherlands),
1 70(8):113–117(1973)—Illustrations.

4776 Jansen, M. T. *and* Mennema, J.
Verslag van de 'Floristenexcursie' naar Dronten in
1 1969.*Gorteria*, 6(7):117–121(1973)—(En); map.

4777 Menke, H. L.
Vossenbroek, een te laat ontdekt natuurgebied. *Natura* (Netherlands), 70(9):131–134(1973)—
1 Illustrations.

4778 Mennema, J. *and* Urk, R. M. van
De huidige stand van de tweede inventarisatie-ronde van het IVON. *Gorteria*, 6(9):153–155(1973)—
1 (En); dot map.

4779 Ogg, W.
De bedreigde schinveldse bossen door de Werkgroep tot behoud van de Schinveldse bossen. *Natuurhist.*
1 *Maandblad*, 62(2–3):13–36(1973)—Illustrations, maps.

4780 Ooststroom, S. J. van *and* Mennema, J.
Nieuwe vondsten van zeldzame planten in Nederland, hoofdzakelijk in 1972. *Gorteria*, 6(12):
1 181–198(1973)—(En).

4781 Ploeg, D. T. E. van der
De waarden langs de Friese Ijsselmeerkust. *Natura* (Netherlands), 69(10):166–169(1972)—Illustrations,
1 map.

4782 Weeda, E. J.
Botanische notities uit Twente. *Natura* (Nether-
1 lands), 70(2):25–29(1973)—Illustrations.

4783 Weeda, E. J.
Het Faisantenbos bij Ellecom. *Natura* (Netherlands),
1 70(5):69–71(1973)—Illustrations.

4784 Westhoff, V.
Vegetatie en bodem op de Beekdalhellingen van het Krijtdistrict. *Natuurhist. Maandblad*, 62(10):124–132
1 (1973)—Illustration.

NORWAY

4785 Elven, R.
Noen plantefunn fra Femundsmarka, en plante-geografisk. (Some plant records from Femundsmarka, central Norway, a phytogeographical discussion).
1 *Blyttia*, 31(4):229–248(1973)—(En); maps, dot maps.

4786 Hygen, K. *and* Hygen, G.
Noen nye plantefunn i Rondane nasjonalpark. (Vascular plants new to Rondane National Park in Central Norway). *Blyttia*, 31(2–3):129–135(1973)—
1 (En).

4787 Lye, K. A.
The vascular plants on alpine peaks at Fillefjell, south Norway. *Norw. J. Bot.*, 20(1):51–55(1973)—
1 Map.

4788 Naustdal, J.
Karplantar på fjella mellom Myrdal og Hallingskeid. (Vascular plants in the mountains between Myrdal and Hallingskeid, Central South Norway). *Blyttia*,
1 31(2–3):137–147(1973)—(En).

4789 Østhagen, H.
Det xeroterme økoelement i Ringerikes flora. (The xerothermic ecological element in the flora of Ringerike, S. E. Norway). *Blyttia*, 31(4):221–228
1 (1973)—(En); map.

POLAND

4790 Browicz, K., *ed.*
Atlas rozmieszczenia drzew i krzewów w Polsce: 13–14. (Atlas of distribution of trees and shrubs in Poland). Warszawa, etc., Polska Akademia Nauk,
1 1973. 2v.—Pol,En,Rus; maps, dot maps.

4791 Czyzewska, K.
O interesujących roślinach Masłowice w województwie łódzkim. (Some more interesting plants of the Masłowice primeval sylvan range in the province of Łódź). *Chrońmy Przyr. Ojczystą*, 29(1):80–83
1 (1973)—Illustration.

4792 Dąbrowska, L.
Występowanie niektórych gatunków roślin naczyniowych w Małych Pieninach. (The occurrence of some vascular plant species in the Małe Pieniny range). *Fragm. Flor. Geobot.*, 19(3):309–321(1973)—
1 (En).

4793 Dzwonko, Z. *and* Zemanek, B.
Rzadkie i interesujące rośliny naczyniowe Gór Słonnych. (Rare and interesting species of vascular plants of the Słonne Mts. (Polish eastern Carpathians)).
1 *Fragm. Flor. Geobot.*, 19(3):323–325(1973)—(En).

4794 Faliński, J. B.
Podstawy i formy eksploracji naukowej Białowieskiego Parku Narodowego. (Basis and forms of scientific exploration of the Białowieza National
1 Park). *Ochr. Přzyr.*, 37:7–55(1972)—(En).

4795 Fijałkowski, D. *and* Izdebski, K.
Projekt Roztoczańskiego Parku Narodowego. (The project of the Roztocze National Park). *Chrońmy*
1 *Przyr. Ojczystą*, 28(5–6):56–66(1972)—Map.

4796 Fijałkowski, D. *and* Pękala, M.
Rzadkie rośliny naczyniowe Nadleśnictwa Strzelce. (Rare vascular plants of the Chief Forestry District Strzelce). *Ann. Univ. Mariae Curie-Skłodowska, C*, 28:139–143(1973)—(Rus,En).
I

4797 Jasnowska, J.
Wąwóz Kamieniec: projektowany rezerwat florystyczny w powiecie szczecińskim. (The Kameniec Gorge: a projected botanical reserve in the district of Szczecin). *Chrońmy Przyr. Ojczystą*, 29(4):31–37(1973)—(En); map.
I

4798 Jelinowski, T.
Rzadziej spotykane rośliny naczyniowe w lasach Lęborszczyzny: część 2. (Rare vascular plant species in the forests of Lebork district (northern Poland): part 2). *Fragm. Flor. Geobot.*, 19(3):279–283(1973)—(En).
I

4799 Kępczyński, K.
Dalsze stanowiska Wolfii bezkorzeniowej *Wolffia arrhiza* (L.) Wimm. na Wysoczyźnie Dobrzyńskiej oraz jej udział w różnych zespolach roślinnych. (Weitere Standorte von *Wolffia arrhiza* (L.) Wimm. auf dem Dobrzyń-Diluvialplateau und ihre Teilnahme in Verschiedenen Pflanzengesellschaften). *Zesz. Nauk. Uniw. Mikol. Kopernika Toruniu, Biol.*, 30: 11–18(1972)—(Ge).
I

4800 Kępczyński, K. *and* Noryśkiewicz, A.
Materiały do flory powiatu wąbrzeskiego: część 1. (Beitrage zur Flora des Kreises Wąbrzeźno: Teil 1). *Zesz. Nauk. Uniw. Mikol. Kopernika Toruniu, Biol.*, 30:91–99(1972)—(Ge).
I

4801 Kępczyński, K. *and* Zielski, A.
Materiały do flory Pojezierza Brodnickiego: część 1. (Beiträge zur Flora der Brodnica-Seenplatte: Teil 1). *Zesz. Nauk. Uniw. Mikol. Kopernika Toruniu, Biol.*, 30:45–62(1972)—(Ge); map.
I

4802 Kostrakiewicz, K. *and* Popek, R.
Góry Pieprzowe jako przyszły rezerwat przyrody. (Góry Pieprzowe (Pepper Hills) as a future nature reserve). *Chrońmy Przyr. Ojczystą*, 28(5–6):11–18 (1972)—(En); illustrations.
I

4803 Mądalski, J.
Atlas flory Polskiej i ziem ościennych. (Florae Polonicae terrarumque adiancentium iconographia) tom 17, zeszyt 1. Scrophulariaceae (pars 4), Orobanchaceae. Warszawa, etc., Wydawnictwo Naukowe, 1973. 77p.—Illustrations.
I

4804 Michalik, S.
Niektóre interesujące gatunki synantropijne z terenu województwa opolskiego. (Some interesting synantropic plants in the Province of Opole (Silesia)). *Fragm. Flor. Geobot.*, 19(3):271–278(1973)—(En).
I

4805 Mizianty, M. *and* Frey, L.
Chromosome numbers of some vascular plants in the western Bieszczady Mts. (south-eastern Poland). (Liczby chromosomów kilku roślin naczyniowych z Bieszczadów Zachodnich). *Fragm. Flor. Geobot.*, 19(3):265–270(1973)—(Pol); chromosome numbers.
I

4806 Myczkowski, S.
Native tree species of the Tatra National Park. *Int. Dendrol. Soc. Year Book 1972*: 70–80(1973)—Illustrations.
I

4807 Nowaczyk, C. *and* Tychowska, W.
Szata leśna i sędziwe dęby rezerwatu 'Kolno Międzychodzkie' w województwie poznańskim. (The sylvan vegetation and old oak trees in the 'Kolno Międzychodskie' nature reserve (Province of Poznan)). *Chrońmy Przyr. Ojczystą*, 29(3):51–59 (1973)—Map.
I

4808 Piękoś, H. *and* Tacik, T., *eds*
Flora Poloniae exsiccata. *Fragm. Flor. Geobot.*, 19(3) suppl.:369–391(1973).
I

4809 Piotrowska, H. *and* Chojnacki, W.
Materiały do rozmieszczenia roślin naczyniowych na Polskim wybrzeżu Bałtyku: część 1. (Materials concerning distribution of vascular plants on the Polish shore of the Baltic sea: part 1). *Badan. Fizjogr. Pol. Zachod., B*, 25:169–175(1972)—(En).
I

4810 Polakowski, B.
W sprawie utworzenia Mazurskiego Parku Narodowego. (Notes on the establishment of the Mazurian National Park). *Chrońmy Przyr. Ojczystą*, 29(4):26–30(1973)—(En); maps.
I

4811 Sokołowski, A. W.
Roślinność rezerwatu 'Perkuć' w Puszczy Augustowskiej. (The vegetation of the Perkuć nature reserve in the Augustów Primeval Forest). *Chrońmy Przyr. Ojczystą*, 28(5–6):68–73(1972)—Illustrations, map.
I

4812 Weretelnik, E.
Flora starych murów Lubania Śląskiego. (The flora of the old walls of the town of Lubań Sląski). *Chrońmy Przyr. Ojczystą*, 29(1):41–45(1973)—(En); illustrations, map.
I

4813 Wójcík, Z.
The plant communities of root-crop fields in lowlands and highlands of Poland: floristic, ecologic and regional differentiation. *Feddes Repert.*, 84(7–8): 573–588(1973)—(Ge); dot maps.
I

4814 Zabawski, J. *and* Matuła, J.
Grzybienie północne *Nymphaea candida* w powiecie Zgorzeleckim na Dolnym Slązku. (The water-lilies, *Nymphaea candida* in the district of Zgorzelec in Lower Silesia). *Chrońmy Przyr. Ojczystą*, 29(1): 76–80 (1973) Illustrations, map.
I

PORTUGAL

4815 Malato-Beliz, J.
Novas especies para a flora de Portugal: 2. *Lagascalia*, 3(1):61–69(1973)—(En); illustrations.
I

4816 Lopes, M. H. Ramos, Martins, J. M., Rainha, B. V., Silva, A. R. Pinto da, Silva, M. da, Silva, Q. de J. G. Pinto da *and* Teles, A. N.
Plantas novas e novas áreas para a flora de Portugal: 10. *Agron. Lusit.*, 34(3):177–196(1972 publ. 1973)—(En).
I

4817 Silva, A. R. Pinto da *and* Lopes, M. H. Ramos
Famílias e géneros da flora Lusitanica, Azorica and Madeirensis. *Anuár. Soc. Brot.*, 38:13–52(1972).
I

RUMANIA

4818 Andrei, M.
Flora cormofitelor cu unele considerații ecologice din complexul de Bălți Crapina-Jijila (2). (La flore des cormophytes avec quelques considérations écologiques du complexe des lacs Crapina-Jijila (2)). *An. Univ. București, Biol. Veg.*, 21:111–123(1972)—(Rus,Fr); illustrations.
I

4819 Bîrjoveanu, C.
Contribuţii la studiul florei împrejurimilor oraşului
Moineşti (jud. Bacău). (Contributions à l'étude de la
flore des environs de la ville de Moineşti (Départ.
1 Bacău)). *Stud. Comun., Bot.*: 105–111(1972)—(Fr).

4820 Cîrţu, D. *and* Cîrţu, M.
Date noi pentru flora cormofită din Oltenia.
(Dates nouvelles pour la flore cormofite d'Oltenie).
1 *Stud. Comun., Bot.*: 113–118(1972)—(Fr).

4821 Ciufudeanu, F.
Aspecte de floră si de vegetaţie din rezervaţia
calcarele de la ampoiţa. (Piatra ampoiţei) (Jud. alba).
(Aspects de la flore et de la végétation de la réserve les
calcaires d'Ampoiţa (Piatra Ampoitei) Dép. d'Alba).
1 *Ocrot. Nat.*, 16(2):217–219(1972)—(Fr).

4822 Diaconescu, F.
Materiale pentru flora vasculară a Masivului
Leaota: 3. *Ann. Şti. Univ. "Al. I. Cuza" Iaşi, Biol.*,
1 18(2):459–461(1973).

4823 Horeanu, Cl.
Aspecte floristice de la Cheia (Dobrogea).
(Quelques aspects floristiques de Cheia (Dobrogea)).
1 *Ocrot. Nat.*, 17(1):83–88(1973)—(Fr); illustrations.

4824 Horeanu, C.
Contribuţii la flora Dobrogei: 3. *Ann. Şti. Univ. "Al.*
1 *I. Cuza" Iaşi, Biol.*, 19(2):475–477(1973).

4825 Mihai, G. *and* Sârbu, I.
Cercetari fitocenologice asupra vegetaţiei de pe
'Valea Lupului'—Iaşi. *Ann. Şti. Univ. "Al. I. Cuza"*
1 *Iaşi, Biol.*, 18(2):469–473(1973).

4826 Mititelu, D. *and* Barabaş, N.
Completări da răspîndirea unor plante în Moldova.
(Complètements à la flore de la Moldavie). *Stud.*
1 *Comun., Bot.*, 103–104(1972)—(Fr).

4827 Mititelu, D. *and* Barabaş, N.
Contribuţie la studiul vegetaţiei lemnoase din lunca
Prutului. *Ann. Şti. Univ. "Al. I. Cuza" Iaşi, Biol.*,
1 18(2):463–468(1973)—Illustrations.

4828 Puşcariu, V.
Păduri, parcuri nationale şi rezervaţii naturale din
România. (Forêts, parcs nationaux et réserves
naturelles de Roumanie). *Ocrot. Nat.*, 16(2):151–166
1 (1972)—(Fr); illustrations, coloured illustrations.

4829 Raclaru, P.
Evoluţia şi succesiunea vegetaţiei din munţii
Rarău. (Evolution et la succession de la végétation
des monts Rarău). *Stud. Comun., Bot.*, 205–211(1972)—
1 (Fr).

4830 Raţiu, O.
Consideraţii floristico-ecologice asupra unor
Pteridofite şi Spermatofite noi pentru vegetaţia
bazinului Stîna de Vale. (Floristical and ecological
considerations on some Pteridophytes and Sperma-
tophytes, new for the vegetation of the Stîna de Vale
1 basin). *Contrib. Bot.*, 119–125(1973)—(En).

4831 Resmeriţă, I.
Cercetări privind completarea inventarului
floristic din Maramureş cu taxoni, noi sau rari.
(Recherches en vue de compléterl'inventaire floristique
de Maramures avec des taxons nouveaux ou rarissimes).
1 *Contrib. Bot.*, 127–132(1973)—(Fr); illustrations.

4832 Resmeriţă, I.
Signalement de marais dans les montagnes de
Maramures avec flore et végétation significatives.
1 *Rev. Roum. Biol., Bot.*, 18(3):137–143(1973).

4833 Roumania: Academia Republicii Socialiste România
Flora Republicii Socialiste România ... 12. Flora
Reipublicae Socialisticae Romania ... [Bucureşti],
Acad. Rep. Social. România, 1972. 810p.—Illustra-
1 tions.

4834 Sanda, V. *and* Popescu, A.
Contribuţii la cunoaşterea vegetaţiei lemnoase din
Cîmpia Română. *Stud. Cerc. Biol., Bot.*, 24(5):381–
1 390(1972)—(Rus); illustrations.

4835 Slonovschi, V.
Materiale pentru flora văii Siretului în sectorul
Mirceşti (Jud. Iaşi). (Materiaux pour la flore de la
vallée de Sireth dans le secteur de Mirceşti (District
1 Jassy). *Stud. Comun., Bot.*, 119–126(1972)—(Fr).

4836 Váczy, C.
Indreptări nomenclaturale la volumele 7–12. Flora
R.S.R. (1960–1972). (Nomenklatorische Berich-
tigungen für die Bände 7–12 der "Flora R.S.R.").
1 *Contrib. Bot.*, 287–305(1973)—(Ge).

SCOTLAND *see* GREAT BRITAIN

SICILY

4837 Maugeri, G.
Entità nuové o interessanti della flora siciliana.
(New or interesting species of the Sicilian flora).
1 *Inform. Bot. Ital.*, 5(2):156(1973)—(En).

SPAIN

4838 Alvarez Diaz, R.
Notas sobre flora gallega. *Trab. Dep. Bot. Fisiol.*
1 *Veg. Univ. Madrid*, 6:7–14(1973).

4839 Castroviejo, S.
Sobre la flora gallega: 2. *Trab. Dep. Bot. Fisiol. Veg.*
1 *Univ. Madrid*, 6:15–22(1973).

4840 Dendaletche, C.
Le peuplement végétal des montagnes entre les
Pics d'Anie et de l'Ohry (Pyrénées occidentales):
notes écologiques, floristiques et phytocénotiques.
1 *Pirineos*, no.105:11–26(1972).

4841 Galiano, E. F.
Una iniciativa que nos preocupa. *Lagascalia*, 3(1):
1 143–145(1973)—(En).

4842 Galiano, E. F. *and* Valdés, B.
Catalogo de las plantas vasculares de la Provincia de
Sevilla: 5. Ranunculales, Aristolochiales. *Lagascalia*,
1 3(2):223–237(1973).

4843 Gaussen, H.
Les cartes de végétation dans le Bassin de l'Ebre.
1 *Pirineos*, no.105:69–83(1972).

4844 Gonzalez Campo, C.
Veinte mil especies vegetales en peligro. *Jard. y*
Paisaje, no.4:9–12(1973)—Illustration, coloured
1 illustrations.

4845 Laínz, M. *and others*
Aportaciones al conocimiento de la flora Cántabro-
Astur: 10(1). *Bol. Inst. Estud. Astur., Supl. Cienc.*,
1 16:159–206(1973)—Illustrations.

4846 Lopez Fernandez, M. L.
Aportacion al conocimiento floristico de la Navarra media occidental. *An. Inst. Bot. A. J. Cavanilles*, 30:
1 183–196(1973)—(Fr,En,Ge).

4847 López Fernández, M. L.
Estudios de flora Navarra: 4. Dicotyledóneas eurosiberianas o de más amplia área, observadas en la montaña media occidental de la provincia. *Pirineos*,
1 no.105:27–46(1972)—(Fr,En,Ge).

4848 Magallon, A. R.
Flora y vegetacion de la Provincia de Alicante (el paisaje vegetal alicantino). Alicante, Inst. Estud. Alicantinos, 1972. xx,403p.—Illustrations, coloured
1 illustrations, map.

4849 Montserrat, P.
Estudios florísticos en el Pirineo occidental.
1 *Pirineos*, no.108:49–64(1973).

4850 Rivas-Martinez, S.
Ensayo sintaxonomico de la vegetacion cormofitica de la Peninsula Iberica, Baleares y Canarias hasta el rango de subalianza: 1. Vegetación acuática, helofítica y turfófila. *Trab. Dep. Bot. Fisiol. Veg. Univ.*
1 *Madrid*, 6:31–43(1973)—(En).

4851 Segura Zubizarreta, A.
De flora soriana y circumsoriana. *Pirineos*, no.109: 35–49(1973). Includes four new taxa and *Ulmus*
1 *laevis* and *Cytisus decumbens* new to Spain.

4852 Villar-Pérez, L.
Notas florísticas del pirineo occidental. *Pirineos*, 103:5–25(1972)—(Fr,En). Includes *Lithospermum*
1 *gastonis* and *Androsace hirtella* new to Spain.

SVALBARD

4853 Hofmann, W. *and* Thannheiser, D. *
Floristische Neufunde von Kross- und Kongsfjord, Spitzbergen. (Nouvelles découvertes floristiques dans les fjords Kross et King au Spitzberg). *Polarforschung*, 7(2):122–124(1972)—(En). Ref. in *Bull. Signal., Biol.*
1 *Physiol. Vég.*, 34(10):9725(1972).

4854 Rønning, O. I.
The distribution of the vascular cryptogams and monocotyledons in Svalbard. *Kong. Norske Vidensk. Selsk. Skr.*, no.24. 63p.(1972)—Dot maps;
1 chromosome numbers.

SWEDEN

4855 Bohlin, A.
Något om floran på Hunneberg. (On the flora of Mt. Hunneberg, Västergötland, SW Sweden).
1 *Svensk Bot. Tidskr.*, 67(4):458(1973)—(En).

4856 Clemedson, C.-J.
Vegetation och flora på Gorsingeholmskullarna vid Strängnäs. (Vegetation and flora of the hills at Gorsingeholm near Strängnäs, Södermanland, Sweden). *Svensk Bot. Tidskr.*, 67(4):423–437(1973)—
1 (En).

4857 Genberg, E.
Kärlväxternas utbredning i nordöstra Östergötland. (The distribution of vascular plants in N.E. Östergötland). *Svensk Bot. Tidskr.*, 67(2):142–171(1973)—
1 (En); map, dot map.

4858 Staav, R.
Den maritima björkskogszonen i Stockholms norra skärgård: nagra reflexioner kring en unik naturtyp.
1 *Fauna Flora*, 67(1):17–22(1972)—Illustrations, map.

SWITZERLAND

4859 Baumgartner, W.
Die Adventivflora des Rheinhafens Basel-Kleinhüngen in den Jahren 1950–1971. *Bauhinia*,
1 5(1):21–27(1973).

4860 Becherer, A.
Erloschene Arten der Schweizer Flora. *Ber.*
1 *Schweiz. Bot. Ges.*, 82(4):300–301(1972).

4861 Becherer, A.
Fortschritte in der Systematik und Floristik der Schweizerflora (Gefässpflanzen) in den Jahren 1970 und 1971. *Ber. Schweiz. Bot. Ges.*, 82(2):159–201
1 (1972).

4862 Becherer, A.
Führer durch die Flora der Schweiz, mit Berücksichtigung der Grenzgebiete. Basel, etc., Schwabe,
1 1972. 207p.

4863 Becherer, A.
Zur Flora des Münstertals. *Jahresber. Naturforsch.*
1 *Ges. Graubündens*, 94:180–192(1969–1971).

4864 Becherer, A. *and* Eckardt, T.
Zur Flora des Puschlav. *Bauhinia*, 5(1):33–56
1 (1973)—Illustrations.

4865 Keller, M.
Kleinräumige Verbreitung von Pflanzenarten im Luzerner Seetal im Vergleich zu Gesamtverbreitung und Umwelt. *Mitt. Naturforsch. Ges. Luzern*, 23.
1 189p. (1972)—Maps, dot maps.

4866 Müller, P. *
Flora des Suhren- und Ruedertales. *Mitt. Aargauischen Naturforsch. Ges.*, no.28:5–233(1971). Abstr. in
1 *Ref. Zhurn. Biol.*, 2(1):V12(1972).

4867 Richard, J.-L.
La vegetation des crêtes rocheuses du Jura. *Ber. Schweiz. Bot. Ges.*, 82(1):68–112(1972)—(Ge,En);
1 illustrations, map.

4868 Schlittler, J.
Beitrag zur Flora des Val d'Hérens. *Ber. Schweiz.*
1 *Bot. Ges.*, 82(1):64–67(1972).

4869 Schmalz, K. L.
Naturschutztätigkeit im Kanton Bern: 1971. *Mitt. Naturforsch. Ges. Bern*, 29:65–90(1972)—
1 Illustrations, maps.

4870 Villaret, P.
Flore et végétation d'Anzeinde. *Mém. Soc. Vaud.*
1 *Sci. Nat.*, 15(3):139–146(1973)—Illustrations.

WALES *see* GREAT BRITAIN

YUGOSLAVIA *see* JUGOSLAVIA

U. S. S. R.

4871 Cherepanov, S. K.
Svod dopolnenii i izmenenii 'Flore SSSR' (tt 1–30). (Additamenta et corrigenda ad 'Floram URSS'
1 (tomi 1–30)). Leningrad, Nauka, 1973. 668p.

4872 Dorogostajskaya, E. V. *
(La végétation du Grand Nord de l'URSS et sa mise en valeur. Les mauvaises herbes du Grand Nord de l'U.R.S.S.). *Rastit. Krajn. Severa SSSR. Osvoen*, 13:1–172(1972)—Rus(En). Abstr. in *Bull. Signal.*,
1 *Biol. Physiol. Vég.*, 34(1):460(1973).

4873 Lithuania: Akademiya Nauk Litovskoĭ SSR: Institut Botaniki
Poleznȳe rasteniya pribaltiĭskikh Respublik i Belorussiĭ . . . 26–27 aprelya 1973 g. [Useful plants of Baltic Republics and Belorussia: data on the II Scientific Conference on the study and enrichment of vegetation of Baltic Republics and Belorussia. (Conference . . . 26–27 April 1973)]. Vilnius, 1973.
1 340p.

4874 Palamarchuk, A. S. *and* Palamarchuk, G. L.
Deyaki florȳstȳchni osoblȳvosti ridkisnȳkh vȳdiv Bilorus'koho Polissya u zv'yazku z poshȳrennyam yikh na Ukrayini. (Some floristic peculiarities of rare species in the Byelorussian Polesye in connection with areas in their distribution in the Ukraine). *Ukr. Bot.*
1 *Zhurn.*, 30(2):238–241(1973)—Uk(En).

4875 Pobedimova, E. G.
Novȳe taksonȳ florȳ SSR, 3. (Taxa nova florae URSS, 3). *Novosti Sist. Vȳssh. Rast.*, 10:339–341
2 (1973).

BALTIC STATES

4876 Issako, L.
Meie looduslikke maitsetaimi: 2. (Our native spice plants: 2). *Eesti Loodus*, 16(6):344–346(1973)—
1 Illustrations.

4877 Issako, L.
Meie looduslikke maitsetaimi: 3. (Our native spice
1 plants: 3). *Eesti Loodus*, 16(8):478–479(1973).

4878 Kask, M. *and* Viljasoo, L.
Huvitavamaid taimeleide Tubiflorae seltsist Eestis. (Some interesting species of Tubiflorae in the Estonian SSR). *Flor. Märkmed*, 1(5):294–300
1 (1973).

4879 Krall, H., Pork, K. *and* Rebasoo, H.-E.
Eesti niitude floora. (The flora of the Estonian meadows). *Flor. Märkmed*, 1(5):315–337(1973).
1 Alphabetic species list.

4880 Lietuvos TSR flora. 4. Red. (Flora Litovskoi *
SSR. T.4.). Vilnius, LietTSR Mokslu Akad. Bot. Inst. Vilnius, 1971. 879p. Note in *Ref. Zhurn. Biol.*,
1 2(1):v411(1972).

4881 Maavara, V.
Kohtumised Tšatkalis: 1. (The Chotkal Nature Reserve: 1). *Eesti Loodus*, 16(8):462–469(1973)—
1 Illustrations.

4882 Mäemets, A. *and* Pihu, E.
Peipsi elustik järvetüüp. (The biota and type of Lake Peipsi). *Eesti Loodus*, 16(12):732–738(1973)—(En,
1 Rus); illustrations.

4883 Rebasoo, H.
Huvitavamaid taimeleide Väinamere laidudelt. (Some interesting plant records from the islets of Muhu Sound). *Flor. Märkmed*, 1(5):306–309
1 (1973).

4884 Reitalu, M.
Haruldased taimed, kenad lilled ja . . . (Rare plants, pretty flowers and . . .). *Eesti Loodus*, 16(7):401–403
1 (1973)—(Rus,En); illustrations.

4885 Snarskis, P.
Lietuvos TSR ažuolynai (floristiniu ir geobotaniniu požiūriu). (Dubrabȳ Litovskoĭ SSR (floristicheskie i geobotanicheskie issledovaniya)). Vilnius, 'Mintis',
1 1972. 94p.—Li(Rus).

CAUCASUS

4886 Aliev, D. A., Gadzhiev, V. D., Vagabov, Z. V. *and* Shikhémirov, M. G.
Materialȳ k rastitel'nosti Lachinskogo raiona Azerbaĭdzhanskoĭ SSR. *Izv. Akad. Nauk Az. SSR*,
2 *Biol. Nauk*, no.3:3–9(1973)—(Az).

4887 Arushanyan, R. I.
Reliktȳ Nagornogo Karabakha i novȳe mestonakhozhdeniya ikh. (Relicts of the mountain Karabakh and their new findings). *Bot. Zhurn.*,
2 58(5):700–706(1973)—Dot map.

4888 Gabriélyan, É. Ts. *and* Gambaryan, P. P.
Novȳe i redkie floristicheskie nakhodki v Armenii.
2 *Biol. Zhurn. Arm.*, 26(11):56–60(1973)—(Arm).

4889 Galushko, A. I.
Novȳe i redkie rasteniya severnogo Kavkaza. (Plantae novae et rarae Caucasi borealis). *Novosti Sist.*
2 *Vȳssh. Rast.*, 10:324–330(1973)—Illustrations.

4890 L'vov, P. L.
Bereznyaki vneshnegornogo Dagestana. (The birchwoods of mountain Daghestan). *Bot. Zhurn.*,
2 58(1):106–113(1973).

4891 Prilipko, L. I.
Floristicheskie nakhodki v gorakh vostochnoĭ chasti bol'shogo Kavkaza. (Novitates florae montium partis orientalis Caucasi majoris). *Novosti Sist. Vȳssh. Rast.*, 10:341–345(1973)—Illustration, map, dot
2 map.

4892 U.S.S.R.: Akademiya Nauk Gruzinskoĭ SSR: Institut Botaniki
Flora Gruzii: 2. Izdanie vtoroe: Paeoniaceae—Salicaceae. Tbilisi, Metsniereba, 1973. 210p.—Geo;
2 illustrations.

CRIMEA

4893 Kovács, A.
A Krím-félsziget flórájának és vegetációjának rövid áttekintése. (O flore i rastitel'nosti Krȳmskogo-poluostrova). *Bot. Közlem.*, 60(3):199–207(1973)—
1 Illustrations.

4894 Luks, Yu. A. *and* Kryukova, I. V.
Tsennȳe redkie i ischezayushchie rasteniya florȳ Krȳma, podlezhashchie zapovednoĭ okhrane. (Valuable, rare and disappearing plants of the Crimea flora, urgently requiring reservation and protection).
1 *Bot. Zhurn.*, 58(1):97–106(1973).

4895 U.S.S.R.: Akademiya Nauk SSSR
Opredelitel' vȳsshikh rasteniĭ Krȳma. [ed.] N.I. Rubtsov. Leningrad, Nauka, 1972. 550p.—
1 Illustrations; keys.

EUROPEAN RUSSIA

4896 Baranova, E. V.
Materialy̆ k analizu konkretny̆kh flor Pskovskoĭ
Oblasti. (Analysis of some concrete floras of Pskov
region). *Vestn. Leningrad. Univ., Biol.*, no.15(3):30–37
1 (1973) —(En); maps.

4897 Belozerov, P. I.
O novy̆kh i podlezhashchikh isklyucheniyu vidakh
rasteniĭ flory̆ Kostromskoĭ oblasti. (On the new and
liable to exclusion plant species of the Kostroma
1 district flora). *Bot. Zhurn.*, 58(5):691–694(1973).

4898 Derviz-Sokolova, T. G. *and* Elenevskiĭ, A. G.
Redkie rasteniya okrestnosteĭ Pavlovskoĭ Slobody̆.
(Rare plants in the environs of Pavlovskaya Sloboda).
Byull. Mosk. Obshch. Ispy̆t. Prir., Biol., 78(2):146–147
1 (1973).

4899 Gusev, Yu. D.
Dopolneniya k adventivnoĭ flore severo-zapadny̆kh
oblasteĭ evropeĭskoĭ Rossii. (Additions to the
adventive flora of north-western regions of the
European Russia). *Bot. Zhurn.*, 58(6):904–909(1973)—
1 Dot map.

4900 Gushchina, E. G.
K flore Ryazanskoĭ Oblasti. (To the flora of
Ryasank Oblast). *Byull. Mosk. Obshch. Ispy̆t. Prir.*,
1 *Biol.*, 78(5):150–151(1973).

4901 Khokhryakov, A. P. *
Sostav i geneticheskie élementy̆ flory̆ kamenisty̆kh
obrazheniĭ Bashkirii. *In* Dikorastushchie i introdukt-
siruemy̆e Polezn. Rast. v. Bashkirii, vy̆p. 3:274–283
1 (1971). Abstr. in *Ref. Zhurn. Biol.*, 2(1):v408(1972).

4902 Khomutova, M. S. *and* Peshkova, G. I.
O novy̆kh i redkikh rasteniyakh yugo-zapada
Kaluzhskoĭ Oblasti. (On new and rare plants in the
south-western part of Kaluzhskaya Oblast). *Byull.*
1 *Mosk. Obshch. Ispy̆t. Prir., Biol.*, 78(5):145–147(1973).

4903 Pchëlkin, Yu. A. *and* Rakov, N. S.
Boreal'no-khvoĭny̆ĭ élement vo flore Ul'yanovskoĭ
oblasti. (Boreal element in the flora of Uljanov
1 region). *Bot. Zhurn.*, 58(6):885–890(1973).

4904 Sergienko, V. G.
K poznaniyu flory̆ yuzhnoĭ chasti Kanina poluo-
strova. (Contribution to the flora of the south part of
Kanin peninsula). *Vestn. Leningrad. Univ., Biol.*,
1 21(4):54–59(1973)—(En).

4905 Shlyakova, E. V.
Rasprostranenie sorny̆kh rasteniĭ na polyakh
Arkhangel'skoĭ oblasti. (The distribution of weed
plants on fields of Archangelsk district). *Bot. Zhurn.*,
1 58(12): 1824–1829(1973).

4906 Shmidt, V. M., Spasskaya, N. A. *and* Val'ma, V. P.
Konkretny̆e flory̆ okrestnosteĭ pos. Lyuby̆tino i
G. Kholma Novgorodskoĭ Oblasti. (The concrete
floras of vicinity of Ljubitino and Holm, Novgorod
region). *Vestn. Leningrad. Univ., Biol.*, no.3(1):
1 41–52(1973)—(En).

4907 Skvortsov, A. K.
Novy̆e danny̆e ob adventivnoĭ flore Moskovskoĭ
oblasti: 1. *Byull. Glavn. Bot. Sada* (Moscow), 87:5–11
1 (1973)—Illustrations.

4908 Skvortsov, A. K.
Novy̆e danny̆e ob adventivnoĭ flore Moskovskoĭ
oblasti: 2. *Byull. Glavn. Bot. Sada* (Moscow), 88:30–
1 35(1973).

4909 Tikhomirov, V. N., Gubanov, I. A., Novikov, V. S.,*
Oktyabreva, N. B. *and* Glazunova, K. P.
(Nouvelles données sur la flore de la région de
Vladimir). *Nauch. Dokl. Vy̆ssh. Shkoly̆., Biol. Nauki*,
no.1:60–66(1973)—Rus. Ref. in *Bull. Signal., Biol.*
1 *Physiol. Vég.*, 34(10):9531(1973).

4910 Tikhomirov, V. N., Novikov, V. S. *and*
Oktyabreva, N. B.
Novy̆e i redkie vidy̆ flory̆ Oksko-Klyaz' minskogo
mezhdurech'ya. (New and rare plants of the Oka-
Kljasma interfluve). *Byull. Mosk. Obshch. Ispy̆t. Prir.*,
1 *Biol.*, 78(5):147–150(1973).

4911 Tsvelev, N. N.
O nekotory̆kh bolee redkikh rasteniyakh Lenin-
gradskoĭ oblasti. (De plantes nonnullis rarioribus in
provincia Leningrad crescentibus). *Novosti Sist.*
1 *Vy̆ssh. Rast.*, 10:361–365(1973).

4912 Yurova, É. A.
Izuchenie sinuziĭ travyanogo pokrova listvenny̆kh
lesov Vladimirskogo opol'ya. (Study of grass cover
synusiae in deciduous forests of Vladimir region).
1 *Bot. Zhurn.*, 58(10):1464–1474(1973).

4913 Zaĭkova, V. A.
Opy̆t izucheniya allelopaticheskogo rezhima v
lugovy̆kh fitotsenozakh Karelii. (The result of
allelopathic regime study in grassland phytocoenoses
1 of Karelia). *Bot. Zhurn.*, 58(12):1753–1760(1973).

KURIL ISLANDS

4914 Chernyaeva, A. M.
K flore ostrova Onekotana. *Byull. Glavn. Bot. Sada*
2 (Moscow), 87:21–28(1973).

MIDDLE ASIA

4915 Chukavina, A. G.
O rastitel'nosti poyasa shiblyaka yugo-vostochno
Khrebta Surkho. *Izv. Akad. Nauk Tadzh. SSR, Biol.*
2 *Nauk*, 1(50):9–13(1973)—(Ta).

4916 Dzhuraev, A.
Rastitelnost' pervichny̆kh osy̆peĭ yshchel'ya Takob
(Gissarskiĭ khrebet). (The vegetation of boulder
stones of the Takob canyon (Giscar mountain ridge)).
Dokl. Akad. Nauk Tadzh. SSR, 15(12):43–46(1972)—
2 (Ta).

4917 Ikonnikov, S. S.
Zametki o flore Badakhshana, 2. (Notae de flora
Badachschanica, 2). *Novosti Sist. Vy̆ssh. Rast.*, 10:331–
2 332(1973).

4918 Ishchenko, L. E. *and* Tochilina, L. A.
O perezimovke nekotory̆kh vidov flory̆ Turkmenii
v 1968/69 i v 1971/72 godakh v botanicheskom sadu
Akademii Nauk Turkmenskoĭ SSR (Soobshchenie
1-e). (About wintering of some plant species referring
to the flora of Turkmenistan in 1968/1969 and 1971/
1972 in the Ashkhabad Botany Garden). *Izv. Akad.*
2 *Nauk Turkm. SSR, Biol. Nauk*, (1):26–30(1973).

4919 Kalenov, G. S.
Ocherk rastitel'nosti Meshedskogo peschanogo
massiva. (An essay of the Meshedsk sand massif
2 vegetation). *Bot. Zhurn.*, 58(5):658–665(1973)—Map.

4920 Kamelin, R. V.
K poznaniyu florȳ Nuratinskikh gor. (An account to the knowledge of the flora of Nuratavian Mountains). *Bot. Zhurn.*, 58(5):625–638(1973)—
2 (En).

4921 Karmȳsheva, N. Kh.
Flora i rastitel'nost zapovednika Aksu-Dzhabaglȳ (Talasskiĭ Alatau). Alma-Ata, Izd. 'Nauka'
2 Kazakhskoĭ SSR, 1973. 177p.—Maps.

4922 Karrȳev, M. O.
Farmakokhimiya nekotorȳkh éfirnomaslichnȳkh rasteniĭ florȳ Turkmenii. Ashgabat, Ȳlȳm, 1973.
2 156p.—Illustrations.

4923 Konnov, A. A.
Flora archovnikov Shakhristana. [Juniper flora of Shakhristan]. Dushanbe, 'Donish', 1973. 177p.—
2 map.

4924 Kovács-Láng, E. *and* Isépy, I.
Botanikai tanulmányút a nyugati Tien-Sanban. (Botanicheskaya ékspeditsiya v zapadnoĭ Tyan'-Shane). *Bot. Közlem.*, 59(3):197–206(1972 publ.
2 1973)—(Rus); maps.

4925 Rakhimova, T.
Materialȳ k florȳ Chartanskikh Adȳrov Ferganskoĭ dolinȳ. *Dokl. Akad. Nauk Uzb. SSR*, no.5:60–
2 61(1973)—(Uz).

4926 Sherbaev, B.
Flora ostantsovȳkh vosvȳshennosteĭ i nizkogoriĭ Karakalpakii. (Flora of relict hills and low plains of
2 Karakalpakia). *Bot. Zhurn.*, 58(8):1216–1222(1973).

4927 Shukurov, A. Sh., Eremina, N. K. *and* Stanyukovich, K. V.
Osnovnȳe poleznȳe dikorastushchie rasteniya florȳ Tadzhikistana. (The main useful plants of wild flora of Tadshikistan). *Dokl. Akad. Nauk Tadzh. SSR*,
2 15(10):62–66(1972)—(Ta).

4928 Shukurov, A. Sh. *and* Stanyukovich, M. B.
Osnovnȳe krasil'nȳe rasteniya Tadzhikistana. *Izv. Akad. Nauk Tadzh. SSR, Biol. Nauk*, 2(47):
2 19–24(1972)—(Ta).

4929 Uzbekistan: Akademiya Nauk Uzbekskoĭ SSR: Institut Botaniki *and* Tashkent: Gosudarstvennȳĭ Universitet im. V. I. Lenina
Opredelitel' rasteniĭ sredneĭ Azii: kriticheskiĭ konspekt florȳ sredneĭ Azii: tom 1–3. (Conspectus florae Asiae Mediae: tomus 1–3). Tashkent, Fan,
2 1968–1972. 3v.—Rus,La; keys.

SIBERIA

4930 Bukhalo, M. A.
Oseredkȳ stepovoyi roslȳnnosti dolȳnȳ r. Lenȳ u tsentral'niĭ Yakutiyi. (Locations of steppe plants of the Lena river valley in central Yakutia). *Ukr. Bot. Zhurn.*,
2 30(3):387–390(1973)—Uk(En).

4931 Doronina, Yu. A. *and* Kurentsova, G. E. *
(Flore et végétation du bassin du fleuve Onda). Novosibirsk, Nauka, 1973. 149p.—Illustrations. Ref.
2 in *Bull. Signal., Biol. Physiol. Vég.*, 34(10):9647 (1973).

4932 Ermolaev, V. I. *
(Les plantes vasculaires aquatiques et de rivages des environs du village de Tareya (Taymir occidental)). *Izv. Sibir. Otdel. Akad. Nauk SSSR, Biol. Nauk*, no.5:20–24(1973)—Rus(En). Ref. in *Bull. Signal.*,
2 *Biol. Physiol. Vég.*, 34(10):9665(1973).

4933 Karavaev, M. N. *and* Skryabin, S. Z. *
Rastitel'nȳi mir Yakutii. Yakutsk, Knigoizdat, 1971. 128p.—Illustrations. Abstr. in *Ref. Zhurn. Biol.*,
2 3(1):v598(1972).

4934 Krasnoborov, I. M., Pen'kovskaya, E. F., Lomonosova, M. N. *and* Khanminchun, V. M.
Novȳe dlya Tuvinskoĭ ASSR rasteniya. (Plants new for the Tuvinsk Autonomous Soviet Socialist
2 Republic). *Bot. Zhurn.*, 58(8):1149–1155(1973).

4935 Siplivinskiĭ, V. N.
Zametki o baĭkal'skoĭ flore, 1. (Notulae de flora baicalensi, 1). *Novosti Sist. Vȳssh. Rast.*, 10:345–361
2 (1973)—Illustrations.

4936 Sobolevskaya, K. A.
Ekologo-istoricheskie spektrȳ florȳ Altaya i ikh znachenie v introduktsii. (Ecologic historical spectra of the Altai flora and their importance for introduction). *In* U.S.S.R.: Akademiya Nauk SSSR: Sovet Botanicheskikh Sadov. Uspekhi introduktsii rasteniĭ:
2 43–65(1973)—Maps.

4937 Soldatenkova, Yu. P.
O rastitel'nosti i temperature poch' okresnosteĭ Yar-Sale. (On vegetation and soil temperatures in the neighbourhood of Yar-Sale (Yamal)). *Vestn. Mosk.*
2 *Univ. Nauch. Zhurn.*, *6*, 27(4):51–57(1972)—(En).

4938 Sviridonov, G. M. *
(Répartition typologique et rendement des plantes utiles dans les forêts du nord de l'Altaï). *Izv. Sibir. Otdel. Akad. Nauk SSSR, Biol. Nauk*, no.5:24–31 (1973)—Rus(En). Ref. in *Bull. Signal., Biol. Physiol.*
2 *Vég.*, 34(10):9664(1973).

4939 Vodop'yanova, N. S., Ivanova, M. M., Kiseleva, A. A. *and* Petrochenko, Yu. N.
Redkie rasteniya s gor Putoranȳ, 2. (Plantae rarae e montibus Putoranae, 2). *Novosti Sist. Vȳssh. Rast.*,
2 10:277–283(1973).

SOVIET FAR EAST

4940 Khokhryakov, A. P.
K flore yuzhnoĭ chasti Magadanskoĭ oblasti. *Byull.*
2 *Glavn. Bot. Sada* (Moscow), 88:43–48(1973).

4941 Khokhryakov, A. P.
Redkie vidȳ rasteniĭ iz yuzhnoĭ chasti Magadanskoĭ oblasti. (Rare plant species from the southern part of Magadan district). *Bot. Zhurn.*, 58(12):1808–1815
2 (1973)—Maps.

4942 Kozhevnikov, Yu. P.
Botaniko-geograficheskie nablyudeniya na zapade Chukotskogo poluostrova v1971–1972gg. (Botanical-geographic observations in the westernmost Chukotka Peninsula in 1971–1972. *Bot. Zhurn.*,
2 58(7):965(1973)—(En).

4943 Kozhevnikov, Yu. P.
Novȳe floristicheskie nakhodki na zapade Chukotskogo poluostrova. (New floristic finds in the southwestern part of Chukotka Peninsula). *Bot. Zhurn.*,
2 58(2):294–300(1973)—Map.

4944 Nechaev, A. P. *and* Nechaev, A. A.
K flore Nizhnego Priamur'ya. *Byull. Glavn. Bot.*
2 *Sada* (Moscow), 88:48–51(1973).

4945 Nechaeva, T. I.
O nekotorȳkh redkikh rasteniyakh yuzhnoĭ chasti dal'nego vostoka. (De plantis nonnullis rarioribus orientis extremi partis australis). *Novosti Sist. Vȳssh. Rast.*, 10:337–339(1973).
2

4946 Petrovskiĭ, V. V.
Spisok sosudistȳkh rasteniĭ O. Vrangelya. (A list of vascular plants of the Wrangel Island). *Bot. Zhurn.*, 58(1):113–127(1973).
2

4947 Pimenov, M. G. *and* Pimenova, M. E.
O nekotorȳkh novȳkh i redkikh rasteniyakh florȳ Priamur'ya. *Byull. Glavn. Bot. Sada* (Moscow), 88:52–53(1973).
2

4948 Shaga, N. I.
Redkie i novȳe vidȳ dlya florȳ nizhnego Amura. *Byull. Glavn. Bot. Sada* (Moscow), 87:30–31 (1973).
2

4949 Yurtsev, B. A.
Botaniko-geograficheskaya zonal'nost' i floristicheskoe raĭonirovanie Chukotskoĭ Tundrȳ. (Phytogeographic zonality and floristic subdivisions of Chukotka). *Bot. Zhurn.*, 58(7):945–965(1973)—(En); maps, dot maps.
2

4950 Yurtsev, B. A., Galanin, A.V., Derviz-Sokolova, T.G., Katenin, A. E., Kozhevnikov, Yu. P., Korobkov, A. A., Petrovskiĭ, V. V., Plieva, T. V., Razzhivin, V. Yu. *and* Taraskina, N. N.
Floristicheskie nakhodki v Chukotskoĭ tundre, 1. (Novitates pro flora terrae Tschuktschorum, 1). *Novosti Sist. Vȳssh. Rast.*, 10:283–324(1973)—Map, dot maps; chromosome numbers.
2

4951 Yurtsev, B. A., Sȳtin, A. K. *and* Sekretareva, N. A.
Interesnȳe floristicheskie nakhodki na vostoke Chukotskogo poluostrova: 2. (Interesting floristic finds in the easternmost Chukotka Peninsula: 2). *Bot. Zhurn.*, 58(12):1742–1753(1973)—(En); chromosome numbers.
2

UKRAINE

4952 Krasnova, A. M.
Ekolohichnȳĭ ta arealohichnȳĭ analiz florȳ pivnichnoho prȳazov'ya. (Ecological and arealogical analysis of the northern Azov sea area flora). *Ukr. Bot. Zhurn.*, 30(5):578–583(1973)—Uk(En).
1

4953 Loskot, N. P.
Analiz florȳ Arabats'koyi strilkȳ. (Analysis of flora in Arabatskaya Strelka). *Ukr. Bot. Zhurn.*, 30(6): 709–714(1973)—Uk(En).
1

4954 Loskot, N. P.
Do florȳ Prȳsȳvashshya. (On the flora of the Sivash area). *Ukr. Bot. Zhurn.*, 30(4):526–529(1973)—Uk(En).
1

4955 Moroz, I. I. *and* Kharkevich, S. S.
Floristicheskie osobennosti Toltrovogo kryazha Podolii. (Floristic peculiarities of the Toltrovy ridge in Podolia). *Bot. Zhurn.*, 58(12):1799–1806(1973).
1

4956 U.S.S.R.: Akademiya Nauk SSSR: Vsesoyuznoe Botanicheskoe Obshchestvo *and* U.S.S.R.: Akademiya Nauk USSR: Ukrainskoe Botanicheskoe Obshchestvo
Tesisȳ dokladov v delegatskogo s'ezda Vsesoyuznogo Botanicheskogo Obshchestva. Kiev, 1973. 383p.—Chromosome numbers.
1

WHITE RUSSIA

4957 Parfenov, V. I., Blazhevich, R. Yu. *and* Kozlovskaya, N. V.
Ekologo-geograficheskie osobennosti proizrastaniya gornȳkh vidov florȳ Belorussii. *Botanika* (Minsk), vȳp 15:82–95(1973)—Map.
1

SOUTH WEST ASIA

AFGHANISTAN

4958 Grey-Wilson, C.
A botanical trek in the Wakham Corridor. *Quart. Bull. Alp. Gard. Soc.*, 41(4):315–323(1973)—Illustration, maps.
2

4959 Hewer, T. F.
A botanical expedition to Iran and Afghanistan, 1971: parts 1–2. *J. Roy. Hort. Soc.*, 98(7):288–294, 98(8):341–348(1973)—Coloured illustrations, maps.
2

EGYPT

4960 Danin, A.
Contribution to the flora of Sinai: 2. New records. *Israel J. Bot.*, 22(1):18–31(1973)—Map, dot map.
2

IRAN

4961 Hewer, T. F.
A botanical expedition to Iran and Afghanistan, 1971: parts 1–2. *J. Roy. Hort. Soc.*, 98(7):288–294, 98(8):341–348(1973)—Coloured illustrations, maps.
2

ISRAEL

4962 Danin, A. *and* Hedge, I. C.
Contributions to the flora of Sinai: 1. New and confused taxa. *Notes Roy. Bot. Gard. Edinburgh*, 32(2): 259–271(1973)—Illustration, dot map, keys.
2

TURKEY

4963 Demiriz, H.
Endemics of the vicinity of Istanbul (Provisory list). (Endemiten der Umgebung von Istanbul (Vorläufige Liste)). Istanbul, 1973. 1p. Unpublished.
1

4964 Huber-Morath, A.
Ergänzungen zur Flora der Türkei. *Verh. Naturforsch. Ges. Basel*, 83(2):193–318(1973).
1, 2

4965 Ponert, J.
Neue taxonomische Kombination Kategorien und Taxa vor allem der türkischen Arten. *Feddes Repert.*, 83(9–10):617–644(1972 publ. 1973)—(En).
2

NORTH AFRICA

LIBYA

4966 Scholz, H. *and* Gabriel, B.
Neue Florenliste aus der libyschen Sahara. (A new list of plants to the flora of the Libyan Sahara). *Willdenowia*, 7(1):169–191(1973)—(En); illustration.
1, 2

MOROCCO

4967 Delay, J. *and* Petit, D.
Littoral atlantique du Maroc Région de Rabat. *Inform. Ann. Caryosyst. Cytogénét.*, no.5:1–16(1971 publ. 1973)—Illustrations, map; chromosome numbers.
1

4968 Peltier, J. P.
Endémiques macaronésiennes au Maroc: inventaire bibliographique et problèmes taxinomiques. *Monogr.*
1 *Biol. Canar.*, no.4:134–142(1973)—(En).

4969 Rauh, W.
Remarks on the succulent vegetation of Morocco. *Cact. Succ. J.* (U.S.A.), 44(5):187–195(1972)—
1 Illustrations, map.

MACARONESIA

4970 Bramwell, D. *and* Moore, D. M.
Flora of Macaronesia project: a progress report.
1, 5 *Monogr. Biol. Canar.*, no.4:165–168(1973).

4971 Bramwell, D. *and* Richardson, I. B. K.
Floristic connections between Macaronesia and the east Mediterranean region. *Monogr. Biol. Canar.*,
1, 2 no.4:118–125(1973)—Maps.

4972 Silva, A. R. Pinto da *and* Lopes, M. H. Ramos
Famílias e géneros da flora Lusitanica, Azorica
1 and Madeirensis. *Anuár. Soc. Brot.*, 38:13–52(1972).

4973 Steinberg, C. H.
Macaronesian collections of phanerogams in the Herbarium Universitatis Florentinae. *Monogr. Biol.*
1 *Canar.*, no.4:30–48(1973).

AZORES

4974 Hansen, A.
Contributions to the flora of the Azores: 2. *Bol.*
1 *Soc. Brot.*, 46:219–238(1972).

4975 Hansen, A.
Contributions to the flora of the Azores: 3. *Anuár.*
1 *Soc. Brot.*, 39:25–38(1973).

4976 Ormonde, J.
Additiones et adnotationes florae azoricae: 3.
1 *Anuár. Soc. Brot.*, 39:65–73(1973).

4977 Ormonde, J. *and* Paiva, J. A. R.
Additiones et adnotationes florae azoricae: 1–2.
1 *Anuár. Soc. Brot.*, 39:39–64(1973)—Illustrations.

4978 Sjögren, E.
Plant communities of the natural vegetation of Madeira and the Azores. *Monogr. Biol. Canar.*, no. 4:
1 107–111(1973).

4979 Sjögren, E.
Recent changes in the vascular flora and vegetation of the Azores Islands. *Mem. Soc. Brot.*, 22. 453p.
1 (1973)—Illustrations, dot maps.

CANARY ISLANDS

4980 Bellot, F. *and* Casaseca, B.
Herborizaciones en Canarias (primera nota). *An.*
1 *Inst. Bot. A. J. Cavanilles*, 30:109–116(1973).

4981 Esteve Chueca, F.
Sinopsis de las alianzas y asociaciones en la clase Cytiso-Pinetea y orden Cytiso Pinetalia. (Synopsis of the Cytiso-Pinetea class in the Canary Islands).
1 *Monogr. Biol. Canar.*, no.4:89–92(1973)—(En).

4982 Gagnieu, A., Linder, R. *and* Voggenreiter, V.
Caryotypes de la flore insulaire de Tenerife. *Monogr. Biol. Canar.*, no.4:126–133(1973)—Map;
1 chromosome numbers.

4983 Hansen, A.
Floristic notes from the Canary Islands (especially La Palma). *Cuad. Bot. Canar.*, no.18–19:5–14(1973)—
1 Illustrations.

4984 Hansen, A.
Some floristic notes from Gran Canaria. *Cuad.*
1 *Bot. Canar.*, no.18–19:39–41(1973)—Illustrations.

4985 Kruess, J.
Canarian common names of wild plants and their meaning in botany and linguistics. *Monogr. Biol.*
1 *Canar.*, no.4:49–59(1973).

4986 Kunkel, G.
La Palma: notas sobre algunas especies introducidas.
1 *Cuad. Bot. Canar.*, 17:15–17(1973).

4987 Kunkel, G.
A new publication on the Canary Islands flora. *Monogr. Biol. Canar.*, no.4:161–164(1973)—(Sp);
1 illustrations.

4988 Kunkel, G.
Novedades en la flora de las islas orientales, con una nota sobre *Dracaena draco* en Gran Canaria. *Cuad. Bot.*
1 *Canar.*, no.17:35–37(1973)—(En).

4989 Kunkel, G.
Nuevas adiciones para la flora de las Islas Orientales (incluyendo Gran Canaria). *Cuad. Bot. Canar.*, no.
1 18–19:25–31(1973)—(En); illustrations.

4990 Kunkel, G.
The role of adventitious plants in the vegetation of the Canary Islands. *Monogr. Biol. Canar.*, no.4:
1 103–106(1973).

4991 Ron Alvarez, M. E.
Plantas canarias en el herbario del Jardin Botanico de Madrid (parte 1). *An. Inst. Bot. A. J. Cavanilles*, 30:
1 117–181(1973)—Illustration.

4992 Sanchez Garcia, I.
Regeneración del bosque subtropical de Laurisilva.
1 *Monogr. Biol. Canar.*, no.4:96–102(1973).

4993 Santos Guerra, A.
Algunos aspectos de la vegetación de la isla de La
1 Palma. *Monogr. Biol. Canar.*, no.4:93–95(1973).

4994 Whicher, A. *and* Whicher, P.
Some interesting plants from Tenerife. *Cact. Succ.*
1 *J. Gr. Brit.*, 35(2):28–32(1973)—Illustrations.

MADEIRA

4995 Silva, A. R. Pinto da *and* Lopes, M. H. Ramos
Famílias e géneros da flora Lusitanica, Azorica and Madeirensis. *Anuár. Soc. Brot.*, 38:13–52
1 (1972).

4996 Sjögren, E.
Plant communities of the natural vegetation of Madeira and the Azores. *Monogr. Biol. Canar.*, no.4:
1 107–111(1973).

EASTERN ASIA

4997 Masamune, G.
Icones Plantarum Asiaticarum: 45–48. *J. Geobot.*, 20(1):i–ii, 20(2):v–vii, 20(3):ix–xi, 20(4):xii–xv
2 (1972)—Ja,En; illustrations.

4998 Ohwi, J.
(Taxonomic notes from eastern Asia). *J. Geobot.*,
2 20(3):58–60(1972)—Ja.

CHINA

4999 Parsons, L. M. H., *6th earl of Rosse*
Dr H. H. Hu's plant-hunting expeditions 1937–40:
2 part 2. *Int. Dendrol. Soc. Year Book 1972*:7–12(1973).

5000 [Tibet: People's Press]
Xizang changyong zhongeaoyao. [Commonly used
Chinese herbal medicines of Tibet]. [s.l.], People's
6 P., 1971. 911p.—Ch; coloured illustrations.

JAPAN

5001 Anon.
(Materials for the distribution of vascular plants in
2 Japan: 81). *J. Jap. Bot.*, 48(7):204(1973)—Ja.

5002 Anon.
(Materials for the distribution of vascular plants in
2 Japan: 82). *J. Jap. Bot.*, 48(8):246(1973)—Ja.

5003 Asai, Y.
(Miscellaneous notes on the naturalized flora of
Japan: 3). *J. Jap. Bot.*, 48(3):67–72(1973)—Ja(En);
illustrations. *Orobanche minor*, var. *flava*, *Amaranthus*
galii, *Heliotropium curassavicum*, *Chenopodium lepto-*
2 *phyllum*, *Hypochoeris glabra*.

5004 DeBelder, R.
Visit to Yakushima. *Int. Dendrol. Soc. Year Book*
2 *1972*:22–30(1973)—Illustrations, map.

5005 Inobe, T.
(Materials for the distribution of vascular plants in
Tokushima Prefecture: 1–2). *J. Geobot.*, 20(2):37,
2 20(3):69–70(1972)—Ja.

5006 Minami, A.
(Comparison with both flora of Hôyo Island and
Sadamisaki Peninsula). *J. Geobot.*, 20(1):26–30(1972)—
2 Ja; maps.

5007 Minami, A.
(Subtropical plants growing in Hakali City
Yamaguchi Pref.). *J. Geobot.*, 20(4):110–112(1972)—
2 Ja.

5008 Numata, M., Miyawaki, A. *and* Itow, D.
Natural and semi-natural vegetation in Japan.
Blumea, 20(2):435–496(1972 publ. 1973)—
2 Illustrations, maps.

5009 Satomi, N.
Icones plantarum Hokurikuensium: 19. *J. Geobot.*,
2 20(1):4(1972)—Ja.

5010 Satomi, N.
Icones plantarum Hokurikuensium: 20. *J. Geobot.*,
2 20(2):8(1972)—Ja.

5011 Satomi, N.
Icones plantarum Hokurikuensium: 21. *J. Geobot.*,
2 20(3):12(1972)—Ja.

5012 Satomi, N.
Icones plantarum Hokurikuensium: 22. *J. Geobot.*,
2 20(4):16(1972)—Ja.

5013 Takahashi, H.
Notes on Japanese alpine plants: 2. *Bull. Kanagawa*
Prefect. Mus., Nat. Hist., no.6:95–111(1973)—Ja,En;
2 illustrations.

5014 Takahashi, H.
(Notes on the plants collected in Izu Islands: 1).
2 *J. Geobot.*, 20(1):14–16(1972)—Ja,En; illustrations.

5015 Takahashi, H.
(Notes on the plants collected in Izu Islands: 2).
2 *J. Geobot.*, 20(2):42–44(1972)—Ja,En; illustrations.

5016 Tawada, S.
(Illustrated important Okinawan medicinal plants).
2 [Japan, 1972]. [vi]176p.—Ja; illustrations.

5017 Uematsu, H.
A list of the plants distributed on the southern
2 Japanese Alps: 10. *J. Geobot.*, 20(1):17–19(1972).

5018 Uematsu, H.
A list of the plants distributed on the southern
2 Japanese Alps: 11. *J. Geobot.*, 20(2):44–48(1972).

5019 Uematsu, H.
A list of the plants distributed on the southern
2 Japanese Alps: 12. *J. Geobot.*, 20(4):103–108(1972).

MONGOLIA

5020 Vasak, V.
Plant hunting in Mongolia: 1. Alpines of Shiir
Chairchan. *J. Scott. Rock Gard. Club*, 13(4):322–326
2 (1973)—Illustrations.

RYUKYU ISLANDS

5021 Nackejima, C.
(On the newly naturalized plants found from
Ryukyus: 4). *J. Geobot.*, 20(1):20–25(1972)—Ja;
2 illustrations.

5022 Walker, E. H.
The making of a regional flora. *Taiwania*, 18(1):
57–61(1973). History of preparation of Okinawa
2 Flora.

5023 Yamada, T.
(A list of plants collected in the Islands of Iheya and
Izena, Okinawa Prefect). *J. Geobot.*, 20(3):70–81
2 (1972)—Ja.

5024 Yamazaki, T.
Supplement of the flora of Ryukyu and Formosa
(10). *J. Jap. Bot.*, 48(4):128(1973)—Ja,En; illustra-
2 tions.

5025 Yamazaki, T.
Supplement of the flora of Ryukyu and Formosa
2 (11). *J. Jap. Bot.*, 48(5):141–142(1973)—Ja,En.

TAIWAN

5026 Hsu, C.-C.
Some noteworthy plants found in Taiwan.
2 *Taiwania*, 18(1):62–72(1973)—Illustrations; keys.

5027 Huang, T. C. *and* Chung, T.-F.
Aeropalynological study of Taiwan: (1)—Chu-shan
2 station. *Taiwania*, 18(2):160–168(1973)—map.

5028 Yamazaki, T.
Supplement of the flora of Ryukyu and Formosa
(10). *J. Jap. Bot.*, 48(4):128(1973)—Ja,En;
2 Illustrations.

5029 Yamazaki, T.
Supplement of the flora of Ryukyu and Formosa
2 (11). *J. Jap. Bot.*, 48(5):141–142(1973)—Ja,En.

INDIAN SUBCONTINENT

5030 Meher-Homji, V. M.
A phytoclimatic approach to the problem of mediterraneity in the Indo-Pakistan sub-continent. *Feddes Repert.*, 83(9–10):757–758(1972 publ. 1973)—
6 (Ge); maps.

BURMA

5031 K., J.
Sino-Himalayan plants (North Burma). *Gard.*
6 *Sept., Roy. Bot. Gard. Edinburgh*, Sept.:6–7(1973).

INDIA

5032 Bhatt, D. C. *and* Inamdar, J. A.
Addition to the vascular flora of Saurashtra. *Proc.*
6 *Indian Sci. Congr.*, 60(3):335–336(1973). Abstract only.

5033 Deb, D. B. *and* Dutta, R. M.
Contribution to the flora of Tirap frontier division. *J. Bombay Nat. Hist. Soc.*, 69(3):547–573(1972 publ.
6 1973).

5034 Gaussen, H. *and others*
Notice de la feuille Rajasthan. *Trav. Sect. Sci. Techn. Inst. Franç. Pondichéry*, Hors sér. no.12:1–154(1972)—
6 Illustrations, maps.

5035 India: Botanical Survey
Botanical Survey of India, Calcutta. Calcutta,
6 Botanical Survey of India, 1973. 26p.—Illustrations.

5036 Kapoor, S. L., Sharma, P. C. *and* Kapoor, L. D.
Additions to the flora of Badrinath and neighbourhood. *Proc. Nation. Acad. Sci., India, B*, 41(3):221–230
6 (1971 publ. 1972).

5037 Koul, A. K. *and* Gohil, R. N.
Cytotaxonomical conspectus of the flora of Kashmir: 1. Chromosome numbers of some common
6 plants. *Phyton* (Austria), 15(1–2):57–66(1973)—(Ge).

5038 Kumar, A. *and* Hingorani, G. R.
Correct botanical identity of some established old exotics in Himachal Pradesh. *Indian Forester*, 99(6):
6 380–390(1973).

5039 Mishra, K. K.
A contribution to our knowledge of the angiospermic flora of the Parasnath Hill. *J. Bihar Bot. Soc.*,
6 1(1–2):48–53(1972).

5040 Misra, A. *and* Jha, D.
Floristics in Darbhanga. *Planto*, 2(1):136–143
6 (1972).

5041 Sanyal, M. N.
A contribution to the shrubs and woody climbers of the forests of Bankura district, West Bengal. *Indian*
6 *Forester*, 99(3):152–158(1973)—Map.

5042 Singh, G. *and* Wafai, B. A.
An exploration of phanerogamic elements along Kishtwar-Duksum route, Jammu and Kashmir State.
6 *Indian Forester*, 99(5):310–320(1973)—Map.

5043 Singh, V.
Recently introduced exotics in the flora of Meerut
6 District. *Curr. Sci.*, 42(20):734–735(1973).

5044 Sinha, S. S. N. *and* Prasad, R.
Cytological studies in some medicinal plants of Chhotanagpur: 2. *J. Bihar Bot. Soc.*, 1(1–2):69(1972)—
6 Chromosome numbers. Abstract.

5045 Thakur, V.
Problems of the flora of Bihar. *J. Bihar Bot. Soc.*,
6 1(1–2):35–44(1972).

5046 Vaid, K. M.
Flora in a bee-dropping—'Ocean in a drop'.
6 *Indian Forester*, 99(11):676–677(1973).

5047 Verma, B. K.
Further studies on the flora of Allahabad. *Curr. Sci.*,
6 42(9):331(1973).

NEPAL *and* HIMALAYA

5048 Barclay, C.
Plant collecting in the Anapurna Himal. *Quart. Bull.*
6 *Alp. Gard. Soc.*, 41(1):64–73(1973)—Popular notes.

5049 Beer, L.
Plants of east Nepal. *Quart. Bull. Alp. Gard. Soc.*,
6 41(3):256–263(1973).

5050 Beer, L.
A walk in the rain. *Cambridge Univ. Bot. Gard. Ass. Newsl.*, 7–9(1972–1973)—Plant collecting
6 expedition to East Nepal in 1971.

5051 Chang, King Wai *and* Chiang, Shu
A primary study on the vertical vegetation belt of Mt. Jolmo-Lungma (Everest) region and its relationship with horizontal zone. *Acta Bot. Sin.*, 15(2):
6 235–252(1973)—Ch(En); illustrations.

5052 Debelmas, A. M., Dobremez, J. F., Michel, S. *and* Benarroche, L.
Plantes médicinales du Nepal. *Pl. Méd. Phytothér.*,
6 7(2):104–113(1973)—(En).

5053 Faugeras, G., Dobremez, J. F., Bourgeois, J. *and* Paris, R. R.
Sur les aconits du Népal: étude préliminaire de six espèces récoltées en 1972. *Pl. Méd. Phytothér.*, 7(2):
6 151–162(1973)—(En); illustrations.

5054 Hara, H.
New or noteworthy flowering plants from Eastern Himalaya. *J. Jap. Bot.*, 48(4):97–103(1973)—(Ja);
6 illustrations, coloured illustrations.

5055 Hara, H.
New or noteworthy flowering plants from Eastern Himalaya. *J. Jap. Bot.*, 48(12):353–369(1973)—
6 Illustrations.

5056 Javeid, G. N.
New records of flowering plants for the Himalaya—
6 1. *Indian Forester*, 99(10):628–631(1973).

5057 Murata, G.
A list of plants collected by the Chiba University Rolwaling Himal Expedition, 1963: 1. *Acta Phytotax.*
6 *Geobot.*, 25(4–6):107–118(1973).

5058 Stainton, A.
To Nepal for flowers. *J. Roy. Hort. Soc.*, 98(2):
6 96–100(1973)—Coloured illustrations.

NICOBAR ISLANDS

5059 Thothathri, K.
Botany of the joint scientific expedition to the Great Nicobar Island. *Proc. Indian Sci. Congr.*, 60(3):
6 341(1973). Abstract only.

WEST PAKISTAN

5060 Kazmi, S. M. A.
Bibliography on the botany of West Pakistan and Kashmir: 4. Miami, Field Research Projects, 1973.
6 vi,111p.

5061 Nasir, E. *and* Ali, S. I., *eds*
Flora of West Pakistan: 25,26,29–43,45–47,49–53. [Rawalpindi], 1972–1973—Illustrations, maps; keys. 25. Sphenocleaceae 3p.; 26. Burseraceae 6p.; 29. Salvadoraceae 4p.; 30. Goodeniaceae 3p.; 31. Parnassiaceae 5p.; 32. Guttiferae 12p.; 33. Pedaliaceae 4p.; 34. Capparidaceae 35p.; 35. Loranthaceae 9p.; 36. Mimosaceae 41p.; 37. Datiscaceae 3p.; 38. Moringaceae 4p.; 39. Sapindaceae 10p.; 40. Molluginaceae 6p.; 41. Aizoaceae 12p.; 42. Dilleniaceae 4p.; 43. Coriariaceae 3p.; 45. Malpighiaceae 4p.; 46. Phrymaceae 3p.; 47. Illecebraceae 12p.; 49. Avicenniaceae 4p.; 50. Alangiaceae 3p.; 51. Portulacaceae 8p.; 52. Polygalaceae 11p.; 53.
6 Dioscoreaceae 6p.

5062 Quraishi, M. A. *and* Khan, S. A.
Flora of Peshawar District and Khyber Agency: volume 2 part I (B). *Pakistan J. Forest.*, 22(3):323–383
6 (1972)—Illustrations.

5063 Stewart, R. R.
Flora of West Pakistan: an annotated catalogue of the vascular plants of West Pakistan and Kashmir.
6 [Rawalpindi], 1972. xx,1028p.—Map.

SOUTH-EAST ASIA

5064 Flora Malesiana Foundation
Identification lists of Malaysian specimens: 42–46. Leyden, Foundation Flora Malesiana, 1972. pp. 727–811. 42. Apocynaceae–2 pp.727–734; 43. *Ficus* (Moraceae) pp.735–784; 44. Taccaceae pp.785–796;
6 45. Lemnaceae pp.797–800; 46. *Hypericum* pp.801–811.

5065 Flore du Cambodge, du Laos et du Viêt-nam. (Révision de la flore générale de l'Indochine): fasc. 14. Paris, Mus. Nation. Hist. Nat., 1973. 91p.—Illustrations, map; keys. Ochnaceae pp.3–16, Onagraceae pp.17–39, Trapaceae pp.40–47, Balanophoraceae pp. 49–58, Rafflesiaceae pp.59–64, Podostemaceae pp.
6 65–74, Tristichaceae pp.75–79.

5066 Murata, G.
New or interesting plants from southeast Asia: 1. *Acta Phytotax. Geobot.*, 25(4–6):97–106(1973)—
2, 6 Illustrations.

5067 Vidal, J. E.
Contribution à l'ethnobotanique du Laos et du Cambodge (Noms vernaculaires, usages). *J. Agric. Trop. Bot. Appl.*, cahier no.2. 88p.(1972)—
6 Illustrations, map.

BALI

5068 Bagian Botani Hutan
Daftar nama pohon-pohonan Bali dan Lombox. (List of tree species Bali and Lombok). *Laporan Lembaga Penelitian Hutan Bogor*, no.145.24p.(1973)—
6 (En); illustrations.

BORNEO

5069 Meijer, W.
Plant life in Kinabalu National Park. *Malayan Nat.*
6 *J.*, 24(3–4):184–189(1971)—Illustrations.

INDONESIA

5070 Indonesia: Bagian Botani Hutan
Daftar nama pohon-pohonan Menado (Sulawesti utara). (List of tree-species Menado (north Sulawesi)). *Laporan Lembaga Penelitian Hutan Bogor*, no.156.
6 63[2]p.(1972)—(En); illustrations.

5071 Prawira, S. A.
Tumbuh-Tumbuhan Bawah Hutan industri dan inventari sasi Cagar Alam. *Laporan Lembaga Penelitian*
6 *Hutan Bogor*, no.163. 48p.(1973)—Map.

JAVA

5072 Backer, C. A.
Atlas of 220 weeds of the sugar-cane fields in Java. Deventer Ysel P 1973. 240p. (Handbook for the cultivation of sugar-cane and manufacturing of cane-sugar in Java, 7, Atlas (final instalment))—Illustrations. Plates 481–747; plates 1–480 published
6 1928–1941.

KHMER REPUBLIC

5073 Legris, P. *and* Blasco, F.
Notice de la carte Cambodge. *Trav. Sect. Sci. Techn. Inst. Franç. Pondichéry*, Hors sér. no.11:1–240
6 (1972)—Illustrations, maps.

5074 Matras, J. *and* Martin, M.
Contribution à l'ethnobotanique des Brou (Cambodge—Province de Ratanakiri). *J. Agric. Trop. Bot. Appl.*, 19(4–5):93–139(1972 publ. 1973)—
6 Illustrations.

MALAYA

5075 Anon.
Identification of plants on Malayan rubber estates: plates 121–128, shrubs. *Pl. Bull.* (Malaya), no.124:
6 14–26(1973)—Illustrations.

5076 Burgess, P. F.
Studies on the regeneration of the hill forests of the Malay Peninsula. *Malayan Forester*, 35(2):103–123
6 (1972)—Map. *Shorea* species.

5077 Whitmore, T. C., *ed.*
Tree flora of Malaya: a manual for foresters. v.1. London, etc. Longman, 1973. 444p. (Malayan Forest
6 Records no.26)—Illustrations, maps; keys.

NEW GUINEA

5078 Henty, E. E. *and* Pritchard, G. H.
Weeds of New Guinea and their control. *Bot. Bull. Dep. Forests Papua New Guinea*, no.7. 195p.(1973)—
6 Illustrations.

SUMATRA

5079 Bagian Botani Hutan
Daftar nama pohon-pohonan Lampung (Sumatera). (List of tree species Lampung (Sumatera)). *Laporan Lembaga Penelitian Hutan Bogor*, no.143. 29p.(1973)—
6 (En); illustrations.

5080 Bagian Botani Hutan
Daftar nama pohon-pohonan Palembang (Sumatera-Selatan). (List of tree species Palembang South-Sumatera). *Laporan Lembaga Penelitian Hutan*
6 *Bogor*, no.141. 91p.(1972)—(En); illustrations.

VIETNAM

5081 Jeanplong, J.
Étude de la végétation des mauvaises herbes du Nord-Viêt-nam. (Gyomvegetácio-tanulmányok Eszak-Vietnamban). *Bot. Közlem.*, 60(3):167–175
6 (1973)—(Hu).

AUSTRALIA

5082 Aston, H. I.
Aquatic plants of Australia: a guide to the identification of the aquatic ferns and flowering plants of Australia, both native and naturalized. Melbourne,
7 U.P., 1973. xvi, 368p.—Illustrations, maps; keys.

5083 Ives, W. A., *comp.*
The Mallee of south-eastern Australia: a short bibliography. Clayton, Victoria, Monash Univ.,
7 1973. xviii, 66p.—Map.

5084 Johnson, L. A. S.
Recent research on the classification of Australian plants. *Austral. Pl.*, 7(56):173–174(1973)—
7 Illustration.

5085 Kratochvil, M., Hannon, N. J. *and* Clarke, L. D.
Mangrove swamp and salt marsh communities in southern Australia. *Proc. Linn. Soc. New S. Wales*,
7 97(4):262–274(1973)—Illustrations, maps.

5086 McGillivray, D. J.
Michel Gandoger's names of Australian plants. *Contrib. New S. Wales Nation. Herb.*, 4(6):319–365
7 (1973).

5087 Melville, R.
Relict plants in the Australian flora and their conservation. *In* Costin, A. B. *and* Groves, R. N., Nature conservation in the Pacific: 83–90(1973)—
7 Illustrations.

5088 Millett, M.
Native trees of Australia. Melbourne, Lansdowne Press, 1971. 112p.—Illustrations, coloured
7 illustrations, portraits.

5089 Telford, I.
Australian native ferns. *Austral. Pl.*, 7(56):159–167,
7 188(1973)—Illustrations, coloured illustrations.

NEW SOUTH WALES

5090 Beadle, N. C. W.
Students flora of north eastern New South Wales: Pt. 1. Pteridophytes. Armidale, University of New
7 England, 1971. 69p.—Illustrations, map; keys.

5091 Beadle, N. C. W.
Students flora of north eastern New South Wales: Pt. 2. Gymnosperms, key to Angiosperm families; Angiosperms: families 37–83. Armidale, University of
7 New England, 1972. 249p.—Illustrations; keys.

5092 Beadle, N. C. W., Evans, O. D., Carolin, R. C. *and* Tindale, M. D.
Flora of the Sydney region. rev. enl. ed. Sydney, etc., Reed, 1972. 724p.—Illustrations, coloured
7 illustrations, map; keys.

5093 Blombery, A. M.
Stony Range Flora Reserve. *Austral. Pl.*, 7(55): 95–122, 127–156(1973)—Illustrations, coloured
7 illustrations. Sydney.

5094 Siddiqi, M. Y., Carolin, R. C. *and* Anderson, D. J.
Studies in the ecology of coastal heath in New South Wales: 1. Vegetation structure. *Proc. Linn. Soc. New S.*
7 *Wales*, 97(3):211–224(1973).

SOUTH AUSTRALIA

5095 Anon.
Botany report on Finniss Scrub. *S. Austral. Nat.*,
7 47(3):54–63(1973).

5096 Spooner, A. G.
Some notes on native grasses in the north eastern suburbs of Adelaide. *S. Austral. Nat.*, 48(1):10–15
7 (1973)—Map.

TASMANIA

5097 Kirkpatrick, J. B.
The vegetation of Sloping Island, Tasmania. *Victorian Nat.*, 90(11):312–321(1973)—Illustrations,
7 maps, dot map.

5098 Stones, M., *illus.*
The endemic flora of Tasmania: botanical and ecological text by Winifred Curtis: part 4. London, Ariel Press, 1973. pp.221–301.—Coloured
7 illustrations, map.

VICTORIA

5099 Baines, J. A.
The origin of generic names of the Victorian flora: part 1. The person behind the name. *Victorian Nat.*, 90(3):72–73, (9):248–250, (10): 279, (11):321–324
7 (1973).

5100 Baines, J. A.
The origin of generic names of the Victorian flora: part 2. Latin, Greek and miscellaneous. *Victorian Nat.*,
7 90(12):349–350, 91(2):38–40(1973).

5101 Timms, Brian *and* Timms, Beverley
The wildflowers of the flat rock area Northern Grampians. *Victorian Nat.*, 90(1):17–26(1973)—Map.
7 Includes species list.

WESTERN AUSTRALIA

5102 Erickson, R., George, A. S., Marchant, N. G. *and* Morcombe, M. K.
Australian flora in colour: flowers and plants of Western Australia. Sydney, A. H. & A. W. Read,
7 1973. 216p.—Coloured illustrations, maps.

5103 Marchant, N. G.
Species diversity in the southwestern flora. *J. Roy.*
7 *Soc. West. Austral.*, 56(1–2):23–30(1973)—Map.

5104 Paterson, H. E. *and* James, S. H.
Animal and plant speciation studies in Western Australia. *J. Roy. Soc. West. Austral.*, 56(1–2):31–43
7 (1973)—Chromosome numbers.

5105 Schmidt, W. E.
A thousand miles of wild flowers in southwestern Australia: part 2. *Calif. Hort. J.*, 34(2):55–60,79
7 (1973)—Illustrations.

5106 Smith, G. G.
A guide to the coastal flora of south-western Australia. Perth, Western Australia Naturalists' Club, 1973. 60p.(Handbook no.10)—Illustrations, coloured
7 illustrations; key.

NEW ZEALAND

5107 Brockie, W. B.
Botanical notes. *New Zealand J. Bot.*, 11(3):581–584(1973)—Chromosome numbers. Annotated by
7 M. J. A. Simpson.

5108 Given, D. R.
Naturalised flowering plants in south-west Fiordland. *New Zealand J. Bot.*, 11(2):247–250(1973)—
7 Maps.

5109 Metcalf, L. J.
The cultivation of New Zealand trees and shrubs. Wellington etc., A. H. & A. W. Reed, 1972. xx,292p.
7 —Illustrations, coloured illustrations.

5110 Raven, P. H.
Evolution of subalpine and alpine plant groups in New Zealand. *New Zealand J. Bot.*, 11(2):177–200
7 (1973)—Maps.

PACIFIC

5111 Fosberg, F. R. *and* Sachet, M.-H.
Plants of southeastern Polynesia: 2. *Micronesica*,
6 8(1–2):43–49(1973).

HAWAIIAN ISLANDS

5112 Degener, O. *and* Degener, I.
Flora Hawaiiensis: new illustrated flora of the Hawaiian Islands. Joinvilleaceae: *Joinvillea*; Thy-
6 melaeaceae: *Wikstroemia*. 1973—Illustrations.

5113 Fosberg, F. R.
Derivation of the flora of the Hawaiian Islands. *In* Kay, E. A., *ed.* A natural history of the Hawaiian
6 Islands: 396–408(1972).

5114 Kay, E. A.
Hawaiian natural history: 1778–1900. *In* Kay, E. A., *ed.* A natural history of the Hawaiian Islands: 604–653,
6 (1972).

5115 Kay, E. A., *ed.*
A natural history of the Hawaiian Islands: selected readings. Honolulu, Hawaii University Press, 1972.
6 xii, 653p.—Illustrations, maps, dot maps; key.

5116 Stewart, M.
Bogs in Hawaii. *Bull. Pac. Trop. Bot. Gard.*, 3(3):
6 55–58(1973)—Illustrations.

MARIANAS ISLANDS

5117 Moore, P. H.
Composition of a limestone forest community on
6 Guam. *Micronesica*, 9(1):45–58(1973)—Map.

NEW CALEDONIA

5118 Dubois, M. J.
Les plantes de la forêt de Maré Iles Loyauté, Nouvelle-Calédonie. *J. Agric. Trop. Bot. Appl.*,
6 19(6–7):249–252(1972 publ. 1973).

NEW HEBRIDES

5119 Marshall, A. G.
A start to nature conservation in the New Hebrides.
6 *Biol. Conserv.*, 5(1):67–69(1973)—Map.

TROPICAL AFRICA

5120 Cremers, G.
Architecture de quelques lianes d'Afrique tropicale. *Candollea*, 28(2):249–280;(1973)—(En,Ge);
5 illustrations.

5121 Exell, A. W.
Angiosperms of the islands of the Gulf of Guinea (Fernando Po, Príncipe, S. Tomé, and Annobon).
5 *Bull. Brit. Mus. (Nat. Hist.), Bot.*, 4(8):325–411(1973).

5122 Exell, A. W.
Relações florísticas entre as ilhas do golfo da Guiné e destas com o continente africano. *Garcia de Orta,*
5 *Ser. Bot.*, 1(1–2):3–9(1973)—(En). Ecological.

5123 Flora of tropical East Africa. 1973. 2v.—Illustrations,
5 map; keys. Combretaceae 99p., Lemnaceae 9p.

5124 Greenway, P. J.
A classification of the vegetationof East Africa. *Kirkia*, 9(1):1–68(1973)—Map. Article written in
5 1943.

5125 Hepper, F. N.
Botanical explorers in West Africa. *Geogr. Mag.*, 45(10):740–742(1973)—Coloured illustrations,
5 map.

5126 Hepper, F. N., *ed.*
Tropical African plants: 33. *Kew Bull.*, 28(2):319–
5 326(1973)—Illustrations.

5127 Keay, R. W. J.
Change in African vegetation. *Afr. Res. Doc.*,
5 no.1:5–10(1973).

5128 Richards, P. W.
Africa, the 'Odd Man Out'. *In* Meggers, B. J., Ayensu, E. S. *and* Duckworth, W. D., *eds.* Tropical forest ecosystems in Africa and South America:
5 21–26(1973).

5129 Weiss, E. A.
Some indigenous trees and shrubs used by local fishermen on the East African coast. *Econ. Bot.*, 27(2):
5 174–192(1973)—Illustrations, map.

ANGOLA

5130 Exell, A. W. *and* Gonçalves, M. L.
A statistical analysis of a sample of the flora of Angola. *Garcia de Orta, Sér. Bot.*, 1(1–2):105–119
5 (1973)—(Por); maps.

BOTSWANA

5131 Banks, C. H.
Ngamiland expedition. *Trees S. Afr.*, 25(1):11–20
5 (1973)—Illustrations, portraits.

CAMEROUN

5132 Aubréville, A. *and* Leroy, J.-F.
Flore du Cameroun: 15. Paris, Mus. Nation. Hist. Nat., 1973. 199p.—Illustrations; keys. Icacinaceae, Olacaceae, Pentadiplandraceae, Opiliaceae, Octoknemaceae. Also published as Flore du Gabon:
5 20(1973).

5133 Aubréville, A. *and* Leroy, J.-F.
Flore du Cameroun: 16. Sapindacées. Paris, Mus. Nation. Hist. Nat., 1973. 202p.—Illustrations; keys. Also published as Flore du Gabon: 23. Sapindacées.
5 1973.

CAPE VERDE ISLANDS

5134 Sunding, P.
Check-list of the vascular plants of the Cape Verde Islands. Oslo, Bot. Gard. Univ. Oslo, 1973. 36p.

5135 Sunding, P.
Endemism in the flora of the Cape Verde Islands, with special emphasis on the Macaronesian flora element. *Monogr. Biol. Canar.*, no.4:112–117(1973).

ETHIOPIA

5136 Arena, M. R. *and* Orsino, F.
Adumbratio Florae Aethiopicae: 24. Rhizophoraceae. *Webbia*, 28(1):135–159(1973)—It(En); illustrations, dot maps.

5137 Boudet, C., Lebrun, J.-P. *and* Rippstein, G.
Southern rangeland livestock development project ecological range survey: botanical appendix. Maisons-Alfort, Inst. d'Élevage Méd. Vét. Pays Trop., 1973. unpaged—Illustrations.

5138 Longhitano, N. *and* Bavazzano, R.
Researches on the vegetation of the Middle Valley of Awash River (Ethiopia). *Webbia*, 28(2):543–565 (1973)—(It); illustrations.

GABON

5139 Aubréville, A. *and* Leroy, J.-F., *eds*
Flore du Gabon: 20. Icacinacées, Olacacées, Pentadiplandracées, Opiliacées, Octoknémacées. Paris, Mus. Nation. Hist. Nat., 1973. 199p.—Illustrations; keys. Also published as Flore du Cameroun: 15(1973).

5140 Aubréville, A. *and* Leroy, J.-F. *eds*
Flore du Gabon: 21. Malpighiacées, Nectaropétalacées, Linacées, Lépidobotryacées, Cténolophonacées, Humiriacées, Erythroxylacées, Ixonanthacées, Santalacées. Paris, Mus. Nation. Hist. Nat., 1973. 70p.—Illustrations; keys.

5141 Aubréville, A. *and* Leroy, J.-F. *eds*
Flore du Gabon: 22. Célastracées, Pandacées, Bombacacées, Cannabacées, Bixacées, Avicenniacées. Paris, Mus. Nation. Hist. Nat., 1973. 71p.—Illustrations; keys.

5142 Aubréville, A. *and* Leroy, J.-F. *eds*
Flore du Gabon: 23. Sapindacées. Paris, Mus. Nation. Hist. Nat., 1973. 202p.—Illustrations; keys Also published as Flore du Cameroun: 16. Sapindacées. 1973.

GUINEA-BISSAU

5143 Lisbon: Jardim e Museu Agricola do Ultramar
Flora da Guiné Portuguesa: Butomaceae. Lisbon, Ministerio do Ultramar, 1972. 5p.

5144 Lisbon: Jardim e Museu Agricola do Ultramar
Flora da Guiné Portuguesa: Caesalpinaceae. Lisbon, Ministerio do Ultramar, 1973. 46p.—Illustrations; chromosome numbers; keys.

5145 Lisbon: Jardim e Museu Agricola do Ultramar
Flora da Guiné Portuguesa: Mimosaceae. Lisbon, Ministerio do Ultramar, 1972. 38p.—Keys.

5146 Lisbon: Jardim e Museu Agricola do Ultramar
Flora da Guiné Portuguesa: Papilionaceae: 1° Apêndice. Lisbon, Ministerio do Ultr amar, 1972. 27p.

GUINEA REPUBLIC

5147 Bazilevskaya, N. A.
Pyat'let v Gvinee: 2. Verkhn-yaya i lesnaya Gvineya. (Five years in Guinea: 2. The upper and forest Guinea). *Bot. Zhurn.*, 58(5):746–753 (1973).

5148 Bazilevskaya, N. A.
Pyat' let v Gvineĭskoĭ Respublike: 1. Nizhnyaya i srednyaya Gvineya. (Five years in Guinea: 1. Lower and middle Guinea). *Bot. Zhurn.*, 58(4):614–622 (1973)—Map.

IVORY COAST

5149 Adjanohoun, E. *and* Aké Assi, L.
Plantes pharmaceutiques de Côte d'Ivoire. Abidjan, République de Côte d'Ivoire, 1970 publ. 1972. 358p.

MALAWI

5150 Brummitt, R. K.
Systematic list of Nyika botanical collections. *In* Wye College 1972 Malawi Project. Final report: 47–77(1973).

5151 [Synge, A. H. M.]
Botanical report. *In* Wye College 1972 Malawi Project. Final report: 7–46,79–85(1973).

MOZAMBIQUE

5152 Fernandes, A., *ed.*
Flora de Moçambique: 4,5,7,9,11,12,13,16,17, 19,20,21,22,37,40,47,49,51. Lisboa, Junta de Investigacao do Ultramar, 1973—Illustrations; chromosome numbers; keys. 4. Ranunculaceae 17p.; 5. Dilleniaceae 3p.; 7. Menispermaceae 27p.; 8. Nymphaeaceae 7p.; 9. Papaveraceae 5p.; 11. Cruciferae 11p.; 12. Capparaceae 66p.; 13. Violaceae 19p.; 16. Flacourtiaceae 41p.; 17. Pittosporaceae 5p.; 19. Caryophyllaceae 19p.; 20. Illecebraceae 7p.; 21. Portulacaceae 15p.; 22. Elatinaceae 7p.; 37. Rutaceae 29p.; 40. Ochnaceae 29p.; 47. Aquifoliaceae 3p.; 49. Rhamnaceae 23p.; 51. Sapindaceae 48p.

5153 Jardim, M. J. H. G.
Alargamento da área de distribuição de algumas espécies de Moçambique. *Rev. Ciênc. Biol.* (Lourenço Marques), 5:1–7(1972)—(En).

RHODESIA

5154 Drummond, R. B. *and* Palgrave, K. C.
Common trees of the Highveld. Salisbury, Longman, 1973. [iv]99p.—Illustrations, coloured illustrations.

5155 Jacobsen, W. B. G.
A checklist and discussion of the flora of a portion of the Lomagundi District, Rhodesia. *Kirkia*, 9(1): 139–207(1973)—Maps.

SAO TOMÉ

5156 Lisbon: Jardim e Museu Agrícola do Ultramar
Flora de S. Tomé e Príncipe: Alangiaceae. Lisboa, Ministerio do Ultramar, 1973. 5p.

5157 Lisbon: Jardim e Museu Agrícola do Ultramar
Flora de S. Tomé e Príncipe: Aquifoliaceae. Lisboa, Ministerio do Ultramar, 1973. 5p.

5158 Lisbon: Jardim e Museu Agrícola do Ultramar
Flora de S. Tomé e Príncipe: Mimosaceae. Lisboa, Ministerio do Ultramar, 1973. 29p.—Keys.

5159 Lisbon: Jardim e Museu Agrícola do Ultramar

Flora de S. Tomé e Príncipe: Papilionaceae. Lisboa, Ministerio do Ultramar, 1972. 90p.—Keys.

SENEGAL

5160 Berhaut, J.

Flore illustrée du Sénégal . . . dicotyledones: tome 1. Acanthacées à Avicenniacées. Dakar, Ministère du Développement Rural, 1971. 626p.—Illustrations.

5161 Lebrun, J.-P.

Énumération des plantes vasculaires du Sénégal. Maisons Alfort, Inst. d'Élevage et de Médicine Vétérinaire des Pays Tropicaux, 1973. 209p. (Etude botanique no.2)—Illustrations, map.

5162 Lebrun, J.-P.

Plantes nouvelles pour le Sénégal. *Adansonia*, 12(4):603–607(1972 publ. 1973)—(En); map, dot map.

SUDAN

5163 Ahti, T., Hämet-Ahti, L. *and* Pettersson, B.

Flora of the inundated Wadi Halfa reach of the Nile, Sudanese Nubia with notes on adjacent areas. *Ann. Bot. Fenn.*, 10(2):131–161(1973)—Illustrations, dot maps.

TANZANIA

5164 Caprotti, E.

Il Kilimangiaro sud-orientale nei suoi caratteri fitogeografici. (A phytogeographical outline of south eastern Kilimanjaro). *Natura* (Italy), 63(3):321–340 (1972)—(En,Fr); illustrations.

ZAÏRE

5165 Flore d'Afrique centrale (Zaïre-Rwanda-Burundi): Spermatophytes. Amaryllidaceae. Bruxelles, Jardin Botanique, 1973. 23p.—Illustrations.

5166 Flore d'Afrique centrale (Zaïre-Rwanda-Burundi): Spermatophytes. Juncaceae. Bruxelles, Jardin Botanique, 1973. 13p.—Illustrations.

5167 Flore d'Afrique centrale (Zaïre-Rwanda-Burundi): Spermatophytes. Oleaceae. Bruxelles, Jardin Botanique, 1973. 36p.—Illustrations; keys.

5168 Flore d'Afrique centrale (Zaïre-Rwanda-Burundi): Spermatophytes. Oliniaceae. Bruxelles, Jardin Botanique, 1973. 5p.—Illustration.

5169 Flore d'Afrique centrale (Zaïre-Rwanda-Burundi): Pteridophytes. Osmundaceae. Bruxelles, Jardin Botanique, 1973. 4p.—Illustration, dot map.

5170 Flore d'Afrique centrale (Zaïre-Rwanda-Burundi): Spermatophytes. Sphenocleaceae. Bruxelles, Jardin Botanique, 1973. 5p.—Illustrations.

SOUTHERN AFRICA

5171 Lucas, A.

Spring on the highveld. *Veld Flora*, 3(3):53–57 (1973)—Illustrations.

5172 Van der Spuy, U.

South African shrubs and trees for the garden. Johannesburg, etc., H. Keartland, 1971. 222p.—Coloured illustrations.

5173 Van der Spuy, U.

Wild flowers of South Africa for the garden. Johannesburg, etc., H. Keartland, 1971. 286p.—Coloured illustrations.

5174 Winter, B. de, Winter, M. de *and* Killick, D. J. B.

A selection of indigenous South African trees. Parow, Reader's Digest, 1973. 32p.—Coloured illustrations, dot maps.

ORANGE FREE STATE

5175 Werger, M. J. A.

An account of plant communities of Tussen die Riviere Game Farm, Orange Free State. *Bothalia*, 11(1–2):165–176(1973)—Illustrations, map.

SOUTH AFRICA (CAPE PROVINCE)

5176 Cone, G. B.

The floras of the south-western part of South Africa: a New Zealand point of view and a suggested explanation of the high speciation. *Tuatara*, 20(3): 160–164(1973).

5177 Eliovson, S.

Namaqualand in flower. Johannesburg, Macmillan, 1972. [xvi]128p.—Coloured illustrations, map.

5178 Kidd, M. M.

Wild flowers of the Cape Peninsula. 2nd ed. Cape Town, etc., Oxford University Press, 1973. Unpaged —Coloured illustrations, map.

5179 Lighton, C.

Cape floral kingdom. The story of South Africa's wild flowers, and the people who found, named and made them famous the world over. [rev. ed.] Cape Town, etc., Juta & Co. Ltd, 1973. 221p.—Illustrations, coloured illustrations, maps.

5180 Mason, H.

Western Cape sandveld flowers. Cape Town, Struik, 1972. 203p.—Illustrations, coloured illustrations.

5181 Werger, M. J. A.

Notes on the phytogeographical affinities of the southern Kalahari. *Bothalia*, 11(1–2):177–180(1973)—Maps.

SOUTH AFRICA (NATAL)

5182 Hilliard, O. M. *and* Burtt, B. L.

Notes on some plants of Southern Africa chiefly from Natal: 3. *Notes Roy. Bot. Gard. Edinburgh*, 32(3):303–387(1973)—Illustrations, dot maps; key.

5183 Musil, C. F.

Water plants of Natal: a guide to the important species. [South Africa], Wildlife Protection and Conservation Society of S. Africa, 1973. [ii]62p.—Illustrations, maps, dot maps; keys.

5184 Ross, J. H.

An analysis of the *Flora of Natal*. *Bothalia*, 11(1–2): 103–106(1973).

5185 Ross, J. H.

The flora of Natal. Pretoria, South Africa, Dep. Agric. Techn. Services, 1973. [ii]418p. (Botanical Survey Memoir no.39)—Maps; keys.

SOUTH AFRICA (TRANSVAAL)

5186 Anon.

Flora conservation. *Transvaal Nat. Conserv. Div. Ann. Rep.*, 7:36(1971–1972 publ. 1972).

5187 Lucas, A.
Wild flowers of the Witwatersrand. Cape Town, Purnell, 1971. viii,113p.—Coloured illustrations by
5 B. Pike.

SOUTH WEST AFRICA

5188 Guillarmod, A. J.
Fish River vegetation. *Veld Flora*, 3(3):47–49,52
5 (1973)—Illustrations.

MALAGASY AND MASCARENE ISLANDS

5189 Dejardin, J., Guillaumet, J.-L. *and* Mangenot, G.
Contribution à la connaissance de l'élément non endémique de la flore malgache (végétaux vasculaires).
5 *Candollea*, 28(2):325–391(1973)—(En,Ge).

5190 Leroy, J. F., *ed.*
Flore de Madagascar et des Comores: 13e Famille-Lycopodiacées; Famille 13bis-Huperziacées, by M. L. Tardieu-Blot. Paris, Musée National de l' Histoire Naturelle, 1971. 48p.—Keys. Lycopodiaceae pp. 3–13;
5 Huperziaceae pp.15–48.

5191 Leroy, J. F., *ed.*
Flore de Madagascar et des Comores: Gymnospermes: 17e Famille-Cycadacées; 18e Famille-Podocarpacées, by D. J. de Laubenfels. Paris, Musée National de l'Histoire Naturelle, 1972. 25p.—Keys.
5 Cycadaceae pp.3–7; Podocarpaceae pp. 9–22.

5192 Markgraf, F. *and* Boiteau, P.
Les 'Bois de Lait' des Mascareignes. *Adansonia*,
5 13(2):241–248(1973)—Illustrations.

MADAGASCAR

5193 Thomasson, M.
Remarques sur la végétation des environs de Tuléar (sud-ouest malgache). 1: Spectres biologiques foliaires. *Bull. Soc. Bot. France*, 118(9):749–752(1971 publ.
5 1973)—(En); map.

REUNION

5194 Melville, R.
L'importance scientifique et le potentiel économique de la flore de la Réunion. *Info-Nat.*, Numéro special:
5 56–59(1973).

NORTH AMERICA

5195 Krauss, H. M.
The information system design for the Flora North America Program. *Brittonia*, 25(2):119–134
3 (1973).

5196 Werger, M. J. A.
Las disyunciones anfitrópicas en las floras xerofiticas norte y sudamericanas. *Darwiniana*, 18
3 (1–2):9–18(1973)—(Ge).

5197 Woodward, B.
Plants of North America. *Quart. Bull. Alp. Gard.*
3 *Soc.*, 41(3):246–255(1973).

ALASKA

5198 Hultén, E.
Supplement to *Flora of Alaska and neighboring territories*: a study in the flora of Alaska and the Transberingian connection. *Bot. Notiser*, 126(4):
1 459–512(1973).

CANADA

5199 Hudson, J. H.
Additions to the floral list of Saskatchewan—
3 1972. *Blue Jay*, 31(2):112–114(1973).

5200 Hudson, J. H.
1971 Botanical records for Saskatchewan. *Blue*
3 *Jay*, 30(2):122–124(1972).

5201 Kuijt, J.
New plant records in Waterton Lakes National
3 Park, Alberta. *Canad. Field-Nat.*, 87(1):67–69(1973).

5202 Lewicki, A. *and* Donahue, P.
A catalogue of plant species from central British
3 Columbia. *Syesis*, 6:183–192(1973)—Dot map.

5203 McCain, J., Pike, R. B. *and* Hodgdon, A. R.
The vascular flora of Kent Island, Grand Manon,
3 New Brunswick. *Rhodora*, 75(802):311–322(1973).

5204 Pinel, H. W. *and* Wallis, C. A.
A botanical investigation in the Drumheller area, Alberta. *Blue Jay*, 30(3):169–194(1972)—Illustrations,
3 map, dot maps.

5205 Shchepanek, M. J.
Botanical investigation of the Otish Mountains, Québec. *Syllogeus*, no.2:1–23(1973)—Illustrations,
3 maps.

5206 Taylor, R. L.
Flora of British Columbia: a new botanical garden program. *Davidsonia*, 4(4):37–40(1973)—
3 Illustrations.

5207 Thaler, G. R. *and* Plowright, R. C.
An examination of the floristic zone concept with special reference to the northern limit of the Carolinian zone in southern Ontario. *Canad. J. Bot.*, 51(10):
3 1765–1789(1973)—(En); maps, dot maps.

5208 Turner, N. C. *and* Bell, M. A. M.
The ethnobotany of the Southern Kwakiutl Indians of British Columbia. *Econ. Bot.*, 27(3):257–
3 310(1973)—Map.

GREENLAND

5209 Fredskild, B.
Studies in the vegetational history of Greenland: palaeobotanical investigations of some holocene lake and bog deposits. *Medd. Grønland*, 198(4).245p.
3 (1973)—Illustrations, map.

5210 Hansen, K.
Microthermic observations in Arctic vegetation.
3 *Medd. Grønland*, 194(6).32p.(1973)—Illustrations.

U.S.A.

5211 Allen, C. M.
Additions to the Louisiana flora. *Sida*, 5(2):130–131
3 (1973).

5212 Barlow, E.
The forests and wetlands of New York City. Boston, Toronto, Little, Brown & Co., 1971. xxiv,
3 160p.—Illustrations, maps.

5213 Boufford, D. E.
New or noteworthy New Hampshire plants.
3 *Rhodora*, 75(801):158–161(1973).

5214 Bradley, T. R.
Plant records for the northern Virginia Fall Belt
3 counties. *Castanea*, 38(2):175–182(1973).

5215 Brown, C. A.
Wildflowers of Louisiana and adjoining States. Baton Rouge, Louisiana State University Press,
3 1972. xl,245p.—Coloured illustrations, map.

5216 Chambers, K. L.
Floristic relationships: Onion Peak with Saddle Mountain, Clatsop County, Oregon. *Madroño*,
3 22(3):105–114(1973)—Map.

5217 Chambers, K. L. *and* Dennis, L. J.
Additional plant records for Oregon. *Madroño*,
3 22(3):149–150(1973).

5218 Chuey, C. F.
The herbaceous angiosperm flora of Mahoning
3 County, Ohio. *Ohio J. Sci.*, 72(1):30–49(1972)—Map.

5219 Clark, R. C., Powell, R. W. *and* Childress, C. G.
Vascular plants in Spartanburg County, South Carolina, new to the State, the Carolinas or the southeastern United States. *Castanea*, 38(3):297–302
3 (1973).

5220 Crum, G. H. *and* Bachmann, R. W.
Submersed aquatic macrophytes of the Iowa Great Lakes region. *Iowa State J. Res.*, 48(2):147–173
3 (1973)—Maps.

5221 Davidson, D. W.
Vascular plants of mature upland forests of northern New Jersey. *Bull. Torrey Bot. Club*, 100(1):44–55
3 (1973).

5222 Del Moral, R.
The vegetation of Findley Lake Basin. *Amer. Midl.*
3 *Nat.*, 89(1):26–40(1973)—Illustration, map.

5223 Douglas, G. W., Naas, D. B. *and* Naas, R. W.
New plant records and ranges for Washington.
3 *Northwest Sci.* (U.S.A.), 47(2):105–108(1973).

5224 Eastman, L. M.
Some new locals in the Maine flora. *Rhodora*,
3 75(801):166–167(1973).

5225 Focht, J.
Additions to the flora of Rockland County, New
3 York. *Bull. Torrey Bot. Club*, 99(5):249–250(1972).

5226 Goodrich, R.
Fall color in Southern California. *Lasca Leaves*,
3 23(3):107–109(1973)—Illustration.

5227 Grimshaw, S. *and* Bradley, T. R.
The vascular flora of Great Falls National Park, Fairfax County, Virginia. *Castanea*, 38(3):229–261
3 (1973)—Illustrations, maps.

5228 Hartman, R. L.
New plant records for New Mexico. *Southwest.*
3 *Nat.*, 18(2):241–242(1973).

5229 Hathaway, W. T. *and* Ramsey, G. W.
The flora of Pittsylvania County, Virginia. *Castanea*,
3 38(1):38–78(1973).

5230 Heckman, H.
Island year. Seattle, London, Washington University Press, 1972. xiv, 255p.—Illustrations. Anderson Island, Puget Sound. Plant names pp.231–
3 238.

5231 Herman, T. *and* See, M. G.
Secondary succession following fire in 'Tight Holler', Kentucky. *Castanea*, 38(3):275–285(1973)—
3 Map.

5232 Heusser, C. J.
Modern pollen spectra from Mount Rainier, Washington. *Northwest Sci.* (U.S.A.), 47(1):1–8
3 (1973)—Map.

5233 Higgins, L. C.
Plants new to the Utah flora. *Great Basin Nat.*,
3 32(2):117–119(1972).

5234 Howell, J. T.
Miscellaneous notes on Munz' *A Californian flora* and its supplement. *Wasmann J. Biol.*, 30(1–2):97–
3 107(1972)—Keys.

5235 Howell, J. T.
A statistical estimate of Munz' supplement to *A Californian flora. Wasmann J. Biol.*, 30(1–2):93–96
3 (1972).

5236 Howitt, B. F. *and* Howell, J. T.
Supplement to the vascular plants of Monterey County, California. Pacific Grove, Museum of Natural
3 History Association, 1973. 60p.—Illustration; keys.

5238 Johnson, M. J.
Naturalists and plant explorers in Georgia, until 1850 (with illustrations of some native plants
3 observed). *Diss. Abstr. Int., B*, 33(9):4150(1973).

5239 Keil, D. J.
Vegetation and flora of the White Tank Mountains Regional Park, Maricopa County, Arizona. *J. Ariz.*
3 *Acad. Sci.*, 8(1):35–48(1973)—Map.

5240 Kral, R.
Some notes on the flora of the southern states, particularly Alabama and middle Tennessee. *Rhodora*,
3 75(803):366–410(1973).

5241 Leonard, S. W.
New records and notes on the flora of the Carolinas.
3 *J. Elisha Mitchell Sci. Soc.*, 88(4):265–266(1973).

5242 Munz, P. A.
Some remarks on the vegetation of southern
3 California. *Aliso*, 8(1):7–12(1973).

5243 Rogers, C. L., Mullens, N. E. *and* Shiflet, G. W.
State records for South Carolina. *Castanea*, 38(1):
3 114–116, (3):302–304(1973).

5244 Rogers, K. E.
Notes on plants of Mississippi: 1. *Castanea*, 38(2):
3 199–203(1973).

5245 Savage, W.
Annotated check-list of vascular plants of Sagehen Creek Drainage Basin, Nevada County, California.
3 *Madroño*, 22(3):115–139(1973)—Map.

5246 Schwegman, J. E.
Additions to the flora of Illinois. *Trans. Illin. Acad.*
3 *Sci.*, 65(3–4):42–44(1972 publ. 1973).

NORTH AMERICA

5247 Shapiro, A. M. *
Notes on some Staten Island weeds. *Proc. Staten Island Inst.*, 26:67–70(1972). Ref. in *Bull. Torrey Bot.*
3 *Club*, 99(5):262(1972).

5248 Smith, G. L.
A flora of Tahoe Basin and neighboring areas. *Wasmann J. Biol.*, 31(1). 231p.(1973)—Illustrations,
3 maps; keys.

5249 Smith, H. V.
Winter wildflowers. Ann Arbor, Michigan Botany Club, 1973. 64p. (Special Publication no.2)—
3 Illustrations.

5250 Stalter, R.
Factors influencing the distribution of vegetation of the Cooper River Estuary. *Castanea*, 38(1):18–24
3 (1973)—Map.

5251 Stalter, R.
The flora of Turtle Island, Jasper Co., South
3 Carolina. *Castanea*, 38(1):35–37(1973).

5252 Stalter, R.
Transplantation of salt marsh vegetation: 2. Georgetown, South Carolina. *Castanea*, 38(2):132–
3 139(1973).

5253 Stanford, J. W.
Vascular plants of the three Central Texas counties of Brown, Comanche and Hamilton. *Diss. Abstr.*
3 *Int., B*, 33(2):601(1972).

5254 Stuckey, R. L.
Changes of vascular aquatic flowering plants during 70 years in Put-in-Bay Harbor, Lake Erie, Ohio. *Ohio J. Sci.*, 71(6):321–342(1971)—Illustrations,
3 map.

5255 Swedberg, K. C.
A transition coniferous forest in the Cascade Mountains of Northern Oregon. *Amer. Midl. Nat.*,
3 89(1):1–25(1973)—Illustrations, maps.

5256 Taylor, R. J., Douglas, G. W. *and* Sundquist, L. M.
Contributions to the flora of Washington: 2.
3 *Northwest Sci.* (U.S.A.), 47(3):169–179(1973).

5257 Todsen, T. K.
Comments on New Mexico flora. *Sida*, 5(2):135
3 (1973).

5258 Uttal, L. J.
Floristic notes from Virginia. *Castanea*, 38(2):194–
3 195(1973).

5259 [Voss, E. G.]
Michigan plants in print: new literature relating to
3 Michigan botany. *Mich. Bot.*, 12(1):73–78(1973).

5260 Weber, W. A.
Additions to the Colorado flora: 5. With nomenclature revisions. *Southwest. Nat.*, 18(3):317–329
3 (1973).

5261 Weedon, R. R.
New and interesting vascular plants from the Great
3 Plains. *Southwest. Nat.*, 18(3):341–342(1973).

5262 Wharton, M. E. *and* Barbour, R. W.
A guide to the wildflowers and ferns of Kentucky. Kentucky University Press, 1971. viii,344p.—
3 Illustrations. coloured illustrations, maps.

5263 Wilcox, W. H.
A survey of the vascular flora of Crittenden County,
3 Arkansas. *Castanea*, 38(3):286–297(1973).

CENTRAL AMERICA

BELIZE

5264 Kellman, M. C.
Dry season weed communities in the upper Belize valley. *J. Appl. Ecol.*, 10(3):683–694(1973)—Map,
4 dot map.

GUATEMALA

5265 Lowden, R. M.
Kellerman's Guatemalan expeditions. *Taxon*,
4 22(2–3):324(1973).

5266 Standley, P. C. *and* Williams, L. O.
Flora of Guatemala: part 9 nos. 3–4. *Fieldiana, Bot.*, 24(9 nos. 3–4):237–418(1973)—Illustrations; keys.
4 Labiatae, Scrophulariaceae.

MEXICO

5267 Rauh, W. *and* Rauh, H.
Baja California: its cacti and succulents: a travelogue. *Nation. Cact. Succ. J.* (U.K.), 28(3):69–73(1973)—
3 Illustrations, map.

NICARAGUA

5268 Seymour, F. C.
Fourth expedition to Nicaragua. *Phytologia*, 23(3):
4 86–96(1973)—Map.

5269 Seymour, F. C.
Third expedition to Nicaragua. *Phytologia*, 25(7):
4 447–457(1973)—Map.

WEST INDIES

5270 Liogier, A. H.
Novitates Antillanae: 6(1). *Phytologia*, 25(5):265–
4 280(1973)—Keys.

BAHAMA ISLANDS

5271 Gillis, W. T., Howard, R. A. *and* Proctor, G. R.
Additions to the Bahama flora since Britton and
4 Millspaugh: 1. *Rhodora*, 75(803):411–425(1973).

5272 Wilcox, M. S., Forbes, A., Shure, S. *and* *
Wilcox, L. V.
A field key to Bahamian mangroves. *Carib. J. Sci.*, 11:155–157(1971 publ. 1972). Ref. in *Bull. Torrey Bot.*
4 *Club*, 99(6):328(1972).

CUBA

5273 Acuña, J., Grudzinskaya, I. A. *and* Muñiz, O.
O kubinskikh polulistopadnȳkh lesakh s *Celtis trinervia* Lam. (On the Cuban semideciduous forests with *Celtis trinervia* Lam.). *Bot. Zhurn.*, 58(12):1693–
4 1700(1973)—(En).

5274 Borhidi, A., Acuña, J. *and* Muñiz, O.
New plants in Cuba: 3. *Acta Bot. Acad. Sci. Hung.*,
4 19(1–4):37–45(1973)—Illustration.

5275 Borhidi, A. *and* Muñiz, O.
New plants in Cuba: 2. *Acta Bot. Acad. Sci. Hung.*
4 18(1–2):29–48(1973)—Illustrations.

5276 Borhidi, A. *and* Muñiz, O.
Új florisztikai adatok Kuba flórájához: 1. (Novedades floristicas en la Flora de Cuba: 1). *Bot. Közlem.*, 59(3):187–192(1972 publ. 1973)—(Sp). 4

JAMAICA

5277 Powell, D.
The voyage of the plant nursery, H.M.S. Providence, 1791–1793. Kingston, Institute of Jamaica, 1973. 80p. (Bulletin of the Institute of Jamaica, Science Series, no.15 pt.2)—Coloured illustrations. 4

SOUTH AMERICA

5278 Agostini, G.
Taxonomic bibliography for the neotropical flora: introduction. *Act. Bot. Venezuel.*, 9(1–4):253–254 (1974 publ. 1973). 4

5279 Cuatrecasas, J.
Miscelanea sobre flora neotropica: 3. *Ciencia*, 27(6): 171–184(1972)—(En); illustrations. 4

5280 Hueck, K. *and* Seibert, P.
Vegetationskarte von Südamerika: (mapa de la vegetación de America del Sur). Stuttgart, G. Fischer, 1972. [viii]71p.—(Vegetations-monographien der einzelnen Grossräume, Bd IIa)—Ge. 4

5281 Maguire, B.
The organization for 'Flora Neotropica'. *Nat. Resources*, 9(3):18–21(1973). 4

5282 Werger, M. J. A.
Las disyunciones anfitrópicas en las floras xerofíticas norte y sudamericanas. *Darwiniana*, 18(1–2):9–18 (1973)—(Ge). 4

5283 Williams, L. O.
Tropical American plants: 13. *Phytologia*, 25(7): 458–464(1973)—Illustration. 4

ARGENTINA

5284 Anon.
Nuevos taxa para la Republica Argentina y regiones limitrofes. *Bol. Soc. Argent. Bot.*, 14(4):381–388 (1972). 4

5285 Anon.
Nuevos taxa para la Republica Argentina y regiones limitrofes. *Bol. Soc. Argent. Bot.*, 15(1):141–146 (1973). 4

5286 Adamoli, J.
Frecuencia, confinamiento y transgresividad en especies del Chaco Argentina. (Frequency, confinement and transgresivity of species from the Argentine Chaco). *Bol. Soc. Argent. Bot.*, 15(1):1–11(1973)—(En). 4

5287 Argentina: Ministerio de Agricultura y Ganaderia de la Nacion
Notas preliminares para la flora Chaqueña (Formosa, Chaco y Santiago de Estero): no. 3. Castelar, 1972. 39p.—Illustrations; keys. 4

5288 Leal, A. R.
Los confines boreal y austral de las provincias Patagonica y Central respectivamente. *Bol. Soc. Argent. Bot.*, 13 Supl.:89–118(1972)—Illustrations, maps. 4

5289 Roig, F. A.
Bosqueja fisionomico de la vegetacion de la Provincia de Mendoza. *Bol. Soc. Argent. Bot.*, 13 Supl.: 49–80(1972)—Map. 4

5290 Toursarkissian, M., Mendiondo, M. E., Rondina, R. V. D. *and* Coussio, J. D.
Tercera contribucion a la bibliografia fitoquimica: catalogo por genero y especie: 1. *IDIA*, no.295. 67p. (1972). 4

BRAZIL

5291 Assumpção, W. R. C. d', *and others*
Levantamento des plantas tóxicas do estado de Minas Gerais. *Oréades*, 3(5):12–19(1972). 4

5292 Baptista, L. R. de Moura *and* Irgang, B. E.
Nota sôbre a composição florística de uma comunidade florestal dos arredores de Pôrto Alegre. *Iheringia*, no.16:3–8(1972)—(Ge). 4

5293 Brazil: Instituto Nacional de Pesquisas da Amazonia: Setor de Fitoquímica.
The chemical composition of Amazonian plants: a catalogue. *Acta Amazonica*, 2(1):37–40,(2):91–92 (1972). 4

5294 Duarte, A. Pereira
Duas novas espécies da flora dos Estados do Espírito Santo e Goiás. *Arq. Jard. Bot. Rio de Janeiro*, 19:217–222(1973)—Illustrations. 4

5295 Ghillány, *Barão* A. de
Uma aventura atrás de 19 porteiras. *Orquídea*, 30(2):61–64(1973)—Illustration, portrait. 4

5296 Gomes, B. A.
Plantas medicinais do Brasil. São Paulo, 1972. 32, ccxxviiip. (Brasiliensia Documenta, 5).—Sp; illustrations, portrait. Facsimile reprint of works, with introduction. 4

5297 Martins Júnior, A. Guedes
Fanerógamos parasitos da flora do Brasil. *Leandra*, 2(2):73–104(1972)—Illustrations. 4

5298 Occhioni, P. *and* Hatschbach, G.
A vegetação arbórea dos ervais do Paraná: 1. *Leandra*, 2(3):5–59(1972)—(En); illustrations, map. 4

5299 Pedersoli, J. L. *and* Martins, J. L.
A vegetação dos afloramentos de calcáreo. *Oréades*, 3(5):27–29(1972)—Illustrations. 4

5300 Ratter, J. A., Richards, P. W., Argent, G. *and* Gifford, D. R.
Observations on the vegetation of northeastern Mato Grosso: 1. The woody vegetation types of the Xavantina-Cachimbo Expedition area. *Phil. Trans. Roy. Soc. London, B*, 266(880):449–492(1973)—Illustrations, maps. 4

5301 Reitz, P. R., *ed.*
Flora ilustrada catarinense. Santa Catarina, 1972. 9v.—Illustrations, dot maps; keys. Amarantáceas 110p., Canáceas 39p., Hidrifiláceas 12p., Mirtáceas 17p., Pteridáceas 244p., Raflesiáceas 16p., Ramnáceas 50p., Tropeoláceas 26p., Umbelíferas 205p. 4

5302 Ribeiro, H. F.
Plantas "medicinais" dos ervanários do Estado da Guanabara. *Leandra*, 2(3):91–94(1972). 4

5303 Santa Catarina: Universidade Federal
Flórula da Ilha de Santa Catarina. Florianópolis, Universidade Federal de Santa Catarina, 1972. 3v.—Illustrations, coloured illustration; key. Chloranthaceae 1p., Goodeniaceae 10p., Myristicaceae 16p.
4

CHILE

5304 Ragonese, A. E. *and* Piccinini, B. G. *
[Vegetation of Isla de los Pájaros (Chubut Province)]. *IDIA*, no.291:30–50(1972)—Sp. Abstr. in *Herb. Abstr.*, 43(9):290(1973).
4

5305 Sparre, B.
'Plants from Robinson Crusoe's Island'. *Taxon*, 22(1):171(1973).
4

5306 Zollner S., O. *
[Natural vegetation of the Azapa Valley]. *Idesia*, no.2:117–125(1972)—Sp(En). Abstr. in *Herb. Abstr.*, 43(9):290(1973).
4

COLOMBIA

5307 Colombia: Universidad Nacional: Instituto de Ciencias Naturales
Catalogo ilustrado de las plantas de Cundinamarca: vol.5. Bogota, Inst. de Ciencias Naturales, 1972. 165p.—Keys. Passifloraceae, Begoniaceae, Melastomataceae.
4

ECUADOR

5308 Harling, G. *and* Sparre, B., *eds*
Flora of Ecuador: 89. Tropaeolaceae. *Op. Bot. Ser. B.*, no. 2. 30p.(1973)—Illustrations; keys.
4

5309 Harling, G. *and* Sparre, B., *eds*
Flora of Ecuador: 216. Cyclanthaceae. *Op. Bot. Ser. B.*, no.1. 47p.—Illustrations; keys.
4

5310 Little, E. L.
New tree species from Esmeraldas, Ecuador (concluded). *Phytologia*, 25(2):59–64(1972)—Illustration.
4

GALAPAGOS ISLANDS

5311 Eckhardt, R. C.
Introduced plants and animals in the Galapagos Islands. *BioScience*, 22(10):585–590(1972)—Maps.
4

5312 Hootele, C. *and* Moorleghem, C. van
Etude des alcaloides d'espèces végétales originaires des Iles Galapagos. *Notic. Galapagos*, no.19/20: 8–9(1972).
4

5313 Houvenaghel, G.
Contribution à l'étude de l'écologie marine des Iles Galapagos. *Mém. Acad. Roy. Sci. Outre-Mer, Sci. Nat. Nouv. Sér.*, 19(1).102p.(1973)—(Du,Sp,En); maps.
4

5314 Schofield, E. K.
Annotated bibliography of Galápagos botany, 1836–1971. *Ann. Missouri Bot. Gard.*, 60(2):461–477 (1973).
4

5315 Schofield, E. K.
Field guide to some common Galapagos plants, with additions. Columbus, Ohio State University, 1971. unpaged—Illustrations, map.
4

5316 Schofield, E. K.
Galápagos flora: the threat of introduced plants. *Biol. Conserv.*, 5(1):48–51(1973)—Illustrations, maps.
4

5317 Schofield, E. K.
A unique and threatened flora. *Gard. J.*, 23(3): 68–73(1973)—Illustrations.
4

GUYANA

5318 Guppy, N.
A young man's journey. London, Murray, 1973. xiv,276p.—Illustrations, map.
4

5319 Warren, A.
Report of the 1971 British Expedition to Mt. Roraima in South America. Oxford, Seacourt Press, [1973]. 152p.—Illustrations, maps. Botanical report by D. Philcox pp.69–72.
4

PANAMA

5320 Dressler, R. L.
Terrestrial plants of Panamá. *In* Jones, M. L. *ed.* The Panamic biota: some observations prior to a sea-level canal: 179–186(1972).
4

5321 Dwyer, J. D.
Notes on Panamanian trees and shrubs collected in 1971 by L. R. Holdridge and others. *Ann. Missouri Bot. Gard.*, 59(2):247–261(1972 publ. 1973).
4

5322 Earle, S. A.
A review of the marine plants of Panamá. *In* Jones, M. L., *ed.* The Panamic biota: some observations prior to a sea-level canal: 69–87(1972).
4

5323 Jones, M. L., *ed.*
The Panamic biota: some observations prior to a sea-level canal. A symposium sponsored by the Biological Society of Washington, The Conservation Foundation, The National Museum of Natural History, The Smithsonian Institution. *Bull. Biol. Soc. Wash.*, no.2. 269p.(1972)—Illustrations, maps.
4

5324 Woodson, R. E. *and* Schery, R. W.
Flora of Panama: parts: 2,4,6,9. *Ann. Missouri Bot. Gard.*, 60(1):1–167(1973)—Illustrations; keys. Najadaceae, Hydrocharitaceae, Rafflesiaceae, Simaroubaceae, Verbenaceae, Avicenniaceae, Caprifoliaceae.
4

PERU

5325 Guillén, J. E. L. *and* Cornelio, I. K. de
Plantas medicinales del Peru: contribución al conocimiento farmacobotánico de especies vegetales, que se emplean vernacularmente como drogas en el Perú. *Biota*, 9(74):283–313, (75):347–376(1973)—Illustrations.
4

5326 Soukup, J.
Las Cletraceas, Ericaceas, Teofrastaceas, Mirsinaceas, Primulaceas y Plumbaginaceas del Peru, sus generos y lista de especies. *Biota*, 9(73):223–256(1972)—Keys.
4

5327 Soukup, J.
Las Ebenaceas, Estiracaceas y Sapotaceas del Peru, sus generos y lista de especies. *Biota*, 9(73):258–264 (1972)—Key.
4

5328 Soukup, J.
[Las Eleocarpaceas, Tiliaceas y Malvaceas del Peru, sus generos y lista de especies]. *Biota*, 9(73):217–222 (1972).
4

5329 Weberbauer, A.
Floral geography of the Peruvian Andes: 2. *Chileans*, 7(26):83–89(1973). Abstracted from 'The Flora of Peru' by J. F. McBride.
4

SURINAM

5330 Donselaar, J. van
Additional data on the lianes of the Brokopondo district, Surinam. *Acta Bot. Neerl.*, 22(2):164–165
4 (1973).

URUGUAY

5331 Klein, R. M.
Árvores nativas da floresta subtropical do Alto
4 Uruguai. *Sellowia*, no.24:9–62(1972)—(Ge).

VENEZUELA

5332 Badillo, V. M. *and* Schnee, L.
Clave de las familias de plantas superiores de Venezuela. (Clé de détermination des familles de plantes supérieures du Venezuela). *Rev. Fac. Agron. Univ. Centr. Venez., Alcance*, no.18:1–246(1972). Ref. in *Bull. Signal., Biol. Physiol. Vég.*, 34(10):8918
4 (1973).

5333 Brewer-Carias, C.
Proyecto de la exploración de las Mesetas de Jaua, Guanacoco y Sarisarinâma. *Bol. Soc. Venezol. Cienc. Nat.*, 30(127) *Suppl.* unpaged.(1973)—(En);
4 illustrations, maps.

5334 Oliva-Esteva, F.
12 arboles de Venezuela. Caracas, Monte Avila Editores, 1972. 71p.—Illustrations, coloured
4 illustrations.

5335 Steyermark, J. A.
Relacion floristica entre la Cordillera de la Costa y la zone de Guayana y Amazonas. *Acta Bot. Venezuel.*,
4 9(1–4):245–252(1974 publ. 1973)—(En).

5336 Steyermark, J. A.
Situacion actual de las exploraciones botanicas en Venezuela. *Acta Bot. Venezuel.*, 9(1–4):241–243(1974
4 publ. 1973).

ANTARCTICA

5337 Allison, J. S. *and* Smith, R. I. L.
The vegetation of Elephant Island, South Shetland Islands. *Brit. Antarct. Surv. Bull.*, nos.33–34:185–212
4 (1973)—Illustrations, dot map.

5338 Greene, D. M.
The herbarium of the British Antarctic Survey.
4 *Brit. Antarct. Surv. Bull.*, no.31:107–109(1972).

5339 Smith, R. I. L. *and* Corner, R. W. M.
Vegetation of the Arthur Harbour–Argentine Islands region of the Antarctic Peninsula. *Brit. Antarct.*
4 *Surv. Bull.*, nos.33–34:89–122(1973)—Map.

5340 Smith, R. I. L.
Vegetation of the South Orkney Islands with particular reference to Signy Island. London, British Antarctic Survey, 1972. 124p. (Scientific Reports,
4 no.68)—Illustrations, maps.

Nomenclature

5341 Bourdoux, P.
Considerations sur la nomenclature des cactées. *Cactus* (Brussels), 4(2):34–40(1972)—Illustrations.

5342 Brickell, C. D.
Problems of horticultural nomenclature. *In* Green, P. S., *ed.* Plants: wild and cultivated: 102–113 (1973).

5343 Compère, P.
Citation des auteurs antérieurs au point de départ: *ex* ou crochets ? *Taxon*, 22(5–6):702–703(1973). Nomenclature proposal 20 to the Leningrad Congress.

5344 Jansonius, J.
Articles 41 and 42 of the I.C.B.N. and fossil spores:
* a proposal. *Taxon*, 22(2–3):259–266(1973). Proposals to the Leningrad Botanical Congress 19.

5345 Jeffrey, C.
Biological nomenclature. London, Arnold, 1973. ix, 69p. (Special Topics in Biology Series).

5346 Jesurún, A.
Principios generales de nomenclatura de plantas. *Orquidea* (Mexico), 1(10):15–20(1971).

5347 MacBryde, B.
A generic homonym gap. *Taxon*, 22(1):161–162 (1973).

5348 Misra, A.
Linne's binomy. *Planto*, 2(1):83–85(1972).

5349 Reda, H.
The international convention for the protection of new varieties of plants of 1961 and the international code of nomenclature of cultivated plants. *Taxon*, 22(1):51–55(1973).

5350 Stearn, W. T.
The principles of botanical nomenclature, their basis and history. *In* Green, P. S. *ed.* Plants: wild and cultivated: 86–101(1973).

5351 Voss, E. G.
Errata in Seattle Code. *Taxon*, 22(4):503(1973). Brief note.

5352 Wuesthoff, F.
Cultivated plant nomenclature and Plant Variety Rights. *Taxon*, 22(4):455–458(1973).

Chromosome Surveys

5353 Anon.
Atlantiques. *Inform. Ann. Caryosyst. Cytogénét.*, no.5:33–36(1971 publ. 1973)—Illustrations;
1 chromosome numbers.

5354 Bawa, K. S.
Chromosome numbers of tree species of a lowland tropical community. *J. Arnold Arbor.*, 54(4):422–434
4 (1973). In Costa Rica.

5355 Darlington, C. D.
Chromosome botany and the origins of cultivated plants. 3rd (rev.) ed. London, Allen & Unwin, 1973.
1–7 xviii,237p.—Maps; chromosome numbers.

5356 Delay, J.
Halophytes et écotypes maritimes. *Inform. Ann. Caryosyst. Cytogénét.*, no.5:29–32(1971 publ. 1973)—
1 Illustrations: chromosome numbers.

5357 Delay, J.
Orophytes. *Inform. Ann. Caryosyst. Cytogénét.*, no.5:17–28(1971 publ. 1973)—Illustrations; chromo-
1 some numbers.

5358 Delay, J.
Région méditerranéenne. *Inform. Ann. Caryosyst. Cytogénét.*, no.5:37–40(1971 publ. 1973)—
1 Chromosome numbers.

5359 Fernandez Casas, J.
Numeros cromosomicos de plantas españolas, 1. *Cuad. Cienc. Biol.* (Granada), 2(1):39–41(1973)—
1 (En).

5360 Fulvio, T. E. di
Recuentos cromosomicos en Angiospermas argentinas: 2. *Kurtziana*, 7:39–42(1973)—
4 Chromosome number.

5361 Garbari, F., Tornadore, N. *and* Pecori, E.
Numeri cromosomici per la flora italiana: 151–
1 161. *Inform. Bot. Ital.*, 5(2):161–169(1973).

5362 Kamari, G. *and* Papatsou, S.
Chromosome studies in some Mediterranean angiosperms. *Bot. Notiser*, 126(2):266–268(1973)—
1 Chromosome numbers.

5363 Leute, G.-H.
Chromosomenzahlen dreier Apiaceen aus Österreich. *Österr. Bot. Zeitschr.*, 121(1–2):143–144
1 (1973).

5364 Löve, Á.
IOPB chromosome number reports: 39. *Taxon*,
1–3, 6, 7 22(1):115–118(1973).

5365 Löve, Á.
IOPB chromosome number report: 40. *Taxon*,
1–3, 6, 7 22(2–3):285–291(1973).

5366 Löve, Á.
IOPB chromosome number reports: 41. *Taxon*,
1–3, 6, 7 22(4):459–464(1973).

5367 Löve, Á.
IOPB chromosome number reports: 42. *Taxon*,
1–3, 6, 7 22(5–6):647–654(1973).

5368 Löve, Á. and Kjellqvist, E.
Cytotaxonomy of Spanish plants: 2. Monooctyledons. *Lagascalia*, 3(2): 147-182(1973)—(Sp);
1 chromosome number.

5369 Marchi, P. *and* D'Amato, G.
Numeri cromosomica per la flora Italiana: 145-150. *Inform. Bot. Ital.*, 5(1): 93-100(1973).
1 *Santolina* spp.

5370 Miège, J. *and* Greuter, W.
Nombres chromosomiques de quelques plantes récoltées en Crète. (Chromosome numbers of some plants collected in Crete). *Ann. Mus. Goulandris*,
1 1:105–112(1973)—(En,Gre).

5371 Moore, R. J., *ed.*
Index to plant chromosome numbers 1967–1971. Utrecht, International Association for Plant
1–7 Taxonomy, 1973. 539p. (Regnum Vegetabile v.90).

5372 Pojar, J.
Levels of polyploids in four vegetation types of southwestern British Columbia. *Canad. J. Bot.*,
3 51(3):621–628(1973)—Chromosome numbers.

5373 Ratter, J. A. *and* Milne, C.
Chromosome numbers of some primitive angiosperms. *Notes Roy. Bot. Gard. Edinburgh*, 32(3):423–
1–7 428(1973).

5374 Ratter, J. A. *and* Milne, C.
Some angiosperm chromosome numbers. *Notes*
1–7 *Roy. Bot. Gard. Edinburgh*, 32(3):429–438(1973).

5375 Reveal, J. L. *and* Styer, E. L.
Miscellaneous chromosome counts of western American plants: 2. *Great Basin Nat.*, 33(1):19–25
3 (1973).

5376 Ritter, J.
Remarques caryologiques et phytosociologiques sur quelques taxons du Jura et des Alpes. *Rev. Cytol. Biol. Vég.*, 35(3–4):281–294(1972 publ. 1973)—
1 (En,Ge); chromosome numbers.

5377 Sharma, A. K., *comp.*
Chromosome census of the plant kingdom: 1. Monocotyledons: part 1. Butomales to Zingiberales.
1–7 *Nucleus*, 15(2). [ii] 20p.(1972).

5378 Valdés, B.
Numeros cromosomicos de algunas plantas españolas: 1. *Lagascalia*, 3(2):211–217(1973)—
1 (En).

5379 Watson, T. J.
Chromosome numbers in Compositae from the southwestern United States. *Southwest. Nat.*, 18(2):
3 117–124(1973).

5380 Zhukova, P. G., Petrovskiĭ, V. V. *and* Ilieva, T. V.
Khromasomnȳe chisla i taksonomiya nekotorȳkh vidov rasteniĭ Sibiri i dal'nego vostoka. (The chromosome numbers and taxonomy of some plant species from Siberia and far east). *Bot. Zhurn.*, 58(9):
2 1331–1342(1973).

5381 Zhukova, P. G. *and* Tikhonova, A. D.
Khromosomnȳe chisla nekotorȳkh vidov rasteniĭ Chukotki: 2. (The chromosome numbers of some Chukotka plant species: 2). *Bot. Zhurn.*, 58(3):395–
2 402(1973).

Chemotaxonomy

5382 Bate-Smith, E. C.
Chemical phylogeny and classification of the angiosperms. *In* Nobel Symposium 25: chemistry in botanical classification, August 20–25, 1973. (1973). 1p. Abstract.

5383 Boudet, A.
Les acides quinique et shikimique chez les angiospermes arborescentes. *Phytochemistry*, 12(2):363–370 (1973)—(En).

5384 Boulter, D.
The use of amino acid sequence data in the classification of higher plants. *In* Nobel Symposium 25: chemistry in botanical classification, August 20–25, 1973. (1973). 1p. Abstract.

5385 Boutard, B., Bouillant, M.-L., Chopin, J. *and* Lebreton, P.
Chimiotaxinomie flaxonique des Fluviales. *Biochem. Syst.*, 1(3):133–140(1973)—(En).

5386 Cronquist, A.
[Morphology]. *In* Nobel Symposium 25: chemistry in botanical classification, August 20–25, 1973. (1973). 2p. Abstract.

5387 Harborne, J. B.
Flavonoids as systematic markers in the angiosperms. *In* Nobel Symposium 25: chemistry in botanical classification, August 20–25, 1973. (1973). 1p. Abstract.

5388 Hegnauer, R.
Die cyanogenen Verbindungen der Liliatae und Magnoliatae–Magnoliidae: zur systematischen Bedeutung des Merkmals der Cyanogenese. *Biochem. Syst.*, 1(4):191–197(1973)—(En).

5389 Herout, V.
A chemical compound as a taxonomic character. *In* Nobel Symposium 25: chemistry in botanical classification, August 20–25, 1973.(1973).1p. Abstract.

5390 Heywood, V. H.
Chemosystematics—an artificial discipline? *In* Nobel Symposium 25: chemistry in botanical classification, August 20–25, 1973. (1973).1p.

5391 Jensen, U.
The interpretation of comparative serological results. *In* Nobel Symposium 25: chemistry in botanical classification, August 20–25, 1973. (1973). 1p. Abstract.

5392 Kjaer, A.
The natural distribution of glucosinolates: a uniform group of sulfur-containing glucosides. *In* Nobel Symposium 25: chemistry in botanical classification, August 20–25,1973.(1973). 1p. Abstract.

5393 Mabry, T. J.
The chemistry of disjunct taxa. *In* Nobel Symposium 25: chemistry in botanical classification, August 20–25, 1973. (1973). 1p. Abstract.

5394 Mabry, T. J.
Is the Centrospermae monophyletic? Betalain and DNA–RNA hybridization data. *In* Nobel Symposium 25: chemistry in botanical classification, August 20–25, 1973. (1973).1p. Abstract.

5395 Ourisson, G.
Some aspects of the distribution of diterpenes in plants. *In* Nobel Symposium 25: chemistry in botanical classification, August 20–25, 1973. (1973). 1p. Abstract.

5396 Sandberg, F.
Biological and chemical screening of plant materials. *In* Nobel Symposium 25: chemistry in botanical classification, August 20–25, 1973. (1973). 1p. Abstract.

5397 Schwarting, A. E.
The structural and biosynthetic heterogeneity and the taxonomic significance of the quinolizidine alkaloids. *In* Nobel Symposium 25: chemistry in botanical classification, August 20–25, 1973. (1973). 1p. Abstract.

5398 Swain, T.
Flavonoids as evolutionary markers in primitive tracheophytes. *In* Nobel Symposium 25: chemistry in botanical classification, August 20–25, 1973. (1973). 1p. Abstract.

5399 Takhtadzhyan, A. L.
The chemical approach to plant classification with the special reference to the higher taxa of Magnoliophyta. *In* Nobel Symposium 25: chemistry in botanical classification, August 20–25, 1973. (1973). 2p. Abstract.

5400 Tétényi, P.
Homology of biosynthetic routes: the base in chemotaxonomy. *In* Nobel Symposium 25: chemistry in botanical classification, August 20–25, 1973. (1973). 1p. Abstract.

5401 Turner, B. L.
Chemosystematic data: their use in the study of disjunctions. *Ann. Missouri Bot. Gard.*, 59(2):152–164(1972 publ. 1973)—Map; chromosome
1–7 numbers.

5402 Turner, B. L. *and* Flake, R.
Volatile constituents, especially terpenes, and their utility and potential as taxonomic characters. *In* Nobel Symposium 25: chemistry in botanical classification, August 20–25, 1973. (1973). 1p. Abstract.

5403 Yeim Yok Siv *and* Paris, R. R.
Sur quelques plantes à flavonoïdes du Cambodge appartenant aux genres *Cananga*, *Colona*, *Grewia*, *Leea* et *Melastoma*. *Pl. Méd. Phytothér.*, 6(4):299–305 (1972)—also publ. in *Trav. Lab. Matière Méd. Pharm.*
5 *Galénique Fac. Pharm. Paris*, 57(2) (1973).

Anatomy and Morphology

5404 Asby, G.
Development of root buds in *Cnidium dubium* and in some other herbaceous plants. *Bot. Tidsskr.*, 68(1): 30–40(1973)—Illustrations.

5405 Behnke, H.-D.
Sieve-tube plastids of Hamamelididae. Electron microscopic investigations with special reference to Urticales. *Taxon*, 22(2–3):205–210(1973).

5406 Blagoveshchenskiĭ, V. V.
O ponyatii ginetseya. (On the concept of gynoecium). *Bot. Zhurn.*, 58(7):1043–1044(1973).

5407 Croizat, L.
Quelques réflexions sur la morphologie, la morphogénèse et la symétrie. *Adansonia*, 13(3):351–382(1973)—Illustrations.

5408 Danilova, M. F. *and* Sokolovskaya, T. B.
Anatomiya prorostka nekotorȳkh vidov zlakov i vopros o prirode odnodol'nosti. (Seedling anatomy of some grass species and the problem of the origin of monocotyledonity). *Bot. Zhurn.*, 58(3):337–349 (1973)—(En).

5409 Doyle, J. A.
The monocotyledons: their evolution and comparative biology: 5. Fossil evidence on early evolution of the monocotyledons. *Quart. Rev. Biol.*, 48(3): 399–413(1973)—Illustrations.

5410 Endress, P. K.
Arils and aril-like structures in woody Ranales. *New Phytol.*, 72(5):1159–1171(1973).

5411 Fagerlind, F. *and* Dunbar, A.
Some electron microscopical methods for solving wood anatomical problems. *Bot. Notiser*, 126(4): 519–533(1973)—Illustrations.

5412 Gegov, A.
Das morphologische Wesen der pflanzlichen Teraten. (Morphological essence of vegetative terrates). *Dokl. Bolg. Akad. Nauk*, 26(4):545–548 (1973).

5413 Gracza, P.
Van-e összefüggés a magház állása és a virágdudor kezdeti szervezödése között? (Is there a connection between the position of the ovary and the original organization of the primordium?). *Bot. Közlem.*, 60(2):129–133(1973)—(En).

5414 Hagemann, W.
The organisation of shoot development. *Rev. Biol.*, 9(1–4):43–67(1973)—Illustrations.

5415 Hickey, L. J.
Classification of the architecture of dicotyledonous leaves. *Amer. J. Bot.*, 60(1):17–33(1973).

5416 Inamdar, J. A. *and* Patel, R. C.
Structure, ontogeny and classification of trichomes in some Polemoniales. *Feddes Repert.*, 83(7–8):473–488(1972 publ. 1973)—(Ge).

5417 Kaplan, D. R.
The monocotyledons: their evolution and comparative biology: 7. The problem of leaf morphology and evolution in the monocotyledons. *Quart. Rev. Biol.*, 48(3):437–457(1973)—Illustrations.

5418 Knobloch, I. W.
The value of plant anatomy. *Taiwania*, 18(1): 42–44(1973).

5419 Levin, D. A.
The role of trichomes in plant defense. *Quart. Rev. Biol.*, 48(1):3–15(1973).

5420 Macdonald, A. D.
Floral development in the Amentiferae. *Diss. Abstr. Int., B*, 33(7):2969(1972)—*Myrica, Comptonia, Juglans, Populus, Laportea, Fagus, Quercus, Betula, Alnus, Ostrya.*

5421 Mogollón, A. P.
Estructura anatomica de 37 maderas de la Guayana venezolana y clave para su. *Acta Bot. Venezuel.*, 8(1–4):9–109(1973)—Key.

5422 Moore, H. E. *and* Uhl, N. W.
The monocotyledons: their evolution and comparative biology: 6. Palms and the origin and evolution of monocotyledons. *Quart. Rev. Biol.*, 48(3):414–436(1973)—Illustrations; chromosome numbers.

5423 Padmanabhan, D., Lakshmanan, M. *and* Veerasamy, S.
Morphological and anatomical studies on little leaf in some flowering plants. *Proc. Indian Acad. Sci., B*, 78(3):130–142(1973)—Illustrations.

5424 Rao, T. A. *and* Bhupal, O. P.
Typology of sclereids. *Proc. Indian Acad. Sci., B*, 77(2):41–55(1973)—Illustrations.

5425 Roth, I.
Estructura anatómica de la corteza de seis especies arbóreas de las familias Araliaceae, Dichapetalaceae, Lacistemaceae, Onacaceae, Oriliaceae y Quiinaceae. (Anatomical structure of the cortex of six arboreous species of the families Araliacae, Dichapetalaceae, Lacistemaceae, Olacaceae, Opiliaceae, and Quiinaceae.) *Acta Biol. Venezuel.*, 8(2):103–129 (1973)—(En).

5426 Sampath, S.
New patterns of symbiosis in some weed species. *Curr. Sci.*, 42(3):100–101(1973)—Illustrations.

5427 Sands, M. J. S.
New aspects of the floral vascular anatomy in some members of the order Rhoeadales sensu Hutch. *Kew Bull.*, 28(2):211–256(1973)—Illustrations.

5428 Scurfield, G., Michell, A. J. *and* Silva, S. R.
Crystals in woody stems. *Bot. J. Linn. Soc.*, 66(4): 277–289(1973).

5429 Speta, F.
Entwicklungsgeschichte und Karyologie von Elaiosomen an Samen und Früchten. *Naturk. Jahrb. Linz*, 9–65(1972)—(En).

5430 Stevenson, D. W.
Phyllode theory in relation to leaf ontogeny in *Sansevieria trifasciata. Amer. J. Bot.*, 60(4):387–395 (1973).

5431 Tixier, P.
Les fougères-nids. *Sci. Nat.*, no.116:13–17(1973)—Illustrations.

5432 Tomlinson, P. B.
The monocotyledons: their evolution and comparative biology: 8. Branching in monocotyledons. *Quart. Rev. Biol.*, 48(3):458–466(1973).

5433 Tomlinson, P. B. *and* Gill, A. M.
Growth habits of tropical trees: some guiding principles. *In* Meggers, B. J., Ayensu, E. S. *and* Duckworth, W. D., *eds.* Tropical forest ecosystems in Africa and South America: 129–143(1973)—Illustrations.

5434 Untawale, A. G. *and* Bhasin, R. K.
On endothecial thickenings in some monocotyledonous families. *Curr. Sci.*, 42(11):398–400 (1973)—Illustrations.

Palynology

5435 Accorsi, C. A.
Presentazione di schede per una flora palinologica italiana. (Introduction of cards for an Italian palynological flora). *Inform. Bot. Ital.*, 5(1):76(1973)—(En).

5436 Ascenso, J. C.
A conservação do pólen e o melhoramento das plantas. *Genet. Iber.*, 24(1–2):107–122(1972)—En. Pollen storage and longevity.

5437 Barth, O. M.
Catálogo sistemático dos pólens das plantas arbóreas do Brasil meridional: 11. Loranthaceae e Olacaceae (Catalogue of southern Brazilian pollen grains). *Mem. Inst. Oswaldo Cruz* (Rio de Janeiro), 70(1):49–61(1972)—(En); illustrations.

5438 Barth, O. M.
Catálogo sistemático dos pólens das plantas arbóreas do Brasil meridional: 13. Flacourtiaceae e Caricaceae. (Catalogue of southern Brazilian pollen grains). *Mem. Inst. Oswaldo Cruz* (Rio de Janeiro), 70(1):1–13 (1972)—(En); illustrations, key.

5439 Barth, O. M. *and* Barbosa, A. F.
Catálogo sistemático dos pólens das plantas arbóreas do Brasil meridional: 14. Nyctaginaceae e Phytolaccaceae. (Catalogue of southern Brazilian pollen grains). *Mem. Inst. Oswaldo Cruz* (Rio de Janeiro), 70(3):241–259(1972)—(En); illustrations.

5440 Barth, O. M. *and* Barbosa, A. F.
Catálogo sistemático dos pólens das plantas arbóreas do Brasil meridional: 15. Myrtaceae. (Catalogue of southern Brazilian pollen grains). *Mem. Inst. Oswaldo Cruz* (Rio de Janeiro), 70(4):467–496(1972)—(En); illustrations.

5441 Barth, O. M. *and* Barbosa, A. F.
Catálogo sistemático dos pólens das plantas arbóreas do Brasil meridional: 16. Ericaceae. (Catalogue of southern Brazilian pollen grains). *Mem. Inst. Oswaldo Cruz* (Rio de Janeiro), 70(4):555–569(1972)—(En); illustrations, key.

5442 Barth, O. M. *and* Barbosa, A. F.
Catálogo sistemático dos pólens das plantas arbóreas do Brasil meridional: 17. Elaeocarpaceae e Tiliaceae. (Catalogue of southern Brazilian pollen grains). *Mem. Inst. Oswaldo Cruz* (Rio de Janeiro), 71(3):203–218(1973)—(En); illustrations.

5443 Cerceau-Larrival, M. T.
Le pollen et la microscopie électronique à balayage. *Sci. Nat.*, no.117:2–13(1973)—Illustrations.

5444 Dexheimer, J.
Etude experimentale des grains de pollen des Angiospermes. *Rev. Cytol. Biol. Vég.*, 35(1–2):17–40 (1972).

5445 Kupriyanova, L. A. *and* Aleshina, L. A.
Pȳl'tsa i sporȳ rasteniĭ florȳ Evropeĭskoĭ chastĭ SSSR: tom 1. (Pollen and spores of plants from the flora of European part of the USSR: 1). Leningrad, "Nauka", 1972. 171p.—Illustrations.

5446 Myachina, A. I., Kazachikhina, L. L., Mamontova, I. B. *and* Kalinina, V. S.
Atlas spori pȳltsȳ nekotorȳkh sovremennȳkh rasteniĭ Dal'nego Vostoka. Khabarovsk, Inst. DVNTs AN SSSR, 1971. 85p.—Illustrations.

5447 Nair, P. K. K.
A palynological basis for plant classification. *J. Palynol.*, 6:111–112(1970 publ. 1972).

5448 Nair, P. K. K.
Pollen evolution in relation to floral evolution with particular reference to the Sympetalae: 2. Pentacyclideae and Tetracyclideae. *J. Palynol.*, 8:105–122 (1972 publ. 1973).

5449 Nekrasov, V. I. *and* Knyazeva, O. M.
Izuchenie kachestva pȳl'tsȳ drevesnȳkh rasteniĭ metodom prorashchivaniya na tsellofane. *Byull. Glavn. Bot. Sada* (Moscow), 88:61–66(1973).

5450 Nillsson, S., *ed.*
World pollen and spore flora: 1–2. Angiospermae. Stockholm, Almqvist & Wiksell, 1973. 2v.—Chromosome numbers; key. Fouqieriaceae 13p., Menyanthaceae 20p.

5451 Saad, S. I.
Pollen structure in relation to phylogeny. *J. Palynol.*, 8: 37–53(1972 publ. 1973).

5452 U.S.S.R.: Akademiya Nauk SSSR: Botanicheskii Institut im. V. L. Komarova
Morfologiya pȳl'tsȳ i spor sovremennȳkh rasteniĭ: trudȳ III Mezhdunarodnoĭ Palinologicheskoĭ Konferentsii SSSR, Novosibirsk, 1971. (Pollen and spores morphology of the recent plants: proceedings of the IIIrd International Palynological Conference Novosibirsk, USSR, 1971). Leningrad, "Nauka", 1973. 112p.—(En); illustrations.

5453 Vasil, I. K.
The new biology of pollen. *Naturwissenschaften*, 60(5):247–253(1973).

5454 Whitehead, D. R.
Approaches to disjunct populations: the contribution of palynology. *Ann. Missouri Bot. Gard.*, 59(2): 125–137(1972 publ. 1973).

5455 Zinderen Bakker, E. M. van, *ed.*
Palaeoecology of Africa: and of the surrounding islands and Antarctica: vol.7. Covering the years 1969–1971. Cape Town, Balkema, 1972. [iv]216p.—Illustrations, maps.

Embryology

5456 Sporne, K. R.
A note on the evolutionary status of tapetal types in dicotyledons. *New Phytol.*, 72(5):1173–1174(1973).

5457 Vakar, B. A.
O mnogokratnom vozniknovenii tipichnogo zarodȳshevogo meshka i dvoĭnogo oplodotvoreniya v svyazi s voprosom o polifileticheskom proiskhozhdenii pokrȳtosemennȳkh rasteniĭ. (On the repeated formation of typical embryo sac and double fertilisation in connection with the problem of a polyphyletic origin of Angiospermae). *Byull. Mosk. Obshch. Ispȳt. Prir., Biol.*, 78(3):89–93(1973)—(En).

Reproductive Biology

5458 Baker, H. G. *and* Baker, I.
Some anthecological aspects of the evolution of nectar-producing flowers particularly amino acid production in nectar. *In* Heywood, V. H., *ed.* Taxonomy and ecology: 243–264(1973).

5459 Baskin, J. M. *and* Baskin, C. C.
Plant population differences in dormancy and germination characteristics of seeds: heredity or environment. *Amer. Midl. Nat.*, 90(2):493–498(1973).

5460 Eisikowitch, D.
Mode of pollination as a consequence of ecological factors. *In* Heywood, V. H. *ed.* Taxonomy and ecology: 283–288(1973).

5461 Gleaves, J. T.
Gene flow mediated by wind-borne pollen. *Heredity*, 31(3):355–366(1973).

5462 Kevan, P. G.
Heliotropism in some Arctic flowers. *Canad.*
3 *Field-Nat.*, 86(1):41–45(1972)—Illustrations.

5463 Kipping, J.
Pollination: 2. *Calif. Hort. J.*, 34(2):76–79(1973)—Illustration.

5464 Melville, R.
Pollination in the flowering plants. Ilford, Central Association of Bee-Keepers, 1973. 8p.

5465 Mulligan, G. A. *and* Kevan, P. G.
Color, brightness and other floral characteristics attracting insects to the blossoms of some Canadian weeds. *Canad. J. Bot.*, 51(10):1939–1952(1973)—Illustrations.

5466 Pijl, L. van der
Principles of dispersal in higher plants. 2nd ed. Berlin etc., Springer-Verlag, 1972. xii,162p.—Illustrations.

5467 Pojar, J.
Pollination of typically anemophilous salt marsh plants by bumble bees *Bombus terricola occidentalis* Grne. *Amer. Midl. Nat.*, 89(2):448–451(1973).

5468 Proctor, M. C. F. *and* Yeo, P. F.
The pollination of flowers. London, Collins, 1973. 418p. (New Naturalist, 54)—Illustrations, coloured illustrations.

5469 Thieret, J. W.
Sex and the angiosperms. *Sida*, 5(2):59–60(1973).

5470 Thompson, P. A.
Seed germination in relation to ecological and geographical distribution. *In* Heywood, V. H. *ed.* Taxonomy and ecology: 93–119(1973)—Maps.

5471 Wycherley, P. R.
The phenology of plants in the humid tropics. *Micronesica*, 9(1):75–96(1973)—Maps.

Bibliography

5472 Anon.
Bibliografia botanica para America Latina. *Bol. Soc. Argent. Bot.*, 15(1):149–153(1973).

5473 Amengual, B. M. *and* Villa Carenzo, M.
Catalogo bibliografico fitoquimico Argentino: 2. Tucuman, Univ. Nac., 1971. 121p. (Miscelanea no. 36)—(En).

5474 Association pour l'Étude Taxonomique de la Flore d'Afrique Tropicale
A.E.T.F.A.T. index 1972 . . . Index of the papers on systematic phanerogamy and of the new taxa concerning Africa south of the Sahara and Madagascar. Bruxelles, A.E.T.F.A.T., 1973. 57p.

5475 Beaulieu sur Mer: Bibliothèque Arpad Plesch
Mille et un livres botaniques: répertoire bibliographique de la Bibliothèque Arpad Plesch. Bruxelles, Arcade, 1973. 517p.—(En); illustrations, coloured illustrations, portraits.

5476 Cullen, J.
William Roscoe's *Monandrian Plants of the order Scitamineae. Notes Roy. Bot. Gard. Edinburgh*, 32(3): 417–421(1973).

5477 Desmond, R.
Literature of plants. *In* Green, P. S. *ed.* Plants: wild and cultivated: 152–163(1973).

5478 Dolezal, H.
Zur Editionsgeschichte der *Icones Plantarum Rariorum* von Nicolaus Joseph von Jacquin. *In* Geck, E. *and* Pressler, G., *eds.* Festschrift für Claus Nissen zum siebzigsten Geburtstag 1971:136–172(1973)—Illustrations.

5479 Duveen, D. I.
Classics of orchid literature: *A century of Indian orchids* by Sir Joseph Hooker. *Amer. Orchid Soc. Bull.*, 42(2):150–151(1973).

5480 Duveen, D. I.
Classics of orchid literature: *A century of orchidaceous plants. Amer. Orchid Soc. Bull.*, 42(5):416–418(1973)—Coloured illustration.

5481 Duveen, D. I.
Classics of orchid literature: *A second century of orchidaceous plants*, edited by James Bateman. *Amer. Orchid Soc. Bull.*, 42(6):508–509(1973).

5482 Duveen, D. I.
Classics of orchid literature: *Le livre des orchidées* by Count Oswald de Kerchove de Denterghem. *Amer. Orchid Soc. Bull.*, 42(11):1013–1015(1973)—Coloured illustration.

5483 Duveen, D. I.
Classics of orchid literature: *Les orchidées: histoire iconographique*, by E. de Puydt. *Amer. Orchid Soc. Bull.*, 42(7):627–629(1973)—Coloured illustration.

5484 Duveen, D. I.
Classics of orchid literature: *Orchids and how to grow them in India and other tropical climates. Amer. Orchid Soc. Bull.*, 42(8):693(1973).

5485 Duveen, D. I.
Classics of Orchid literature: *Pescatorea: iconographie des Orchidées, volume 1*, by Jean Linden. *Amer. Orchid Soc. Bull.*, 42(9):822–824(1973)—Coloured illustration.

5486 Duveen, D. I.
Classics of orchid literature: *Select orchidaceous plants. Amer. Orchid Soc. Bull.*, 42(3):244–246(1973)—Coloured illustration.

5487 Duvigneaud, J. *and* Lambinon, J.
Bibliographie de l'histoire naturelle en Belgique: B. Botanique 1970. *Nat. Mosana, Suppl.*, 25(1–2). 51p.(1972).

5488 Duvigneaud, J. *and* Lambinon, J.
Bibliographie de l'histoire naturelle en Belgique: B. Botanique 1971. *Nat. Mosana, Suppl.*, 26(1–2) 56p.(1973).

5489 Fenton, M. O.
Publication dates of *Notes on Mesembryanthemum & allied genera* by H. M. L. Bolus. *Taxon*, 22(2–3): 323(1973).

5490 Geck, E. *and* Pressler, G., *eds*
Festschrift für Claus Nissen zum siebzigsten Geburtstag 2. September 1971. Wiesbaden, G. Pressler, 1973. 686p.

5491 Goff, F. R.
Johann Petri of Passau and his *Herbarius Latinus*. *In* Geck, E. *and* Pressler, G., *eds*. Festschrift für Claus Nissen zum siebzigsten Geburtstag 1971:302–310 (1973).

5492 Hansen, A.
Dansk botanisk litteratur i 1969, 1970 og 1971. *Bot. Tidsskr.*, 67(4):344–384(1973).

5493 Hansen, B.
Bibliography of Thai botany. *Nat. Hist. Bull. Siam Soc.*, 24(3–4):319–408(1973).

5494 Hansen, P. *and* Maule, A. F.
Pehr Osbeck's collections and Linnaeus's *Species Plantarum* (1753). *Bot. J. Linn. Soc.*, 67(3):189–212 (1973)—Illustrations.

5495 Hope, J. H., Murray-Smith, S. *and* Norman, F. I.
A bibliography of Bass Strait 1797–1971. *Proc. Roy. Soc. Victoria*, 85(2):79–116(1973)—Map.

5496 Howard, R. A.
The *Enumeratio* and *Selectarum* of Nicolaus von Jacquin. *J. Arnold Arbor.*, 54(4):435–470(1973).

5497 Krauss, N. L. H.
Bibliography of Mortlock Atoll (Tauu), Melanesia. *Pac. Isl. Stud. Notes*, no.10.6p.(1973).

5498 Krauss, N. L. H.
Bibliography of Tasman Atoll (Nukumanu) Melanesia. *Pac. Isl. Stud. Notes*, no.9. 4p.(1973).

5499 Kristinsson, H.
Recent literature on the botany of Iceland: 1. *Acta Bot. Islandica*, 2:67–76(1973).

5500 Mazrimas, J. A.
Edgar T. Wherry's *Sarracenia* publications. *Carniv. Pl. Newsl.*, 2(3):37–39(1973)—Bibliography.

5501 Metcalfe, C. R.
Metcalfe and Chalk's *Anatomy of the Dicotyledons* and its revision. *Taxon*, 22(5–6):659–668(1973).

5502 Moskovskoe Obshchestvo Ispȳtateleĭ Prirodȳ
Flora i rastitel'nost' Moskovskoĭ Oblasti (istoriya izucheniya i annotirovannaya bibliografiya). (Flora and vegetation of Moscow Region (history of research and annotated bibliography)). Moskva, Izd. Moskovsk. Univ., 1972. 288p.

5503 New York: Botanical Garden: Library
Catalog of the manuscript and archival collections and index to the correspondence of John Torrey. Boston, Mass., Hall, 1973. viii,473p.

5504 Opsomer, J. E.
Cinq cents ans de botanique exotique en Belgique: (13e–17e siècles). *Bull. Séances Acad. Roy. Sci. Outre-Mer*, 4:446–479(1972 publ. 1973).

5505 St. John, H.
The scientific names in the German edition of Parkinson's plants of use for food, medicine etc., in Otaheite. *Biol. J. Linn. Soc.*, 4(4):305–310(1972 publ. 1973).

5506 Toursarkissian, M., Mendiondo, M. E., Rondina, R. V. D. *and* Coussio, J. D.
Tercera contribución a la bibliografía fitoquímica argentina: catalogo por género y especie: 1. *IDIA*, no.295.67p.(1972).

5507 Tralau, H., *ed.*
Index Holmensis: 3. A world index of plant distribution maps. Monocotyledoneae J-Z. Zürich, Scientific Publishers, 1973. 224p.

5508 Sunding, P., *comp.*
A botanical bibliography of the Canary Islands. 2nd ed. Oslo, Botanic Garden, University of Oslo, 1973. 46p.

5509 Van Royen, P.
Publication dates of and index to taxa in reports by R. Kanehira on plants collected during the Kanehira-Hatusima 1940 collection in New Guinea. *Bot. Mag. Tokyo*, 86(1002):143–154(1973).

5510 Vidal, J. E.
Bibliographie botanique Indochinoise de 1955–1969. *Bull. Soc. Étude Indoch.*, 47(4):657–748(1972).

Personalia

5511 Anon.
Directório dos botânicos taxonomistas portugueses (1971). *Bol. Soc. Port. Cienc. Nat.*, 14:143–147
1 (1972).

5512 Anon.
South American explorers: 3. *Chileans*, 7(25):
4 26–28(1973).

5513 Baines, J. A.
The origin of generic names of Victorian flora: part 1. The person behind the name. *Victorian Nat.*, 90(3):72–73, (5):122–124, (6):170–172, (7):186–187, (8):227–229, (9):248–250, (10):279, (11):321–324
7 (1973).

5514 Burdet, H. M.
Cartulae ad botanicorum graphicem: 2. *Candollea*, 28(1):137–170(1973)—Ge(En,Fr). Buckley to
1–3 R. E. A. de Candolle.

5515 Burdet, H. M.
Cartulae ad botanicorum graphicem: 3. *Candollea*,
1–7 28(2):407–440(1973)—(En,Ge). Capus to Chabert.

5516 Gomes e Sousa, A.
Exploradores e naturalistas da flora de Moçambique. *Bol. Soc. Estud. Moçambique*, 40(166–167):99–115
5 (1971)—Portraits.

5517 Howard, R. A.
A partial history of the New England Botanical
3 Club. *Rhodora*, 75(804):493–516(1973).

5518 Shaw, G. A.
Some Yorkshire bryologists and bryophytes.
1 *Naturalist*, no.924:1–12(1973).

5519 Tamlyn, P.
Some early Jamaican botanists. *West Indian Rev.*,
4 31,51(1973 ?).

ABBOT, John (c.1776–c.1840)

5520 Johnson, M. J.
Naturalists and plant explorers in Georgia, until 1850 (with illustrations of some native plants
3 observed). *Diss. Abstr. Int.*, *B*, 33(9):4150(1973).

AELLEN, Paul (1896–1973)

5521 Anon.
1 Paul Aellen. *Taxon*, 22(5–6):708(1973)—En.

5522 Dörr, E.
Zum Tode von Paul Aellen. *Mitt. Naturwiss.*
1 *Arbeitskr. Kempten (Allgäu)*, 17(3):1(1973).

AFANAS'EV, Danilo Yakovich (1902–)

5523 Bilik, G. I.
Danȳlo Yakovȳch Afanas'yev (do 70-richchya narodzhennya). (Danylo Yakovych Afanasyev (on his 70th birthday.)) *Ukr. Bot. Zhurn.*, 30(1):122–124
1 (1973)—Uk; portrait.

ALCOCK, Nora (c.1875–1972)

5524 Foister, C. E.
Mrs. Nora Alcock. *B.S.E. News*, no.10:10–11
1 (1973).

ALEXANDER, H. G. (c.1876–1972)

5525 Black, P.
H. G. Alexander V.M.H. *J. Roy. Hort. Soc.*, 98(6):
1 277(1973)—En.

ALTEHAGE, Carl (1899–1970)

5526 Stredelmann, W.
Carl Altehage 1.4.1899–12.12.1970. *Osnabrücker*
1 *Naturwiss. Mitt.*, 1:9–12(1972)—Portrait, bibliography.

ANDERSON, Edgar (1897–1969)

5527 Eisendrath, E. R.
The publications of Edgar Anderson. *Ann. Missouri Bot. Gard.*, 59(3):346–361(1972 publ. 1973)—
3 Bibliography.

5528 Finan, J. J.
Edgar Anderson 1897–1969. *Ann. Missouri Bot. Gard.*, 59(3):325–345(1972 publ. 1973)—En;
3 portraits.

5529 Gamble, M. A.
The Edgar Anderson Boxwood Garden. *Herbarist*,
3 no.39:24–32(1973)—Portrait.

5530 Heiser, C. B.
Student days with Edgar Anderson or how I came to study sunflowers. *Ann. Missouri Bot. Gard.*, 59(3):
3 362–372(1972 publ. 1973)—Portraits.

5531 Porter, D. M.
Hybridization, evolution, and systematics. *Ann.*
3 *Missouri Bot. Gard.*, 59(3):380–381(1972 publ. 1973).

5532 Stebbins, G. L.
Edgar Anderson: recollections of a long friendship. *Ann. Missouri Bot. Gard.*, 59(3):373–379(1972 publ.
3 1973).

ANDERSON, Edward Bertram (c.1886–1971)

5533 Bellia, G. G.
Due amici scomparsi: Aymon Correvon ed E. Bertram Anderson. *Rea*, no.5:41–44(1972)—(En);
1 portrait.

5534 Synge, P. M.
E. B. Anderson Memorial Lecture, 1972. *Quart. Bull. Alp. Gard. Soc.*, 41(2):138–148(1973)—
1 Illustrations.

ANDERSON, Ernest Gustav (1891–1973)

5535 Emerson, S.
Ernest Gustav Anderson March 3, 1891–January 30,
3 1973. *Genetics*, 74 *Suppl.* 1(2):s19–s20(1973).

ANDREWS, E. Wyllys (c.1916–c.1971)

5536 Andrews, J. M. de
In memoriam Dr E. Wyllys Andrews. *Orquídea*
4 (Mexico), 1(12):18–20(1971 publ. 1972)—Portrait.

ANDREWS, William George (1893–1972)

5537 Cheek, A. J. W.
William George Andrews. *J. Kew Guild*, 9(77):
1 156–157(1973)—En; portrait.

ARBER, Agnes (1879–1960)

5538 Misra, J.
Agnes Arber (1879–1960). *Planto*, 2(1):89–90
1 (1972)—En.

ARCHER, William Andrew (1894–1973)

5539 Anon.
[W. Andrew Archer]. *Taxon*, 22(4):523(1973)—
3 En.

ARNOLD, Joseph (1782–1818)

5540 Bastin, J.
Dr Joseph Arnold and the discovery of *Rafflesia arnoldi* in west Sumatra in 1818. *J. Soc. Bibliogr. Nat.*
1 *Hist.*, 6(5):305–372(1973)—Illustration, portraits.

ARROJADOA LISBOA, Miguel

5541 Buxbaum, F.
The genus *Arrojadoa*. *Chileans*, 6(23):62–65(1972)—Illustrations. Translated by H. Middleditch from
4 *Kakteen*, 1.vii.1968.

ARZT, Theodor (1905–1973)

5542 Eberle, G.
In memoriam Theodor Arzt. *Hess. Flor. Briefe*,
1 262–264:50–52(1973)—Ge.

AUSTIN, Coe Finch (1831–1880)

5543 Meyer, F. G. *and* Elsasser, S.
The 19th century Herbarium of Isaac C. Martindale.
3 *Taxon*, 22(4):375–404(1973).

BAGCHEE, Krishnadas (1898–1973)

5544 Anon.
Obituary Dr. K. Bagchee, D.Sc. *Indian Forester*,
6 99(4):[2](1973)—Portrait.

5545 Bakshi, B. K.
Obituary. *Indian Phytopathol.*, 26(1):1(1973)—
6 Portrait.

BAILLON, Henri (1827–1895)

5546 Stafleu, F. A.
Baillon and his *Dictionnaire de Botanique*. *Acta Bot. Acad. Sci. Hung.*, 19(1–4):335–341(1973)—
1 Illustration.

BAINES, Thomas (1822–1875)

5547 Winfield, J. J.
Thomas Baines Nature Reserve. *East. Cape Nat.*,
1, 5 no.49:19–20(1973).

BALFOUR, *Sir* Isaac Bayley (1853–1922)

5548 [Portrait only]. *J. Roy. Hort. Soc.*, 98(3):(1973).

5549 Matthews, J. R.
Examinations: the Balfour way. *B. S. E. News*,
1 no.11:5–7(1973).

5550 Matthews, J. R.
Presidents and professors. *B.S.E. News*, no.10:
1 4–6(1973).

BALLAND, René (1905–1970)

5551 Dupuis, C.
René Balland (1905–1970), président de la Fédération Française des Sociétés de Sciences Naturelles.
Rev. Féd. Franç. Soc. Sci. Nat., 11(49):120–121
1 (1972).

BANKS, *Sir* Joseph (1743–1820)

5552 Cannon, J. F. M. *and* Lysaght, A. M.
The British herbarium of Sir Joseph Banks. *Taxon*,
1 22(2–3):323–324(1973).

5553 Lysaght, A. M. *and* Cannon, J. F. M.
Sir Joseph Banks's British herbarium. *B.S.B.I. News*,
1 2(1):15–16(1973).

BAPTIST, John Thomas

5554 Hazelwood, W. G.
Australia's first nurserymen: John Thomas Baptist.
7 *Camellia News*, no.51:31(1973).

BARBARICH, Andreĭ Ivanovich (1903–)

5555 Morozyuk, S. S. *and* Kotov, M. I.
Andriĭ Ivanovich Barbarȳch (do 70-richchya vid dnya narodzhennya). (Andrei Ivanovich Barbarich (on his 70th birthday)). *Ukr. Bot. Zhurn.*, 30(4):530–
1 531(1973)—Uk; portrait.

BARNETT, Euphemia Cowan (1890–1970)

5556 Santisuk, T.
[Commemorative name *Barnettia* (Bignoniaceae)].
1 *Kew Bull.*, 28(2):172(1973).

BARROS, Manuel (1880–1973)

5557 Burkart, A.
Manuel Barros su fallecimiento. *Darwiniana*, 18
1 (1–2):282(1973).

BERNARDI, Luciano

5558 Geneva: Conservatoire et Jardin Botaniques
Catalogus plantarum a Luciano Bernardi lectarum, loco et die manifestis in schedulis annexis. Geneva,
1 197–? var. pag.

BÍLÝ, Julius (1895–1970)

5559 Siemińska, J.
Wspomnienie o Juliuszu Bílým. *Wiad. Bot.*, 16(4):
1 227–228(1972)—Pol.

BLACKWELL, Elizabeth Marianne (1889–1973)

5560 Waterhouse, G. M.
Obituary: Elizabeth Marianne Blackwell
8 January 1889–25 May 1973. *Trans. Brit. Mycol. Soc.*,
1 61(3):610–614(1973)—Portrait.

5561 Watling, R.
Elizabeth M. Blackwell. 1889–1973. *Naturalist*,
1 no.926:113–114(1973)—En.

BLAKE, Stanley Thatcher (1910–1973)

5562 Anon.
7 [S. T. Blake]. *Taxon*, 22(4):523(1973)—En.

5563 Anon.
Australia loses a famous botanist. *Austral. Pl.*,
7 7(54):60(1973).

BLANC, Albert A. (1850–1928)

5564 Mitich, L. W.
The world of A. Blanc. *Cact. Succ. J.* (U.S.A.),
1, 3 45(4):158–170(1973)—Illustrations.

5565 Mitich, L. W.
The world of A. Blanc, part 2. *Cact. Succ. J.*
3 (U.S.A.), 45(5):203–213(1973)—Illustrations.

5566 Mitich, L. W.
The world of A. Blanc: pt.3. *Cact. Succ. J.* (U.S.A.),
3 45(6):259–269(1973)—Illustrations.

BLAŠKOVIČ, Dionýz (c.1913–)

5567 Borecký, L.
K šestdesiatym narodeninám akademika D. Blaškoviča. (To the sixtieth birthday of the academician D. Blaškovič). *Biológia, A* (Czecho-
1 slovakia), 28(9):793–794(1973).

BLUME, Karl Ludwig (1796–1862)

5568 Reinikka, M. A.
Karl Ludwig Blume (1796–1862). *In* A history of
1, 6 the Orchid:145–148(1972)—Portrait.

BÖDEKER, Friedrich (1867–1937)

5569 Hunt, D. R.
Who was who: 2. Friedrich Bödeker (1867–1937).
1 *J. Mammillaria Soc.*, 13(1):5–6(1973).

BOLIN, Rolf L. (–1973)

5570 Gripp, P.
In memoriam—Rolf L. Bolin. *Amer. Orchid Soc.*
3 *Bull.*, 42(11):968(1973).

BOR, Norman Loftus (1893–1972)

5571 Anon.
Dr Norman Bor: authority on grasses. *The Times*,
1 1 January,(1973).

5572 Anon.
[Norman Loftus Bor]. *Taxon*, 22(4):523(1973)—
1 En.

5573 Anon.
Norman Loftus Bor, C.I.E., M.A., D.SC. *Biol. J.*
1 *Linn. Soc.*, 5(4):391(1973)—En. Short note.

5574 Clayton, D.
Dr Norman Loftus Bor. *J. Kew Guild*, 9(77):151
1 (1973)—En.

5575 Rowntree, J.
H. P. Smith; N. L. Bor. *Commonw. Forest. Rev.*,
1, 6 52(2):116–117(1973)—En.

5576 S., K. C.
Obituary Dr Norman Loftus Bor, C.I.E., M.A. (Dublin), D.SC. (Edin.), SC.D., F.R.S.E., F.N.I., I.F.S. *Indian Forester*, 99(3). [2p.] (1973)—
1, 6 Portrait.

BOROS, Ádám (1900–1973)

5577 Jánossy, A.
Boros Ádám. *Agrobotanika*, no.14:5–8(1972 publ.
1 1973)—Hu; portrait, bibliography.

5578 Simon, T.
Megemlékezés Boros Ádámról (1900–1973).
1 *Bot. Közlem.*, 60(1):3–5(1973)—Portrait.

5579 Vajda, L.
Prof. Dr. Adám Boros. *Rev. Bryol. Lichénol.*,
1 39(3):495–496(1973)—Fr; portrait.

BOUILLENNE, Raymond (c.1896–1972)

5580 Homès, M.
Hommage à Raymond Bouillenne. *Bull. Classe Sci.*
1 *Acad. Roy. Belg.*, 58(7):811–814(1972).

5581 BRAARUD, Trygve (1903–)
1 [Portrait only] *Norw. J. Bot.*, 20(2–3):75(1973).

BRADE, Alexandre Curt (1881–1971)

5582 Anon.
[Commemorative name × *Bradeara* (Orchidaceae)].
1, 4 *Orchid Rev.*, 81(961):centre page pull-out 8(1973).

5583 Markgraf, F.
Alexander Curt Brade. *Bot. Jahrb.*, 93(1):1–8
1, 4 (1973)—Ge; map.

BRADLEY, Richard (1688–1732)

5584 Tjaden, W.
Richard Bradley, F.R.S. 1688–1732, succulent plant
pioneer: part 1. *Bull. Afr. Succ. Pl. Soc.*, 8(4):132–
1 135(1973).

5585 Tjaden, W.
Richard Bradley, F.R.S. 1688–1732, succulent plant
pioneer: part 2. *Bull. Afr. Succ. Pl. Soc.*, 8(5):182–185
1 (1973).

BRÂNDZĂ, Dimitrie (1846–1895)

5586 Pop, E.
Viața și personalitatea lui Dimitrie Brândză
(22 Oct. 1846–3 Aug. 1895). (La vie et la personnalité
de Dimitrie Brândză (1846–1895)). *Contrib. Bot.*,
1 7–15(1973)—(Fr).

BRAUN, Emma Lucy (1889–1971)

5587 Stuckey, R. L.
E. Lucy Braun (1889–1971) outstanding botanist
and conservationist: a biographical sketch, with
bibliography. *Mich. Bot.*, 12(2):83–106(1973)—
3 Portraits, bibliography.

BRENAN, John Patrick Micklethwait (1917–)

5588 Anon.
John Patrick Micklethwait Brenan: President
1972–73. *J. Kew Guild*, 9(77):108–109(1973)—En;
1 portrait.

BRIGGS, Mary

5589 [Portrait only]. *Watsonia*, 9(4):421(1973).

BROWN, Lancelot (1716–1783)

5590 Bilbrough, H.
Documents in record offices which might affect
the assessment of the achievement of 'Capability'
Brown. *J. Gard. Hist. Soc.*, 1(3):9–22(1973)—
1 Illustration.

BROWN, Nicholas Edward (1849–1934)

5591 Squire, A. W. L.
Qui était N.E. Br. ? *Cactus* (Brussels), 4(6):119–121
1 (1972)—Illustrations.

BUBANI, Pietro (1806–1888)

5592 Burdet, H. M.
Cartulae ad botanicorum graphicem. *Candollea*,
1 27(2):338–340(1972)—Fr.

BUINING, Albert Frederik Hendrik (1901–)

5593 Anon.
Who's who in succulents: A. F. H. Buining. *Cact.*
1 *Succ. J.* (U.S.A.), 44(3):99,110(1972)—Portrait.

BUISSON, J. P.

5594 Guédès, M.
Duchesne, Buisson, Durande, early followers of
the natural method of the Jussieus. *Taxon*, 22(2–3):
1 211–219(1973).

BURBIDGE, Frederick William (1847–1905)

5595 Reinikka, M. A.
Frederick William Burbidge (1847–1905). *In*
1 A history of the Orchid: 258–259(1972).

BURDON-SANDERSON, *Sir* John Scott
(1828–1905)

5596 Williams, S. E.
A salute to Sir John Burdon-Sanderson and Mr
Charles Darwin on the centennial of the discovery of
the nerve-like activity in the Venus' Flytrap. *Carniv.*
1 *Pl. Newsl.*, 2(3):41–42(1973).

BURKE, David (1854–1897)

5597 Reinikka, M. A.
David Burke (1854–1897). *In* A history of the
1 Orchid: 269–270(1972).

BURRAGE, Albert Cameron (1859–1931)

5598 Reinikka, M. A.
Albert Cameron Burrage (1859–1931). *In* A history
3 of the Orchid: 287–289(1972)—Portrait.

BUȘ, Nikolai Adolfovici (1869–1941)

5599 Ștefureac, T. I.
Viata și opera botanistului Nikolai Adolfocici
Buș (1869–1941). (Vie et oeuvre du botaniste Nikolai
Adolfovici Buș (1869–1941)). *Stud. Comun., Bot.*,
1 285–299(1972)—(Fr); portrait.

CABEJSZEK, Irena (1910–1972)

5600 Siemińska, J.
Wspomnienie o Prof. Dr. Irenie Cabejszek. *Wiad.*
1 *Bot.*, 17(3):135–140(1973)—Portrait, bibliography.

CALDERON, Cesar Vargas (1905–)

5601 Calderon, C. V.
Cesar Vargas Calderon: an autobiography. *Pl. Life*,
4 29(1–4):6–7,9(1973)—Portrait.

CÂMARA, António de Sousa da (1901–)

5602 Fernandes, A.
Prof. Eng°. António de Sousa da Câmara o profes-
sor e o investigador. *Bol. Soc. Brot.*, 46:vii–xli(1972)—
1 Portrait, bibliography; chromosome numbers.

CARDENAS HERMOSA, Martin (1899–1973)

5603 Anon.
Dr. Martin Cardenas H. *Cact. Succ. J.* (U.S.A.),
4 45(2):85(1973)—En.

5604 Anon.
Martin Cardenas. *Biol. J. Linn. Soc.*, 5(4):393(1973)
4 —En.

5605 Anon.
Martin Cárdenas Hermosa. *Biota*, 9(74):314(1973)—
4 Sp.

5606 Anon.
Martin Cardenas Hermosa. *Taxon*, 22(5–6):708
4 (1973)—En.

5607 Anon.
Le professeur Martin Cardenas est mort. *Cactus*
4 (Brussels), no.5:90(1973).

5608 Diers, L.
Prof. Dr. Martin Cárdenas. *Kakt. And. Sukk.*,
4 24(8):175(1973)—Ge.

5609 Vasquez Ch., R.
Dr. Martin Cardenas Hermosa. *Cact. Succ. J.*
4 (U.S.A.), 45(3):124(1973)—En; portrait.

CARRINGTON, Benjamin (1827–1893)

5610 Shaw, G. A.
Some Yorkshire bryologists and bryophytes.
1 *Naturalist*, no.924:6–7(1973).

CASTELL, Cyril Philip (1907–1972)

5611 Beven, G.
Cyril Philip Castell, 1907–1972. *London Nat.*,
1 no.52:121–129(1973)—En; portrait, bibliography.

CASTELLANOS, Alberto (1896–1968)

5612 Castellanos, A.
Cactaceas. [Buenos Aires, Librart], 1972. xxiv, 385p.
—Illustrations, portrait, dot maps; keys. Reprints
4 from various journals.

CATESBY, Mark (c.1682–c.1750)

5613 Anderson, F. J.
Mark Catesby: pioneer naturalist of the American
Colonies. *Gard. J.*, 23(3):77–81(1973)—
3 Illustrations.

5614 Johnson, M. J.
Naturalists and plant explorers in Georgia, until
1850 (with illustrations of some native plants
3 observed). *Diss. Abstr. Int.*, *B*, 33(9):4150(1973).

CHARPENTIER, Jean de (1786–1855)

5615 Meyer, F. G. *and* Elsasser, S.
The 19th century Herbarium of Isaac C. Martindale.
1 *Taxon*, 22(4):375–404(1973).

CHEETHAM, Christopher Arthington (1875–1954)

5616 Shaw, G. A.
Some Yorkshire bryologists and bryophytes.
1 *Naturalist*, no.924:7–8(1973)—Portrait.

CHRISTENSEN, Carl (1872–1942)

5617 Stafleu, F. A.
Christensen and his *Index Filicum*. *Taxon*, 22(4):
1 465–466(1973).

CINQ-MARS, Lionel (1919–1973)

5618 Grandtner, M. M.
Lionel Cinq-Mars 12 juin 1919–6 août 1973. *Nat.*
3 *Canad.*, 100(5):521–523(1973)—Fr; portrait.

5619 Morton, J. K.
Dr. Lionel Cinq-Mars, an obituary. *Int. Organ. Pl.*
3 *Biosyst. Newsl.*, no.7:23(1973).

CIPRA, Václav (1892–1973)

5620 Jedlička, V.
Za reditelem Václavem Ciprou (1892–1973).
1 *Česká Mykol.*, 27(4):247(1973).

CLAVAUD, Armand (1828–1890)

5621 Schotsman, H. D.
Clavaud: sa vie son oeuvre. *Bull. Centre Etudes*
Rech. Sci. Biarritz, 8(4):708–791(1971 publ. 1972)—
1 Illustrations, portrait, bibliography.

CODD, Leslie Edward Wostall (1908–)

5622 Anon.
Dr. L. E. W. Codd. *Forum Bot.*, 11(11):81–83
5 (1973)—En.

COMAN, Artur (1881–1972)

5623 Ţopa, E. *and* Toma, C.
Artur Coman (15.iv.1881–6.viii.1972). *An. Şti.*
Univ. Al. I. Cusa Iaşi, Biol., 19(1):233–236(1973)—
1 Rum; portrait.

COMBER, Harold F. (1897–1969)

5624 Woodward, B.
H. F. Comber and plants of the Andes. *Quart. Bull.*
Alp. Gard. Soc., 41(2):98–121(1973)—Illustrations,
1, 4 portrait.

COMMERSON, Philibert (1727–1773)

5625 Bernardi, L.
Philibert Commerson 'Ab Humbertis' (1727–
1773) ou de l'intrépidité. *Mus. Genève*, 140:10–21
1 (1973)—Illustrations, portrait.

CONARD, Henry Shoemaker (1874–1971)

5626 Anon.
In memoriam Henry S. Conard (1874–1971).
3 *Proc. Iowa Acad. Sci.*, 79(2):96(1972)—En; portrait.

CONDRY, William

5627 Soan, H.
William Condry. *Nat. Wales*, 13(3):182–185
1 (1973)—En; portrait.

COOK, James (1728–1779)

5628 Hallé, N.
[Commemorative name *Captaincookia* (Rubiaceae)].
1 *Adansonia*, 13(2):197(1973).

COOKE, Randle Blair (1880–1973)

5629 Anon.
1 R. B. Cooke. *Vasculum*, 58(4):51(1973)—En.

COPERNICUS (1473–1543)

5630 Karpowiczowa, L.
Mikołaj Kopernik a botanika. *Wiad. Bot.*, 17(3):
1 169–172(1973)—Illustrations, portrait.

CORRELL, Donovan Stewart (1908–)

5631 D'Arcy, W. G.
[Commemorative name *Correlliana* (Myrsinaceae)]
3 *Ann. Missouri Bot. Gard.*, 60(2):442(1973).

5632 Powell, A. M.
[Commemorative name *Correllia* (Compositae)].
3 *Brittonia*, 25(2):116(1973).

CORREVON, Aymon

5633 Bellia, G. G.
Due amici scomparsi: Aymon Correvon ed E. Bertram Anderson. *Rea*, no.5:41–44(1972)—(En);
1 portrait.

5634 Page, X.
Aymon Correvon: hommage à un ami disparu. *Bull. Mens. Soc. Linn. Lyon*, 42(2):xxi–xxii(1973)—
1 Portrait.

CROSSLAND, Charles (1854–1917)

5635 Shaw, G. A.
Some Yorkshire bryologists and bryophytes.
1 *Naturalist*, no.924:2(1973)—Portrait.

CUMING, Hugh (1791–1865)

5636 Reinikka, M. A.
Hugh Cuming (1791–1865). *In* A history of the
1 Orchid: 141–144(1972)—Portrait.

CURTIS, Charles (1853–1928)

5637 Reinikka, M. A.
Charles Curtis (1853–1928). *In* A history of the
1, 6 Orchid: 266–268(1972).

CUTAK, Ladislaus (1908–1973)

5638 Anon.
Ladislaus Cutak. *Missouri Bot. Gard. Bull.*, 62(2):
3 2(1973)—En; portrait.

5639 Pring, G.
Lad Cutak, the horticulturist. *Missouri Bot. Gard.*
3 *Bull.*, 62(2):2–3(1973).

DAHNKE, Walter (1890–1972)

5640 Kintzel, W.
In memoriam Walter Dahnke (1890–1972).
1 *Mykol. Mitteilungsbl.*, 17(1):23–26(1973)—Portrait.

DARWIN, Charles Robert (1809–1882)

5641 Hull, D. L.
Darwin and his critics. The reception of Darwin's theory of evolution by the scientific community. Cambridge, Mass., Harvard University Press, 1973.
1 xii,473p.

5642 Williams, S. E.
A salute to Sir John Burdon-Sanderson and Mr. Charles Darwin on the centennial of the discovery of nerve-like activity in the Venus' Flytrap. *Carniv.*
1 *Pl. Newsl.*, 2(3):41–42(1973).

DARWIN, Erasmus (1731–1802)

5643 King-Hele, D.
Erasmus Darwin, master of many crafts. *Nature*,
1 247(5436):87–91(1973)—Portrait.

DAVALL, Edmund (1763–1798)

5644 Meyer, F. G. *and* Elsasser, S.
The 19th century Herbarium of Isaac C. Martindale.
1 *Taxon*, 22(4):375–404(1973).

DAVIS, Ralph

5645 Gall, I.
The work of Ralph Davis. *J. Bromeliad Soc.*, 23(5):
3 182–184(1973)—Portrait.

DAVIS, Walter (1847–1930)

5646 Reinikka, M. A.
Walter Davis (1847–1930). *In* A history of the
1 Orchid: 264–265(1972).

DEVOL, Charles E. (1903–)

5647 Huang, T.-C.
Dr Charles E. DeVol. *Taiwania*, 18(1):i–ii(1973)—
2 En; portrait; bibliography.

DIMOCK, Arthur Watson (1908–1972)

5648 Bateman, D. F.
Arthur Watson Dimock, 1908–1972. *Phyto-*
3 *pathology*, 63(8):945(1973)—En; portrait.

DIMOND, Albert E. (1914–1972)

5649 Rich, S. *and* Horsfall, J. G.
A. E. Dimond 1914–1972. *Phytopathology*, 63(6):
3 657(1973)—Portrait.

DIXIE, Ethel May (1876–1973)

5650 Anon.
Ethel May Dixie. *Forum Bot.*, 11(11):87–88(1973)—
5 En.

DOMIN, Karel (1882–1953)

5651 McGillivray, D. J.
Domin's *Beiträge zur Flora und Pflanzengeographie Australiens. Contrib. New S. Wales Nation. Herb.*,
1, 7 4(6):366–368(1973).

DOMINY, John (1816–1891)

5652 Reinikka, M. A.
John Dominy (1816–1891). *In* A history of the
1 Orchid: 196–198(1972).

DONK, Marinus Anton (1908–1972)

5653 Kotlaba, F.
Za dr M. A. Donkem (14.8.1908–2.9.1972). (Dr M. A. Donk in memoriam (14.8.1908–2.9.1972)).
1 *Česká Mykol.*, 27(2):121–122(1973)—Portrait.

5654 Maas Geesteranus, R. A.
Dr. M. A. Donk, his life and work. *Persoonia*,
1, 6 7(2):119–126(1973)—Portrait; bibliography.

5655 N., M.
Obituary: Dr. Marinus Anton Donk, 1908–1972.
1, 6 *Trans. Brit. Mycol. Soc.*, 61(1):201(1973).

5656 Singer, R.
Marinus Anton (1908–1972). *Mycologia*, 65(3):
1, 6 503–506(1973)—En; portrait.

DORE, William George (1912–)

5657 Anon.
New honorary members of the Ottawa Field-Naturalist's Club Dr. William G. Dore. *Canad.*
3 *Field-Nat.*, 87(1):77(1973).

DOSTÁL, Rudolf (1885–1973)

5658 Kolek, J.
Prof. PhDr MVDr. h.c. Rudolf Dostál, Dr Sc. zomrel. (Prof. PhDr MVDr. h.c. Rudolf Dostál has
1 died). *Biológia, A* (Czechoslovakia), 28(10):855(1973).

DOYLE, Conrad B. (c.1885–1973)

5659 Anon.
3, 4 [Conrad B. Doyle]. *Taxon*, 22(4):523(1973)—En.

DUCHESNE, Antoine Nicolas (1747–1827)

5660 Guédès, M.
Duchesne, Buisson, Durande, early followers of the natural method of the Jussieus. *Taxon*, 22(2–3):
1 211–219(1973).

DUERINCK, Jean-Baptiste (1809–1859)

5661 Robyns, A.
Les collections du Père Jean-Baptiste Duerinck S. J. conservées au Jardin botanique national de Belgique. *Bull. Jard. Bot. Nation. Belg.*, 43(1–2):15–16
1, 3 (1973).

DUGAND, Armando (1906–c.1972)

5662 Cabrera, M. E.
Armando Dugand. *Bol. Soc. Argent. Bot.*, 14(4):
4 378–379(1972)—Sp.

DUHAMEL du MONCEAU, Henry Louis (1700–1782)

5663 Pacher, J.
Die Bedeutung von Henry Louis Duhamel du Monceau für die Entwicklung der Forstbotanik. (The importance of Henry Louis Duhamel du Monceau for the development of the science of forest botany during the late 18th century). *Allg. Forst JagdZeitung*, 144(3):49–55(1973)—(En,Fr); biblio-
1 graphy.

DUNCAN, Ursula Katharine (1910–)

5664 Anon.
Dr. Ursula Katharine Duncan, M.A., LL.D., F.L.S.
1 *Biol. J. Linn. Soc.*, 5(4):369–370(1973)—En.

DURANDE, Jean François (1732–1794)

5665 Guédès, M.
Duchesne, Buisson, Durande, early followers of the natural method of the Jussieus. *Taxon*, 22(2–3):211–
1 219(1973).

DUTILLY, *Père* Arthème (1896–1973)

5666 Lepage, E.
Le père Arthème Dutilly 22 novembre 1896–31 janvier 1973. *Nat. Canad.*, 100(3):327–329(1973)—
3 Portrait.

DVOŘÁK, Josef (1913–)

5667 Smejkal, M.
Josef Dvořák—60 let. *Preslia*, 45(3):278–279(1973)—
1 Cz; portrait; bibliography.

EDISON, Thomas Alva

5668 Burt, L. S.
The Herbarium of Thomas Alva Edison. *Herbarist*,
3 no.39:8–11(1973).

EHRET, Georg Dionysius (1708–1770)

5669 Daniels, G. S. *and* Stafleu, F. A.
Georg Dionysius Ehret. *Taxon*, 22(2–3):292(1973)
1 (HI-IAPT Portraits of Botanists no.41)—En.

EICHHORN, André (1902–1972)

5670 Chardard, R.
André Eichhorn 1902–1972. *Rev. Gén. Bot.*, 80
1 (944–946):75–82(1973)—Fr; portrait.

5671 Hamel, J.-L.
André Eichhorn (17 février 1902–1er mai 1972). *Bull. Soc. Bot. France*, 119(5–6):405–414(1972 publ.
1 1973).

5672 Mangenot, G.
André Eichhorn (1902–1972). *Rev. Cytol. Biol. Vég.*, 35(3–4):331–343(1972 publ. 1973)—Fr; portrait,
1 bibliography.

ELSLEY, John E.

5673 Anon.
John Elsley appointed curator of hardy plants.
1, 3 *Missouri Bot. Gard. Bull.*, 62(8):3,5(1973)—Portrait.

ENDRES, Philippe (1806–1831)

5674 Meyer, F. G. *and* Elsasser, S.
The 19th century Herbarium of Isaac C. Martindale.
1 *Taxon*, 22(4):375–404(1973).

EPLING, Carl Clawson (1894–1968)

5675 Williams, L. O.
[Commemorative name *Eplingia* (Labiatae)].
3 *Fieldiana, Bot.*, 26(3):18(1973).

ERDTMAN, Otto Gunnar Elias (1897–1973)

5676 Anon.
1 [Gunnar Erdtman]. *Taxon*, 22(4):523(1973)—En.

5677 Anon.
Obituary: Professor Gunnar Erdtman. *Sci. Cult.*,
1 39(9):387–388(1973)—Portrait.

5678 Anon.
[Professor Gunnar Erdtman]. *Botanica* (Delhi),
1 23(2–3):63(1973)—En.

5679 Burkart, A.
Gunnar Erdtman su fallecimiento. *Darwiniana*,
1 18(1–2):281–282(1973).

5680 Faegri, K.
In memoriam O. Gunnar E. Erdtman 1897–1973.
1 *Pollen Spores*, 15(1):5–12(1973)—En; portrait.

5681 Gullvåg, B.
Gunnar Erdtman 28.11.1897–18.2.1973. *Grana*,
1 12(3):129–130(1972 publ. 1973)—En; portrait.

5682 Henderson, D.
Professor Gunnar Erdtman. *B.S.E. News*, no.10:
1 9–10(1973).

5683 Nilsson, S. *and* Praglowski, J.
In memoriam O.G.E. (Gunnar) Erdtman (1897–1973). *Rev. Palaeobot. Palynol.*, 15(1):1–2(1973)—En;
1 portrait.

ERNST, Wallace Roy (1928–1971)

5684 Thomas, J. H. *and* Shetler, S. G.
Wallace Roy Ernst, 1928–1971. *Madroño*, 22(4):
3 207–213(1973)—En; bibliography.

FALK, Peter

5685 Anon.
An article on P. Falk. *Rep. Rugby Sch. Nat. Hist. Soc.*,
1 10–11(1973).

FARWELL, Oliver Atkins (1867–1944)

5686 Wells, J. R. *and* Thompson, P. W.
Oliver Atkins Farwell—a reflection. *Mich. Bot.*,
3 12(1):69–73(1973)—Portrait.

FAWCETT, Benjamin (1808–1893)

5687 Desmond, R.
Benjamin Fawcett (1808–1893): printer and engraver of natural history books. *In* Geck, E. *and* Pressler, G., *eds*. Festschrift für Claus Nissen zum siebzigsten
1 Geburtstag 1971:91–135(1973)—Bibliography.

FAWDRY, Dallas (1932–1972)

5688 Collett, T. G.
Dallas Fawdry, 1932–1972. *London Nat.*, no.52:
1 130(1973)—En.

FEDCHENKO, Boris Alekseevich (1872–1947)

5689 Bobrov, E. G.
Pamyati B. A. Fedchenko (k100—letiyu so dnya rozhdeniya) (To the memory B. A. Fedtchenko (to the 100th birth anniversary)). *Bot. Zhurn.*, 58(5):
1 754–757(1973).

5690 Vasil'chenko, I. T.
Nauchnaya sessiya posvyashchennaya 100-letiyu so dnya rozhdeniya Borisa Alekseevicha Fedchenko (1872–1972). (Scientific session, dedicated to the 100th anniversary of B. A. Fedchenko (1872–1972)).
2 *Bot. Zhurn.*, 58(3):471(1973).

FORREST, George (1873–1932)

5691 Forrest, G.
George Forrest: 'The man'. *J. Scott. Rock Gard.*
1, 2 *Club*, 13(3):169–175(1973)—Portrait.

5692 Keenan, J.
George Forrest, 1873–1932. *J. Roy. Hort. Soc.*, 98(3): 112–117(1973)—En; illustrations, coloured illustra-
1 tions, portrait.

5693 Lancaster, R.
George Forrest of China. *Gard. Chron.*, 174(17):
1 30–33(1973)—Illustrations.

5694 Taylor, D.
The George Forrest adventure. *Amat. Gard.*,
1 4624:22–23(1973)—Illustration, portrait.

FOSTER, Adriance Sherwood (1901–1973)

5695 Anon.
[Adriance Sherwood Foster]. *Taxon*, 22(4):523–
3 524(1973)—En.

FOSTER, Mulford Bateman (1888–)

5696 Nally, J.
The men behind the scene—Mulford B. Foster. *J. Bromeliad Soc.*, 23(5):168–171(1973)—Illustration,
3 portrait.

FRAGNER, Petr (1923–)

5697 Herink, J.
K padesátinám Dr. Petra Fragnera. (Dr. Petr Fragner quinquagenarius). *Česká Mykol.*, 27(4):241–
1 247(1973)—Portrait, bibliography.

FRIES, Elias Magnus (1794–1878)

5698 Daniels, G. S. *and* Stafleu, F. A.
Elias Magnus Fries. *Taxon*, 22(4):374–(1973) (HI-
1 IAPT Portraits of Botanists no.42)—En.

FRIVALDSZKY VON FRIVALD, Emerich (1799–1870)

5699 Stanev, S.
Prouchvaniyata na E. Fon Frivaldski vŭrkhu Bŭlgarskata flora. (Studies of E. Fon Frivaldszky on Bulgarian flora). *Izv. Prir. Mus. Plovdiv*, 2:17–30
1 (1973)—(Rus,En); portrait, map.

FURBER, William

5700 Hazelwood, W. G.
Australia's first nurserymen: William Furber.
7 *Camellia News*, no.51:28(1973).

FUTÔ, S.

5701 Mizuno, M.
(Obituary of the late Dr. S. Futô). *J. Geobot.*,
2 20(4):93(1972)—Ja.

GANDOGER, Michel (1850–1926)

5702 McGillivray, D. J.
Michel Gandoger's names of Australian plants. *Contrib. New S. Wales Nation. Herb.*, 4(6):319–365
1, 7 (1973).

GAUDIN, Jean François (1766–1833)

5703 Meyer, F. G. *and* Elsasser, S.
The 19th century Herbarium of Isaac C. Martindale.
1 *Taxon*, 22(4):375–404(1973).

GAUSSEN, Henri Marcel (1891–)

5704 Anon.
Travaux de H. Gaussen. *Trav. Lab. Forest. Toulouse*,
1 tome bibliographie. 124p.(1971)—Portrait.

GAY, Jacques (1786–1864)

5705 Meyer, F. G. *and* Elsasser, S.
The 19th century Herbarium of Isaac C. Martindale.
1 *Taxon*, 22(4):375–404(1973).

GAYTHORNE-HARDY, Robert (1902–1973)

5706 Anon.
Hon. Robert Gaythorne-Hardy, B. A. *Biol. J.*
1 *Linn. Soc.*, 5(4):391(1973)—En. Short note.

GEORGI, Johann Gottlieb (1729–1802)

5707 Karavaev, M. N.
Zametki o gerbarii I. I. Georgi. (Animadversiones ex herbario J. G. Georgiano). *Novosti Sist. Vyssh.*
1 *Rast.*, 10:333–336(1973).

GIELSDORF, Karl (–1973)

5708 Anon.
Karl Gielsdorf. *Kakt. And. Sukk.*, 24(10):225
1 (1973)—Ge.

GILKEY, Helen Margaret (1886–1972)

5709 Anon.
Helen Margaret Gilkey. *Taxon*, 22(5–6):708
3 (1973)—En.

GODNEĬ, Tsikhan Mikalaevich (1893–)

5710 Shlyk, A. A. *and* Gancharyk, M. M.
Tsikhan Mikalaevich Hodneĭ [Godneĭ] (da 80-goddzya dnya naradzhénnya). *Vestsi Akad. Navuk BSSR, Biyal. Navuk*, no.2:123–126(1973)—Be;
1 portrait.

GOETHE, Johann Wolfgang von (1749–1832)

5711 Winter, L.
Le *Ginkgo biloba* L. et le poète Johann Wolfgang von Goetbe. *Jard. France*, no.7:13–14(1973)—
1 Illustration.

GOFF, Marvin E.

5712 Anon.
[Commemorative name × *Goffara* (Orchidaceae)].
3 *Orchid Rev.*, 81(961):centre page pull-out 8 (1973).

GOMES, Bernardino António (1768–1823)

5713 Fonseca, O. da
Bernardino António Gomes. *In* Gomes, B. A. Plantas medicinais do Brasil: 11–20(1972)—Sp;
4 portrait.

GORSKI, Stanisław Batys (1802–1864)

5714 Mowszowicz, J.
Stanisław Batys Gorski (1802–1864) w 170-lecie urodzin. *Wiad. Bot.*, 17(3):141–143(1973)—
1 Portrait.

GOTTLIEB, Heinrich (1880–1970)

5715 Koppe, F.
Heinrich Gottlieb 1880–1970. *Ber. Naturwiss. Ver. Bielefeld*, 21:265–267(1971–1972 publ. 1973)—
1 Portrait, bibliography.

GRAY, Asa (1810–1888)

5716 Meyer, F. G. *and* Elsasser, S.
The 19th century Herbarium of Isaac C. Martindale.
1 *Taxon*, 22(4):375–404(1973).

GREENE, Edward Lee (1843–1915)

5717 Crovello, T. J.
The computerization of the Edward Lee Greene Herbarium (NDG) at Notre Dame. *Proc. Indiana*
3 *Acad. Sci.*, 81:275(1971 publ. 1972). Abstract.

GREICHEN, Frank (c.1918–1973)

5718 Higgins, T. R.
In memoriam—Frank Greichen. *Amer. Orchid Soc.*
3 *Bull.*, 42(12):1070(1973).

GRIFFITH, William (1810–1845)

5719 Daniels, G. S. *and* Stafleu, F. A.
William Griffith. *Taxon*, 22(4):424(1973) (HI-IAPT
1, 6 Portraits of Botanists no.43)—En.

GROSE, Donald (c.1901–1973)

5720 S., R. E.
Donald Grose. *Wiltshire Archaeol. Nat. Hist. Mag.*,
1 *A*, 68:35–36(1973)—En; portrait.

GUILFOYLE, Michael (1809–1884)

5721 Hazelwood, W. G.
Australia's first nurserymen: Michael Guilfoyle.
7 1809–1884. *Camellia News*, no.51:31(1973).

GUILFOYLE, William Robert (1840–1912)

5722 Hazelwood, W. G.
Australia's first nurserymen: William Robert Guilfoyle. 1840–1912. *Camellia News*, no.51:31–32
7 (1973).

GUNNERUS, Johan Ernst (1718–1773)

5723 Engegard, G.
Biskop Gunnerus og '*Flora Norvegica*'. *Blyttia*,
1 31(1):3–15(1973).

HALÁSZ, Márta (1905–1971)

5724 Komáromy, P. Zs.
In memoriam Dr. Márta Halász (1905–1971). *Ann. Hist.-Nat. Mus. Nation. Hung.*, 65:15–17(1973)—Ge;
1 bibliography.

HARDING, Charles B. (1899–)

5725 [Portrait only]. *New York Bot. Gard. Ann. Rep.*,
3 *1971–1972*. [1]p. [1973].

HAYNE, Friedrich Gottlieb (1763–1832)

5726 Daniels, G. S. *and* Stafleu, F. A.
Friedrich Gottlieb Hayne. *Taxon*, 22(2–3):240
1 (1973) (HI-IAPT Portraits of Botanists no.40)—En.

HEINIS, Fritz (1883–1970) *

5727 Althaus, H., Plattner, W., Schmassmann, W. *and* Meyer, A.
Aus dem Leben von Dr. phil. Fritz Heinis (1883–1970). *Tätigkeitsber. Naturforsch. Ges. Baselland*, 28: 12–21(1972). Abstract in *Excerpta Bot.*, *A*, 22(2):128
1 (1973).

HELLER, Alfonse Henry

5728 Williams, L. O.
In memoriam—Alfonse Henry Heller. *Amer.*
3, 4 *Orchid Soc. Bull.*, 42(11):987(1973).

HENNIG, Bruno (c.1894–1972)

5729 Moser, M.
Zum Tode von Bruno Hennig. *Zeitschr. Pilzk.*,
1 38(1–4):180–182(1972)—Portrait.

HERRE, Hans

5730 Anon.
5 Mr. Hans Herre. *Forum Bot.*, 11(12):92(1973).

HODGKIN, Eliot (–1973)

5731 Synge, P. M.
Obituary: Eliot Hodgkin, V. M. H. *Quart. Bull.*
1 *Alp. Gard. Soc.*, 41(2):156–157(1973)—Portrait.

5732 HODNEĬ, Tsikhan Mikalaevich *see* GODNEĬ, Tsikan Mikalaevich

HOLGATE, Jeanne (1920–)

5733 Brindle, J. V., *comp.*
Jeanne Holgate: paintings and drawings June 4–September 14, 1973. Pittsburgh, Hunt Inst. Bot.
1, 3 Documentation, 1973. 16p.—Illustrations, portrait.

HOLMES, Edward Morell (1843–1930)

5734 Shellard, E. J.
E. M. Holmes (1843–1930)—Curator of the Materia Medica Museum of the Pharmaceutical Society (1872–1922). *Bot. J. Linn. Soc.*, 67(1):1–2
1 (1973).

HOLTTUM, Richard Eric (1895–)

5735 [Portrait only]. *J. Roy. Hort. Soc.*, 98(2) (1973).

HOOKER, *Sir* Joseph Dalton (1817–1911)

5736 [Portrait only]. *Arnoldia*, 33(2):156
1 (1973).

HOPPE, Paul Eric (1896–1972)

5737 Hanson, E. W. and Arny, D. C.
Paul Eric Hoppe, 1896–1972. *Phytopathology*,
3 63(8):946(1973)—En; portrait.

HOUGHTON, Douglass (1809–1845)

5738 Meyer, F. G. *and* Elsasser, S.
The 19th century Herbarium of Isaac C. Martindale.
3 *Taxon*, 22(4):375–404(1973).

HOUTTUYN, Martin (1720–1798)

5739 Steenis, C. G. G. J. van *and* Veldkamp, J. F.
Note on Houttuyn's Herbarium. *Blumea*, 21(1):
1 151–152(1973).

HUMBOLDT, Friedrich Heinrich Alexander von (1769–1859)

5740 Jahn, I. *
Über die Einwirkung Alexander von Humboldts auf die Entwicklung der Naturwissenschaften an der Berliner Universität. *Wiss. Zeitschr. Univ. Berlin, Math.-Nat. R.*, 21:131–144(1972). Abstract in *Excerpta*
1 *Bot., A*, 22(2):134(1973).

5741 Ogden, M. H.
A. von Humboldt. *Explorers J.*, 50:52–56(1972)—
1 Illustration, portrait.

HUMMELBERG, Gabriel (c.1490–1544)

5742 Burmeister, K. H.
Der Humanist und Botaniker Gabriel Hummelberg (ca 1490–1544). *In* Geck, E. *and* Pressler, G., *eds.* Festschrift für Claus Nissen zum siebzigsten
1 Geburtstag 1971: 43–71(1973)—Bibliography.

HUTCHINSON, John (1884–1972)

5743 Anon.
John Hutchinson 1884–1972. *Pl. Sci. Bull.*, 18(4):
1 43–44(1972)—En.

5744 Anon.
John Hutchinson, O.B.E., F.R.S., LL.D., V.M.H.
1 *Biol. J. Linn. Soc.*, 5(4):392(1973)—En. Short note.

5745 Bramwell, D.
Dr. John Hutchinson F.R.S., O.B.E., 1884–1972.
1 *Cuad. Bot. Canar.*, 17:3–4(1973)—En; bibliography.

5746 Brenan, J. P. M.
Dr. John Hutchinson (1884–1972). *Taxon*, 22(5–6):
1 705–706(1973).

5747 Gilbert, J. L.
John Hutchinson (1884–1972). *Watsonia*, 9(4):
1 408(1973)—En.

5748 Gunn, M. D.
John Hutchinson, O.B.E., F.R.S., LL.D., V.M.H., F.L.S. *Bothalia*, 11(1–2):1–3(1973)—En;
1 portrait.

5749 Hepper, F. N.
Dr. John Hutchinson. *J. Kew Guild*, 9(77):159
1 (1973)—En.

5750 Taylor, *Sir* G.
John Hutchinson. *J. Roy. Hort. Soc.*, 98(1):30–31
1 (1973)—En.

HYDE, Harold Augustus (c.1892–1973)

5751 Anon.
1 [H. A. Hyde]. *Taxon*, 22(4):524(1973).

5752 Anon.
Harold Augustus Hyde, M.A., Ph.D. *Biol. J. Linn.*
1 *Soc.*, 5(4):392(1973)—En. Short note.

5753 H., S. G.
Harold Augustus Hyde, M.A., Ph.D., F.L.S.,
1 F.M.A. *Mus. Ass. Mon. Bull.*, 13(2):19–20(1973).

INGEN-HOUSZ, Jan (1730–1799)

5754 Pas, P. W. van der
In Gillispie, C. C., *ed.* Dictionary of scientific
1 biography, 7:11–16(1973)—Bibliography.

INGHAM, William (1854–1923)

5755 Shaw, G. A.
Some Yorkshire bryologists and bryophytes.
1 *Naturalist*, no.924:2–3(1973).

INGRAM, Collingwood (c.1880–)

5756 Palmer, A.
Collingwood Ingram, F.L.S., V.M.H. *Int. Dendrol.*
1 *Soc. Year Book 1972*: 20–21(1973)—Portrait.

IRWIN, Howard Samuel (1928–)

5757 [Portrait only]. *New York Bot. Gard. Ann. Rep.*
3 *1971–1972*, [1]p.[1973].

IVANOĬ, Alyakseĭ Fyadosavich (c.1910–)

5758 Rakhtseenka, I. N. *and* Smolyak, L. P.
Alyakseĭ Fyadosavich Ivanoĭ (da 60-goddzya z dnya naradzhennya). *Vestsi Akad. Navuk BSSR, Biyal.*
1 *Navuk*, no.2:127–129(1973)—Be; portrait.

IVANOVSKY, Dmitri Iosifovich (1864–1920)

5759 Gutina, V.
In Gillispie, C. C., *ed.* Dictionary of scientific
1 biography, 7:34–36(1973)—Bibliography.

IVERSEN, Johannes (1904–1971)

5760 Godwin, H.
Johannes Iversen (26 December 1904 to 17 October 1971). *New Phytol.*, 72(5):1245–1246(1973)—
1 Portrait.

IWATSUKI, Kunio (1934–)

5761 Pichi Sermolli, R. E. G.
[Commemorative name *Kuniwatsukia* (Pterido-
2 phyta)]. *Webbia*, 28(2):455(1973).

JACQUIN, Nicolaus Joseph (1727–1817)

5762 Oberhummer, W.
In Gillispie, C. C., *ed.* Dictionary of scientific
1 biography, 7:57–59(1973)—Bibliography.

JEFFREY, Edward Charles (1866–1952)

5763 Wetmore, R. H.
In Gillispie, C. C. *ed.* Dictionary of scientific
3 biography, 7:90–91(1973)—Bibliography.

JENNINGS, Otto Emery (1877–1964)

5764 Pearth, D. L.
Bibliography, Otto Emery Jennings. *Ann. Carnegie*
3 *Mus. Pittsburgh*, 44(6):59–69(1973).

JEPSON, Willis Linn (1867–1946)

5765 Ewan, J.
In Gillispie, C. C., *ed.* Dictionary of scientific
3 biography, 7:102–103(1973)—Bibliography.

JESSEN, Knud (1884–1971)

5766 Godwin, H.
Knud Jessen (29 November 1884 to 14 April 1971). *New Phytol.*, 72(5):1247–1248(1973)—
1 Portrait.

JIMENEZ, Jose de Jesus (1905–)

5767 Anon.
[Commemorative name × *Jimenezara* (Orchidaceae)]. *Orchid Rev.*, 81(964): centre page pull-out 8
4 (1973).

JOHNSON, Thomas (c.1600–1644)

5768 Coombe, D. E.
In Gillispie, C. C., *ed.* Dictionary of scientific
1 biography, 7:146–148(1973)—Bibliography.

JOHNSTON, *Mrs.* Earl M. (–1973)

5769 Anon.
Death of Mrs. Earl M. Johnston. *Bull. Missouri*
3 *Bot. Gard.*, 62(1):6(1973)—Illustration.

JÓNASSON, Helgi (1887–1972)

5770 Kristinsson, H.
Helgi Jónasson frá Gvendarstödum. Sept. 26, 1887–April 13, 1972. *Acta Bot. Islandica*, 2:3–6(1973)—
1 Portrait, map, bibliography.

JORDAN, (Claude Thomas) Alexis (1814–1897)

5771 Piquemal, J.
In Gillispie, C. C., *ed.* Dictionary of scientific
1 biography, 7:165–167(1973)—Bibliography.

JUHÁSZ, Peter Melius (1536?–1572)

5772 Soo, R.
Peter Melius Juhász (1536?–1572). *Acta Bot. Acad.*
1 *Sci. Hung.*, 18(1–2):1–6(1973)—Ge(En).

JURATZKA, Jakob (1821–1878)

5773 Daniels, G. S. *and* Stafleu, F. A.
Jakob Juratzka. *Taxon*, 22(5–6):556(1973) (HI-
1 IAPT portraits of Botanists no.46)—En.

JUSSIEU, Adrien Henri Laurent de (1797–1853)

5774 Stafleu, F. A.
In Gillispie, C. C., *ed.* Dictionary of scientific
1 biography, 7:196–197(1973)—Bibliography.

JUSSIEU, Antoine de (1686–1758)

5775 Stafleu, F. A.
In Gillispie, C. C., *ed.* Dictionary of scientific
1 biography, 7:197–198(1973)—Bibliography.

JUSSIEU, Antoine-Laurent de (1748–1836)

5776 Stafleu, F. A.
In Gillispie, C. C., *ed.* Dictionary of scientific
1 biography, 7:198–199(1973)—Bibliography.

JUSSIEU, Bernard de (1699–1777)

5777 Stafleu, F. A.
In Gillispie, C. C., *ed.* Dictionary of scientific
1 biography, 7:199–200(1973)—Bibliography.

JUSTICE, James (1698–1763)

5778 Minay, P.
James Justice (1698–1763): eighteenth-century Scots horticulturalist and botanist: 1. *Gard. Hist.*,
1 1(2):41–62(1973)—Illustrations, map.

KAEMPFER, Engelbert (1651–1716)

5779 Pas, P. W. van der
In Gillispie, C. C., *ed.* Dictionary of scientific
1, 6 biography, 7:204–206(1973)—Bibliography.

KAISLER, Jaroslav (1913–)

5780 Morávek, A.
Jubilea členů ČSBS: Jaroslav Kaisler. *Preslia*, 45(4):
1 374(1973).

KARAVAEV, Mikhail Nikolaevich (1903–)

5781 Skryabin, S. Z. *and* Sheludyakova, V. A.
Mikhail Nikolaevich Karavaev (k 70-letiyu so dnya rozhdeniya) (M. N. Karavaev (on the 70th anniversary of his birth)). *Byull. Mosk. Obshch. Ispȳt.*
1 *Prir., Biol.*, 78(3):147–149(1973).

KÁRPATI, Zoltan (1909–1972)

5782 András, T.
Kárpati Zoltán emlékezete (1909–1972). *Bot.*
1 *Közlem.*, 60(3):145–148(1973)—Portrait.

KAUFMANN, Guido (1897–1972)

5783 Panzera, O.
Guido Kaufmann 1897–1972. *Boll. Soc. Ticin. Sci. Nat.*, 63:9–13(1972–73 publ. 1973)—It; portrait,
1 bibliography.

KAUSEL, Eberhard (1910–1972)

5784 Looser, G.
El botanico chileno Dr. Eberhard Kausel. *Bol. Soc.*
4 *Argent. Bot.*, 15(1):137–139(1973)—Portrait.

KELLOGG, Albert (1813–1887)

5785 Ewan, J.
In Gillispie, C. C., *ed.* Dictionary of scientific
3 biography, 7:285(1973)—Bibliography.

KELSEY, Seth Low (1906–1972)

5786 Gavitt, B.
Seth Low Kelsey. *Arbor. Bot. Gard. Bull.*, 7(2):
3 62–63(1973)—Portrait.

KEYSERLING, Alexander Andreevich

5787 Tikhomirov, V. V.
In Gillispie, C. C., *ed.* Dictionary of scientific
1 biography, 7:319–320(1973)—Bibliography.

KHOLODNȲĬ, Nikolai Gregor'evich (1882–)

5788 Manoĭlenko, K. V.
(N. G. Kholodny and evolutionary physiology of plants). *Ukr. Bot. Zhurn.*, 30(3):279–285(1973)—
1 Uk(En).

5789 Rubenchik, L. Ĭ.
(Some reminiscences about N. G. Kholodny).
1 *Ukr. Bot. Zhurn.*, 30(3):300–301(1973)—Uk(En).

5790 Velokon, I. P.
N. G. Kholodnȳĭ o S. G. Navashine. *Byull. Glavn.*
1 *Bot. Sada* (Moscow), 89:4(1973).

KIRK, Thomas (1828–1898)

5791 Moore, L. B.
Thomas Kirk, botanist. *Tuatara*, 20(2):51–56
1, 7 (1973).

KLEBS, Georg Albrecht (1857–1918)

5792 Bopp, M.
In Gillispie, C. C., *ed.* Dictionary of scientific
1 biography 7:395–396(1973)—Bibliography.

KLINE, David McKendree (1923–1972)

5793 Hebert, T. T. *and* Ellis, D. E.
David McKendree Kline, 1923–1972. *Phytopathology*,
3 63(12):1434(1973)—En; portrait.

KNIGHT, Thomas Andrew (1759–1838)

5794 Simpkins, D. M.
In Gillispie, C. C., *ed.* Dictionary of scientific
1 biography, 7:408–410(1973)—Bibliography.

KNUDSON, Lewis (1884–1958)

5795 Arditti, J.
El Profesor Lewis Knudson y la germinacion de semillas de Orquideas cincuentenario. *Orquidea*
3 (Mexico), 2(9):229–236(1972)—(En).

KNUTH, Paul Erich Otto Wilhelm (1854–1900)

5796 Pas, P. W. van der
In Gillispie, C. C., *ed.* Dictionary of scientific
1 biography, 7:417–418(1973)—Bibliography.

KODÁLY, Zoltán

5797 Borhidi, A. *and* Acuña, J.
[Commemorative name *Kodalyodendron* (Rutaceae)].
1 *Acta Bot. Acad. Sci. Hung.*, 19(1–4):42(1973).

KOELREUTER, Joseph Gottlieb (1733–1806)

5798 Olby, R.
In Gillispie, C. C., *ed.* Dictionary of scientific
1 biography, 7:440–442(1973)—Bibliography.

KOMAROV, Vladimir Leont'evich (1869–)

5799 U.S.S.R. Akademiya Nauk SSSR: Dal'nevostochnȳĭ Nauchnȳĭ Tsentr Biologo-Pochvennȳĭ Institut
Komarovskie chteniya vyp. 18. [Komarov's
1 lectures]. Vladivostok, 1971. 51p.—Portrait.

KOWNAS, Stefan (1898–)

5800 Głazek, T., Grinn, U. *and* Stachak, A.
Prof. Dr. Stefan Kownas zasłużony pedagog, naukowiec i społecznik. *Wiad. Bot.*, 17(1):3–7(1973)—
1 Pol; portrait.

KOYAMA, Tetsuo Micheal (1933–)

5801 Robinson, H. *and* Brettell, R. D.
[Commemorative name *Koyamacalia* (Compositae)].
2, 3 *Phytologia*, 27(4):270(1973).

KRÄNZLIN, Friedrich Wilhelm (1847–1934)

5802 [Portrait only]. *Taxon*, 22(2–3):233
1 (1973).

KRAEPELIN, Emil (1856–1926)

5803 Weidner, H.
Karl und Emil Kraepelins Studienreise nach Indien und Java im Jahr 1904. *Abh. Verh. Naturwiss. Ver.*
1, 6 *Hamburg*, 16:19–72(1972).

KRAEPELIN, Karl (1848–1915)

5804 Weidner, H.
Karl und Emil Kraepelins Studienreise nach Indien und Java im Jahr 1904. *Abh. Verh. Naturwiss. Ver.*
1, 6 *Hamburg*, 16:19–72(1972).

KUBIČKA, Jiřího (1913–)

5805 Svrček, M.
Kšedesátinám MUDr. Jiřího Kubičky. (MUDr. Georgius Kubička sexagenario ad salutum). *Česká*
1 *Mykol.*, 27(2):123–127(1973)—Portrait, bibliography.

KUBISCH, Fritz

5806 Goddard, W. E.
The men behind the scene—Fritz Kubisch. *J.*
1, 3 *Bromeliad Soc.*, 23(6):227–228(1973).

KÜTZING, Friedrich Traugott (1807–1893)

5807 Papenfuss, G. F.
In Gillispie, C. C., *ed.* Dictionary of scientific
1 biography, 7:533–534(1973)—Bibliography.

KRUPTSOV, Aleksandr Ivanovich (1900–)

5808 Ioganzen, V. G.
Aleksandr Ivanovich Kruptsov (k 70- letiyu so dnya rozhdeniya). *Rast. Resursȳ*, 9(2):313–318(1973)—
1 Rus; bibliography.

KYLIN, Johann Harald (1879–1949)

5809 Papenfuss, G. F.
In Gillispie, C. C., *ed.* Dictionary of scientific
1 biography, 7:534–536(1973)—Bibliography.

LA BROSSE, Guy de (c.1586–1641)

5810 Guerlac, H.
In Gillispie, C. C., *ed.* Dictionary of scientific
1 biography, 7:536–541(1973)—Bibliography.

LAFLIN, Tom (1914–1972)

5811 Clark, M. C.
Obituary: Tom Laflin. *Proc. Birmingham Nat.*
1 *Hist. Soc.*, 22(3):159–160(1973).

5812 Hawkes, J. G.
Tom Laflin. (1914–1972). *J. Bryol.*, 7(3):468–469
1 (1973)—En.

LAMARCK, Jean Baptiste Pierre Antoine de Monet de (1744–1829)

5813 Burlingame, L. J.
In Gillispie, C. C., *ed.* Dictionary of scientific
1 biography, 7:584–594(1973)—Bibliography.

LANKESTER, Charles Herbert (1879–1969)

5814 Northen, R. T.
Dedication of the Lankester Garden. *Amer. Orchid Soc. Bull.*, 42(7):609–614(1973)—Coloured illustra-
4 tions, portrait.

LAPEYROUSE, Philippe Picot de (1744–1818)

5815 Laissus, Y.
Philippe Picot de Lapeyrouse (1744–1818), d'après les documents conservés à la Bibliothèque du Muséum National d'Histoire Naturelle. *Bull. Soc. Hist. Nat. Toulouse*, 108(3–4):365–385(1972)—Portrait,
1 bibliography.

LARIN, Ivan Vasil'evich (1889–1972)

5816 Levina, F. Ya.
Pamyati Ivana Vasil'evicha Larina (10 vi 1889 n. st. – 4 v 1972). (To the memory of Ivan Vasiljevich Larin (10 VI 1889 – 4 V 1972)). *Bot. Zhurn.*, 58(6):931–
2 936(1973)—Portrait.

LAUTERBACH, Charles Adolph George (1864–1937)

5817 Rostański, K. *and* Weretelnik, E.
Catalogus plantarum quae in herbario Instituti Botanici Universitatis Wratislaviensis asservantur: 3. Herbarium Lauterbach pars 1 (The Lauterbach's herbarium part 1). *Prace Bot.* (Wrocław), 16:77–90
1 (1973 publ. 1972).

LEANDRO DO SACRAMENTO, *Brother* (1778–1829)

5818 Triaini, H. *and* Campos, V. R. de Paiva
Frei Leandro do Sacramento: 1778–1829. *Leandra*,
4 2(2):5–9(1972)—Por.

LEEDALE Gordon Frank

5819 [Cave A. J. E.]
Dr Gordon Frank Leedale. *Biol. J. Linn. Soc.*, 4(4):
1 339(1972 publ. 1973)—En.

LEES, Frederic Arnold (1847–1921)

5820 Seaward, M. R. D.
F. A. Lees' botanical collections: part 4.
1 *Naturalist*, no.924:35–36(1973).

LESCHKE, Hans (1883–1973)

5821 Anon.
1, 3 Hans Leschke. *Taxon*, 22(5–6):708(1973)—En.

L'ESCLUSE, Jules Charles (1526–1609)

5822 Traxler, G.
Die burgenländischen Pflanzenstandorte bei Carolus Clusius. *Burgenländ. Heimatbl.*, 35(2):49–59
1 (1973).

LID, Dagny Tande (1903–)

5823 Löve, A.
Art that immortalizes flowers. *Taxon*, 22(2–3):
1 279–280(1973).

LINDEN, Jean (1817–1898)

5824 Reinikka, M. A.
Jean Linden (1817–1898). *In* A history of the
1 Orchid: 204–207(1972)—Illustration, portrait.

LINDLEY, John (1799–1865)

5825 [Portrait only] *Taxon*, 22(2–3):230
1 (1973).

LINNÉ, Carl von (1707–1778)

5826 Åsberg, M. *and* Stearn, W. T.
Linnaeus's Öland and Gotland journey 1741; translated from the Swedish edition 1745. *Biol. J. Linn. Soc.*, 5(1):1–107(1973)—Map. Introduction by
1 W. T. Stearn pp.1–17.

5827 Åsberg, M. *and* Stearn, W. T.
Linnaeus's Öland and Gotland journey 1741; translated from the Swedish edition 1745. *Biol. J.*
1 *Linn. Soc.*, 5(2):108–220,viii(1973)—Maps.

5828 Hunt Botanical Library
A Linnaean keepsake: issued to commemorate the opening of the Strandell collection of Linnaeana at the Hunt Botanical Library Carnegie-Mellon University. Pittsburgh, Hunt Botanical Library, 1973. 70p.—
1 Illustrations, portraits.

5829 Kukkonen, I. *and* Viljamaa, K.
Linnaean specimens in the Botanical Museum of the University of Helsinki. *Ann. Bot. Fenn.*, 10(4):309–
1 336(1973)—Illustrations.

5830 Minelli, A.
Un'escursione con il prof. Carlo Linneo. *Nat.*
1 *Montagna*, 20(1):53–60(1973).

5831 Strandell, B.
Ett par tidiga biografier ïver Linné och Rosén. (Two early biographies of Linnaeus and Rosén).
1 Uppsala, Almqvist & Wiksell, 1973. 8p.—(En).

5832 Sydow, C. von, *ed.*
[Whole issue devoted to Linnaeus]. *Svenska Linné-Sällsk. Årsskr.*, 1970–1971, 138p.(1973)—Sw(En);
1 illustrations, coloured illustration.

LOBB, Thomas (1820–1894)

5833 Reinikka, M. A.
Thomas Lobb (1820–1894). *In* A history of the
1, 6 Orchid: 208–210(1972).

LOBB, William (1809–1864)

5834 Ewan, J.
William Lobb, plant hunter for Veitch and messenger of the Big Tree. *Univ. Calif. Publ. Bot.*,
1 67. 40p.(1973)—Illustration.

LÜCKHOFF, James (1875–1973)

5835 Anon.
5 Dr. James Lückhoff. *Forum Bot.*, 11(12):93(1973).

LYUBIMENKO, Volodimir Mikolaïovich (1873–1937)

5836 Kotov, M. I. *and* Sud'ina, E. G.
Zhȳttya diyal'nist V. M. Lyubȳmenka ta rozvȳtok ïoho ideï u suchasniï nautsi (do 100-richchya vid dnya narodzhennya). *Ukr. Bot. Zhurn.*, 30(6):787–790
1 (1973)—Uk(En).

McCLINTOCK, David

5837 [Portrait only]. *Watsonia*, 9(4):421(1973)—
1 Portrait.

McCLURE, Floyd A. (1897–1970)

5838 Calderón, C. E. *and* Soderstrom, T. R.
[Commemorative name *Maclurolyra* (Gramineae)].
3, 4 *Smithsonian Contrib. Bot.*, 11:6(1973).

MACDOUGALL, Thomas Baille (c.1898–1973)

5839 Anon.
Tom Macdougall. *Cact. Succ. J.* (U.S.A.), 45(2):
3, 4 55(1973)—En.

5840 Bravo H., H.
Thomas Macdougall. *Orquídea* (Mexico), 3(6): 169–172(1973)—Sp,En. Reprinted from *Macpalxochitl*,
1, 4 47:6(1973).

5841 Richardson, D.
In memoriam—Thomas Baille MacDougall. *Amer.*
1, 4 *Orchid Soc. Bull.*, 42(7):626(1973).

5842 Ziesenhenne, R.
Thomas MacDougall. *Begonian*, 40(6):126–127
3 (1973)—En.

MAGNOL, Pierre (1638–1715)

5843 Stearn, W. T.
Magnol's *Botanicum monspeliense* and Linnaeus's *Flora monspeliensis*. *In* Geck, E. *and* Pressler, G., *eds.* Festschrift für Claus Nissen zum siebzigsten Geburt-
1 stag 1971: 612–650(1973)—Illustration, portrait.

MANDENOVA, Ida P. (1907–)

5844 Alava, R.
[Commemorative name *Mandenovia* (Umbelliferae)]. *Notes Roy. Bot. Gard. Edinburgh*, 32(2):191
1 (1973).

MARIÉTAN, I. (1882–1971)

5845 Chastellain, F. *
A la mémoire de l'abbé I. Mariétan (1882–1971). *Bull. Murith.*, 88:3–7(1971). Abstract in *Excerpta Bot.*,
1 *A*, 22(2):130(1973).

MARSCHALL VON BIEBERSTEIN, Friedrich August *Freiherr* (1768–1826)

5846 Stafleu, F. A.
Marschall von Bieberstein and his *Flora Taurico-*
1, 2 *Caucasica. Taxon*, 22(1):126–128(1973).

MARTIN, George Willard (1886–1971)

5847 Hulbary, R. L.
George Willard Martin 1886–1971. *Pl. Sci. Bull.*,
3 18(4):44(1972)—En.

5848 Wells, K. *and* Lentz, P. L.
George Willard Martin (1886–1971). *Mycologia*,
3 65(5):985–998(1973)—En; portrait.

MARTINDALE, Isaac Comly (1842–1893)

5849 Meyer, F. G. *and* Elsasser, S.
The 19th century Herbarium of Isaac C. Martindale. *Taxon*, 22(4):375–404(1973)—Illustrations, portrait,
3 bibliography.

MARTINEZ, Argentino (1914–1972)

5850 Wright, J. E.
In memoriam Dr. Argentino Martinez 3.6.1914–11.2.1972. *Bol. Soc. Argent. Bot.*, 14(4):375–378
4 (1972)—Sp; portrait, bibliography.

MARTINOVSKÝ, Jan Otakar (1903–)

5851 Morávek, A.
Jubilea členů ČSBS: RNDR. Jan Otakar
1 Martinovský. *Preslia*, 45(4):375(1973).

MARTIUS, Carl Friedrich Philipp von (1794–1868)

5852 Campos, P. M. *
Um naturalista e a história. *Rev. Hist.*, 43:241–248(1971). Abstract in *Excerpta Bot., A*, 22(2):130
1 (1973).

MASON, William (1724–1797)

5853 Batey, M.
William Mason, English gardener. *Gard. Hist.*,
1 1(2):11–25(1973)—Illustrations, portrait.

5854 [Lemmon, K.]
William Mason 1724–1797. *North. Gard.*, 27(4):
1 101–103,107(1973)—En; illustration.

MATOUŠEK, Václav (1891–)

5855 Stepán, J.
Vlastivědný pracovník Václav Matoušek. (Heimatkundlicher Forscher Václav Matoušek). *Zprávy Česk. Bot. Spol.*, 8(1):75–76(1973)—
1 Bibliography.

MATUDA, Eizi (1894–)

5856 King, R. M. *and* Robinson, H.
[Commemorative name *Matudina* (Compositae)].
2, 4 *Phytologia*, 26(3):171(1973).

MAYER, Johann (1784–1807)

5857 Eisnerova, V. *
[Zur botanischen Tätigkeit Johann Mayers.] *Comm. Czech. Soc. Hist. Sci. Technol.*, 1:20–24(1971)—Cz.
1 Abstract in *Excerpta Bot., A*, 22(2):131(1973).

MENDEL, Gregor (1822–1884)

5858 Musil, R., *ed.*
Papers relating to Mendel and to the early development of genetics. *Folia Mendeliana*, 7. 53p.
1 (1972)—Ge,En; portraits.

MENDES, Eduardo José Santos Moreira (1924–)

5859 Gomes e Sousa, A.
Exploradores e naturalistas da flora de Moçambique. *Bol. Soc. Estud. Moçambique*, 40(168–169):93–97
1, 5 (1971)—Portrait, bibliography.

MESSERI, Albina (1904–1972)

5860 Corti, E. F.
Albina Messeri 25 Agosto 1904–19 Gennaio 1972. *Inform. Bot. Ital.*, 5(1):40–53(1973)—Portrait; biblio-
1 graphy.

MEYER, Teodoro (1910–1972)

5861 Cabrera, A. L.
Teodoro Meyer 1910–1972. *Bol. Soc. Argent. Bot.*,
4 14(4):374–375(1972)—Sp; portrait.

MICHOLITZ, Wilhelm (1854–1932)

5862 Reinikka, M. A.
Wilhelm Micholitz (1854–1953). *In* A history of the
1 Orchid: 271–274(1972)—Illustration.

MILL, John Stuart (1806–1873)

5863 Meyer, F. G. *and* Elsasser, S.
The 19th century Herbarium of Isaac C. Martindale.
1 *Taxon*, 22(4):375–404(1973).

MILLER, Philip (1691–1771)

5864 Squire, A. W. L.
Qui était Mill.? *Cactus* (Brussels), 4(6):121–122
1 (1972).

MISRA, Jo Ann K.

5865 Anon.
[Commemorative name × Joannara (Orchidaceae)].
6 *Orchid Rev.*, 81(964): centre page pull-out 8(1973).

MIZUSHIMA, M.

5866 Satomi, N.
(Obituary of the late Dr. M. Mizushima). *J. Geobot.*,
2 20(3):82–83(1972)—Ja; portrait.

MOBBS, Eric Charles

5867 C., H. G.
Obituary: Professor Eric Charles Mobbs, C.B.E.
1, 6 *Commonw. Forest Rev.*, 52(1):12–13(1973).

MONTARNAL, Pierre (1897–1971)

5868 Heim, R.
Pierre Montarnal (1897–1971) un esprit, un coeur.
1 *Rev. Mycol.*, 37(1–2):111–116(1973)—Portrait.

MOON, Henry George (1857–1905)

5869 Reinikka, M. A.
Henry George Moon (1857–1905). *In* A history of
1 the Orchid: 283–286(1972).

MOORE, Lucy Beatrice (1906–)

5870 [Portrait only]. *In* New Zealand: Department of Scientific and Industrial Research: Botany Division.
7 1969–71 triennial report (1973?).

MORISON, Robert (1620–1683)

5871 Hedge, I. C.
Umbelliferae in 1672 and 1972. *Notes Roy. Bot.*
1 *Gard. Edinburgh*, 32(2):151–158(1973).

MORITZ, Wolfram (1942–1973)

5872 Anon.
In memoriam. Dr. Wolfram Moritz, 15.2.1942–3.2.1973. *Kulturpflanze*, 21:331(1973)—Portrait.
1 Portrait only.

MORREN, Charles Jacques Edouard (1833–1886)

5873 Padilla, V.
The legacy of C. Jacques Edouard Morren. *J. Bromeliad Soc.*, 22(4):161–163(1972)—Coloured
1 illustrations, portrait.

MORTON, Conrad V. (1905–1972)

5874 Anon.
Conrad V. Morton, 1905–1972. *Madroño*, 22(1):
3 31(1973)—En.

5875 Anon.
Conrad Vernon Morton. *Biota*, 9(73):257(1972)—
3 Sp; portrait.

5876 Burkart, A.
Conrad Vernon Morton su fallecimiento.
3 *Darwiniana*, 18(1–2):281(1973).

5877 Ewan, J.
Conrad Vernon Morton 1905–1972. *Taxon*, 22
3 (2–3):271–274(1973)—En.

5878 Lellinger, D. B.
Conrad Vernon Morton (1905–1972). *Amer. Fern J.*,
3 63(3):49–60(1973)—En; portrait, bibliography.

5879 Rosenbaum, I.
In memoriam: Conrad Vernon Morton. *Gloxinian*,
3 23(1):3(1973)—En.

MOTHES, Kurt A. (1900–)

5880 Jacob, F. *
Kurt Mothes—70 Jahre. *Hercynia*, 8:233–234
1 (1971). Abstract in *Excerpta Bot., A*, 22(2):134(1973).

MRKOS, Otto (c.1903–)

5881 Bednářová, J.
Prof. RNDR. Otto Mrkos—70 let. *Preslia*, 45(4):
1 369–373(1973)—Cz; portrait, bibliography.

MUNRO SMITH, Denys D. (1890–1971)

5882 Willis, A. J.
Denys D. Munro Smith. *Watsonia*, 9(4):408–409
1 (1973)—En.

MURITH, Laurent Joseph (1742–1818)

5883 Meyer, F. G. *and* Elsasser, S.
The 19th century Herbarium of Isaac C. Martindale.
1 *Taxon*, 22(4):376–404(1973).

NAGEL, Otto (1894–1972)

5884 Halbinger, F.
In memoriam—Otto Nagel. *Amer. Orchid Soc.*
1, 4 *Bull.*, 42(9):784(1973)—En.

NAPPER, Diana Margaret (1930–1972)

5885 Ballard, F.
Diana Margaret Napper. *J. Kew Guild*, 9(77):
1, 5 165–166(1973)—En; portrait.

5886 Verdcourt, B.
Miss D. M. Napper. *Kew Bull.*, 28(1):1–4(1973)—
1, 5 Portrait; bibliography.

NARSIMHAN, M. J. (1891–1970)

5887 Anon.
Narsimhan, M. J. 1891–1970. *Planto*, 2(1):92
6 (1972)—En.

NAVASHIN, Sergeĭ Gavrilovich (1857–1930)

5888 Kholodnȳĭ, N. G.
Chemu uchil S. G. Navashin kak estestvoispȳtatel.
1 *Byull. Glavn. Bot. Sada* (Moscow), 89:5–10(1973).

5889 Tsitsin, N. V.
K 75-letiyu otkrȳtiya S. G. Navashinȳm dvoĭnogo oplodotvoreniya u pokrȳtosemennȳkh rasteniĭ. *Byull. Glavn. Bot. Sada* (Moscow), 89:3(1973)—
1 Portrait.

5890 Velokon, I. P.
N. G. Kholodnȳi o S. G. Navashine. *Byull. Glavn.*
1 *Bot. Sada* (Moscow), 89:4(1973).

NECKER, Noel Martin Joseph de (1730–1793)

5891 Daniels, G. S. *and* Stafleu, F. A.
Noel Martin Joseph de Necker. *Taxon*, 22(1):56 (1973) (HI-IAPT Portraits of Botanists no. 34)—
1 En.

NEEDHAM, James (1849–1913)

5892 Shaw, G. A.
Some Yorkshire bryologists and bryophytes.
1 *Naturalist*, no.924:4(1973)—Portrait.

NELSON, E. W. (1855–1934)

5893 Robinson, H. *and* Brettell, R. D.
[Commemorative name *Nelsonianthus* (Compo-
3 sitae)]. *Phytologia*, 27(1):54(1973).

NĚMEC, Bohumil Řehoř (1873–1966)

5894 Blattný, C.
Fragments of reminiscences on Academician
1 Bohumil Němec. *Biol. Pl.*, 15(1):12–16(1973).

5895 Dostál, R.
On the occasion of the 100th anniversary of the birth of Academician Bohumil Němec. *Biol. Pl.*,
1 15(1):9–11(1973).

5896 Kloz, J.
Die Jugend Bohumil Němec's. *Biol. Pl.*, 15(1):
1 3–8(1973)—Portrait.

5897 Kloz, J.
Einige Gedanken von Bohumil Němec. *Biol. Pl.*,
1 15(1):1–2(1973)—Portraits.

5898 Luxová, M.
100 výročie narodenín akademika Bohumila Němca. (Centenary of the birth of the academician Bohumil Němec). *Biológia, A*, (Czechoslovakia),
1 28(4):225–226(1973)—Portrait.

5899 Prát, S.
100 anniversary of the birth of Professor Němec.
1 *Biol. Pl.*, 15(1). [7]p.(1973)—Portraits.

5900 Sekerka, V.
Niektoré všeobecnobiologicke aspekty v diele Bohumila Němca (prispevok k 100. výročiu narodenia). (Some general aspects on biology in the work of Bohumil Němec (a contribution to the centenary of his birth)). *Biológia, A*, (Czechoslovakia), 28(10):
1 845–849(1973).

NISSEN, Claus (1901–)

5901 Geck, E. *and* Pressler, G., *eds*
Festschrift für Claus Nissen zum siebzigsten
Geburtstag 2. September 1971. Wiesbaden, G.
1 Pressler, 1973. 686p.

NITSCH, Jean (1921–1971)

5902 Maresquelle, H. J.
Jean Nitsch. *Bull. Soc. Bot. France, Mém.*, 11–22
1 (1973)—Fr; portrait, bibliography.

NOVÁK, František A. (1892–1964)

5903 Stary, F.
K nedožitým osmdesátinám univ. prof. PhDr.
Frant. A. Nováka. *Preslia*, 45(2):183–185(1973)—
1 Bibliography from 1965.

NOVOTNÝ, Čeněk (1903–1972)

5904 Houfek, J. *and* Slavík, B.
Dr Čeněk Novotný zemřel. *Preslia*, 45(2):186–190
1 (1973)—Portrait, bibliography.

NUTTALL, Thomas (1786–1859)

5905 Daniels, G. S. *and* Stafleu, F. A.
Thomas Nuttall. *Taxon*, 22(1):92(1973) (HI-IAPT
1, 3 Portraits of Botanists no. 34)—En.

OBERMEYER-MAUVE, Anna Amelia (1907–)

5906 Anon.
Anna Amelia Obermeyer-Mauve. *Flow. Pl. Afr.*,
5 42(3–4). 2p.(1973)—En; portrait.

OEDER, Georg Christian von (1728–1791)

5907 Daniels, G. S. *and* Stafleu, F. A.
Georg Christian von Oeder. *Taxon*, 22(1):164
(1973) (HI-IAPT Portraits of Botanists no. 34)—
1 En.

OESCU, Constantin V. (1891–1973)

5908 Cărăuşu, D., Zolyneak, C. *and* Constantinescu, N. M.
Un înaintaş al hibridologiei ovezelor (*Avena*) în
România: Professor Constantin V. Oescu (1891–1973).
An. Şti. Univ. Al. I. Cuza Iaşi, Biol., 19(1):227–232
1 (1973)—Portrait.

5909 OGOLEVETS, Georgiĭ Stepanovich (1897–1973)

Tsitsin, N. V., *and others*
Pamyati Georgiya Stepanovicha Ogolevtsa
(1897–1973). *Byull. Glavn. Bot. Sada* (Moscow), 90:
2 78–79(1973)—Portrait.

OSBECK, Pehr (1723–1805)

5910 Hansen, C. *and* Maule, A. F.
Pehr Osbeck's collections and Linnaeus's *Species
Plantarum* (1753). *Bot. J. Linn. Soc.*, 67(3):189–212
1 (1973)—Illustrations.

OSBORN, Theodore George Bentley (1887–1973)

5911 Anon.
Obituary: T. G. B. Osborn. *Nature*, 244(5415):
1, 7 377(1973).

5912 Anon.
1, 7 Obituary. *S. Austral. Nat.*, 48(1):9(1973)—En.

5913 Robertson, R. N. *and* Eardley, C. M.
Theodore George Bentley Osborn, D.SC., M.A.,
F.L.S. 2.10.1887–3.6.1973. *Trans. Roy. Soc. S.
1, 7 Austral.*, 97(4):316–320(1973)—En; portrait.

OSVALD, Hugo (1892–1970)

5914 Godwin, H.
Hugo Osvald (21 November 1892 to 2 November
1970). *New Phytol.*, 72(5):1246–1247(1973)—
1 Portrait.

OVCHINNIKOV, Pavel Nikolaevich (1903–)

5915 Asimov, M. S., Maksumov, A. N., Narzikulov,
M. N. *and* Rasulova, M. R.
Pavel Nikolaevich Ovchinnikov (k letiyu so dnya
rozhdeniya). *Izv. Akad. Nauk Tadzh. SSR, Biol. Nauk*,
2 1(50):3–8(1973)—Rus; portrait.

5916 Korovin, E. P.
[Commemorative name *Paulia* (Umbelliferae)].
Izv. Akad. Nauk Tadzh. SSR, Biol. Nauk, 1(50):
2 14(1973).

PABST, Guido F. J. (1914–)

5917 Garay, L. A.
[Commemorative name *Pabstia* (Orchidaceae)].
4 *Bradea*, 1(27):306(1973).

PAL, Benjamin Peary (1906–)

5918 Kallarackal, J.
Great biologists: B. P. Pal. *Botanica* (Delhi),
6 23(2–3):61–63(1973)—Portrait.

PALLAS, Peter Simon (1741–1811)

5919 Daniels, G. S. *and* Stafleu, F. A.
Peter Simon Pallas. *Taxon*, 22(4):438(1973) (HI-
1 IAPT Portraits of Botanists no. 44)—En.

PALMITER, De Forest Harold (1904–1972)

5920 Braun, A. J. *and* Szkolnik, M.
De Forest H. Palmiter, 1904–1972. *Phytopathology*,
3 63(12):1433(1973)—En; portrait.

PAPP, Constantin (1896–1972)

5921 Mihai, G.
Prof. Dr. Doc. Constantin Papp 1896–1972. *An.
Şti. Univ. Al. I. Cuza Iaşi, Biol.*, 19(2):479–481(1973)—
1 Rum; portrait.

5922 Ştefureac, T. I.
Profesorul şi briologul Constantin Papp. (1.1.1896–
17.8.1972). *Stud. Circ. Biol., Bot.*, 25(5):467–472
1 (1973)—Portrait.

PARKER, Charles F. (1820–1883)

5923 Meyer, F. G. *and* Elsasser, S.
The 19th century Herbarium of Isaac C. Martindale.
3 *Taxon*, 22(4):375–404(1973).

PARKMAN, Francis (1823–1893)

5924 Whitehill, W. M.
Francis Parkman as horticulturist. *Arnoldia*, 33(3):
3 169–183(1973)—Illustration, portraits.

PARRY, William Edward (1790–1855)

5925 Meyer, F. G. *and* Elsasser, S.
The 19th century Herbarium of Isaac C. Martindale.
1 *Taxon*, 22(4):375–404(1973).

PASTÝRIK, L'udovít (c.1913–)

5926 Sekerka, V. *and* Erdelský, K.
Prof. RNDr. L'udovít Pastýrik, DR.SC. *Acta Fac.
Rerum Nat. Univ. Comen. Physiol. Pl.*, 6:1–12(1973)—
1 En; portrait, bibliography.

PAULLI, Simon (1603–1680)

5927 Cockx-Indestege, E.
De *Flora Danica* von Simon Paulli: bijdrage van Balthasar Moretus tot uitgawe 1647–1648. *In* Geck, E. *and* Pressler, G., *eds*. Festschrift für Claus Nissen zum siebzigsten Geburtstag 1971:72–90(1973)—
1 Bibliography.

PAWŁOWSKI, Bogumił (1898–1971)

5928 Zarzycki, K., Guzik, J. *and* Tacik, T.
Bogumił Pawłowski (1898–1971) i jego dorobek w zakresie systematyki roślin. (Bogumił Pawłowski (1898–1971) and his scientific work in the scope of plant taxonomy). *Acta Soc. Bot. Pol.*, 42(1):3–20
1 (1973)—(En); portrait; bibliography.

PAYLEN, William R.

5929 Anon.
Introducing our President. *J. Bromeliad Soc.*, 23(6):
3, 6 207(1973)—Portrait.

5930 PEARCE, Sidney A. (1906–1972)
Brown, G. E.
Sidney A. Pearce. *J. Kew Guild*, 9(77):150–151
1 (1973)—En.

PEMBERTON, Joseph Hardwick (1852–1926)

5931 Anon.
Rev. Joseph Hardwick Pemberton. *Rose Bull.*,
1 32–33(1973)—En.

PERCY-LANCASTER, Sydney (1886–1972)

5932 Anon.
Sydney Percy-Lancaster. *Biol. J. Linn. Soc.*, 5(4):
6 392(1973)—En. Short note.

PERRY, Ralph F.

5933 Spongberg, H. C.
The Ralph F. Perry wood collection. *Arnoldia*,
3 33(4):231–234(1973)—Portrait.

PFITZER, Ernst Hugo Heinrich (1846–1906)

5934 Reinikka, M. A.
Ernst Hugo Heinrich Pfitzer (1846–1906). *In* A
1 history of the Orchid: 255–257(1972)—Portrait.

PIANE, Victor (1900–1972)

5935 M., J.
Victor Piane (1900–1972). *Bull. Trim. Soc. Mycol.*
1 *France*, 89(3–4):225–227(1973)—Fr; portrait.

PILÁT, Albert (1903–)

5936 Herink, J.
Sedmdesát let doc. dr. Alberta Piláta, Dr. Sc., člena korespondenta ČSAV. (Ad septuagesimum diem natalem Alberti Pilatii). *Česk á Mykol.*, 27(4):193–
1 200(1973)—Portrait; bibliography.

5937 Neuhäuslová-Novotna, Z.
Člen korespondent ČSAV Albert Pilát—70 let.
1 *Preslia*, 45(4):366–368(1973)—Bibliography.

5938 Zerova, M. Ya. *and* Vasser, S. P.
Al'bertu Pilatu—70 rokiv. (Albert Pilat is 70 years
1 old). *Ukr. Bot. Zhurn.*, 30(4):532–533(1973)—Uk.

PITCHER, Zina (1797–1872)

5939 Meyer, F. G. *and* Elsasser, S.
The 19th century Herbarium of Isaac C. Martindale.
3 *Taxon*, 22(4):375–404(1973).

PITTIER, Henri (1857–1950)

5940 Dwyer, J. D.
Henri Pittier's botanical activity in Panama. *Taxon*,
1, 4 22(5–6):557–576(1973).

POEPPIG, Eduard Friedrich (1796–1868)

5941 Daniels, G. S. *and* Stafleu, F. A.
Eduard Friedrich Poeppig. *Taxon*, 22(4):444(1973)
1, 4 (HI-IAPT Portraits of Botanists no.45)—En.

POIRET, Jean Louis Marie (1755–1834)

5942 Daniels, G. S. *and* Stafleu, F. A.
Jean Louis Marie Poiret. *Taxon*, 22(5–6):590
1 (1973) (HI-IAPT Portraits of Botanists no.4)—En.

PRANCE, Ghillean Tolmie (1937–)

5943 Anon.
Biographical sketch. *In* Prance, G. T., Flora neotropica monograph no.11. Rhabdodendraceae: 23
1, 3 (1972)—Portrait.

PRINGLE, Cyrus Guernsey (1838–1911)

5944 Seymour, F. C.
The Pringle Herbarium. *Rhodora*, 75(802):297–303
3 (1973).

PRINGSHEIM, Ernst Georg (1881–1970)

5945 Pirson, A.
Ernst Georg Pringsheim 1881–1970. *Ber. Deutsch. Bot. Ges.*, 85(10–12):651–659(1972 publ. 1973)—
1 Ge; portrait, bibliography.

PRITZEL, Georg August (1815–1874)

5946 Stafleu, F. A.
Pritzel and his Thesaurus. *Taxon*, 22(1):119–126
1 (1973)—Portrait.

RAFINESQUE, Constantine Samuel (1783–1840)

5947 Daniels, G. S. *and* Stafleu, F. A.
Constantine Samuel Rafinesque. *Taxon*, 22(2–3): 228(1973) (HI-IAPT Portraits of Botanists no.39)—
2, 3 En.

5948 Meijer, W.
The contribution by Rafinesque to the early botanical exploration of Kentucky. *Castanea*, 38(3):
2, 3 261–265(1973).

RÁSKY, Klára (1908–1971)

5949 Szujko-Lacza, J.
In memoriam Dr. Klára Rásky (1908–1971). *Ann. Hist.-Nat. Mus. Nation. Hung.*, 65:11–13(1973)—
1 Ge; bibliography.

RAUH, Werner (1913–)

5950 Padilla, V.
The men behind the scene—Dr. Werner Rauh. *J. Bromeliad Soc.*, 23(3):83–85(1973)—Illustration,
1 portraits.

RAVEN, Peter Hamilton (1936–)

5951 [Portrait only]. *Missouri Bot. Gard. Bull.*, 62(4):
3 [1]p.(1973).

RAWLINGS, Stanley W. (1912–)

5952 Anon.
S. W. Rawlings Vice-President 1972–74. *J. Kew*
1 *Guild*, 9(77):110–112(1973)—En; portrait.

RAY, John (1627–1705)

5953 Daniels, G. S. *and* Stafleu, F. A.
John Ray. *Taxon*, 22(2–3):204(1973) (HI-IAPT
1 Portraits of Botanists no.37)—En.

RAYMOND, Louis Florent Marcel (1915–1972)

5954 Batcheller, F. N.
In memoriam Louis Florent Marcel Raymond.
3 *Gloxinian*, 23(3):4–5(1973)—En.

5955 Boivin, B.
Marcel Raymond 1915–1972. *Taxon*, 22(2–3):275–
3 278(1973)—En; portrait.

REICHENBACH, Heinrich Gottlieb Ludwig (1793–1879)

5956 Helm, J.
Die Zeichner der botanischen Iconographien von Ludwig Reichenbach im Zeitraum 1820–1850. *In* Geck, E. *and* Pressler, G., *eds.* Festschrift für Claus Nissen zum siebzigsten Geburtstag 1971: 351–370
1 (1973)—Bibliography.

REICHENBACH, Heinrich Gustav (1824–1889)

5957 [Portrait only]. *Taxon*, 22(2–3):231
1 (1973).

REITZ, Raulino (1919–)

5958 Padilla, V.
The men behind the scene—P. Raulino Reitz. *J.*
4 *Bromeliad Soc.*, 23(4):130–131(1973)—Portrait.

RENNÓ, Lair Remusat (1909–1972)

5959 Anon.
Professor Lair Remusat Rennó. *Oréades*, 3(5):3–4
4 (1972)—Por; portrait.

5960 Costa, O. de A.
Lair Remusat Rennó (1.11.1909–6.2.1972).
4 *Leandra*, 2(3):105–108(1972).

RETZIUS, Anders Jahan (1742–1821)

5961 Daniels, G. S. *and* Stafleu, F. A.
Anders Jahan Retzius. *Taxon*, 22(2–3):220(1973)
1 (HI-IAPT Portraits of Botanists no.38)—En.

RHOADES, Marcus Morton (1903–)

5962 Dempsey, E.
Random observations on a distinguished professor.
3 *Theor. Appl. Genet.*, 43(3–4):97–100(1973)—Portraits.

5963 Peterson, P. A. *and* Peterson, S. R.
Marcus M. Rhoades. *Theor. Appl. Genet.*, 43(3–4):
3 93–96(1973)—En; portrait.

RICHARD, Louis Claude Marie (1754–1821)

5964 Daniels, G. S. *and* Stafleu, F. A.
Louis Claude Marie Richard. *Taxon*, 22(5–6):658
(1973) (HI-IAPT Portraits of Botanists no.48)—
1, 4 En.

RICKER, Percy L. (c.1879–1973)

5965 Anon.
[Percy L. Ricker]. *Taxon*, 22(2–3):328(1973)—
3 En.

RIVAS GODAY, Salvador (1905–)

5966 Esteve Chueca, F.
[Commemorative name *Rivasgodaya* (Leguminosae-Papilionoideae)]. *Trab. Dep. Bot. Univ. Granada*
1 2(2):81(1973).

ROBB, Mary Ann (1829–1912)

5967 Stearn, W. T.
Mrs Robb and 'Mrs Robb's bonnet' (*Euphorbia robbiae*). *J. Roy. Hort. Soc.*, 98(7):306–310(1973)—
1 Portrait.

ROBBIE, James (1904–1972)

5968 Anon.
James Robbie. *J. Kew Guild*, 9(77):155(1973)—En;
1 portrait.

ROEZL, J. Benedict (1823–1885)

5969 Anon.
The celebrated collector J. Benedict Roezl (1823–1885). *Orchid Dig.*, 37(6):221–222(1973)—Portrait. Reprinted from *L'Orchidophile*, pp.140–143, 1885,
1, 3, 4 translated by Nico Lek De Tachinville.

5970 Reinikka, M. A.
Benedict Roezl (1823–1885). *In* A history of the
1, 3, 4 Orchid: 219–222(1972)—Illustration, portrait.

ROLFE, Robert Allen (1855–1921)

5971 [Portrait only]. *Taxon*, 22(2–3):234
1 (1973).

5972 Reinikka, M. A.
Robert Allen Rolfe (1855–1921). *In* A history of the
1 Orchid: 275–278(1972)—Portrait.

ROLLAND-GERMAIN, *Brother* (1881–1972)

5973 Anon.
[Brother Rolland-Germain]. *Taxon*, 22(2–3):328
3 (1973)—En.

ROMERO-CASTAÑEDA, Rafael (1910–1973)

5974 Anon.
[Rafael Romero-Castañeda]. *Taxon*, 22(4):523
4 (1973)—En.

5975 Forero G., E.
Rafael Romero Castañeda. *Taxon*, 22(4):520–521
4 (1973)—En.

RONALDS, Hugh (1759–1833)

5976 Gilbert, J. L.
Hugh Ronalds (1759–1833)—a friend of Kew. *J.*
1 *Kew Guild*, 9(77):127–129(1973).

ROSCOE, William (1753–1831)

5977 S., R. M.
What the label doesn't tell. *Gard. July, Roy. Bot.*
1 *Gard. Edinburgh*, 4–5(1973)—Illustration.

ROSÉN von ROSENSTEIN, Nils (1706–1773)

5978 Strandell, B.
Ett par tidiga biografier över Linné och Rosén. (Two early biographies of Linnaeus and Rosén).
1 Uppsala, Almqvist & Wiksell, 1973. 8p.—(En).

ROTREKL, Bohuslav (1913–)

5979 Morávek, A.
Jubilea členu ČSBS: Bohuslav Rotrekl. *Preslia*,
1 45(4):374(1973).

ROUBAL, Jan (1880–1971)

5980 Likovský, Z.
23 října zemřel profesor Jan Roubal. *Acta Mus. Reginaehrad., A*, 12:199–200(1971 publ. 1972)—Cz;
1 portrait.

5981 Pfeffer, A. *and* Štys, P.
K úmrtí profesora Jana Roubala. (Zum Tode Professors Jan Roubal). *Sborn. Jihočesk. Muz. Česk.*
1 *Budějovicich Přír. Ved.*, 12(3):156–158(1972).

ROY, Ram Parikshan (1921–)

5982 Thakur, V.
Prof. R. P. Roy. *J. Bihar Bot. Soc.*, 1(1–2):57–59
6 (1972)—En; portrait.

RUGEL, Ferdinand (1806–1879)

5983 Meyer, F. G. *and* Elsasser, S.
The 19th century Herbarium of Isaac C. Martindale.
1, 3 *Taxon*, 22(4):375–404(1973).

RYPÁČEK, Vladimír (1910–)

5984 Penka, M.
Professor Vladimír Rypáček sixty years old. *Acta Univ. Agric., Ser. C* (Brno), 40(1):79–82(1971)—
1 Cz,En; portrait.

SALISBURY, Richard (1761–1829)

5985 McGillivray, D. J.
The handwriting of Richard Salisbury (1761–1829) (and some incidental notes). *Contrib. New S. Wales*
1, 7 *Nation. Herb.*, 4(6):369–372(1973).

SANTAPAU, Hermenegild (1903–1970)

5986 Ratnaparkhi, S.
Great biologists: Rev. Fr. Hermenegild Santapau.
1, 6 *Botanica* (Delhi), 22(3):100–103(1972).

SASAKI, Yoshiyuki (1926–1972)

5987 Suzuki, H.
Dr. Yoshiyuki Sasaki (1926–1972). *Hikobia*, 16
2 (3–4):293–298(1973)—Ja; portrait, bibliography.

SAX, Karl (c.1892–1973)

5988 Anon.
Karl Sax. *HortScience*, 8(6):519(1973)—En;
3 portrait.

SAXTON, Walter Theodore (1882–1973)

5989 Anon.
Walter Theodore Saxton, M.A. *Biol. J. Linn. Soc.*,
1 5(4):392(1973)—En. Short note.

SCHAERER, Ludwig Emanuel (1785–1853)

5990 Daniels, G. S. *and* Stafleu, F. A.
Ludwig Emanuel Schaerer. *Taxon*, 22(5–6):704 (1973) (HI-IAPT Portraits of Botanists no.49)—
1 En.

SCHLECHTER, Friedrich Richard Rudolf (1872–1925)

5991 [Portrait only]. *Taxon*, 22(2–3):235
1 (1973).

SCHMIDEL, Casimir Christoph (1718–1792)

5992 Fuchs-Eckert, H. P.
Die *Icones plantarum et analyses partium* des Casimir Christoph Schmidel. *In* Geck, E. *and* Pressler, G., *eds.* Festschrift für Claus Nissen zum siebzigsten Geburtstag 1971: 185–276(1973)—Illustrations,
1 bibliography.

SCHROETER, Otto Nagel (1894–1972)

5993 Halbinger, F.
In memoriam Otto Nagel Schroeter. *Orquídea*
4 (Mexico), 2(9):255–258(1972)—Sp,En; portrait.

SCHWARTEN, Lazella Harenberg (1900–1973)

5994 Howard, R. A.
Lazella Harenberg Schwarten 1900–1973. *J. Arnold*
3 *Arbor.*, 54(4):419–421(1973)—En.

SCOPOLI, Johannes Antonius (1723–1788)

5995 Mayer, E. *
[Ein Brief von Scopoli an Linné]. *Proteus* (Ljubljana), 34:307–308(1972)—Sl. Abstract in *Excerpta Bot., A*,
1 22(2):135(1973).

5996 Mayer, E. *
[Zweihundert Jahre 'Flora carniolica' von J. A. Scopoli]. *Proteus* (Ljubljana),34:299–305(1972)—Sl.
1 Abstract in *Excerpta Bot., A*, 22(2):135(1973).

SEAGRIEF, Stanley Charles (1927–)

5997 Anon.
Stanley Charles Seagrief. *Veld Flora*, 3(4):62(1973)—
5 En.

SEAVER, Fred Jay (1877–1970)

5998 Degener, O.
Fred Jay Seaver (1877–1970) *Phytologia*, 25(4):218–
3 219(1973)—En; portrait.

5999 Rogerson, C. T.
Fred Jay Seaver 1877–1970. *Mycologia*, 65(4):721–
3 724(1973)—En; portrait.

SHEBALINA, Mariya Aleksandrovna (c.1900–1972)

6000 Alekseeva, N. P., *and others*
Pamyati tovarishcha Mariya Aleksandrovna Shebalina. *Byull. Vses. Ord. Lenina Inst. Rast. N.I.*
1 *Vavilova*, 31:87–88(1973).

SHEPHERD, Thomas (1780–1835)

6001 Hazelwood, W. G.
Australia's first nurserymen: Thomas Shepherd
7 1780–1835. *Camellia News*, no.51:28–31(1973).

SHIMADA, Y.

6002 Satomi, N.
(Obituary of the late Mr. Y. Shimada). *J. Geobot.*,
2 20(1):4(1972)—Ja.

SHINNERS, Lloyd Herbert (1918–1971)

6003 Flook, J. M.
Guide to the botanical contributions of Lloyd H. Shinners (1918–1971). *Sida*, 5(3):137–180(1973)—
3 Bibliography.

6004 Tomb, A. S.
[Commemorative name *Shinnersoseris* (Composi-
3 tae)]. *Sida*, 5(3):186(1973).

SKINNER, George Ure (1804–1867)

6005 Reinikka, M. A.
George Ure Skinner (1804–1867). *In* A history of the Orchid: 169–173(1972)—Illustration,
1, 4 portrait.

SLAVÍK, Miloslav (1903–)

6006 Morávek, A.
Jubilea členů CSBS: Miloslav Slavík. *Preslia*,
1 45(4):374(1973).

SMITH, Harry Percival (–1972)

6007 Rowntree, J.
H. P. Smith; N. L. Bor. *Commonw. Forest. Rev.*,
1, 6 52(2):116–117(1973)—En.

SMITH, Lyman Bradford (1904–)

6008 Padilla, V.
The men behind the scene—Lyman B. Smith. *J.*
3 *Bromeliad Soc.*, 23(2):43–45(1973)—Portraits.

SMITINAND, Tem (1920–)

6009 Kostermans, A. J. G. H.
[Commemorative name *Temmodaphne* (Lauraceae)].
6 *Nat. Hist. Bull. Siam Soc.*, 24(3–4):479(1973).

SNELLING, Lilian (1879–1972)

6010 Douglas, S. C. J. H., *21st Earl of Morton*
Miss Lilian Snelling, V.M.H. *J. Roy. Hort. Soc.*,
1 98(3):139(1973).

SØRENSEN, Thorwald Julius (1902–1973)

6011 Anon.
Thorwald J. Sørensen. *Taxon*, 22(5–6):708(1973)—
1 En.

6012 Jakobsen, K.
Dansk botanisk forening: Thorvald Sørensen
4. juli 1902–21 juni 1973. *Bot. Tidsskr.*, 68(2):165–168
1 (1973)—Portraits.

SOKOLOVS'KIĬ, O. I. (1892–)

6013 Kharkevich, S. S.
O. I. Sokolovs'kȳi (do 80-richchya vid dnya
narodzhennya i 55-richchya naukovoyi ta pedaho-
hichnoyi diya-l'nosti). (O.I. Sokolovsky (on his
80th birthday and the 55th anniversary of his scientific
and pedagogical activity)). *Ukr. Bot. Zhurn.*, 30(1):
1 119–121(1973)—Uk; portrait.

SOÓ, Rezsö (1903–)

6014 [Portrait only]. *Bot. Közlem.*, 60(1):1
1 (1973).

6015 Editorial Staff, *Acta Bot. Hung.*
Prof. Rezsö Soó, Academician on his seventieth
birthday. *Acta Bot. Acad. Sci. Hung.*, 19(1–4):III-XI
1 (1973)—Illustrations, portrait, bibliography.

6016 Michalko, J.
Profesor Soó Rezsó člen madarskej akadémie vied,
sedemdesiatročný. (Professor Soó Rezsö, member of
the Hungarian Academy of Sciences, now seventy).
1 *Biológia, A* (Czechoslovakia), 28(10):853–854(1973).

6017 Pócs, T.
[Commemorative name *Sooia* (Acanthaceae)].
1 *Acta Bot. Acad. Sci. Hung.*, 19(1–4):461(1973).

SOWERBY FAMILY

6018 Collins, J.
A note on the history of the Sowerby family
archive; together with a short title catalogue of
Natural History works written or illustrated by
members of the family. London, Seaton, 1973. 22p.—
1 Bibliographies.

SOWTER, Frederick Archibald (1899–1972)

6019 Anon.
Frederick Archibald Sowter. *Biol. J. Linn. Soc.*,
1 5(4):392(1973)—En. Short notes.

6020 Hawksworth, D. L.
Frederick Archibald Sowter 1899–1972.
1 *Lichenologist*, 5:345–348(1973)—En; portrait.

6021 Jones, E. W.
Frederick Archibald Sowter 1899–1972. *J. Bryol.*,
1 7(3):465–467(1973)—En; portrait; bibliography.

STEBBINS, George Ledyard (1906–)

6022 Anon.
Professor George Ledyard Stebbins, F.M.L.S.
3 *Biol. J. Linn. Soc.*, 5(4):366–368(1973).

STEENIS, Cornelius Gijsbert Gerrit Jan van
(1901–)

6023 Leenhouts, P. W. *
Professor C. G. G. J. van Steenis. *Trop. Ecol.*,
13:1–4(1972). Abstract in *Excerpta Bot.*, *A*, 22(2):
1 134(1973).

STEERE, William Campbell (1907–)

6024 [Portrait only]. *New York Bot. Gard. Ann. Rep.*,
3 *1971–1972*. [1]p. [1973].

STELFOX, Arthur Wilson (1883–1972)

6025 Anon.
Arthur Wilson Stelfox, 1883–1972. *Irish Nat. J.*,
1 17(9):285–291,295(1973)—En; portraits.

6026 Anon.
Stelfox—botanist. *Irish Nat. J.*, 17(9):292–293
1 (1973).

6027 Meikle, R. D.
The Stelfox home in Dublin. *Irish Nat. J.*, 17(9):
1 298–301(1973).

STONE, Witmer (1866–1939)

6028 Stafleu, F. A.
Witmer Stone and his *Plants of Southern New Jersey*.
3 *Taxon*, 22(4):467–468(1973).

STROH, Rudolf (1897–1971)

6029 Sunderman, H. *
Dr. Rudolf Stroh. *Jahresber. Naturwiss. Ver.
Wuppertal*, 25:213(1972). Abstract in *Excerpta Bot.*, *A*,
1 22(2):139(1973).

SUTTOR, George (1774–1859)

6030 Hazelwood, W. G.
Australia's first nurserymen: George Suttor 1774–
7 1859. *Camellia News*, no.51:27(1973).

SVENTENIUS, Eric R. S. (1910–1973)

6031 Kunkel, G.
In memoriam: Eric R. S. Sventenius 1910–1973.
Cuad. Bot. Canar., 18–19:1–4(1973)—Sp; portrait,
1 bibliography.

SWAMINATHAN, Moncombu Sambasiva (1925–)

6032 Behera, N.
Great biologists: M. S. Swaminathan. *Botanica*
6 (Delhi), 23(4):139–140(1973).

SWARTZ, Olof (1760–1818)

6033 Reinikka, M. A.
Olof Swartz (1760–1818). *In* A history of the
1 Orchid: 127–129(1972).

SZAFER, Władysław (1886–1970)

6034 Godwin, H.
Władysław Szafer (23 July 1886 to 16 November
1970). *New Phytol.*, 72(5):1249–1250(1973)—
1 Portrait.

SZOVITS, Johann Nepomuk (–1830)

6035 Lamond, J. M.
The Transcaucasian and Iranian collections of J. N. Szovits. *Notes Roy. Bot. Gard. Edinburgh*, 32(2):
1, 2 239–245(1973)—Map.

TALIEV, Valeriĭ Ivanovich (1872–c.1957)

6036 Kotov, M. I.
Vklad V. I. Taliyeva v rozvȳtok botanikȳ. (do 100-richchya z dnya narodzhennya) (Contribution of V. I. Taliev to development of botany (on a centenary of his birth)). *Ukr. Bot. Zhurn.*, 30(2):251–252(1973)—
1 Uk.

TALLON, Gabriel (1890–1972)

6037 Molinier, R.
Nécrologie: Gabriel Tallon (1890–1972). *Terre Vie*,
1 27(1):147–151(1973)—Portrait, bibliography.

TAVARES, Carlos das Neves (1914–1972)

6038 Ferri, M. G.
Carlos das Neves Tavares sua vida e sua obra: homenagem. *Rev. Biol.*, 9(1–4): 1-x(1973)—
1 Portrait.

6039 Telles-Antunes, M. S. S. P.
Carlos das Neves Tavares. *Rev. Bryol. Lichénol.*,
1 39(3):493–494(1973)—Portrait.

TAYLOR, *Sir* George (1904–)

6040 [Portrait only]. *Amer. Orchid Soc. Bull.*, 42(7):611
1 (1973).

THOMAS, Abraham (1740–1824)

6041 Meyer, F. G. *and* Elsasser, S.
The 19th century Herbarium of Isaac C. Martindale.
1 *Taxon*, 22(4):375–404(1973).

THOMAS, Philippe (1782–1831)

6042 Meyer, F. G. *and* Elsasser, S.
The 19th century Herbarium of Isaac C. Martindale.
1 *Taxon*, 22(4):375–404(1973).

THWAITES, W. A.

6043 Shaw, G. A.
Some Yorkshire bryologists and bryophytes.
1 *Naturalist*, no.924:3(1973)—Portrait only.

TISDALE, Wendell H. (1892–1973)

6044 Ryker, T. C.
Wendell H. Tisdale, 1892–1973. *Phytopathology*,
3 63(11):1323(1973)—En; portrait.

TITIUS, Pius Vendel (1801–1884)

6045 Vöröss, L. Z.
Titius Pius Vendel (1801–1884) élete és munkássága. (Das Leben und die Tätigkeit von Pius Vendel Titius (1801–1884)). *Bot. Közlem.*, 59(3):141–143
1 (1972 publ. 1973)—(Ge); map.

TOLMACHEV, Aleksandr Innokent'evich (1903–)

6046 Vasilevskaya, V. K., Minyaev, N. A., Orlova, N. I. *and* Dobryakov, P. M.
Sistematik, geograf, geobotanik (k 70-letiyu so dnya rozhdennya Aleksandra Innokent'evicha Tolmacheva). (Systematizator, geographer, geobotanic (to the 70 anniversary of A. I. Tolmachev's birthday)). *Vestn. Leningrad. Univ., Biol.*, 21(4):152–153(1973)—
1 Portrait.

TORREY, John (1796–1873)

6047 New York: Botanical Garden: Library
Catalog of the manuscript and archival collections and index to the correspondence of John Torrey.
3 Boston, Mass., Hall, 1973. viii,473p.

TOURNAY, Roland Louis Jules Alfred (1925–1972)

6048 Anon.
[Roland Louis Jules Alfred Tournay]. *Taxon*,
1 22 (2–3):328(1973).

6049 Demaret, F.
Roland Tournay (1925–1972). *Bull. Jard. Bot.*
1 *Nation. Belg.*, 43(1–2):1–5(1973)—Fr; portrait.

6050 Dendal, R.
In memoriam Roland Tournay (1925–1972). *Nat.*
1 *Mosana*, 26(1–2):21–24(1973)—Fr.

6051 Lawalrée, A.
Roland Tournay (1925–1972). *Bull. Soc. Roy. Bot.*
1 *Belg.*, 106(3):4–16(1973)—Fr; bibliography.

TOUTON, Jean-Baptiste (1881–1972)

6052 Bonnot, E.-J.
Jean-Baptiste Touton, botaniste 1881–1972. *Bull. Soc. Bot. France*, 120(5–6):247–252(1973)—Illustrations,
1 portrait.

TROFIMOVSKOĬ, Aleksandra Yakovlevna (1903–)

6053 Anon.
Slavnȳĭ put uchenogo (k 70-letiyu A. Ya. Trofimovskoĭ). *Trudȳ Prikl. Bot. Genet. Selek.*, 50(1):
1 313–314(1973)—Portrait.

TRONCHET, A.

6054 Anon.
Liste des travaux de A. Tronchet, par ordre chronologique. *Ann. Sci. Univ. Besançon, Bot.*, 12:5–9
1 (1972)—Portrait.

TSWETT, Michel (1872–1919)

6055 Miège, J.
Voici cent ans naissait Michel Tswett, le fondateur de la chromatographie. *Mus. Genève*, 133:18–22
1 (1973)—Portraits.

TUCK, Gillian

6056 Swann, E. L.
Profile: Miss Gillian Tuck. *B.S.B.I. News*, 2(2):
1 30–31(1973).

TUOMIKOSKI, Risto (1911–)

6057 Anon. *
[Risto Tuomikoski 60 years.] *Luonnon Tutkija*, 75:56(1971)—Fi. Abstract in *Excerpta Bot., A*, 22(2):
1 128(1973).

TURESSON, Göte (1892–1970)

6058 [Roger, J. A.]
Göte Turesson 1892–1970. *B.S.E. News*, no.6:
1 11–12(1972)—En.

TWISSELMANN, Ernest Christian (1917–1972)

6059 Anon.
Ernest Christian Twisselmann. *Taxon*, 22(5–6):708
3 (1973)—En.

6060 Anon.
Ernest Twisselmann 1917–1972. *Four Seasons*, 4(3):
3 13–14(1973)—En; portrait.

UENO, H.

6061 Satomi, N.
(Obituary of the late Mr. H. Ueno). *J. Geobot.*,
2 20(4):109(1972)—Ja.

VASIL'CHENKO, Ivan Tikhonovich (c.1903–)

6062 Zakirov, K. Z., Granitov, I. I., Pyataeva, A. D., Khasanov, O. Kh. *and* Vernik, R. S.
Ivan Tikhonovich Vasil'chenko (k 70-letiyu so dnya rozhdeniya). *Uzb. Biol. Zhurn.*, no.4:76–77
1 (1973)—Portrait.

VASILEVSKAYA, Veronika Kazimirovna (1912–)

6063 Kondrat'eva-Mel'vil', E. A. *and* Tolmachev, A. I.
Krupnÿĭ anatom rasteniĭ: (k 60-letiyu Veroniki Kazimirovnÿ Vasilevskoĭ). (The greatest anatomist of plants (to the 60th birthday of Veronika Kazimirovna Vasilevskaya)). *Vestn. Leningrad. Univ., Biol.*,
2 no.3(1):136–138(1973)—Portrait.

VEITCH, *Sir* Harry James (1840–1924)

6064 Fuller, K. A. P. *and* Langdon, J. M.
The house of Veitch. *Int. Dendrol. Soc. Year Book*
1 *1972*:63–69(1973)—Portrait.

VEITCH, James (1792–1863)

6065 Fuller, K. A. P. *and* Langdon, J. M.
The house of Veitch. *Int. Dendrol. Soc. Year Book*
1 *1972*:63–69(1973)—Portrait.

VEITCH, James (1815–1869)

6066 Fuller, K. A. P. *and* Langdon, J. M.
The house of Veitch. *Int. Dendrol. Soc. Year Book*
1 *1972*: 63–69(1973)—Portrait. James Veitch junior.

VEITCH, John (1752–1839)

6067 Fuller, K. A. P. *and* Langdon, J. M.
The house of Veitch. *Int. Dendrol. Soc. Year Book,*
1 *1972*: 63–69(1973)—Portrait.

VELASCO, José María (1840–1912)

6068 Campo, R. M. del
Pericácticas 6—José María Velasco, pintor y naturalista. *Cact. Suc. Mex.*, 18(3):69–72(1973)—(En);
4 coloured illustration, portrait.

VELLOZO, José Mariano da Conceiçao (1742–1811)

6069 Carauta, J. P. P.
The text of Vellozo's *Flora Fluminensis* and its effective date of publication. *Taxon*, 22(2–3):281–284
4 (1973).

VENKATA RAO, Chellapilla (1910–1971)

6070 Steeramulu, T.
C. Venkata Rao (1910–1971). *Phytomorphology*,
6 22(1):97(1972 publ. 1973).

VISYULINA, Olena Dmitrivna (1898–1972)

6071 Kotov, M. I. *and* Morozyuk, S. S.
Pam'ya ti Olenÿ Dmÿtrivnÿ Visyulinoyi (1898–1972). (In memory of Olena Dmytrivna Visyulina). *Ukr. Bot. Zhurn.*, 30(2):246–250(1973)—Uk;
1 portrait; bibliography.

VŸKHODTSEV, Ivan Vasil'evich (1893–1971)

6072 Sharashova, V. S., Popova, L. I. *and* Lebedeva, L. P.
Pamyati Ivana Vasil'evicha Vÿkhodtseva (3 IX 1893–31 III 1971). (To the memory of I. V. Vykhodtsev). *Bot. Zhurn.*, 58(2):307–312(1973)—
1 Portrait, bibliography.

WAGENITZ, Gerhard (1927–)

6073 Dostál, J.
[Commemorative name *Wagenitzia* (Compositae)]. *Acta Bot. Acad. Sci. Hung.*, 19(1–4):76
1 (1973).

WAKEFIELD, Elsie Maud (1886–1972)

6074 Anon.
Elsie Wakefield, O.B.E., M.A. *Biol. J. Linn. Soc.*,
1 5(4):392–393(1973)—En.

6075 Blackwell, E. M.
Obituary: Elsie M. Wakefield, O.B.E., M.A. 1886–1972. *Trans. Brit. Mycol. Soc.*, 60(1):166–170
1 (1973)—Portrait.

6076 Dennis, R. W. G.
Miss E. M. Wakefield. *J. Kew Guild*, 9(77):163–164
1 (1973)—En.

6077 Dring, D. M. *and* Forwood, S.
List of publications. *Trans. Brit. Mycol. Soc.*, 60(1):
1 170–174(1973).

WAKEFIELD, Norman Arthur (–1972)

6078 Willis, J. H.
Vale, Norman Arthur Wakefield. *Victorian Nat.*,
7 90(4):103–105(1973).

WALLIS, Gustave (1830–1878)

6079 Reinikka, M. A.
Gustave Wallis (1830–1879). *In* A history of the
1, 4 Orchid: 234–235(1972).

WALPOLE, Frederick A. (1861–1904)

6080 Brindle, J. V. *comp.*
Paintings and drawings by Frederick A. Walpole: October 1, 1973—February 28, 1974. Hunt Institute,
3 Pittsburgh, 1973. 20p.—Illustrations, portrait.

WALTHER, Eric (1892–1959)

6081 Stafleu, F. A.
Eric Walther and his Echeverias. *Taxon*, 22(4):
1, 3 472–473(1973).

WALTON, John (1895–1971)

6082 Braid, K. W.
Obituary: John Walton (1895–1971). *Glasgow Nat.*,
1 19(1):70–71(1973)—Portrait.

WARD, Nathaniel Bagshaw (1791–1868)

6083 Etter, H.
Out of a chrysalis a Wardian case. *Morton Arbor.*
1 *Quart.*, 9(4):49–55(1973)—Illustrations.

WATERFALL, Umaldy Theodore (1910–1971)

6084 Stanford, J. W.
Umaldy T. Waterfall (1910–1971). *Rhodora*, 75
3 (801):146–149(1973)—En.

WATKINS, Ken

6085 Mercer, I.
[Ken Watkins]. *J. Devon Trust Nat. Conserv.*, 5(3): 110–111(1973)—Portrait. Former editor of the
1 Journal.

WEBB, Philip Barker (1793–1854)

6086 Stearn, W. T.
Philip Barker Webb and Canarian botany.
1 *Monogr. Biol. Canar.*, no.4:15–29(1973)—Portraits.

WELCH, Donald Stuart (1894–1972)

6087 Sinclair, W. A.
Donald Stuart Welch 1894–1972. *Phytopathology*,
1, 3 63(4):437(1973)—En; portrait.

WELWITSCH, Friedrich Martin Josef (1806–1872)

6088 Edwards, P. I.
Friedrich Welwitsch, 1806–72: his manuscripts and correspondence in the Departments of Botany and Zoology, British Museum (Natural History), the Linnean Society of London, and the Royal Botanic Gardens, Kew. *Biol. J. Linn. Soc.*, 4(4):291–303(1972
1, 5 publ. 1973)—Illustrations.

6089 Mendes, E. J.
A Friedrich Welwitsch. *Garcia de Orta, Sér. Bot.*,
1, 5 1(1–2):1–2(1973)—Por; portrait.

6090 Stearn, W. T.
The catalogue of the African Plants collected by Dr. Friedrich Welwitsch (1896–1901) and its litigious background. *Garcia de Orta, Sér. Bot.*, 1(1–2):101–
1, 5 104(1973)—(Por).

6091 Swinscow, T. D. V.
Friedrich Welwitsch, 1806–72: a centennial memoir. *Biol. J. Linn. Soc.*, 4(4):269–289(1972 publ.
1, 5 1973)—Illustrations, portrait.

WEST, Oliver (1910–)

6092 Anon.
Dr. Oliver West retires. *Rhodesia Agric. J.*, 70(1):
5 19–20(1973)—Portrait.

WEST, William (1848–1914)

6093 Shaw, G. A.
Some Yorkshire bryologists and bryophytes.
1 *Naturalist*, no.924:4–5(1973).

WETMORE, Ralph Hartley (1892–)

6094 De Maggio, A. E. *and* Steeves, T. A.
Ralph H. Wetmore: dedication. *Bot. Gaz.*, 133(3):
3 197(1972)—Portrait.

6095 Torrey, J. G. *and* Thimann, K. V.
On the beginning of modern studies of plant morphogenesis: a tribute to Ralph H. Wetmore's
3 contributions. *Bot. Gaz.*, 133(3):199–206(1972).

WHERRY, Edgar Theodore (1885–)

6096 Mazrimas, J. A.
Edgar T. Wherry's *Sarracenia* publications. *Carniv.*
3 *Pl. Newsl.*, 2(3):37–39(1973)—Bibliography.

WHITTET, John N. (c.1888–1973)

6097 Anon.
. . . Whittet kikuyu obituary . . . J. N. Whittet.
7 *Agric. Gaz. New S. Wales*, 84(5):302(1973).

WINTERL, Jakob Joseph (1739–1809)

6098 Priszter, S.
A pesti Egyetemi Botanikus Kert történetéhez: 7. Néhány újabb dokumentum Winterl életéhez és müködéséhez. (Zur Geschichte des Botanischen Gartens der Universität zu Pest: 7. Einige neuere Dokumente zum Leben und Wirken von Winterl). *Bot. Közlem.*, 60(1):53–64(1973)—Illustrations,
1 portraits.

6099 Priszter, S.
Zur Geschichte des Botanischen Gartens der Universität zu Pest. 6. Winterls handschriftlicher Pflanzenkatalog (1785). *Ann. Univ. Sci. Budapest.*
1 *Rolando Eötvos Nom., Biol.*, 15:93–101(1973).

WIŚNIEWSKI, Piotr (1881–1971)

6100 Mowszowicz, J.
Prof. Dr. Piotr Wiśniewski (1881–1971). *Wiad. Bot.*,
1 17(2):75–77(1973)—Pol.

WISTAR, Caspar (1761–1818)

6101 McClintock, E.
The generic name *Wisteria*. *Calif. Hort. J.*, 34(4):
3 159(1973).

WITHERING, William (1741–1799)

6102 Cooper, G.
A famous Warwickshire botanist: William Withering. *Ann. Rep. Warwick. Nat. Hist. Soc.*, 19:
1 24–26(1973).

WRABER, Maks (1905–1972)

6103 Horvatić, S.
Prof. Dr. Maks Wraber: in memoriam. *Vegetatio*,
1 28(1–2):125–128(1973)—Bibliography.

6104 Zupančič, M.
V slovo profesorju Dr. Maksu Wrabru. *Biol.*
1 *Vestn.*, 20:7–10(1972)—Portrait.

WRIGHT, John (1811–1846)

6105 Meyer, F. G. *and* Elsasser, S.
The 19th century Herbarium of Isaac C. Martindale.
3 *Taxon*, 22(4):375–404(1973).

WYATT, Oliver E. P. (–1973)

6106 Synge, P. M.
Obituary: Oliver E. P. Wyatt, M. C., V.M.H.
1 *Quart. Bull. Alp. Gard. Soc.*, 41(2):185(1973).

YAJNVALKYA, Bharadvaja (1895–)

6107 Chaurasia, A.
Bharadvaja Yajnvalkya (1895–). *Planto*, 2(1):
6 91(1972)—En.

YANG, Bao-Yu (1902–1973)

6108 Huang, T-C.
Bao-Yu Yang. *Taiwania*, 18(2):i–iv(1973)—
2 Portrait, bibliography.

YOUNG, Donald Peter (1917–1972)

6109 Prime, C. T.
Donald Peter Young (1917–1972). *Watsonia*, 9(3):
1 293–295(1973)—En; bibliography.

YURKEVICH, Ivan Danilovich (c.1903–)

6110 Nesterovich, N. D. *and* Gel'tman, V.S.
Ivan Danilovich Yurkevich (K 70-letiyu so dnya rozhdeniya). (Ivan Danilovich Yurkevich (towards the 70th birthday anniversary)). *Bot. Zhurn.*, 58(1):152–
1 154(1973)—Portrait.

ZANDER, Robert (1892–1969)

6111 Stafleu, F. A.
Zander's *Handwörterbuch. Taxon*, 22(1):129–130
1 (1973)—En.

ZAPRYAGAEVA, Vera Ivanovoa (c.1912–)

6112 Anon.
Vera Ivanovna Zapryagaeva (k 60- letiyu so dnya rozhdeniya). *Izv. Akad. Nauk Tadzh. SSR, Biol. Nauk*,
2 2(47):100–102(1972)—Rus; portrait.

ZAWADSKI, Alexander (1798–1868)

6113 Orel, V.
Professor Alexander Zawadski (1798–1868). Mendel's superior at the Technical Modern School in Brno. *Folia Mendeliana*, 7:13–20(1972)—(Ge);
1 portraits.

ZEROV, Dmitriĭ Konstantinovich (1895–1971)

6114 Belokon, I. P.
Dmitriĭ Konstantinovich Zerov (20 ix 1895–20 xii 1971). (Dmitriy Konstantinovich Zerov (20 ix 1895–20 xii 1971)). *Bot. Zhurn.*, 58(8):1228–1235(1973)—
1 Portrait; bibliography.

ZHUKOVSKIĬ, Petr Mukhaĭlovich (1888–)

6115 Anon.
Entuziast nauki (k 85-letiyu P. M. Zhukovskogo). *Byull. Vses. Ord. Lenina Inst. Rast. N.I. Vavilova*, 30:
1 88–90(1973)—Portrait.

ZIRKLE, Conway (1895–1972)

6116 Anon.
Conway Zirkle 1895–1972. *Stain Techn.*, 48(1):2–4
3 (1973)—En; portrait.

Botanical Institutions

6117 Apel, J.
Die Botanischen Gärten sind im Kommen! *Gärtn.-*
1 *Bot. Brief*, 41:1245–1264(1973)—Maps.

6118 Apel, J.
Das 'System' im Botanischen Garten. Fragen einer zweckmässigen Gestaltung. *Gärtn.-Bot. Brief*, 41:
1 1215–1243(1973)—Illustrations.

6119 Bryan, J. E.
The purpose of an arboretum. *Calif. Hort. J.*, 34(2):
3 70–74(1973).

6120 Henderson, D. M.
The role of botanic gardens in horticulture and field botany. *In* Green, P. S. *ed.* Plants: wild and cultivated:
1 66–71(1973).

6121 Lavrenko, E. M., Skvortsov, A. K., Takhtadzhyan, A. L., Tikhomirov, V. N. *and* Lyurtsev, B. A.
Gerbarii: znachenie dlya obshchestva, sovremennoe sostoyanie, perspektivȳ. (Herbaria: their significance for society, contemporary state, perspectives). *Izv.*
1–7 *Akad. Nauk SSSR, Biol.*, no.1:5–11(1973)—(En).

6122 Walters, S. M.
The role of botanical gardens in conservation. *J. Roy. Hort. Soc.*, 98(7):311–315(1973)—Coloured
1–7 illustrations.

COSTA RICA

6123 Northen, R. T.
Dedication of the Lankester Garden. *Amer. Orchid Soc. Bull.*, 42(7):609–614(1973)—Coloured illustra-
4 tions, portraits.

6124 Northen, R. T.
Salvando los Jardines Lankester. *Orquídea* (Mexico),
4 2(1):20–23(1972).

KAZAKHSTAN

6125 Uzenbaev, E. Kh., Vintergoller, B. A. *and* Dzhadaĭbaev, Zh. M.
Botanicheskie sadȳ Kazakhstana Alma-Ata, Izd.
2 'Nauka' Kazakhskoĭ SSR, 1973. 101p.—Illustrations.

MONACO

6127 Funke, G.
Die exotischen Gärten zu Monaco. *Palmengarten*,
1 37(2):41–42(1973)—Illustration.

SPAIN

6128 Ozores y Saavedra, M. T. de, *Marquesa de Casa Valdés*
Jardines de España. Madrid, Aguilar, 1973. 229p.—
1 Illustrations, coloured illustrations, maps.

U.S.S.R.

6129 U.S.S.R.: Akademiya Nauk SSSR: Sovet Botanicheskikh Sadov
Uspekhi introduktsii rasteniĭ: k 75- letiyu so dnya rozhdeniya akademika Nikolaya Vasil'evicha Tsitsina. (Advances of plant introduction, dedicated to academician N. V. Tsitsin's 75th birthday). Moskva, Nauka, 1973. 334p.—Illustrations, coloured
2 illustrations, maps.

ADIOPODOUMÉ

6130 Binder, E.
Le Centre Suisse de Recherche Scientifique en Côte-d'Ivoire. *Mus. Genève*, 132:11–14(1973)—
5 Illustrations.

ALTON

6131 Hellyer, A. G. L.
The garden of a great romantic. *Amat. Gard.*, 90 (4622):1,16–17(1973)—Illustrations, coloured
1 illustrations.

ARCADIA

6132 Spalding, G. H.
A history of Los Angeles State and County Arboretum: the first 25 years. *Lasca Leaves*, 23(1): 11–56(1973)—Illustrations, coloured illustrations,
3 portraits.

AUSTIN

6133 Turner, B. L.
Transfer of the Lundell Herbarium (LL) to the University of Texas, Austin (Tex). *Taxon*, 22(1):
3 172(1973).

BALTIMORE

6134 Windler, D. R. *and* Windler, B. K.
The Towson State Herbarium family index.
3 *Phytologia*, 26(2):103–115(1973).

BELFAST

6135 Hackney, P.
Additional notes on the herbarium of the Ulster
1 Museum. *Irish Nat. J.*, 17(9):318(1973).

BERLIN

6136 Kraft, H.
Der Botanische Garten. *In* Berlin und seine Bauten.
1 9p.(1972)—Illustrations, maps.

BODNANT

6137 Hellyer, A. G. L.
Bodnant: a gardener's paradise. *Amat. Gard.*, 90
(4611):1,34(1973)—Illustrations, coloured
1 illustration.

BOGOR

6138 Lamoureux, C. *and* Arditti, J.
Orchids notes from U.C.I.: orchids at Kebun Raya.
Amer. Orchid Soc. Bull., 42(11):969–971(1973)—
6 Coloured illustrations.

BRATISLAVA

6139 Gorchakovskiĭ, P. L.
Floristicheskie i geobotanicheskie issledovaniya v
Botanicheskom Institute Slovatskoĭ Akademii Nauk.
(Floristical and geobotanical research at the Botanical
Institute of the Slovak Academy of Sciences). *Bot.*
1 *Zhurn.*, 58(10):1570–1573(1973).

6140 Kolek, J. *and* Michalko, J.
Dvadsat rokov botanickeho výskumu v Slovenskej
Akadémii vied. (Twenty years of botanical research
by the Slovak Academy of Sciences). *Biológia, A*
1 (Czechoslovakia), 28(7):604–607(1973).

BRESSINGHAM

6141 Anon.
Un établissement anglais spécialisé dans la production des plantes vivaces: 'Bressingham Gardens'. *Jard.*
1 *France*, no.1:18–20,29–33(1973)—Illustrations.

BRUSSELS

6142 Bamps, P.
Collections botaniques en Angola déposées dans
l'herbier de Bruxelles. *Garcia de Orta, Sér. Bot.*, 1(1–2):
1, 5 43–44(1973).

BUDAPEST

6143 Priszter, S.
A pesti egyetemi Botanikus Kert történetéhez: 5. A
közép-magyarországi flórakutatás megindulása. (Zur
Geschichte des Botanischen Gartens der Universität zu
Pest: 5. Beginn der Florenforschung in Mittelungarn).
Bot. Közlem., 59(3):145–154(1972 publ. 1973)—
1 (Ge).

6144 Priszter, S.
A pesti Egyetemi Botanikus Kert történetéhez: 7.
Néhány újabb dokumentum Winterl életéhez és
müködeséhez. (Zur Geschichte des Botanischen
Gartens der Universität zu Pest: 7. Einige neuere
Dokumente zum Leben und Wirken von Winterl).
Bot. Közlem., 60(1):53–64(1973)—Illustrations,
1 portraits.

6145 Priszter, Sz.
Jubiläum des 200 jährigen Botanischen Gartens der
Universität. *Ann. Univ. Budapest. Rolando Eötvos Nom.*,
1 *Biol.*, 14:127–132(1972 publ. 1973).

6146 Priszter, Sz.
Zur Geschichte des Botanischen Gartens der
Universität zu Pest. 6. Winterls handschriftlicher
Pflanzenkatalog (1785). *Ann. Univ. Sci. Budapest.*
1 *Rolando Eötvos Nom., Biol.*, 15:93–101(1973).

6147 Soó, R.
Növénygyujtéseim, 1917–1968. (Meine Pflanzensammeltätigkeit, 1917–1968). *Bot. Közlem.*, 59(3):
1 211–214(1972 publ. 1973).

6148 Szujkó Lacza, J.
A 100 éves Növénytár herbáriumainak története. 1.
Általános rész. (Ungarische Herbarien: 10. Die
Geschichte der Herbarien der 100-jährigen Botanischen
Abteilung des Museums: 1. Allgemeiner Teil). *Bot.*
1 *Közlem.*, 60(1):49–52(1973)—(Ge).

BUENA PARK, CALIFORNIA

6149 Woodson, W. D.
Ghost Town Arboretum. *Trees*, 36(2):3–9(1973)—
3 Illustrations.

BURLINGTON

6150 Seymour, F. C.
The Pringle Herbarium. *Rhodora*, 75(802):297–
3 303(1973).

CAERHAYS

6151 Williams, J.
George Forrest's plants at Caerhays. *J. Scott. Rock*
1 *Gard. Club*, 13(3):175–179(1973).

CAMBRIDGE

6152 Orriss, P. T.
The Winter Garden at Cambridge. *J. Roy. Hort.*
1 *Soc.*, 98(1):8–13(1973)—Coloured illustrations.

CAMBRIDGE

6153 Sander, D. F.
The Cambridge Botanic Gardens. *Orchid Rev.*,
1 81(957):76–78(1973)—Illustrations.

CANBERRA

6154 Sinadskiĭ, Yu. V.
Botanicheskiĭ Sad g Kanberra. *Byull. Glavn. Bot.*
7 *Sada* (Moscow), 90:80–82(1973).

CARACAS

6155 Lasser, T., Braun, A. *and* Steyermark, J. A.
Catalogo de las plantas que crecen en el Jardin
Botanico del Ministerio de Agricultura y Cria,
Caracas: catalogo de las plantas del Jardin Botanico.
4 *Acta Bot. Venezuel.*, 9(1–4):9–61(1974 publ. 1973).

CHISWICK

6156 Carré, J.
Lord Burlington's garden at Chiswick. *J. Gard.*
1 *Hist. Soc.*, 1(3):23–30(1973)—Illustrations.

DUNEDIN

6157 Anon.
Botanical Garden, city of Dunedin. *Rhododendron*,
7 12(4):2–6(1973)—Illustration.

6158 Paterson, G.
The Dunedin Botanic Garden. *Roy. New Zealand*
7 *Inst. Hort. Ann. J.*, no.1:30–42(1973)—Illustrations.

EAST BERGHOLT

6159 Eley, M.
East Bergholt Place, Suffolk. *Int. Dendrol. Soc. Year*
1 *Book 1972*: 30–39(1973)—Illustrations.

EDINBURGH

6160 Keenan, J.
Forrest and the Royal Botanic Garden, Edinburgh.
1 *J. Scott. Rock Gard. Club*, 13(3):212–215(1973).

FLORENCE

6161 Innamorati, T. F.
Notizie di medicina popolare africana nell'Erbario
Tropicale di Firenze. *Webbia*, 28(1):81–134(1973)—
1 (En); illustrations.

6162 Steinberg, C. H.
Macaronesian collections of phanerogams in the
Herbarium Universitatis Florentinae. *Monogr. Biol.*
1 *Canar.*, no.4:30–48(1973).

FONTAINEBLEAU

6163 Boissière, J. C.
Un jardin botanique au Laboratoire de biologie
végétale de Fontainebleau. *Bull. Ass. Nat. Vallée*
Loing Massif Fontainebleau, 49(3–4):33(1973)—Short
1 note on a garden to be established.

FRUNZE

6164 Kirgizia: Akademiya Nauk Kirgizskoĭ SSR:
Botanicheskiĭ Sad
Tsvetȳ Botanicheskogo Sada Akademii Nauk
Kirgizskoĭ SSR. [Flowers of the Botanical Garden of
the Academy of Sciences Kirgiz SSR]. Frunze, ILIM,
2 1973. 79p.—Illustrations.

GENEVA

6165 Miège, J., *ed.*
Exsiccatorum genavensium a conservatorio
botanico distributorum. 3. Geneva, 1972. pp.
1 33–52.

6166 Miège, J. *ed.*
Exsiccatorum genavensium a conservatorio botanico
1 distributorum: 4. Geneva, 1973. pp. 53–71.

GLINNA

6167 Tumiłowicz, J.
Ogród dendrologiczny w Glinnejstan obecny i
plany rozwoju. (The Dendrological Garden at
Glinna—its present condition and plans for future
development). *Roczn. Sekc. Dendrol. Pol. Tow. Bot.*,
1 27:175–185(1973)—(En); illustrations.

HONOMAELE, HAWAII

6168 Stewart, W. S.
Work is initiated at Kahanu Gardens. *Bull. Pac.*
6 *Trop. Bot. Gard.*, 3(1):9–12(1973)—Illustrations.

JAMAICA PLAIN

6169 Anon. *
The Arnold Arboretum—100 growing years.
Horticulture, 50(5):46–47(1972). Abstract in *Excerpta*
3 *Bot., A*, 22(2):112(1973).

6170 Sutton, S. B. *
The Arnold Arboretum: the first century.
Cambridge (Mass.), Harvard University, 1971. 72p.
3 Abstract in *Excerpta Bot., A*, 22(2):124(1973).

KANAGAWA

6171 Kanagawa: Botanical Garden
List of plants cultivated in Kanagawa Prefectural
Ofuna Botanical Garden. Kamakura, Botanical
2 Garden, 1972. 65p.

KEPONG

6172 Ng, F. S. P.
100,000 specimens in the Kepong Forest Herbarium.
6 *Malayan Forester*, 35(2):139(1972).

6173 Wong, T. M.
The catalogue to the Kepong wood collection. *For.*
Res. Inst. Res. Pamph. (W. Malaysia), no.61. 97p.
6 (1972).

KEW

6174 Christopher, B.
Irises at Kew. *Iris Year Book 1972*: 112–114
1, 6 (1973).

6175 King, R. W.
The Queen's Garden at Kew. *Amat. Gard.*, 90
(4610):1,30(1973)—Illustrations, coloured
1 illustration.

KIEV

6176 Kharkevich, S. S.
Gerbariĭ Tsentral'nogo Respublikanskogo Botani-
cheskogo Sada Akademii Nauk Ukrainiskoĭ SSR v
Kieve. (The Herbarium of the Central Republican
Botanical Garden of the Academy of Sciences of the
1 Ukrainian SSR). *Bot. Zhurn.*, 58(11):1680–1682(1973).

KIRSTENBOSCH

6177 Anon.
Conservation on the slopes of Table Mountain.
Amat. Gard., 90(4614):1,34–35(1973)—Illustrations,
5 coloured illustration.

KORNIK

6178 Bojarczuk, K. *and* Bojarczuk, T.
Wykaz drzew i krzewów w Arboretum Kórnickim:
3. (*Callicarpa–Ostrya*). (List of trees and shrubs in the
Kórnik Arboretum). *Arbor. Kórnickie*, 18:237–263
1 (1973).

KYOTO

6179 Hatakeyama, I., Murata, G. *and* Tabata, H.
A list of plants in the botanical garden of Kyoto
University and some ecological data. *Mem. Fac. Sci.*
Kyoto Univ., Biol., 6(2):91–148(1973)—En,Ja;
2 map.

LAUNCESTON, TASMANIA

6180 Anon.
The Queen Victoria Museum Herbarium.
7 *Launceston Nat.*, 5(4):2(1972).

LAWAI, HAWAII

6181 Woolliams, K.
Summary of the native species in the Pacific Garden.
6 *Bull. Pac. Trop. Bot. Gard.*, 3(1):14–15(1973).

LEEUWARDEN

6182 Stobbe, G. *and* Veen, B. van der
Het Fries Natuurhistorisch Museum 50 jaar. *Natura*
1 (Netherlands), 70(10): 152–154(1973)—Illustrations.

LEICESTER

6183 Balshaw, W. J.
Leicester arboretum. *Arboricult. Ass. J.*, 2(4):110–
1 113(1973).

LEIPZIG

6184 Clos, E. C.
El jardin botanico de Leipzig (Alemania Oriental) cumple 430 años. *Rev. Fac. Agron. Univ. Nac. La Plata*,
1 49(1):101(1973).

LENINGRAD

6185 Levina, F. Ya. *and* Rodin, L. E.
50-letie otdela geobotaniki Botanicheskogo Instituta im. V. L. Komarova Akademiĭ Nauk SSSR (1922–1972). (The 50th anniversary of the department of geobotany of the V.L. Komarov Botanical Institute Academy of Sciences of the U.S.S.R. (1922–
2 1972)). *Bot. Zhurn.*, 58(12):1830–1845(1973).

6186 U.S.S.R.: Akademiya Nauk SSSR: Botanicheskiĭ Institut [V. L. Komarova] *
Tropicheskie i subtropicheskie rasteniya v oranzhereyakh Botanicheskogo Instituta AN SSSR. (Les plantes tropicales et subtropicales dans les serres de l'Institut Botanique de l'Académie des Sciences de l'U.R.S.S.). Leningrad, Nauka, 1973. 275p.—Illustrations. Ref. in *Bull. Signal., Biol. Physiol. Vég.*,
2 34(10):9742(1973).

LINZ

6187 Schatzl,
New and old in flower in the Botanic Garden. *Chileans*, 7(25):37–38(1973). Translation by E. W.
1 Bentley from the *G.O.K. Bull.* for Jan. 1971.

LISLE

6188 MacPhail, I.
Notes on the Morton Arboretum Library. *Morton*
3 *Arbor. Quart.*, 9(4):56–61(1973)—Illustrations.

6189 Schulenberg, R.
Notes on the Morton Arboretum Herbarium. *Morton Arbor. Quart.*, 9(2):17–26(1973)—
3 Illustrations.

MANAUS

6190 Silva, M. Freitas da *
Os tipos do herbário do Instituto Nacional de Pesquisas da Amazonia. *Acta Amazonica*, 1(2):19–32 (1971 publ. 1972). Ref. in *Bull. Torrey Bot. Club*, 99(6):
4 325(1972).

MANCHESTER

6191 Franks, J. W.
Herb. Manch.: a guide to the contents of the Herbarium of Manchester Museum. Manchester Museum, 1973. 29p. (Manchester Museum publication
1 no. N.S.1, 73).

MARBURG

6192 Kellner, K. *
Der Marburger botanische Garten. *Hess. Heimat*, 22:50–52(1972). Abstract in *Excerpta Bot.*, *A*, 22(2):
1 119(1973).

MIAMI

6193 Stephens, J. A.
Fairchild Tropical Garden. *J. Bromeliad Soc.*, 23(5):
3 178–179(1973).

MISSOURI

6194 Goldblatt, P.
The African collection at Missouri Botanical
3, 5 Garden. *Taxon*, 22(2–3):325(1973).

MOSCOW

6195 Skvortsov, A. K. *and* Proskuryakova, G. M.
Gerbariĭ Glavnogo Botanicheskogo Sada Akademii Nauk SSSR. (The herbarium of the Central Botanical Garden of the Academy of Sciences of the USSR).
1 *Bot. Zhurn.*, 58(1):155–162(1973).

NEW YORK

6196 Brown, D. A.
The New York Botanical Garden: a natural sanctuary amid Gotham City's concrete and steel. *Amer. Nurseryman*, 138(9):9,79,82–84(1973)—
3 Illustrations.

6197 Clos, E. C.
L. H. Bailey Hortorium. *Rev. Fac. Agron. Univ.*
3 *Nac. La Plata*, 49(1):101(1973)—Sp.

OXFORD

6198 Paterson, A.
Oxford Botanic Garden. *Amat. Gard.*, 90(4642):1,
1 18–19(1973)—Illustrations, coloured illustration.

PRETORIA

6199 South Africa: Department of Agricultural Technical Services
Botanical Research Institute. Pretoria, D.A.T.S., [1973 ?] 47p.—Illustrations, coloured illustrations,
5 map.

RICEBORO

6200 Anon.
Restoration of LeConte Botanical Garden. *Amer. Camellia Yearbook 1973*: 178–184(1972)—
3 Illustrations, coloured illustrations.

RIO DE JANEIRO

6201 Occhioni, P. *
Catálogo do Herbario da Faculdade Nacional de Farmacia. *An. Fac. Nac. Farm.*, (Rio de Janeiro), 7:93–222(1972). Abstract in *Excerpta Bot.*, *A*, 22(2):121
4 (1973).

6202 Reitz, R.
The Bromeliario in the Jardim Botanico in Rio. *J. Bromeliad Soc.*, 23(4):132–137(1973)—Illustrations,
4 coloured illustrations.

ROCCA TRAMONTI

6203 Trucchi, M.
Un giardino botanico appenninico nel Parco
1 Nazionale d'Abruzzo. *Rea*, no.5:21–23(1972)—(En).

SALISBURY

6204 Kimberley, M. J.
History of Rhodesia's National Herbarium and National Botanic Garden. *Aloe, Cact. Succ. Soc. Rhodesia Quart. Newsl.*, no.19:31–32(1973). First
5 published in 1969 by the Rhodesian Government.

SKRIVERI

6205 Zvirgzd, A. V., Maurin', A. M. *and* Tsinovskis, R. E.
Skriverskiĭ Dendrariĭ. Riga, Zinatne, 1972. 171p.—
1 Rus(Ge); illustrations.

STIRLING

6206 Dunn, J. M. *and* Milne, H. H.
University of Stirling—George Forrest Walk.
1 *J. Scott. Rock Gard. Club*, 13(3):255–256(1973).

STOCKPORT

6207 Messenger, K. G.
A newly-discovered hortus siccus dating from the 1820s. *Watsonia*, 9(4):430(1973). Collection of
1 unknown origin found in Stockport Museum in 1972.

SZEGED

6208 Wróbel-Stermińska, W.
50-lecie ogrodu Botanicznego w Szegedzie. *Wiad.*
1 *Bot.*, 17(3):172–173(1973).

TELČ

6209 Diener, J.
Nález zajímavéno herbáře v Telči. (Ein interessantes Herbarium wurde in der Stadt Telč gefunden).
1 *Zprávy Česk. Bot. Spol.*, 8(3):185–187(1973).

THREAVE

6210 Taylor, D.
Looking and learning at Threave. *Amat. Gard.*, 90(4606):1,30(1973)—Illustrations, coloured
1 illustration.

TOMSK

6211 Polozhii, A. V.
V gerbarii imeni P. N. Krȳlova pri Tomskom Universitete. (In the P. N. Krȳlov Herbarium at the Tomsk University). *Bot. Zhurn.*, 58(10):1574–1578
1 (1973).

URBINO

6212 Petri, P. S.
L'Orto Botanico di Urbino. *Nat. Montagna*, 20(1):
1 53–60(1973)—Illustrations.

UTRECHT

6213 Lanjouw, J.
De botanische tuinen van de rijksuniversiteit te Utrecht. *Wachendorffia*, 1(2–3):13–16,25–28(1973)—
1 Illustrations.

6214 Nieuman, W.
Enige verwilderde en wilde planten op 'Sandwijck'.
1 *Wachendorffia*, 1(3):36–40(1973)—Illustrations.

VERSAILLES

6215 Callen, G.
L'Arboretum de Chèvreloup. *Jard. France*, no.5:
1 18–24(1973)—Illustrations.

WASHINGTON

6216 Meyer, F. G. *and* Elsasser, S.
The 19th century Herbarium of Isaac C. Martindale. *Taxon*, 22(4):375–404(1973)—Illustrations, portrait,
3 bibliography. In the National Arboretum.

WAU

6217 Wau: Ecology Institute
Wau Ecology Institute: dedicated to research and education in tropical ecology and to conservation of the unique biota of Papua New Guinea. Wau,
6 Ecology Institute, [1973] [5]p.—Illustration, maps.

WINDHOEK

6218 Herne, H.
The first botanic garden for succulents in South West Africa. *Cact. Succ. J.* (U.S.A.), 44(2):79(1972)—
5 Illustration.

Periodical abbreviations

The following list represents additions to the lists of periodical abbreviations which were published in the *Kew Record of Taxonomic Literature 1971*, and the *Kew Record of Taxonomic Literature 1972*. Full titles of periodicals cited in the *Kew Record of Taxonomic Literature 1973* are given, with addresses of publishers where it has been possible to ascertain them.

Abh. Ber. Naturk. Mus. 'Mauritianum' Altenburg
Abhandlungen und Berichte des Naturkundlichen Museums 'Mauritianum' Altenburg. Altenburg, Austria.

Abh. Staatl. Mus. Mineral.Geol., Dresden
Abhandlungen: Staatlichen Museums für Mineralogie und Geologie zu Dresden. Dresden, German Democratic Republic.

Acta Agron. Acad. Sci. Hung.
Acta Agronomica Academiae Scientiarum Hungaricae. Magyar Tudomanyos Akademia, Budapest, Hungary.

Acta Biol. Parana.
Acta Biológica Paranaense: Revista do Instituto de Biologia da Universidade Federal do Paraná. Curitiba, Brazil.

Acta Bot. Fenn.
Acta Botanica Fennica. Societas pro Fauna et Flora Fennica, Helsinki, Finland.

Acta Bot. Sin.
Acta Botanica Sinica. Institutum Botanicum, Academia Sinica, Peking, China.

Acta Univ. Agric., Ser. C (Brno)
Acta Universitatis Agriculturae: Ser. C: Facultas Silviculturae. Brno, Czechoslovakia.

Afr. Res. Doc.
African Research and Documentation. African Studies Association of the U.K., Centre for West African Studies, Birmingham, U.K.

Allg. Forstzeitung (Vienna)
Allgemeine Forstzeitung. Wien, Austria.

Aloe, Cact. Succ. Soc. Rhodesia Quart. Newsl.
The Aloe, Cactus and Succulent Society of Rhodesia Quarterly Newsletter. P.O. Box 922, Salisbury, Rhodesia.

An. Estac. Exp. Aula Dei
Anales de la Estacion Experimental de Aula Dei. Consejo Superior de Investigaciones Cientificas, Estacion Experimental de Aula Dei, Zaragoza, Spain.

An. Mus. Hist. Nat.Valparaiso
Anales del Museo de Historia Natural de Valparaiso. Valparaiso, Chile.

An. Soc. Bot. Brasil
Anais da Sociedade Botânica do Brasil. Jardim Botânico, Rio de Janeiro, Brazil.

Ann. Mus.Goulandris
Annales Musei Goulandris. The Goulandris Natural History Museum, Athens, Greece.

Ann. Rep. North. Nut Growers Ass.
Annual Report of the Northern Nut Growers Association. Iowa State University, Ames, Iowa, U.S.A.

Ann. Rep. Warwick Nat. Hist. Soc.
Annual Report of the Warwick Natural History Society. County Museum, Market Place, Warwick, U.K.

Ann. Rep. York. Nat. Union
Annual Report: Yorkshire Naturalists' Union. Leeds, Yorkshire, U.K.

Ann. Sci. Univ. Besançon, Bot.
Annales Scientifiques de l'Université de Besançon. 3e serie: Botanique. Institut des Sciences Naturelles, Besançon, France.

Ann. Univ. Budapest. Rolando Eötvos Nom., Biol.
Annales Universitatis Scientiarum Budapestinensis de Rolando Eötvos nominatae, Sectio Biologica. Budapest, Hungary.

Anz. Österr. Akad. Wiss. Mat.-Naturwiss.
Anzeiger Österreichische Akademie der Wissenschaften Mathematisch-Naturwissenschaftliche. Springer Verlag, Berlin, Germany.

Arch. Sci.
Archives des Sciences. Société de Physique et d'Histoire Naturelle de Genève, Genève, Switzerland.

Arq. Jard. Bot. Rio de Janeiro
Arquivos do Jardim Botânico do Rio de Janeiro. Ministério da Agricultura, Rio de Janeiro, Brazil.

Ashingtonia
Ashingtonia. Hollygate Nurseries Ltd., Ashington, Sussex, U.K.

Atti Ist. Veneto Sci. Lett. Arti, Sci. Mat. Nat.
Atti Istituto Veneto di Scienze, Lettere ed Arti. Classe di Scienze Matematiche e Naturali. Venezia, Italy.

Atti Mus. Civ. Stor. Nat. Trieste
Atti del Museo Civico di Storia Naturali di Trieste. Trieste.

Austral. Forest. Timber Bur. Forest Tree Ser.
Australian Forestry & Timber Bureau Forest Tree Series. Canberra, Australia.

Austral. Nat. Hist.
Australian Natural History. Australian Museum, College Street, Sydney.

B.S.B.I. Welsh Region Bull.
Botanical Society of the British Isles: Welsh Region Bulletin. Cardiff, U.K.

B.S.E. News
B.S.E. News: Newsletter of the Botanical Society of Edinburgh. Royal Botanic Garden, Inverleith Row, Edinburgh, U.K.

Ber. Naturwiss. Ver. Bielefeld
Bericht des Naturwissenschaftlichen Vereins für Bielefeld und Umgegend. Bielefeld, Federal Republic of Germany.

Biochem. Syst.
Biochemical Systematics. Pergamon Press, Oxford, U.K.

Biol. Práce
Biologicke Práce. Slovenskej Akadémie Vied, Bratislava, Czechoslovakia.

Blue Jay
Blue Jay: A Journal of Natural History and Conservation for Saskatchewan and Adjacent Regions. Saskatchewan Natural History Society, Regina, Saskatchewan, Canada.

Bocagiana
Bocagiana. Museu Municipal do Funchal, Madeira.

Bol. Estac. Central Ecol.
Boletin de la Estación Central de Ecología. Instituto Nacional para la Conservación de la Naturaleza. Madrid, Spain.

Bol. Fac. Agron. (Montevideo)
Boletín: Facultad de Agronomia. Universidad de la Republica, Montevideo, Uruguay.

Bol. Inst. Estud. Astur., Supl. Cienc.
Boletin del Instituto de Estudios Asturianos, Suplemento de Ciencias. Oviedo, Spain.

Bol. Soc. Estud. Moçambique
Boletim da Sociedade de Estudos de Moçambique. Lourenço Marques, Mozambique.

Boll. Soc. Ticin. Sci. Nat.
Bollettino della Società Ticinese di Scienze Naturali. Lugano, Switzerland.

Brenesia
Brenesia. Departamento de Historia Natural, Museo Nacional de Costa Rica.

Broteria, Sci. Nat.
Broteria: Série de Ciencias Naturais. Lisbon, Portugal.

Bul. Şti., B (Baiă-Mare)
Buletin Ştiinţific Seria B: Biologie, Fizică-Chimie, Matematică. Institutul Pedagogic, Baiă-Mare, Romania.

Bull. Biol. Soc. Wash.
Bulletin of the Biological Society of Washington. Washington, D.C., U.S.A.

Bull. Classe Sci. Acad. Roy. Belg.
Bulletin de la Classe des Sciences de l'Académie Royale de Belgique. Palais des Académies, Rue Ducale, Bruxelles, Belgium.

Bull. Féd. Soc. Hist. Nat. Franche-Comté
Bulletin de la Fédération des Sociétés d'Histoire Naturelle de Franche-Comté. Institut des Sciences Naturelles, Place Leclerc, Besançon, France.

Bull. Murith.
Bulletin de la Murithienne. Société Valaisanne des Sciences Naturelles, Sion, Switzerland.

Bull. Orchid Soc. S.E. Asia
Bulletin of the Orchid Society of South East Asia. P.O. Box. 2363, Singapore, Malaysia.

Bull. Sci. Bourgogne
Bulletin Scientifique de Bourgogne. Ville et Université de Dijon, Dijon, France.

Bull. Séances Acad. Roy. Sci. Outre-Mer
Bulletin des Séances de l'Académie Royale des Sciences d'Outre Mer. Bruxelles, Belgium.

Bull. Soc. Sci. Bretagne
Bulletin de la Société Scientifique de Bretagne. Rennes, France.

Buxtonian
The Buxtonian. Department of Biology, Northeastern University, Boston, Massachusetts.

Cact. Comments
Cactus Comments: Monthly News Letter. The New York Cactus and Succulent Society Inc., 44 West 75 St., New York, N.Y., U.S.A.

Cactus (France)
Cactus. Association française des Amateurs de Cactées et Plantes Grasses, 84, Rue de Grenelle, Paris, France.

Cah. O.R.S.T.O.M., Biol.
Cahiers O.R.S.T.O.M, Série Biologie. Office de la Recherche Scientifique et Technique Outre-Mer, Service Central de Documentation, 70–74, Route d'Aulnay, Paris, France.

Cambridge Univ. Bot. Gard. Ass. Newsl.
Cambridge University Botanic Garden Association Newsletter. Cambridge, U.K.

Carib. J. Sci.
Caribbean Journal of Science. Institute of Caribbean Science, University of Puerto Rico, Mayaguez, Puerto Rico.

Chileans
The Chileans. 5, Oakfield Road, Bristol, U.K.

Compt. R. Acad. Agric. Georgi Dimitrov
Comptes Rendus de l'Académie Agricole Georgi Dimitrov. Sofia, Bulgaria.

Dendroflora
Dendroflora: Koninklijke Vereniging voor Boskoopse Culturen. Boskoop, Netherlands.

Dendrol. Uzb.
Dendrologiya Uzbekistana. Izdatel'stvo 'Fan' Uzbekskoi SSR, Tashkent, U.S.S.R.

Eivissa
Eivissa: Revista de l'Institut d'Estudis Eivissencs. Pati d'Armes, Eivissa [Ibiza], Spain.

Ekol. Pol.
Ekologia Polska. Institute of Ecology, Polish Academy of Sciences, Warszawa, Poland.

Environment Southwest
Environment Southwest. San Diego Society of Natural History, P.O. Box 1390, San Diego, California, U.S.A.

Excerpta Bot., A
Excerpta Botanica: Sectio A: Taxonomica et Chorologica. Gustav Fischer Verlag, Stuttgart, Federal Republic of Germany.

Faun.-Flor. Notizen Saarland
Faunistisch-Floristiche Notizen aus dem Saarland. Arbeitsgemeinschaft für Tier- und Pflanzengeophysische Heimatforschung im Saarland, Saarbrucken, Federal Republic of Germany.

Flor. Märkm.
Floristilised Märkmed. Loodusuurijate Selts Eesti NSV Teaduste Akadeemia Juures, Tallinn, Estonian SSR, U.S.S.R.

Folia Mus. Rer. Nat. Bohem. Occid. Bot.
Folia Musei Rerum Naturalium Bohemiae Occidentalis, Botanica. Západoceské Muzeum, Plzeň, Czechoslovakia.

Forest. Commis. N.S.W. Res. Note
Forestry Commission of New South Wales Research Note. Sydney, N.S.W., Australia.

Forest Res. Inst. Res. Pamph. (W. Malaysia)
Forest Research Institute West Malaysia Research Pamphlet. Forest Department, West Malaysia.

Garcia de Orta, Sér. Bot.
Garcia de Orta, Série de Botânica. Revista da Junta de Investigaçoes do Ultramar, Lisboa, Portugal. Formerly *Garcia de Orta.*

Geigy Weed Tables
Geigy Weed Tables. Geigy Ltd: Agricultural Chemicals Division, Manchester, U.K.

Genet. Selekts. (Bulgaria)
Genetika i Selektsiya. Akademiya Georgi Dimitrov, Sofiya, Bulgaria.

Gentes Herb. (Ithaca)
Gentes Herbarum. L. H. Bailey Hortorium of the New York State College of Agriculture and Life Sciences, Cornell University, Ithaca, New York, U.S.A.

Geoforum
Geoforum: Journal of Physical, Human and Regional Geosciences. Pergamon Press, New York, U.S.A., and Vieweg GmbH, Braunschweig, Federal Republic of Germany.

Geogr. Mag.
The Geographical Magazine. New Science Publications, London, U.K.

Ginkgoana
Ginkgoana: Contributions to the Flora of Asia and the Pacific Region. Academic Scientific Book Co. Inc., Tokyo, Japan.

Glasn. Prirodnj. Muz. Beogradu, B
Glasnik Prirodnjačkog Muzeja u Beogradu. Beograd, Yugoslavia.

Herba Pol.
Herba Polonica. Instytut Przemyslu Zielarskiego, Poznan, Poland.

Horticulture
Horticulture. Massachusetts Horticultural Society, Horticultural Hall, Boston, Massachusetts, U.S.A.

IDIA
IDIA: Informativo de investigacones agricolas. Buenos Aires, Argentina.

IUCN Occ. Pap.
IUCN Occasional Paper. International Union for the Conservation of Nature and Natural Resources, Morges, Switzerland.

Idesia
Idesia. Biblioteca Centrale de la Universidade de Norte, Arica, Chile.

Incompat. Newsl.
Incompatibility Newsletter. Association Euratom-Ital, Wageningen, Netherlands.

Info-Nat.
Info-Nature: Bulletin de Liaison de la Société Réunionnaise pour l'étude et la Protection de la Nature. Muséum d'Histoire Naturelle, St. Denis, Réunion.

Inform. ZAG Wasserpfl.
Informationen ZAG Wasserpflanzen. Ingo Hertel, 582 Bad Langensalza, German Democratic Republic.

Inverness. Surv. 1970–74 Newsl.
Inverness-shire Survey 1970–74: Newsletter. Committee for the Study of the Scottish Flora, Glasgow, U.K.

Istanbul Univ. Orman Fak. Yayinl.
Istanbul Universitesi Orman Fakültesi Yayinlari. Istanbul, Turkey.

J. Barbados Mus. Hist. Soc.
Journal of the Barbados Museum and Historical Society. St. Ann's Garrison, Barbados, West Indies.

J. Bihar Bot. Soc.
Journal of the Bihar Botanical Society. Bihar, India.

J. Bromeliad Soc.
Journal of the Bromeliad Society. Los Angeles, California, U.S.A.

J. Devon Trust Nat. Conserv.
Journal of the Devon Trust for Nature Conservation. 2, Pennsylvania Road, Exeter, Devon, U.K.

J. Genet.
Journal of Genetics. Cambridge University Press, Cambridge, U.K.

J. Nation. Inst. Agric. Bot.
The Journal of the National Institute of Agricultural Botany. The National Institute of Agricultural Botany, Huntingdon Road, Cambridge, U.K.

J. Sci. Food Agric.
Journal of the Science of Food and Agriculture. Society of Chemical Industry, London, U.K.

Jahrb. Oberösterr. Musealver.
Jahrbuch des Oberösterreichischen Musealvereines. Linz, Austria.

Jahresber. Naturforsch. Ges. Graubündens
Jahresbericht der Naturforschenden Gesellschaft Graubündens. Bischofberger, Chur, Switzerland.

Jahresber. Naturwiss. Ver. Wuppertal
Jahresbericht des Naturwissenschaftlichen Vereins zu Wuppertal. Wuppertal, Federal Republic of Germany.

John Innes Inst. Ann. Rep.
John Innes Institute Annual Report. Colney Lane, Norwich, U.K.

Laporan Lembaga Penelitian Hutan
Laporan Lembaga Penelitian Hutan. Bogor, Indonesia.

Launceston Nat.
The Launceston Naturalist. Launceston Field Naturalists' Club, Launceston, Tasmania.

Lethaia
Lethaia: an International Journal of Palaeontology and Stratigraphy. Academic Press, New York, U.S.A.

Lichenologist
The Lichenologist. British Lichen Society, c/o Department of Botany, British Museum (Natural History), Cromwell Road, London, S.W. 7. Publ. for the Soc. by Blackwell Scientific Publications Ltd., 5, Alfred Street, Oxford.

Loodusuur. Seltsi Aastar.
Loodusuurijate Seltsi Aastaraamat. Loodusuurijate Selts Eesti NSV Teaduste Akadeemia Juures, Kirjastus 'Valgur', Tallinn, Estonia, U.S.S.R.

Lorentzia
Lorentzia. Museo Botanico, Facultad de Ciencias Exactas, Fisicas y Naturales, Cordoba, Argentina.

Malayan Forester
The Malayan Forester. Officers of the Forest Department, Malaysia, Kuala Lumpur, Malaysia.

Malaysian J. Sci.
Malaysian Journal of Science. Kuala Lumpur, Malaysia.

Mém. Acad. Roy. Sci. Outre-Mer, Sci. Nat. Nouv. Sér.
Memoire de l'Academie Royale des Sciences d'Outre-Mer, Classe des Sciences Naturelles et Medicales, Nouvelle Série. Bruxelles, Belgium.

Mem. Fac. Agric. Hokkaido Univ.
Memoirs of the Faculty of Agriculture Hokkaido University. Sapporo, Japan.

Mich. Academician
Michigan Academician. Michigan Academy of Science, Arts and Letters, Ann Arbor, Michigan, U.S.A.

Mitt. Abteil. Bot. Landesmus. 'Joanneum' Graz
Mitteilungen der Abteilung für Botanik am Landesmuseum 'Joanneum' Graz. Graz, Austria.

Mitt. Bad. Landesver. Naturk. Naturschutz
Mitteilungen des Badischen Landesvereins für Naturkunde und Naturschutz. Freiburg, Federal Republic of Germany.

Mitt. Naturforsch. Ges. Schaffhausen
Mitteilungen der Naturforschenden Gesellschaft Schaffhausen. Schaffhausen, Switzerland.

Mus. Ass. Mon. Bull.
Museums Association Monthly Bulletin. London, U.K.

Mus. Nac. Hist. Nat. Not. Mens. (Chile)
Museo Nacional de Historia Natural: Noticiario Mensual. Santiago, Chile.

Mycologia
Mycologia. The Mycological Society of America, New York Botanical Garden, Bronx, New York, U.S.A.

Nat. Landschap
Natuur en Landschap: Tijdschrift voor Natuurbehoud en Milieubekeer. Herengracht 540, Amsterdam, Netherlands.

Nat. Mensch
Natur und Mensch: Jahresmitteilung der Naturhistorischen Gesellschaft Nürnberg. Nürnberg, Federal Republic of Germany.

Nat. Resources
Nature and Resources. UNESCO, Paris, France.

Nature New Biol.
Nature New Biology. Macmillan Journals Ltd., Brunel Road, Basingstoke, Hampshire, U.K.

Naturk. Jahrb. Linz
Naturkundliches Jahrbuch der Stadt Linz. Stadt Museum Linz, Linz/Donau, Austria.

Nauch. Trud. Plovdiv. Univ. 'Paisii Khilendarski'
Nauchni Trudove na Plovdivski Universitet 'Paisiĭ Khilendarski'. Izdatel'stvo Khristo G. Panov, Plovdiv, Bulgaria.

Neujahrsbl. Naturforsch. Ges. Schaffhausen
Neujahrsblatt der Naturforschenden Gesellschaft Schaffhausen, Schaffhausen, Switzerland.

New York Bot. Gard. Ann. Rep.
New York Botanical Garden Annual Report. Bronx, New York, U.S.A.

Norsk. Polarinst. Årb.
Norsk Polarinstitut Årbok. Oslo, Norway.

Oléagineux
Oléagineux: Revue générale des Corps Gras et Dérivés. SETCO, Paris, France.

Op. Bot. Ser. B.
Opera Botanica Series B: Flora of Ecuador. Department of Systematic Botany, University of Göteborg, and Section for Botany, Ritsmuseum, Stockholm, Sweden.

Orquidea (Mexico)
Orquidea. Asociacion Mexicana de Orquideologia, México, D.F., Mexico.

Osnabrücker Naturwiss. Mitt.
Osnabrücker Naturwissenschaftliche Mitteilungen. Naturwissenschaftlichen Museum Osnabrück und Naturwissenschaftlichen Verein Osnabrück, Osnabrück, Federal Republic of Germany.

Pac. Isl. Stud. Notes
Pacific Island Studies and Notes. Honolulu, Hawaii, U.S.A.

Paphioped. World
Paphiopedilum World: The International Magazine of Cypripedieae. J. E. Wilson, 2865 Sheffield Road, San Marino, California. U.S.A.

Persoonia
Persoonia: A Mycological Journal. Rijksherbarium, Leiden. Netherlands.

Pl. Sci. (Lucknow)
Plant Science. Association for the Advancement of Plant Sciences, National Botanic Gardens, Lucknow, India.

Planto
Planto: Bulletin of Darbhanga Botanical Society. Botany Department, Mithila University, Darbhanga, India.

Prace Bot. (Wrocław)
Prace Botaniczne: Acta Universitatis Wratislaviensis. Wrocław, Poland.

Proc. Indian Sci. Congr.
Proceedings of the Indian Science Congress. Calcutta, India.

Proc. Indiana Acad. Sci.
Proceedings: Indiana Academy of Science. Indiana State Library, Indianapolis, Indiana, U.S.A.

Proc. Iowa Acad. Sci.
Proceedings: Iowa Academy of Science. University of N. Iowa, Cedar Falls, Iowa, U.S.A.

Proc. Kon. Nederl. Akad. Wetensch., B
Proceedings: Koninklijke Nederlandse Akademie van Wetenschappen Series B. North-Holland Publishing Company, Amsterdam, Netherlands.

Proteus (Ljubljana)
Proteus. Prirodoslovnodrustvo Slovenije v Ljubljani, Ljubljana, Jugoslavia.

Publ. Mus. Hist. Nat. Javier Prado, B
Publicaciones del Museo de Historia Natural "Javier Prado". Series B: Botanica. Lima, Peru.

Quart. Amer. Primrose Soc.
Quarterly of the American Primrose Society. 7213 South 88th Street, Seattle, Washington, U.S.A.

Ras. Nauki
Rastenievŭdni Nauki. Selskostopanska Akademiya Georgi Dimitrov, Sofiya, Bulgaria.

Rev. Fac. Nac. Agron. (Medellin)
Revista Facultad Nacional de Agronomia. Medellin, Colombia.

Rev. Hist.
Revue Historique. Presses Universitaires de France, Paris, France.

Rev. Mus. Argent. Cienc. Nat. Bernardino Rivadavia, Bot.
Revista del Museo Argentino de Ciencias Naturales "Bernardino Rivadavia" Ciencias Botanicas. Buenos Aires, Argentina.

Rev. Mycol.
Revue de Mycologie. Laboratoire de Cryptogamie du Museum National d'Histoire Naturelle, 12, Rue de Buffon, Paris.

Rev. Soc. Sav. Haute Normandie
Revue: Sociétés Savantes de Haute Normandie. Rouen, France.

Rev. Univ. Burundi
Revue Universitaire du Burundi. The University, Bujumbura, Burundi.

Roy. New Zealand Inst. Hort. Ann. J.
Royal New Zealand Institute of Horticulture Annual Journal. Christchurch, New Zealand.

Sri-Lanka Forest.
Sri-Lanka Forester: Journal of the Sri-Lanka Forest Department. Colombo, Sri Lanka. Formerly *Ceylon Forester.*

Stachelpost
Stachelpost: Mitteilungsblätter für Kakteenfreunde. C. Kilian, 6502 Mainz-Kostheim, Wallufer Str. 30, Federal Republic of Germany.

Stud. Bot. Hung.
Studia Botanica Hungarica. Természettudományi Múzeum, Budapest, Hungary. Formerly *Fragmenta Botanica.*

Stud. Hist. Phil. Sci.
Studies in History and Philosophy of Science. Macmillan Journals Ltd., Little Essex Street, London, U.K.

Stuttgarter Beitr. Naturk., A
Stuttgarter Beiträge zur Naturkunde, Serie A (Biologie). Staatliche Museum für Naturkunde in Stuttgart, Stuttgart, Federal Republic of Germany.

Svenska Linné-Sällsk. Årsskr.
Svenska Linné-Sällskapets Årsskrift. Uppsala, Sweden.

Syllogeus
Syllogeus. National Museum of Natural Sciences (Canada), Ottawa, Canada.

Tätigkeitsber. Naturforsch. Ges. Baselland
Tätigkeitsberichte der Naturforschenden Gesellschaft Baselland. Buchdruckerei Lüdin, Liestal, Switzerland.

Terre Vie
La Terre et la Vie: Revue d'Ecologie appliquée à la Conservation de la Nature. Société Nationale de Protection de la Nature et d'Acclimatation de France, Paris, France.

Trans. Kent Field Club
Transactions of the Kent Field Club. Maidstone, Kent, U.K.

Transvaal Nat. Conserv. Div. Ann. Rep.
Transvaal Nature Conservation Division: Annual Report. Transvaal, South Africa.

Trees
Trees: Journal of the Men of the Trees. Crawley Down, Crawley, Sussex, U.K.

Trop. Grasslands
Tropical Grasslands. Tropical Grasslands Society of Australia, Cunningham Laboratory, St. Lucia, Queensland, Australia.

Univ. Calif. Publ. Entomol.
University of California Publications in Entomology. Berkeley, California, U.S.A.

Univ. Waterloo Biol. Ser.
University of Waterloo Biology Series. Waterloo, Ontario, Canada.

West Indian Rev.
West Indian Review. Dorking, Surrey, U.K.

Wiss. Zeitschr. Friedrich-Schiller-Univ. Jena, Mat.-Naturwiss.
Wissenschaftliche Zeitschrift: Friedrich-Schiller-Universität Jena, Mathematisch-Naturwissenschatiche Reihe. Jena, German Democratic Republic.

Wiss. Zeitschr. Univ. Berlin, Math.-Nat.
Wissenschaftliche Zeitschrift der Humboldt Universität zu Berlin, Mathematisch-Naturwissenschaftliche Reihe. Berlin, German Democratic Republic.

Zesz. Nauk Uniw. Mikol. Kopernika Toruniu, Biol.
Zeszyty Naukowe Uniwersytet Mikolaja Kopernika w. Toruniu, Torun, Poland.

Zprávy Geogr. Uśt., CSAV
Zprávy Geografického Uśtavu CSAV. Brno, Czechoslovakia.

Abbreviations of Floras

Fl. France
Flore de France. Centre National de la Recherche Scientifique, Paris, France.

Fl. Moçamb.
Flore de Moçambique. Centro de Botanica, Junta de Investigacões do Ultramar, Lisboa, Portugal.

Fl. Nar. Republ. Bŭlgariya
Flora na Narodna Republika Bŭlgariya. Bulgarskata Akademiya na Naukite, Sofiya, Bulgaria.

Fl. Republ. Social. Roman.
Flora Republicii Socialiste România. Academiei Republicii Socialiste România, Bucuresti, Rumania.

Nouv. Fl. Belg., Luxemb., N. France etc.
Nouvelle Flore de la Belgique, du Grand-Duché de Luxembourg, du Nord de la France et des Régions Voisines (Pteridophytes et Spermatophytes). Jardin Botanique National de Belgique, Bruxelles, Belgium.

Tree Fl. Malaya
Tree Flora of Malaya: a Manual for Foresters. Longman, London, U.K.

Author Index

C

E

F

H

I

L

M

N

O

P

T

U

V

W

X

Y

Z

Printed in England for Her Majesty's Stationery Office
By McCorquodale Printers Ltd., London
HM 7527 Dd. 506749 K9 11/75 McC3309